Discovering Geometry

An Investigative Approach

Third Edition

Michael Serra

Project Editor
Ladie Malek

Project Administrator
Shannon Miller

Editors
Christian Aviles-Scott, Dan Bennett, Mary Jo Cittadino, Curt Gebhard

Editorial Assistants
Halo Golden, Erin Gray, Susan Minarcin, Laura Schattschneider, Jason Taylor

Editorial Consultants
Cavan Fang, David Hoppe, Stacey Miceli, Davia Schmidt

Mathematics Reviewers
Michael de Villiers, Ph.D., University of Durban, Westville, Pinetown, South Africa

David Rasmussen, Neil's Harbour, Nova Scotia

Multicultural and Equity Reviewers
David Keiser, Montclair State University, Upper Montclair, New Jersey

Swapna Mukhopadhyay, Ph.D., San Diego State University, San Diego, California

Accuracy Checkers
Dudley Brooks, Marcia Ellen Olmstead

Editorial Production Manager
Deborah Cogan

Production Editor
Kristin Ferraioli

Copyeditor
Margaret Moore

Production Director
Diana Jean Parks

Production Coordinator
Ann Rothenbuhler

Cover Designer
Jill Kongabel

Text Designer
Marilyn Perry

Art Editor
Jason Luz

Photo Editor
Margee Robinson

Art and Design Coordinator
Caroline Ayres

Illustrators
Juan Alvarez, Andy Levine, Claudia Newell, Bill Pasini, William Rieser, Sue Todd, Rose Zgodzinski

Technical Art
Precision Graphics

Compositor and Prepress
TSI Graphics

Printer
Von Hoffman Press

Executive Editor
Casey FitzSimons

Publisher
Steven Rasmussen

Key Curriculum Press
1150 65th Street
Emeryville, CA 94608
editorial@keypress.com
http://www.keypress.com

Printed in the United States of America

10 9 8 7 6 5 4 3 2 06 05 04 03

ISBN 1-55953-459-1

Acknowledgments

First, to all the teachers who have used *Discovering Geometry,* a sincere thank you for your wonderful support and encouragement. I wish to thank my always delightful and ever-so-patient students for their insight, humor, and hard work. I also wish to thank the many students across the country who have written to me with their kind words, comments, and suggestions. And thanks to Kelvin Taylor and the rest of the marketing and sales staff at Key Curriculum Press for their successful efforts in bringing the first two editions into so many classrooms.

There are two people who have added their touch to earlier editions of *Discovering Geometry*: Steve Rasmussen was editor on the first edition and Dan Bennett was editor on the second edition. Thank you, Steve and Dan.

This third edition, as you can see from the credits, involved a much larger team. While working on this edition of *Discovering Geometry,* I was fortunate to have the assistance of Ladie Malek as project editor. Thank you, Ladie, for your creativity, dedication, and especially your patience. To the editorial and production staff and managers at Key, the field testers, the advisors, the consultants, and the reviewers, I am grateful for your quality work.

Michael Serra

A Note from the Publisher

When Key Curriculum Press first published *Discovering Geometry* in 1989, it was unique among high school geometry books because of its discovery approach. *Discovering Geometry* still presents concepts visually, and students explore ideas analytically, then inductively, and finally deductively—developing insight, confidence, and increasingly sophisticated mathematical understanding. As J. Michael Shaughnessy, mathematics professor at Portland State University, said, "This is a book for 'doers.' Students constantly *do* things in this book, both alone and in groups. If you want your students to become actively involved in the process of learning and creating geometry, then this is the book for you."

The mathematics we learn and teach in school changes over time, driven by new scientific discoveries, new research in education, changing societal needs, and by the use of new technology in work and in education. The effectiveness of *Discovering Geometry*'s investigative approach has been substantiated in many thousands of classrooms and is reflected in the *Principles and Standards for School Mathematics,* the guiding document of the National Council of Teachers of Mathematics (NCTM). In this, the third edition, you will find many of the text's original and hallmark features—plus a host of improvements. The layout is easier to follow, and the additional examples from art and science will be a motivating complement to your curriculum. We have carefully analyzed the exercises for optimal practice and real-world interest. There are more opportunities to review algebra and more ways to use technology, especially The Geometer's Sketchpad® software, in the projects and homework assignments. These changes will give you—the student, parent, or teacher—greater flexibility in attaining your educational goals. They will enable more students to succeed in high school geometry and achieve continued success in future mathematics courses, other areas of education, and eventual careers.

Experience as well as sound educational research on how geometric thinking develops during adolescence tells us that, regardless of the subject or level, students learn mathematics best when they understand the concepts. The positive feedback we have received over the years for *Discovering Geometry* has inspired us to create an entire series. Key Curriculum Press now offers the *Discovering Mathematics* series, a complete program of algebra, geometry, and advanced algebra. Through the investigations that are the heart of the series, students discover many important mathematical principles. In the process, they come to believe in their ability to succeed at mathematics, they understand the course content more deeply, and they realize that they can re-create their discoveries if they need to.

If you are a student, we hope that as you work through this course you gain knowledge for a lifetime. If you are a parent, we hope you enjoy watching your student develop mathematical power. If you are a teacher, we hope you find that *Discovering Geometry* makes a significant positive impact in your classroom. Whether you are learning, guiding, or teaching, please share your trials and successes with the professional team at Key Curriculum Press.

Steven Rasmussen, President
Key Curriculum Press

Contents

A Note to Students from the Author xiv

CHAPTER 0 Geometric Art 1

0.1 Geometry in Nature and in Art 2
0.2 Line Designs 7
0.3 Circle Designs 10
0.4 Op Art 13
0.5 Knot Designs 16
Project: Symbolic Art 19
0.6 Islamic Tile Designs 20
Project: Photo or Video Safari 23
Chapter 0 Review 24
Assessing What You've Learned 26

CHAPTER 1 Introducing Geometry 27

1.1 Building Blocks of Geometry 28
Investigation: Mathematical Models 29
Project: Spiral Designs 35
Using Your Algebra Skills 1: Midpoint 36
1.2 Poolroom Math 38
Investigation: Virtual Pool 41
1.3 What's a Widget? 47
Investigation: Defining Angles 49
1.4 Polygons 54
1.5 Triangles and Special Quadrilaterals 59
Investigation: Triangles and Special Quadrilaterals 60
Project: Drawing the Impossible 66
1.6 Circles 67
Investigation: Defining Circle Terms 69
1.7 A Picture Is Worth a Thousand Words 73
1.8 Space Geometry 80
Investigation: Space Geometry 82
Exploration: Geometric Probability I 86
Activity: Chances Are 86
Chapter 1 Review 88
Assessing What You've Learned 92

CHAPTER 2 **Reasoning in Geometry** 93

2.1 **Inductive Reasoning** 94
Investigation: Shape Shifters 96
2.2 **Deductive Reasoning** 100
Investigation: Overlapping Segments 102
2.3 **Finding the *nth* Term** 106
Investigation: Finding the Rule 106
Project: Best-Fit Lines 111
2.4 **Mathematical Modeling** 112
Investigation: Party Handshakes 112
Exploration: The Seven Bridges of Königsberg 118
Activity: Traveling Networks 118
2.5 **Angle Relationships** 120
Investigation 1: The Linear Pair Conjecture 120
Investigation 2: Vertical Angles Conjecture 121
2.6 **Special Angles on Parallel Lines** 126
Investigation 1: Which Angles Are Congruent? 126
Investigation 2: Is the Converse True? 128
Project: Line Designs 132
Using Your Algebra Skills 2: Slope 133
Exploration: Patterns in Fractals 135
Activity: The Sierpiński Triangle 136
Chapter 2 Review 138
Assessing What You've Learned 140

CHAPTER 3 **Using Tools of Geometry** 141

3.1 **Duplicating Segments and Angles** 142
Investigation 1: Copying a Segment 143
Investigation 2: Copying an Angle 144
3.2 **Constructing Perpendicular Bisectors** 147
Investigation 1: Finding the Right Bisector 147
Investigation 2: Right Down the Middle 148
3.3 **Constructing Perpendiculars to a Line** 152
Investigation 1: Finding the Right Line 152
Investigation 2: Patty-Paper Perpendiculars 153
3.4 **Constructing Angle Bisectors** 157
Investigation 1: Angle Bisecting by Folding 157
Investigation 2: Angle Bisecting with Compass 158
3.5 **Constructing Parallel Lines** 161
Investigation: Constructing Parallel Lines by Folding 161
Using Your Algebra Skills 3: Slopes of Parallel and Perpendicular Lines 165
3.6 **Construction Problems** 168

Exploration: Perspective Drawing 172
Activity: Boxes in Space 173
3.7 Constructing Points of Concurrency 176
Investigation 1: Concurrence 176
Investigation 2: Incenter and Circumcenter 177
3.8 The Centroid 183
Investigation 1: Are Medians Concurrent? 183
Investigation 2: Balancing Act 184
Exploration: The Euler Line 189
Activity: Three Out of Four 189
Project: Is There More to the Orthocenter? 190
Chapter 3 Review 191
Mixed Review 194
Assessing What You've Learned 196

CHAPTER 4 Discovering and Proving Triangle Properties **197**

4.1 Triangle Sum Conjecture 198
Investigation: The Triangle Sum 199
4.2 Properties of Special Triangles 204
Investigation 1: Base Angles in an Isosceles Triangle 205
Investigation 2: Is the Converse True? 206
Using Your Algebra Skills 4: Writing Linear Equations 210
4.3 Triangle Inequalities 213
Investigation 1: What Is the Shortest Path from *A* to *B*? 214
Investigation 2: Where Are the Largest and Smallest Angles? 215
Investigation 3: Exterior Angles of a Triangle 215
Project: Random Triangles 218
4.4 Are There Congruence Shortcuts? 219
Investigation 1: Is SSS a Congruence Shortcut? 220
Investigation 2: Is SAS a Congruence Shortcut? 221
4.5 Are There Other Congruence Shortcuts? 225
Investigation: Is ASA a Congruence Shortcut? 225
4.6 Corresponding Parts of Congruent Triangles 230
Project: Polya's Problem 234
4.7 Flowchart Thinking 235
4.8 Proving Isosceles Triangle Conjectures 241
Investigation: The Symmetry Line in an Isosceles Triangle 242
Exploration: Napoleon's Theorem 247
Activity: Napoleon Triangles 247
Project: Lines and Isosceles Triangles 248
Chapter 4 Review 249
Take Another Look 253
Assessing What You've Learned 254

CHAPTER 5 Discovering and Proving Polygon Properties 255

5.1 **Polygon Sum Conjecture** 256
Investigation: Is There a Polygon Sum Formula? 256
5.2 **Exterior Angles of a Polygon** 260
Investigation: Is There an Exterior Angle Sum? 260
Exploration: Star Polygons 264
Activity: Exploring Star Polygons 264
5.3 **Kite and Trapezoid Properties** 266
Investigation 1: What Are Some Properties of Kites? 266
Investigation 2: What Are Some Properties of Trapezoids? 268
Project: Drawing Regular Polygons 272
5.4 **Properties of Midsegments** 273
Investigation 1: Triangle Midsegment Properties 273
Investigation 2: Trapezoid Midsegment Properties 274
Project: Building an Arch 278
5.5 **Properties of Parallelograms** 279
Investigation: Four Parallelogram Properties 279
Using Your Algebra Skills 5: Solving Systems of Linear Equations 285
5.6 **Properties of Special Parallelograms** 287
Investigation 1: What Can You Draw with the Double-Edged Straightedge? 287
Investigation 2: Do Rhombus Diagonals Have Special Properties? 288
Investigation 3: Do Rectangle Diagonals Have Special Properties? 289
5.7 **Proving Quadrilateral Properties** 294
Project: Japanese Puzzle Quilts 299
Chapter 5 Review 300
Take Another Look 303
Assessing What You've Learned 304

CHAPTER 6 Discovering and Proving Circle Properties 305

6.1 **Chord Properties** 306
Investigation 1: How Do We Define Angles in a Circle? 307
Investigation 2: Chords and Their Central Angles 307
Investigation 3: Chords and the Center of the Circle 309
Investigation 4: Perpendicular Bisector of a Chord 309
6.2 **Tangent Properties** 313
Investigation 1: Going Off on a Tangent 313
Investigation 2: Tangent Segments 314
6.3 **Arcs and Angles** 319
Investigation 1: Inscribed Angle Properties 319
Investigation 2: Inscribed Angles Intercepting the Same Arc 320

Investigation 3: Angles Inscribed in a Semicircle 320
Investigation 4: Cyclic Quadrilaterals 321
Investigation 5: Arcs by Parallel Lines 321
6.4 Proving Circle Conjectures 325
Using Your Algebra Skills 6: Finding the Circumcenter 329
6.5 The Circumference/Diameter Ratio 331
Investigation: A Taste of Pi 332
Project: Needle Toss 336
6.6 Around the World 337
6.7 Arc Length 341
Investigation: Finding the Arcs 342
Project: Racetrack Geometry 345
Exploration: Cycloids 346
Activity: Turning Wheels 346
Chapter 6 Review 349
Mixed Review 352
Take Another Look 355
Assessing What You've Learned 356

CHAPTER 7 Transformations and Tessellations 357

7.1 Transformations and Symmetry 358
Investigation: The Basic Property of a Reflection 360
7.2 Properties of Isometries 366
Investigation 1: Transformations on a Coordinate Plane 367
Investigation 2: Finding a Minimal Path 367
7.3 Compositions of Transformations 373
Investigation 1: Reflections over Two Parallel Lines 374
Investigation 2: Reflections over Two Intersecting Lines 375
Project: Kaleidoscopes 378
7.4 Tessellations with Regular Polygons 379
Investigation: The Semiregular Tessellations 380
7.5 Tessellations with Nonregular Polygons 384
Investigation 1: Do All Triangles Tessellate? 384
Investigation 2: Do All Quadrilaterals Tessellate? 385
Project: Penrose Tilings 388
7.6 Tessellations Using Only Translations 389
7.7 Tessellations That Use Rotations 393
7.8 Tessellations That Use Glide Reflections 398
Using Your Algebra Skills 7: Finding the Orthocenter and Centroid 401
Chapter 7 Review 404
Assessing What You've Learned 408

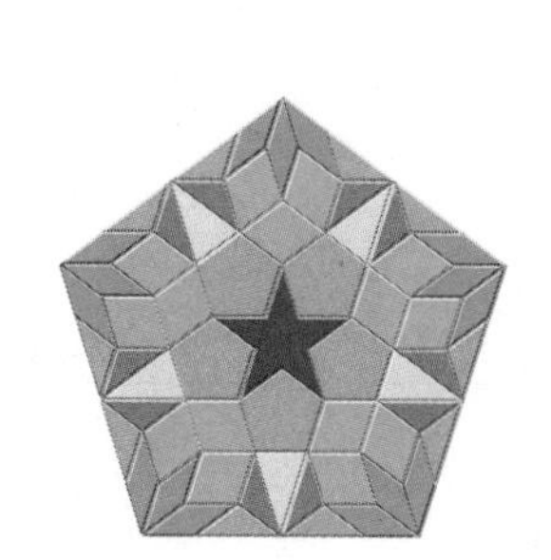

CHAPTER 8 Area 409

8.1 **Areas of Rectangles and Parallelograms** 410
Investigation: Area Formula for Parallelograms 412
Project: Random Rectangles 416
8.2 **Areas of Triangles, Trapezoids, and Kites** 417
Investigation 1: Area Formula for Triangles 417
Investigation 2: Area Formula for Trapezoids 417
Investigation 3: Area Formula for Kites 418
Project: Maximizing Area 421
8.3 **Area Problems** 422
Investigation: Solving Problems with Area Formulas 422
8.4 **Areas of Regular Polygons** 426
Investigation: Area Formula for Regular Polygons 426
Exploration: Pick's Formula for Area 430
Activity: Dinosaur Footprints and Other Shapes 431
8.5 **Areas of Circles** 433
Investigation: Area Formula for Circles 433
8.6 **Any Way You Slice It** 437
Exploration: Geometric Probability II 442
Activity: Where the Chips Fall 442
Project: Different Dice 444
8.7 **Surface Area** 445
Investigation 1: Surface Area of a Regular Pyramid 448
Investigation 2: Surface Area of a Cone 449
Exploration: Alternative Area Formulas 453
Activity: Calculating Area in Ancient Egypt 453
Chapter 8 Review 455
Take Another Look 459
Assessing What You've Learned 460

CHAPTER 9 The Pythagorean Theorem 461

9.1 **The Theorem of Pythagoras** 462
Investigation: The Three Sides of a Right Triangle 462
Project: Creating a Geometry Flip Book 467
9.2 **The Converse of the Pythagorean Theorem** 468
Investigation: Is the Converse True? 468
Using Your Algebra Skills 8: Radical Expressions 473
9.3 **Two Special Right Triangles** 475
Investigation 1: Isosceles Right Triangles 475
Investigation 2: 30°-60°-90° Triangles 476
Exploration: A Pythagorean Fractal 480
Activity: The Right Triangle Fractal 481
9.4 **Story Problems** 482

9.5 **Distance in Coordinate Geometry** 486
Investigation 1: The Distance Formula 486
Investigation 2: The Equation of a Circle 488
Exploration: Ladder Climb 491
Activity: Climbing the Wall 491
9.6 **Circles and the Pythagorean Theorem** 492
Chapter 9 Review 496
Mixed Review 499
Take Another Look 501
Assessing What You've Learned 502

CHAPTER 10 Volume 503

10.1 **The Geometry of Solids** 504
Exploration: Euler's Formula for Polyhedrons 512
Activity: Toothpick Polyhedrons 512
10.2 **Volume of Prisms and Cylinders** 514
Investigation: The Volume Formula for Prisms and Cylinders 515
Project: The Soma Cube 521
10.3 **Volume of Pyramids and Cones** 522
Investigation: The Volume Formula for Pyramids and Cones 522
Project: The World's Largest Pyramid 527
Exploration: The Five Platonic Solids 528
Activity: Modeling the Platonic Solids 528
10.4 **Volume Problems** 531
10.5 **Displacement and Density** 535
Project: Maximizing Volume 538
Exploration: Orthographic Drawing 539
Activity: Isometric and Orthographic Drawings 541
10.6 **Volume of a Sphere** 542
Investigation: The Formula for the Volume of a Sphere 542
10.7 **Surface Area of a Sphere** 546
Investigation: The Formula for the Surface Area of a Sphere 546
Exploration: Sherlock Holmes and Forms of Valid Reasoning 551
Activity: It's Elementary! 552
Chapter 10 Review 554
Take Another Look 557
Assessing What You've Learned 558

CHAPTER 11 Similarity 559

Using Your Algebra Skills 9: Proportion and Reasoning 560
11.1 Similar Polygons 563
Investigation 1: What Makes Polygons Similar? 564
Investigation 2: Dilations on a Coordinate Plane 566
Project: Making a Mural 571
11.2 Similar Triangles 572
Investigation 1: Is AA a Similarity Shortcut? 572
Investigation 2: Is SSS a Similarity Shortcut? 573
Investigation 3: Is SAS a Similarity Shortcut? 574
Exploration: Constructing a Dilation Design 578
Activity: Dilation Creations 578
11.3 Indirect Measurement with Similar Triangles 581
Investigation: Mirror, Mirror 581
11.4 Corresponding Parts of Similar Triangles 586
Investigation 1: Corresponding Parts 586
Investigation 2: Opposite Side Ratios 588
11.5 Proportions with Area and Volume 592
Investigation 1: Area Ratios 592
Investigation 2: Volume Ratios 594
Project: In Search of the Perfect Rectangle 598
Exploration: Why Elephants Have Big Ears 599
Activity: Convenient Sizes 599
11.6 Proportional Segments Between Parallel Lines 603
Investigation 1: Parallels and Proportionality 604
Investigation 2: Extended Parallel/Proportionality 606
Exploration: Two More Forms of Valid Reasoning 611
Activity: Symbolic Proofs 613
Chapter 11 Review 614
Take Another Look 617
Assessing What You've Learned 618

CHAPTER 12 Trigonometry 619

12.1 Trigonometric Ratios 620
Investigation: Trigonometric Tables 622
12.2 Problem Solving with Right Triangles 627
Project: Light for All Seasons 631
Exploration: Indirect Measurement 632
Activity: Using a Clinometer 632
12.3 The Law of Sines 634
Investigation 1: Area of a Triangle 634
Investigation 2: The Law of Sines 635
12.4 The Law of Cosines 641
Investigation: A Pythagorean Identity 641

Project: Japanese Temple Tablets 646
12.5 Problem Solving with Trigonometry 647
Exploration: Trigonometric Ratios and the Unit Circle 651
Activity: The Unit Circle 651
Project: Trigonometric Functions 654
Exploration: Three Types of Proofs 655
Activity: Prove It! 657
Chapter 12 Review 659
Mixed Review 662
Take Another Look 665
Assessing What You've Learned 666

CHAPTER 13 Geometry as a Mathematical System **667**

13.1 The Premises of Geometry 668
13.2 Planning a Geometry Proof 679
13.3 Triangle Proofs 686
13.4 Quadrilateral Proofs 692
Investigation: Proving Parallelogram Conjectures 692
Exploration: Proof as Challenge and Discovery 696
Activity: Exploring Properties of Special Constructions 696
13.5 Indirect Proof 698
Investigation: Proving the Tangent Conjecture 699
13.6 Circle Proofs 703
13.7 Similarity Proofs 706
Investigation: Can You Prove the SSS Similarity Conjecture? 708
Using Your Algebra Skills 10: Coordinate Proof 712
Project: Special Proofs of Special Conjectures 717
Exploration: Non-Euclidean Geometries 718
Activity: Elliptic Geometry 719
Chapter 13 Review 721
Assessing What You've Learned 723

Hints for Selected Exercises 725
Index 747
Photo Credits 765

A Note to Students from the Author

Michael Serra

What Makes *Discovering Geometry* Different?

Discovering Geometry was designed so that you can be actively engaged as you learn geometry. In this book you "learn by doing." You will learn to use the tools of geometry and to perform geometry **investigations** with them. Many of the geometry investigations are carried out in small **cooperative groups** in which you jointly plan and find solutions with other students. Your investigations will lead you to the discovery of geometry properties. In addition, you will gradually learn about proof, a form of reasoning that will help explain why your discoveries are true.

Discovering Geometry was designed so that you and your teacher can have fun while learning geometry. It has a lot of "extras." Each lesson begins with a **quote** that I hope you will find funny or thought provoking. I think you'll enjoy the extra challenges in the **Improving Your...Skills** puzzles at the end of most lessons. To solve each puzzle, you'll need clever visual thinking skills or sharp reasoning skills or both. I hope you will find some of the **illustrated word problems** humorous. I created them in the hope of reducing any anxiety you might have about word problems. In the **explorations** you will learn about geometric probability, build geometric solids, find the height of your school building, and discover why elephants have big ears. In the **projects** you will draw the impossible, make kaleidoscopes, design a racetrack, and create a mural. There are also several **graphing calculator projects, Fathom Dynamic Statistics™ projects, The Geometer's Sketchpad explorations,** and **web links** that will allow you to practice and improve your skills with the latest educational technology.

You can do the projects, the puzzles, and the calculator and computer activities independently, whether or not your class tackles them as a group. Read through them as you proceed through the book.

Suggestions for Success

It is important to be organized. Keep a notebook with a section for definitions, a section for your geometry investigations, a section for discoveries, and a section for daily notes and exercises. Develop the habit of writing a summary page when you have completed each chapter. Study your notebook regularly.

You will need four tools for the investigations: a compass, a protractor, a straightedge, and a ruler. Some investigations use waxed "patty paper" that can be used as a unique geometry tool. Keep a graphing calculator handy, too.

You will find hints for some exercises in the back of the book. Those exercises are marked with an ⓗ. Try to solve the problems on your own first. Refer to the hints to check your method or as a last resort if you can't solve a problem.

Discovering Geometry will ask you to work cooperatively with your classmates. When you are working cooperatively, always be willing to listen to each other, to actively participate, to ask each other questions, and to help each other when asked. You can accomplish much more cooperatively than you can individually. And, best of all, you'll experience less frustration and have much more fun.

Michael Serra

CHAPTER

0 Geometric Art

A work by Dutch graphic artist M. C. Escher (1898–1972) opens each chapter in this book. Escher used geometry in creative ways to make his interesting and unusual works of art. As you come to each new chapter, see if you can connect the Escher work to the content of the chapter.

My subjects are often playful. . . . It is, for example, a pleasure to deliberately mix together objects of two and of three dimensions, surface and spatial relationships, and to make fun of gravity.

M. C. ESCHER

Print Gallery, M. C. Escher, 1956

OBJECTIVES

In this chapter you will

- see examples of geometry in nature
- study geometric art forms of cultures around the world
- study the symmetry in flowers, crystals, and animals
- see geometry as a way of thinking and of looking at the world
- practice using a compass and straightedge

LESSON 0.1

Geometry in Nature and in Art

There is one art,
no more no less,
To do all things
with artlessness.
PIET HEIN

Nature displays a seemingly infinite variety of geometric shapes, from tiny atoms to great galaxies. Crystals, honeycombs, snowflakes, spiral shells, spiderwebs, and seed arrangements on sunflowers and pinecones are just a few of nature's geometric masterpieces.

Circle

Hexagon

Pentagon

Geometry includes the study of the properties of shapes such as circles, hexagons, and pentagons. Outlines of the sun, the moon, and the planets appear as circles. Snowflakes, honeycombs, and many crystals are hexagonal (6-sided). Many living things, such as flowers and starfish, are pentagonal (5-sided).

People observe geometric patterns in nature and use them in a variety of art forms. Basket weavers, woodworkers, and other artisans often use geometric designs to make their works more interesting and beautiful. You will learn some of their techniques in this chapter.

In the Celtic knot design above, the curves seem to weave together.

This Islamic design from Egypt uses 4-sided and 6-sided shapes, as well as 5-pointed and 12-pointed stars.

Artists rely on geometry to show perspective and proportion, and to produce certain optical effects. Using their understanding of lines, artists can give depth to their drawings. Or they can use lines and curves to create designs that seem to pop out of the page. You will create your own optical designs in Lesson 0.4.

Hungarian artist Victor Vasarely (1908–1997) had a strong interest in geometry, which was reflected in his work. In this series, he used curved lines to produce the illusion of three spheres.

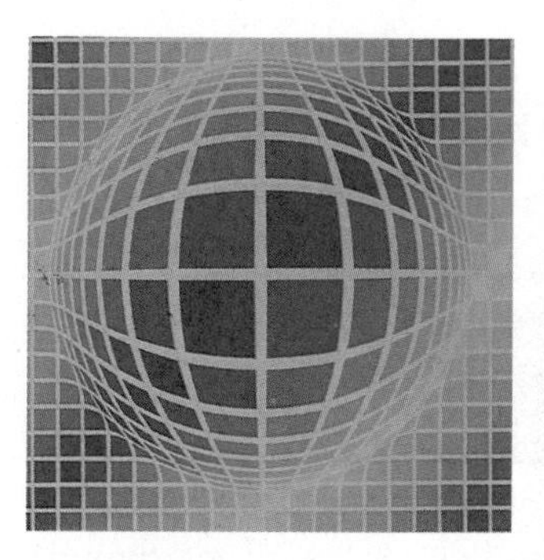

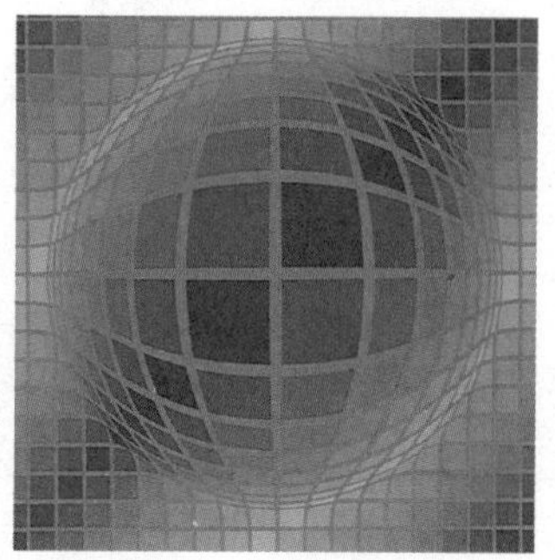

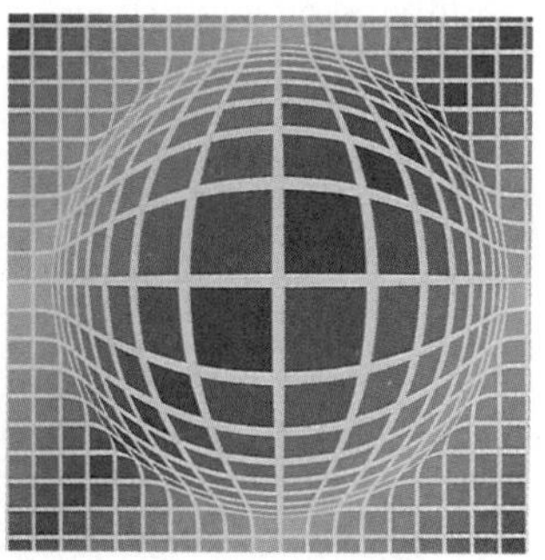

Tsiga I, II, III (1991), Victor Vasarely, courtesy of the artist.

Symmetry is a geometric characteristic of both nature and art. You may already know the two basic types of symmetry, reflectional symmetry and rotational symmetry. A design has **reflectional symmetry** if you can fold it along a **line of symmetry** so that all the points on one side of the line exactly coincide with (or match) all the points on the other side of the line.

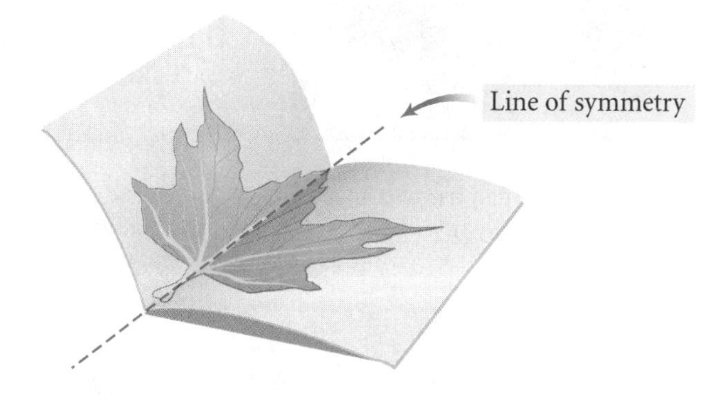

This leaf has reflectional symmetry.

You can place a mirror on the line of symmetry so that half the figure and its mirror image re-create the original figure. So, reflectional symmetry is also called *line symmetry* or *mirror symmetry.*

An object with reflectional symmetry looks balanced. And an object with just one line of symmetry, like the human body or a butterfly, has **bilateral symmetry.**

A butterfly has one line of reflectional symmetry.

A design has **rotational symmetry** if it looks the same after you turn it around a point by less than a full circle. The number of times that the design looks the same as you turn it through a complete 360° circle determines the type of rotational symmetry. The Apache basket has 3-fold rotational symmetry because it looks the same after you rotate it 120° (a third of a circle), 240° (two-thirds of a circle), and 360° (one full circle).

This Apache basket has 3-fold rotational symmetry.

A starfish has 5-fold symmetry. It looks the same after you rotate it 72°, 144°, 216°, 288°, or 360°.

The square fabric has 4-fold rotational symmetry and a starfish has 5-fold rotational symmetry. What type of rotational symmetry does a circular plate have?

Countries throughout the world use symmetry in their national flags. Notice that the Jamaican flag has rotational symmetry in addition to two lines of reflectional symmetry. You can rotate the flag 180° without changing its appearance. The origami boxes, however, have rotational symmetry, but not reflectional symmetry. (The Apache basket on page 3 *almost* has reflectional symmetry. Can you see why it doesn't?)

The Jamaican flag has two lines of reflectional symmetry.

If you ignore colors, the Japanese origami box on the left has 3-fold rotational symmetry. What type of symmetry does the other box have?

Consumer

Many products have eye-catching labels, logos, and designs. Have you ever paid more attention to a product because the geometric design of its logo was familiar or attractive to you?

EXERCISES

1. Name two objects from nature whose shapes are hexagonal. Name two living organisms whose shapes are pentagonal.

2. Describe some ways that artists use geometry.

3. Name some objects with only one line of symmetry. What is the name for this type of symmetry?

4. Which of these objects have reflectional symmetry (or approximate reflectional symmetry)?

A.

B.

C.

D.

E.

F.

5. Which of the objects in Exercise 4 have rotational symmetry (or approximate rotational symmetry)?

6. Which of these playing cards have rotational symmetry? Which ones have reflectional symmetry?

7. British artist Andy Goldsworthy (b 1956) uses materials from nature to create beautiful outdoor sculptures. The artful arrangement of sticks below might appear to have rotational symmetry, but instead it has one line of reflectional symmetry. Can you find the line of symmetry? ⓗ

If an exercise has an ⓗ at the end, you can find a hint to help you in Hints for Selected Exercises at the back of the book.

For the title of this outdoor sculpture by Andy Goldsworthy, see the hint to Exercise 7 in the Hints section.

Courtesy of the artist and Galerie Lelong, New York.

8. Shah Jahan, Mughal emperor of India from 1628 to 1658, had the beautiful Taj Mahal built in memory of his wife, Mumtaz Mahal. Its architect, Ustad Ahmad Lahori, designed it with perfect symmetry. Describe two lines of symmetry in this photo. How does the design of the building's grounds give this view of the Taj Mahal even more symmetry than the building itself has?

The Taj Mahal in Agra, India, was described by the poet Rabindranath Tagore as "rising above the banks of the river like a solitary tear suspended on the cheek of time."

9. Create a simple design that has two lines of reflectional symmetry. Does it have rotational symmetry? Next, try to create another design with two lines of reflectional symmetry, but without rotational symmetry. Any luck?

10. Bring to class an object from nature that shows geometry. Describe the geometry that you find in the object as well as any symmetry the object has.

11. Bring an object to school or wear something that displays a form of handmade or manufactured geometric art. Describe any symmetry the object has.

IMPROVING YOUR VISUAL THINKING SKILLS

Pickup Sticks

Pickup sticks is a good game for developing motor skills, but you can turn it into a challenging visual puzzle. In what order should you pick up the sticks so that you are always removing the top stick?

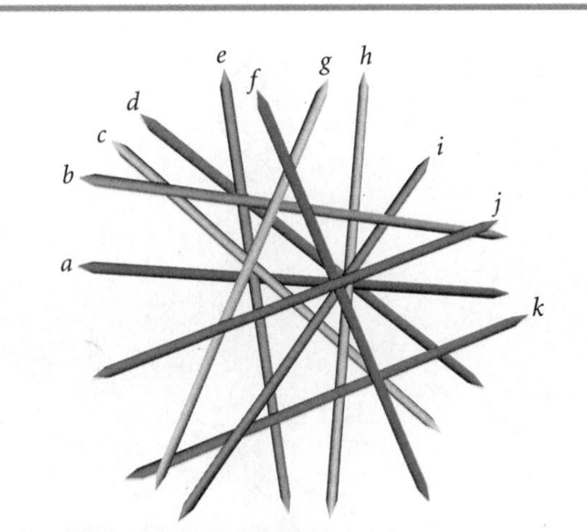

LESSON

0.2

Line Designs

We especially need imagination in science. It is not all mathematics, nor all logic, but it is somewhat beauty and poetry.

MARIA MITCHELL

The symmetry and patterns in geometric designs make them very appealing. You can make many designs using the basic tools of geometry—**compass** and **straightedge.**

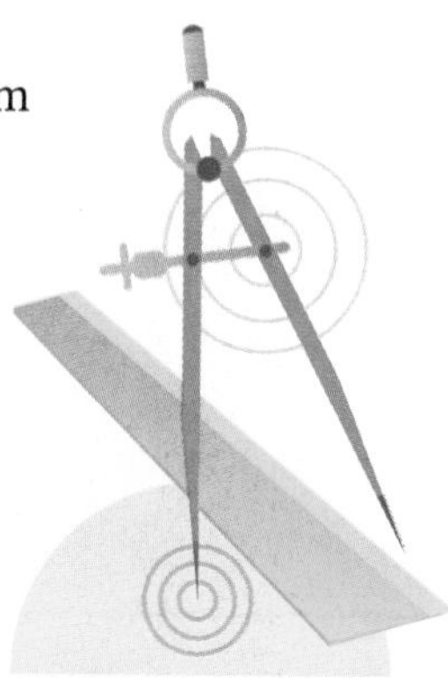

You'll use a straightedge to construct straight lines and a compass to construct circles and to mark off equal distances. A straightedge is like a ruler but it has no marks. You can use the edge of a ruler as a straightedge. The straightedge and the compass are the classical construction tools used by the ancient Greeks, who laid the foundations of the geometry that you are studying.

Japanese design is known for its simple, clean lines.

The complementary line designs on the arched ceiling and tile floor make this building lobby look grandiose.

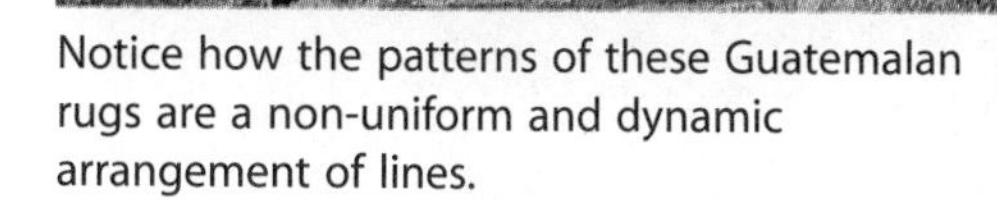

Notice how the patterns of these Guatemalan rugs are a non-uniform and dynamic arrangement of lines.

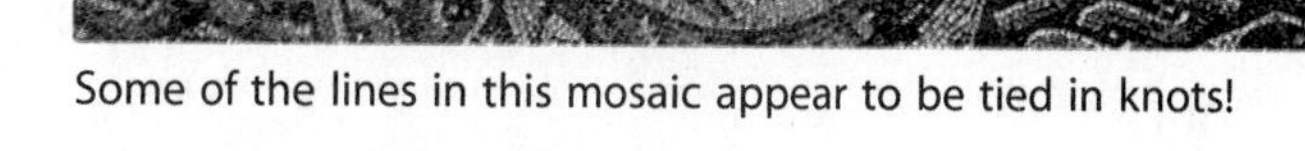

Some of the lines in this mosaic appear to be tied in knots!

You can create many types of designs using only straight lines. Here are two line designs and the steps for creating each one.

The Astrid **The 8-pointed Star**

The Astrid

Step 1 Step 2 Step 3 Step 4

The 8-pointed Star

Step 1 Step 2 Step 3 Step 4

EXERCISES

1. What are the classical construction tools of geometry?

2. Create a line design from this lesson. Color your design.

3. Each of these line designs uses straight lines only. Select one design and re-create it on a sheet of paper. ⓗ

4. Describe the symmetries of the three designs in Exercise 3. For the third design, does color matter?

5. Many quilt designers create beautiful geometric patterns with reflectional symmetry. One-fourth of a 4-by-4 quilt pattern and its reflection are shown at right. Copy the designs onto graph paper, and complete the 4-by-4 pattern so that it has two lines of reflectional symmetry. Color your quilt.

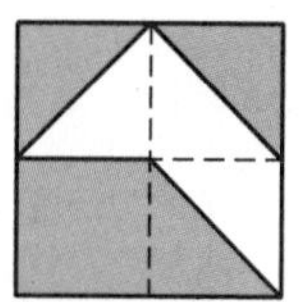

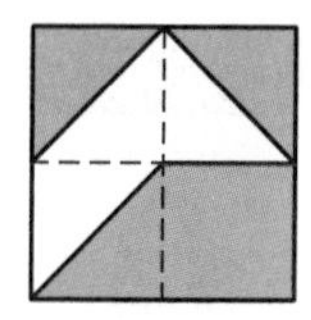

6. Geometric patterns seem to be in motion in a quilt design with rotational symmetry. Copy the 4-by-4 quilt piece shown in Exercise 5 onto graph paper, and complete the quilt pattern so that it has 4-fold rotational symmetry. Color your quilt.

7. Organic molecules have geometric shapes. How many different lines of reflectional symmetry does this benzene molecule have? How about rotational symmetry? Sketch your answers.

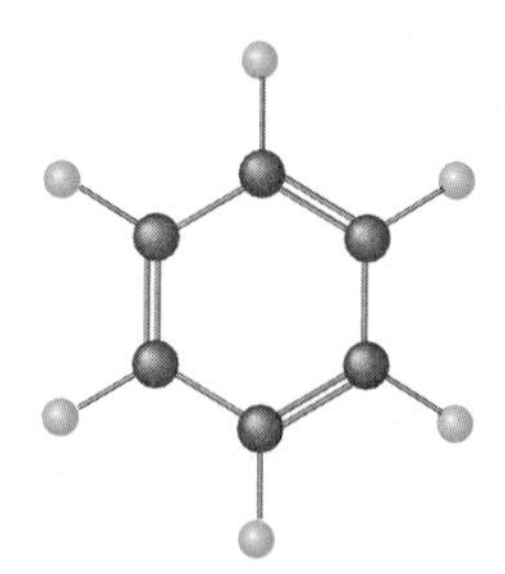

Benzene molecule

Architecture CONNECTION

Frank Lloyd Wright (1867–1959) is often called America's favorite architect. He built homes in 36 states—sometimes in unusual settings. *Fallingwater,* located in Pennsylvania, is a building designed by Wright that displays his obvious love of geometry. Can you describe the geometry you see? Find more information on Frank Lloyd Wright at www.keymath.com/DG .

IMPROVING YOUR ALGEBRA SKILLS

Pyramid Puzzle I

Place four different numbers in the bubbles at the vertices of the pyramid so that the two numbers at the ends of each edge or diagonal add up to the number on that edge.

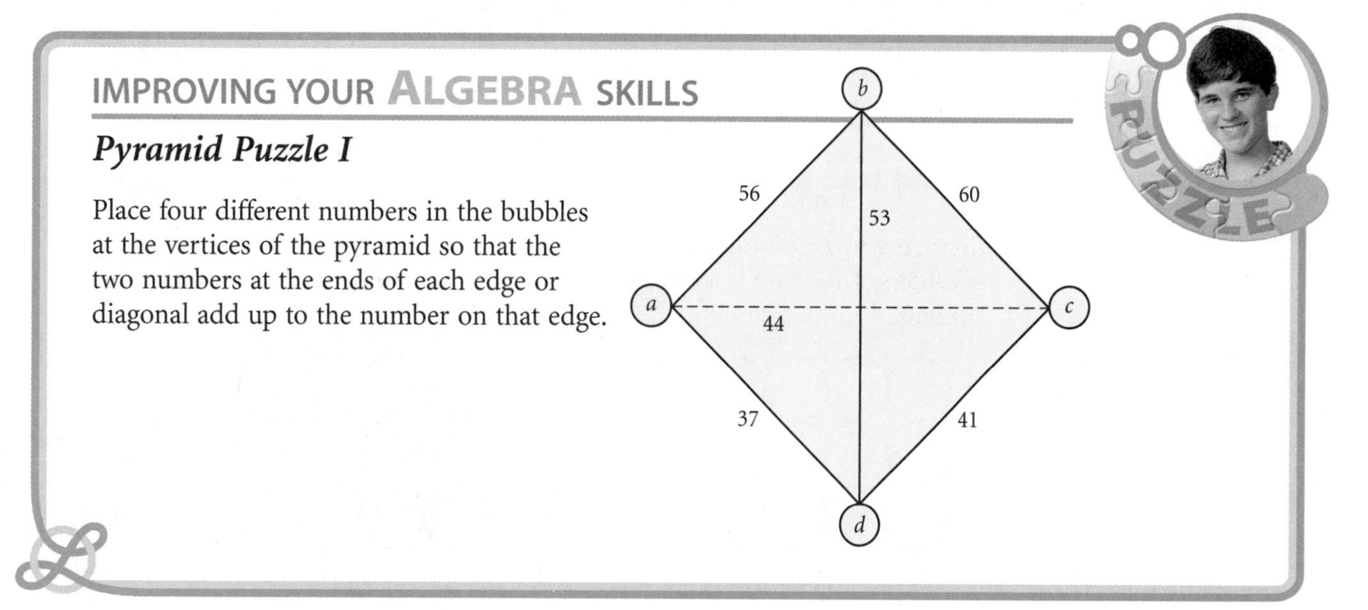

LESSON

0.3

Circle Designs

It's where we go, and what we do when we get there, that tells us who we are.

JOYCE CAROL OATES

People have always been fascinated by circles. Circles are used in the design of mosaics, baskets, and ceramics, as well as in the architectural design of buildings.

Chinese pottery

Palestinian cloth

Circular window

You can make circle designs with a compass as your primary tool. For example, here is a design you can make on a square dot grid.

Begin with a 7-by-9 square dot grid. Construct three rows of four circles.

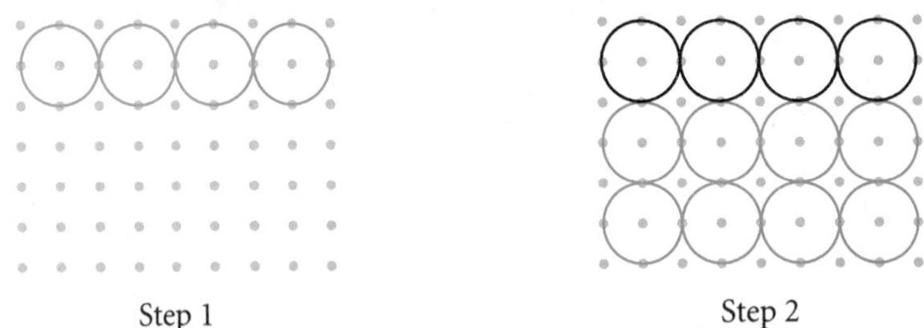

Step 1

Step 2

Construct two rows of three circles using the points between the first set of circles as centers. The result is a set of six circles overlapping the original 12 circles. Decorate your design.

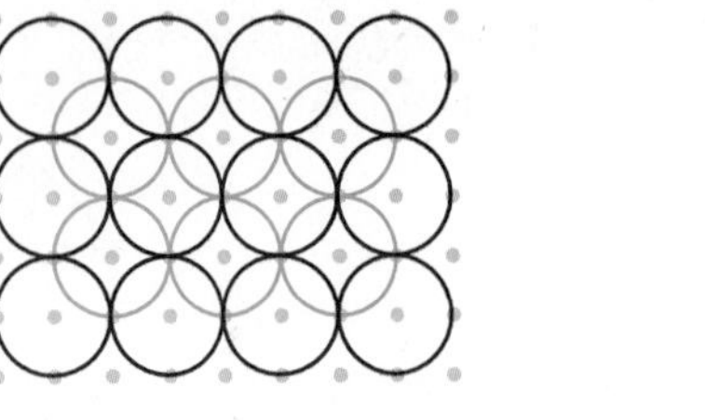

Step 3

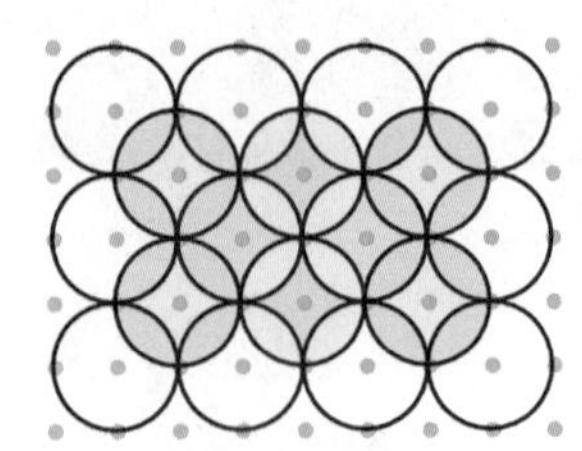

Step 4

Here is another design that you can make using only a compass. Start by constructing a circle, then select any point on it. Without changing your compass setting, swing an arc centered at the selected point. Swing an arc with each of the two new points as centers, and so on.

The Daisy

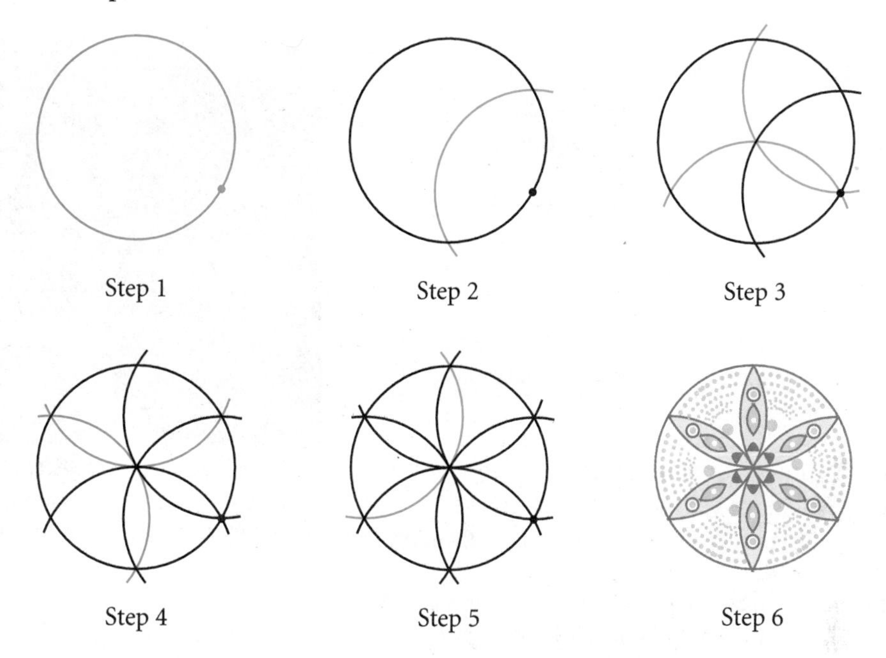

Instead of stopping at the perimeter of the first circle, you can continue to swing full circles. Then you get a "field of daisies," as shown above.

Notice the shape you get by connecting the six petal tips of the daisy. This is a **regular hexagon,** a 6-sided figure whose sides are the same length and whose angles are all the same size.

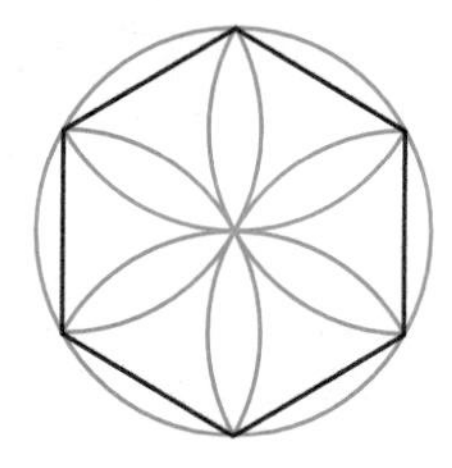

You can do many variations on a daisy design.

12-petal daisy

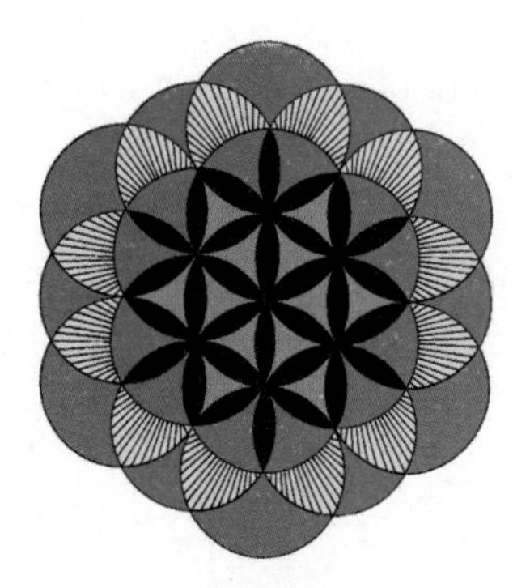

Field of daisies

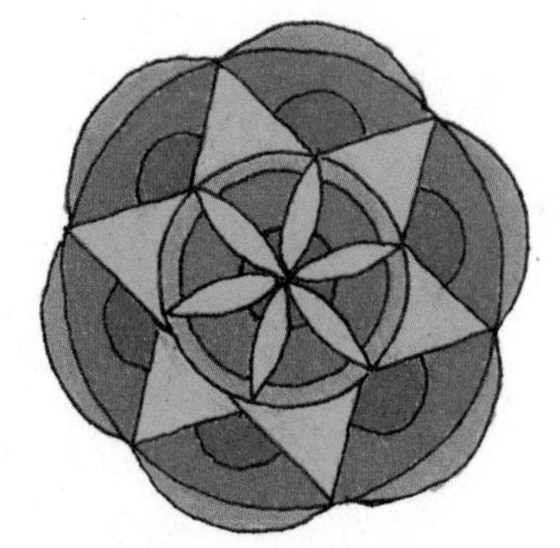

Combination line and circle design (Can you see how it was made?)
Schuyler Smith, geometry student

EXERCISES

You will need

Construction tools for Exercises **1–5**

For Exercises 1–5, use your construction tools.

1. Use square dot paper to create a 4-by-5 grid of 20 circles, and 12 circles overlapping them. Color or shade the design so that it has reflectional symmetry.

2. Use your compass to create a set of seven identical circles that touch but do not overlap. Draw a larger circle that encloses the seven circles. Color or shade your design so that it has rotational symmetry. ⓗ

This rose window at the National Cathedral in Washington, D.C. has a central design of seven circles enclosed in a larger circle.

3. Create a 6-petal daisy design and color or shade it so that it has rotational symmetry, but not reflectional symmetry.

4. Make a 12-petal daisy by drawing a second 6-petal daisy between the petals of the first 6-petal daisy. Color or shade the design so that it has reflectional symmetry, but not rotational symmetry.

5. Using a 1-inch setting for your compass, construct a central regular hexagon and six regular hexagons that each share one side with the original hexagon. Your hexagon design should look similar to, but larger than, the figure at right. This design is called a tessellation, or tiling, of regular hexagons.

IMPROVING YOUR ALGEGRA SKILLS

PUZZLE

Algebraic Magic Squares I

A magic square is an arrangement of numbers in a square grid. The numbers in every row, column, or diagonal add up to the same number. For example, in the magic square on the left, the sum of each row, column, and diagonal is 18.

Complete the 5-by-5 magic square on the right. Use only the numbers in this list: 6, 7, 9, 13, 17, 21, 23, 24, 27, and 28.

5	10	3
4	6	8
9	2	7

20			8	14
	19	25	26	
	12	18		30
29	10	11		
22			15	16

LESSON

0.4

Op Art

Everything is an illusion, including this notion.

STANISLAW J. LEC

Op art, or optical art, is a form of abstract art that uses lines or geometric patterns to create a special visual effect. The contrasting dark and light regions sometimes appear to be in motion or to represent a change in surface, direction, and dimension. Victor Vasarely was one artist who transformed grids so that spheres seem to bulge from them. Recall the series *Tsiga I, II,* and *III* that appears in Lesson 0.1. *Harlequin,* shown at right, is a rare Vasarely work that includes a human form. Still, you can see Vasarely's trademark sphere in the clown's bulging belly.

In *Hesitate,* by contemporary op artist Bridget Riley (b 1931), what effect do the changing dots produce?

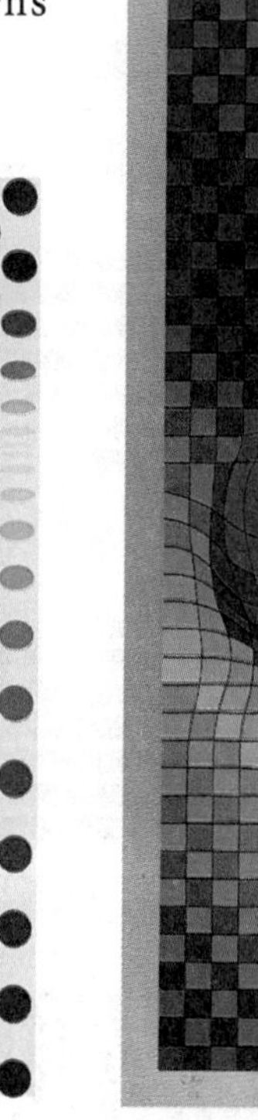

Harlequin, Victor Vasarely, courtesy of the artist.

In *Harlequin,* Victor Vasarely used curved lines and shading to create the form of a clown in motion.

Op art is fun and easy to create. To create one kind of op art design, first make a design in outline. Next, draw horizontal or vertical lines, gradually varying the space between the lines to create an illusion of hills and valleys. Finally, color in alternating spaces.

The Wavy Letter

Step 1

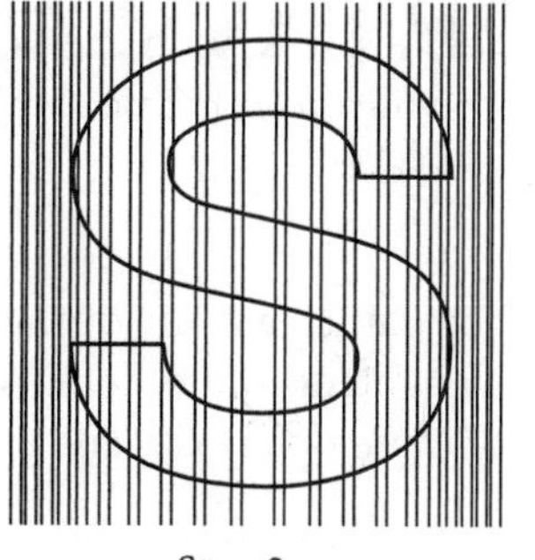

Step 2

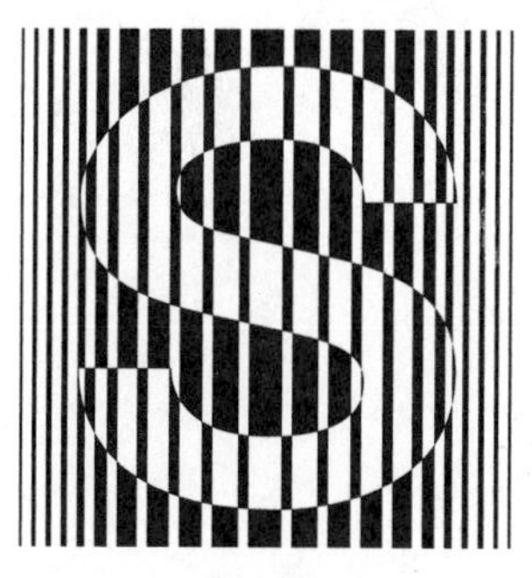

Step 3

To create the next design, first locate a point on each of the four sides of a square. Each point should be the same distance from a corner, as shown. Your compass is a good tool for measuring equal lengths. Connect these four points to create another square within the first. Repeat the process until the squares appear to converge on the center. Be careful that you don't fall in!

The Square Spiral

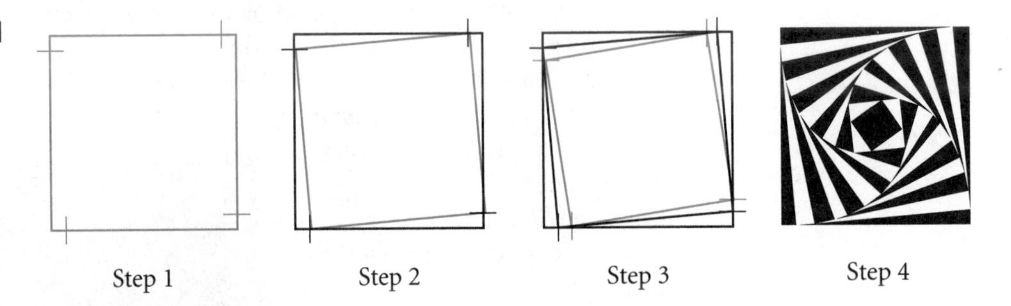

Here are some other examples of op art.

Square tunnel or top of pyramid?

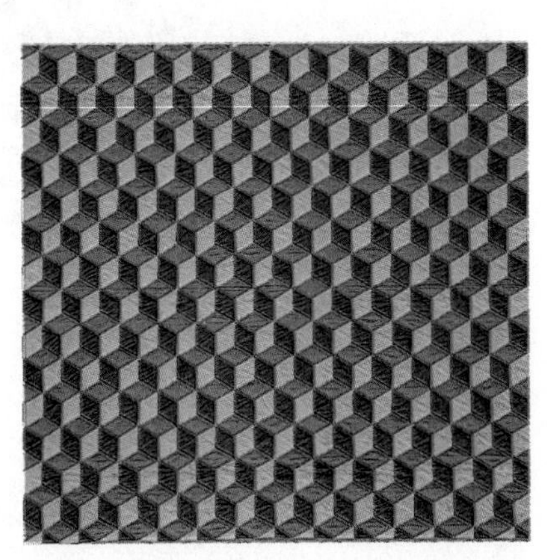

Amish quilt, tumbling block design

Japanese Op Art, Hajime Juchi, Dover Publications

Op art by Carmen Apodaca, geometry student

You can create any of the designs on this page using just a compass and straightedge (and doing some careful coloring). Can you figure out how each of these op art designs was created?

EXERCISES

1. What is the optical effect in each piece of art in this lesson?

2. Nature creates its own optical art. At first the black and white stripes of a zebra appear to work against it, standing out against the golden brown grasses of the African plain. However, the stripes do provide the zebras with very effective protection from predators. When and how?

3. Select one type of op art design from this lesson and create your own version of it.

4. Create an op art design that has reflectional symmetry, but not rotational symmetry.

5. Antoni Gaudí (1852–1926) designed the Bishop's Palace in Astorga, Spain. List as many geometric shapes as you can recognize on the palace (flat, two-dimensional shapes such as rectangles as well as solid, three-dimensional shapes such as cylinders). What type of symmetry do you see on the palace?

Bishop's Palace, Astorga, Spain

IMPROVING YOUR REASONING SKILLS

Bagels

In the original computer game of bagels, a player determines a three-digit number (no digit repeated) by making educated guesses. After each guess, the computer gives a clue about the guess. Here are the clues.

bagels: no digit is correct

pico: one digit is correct but in the wrong position

fermi: one digit is correct and in the correct position

In each of the games below, a number of guesses have been made, with the clue for each guess shown to its right. From the given set of guesses and clues, determine the three-digit number. If there is more than one solution, find them all.

Game 1:			**Game 2:**		
	1 2 3	*bagels*		9 0 8	*bagels*
	4 5 6	*pico*		1 3 4	*pico*
	7 8 9	*pico*		3 8 7	*pico fermi*
	0 7 5	*pico fermi*		2 5 6	*fermi*
	0 8 7	*pico*		2 3 7	*pico pico*
	? ? ?			? ? ?	

LESSON

0.5

Knot Designs

In the old days, a love-sick sailor might send his sweetheart a length of fishline loosely tied in a love knot. If the knot was returned pulled tight it meant the passion was strong. But if the knot was returned untied—ah, matey, time to ship out.

OLD SAILOR'S TALE

Knot designs are geometric designs that appear to weave or to interlace like a knot. Some of the earliest known designs are found in Celtic art from the northern regions of England and Scotland. In their carved stone designs, the artists imitated the rich geometric patterns of three-dimensional crafts such as weaving and basketry. The *Book of Kells* (8th and 9th centuries) is the most famous collection of Celtic knot designs.

Celtic knot design

Carved knot pattern from Nigeria

Today a very familiar knot design is the set of interconnected rings (shown at right) used as the logo for the Olympic Games.

Here are the steps for creating two examples of knot designs. Look them over before you begin the exercises.

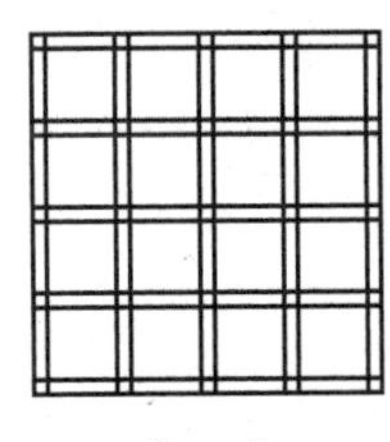

Step 1

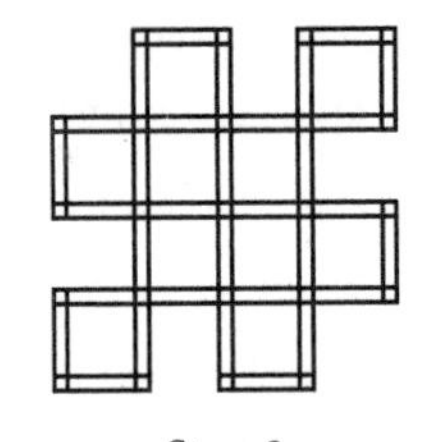

Step 2

Step 3

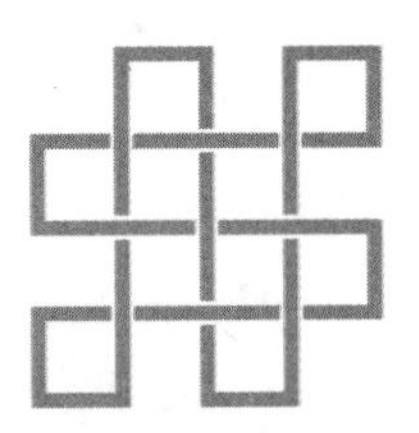

Step 4

You can use a similar approach to create a knot design with rings.

Step 1

Step 2

Step 3

Step 4

Here are some more examples of knot designs.

Knot design by Scott Shanks, geometry student

Tiger Tail, Diane Cassell, parent of geometry student

Medieval Russian knot design

Japanese knot design

The last woodcut made by M. C. Escher is a knot design called *Snakes.* The rings and the snakes interlace, and the design has 3-fold rotational symmetry.

Snakes, M. C. Escher, 1969/

EXERCISES

You will need

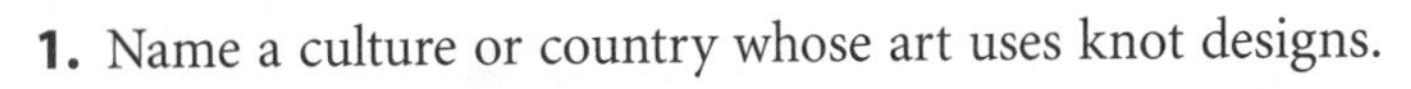

1. Name a culture or country whose art uses knot designs.

2. Create a knot design of your own, using only straight lines on graph paper.

3. Create a knot design of your own with rotational symmetry, using a compass or a circle template.

4. Sketch five rings linked together so that you could separate all five by cutting open one ring.

5. The coat of arms of the Borromeo family, who lived during the Italian Renaissance (ca. 15th century), showed a very interesting knot design known as the Borromean Rings. In it, three rings are linked together so that if any one ring is removed the remaining two rings are no longer connected. Got that? Good. Sketch the Borromean Rings.

6. The Chokwe storytellers of northeastern Angola are called *Akwa kuta sona* ("those who know how to draw"). When they sit down to draw and to tell their stories, they clear the ground and set up a grid of points in the sand with their fingertips, as shown below left. Then they begin to tell a story and, at the same time, trace a finger through the sand to create a *lusona* design with one smooth, continuous motion. Try your hand at creating *sona* (plural of *lusona*). Begin with the correct number of dots. Then, in one motion, re-create one of the *sona* below. The initial dot grid is shown for the rat.

7. In Greek mythology, the Gordian knot was such a complicated knot that no one could undo it. Oracles claimed that whoever could undo the knot would become the ruler of Gordium. When Alexander the Great (356–323 B.C.E.) came upon the knot, he simply cut it with his sword and claimed he had fulfilled the prophecy, so the throne was his. The expression "cutting the Gordian knot" is still used today. What do you think it means?

Science CONNECTION

Mathematician DeWitt Sumners at Florida State University and biophysicist Sylvia Spenger at the University of California, Berkeley, have discovered that when a virus attacks DNA, it creates a knot on the DNA.

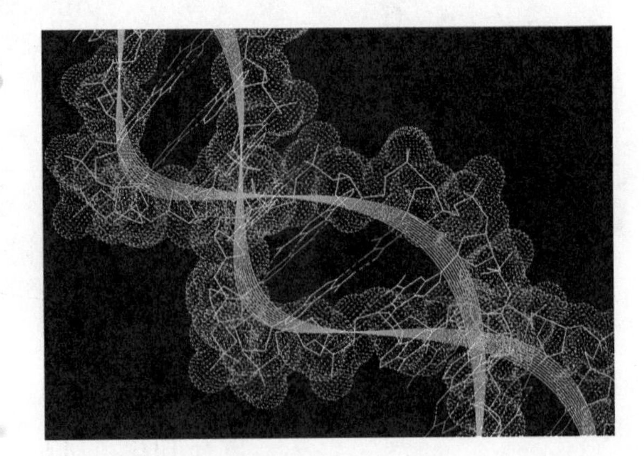

8. The square knot and granny knot are very similar but do very different things. Compare their symmetries. Use string to re-create the two knots and explain their differences.

Square knot

Granny knot

9. Cut a long strip of paper from a sheet of lined paper or graph paper. Tie the strip of paper snugly, but without wrinkles, into a simple knot. What shape does the knot create? Sketch your knot.

project

SYMBOLIC ART

Japanese artist Kunito Nagaoka (b 1940) uses geometry in his work. Nagaoka was born in Nagano, Japan, and was raised near the active volcano Asama. In Japan, he experienced earthquakes and typhoons as well as the human tragedies of Hiroshima and Nagasaki. In 1966, he moved to Berlin, Germany, a city rebuilt in concrete from the ruins of World War II. These experiences clearly influenced his work.

You can find other examples of symbolic art at www.keymath.com/DG .

- Look at the etching shown here, or another piece of symbolic art. Write a paragraph describing what you think might have happened in the scene or what you think it might represent. What types of geometric figures do you find?
- Use geometric shapes in your own sketch or painting to evoke a feeling or to tell a story. Write a one- or two-page story related to your art.

ISEKI/PY XVIII (1978), Kunito Nagaoka

LESSON

0.6

Islamic Tile Designs

Islamic art is rich in geometric forms. Early Islamic, or Muslim, artists became familiar with geometry through the works of Euclid, Pythagoras, and other mathematicians of antiquity, and they used geometric patterns extensively in their art and architecture.

Patience with small details makes perfect a large work, like the universe.

JALALUDDIN RUMI

An exterior wall of the Dome of the Rock (660–750 C.E.) mosque in Jerusalem

Alcove in the Hall of Ambassadors, the Alhambra, a Moorish palace in Granada, Spain

Islam forbids the representation of humans or animals in religious art. So, instead, the artists use intricate geometric patterns.

One of the most striking examples of Islamic architecture is the Alhambra Palace, in Granada, Spain. Built over 600 years ago by Moors and Spaniards, the Alhambra is filled from floor to ceiling with marvelous geometric patterns. The designs you see on this page are but a few of the hundreds of intricate geometric patterns found in the tile work and the inlaid wood ceilings of buildings like the Alhambra and the Dome of the Rock.

Carpets and hand-tooled bronze plates from the Islamic world also show geometric designs. The patterns often elaborate on basic grids of regular hexagons, equilateral triangles, or squares. These complex Islamic patterns were constructed with no more than a compass and a straightedge. Repeating patterns like these are called **tessellations.** You'll learn more about tessellations in Chapter 7.

The two examples below show how to create one tile in a square-based and a hexagon-based design. The hexagon-based pattern is also a knot design.

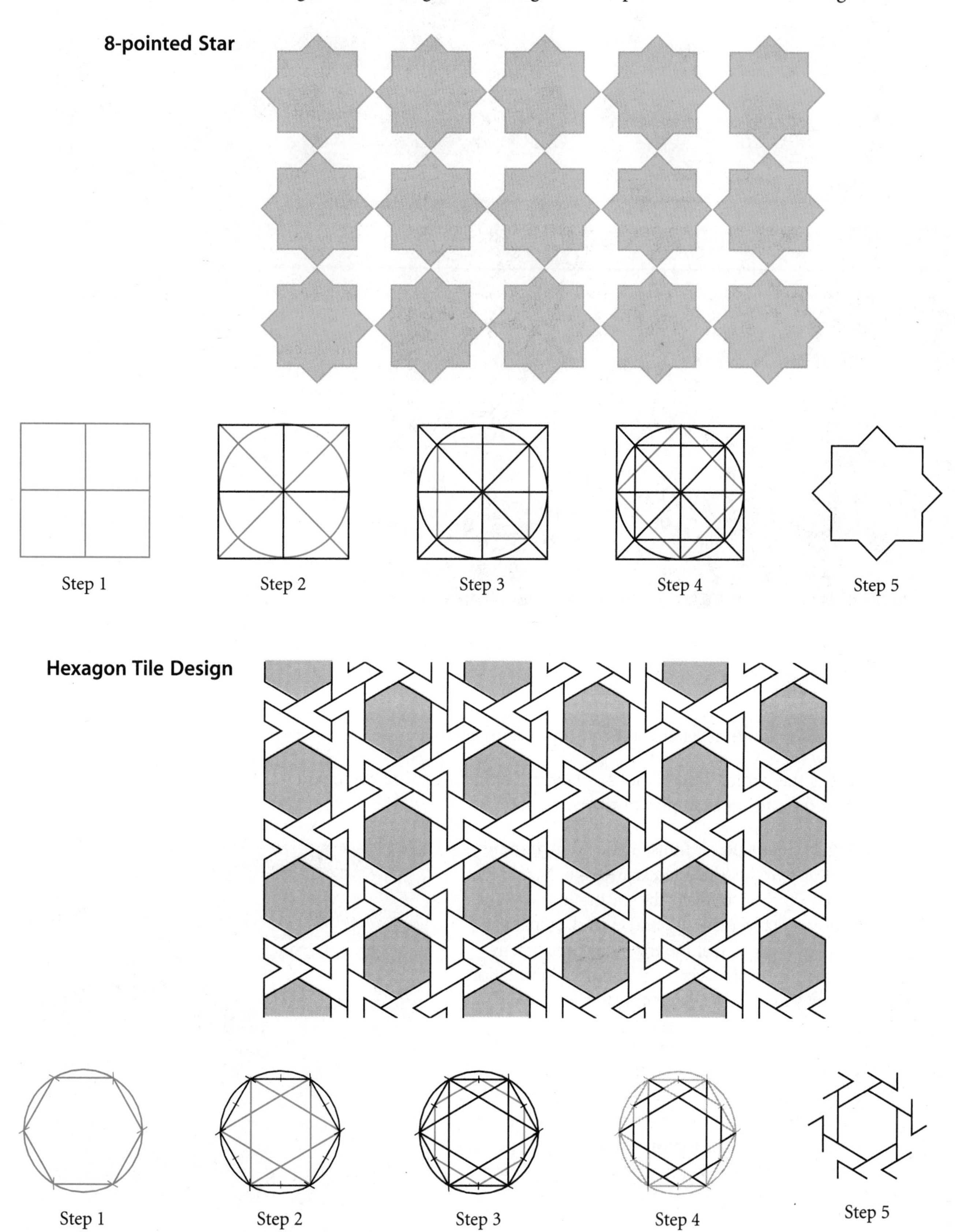

In Morocco, *zillij,* the art of using glazed tiles to form geometric patterns, is the most common practice for making mosaics. *Zillij* artists cut stars, octagons, and other shapes from clay tiles and place them upside down into the lines of their design. When the tiling is complete, artists pour concrete over the tiles to form a slab. When the concrete dries, they lift the whole mosaic, displaying the colors and connected shapes, and mount it against a fountain, palace, or other building.

EXERCISES

You will need

Construction tools for Exercises **5–7**

1. Name two countries where you can find Islamic architecture.

2. What is the name of the famous palace in Granada, Spain, where you can find beautiful examples of tile patterns?

3. Using tracing paper or transparency film, trace a few tiles from the 8-pointed star design. Notice that you can slide, or translate, the tracing in a straight line horizontally, vertically, and even diagonally to other positions so that the tracing will fit exactly onto the tiles again. What is the shortest translation distance you can find, in centimeters?

4. Notice that when you rotate your tracing from Exercise 3 about certain points in the tessellation, the tracing fits exactly onto the tiles again. Find two different points of rotation. (Put your pencil on the point and try rotating the tracing paper or transparency.) How many times in one rotation can you make the tiles match up again?

Architecture
CONNECTION

After studying buildings in other Muslim countries, the architect of the Petronas Twin Towers, Cesar Pelli (b 1926), decided that geometric tiling patterns would be key to the design. For the floor plan, his team used a very traditional tile design, the 8-pointed star—two intersecting squares. To add space and connect the design to the traditional "arabesques," the design team added arcs of circles between the eight points.

5. Currently the tallest buildings in the world are the Petronas Twin Towers in Kuala Lumpur, Malaysia. Notice that the floor plans of the towers have the shape of Islamic designs. Use your compass and straightedge to re-create the design of the base of the Petronas Twin Towers, shown at right. ⓗ

6. Use your protractor and ruler to draw a square tile. Use your compass, straightedge, and eraser to modify and decorate it. See the example in this lesson for ideas, but yours can be different. Be creative!

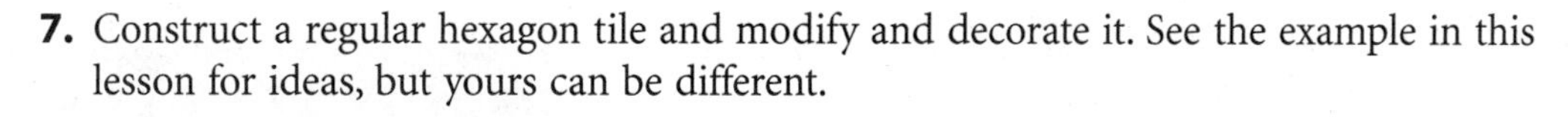

7. Construct a regular hexagon tile and modify and decorate it. See the example in this lesson for ideas, but yours can be different.

8. Create a tessellation with one of the designs you made in Exercises 6 and 7. Trace or photocopy several copies and paste them together in a tile pattern. (You can also create your tessellation using geometry software and print out a copy.) Add finishing touches to your tessellation by adding, erasing, or whiting out lines as desired. If you want, see if you can interweave a knot design within your tessellation. Color your tessellation.

PHOTO OR VIDEO SAFARI

In Lesson 0.1, you saw a few examples of geometry and symmetry in nature and art. Now go out with your group and document examples of geometry in nature and art. Use a camera or video camera to take pictures of as many examples of geometry in nature and art as you can. Look for many different types of symmetry, and try to photograph art and crafts from many different cultures. Consider visiting museums and art galleries, but make sure it's okay to take pictures when you visit. You might find examples in your home or in the homes of friends and neighbors.

If you take photographs, write captions for them that describe the geometry and the types of symmetries you find. If you record video, record your commentary on the soundtrack.

CHAPTER 0 REVIEW

In this chapter, you described the geometric shapes and symmetries you see in nature, in everyday objects, in art, and in architecture. You learned that geometry appears in many types of art—ancient and modern, from every culture—and you learned specific ways in which some cultures use geometry in their art. You also used a compass and straightedge to create your own works of geometric art.

The end of a chapter is a good time to review and organize your work. Each chapter in this book will end with a review lesson.

EXERCISES

You will need

Construction tools for Exercises **4**, **5**, and **10**

1. List three cultures that use geometry in their art.
2. What is the optical effect of the op art design at right?
3. Name the basic tools of geometry you used in this chapter and describe their uses.
4. With a compass, draw a 12-petal daisy.
5. *Construction* With a compass and straightedge, construct a regular hexagon.
6. List three things in nature that have geometric shapes. Name their shapes.
7. Draw an original knot design.
8. Which of the wheels below have reflectional symmetry? How many lines of symmetry does each have?

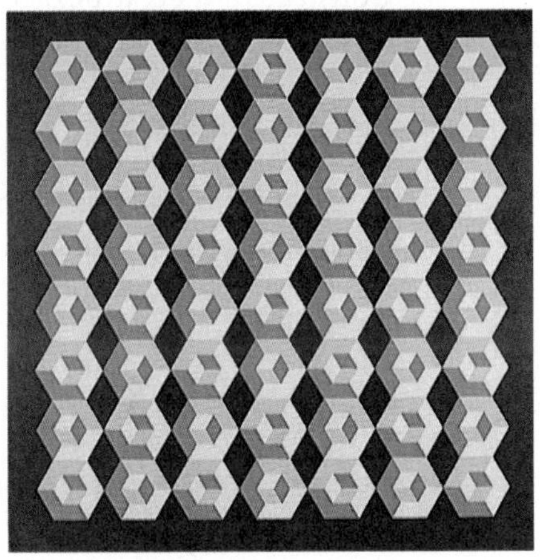

Hot Blocks (1966–67), Edna Andrade

Wheel A   Wheel B

Wheel C

Wheel D

9. Which of the wheels in Exercise 8 have *only* rotational symmetry? What kind of rotational symmetry does each have?

10. A *mandala* is a circular design arranged in rings that radiate from the center. (See the Cultural Connection below.) Use your compass and straightedge to create a mandala. Draw several circles using the same point as the center. Create a geometric design in the center circle, and decorate each ring with a symmetric geometric design. Color or decorate your mandala. Two examples are shown below.

The first mandala uses daisy designs. The second mandala is a combination knot and Islamic design by Scott Shanks, geometry student.

11. Create your own personal mandala. You might include your name, cultural symbols, photos of friends and relatives, and symbols that have personal meaning for you. Color it.

12. Create one mandala that uses techniques from Islamic art, is a knot design, and also has optical effects.

Cultural CONNECTION

The word *mandala* comes from Sanskrit, the classical language of India, and means "circle" or "center." Hindus use mandala designs for meditation. The Aztec calendar stone below left is an example of a mandala. In the center is the mask of the sun god. Notice the symbols are arranged symmetrically within each circle. The rose windows in many gothic cathedrals, like the one below right from the Chartres Cathedral in France, are also mandalas. Notice all the circles within circles, each one filled with a design or picture.

13. Did you know that "flags" is the most widely read topic of the *World Book Encyclopedia*? Research answers to these questions. More information about flags is available at www.keymath.com/DG .

a. Is the flag of Puerto Rico symmetric? Explain.

b. Does the flag of Kenya have rotational symmetry? Explain.

c. Name a country whose flag has both rotational and reflectional symmetry. Sketch the flag.

This section suggests how you might review, organize, and communicate to others what you've learned. Whether you follow these suggestions or directions from your teacher, or use study strategies of your own, be sure to reflect on all you've learned.

Assessing What You've Learned

KEEPING A PORTFOLIO

An essential part of learning is being able to show yourself and others how much you know and what you can do. Assessment isn't limited to tests and quizzes. Assessment isn't even limited to what your teacher sees or what makes up your grade. Every piece of art you make, and every project or exercise you complete, gives you a chance to demonstrate to somebody—yourself, at least—what you're capable of.

BEGIN A PORTFOLIO This chapter is primarily about art, so you might organize your work the way a professional artist does—in a portfolio. A portfolio is different from a notebook, both for an artist and for a geometry student. An artist's notebook might contain everything from scratch work to practice sketches to random ideas jotted down. A portfolio is reserved for an artist's most significant or *best* work. It's his or her portfolio that an artist presents to the world to demonstrate what he or she is capable of doing. The portfolio can also show how an artist's work has changed over time.

Review all the work you've done so far and choose one or more examples of your best art projects to include in your portfolio. Write a paragraph or two about each piece, addressing these questions:

- What is the piece an example of?
- Does this piece represent your best work? Why else did you choose it?
- What mathematics did you learn or apply in this piece?
- How would you improve the piece if you redid or revised it?

Portfolios are an ongoing and ever-changing display of your work and growth. As you finish each chapter, update your portfolio by adding new work.

CHAPTER

1 Introducing Geometry

Although I am absolutely without training or knowledge in the exact sciences, I often seem to have more in common with mathematicians than with my fellow artists.

M. C. ESCHER

Three Worlds, M. C. Escher, 1955

OBJECTIVES

In this chapter you will

- write your own definitions of many geometry terms and geometric figures
- start a notebook with a list of all the terms and their definitions
- develop very useful visual thinking skills

LESSON

1.1

Building Blocks of Geometry

Nature's Great Book is written in mathematical symbols.

GALILEO GALILEI

Three building blocks of geometry are points, lines, and planes. A **point** is the most basic building block of geometry. It has no size. It has only location. You represent a point with a dot, and you name it with a capital letter. The point shown below is called P.

P

Mathematical model of a point

A tiny seed is a physical model of a point.

A **line** is a straight, continuous arrangement of infinitely many points. It has infinite length but no thickness. It extends forever in two directions. You name a line by giving the letter names of any two points on the line and by placing the line symbol above the letters, for example, $\overleftrightarrow{AB}$ or $\overleftrightarrow{BA}$.

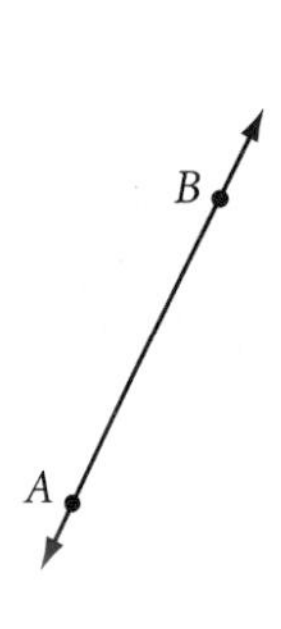

Mathematical model of a line

A piece of spaghetti is a physical model of a line. A line, however, is longer, straighter, and thinner than any piece of spaghetti ever made.

A **plane** has length and width but no thickness. It is like a flat surface that extends infinitely along its length and width. You represent a plane with a four-sided figure, like a tilted piece of paper, drawn in perspective. Of course, this actually illustrates only part of a plane. You name a plane with a script capital letter, such as $\mathcal{P}$.

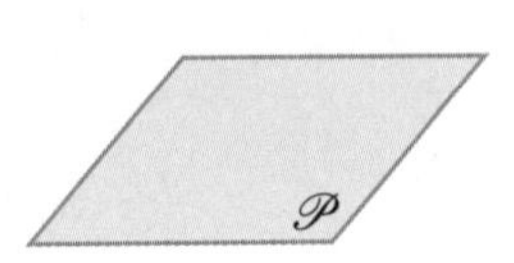

Mathematical model of a plane

A flat piece of rolled-out dough is a model of a plane, but a plane is broader, wider, and thinner than any piece of dough you could roll.

Investigation

Mathematical Models

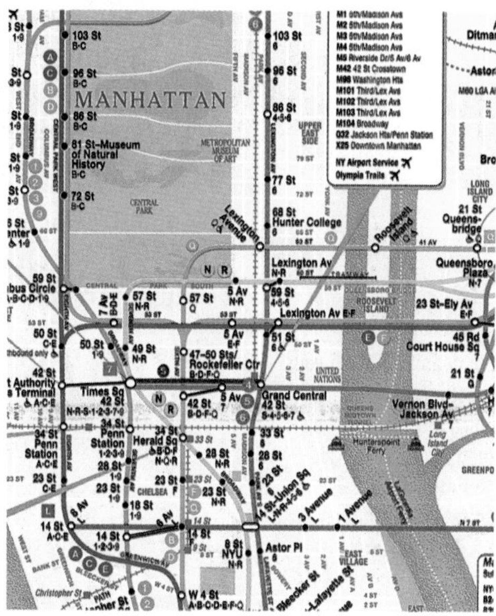

Step 1 | Identify examples of points, lines, and planes in these pictures.

Step 2 | Explain in your own words what point, line, and plane mean.

It can be difficult to explain what points, lines, and planes are. Yet, you probably recognized several models of each in the investigation. Early mathematicians tried to define these terms.

By permission of Johnny Hart and Creators Syndicate, Inc.

The ancient Greeks said, "A point is that which has no part. A line is breadthless length." The Mohist philosophers of ancient China said, "The line is divided into parts, and that part which has no remaining part is a point." Those definitions don't help much, do they?

A **definition** is a statement that clarifies or explains the meaning of a word or a phrase. However, it is impossible to define point, line, and plane without using words or phrases that themselves need definition. So these terms remain undefined. Yet, they are the basis for all of geometry.

Using the undefined terms *point, line,* and *plane,* you can define all other geometry terms and geometric figures. Many are defined in this book, and others will be defined by you and your classmates.

Keep a definition list in your notebook, and each time you encounter new geometry vocabulary, add the term to your list. Illustrate each definition with a simple sketch.

Here are your first definitions. Begin your list and draw sketches for all definitions.

Collinear means on the same line.

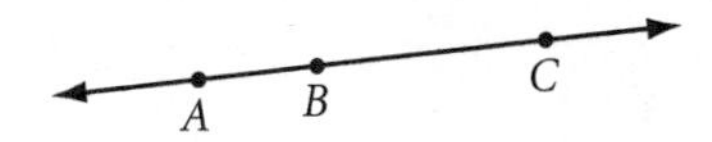

Points *A, B,* and *C* are collinear.

Coplanar means on the same plane.

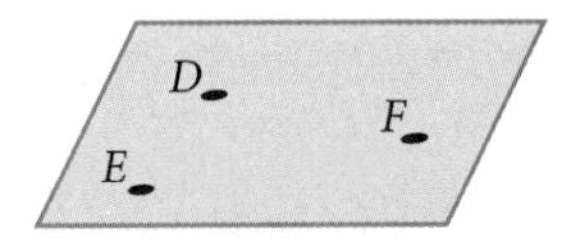

Points *D, E,* and *F* are coplanar.

Name three balls that are collinear. Name three balls that are coplanar but not collinear. Name four balls that are not coplanar.

A **line segment** consists of two points called the **endpoints** of the segment and all the points between them that are collinear with the two points.

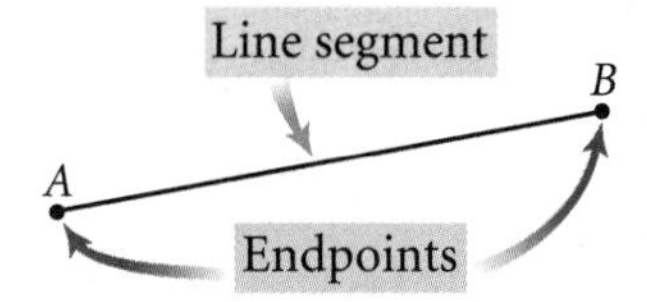

You can write line segment AB, using a segment symbol, as $\overline{AB}$ or $\overline{BA}$. There are two ways to write the length of a segment. You can write $AB = 2$ in., meaning the distance from A to B is 2 inches. You can also use an m for "measure" in front of the segment name, and write the distance as $m\overline{AB} = 2$ in. If no measurement units are used for the length of a segment, it is understood that the choice of units is not important, or is based on the length of the smallest square in the grid.

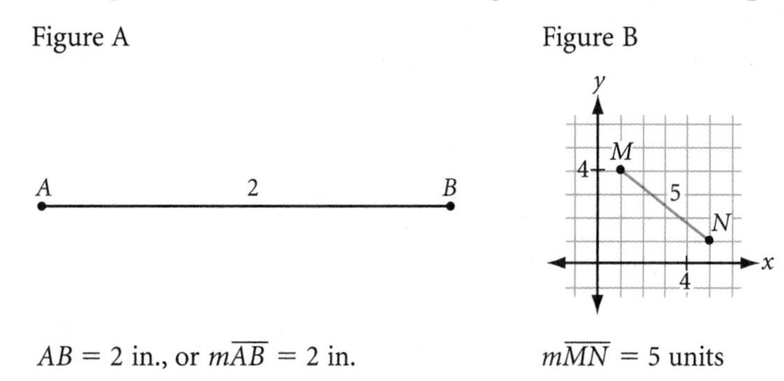

$AB = 2$ in., or $m\overline{AB} = 2$ in.

$m\overline{MN} = 5$ units

Two segments are **congruent segments** if and only if they have the same measure or length. The symbol for congruence is $\cong$, and you say it as "is congruent to." You use the equals symbol, $=$, between equal numbers and the congruence symbol, $\cong$, between congruent figures.

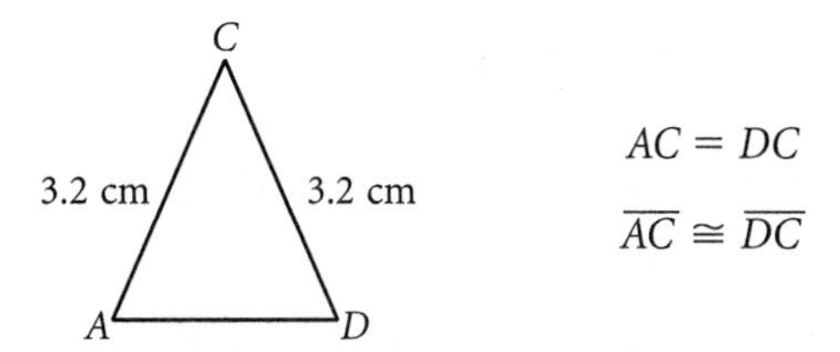

$AC = DC$

$\overline{AC} \cong \overline{DC}$

When drawing figures, you show congruent segments by making identical markings.

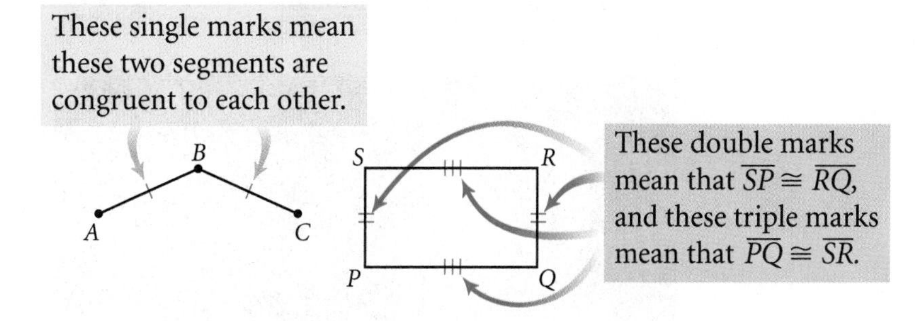

The **midpoint** of a segment is the point on the segment that is the same distance from both endpoints. The midpoint **bisects** the segment, or divides the segment into two congruent segments.

EXAMPLE

Study the diagrams below.

a. Name each midpoint and the segment it bisects.

b. Name all the congruent segments. Use the congruence symbol to write your answers.

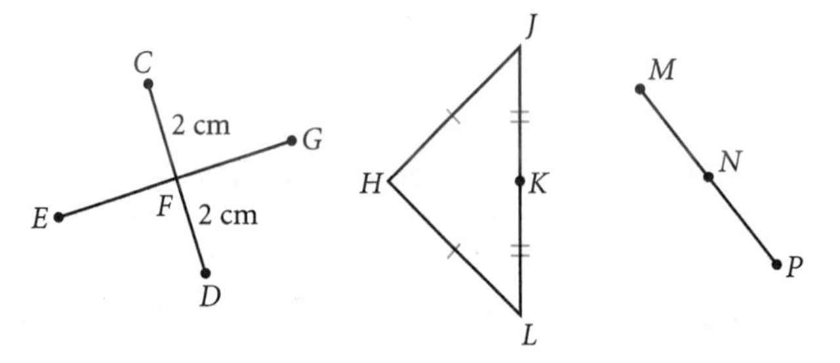

▸ Solution

Look carefully at the markings and apply the midpoint definition.

a. $CF \cong FD$, so F is the midpoint of $\overline{CD}$; $\overline{JK} \cong \overline{KL}$, so K is the midpoint of $\overline{JL}$.

b. $\overline{CF} \cong \overline{FD}$, $\overline{HJ} \cong \overline{HL}$, and $\overline{JK} \cong \overline{KL}$.

Even though $\overline{EF}$ and $\overline{FG}$ appear to have the same length, you cannot assume they are congruent without the markings. The same is true for $\overline{MN}$ and $\overline{NP}$.

Ray AB is the part of $\overleftrightarrow{AB}$ that contains point A and all the points on $\overleftrightarrow{AB}$ that are on the same side of point A as point B. Imagine cutting off all the points to the left of point A.

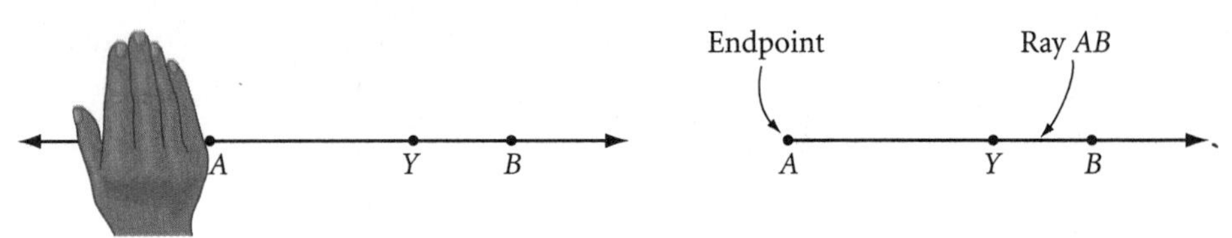

In the figure above, $\overrightarrow{AY}$ and $\overrightarrow{AB}$ are two ways to name the same ray. Note that $\overrightarrow{AB}$ is not the same as $\overrightarrow{BA}$!

A ray begins at a point and extends infinitely in one direction. You need two letters to name a ray. The first letter is the endpoint of the ray, and the second letter is any other point that the ray passes through.

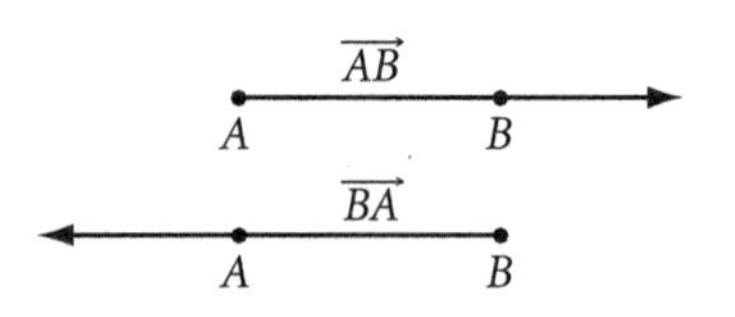

Physical model of a ray: beams of light

EXERCISES

1. Identify the models in the photos below for point, segment, plane, collinear points, and coplanar points.

For Exercises 2–4, name each line in two different ways.

2. P T

3. ⓗ

4. y 5 M A S x 8

For Exercises 5–7, draw two points and label them. Then use a ruler to draw each line. Don't forget to use arrowheads to show that it extends indefinitely.

5. $\overleftrightarrow{AB}$

6. $\overleftrightarrow{KL}$

7. $\overleftrightarrow{DE}$ with $D(-3, 0)$ and $E(0, -3)$

For Exercises 8–10, name each line segment.

8. A C

9.

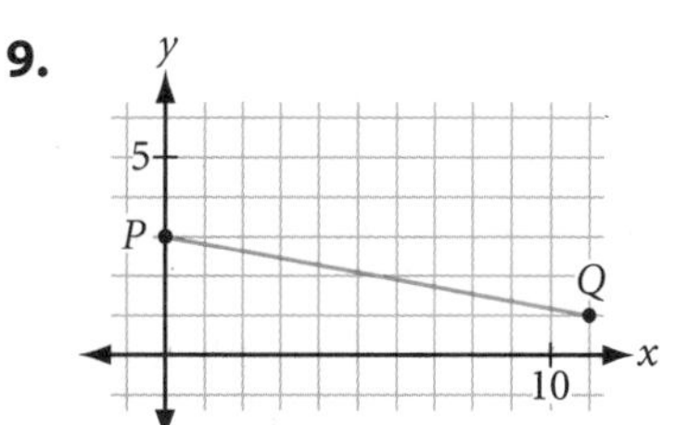

10.

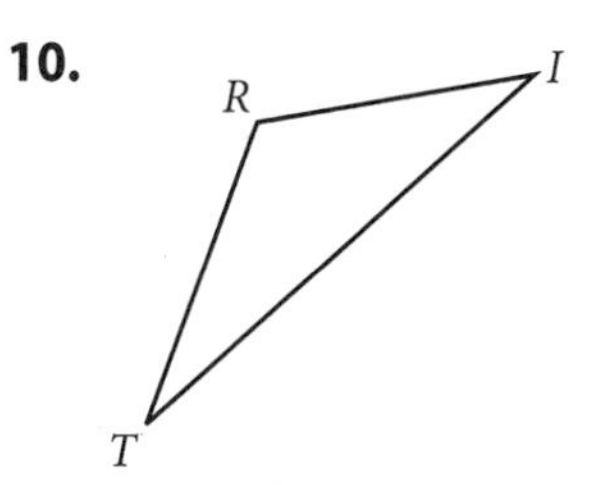

For Exercises 11 and 12, draw and label each line segment.

11. $\overline{AB}$

12. $\overline{RS}$ with $R(0, 3)$ and $S(-2, 11)$

For Exercises 13 and 14, use your ruler to find the length of each line segment to the nearest tenth of a centimeter. Write your answer in the form $m\overline{AB} = \underline{\ ?\ }$.

13. A B

14. C D

For Exercises 15–17, use your ruler to draw each segment as accurately as you can. Label each segment.

15. $AB = 4.5$ cm

16. $CD = 3$ in.

17. $EF = 24.8$ cm

18. Name each midpoint and the segment it bisects.

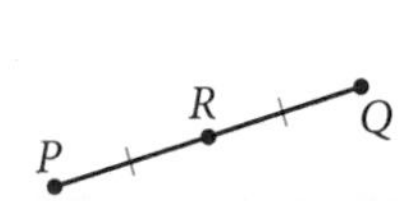

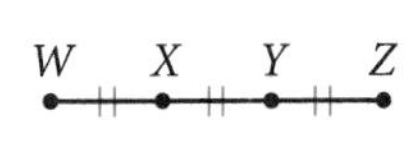

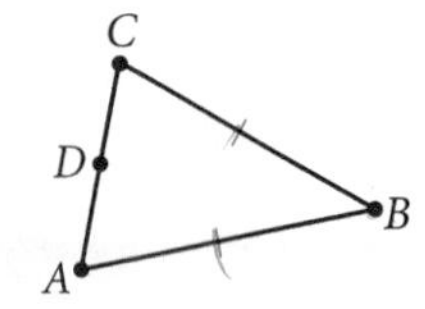

19. Draw two segments that have the same midpoint. Mark your drawing to show congruent segments.

20. Draw and mark a figure in which M is the midpoint of $\overline{ST}$, $SP = PT$, and T is the midpoint of $\overline{PQ}$.

For Exercises 21–23, name the ray in two different ways.

21. ⓗ A B C

22. M N P

23. Z Y X

For Exercises 24–26, draw and label each ray.

24. $\overrightarrow{AB}$

25. $\overrightarrow{YX}$

26. $\overrightarrow{MN}$

27. Draw a plane containing four coplanar points A, B, C, and D, with exactly three collinear points A, B, and D.

28. Given two points A and B, there is only one segment that you can name: $\overline{AB}$. With three collinear points A, B, and C, there are three different segments that you can name: $\overline{AB}$, $\overline{AC}$, and $\overline{BC}$. With five collinear points A, B, C, D, and E, how many different segments can you name?

For Exercises 29–31, draw axes onto graph paper and locate point $A(4, 0)$ as shown.

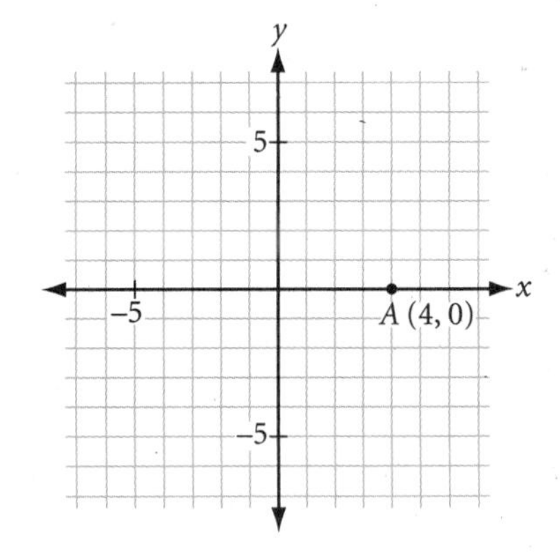

29. Draw $\overline{AB}$, where point B has coordinates $(2, -6)$.

30. Draw $\overrightarrow{OM}$ with endpoint $(0, 0)$ that goes through point $M(2, 2)$.

31. Draw $\overleftrightarrow{CD}$ through points $C(-2, 1)$ and $D(-2, -3)$.

Career CONNECTION

Woodworkers use a tool called a plane to shave a rough wooden surface to create a perfectly smooth planar surface. The smooth board can then be made into a tabletop, a door, or a cabinet.

Woodworking is a very precise process. Producing high-quality pieces requires an understanding of lines, planes, and angles as well as careful measurements.

32. If the signs of the coordinates of collinear points $P(-6, -2)$, $Q(-5, 2)$, and $R(-4, 6)$ are reversed, are the three new points still collinear? Draw a picture and explain why.

33. Draw a segment with midpoint $N(-3, 2)$. Label it $\overline{PQ}$.

34. Copy triangle *TRY* shown at right. Use your ruler to find the midpoint *A* of side $\overline{TR}$ and the midpoint *G* of side $\overline{TY}$. Draw $\overline{AG}$.

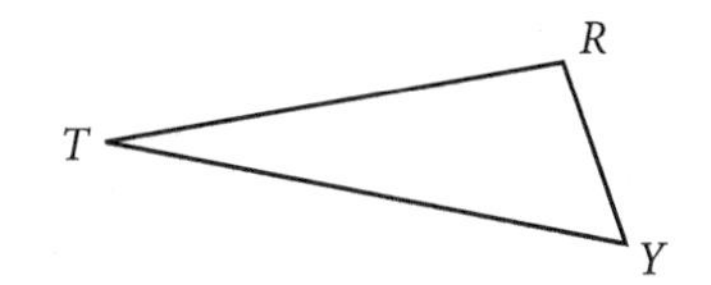

project

SPIRAL DESIGNS

The circle design shown below is used in a variety of cultures to create mosaic decorations. The spiral design may have been inspired by patterns in nature. Notice that the seeds on the sunflower also spiral out from the center.

Here are the steps to make the spirals.

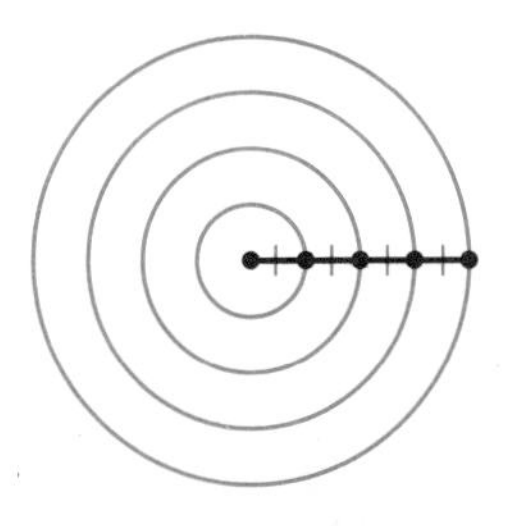

Step 1

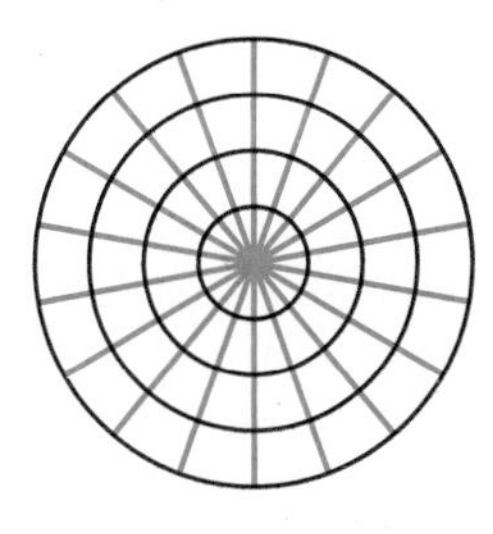

Step 2

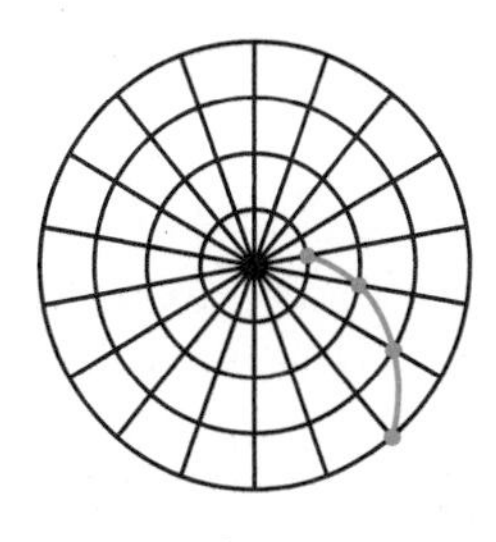

Step 3

Step 4

The more circles and radii you draw, the more detailed your design will be. Create and decorate your own spiral design.

Midpoint

A midpoint is the point on a line segment that is the same distance from both endpoints.

You can think of a midpoint as being halfway between two locations. You know how to mark a midpoint. But when the position and location matter, such as in navigation and geography, you can use a coordinate grid and some algebra to find the exact location of the midpoint. You can calculate the coordinates of the midpoint of a segment on a coordinate grid using a formula.

Coordinate Midpoint Property

If (x_1, y_1) and (x_2, y_2) are the coordinates of the endpoints of a segment, then the coordinates of the midpoint are

$$\left(\frac{x_1 + x_2}{2}, \frac{y_1 + y_2}{2}\right)$$

History CONNECTION

Surveyors and mapmakers of ancient Egypt, China, Greece, and Rome used various coordinate systems to locate points. Egyptians made extensive use of square grids and used the first known rectangular coordinates at Saqqara around 2650 B.C.E. By the seventeenth century, the age of European exploration, the need for accurate maps and the development of easy-to-use algebraic symbols gave rise to modern coordinate geometry. Notice the lines of latitude and longitude in this seventeenth-century map.

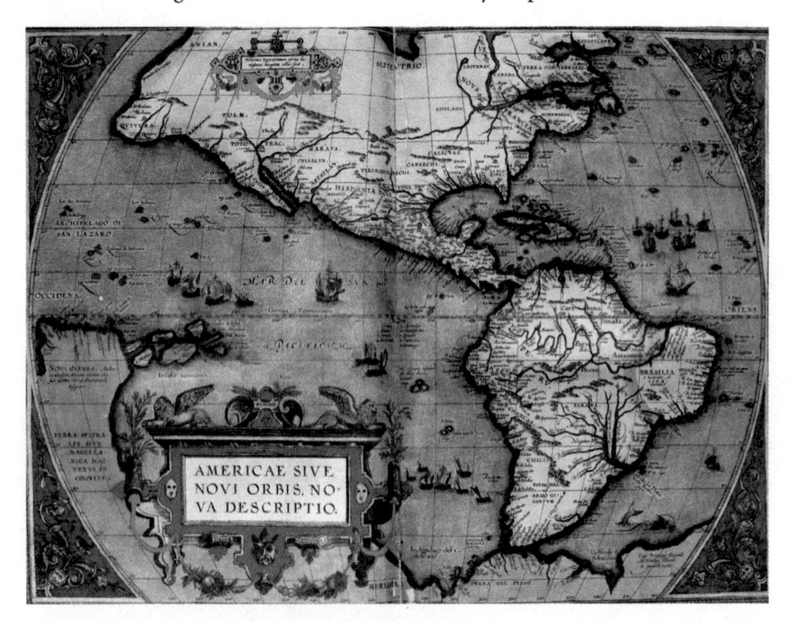

EXAMPLE Segment AB has endpoints $(-8, 5)$ and $(3, -6)$. Find the coordinates of the midpoint of $\overline{AB}$.

▶ Solution The midpoint is not on a grid intersection point, so we can use the coordinate midpoint property.

$$x = \frac{x_1 + x_2}{2} = \frac{(-8 + 3)}{2} = -2.5$$

$$y = \frac{y_1 + y_2}{2} = \frac{(5 + -6)}{2} = -0.5$$

The midpoint of $\overline{AB}$ is $(-2.5, -0.5)$.

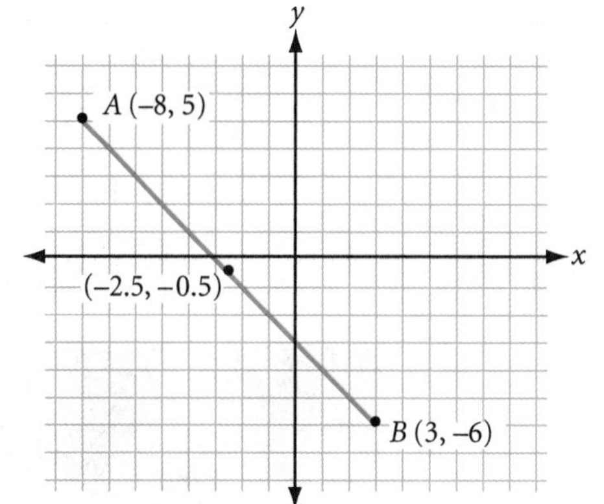

EXERCISES

For Exercises 1–3, find the coordinates of the midpoint of the segment with each pair of endpoints.

1. $(12, -7)$ and $(-6, 15)$ **2.** $(-17, -8)$ and $(-1, 11)$ **3.** $(14, -7)$ and $(-3, 18)$

4. One endpoint of a segment is $(12, -8)$. The midpoint is $(3, 18)$. Find the coordinates of the other endpoint.

5. A classmate tells you, "Finding the coordinates of a midpoint is easy. You just find the averages." Is there any truth to it? Explain what you think your classmate means.

6. Find the two points on $\overline{AB}$ that divide the segment into three congruent parts. Point A has coordinates $(0, 0)$ and point B has coordinates $(9, 6)$. Explain your method.

7. Describe a way to find points that divide a segment into fourths.

8. In each figure below, imagine drawing the diagonals $\overline{AC}$ and $\overline{BD}$.

a. Find the midpoint of $\overline{AC}$ and the midpoint of $\overline{BD}$ in each figure.

b. What do you notice about the midpoints?

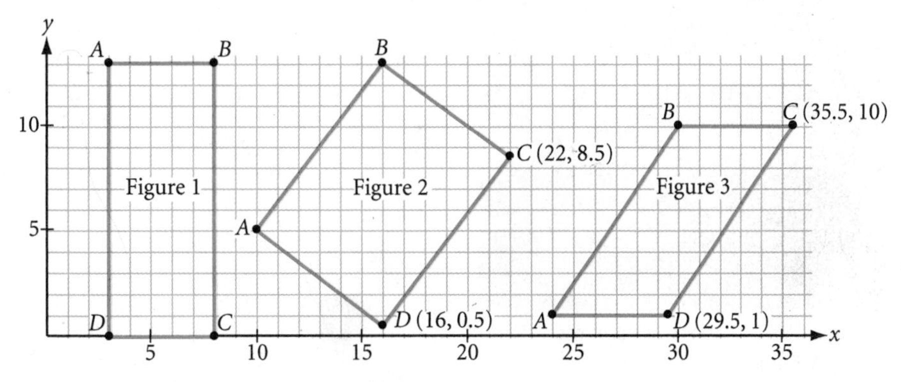

LESSON 1.2

Poolroom Math

Inspiration is needed in geometry, just as much as in poetry.

ALEKSANDR PUSHKIN

People use angles every day. Plumbers measure the angle between connecting pipes to make a good fitting. Woodworkers adjust their saw blades to cut wood at just the correct angle. Air traffic controllers use angles to direct planes. And good pool players must know their angles to plan their shots.

Is the angle between the two hands of the wristwatch smaller than the angle between the hands of the large clock?

"Little Benji," the wristwatch

Big Ben at the Houses of Parliament in London, England

You can use the terms that you defined in Lesson 1.1 to write a precise definition of angle. An **angle** is formed by two rays that share a common endpoint, provided that the two rays are noncollinear. In other words, the rays cannot lie on the same line. The common endpoint of the two rays is the **vertex** of the angle. The two rays are the **sides** of the angle.

You can name the angle in the figure below angle *TAP* or angle *PAT*, or use the angle symbol and write $\angle TAP$ or $\angle PAT$. Notice that the vertex must be the middle letter, and the first and last letters each name a point on a different ray. Since there are no other angles with vertex *A*, you can also simply call this $\angle A$.

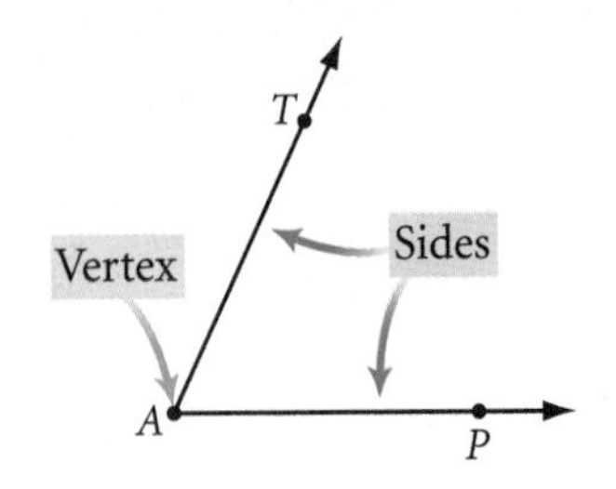

Career CONNECTION

In sports medicine, specialists may examine the healing rate of an injured joint by its angle of recovery. For example, a physician may assess how much physical therapy a patient needs by measuring the degree to which a patient can bend his or her ankle from the floor.

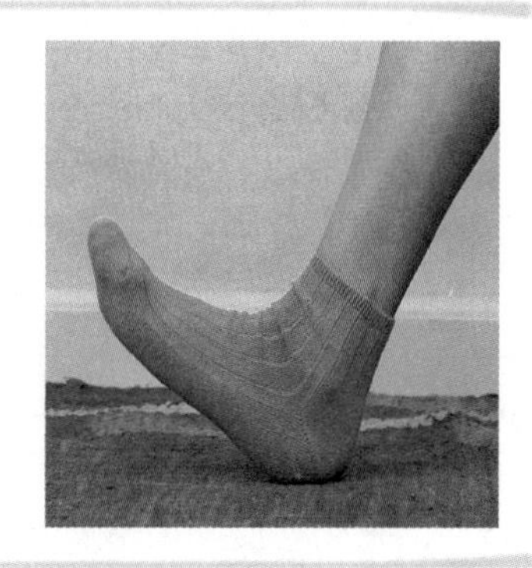

EXAMPLE A Name all the angles in these drawings.

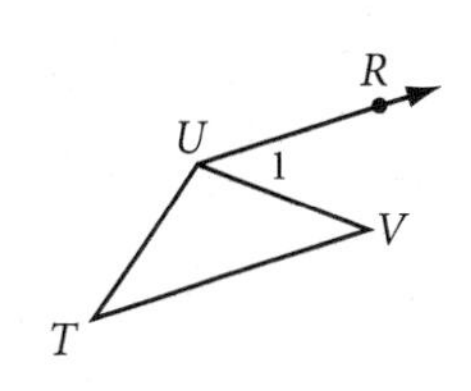

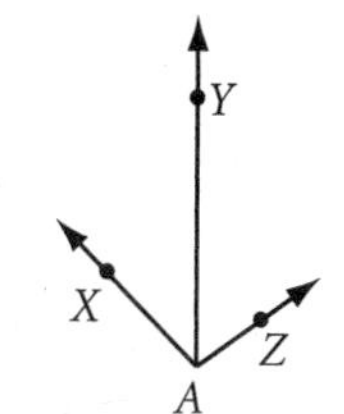

▶ Solution The angles are $\angle T$, $\angle V$, $\angle TUV$, $\angle 1$, $\angle TUR$, $\angle XAY$, $\angle YAZ$, and $\angle XAZ$. (Did you get them all?) Notice that $\angle 1$ is a shorter way to name $\angle RUV$.

Which angles in Example A seem big to you? Which seem small? The **measure of an angle** is the smallest amount of rotation about the vertex from one ray to the other, measured in **degrees.** According to this definition, the measure of an angle can be any value between 0° and 180°.

The smallest amount of rotation from $\overrightarrow{PG}$ to $\overrightarrow{PS}$ is 136°.

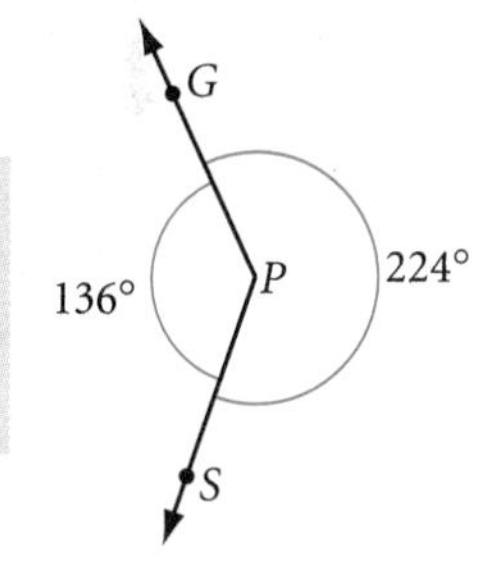

The geometry tool you use to measure an angle is a **protractor.** Here's how you use it.

Step 1: Place the center mark of the protractor on the vertex.

Step 2: Line up the 0-mark with one side of the angle.

Step 3: Read the measure on the protractor scale.

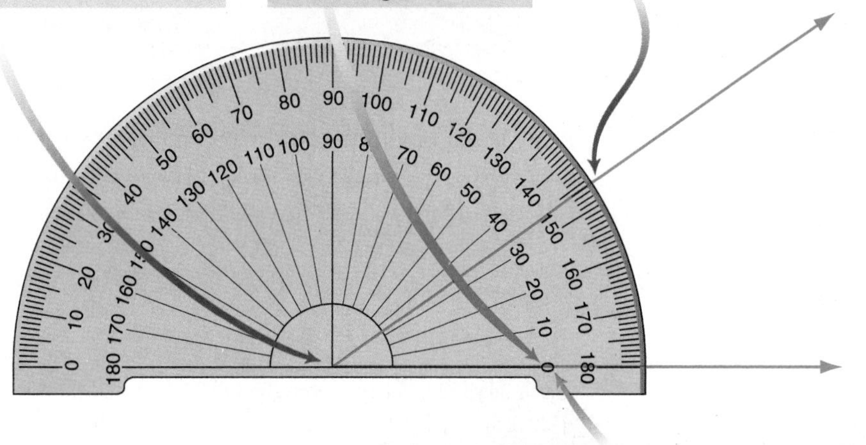

Step 4: Be sure you read the scale that has the 0-mark you are using! The angle in the diagram measures 34° and not 146°.

A television antenna is a physical model of an angle. Note that changing the length of the antenna doesn't change the angle.

To show the measure of an angle, use an *m* before the angle symbol. For example, $m\angle ZAP = 34°$ means the measure of $\angle ZAP$ is 34 degrees.

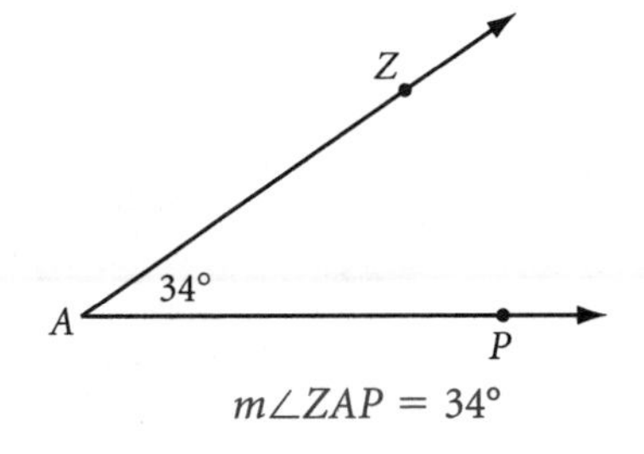

EXAMPLE B Use your protractor to measure these angles as accurately as you can. Which ones measure more than 90°?

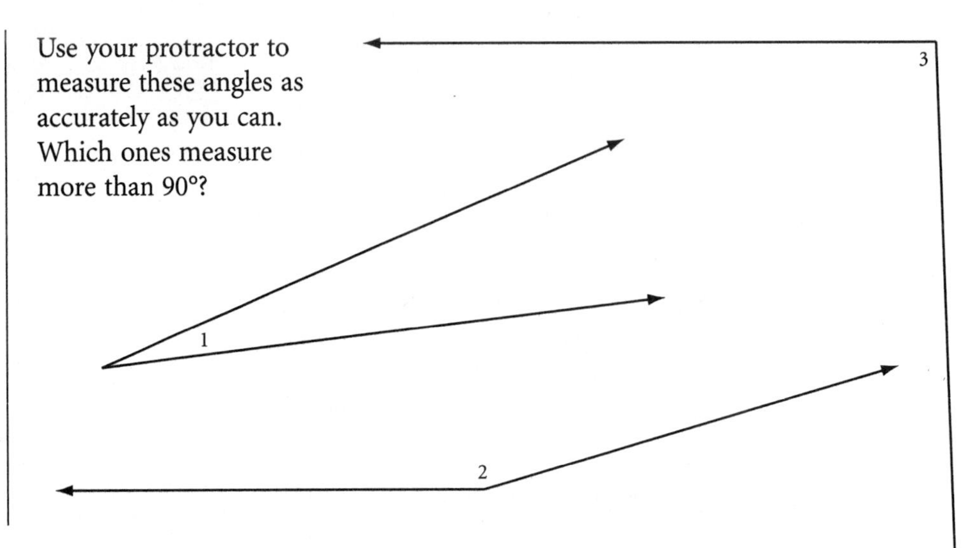

▶ Solution Measuring to the nearest degree, you should get these approximate answers. (The symbol ≈ means "is approximately equal to.")

$m\angle 1 \approx 16°$ $\quad$ $m\angle 3 \approx 92°$

$m\angle 2 \approx 164°$ $\quad$ $\angle 2$ and $\angle 3$ measure more than 90°.

Two angles are **congruent angles** if and only if they have the same measure. You use identical markings to show that two angles in a figure are congruent.

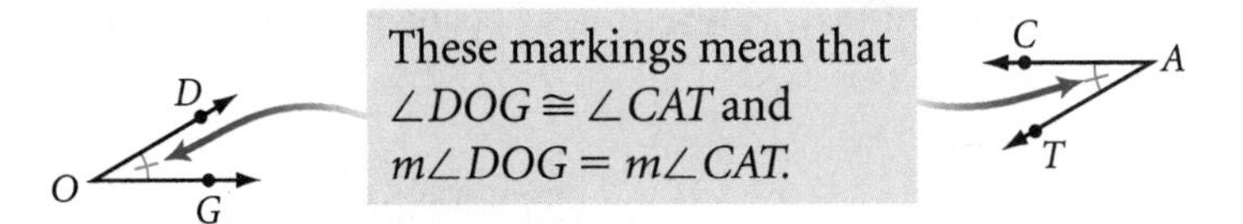

A ray is the **angle bisector** if it contains the vertex and divides the angle into two congruent angles. In the figure at right, $\overrightarrow{CD}$ bisects $\angle ACB$ so that $\angle ACD \cong \angle BCD$.

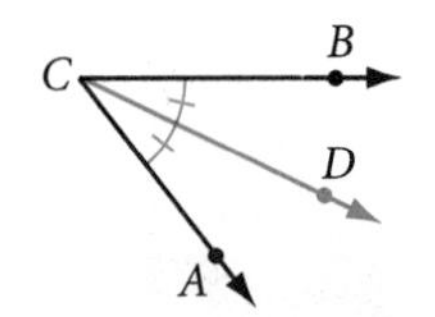

Science CONNECTION

Earth takes 365.25 days to travel a full 360° around the Sun. That means that each day, Earth travels a little less than 1° in its orbit around the Sun. Meanwhile, Earth also completes one full rotation each day, making the Sun appear to rise and set. By how many degrees does the Sun's position in the sky change every hour?

EXAMPLE C Look for angle bisectors and congruent angles in the figures below.

a. Name each angle bisector and the angle it bisects.

b. Name all the congruent angles in the figure. Use the congruence symbol and name the angles so there is no confusion about which angle you mean.

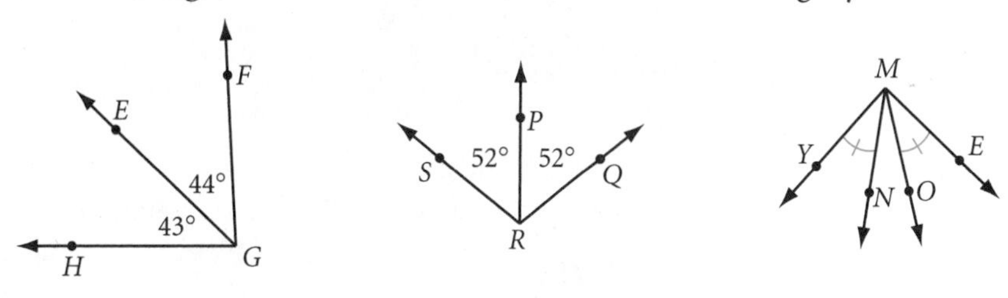

▸ Solution

a. Use the angle bisector definition. $\angle SRP \cong \angle PRQ$, so $\overrightarrow{RP}$ bisects $\angle SRQ$.

b. $\angle SRP \cong \angle PRQ$ and $\angle YMN \cong \angle OME$.

Investigation
Virtual Pool

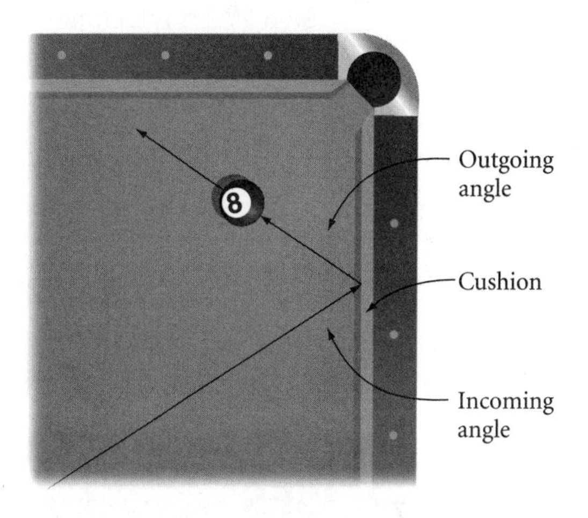

Pocket billiards, or pool, is a game of angles. When a ball bounces off the pool table's cushion, its path forms two angles with the edge of the cushion. The **incoming angle** is formed by the cushion and the path of the ball approaching the cushion.

The **outgoing angle** is formed by the cushion and the path of the ball leaving the cushion. As it turns out, the measure of the outgoing angle equals the measure of the incoming angle.

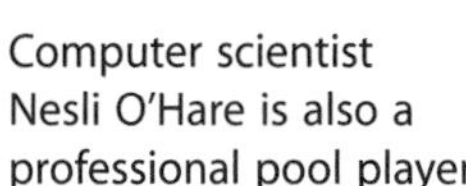

Computer scientist Nesli O'Hare is also a professional pool player.

Use your protractor to study these shots.

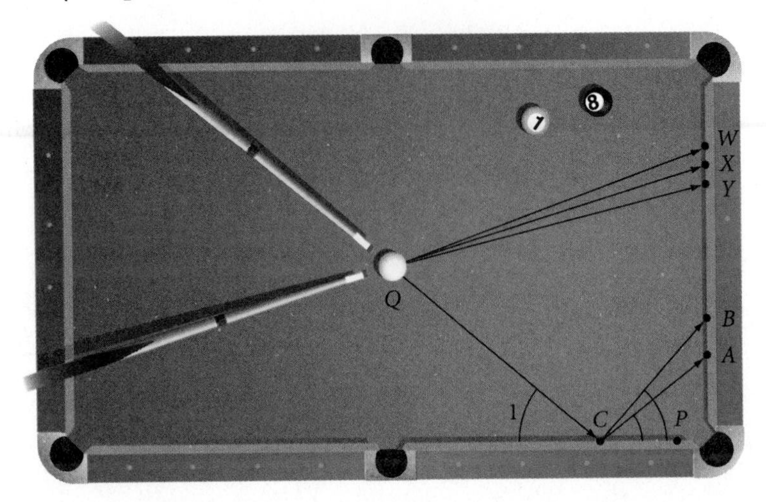

Step 1 Use your protractor to find the measure of $\angle 1$. Which is the correct outgoing angle? Which point—*A* or *B*—will the ball hit?

Step 2 Which point on the cushion—*W*, *X*, or *Y*—should the white ball hit so that the ray of the outgoing angle passes through the center of the 8-ball?

Step 3 Compare your results with your group members' results. Does everyone agree?

Step 4 How would you hit the white ball against the cushion so that the ball passes over the same spot on the way back?

Step 5 How would you hit the ball so that it bounces off three different points on the cushions, without ever touching cushion $\overleftrightarrow{CP}$?

EXERCISES

1. Name each angle in three different ways.

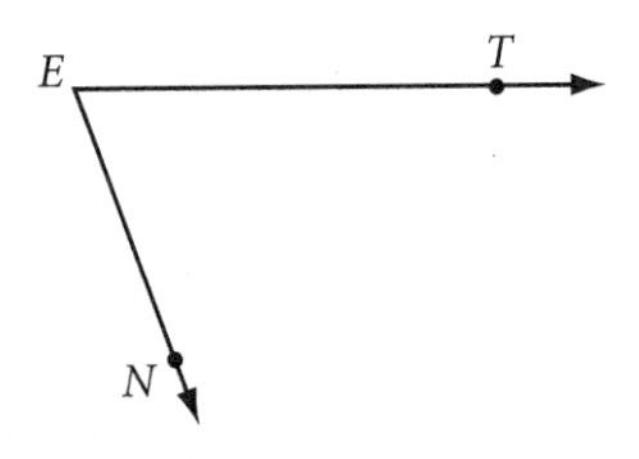

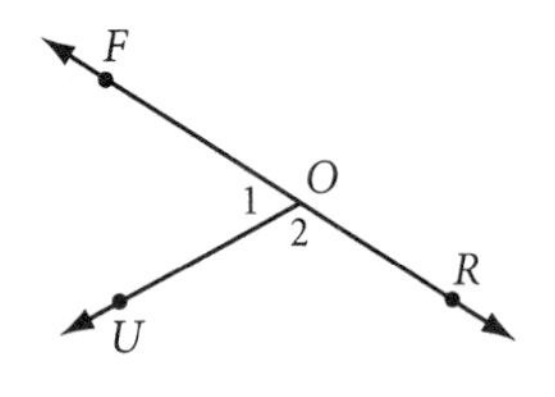

For Exercises 2–4, draw and label each angle.

2. $\angle TAN$

3. $\angle BIG$

4. $\angle SML$

5. For each figure at right, list the angles that you can name using only the vertex letter.

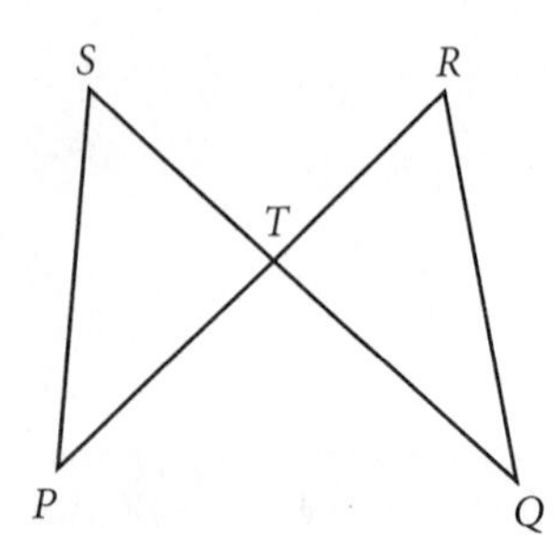

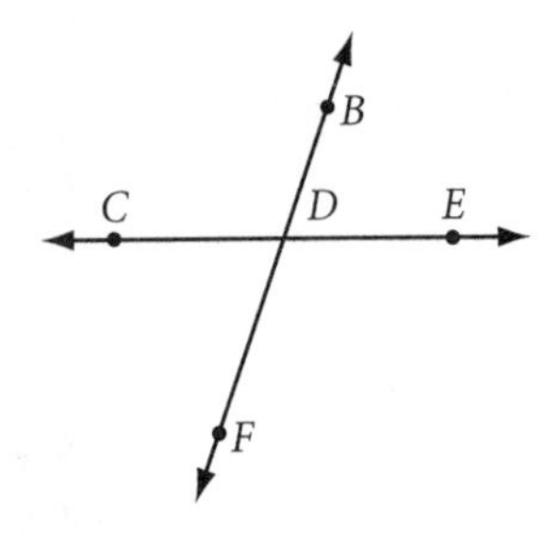

6. Draw a figure that contains at least three angles and requires three letters to name each angle.

For Exercises 7–14, find the measure of each angle.

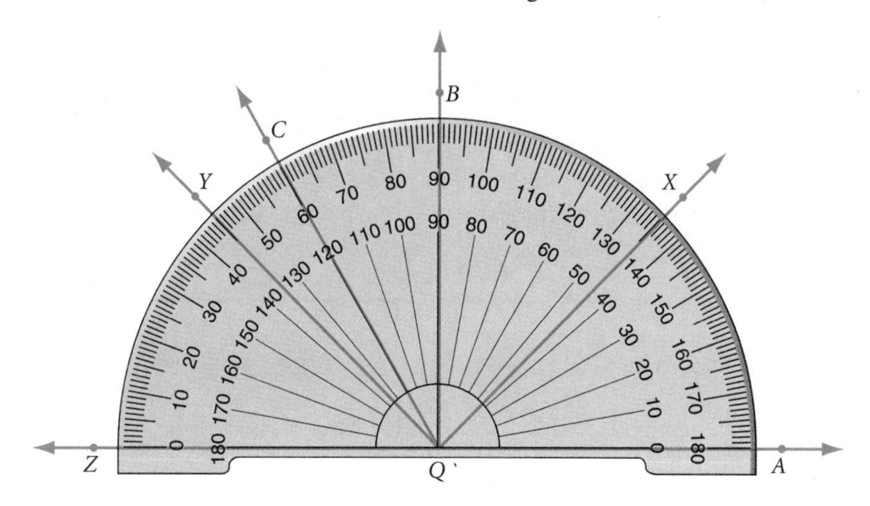

7. $m\angle AQB \approx$ _?_ **8.** $m\angle AQC \approx$ _?_ **9.** $m\angle XQA \approx$ _?_ **10.** $m\angle AQY \approx$ _?_

11. $m\angle ZQY \approx$ _?_ **12.** $m\angle ZQX \approx$ _?_ **13.** $m\angle CQB \approx$ _?_ ⓗ **14.** $m\angle XQY \approx$ _?_

For Exercises 15–19, use your protractor to find the measure of the angle to the nearest degree.

15. $m\angle MAC \approx$ _?_

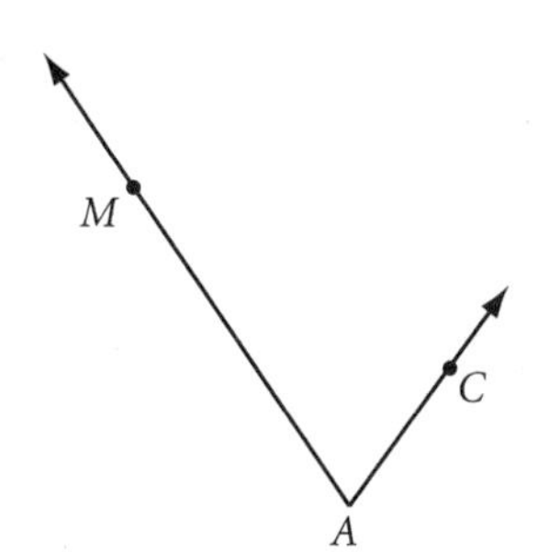

16. $m\angle IBM \approx$ _?_

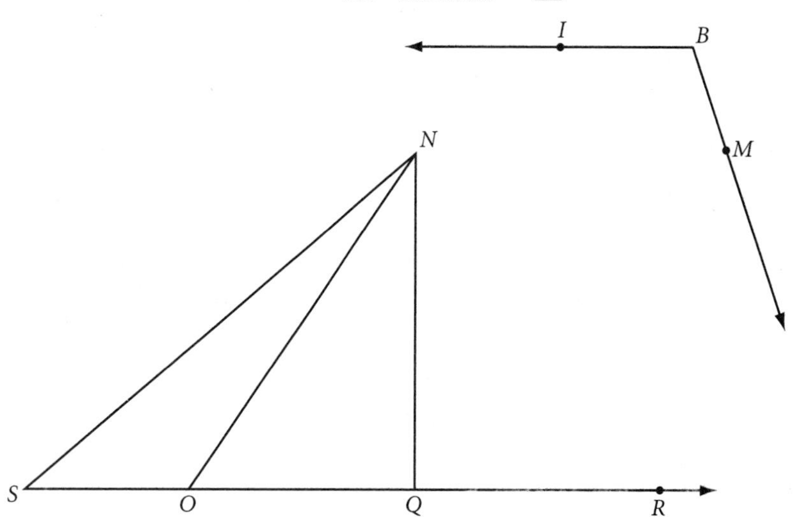

17. $m\angle S \approx$ _?_

18. $m\angle SON \approx$ _?_

19. $m\angle NOR \approx$ _?_

20. Which angle below has the greater measure, $\angle SML$ or $\angle BIG$? Why?

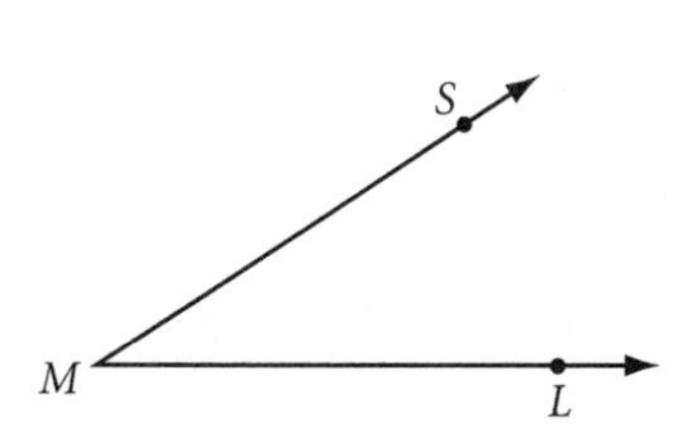

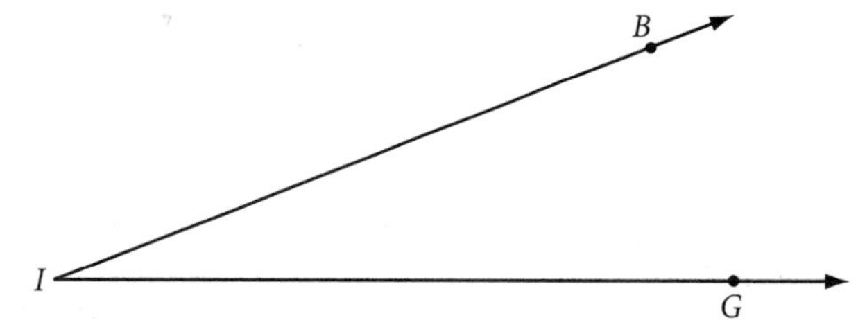

For Exercises 21–23, use your protractor to draw angles with these measures. Label them.

21. $m\angle A = 44°$ **22.** $m\angle B = 90°$ **23.** $m\angle CDE = 135°$

24. Use your protractor to draw the angle bisector of $\angle A$ in Exercise 21 and the angle bisector of $\angle D$ in Exercise 23. Use markings to show that the two halves are congruent.

25. Copy triangle *CAN* shown at right. Use your protractor to find the angle bisector of $\angle A$. Label the point where it crosses $\overline{CN}$ point *Y*. Use your ruler to find the midpoint of $\overline{CN}$ and label it *D*. Are *D* and *Y* the same point?

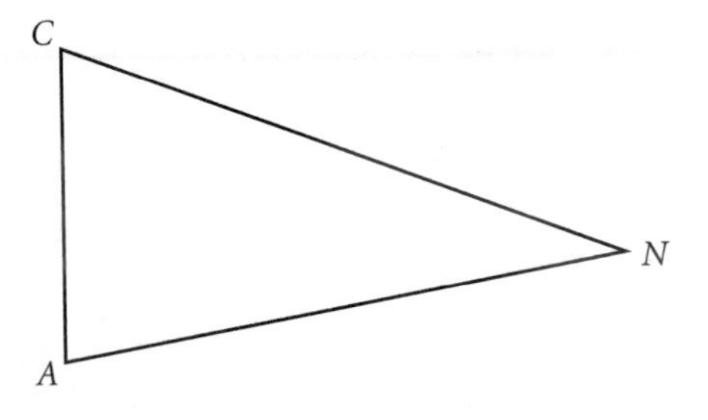

For Exercises 26–28, draw a clock face with hands to show these times.

26. 3:30 ⓗ **27.** 3:40 **28.** 3:15

29. Give an example of a time when the angle made by the hands of the clock will be greater than 90°.

For Exercises 30–33, copy each figure and mark it with all the given information.

30. $TH = 6$
$m\angle THO = 90°$
$OH = 8$

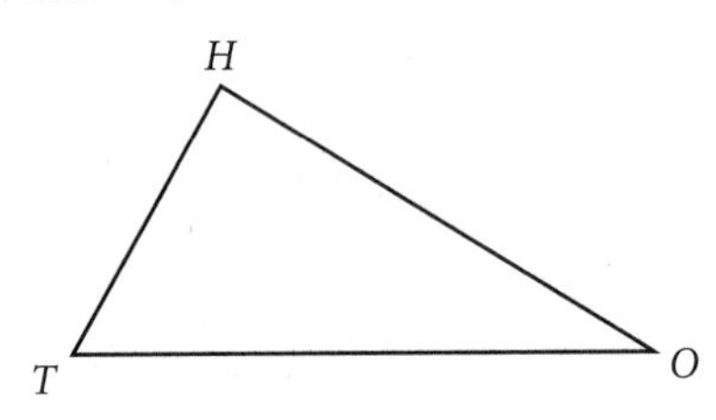

31. $RA = SA$
$m\angle T = m\angle H$
$RT = SH$

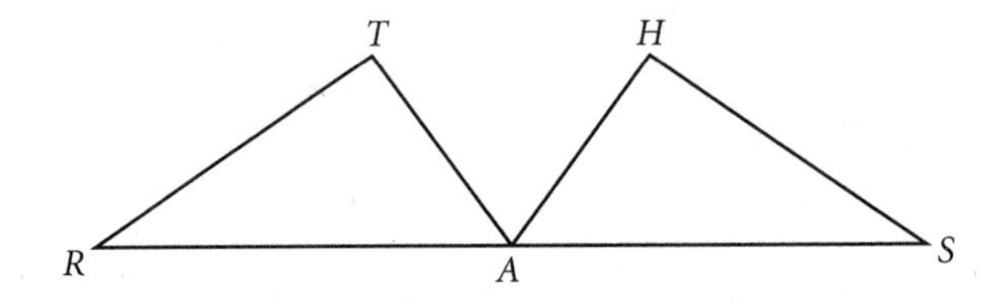

32. $AT = AG$ $\angle AGT \cong \angle ATG$
$AI = AN$ $GI = TN$

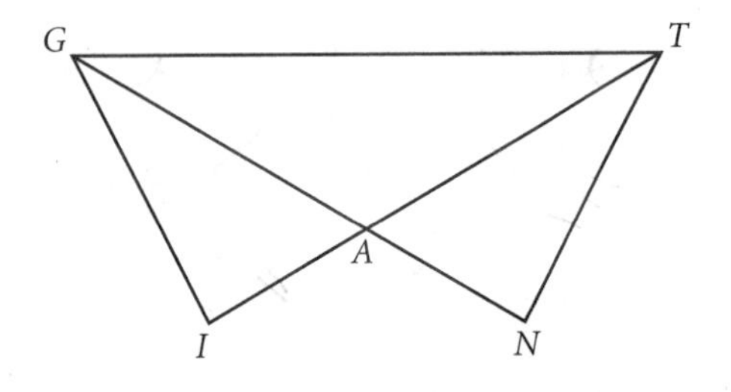

33. $\overline{BW} \cong \overline{TI}$ $\angle WBT \cong \angle ITB$
$\overline{WO} \cong \overline{IO}$ $\angle BWO \cong \angle TIO$

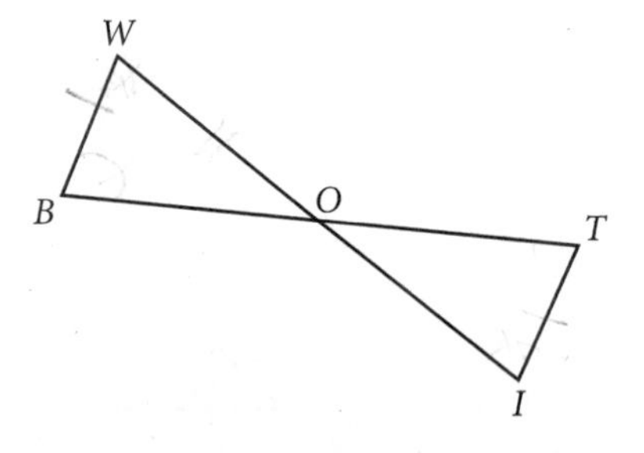

For Exercises 34 and 35, write down what you know from the markings. Do not use your protractor or your ruler.

34. $MI =$?
$IC =$?
$m\angle M =$?

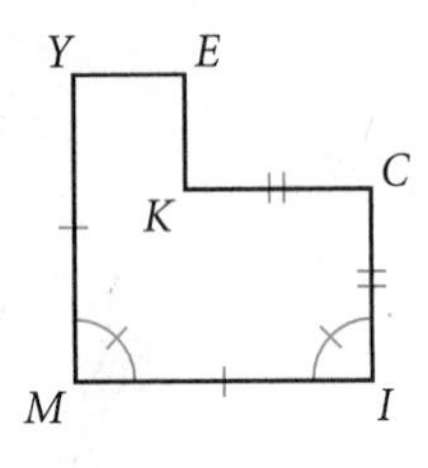

35. $\angle MEO \cong$?
$\angle SUE \cong$?
$OU =$?

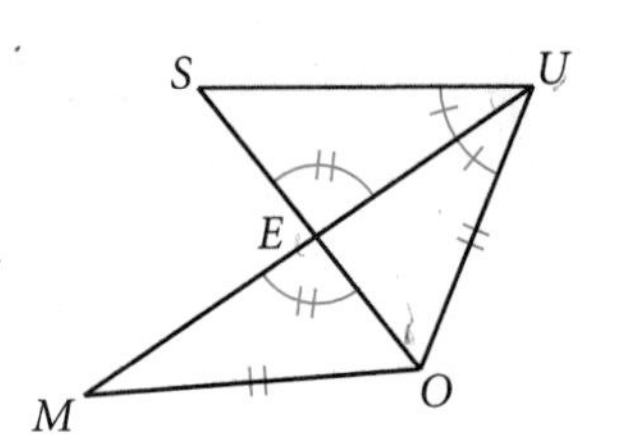

For Exercises 36–38, do not use a protractor. Recall from Chapter 0 that a complete rotation around a point is 360°. Find the angle measures represented by each letter.

36. ⓗ

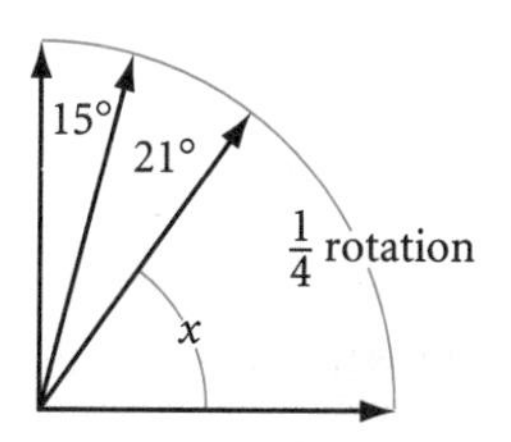

37.

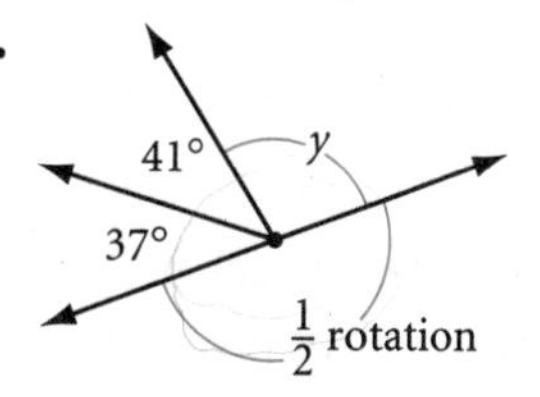

38.

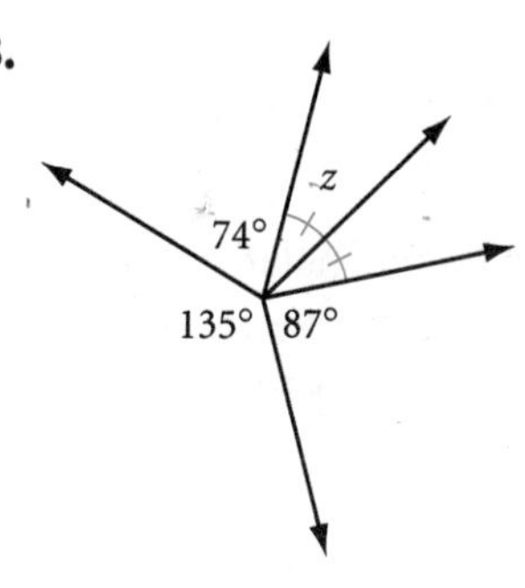

39. If the 4-ball is hit as shown, will it go into the corner pocket? Find the path of the ball using only your protractor and straightedge.

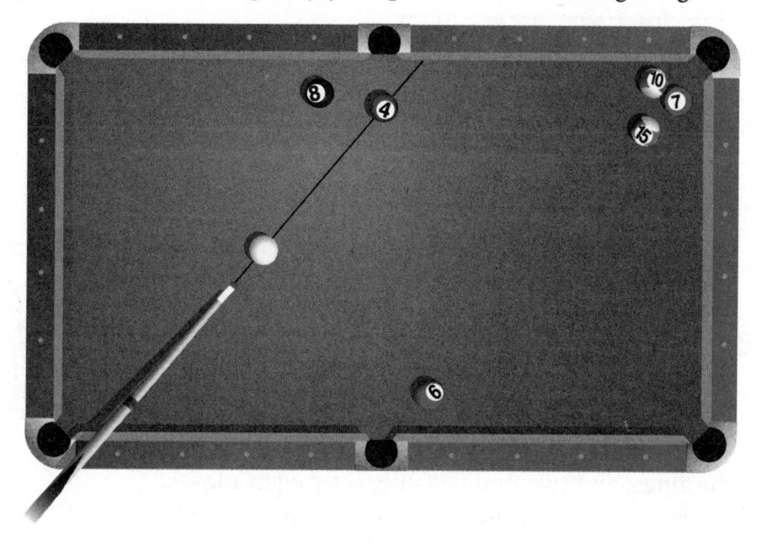

40. What is the measure of the incoming angle? Which point will the ball pass through? Use your protractor to find out.

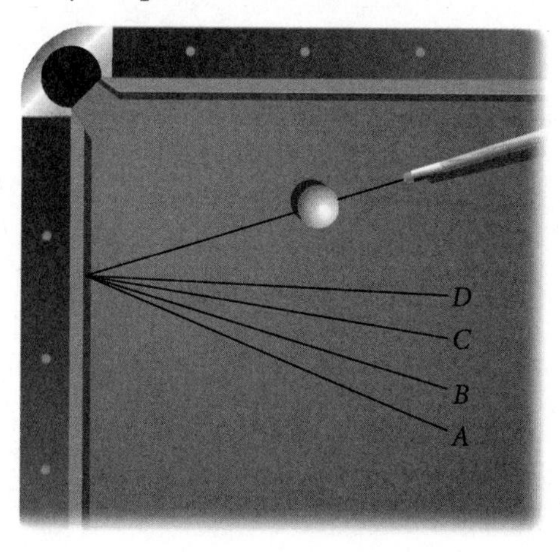

41. The principle you just learned for billiard balls is also true for a ray of light reflecting from a mirror. What you "see" in the mirror is actually light from an object bouncing off the mirror and traveling to your eye.

Will you be able to see your shoes in this mirror? Copy the illustration and draw rays to show the light traveling from your shoes to the mirror and back to your eye.

Review

42. Use your ruler to draw a segment with length 12 cm. Then use your ruler to locate the midpoint. Label and mark the figure.

43. The balancing point of an object is called its *center of gravity*. Where is the center of gravity of a thin, rodlike piece of wire or tubing? Copy the thin wire shown below onto your paper. Mark the balance point or center of gravity.

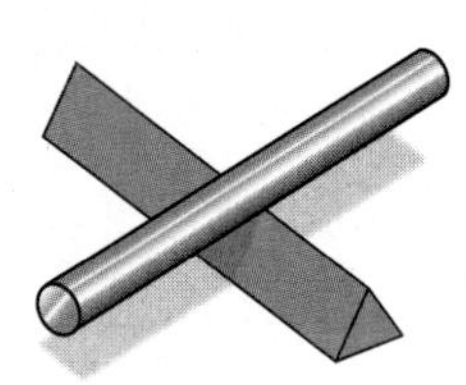

44. Explain the difference between $MS = DG$ and $\overline{MS} \cong \overline{DG}$.

IMPROVING YOUR VISUAL THINKING SKILLS

PUZZLE

Coin Swap I and II

1. Arrange two dimes and two pennies on a grid of five squares, as shown. Your task is to switch the position of the two dimes and two pennies in exactly eight moves. A coin can slide into an empty square next to it, or it can jump over one coin into an empty space. Record your solution by drawing eight diagrams that show the moves.

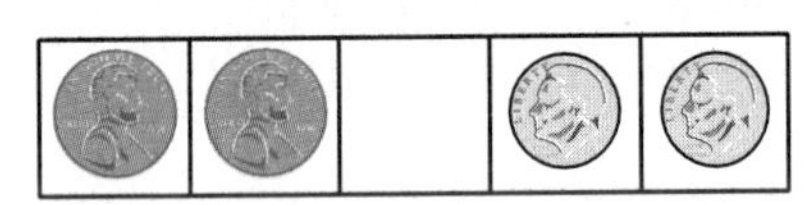

2. Arrange three dimes and three pennies on a grid of seven squares, as shown. Follow the same rules as above to switch the position of the three dimes and three pennies in exactly 15 moves. Record your solution by listing in order which coin is moved. For example, your list might begin PDP. . . .

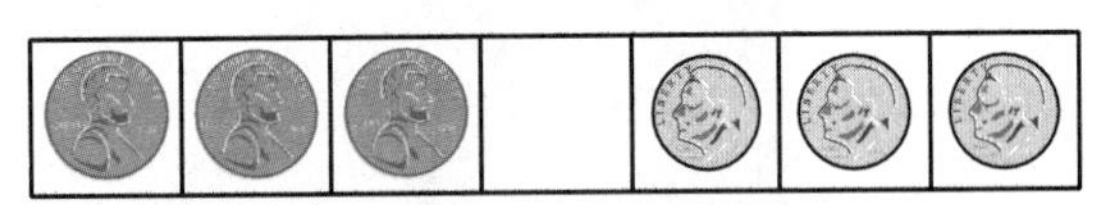

LESSON

1.3

What's a Widget?

"When I use a word," Humpty replied in a scornful tone, "it means just what I choose it to mean—neither more nor less." "The question is," said Alice, "whether you can make a word mean so many different things."

LEWIS CARROLL

Good definitions are very important in geometry. In this lesson you will write your own geometry definitions.

Which creatures in the last group are Widgets?

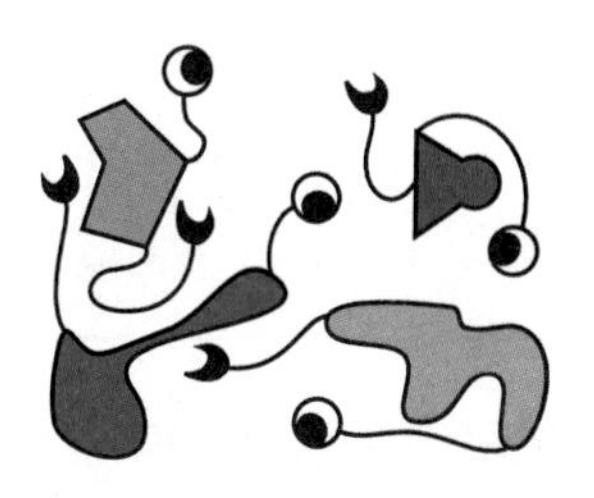

Widgets

Not Widgets

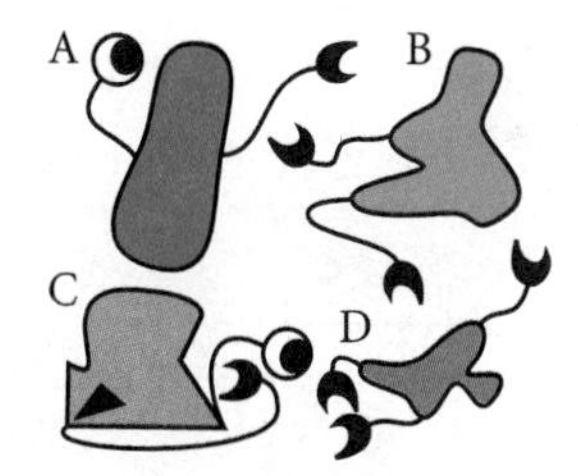

Who are Widgets?

You might have asked yourself, "What things do all the Widgets have in common, and what things do Widgets have that others do not have?" In other words, what characteristics make a Widget a Widget? They all have colorful bodies with nothing else inside; two tails—one like a crescent moon, the other like an eyeball.

By observing what a Widget is and what a Widget isn't, you identified the characteristics that distinguish a Widget from a non-Widget. Based on these characteristics, you should have selected A as the only Widget in the last group. This same process can help you write good definitions of geometric figures.

This statement defines a protractor: "A protractor is a geometry tool used to measure angles." First, you classify what it is (a geometry tool), then you say how it differs from other geometry tools (it is the one you use to measure angles). What should go in the blanks to define a square?

A square is a ________ that ________.

Classify it. What is it? How does it differ from others?

Once you've written a definition, you should test it. To do this, you look for a **counterexample.** That is, try to create a figure that fits your definition but *isn't* what you're trying to define. If you can come up with a counterexample for your definition, you don't have a good definition.

EXAMPLE A

Everyone knows, "A square is a figure with four equal sides." What's wrong with this definition?

a. Sketch a counterexample. (You can probably find more than one!)

b. Write a better definition for a square.

▶ Solution

You probably noticed that "figure" is not specific enough to classify a square, and that "four equal sides" does not specify how it differs from the first counterexample shown below.

a. Three counterexamples are shown here, and you may have found others, too.

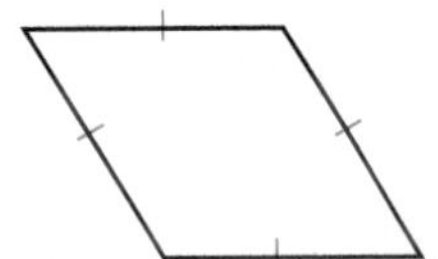

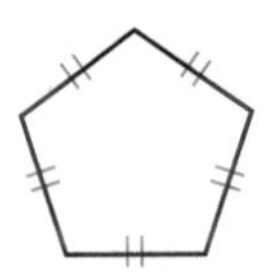

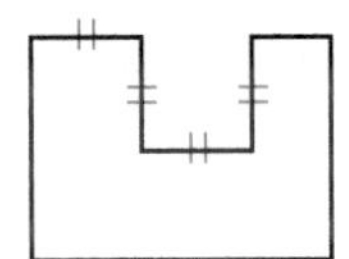

b. One better definition is "A square is a 4-sided figure that has all sides congruent and all angles measuring 90 degrees."

A restaurant counter example

Beginning Steps to Creating a Good Definition

1. **Classify** your term. What is it? ("A square is a 4-sided figure . . .")
2. **Differentiate** your term. How does it differ from others in that class? (". . . that has four congruent sides and four right angles.")
3. **Test** your definition by looking for a counterexample.

Ready to write a couple of definitions? First, here are two more types of markings that are very important in geometry.

The same number of arrow marks indicates that lines are parallel. The symbol $\parallel$ means "is parallel to." A small square in the corner of an angle indicates that it measures 90°. The symbol $\perp$ means "is perpendicular to."

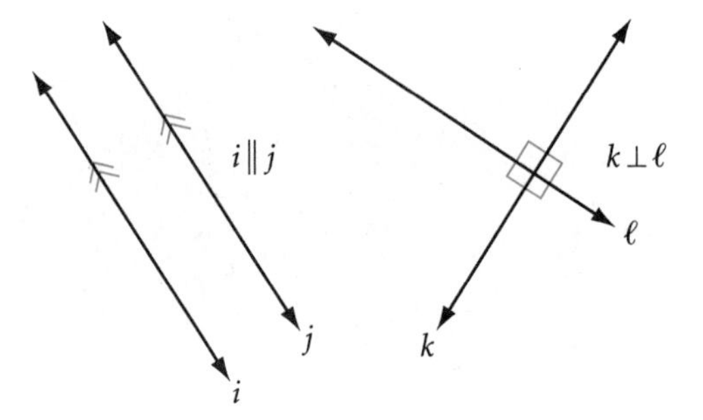

EXAMPLE B

Define these terms:

a. Parallel lines

b. Perpendicular lines

▶ Solution

Following these steps, classify and differentiate each term.

Classify. Differentiate.

a. Parallel lines are lines in the same plane that never meet.

b. Perpendicular lines are lines that meet at 90° angles.

Why do you need to say "in the same plane" for parallel lines but not for perpendicular lines? Sketch or demonstrate a counterexample to show the following definition is incomplete: "Parallel lines are lines that never meet." (Two lines that do not intersect and are noncoplanar are **skew** lines.)

Investigation
Defining Angles

Here are some examples and non-examples of special types of angles.

Step 1 Write a definition for each boldfaced term. Make sure your definitions highlight important differences.

Step 2 Trade definitions and test each other's definitions by looking for counterexamples.

Step 3 If another group member finds a counterexample to one of your definitions, write a better definition. As a group, decide on the best definition for each term.

Step 4 As a class, agree on common definitions. Add these to your notebook. Draw and label a picture to illustrate each definition.

Right Angle

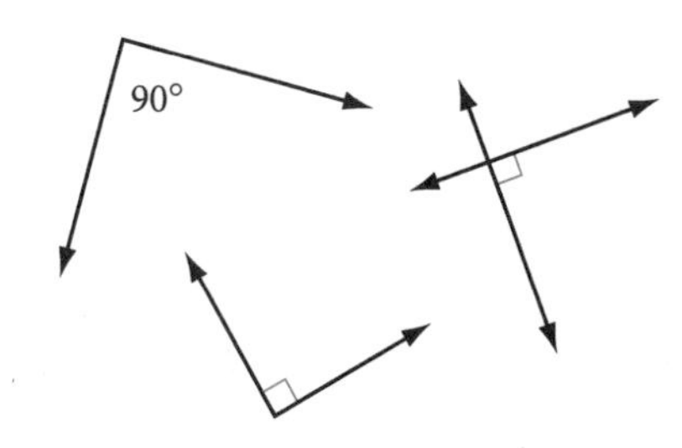

Right angles

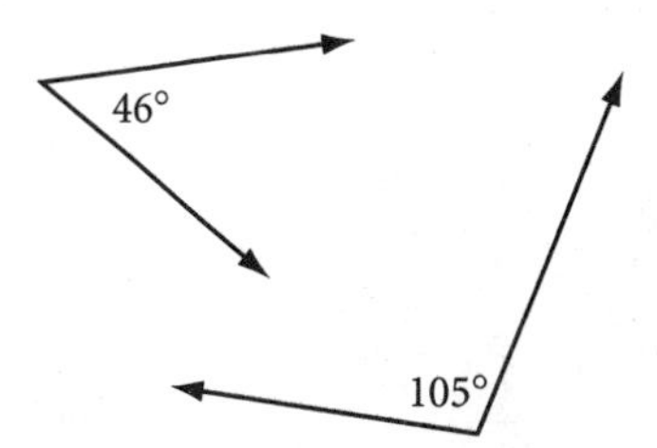

Not right angles

Acute Angle

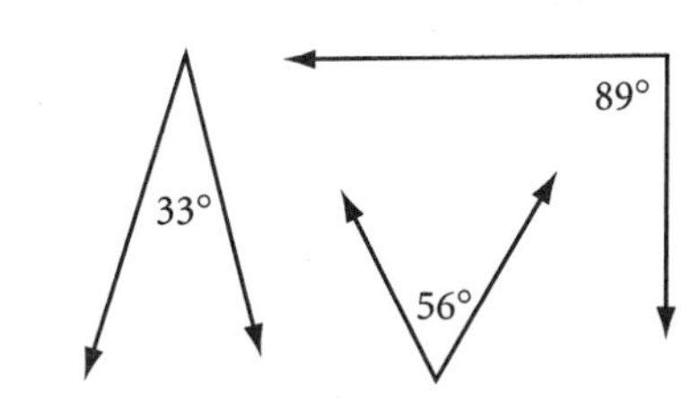

Acute angles

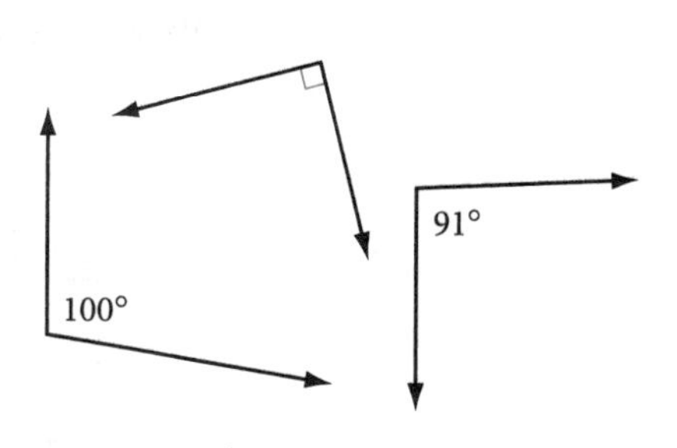

Not acute angles

Obtuse Angle

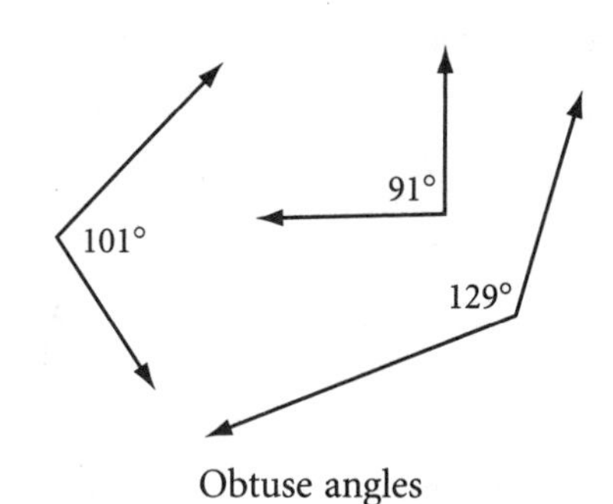

Obtuse angles

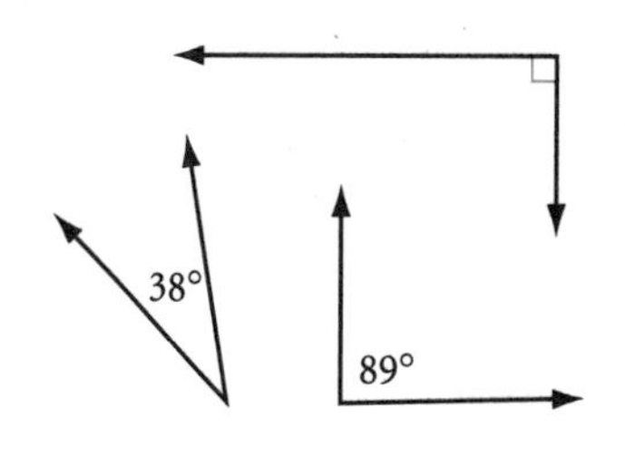

Not obtuse angles

Notice the many congruent angles in this Navajo transitional Wedgeweave blanket. Are they right, acute, or obtuse angles?

Pair of Vertical Angles

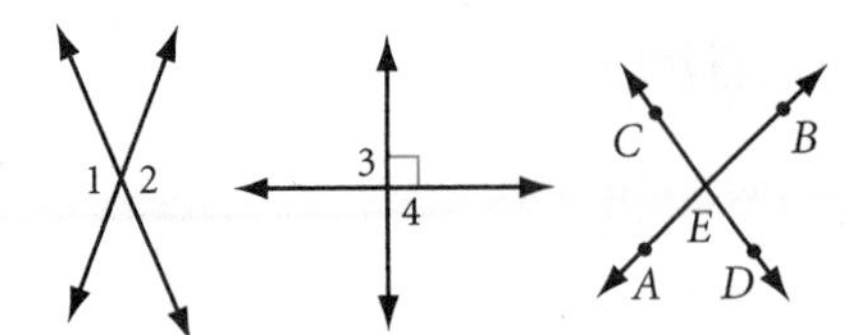

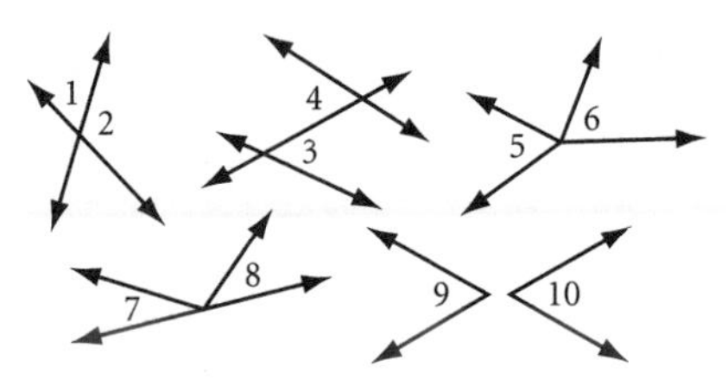

Pairs of vertical angles:

∠1 and ∠2
∠3 and ∠4
∠*AED* and ∠*BEC*
∠*AEC* and ∠*DEB*

Not pairs of vertical angles:

∠1 and ∠2
∠3 and ∠4
∠5 and∠6
∠7 and ∠8
∠9 and ∠10

What types of angles or angle pairs do you see in this magnified view of a computer chip?

Linear Pair of Angles

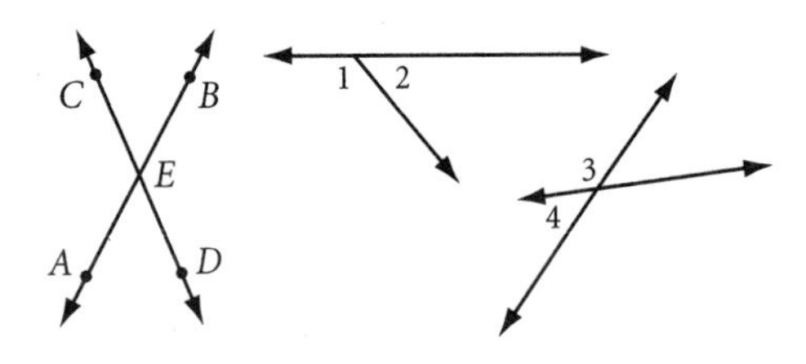

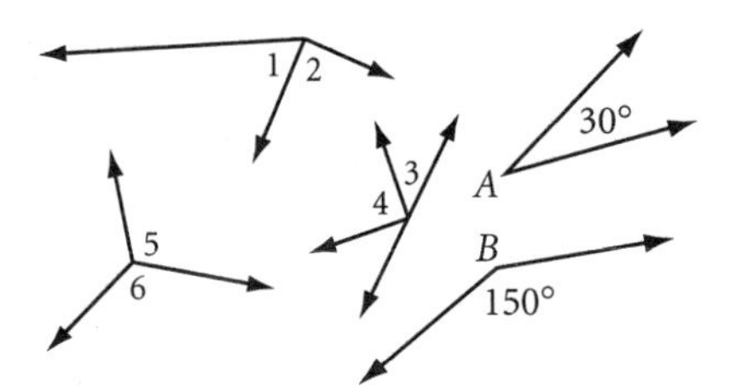

Linear pairs of angles:

∠1 and ∠2
∠3 and ∠4
∠*AED* and ∠*AEC*
∠*BED* and ∠*DEA*

Not linear pairs of angles:

∠1 and ∠2
∠3 and ∠4
∠5 and ∠6
∠*A* and ∠*B*

Pair of Complementary Angles

$m\angle 1 + m\angle 2 = 90°$

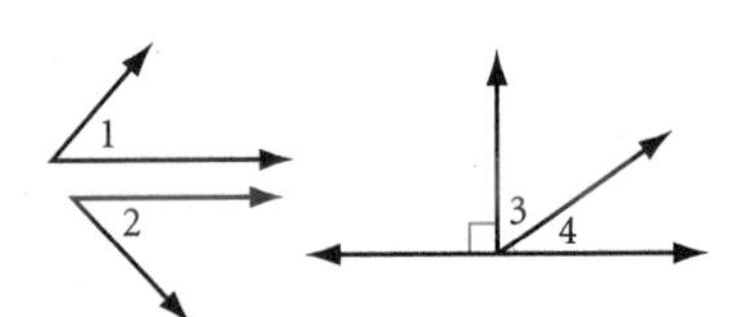

$m\angle 1 + m\angle 2 \neq 90°$

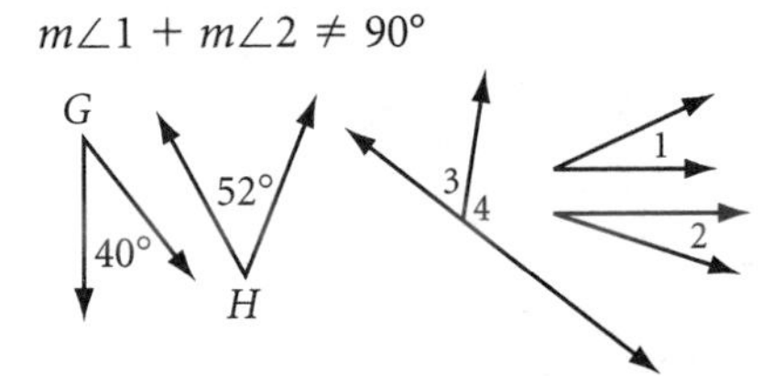

Pairs of complementary angles:

∠1 and ∠2
∠3 and ∠4

Not pairs of complementary angles:

∠*G* and ∠*H* ∠1 and ∠2
∠3 and ∠4

Pair of Supplementary Angles

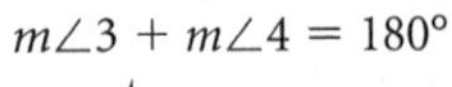

$m\angle 3 + m\angle 4 = 180°$

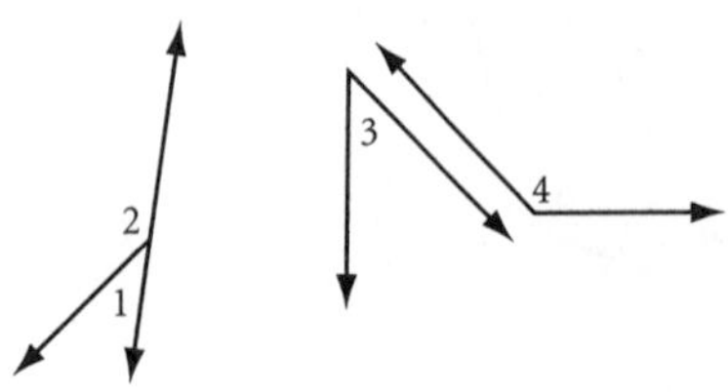

$m\angle 4 + m\angle 5 > 180°$

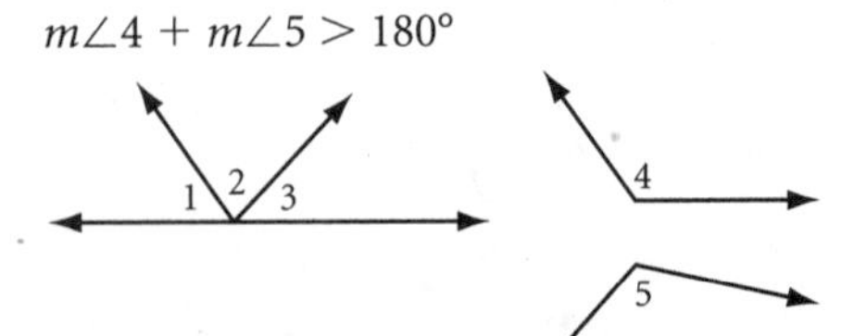

Pairs of supplementary angles:

∠1 and ∠2
∠3 and ∠4

Not pairs of supplementary angles:

∠1, ∠2, and ∠3
∠4 and ∠5

How did you do? Did you notice the difference between a supplementary pair and a linear pair? Did you make it clear which is which in your definitions? The more you practice writing geometry definitions, the better you will get at it.

The design of this African Kente cloth contains examples of parallel and perpendicular lines, obtuse and acute angles, and complementary and supplementary angle pairs. To learn about the significance of Kente cloth designs, visit www.keymath.com/DG .

EXERCISES

For Exercises 1–8, draw and carefully label the figures. Use the appropriate marks to indicate right angles, parallel lines, congruent segments, and congruent angles. Use a protractor and a ruler when you need to.

1. Acute angle DOG with a measure of 45°

2. Right angle RTE

3. Obtuse angle BIG with angle bisector $\overrightarrow{IE}$

4. $\overline{DG} \parallel \overleftrightarrow{MS}$

5. $\overline{PE} \perp \overrightarrow{AR}$

6. Vertical angles ABC and DBE

7. Complementary angles $\angle A$ and $\angle B$ with $m\angle A = 40°$

8. Supplementary angles $\angle C$ and $\angle D$ with $m\angle D = 40°$

9. Which creatures in the last group below are Zoids? What makes a Zoid a Zoid?

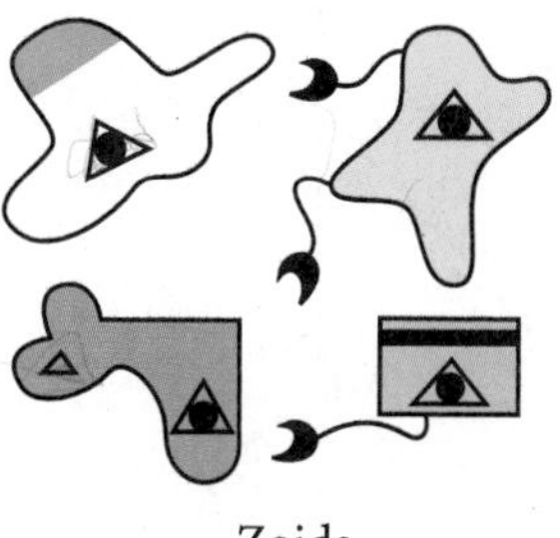

Zoids

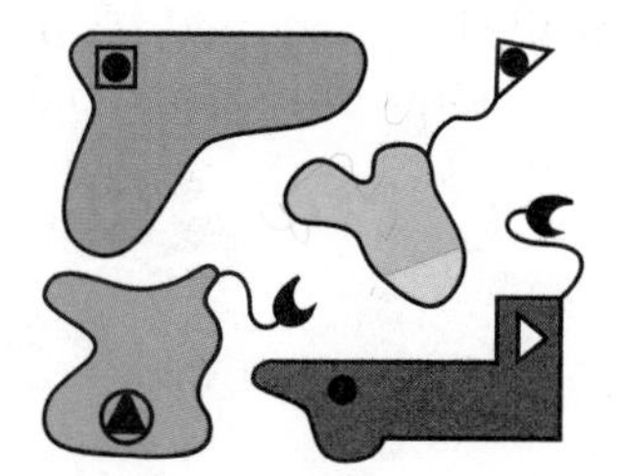

Not Zoids

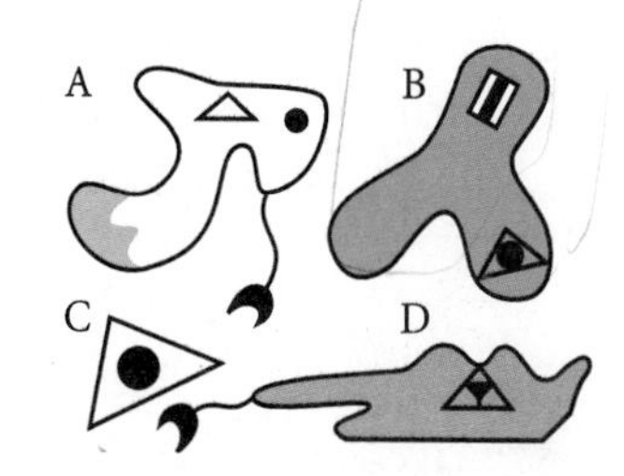

Who are Zoids?

10. What are the characteristics of a good definition?

For Exercises 11–20, four of the statements are true. Make a sketch or demonstrate each true statement. For each false statement, draw a counterexample.

11. For every line segment there is exactly one midpoint.

12. For every angle there is exactly one angle bisector.

13. If two different lines intersect, then they intersect at one and only one point.

14. If two different circles intersect, then they intersect at one and only one point.

15. Through a given point on a line there is one and only one line perpendicular to the given line. ⓗ

16. In every triangle there is exactly one right angle.

17. Through a point not on a line, one and only one line can be constructed parallel to the given line.

18. If $CA = AT$, then A is the midpoint of $\overline{CT}$.

19. If $m\angle D = 40°$ and $m\angle C = 140°$, then angles C and D are a linear pair.

20. If point A is not the midpoint of $\overline{CT}$, then $CA \neq AT$.

21. There is something wrong with this definition for a pair of vertical angles: "If $\overleftrightarrow{AB}$ and $\overleftrightarrow{CD}$ intersect at point P, then $\angle APC$ and $\angle BPD$ are a pair of vertical angles." Sketch a counterexample to show why it is not correct. Can you add a phrase to correct it?

Review

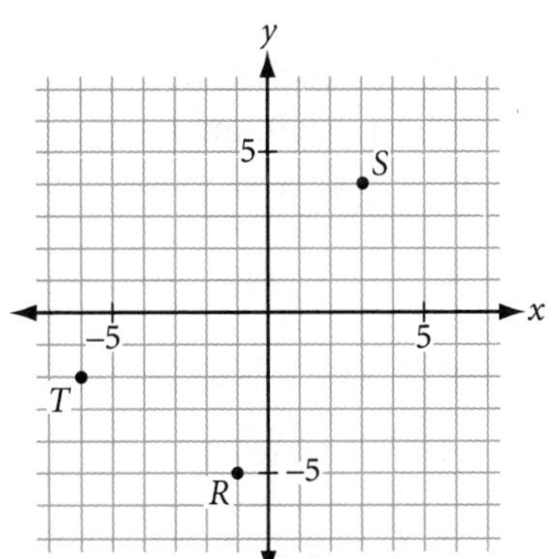

For Exercises 22 and 23, refer to the graph at right.

22. Find possible coordinates of a point P so that points P, T, and S are collinear.

23. Find possible coordinates of a point Q so that $\overleftrightarrow{QR} \parallel \overleftrightarrow{TS}$.

24. A *partial mirror* reflects some light and lets the rest of the light pass through. In the figure at right, half the light from point A passes through the partial mirror to point B. Copy the figure, then draw the outgoing angle for the light reflected from the mirror. What do you notice about the ray of reflected light and the ray of light that passes through? ⓗ

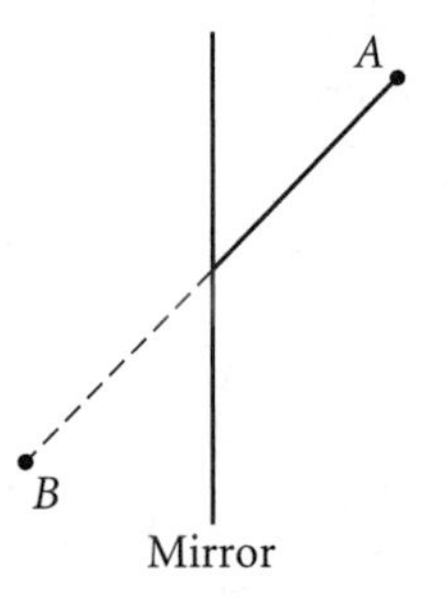

Science CONNECTION

Albert Abraham Michelson (1852–1931) designed the Michelson Interferometer to find the wavelength of light. A modern version of the experiment uses a partial mirror to split a laser beam so that it travels in two different directions, and mirrors to recombine the separated beams.

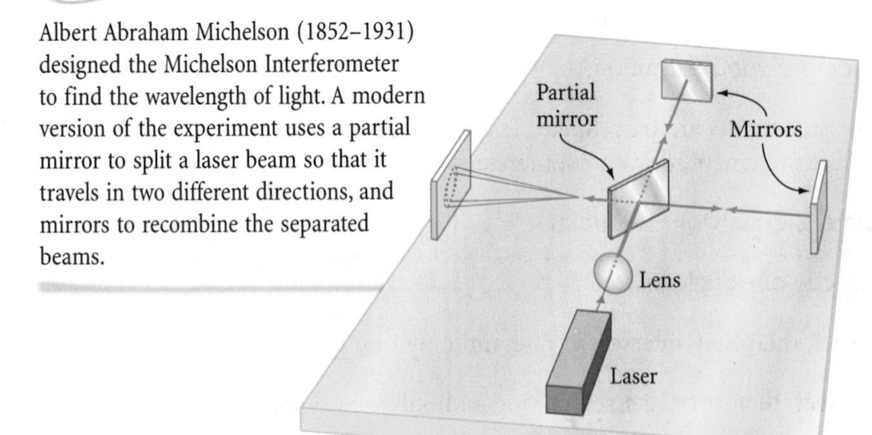

25. Find possible coordinates of points A, B, and C so that $\angle BAC$ is a right angle, $\angle BAT$ is an acute angle, $\angle ABS$ is an obtuse angle, and the points C, T, and R are collinear. ⓗ

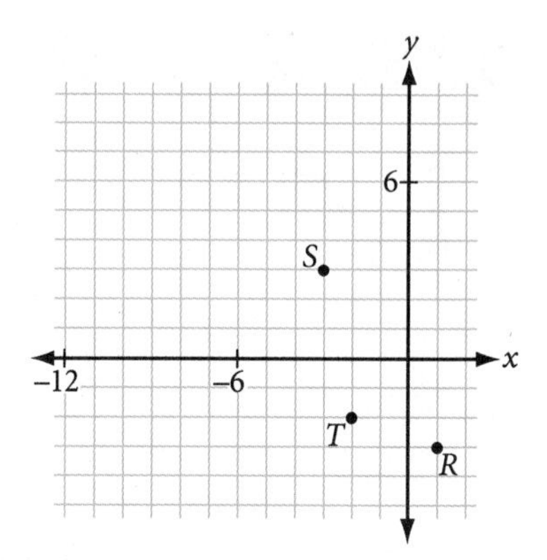

26. If D is the midpoint of $\overline{AC}$ and C is the midpoint of $\overline{AB}$, and $AD = 3$ cm, what is the length of $\overline{AB}$?

27. If $\overrightarrow{BD}$ is the angle bisector of $\angle ABC$, $\overrightarrow{BE}$ is the angle bisector of $\angle ABD$, and $m\angle DBC = 24°$, what is $m\angle EBC$?

28. Draw and label a figure that has two congruent segments and three congruent angles. Mark the congruent angles and congruent segments.

29. Show how three lines in a plane can intersect in no points, exactly one point, exactly two points, or exactly three points.

30. Show how it is possible for two triangles to intersect in one point, two points, three points, four points, five points, or six points, but not seven points. Show how they can intersect in infinitely many points.

31. Each pizza is cut into slices from the center.

a. What fraction of the pizza is left?

b. What fraction of the pizza is missing?

c. If the pizza is cut into nine equal slices, how many degrees is each angle at the center of the pizza?

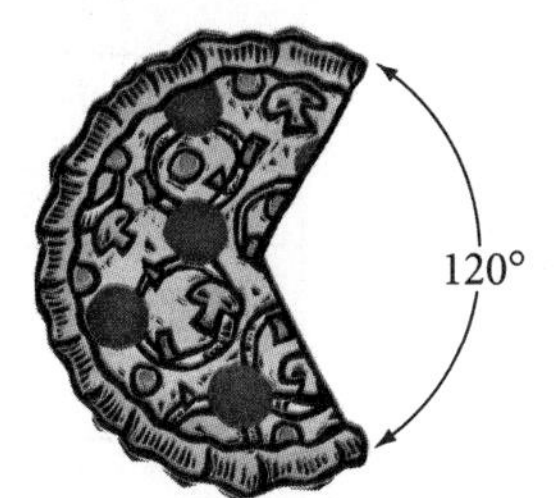

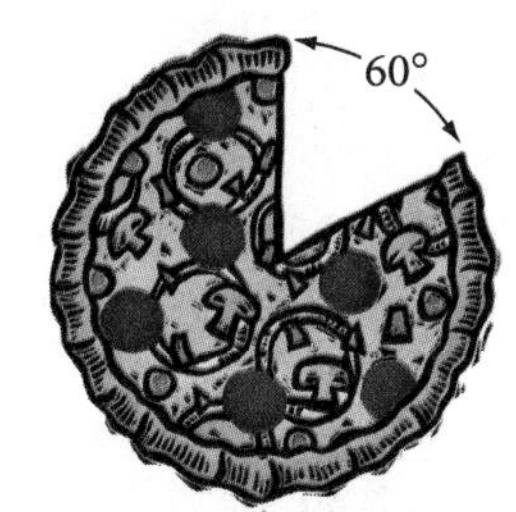

IMPROVING YOUR VISUAL THINKING SKILLS

PUZZLE

Polyominoes

In 1953, United States mathematician Solomon Golomb introduced polyominoes at the Harvard Mathematics Club, and they have been played with and enjoyed throughout the world ever since. Polyominoes are shapes made by connecting congruent squares. The squares are joined together side to side. (A complete side must touch a complete side.) Some of the smaller polyominoes are shown below. There is only one monomino and only one domino, but there are two trominoes, as shown. There are five tetrominoes—one is shown. Sketch the other four.

Monomino

Domino

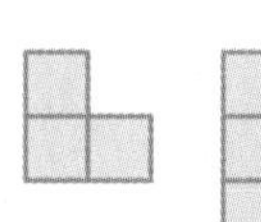

Trominoes

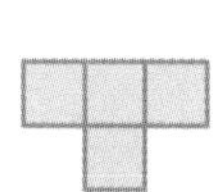

Tetromino

LESSON

1.4

Polygons

There are two kinds of people in this world: those who divide everything into two groups, and those who don't.

KENNETH BOULDING

A **polygon** is a closed figure in a plane, formed by connecting line segments endpoint to endpoint with each segment intersecting exactly two others. Each line segment is called a **side** of the polygon. Each endpoint where the sides meet is called a **vertex** of the polygon.

Polygons

Not polygons

You classify a polygon by the number of sides it has. Familiar polygons have specific names, listed in this table. The ones without specific names are called n-sided polygons, or n-gons. For instance, you call a 25-sided polygon a 25-gon.

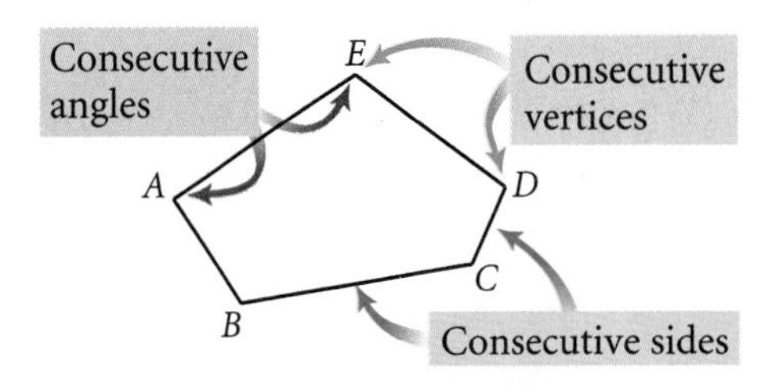

Sides	Name
3	Triangle
4	Quadrilateral
5	Pentagon
6	Hexagon
7	Heptagon
8	Octagon
9	Nonagon
10	Decagon
11	Undecagon
12	Dodecagon
n	n-gon

To name a polygon, list the vertices in consecutive order. You can name the pentagon above pentagon $ABCDE$. You can also call it $DCBAE$, but not $BCAED$. When the polygon is a triangle, you use the triangle symbol. For example, $\triangle ABC$ means triangle ABC.

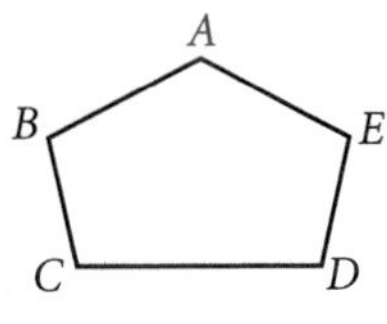

Pentagon $ABCDE$

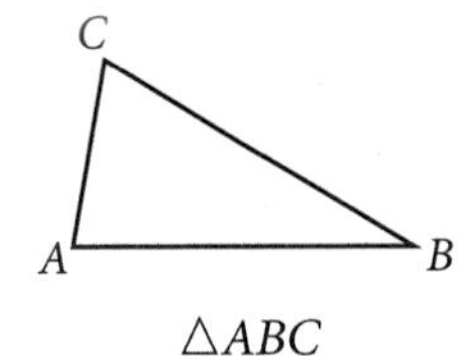

$\triangle ABC$

A **diagonal** of a polygon is a line segment that connects two nonconsecutive vertices.

Diagonal

E

F

D

G

A polygon is **convex** if no diagonal is outside the polygon. A polygon is **concave** if at least one diagonal is outside the polygon.

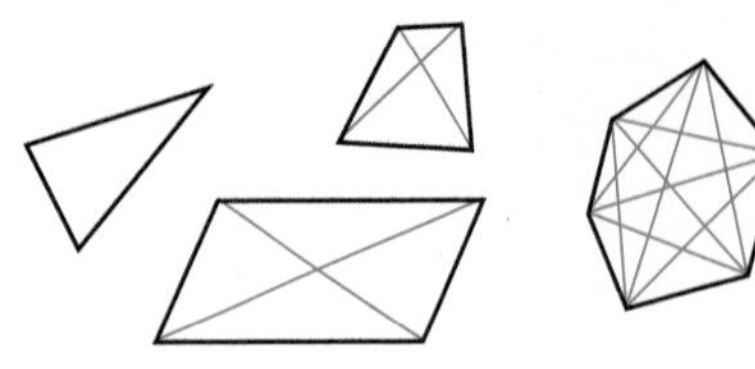

Convex polygons: All diagonals are inside

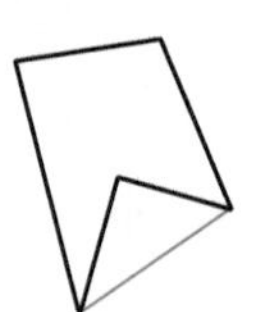

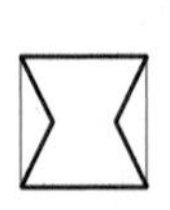

Concave polygons: One or more diagonals are outside

Recall that two segments or two angles are congruent if and only if they have the same measures. Two polygons are **congruent polygons** if and only if they are exactly the same size and shape. "If and only if" means that the statements work both ways.

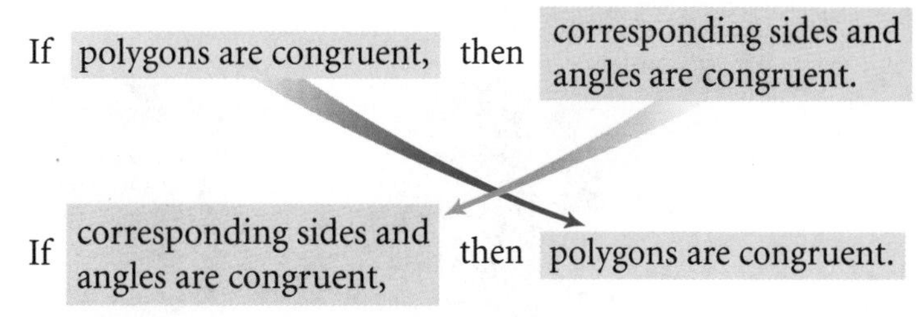

How does the shape of the framework of this Marc Chagall (1887–1985) stained glass window support the various shapes of the design?

For example, if quadrilateral *CAMP* is congruent to quadrilateral *SITE*, then their four pairs of corresponding angles and four pairs of corresponding sides are also congruent. When you write a statement of congruence, always write the letters of the corresponding vertices in an order that shows the correspondences.

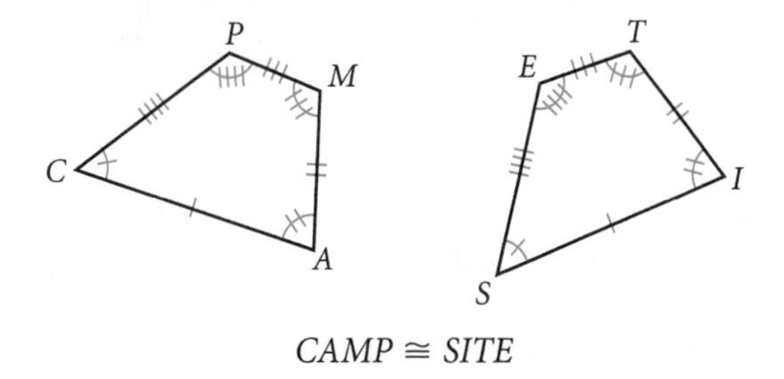

$CAMP \cong SITE$

EXAMPLE

Which polygon is congruent to *ABCDE*?
$ABCDE \cong$?

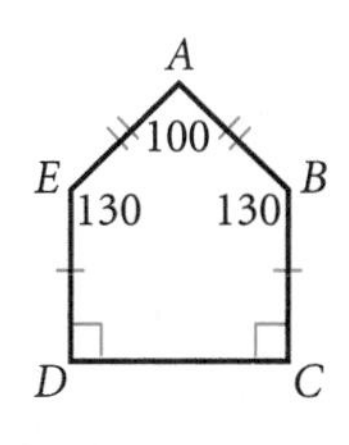

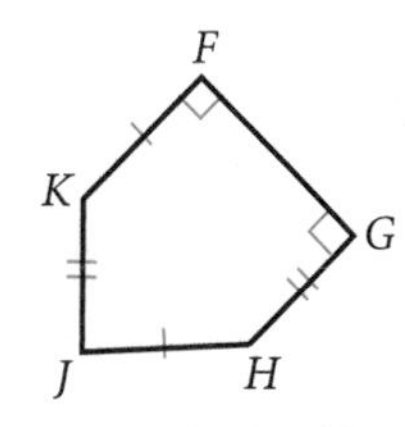

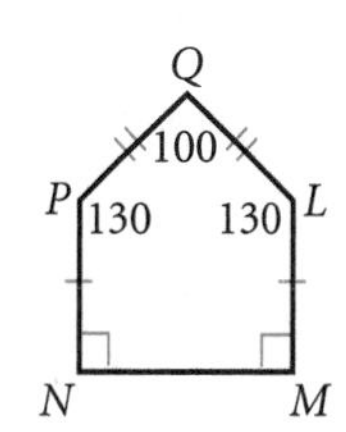

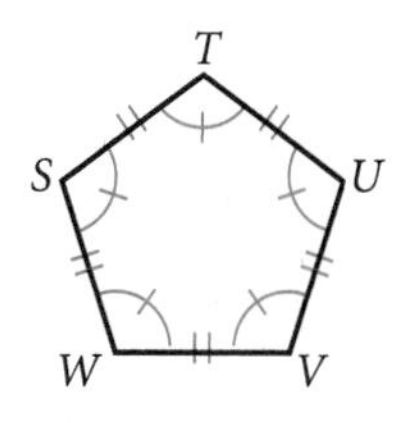

▶ Solution

All corresponding sides and angles must be congruent, so polygon $ABCDE \cong$ polygon *QLMNP*. You could also say $ABCDE \cong QPNML$, because all the congruent parts would still match.

In an **equilateral polygon,** all the sides have equal length. In an **equiangular polygon,** all the angles have equal measure. A **regular polygon** is both equilateral and equiangular.

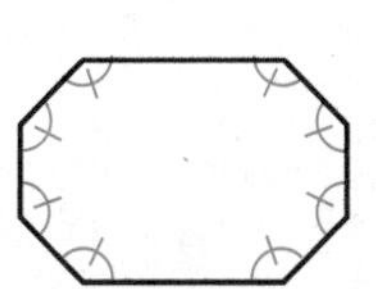
Equiangular octagon

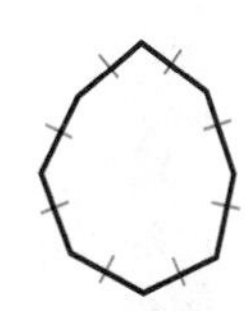
Equilateral octagon

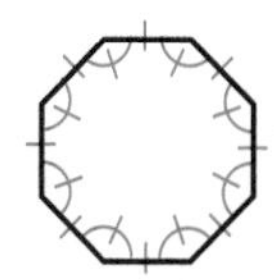
Regular octagon

EXERCISES

For Exercises 1–8, classify each polygon. Assume that the sides of the chips and crackers are straight.

1. 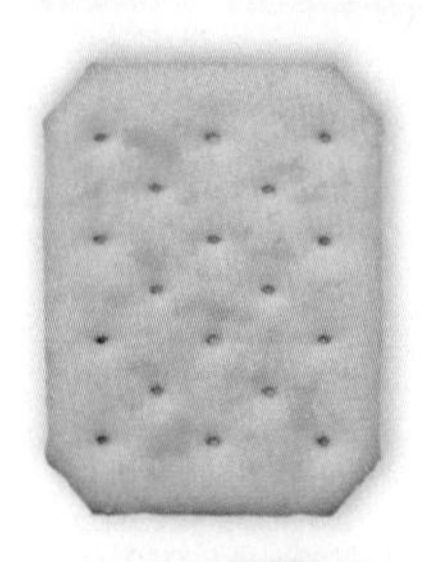

2. 

3. 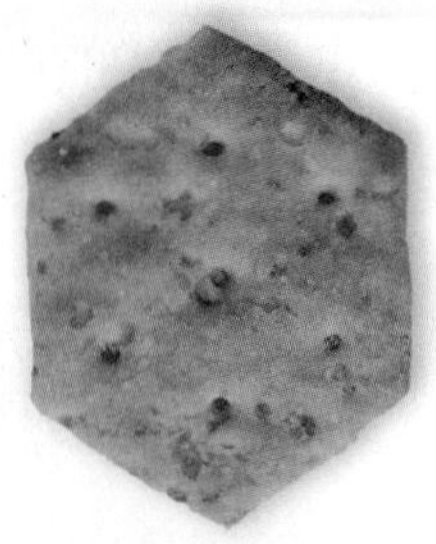

4. 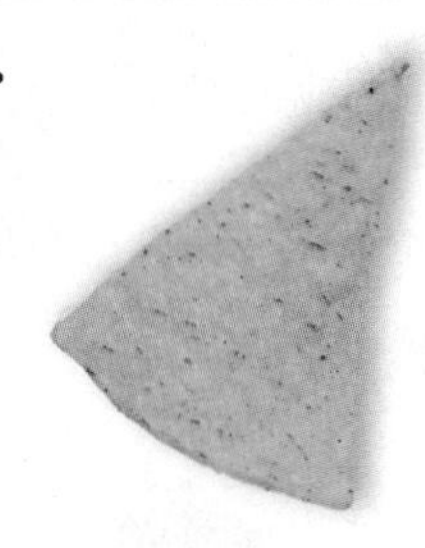

5. 

6. 

7. 

8. 

For Exercises 9–11, draw an example of each polygon.

9. Quadrilateral

10. Dodecagon

11. Octagon

For Exercises 12 and 13, give one possible name for each polygon.

12. 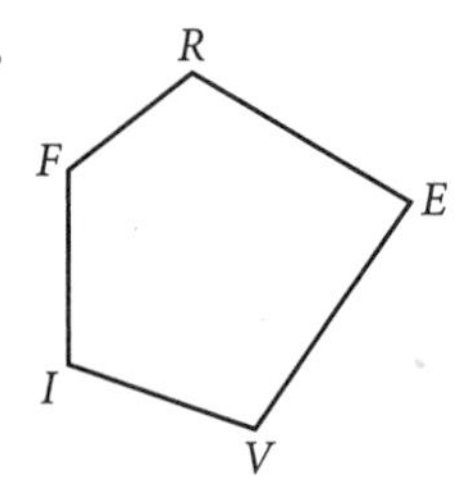

13. 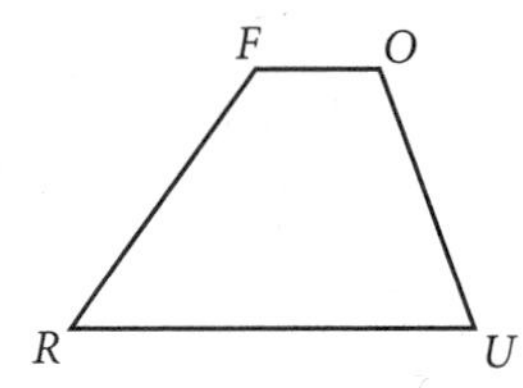

14. 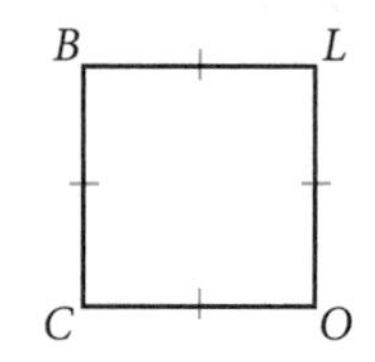

15. Name a pair of consecutive angles and a pair of consecutive sides in the figure below.

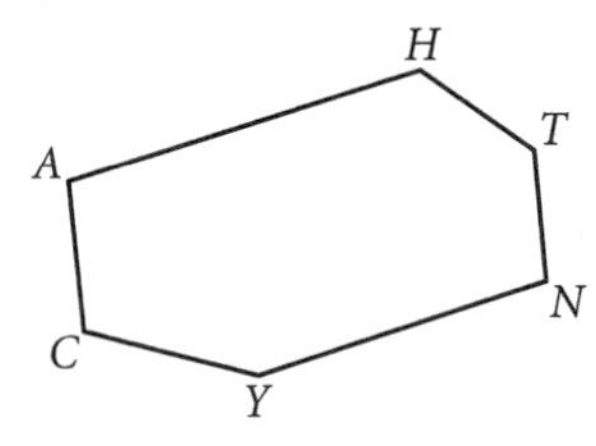

16. Draw a concave hexagon. How many diagonals does it have?

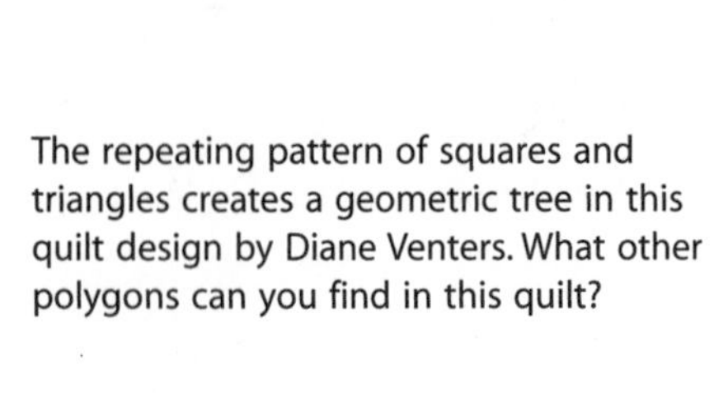

The repeating pattern of squares and triangles creates a geometric tree in this quilt design by Diane Venters. What other polygons can you find in this quilt?

17. Name the diagonals of pentagon *ABCDE.*

For Exercises 18 and 19, use the information given to name the triangle that is congruent to the first one.

18. $\triangle EAR \cong \triangle$ $\underline{?}$ ⓗ

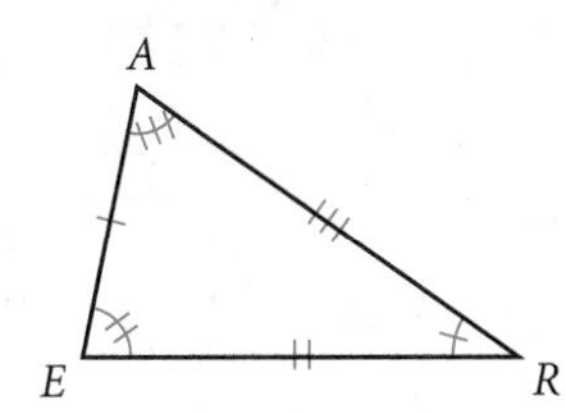

19. $\triangle OLD \cong \triangle$ $\underline{?}$

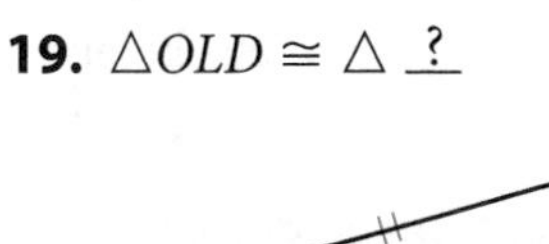

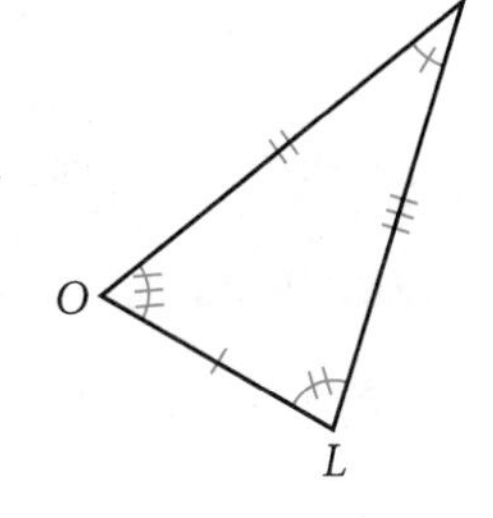

20. In the figure at right, $THINK \cong POWER$.

a. Find the missing measures.

b. If $m\angle P = 87°$ and $m\angle W = 165°$, which angles in *THINK* do you know? Write their measures.

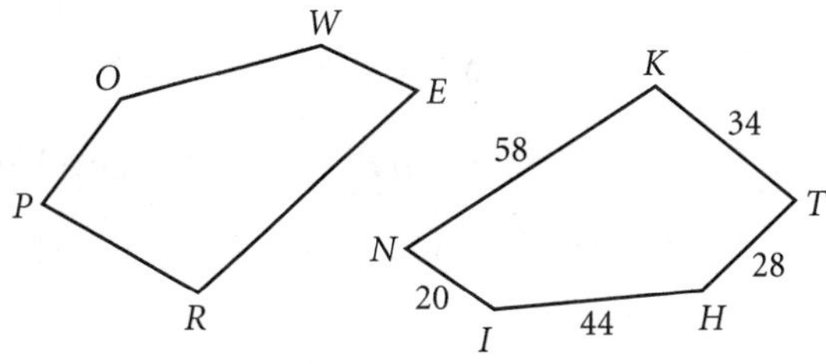

21. If pentagon *FIVER* is congruent to pentagon *PANCH,* then which side in pentagon *FIVER* is congruent to side $\overline{PA}$? Which angle in pentagon *PANCH* is congruent to $\angle IVE$?

22. Draw an equilateral concave pentagon. Then draw an equiangular concave pentagon.

For Exercises 23–26, copy the given polygon and segment onto graph paper. Give the coordinates of the missing points.

23. $\triangle CAR \cong \triangle PET$

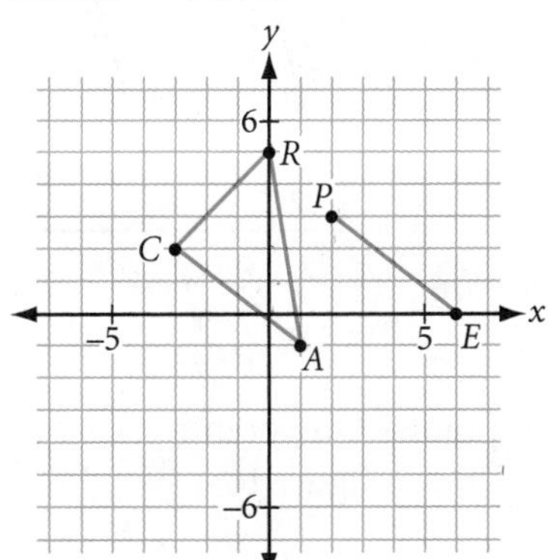

24. $TUNA \cong FISH$

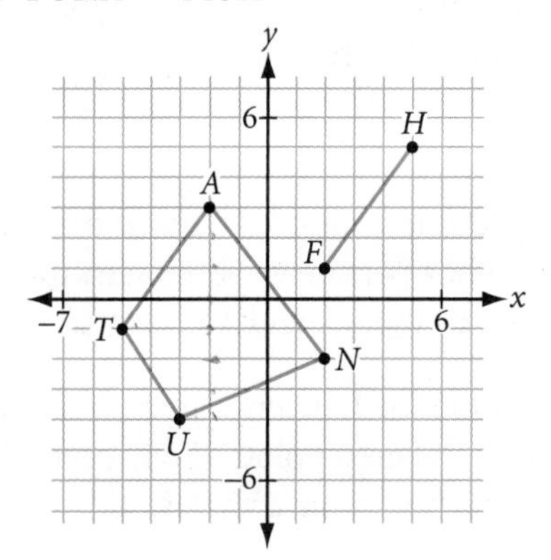

25. $BLUE \cong FISH$

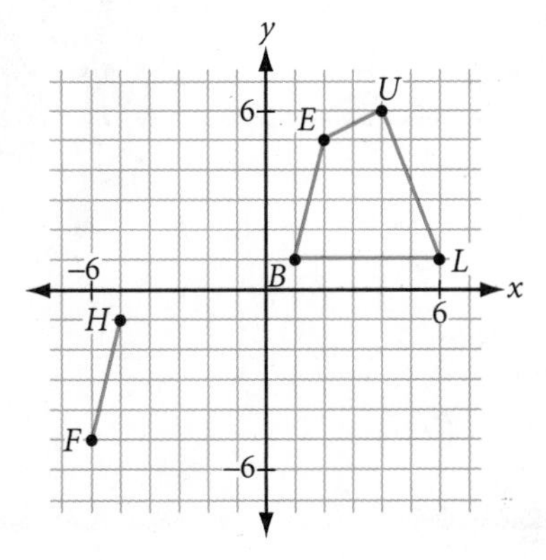

26. $RECT \cong ANGL$

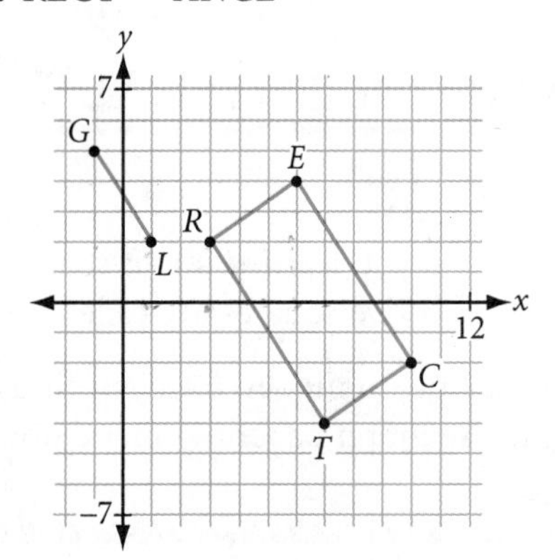

27. Draw an equilateral octagon $ABCDEFGH$ with $A(5, 0)$, $B(4,4)$, and $C(0, 5)$ as three of its vertices. Is it regular?

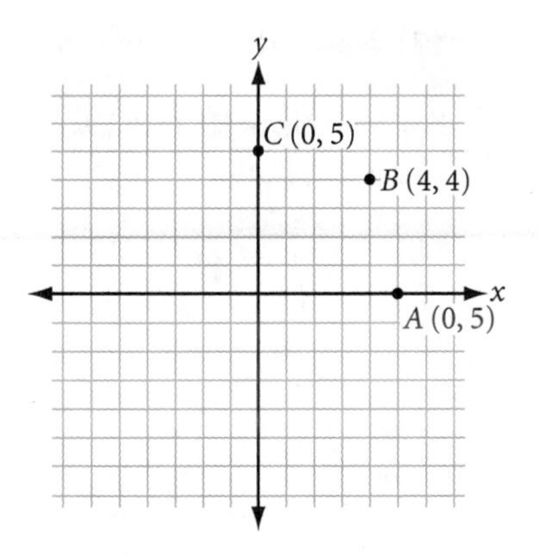

For Exercises 28–32, sketch and carefully label the figure. Mark the congruences.

28. Pentagon $PENTA$ with $PE = EN$

29. Hexagon $NGAXEH$ with $\angle HEX \cong \angle EXA$

30. Equiangular quadrilateral $QUAD$ with $QU \neq QD$

31. A hexagon with exactly one line of reflectional symmetry ⓗ

32. Two different equilateral pentagons with perimeter 25 cm

33. Use your compass, protractor, and straightedge to draw a regular pentagon.

34. A rectangle with perimeter 198 cm is divided into five congruent rectangles as shown in the diagram at right. What is the perimeter of one of the five congruent rectangles? ⓗ

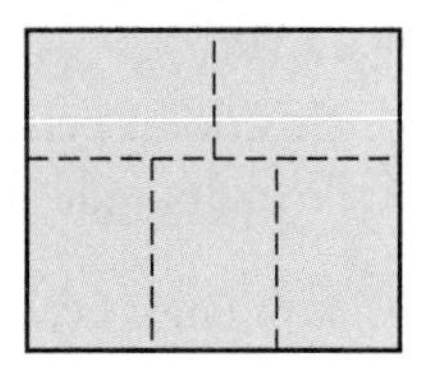

Review

35. Name a pair of complementary angles and a pair of vertical angles in the figure at right.

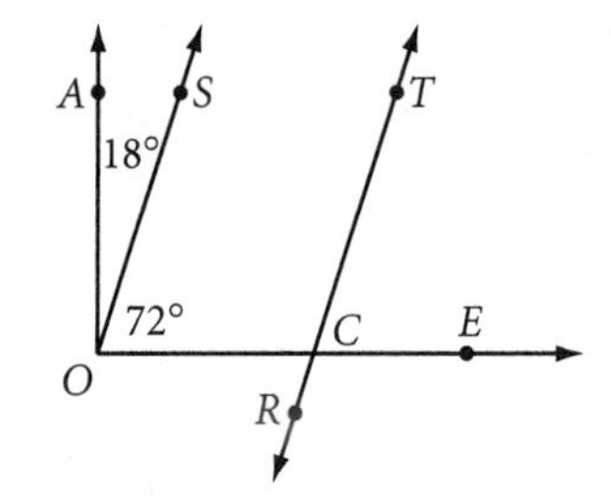

36. Draw $\overleftrightarrow{AB}$, $\overleftrightarrow{CD}$, and $\overleftrightarrow{EF}$ with $\overleftrightarrow{AB} \parallel \overleftrightarrow{CD}$ and $\overleftrightarrow{CD} \perp \overleftrightarrow{EF}$.

37. Draw a counterexample to show that this statement is false: "If a rectangle has perimeter 50 meters, then a pair of adjacent sides measures 10 meters and 15 meters."

38. Is it possible for four lines in a plane to intersect in exactly zero points? One point? Two points? Three points? Four points? Five points? Six points? Draw a figure to support each of your answers. ⓗ

IMPROVING YOUR VISUAL THINKING SKILLS

Pentominoes I

In Polyominoes I, you learned about shapes called polyominoes. Polyominoes with five squares are called pentominoes.

There are 12 pentominoes. Can you find them all? One is shown at right. Use graph paper or square dot paper to sketch all 12.

LESSON

1.5

Triangles and Special Quadrilaterals

The difference between the right word and the almost right word is the difference between lightning and the lightning bug.

MARK TWAIN

You have learned to be careful with geometry definitions. It turns out that you also have to be careful with diagrams.

When you look at a diagram, be careful not to assume too much from it. To **assume** something is to accept it as true without facts or proof.

Lightning

Not lightning

Things you can assume:

You may assume that lines are straight, and if two lines intersect, they intersect at one point.

You may assume that points on a line are collinear and that all points shown in a diagram are coplanar unless planes are drawn to show that they are noncoplanar.

Things you can't assume:

You may not assume that just because two lines or segments *look* parallel that they *are* parallel—they must be *marked* parallel!

You may not assume that two lines *are* perpendicular just because they *look* perpendicular—they must be *marked* perpendicular!

Pairs of angles, segments, or polygons are not necessarily congruent, unless they are *marked* with information that tells you they must be congruent!

EXAMPLE

In the diagrams below, which pairs of lines are perpendicular? Which pairs of lines are parallel? Which pair of triangles is congruent?

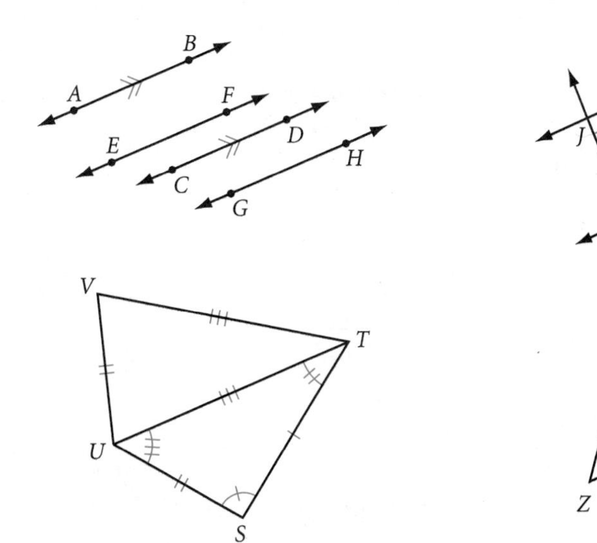

▸ Solution By studying the markings, you can tell that $\overleftrightarrow{AB} \parallel \overleftrightarrow{CD}$, $\overleftrightarrow{JK} \perp \overleftrightarrow{JM}$, and $\triangle STU \cong \triangle XYZ$.

In this lesson you will write definitions that classify different kinds of triangles and special quadrilaterals, based on relationships among their sides and angles.

Investigation
Triangles and Special Quadrilaterals

Write a good definition of each boldfaced term. Discuss your definitions with others in your group. Agree on a common set of definitions for your class and add them to your definition list. In your notebook, draw and label a figure to illustrate each definition.

What shape is the basis for the design on this textile from Uzbekistan?

Right Triangle

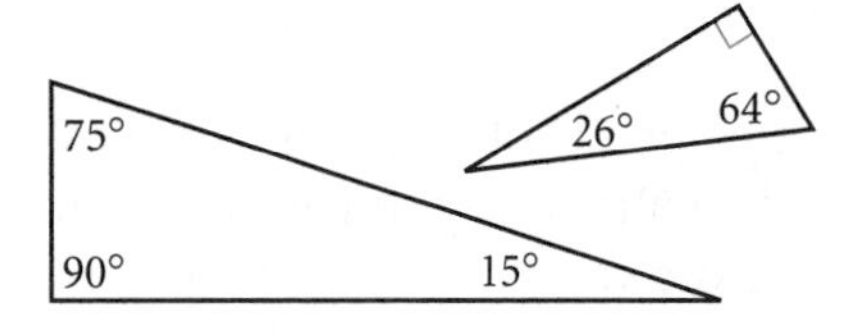

Right triangles

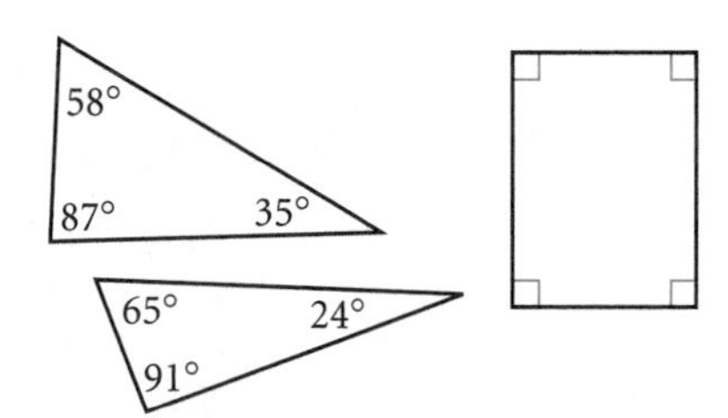

Not right triangles

Acute Triangle

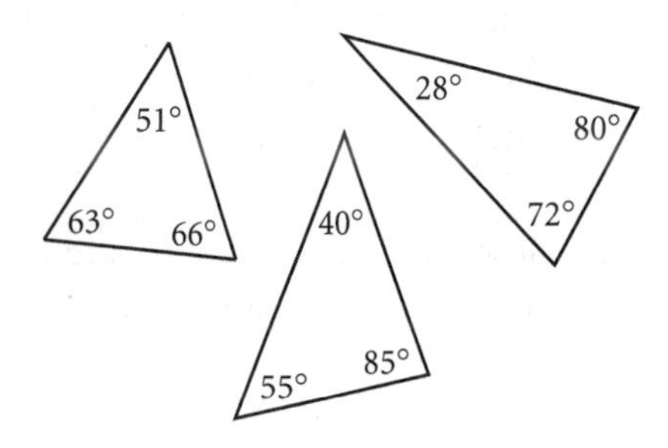

Acute triangles

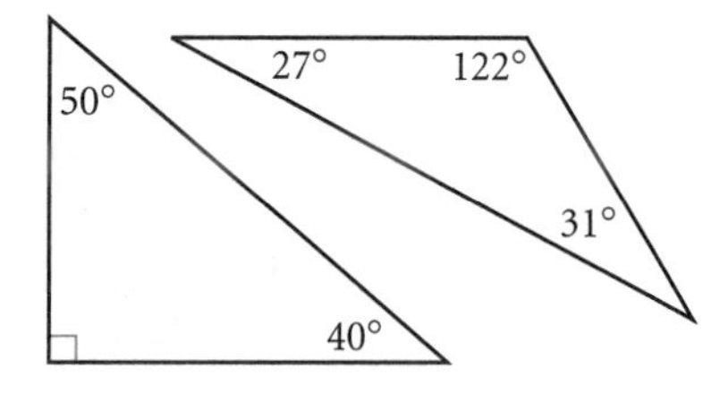

Not acute triangles

Obtuse Triangle

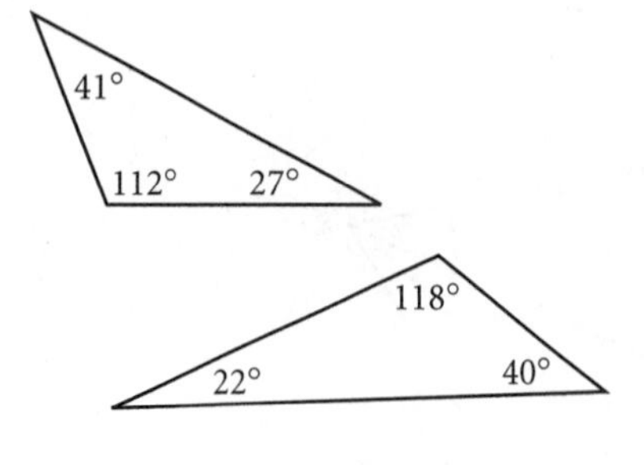

Obtuse triangles

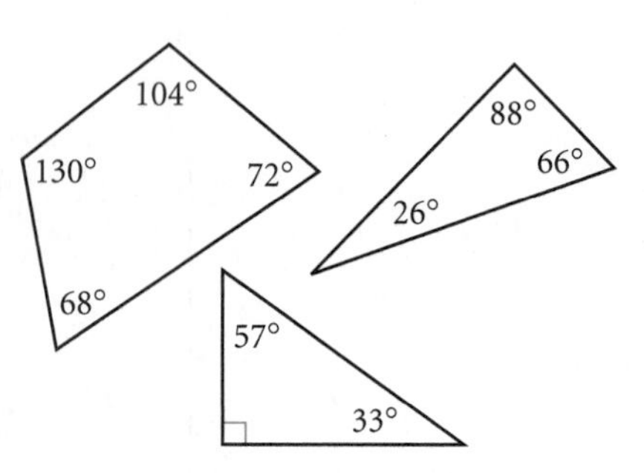

Not obtuse triangles

The Sol LeWitt (b 1928, United States) design inside this art museum uses triangles and quadrilaterals to create a painting the size of an entire room.

Sol LeWitt, *Wall Drawing #652—On three walls, continuous forms with color ink washes superimposed, color in wash.* Collection: Indianapolis Museum of Art, Indianapolis, IN. September, 1990. Courtesy of the artist.

Scalene Triangle

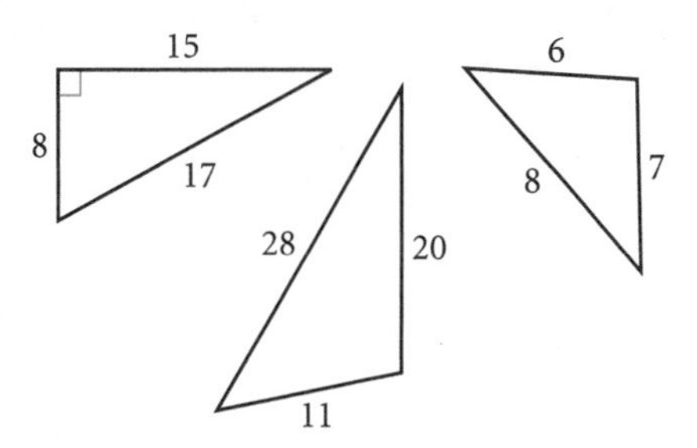

Scalene triangles

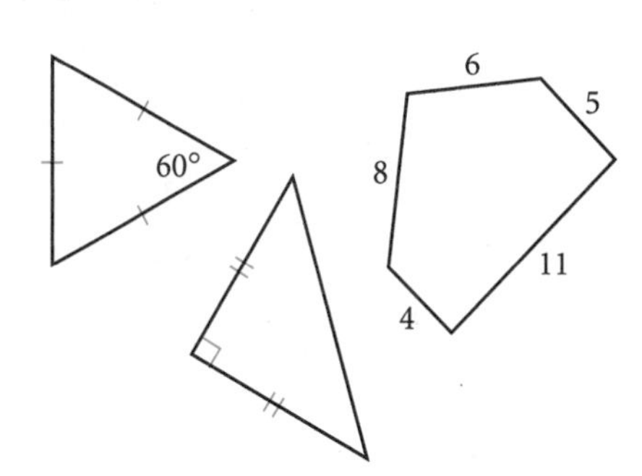

Not scalene triangles

Equilateral Triangle

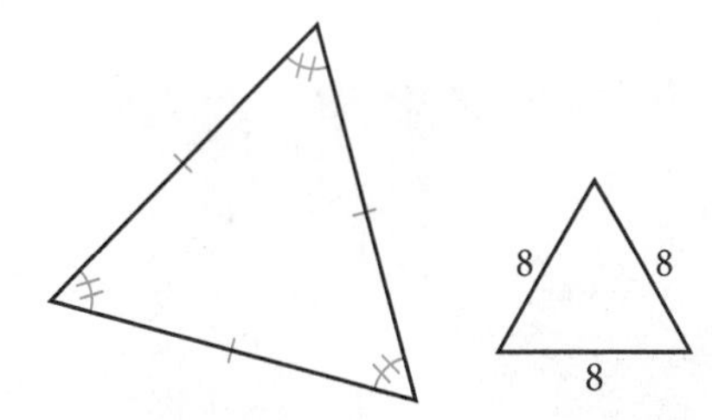

Equilateral triangles

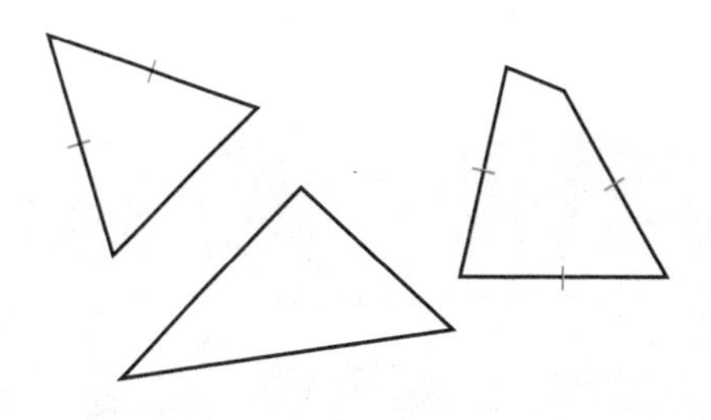

Not equilateral triangles

Isosceles Triangle

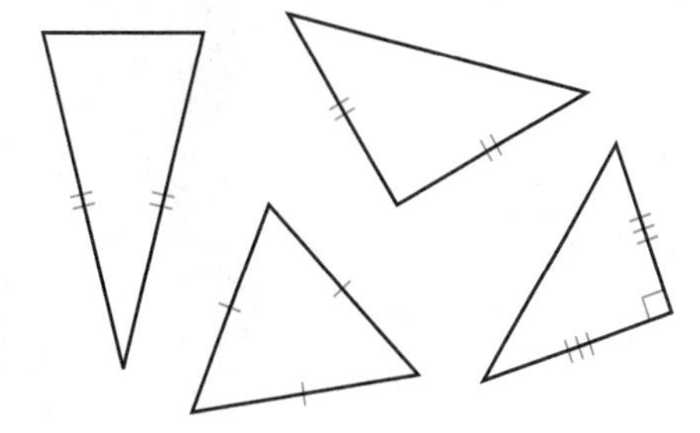

Isosceles triangles

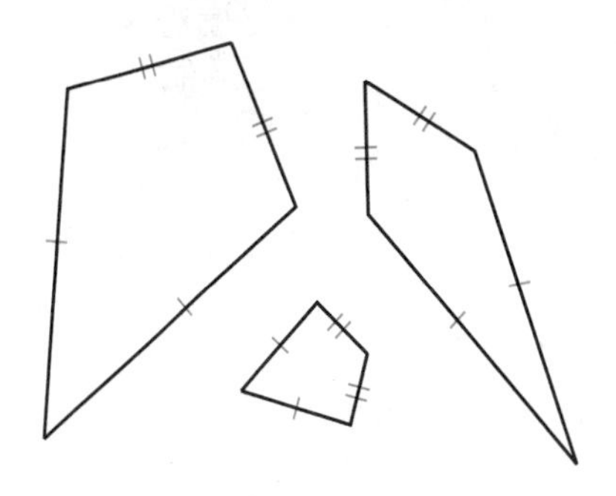

Not isosceles triangles

In an isosceles triangle, the angle between the two sides of equal length is called the **vertex angle.** The side opposite the vertex angle is called the **base** of the isosceles triangle. The two angles opposite the two sides of equal length are called the **base angles** of the isosceles triangle.

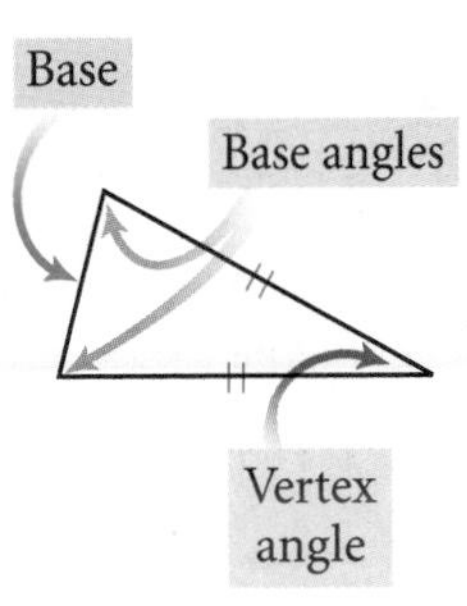

Now let's write definitions for quadrilaterals.

Trapezoid

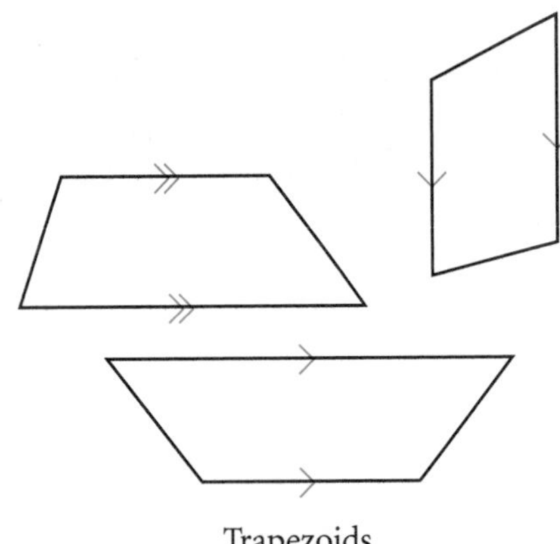

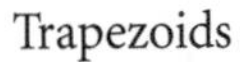

Trapezoids

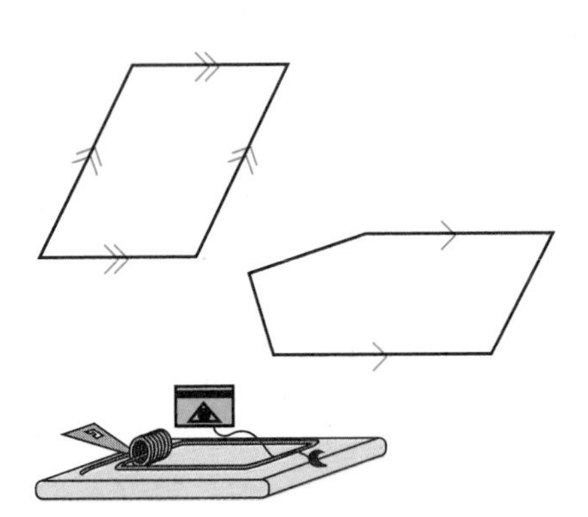

Not trapezoids

How does this ancient pyramid in Mexico use shapes for practical purposes and also for its overall attractiveness?

At the Acoma Pueblo Dwellings in New Mexico, how do sunlight and shadows enhance the existing shape formations?

How many shapes make up the overall triangular shapes of these pyramids at the Louvre in Paris?

Kite

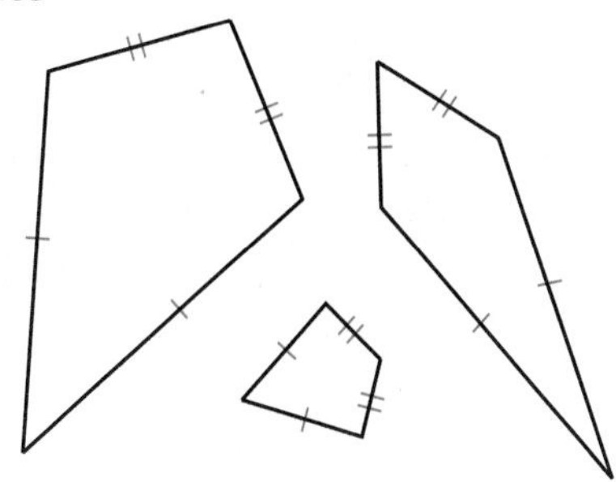

Kites

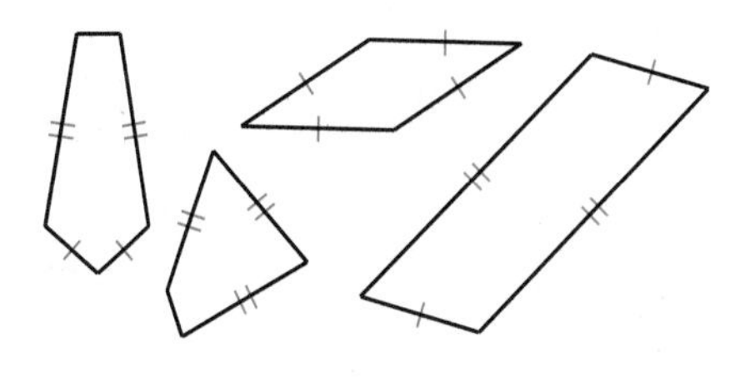

Not kites

Today's kite designers use lightweight plastics, synthetic fabrics, and complex shapes to sustain kites in the air longer than earlier kites that were made of wood, cloth, and had the basic "kite" shape. Many countries even hold annual kite festivals where contestants fly flat kites, box kites, and fighter kites. The design will determine the fastest and most durable kite in the festival.

Parallelogram

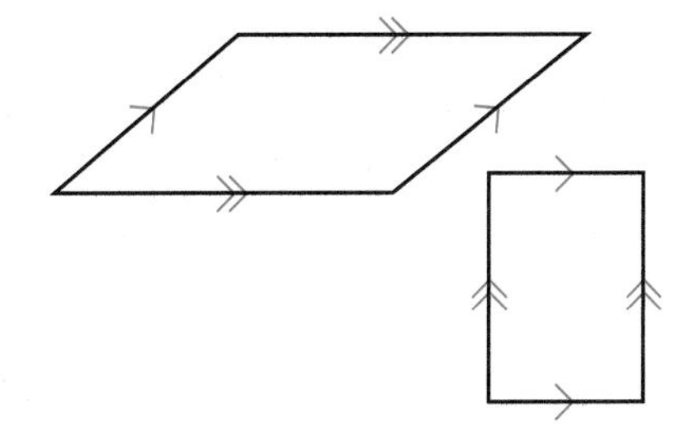

Parallelograms

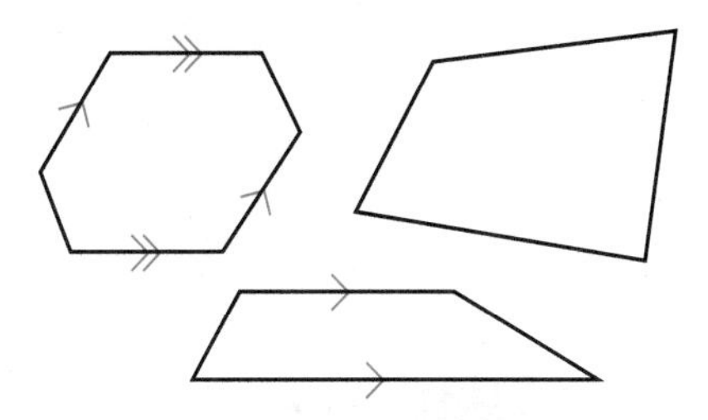

Not parallelograms

Rhombus

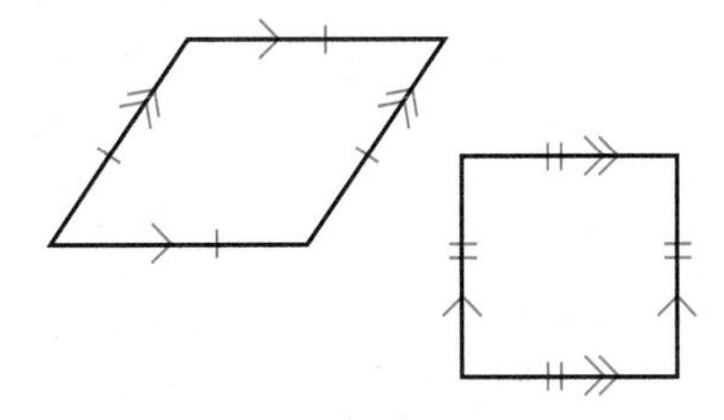

Rhombuses

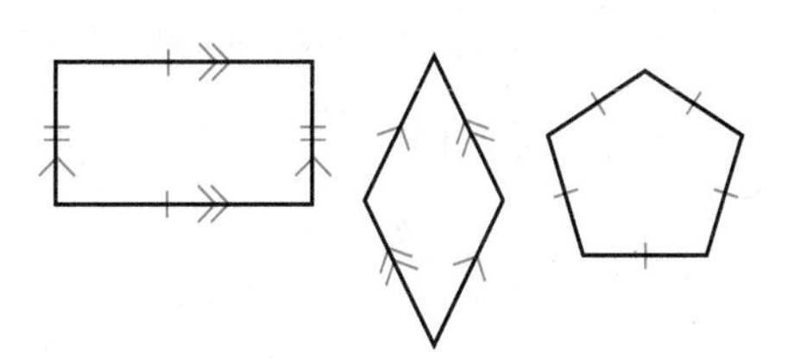

Not rhombuses

Rectangle

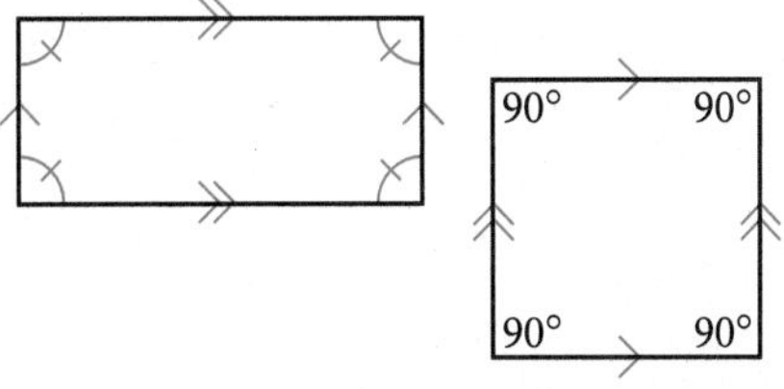

Rectangles

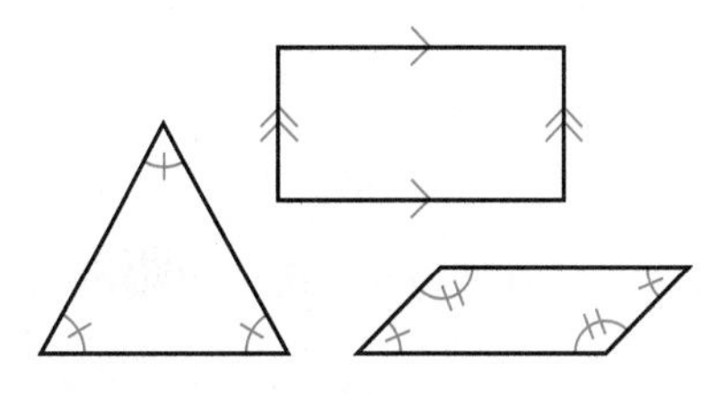

Not rectangles

Square

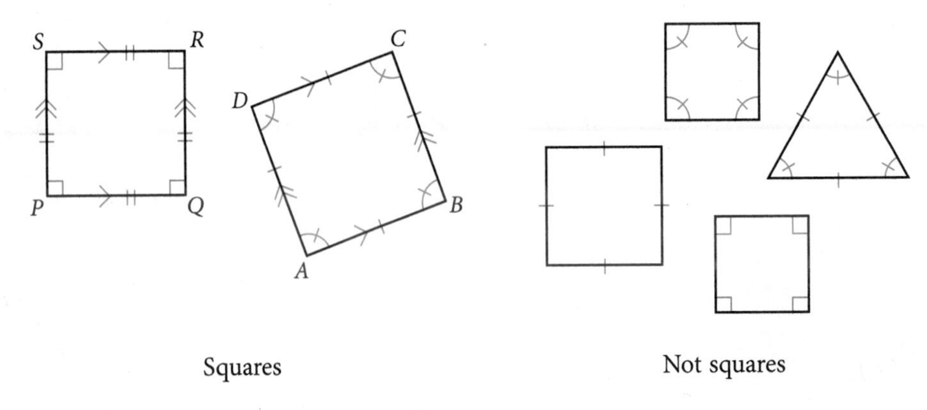

Squares

Not squares

As you learned in the investigation, a square is not a square unless it has the proper markings. Keep this in mind as you work on the exercises.

EXERCISES

1. Based on the marks, what can you assume to be true in each figure?

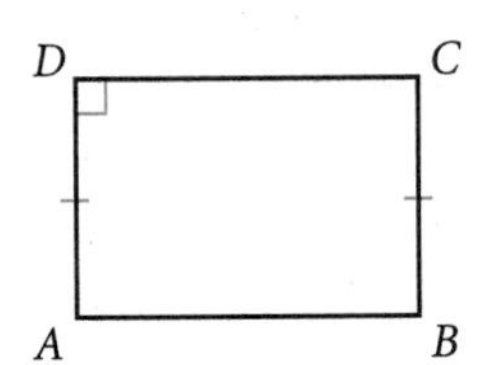

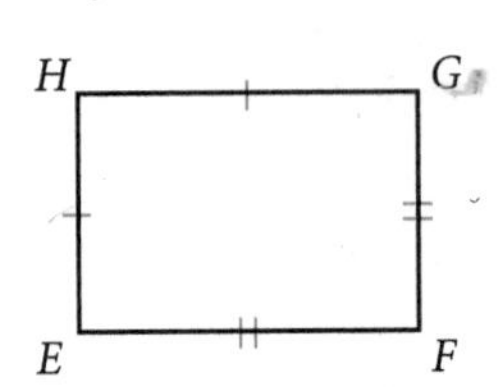

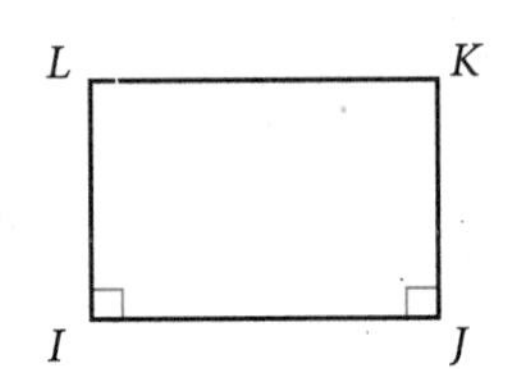

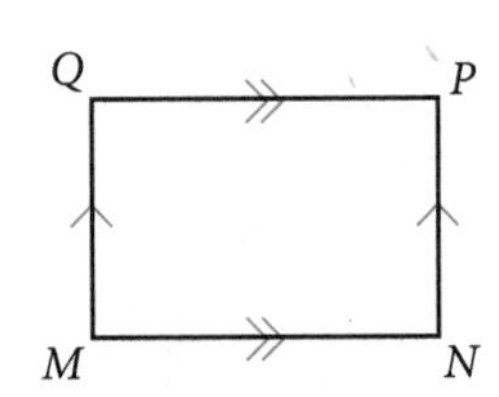

For Exercises 2–9, match the term on the left with its figure on the right.

2. Scalene right triangle
3. Isosceles right triangle
4. Isosceles obtuse triangle
5. Trapezoid
6. Rhombus
7. Rectangle
8. Kite
9. Parallelogram

A.

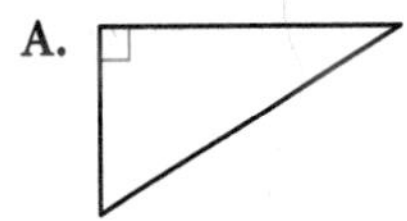

B.

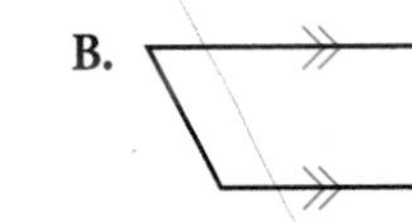

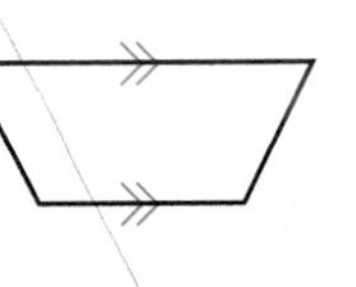

C.

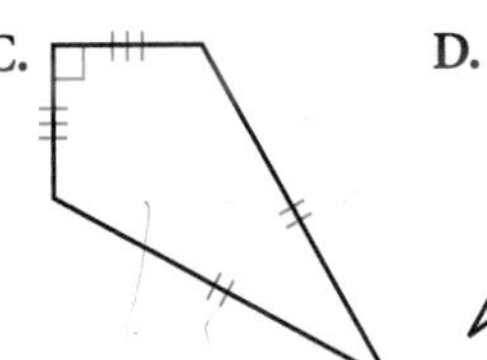

D.

E.

F.

G.

H.

I.

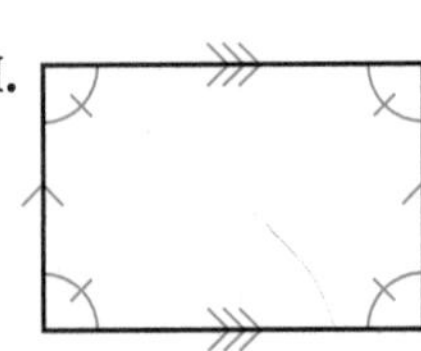

For Exercises 10–18, sketch and label the figure. Mark the figures.

10. Isosceles acute triangle ACT with $AC = CT$

11. Scalene triangle SCL with angle bisector $\overline{CM}$

12. Isosceles right triangle CAR with $m\angle CRA = 90°$

13. Trapezoid $ZOID$ with $ZO \parallel ID$

14. Kite $BENF$ with $BE = EN$

15. Rhombus $EQUL$ with diagonals EU and QL intersecting at A

16. Rectangle $RGHT$ with diagonals RH and GT intersecting at I

17. Two different isosceles triangles with perimeter $4a + b$

18. Two noncongruent triangles, each with side 6 cm and an angle measuring 40°

19. Draw a hexagon with exactly two outside diagonals.

20. Draw a regular quadrilateral. What is another name for this shape?

21. Find the other two vertices of a square with one vertex (0, 0) and another vertex (4, 2). Can you find another answer?

Austrian architect and artist Friedensreich Hundertwasser (1928–2000) designed the apartment house Hundertwasser-House (1986) with a square spiral staircase in its center. What other shapes do you see?

For Exercises 22–24, use the graphs below. Can you find more than one answer?

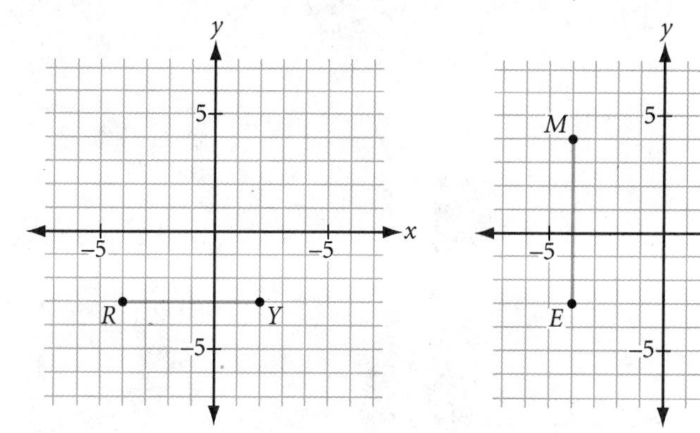

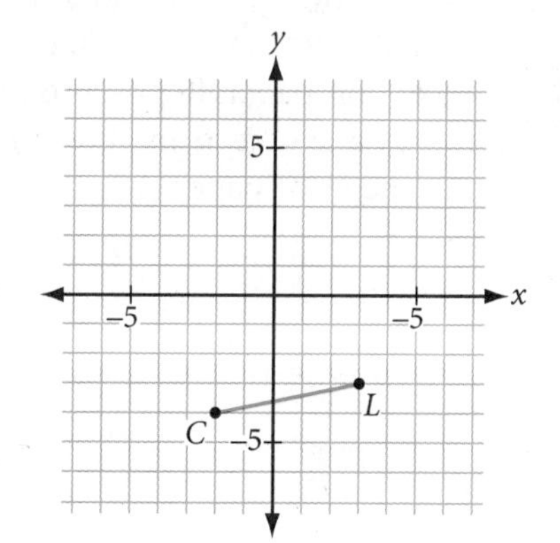

22. Locate a point L so that $\triangle LRY$ is an isosceles triangle.

23. Locate a point O so that $\triangle MOE$ is an isosceles right triangle.

24. Locate a point R so that $\triangle CRL$ is an isosceles right triangle. ⓗ

Review

For Exercises 25–29, tell whether the statement is true or false. For each false statement, sketch a counterexample or explain why the statement is false.

25. An acute angle is an angle whose measure is less than 90°.

26. If two lines intersect to form a right angle, then the lines are perpendicular.

27. A diagonal is a line segment that connects any two vertices of a polygon.

28. A ray that divides the angle into two angles is the angle bisector.

29. An obtuse triangle has exactly one angle whose measure is greater than 90°.

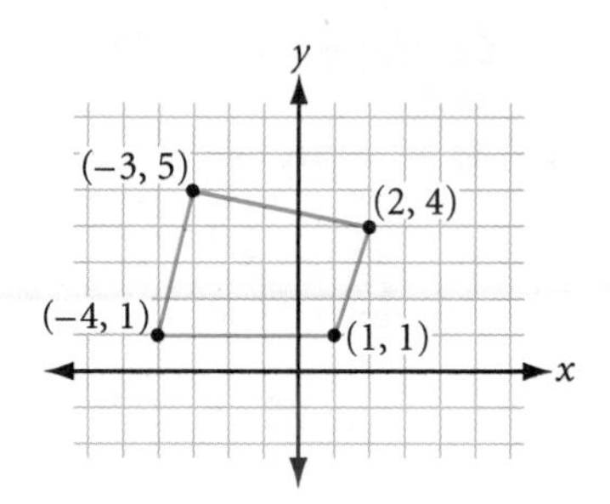

30. Use the ordered pair rule $(x, y) \rightarrow (x + 1, y - 3)$ to relocate the four vertices of the given quadrilateral. Connect the four new points to create a new quadrilateral. Do the two quadrilaterals appear congruent? Check your guess with tracing paper or patty paper.

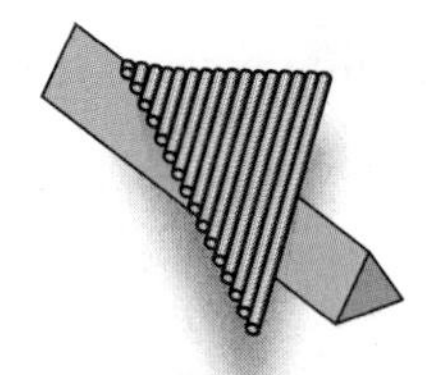

31. Suppose a set of thin rods is glued together into a triangle as shown. How would you place the triangular arrangement of rods onto the edge of a ruler so that they balance? Explain why. ⓗ

project

DRAWING THE IMPOSSIBLE

You experienced some optical illusions with op art in Chapter 0. Some optical illusions are tricks—they at first appear to be drawings of real objects, but actually they are impossible to make, except on paper. Reproduce the two impossible objects by drawing them on full sheets of paper.

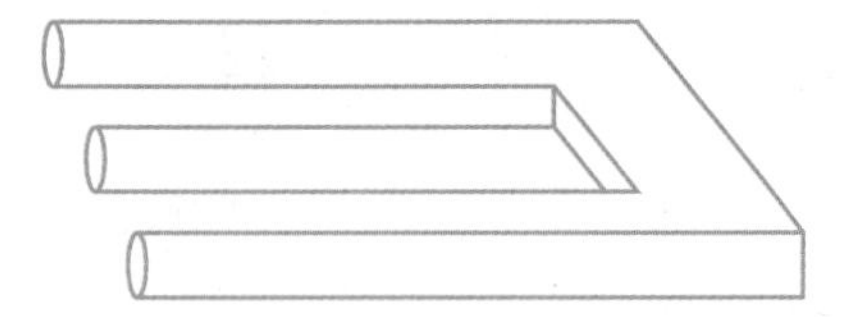

Can you create an impossible object of your own? Try it.

From WALTER WICK'S OPTICAL TRICKS. Published by Cartwheel Books, a division of Scholastic Inc. ©1998 by Walter Wick. Reprinted by permission.

LESSON

1.6

Circles

I can never remember things I didn't understand in the first place.

AMY TAN

Unless you walked to school this morning, you arrived on a vehicle with circular wheels.

A **circle** is the set of all points in a plane at a given distance (radius) from a given point (center) in the plane. You name a circle by its center. The circle on the bicycle wheel, with center O, is called circle O. When you see a dot at the center of a circle, you can assume that it represents the center point.

A segment from the center to a point on the edge of the circle is called a **radius.** Its length is also called the radius.

The **diameter** is a line segment containing the center, with its endpoints on the circle. The length of this segment is also called the diameter.

If two or more circles have the same radius, they are **congruent circles.** If two or more coplanar circles share the same center, they are **concentric circles.**

Congruent circles

Concentric circles

An **arc of a circle** is two points on the circle and the continuous (unbroken) part of the circle between the two points. The two points are called the **endpoints** of the arc.

You write arc AB as $\overset{\frown}{AB}$ or $\overset{\frown}{BA}$. You classify arcs into three types: semicircles, minor arcs, and major arcs. A **semicircle** is an arc of a circle whose endpoints are the endpoints of a diameter. A **minor arc** is an arc of a circle that is smaller than a semicircle. A **major arc** is an arc of a circle that is larger than a semicircle. You can name minor arcs with the letters of the two endpoints. For semicircles and major arcs, you need three points to make clear which arc you mean—the first and last letters are the endpoints and the middle letter is any other point on the arc.

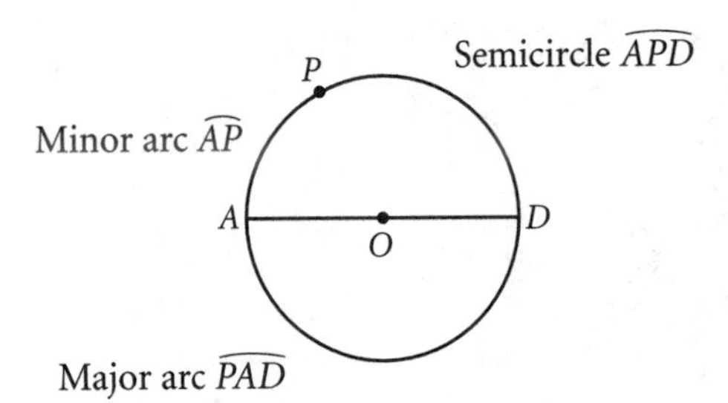

Arcs have a degree measure, just as angles do. A full circle has an arc measure of 360°, a semicircle has an arc measure of 180°, and so on. You find the **arc measure** by measuring the **central angle,** the angle with its vertex at the center of the circle, and sides passing through the endpoints of the arc.

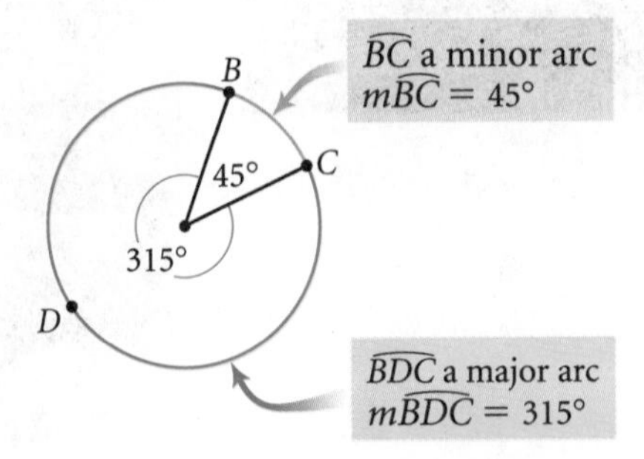

Investigation

Defining Circle Terms

Step 1 Write a good definition of each boldfaced term. Discuss your definitions with others in your group. Agree on a common set of definitions as a class and add them to your definition list. In your notebook, draw and label a figure to illustrate each definition.

Chord

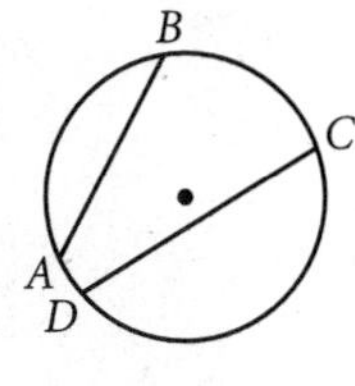

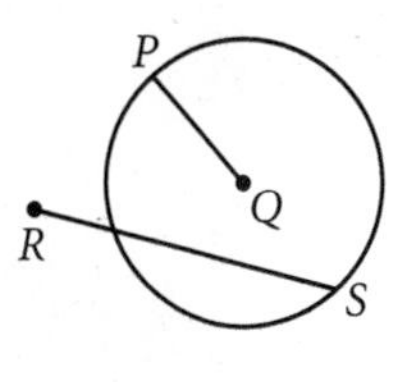

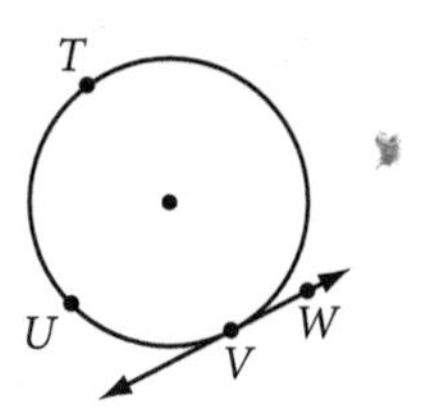

Chords:
$\overline{AB}$, $\overline{CD}$, $\overline{EF}$, $\overline{GH}$, and $\overline{IJ}$

Not chords:
$\overline{PQ}$, $\overline{RS}$, $\overparen{TU}$, and $\overleftrightarrow{VW}$

Diameter

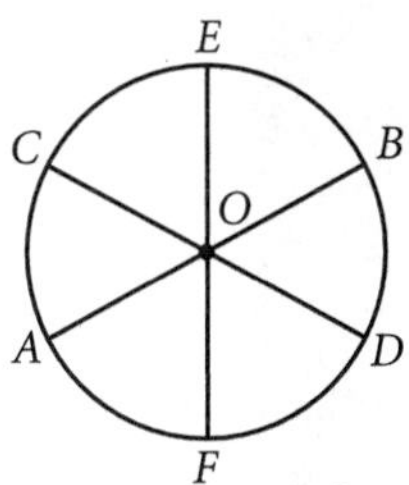

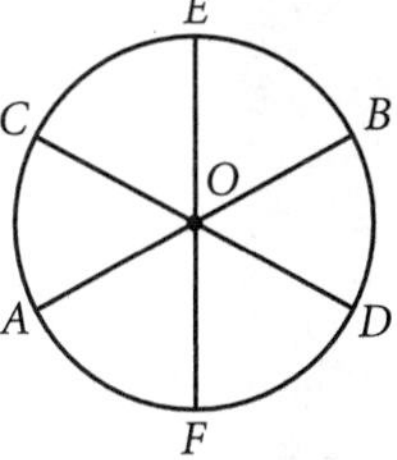

Diameters:
$\overline{AB}$, $\overline{CD}$, and $\overline{EF}$

Not diameters:
$\overline{PQ}$, $\overline{RS}$, $\overleftrightarrow{TU}$, and $\overline{VW}$

Tangent

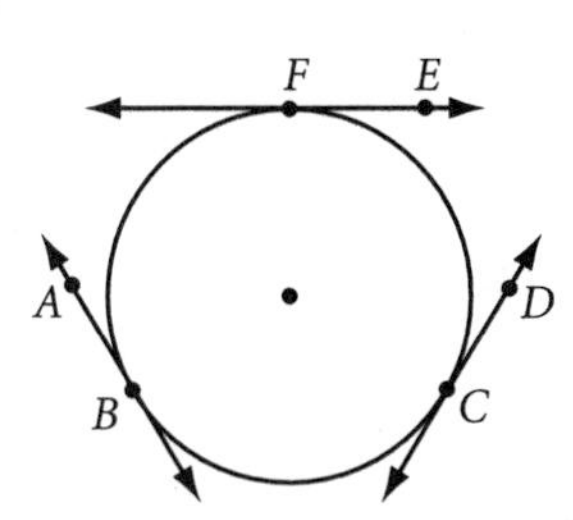

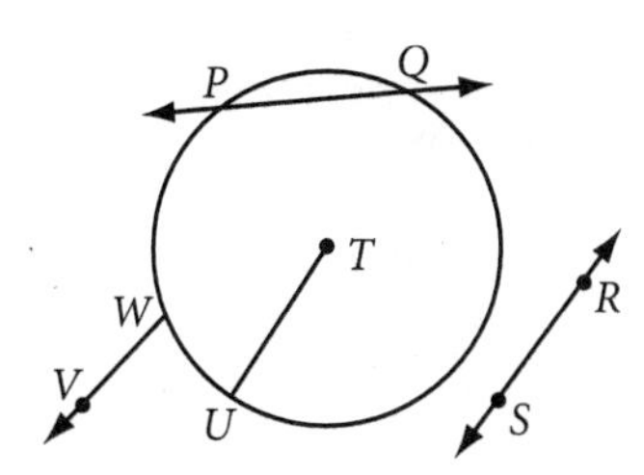

Tangents:
$\overleftrightarrow{AB}$, $\overleftrightarrow{CD}$, and $\overleftrightarrow{EF}$

Not tangents:
$\overleftrightarrow{PQ}$, $\overleftrightarrow{RS}$, $\overline{TU}$, and $\overline{VW}$

Note: You can say $\overleftrightarrow{AB}$ is a tangent, or you can say $\overleftrightarrow{AB}$ is tangent to circle O. The point where the tangent touches the circle is called the **point of tangency.**

Step 2 Can a chord of a circle also be a diameter of the circle? Can it be a tangent? Explain why or why not.

Step 3 Can two circles be tangent to the same line at the same point? Draw a sketch and explain.

In each photo, find examples of the terms introduced in this lesson.

Japanese wood bridge

Circular irrigation on a farm

CAD design of pistons in a car engine

EXERCISES

For Exercises 1–8, use the diagram at right. Points E, P, and C are collinear.

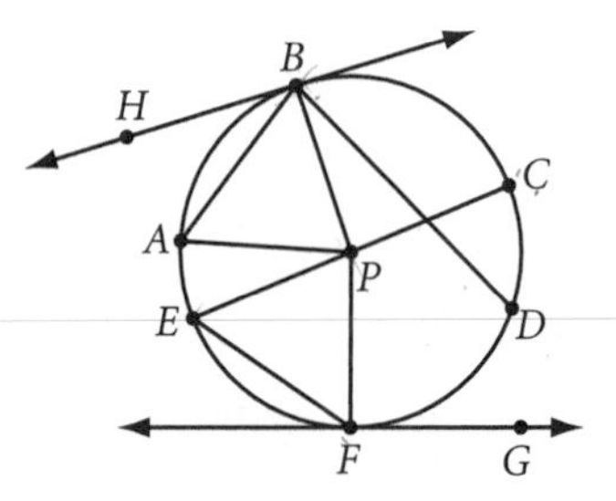

1. Name three chords.

2. Name one diameter.

3. Name five radii.

4. Name five minor arcs.

5. Name two semicircles.

6. Name two major arcs.

7. Name two tangents.

8. Name a point of tangency.

9. Name two types of vehicles that use wheels, two household appliances that use wheels, and two uses of the wheel in the world of entertainment.

10. In the figure at right, what is $m\overset{\frown}{PQ}$? $m\overset{\frown}{PRQ}$?

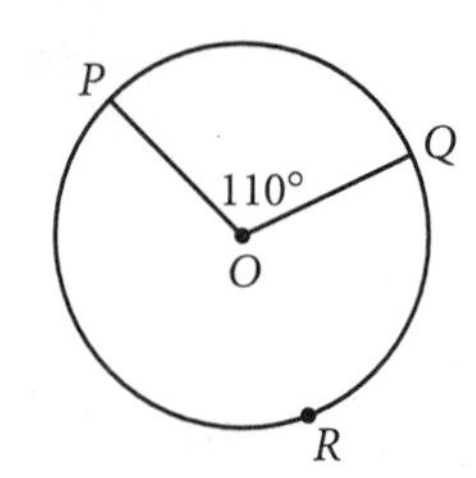

11. Use your compass and protractor to make an arc with measure 65°. Now make an arc with measure 215°. Label each arc with its measure.

12. Name two places or objects where concentric circles appear. Bring an example of a set of concentric circles to class tomorrow. You might look in a magazine for a photo or make a copy of a photo from a book (but not this book!).

13. Sketch two circles that appear to be concentric. Then use your compass to construct a pair of concentric circles.

14. Sketch circle P. Sketch a triangle inside circle P so that the three sides of the triangle are chords of the circle. This triangle is "inscribed" in the circle. Sketch another circle and label it Q. Sketch a triangle in the exterior of circle Q so that the three sides of the triangle are tangents of the circle. This triangle is "circumscribed" about the circle.

15. Use your compass to construct two circles with the same radius intersecting at two points. Label the centers P and Q. Label the points of intersection of the two circles A and B. Construct quadrilateral $PAQB$. What type of quadrilateral is it?

16. Do you remember the daisy construction from Chapter 0? Construct a circle with radius s. With the same compass setting, divide the circle into six congruent arcs. Construct the chords to form a regular hexagon inscribed in the circle. Construct radii to each of the six vertices. What type of triangles are formed? What is the ratio of the perimeter of the hexagon to the diameter of the circle?

17. Sketch the path made by the midpoint of a radius of a circle if the radius is rotated about the center.

For Exercises 18–20, use the ordered pair rule shown to relocate the four points on the given circle. Can the four new points be connected to create a new circle? Does the new figure appear congruent to the original circle?

18. $(x, y) \rightarrow (x - 1, y + 2)$

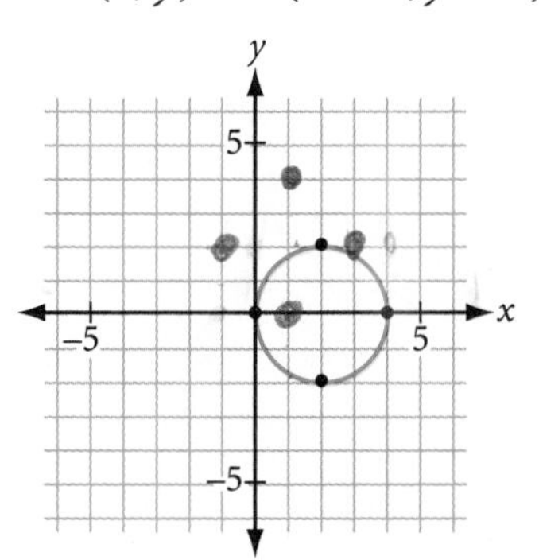

19. $(x, y) \rightarrow (2x, 2y)$ ⓗ

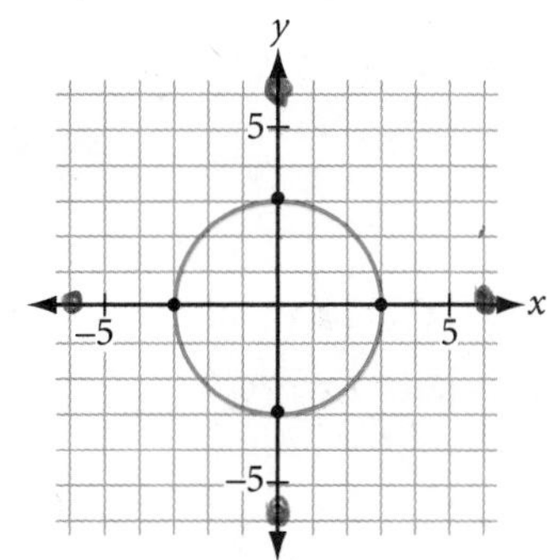

20. $(x, y) \rightarrow (2x, y)$

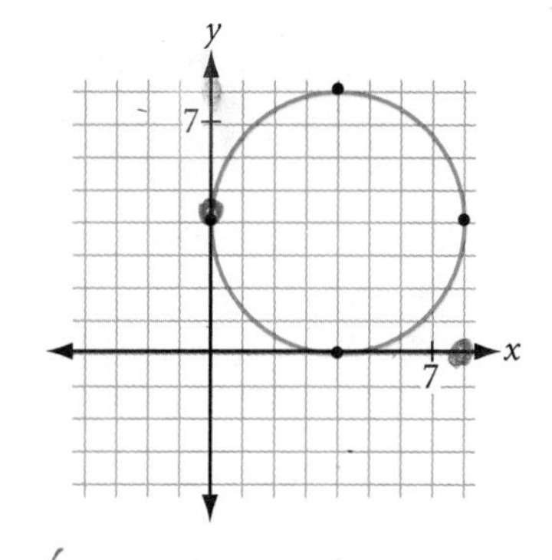

Review

For Exercises 21–24, draw each kind of triangle or write "not possible" and explain why. Use your geometry tools to make your drawings as accurate as possible.

21. Isosceles right triangle

22. Scalene isosceles triangle

23. Scalene obtuse triangle

24. Isosceles obtuse triangle

For Exercises 25–33, sketch, label, and mark the figure.

25. Obtuse scalene triangle FAT with $m\angle FAT = 100°$

26. Trapezoid $TRAP$ with $\overline{TR} \parallel \overline{AP}$ and $\angle TRA$ a right angle

27. Two different (noncongruent) quadrilaterals with angles of 60°, 60°, 120°, and 120°

28. Equilateral right triangle

29. Right isosceles triangle RGT with $RT = GT$ and $m\angle RTG = 90°$

30. An equilateral triangle with perimeter $12a + 6b$

31. Two triangles that are not congruent, each with angles measuring 50° and 70°

32. Rhombus $EQUI$ with perimeter $8p$ and $m\angle IEQ = 55°$

33. Kite $KITE$ with $TE = 2EK$ and $m\angle TEK = 120°$

IMPROVING YOUR REASONING SKILLS

PUZZLE

Checkerboard Puzzle

1. Four checkers—three red and one black—are arranged on the corner of a checkerboard, as shown at right. Any checker can jump any other checker. The checker that was jumped over is then removed. With exactly three horizontal or vertical jumps, remove all three red checkers, leaving the single black checker. Record your solution.

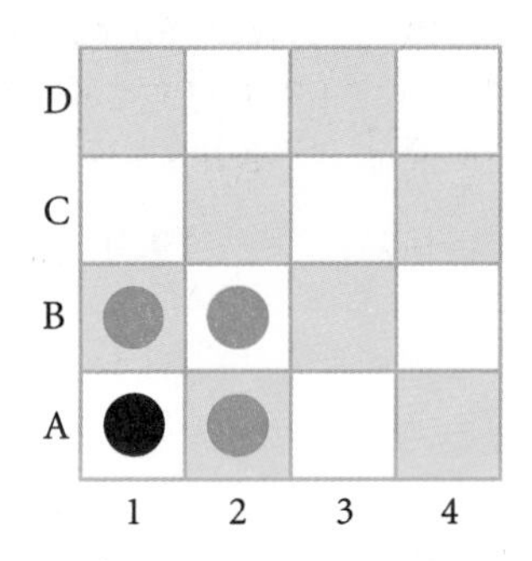

2. Now, with exactly seven horizontal or vertical jumps, remove all seven red checkers, leaving the single black checker. Record your solution.

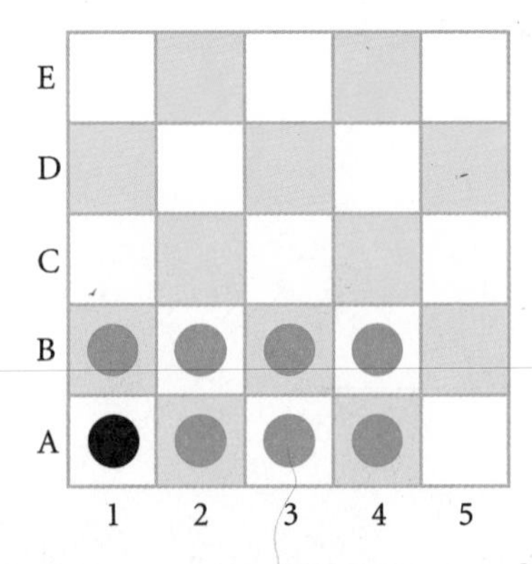

LESSON 1.7

A Picture Is Worth a Thousand Words

You can observe a lot just by watching.

YOGI BERRA

A picture is worth a thousand words! That expression certainly applies to geometry. A drawing of an object often conveys information more quickly than a long written description. People in many occupations use drawings and sketches to communicate ideas. Architects create blueprints. Composers create musical scores. Choreographers visualize and map out sequences of dance steps. Basketball coaches design plays. Interior designers—well, you get the picture.

Visualization skills are extremely important in geometry. So far, you have visualized geometric situations in every lesson. To visualize a plane, you pictured a flat surface extending infinitely. In another lesson, you visualized the number of different ways that four lines can intersect. Can you picture what the hands of a clock look like when it is 3:30?

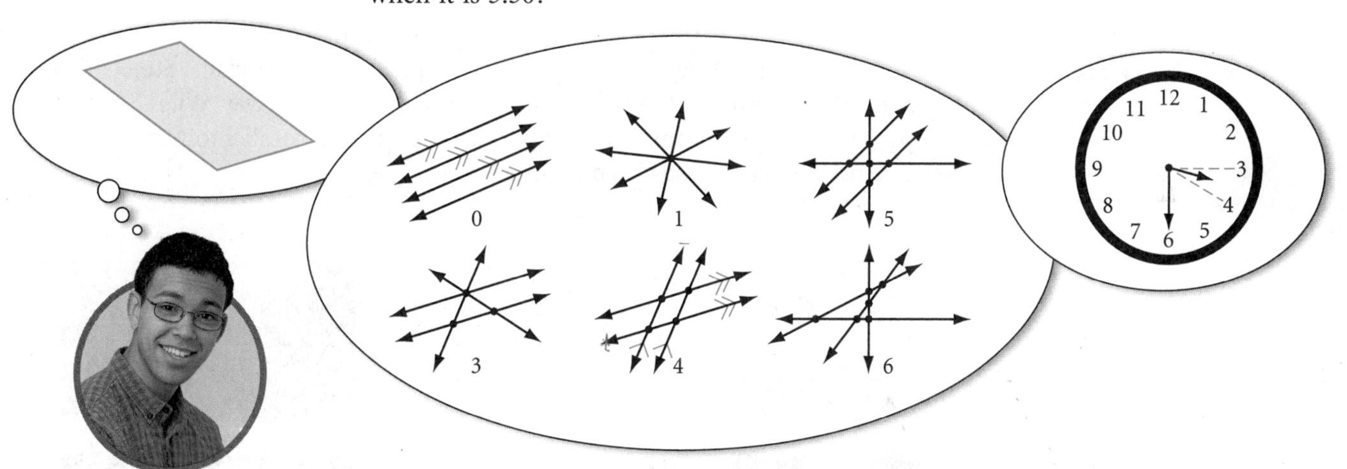

By drawing diagrams, you apply visual thinking to problem solving. Let's look at some examples that show how to use visual thinking to solve word problems.

EXAMPLE A

Volumes 1 and 2 of a two-volume set of math books sit next to each other on a shelf. They sit in their proper order: Volume 1 on the left and Volume 2 on the right. Each front and back cover is $\frac{1}{8}$ inch thick, and the pages portion of each book is 1 inch thick. If a bookworm starts at the first page of Volume 1 and burrows all the way through to the last page of Volume 2, how far will she travel?

Take a moment and try to solve the problem in your head.

Bookplate for Albert Ernst Bosman, M. C. Escher, 1946

▸ Solution

Did you get $2\frac{1}{4}$ inches? It seems reasonable, doesn't it?

Guess what? That's not the answer. Let's get organized. Reread the problem to identify what information you are given and what you are trying to find.

You are given the thickness of each cover, the thickness of the page portion, and the position of the books on the shelf. You are trying to find how far it is from the first page of Volume 1 to the last page of Volume 2. Draw a picture and locate the position of the pages referred to in the problem.

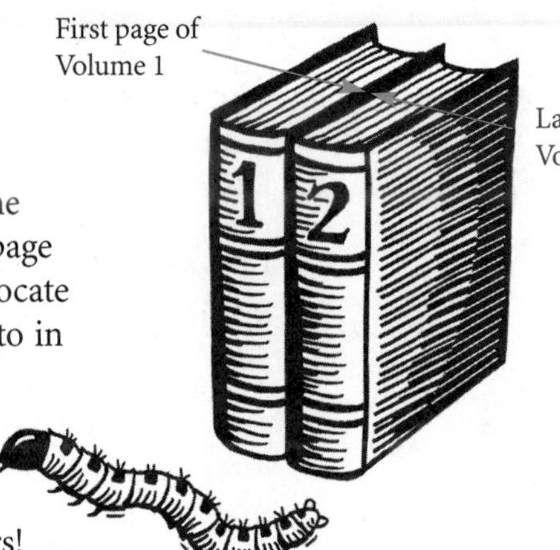

Now "look" how easy it is to solve the problem. She traveled only $\frac{1}{4}$ inch through the two covers!

EXAMPLE B

In Reasonville, many streets are named after famous mathematicians. Streets that end in an "s" run east-west. All other streets might run either way. Wiles Street runs perpendicular to Germain Street. Fermat Street runs parallel to Germain Street. Which direction does Fermat Street run?

Mathematics CONNECTION

The French mathematician Pierre de Fermat (1601–1665) developed analytic geometry. His algebraic approach is what made his influence on geometry so strong.

Sophie Germain (1776–1831), a French mathematician with no formal education, wrote a prize treatise, contributed to many theories, and worked extensively on Fermat's Last Theorem.

Andrew Wiles (b 1953), an English mathematician at Princeton University, began trying to prove Fermat's Last Theorem when he was just 10 years old. In 1993, after spending most of his career working on the theorem, sometimes in complete isolation, he announced a proof of the problem.

▸ Solution

Did you make a diagram? You can start your diagram with the first piece of information. Then you can add to the diagram as new information is added. Wiles Street ends in an "s," so it runs east-west. You are trying to find the direction of Fermat Street.

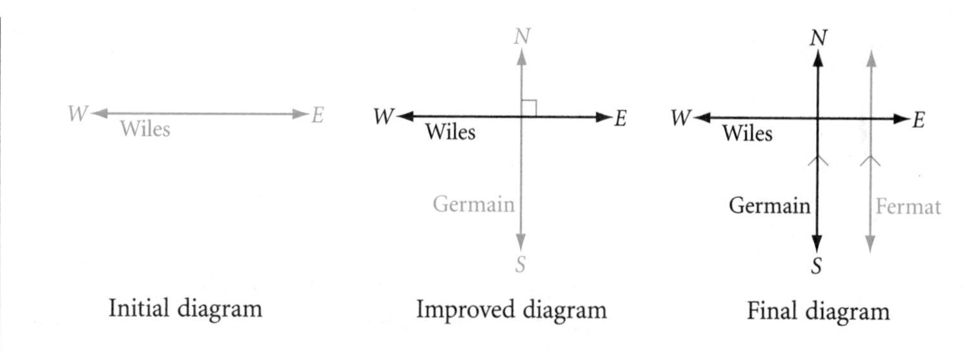

The final diagram reveals the answer. Fermat Street runs north–south.

Sometimes there is more than one point or even many points that satisfy a set of conditions. The set of points is called a **locus** of points. Let's look at an example showing how to solve a locus problem.

EXAMPLE C

Harold, Dina, and Linda are standing on a flat, dry field reading their treasure map. Harold is standing at one of the features marked on the map, a gnarled tree stump, and Dina is standing atop a large black boulder. The map shows that the treasure is buried 60 meters from the tree stump and 40 meters from the large black boulder. Harold and Dina are standing 80 meters apart. What is the locus of points where the treasure might be buried?

▶ Solution

Start by drawing a diagram based on the information given in the first two sentences, then add to the diagram as new information is added. Can you visualize all the points that are 60 meters from the tree stump? Mark them on your diagram. They should lie on a circle. The treasure is also 40 meters from the boulder. All the possible points lie in a circle around the boulder. The two possible spots where the treasure might be buried, or the locus of points, are the points where the two circles intersect.

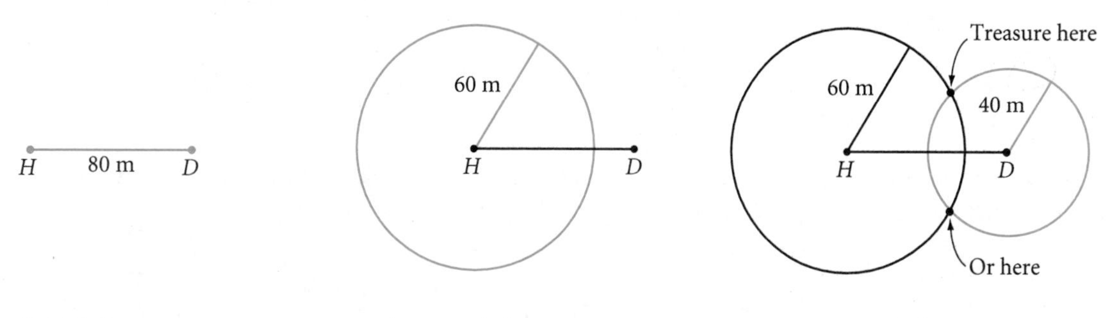

EXERCISES

You will need

Construction tools for Exercises **33** and **34**

1. Surgeons, engineers, carpenters, plumbers, electricians, and furniture movers all rely on trained experience with visual thinking. Describe how one of these tradespeople or someone in another occupation uses visual thinking in his or her work.

Now try your hand at some word problems. Read each problem carefully, determine what you are trying to find, and draw and label a diagram. Finally, solve the problem.

2. In the city of Rectangulus, all the streets running east–west are numbered and those streets running north–south are lettered. The even-numbered streets are one-way east and the odd-numbered streets are one-way west. All the vowel-lettered avenues are one-way north and the rest are two-way. Can a car traveling south on S Street make a legal left turn onto 14th Street?

3. Freddie the Frog is at the bottom of a 30-foot well. Each day he jumps up 3 feet, but then, during the night, he slides back down 2 feet. How many days will it take Freddie to get to the top and out? ⓗ

4. Mary Ann is building a fence around the outer edge of a rectangular garden plot that measures 25 feet by 45 feet. She will set the posts 5 feet apart. How many posts will she need?

5. Midway through a 2000-meter race, a photo is taken of five runners. It shows Meg 20 meters behind Edith. Edith is 50 meters ahead of Wanda, who is 20 meters behind Olivia. Olivia is 40 meters behind Nadine. Who is ahead? In your diagram, use M for Meg, E for Edith, and so on.

6. Here is an exercise taken from Marilyn vos Savant's Ask Marilyn® column in *Parade* magazine. It is a good example of a difficult-sounding problem becoming clear once a diagram has been made. Try it. ⓗ

   A 30-foot cable is suspended between the tops of two 20-foot poles on level ground. The lowest point of the cable is 5 feet above the ground. What is the distance between the two poles?

7. Points A and B lie in a plane. Sketch the locus of points in *the plane* that are equally distant from points A and B. Sketch the locus of points in *space* that are equally distant from points A and B. ⓗ

8. Draw an angle. Label it $\angle A$. Sketch the locus of points in the plane of angle A that are the same distance from the two sides of angle A.

9. Line AB lies in plane $\mathcal{P}$. Sketch the locus of points in plane $\mathcal{P}$ that are 3 cm from $\overleftrightarrow{AB}$. Sketch the locus of points in space that are 3 cm from $\overleftrightarrow{AB}$.

10. Beth Mack and her dog Trouble are exploring in the woods east of Birnam Woods Road, which runs north-south. They begin walking in a zigzag pattern: 1 km south, 1 km west, 1 km south, 2 km west, 1 km south, 3 km west, and so on. They walk at the rate of 4 km/h. If they started 15 km east of Birnam Woods Road at 3:00 P.M., and the sun sets at 7:30 P.M., will they reach Birnam Woods Road before sunset?

In geometry you will use visual thinking all the time. In Exercises 11 and 12 you will be asked to locate and recognize congruent geometric figures even if they are in different positions due to translations (slides), rotations (turns), or reflections (flips).

11. If trapezoid $ABCD$ were rotated 90° counterclockwise about (0, 0), to what (x, y) location would points A, B, C, and D be relocated? ⓗ

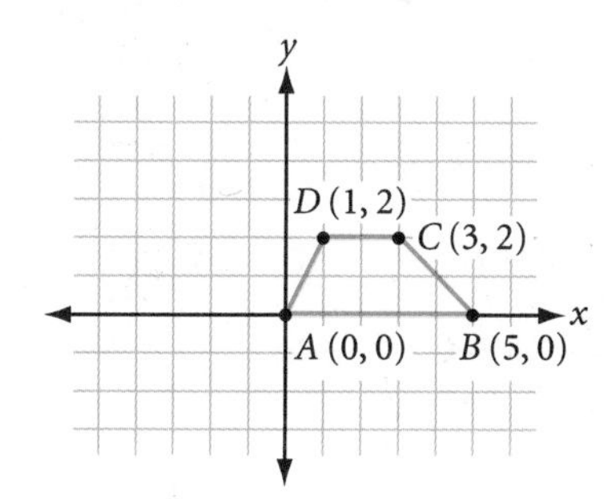

12. If $\triangle CYN$ were reflected over the y-axis, to what location would points C, N, and Y be relocated?

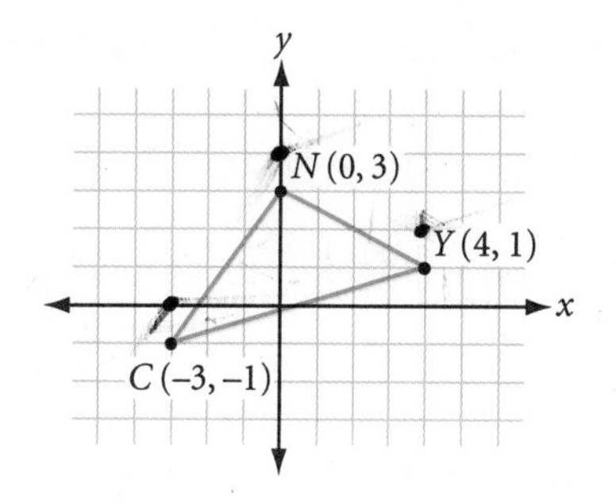

13. What was the ordered pair rule used to relocate the four vertices of $ABCD$ to $A'B'C'D'$?

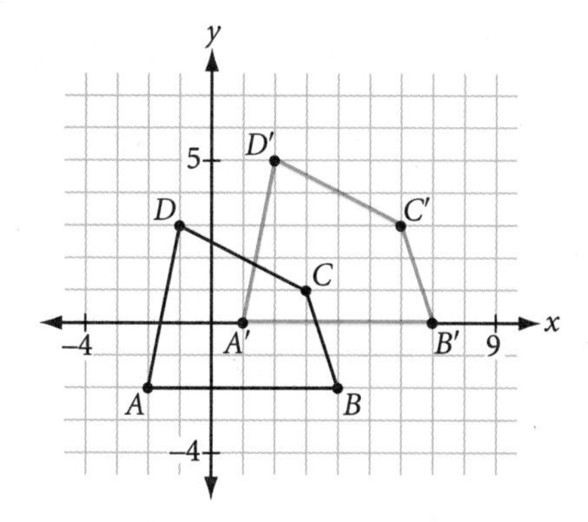

14. Which lines are perpendicular? Which lines are parallel?

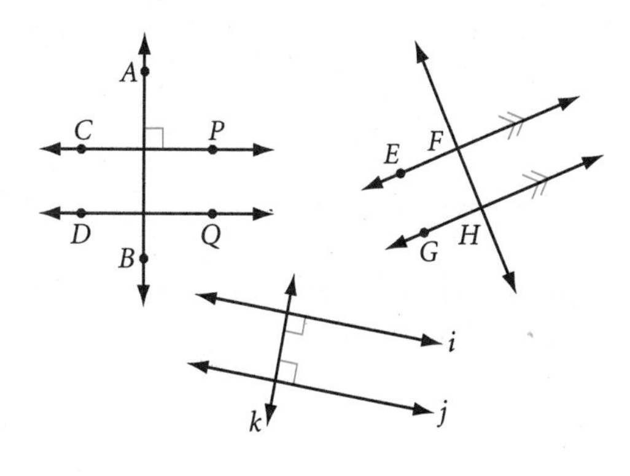

15. Sketch the next two figures in the pattern below. If this pattern were to continue, what would be the perimeter of the eighth figure in the pattern? (Assume the length of each segment is 1 cm.) ⓗ

, , , , . . .

16. Many of the geometric figures you have defined are closely related to one another. A diagram can help you see the relationships among them. At right is a concept map showing the relationships among members of the triangle family. This type of concept map is known as a **tree diagram** because the relationships are shown as branches of a tree. Copy and fill in the missing branches of the tree diagram for triangles.

17. At right is a concept map showing the relationships among some members of the parallelogram family. This type of concept map is known as a **Venn diagram.** Fill in the missing names.

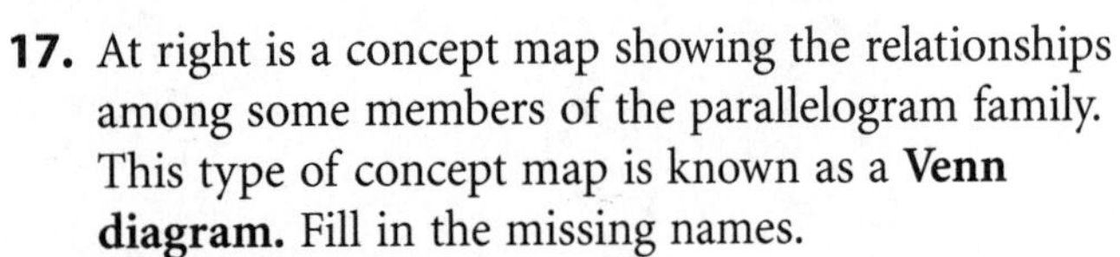

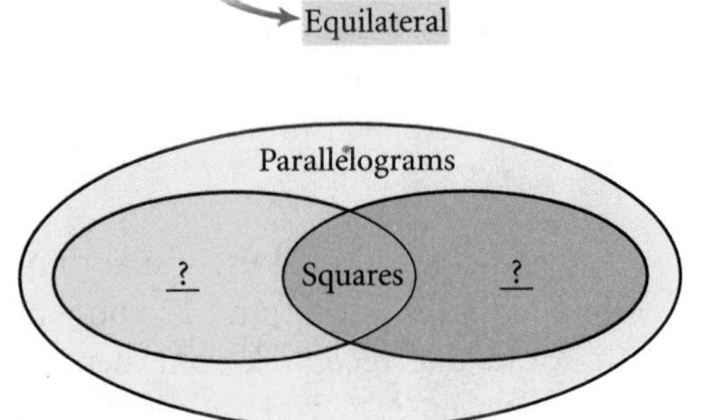

A **net** is a two-dimensional pattern that you can cut and fold to form a three-dimensional figure. Another visual thinking skill you will need is the ability to visualize nets being folded into solid objects and geometric solids being unfolded into nets. The net below left can be folded into a cube and the net below right can be folded into a pyramid.

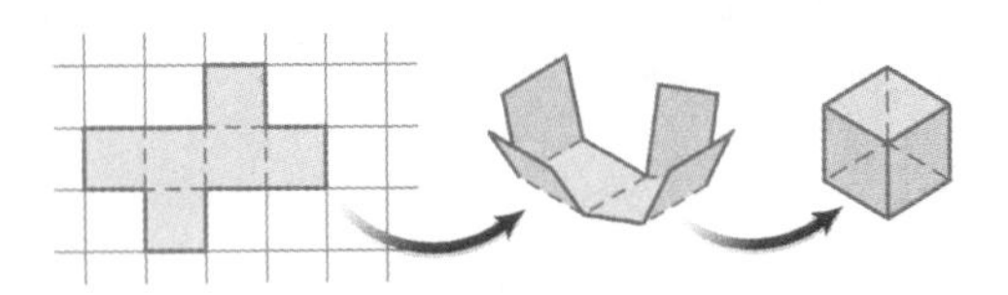

Net for a cube

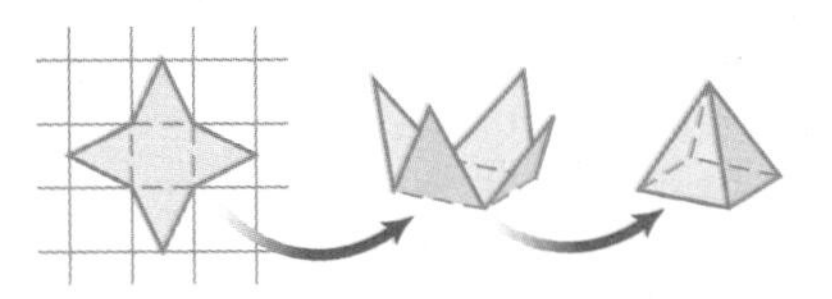

Net for a square-based pyramid

18. Which net(s) will fold to make a cube?

A. 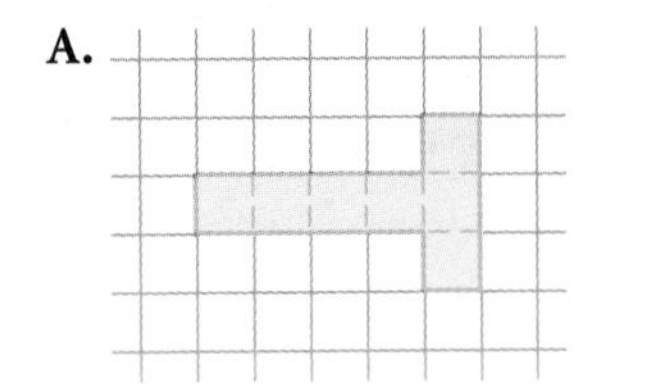

B.

C. 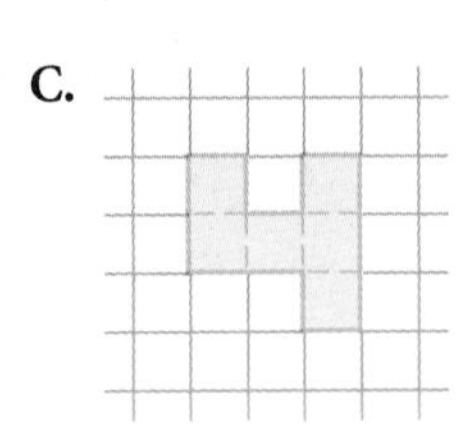

D. 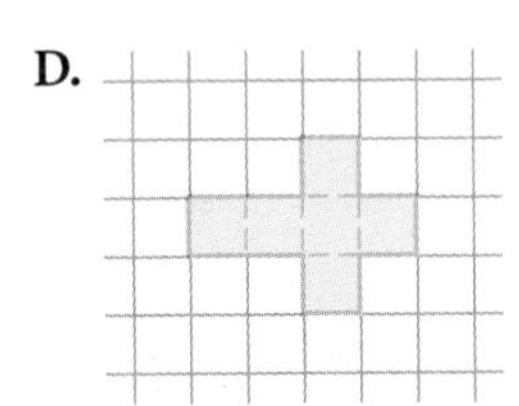

E. 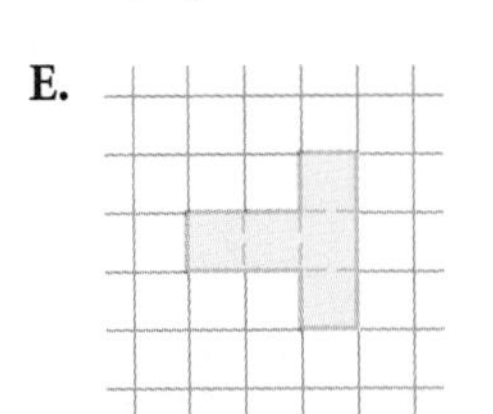

F. 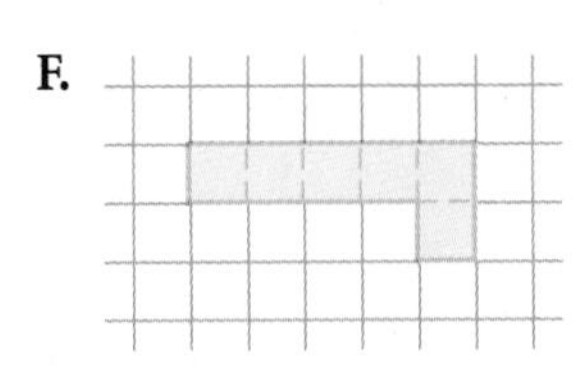

For Exercises 19–22, match the net with its geometric solid.

19. 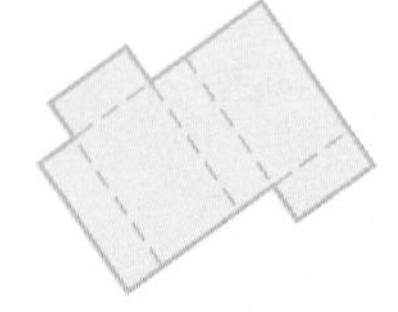

20. 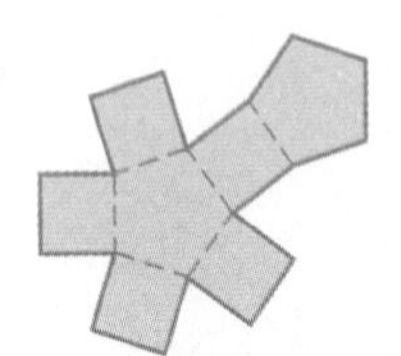

21. 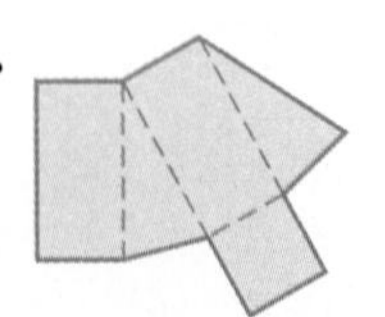

22. 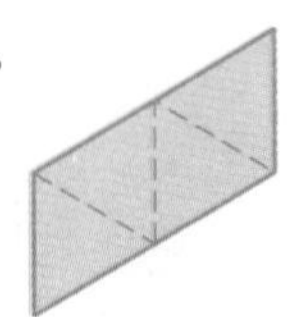

A. 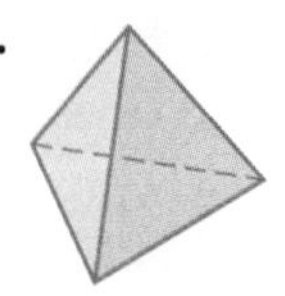

B. 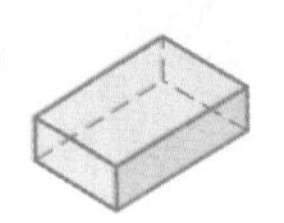

C. 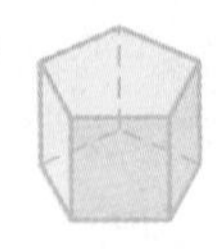

D. 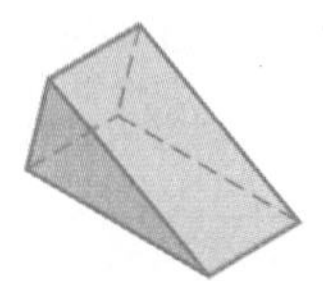

Review

For Exercises 23–32, write the words or the symbols that make the statement true.

William Thomas Williams, DO YOU THINK A IS B, acrylic on canvas, 1975–77, Fisk University Galleries, Nashville, Tennessee.

23. The three undefined terms of geometry are ?, ?, and ?.

24. "Line *AB*" may be written using a symbol as ?.

25. "Arc *AB*" may be written using a symbol as ?.

26. The point where the two sides of an angle meet is the ? of the angle.

27. "Ray *AB*" may be written using a symbol as ?.

28. "Line *AB* is parallel to segment *CD*" is written in symbolic form as ?.

29. The geometry tool you use to measure an angle is a ?.

30. "Angle *ABC*" is written in symbolic form as ?.

31. The sentence "Segment *AB* is perpendicular to line *CD*" is written in symbolic form as ?.

32. The angle formed by a light ray coming into a mirror is ? the angle formed by a light ray leaving the mirror.

33. Use your compass to draw two congruent circles intersecting in exactly one point. How does the distance between the two centers compare with the radius?

34. Use your compass to construct two congruent circles so that each circle passes through the center of the other circle. Label the centers *P* and *Q*. Construct $\overline{PQ}$ connecting the centers. Label the points of intersection of the two circles *A* and *B*. Construct chord $\overline{AB}$. What is the relationship between $\overline{AB}$ and $\overline{PQ}$? (h)

IMPROVING YOUR VISUAL THINKING SKILLS

Hexominoes

Polyominoes with six squares are called hexominoes. There are 35 different hexominoes. There is 1 with a longest string of six squares; there are 3 with a longest string of five squares, 13 with a longest string of four squares, 17 with a longest string of three squares; and there is 1 with a longest string of two squares. Use graph paper to sketch the 35 hexominoes. Which are nets for cubes? Here is one hexomino that does fold into a cube.

LESSON 1.8

Space Geometry

When curiosity turns to serious matters, it's called research.

MARIE VON EBNER-ESCHENBACH

Lesson 1.1 introduced you to point, line, and plane. Throughout this chapter you have used these terms to define a wide range of other geometric figures, from rays to polygons. You did most of your work on a single flat surface, a single plane. Some problems, however, required you to step out of a single plane to visualize geometry in space. In this lesson you will learn more about space geometry, or solid geometry.

Space is the set of all points. Unlike lines and planes, space cannot be contained in a flat surface. Space is three-dimensional, or "3-D."

In an "edge view," you see the front edge of a building as a vertical line, and the other edges as diagonal lines. Isometric dot paper helps you draw these lines, as you can see in the steps below.

Let's practice the visual thinking skill of presenting three-dimensional (3-D) objects in two-dimensional (2-D) drawings.

The geometric solid you are probably most familiar with is a box, or rectangular prism. Below are steps for making a two-dimensional drawing of a rectangular prism. This type of drawing is called an **isometric drawing.** It shows three sides of an object in one view (an edge view). This method works best with isometric dot grid paper. After practicing, you will be able to draw the box without the aid of the dot grid.

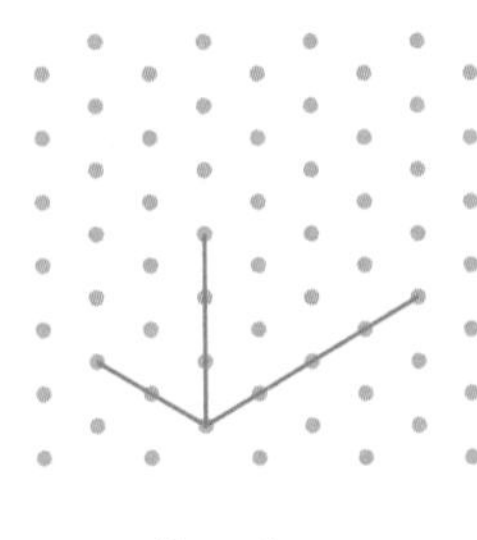

Step 1

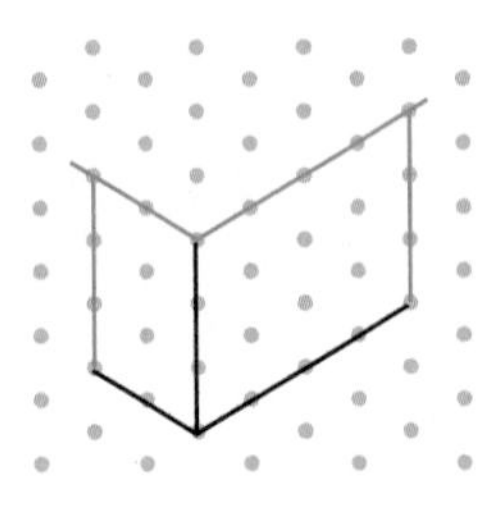

Step 2

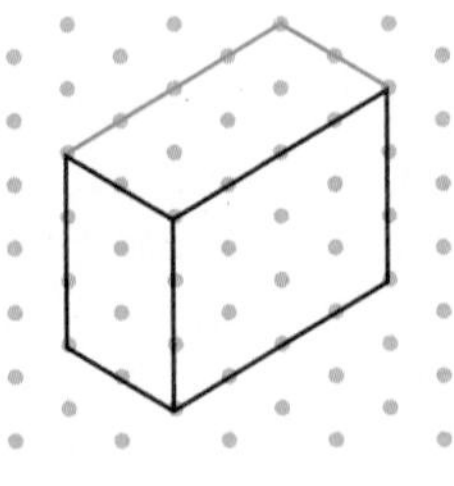

Step 3

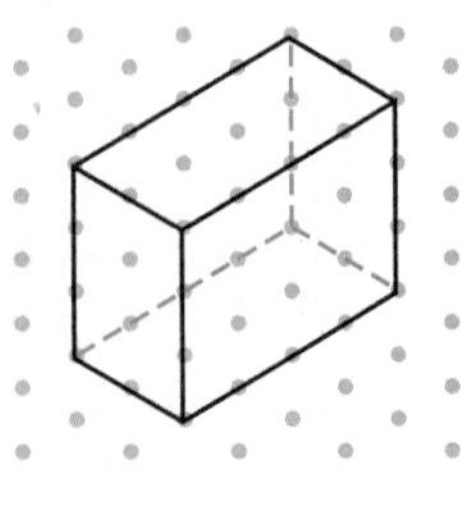

Step 4

Use dashed lines for edges that you couldn't see if the object were solid.

The three-dimensional objects you will study include the six types of geometric solids shown below.

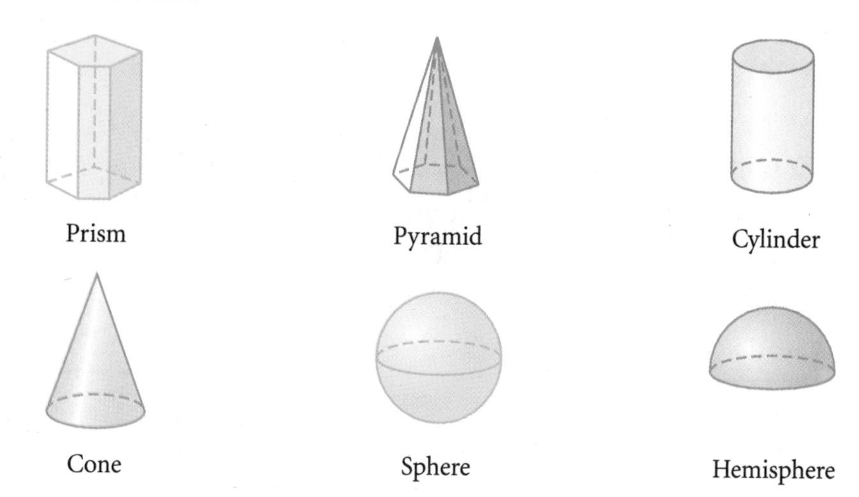

The shapes of these solids are probably already familiar to you even if you are not familiar with their proper names. The ability to draw these geometric solids is an important visual thinking skill. Here are some drawing tips. Remember to use dashes for the hidden lines.

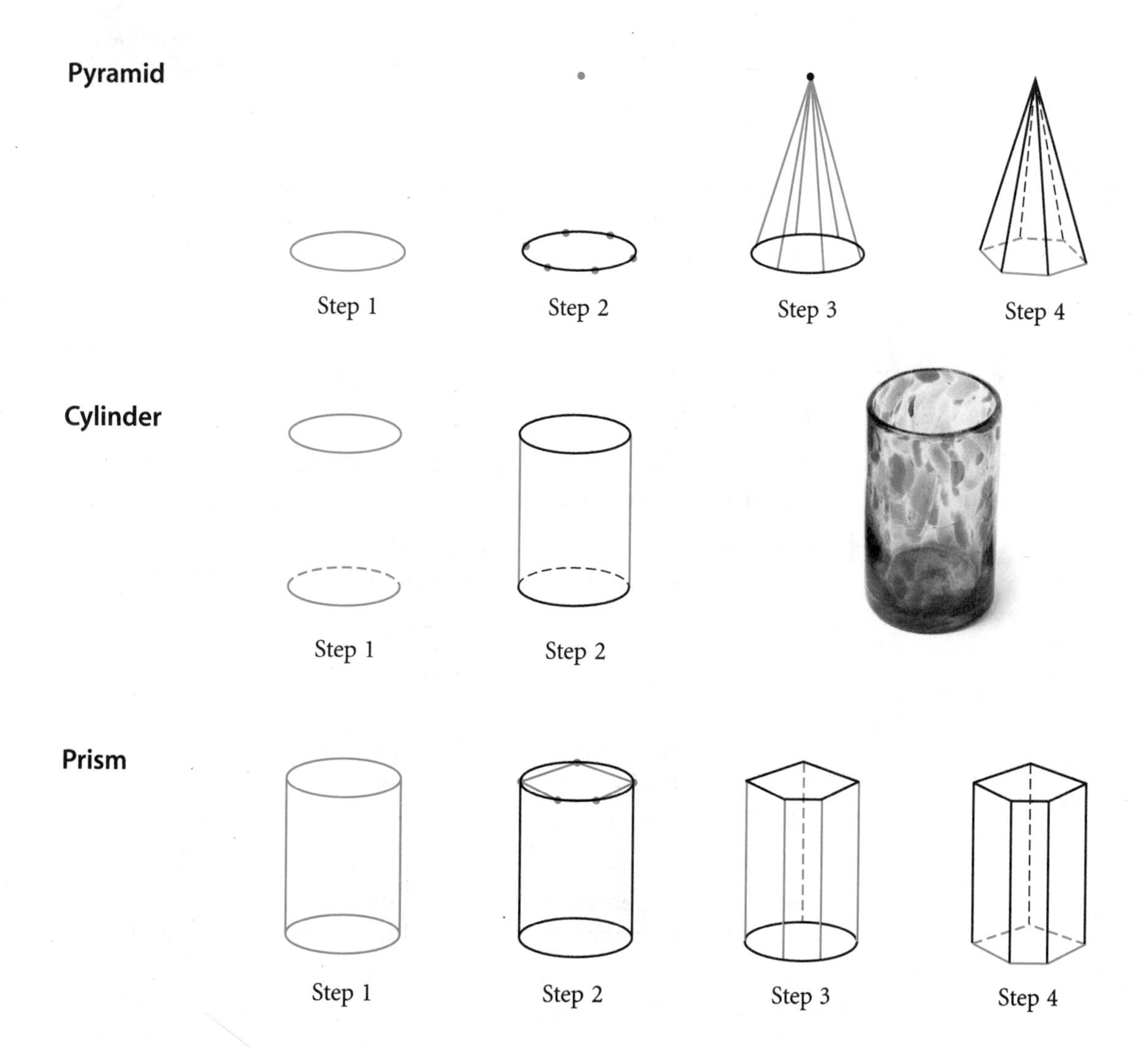

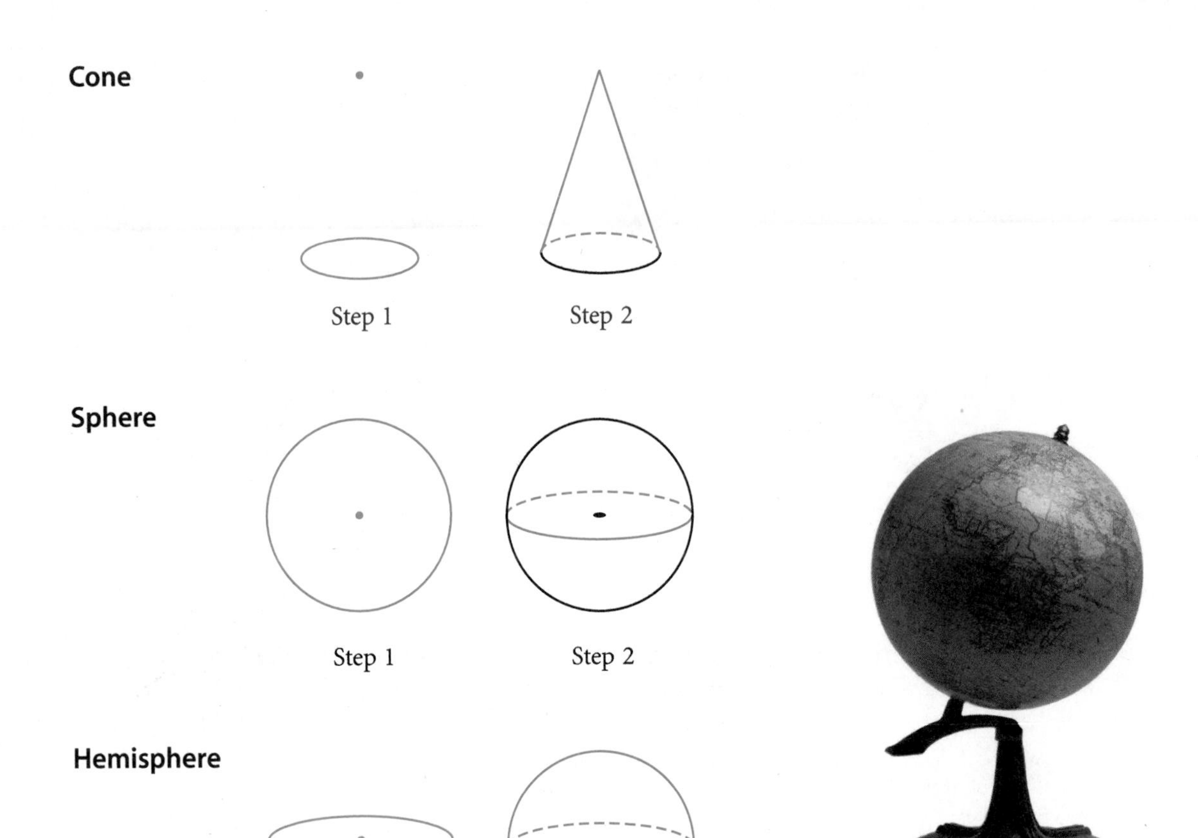

Solid geometry also involves visualizing points and lines in space. In the following investigation, you will have to visualize relationships between geometric figures in a plane and in space.

Investigation
Space Geometry

Step 1 Make a sketch or use physical objects to demonstrate each statement in the list below.

Step 2 Work with your group to determine whether each statement is true or false. If the statement is false, draw a picture and explain why it is false.

1. Only one line can be drawn through two different points.
2. Only one plane can pass through one line and a point that is not on the line.
3. If two coplanar lines are both perpendicular to a third line in the same plane, then the two lines are parallel.
4. If two planes do not intersect, then they are parallel.
5. If two lines do not intersect, then they must be parallel.
6. If a line is perpendicular to two lines in a plane, but the line is not contained in the plane, then the line is perpendicular to the plane.

EXERCISES

For Exercises 1–6, draw each figure. Study the drawing tips provided on the previous page before you start.

A police station, or *koban,* in Tokyo, Japan

1. Cylinder
2. Cone
3. Prism with a hexagonal base
4. Sphere
5. Pyramid with a heptagonal base
6. Hemisphere
7. The photo at right shows a prism-shaped building with a pyramid roof and a cylindrical porch. Draw a cylindrical building with a cone roof and a prism-shaped porch.

For Exercises 8 and 9, make a drawing to scale of each figure. Use isometric dot grid paper. Label each figure. (For example, in Exercise 8, draw the solid so that the dimensions measure 2 units by 3 units by 4 units, then label the figure with meters.)

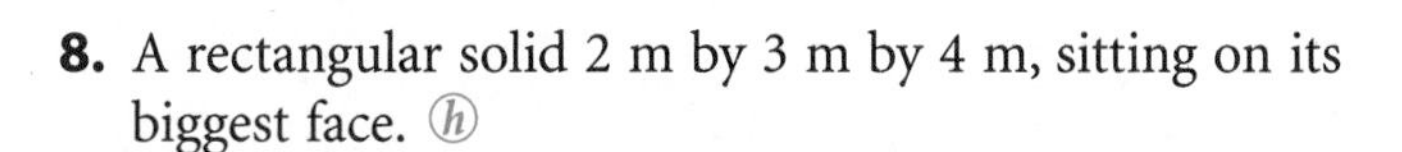

8. A rectangular solid 2 m by 3 m by 4 m, sitting on its biggest face. ⓗ
9. A rectangular solid 3 inches by 4 inches by 5 inches, resting on its smallest face. Draw lines on the three visible surfaces showing how you can divide the solid into cubic-inch boxes. How many such boxes will fit in the solid? ⓗ

For Exercises 10–12, use isometric dot grid paper to draw the figure shown.

10.

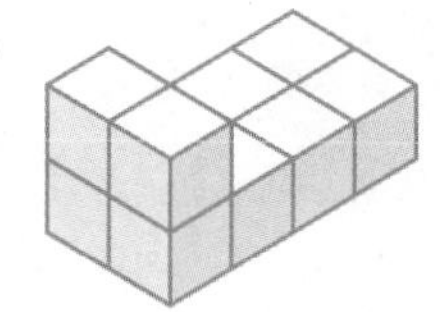

11.

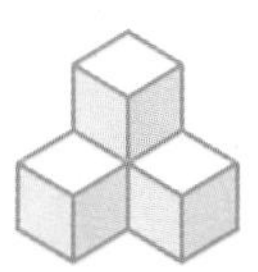

12.

For Exercises 13–15, sketch the three-dimensional figure formed by folding each net into a solid. Name the solid.

13.

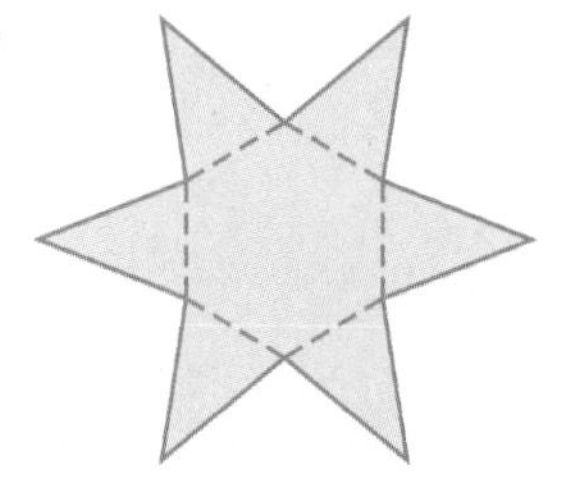

14.

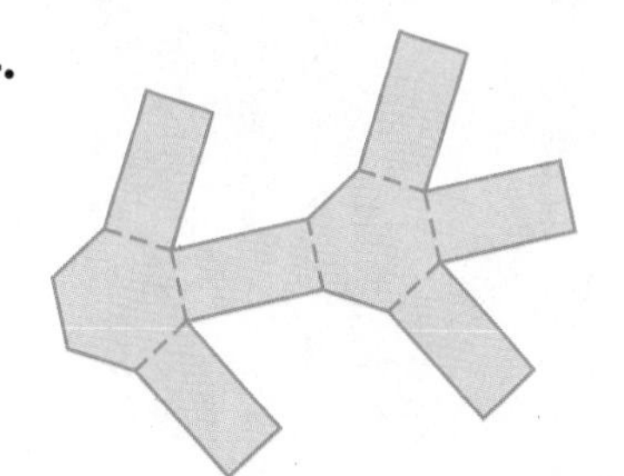

15.

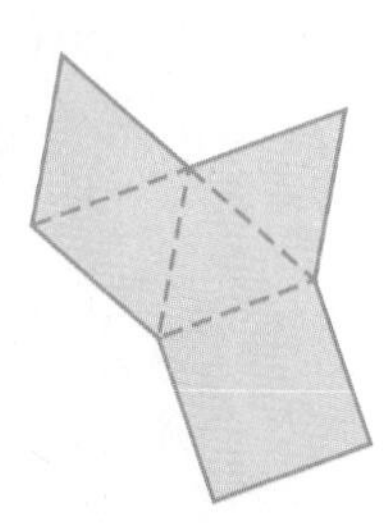

For Exercises 16 and 17, find the lengths x and y. (Every angle on each block is a right angle.)

16.

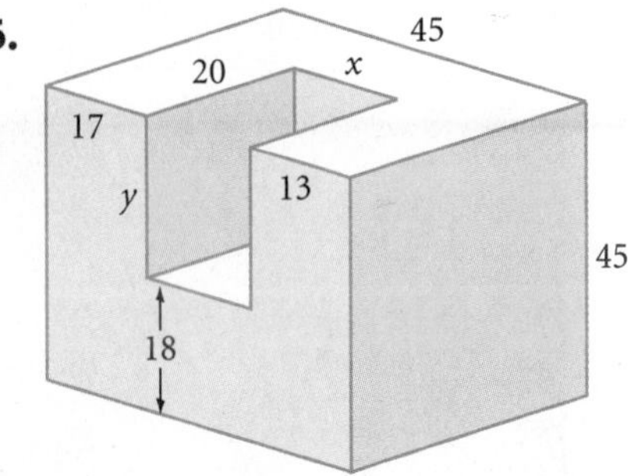

17.

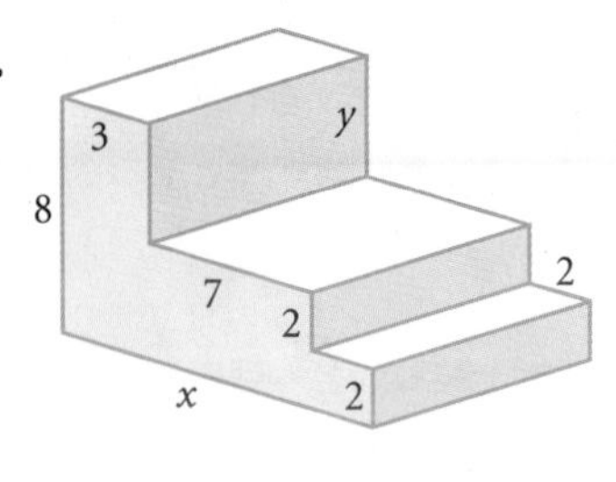

In Exercises 18 and 19, each figure represents a two-dimensional figure with a wire attached. The three-dimensional solid formed by spinning the figure on the wire between your fingers is called a **solid of revolution.** Sketch the solid of revolution formed by each two-dimensional figure.

18. ⓗ

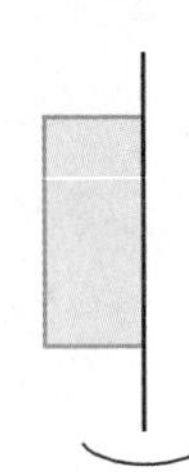

19.

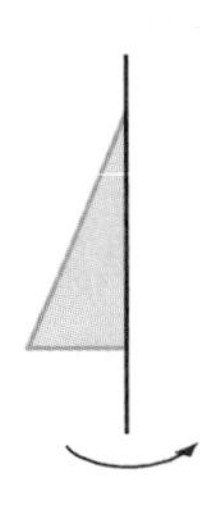

A real-life example of a "solid of revolution" is a clay pot on a potter's wheel.

When a solid is cut by a plane, the resulting two-dimensional figure is called a **section.** For Exercises 20 and 21, sketch the section formed when each solid is sliced by the plane, as shown.

20. ⓗ

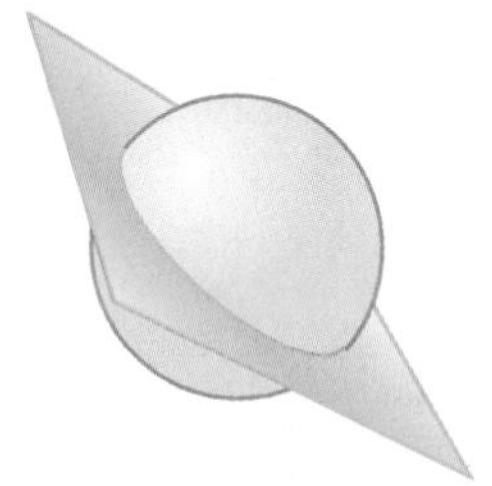

21.

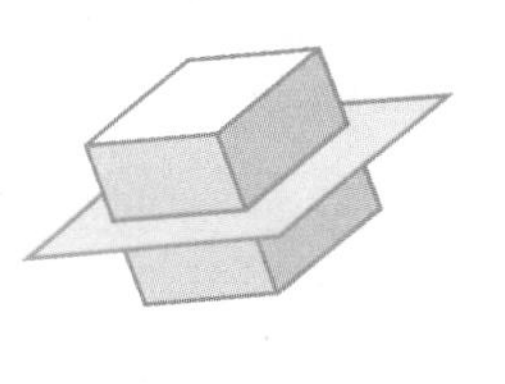

Slicing a block of clay reveals a section of the solid. Here, the section is a rectangle.

All of the statements in Exercises 22–29 are true except for two. Make a sketch to demonstrate each true statement. For each false statement, draw a sketch and explain why it is false.

Physical models can help you visualize the intersections of lines and planes in space. Can you see examples of intersecting lines in this photo? Parallel lines? Planes? Points?

22. Only one plane can pass through three noncollinear points.

23. If a line intersects a plane that does not contain the line, then the intersection is exactly one point.

24. If two lines are perpendicular to the same line, then they are parallel. ⓗ

25. If two different planes intersect, then their intersection is a line.

26. If a line and a plane have no points in common, then they are parallel.

27. If a plane intersects two parallel planes, then the lines of intersection are parallel.

28. If three random planes intersect (no two are parallel and all three do not share the same line), then they divide space into six parts.

29. If two lines are perpendicular to the same plane, then they are parallel to each other.

Review

30. If the kite *DIAN* were rotated 90° clockwise about the origin, to what location would point *A* be relocated?

31. Use your ruler to measure the perimeter of $\triangle WIM$ (in centimeters) and your protractor to measure the largest angle.

32. Use your geometry tools to draw a triangle with two sides of length 8 cm and length 13 cm and the angle between them measuring 120°.

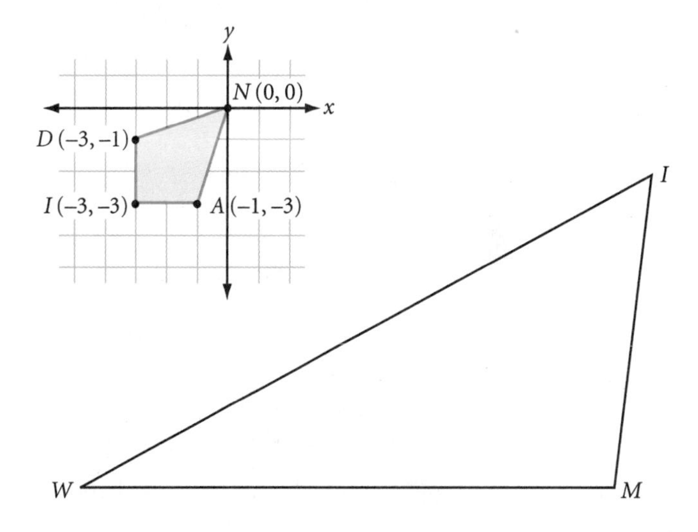

IMPROVING YOUR VISUAL THINKING SKILLS

Equal Distances

Here's a real challenge:

Show four points *A*, *B*, *C*, and *D* so that $AB = BC = AC = AD = BD = CD$.

Exploration

Geometric Probability I

You probably know what probability means. The **probability,** or likelihood, of a particular outcome is the ratio of the number of successful outcomes to the number of possible outcomes. So the probability of rolling a 4 on a 6-sided die is $\frac{1}{6}$. Or, you can name an event that involves more than one outcome, like getting the total 4 on two 6-sided dice. Since each die can come up in six different ways, there are 6×6 or 36 combinations (count 'em!). You can get the total 4 with a 1 and a 3, a 3 and a 1, or a 2 and a 2. So the probability of getting the total 4 is $\frac{3}{36}$ or $\frac{1}{12}$. Anyway, that's the theory.

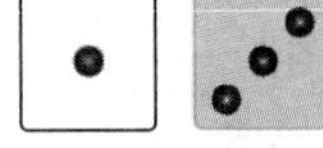

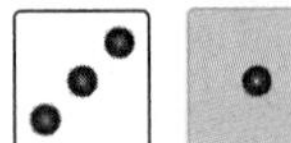

Activity
Chances Are

You will need

- a protractor
- a ruler

In this activity you'll see that you can apply probability theory to geometric figures.

The Spinner

After you've finished your homework and have eaten dinner, you play a game of chance using the spinner at right. Where the spinner lands determines how you'll spend the evening.

A
B
C
D

Sector A: Playing with your younger brother the whole evening

Sector B: Half the evening playing with your younger brother and half the evening watching TV

Sector C: Cleaning the birdcage, the hamster cage, and the aquarium the whole evening

Sector D: Playing in a band in a friend's garage the whole evening

Step 1 What is the probability of landing in each sector?

Step 2 What is the probability that you'll spend at least half the evening with your younger brother? What is the probability that you won't spend any time with him?

The Bridge

A computer programmer who is trying to win money on a TV survival program builds a 120 ft rope bridge across a piranha-infested river 90 ft below.

Step 3 If the rope breaks where he is standing (a random point), but he is able to cling to one end of it, what is the probability that he'll avoid getting wet (or worse)?

Step 4 Suppose the probability that the rope breaks at all is $\frac{1}{2}$. Also suppose that, as long as he doesn't fall more than 30 ft, the probability that he can climb back up is $\frac{3}{4}$. What is the probability that he won't fall at all? What is the probability that if he does, he'll be able to climb back up?

The Bus Stop

Noriko arrives at the bus stop at a random time between 3:00 and 4:30 P.M. each day. Her bus stops there every 20 minutes, including at 3:00 P.M.

Step 5 Draw a number line to show stopping times. (Don't worry about the length of time that the bus is actually stopped. Assume it is 0 minutes.)

Step 6 What is the probability that she will have to wait 5 minutes or more? 10 minutes or more? Hint: What line lengths represent possible waiting time?

Step 7 If the bus stops for exactly 3 minutes, how do your answers to Step 5 change?

Step 8 List the geometric properties you needed in each of the three scenarios above and tell how your answers depended on them.

Step 9 How is geometric probability like the probability you've studied before? How is it different?

Step 10 Create your own geometric probability problem.

CHAPTER 1 REVIEW

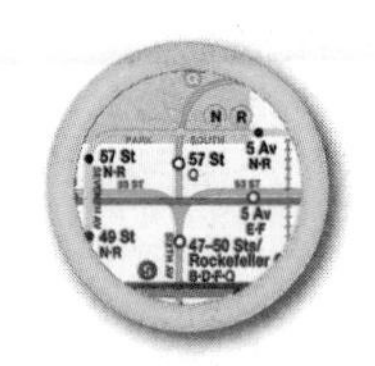

It may seem that there's a lot to memorize in this chapter. But having defined terms yourself, you're more likely to remember and understand them. The key is to practice using these new terms and to be organized. Do the following exercises, then read Assessing What You've Learned for tips on staying organized.

Whether you've been keeping a good list or not, go back now through each lesson in the chapter and double-check that you've completed each definition and that you understand it. For example, if someone mentions a geometry term to you, can you sketch it? If you are shown a geometric figure can you classify it? Compare your list of geometry terms with the lists of your group members.

EXERCISES

For Exercises 1–16, identify the statement as true or false. For each false statement, explain why it is false or sketch a counterexample.

1. The three basic building blocks of geometry are point, line, and plane.

2. "The ray through point P from point Q" is written in symbolic form as $\overrightarrow{PQ}$.

3. "The length of segment PQ"can be written as PQ.

4. The vertex of angle PDQ is point P.

5. The symbol for *perpendicular* is $\perp$.

6. A scalene triangle is a triangle with no two sides the same length.

7. An acute angle is an angle whose measure is more than 90°.

8. If $\overleftrightarrow{AB}$ intersects $\overleftrightarrow{CD}$ at point P, then $\angle APD$ and $\angle APC$ are a pair of vertical angles.

9. A diagonal is a line segment in a polygon connecting any two nonconsecutive vertices.

10. If two lines lie in the same plane and are perpendicular to the same line, then they are parallel.

11. If the sum of the measures of two angles is 180°, then the two angles are complementary.

12. A trapezoid is a quadrilateral having exactly one pair of parallel sides.

13. A polygon with ten sides is a decagon.

A knowledge of parallel lines, planes, arcs, circles, and symmetry is necessary to build durable guitars that sound pleasing.

14. A square is a rectangle with all the sides equal in length.

15. A pentagon has five sides and six diagonals.

16. The largest chord of a circle is a diameter of the circle.

For Exercises 17–25, match each term with its figure below, or write "no match."

17. Isosceles acute triangle

18. Isosceles right triangle

19. Rhombus

20. Trapezoid

21. Pyramid

22. Cylinder

23. Concave polygon

24. Chord

25. Minor arc

A. 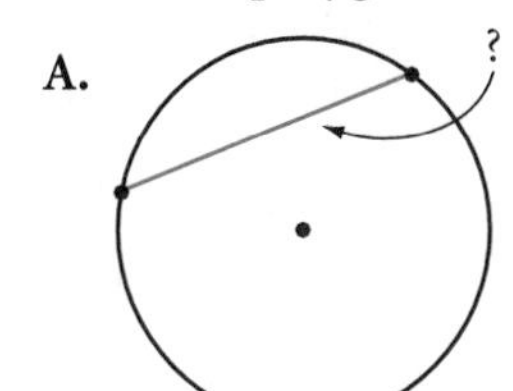

B.

C. 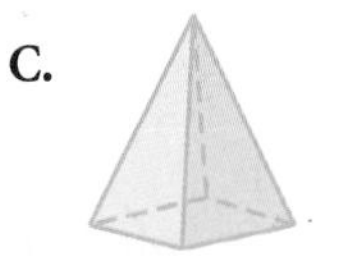

D. 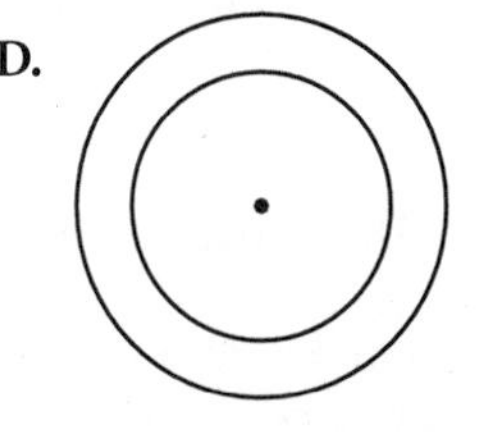

E.

F. 

G.

H. 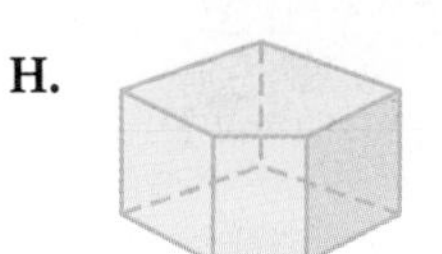

I.

J. 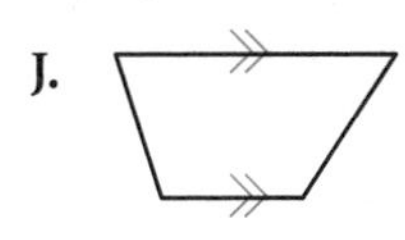

K. 

L. 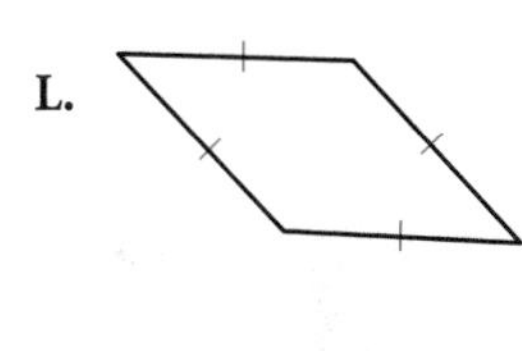

M.

N. 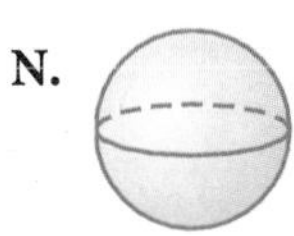

O. 

For Exercises 26–33, sketch, label, and mark each figure.

26. Kite *KYTE* with $\overline{KY} \cong \overline{YT}$

27. Scalene triangle *PTS* with $PS = 3$, $ST = 5$, $PT = 7$, and angle bisector $\overline{SO}$

28. Hexagon *REGINA* with diagonal $\overline{AG}$ parallel to sides $\overline{RE}$ and $\overline{NI}$

29. Trapezoid *TRAP* with $\overline{AR}$ and $\overline{PT}$ the nonparallel sides. Let *E* be the midpoint of $\overline{PT}$ and let *Y* be the midpoint of $\overline{AR}$. Draw $\overline{EY}$.

30. A triangle with exactly one line of reflectional symmetry

31. A circle with center at *P*, radii $\overline{PA}$ and $\overline{PT}$, and chord $\overline{TA}$ creating a minor arc $\overset{\frown}{TA}$

32. A pair of concentric circles with the diameter $\overline{AB}$ of the inner circle perpendicular at B to a chord $\overline{CD}$ of the larger circle

33. A pyramid with a pentagonal base

34. Draw a rectangular prism 2 inches by 3 inches by 5 inches, resting on its largest face. Draw lines on the three visible faces, showing how the solid can be divided into 30 smaller cubes.

35. Use your protractor to draw a 125° angle.

36. Use your protractor, ruler, and compass to draw an isosceles triangle with a vertex angle having a measure of 40°.

37. Use your geomety tools to draw a regular octagon. ⓗ

38. What is the measure of $\angle A$? Use your protractor.

A

For Exercises 39–42, find the lengths x and y.

39.

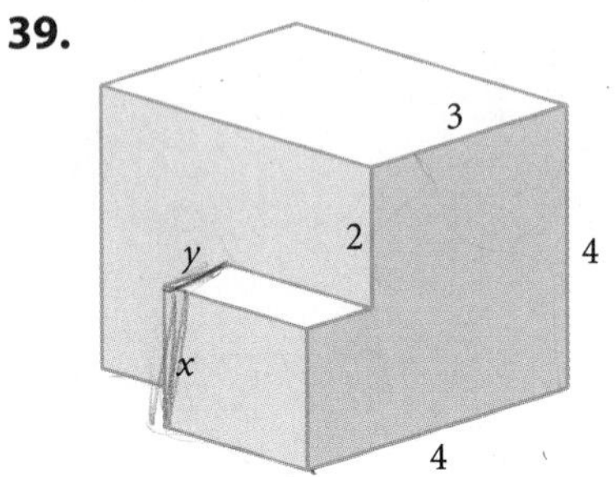

40.

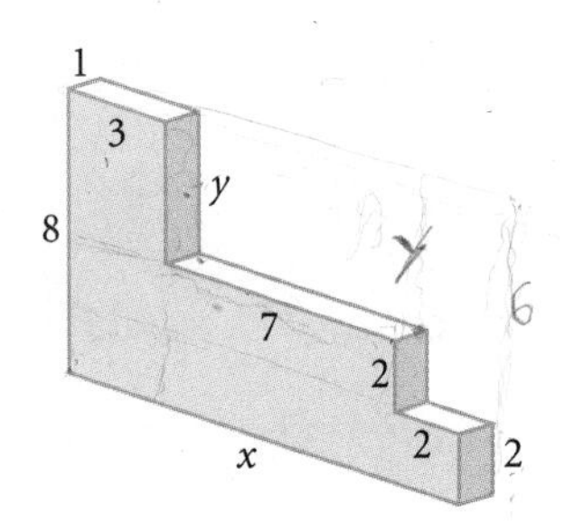

41.

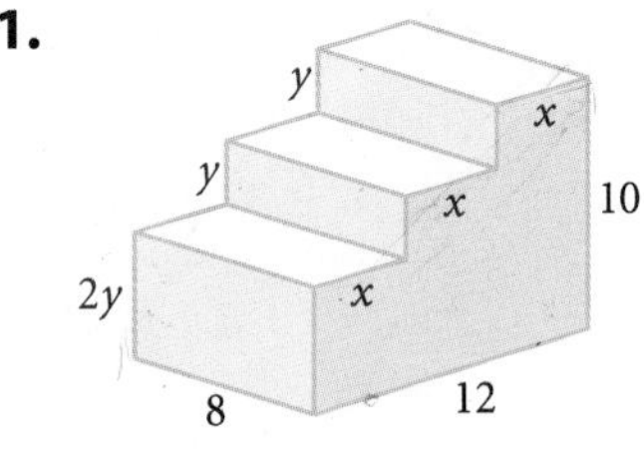

42.

18

20

x y 12

43. If D is the midpoint of $\overline{AC}$, C is the midpoint of $\overline{AB}$, and $BD = 12$ cm, what is the length of $\overline{AB}$?

44. If $\overrightarrow{BD}$ is the angle bisector of $\angle ABC$ and $\overrightarrow{BE}$ is the angle bisector of $\angle DBC$, find $m\angle EBA$ if $m\angle DBE = 32°$?

45. What is the measure of the angle formed by the hands of the clock at 2:30? ⓗ

46. If the pizza is cut into 12 congruent pieces, how many degrees are in each central angle?

47. Make a concept map (a tree diagram or a Venn diagram) to organize these quadrilaterals: rhombus, rectangle, square, trapezoid.

48. The box at right is wrapped with two strips of ribbon, as shown. What is the minimum length of ribbon needed to decorate the box?

9 in.
5 in.
14 in.

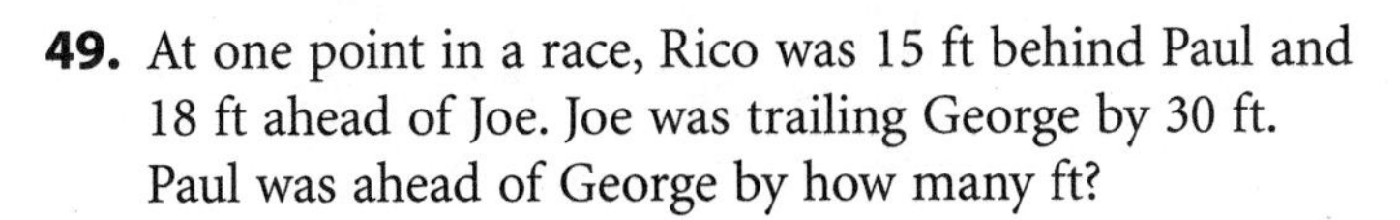
49. At one point in a race, Rico was 15 ft behind Paul and 18 ft ahead of Joe. Joe was trailing George by 30 ft. Paul was ahead of George by how many ft?

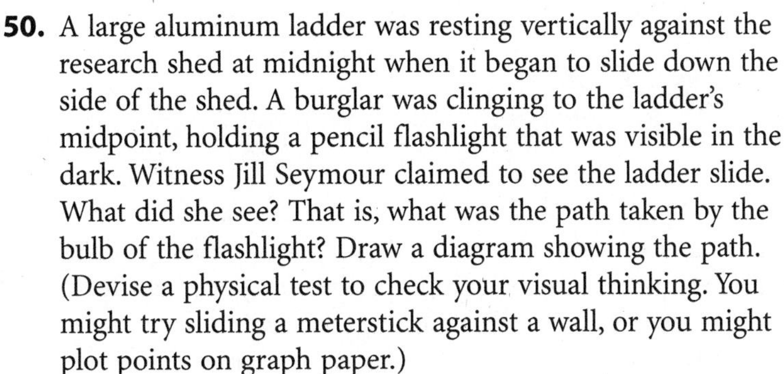
50. A large aluminum ladder was resting vertically against the research shed at midnight when it began to slide down the side of the shed. A burglar was clinging to the ladder's midpoint, holding a pencil flashlight that was visible in the dark. Witness Jill Seymour claimed to see the ladder slide. What did she see? That is, what was the path taken by the bulb of the flashlight? Draw a diagram showing the path. (Devise a physical test to check your visual thinking. You might try sliding a meterstick against a wall, or you might plot points on graph paper.)

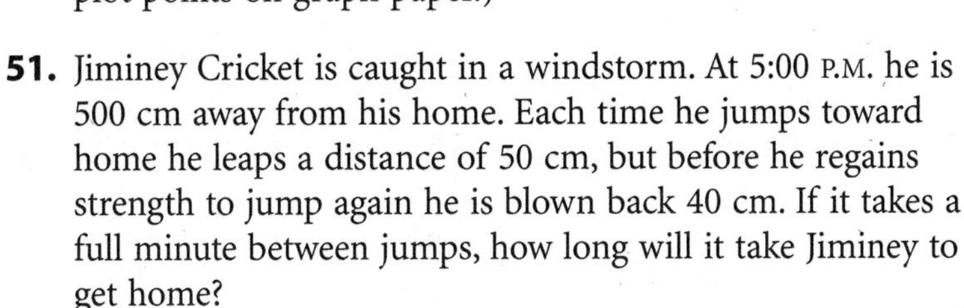
51. Jiminey Cricket is caught in a windstorm. At 5:00 P.M. he is 500 cm away from his home. Each time he jumps toward home he leaps a distance of 50 cm, but before he regains strength to jump again he is blown back 40 cm. If it takes a full minute between jumps, how long will it take Jiminey to get home?

52. If the right triangle *BAR* were rotated 90° clockwise about point *B*, to what location would point *A* be relocated?

y
$R\,(2, 2)$
x
$A\,(-2, -1)$
$B\,(2, -1)$

53. Sketch the three-dimensional figure formed by folding the net below into a solid.

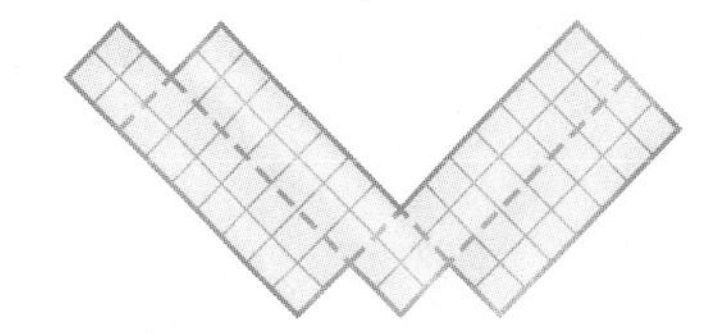

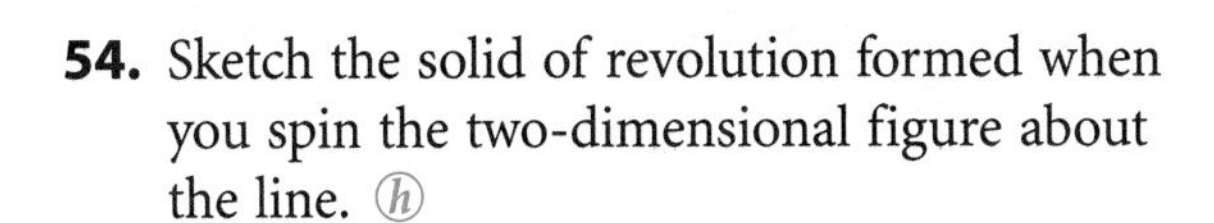
54. Sketch the solid of revolution formed when you spin the two-dimensional figure about the line. ⓗ

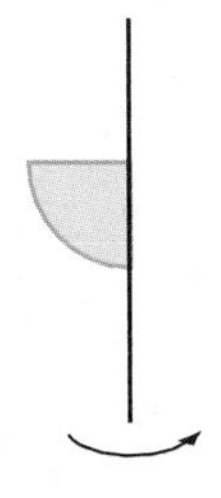

55. Sketch the section formed when the solid is sliced by the plane, as shown.

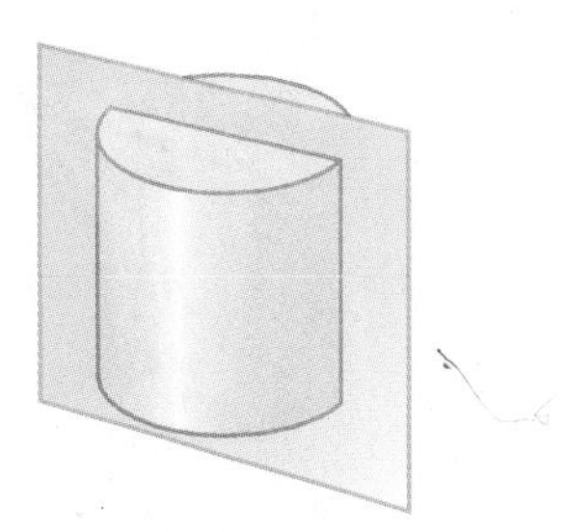

56. Use an isometric dot grid to sketch the figure shown below.

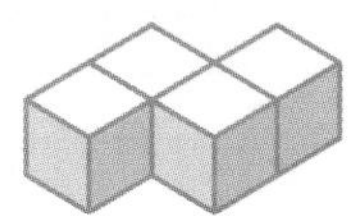

57. Sketch the figure shown with the red edge vertical and facing the viewer. ⓗ

Assessing What You've Learned

ORGANIZE YOUR NOTEBOOK

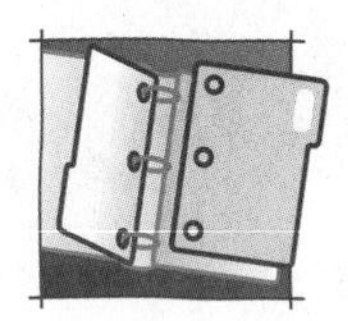

Is this textbook filling up with folded-up papers stuffed between pages? If so, that's a bad sign! But it's not too late to get organized. Keeping a well-organized notebook is one of the best habits you can develop to improve and assess your learning. You should have sections for your classwork, definition list, and homework exercises. There should be room to make corrections and to summarize what you learned and write down questions you still have.

Many books include a definition list (sometimes called a glossary) in the back. This book makes you responsible for your own glossary, so it's essential that, in addition to taking good notes, you keep a complete definition list that you can refer to. You started a definition list in Lesson 1.1. Get help from classmates or your teacher on any definition you don't understand.

As you progress through the course, your notebook will become more and more important. A good way to review a chapter is to read through the chapter and your notes and write a one-page summary of the chapter. And if you create a one-page summary for each chapter, the summaries will be very helpful to you when it comes time for midterms and final exams. You'll find no better learning and study aid than a summary page for each chapter, and your definition list, kept in an organized notebook.

UPDATE YOUR PORTFOLIO

- ▶ If you did the project in this chapter, document your work and add it to your portfolio.
- ▶ Choose one homework assignment that demonstrates your best work in terms of completeness, correctness, and neatness. Add it (or a copy of it) to your portfolio.

CHAPTER

2 Reasoning in Geometry

That which an artist makes is a mirror image of what he sees around him.

M. C. ESCHER

Hand with Reflecting Sphere (Self-Portrait in Spherical Mirror), M. C. Escher

OBJECTIVES

In this chapter you will

- perform geometry investigations and make many discoveries by observing common features or patterns
- use your discoveries to solve problems through a process called inductive reasoning
- use inductive reasoning to discover patterns
- learn to use deductive reasoning
- learn about vertical angles and linear pairs
- make conjectures

LESSON

2.1

Inductive Reasoning

We have to reinvent the wheel every once in a while, not because we need a lot of wheels; but because we need a lot of inventors.

BRUCE JOYCE

As a child you learned by experimenting with the natural world around you. You learned how to walk, to talk, and to ride your first bicycle, all by trial and error. From experience you learned to turn a water faucet on with a counterclockwise motion and to turn it off with a clockwise motion. You achieved most of your learning by a process called **inductive reasoning.** It is the process of observing data, recognizing patterns, and making generalizations about those patterns.

Geometry is rooted in inductive reasoning. In ancient Egypt and Babylonia, geometry began when people developed procedures for measurement after much experience and observation. Assessors and surveyors used these procedures to calculate land areas and to reestablish the boundaries of agricultural fields after floods. Engineers used the procedures to build canals, reservoirs, and the Great Pyramids. Throughout this course you will use inductive reasoning. You will perform investigations, observe similarities and patterns, and make many discoveries that you can use to solve problems.

Language CONNECTION

The word "geometry" means "measure of the earth" and was originally inspired by the ancient Egyptians. The ancient Egyptians devised a complex system of land surveying in order to reestablish land boundaries that were erased each spring by the annual flooding of the Nile River.

Inductive reasoning guides scientists, investors, and business managers. All of these professionals use past experience to assess what is likely to happen in the future.

When you use inductive reasoning to make a generalization, the generalization is called a **conjecture.** Consider the following example from science.

EXAMPLE A

A scientist dips a platinum wire into a solution containing salt (sodium chloride), passes the wire over a flame, and observes that it produces an orange-yellow flame.

She does this with many other solutions that contain salt, finding that they all produce an orange-yellow flame. Make a conjecture based on her findings.

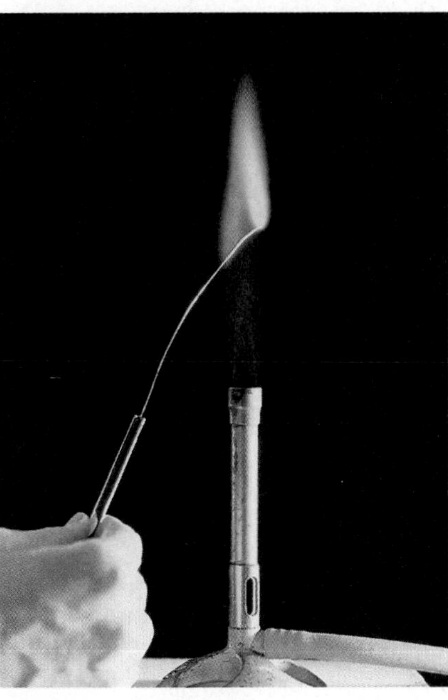

Platinum wire flame test

▶ ***Solution***

The scientist tested many other solutions containing salt, and found no counterexamples. You should conjecture: "If a solution contains sodium chloride, then in a flame test it produces an orange-yellow flame."

Like scientists, mathematicians often use inductive reasoning to make discoveries. For example, a mathematician might use inductive reasoning to find patterns in a number sequence. Once he knows the pattern, he can find the next term.

EXAMPLE B

Consider the sequence

2, 4, 7, 11, . . .

Make a conjecture about the rule for generating the sequence. Then find the next three terms.

▶ ***Solution***

Look at the numbers you add to get each term. The 1st term in the sequence is 2. You add 2 to find the 2nd term. Then you add 3 to find the 3rd term, and so on.

$$2 \xrightarrow{+2} 4 \xrightarrow{+3} 7 \xrightarrow{+4} 11$$

You can conjecture that if the pattern continues, you always add the next counting number to get the next term. The next three terms in the sequence will be 16, 22, and 29.

$$11 \xrightarrow{+5} 16 \xrightarrow{+6} 22 \xrightarrow{+7} 29$$

In the following investigation you will use inductive reasoning to recognize a pattern in a series of drawings and use it to find a term much farther out in a sequence.

Investigation
Shape Shifters

Look at the sequence of shapes below. Pay close attention to the patterns that occur in every other shape.

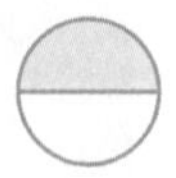  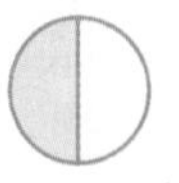 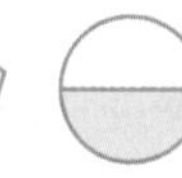 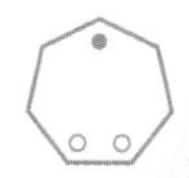

Step 1 What patterns do you notice in the 1st, 3rd, and 5th shapes?

Step 2 What patterns do you notice in the 2nd, 4th, and 6th shapes?

Step 3 Draw the next two shapes in the sequence.

Step 4 Use the patterns you discovered to draw the 25th shape.

Step 5 Describe the 30th shape in the sequence. You do not have to draw it!

Sometimes a conjecture is difficult to find because the data collected are unorganized or the observer is mistaking coincidence with cause and effect. Good use of inductive reasoning depends on the quantity and quality of data. Sometimes not enough information or data have been collected to make a proper conjecture. For example, if you are asked to find the next term in the pattern 3, 5, 7, you might conjecture that the next term is 9—the next odd number. Someone else might notice that the pattern is the consecutive odd primes and say that the next term is 11. If the pattern was 3, 5, 7, 11, 13, what would you be more likely to conjecture?

Exercises

1. On his way to the local Hunting and Gathering Convention, caveperson Stony Grok picks up a rock, drops it into a lake, and notices that it sinks. He picks up a second rock, drops it into the lake, and notices that it also sinks. He does this five more times. Each time, the rock sinks straight to the bottom of the lake. Stony conjectures: "Ura nok seblu," which translates to __?__. What counterexample would Stony Grok need to find to disprove, or at least to refine, his conjecture? ⓗ

2. Sean draws these geometric figures on paper. His sister Courtney measures each angle with a protractor. They add the measures of each pair of angles to form a conjecture. Write their conjecture.

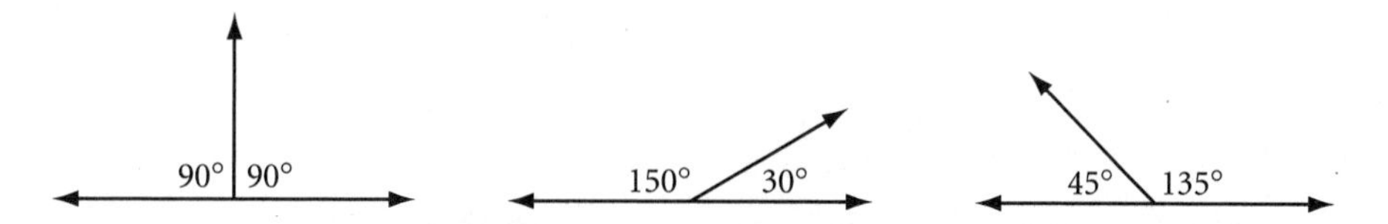

For Exercises 3–10, use inductive reasoning to find the next two terms in each sequence.

3. 1, 10, 100, 1000, ?, ?

4. $\frac{1}{6}, \frac{1}{3}, \frac{1}{2}, \frac{2}{3}$, ?, ? ⓗ

5. 7, 3, −1, −5, −9, −13, ?, ?

6. 1, 3, 6, 10, 15, 21, ?, ?

7. 1, 1, 2, 3, 5, 8, 13, ?, ? ⓗ

8. 1, 4, 9, 16, 25, 36, ?, ? ⓗ

9. 32, 30, 26, 20, 12, 2, ?, ?

10. 1, 2, 4, 8, 16, 32, ?, ?

For Exercises 11–16, use inductive reasoning to draw the next shape in each picture pattern.

11.

12.

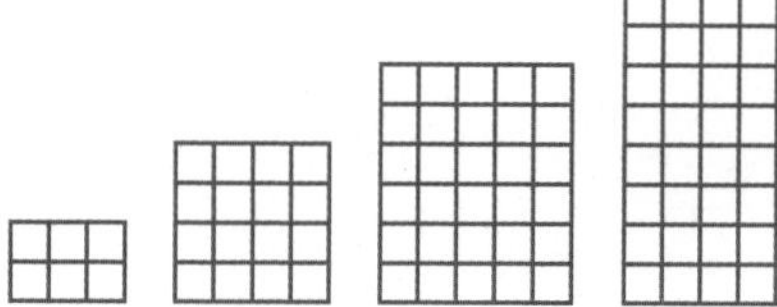

13. ⓗ

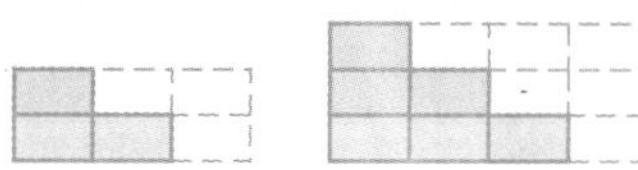

14. ⓗ

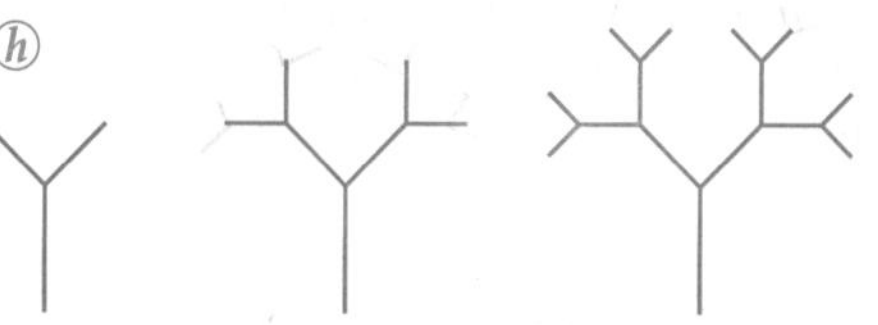

15. ⓗ

16.

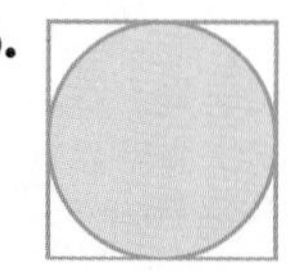

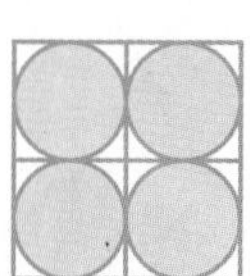

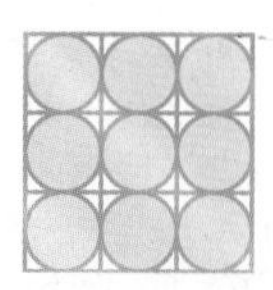

Use the rule provided to generate the first five terms of the sequence in Exercise 17 and the next five terms of the sequence in Exercise 18.

17. $3n - 2$ ⓗ

18. 1, 3, 6, 10, . . . , $\frac{n(n+1)}{2}$, . . .

19. Now it's your turn. Generate the first five terms of a sequence. Give the sequence to a member of your family or to a friend and ask him or her to find the next two terms in the sequence. Can he or she find your pattern?

20. Write the first five terms of two different sequences in which 12 is the 3rd term.

21. Think of a situation in which you have used inductive reasoning. Write a paragraph describing what happened and explaining why you think it was inductive reasoning. ⓗ

22. Look at the pattern in these pairs of equations. Decide if the conjecture is true. If it is not true, find a counterexample.

$12^2 = 144$ and $21^2 = 441$

$13^2 = 169$ and $31^2 = 961$

$103^2 = 10609$ and $301^2 = 90601$

$112^2 = 12544$ and $211^2 = 44521$

Conjecture: If two numbers have the same digits in reverse order, then the squares of those numbers will have identical digits but in reverse order.

Review

23. Sketch the section formed when the cone is sliced by the plane, as shown.

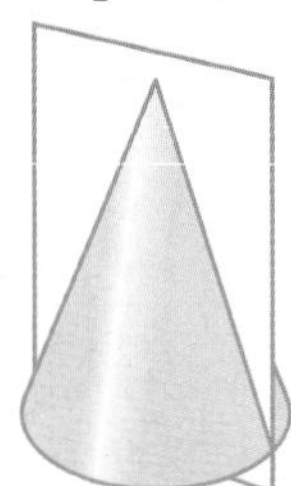

24.

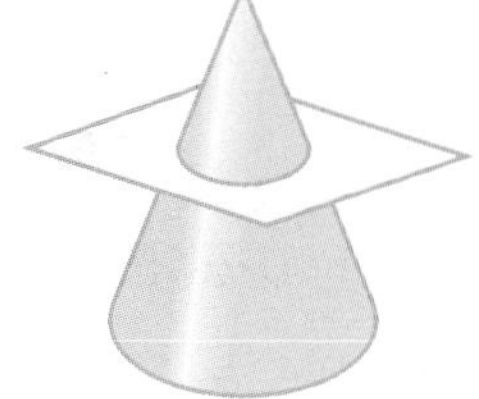

25.

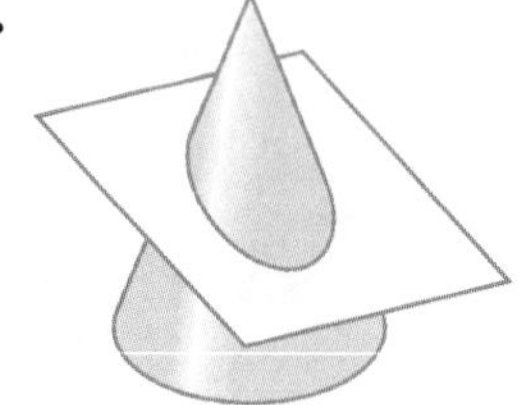

26. Sketch the three-dimensional figure formed by folding the net below into a solid. ⓗ

27. Sketch the figure shown below but with the red edge vertical and facing you. ⓗ

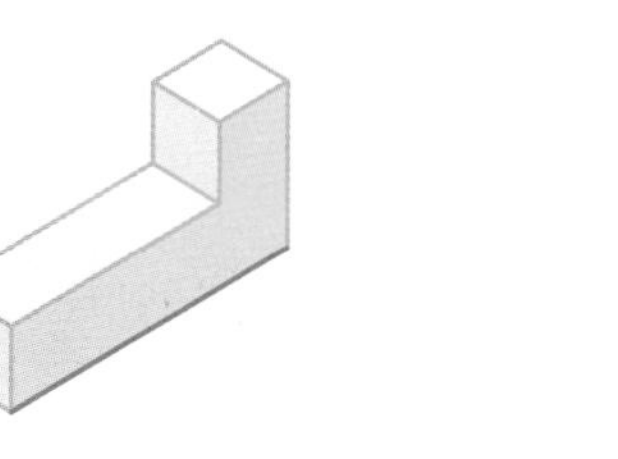

28. Sketch the solid of revolution formed when the two-dimensional figure is rotated about the line. ⓗ

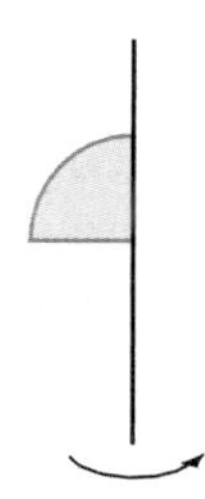

For Exercises 29–38, write the word that makes the statement true.

29. Points are _?_ if they lie on the same line.

30. A triangle with two congruent sides is _?_.

31. The geometry tool used to measure the size of an angle in degrees is called a(n) _?_.

32. A(n) _?_ of a circle connects its center to a point on the circle.

33. A segment connecting any two non-adjacent vertices in a polygon is called a(n) _?_.

34. A polygon with 12 sides is called a(n) _?_.

35. A trapezoid has exactly one pair of _?_ sides.

36. A(n) $\underline{?}$ polygon is both equiangular and equilateral.

37. If angles are complementary, then their measures add to $\underline{?}$.

38. If two lines intersect to form a right angle, then they are $\underline{?}$.

For Exercises 39–42, sketch and label the figure.

39. Pentagon *GIANT* with diagonal $\overline{AG}$ parallel to side $\overline{NT}$

40. A quadrilateral that has reflectional symmetry but not rotational symmetry

41. A prism with a hexagonal base

42. A counterexample to show that the following statement is false: The diagonals of a kite bisect the angles. ⓗ

IMPROVING YOUR REASONING SKILLS

PUZZLE

Puzzling Patterns

These patterns are "different." Your task is to find the next term.

1. 18, 46, 94, 63, 52, 61, $\underline{?}$
2. O, T, T, F, F, S, S, E, N, $\underline{?}$
3. 1, 4, 3, 16, 5, 36, 7, $\underline{?}$
4. 4, 8, 61, 221, 244, 884, $\underline{?}$
5. 6, 8, 5, 10, 3, 14, 1, $\underline{?}$
6. **B, 0, C, 2, D, 0, E, 3, F, 3, G,** $\underline{?}$
7. 2, 3, 6, 1, 8, 6, 8, 4, 8, 4, 8, 3, 2, 3, 2, 3, $\underline{?}$
8. A E F H I K L M N T V W
   B C D G J O P Q R S U
   Where do the X, Y, and Z go?

LESSON

2.2

Deductive Reasoning

That's the way things come clear. All of a sudden. And then you realize how obvious they've been all along.

MADELEINE L'ENGLE

Have you ever noticed that the days are longer in the summer? Or that mosquitoes appear after a summer rain? Over the years you have made conjectures, using inductive reasoning, based on patterns you have observed. When you make a conjecture, the process of discovery may not always help explain *why* the conjecture works. You need another kind of reasoning to help answer this question.

Deductive reasoning is the process of showing that certain statements follow logically from agreed-upon assumptions and proven facts. When you use deductive reasoning, you try to reason in an orderly way to convince yourself or someone else that your conclusion is valid. If your initial statements are true, and you give a logical argument, then you have shown that your conclusion is true. For example, in a trial, lawyers use deductive arguments to show how the evidence that they present proves their case. A lawyer might make a very good argument. But first, the court must believe the evidence and accept it as true.

The success of an attorney's case depends on the jury accepting the evidence as true and following the steps in her deductive reasoning.

You use deductive reasoning in algebra. When you provide a reason for each step in the process of solving an equation, you are using deductive reasoning. Here is an example.

EXAMPLE A

Solve the equation for x. Give a reason for each step in the process.

$$3(2x + 1) + 2(2x + 1) + 7 = 42 - 5x$$

▸ ***Solution***

$3(2x + 1) + 2(2x + 1) + 7 = 42 - 5x$	The original equation.
$5(2x + 1) + 7 = 42 - 5x$	Combining like terms.
$5(2x + 1) = 35 - 5x$	Subtraction property of equality.
$10x + 5 = 35 - 5x$	Distributive property.
$10x = 30 - 5x$	Subtraction property of equality.
$15x = 30$	Addition property of equality.
$x = 2$	Division property of equality.

The next example shows how to use both kinds of reasoning: inductive reasoning to discover the property and deductive reasoning to explain why it works.

EXAMPLE B

In each diagram, $\overrightarrow{AC}$ bisects obtuse angle BAD. Classify $\angle BAD$, $\angle DAC$, and $\angle CAB$ as *acute, right,* or *obtuse*. Then complete the conjecture.

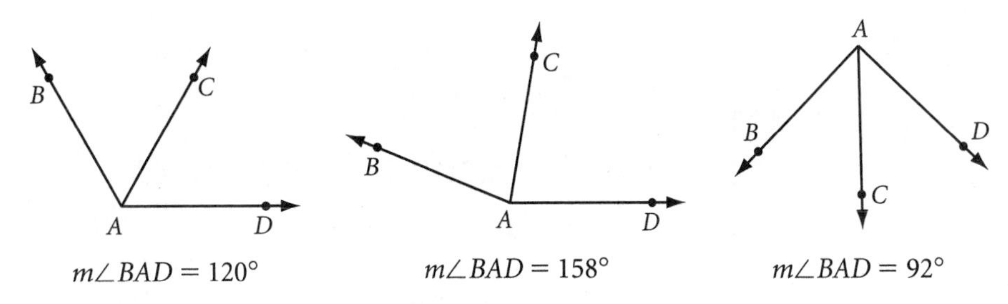

Conjecture: If an obtuse angle is bisected, then the two newly formed congruent angles are _?_.

Justify your answers with a deductive argument.

▶ Solution

In each diagram, $\angle BAD$ is obtuse because $m\angle BAD$ is greater than 90°. In each diagram, the angles formed by the bisector are acute because their measures—60°, 79°, and 46°—are less than 90°. So one possible conjecture is

Conjecture: If an obtuse angle is bisected, then the two newly formed congruent angles are *acute.*

Why? According to our definition of an angle, every angle measure is less than 180°. So, using algebra, if m is the measure of an obtuse angle, then $m < 180°$. When you bisect an angle, the two newly formed angles each measure half of the original angle, or $\frac{1}{2}m$. If $m < 180°$, then $\frac{1}{2}m < \frac{1}{2}(180)$, so $\frac{1}{2}m < 90°$. The two angles are each less than 90°, so they are acute.

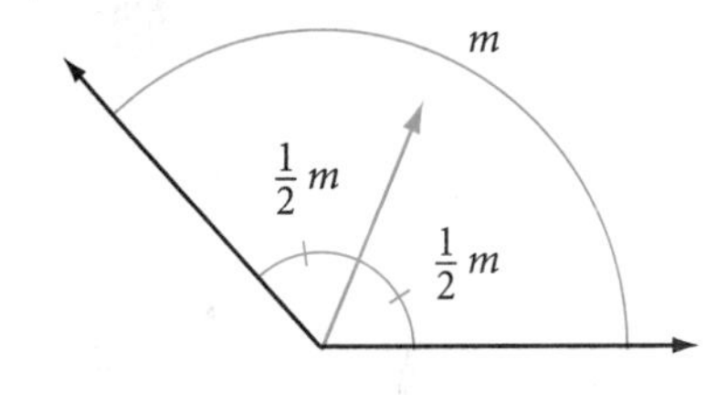

Science CONNECTION

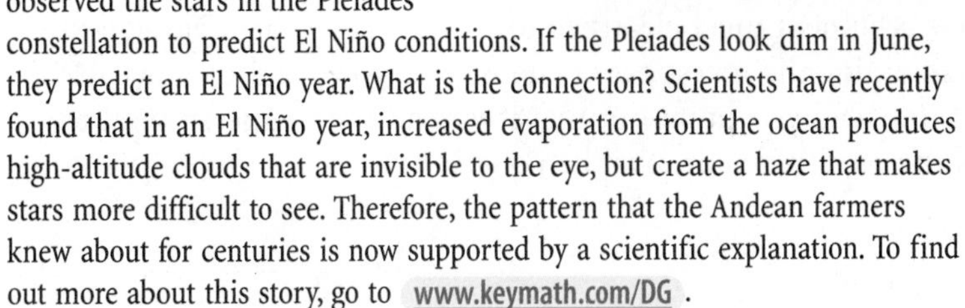

Here is an example of inductive reasoning, supported by deductive reasoning. El Niño is the warming of water in the tropical Pacific Ocean, which produces unusual weather conditions and storms worldwide. For centuries, farmers living in the Andes Mountains of South America have observed the stars in the Pleiades constellation to predict El Niño conditions. If the Pleiades look dim in June, they predict an El Niño year. What is the connection? Scientists have recently found that in an El Niño year, increased evaporation from the ocean produces high-altitude clouds that are invisible to the eye, but create a haze that makes stars more difficult to see. Therefore, the pattern that the Andean farmers knew about for centuries is now supported by a scientific explanation. To find out more about this story, go to www.keymath.com/DG .

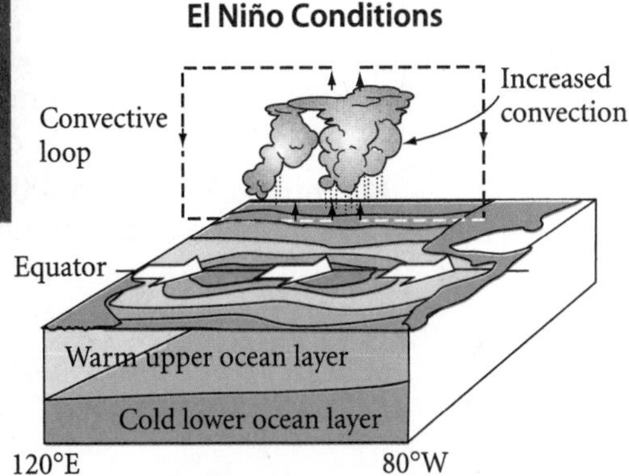

Inductive reasoning allows you to discover new ideas based on observed patterns. Deductive reasoning can help explain why your conjectures are true.

Inductive and deductive reasoning work very well together. In this investigation you will use inductive reasoning to form a conjecture and deductive reasoning to explain why it's true.

Investigation
Overlapping Segments

In each segment, $\overline{AB} \cong \overline{CD}$.

25 cm 75 cm 25 cm
A B C D

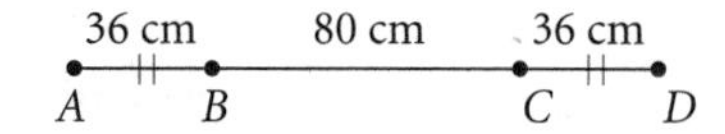

Step 1 From the markings on each diagram, determine the lengths of $\overline{AC}$ and $\overline{BD}$. What do you discover about these segments?

Step 2 Draw a new segment. Label it $\overline{AD}$. Place your own points B and C on $\overline{AD}$ so that $\overline{AB} \cong \overline{CD}$.

A B C D

Step 3 Measure $\overline{AC}$ and $\overline{BD}$. How do these lengths compare?

Step 4 Complete the conclusion of this conjecture:

If $\overline{AD}$ has points A, B, C, and D in that order with $\overline{AB} \cong \overline{CD}$, then $\underline{?}$.

Now you will use deductive reasoning and algebra to explain why the conjecture from Step 4 is true.

Step 5 Use deductive reasoning to convince your group that AC will always equal BD. Take turns explaining to each other. Write your argument algebraically.

In the investigation you used both inductive and deductive reasoning to convince yourself of the overlapping segments property. You will use a similar process in the next lesson to discover and prove the overlapping angles property in Exercise 17.

Good use of deductive reasoning depends on the quality of the argument. Just like the saying, "A chain is only as strong as its weakest link," a deductive argument is only as good (or as true) as the statements used in the argument. A conclusion in a deductive argument is true only if *all* the statements in the argument are true. Also, the statements in your argument must clearly follow from each other. Did you use clear arguments in explaining the investigation steps? Did you point out that $\overline{BC}$ is part of both $\overline{AC}$ and $\overline{BD}$? Did you point out that if you add the same amount to things that are equal the resulting sum must be equal?

Exercises

1. When you use __?__ reasoning you are generalizing from careful observation that something is probably true. When you use __?__ reasoning you are establishing that, if a set of properties is accepted as true, something else must be true.

2. $\angle A$ and $\angle B$ are complementary. $m\angle A = 25°$. What is $m\angle B$? What type of reasoning do you use, inductive or deductive reasoning, when solving this problem?

3. If the pattern continues, what are the next two terms?

   What type of reasoning do you use, inductive or deductive reasoning, when solving this problem?

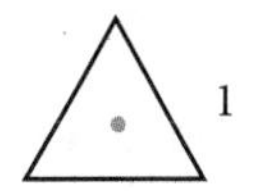

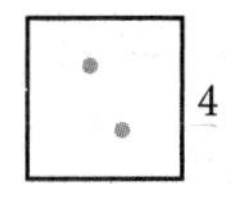

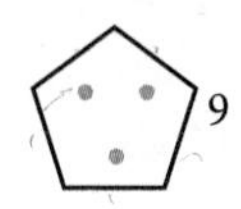

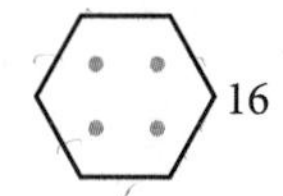

4. $\triangle DGT$ is isosceles with $TD = DG$. If the perimeter of $\triangle DGT$ is 756 cm and $GT = 240$ cm, then $DG =$ __?__. What type of reasoning do you use, inductive or deductive reasoning, when solving this problem?

5. ***Mini-Investigation*** The sum of the measures of the five marked angles in stars A through C is shown below each star. Use your protractor to carefully measure the five marked angles in star D.

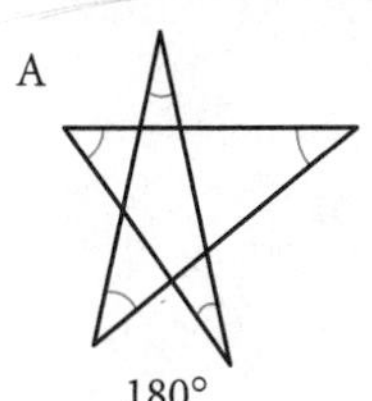

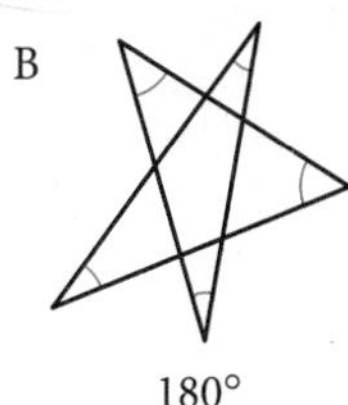

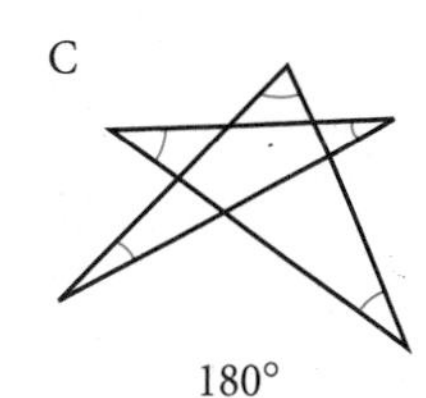

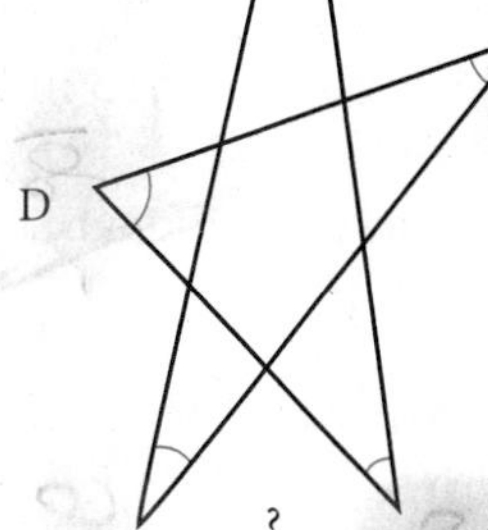

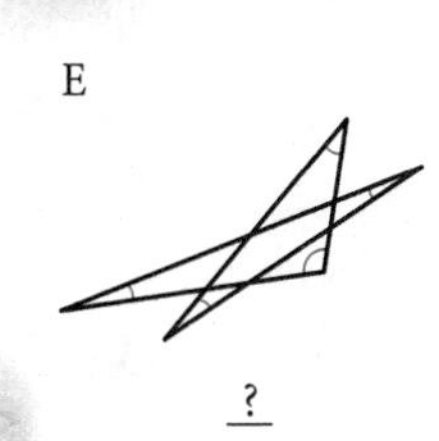

   If this pattern continues, without measuring, what would be the sum of the measures of the marked angles in star E? What type of reasoning do you use, inductive or deductive reasoning, when solving this problem?

6. The definition of a parallelogram says, "If both pairs of opposite sides of a quadrilateral are parallel, then the quadrilateral is a parallelogram." Quadrilateral *LNDA* has both pairs of opposite sides parallel. What conclusion can you make? What type of reasoning did you use?

7. Use the overlapping segments property to complete each statement.

   a. If $AB = 3$, then $CD =$ __?__.

   b. If $AC = 10$, then $BD =$ __?__.

   c. If $BC = 4$ and $CD = 3$, then $AC =$ __?__.

8. In Example B of this lesson you discovered through inductive reasoning that if an obtuse angle is bisected, then the two newly formed congruent angles are acute. You then used deductive reasoning to explain why they were acute. Go back to the example and look at the sizes of the acute angles formed. What is the smallest possible size for the two congruent acute angles formed by the bisector? Can you use deductive reasoning to explain why? ⓗ

9. Study the pattern and make a conjecture by completing the fifth line. What would be the conjecture for the sixth line? The tenth line? (h)

$1 \cdot 1 = 1$

$11 \cdot 11 = 121$

$111 \cdot 111 = 12{,}321$

$1{,}111 \cdot 1{,}111 = 1{,}234{,}321$

$11{,}111 \cdot 11{,}111 =$ _?_

10. Think of a situation you observed outside of school in which deductive reasoning was used correctly. Write a paragraph or two describing what happened and explaining why you think it was deductive reasoning.

Review

11. Mark Twain once observed that the lower Mississippi River is very crooked and that over the years, as the bends and the turns straighten out, the river gets shorter and shorter. Using numerical data about the length of the lower part of the river, he noticed that in the year 1700 the river was more than 1200 miles long, yet by the year 1875 it was only 973 miles long. Twain concluded that any person "can see that 742 years from now the lower Mississippi will be only a mile and three-quarters long." What is wrong with this inductive reasoning?

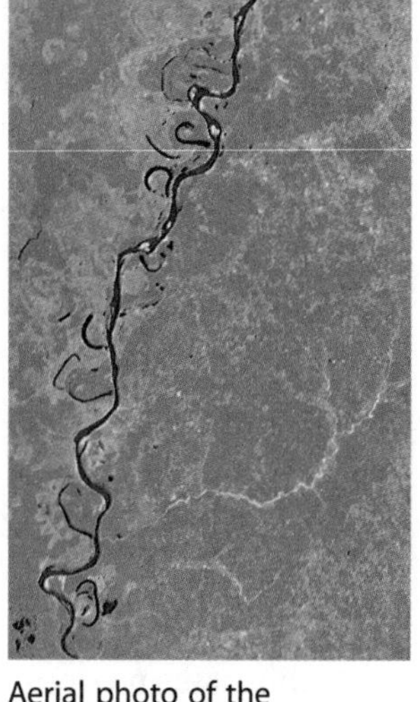

Aerial photo of the Mississippi River

For Exercises 12–14, use inductive reasoning to find the next two terms of the sequence.

12. 180, 360, 540, 720, _?_, _?_ (h)

13. 0, 10, 21, 33, 46, 60, _?_, _?_

14. $\frac{1}{2}$, 9, $\frac{2}{3}$, 10, $\frac{3}{4}$, 11, _?_, _?_

For Exercises 15–18, draw the next shape in each picture pattern.

15.

16. (h)

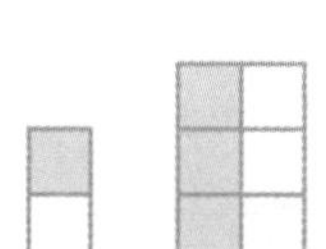
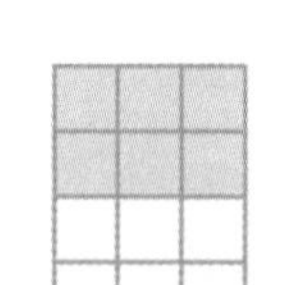
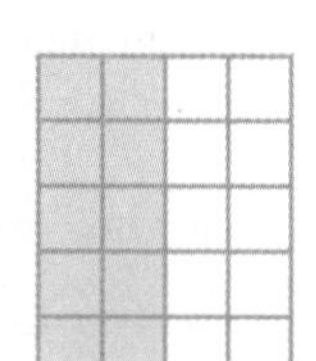

17. (h)

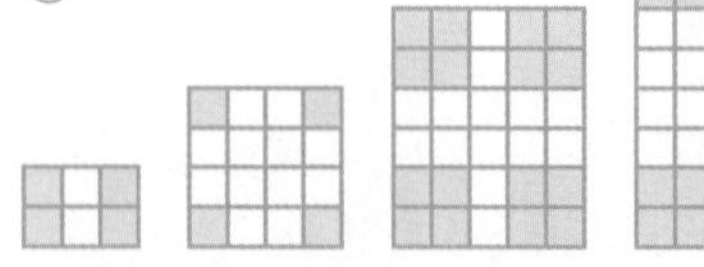

18.

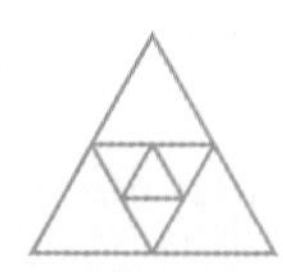
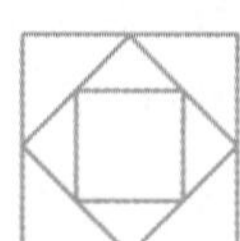
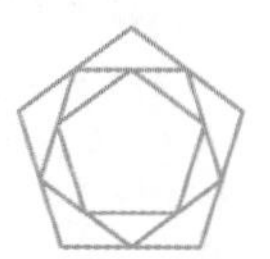

19. Think of a situation you have observed in which inductive reasoning was used incorrectly. Write a paragraph or two describing what happened and explaining why you think it was an incorrect use of inductive reasoning.

Match each term in Exercises 20–29 with one of the figures A–O.

20. Kite

21. Consecutive angles in a polygon

22. Trapezoid

23. Diagonal in a polygon

24. Pair of complementary angles

25. Radius

26. Pair of vertical angles

27. Chord

28. Acute angle

29. Angle bisector in a triangle

A.

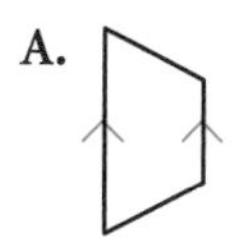

B.

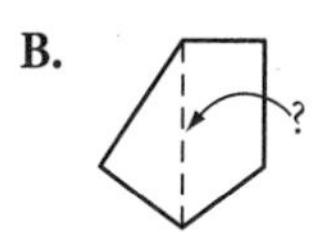

C.

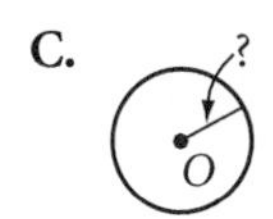

D.

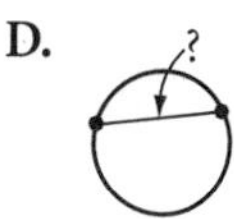

E.

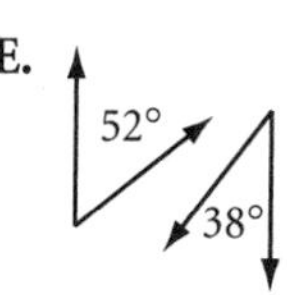

F. 48° 132°

G.

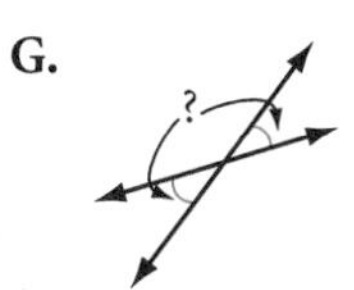

H.

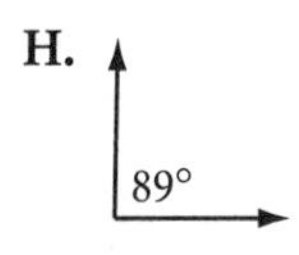

I.

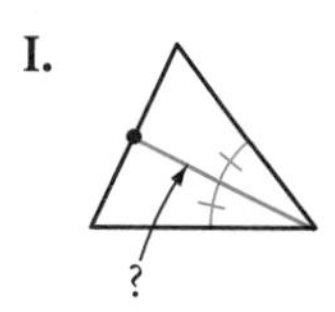

J.

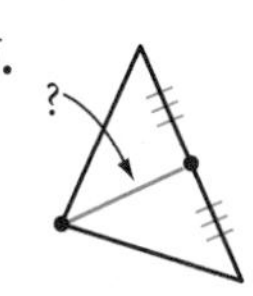

K.

L.

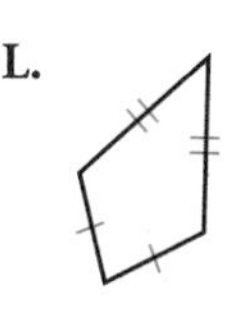

M.

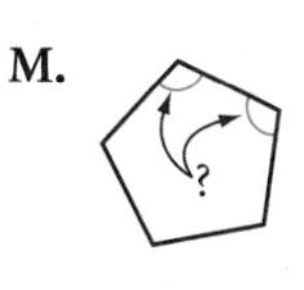

N.

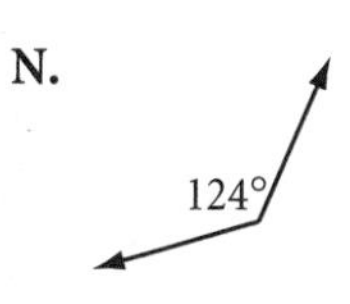

O.

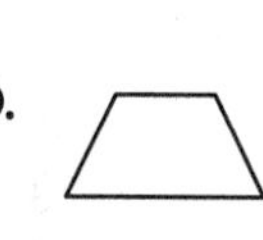

For Exercises 30–33, sketch and carefully label the figure.

30. Pentagon *WILDE* with $\angle ILD \cong \angle LDE$ and $\overline{LD} \cong \overline{DE}$

31. Isosceles obtuse triangle *OBG* with $m\angle BGO = 140°$

32. Circle *O* with a chord $\overline{CD}$ perpendicular to radius $\overline{OT}$

33. Circle *K* with angle *DIN* where *D, I,* and *N* are points on circle *K*

IMPROVING YOUR VISUAL THINKING SKILLS

Rotating Gears

In what direction will gear E rotate if gear A rotates in a counterclockwise direction?

LESSON

2.3

Finding the nth Term

If you do something once, people call it an accident. If you do it twice, they call it a coincidence. But do it a third time and you've just proven a natural law.

GRACE MURRAY HOPPER

What would you do to get the next term in the sequence 20, 27, 34, 41, 48, 55, . . . ? A good strategy would be to find a pattern, using inductive reasoning. Then you would look at the differences between consecutive terms and predict what comes next. In this case there is a constant difference of $+7$. That is, you add 7 each time.

The next term is $55 + 7$, or 62. What if you needed to know the value of the 200th term of the sequence? You certainly don't want to generate the next 193 terms just to get one answer. If you knew a rule for calculating *any* term in a sequence, without having to know the previous term, you could apply it to directly calculate the 200th term. The rule that gives the nth term for a sequence is called the **function rule.**

Let's see how the constant difference can help you find the function rule for some sequences.

DRABBLE reprinted by permission of United Feature Syndicate, Inc.

Investigation
Finding the Rule

Step 1 Copy and complete each table. Find the differences between consecutive values.

a.

n	1	2	3	4	5	6	7	8
$n - 5$	-4	-3	-2					

b.

n	1	2	3	4	5	6	7	8
$4n - 3$	1	5	9					

c.

n	1	2	3	4	5	6	7	8
$-2n + 5$	3	1	-1					

d.

n	1	2	3	4	5	6	7	8
$3n - 2$	1	4	7					

e.

n	1	2	3	4	5	6	7	8
$-5n + 7$	2	-3	-8					

Step 2 Did you spot the pattern? If a sequence has a constant difference of 4, then the number in front of the n (the coefficient of n) is _?_. In general, if the difference between the values of consecutive terms of a sequence is always the same, say m (a constant), then the coefficient of n in the formula is _?_.

Let's return to the sequence at the beginning of the lesson.

Term	1	2	3	4	5	6	7	...	n
Value	20	27	34	41	48	55	62	...	

+7 +7

The constant difference is 7, so you know part of the rule is $7n$. How do you find the rest of the rule?

Step 3 The first term ($n = 1$) of the sequence is 20, but if you apply the part of the rule you have so far using $n = 1$, you get $7n = 7(1) = 7$, not 20. So how should you fix the rule? How can you get from 7 to 20? What is the rule for this sequence?

Step 4 Check your rule by trying the rule with other terms in the sequence.

Let's look at an example of how to find a function rule, for the nth term in a number pattern.

EXAMPLE A

Find the rule for the sequence 7, 2, −3, −8, −13, −18, . . .

▶ *Solution*

Placing the terms and values in a table we get

Term	1	2	3	4	5	6	...	n
Value	7	2	−3	−8	−13	−18	...	

The difference between the terms is always −5. So the rule is

$-5n +$ "something"

Let's use c to stand for the unknown "something." So the rule is

$5n + c$

To find c, replace the n in the rule with a term number. Try $n = 1$ and set the expression equal to 7.

$$-5(1) + c = 7$$
$$c = 12$$

The rule is $-5n + 12$.

You can find the value of any term in the sequence by substituting the term number for n in the function rule. Let's look at an example of how to find the 200th term in a geometric pattern.

EXAMPLE B If you place 200 points on a line, into how many non-overlapping rays and segments does it divide the line?

▶ Solution Wait! don't start placing 200 points on a line. You need to find a rule that relates the number of points placed on a line to the number of parts created by those points. Then you can use your rule to answer the problem.

Start by creating a table.

Points dividing the line	1	2	3	4	5	6	...	n	...	200
Non-overlapping rays							...		...	
Non-overlapping segments							...		...	
Total							...		...	

Sketch one point dividing a line. One point gives you just two rays. Enter that into the table.

Next, sketch two points dividing a line. This gives one segment and the two end rays. Enter the value into your table.

Next, sketch three points dividing a line, then four, then five, and so on. The table completed for one to three points is

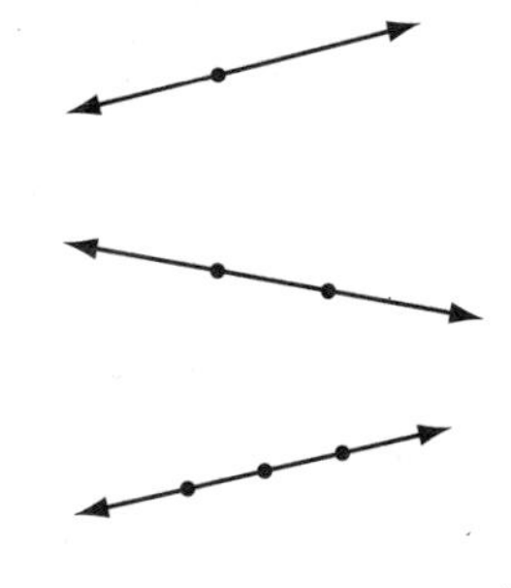

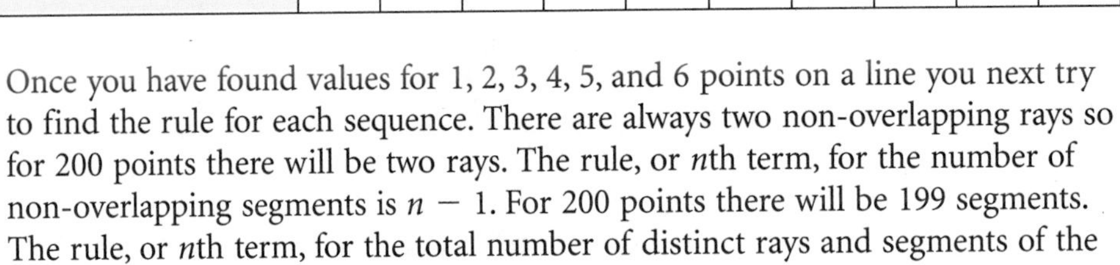

Points dividing the line	1	2	3	4	5	6	...	n	...	200
Non-overlapping rays	2	2	2				...		...	
Non-overlapping segments	0	1	2				...		...	
Total	2	3	4				...		...	

Once you have found values for 1, 2, 3, 4, 5, and 6 points on a line you next try to find the rule for each sequence. There are always two non-overlapping rays so for 200 points there will be two rays. The rule, or nth term, for the number of non-overlapping segments is $n - 1$. For 200 points there will be 199 segments. The rule, or nth term, for the total number of distinct rays and segments of the line is $n + 1$. For 200 points there will be 201 distinct parts of the line.

This process of looking at patterns and generalizing a rule, or nth term, is inductive reasoning. To understand why the rule is what it is, you can turn to deductive reasoning. Notice that adding another point on a line divides a segment into two segments. So each new point adds one more segment to the pattern.

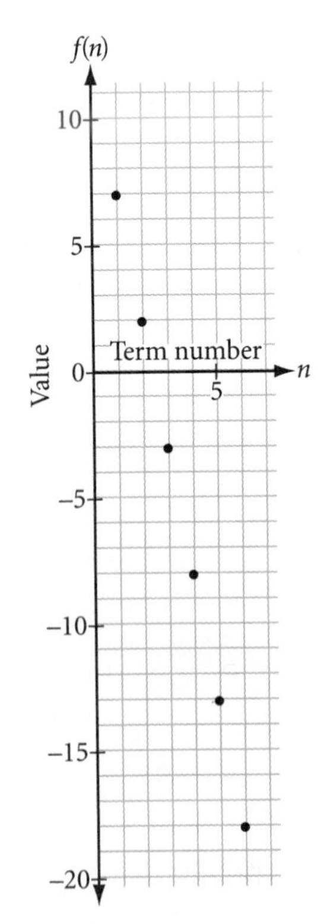

Rules that generate a sequence with a constant difference are **linear functions.** To see why they're called linear, you can graph the term number and the value for the sequence as ordered pairs of the form (*term number, value*) on the coordinate plane. At left is the graph of the sequence from Example A.

Term number n	1	2	3	4	5	6	...	n
Value $f(n)$	7	2	-3	-8	-13	-18	...	$-5n + 12$

EXERCISES

For Exercises 1–3, find the function rule $f(n)$ for each sequence. Then find the 20th term in the sequence.

1. ⓗ

n	1	2	3	4	5	6	...	n	...	20
$f(n)$	3	9	15	21	27	33	...		...	

2.

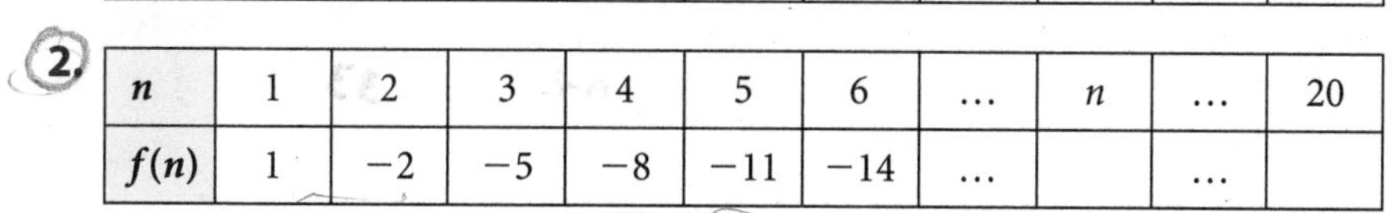

n	1	2	3	4	5	6	...	n	...	20
$f(n)$	1	−2	−5	−8	−11	−14	...		...	

3.

n	1	2	3	4	5	6	...	n	...	20
$f(n)$	−4	4	12	20	28	36	...		...	

For Exercises 4–6, find the rule for the nth figure. Then find the number of colored tiles or matchsticks in the 200th figure.

4. ⓗ

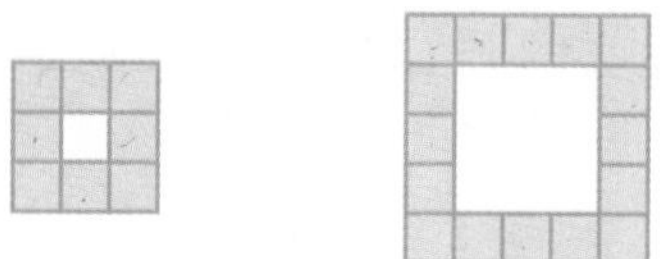
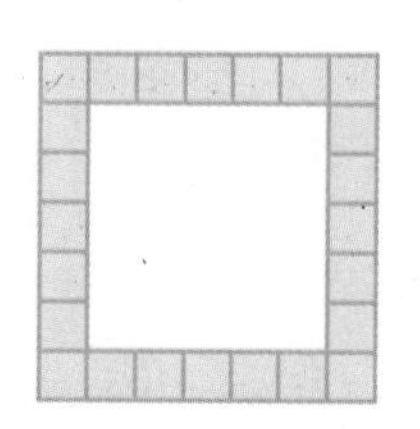
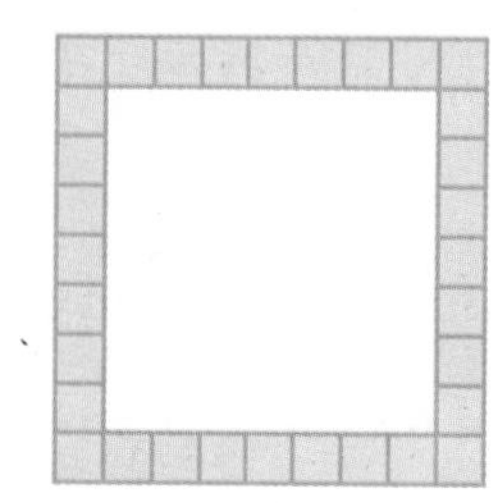

Figure number	1	2	3	4	5	6	...	n	...	200
Number of tiles	8	16	24	32	40	48	...		...	

5.

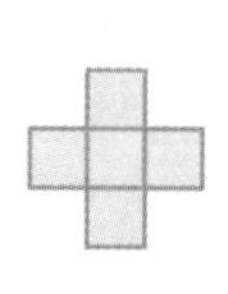
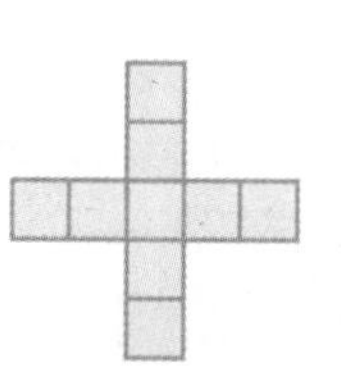
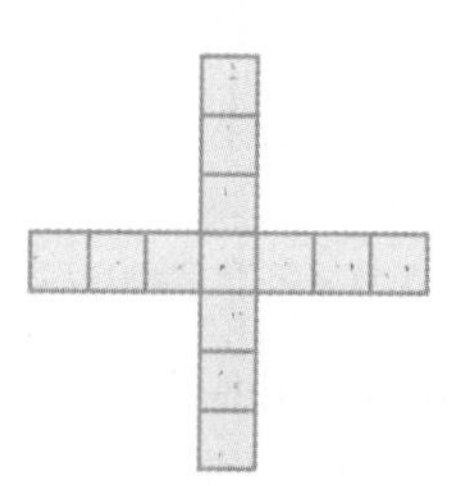

Figure number	1	2	3	4	5	6	...	n	...	200
Number of tiles	1	5	9	13			...		...	

4n-3

6.

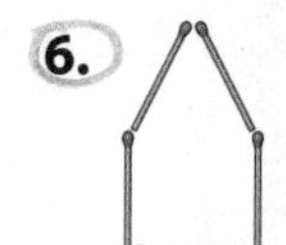
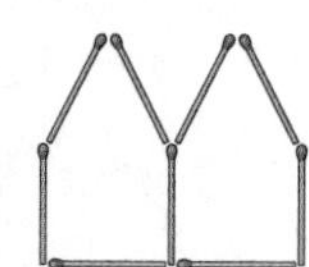
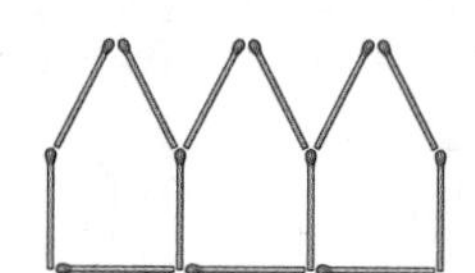
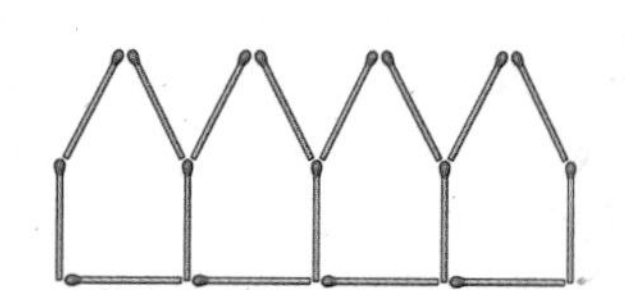

Figure number	1	2	3	4	5	6	...	n	...	200
Number of matchsticks	5	9	13	17	21		...	4n+1	...	
Number of matchsticks in perimeter of figure	5	8	11	14	19		...		...	

7. How many triangles are formed when you draw all the possible diagonals from just one vertex of a 35-gon? ⓗ

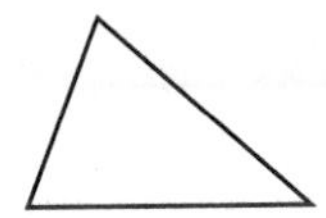 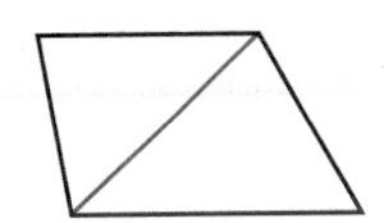 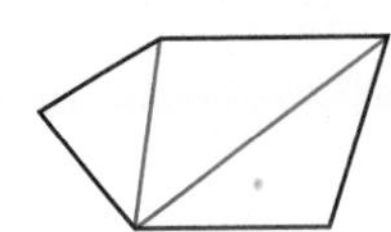 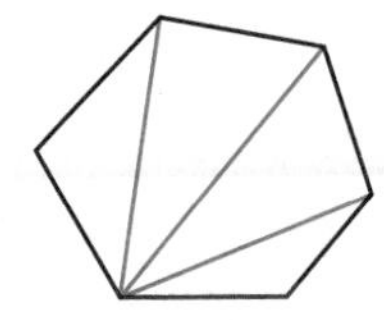

Number of sides	3	4	5	6	...	n	...	35
Number of triangles formed	1	2	3	4	...	n-2	...	33

8. Graph the values in your tables from Exercises 4–6. Which set of points lies on a steeper line? What number in the rule gives a measure of steepness?

9. Find the rule for the set of points in the graph shown at right. Place the x-coordinate of each ordered pair in the top row of your table and the corresponding y-coordinate in the second row. What is the value of y in terms of x?

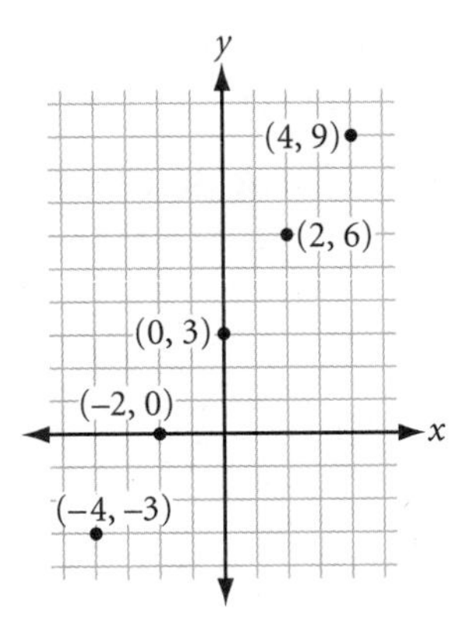

Review

For Exercises 10–13, sketch and carefully label the figure.

10. Equilateral triangle EQL with $\overline{QT}$ where T lies on $\overline{EL}$ and $\overline{QT} \perp \overline{EL}$

11. Isosceles obtuse triangle OLY with $\overline{OL} \cong \overline{YL}$ and angle bisector $\overline{LM}$

12. A cube with a plane passing through it; the cross section is rectangle $RECT$

13. A net for a rectangular solid with the dimensions 1 by 2 by 3 cm

14. Márisol's younger brother José was drawing triangles when he noticed that every triangle he drew turned out to have two sides congruent. José conjectures: "Look, Márisol, all triangles are isosceles." How should Márisol respond?

15. A midpoint divides a segment into two congruent segments. Point M divides segment $\overline{AY}$ into two congruent segments $\overline{AM}$ and $\overline{MY}$. What conclusion can you make? What type of reasoning did you use?

16. Tanya's favorite lunch is peanut butter and jelly on wheat bread with a glass of milk. Lately, she has been getting an allergic reaction after eating this lunch. She is wondering if she might be developing an allergy to peanut butter, wheat, or milk. What experiment could she do to find out which food it is? What type of reasoning would she be using?

17. ***Mini-Investigation*** Do the geometry investigation and make a conjecture.

Given $\angle APB$ with points C and D in its interior and $m\angle APC = m\angle DPB$,

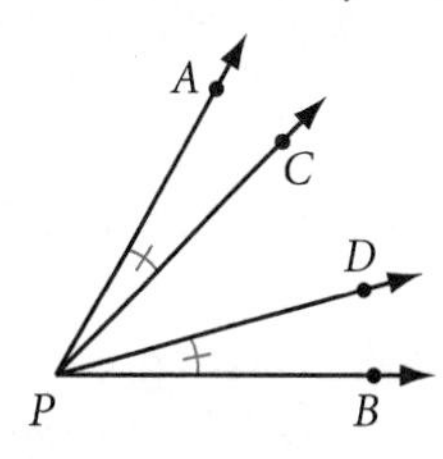

If $m\angle APD = 48°$, then $m\angle CPB = \underline{?}$

If $m\angle CPB = 17°$, then $m\angle APD = \underline{?}$

If $m\angle APD = 62°$, then $m\angle CPB = \underline{?}$

Conjecture: If points C and D lie in the interior of $\angle APB$, and $m\angle APC = m\angle DPB$ then $m\angle APD = \underline{?}$ (Overlapping angles property)

project

BEST-FIT LINES

The following table and graph show the mileage and lowest priced round-trip airfare between New York City and each destination city. Is there a relationship between the money you spend and how far you can travel?

Fathom

With Fathom Dynamic Statistics™ software, you can plot your data points and find the linear equation that best fits your data.

Lowest Round-trip Airfares from New York City on February 25, 2002

Destination City	Distance (miles)	Price ($)
Boston	215	$118
Chicago	784	$178
Atlanta	865	$158
Miami	1286	$170
Denver	1791	$238
Phoenix	2431	$338
Los Angeles	2763	$298

Source: http://www.Expedia.com

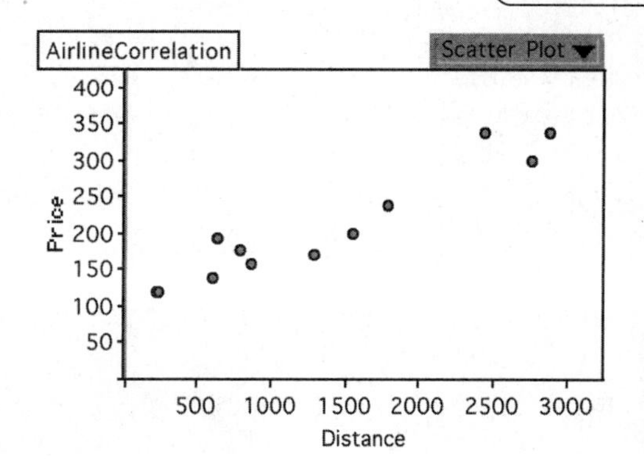

Even though the data are not linear, you can find a linear equation that *approximately* fits the data. The graph of this equation is called the **line of best fit.** How would you use the line of best fit to predict the cost of a round-trip ticket to Seattle (2814 miles)? How would you use it to determine how far you could travel (in miles) with $250? How accurate do you think the answer would be?

Choose a topic and a relationship to explore. You can use data from the census (such as age and income), or data you collect yourself (such as number of ice cubes in a glass and melting time). For more sources and ideas, go to **www.keymath.com/DG** .

Collect pairs of data points. Use Fathom to graph your points and to find the line of best fit. Write a summary of your results.

LESSON

2.4

Mathematical Modeling

It's amazing what one can do when one doesn't know what one can't do.

GARFIELD THE CAT

Physical models have many of the same features as the original object or activity they represent, but are often more convenient to study. For example, building a new airplane and testing it is difficult and expensive. But you can analyze a new airplane design by building a model and testing it in a wind tunnel.

In Chapter 1 you learned that geometry ideas such as points, lines, planes, triangles, polygons, and diagonals are **mathematical models** of physical objects.

When you draw graphs or pictures of situations, or when you write equations that describe a problem, you are creating mathematical models. A physical model of a complicated telecommunications network, for example, might not be practical, but you can draw a mathematical model of the network using points and lines.

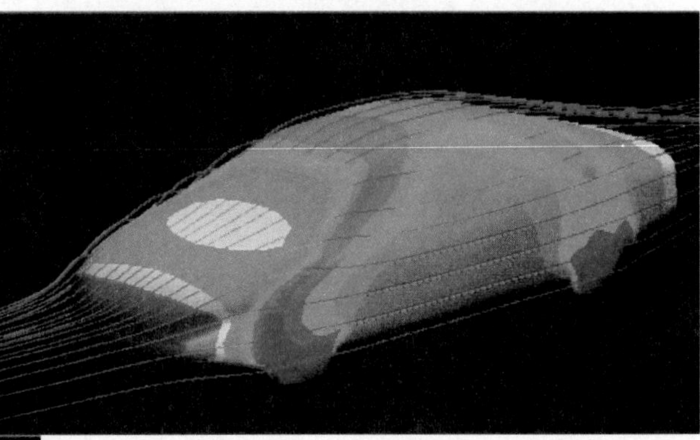

This computer model tests the effectiveness of the car's design for minimizing wind resistance.

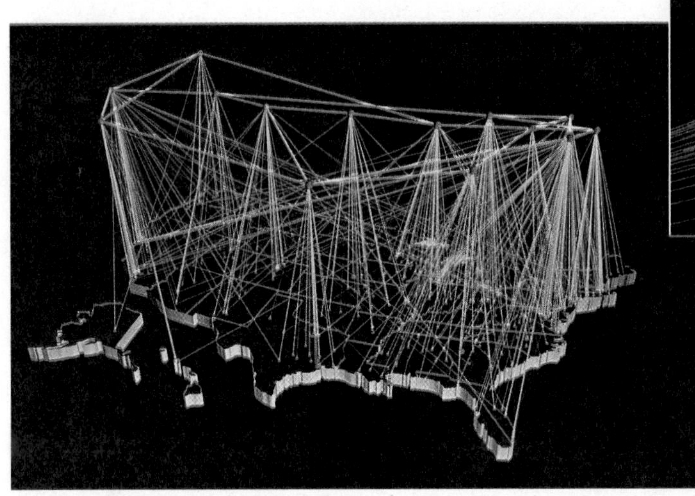

This computer-generated model uses points and line segments to show the volume of data traveling to different locations on the National Science Foundation Network.

In this investigation, you will attempt to solve a problem first by acting it out, then by creating a mathematical model.

Investigation

Party Handshakes

Each of the 30 people at a party shook hands with everyone else. How many handshakes were there altogether?

Step 1 Act out this problem with members of your group. Collect data for "parties" of one, two, three, and four people and record your results in a table.

People	1	2	3	4	...	30
Handshakes	0	1			...	

Step 2 | Look for a pattern. Can you generalize from your pattern to find the 30th term?

Acting out a problem is a powerful problem-solving strategy that can give you important insight into a solution. Were you able to make a generalization from just four terms? If so, how confident are you of your generalization? To collect more data, you can ask more classmates to join your group. You can see, however, that acting out a problem sometimes has its practical limitations. That's where you can use mathematical models.

Step 3 | Model the problem by using points to represent people and line segments connecting the points to represent handshakes.

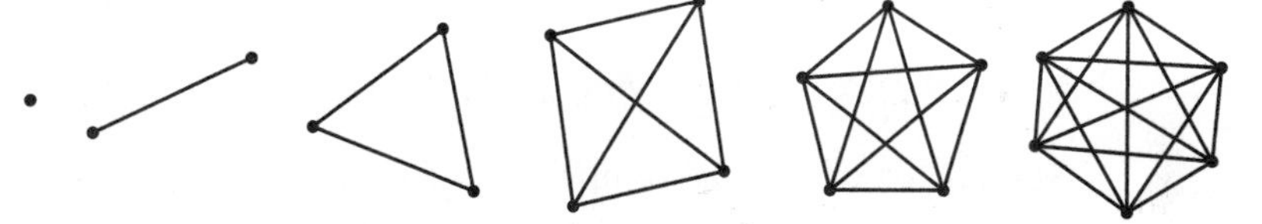

Record your results in a table like this one:

Number of points (people)	1	2	3	4	5	6	...	n	...	30
Number of segments (handshakes)	0	1					...		...	

Notice that the pattern does not have a constant difference. That is, the rule is not a linear function. So we need to look for a different kind of rule.

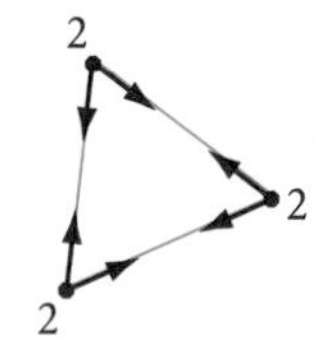

3 points
2 segments per vertex

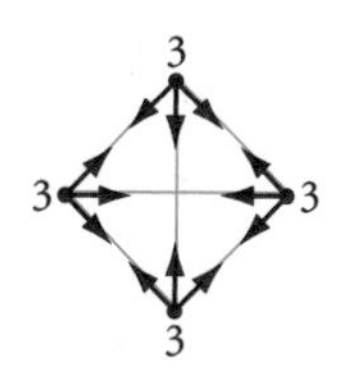

4 points
3 segments per vertex

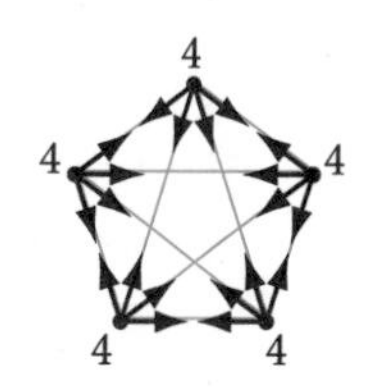

5 points
_?_ segments per vertex

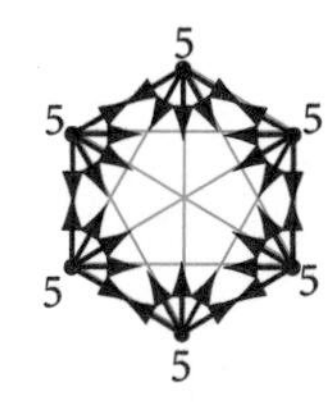

6 points
_?_ segments per vertex

Step 4 | Refer to the table you made for Step 3. The pattern of differences is increasing by one: 1, 2, 3, 4, 5, 6, 7. Read the dialogue between Erin and Stephanie as they attempt to combine inductive and deductive reasoning to find the rule.

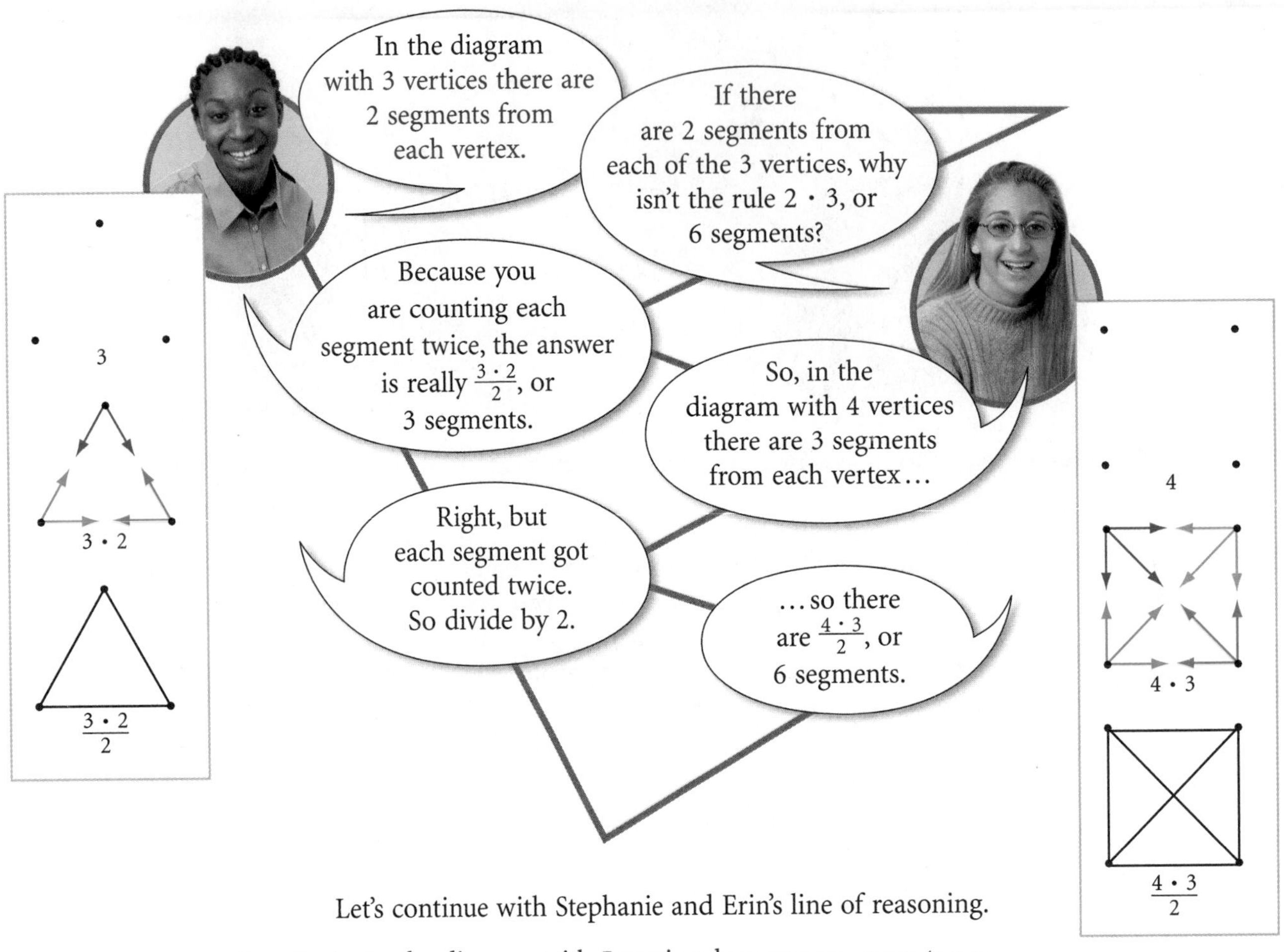

Let's continue with Stephanie and Erin's line of reasoning.

Step 5 | In the diagram with 5 vertices how many segments are there from each vertex? So the total number of segments written in factored form is $\frac{5 \cdot ?}{2}$.

Step 6 | Complete the table below by expressing the total number of segments in factored form.

Number of points (people)	1	2	3	4	5	6	...	n
Number of segments (handshakes)	$\frac{(1)(0)}{2}$	$\frac{(2)(1)}{2}$	$\frac{(3)(2)}{2}$	$\frac{(4)(3)}{2}$	$\frac{(5)(?)}{2}$	$\frac{(6)(?)}{2}$	...	$\frac{(?)(?)}{2}$

Step 7 | The larger of the two factors in the numerator represents the number of points. What does the smaller of the two numbers in the numerator represent? Why do we divide by 2?

Step 8 | Write a function rule. How many handshakes were there at the party?

Fifteen pool balls can be arranged in a triangle, so 15 is a triangular number.

The numbers in the pattern in the previous investigation are called the **triangular numbers** because you can arrange them into a triangular pattern of dots.

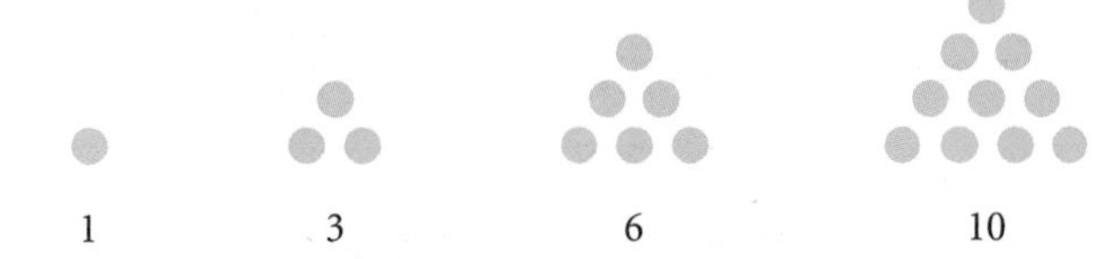

The triangular numbers appear in many geometric situations. You will see some of them in the exercises.

Here is a visual approach to arrive at the rule for this special pattern of numbers. If we arrange the triangular numbers in stacks,

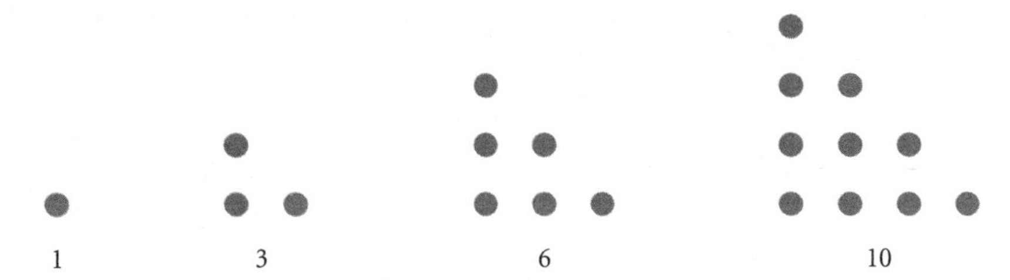

you can see that each is half of a **rectangular number.**

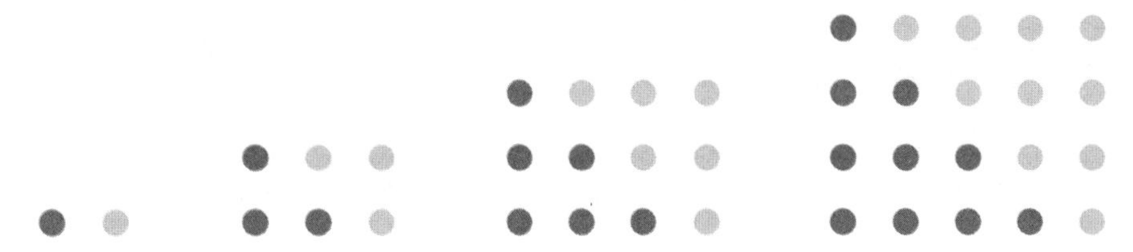

To get the total number of dots in a rectangular array, you multiply the number of rows by the number of dots in each row. In the case of this rectangular array, the nth rectangle has $n(n + 1)$ dots. So, the triangular array has $\frac{n(n+1)}{2}$ dots.

EXERCISES

For Exercises 1–6, draw the next figure. Complete a table and find the function rule. Then find the 35th term.

1. Lines passing through the same point are **concurrent.** Into how many regions do 35 concurrent lines divide the plane?

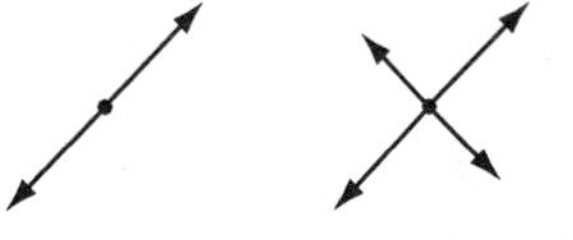

Lines	1	2	3	4	5	...	n	...	35
Regions	2					...		...	

2. Into how many regions do 35 parallel lines in a plane divide that plane?

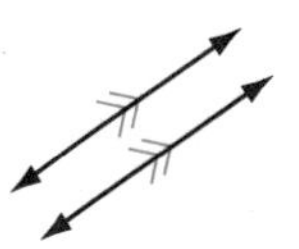

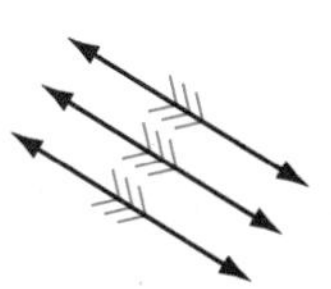

3. How many diagonals can you draw from one vertex in a polygon with 35 sides?

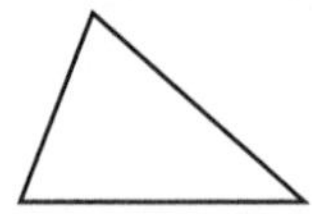 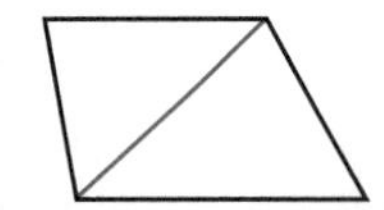 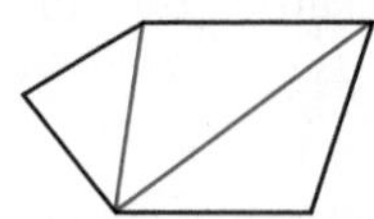

4. What's the total number of diagonals in a 35-sided polygon? ⓗ

 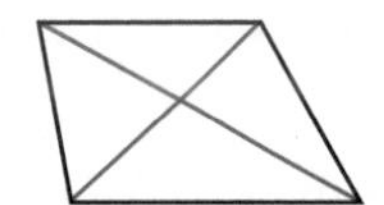 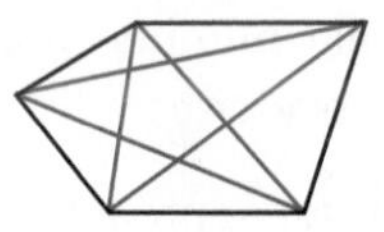

5. If you place 35 points on a piece of paper so that no three points are in a line, how many line segments are necessary to connect each point to all the others? ⓗ

 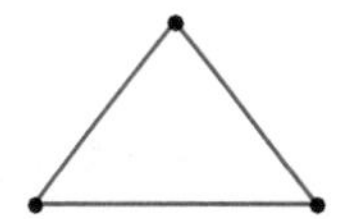 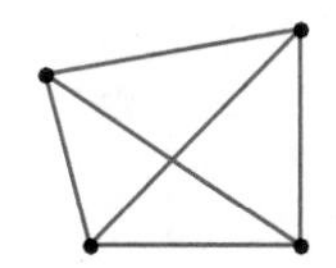

6. If you draw 35 lines on a piece of paper so that no two lines are parallel to each other and no three lines are concurrent, how many times will they intersect? ⓗ

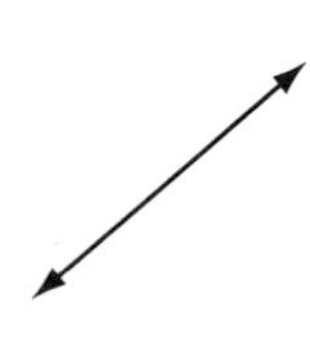 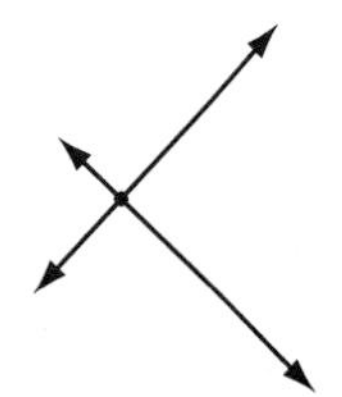 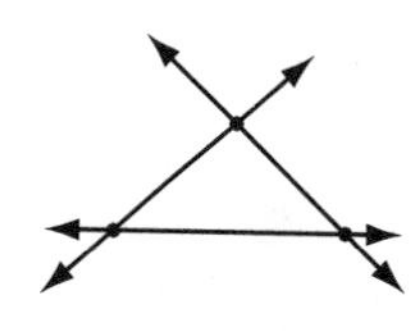

7. Look at the formulas you found in Exercises 4–6. Describe how the formulas are related. Then explain how the three problems are related geometrically. ⓗ

For Exercises 8–10, draw a diagram, find the appropriate geometric model, and solve.

8. If 40 houses in a community all need direct lines to one another in order to have telephone service, how many lines are necessary? Is that practical? Sketch and describe two models: first, model the situation in which direct lines connect every house to every other house and, second, model a more practical alternative.

9. If each team in a ten-team league plays each of the other teams four times in a season, how many league games are played during one season? What geometric figures can you use to model teams and games played? ⓗ

10. Each person at a party shook hands with everyone else exactly once. There were 66 handshakes. How many people were at the party? ⓗ

Review

For Exercises 11–19, identify the statement as true or false. For each false statement, explain why it is false or sketch a counterexample.

11. The largest chord of a circle is a diameter of the circle.

12. The vertex of $\angle TOP$ is point O.

13. An isosceles right triangle is a triangle with an angle measuring 90° and no two sides congruent.

14. If $\overleftrightarrow{AB}$ intersects $\overleftrightarrow{CD}$ in point E, then $\angle AED$ and $\angle BED$ form a linear pair of angles. ⓗ

15. If two lines lie in the same plane and are perpendicular to the same line, they are perpendicular.

16. The opposite sides of a kite are never parallel.

17. A rectangle is a parallelogram with all sides congruent.

18. A line segment that connects any two vertices in a polygon is called a diagonal.

19. To show that two lines are parallel, you mark them with the same number of arrowheads.

20. Hydrocarbons are molecules that consist of carbon (C) and hydrogen (H). Hydrocarbons in which all the bonds between the carbon atoms are single bonds are called *alkanes*. The first four alkanes are modeled below.

Sketch the alkane with eight carbons in the chain. What is the general rule for alkanes $(C_nH_?)$? In other words, if there are n carbon atoms (C), how many hydrogen atoms (H) are in the alkane?

```
    H          H   H          H   H   H          H   H   H   H
    |          |   |          |   |   |          |   |   |   |
H - C - H  H - C - C - H  H - C - C - C - H  H - C - C - C - C - H
    |          |   |          |   |   |          |   |   |   |
    H          H   H          H   H   H          H   H   H   H
```

Methane (CH_4) Ethane (C_2H_6) Propane (C_3H_8) Butane (C_4H_{10})

Science CONNECTION

Organic chemistry is the study of carbon compounds and their reactions. Drugs, vitamins, synthetic fibers, and food all contain organic molecules. Organic chemists continue to improve our quality of life by the advances they make in medicine, nutrition, and manufacturing. To learn about new advances in organic chemistry, go to www.keymath.com/DG .

IMPROVING YOUR VISUAL THINKING SKILLS

Pentominoes II

In Pentominoes I, you found the 12 pentominoes. Which of the 12 pentominoes can you cut along the edges and fold into a box without a lid? Here is an example.

Exploration

The Seven Bridges of Königsberg

The River Pregel runs through the university town of Königsberg (now Kaliningrad in Russia). In the middle of the river are two islands connected to each other and to the rest of the city by seven bridges. Many years ago, a tradition developed among the townspeople of Königsberg. They challenged one another to make a round trip over all seven bridges, walking over each bridge once and only once before returning to the starting point.

Leonhard Euler

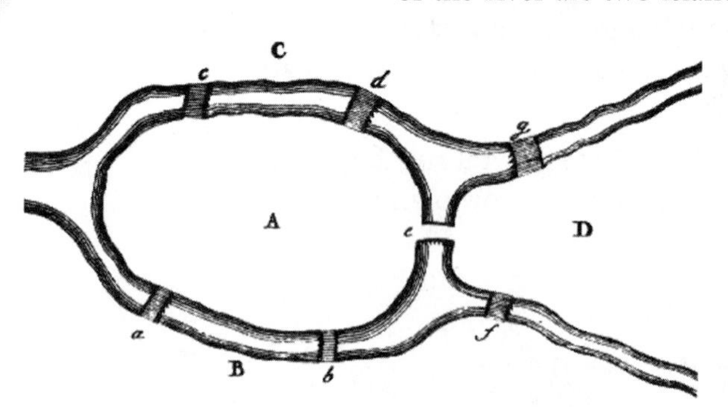

The seven bridges of Königsberg

For a long time no one was able to do it, and yet no one was able to show that it couldn't be done. In 1735, they finally wrote to Leonhard Euler (1707–1783), a Swiss mathematician, asking for his help on the problem. Euler (pronounced "oyler") reduced the problem to a network of paths connecting the two sides of the rivers C and B, and the two islands A and D, as shown in the network at right. Then Euler demonstrated that the task is impossible.

In this activity you will work with a variety of networks to see if you can come up with a rule to find out whether a network can or cannot be "traveled."

Activity

Traveling Networks

A collection of points connected by paths is called a **network.** When we say a network can be traveled, we mean that the network can be drawn with a pencil without lifting the pencil off the paper and without retracing any paths. (Points can be passed over more than once.)

Step 1 Try these networks and see which ones can be traveled and which are impossible to travel.

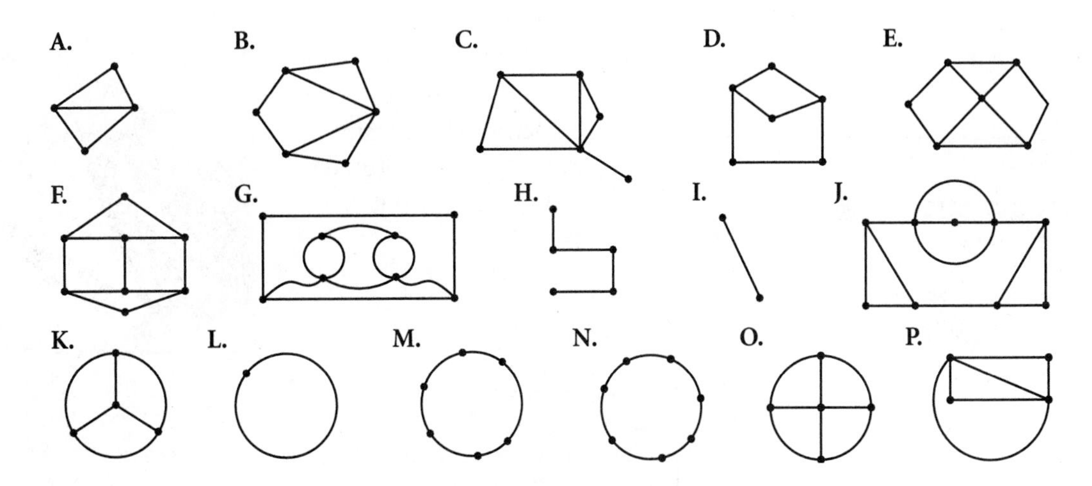

Which networks were impossible to travel? Are they impossible or just difficult? How can you be sure? As you do the next few steps, see if you can find the reason why some networks are impossible to travel.

Step 2 Draw the River Pregel and the two islands shown on the first page of this exploration. Draw an eighth bridge so that you can travel over all the bridges exactly once if you start at point *C* and end at point *B*.

Step 3 Draw the River Pregel and the two islands. Can you draw an eighth bridge so that you can travel over all the bridges exactly once, starting and finishing at the same point? How many solutions can you find?

Step 4 Euler realized that it is the points of intersection that determine whether a network can be traveled. Each point of intersection is either "odd" or "even."

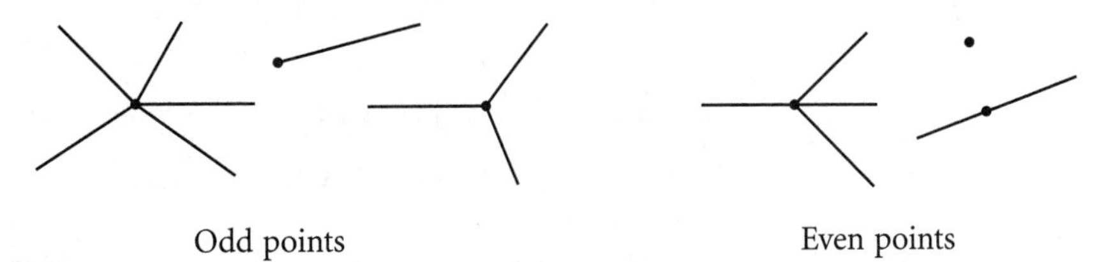

Odd points Even points

Did you find any networks that have only one odd point? Can you draw one? Try it. How about three odd points? Or five odd points? Can you create a network that has an odd number of odd points? Explain why or why not.

Step 5 How does the number of even points and odd points affect whether a network can be traveled?

Conjecture

A network can be traveled if __?__.

LESSON

2.5

Angle Relationships

Discovery consists of looking at the same thing as everyone else and thinking something different.

ALBERT SZENT-GYÖRGYI

Now that you've had experience with inductive reasoning, let's use it to start discovering geometric relationships. This investigation is the first of many investigations you will do using your geometry tools.

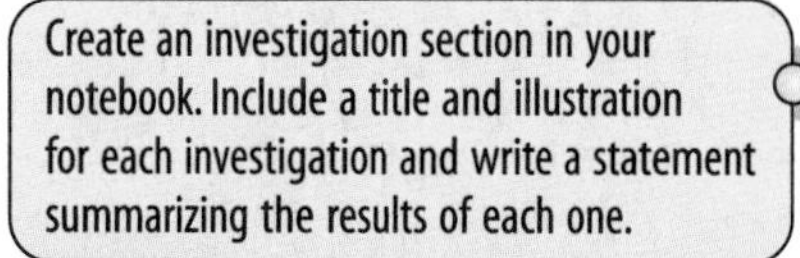

Investigation 1
The Linear Pair Conjecture

You will need

- a protractor

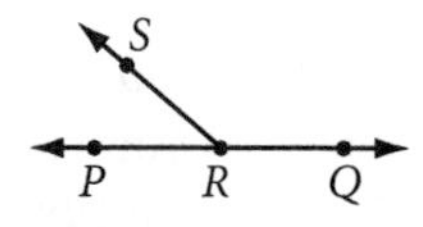

Step 1 On a sheet of paper, draw $\overleftrightarrow{PQ}$ and place a point R between P and Q. Choose another point S not on $\overleftrightarrow{PQ}$ and draw $\overrightarrow{RS}$. You have just created a linear pair of angles. Place the "zero edge" of your protractor along $\overleftrightarrow{PQ}$. What do you notice about the sum of the measures of the linear pair of angles?

Step 2 Compare your results with those of your group. Does everyone make the same observation? Complete the statement.

Linear Pair Conjecture C-1

If two angles form a linear pair, then _?_.

The important conjectures have been given a name and a number. Start a list of them in your notebook. The Linear Pair Conjecture (C-1) and the Vertical Angles Conjecture (C-2) should be the first entries on your list. Make a sketch for each conjecture.

In the previous investigation you discovered the relationship between a linear pair of angles, such as $\angle 1$ and $\angle 2$ in the diagram at right.

You will discover the relationship between vertical angles, such as $\angle 1$ and $\angle 3$, in the next investigation.

Investigation 2
Vertical Angles Conjecture

You will need

- a protractor
- patty paper

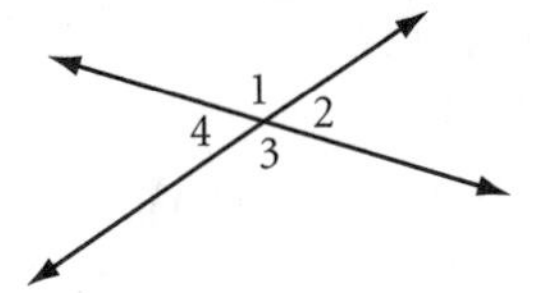

Step 1 Draw two intersecting lines onto patty paper or tracing paper. Label the angles as shown. Which angles are vertical angles?

Step 2 Fold the paper so that the vertical angles lie over each other. What do you notice about their measures?

Step 3 Repeat this investigation with another pair of intersecting lines.

Step 4 Compare your results with the results of others. Complete the statement.

Vertical Angles Conjecture C-2

If two angles are vertical angles, then __?__.

You used inductive reasoning to discover both the Linear Pair Conjecture and the Vertical Angles Conjecture. Are they related in any way? That is, if we accept the Linear Pair Conjecture as true, can we use deductive reasoning to show that the Vertical Angles Conjecture must be true?

EXAMPLE

The Linear Pair Conjecture states that every linear pair adds up to 180°. Using this conjecture and the diagram, write a logical argument explaining why $\angle 1$ must be congruent to $\angle 3$.

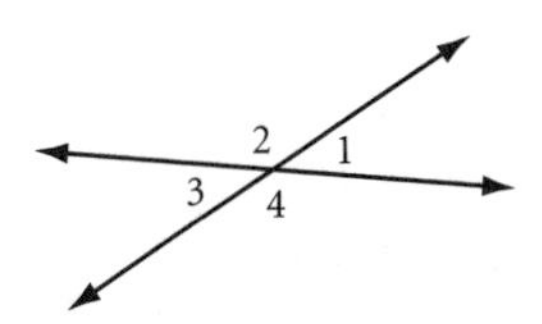

▶ Solution

You can see that the measures of $\angle 1$ and $\angle 2$ add up to 180°, and that the measures of $\angle 3$ and $\angle 2$ also add up to 180°. Using algebra, we can write a logical argument to show that $\angle 1$ and $\angle 3$ must be congruent.

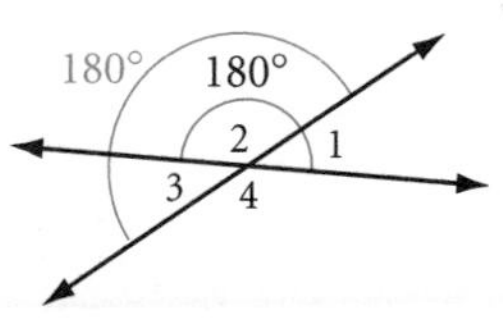

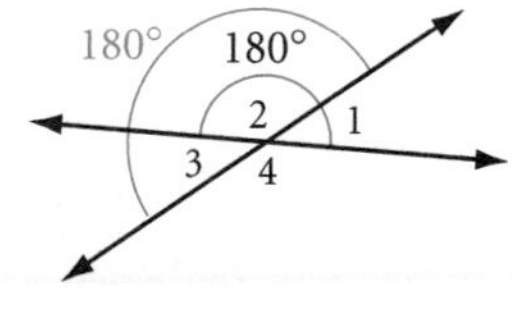

According to the Linear Pair Conjecture, $m\angle 1 + m\angle 2 = 180°$ and $m\angle 2 + m\angle 3 = 180°$. By substituting $m\angle 2 + m\angle 3$ for 180° in the first statement, you get $m\angle 1 + m\angle 2 = m\angle 2 + m\angle 3$. By the subtraction property of equality, you can subtract $m\angle 2$ from both sides of the equation to get $m\angle 1 = m\angle 3$. Therefore, vertical angles 1 and 3 have equal measures and are congruent.

Here are the algebraic steps:

$m\angle 2 + m\angle 3 = 180°$
$m\angle 1 + m\angle 2 = 180°$
$m\angle 1 + m\angle 2 = m\angle 2 + m\angle 3$
thus $m\angle 1 = m\angle 3$
therefore $\angle 1 \cong \angle 3$

This type of logical explanation, written as a paragraph, is called a **paragraph proof.**

Now consider another idea. You discovered the Vertical Angles Conjecture: If two angles are vertical angles, then they are congruent. Does that also mean that all congruent angles are vertical angles? The **converse** of an "if-then" statement switches the "if" and "then" parts. The converse of the Vertical Angles Conjecture may be stated: If two angles are congruent, then they are vertical angles. Is this converse statement true? Remember that if you can find even one counterexample, like the diagram below, then the statement is false.

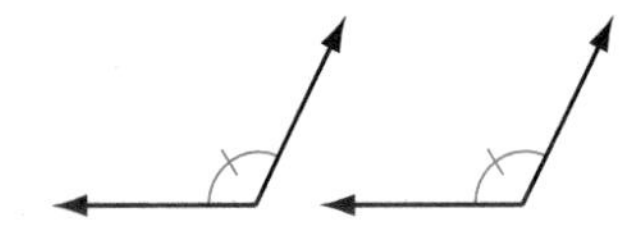

Therefore, the converse of the Vertical Angles Conjecture is false.

EXERCISES

You will need

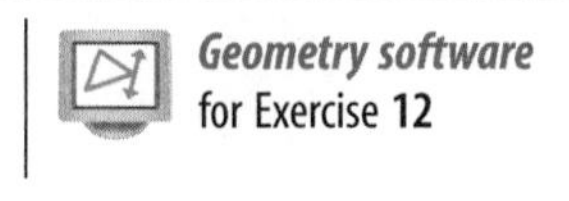

Without using a protractor, but with the aid of your two new conjectures, find the measure of each lettered angle in Exercises 1–5. Copy the diagrams so that you can write on them. List your answers in alphabetical order.

1.

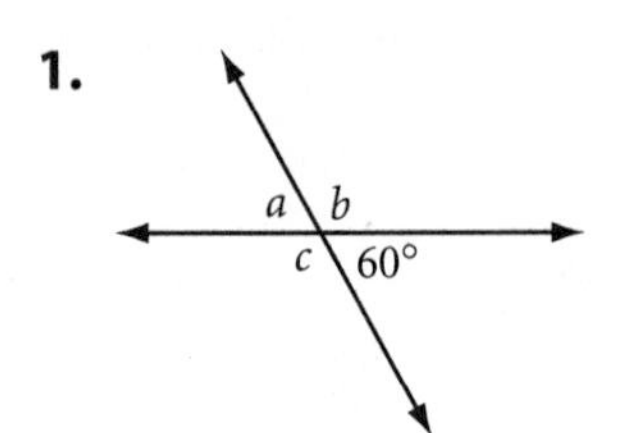

2.

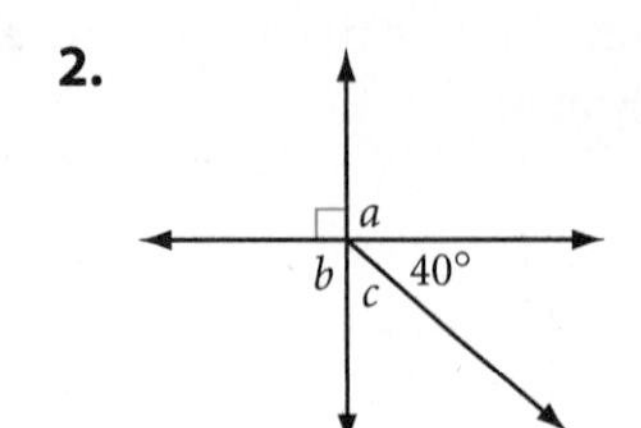

3.

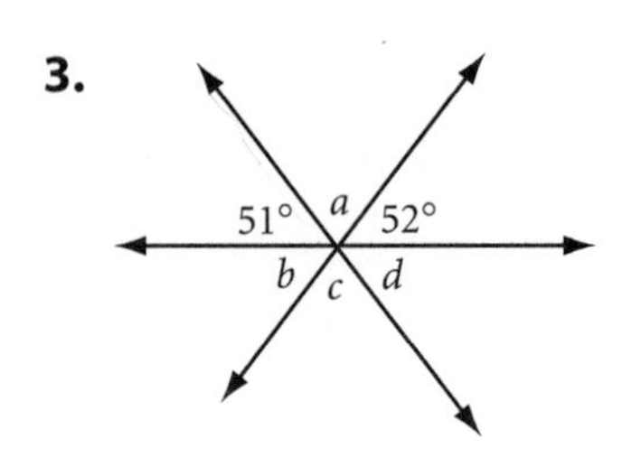

4. (h)

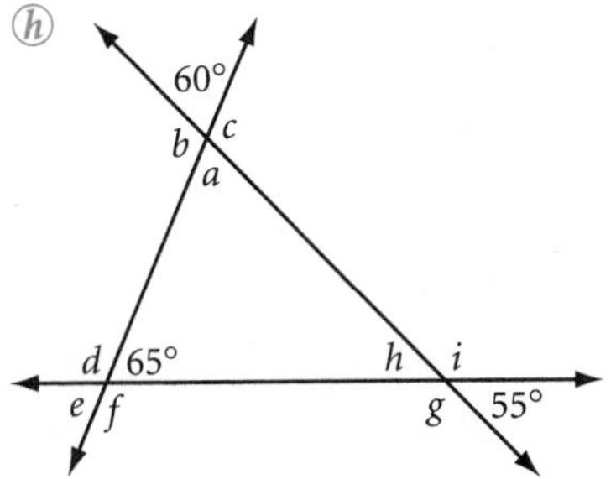

5.

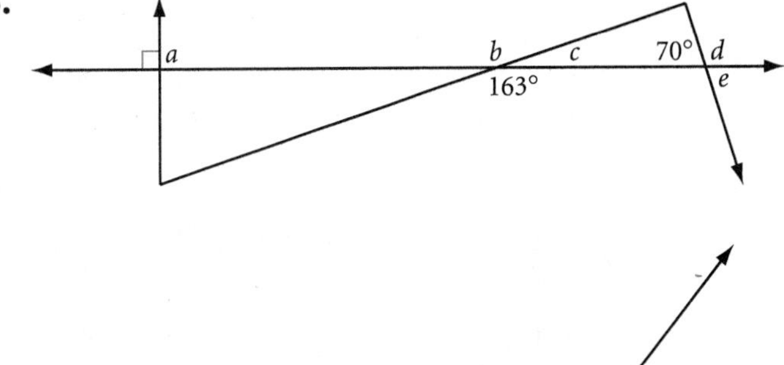

6. Points A, B, and C at right are collinear. What's wrong with this picture?

7. Yoshi is building a cold frame for his plants. He wants to cut two wood strips so that they'll fit together to make a right-angled corner. At what angle should he cut ends of the strips?

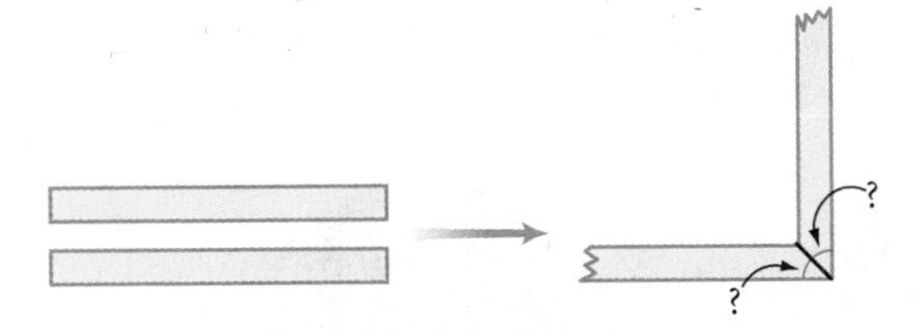

8. A tree on a 30° slope grows straight up. What are the measures of the greatest and smallest angles the tree makes with the hill? Explain.

9. You discovered that if a pair of angles is a linear pair then the angles are supplementary. Does that mean that all supplementary angles form a linear pair of angles? Is the converse true? If not, sketch a counterexample.

10. If two congruent angles are supplementary, what must be true of the two angles? Make a sketch, then complete the following conjecture: If two angles are both congruent and supplementary, then ___?___.

11. Using algebra, write a paragraph proof that explains why the conjecture from Exercise 10 is true.

12. *Technology* Use geometry software to construct two intersecting lines. Measure a pair of vertical angles. Use **calculate** to find the ratio of their measures. What is the ratio? Drag one of the lines. Does the ratio ever change? Does this demonstration convince you that the Vertical Angles Conjecture is true? Does it explain why it is true?

Review

For Exercises 13–17, sketch, label, and mark the figure.

13. Scalene obtuse triangle PAT with $PA = 3$ cm, $AT = 5$ cm, and $\angle A$ an obtuse angle

14. A quadrilateral that has rotational symmetry but not reflectional symmetry

15. A circle with center at O and radii $\overline{OA}$ and $\overline{OT}$ creating a minor arc $\overset{\frown}{AT}$

16. A pyramid with an octagonal base

17. A 3-by-4-by-6-inch rectangular solid rests on its smallest face. Draw lines on the three visible faces, showing how you can divide it into 72 identical smaller cubes.

18. Miriam the Magnificent placed four cards face up (the first four cards shown below). Blindfolded, she asked someone from her audience to come up to the stage and turn one card 180°.

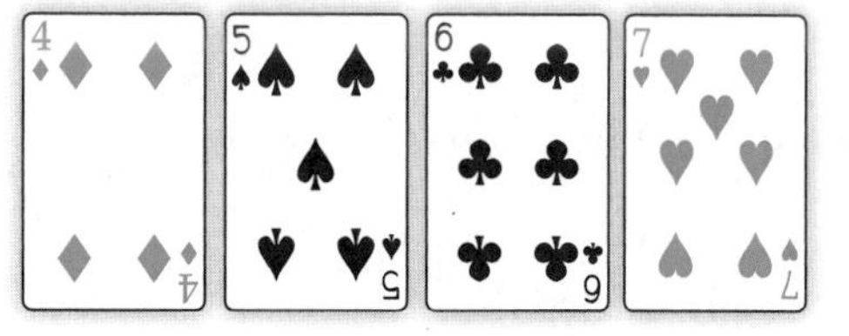

Before turn

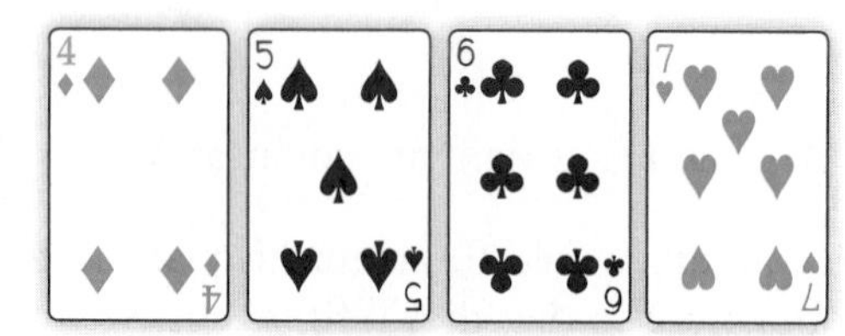

After turn

Miriam removed her blindfold and claimed she was able to determine which card was turned 180°. What is her trick? Can you figure out which card was turned? Explain.

19. If a pizza is cut into 16 congruent pieces, how many degrees are in each angle at the center of the pizza?

20. Paulus Gerdes, a mathematician from Mozambique, uses traditional *lusona* patterns from Angola to practice inductive thinking. Shown below are three *sona* designs. Sketch the fourth *sona* design, assuming the pattern continues.

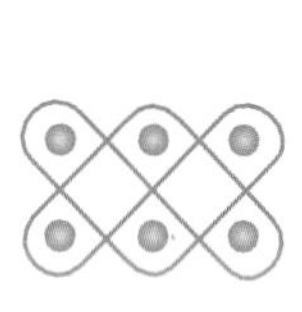

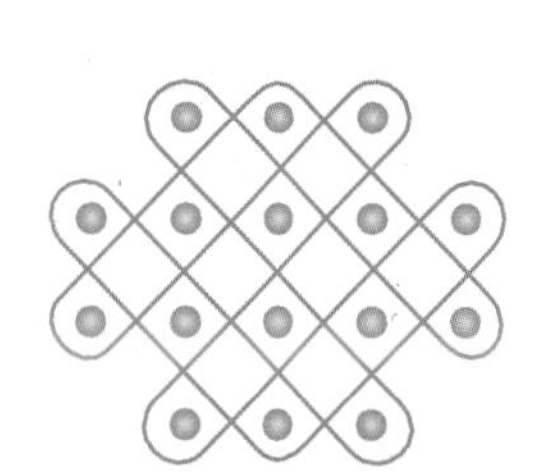

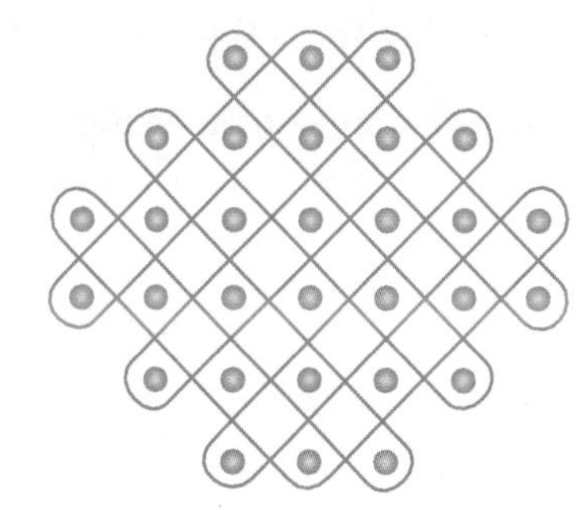

21. Hydrocarbon molecules in which all the bonds between the carbon atoms are single bonds except one double bond are called *alkenes*. The first three alkenes are modeled below.

```
        H          H       H          H   H       H
        |          |       |          |   |       |
  H - C = C    H - C - C = C    H - C - C - C = C
      |   |        |   |   |        |   |   |   |
      H   H        H   H   H        H   H   H   H
```

Ethene (C_2H_4)

Propene (C_3H_6)

Butene (C_4H_8)

Sketch the alkene with eight carbons in the chain. What is the general rule for alkenes $(C_nH_?)$? In other words, if there are n carbon atoms (C), how many hydrogen atoms (H) are in the alkene?

22. If the pattern of rectangles continues, what is the rule for the perimeter of the nth rectangle, and what is the perimeter of the 200th rectangle?

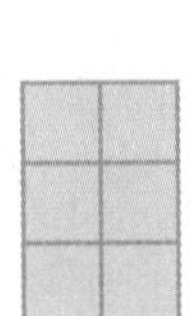
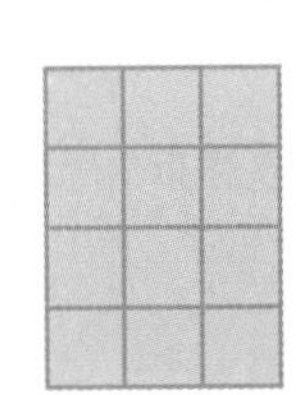
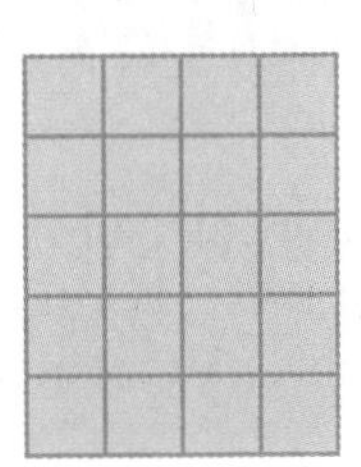
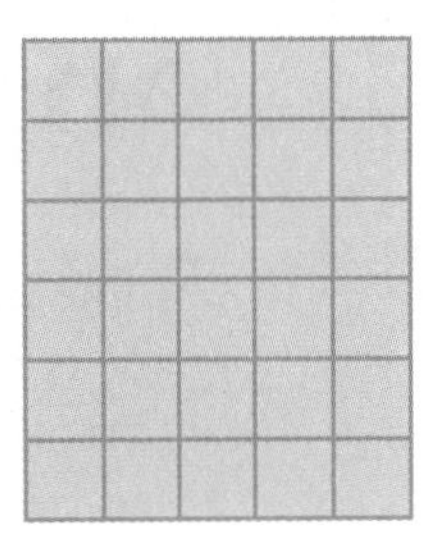

Perimeter in a rectangular pattern

Rectangle	1	2	3	4	5	6	...	n	...	200
Perimeter of rectangle	10	14	18				...		...	

23. The twelfth grade class of 80 students is assembled in a large circle on the football field at halftime. Each student is connected by a string to each of the other class members. How many pieces of string are necessary to connect each student to all the others? ⓗ

24. If you draw 80 lines on a piece of paper so that no 2 lines are parallel to each other and no 3 lines pass through the same point, how many intersections will there be? ⓗ

25. If there are 20 couples at a party, how many different handshakes can there be between pairs of people? Assume that the two people in each couple do not shake hands with each other. ⓗ

26. If a polygon has 24 sides, how many diagonals are there from each vertex? How many diagonals are there in all?

27. If a polygon has a total of 560 diagonals, how many vertices does it have? ⓗ

IMPROVING YOUR ALGEBRA SKILLS

Number Line Diagrams

1. The two segments at right have the same length. Translate the number line diagram into an equation, then solve for the variable x.

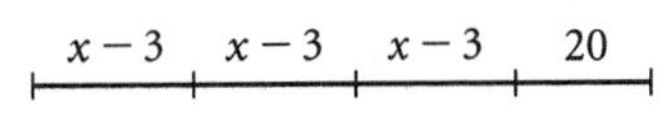

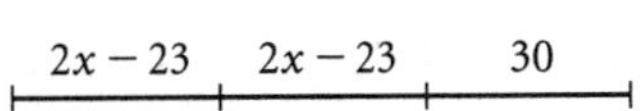

2. Translate this equation into a number line diagram.

$2(x + 3) + 14 = 3(x - 4) + 11$

LESSON 2.6

Special Angles on Parallel Lines

The greatest mistake you can make in life is to be continually fearing that you will make one.

ELLEN HUBBARD

A line intersecting two or more other lines in the plane is called a **transversal.** A transversal creates different types of angle pairs. Three types are listed below.

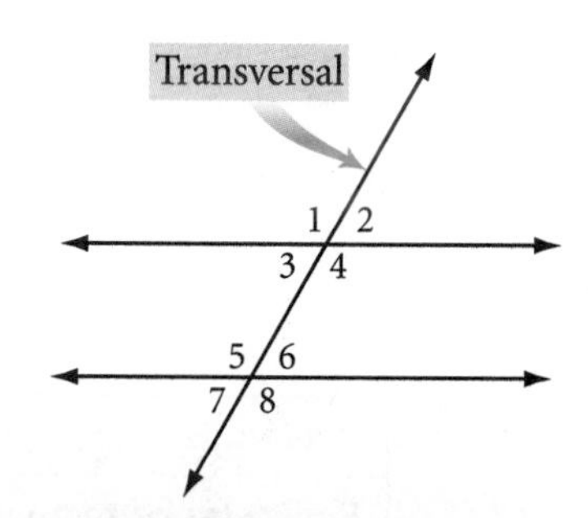

One pair of **corresponding angles** is $\angle 1$ and $\angle 5$. Can you find three more pairs of corresponding angles?

One pair of **alternate interior angles** is $\angle 3$ and $\angle 6$. Do you see another pair of alternate interior angles?

One pair of **alternate exterior angles** is $\angle 2$ and $\angle 7$. Do you see the other pair of alternate exterior angles?

When parallel lines are cut by a transversal, there is a special relationship among the angles. Let's investigate.

Investigation 1
Which Angles Are Congruent?

You will need

- lined paper or a straightedge
- patty paper
- a protractor

Using the lines on your paper as a guide, draw a pair of parallel lines. Or use both edges of your ruler or straightedge to create parallel lines. Label them k and ℓ. Now draw a transversal that intersects the parallel lines. Label the transversal m, and label the angles with numbers, as shown at right.

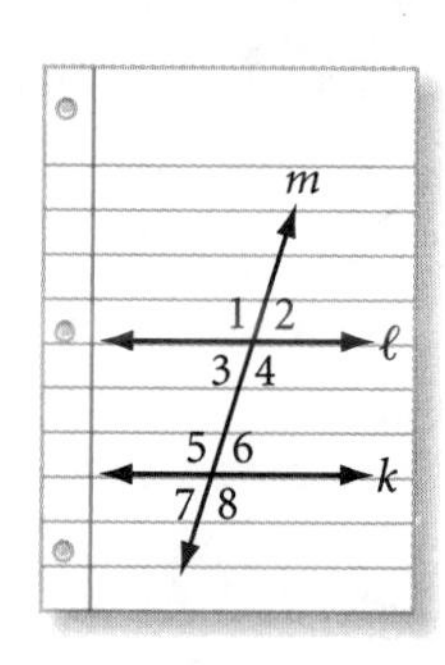

Step 1 Place a piece of patty paper over the set of angles 1, 2, 3, and 4. Copy the two intersecting lines m and ℓ and the four angles onto the patty paper.

Slide the patty paper down to the intersection of lines m and k, and compare angles 1 through 4 with each of the corresponding angles 5 through 8. What is the relationship between corresponding angles? Alternate interior angles? Alternate exterior angles?

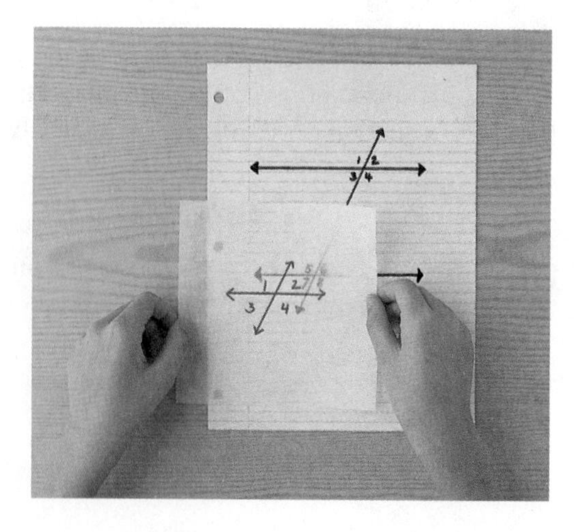

Compare your results with the results of others in your group and complete the three conjectures below.

C-3a

Corresponding Angles Conjecture, or CA Conjecture

If two parallel lines are cut by a transversal, then corresponding angles are _?_.

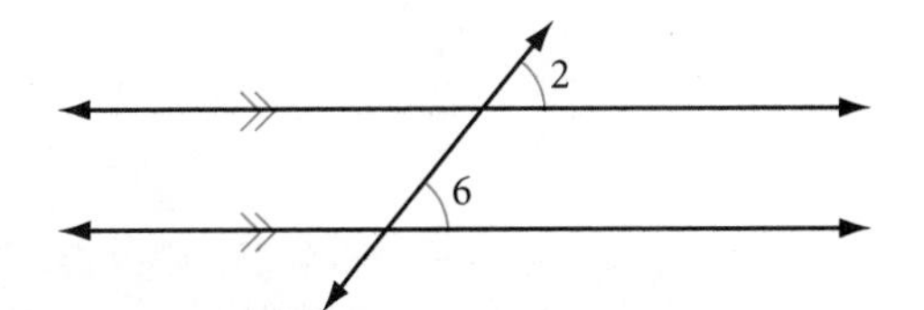

C-3b

Alternate Interior Angles Conjecture, or AIA Conjecture

If two parallel lines are cut by a transversal, then alternate interior angles are _?_.

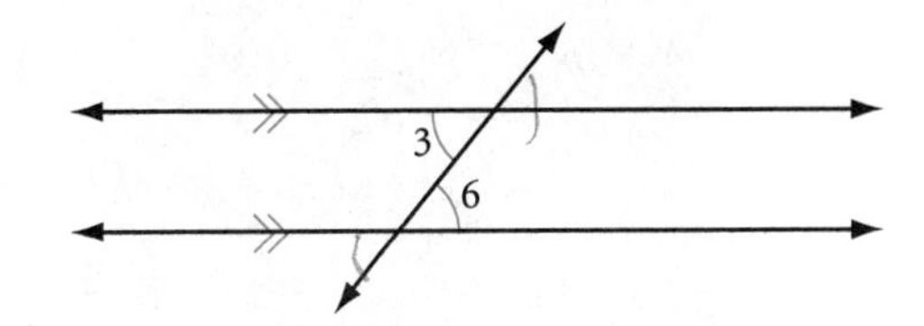

C-3c

Alternate Exterior Angles Conjecture, or AEA Conjecture

If two parallel lines are cut by a transversal, then alternate exterior angles are _?_.

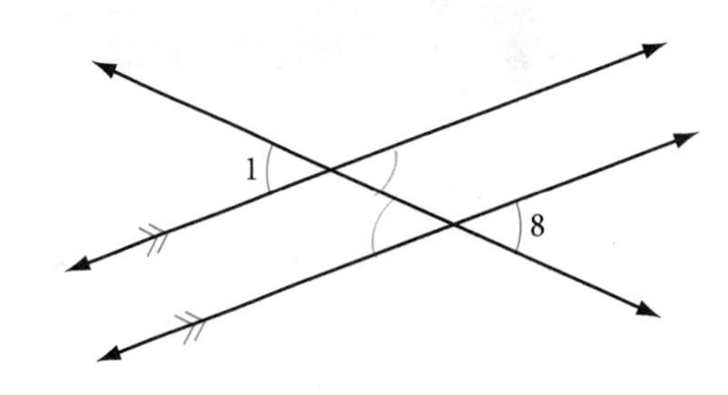

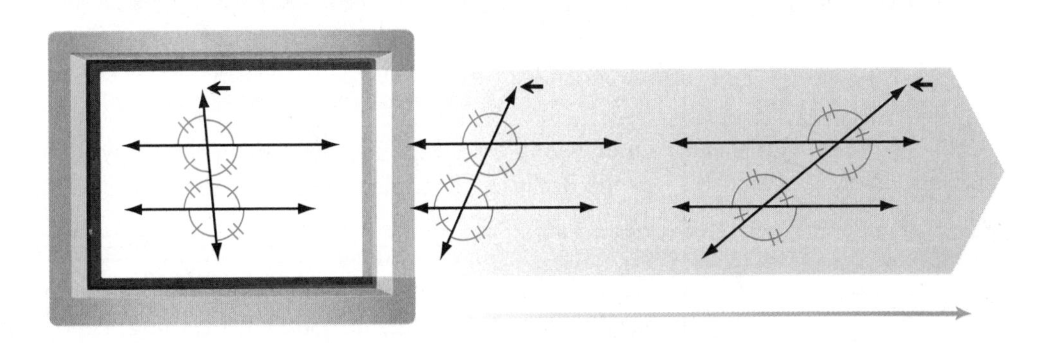

The three conjectures you wrote can all be combined to create a Parallel Lines Conjecture, which is really three conjectures in one.

C-3

Parallel Lines Conjecture

If two parallel lines are cut by a transversal, then corresponding angles are _?_, alternate interior angles are _?_, and alternate exterior angles are _?_.

Step 2 — What happens if the lines you start with are not parallel? Check whether your conjectures will work with nonparallel lines.

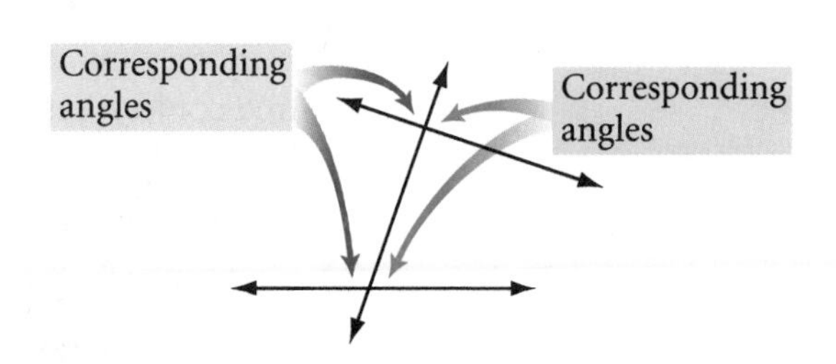

What about the converse of each of your conjectures? Suppose you know that a pair of corresponding angles, or alternate interior angles, is congruent. Will the lines be parallel? Is it possible for the angles to be congruent but for the lines not to be parallel?

Investigation 2
Is the Converse True?

You will need

- lined paper or a straightedge
- patty paper
- a protractor

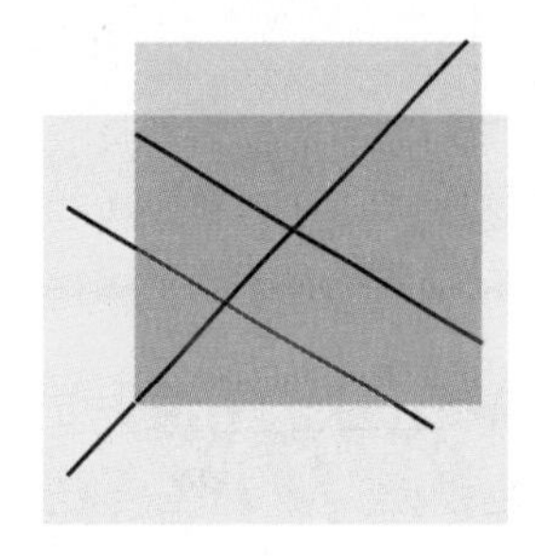

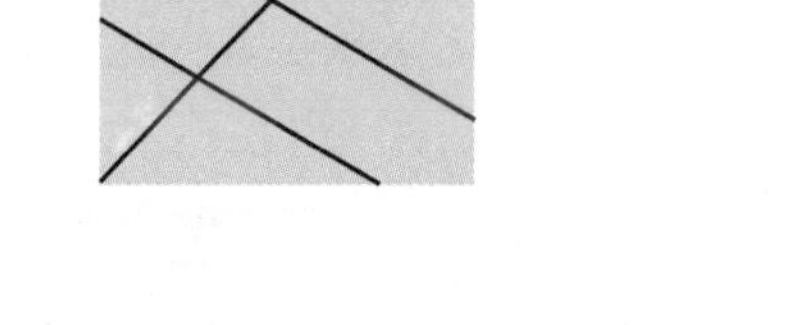

Step 1 — Draw two intersecting lines on your paper. Copy these lines onto a piece of patty paper. Because you copied the angles, the two sets of angles are congruent.

Slide the top copy so that the transversal stays lined up.

Trace the lines and the angles from the bottom original onto the patty paper again. When you do this, you are constructing sets of congruent corresponding angles. Mark the congruent angles.

Are the two lines parallel? You can test to see if the distance between the two lines remains the same, which guarantees that they will never meet.

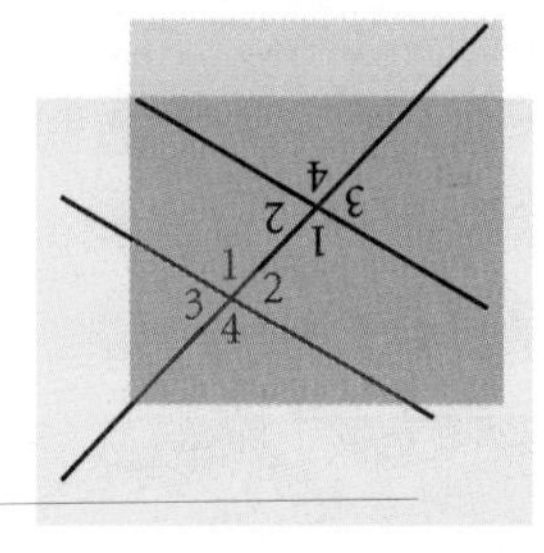

Step 2 — Repeat Step 1, but this time rotate your patty paper 180° so that the transversal lines up again. What kinds of congruent angles have you created? Trace the lines and angles and mark the congruent angles. Are the lines parallel? Check them.

Step 3 | Compare your results with those of your group. If your results do not agree, discuss them until you have convinced each other. Complete the conjecture below and add it to your conjecture list.

Converse of the Parallel Lines Conjecture

C-4

If two lines are cut by a transversal to form pairs of congruent corresponding angles, congruent alternate interior angles, or congruent alternate exterior angles, then the lines are _?_.

You used inductive reasoning to discover all three parts of the Parallel Lines Conjecture. However, if you accept any one of them as true, you can use deductive reasoning to show that the others are true.

EXAMPLE

Suppose we assume that the Vertical Angles Conjecture is true. Write a paragraph proof showing that if corresponding angles are congruent, then the Alternate Interior Angles Conjecture is true.

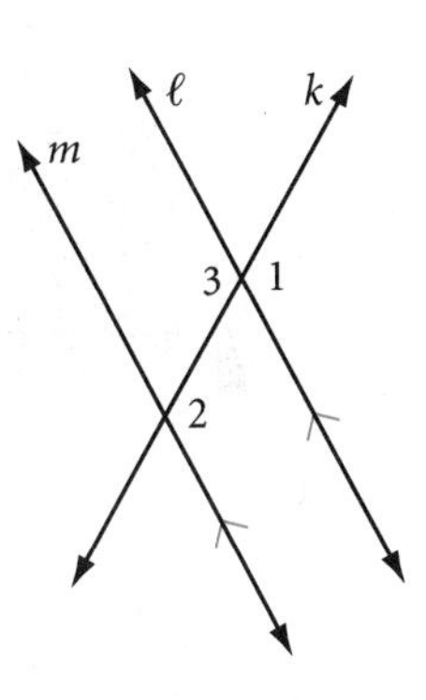

▶ ***Solution***

Paragraph Proof

Lines ℓ and m are parallel and intersected by transversal k. Pick any two alternate interior angles, such as $\angle 2$ and $\angle 3$. According to the Corresponding Angles Conjecture, $\angle 2 \cong \angle 1$. And, according to the Vertical Angles Conjecture, $\angle 1 \cong \angle 3$. Substitute $\angle 3$ for $\angle 1$ in the first statement to get $\angle 2 \cong \angle 3$. But $\angle 2$ and $\angle 3$ are alternate interior angles. Therefore, if the corresponding angles are congruent, then the alternate interior angles are congruent. ■

Here are the algebraic steps:

$\angle 2 \cong \angle 1$

$\angle 3 \cong \angle 1$

So $\angle 2 \cong \angle 3$

It helps to visualize each statement and to mark all congruences you know on your paper.

EXERCISES

Use your new conjectures in Exercises 1–6. A small letter in an angle represents the angle measure.

1. $r \parallel s$
$w =$ _?_

2. $p \parallel q$
$x =$ _?_

3. Is line k parallel to line ℓ?

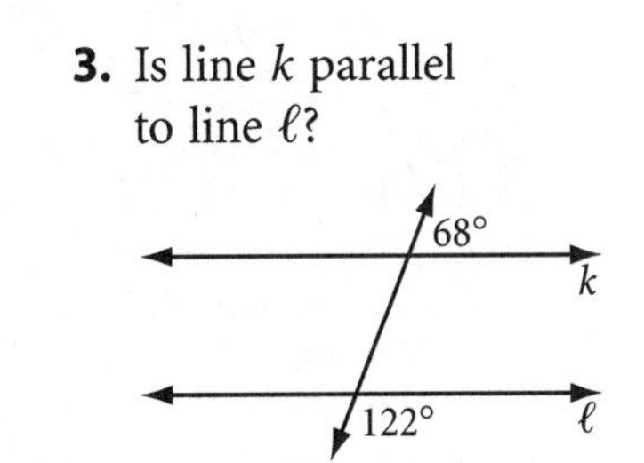

4. Quadrilateral *TUNA* is a parallelogram.
$y = \underline{?}$ ⓗ

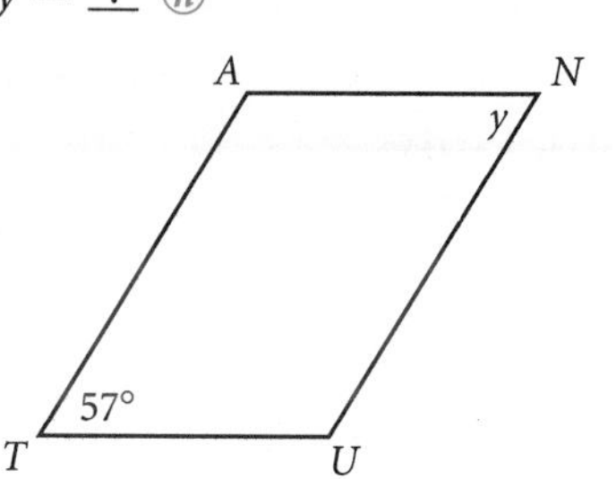

5. Is quadrilateral *FISH* a parallelogram?

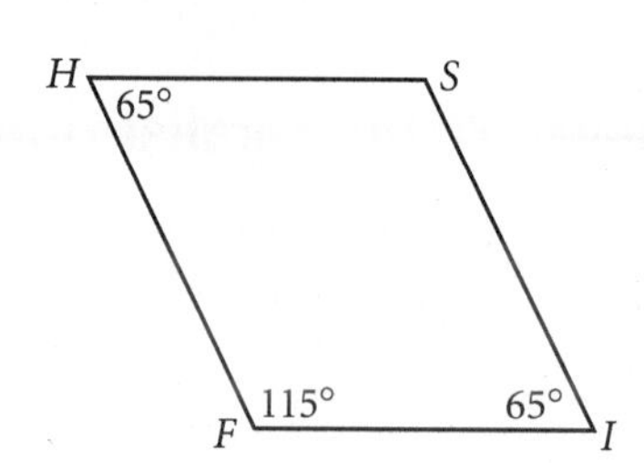

6. $m \parallel n$
$z = \underline{?}$ ⓗ

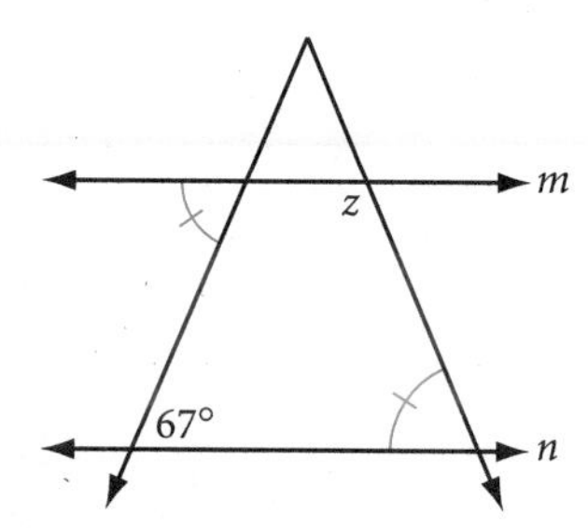

7. Trace the diagram below. Calculate each lettered angle measure. ⓗ

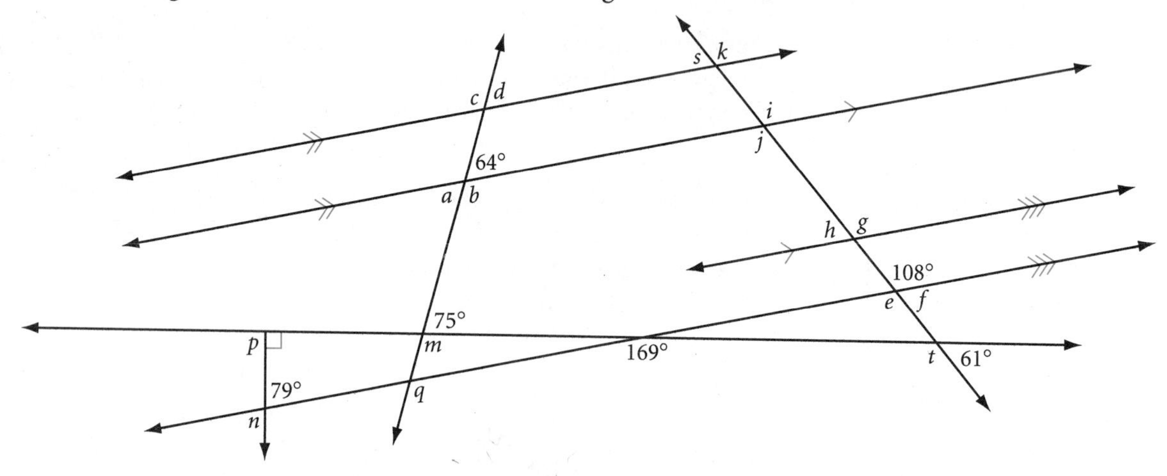

8. You've seen before how parallel lines appear to meet in the distance. Let's look at the converse of this effect: The top and the bottom of the Vietnam Veterans Memorial Wall appear to be parallel because they appear to meet so far in the distance. Consider the diagram of the corner of the memorial, shown below. You know that line ℓ_1 and line ℓ_2 eventually meet. Is the blue shaded portion of the wall a rectangle? Write a paragraph proof explaining why it is or is not a rectangle.

Sculptor Maya Lin designed the Vietnam Veterans Memorial Wall in Washington, D.C. Engraved in the granite wall are the names of United States armed forces service members who died in the Vietnam War or remain missing in action. To learn more about the Memorial Wall and Lin's other projects, visit www.keymath.com/DG

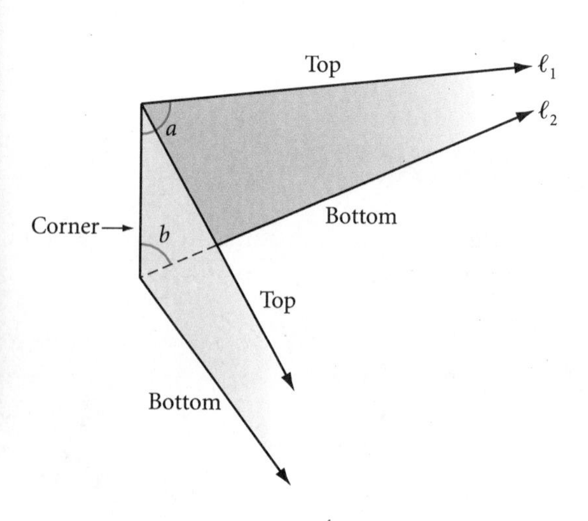

9. What's wrong with this picture?

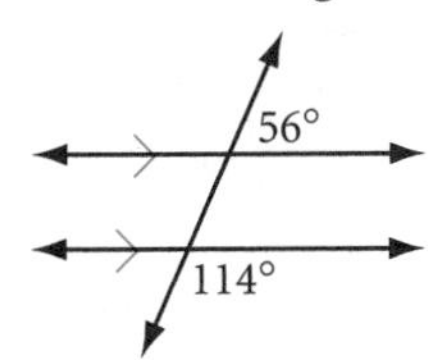

10. What's wrong with this picture?

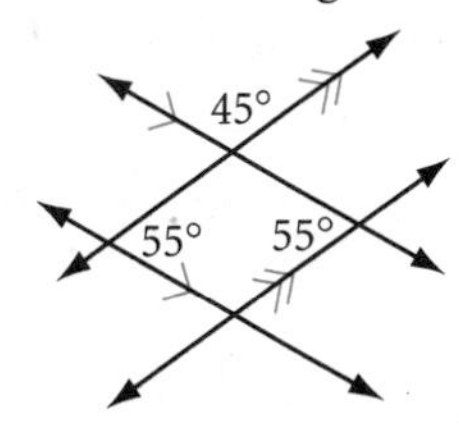

11. A periscope permits a sailor on a submarine to see above the surface of the ocean. This periscope is designed so that the line of sight a is parallel to the light ray b. The middle tube is perpendicular to the top and bottom tubes. What are the measures of the incoming and outgoing angles formed by the light rays and the mirrors in this periscope? Are the surfaces of the mirrors parallel? How do you know?

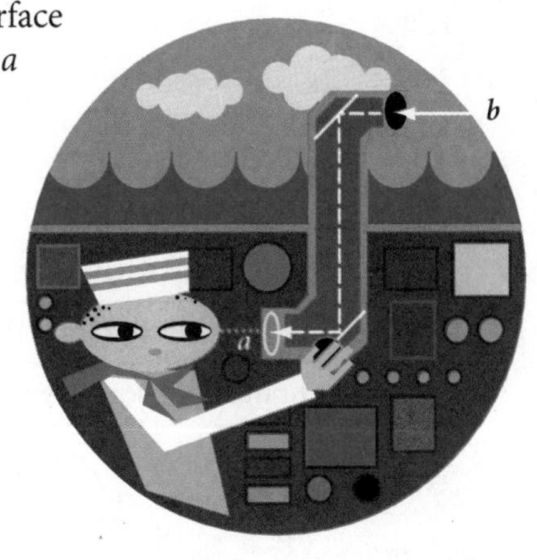

Review

12. What type (or types) of triangle has one or more lines of symmetry?

13. What type (or types) of quadrilateral has only rotational symmetry? ⓗ

14. If D is the midpoint of $\overline{AC}$ and C is the midpoint of $\overline{BD}$, what is the length of $\overline{AB}$ if $BD = 12$ cm?

15. If $\overrightarrow{AI}$ is the angle bisector of $\angle KAN$ and $\overrightarrow{AR}$ is the angle bisector of $\angle KAI$, what is $m\angle RAN$ if $m\angle RAK = 13°$?

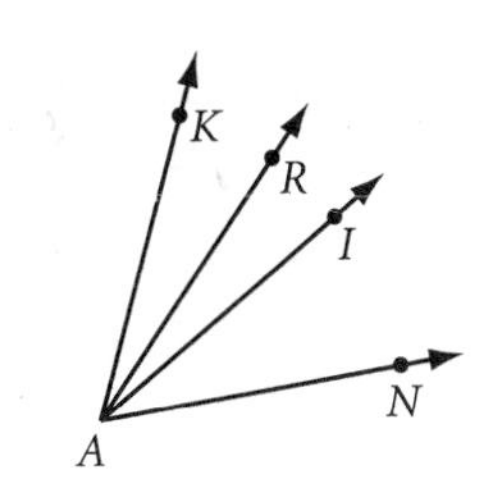

For Exercises 16–18, draw each polygon on graph paper. Relocate the vertices according to the rule. Connect the new points to form a new polygon. Describe what happened to the figure. Is the new polygon congruent to the original?

16. **Rule:** Subtract 1 from each x-coordinate. ⓗ

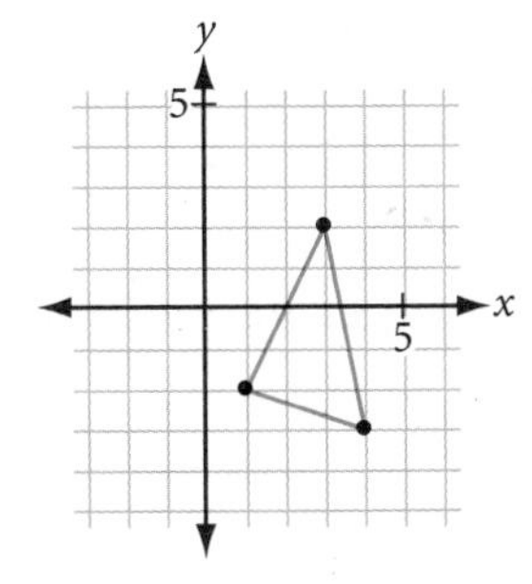

17. **Rule:** Reverse the sign of each x- and y-coordinate.

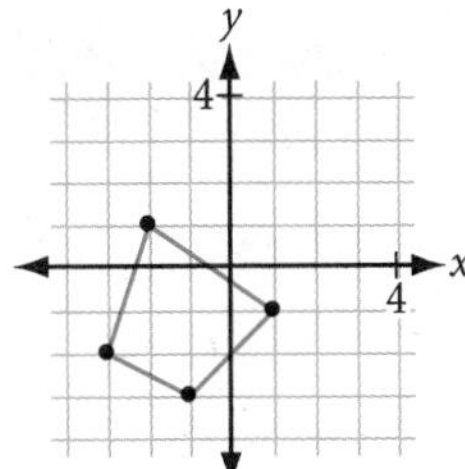

18. **Rule:** Switch the x- and y-coordinates. Pentagon $LEMON$ with vertices:

$L(-4, 2)$
$E(-4, -3)$
$M(0, -5)$
$O(3, 1)$
$N(-1, 4)$

19. If everyone in the town of Skunk's Crossing (population 84) has a telephone, how many different lines are needed to connect all the phones to each other?

20. How many squares of all sizes are in a 4-by-4 grid of squares? (There are more than 16!) ⓗ

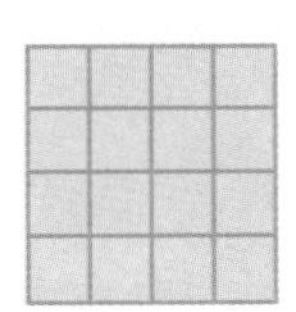

21. Assume the pattern of blue and yellow shaded **T**'s continues. Copy and complete the table for blue shaded and yellow shaded squares and for the total number of squares. ⓗ

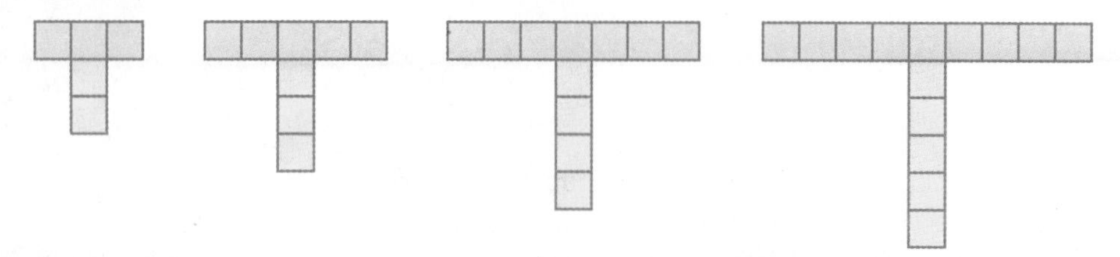

The T-formation

Figure number	1	2	3	4	5	6	...	n	...	35
Number of yellow squares	2						...		...	
Number of blue squares	3						...		...	
Total number of squares	5						...		...	

LINE DESIGNS

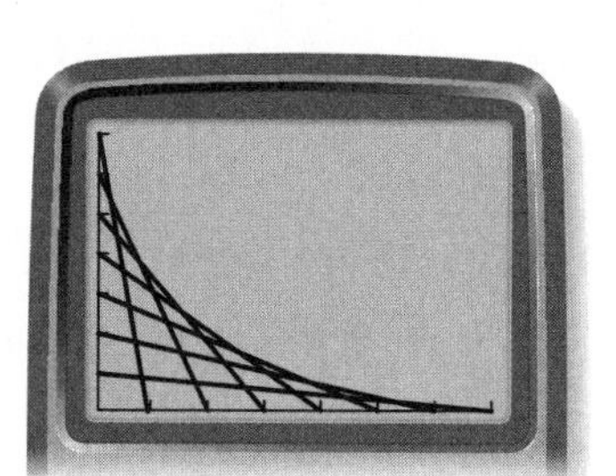

Can you use your graphing calculator to make the line design shown at right? You'll need to recall some algebra. Here are some hints.

1. The x- and y-ranges are set to minimums of 0 and maximums of 7.
2. The design consists of the graphs of seven lines.
3. The equation for one of the lines is $y = -\frac{1}{7}x + 1$.
4. There's a simple pattern in the slopes and y-intercepts of the lines.

You're on your own from here. Experiment! Then create a line design of your own and write the equations for it.

Slope

The slope of a line is a measure of its steepness. Measuring slope tells us the steepness of a hill, the pitch of a roof, or the incline of a ramp. On a graph, slope can tell us the rate of change, or speed.

To calculate slope, you find the ratio of the vertical distance to the horizontal distance traveled, sometimes referred to as "rise over run."

$$\text{slope} = \frac{\text{vertical change}}{\text{horizontal change}}$$

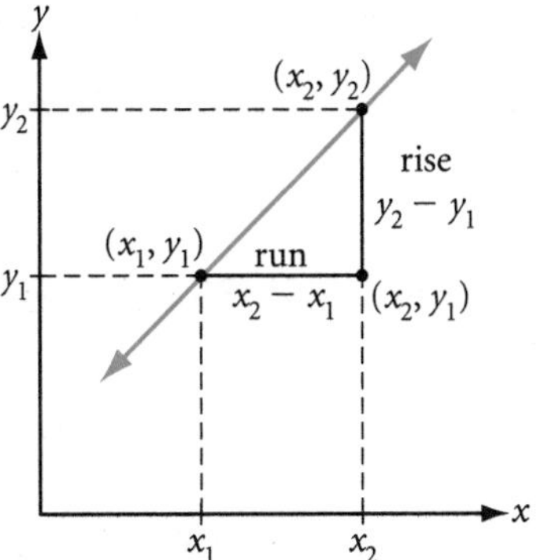

One way to find slope is to use a **slope triangle.** Then use the coordinates of its vertices in the formula.

Slope Formula

The slope m of a line (or segment) through two points with coordinates (x_1, y_1) and (x_2, y_2) is

$$m = \frac{y_2 - y_1}{x_2 - x_1}$$

where $x_2 - x_1 \neq 0$.

EXAMPLE

Draw the slope triangle and find the slope for $\overline{AB}$.

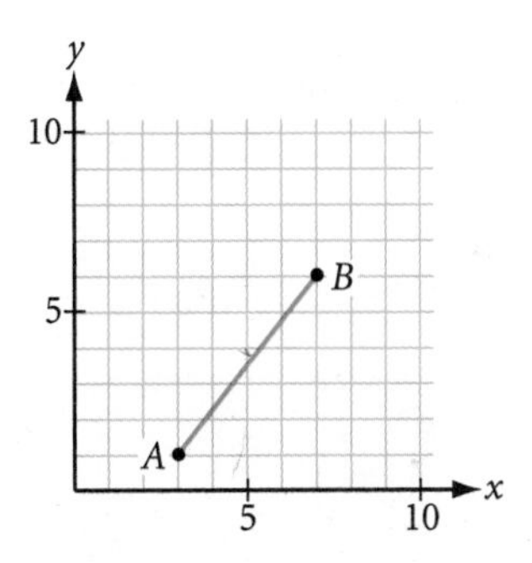

Solution

Draw the horizontal and vertical sides of the slope triangle below the line. Use them to calculate the side lengths.

$$m = \frac{y_2 - y_1}{x_2 - x_1} = \frac{6 - 1}{7 - 3} = \frac{5}{4}$$

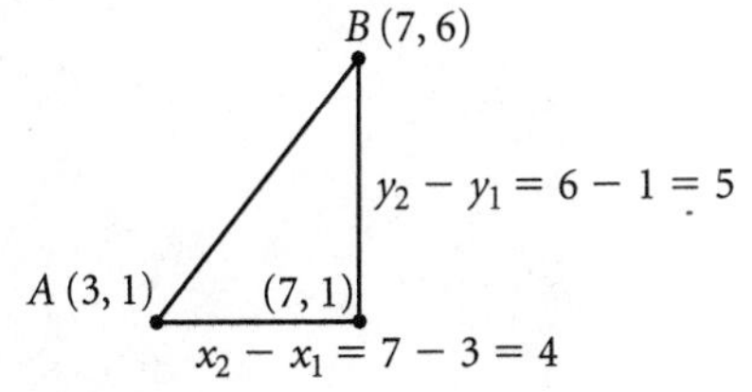

Note that if the slope triangle is *above* the line, you subtract the numbers in reverse order, but still get the same result.

$$m = \frac{1 - 6}{3 - 7} = \frac{-5}{-4} = \frac{5}{4}$$

The slope is positive when the line goes up from left to right. The slope is negative when the line goes down from left to right. When is the slope 0? What is the slope of a vertical line?

EXERCISES

In Exercises 1–3 find the slope of the line through the given points.

1. (16, 0) and (12, 8) **2.** (−3, −4) and (−16, 8) **3.** (5.3, 8.2) and (0.7, −1.5)

4. A line through points (−5, 2) and (2, y) has a slope of 3. Find y.

5. A line through points (x, 2) and (7, 9) has a slope of $\frac{7}{3}$. Find x.

6. Find the coordinates of three more points that lie on the line passing through the points (0, 0) and (3, −4). Explain your method.

7. What is the speed, in miles per hour, represented by Graph A?

8. From Graph B, which in-line skater is faster? How much faster?

9. The grade of a road is its slope, given as a percent. For example, a road with a 6% grade has slope $\frac{6}{100}$. It rises 6 feet for every 100 feet of horizontal run. Describe a 100% grade. Do you think you could drive up it? Could you walk up it? Is it possible for a grade to be greater than 100%?

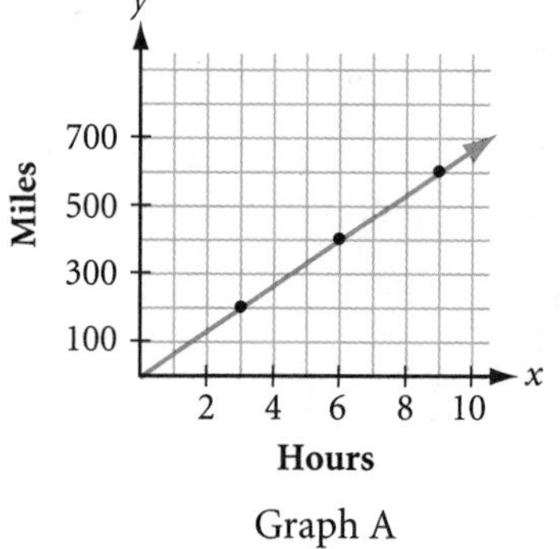

Graph A

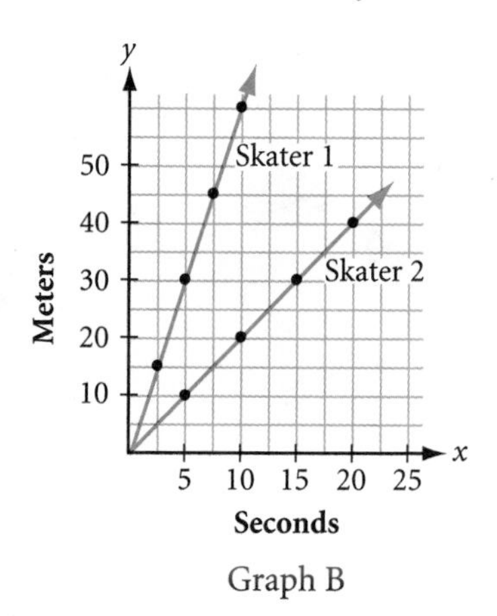

Graph B

10. What's the slope of the roof on the adobe house? Why might a roof in Connecticut be steeper than a roof in the desert?

Adobe house, New Mexico

Pitched-roof house, Connecticut

Exploration

Patterns in Fractals

In Lesson 2.1, you discovered patterns and used them to continue number sequences. In most cases, you found each term by applying a rule to the term before it. Such rules are called **recursive rules.**

Some picture patterns are also generated by recursive rules. You find the next picture in the sequence by looking at the picture before it and comparing that to the picture before it, and so on.

The Geometer's Sketchpad® can repeat a recursive rule on a figure using a command called **Iterate.** Using **Iterate,** you can create the initial stages of fascinating geometric figures called **fractals.** Fractals have **self-similarity,** meaning that if you zoom in on a part of the figure, it looks like the whole. A true fractal would need infinitely many applications of the recursive rule. In this exploration, you'll use **Iterate** to create the first few stages of a fractal called the **Sierpiński triangle.**

In this procedure you will construct a triangle, its interior, and midpoints on its sides. Then you will use **Iterate** to repeat the process on three outer triangles formed by connecting the midpoints and vertices of the original triangle.

This fern frond illustrates self-similarity. Notice how each curled leaf resembles the shape of the entire curled frond.

Activity

The Sierpiński Triangle

Procedure Note

Sierpiński Triangle Iteration

1. Open a new Sketchpad™ sketch.
2. Use the **Segment** tool to draw triangle *ABC*.
3. Select the vertices, and construct the triangle interior.
4. Select $\overline{AB}$, $\overline{BC}$, and $\overline{CA}$, in that order, and construct the midpoints *D*, *E*, and *F*.
5. Select the vertices again, and choose **Iterate** from the Transform menu. An Iterate dialog box will open. Select points *A*, *D*, and *F*. This maps triangle *ABC* onto the smaller triangle *ADF*.
6. In the Iterate dialog box, choose **Add A New Map** from the Structure pop-up menu. Map triangle *ABC* onto triangle *BED*.
7. Repeat Step 6 to map triangle *ABC* onto triangle *CFE*.
8. In the Iterate dialog box, choose **Final Iteration Only** from the Display pop-up menu.
9. In the Iterate dialog box, click Iterate to complete your construction.
10. Click in the center of triangle *ABC* and hide the interior to see your fractal at Stage 3.
11. Use Shift+plus or Shift+minus to increase or decrease the stage of your fractal.

Step 1 Follow the Procedure Note to create the Stage 3 Sierpiński triangle. The original triangle *ABC* is a Stage 0 Sierpiński triangle. Practice the last step of the Procedure Note to see how the fractal grows in successive stages. Write a sentence or two explaining what the Sierpiński triangle shows you about self-similarity.

Notice that the fractal's property of self-similarity does not change as you drag the vertices.

Step 2 What happens to the number of shaded triangles in successive stages of the Sierpiński triangle? Decrease your construction to Stage 1 and investigate. How many triangles would be shaded in a Stage n Sierpiński triangle? Use your construction and look for patterns to complete this table.

Stage number	0	1	2	3	...	n	...	50
Number of triangles	1	3			...		...	

What stage is the Sierpiński triangle shown on page 135?

Step 3 Suppose you start with a Stage 0 triangle (just a plain old triangle) with an area of 1 unit. What would be the shaded area at Stage 1? What happens to the shaded area in successive stages of the Sierpiński triangle? Use your construction and look for patterns to complete this table.

Stage number	0	1	2	3	...	n	...	50
Shaded area	1	$\frac{3}{4}$			...		...	

What would happen to the shaded area if you could infinitely increase the stage number?

Step 4 Suppose you start with a Stage 0 triangle with a perimeter of 6 units. At Stage 1 the perimeter would be 9 units (the sum of the perimeters of the three triangles, each half the size of the original triangle). What happens to the perimeter in successive stages of the Sierpiński triangle? Complete this table.

Stage number	0	1	2	3	...	n	...	50
Perimeter	6	9			...		...	

What would happen to the perimeter if you could infinitely increase the stage number?

Step 5 Increase your fractal to Stage 3 or 4. If you print three copies of your sketch, you can put the copies together to create a larger triangle one stage greater than your original. How many copies would you need to print in order to create a triangle two stages greater than your original? Print the copies you need and combine them into a poster or a bulletin board display.

Step 6 Sketchpad comes with a sample file of interesting fractals. Explore these fractals and see if you can use **Iterate** to create them yourself. You can save any of your fractal constructions by selecting the entire construction and then choosing **Create New Tool** from the Custom Tools menu. When you use your custom tool in the future, the fractal will be created without having to use **Iterate.**

The word *fractal* was coined by Benoit Mandelbrot (b 1924), a pioneering researcher in this new field of mathematics. He was the first to use high-speed computers to create the figure below, called the Mandelbrot set.

Only the black area is part of the set itself. The rainbow colors represent properties of points near the Mandelbrot set. To learn more about different kinds of fractals, visit www.keymath.com/DG.

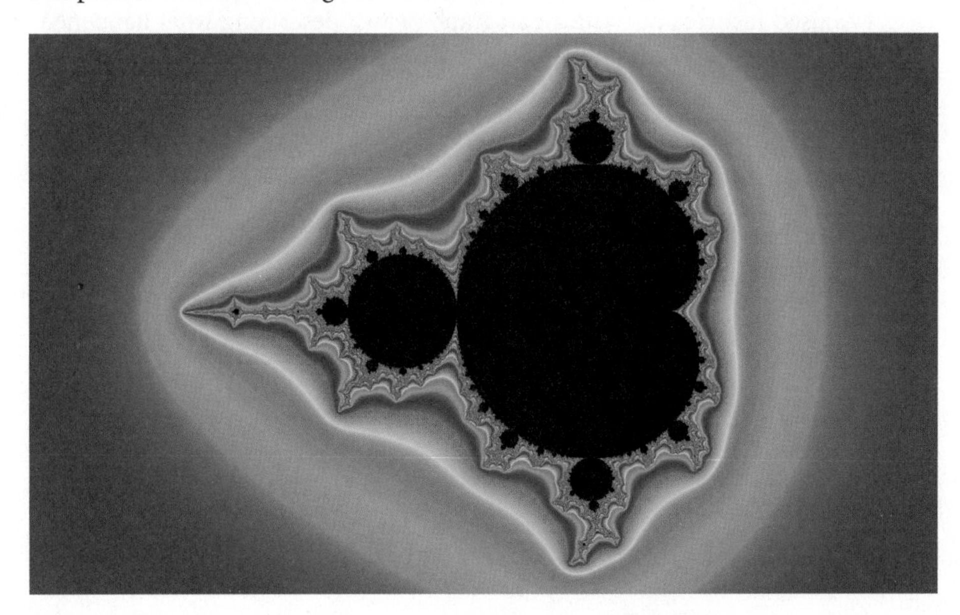

CHAPTER 2 REVIEW

This chapter introduced you to inductive reasoning. You used inductive reasoning to observe patterns and make conjectures. You learned to disprove a conjecture with a counterexample and to explain why a conjecture is true with deductive reasoning. You learned how to predict number sequences with rules and how to use these rules to model application problems. Then you discovered special relationships about angle pairs and made your first geometry conjectures. Finally you explored the properties of corresponding, alternate interior, and alternate exterior angles formed by a transversal across parallel lines. As you review the chapter, be sure you understand all the important terms. Go back to the lesson to review any terms you're unsure of.

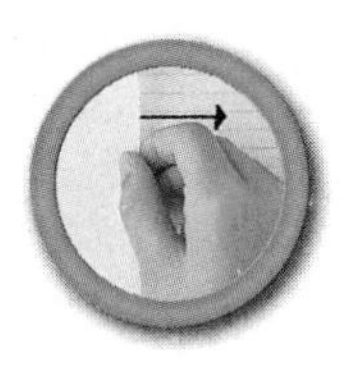

EXERCISES

1. "My dad is in the navy, and he says that food is great on submarines," said Diana. "My mom is a pilot," added Jill, "and she says that airline food is notoriously bad." "My mom is an astronaut trainee," said Julio, "and she says that astronauts' food is the worst imaginable." "You know," concluded Diana, "I bet no life exists beyond Earth! As you move farther and farther from the surface of Earth, food tastes worse and worse. At extreme altitudes, food must taste so bad that no creature could stand to eat. Therefore, no life exists out there." What do you think of Diana's reasoning? Is it inductive or deductive?

2. Think of a situation you observed outside of school in which inductive reasoning was used incorrectly. Write a paragraph or two describing what happened and explaining why you think it was poor inductive reasoning.

3. Think of a situation you observed outside of school in which deductive reasoning was used incorrectly. Write a paragraph or two describing what happened and explaining why you think it was poor deductive reasoning.

For Exercises 4–7, find the next two terms in the sequence.

4. 7, 21, 35, 49, 63, 77, ?, ?

5. Z, 1, Y, 2, X, 4, W, 8, ?, ? (h)

6. 7, 2, 5, −3, 8, −11, ?, ?

7. A, 4, D, 9, H, 16, M, 25, ?, ? (h)

For Exercises 8 and 9, generate the first six terms in the sequence for each function rule.

8. $f(n) = n^2 + 1$

9. $f(n) = 2^{n-1}$ (h)

For Exercises 10 and 11, draw the next shape in the pattern.

10.

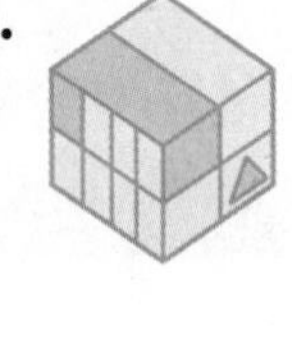

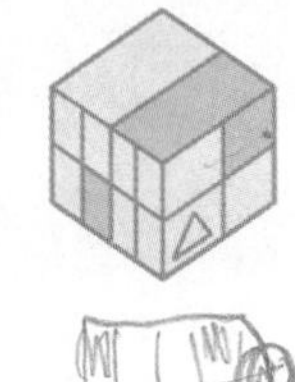

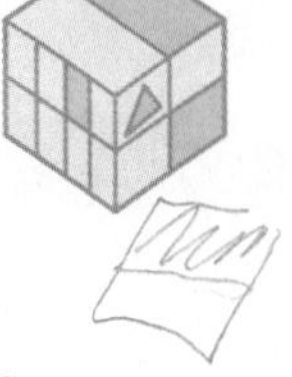

11. (h)

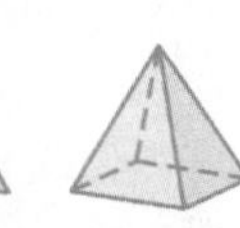

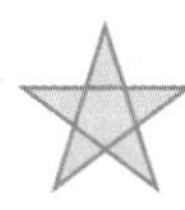

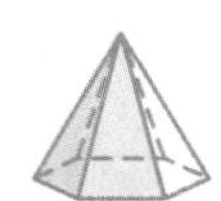

For Exercises 12–15, find the nth term and the 20th term in the sequence.

12.

n	1	2	3	4	5	6	...	n	...	20
$f(n)$	2	−1	−4	−7	−10	−13	...		...	

13.

n	1	2	3	4	5	6	...	n	...	20
$f(n)$	1	4	9	16	25	36	...		...	

14. ⓗ

n	1	2	3	4	5	6	...	n	...	20
$f(n)$	0	1	3	6	10	15	...		...	

15. ⓗ

n	1	2	3	4	5	6	...	n	...	20
$f(n)$	1	3	6	10	15	21	...		...	

For Exercises 16 and 17, find a relationship. Then complete the conjecture.

16. Conjecture: The sum of the first 30 odd whole numbers is _?_.

17. Conjecture: The sum of the first 30 even whole numbers is _?_. ⓗ

18. Viktoriya is a store window designer for Savant Toys. She plans to build a stack of blocks similar to the ones shown below but 30 blocks high. Make a conjecture for the value of the nth term and for the value of the 30th term. How many blocks will she need? ⓗ

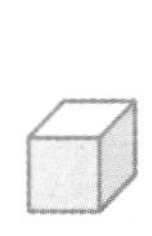
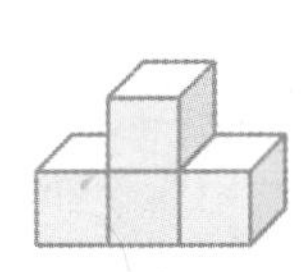
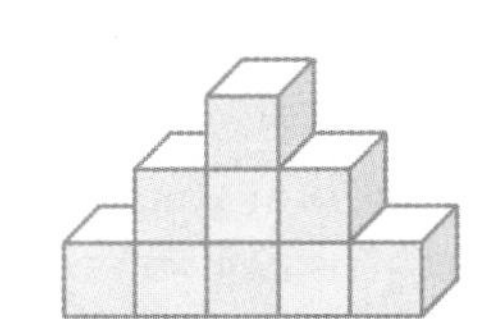
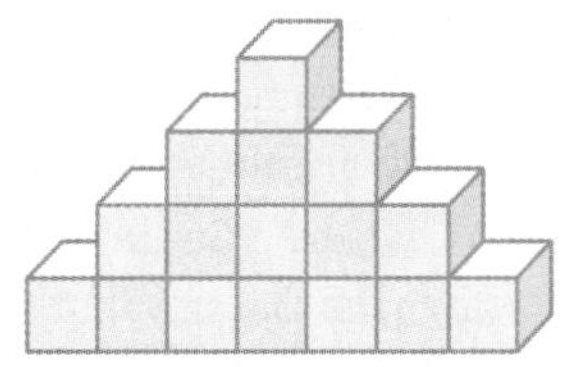

19. The stack of bricks at right is four bricks high. Find the total number of bricks for a stack that is 100 bricks high.

20. For the 4-by-7 rectangular grid, the diagonal passes through 10 squares and 9 interior segments. In an 11-by-101 grid of squares, how many squares will the diagonal pass through? How many interior segments will it pass through? ⓗ

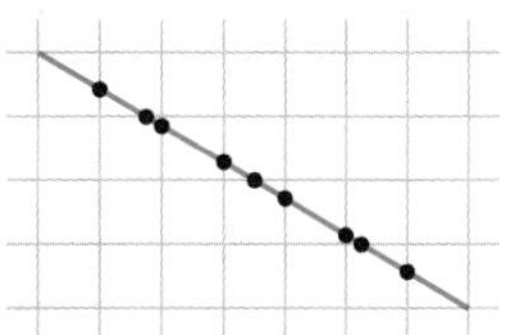

21. If at a party there are a total of 741 handshakes and each person shakes hands with everyone else at the party exactly once, how many people are at the party?

22. If 28 lines are drawn on a plane, what is the maximum number of points of intersection possible?

23. If a whole bunch of lines (no two parallel, no three concurrent) intersect in a plane 2926 times, how many lines are a whole bunch? ⓗ

24. If in a 54-sided polygon all possible diagonals are drawn from one vertex, they divide the interior of the polygon into how many regions?

25. How many sides does the polygon have if all possible diagonals drawn from one vertex divide the interior of the polygon into 54 regions?

26. Trace the diagram at right. Calculate each lettered angle measure.

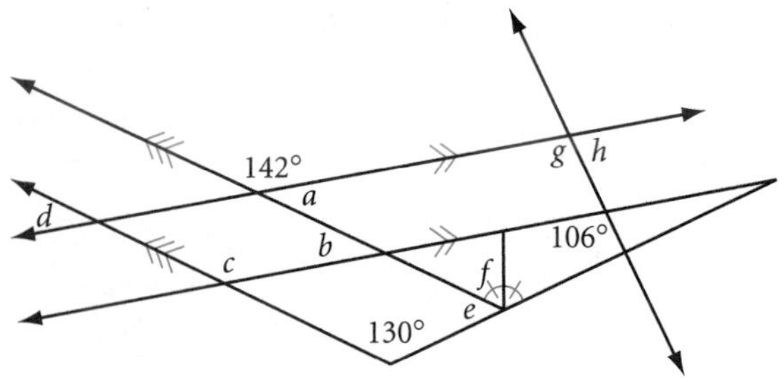

Assessing What You've Learned

WRITE IN YOUR JOURNAL

Many students find it useful to reflect on the mathematics they're learning by keeping a journal. Like a diary or a travel journal, a mathematics journal is a chance for you to reflect on what happens each day and your feelings about it. Unlike a diary, though, your mathematics journal isn't private—your teacher may ask to read it too, and may respond to you in writing. Reflecting on your learning experiences will help you assess your strengths and weaknesses, your preferences, and your learning style. Reading through your journal may help you see what obstacles you have overcome. Or it may help you realize when you need help.

- So far, you have written definitions, looked for patterns, and made conjectures. How does this way of doing mathematics compare to the way you have learned mathematics in the past?
- What are some of the most significant concepts or skills you've learned so far? Why are they significant to you?
- What are you looking forward to in your study of geometry? What are your goals for this class? What specific steps can you take to achieve your goals?
- What are you uncomfortable or concerned about? What are some things you or your teacher can do to help you overcome these obstacles?

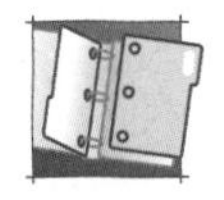

KEEPING A NOTEBOOK You should now have four parts to your notebook: a section for homework and notes, an investigation section, a definition list, and now a conjecture list. Make sure these are up-to-date.

UPDATE YOUR PORTFOLIO Choose one or more pieces of your most significant work in this chapter to add to your portfolio. These could include an investigation, a project, or a complex homework exercise. Make sure your work is complete. Describe why you chose the piece and what you learned from it.

CHAPTER

3 Using Tools of Geometry

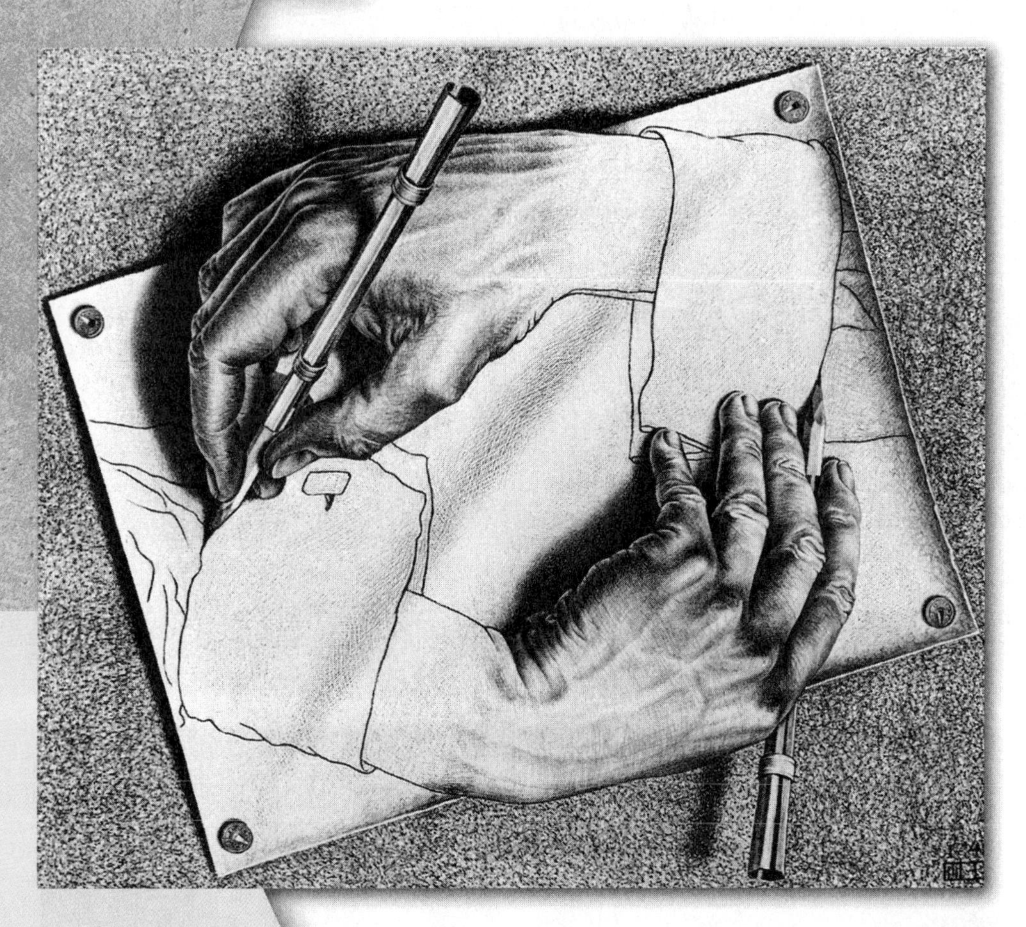

There is indeed great satisfaction in acquiring skill, in coming to thoroughly understand the qualities of the material at hand and in learning to use the instruments we have—in the first place, our hands!—in an effective and controlled way.

M. C. ESCHER

Drawing Hands, M. C. Escher, 1948

OBJECTIVES

In this chapter you will

- learn about the history of geometric constructions
- develop skills using a compass, a straightedge, patty paper, and geometry software
- see how to create complex figures using only a compass, a straightedge, and patty paper

LESSON

3.1 Duplicating Segments and Angles

It is only the first step that is difficult.

MARIE DE VICHY-CHAMROD

The compass, like the straightedge, has been a useful geometry tool for thousands of years. The ancient Egyptians used the compass to mark off distances. During the Golden Age of Greece, Greek mathematicians made a game of geometric constructions. In his work *Elements,* Euclid (325–265 B.C.E.) established the basic rules for constructions using only a compass and a straightedge. In this course you will learn how to construct geometric figures using these tools as well as patty paper.

Constructions with patty paper are a variation on the ancient Greek game of geometric constructions. Almost all the figures that can be constructed with a compass and a straightedge can also be constructed using a straightedge and patty paper, waxed paper, or tracing paper. If you have access to a computer with a geometry software program, you can do constructions electronically.

In the previous chapters, you drew and sketched many figures. In this chapter, however, you'll construct geometric figures. The words *sketch, draw,* and *construct* have specific meanings in geometry.

Mathematics CONNECTION

Euclidean geometry is the study of geometry based on the assumptions of Euclid (325–265 B.C.E.). Euclid established the basic rules for constructions using only a compass and a straightedge. In his work *Elements,* Euclid proposed definitions and constructions about points, lines, angles, surfaces, and solids. He also explained why the constructions were correct with deductive reasoning.

A page from a book on Euclid, above, shows some of his constructions and a translation of his explanations from Greek into Latin.

When you *sketch* an equilateral triangle, you may make a freehand sketch of a triangle that looks equilateral. You don't need to use any geometry tools.

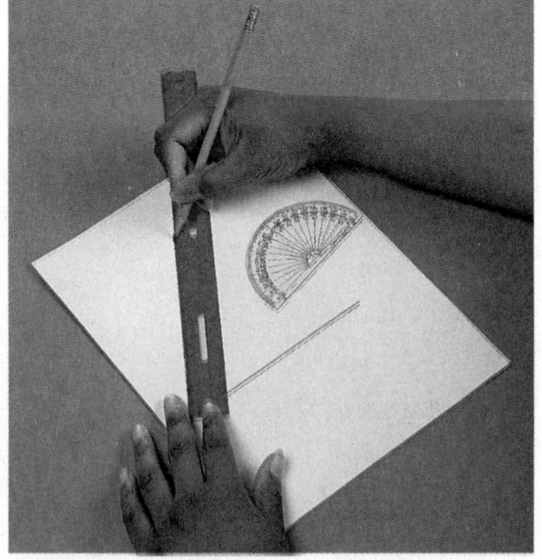

When you *draw* an equilateral triangle, you should draw it carefully and accurately, using your geometry tools. You may use a protractor to measure angles and a ruler to measure the sides to make sure they are equal in measure.

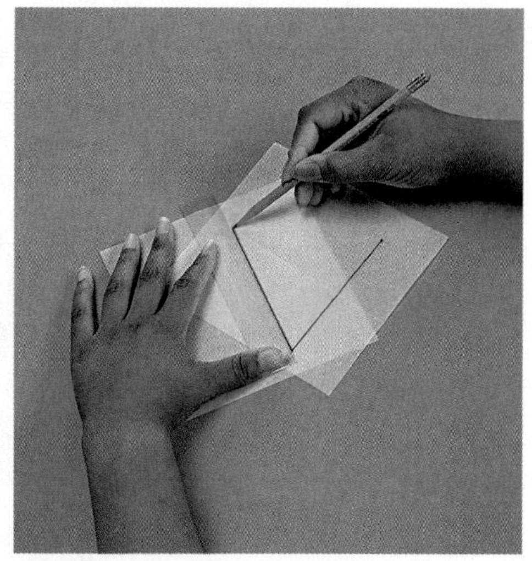

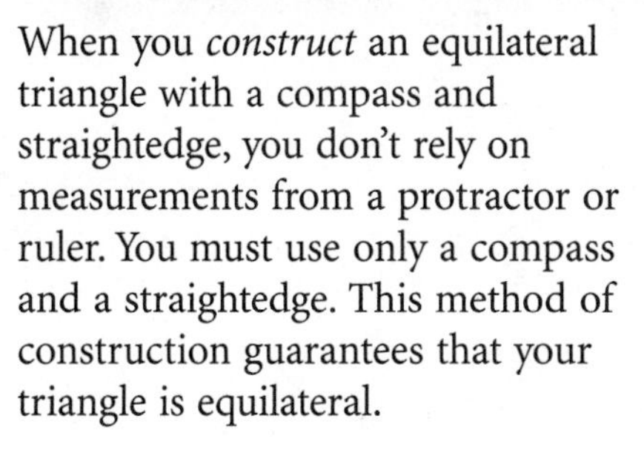

When you *construct* an equilateral triangle with a compass and straightedge, you don't rely on measurements from a protractor or ruler. You must use only a compass and a straightedge. This method of construction guarantees that your triangle is equilateral.

When you *construct* an equilateral triangle with patty paper and straightedge, you fold the paper and trace equal segments. You may use a straightedge to draw a segment, but you may not use a compass or any measuring tools.

When you sketch or draw, use the special marks that indicate right angles, parallel segments, and congruent segments and angles.

By tradition, neither a ruler nor a protractor is ever used to perform geometric constructions. Rulers and protractors are measuring tools, not construction tools. You may use a ruler as a straightedge in constructions, provided you do not use its marks for measuring. In the next two investigations you will discover how to duplicate a line segment using only your compass and straightedge, or using only patty paper.

Investigation 1
Copying a Segment

You will need

- a compass
- a straightedge
- a ruler
- patty paper

A B

C — Stage 1

A B — Stage 2

C D — Stage 3

Step 1 The complete construction for copying a segment, $\overline{AB}$, is shown above. Describe each stage of the process.

Step 2 Use a ruler to measure $\overline{AB}$ and $\overline{CD}$. How do the two segments compare?

Step 3 Describe how to duplicate a segment using patty paper instead of a compass.

Using only a compass and a straightedge, how would you duplicate an angle? In other words, how would you construct an angle that is congruent to a given angle? You may not use your protractor, because a protractor is a measuring tool, not a construction tool.

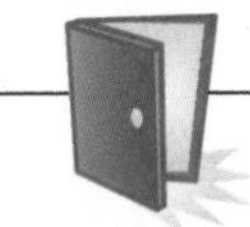

Investigation 2
Copying an Angle

You will need

- a compass
- a straightedge

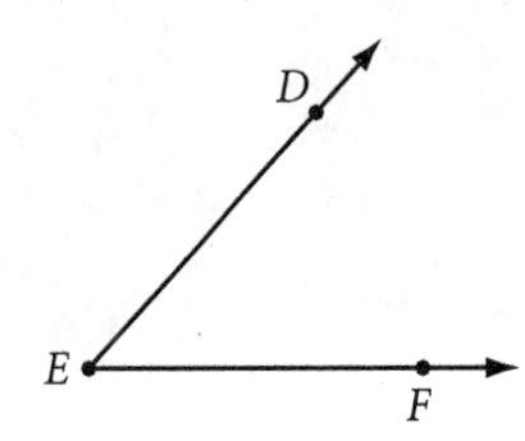

Step 1 | The first two stages for copying $\angle DEF$ are shown below. Describe each stage of the process.

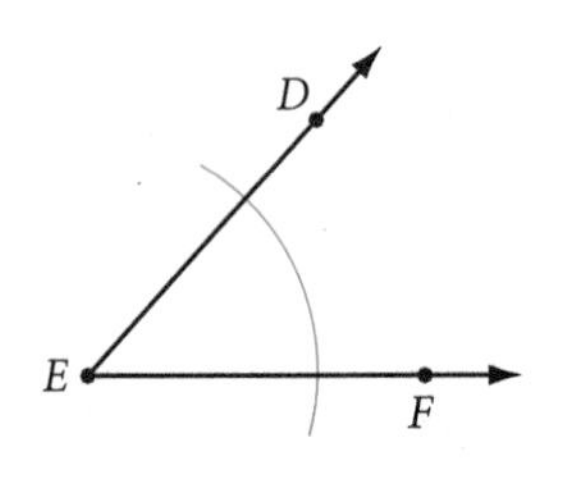

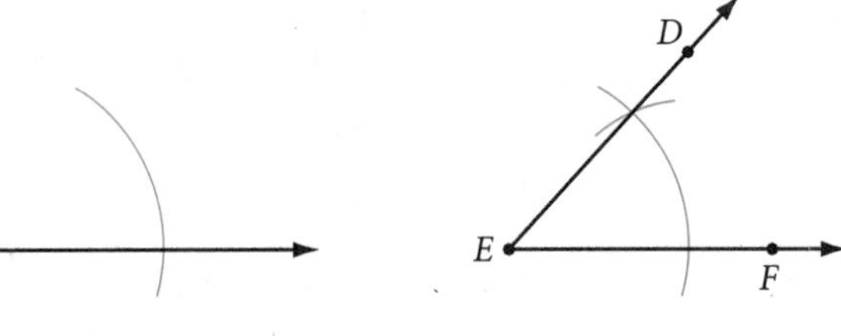

Stage 1

Stage 2

Step 2 | What will be the final stage of the construction?

Step 3 | Use a protractor to measure $\angle DEF$ and $\angle G$. What can you state about these angles?

Step 4 | Describe how to duplicate an angle using patty paper instead of a compass.

You've just discovered how to duplicate segments and angles using a straightedge and compass or patty paper. These are the basic constructions. You will use combinations of these to do many other constructions. You may be surprised that you can construct figures more precisely *without* using a ruler or protractor!

Called *vintas,* these canoes with brightly patterned sails are used for fishing in Zamboanga, Philippines. What angles and segments are duplicated in this photo?

EXERCISES

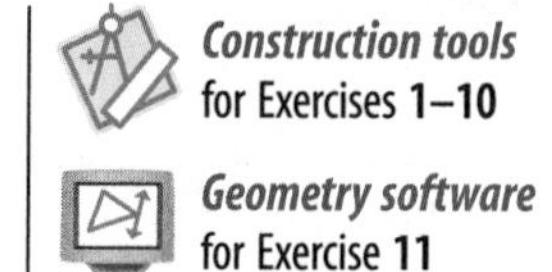

Construction Now that you can duplicate line segments and angles, do the constructions in Exercises 1–10. You will duplicate polygons in Exercises 7 and 10.

1. Using only a compass and a straightedge, duplicate the three line segments shown below. Label them as they're labeled in the figure.

2. Use the segments from Exercise 1 to construct a line segment with length $AB + CD$. ⓗ

3. Use the segments from Exercise 1 to construct a line segment with length $AB + 2EF - CD$.

4. Use a compass and a straightedge to duplicate each angle. There's an arc in each angle to help you.

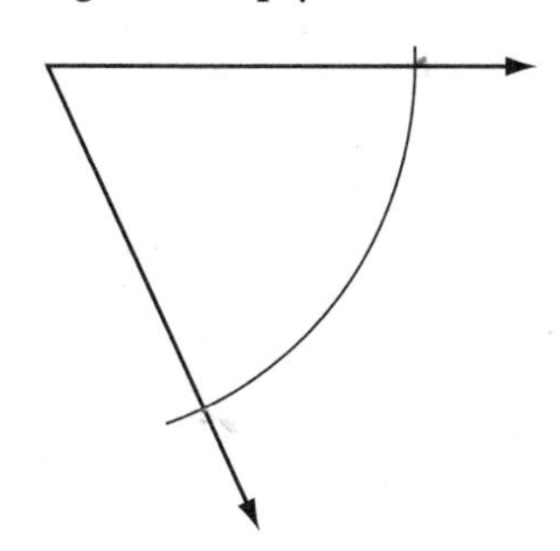

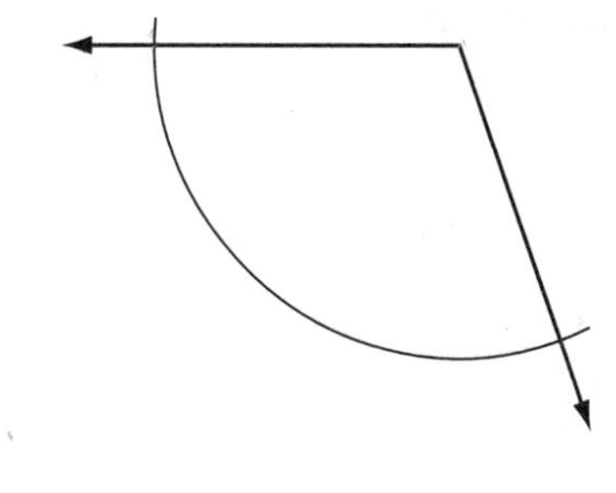

5. Draw an obtuse angle. Label it *LGE*, then duplicate it.

6. Draw two acute angles on your paper. Construct a third angle with a measure equal to the sum of the measures of the first two angles. Remember, you cannot use a protractor—use a compass and a straightedge only.

7. Draw a large acute triangle on the top half of your paper. Duplicate it on the bottom half, using your compass and straightedge. Do not erase your construction marks, so others can see your method.

8. Construct an equilateral triangle. Each side should be the length of this segment.

9. Repeat Exercises 7 and 8 using constructions with patty paper.

10. Draw quadrilateral *QUAD*. Duplicate it, using your compass and straightedge. Label the construction *COPY* so that $QUAD \cong COPY$. ⓗ

11. ***Technology*** Use geometry software to construct an equilateral triangle. Drag each vertex to make sure it remains equilateral.

Review

12. Copy the diagram at right. Use the Vertical Angles Conjecture and the Parallel Lines Conjecture to calculate the measure of each angle.

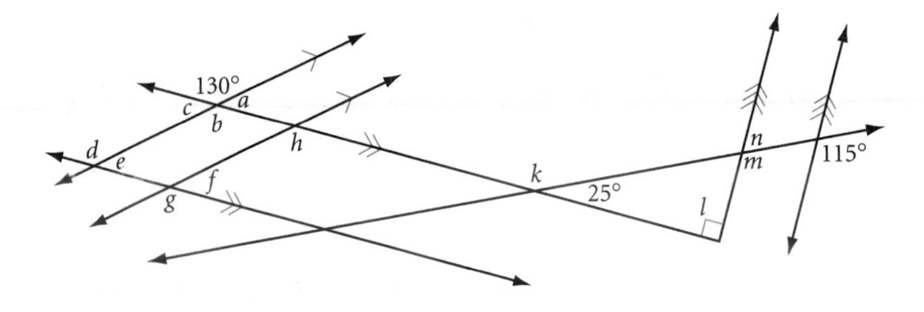

13. Hyacinth is standing on the curb waiting to cross 24th Street. A half block to her left is Avenue J, and Avenue K is a half block to her right. Numbered streets run parallel to one another and are all perpendicular to lettered avenues. If Avenue P is the northernmost avenue, which direction (N, S, E, or W) is she facing?

14. Write a new definition for an isosceles triangle, based on the triangle's reflectional symmetry. Does your definition apply to equilateral triangles? Explain. ⓗ

15. Draw $\triangle DAY$ after it is rotated 90° clockwise about the origin. Label the coordinates of the vertices.

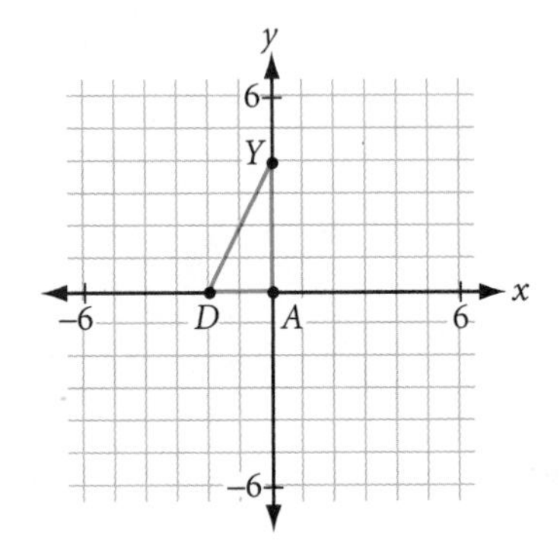

16. Sketch the three-dimensional figure formed by folding this net into a solid.

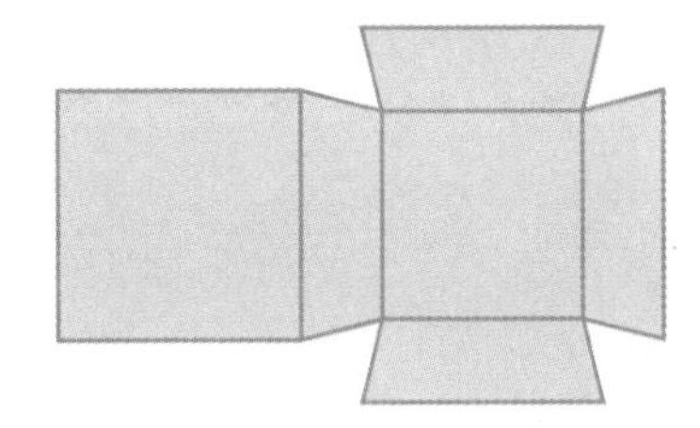

IMPROVING YOUR ALGEBRA SKILLS

Pyramid Puzzle II

Place four different numbers in the bubbles at the vertices of each pyramid so that the two numbers at the ends of each edge add to the number on that edge.

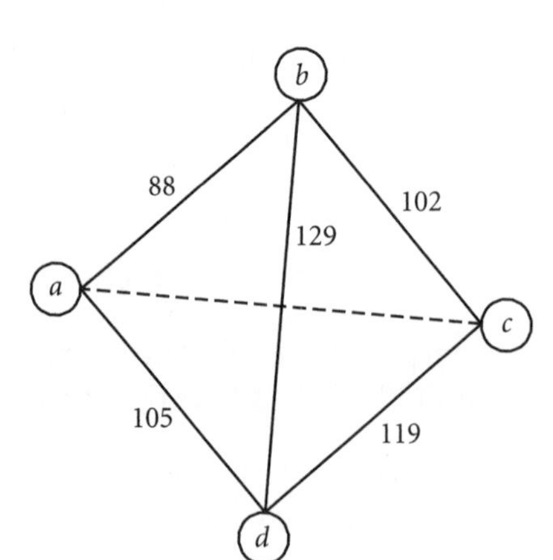

LESSON
3.2 Constructing Perpendicular Bisectors

To be successful, the first thing to do is to fall in love with your work.

SISTER MARY LAURETTA

Each segment has exactly one midpoint. A **segment bisector** is a line, ray, or segment in a plane that passes through the midpoint of a segment in a plane.

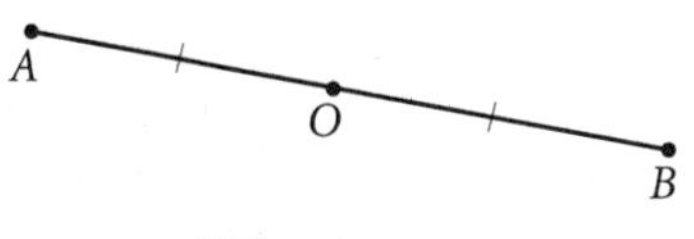

Segment $\overline{AB}$ has a midpoint O.

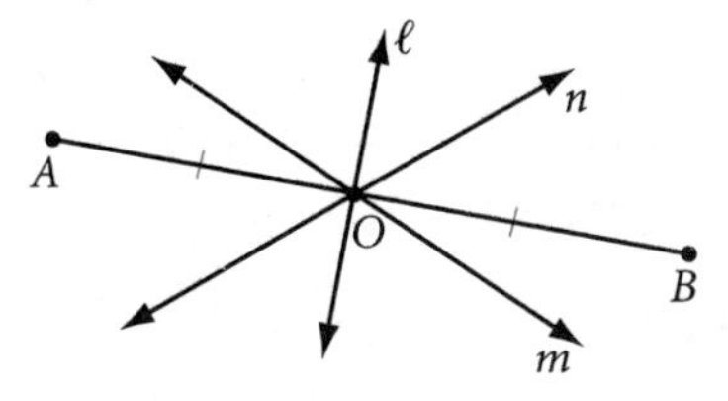

Lines ℓ, m, and n bisect $\overline{AB}$.

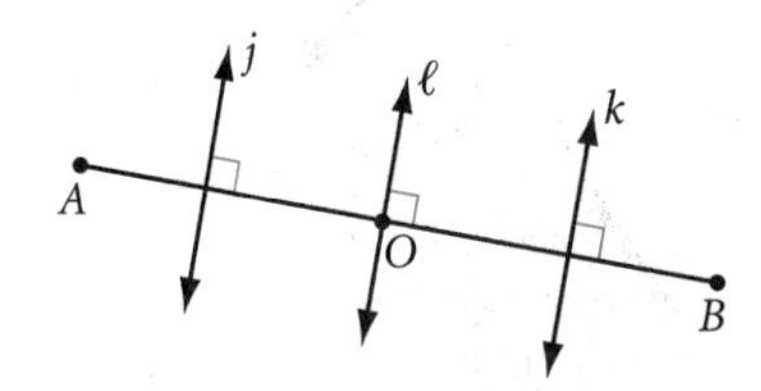

Lines j, k, and ℓ are perpendicular to $\overline{AB}$.

A segment has many perpendiculars and many bisectors, but each segment in a plane has only one bisector that is also perpendicular to the segment. This line is its **perpendicular bisector.**

Line ℓ is the perpendicular bisector of $\overline{AB}$.

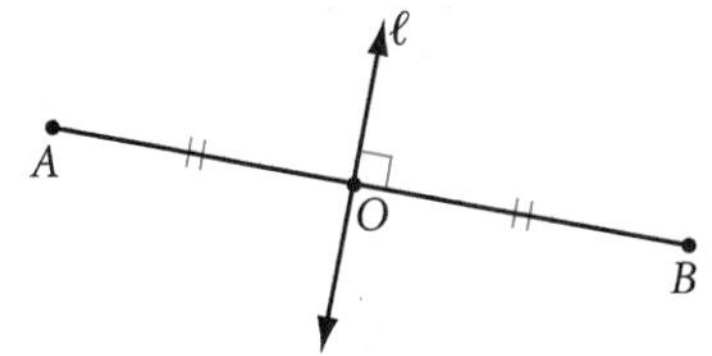

Investigation 1
Finding the Right Bisector

You will need

- patty paper

In this investigation you will discover how to construct the perpendicular bisector of a segment.

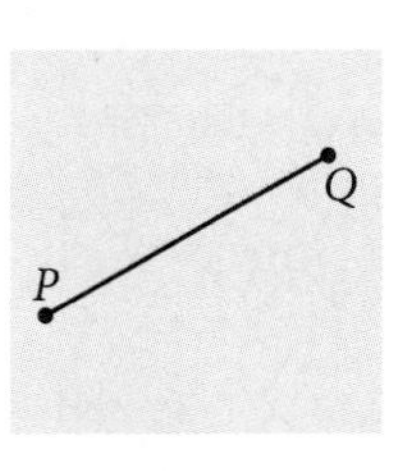

Step 1

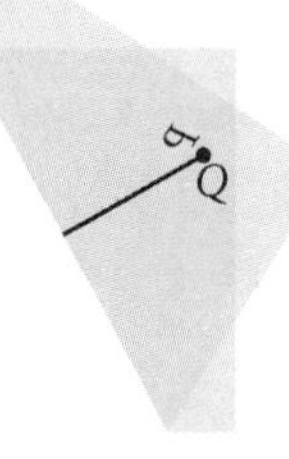

Step 2

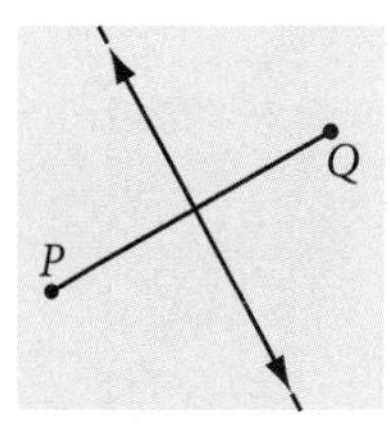

Step 3

Step 1 Draw a segment on patty paper. Label it $\overline{PQ}$.

Step 2 Fold your patty paper so that endpoints P and Q land exactly on top of each other, that is, they **coincide.** Crease your paper along the fold.

Step 3 Unfold your paper. Draw a line in the crease. What is the relationship of this line to $\overline{PQ}$? Check with others in your group. Use your ruler and protractor to verify your observations.

How would you describe the relationship of the points on the perpendicular bisector to the endpoints of the bisected segment? There's one more step in your investigation.

Step 4 Place three points on your perpendicular bisector. Label them *A, B,* and *C.* With your compass, compare the distances *PA* and *QA*. Compare the distances *PB* and *QB*. Compare the distances *PC* and *QC*. What do you notice about the two distances from each point on the perpendicular bisector to the endpoints of the segment? Compare your results with the results of others. Then copy and complete the conjecture.

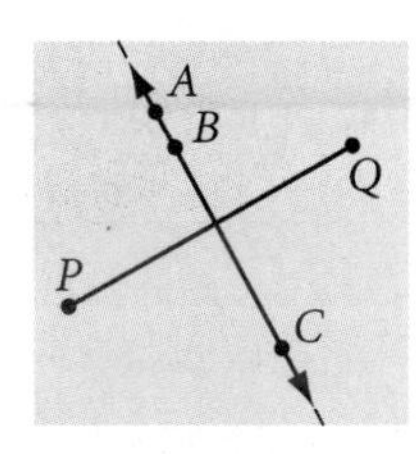

Remember to add each conjecture to your conjecture list and draw a figure for it.

Perpendicular Bisector Conjecture C-5

If a point is on the perpendicular bisector of a segment, then it is _?_ from the endpoints.

You've just completed the Perpendicular Bisector Conjecture. What about the converse of this statement?

Investigation 2
Right Down the Middle

You will need

- a compass
- a straightedge

If a point is **equidistant,** or the same distance, from two endpoints of a line segment in a plane, will it be on the segment's perpendicular bisector? If so, then locating two such points can help you construct the perpendicular bisector.

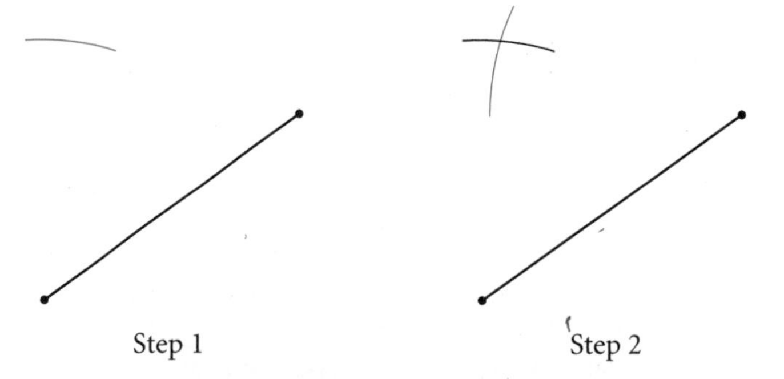

Step 1 Draw a line segment. Using one endpoint as center, swing an arc on one side of the segment.

Step 2 Using the same compass setting, but using the other endpoint as center, swing a second arc intersecting the first.

Step 3 The point where the two arcs intersect is equidistant from the endpoints of your segment. Use your compass to find another such point. Use these points to construct a line. Is this line the perpendicular bisector of the segment? Use the paper-folding technique of Investigation 1 to check.

Step 4 | Complete the conjecture below, and write a summary of what you did in this investigation.

Converse of the Perpendicular Bisector Conjecture C-6

If a point is equidistant from the endpoints of a segment, then it is on the __?__ of the segment.

Notice that constructing the perpendicular bisector also locates the midpoint of a segment. Now that you know how to construct the perpendicular bisector and the midpoint, you can construct rectangles, squares, and right triangles. You can also construct two special segments in any triangle: medians and midsegments.

The segment connecting the vertex of a triangle to the midpoint of its opposite side is a **median.** There are three midpoints and three vertices in every triangle, so every triangle has three medians.

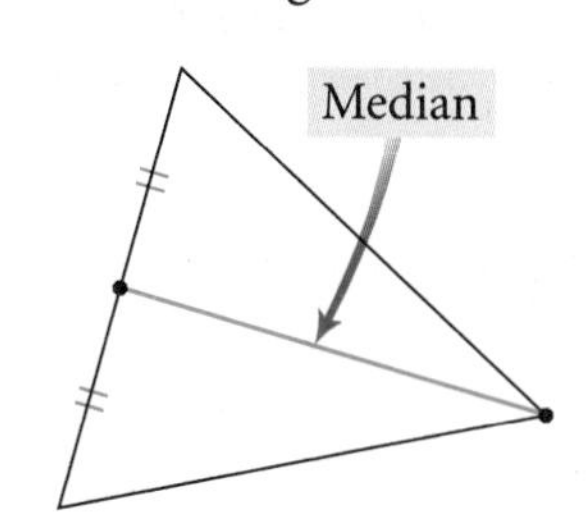

The segment that connects the midpoints of two sides of a triangle is a **midsegment.** A triangle has three sides, each with its own midpoint, so there are three midsegments in every triangle.

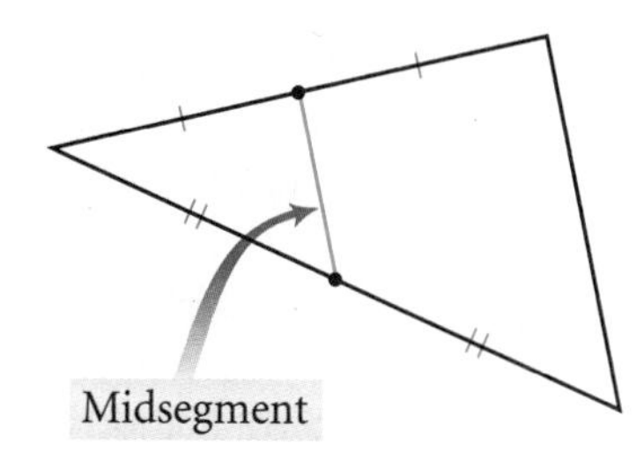

EXERCISES

You will need

Construction tools for Exercises **1–10** and **14**

Geometry software for Exercise **13**

Construction For Exercises 1–5, construct the figures using only a compass and a straightedge.

1. Construct and label $\overline{AB}$. Construct the perpendicular bisector of $\overline{AB}$.

2. Construct and label $\overline{QD}$. Construct perpendicular bisectors to divide $\overline{QD}$ into four congruent segments. ⓗ

3. Construct a line segment so close to the edge of your paper that you can swing arcs on only one side of the segment. Then construct the perpendicular bisector of the segment. ⓗ

4. Using $\overline{AB}$ and $\overline{CD}$, construct a segment with length $2AB - \frac{1}{2}CD$. ⓗ

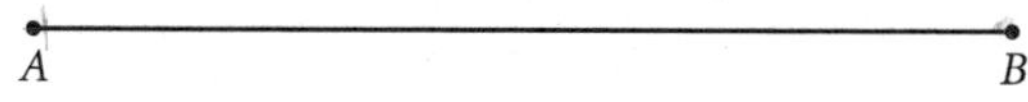

5. Construct $\overline{MN}$ with length equal to the average length of $\overline{AB}$ and $\overline{CD}$ above. ⓗ

6. ***Construction*** Do Exercises 1–5 using patty paper.

Construction For Exercises 7–10, you have your choice of construction tools. Use either a compass and a straightedge or patty paper and a straightedge. Do *not* use patty paper and compass together.

7. Construct $\triangle ALI$. Construct the perpendicular bisector of each side. What do you notice about the three bisectors?

8. Construct $\triangle ABC$. Construct medians $\overline{AM}$, $\overline{BN}$, and $\overline{CL}$. Notice anything special? ⓗ

9. Construct $\triangle DEF$. Construct midsegment $\overline{GH}$ where G is the midpoint of $\overline{DF}$ and H is the midpoint of $\overline{DE}$. What do you notice about the relationship between $\overline{EF}$ and $\overline{GH}$?

10. Copy rectangle $DSOE$ onto your paper. Construct the midpoint of each side. Label the midpoint of $\overline{DS}$ point I, the midpoint of $\overline{SO}$ point C, the midpoint of $\overline{OE}$ point V, and the midpoint of $\overline{ED}$ point R. Construct quadrilateral $RICV$. Describe $RICV$.

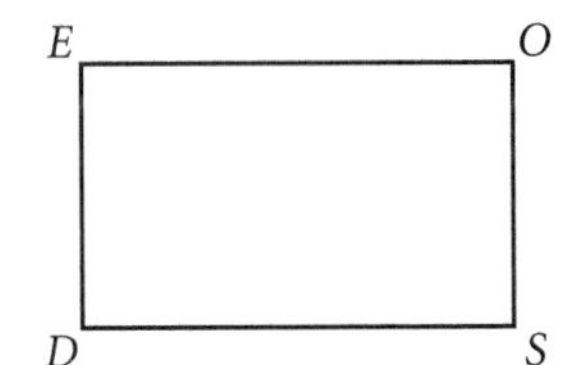

11. The island shown at right has two post offices. The postal service wants to divide the island into two zones so that anyone within each zone is always closer to their own post office than to the other one. Copy the island and the locations of the post offices and locate the dividing line between the two zones. Explain how you know this dividing line solves the problem. Or, pick several points in each zone and make sure they are closer to that zone's post office than they are to the other one.

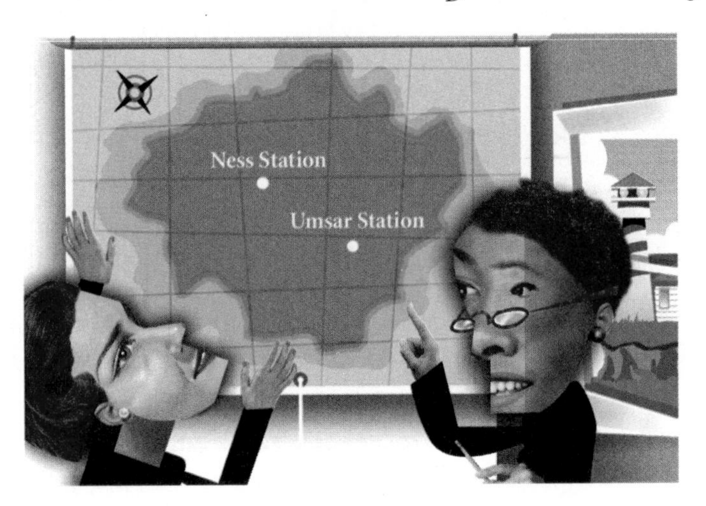

12. Copy parallelogram $FLAT$ onto your paper. Construct the perpendicular bisector of each side. What do you notice about the quadrilateral formed by the four lines?

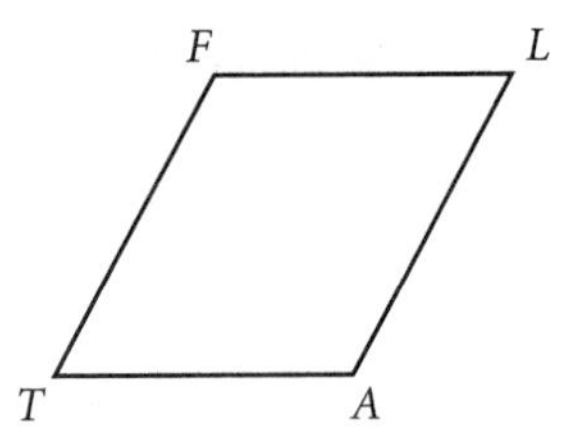

Review

13. ***Technology*** Use geometry software to construct a triangle. Construct a median. Are the two triangles created by the median congruent? Use an area measuring tool in your software program to find the areas of the two triangles. How do they compare? If you made the original triangle from heavy cardboard, and you wanted to balance that cardboard triangle on the edge of a ruler, what would you do?

14. ***Construction*** Construct a very large triangle on a piece of cardboard or mat board and construct its median. Cut out the triangle and see if you can balance it on the edge of a ruler. Sketch how you placed the triangle on the ruler. Cut the triangle into two pieces along the median and weigh the two pieces. Are they the same weight?

In Exercises 15–20, match the term with its figure below.

15. Scalene acute triangle

16. Isosceles obtuse triangle

17. Isosceles right triangle

18. Isosceles acute triangle

19. Scalene obtuse triangle

20. Scalene right triangle

A.

B.

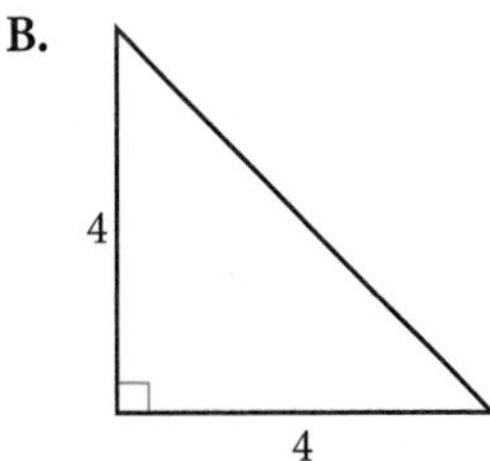

C.

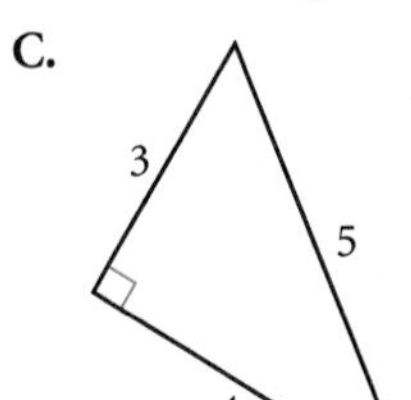
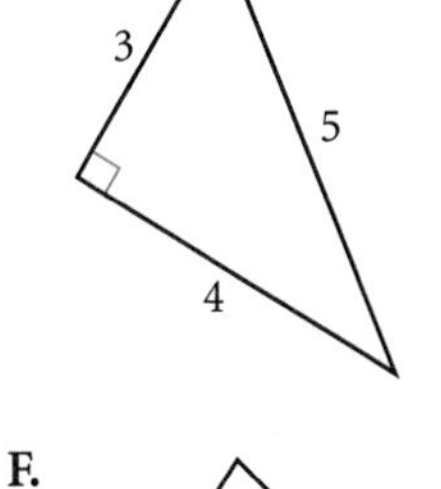

D.

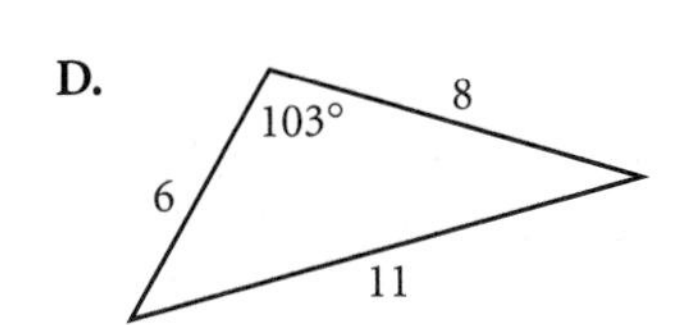

E.

F.

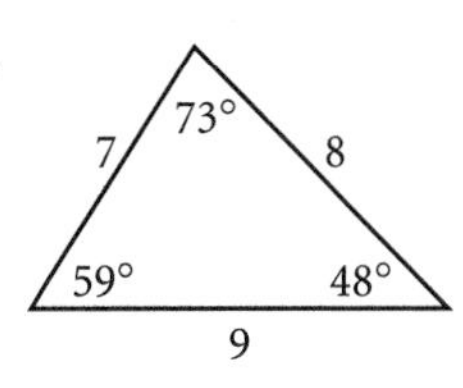

21. Sketch and label a polygon that has exactly three sides of equal length and exactly two angles of equal measure.

22. Sketch two triangles. Each should have one side measuring 5 cm and one side measuring 9 cm, but they should not be congruent.

23. List the letters from the alphabet below that have a horizontal line of symmetry.

A B C D E F G H I J K L M N O P Q R S T U V W X Y Z

IMPROVING YOUR VISUAL THINKING SKILLS

Folding Cubes I

In the problems below, the figure at the left represents the net for a cube. When the net is folded, which cube at the right will it become?

1. 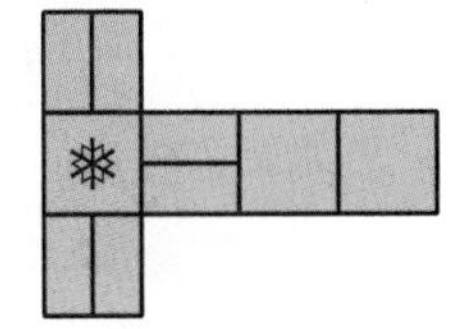

A. 

B. 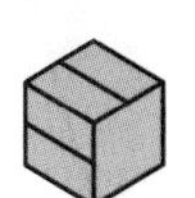

C. 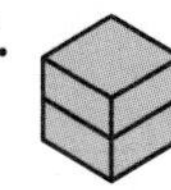

2. 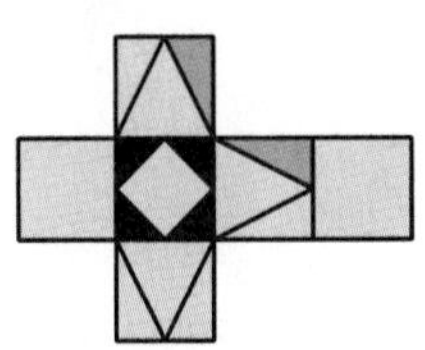

A. 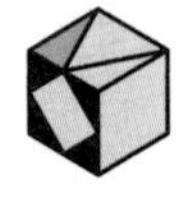

B. 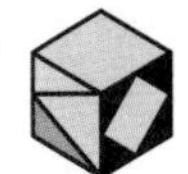

C. 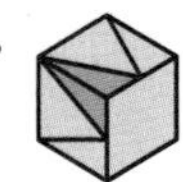

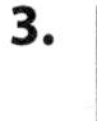

3. 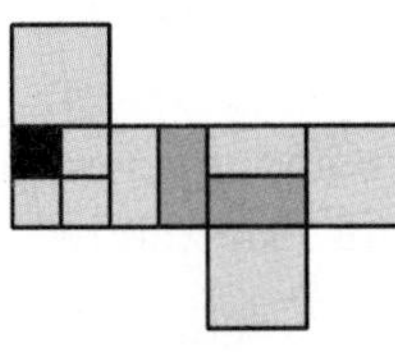

A. 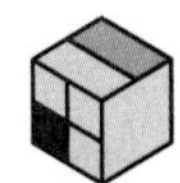

B. 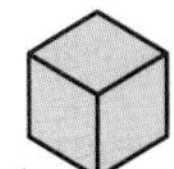

C. 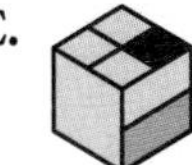

LESSON

3.3

Constructing Perpendiculars to a Line

Intelligence plus character—that is the goal of true education.

MARTIN LUTHER KING, JR.

If you are in a room, look over at one of the walls. What is the distance from where you are to that wall? How would you measure that distance? There are a lot of distances from where you are to the wall, but in geometry when we speak of a distance from a point to a line we mean a particular distance.

The construction of a perpendicular from a point to a line (with the point not on the line) is another of Euclid's constructions, and it has practical applications in many fields including agriculture and engineering. For example, think of a high-speed Internet cable as a line and a building as a point not on the line. Suppose you wanted to connect the building to the Internet cable using the shortest possible length of connecting wire. How can you find out how much wire you need, without buying too much?

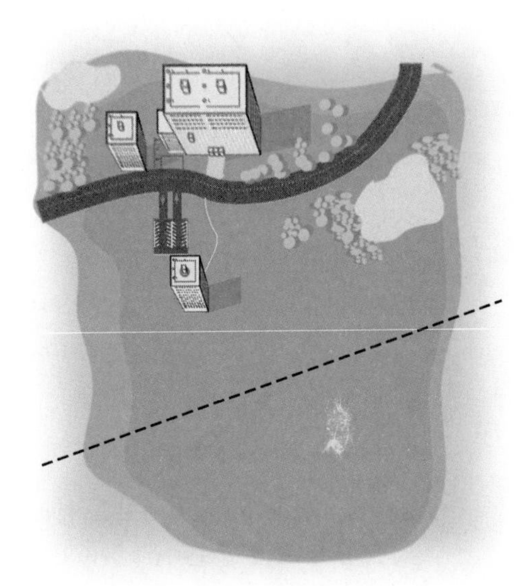

Investigation 1

Finding the Right Line

You will need

- a compass
- a straightedge

You already know how to construct perpendicular bisectors. You can't bisect a line because it is infinitely long, but you can use that know-how to construct a perpendicular from a point to a line.

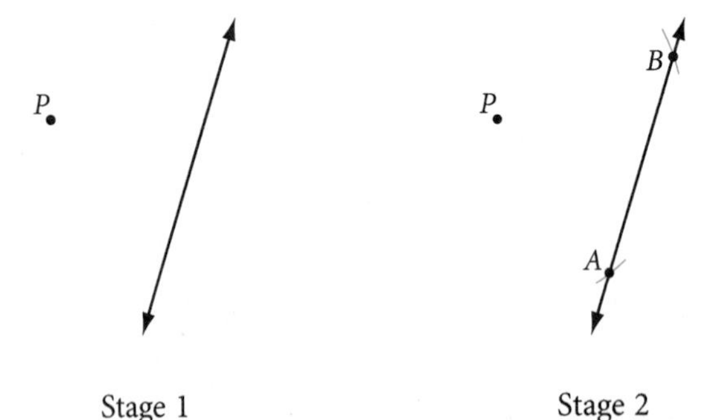

Step 1 Draw a line and a point labeled P not on the line, as shown above.

Step 2 Describe the construction steps you take at Stage 2.

Step 3 How is PA related to PB? What does this answer tell you about where point P lies? Hint: See the Converse of the Perpendicular Bisector Conjecture.

Step 4 Construct the perpendicular bisector of $\overline{AB}$. Label the midpoint M.

You have now constructed a perpendicular through a point not on the line. This is useful for finding the distance to a line.

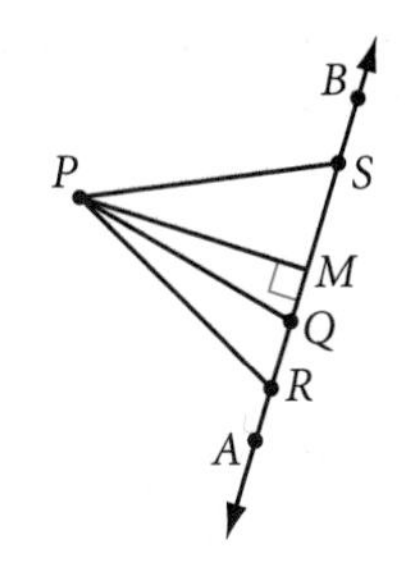

Step 5 Label three randomly placed points on $\overleftrightarrow{AB}$ as Q, R, and S. Measure PQ, PR, PS, and PM. Which distance is shortest? Compare results with others in your group.

You are now ready to state your observations by completing the conjecture.

Shortest Distance Conjecture C-7

The shortest distance from a point to a line is measured along the _?_ from the point to the line.

Let's take another look. How could you use patty paper to do this construction?

Investigation 2
Patty-Paper Perpendiculars

You will need

- patty paper
- a straightedge

In Investigation 1, you constructed a perpendicular from a point to a line. Now let's do the same construction using patty paper.

On a piece of patty paper, perform the steps below.

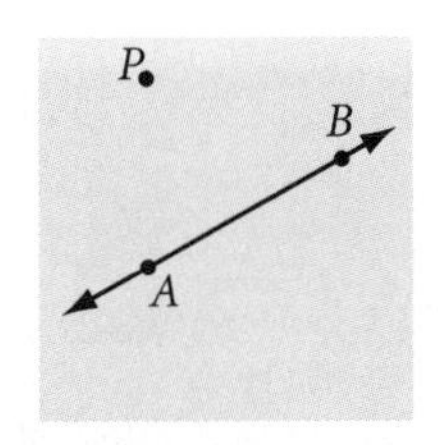

Step 1

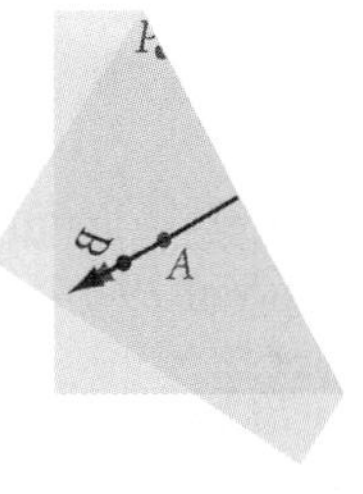

Step 2

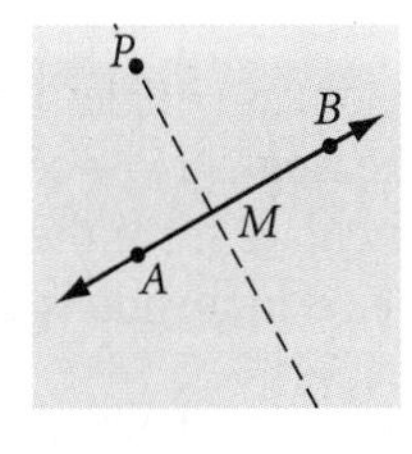

Step 3

Step 1 Draw and label $\overleftrightarrow{AB}$ and a point P not on $\overleftrightarrow{AB}$.

Step 2 Fold the line onto itself, and slide the layers of paper so that point P appears to be on the crease. Is the crease perpendicular to the line? Check it with the corner of a piece of patty paper.

Step 3 Label the point of intersection M. Are $\angle AMP$ and $\angle BMP$ congruent? Supplementary? Why or why not?

In Investigation 2, is M the midpoint of $\overline{AB}$? Do you think it needs to be? Think about the techniques used in the two investigations. How do the techniques differ?

The construction of a perpendicular from a point to a line lets you find the shortest distance from a point to a line. The geometry definition of distance from a point to a line is based on this construction, and it reads, "The **distance from a point to a line** is the length of the perpendicular segment from the point to the line."

You can also use this construction to find the altitude of a triangle. An **altitude** of a triangle is a perpendicular segment from a vertex to the opposite side or to a line containing the opposite side.

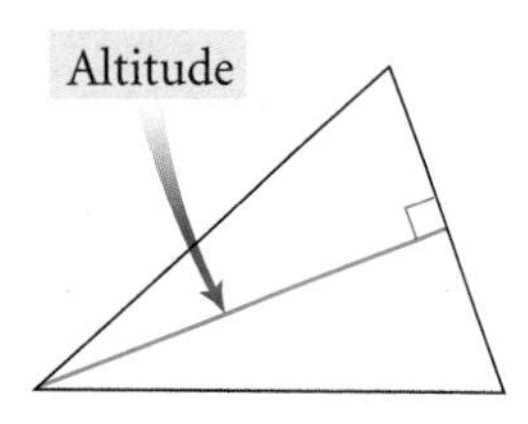

An altitude can be inside the triangle.

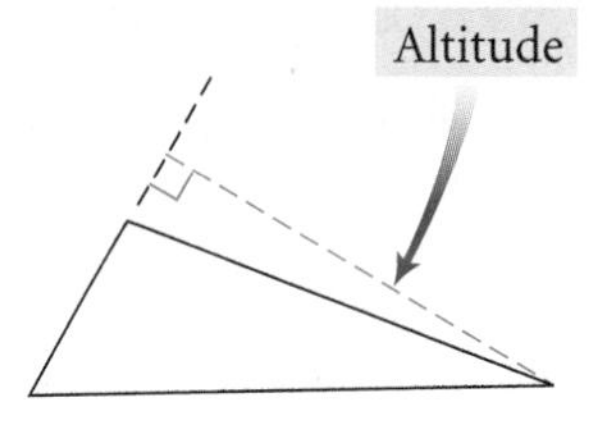

An altitude can be outside the triangle.

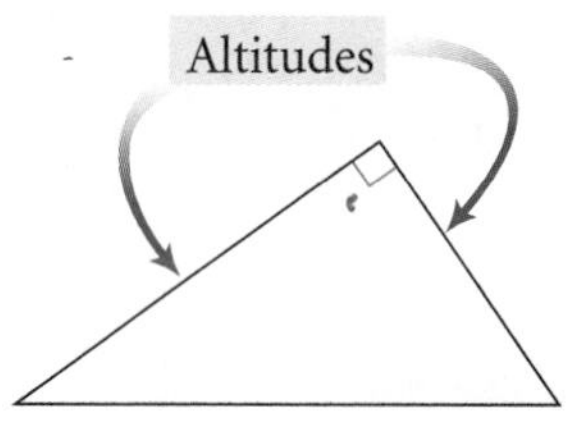

An altitude can be one of the sides of the triangle.

The length of the altitude is the height of the triangle. A triangle has three different altitudes, so it has three different heights.

EXERCISES

You will need

Construction tools for Exercises **1–12**

Construction Use your compass and straightedge and the definition of distance to do Exercises 1–5.

1. Draw an obtuse angle *BIG*. Place a point *P* inside the angle. Now construct perpendiculars from the point to both sides of the angle. Which side is closer to point *P*?

2. Draw an acute triangle. Label it *ABC*. Construct altitude $\overline{CD}$ with point *D* on $\overleftrightarrow{AB}$. (We didn't forget about point *D*. It's at the *foot* of the perpendicular. Your job is to locate it.)

In this futuristic painting, American artist Ralston Crawford (1906–1978) has constructed a set of converging lines and vertical lines to produce an illusion of distance.

3. Draw obtuse triangle *OBT* with obtuse angle *O*. Construct altitude $\overline{BU}$. In an obtuse triangle, an altitude can fall outside the triangle. To construct an altitude from point *B* of your triangle, extend side $\overline{OT}$. In an obtuse triangle, how many altitudes fall outside the triangle and how many fall inside the triangle? ⓗ

4. How can you construct a perpendicular to a line through a point that is on the line? Draw a line. Mark a point on your line. Now experiment. Devise a method to construct a perpendicular to your line at the point. ⓗ

5. Draw a line. Mark two points on the line and label them *Q* and *R*. Now construct a square *SQRE* with $\overline{QR}$ as a side. ⓗ

Construction For Exercises 6–9, use patty paper. (Attach your patty-paper work to your problems.)

6. Draw a line across your patty paper with a straightedge. Place a point P not on the line, and fold the perpendicular to the line through the point P. How would you fold to construct a perpendicular through a point on a line? Place a point Q on the line. Fold a perpendicular to the line through point Q. What do you notice about the two folds?

7. Draw a very large acute triangle on your patty paper. Place a point inside the triangle. Now construct perpendiculars from the point to all three sides of the triangle by folding. Mark your figure. How can you use your construction to decide which side of the triangle your point is closest to?

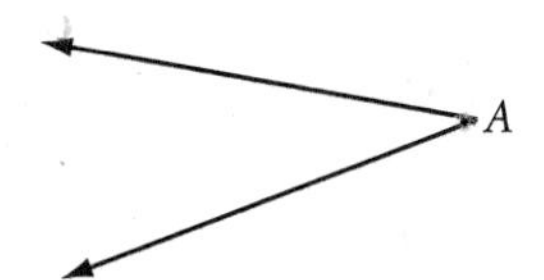

8. Construct an isosceles right triangle. Label its vertices A, B, and C, with point C the right angle. Fold to construct the altitude $\overline{CD}$. What do you notice about this line?

9. Draw obtuse triangle OBT with angle O obtuse. Fold to construct the altitude $\overline{BU}$. (Don't forget, you must extend the side $\overline{OT}$.)

Construction For Exercises 10–12, you may use either patty paper or a compass and a straightedge.

10. Construct a square $ABLE$ given $\overline{AL}$ as a diagonal.

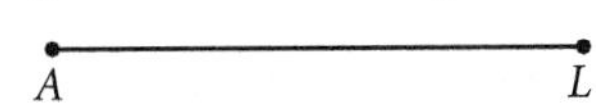

11. Construct a rectangle whose width is half its length.

12. Construct the complement of $\angle A$.

A

Review

13. Copy and complete the table. Make a conjecture for the value of the nth term and for the value of the 35th term. ⓗ

Rectangular pattern with triangles

Rectangle	1	2	3	4	5	6	...	n	...	35
Number of shaded triangles	2	9					...		...	

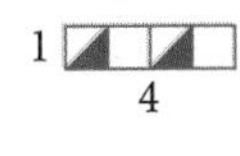

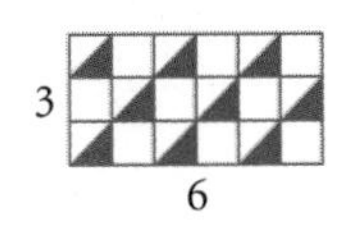

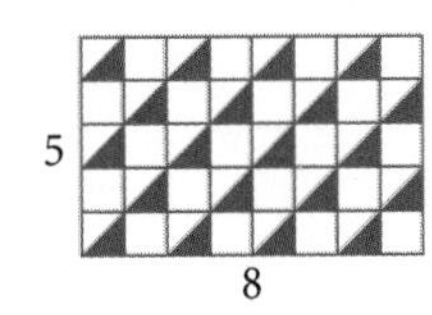

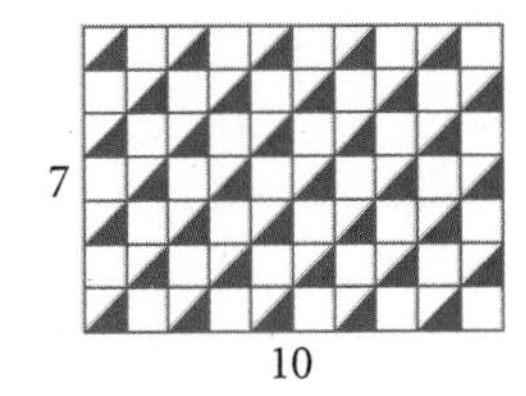

14. Sketch the solid of revolution formed when the two-dimensional figure at right is revolved about the line.

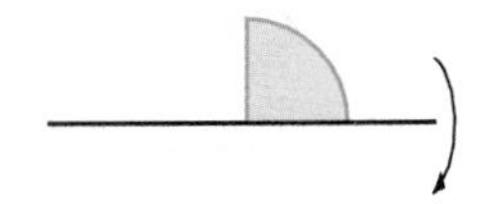

For Exercises 15–18, label the vertices with the appropriate letters. When you sketch or draw, use the special marks that indicate right angles, parallel segments, and congruent segments and angles.

15. Sketch obtuse triangle *FIT* with $m\angle I > 90°$ and median $\overline{IY}$.

16. Sketch $\overline{AB} \perp \overline{CD}$ and $\overline{EF} \perp \overline{CD}$.

17. Use your protractor to *draw* a regular pentagon. Draw all the diagonals. Use your compass to *construct* a regular hexagon. Draw three diagonals connecting alternating vertices. Do the same for the other three vertices. ⓗ

18. Draw a triangle with a 6 cm side and an 8 cm side and the angle between them measuring 40°. Draw a second triangle with a 6 cm side and an 8 cm side and exactly one 40° angle that is not between the two given sides. Are the two triangles congruent?

IMPROVING YOUR VISUAL THINKING SKILLS

PUZZLE

Constructing an Islamic Design

This Islamic design is based on two intersecting squares that form an 8-pointed star. Most Islamic designs of this kind can be constructed using only a compass and a straightedge. Try it. Use your compass and straightedge to re-create this design or to create a design of your own based on an 8-pointed star.

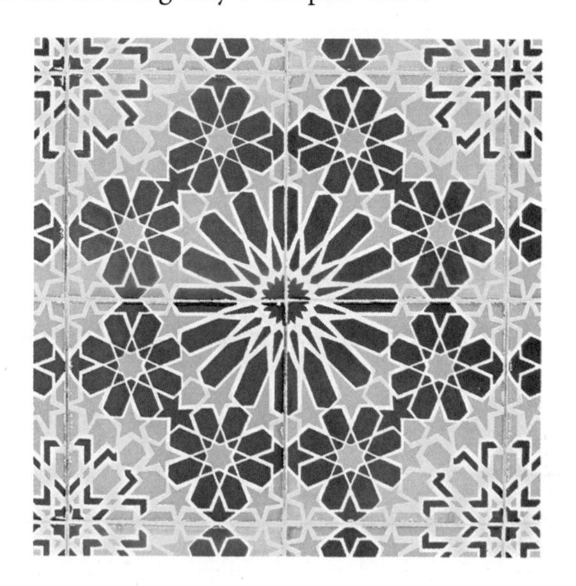

Here are two diagrams to get you started.

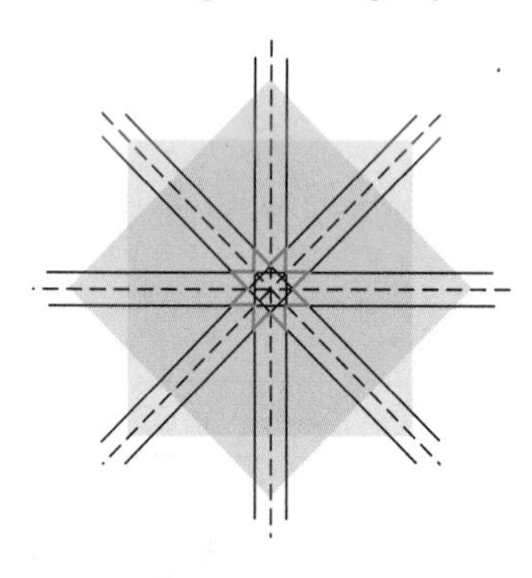

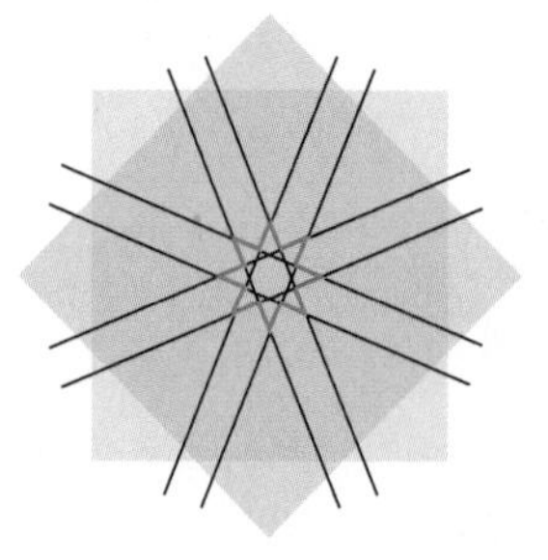

LESSON

3.4

Constructing Angle Bisectors

Challenges make you discover things about yourself that you never really knew.

CICELY TYSON

In Chapter 1, you learned that an **angle bisector** divides an angle into two congruent angles. While the definition states that the bisector of an angle is a ray, you may also refer to a segment as an angle bisector if the segment lies on the ray and passes through the vertex. For example, in this triangle the two angle bisectors are both segments.

In Investigations 1 and 2, you will learn to construct the bisector of an angle.

Investigation 1

Angle Bisecting by Folding

You will need

- patty paper

Each person should draw his or her own acute angle for this investigation.

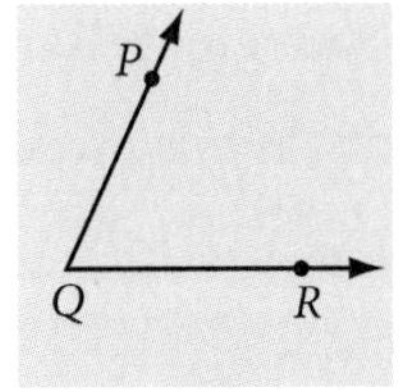

Step 1

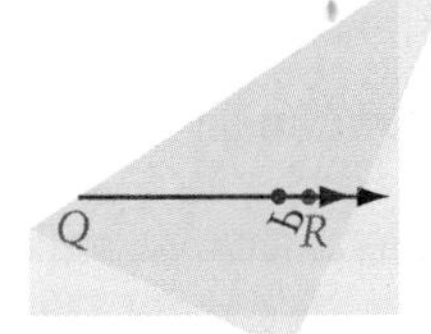

Step 2

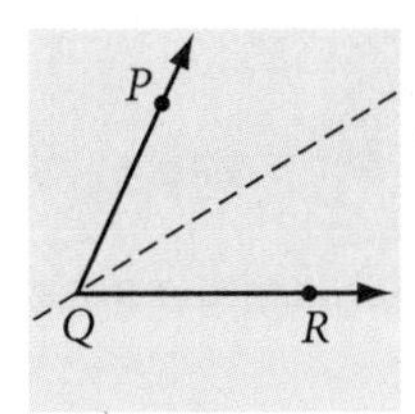

Step 3

Step 1 On patty paper, draw a large-scale angle. Label it PQR.

Step 2 Fold your patty paper so that $\overrightarrow{QP}$ and $\overrightarrow{QR}$ coincide. Crease the paper along the fold.

Step 3 Unfold your patty paper. Draw a ray with endpoint Q along the crease. Does the ray bisect $\angle PQR$? How can you tell?

Step 4 Repeat Steps 1–3 with an obtuse angle. Do you use different methods for finding the bisectors of different kinds of angles?

Step 5 Place a point on your angle bisector. Label it A. Compare the distances from A to each of the two sides. Remember that "distance" means *shortest* distance! Try it with other points on the angle bisector. Compare your results with those of others. Copy and complete the conjecture.

Angle Bisector Conjecture C-8

If a point is on the bisector of an angle, then it is $\underline{?}$ from the sides of the angle.

You've found the bisector of an angle by folding patty paper. Now let's see how you can construct the angle bisector with a compass and a straightedge.

Investigation 2
Angle Bisecting with Compass

You will need

- a compass
- a straightedge

In this investigation, you will find a method for bisecting an angle using a compass and straightedge. Each person in your group should investigate a different angle.

Step 1 Draw an angle.

Step 2 Find a method for constructing the bisector of the angle. Experiment!

Hint: Start by drawing an arc centered at the vertex.

Step 3 Once you think you have constructed the angle bisector, fold your paper to see if the ray you constructed is actually the bisector. Share your method with other students in your group. Agree on a best method.

Step 4 Write a summary of what you did in this investigation.

In earlier lessons, you learned to construct a 90° angle. Now you know how to bisect an angle. What angles can you construct by combining these two skills?

EXERCISES

You will need

Construction tools for Exercises **1–12**

Geometry software for Exercise **17**

Construction For Exercises 1–5, match each geometric construction with its diagram.

1. Construction of an angle bisector

2. Construction of a median

3. Construction of a midsegment

4. Construction of a perpendicular bisector

5. Construction of an altitude

A.
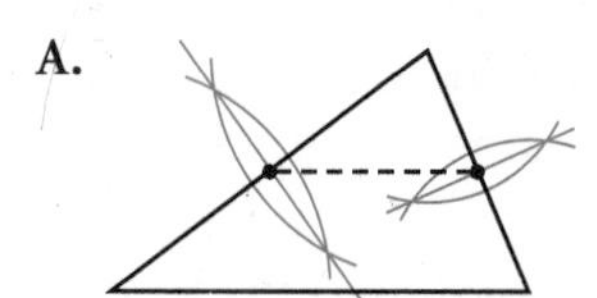

B.
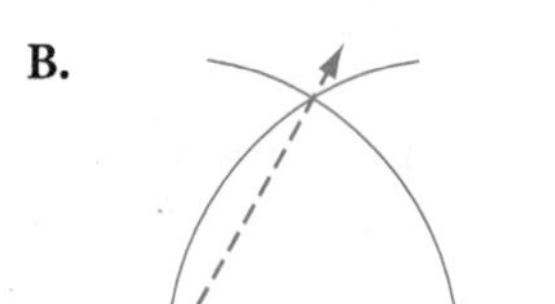

C.
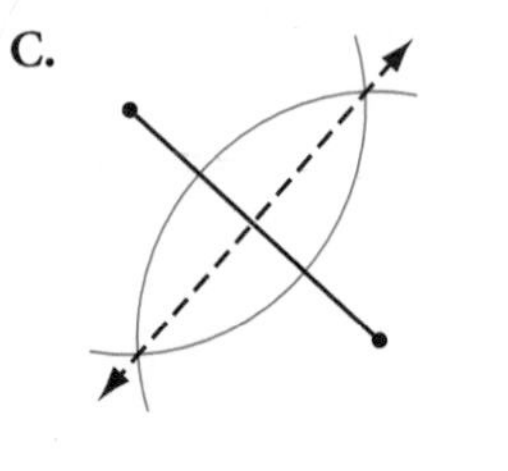

D.
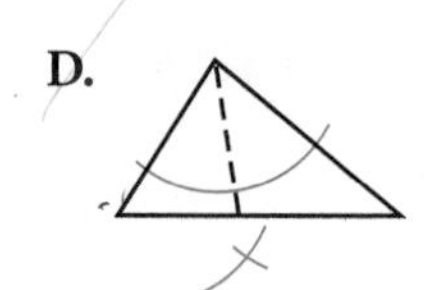

E.
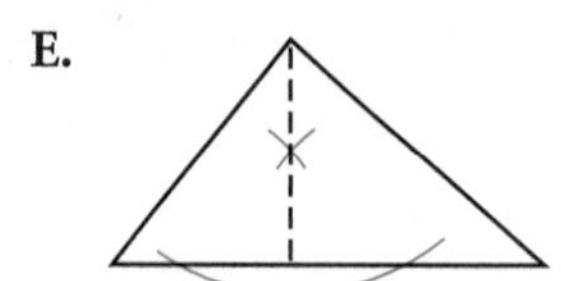

F.
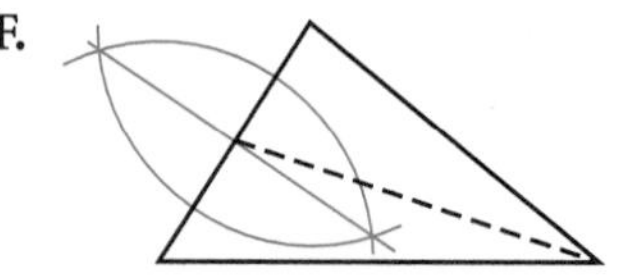

Construction For Exercises 6–12, construct a figure with the given specifications.

6. Given:

z

Construct: An isosceles right triangle with z as the length of each of the two congruent sides

7. Given:

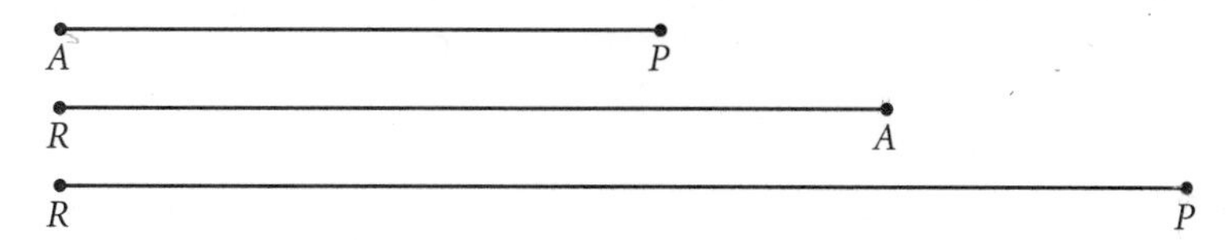

Construct: $\triangle RAP$ with median $\overline{PM}$ and angle bisector $\overline{RB}$

8. Given:

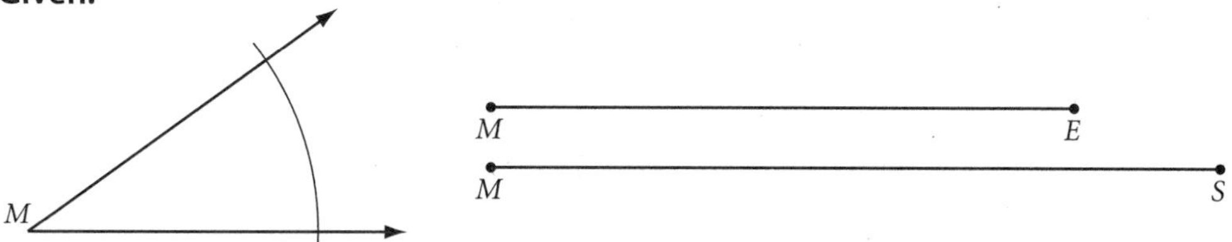

Construct: $\triangle MSE$ with $\overline{OU}$, where O is the midpoint of $\overline{MS}$ and U is the midpoint of $\overline{SE}$

9. Construct an angle with each given measure and label it. Remember, you may use only your compass and straightedge. No protractor!

a. 90° **b.** 45° **c.** 135°

10. Draw a large acute triangle. Bisect one vertex with a compass and a straightedge. Construct an altitude from the second vertex and a median from the third vertex.

11. Repeat Exercise 10 with patty paper. Which set of construction tools do you prefer? Why?

12. Use your straightedge to draw a linear pair of angles. Use your compass to bisect each angle of the linear pair. What do you notice about the two angle bisectors? Can you make a conjecture? Can you think of a way to explain why it is true?

13. In this lesson you discovered the Angle Bisector Conjecture. Write the converse of the Angle Bisector Conjecture. Do you think it's true? Why or why not?

Notice how this mosaic floor at Church of Pomposa in Italy (ca. 850 C.E.) uses many duplicated shapes. What constructions do you see in the square pattern? Are all the triangles in the isosceles triangle pattern identical? How can you tell?

Review

Sketch, draw, or construct each figure in Exercises 14–17. Label the vertices with the appropriate letters.

14. Draw a regular octagon. What traffic sign comes to mind? ⓗ

15. Construct regular octagon *ALTOSIGN*. ⓗ

16. Draw a triangle with a 40° angle, a 60° angle, and a side between the given angles measuring 8 cm. Draw a second triangle with a 40° angle and a 60° angle but with a side measuring 8 cm *opposite* the 60° angle. Are the triangles congruent? ⓗ

17. *Technology* Use geometry software to construct $\overline{AB}$ and $\overline{CD}$, with point C on $\overline{AB}$ and point D not on $\overleftrightarrow{AB}$. Construct the perpendicular bisector of $\overline{CD}$.

a. Trace this perpendicular bisector as you drag point C along $\overline{AB}$. Describe the shape formed by this locus of lines.

b. Erase the tracings from part a. Now trace the midpoint of $\overline{CD}$ as you drag C. Describe the locus of points.

IMPROVING YOUR VISUAL THINKING SKILLS

PUZZLE

Coin Swap III

Arrange four dimes and four pennies in a row of nine squares, as shown. Switch the position of the four dimes and four pennies in exactly 24 moves. A coin can slide into an empty square next to it or can jump over one coin into an empty space. Record your solution by listing, in order, which type of coin is moved. For example, your list might begin PDPDPPDD . . . .

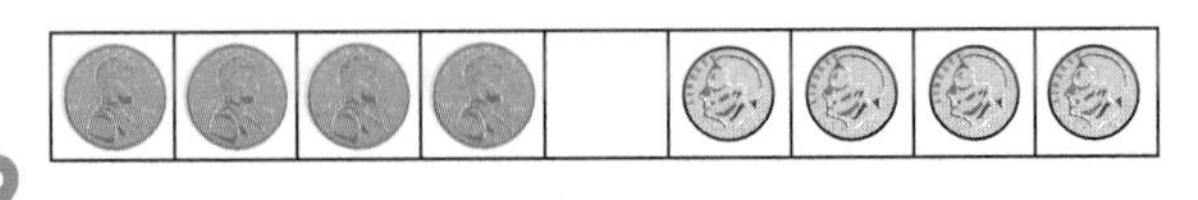

LESSON

3.5

Constructing Parallel Lines

When you stop to think, don't forget to start up again.

ANONYMOUS

Parallel lines are lines that lie in the same plane and do not intersect.

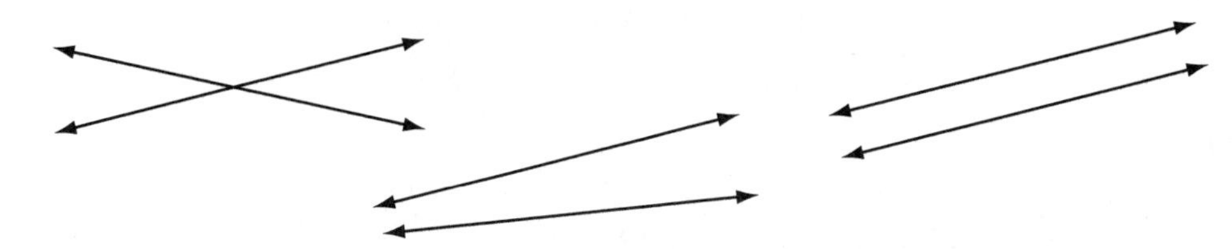

The lines in the first pair shown above intersect. They are clearly not parallel. The lines in the second pair do not meet as drawn. However, if they were extended, they would intersect. Therefore, they are not parallel. The lines in the third pair appear to be parallel, but if you extend them far enough in both directions, can you be sure they won't meet? There are many ways to be sure that the lines are parallel.

Investigation
Constructing Parallel Lines by Folding

You will need

- patty paper

How would you check whether two lines are parallel? One way is to draw a transversal and compare corresponding angles. You can also use this idea to *construct* a pair of parallel lines.

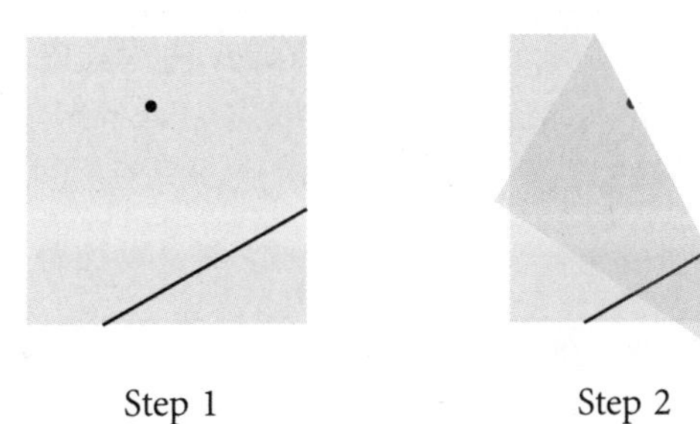

Step 1 Step 2

Step 1 Draw a line and a point on patty paper as shown.

Step 2 Fold the paper to construct a perpendicular so that the crease runs through the point as shown. Describe the four newly formed angles.

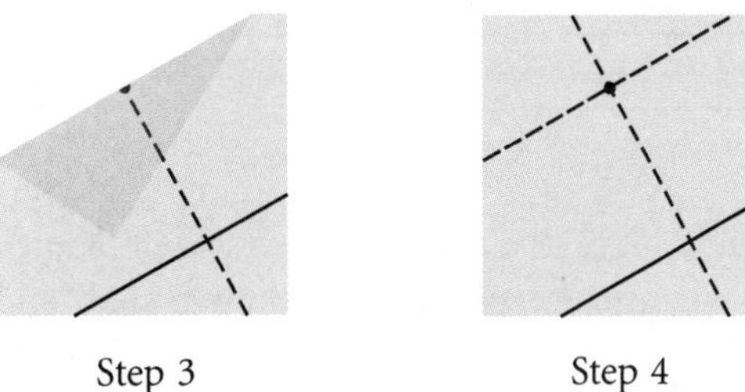

Step 3 Step 4

Step 3 Through the point, make another fold that is perpendicular to the first crease.

Step 4 Match the pairs of corresponding angles created by the folds. Are they all congruent? Why? What conclusion can you make about the lines?

There are many ways to construct parallel lines. You can construct parallel lines much more quickly with patty paper than with compass and straightedge. You can also use properties you discovered in the Parallel Lines Conjecture to construct parallel lines by duplicating corresponding angles, alternate interior angles, or alternate exterior angles. Or you can construct two perpendiculars to the same line. In the exercises you will practice all of these methods.

EXERCISES

You will need

Construction tools for Exercises **1–9**

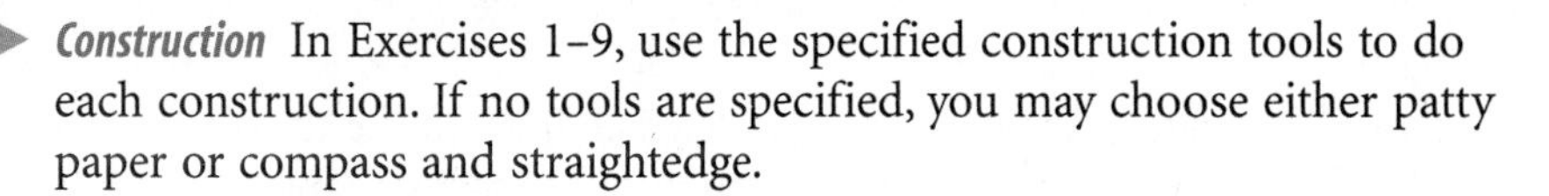

Construction In Exercises 1–9, use the specified construction tools to do each construction. If no tools are specified, you may choose either patty paper or compass and straightedge.

1. Use compass and straightedge. Draw a line and a point not on the line. Construct a second line through the point that is parallel to the first line, by duplicating alternate interior angles.

2. Use compass and straightedge. Draw a line and a point not on the line. Construct a second line through the point that is parallel to the first line, by duplicating corresponding angles.

3. Construct a square with perimeter z. ⓗ

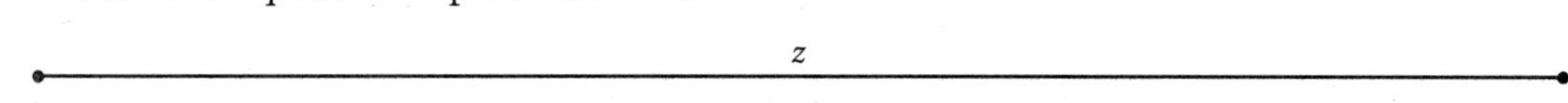

4. Construct a rhombus with x as the length of each side and $\angle A$ as one of the acute angles.

x

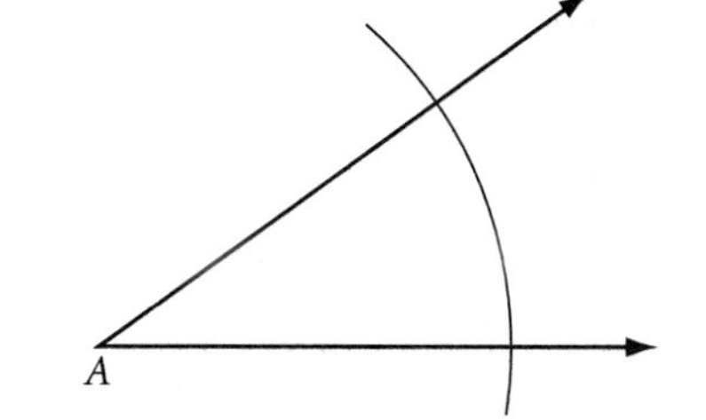

5. Construct trapezoid $TRAP$ with $\overline{TR}$ and $\overline{AP}$ as the two parallel sides and with AP as the distance between them. (There are many solutions!)

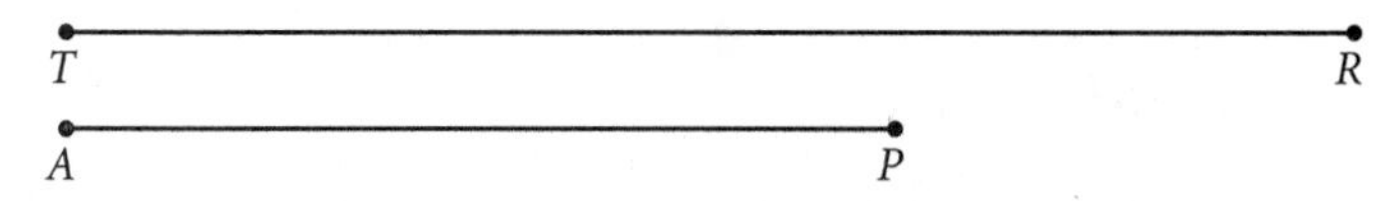

6. Using patty paper and straightedge, or a compass and straightedge, construct parallelogram $GRAM$ with $\overline{RG}$ and $\overline{RA}$ as two consecutive sides and ML as the distance between $\overline{RG}$ and $\overline{AM}$. (How many solutions can you find?)

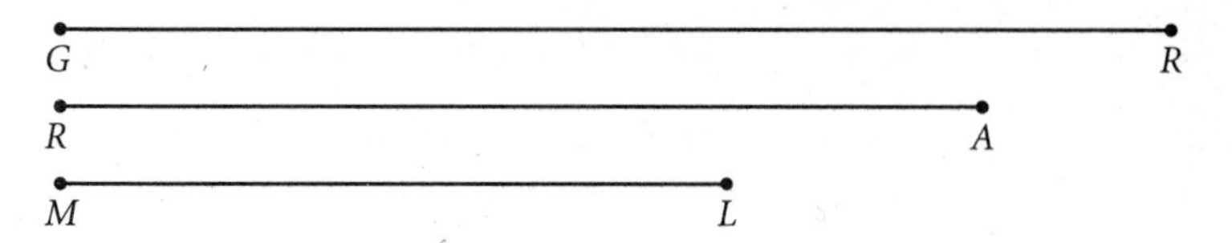

7. *Mini-Investigation* Construct a large scalene acute triangle and label it $\triangle SUM$. Through vertex M construct a line parallel to side $\overline{SU}$ as shown in the diagram. Use your protractor or a piece of patty paper to compare $\angle 1$ and $\angle 2$ with the other two angles of the triangle ($\angle S$ and $\angle U$). Notice anything special? Can you explain why?

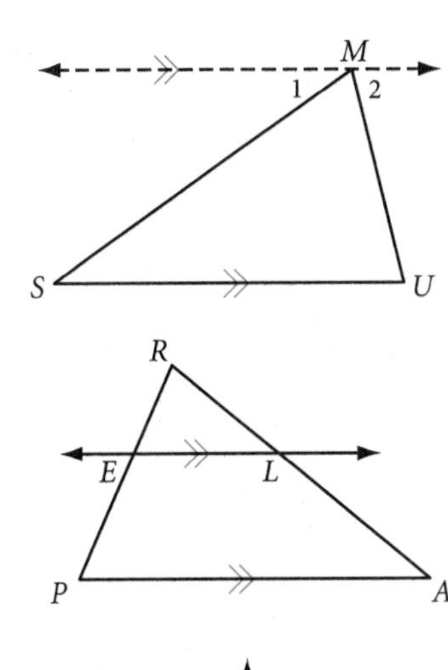

8. *Mini-Investigation* Construct a large scalene acute triangle and label it $\triangle PAR$. Place point E anywhere on side PR, and construct a line $\overleftrightarrow{EL}$ parallel to side $\overline{PA}$ as shown in the diagram. Use your ruler to measure the lengths of the four segments $\overline{AL}$, $\overline{LR}$, $\overline{RE}$, and $\overline{EP}$, and compare ratios $\frac{RL}{LA}$ and $\frac{RE}{EP}$. Notice anything special? (You may also do this problem using geometry software.)

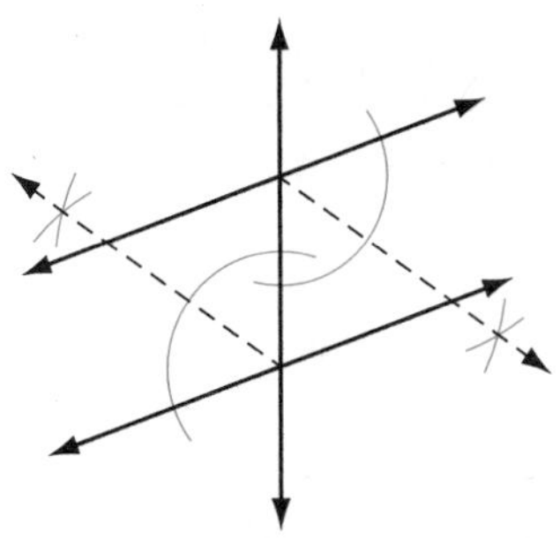

9. *Mini-Investigation* Draw a pair of parallel lines by tracing along both edges of your ruler. Draw a transversal. Use your compass to bisect each angle of a pair of alternate interior angles. What shape is formed? Can you explain why?

Review

10. There are three fire stations in the small county of Dry Lake. County planners need to divide the county into three zones so that fire alarms alert the closest station. Trace the county and the three fire stations onto patty paper, and locate the boundaries of the three zones. Explain how these boundaries solve the problem. ⓗ

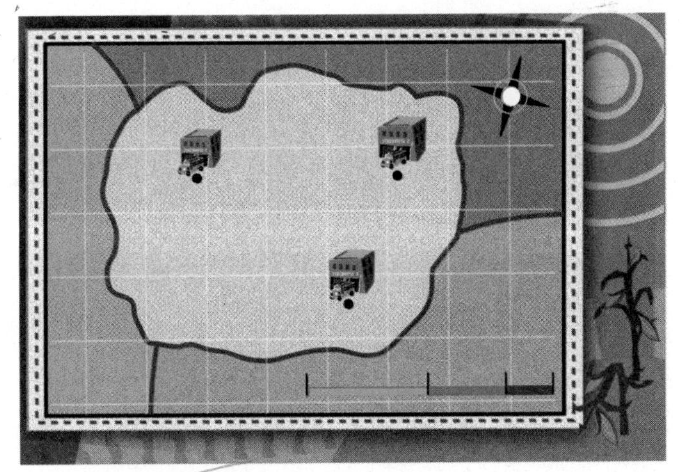

11. Copy the diagram below. Use your conjectures to calculate the measure of each lettered angle.

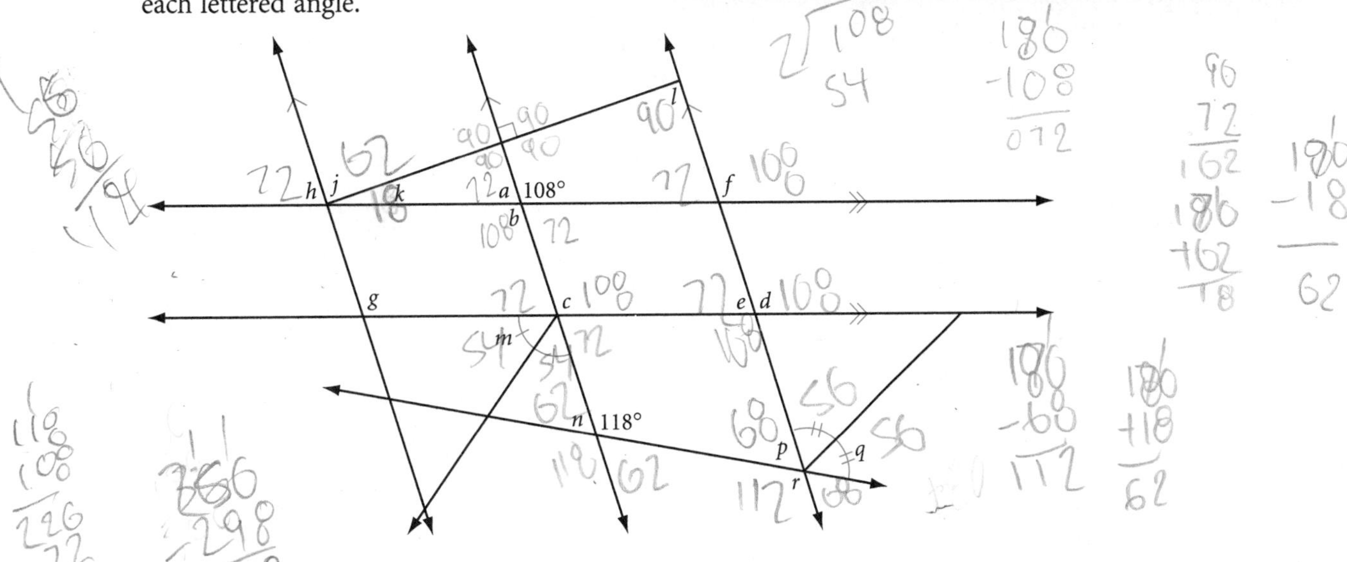

Sketch or draw each figure in Exercises 12–14. Label the vertices with the appropriate letters. Use the special marks that indicate right angles, parallel segments, and congruent segments and angles.

12. Sketch trapezoid *ZOID* with $\overline{ZO} \parallel \overline{ID}$, point *T* the midpoint of $\overline{OI}$, and *R* the midpoint of $\overline{ZD}$. Sketch segment *TR*.

13. Draw rhombus *ROMB* with $m\angle R = 60°$ and diagonal $\overline{OB}$.

14. Draw rectangle *RECK* with diagonals $\overline{RC}$ and $\overline{EK}$ both 8 cm long and intersecting at point *W*.

IMPROVING YOUR VISUAL THINKING SKILLS

PUZZLE

Visual Analogies

Which of the designs at right complete the statements at left? Explain.

1. 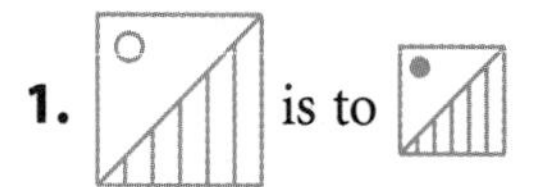

as  is to ?

A. 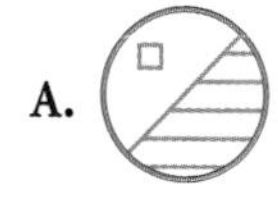

B. 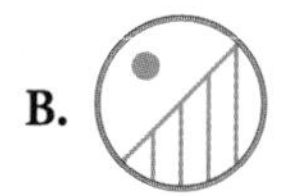

C. 

D. 

2. 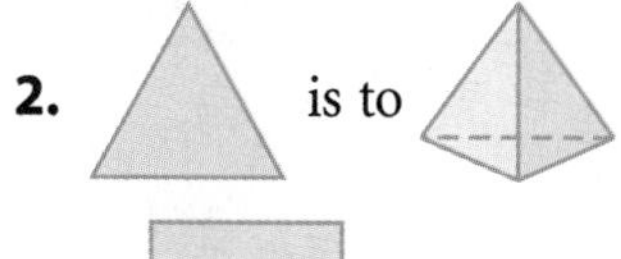

as 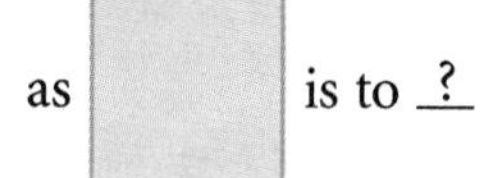 is to ?

A. 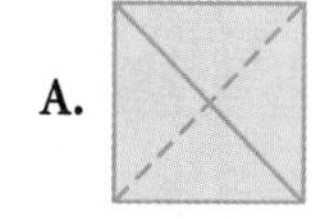

B. 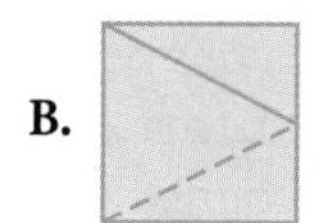

C. 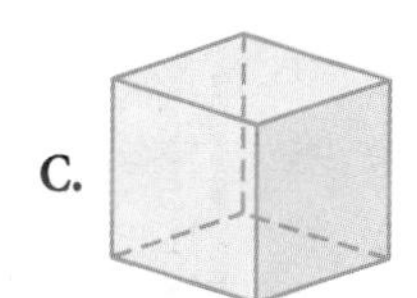

D. 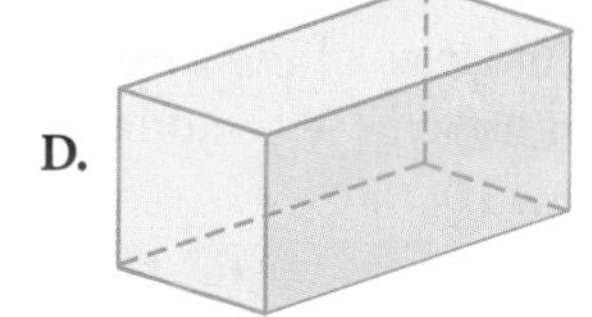

3. 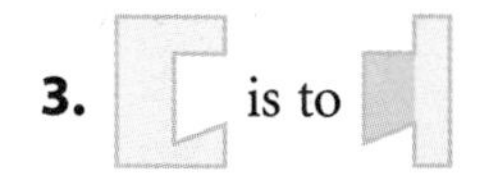

as 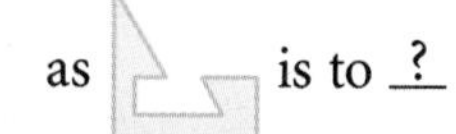 is to ?

A. 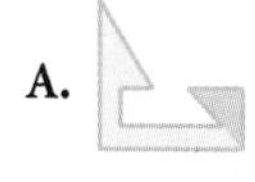

B. 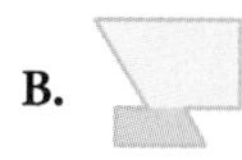

C. 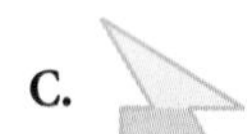

D. 

Slopes of Parallel and Perpendicular Lines

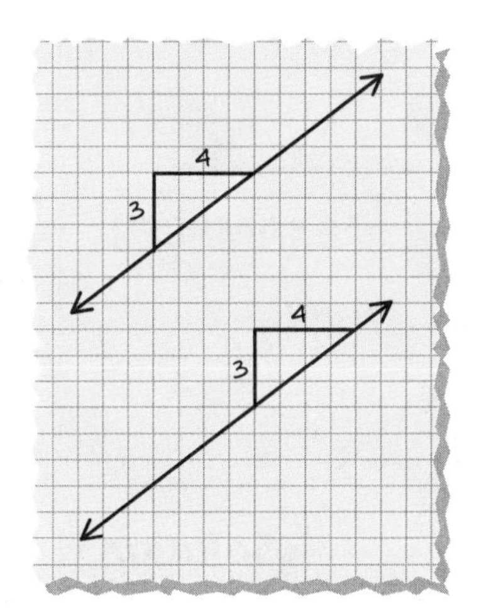

If two lines are parallel, how do their slopes compare? If two lines are perpendicular, how do *their* slopes compare? In this lesson you will review properties of the slopes of parallel and perpendicular lines.

If the slopes of two or more distinct lines are equal, are the lines parallel? To find out, try drawing on graph paper two lines that have the same slope triangle.

Yes, the lines are parallel. In fact, in coordinate geometry, this is the definition of parallel lines. The converse of this is true as well: If two lines are parallel, their slopes must be equal.

Parallel Slope Property

In a coordinate plane, two distinct lines are parallel if and only if their slopes are equal.

If two lines are perpendicular, their slope triangles have a different relationship. Study the slopes of the two perpendicular lines at right.

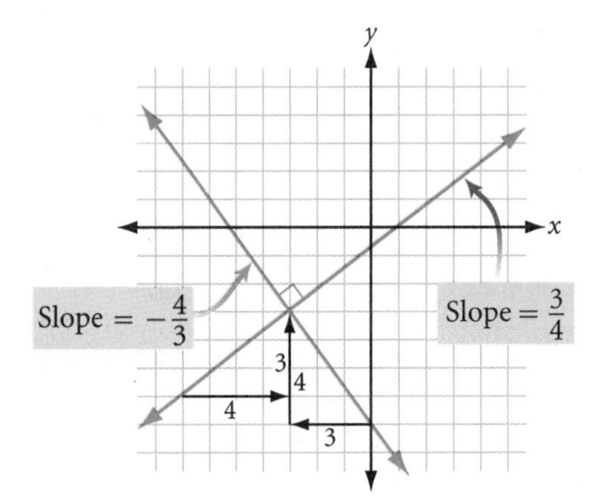

Perpendicular Slope Property

In a coordinate plane, two nonvertical lines are perpendicular if and only if their slopes are negative reciprocals of each other.

Can you explain why the slopes of perpendicular lines would have opposite signs? Can you explain why they would be reciprocals? Why do the lines need to be nonvertical?

EXAMPLE A

Consider $A(-15, -6)$, $B(6, 8)$, $C(4, -2)$, and $D(-4, 10)$. Are $\overleftrightarrow{AB}$ and $\overleftrightarrow{CD}$ parallel, perpendicular, or neither?

▶ Solution

Calculate the slope of each line.

$$\text{slope of } \overleftrightarrow{AB} = \frac{8-(-6)}{6-(-15)} = \frac{2}{3} \qquad \text{slope of } \overleftrightarrow{CD} = \frac{10-(-2)}{-4-4} = -\frac{3}{2}$$

The slopes, $\frac{2}{3}$ and $-\frac{3}{2}$, are negative reciprocals of each other, so $\overleftrightarrow{AB} \perp \overleftrightarrow{CD}$.

EXAMPLE B

Given points $E(-3, 0)$, $F(5, -4)$, and $Q(4, 2)$, find the coordinates of a point P such that $\overleftrightarrow{PQ}$ is parallel to $\overleftrightarrow{EF}$.

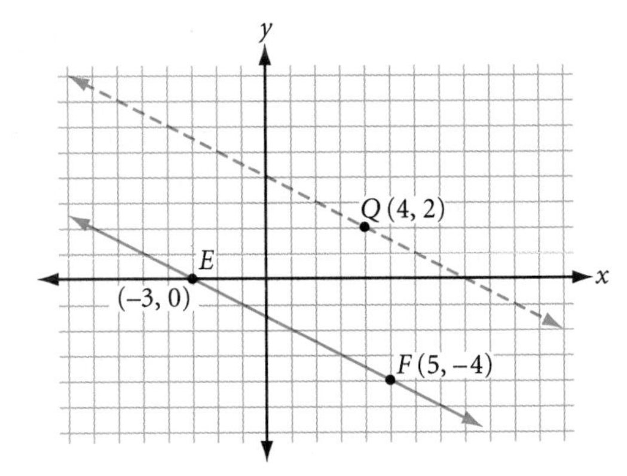

▶ Solution

We know that if $\overleftrightarrow{PQ} \parallel \overleftrightarrow{EF}$, then the slope of $\overleftrightarrow{PQ}$ equals the slope of $\overleftrightarrow{EF}$. First find the slope of $\overleftrightarrow{EF}$.

$$\text{slope of } \overleftrightarrow{EF} = \frac{-4-0}{5-(-3)} = \frac{-4}{8} = -\frac{1}{2}$$

There are many possible ordered pairs (x, y) for P. Use (x, y) as the coordinates of P, and the given coordinates of Q, in the slope formula to get

$$\frac{2-y}{4-x} = -\frac{1}{2}$$

Now you can treat the denominators and numerators as separate equations.

$$\begin{aligned} 4 - x &= 2 \\ -x &= -2 \\ x &= 2 \end{aligned} \qquad \begin{aligned} 2 - y &= -1 \\ -y &= -3 \\ y &= 3 \end{aligned}$$

Thus one possibility is $P(2, 3)$. How could you find another ordered pair for P? Here's a hint: How many different ways can you express $-\frac{1}{2}$?

Language CONNECTION

Coordinate geometry is sometimes called "analytic geometry." This term implies that you can use algebra to further analyze what you see. For example, consider $\overleftrightarrow{AB}$ and $\overleftrightarrow{CD}$. They look parallel, but looks can be deceiving. Only by calculating the slopes will you see that the lines are not truly parallel.

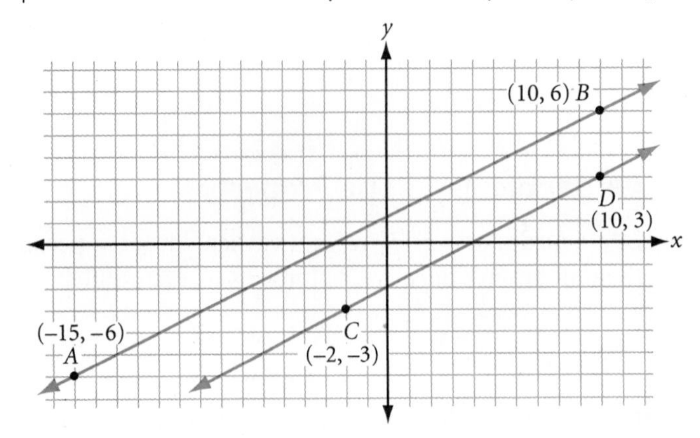

EXERCISES

For Exercises 1–4, determine whether each pair of lines through the points given below is parallel, perpendicular, or neither.

$A(1, 2)$ $B(3, 4)$ $C(5, 2)$ $D(8, 3)$ $E(3, 8)$ $F(-6, 5)$

1. $\overleftrightarrow{AB}$ and $\overleftrightarrow{BC}$ **2.** $\overleftrightarrow{AB}$ and $\overleftrightarrow{CD}$ **3.** $\overleftrightarrow{AB}$ and $\overleftrightarrow{DE}$ **4.** $\overleftrightarrow{CD}$ and $\overleftrightarrow{EF}$

5. Given $A(0, -3)$, $B(5, 3)$, and $Q(-3, -1)$, find two possible locations for a point P such that $\overleftrightarrow{PQ}$ is parallel to $\overleftrightarrow{AB}$.

6. Given $C(-2, -1)$, $D(5, -4)$, and $Q(4, 2)$, find two possible locations for a point P such that $\overleftrightarrow{PQ}$ is perpendicular to $\overleftrightarrow{CD}$.

For Exercises 7–9, find the slope of each side, and then determine whether each figure is a trapezoid, a parallelogram, a rectangle, or just an ordinary quadrilateral. Explain how you know.

7.

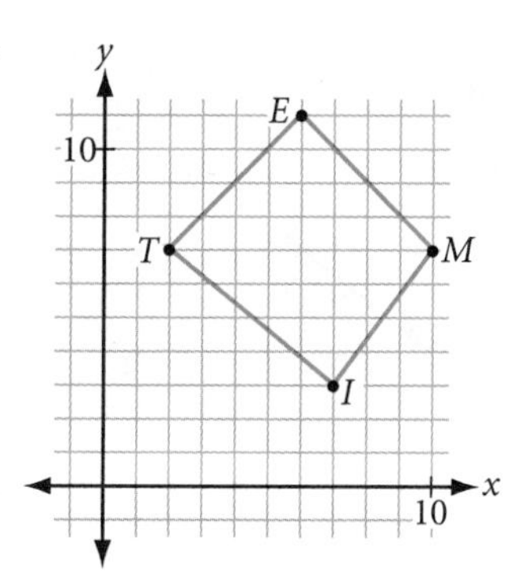

8.

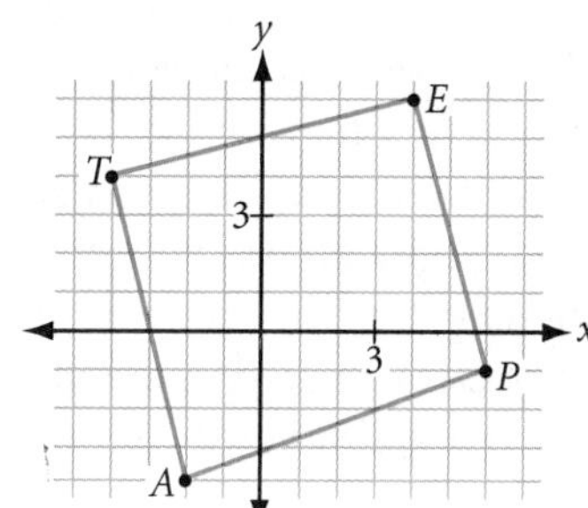

9.

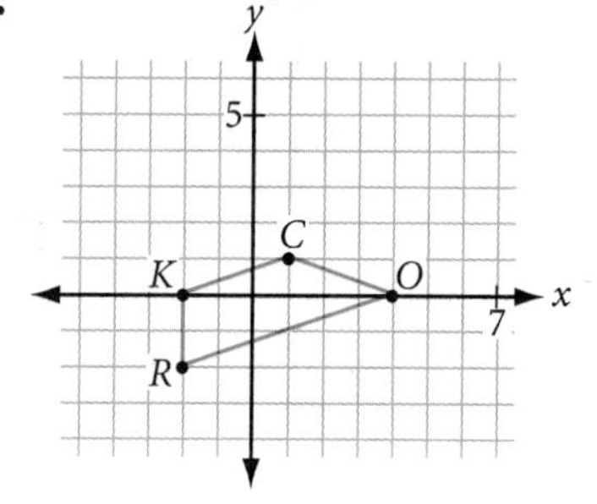

10. Quadrilateral *HAND* has vertices $H(-5, -1)$, $A(7, 1)$, $N(6, 7)$, and $D(-6, 5)$.

a. Is quadrilateral *HAND* a parallelogram? A rectangle? Neither? Explain how you know.

b. Find the midpoint of each diagonal. What can you conjecture?

11. Quadrilateral *OVER* has vertices $O(-4, 2)$, $V(1, 1)$, $E(0, 6)$, and $R(-5, 7)$.

a. Are the diagonals perpendicular? Explain how you know.

b. Find the midpoint of each diagonal. What can you conjecture?

c. What type of quadrilateral does *OVER* appear to be? Explain how you know.

12. Consider the points $A(-5, -2)$, $B(1, 1)$, $C(-1, 0)$, and $D(3, 2)$.

a. Find the slopes of $\overline{AB}$ and $\overline{CD}$.

b. Despite their slopes, $\overline{AB}$ and $\overline{CD}$ are not parallel. Why not?

c. What word in the Parallel Slope Property addresses the problem in 12b?

13. Given $A(-3, 2)$, $B(1, 5)$, and $C(7, -3)$, find point D such that quadrilateral $ABCD$ is a rectangle.

LESSON

3.6

Construction Problems

Once you know the basic constructions, you can create more complex geometric figures.

People who are only good with hammers see every problem as a nail.

ABRAHAM MASLOW

You know how to duplicate segments and angles with a compass and straightedge. If given a triangle, you can use these two constructions to duplicate the triangle by copying each segment and angle. Can you construct a triangle if you are given the parts separately? Would you need all six parts—three segments and three angles—to construct a triangle?

Let's first consider a case in which only three segments are given.

EXAMPLE A

Construct $\triangle ABC$ using the three segments $\overline{AB}$, $\overline{BC}$, and $\overline{CA}$ shown below. How many different-size triangles can be drawn?

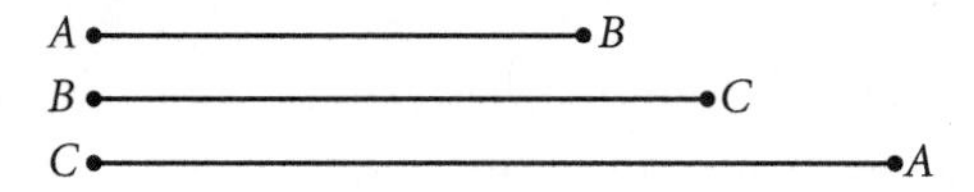

▸ Solution

You can begin by copying one segment, for example $\overline{AC}$. Then adjust your compass to match the length of another segment. Using this length as a radius, draw a circle centered at one endpoint of the first segment. Now use the third segment length as the radius for another circle, this one centered at the other endpoint. The third vertex of the triangle is where the circles intersect.

In the construction above, the segment lengths determine the sizes of the circles and where they intersect. Once the triangle "closes" at the intersection of the arcs, the angles are determined, too. So the lengths of the segments affect the size of the angles. There is only one size of triangle that can be drawn with the segments given, so the segments **determine** the triangle. Does having three angles also determine a triangle?

EXAMPLE B

Construct $\triangle ABC$ with patty paper by copying the three angles $\angle A$, $\angle B$, and $\angle C$ shown below. How many different size triangles can be drawn?

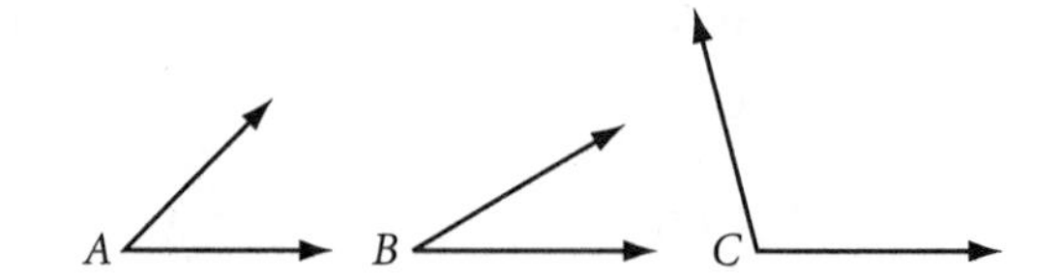

▸ Solution

In this patty-paper construction the angles do not determine the segment length. You can locate the endpoint of a segment anywhere along an angle's side without affecting the angle measures. Infinitely many different triangles can be drawn with the angles given. Here are just a few examples.

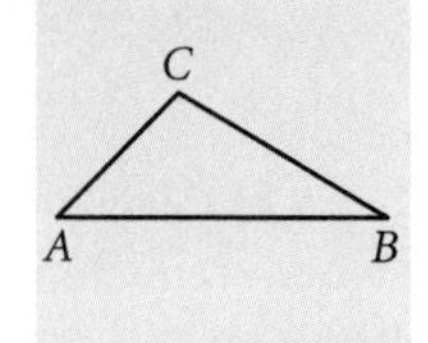

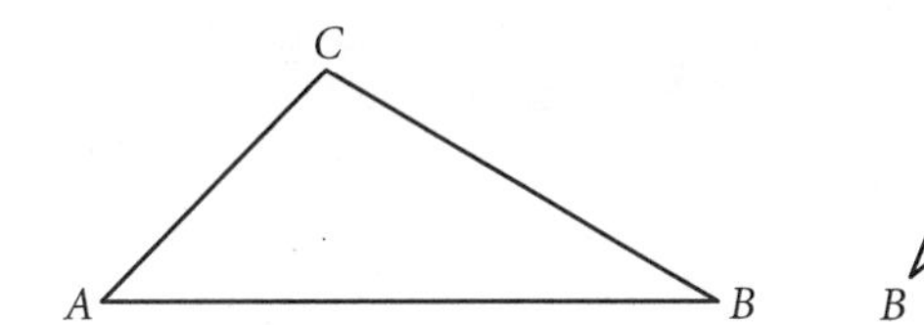

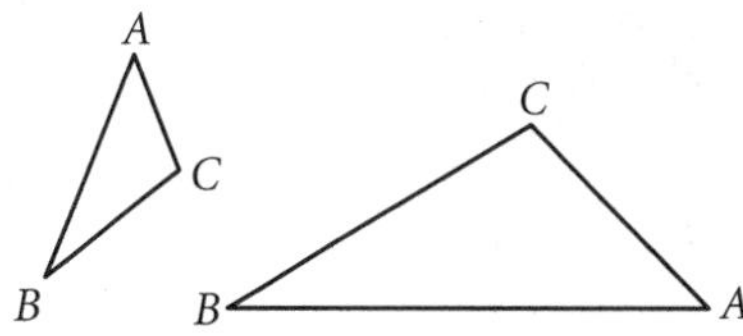

The angles do not determine a particular triangle.

Since a triangle has a total of six parts, there are several combinations of segments and angles that may or may not determine a triangle. Having one or two parts given is not enough to determine a triangle. Is having three parts enough? That answer depends on the combination. In the exercises you will construct triangles and quadrilaterals with various combinations of parts given.

EXERCISES

You will need

Construction tools for Exercises **1–10**

Geometry software for Exercise **11**

Construction In Exercises 1–10, first sketch and label the figure you are going to construct. Second, construct the figure, using either a compass and straightedge, or patty paper and straightedge. Third, describe the steps in your construction in a few sentences.

1. Given:

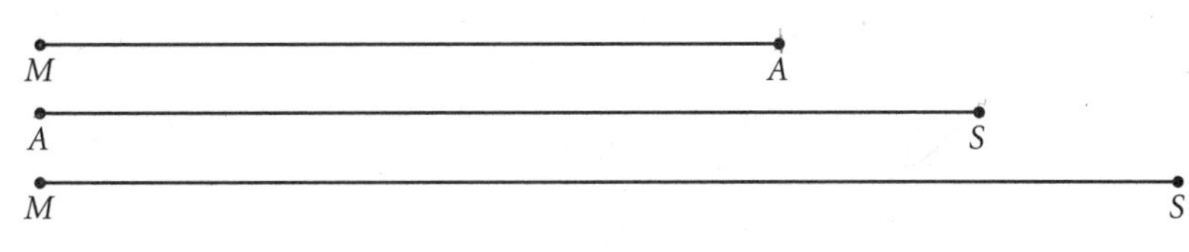

Construct: $\triangle MAS$ ⓗ

2. Given:

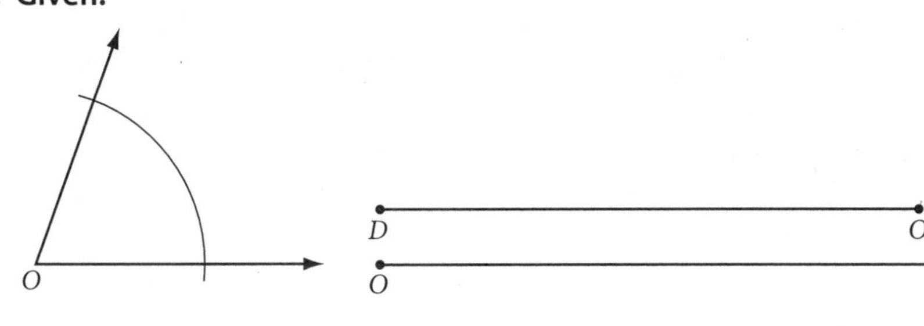

Construct: $\triangle DOT$

3. Given:

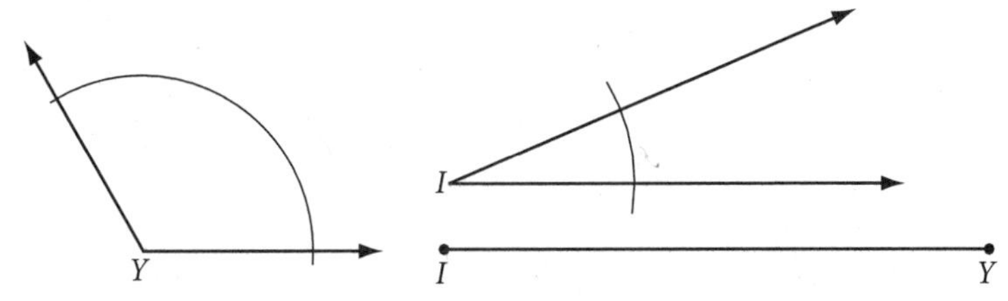

Construct: $\triangle IGY$

4. Given the triangle shown at right, construct another triangle with angles congruent to the given angles but with sides *not* congruent to the given sides. Is there more than one triangle with the same three angles?

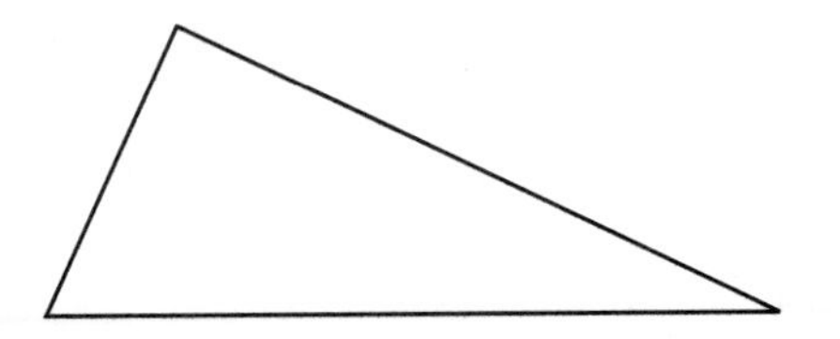

5. The two segments and the angle below do not determine a triangle.

Given:

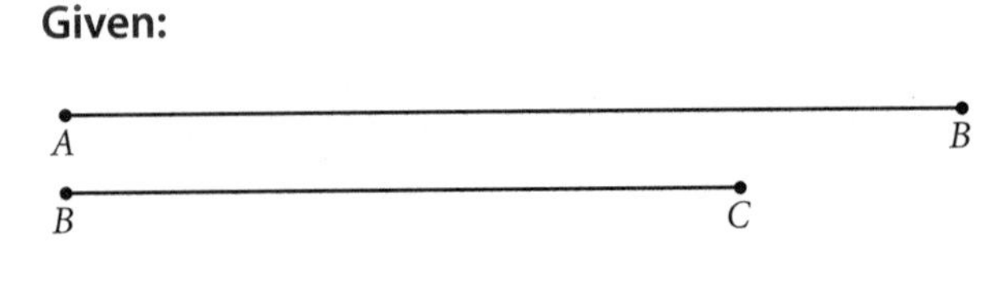

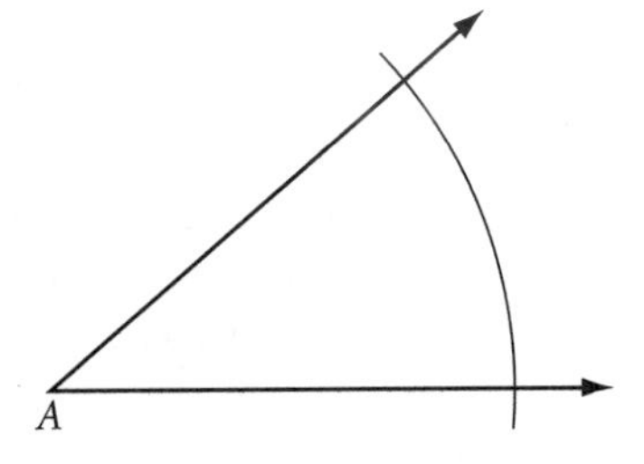

Construct: Two different (noncongruent) triangles named $\triangle ABC$ that have the three given parts ⓗ

6. Given:

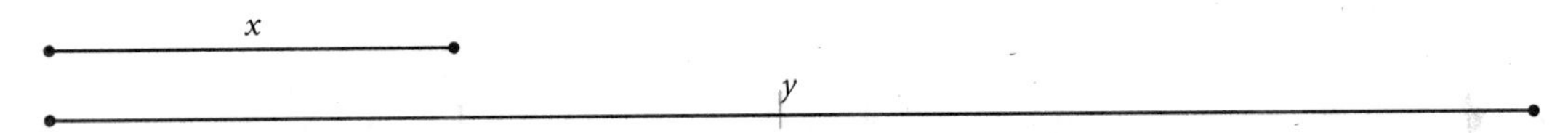

Construct: Isosceles triangle *CAT* with perimeter y and length of the base equal to x ⓗ

7. Construct a kite.

8. Construct a quadrilateral with two pairs of opposite sides of equal length.

9. Construct a quadrilateral with exactly three sides of equal length.

10. Construct a quadrilateral with all four sides of equal length.

11. ***Technology*** Using geometry software, draw a large scalene obtuse triangle ABC with $\angle B$ the obtuse angle. Construct the angle bisector $\overline{BR}$, the median $\overline{BM}$, and the altitude $\overline{BS}$. What is the order of the points on $\overline{AC}$? Drag B. Is the order of points always the same? Write a conjecture.

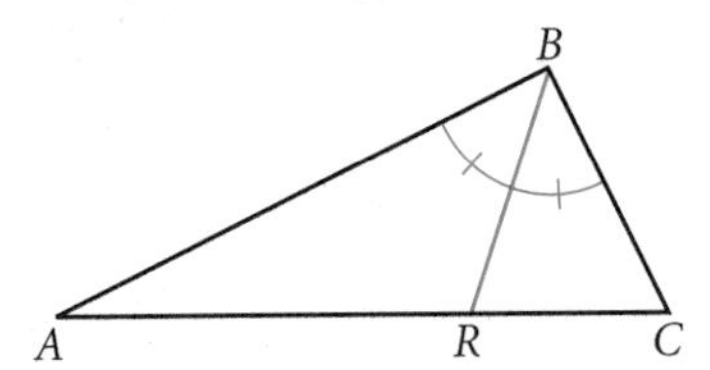

Review

12. Draw each figure and decide how many reflectional and rotational symmetries it has. Copy and complete the table at right.

Figure	Reflectional symmetries	Rotational symmetries
Trapezoid		
Kite		
Parallelogram		
Rhombus		
Rectangle		

13. Draw the new position of $\triangle TEA$ if it is reflected over the dotted line. Label the coordinates of the vertices.

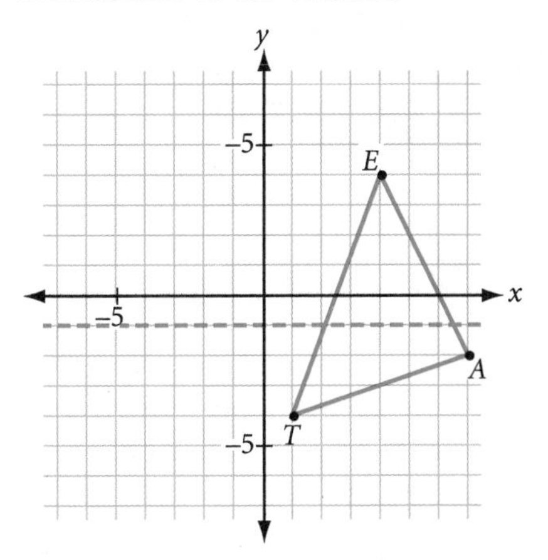

14. Sketch the three-dimensional figure formed by folding the net below into a solid.

15. If a polygon has 500 diagonals from each vertex, how many sides does it have?

IMPROVING YOUR REASONING SKILLS

PUZZLE

Spelling Card Trick

This card trick uses one complete suit (hearts, clubs, spades, or diamonds) from a regular deck of playing cards. How must you arrange the cards so that you can successfully complete the trick? Here is what your audience should see and hear as you perform.

1. As you take the top card and place it at the bottom of the pile, say "A."
2. Then take the second card, place it at the bottom of the pile, and say "C."
3. Take the third card, place it at the bottom, and say "E."
4. You've just spelled *ace.* Now take the fourth card and turn it faceup on the table. The card should be an ace.
5. Continue in this fashion, saying "T," "W," and "O" for the next three cards. Then turn the next card faceup. It should be a 2.
6. Continue spelling *three, four, . . . , jack, queen, king.* Each time you spell a card, the next card turned faceup should be that card.

Exploration

Perspective Drawing

You know from experience that when you look down a long straight road, the parallel edges and the center line seem to meet at a point on the horizon. To show this effect in a drawing, artists use perspective, the technique of portraying solid objects and spatial relationships on a flat surface. Renaissance artists and architects in the fifteenth century developed **perspective,** turning to geometry to make art appear true-to-life.

In a perspective drawing, receding parallel lines (lines that run directly away from the viewer) converge at a **vanishing point** on the **horizon line.** Locate the horizon line, the vanishing point, and converging lines in the perspective study below by Jan Vredeman de Vries.

(Below) Perspective study by Dutch artist Jan Vredeman de Vries (1527–1604)

Activity

Boxes in Space

You will need

- a ruler

In this activity you'll learn to draw a box in perspective.

Perspective drawing is based on the relationships between many parallel and perpendicular lines. The lines that recede to the horizon make you visually think of parallel lines even though they actually intersect at a vanishing point.

First, you'll draw a rectangular solid, or box, in **one-point perspective.** Look at the diagrams below for each step.

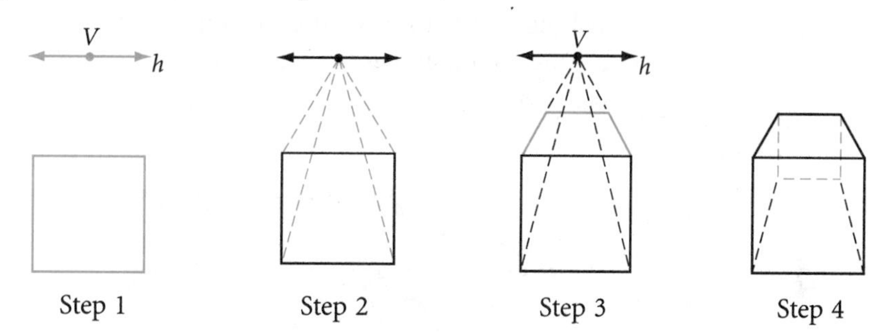

Step 1 Draw a horizon line h and a vanishing point V. Draw the front face of the box with its horizontal edges parallel to h.

Step 2 Connect the corners of the box face to V with dashed lines.

Step 3 Draw the upper rear box edge parallel to h. Its endpoints determine the vertical edges of the back face.

Step 4 Draw the hidden back vertical and horizontal edges with dashed lines. Erase unnecessary lines and dashed segments.

Step 5 Repeat Steps 1–4 several more times, each time placing the first box face in a different position with respect to h and V—above, below, or overlapping h; to the left or right of V or centered on V.

Step 6 Share your drawings in your group. Tell which faces of the box recede and which are parallel to the imaginary window or picture plane that you see through. What is the shape of a receding box face? Think of each drawing as a scene. Where do you seem to be standing to view each box? That is, how is the viewing position affected by placing V to the left or right of the box? Above or below the box?

You can also use perspective to play visual tricks. The Italian architect Francesco Borromini (1599–1667) designed and built a very clever colonnade in the Palazzo Spada. The colonnade is only 12 meters long, but he made it look much longer by designing the sides to get closer and closer to each other and the height of the columns to gradually shrink.

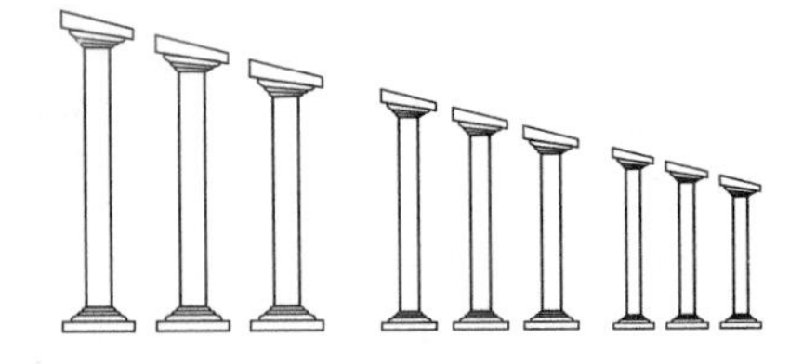

If the front surface of a box is not parallel to the picture plane, then you need two vanishing points to show the two front faces receding from view. This is called **two-point perspective.** Let's look at a rectangular solid with one edge viewed straight on.

Step 7 Draw a horizon line h and select two vanishing points on it, V_1 and V_2. Draw a vertical segment for the nearest box edge.

Step 8 Connect each endpoint of the box edge to V_1 and V_2 with dashed lines.

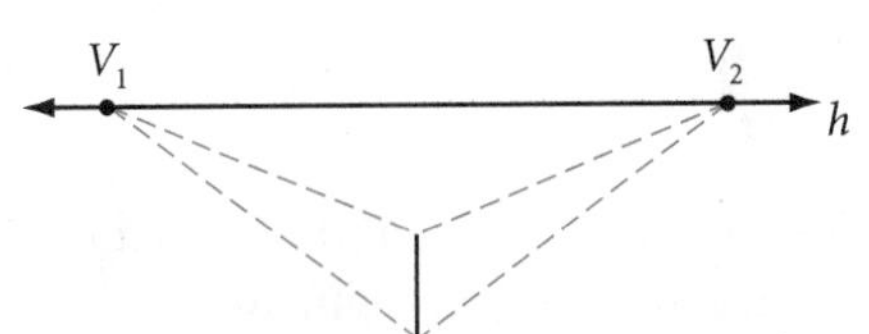

Step 9 Draw two vertical segments within the dashed lines as shown. Connect their endpoints to the endpoints of the front edge along the dashed lines. Now you have determined the position of the hidden back edges that recede from view.

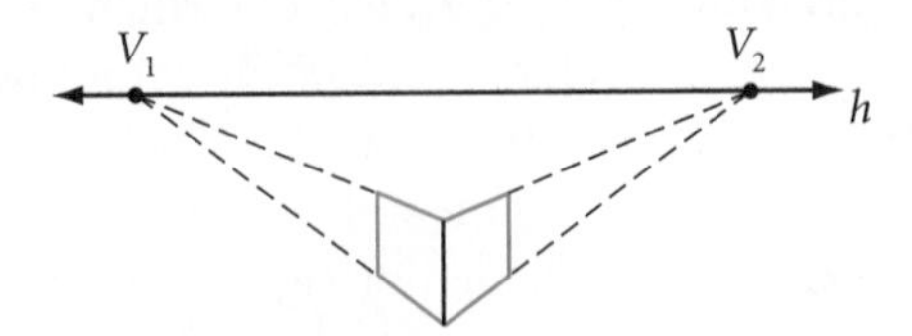

Step 10 Draw the remaining edges along vanishing lines, using dashed lines for hidden edges. Erase unnecessary dashed segments.

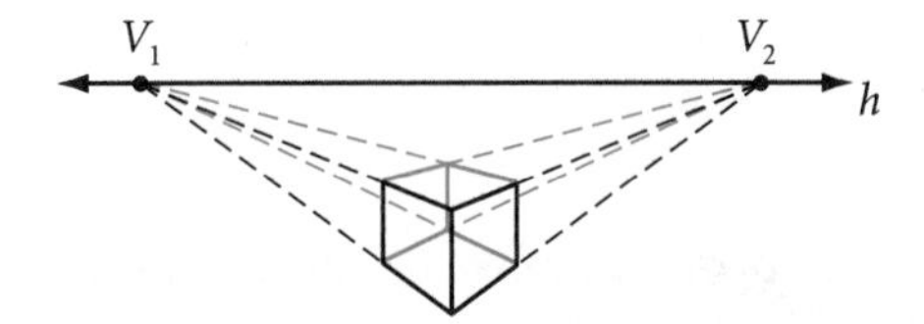

Step 11 Repeat Steps 7–10 several times, each time placing the nearest box edge in a different position with respect to h, V_1, and V_2, and varying the distance between V_1, and V_2. You can also experiment with different-shaped boxes.

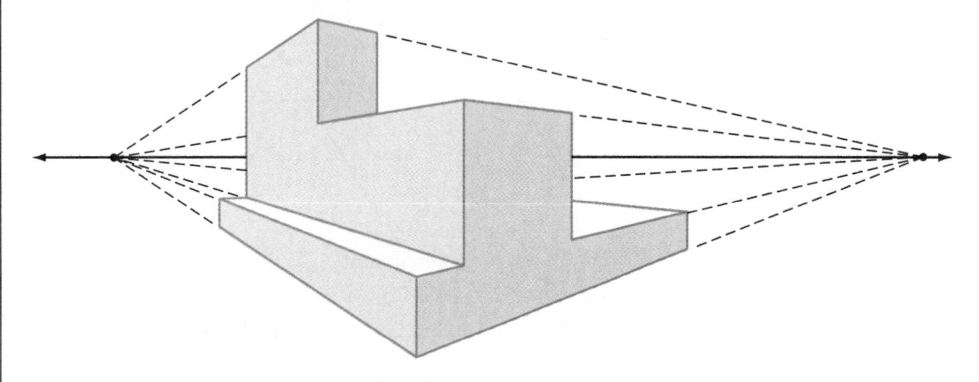

Step 12 Share your drawings in your group. Are any faces of the box parallel to the picture plane? Does each box face have a pair of parallel sides?

Explain how the viewing position is affected by the distance between V_1 and V_2 relative to the size of the box. Must the box be between V_1 and V_2?

Perspective helps in designing the lettering painted on streets. From above, letters appear tall, but from a low angle they appear normal. Tilt the page up to your face. How do the letters look?

LESSON

3.7

Constructing Points of Concurrency

Nothing in life is to be feared, it is only to be understood.

MARIE CURIE

You now can perform a number of constructions in triangles, including angle bisectors, perpendicular bisectors of the sides, medians, and altitudes. In this lesson and the next lesson you will discover special properties of these lines and segments. When three or more lines have a point in common, they are **concurrent.** Segments, rays, and even planes are concurrent if they intersect in a single point.

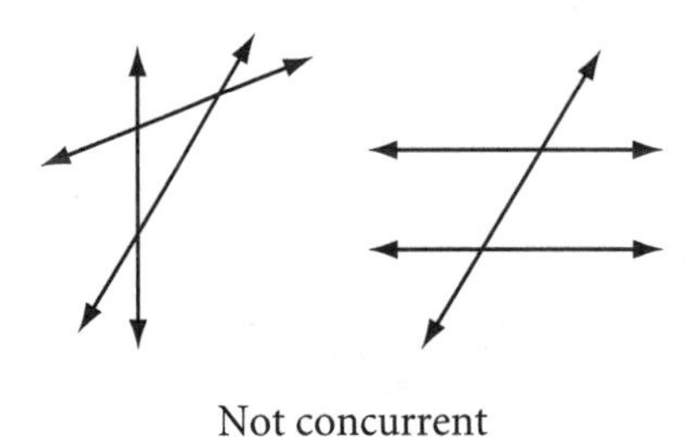

Not concurrent

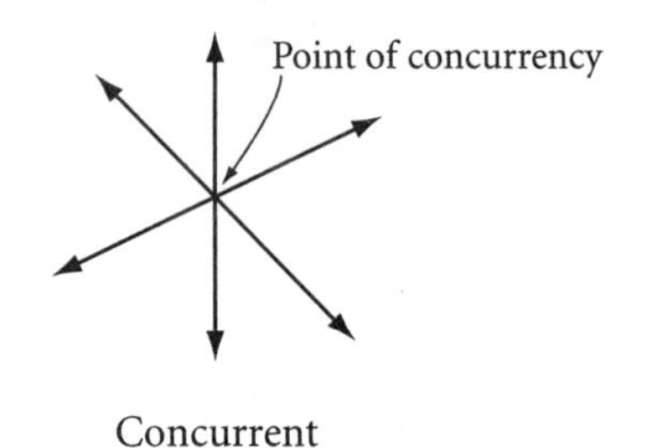

Concurrent

The point of intersection is the **point of concurrency.**

Investigation 1
Concurrence

You will need

- patty paper
- a compass
- a straightedge

In this investigation you will discover that some special lines in a triangle have points of concurrency.

You should investigate each set of lines on an acute triangle, an obtuse triangle, and a right triangle to be sure your conjecture applies to all triangles.

Step 1 Draw a large acute triangle on one patty paper and an obtuse triangle on another. If you're using a compass and a straightedge, draw your triangles on the top and bottom halves of a piece of paper.

Step 2 Construct the three angle bisectors for each triangle. Are they concurrent?

Compare your results with the results of others. State your observations as a conjecture.

Angle Bisector Concurrency Conjecture C-9

The three angle bisectors of a triangle ?.

Step 3 Construct the perpendicular bisector for each side of the triangle and complete the conjecture.

Perpendicular Bisector Concurrency Conjecture

C-10

The three perpendicular bisectors of a triangle _?_.

Step 4 Construct the lines containing the altitudes of your triangle and complete the conjecture.

Altitude Concurrency Conjecture

C-11

The three altitudes (or the lines containing the altitudes) of a triangle _?_.

Step 5 For what kind of triangle will the points of concurrency be the same point?

The point of concurrency for the three angle bisectors is the **incenter.** The point of concurrency for the perpendicular bisectors is the **circumcenter.** The point of concurrency for the three altitudes is called the **orthocenter.** You will investigate a triangle's medians in the next lesson.

Rudolf Bauer (1889–1953) titled this painting *Rounds and Triangles.*

Investigation 2
Incenter and Circumcenter

You will need

- construction tools

In this investigation you will discover special properties of the incenter and the circumcenter.

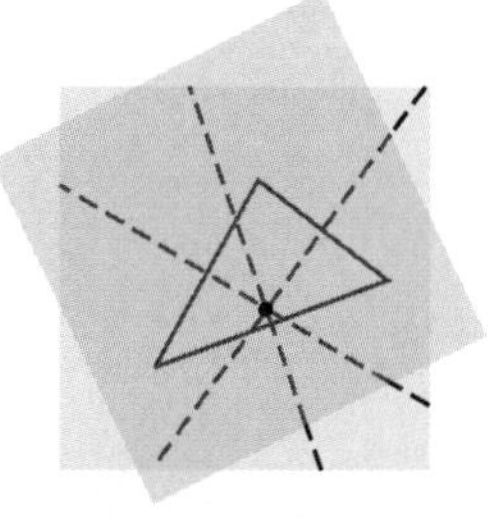

Step 1 Measure and compare the distances from the circumcenter to each of the three vertices. Are they the same? Compare the distances from the circumcenter to each of the three sides. Are they the same? State your observations as your next conjecture.

Circumcenter Conjecture

C-12

The circumcenter of a triangle _?_.

Step 2 Measure and compare the distances from the incenter to each of the three sides. Are they the same? State your observations as your next conjecture.

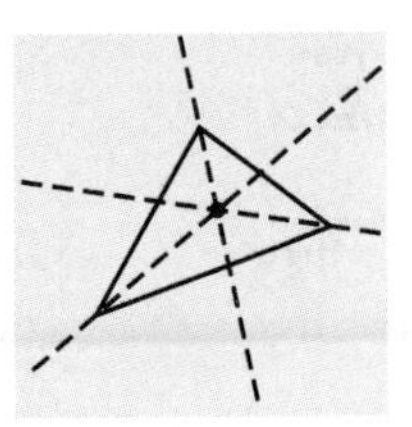

Incenter Conjecture C-13

The incenter of a triangle ___?___.

You just discovered a very useful property of the circumcenter and a very useful property of the incenter. You will see some applications of these properties in the exercises. With earlier conjectures and logical reasoning, you can explain why your conjectures are true. Let's look at a paragraph proof of the Circumcenter Conjecture.

Paragraph Proof of the Circumcenter Conjecture

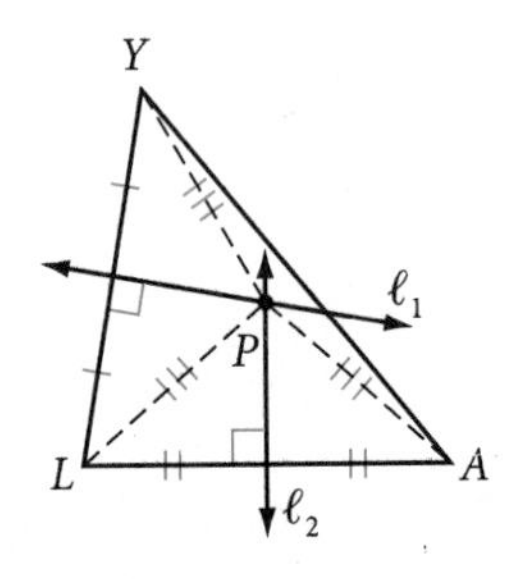

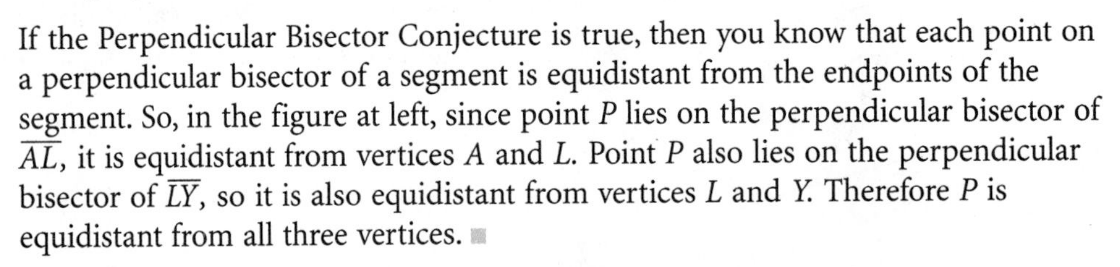

If the Perpendicular Bisector Conjecture is true, then you know that each point on a perpendicular bisector of a segment is equidistant from the endpoints of the segment. So, in the figure at left, since point P lies on the perpendicular bisector of $\overline{AL}$, it is equidistant from vertices A and L. Point P also lies on the perpendicular bisector of $\overline{LY}$, so it is also equidistant from vertices L and Y. Therefore P is equidistant from all three vertices. ■

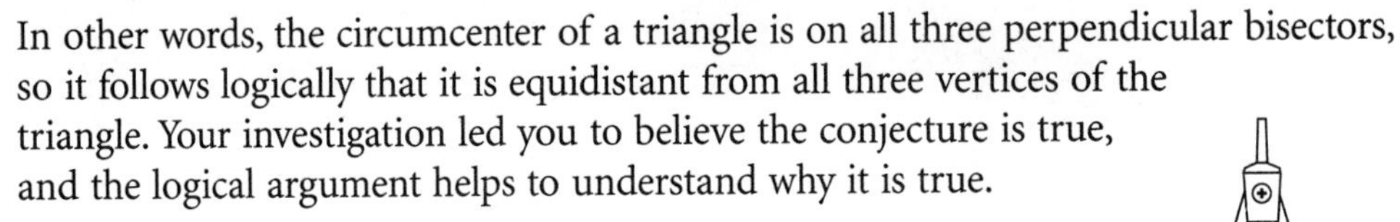

In other words, the circumcenter of a triangle is on all three perpendicular bisectors, so it follows logically that it is equidistant from all three vertices of the triangle. Your investigation led you to believe the conjecture is true, and the logical argument helps to understand why it is true.

You can use your compass to construct a circle that passes through the three vertices of the triangle with the circumcenter as the center of the circle.

So, you can also think of the circumcenter as the center of a circle that passes through the three vertices of a triangle.

You can make a similar logical argument for the Incenter Conjecture.

Paragraph Proof of the Incenter Conjecture

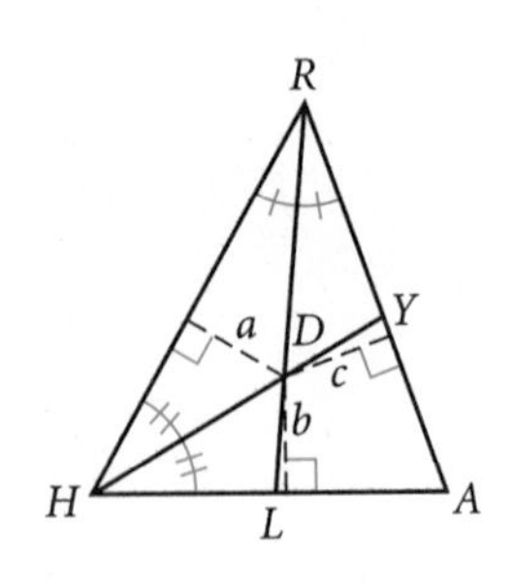

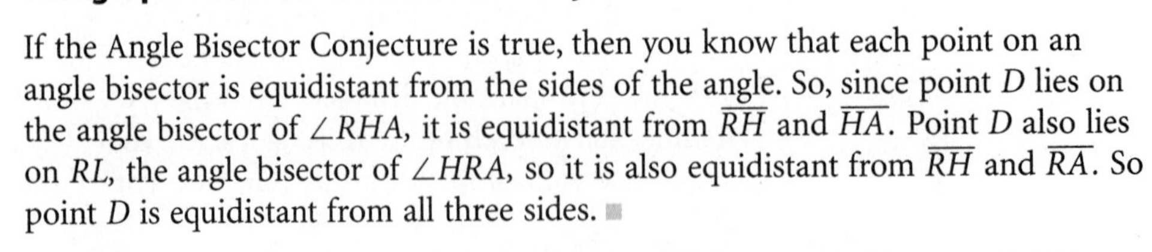

If the Angle Bisector Conjecture is true, then you know that each point on an angle bisector is equidistant from the sides of the angle. So, since point D lies on the angle bisector of $\angle RHA$, it is equidistant from $\overline{RH}$ and $\overline{HA}$. Point D also lies on RL, the angle bisector of $\angle HRA$, so it is also equidistant from $\overline{RH}$ and $\overline{RA}$. So point D is equidistant from all three sides. ■

In other words, the incenter of a triangle is on all three angle bisectors. It follows logically that the incenter is equidistant from all three sides. Again your investigation may have convinced you that this was true, but the logical argument explains why it is true.

You can use your compass to construct a circle that is tangent to the three sides with the incenter as the center of a circle. To construct the circle you need the radius, the shortest distance from the incenter to each side. To get the radius, you construct the perpendicular from the incenter to one of the sides.

The incenter is the center of a circle that touches each side of the triangle. Here are a few vocabulary terms that help describe these geometric situations.

A circle is **circumscribed** about a polygon if and only if it passes through each vertex of the polygon. (The polygon is inscribed in the circle.)

A circle is **inscribed** in a polygon if and only if it touches each side of the polygon at exactly one point. (The polygon is circumscribed about the circle.)

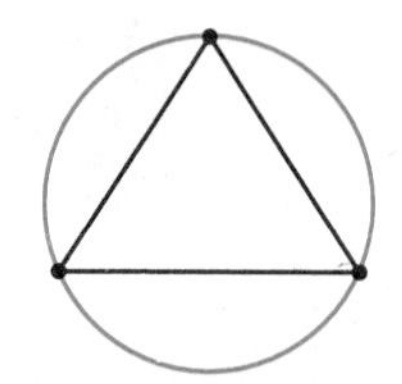

Circumscribed circle
(inscribed triangle)

Inscribed circle
(circumscribed triangle)

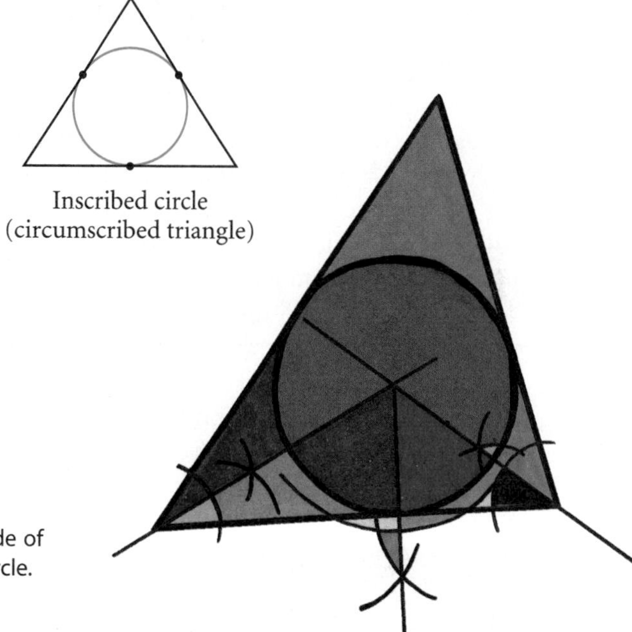

This geometric art by geometry student Ryan Garvin shows the construction of the incenter, its perpendicular distance to one side of the triangle, and the inscribed circle.

EXERCISES

You will need

Construction tools for Exercises **6, 7, 12–15,** and **17**

Geometry software for Exercises **10, 11,** and **18**

For Exercises 1–4, make a sketch and explain how to find the answer.

1. The first-aid center of Mt. Thermopolis State Park needs to be at a point that is equidistant from three bike paths that intersect to form a triangle. Locate this point so that in an emergency medical personnel will be able to get to any one of the paths by the shortest route possible. Which point of concurrency is it?

2. Rosita wants to install a circular sink in her new triangular countertop. She wants to choose the largest sink that will fit. Which point of concurrency must she locate? Explain.

3. Julian Chive wishes to center a butcher-block table at a location equidistant from the refrigerator, stove, and sink. Which point of concurrency does Julian need to locate?

Art

CONNECTION

The designer of stained glass arranges pieces of painted glass to form the elaborate mosaics that you might see in Gothic cathedrals or on Tiffany lampshades. He first organizes the glass pieces by shape and color according to the design. He mounts these pieces into a metal framework that will hold the design. With precision, the designer cuts every glass piece so that it fits against the next one with a strip of cast lead. The result is a pleasing combination of colors and shapes that form a luminous design when viewed against light.

4. A stained-glass artist wishes to circumscribe a circle about a triangle in her latest abstract design. Which point of concurrency does she need to locate?

5. One event at this year's Battle of the Classes will be a pie-eating contest between the sophomores, juniors, and seniors. Five members of each class will be positioned on the football field at the points indicated below. At the whistle, one student from each class will run to the pie table, eat exactly one pie, and run back to his or her group. The next student will then repeat the process. The first class to eat five pies and return to home base will be the winner of the pie-eating contest. Where should the pie table be located so that it will be a fair contest? Describe how the contest planners should find that point.

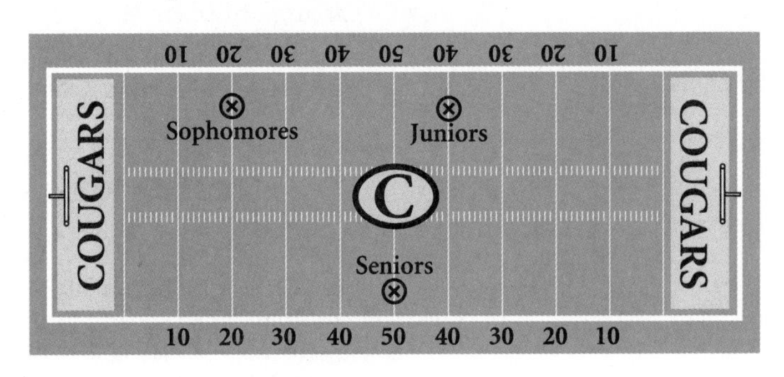

6. ***Construction*** Draw a large triangle. Construct a circle inscribed in the triangle. ⓗ

7. ***Construction*** Draw a triangle. Construct a circle circumscribed about the triangle. ⓗ

8. Is the inscribed circle the greatest circle to fit within a given triangle? Explain. If you think not, give a counterexample. ⓗ

9. Does the circumscribed circle create the smallest circular region that contains a given triangle? Explain. If you think not, give a counterexample. ⓗ

10. ***Technology*** Notice that the circumcenter is in the interior of acute triangles and on the exterior of obtuse triangles. Use geometry software to construct a right triangle. Where is the circumcenter of a right triangle? (This problem can also be done by drawing several right triangles on graph paper.)

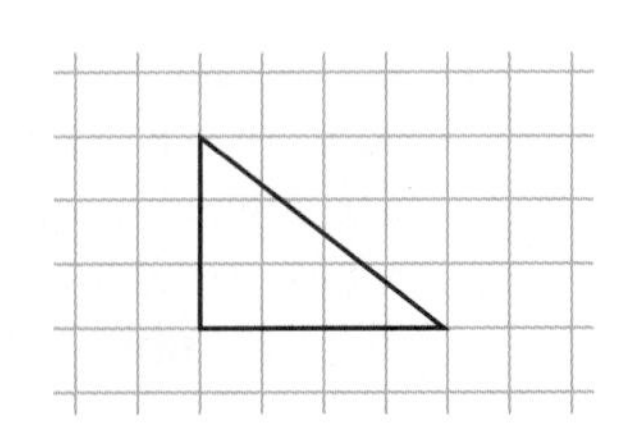

11. ***Technology*** Notice that the orthocenter is in the interior of acute triangles and on the exterior of obtuse triangles. Use geometry software to construct a right triangle. Where is the orthocenter of a right triangle? (This problem can also be done by drawing several right triangles on graph paper.)

Review

Construction Use the segments and angle at right to construct each figure in Exercises 12–15.

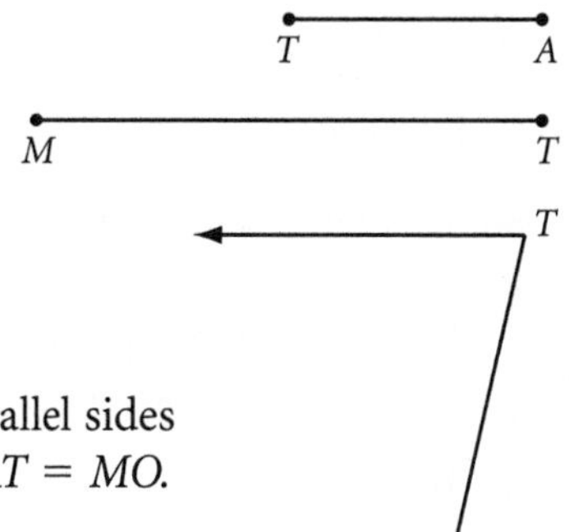

12. ***Mini-Investigation*** Construct $\triangle MAT$. Construct H the midpoint of $\overline{MT}$ and S the midpoint of $\overline{AT}$. Construct the midsegment $\overline{HS}$. Compare the lengths of $\overline{HS}$ and $\overline{MA}$. Notice anything special?

13. ***Mini-Investigation*** An isosceles trapezoid is a trapezoid with the nonparallel sides congruent. Construct isosceles trapezoid $MOAT$ with $\overline{MT} \parallel \overline{OA}$ and $AT = MO$. Use patty paper to compare $\angle T$ and $\angle M$. Notice anything special?

14. ***Mini-Investigation*** Construct a circle with diameter MT. Construct chord $\overline{TA}$. Construct chord $\overline{MA}$ to form $\triangle MTA$. What is the measure of $\angle A$? Notice anything special?

15. ***Mini-Investigation*** Construct a rhombus with TA as the length of a side and $\angle T$ as one of the acute angles. Construct the two diagonals. Notice anything special?

16. Sketch the locus of points on the coordinate plane in which the sum of the x-coordinate and the y-coordinate is 9. ⓗ

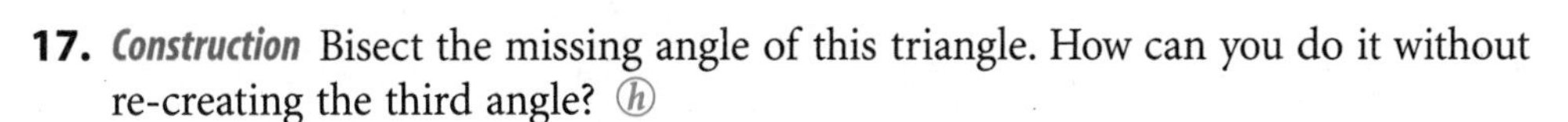

17. ***Construction*** Bisect the missing angle of this triangle. How can you do it without re-creating the third angle? ⓗ

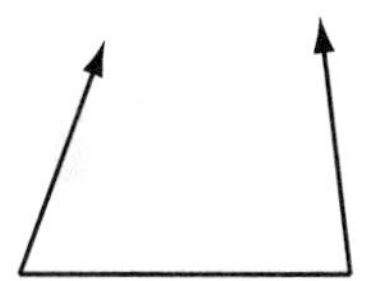

18. ***Technology*** Is it possible for the midpoints of the three altitudes of a triangle to be collinear? Investigate by using geometry software. Write a paragraph describing your findings.

19. Sketch the section formed when the plane slices the cube as shown.

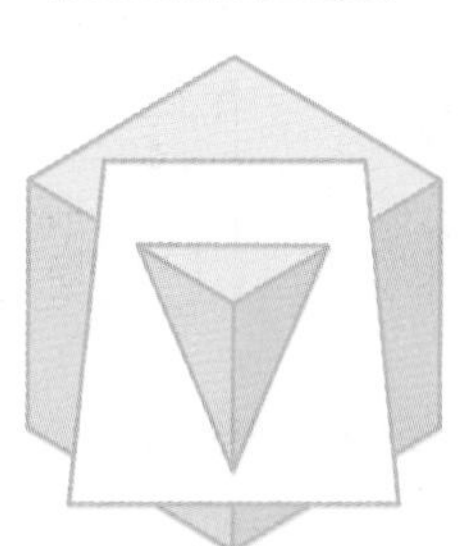

For Exercises 20–24, match each geometric construction with one of the figures below.

20. Construction of a perpendicular bisector

21. Construction of an angle bisector

22. Construction of a perpendicular through a point on a line

23. Construction of a line parallel to a given line through a given point not on the line

24. Construction of an equilateral triangle

A.

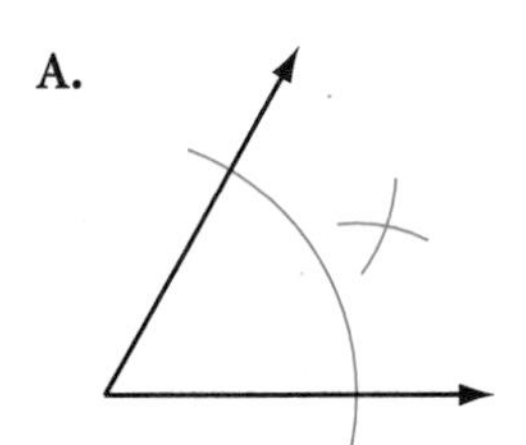

B.

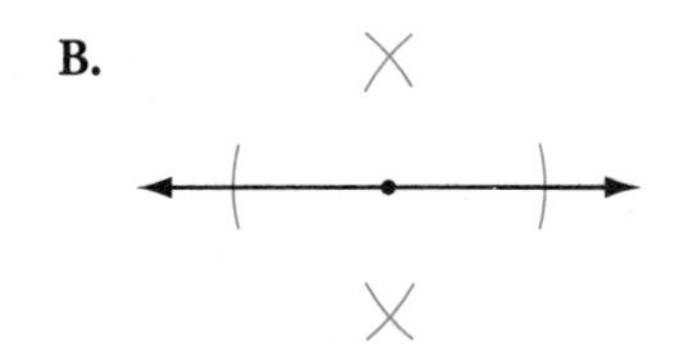

C.

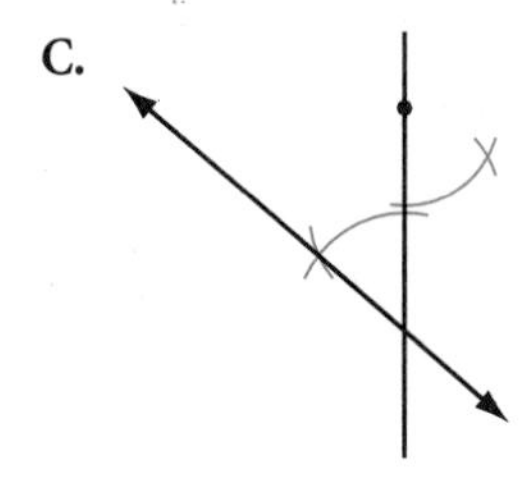

D.

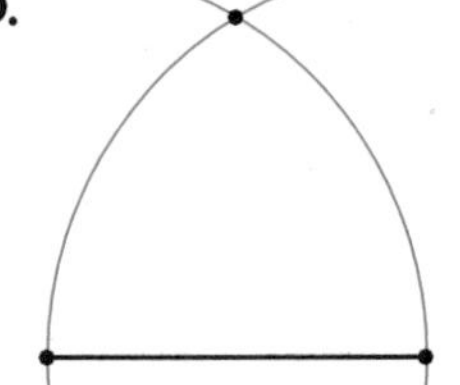

E.

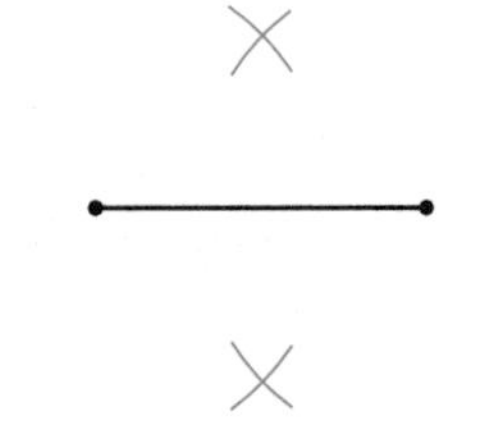

IMPROVING YOUR VISUAL THINKING SKILLS

The Puzzle Lock

This mysterious pattern is a lock that must be solved like a puzzle. Here are the rules:

- You must make eight moves in the proper sequence.
- To make each move (except the last), you place a gold coin onto an empty circle, then slide it along a diagonal to another empty circle.
- You must place the first coin onto circle 1, then slide it to either circle 4 or circle 6.
- You must place the last coin onto circle 5.
- You do not slide the last coin.

Solve the puzzle. Copy and complete the table to show your solution.

Coin Movements

Coin	Placed on		Slid to
First	1	→	
Second		→	
Third		→	
Fourth		→	
Fifth		→	
Sixth		→	
Seventh		→	
Eighth	5		

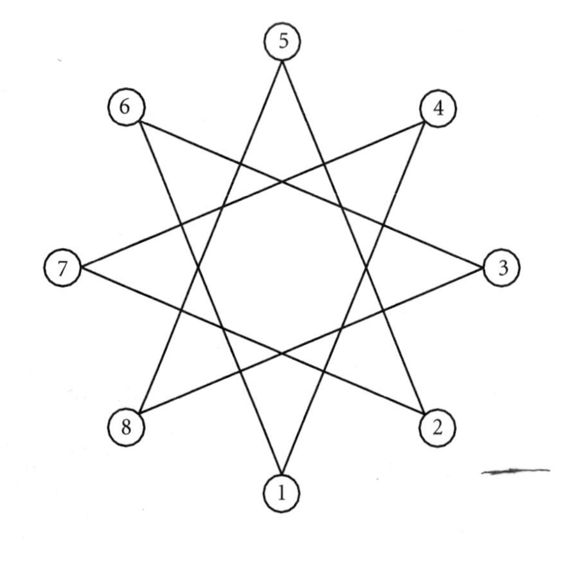

LESSON

3.8

The Centroid

The universe may be as great as they say, but it wouldn't be missed if it didn't exist.

PIET HEIN

In the previous lesson you discovered that the three angle bisectors are concurrent, the three perpendicular bisectors of the sides are concurrent, and the three altitudes in a triangle are concurrent. You also discovered the properties of the incenter and the circumcenter. In this lesson you will investigate the medians of a triangle.

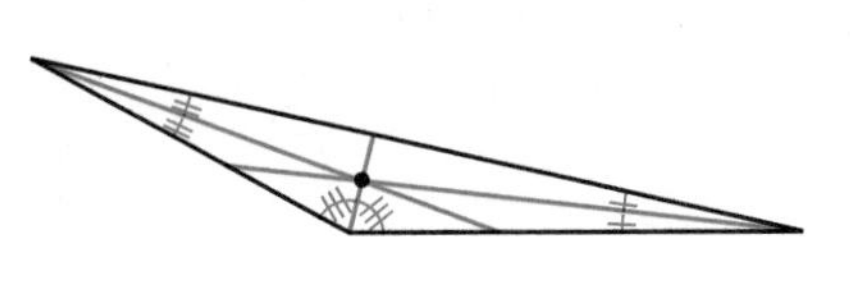

Three angle bisectors (incenter)

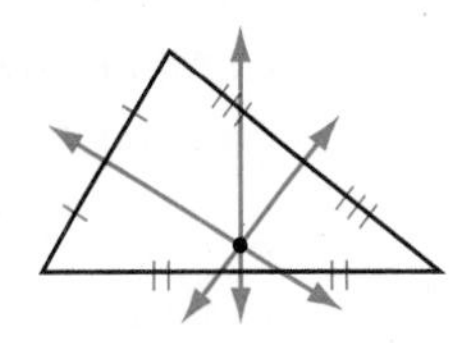

Three perpendicular bisectors (circumcenter)

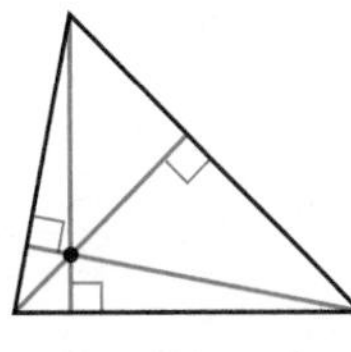

Three altitudes (orthocenter)

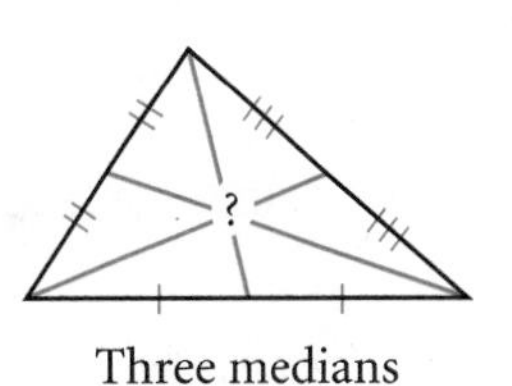

Three medians $\underline{?}$

Investigation 1
Are Medians Concurrent?

You will need

- patty paper
- a straightedge

Each person in your group should draw a different triangle for this investigation. Make sure you have at least one acute triangle, one obtuse triangle, and one right triangle in your group.

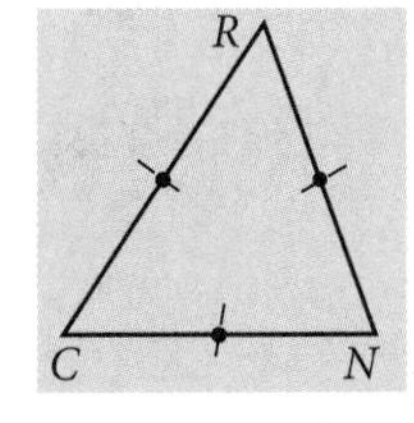

Step 1 On a sheet of patty paper, draw as large a scalene triangle as possible and label it *CNR*, as shown at right. Locate the midpoints of the three sides. Construct the medians and complete the conjecture.

Median Concurrency Conjecture C-14

The three medians of a triangle $\underline{?}$.

The point of concurrency of the three medians is the **centroid.**

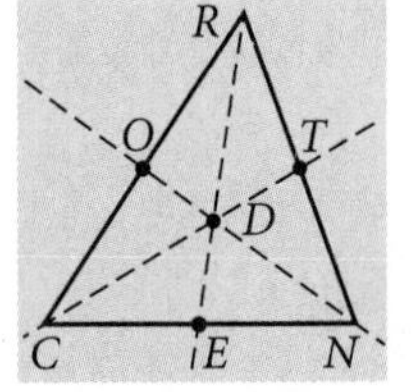

Step 2 Label the three medians $\overline{CT}$, $\overline{NO}$, and $\overline{RE}$. Label the centroid *D*.

Step 3 Use your compass or another sheet of patty paper to investigate whether there is anything special about the centroid. Is the centroid equidistant from the three vertices? From the three sides? Is the centroid the midpoint of each median?

Step 4 The centroid divides a median into two segments. Use your patty paper or compass to compare the length of the longer segment to the length of the shorter segment and find the ratio.

Step 5 Find the ratios of the lengths of the segment parts for the other two medians. Do you get the same ratio for each median?

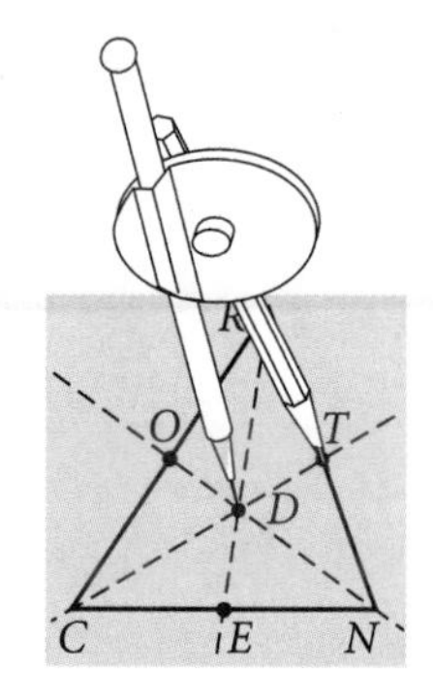

Compare your results with the results of others. State your discovery as a conjecture, and add it to your conjecture list.

Centroid Conjecture

C-15

The centroid of a triangle divides each median into two parts so that the distance from the centroid to the vertex is _?_ the distance from the centroid to the midpoint of the opposite side.

In earlier lessons you discovered that the midpoint of a segment is the balance point or center of gravity. You also saw that when a set of segments is arranged into a triangle, the line through the midpoints of these segments can act as a line of balance for the triangle. Can you then balance a triangle on a median? Let's take a look.

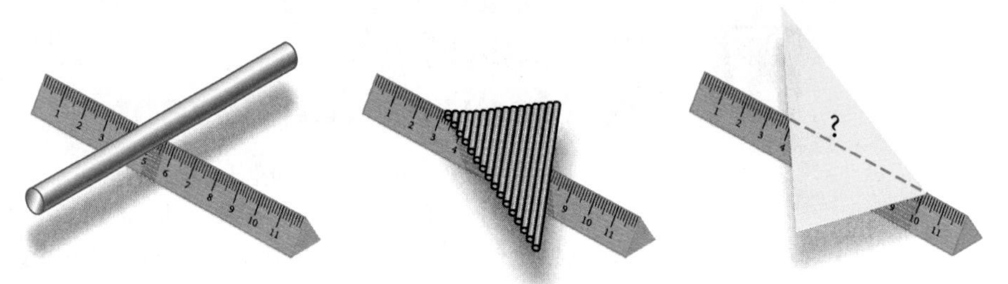

Investigation 2
Balancing Act

You will need

- cardboard

Use your patty paper from Investigation 1 for this investigation.

Step 1 Place your patty paper from the previous investigation on a piece of mat board or cardboard. With a sharp pencil tip or compass tip, mark the three vertices, the three midpoints, and the centroid on the board.

Step 2 Draw in the triangle and medians on the cardboard. Cut out the cardboard triangle.

Step 3 Try balancing the triangle on each of the three medians by placing the median on the edge of a ruler. If you are successful, what does that imply about the areas of the two triangles formed by one median?

Step 4 Is there a single point where you can balance the triangle?

If you have found the balancing point for the triangle, you have found its **center of gravity.** State your discovery as a conjecture, and add it to your conjecture list.

Center of Gravity Conjecture C-16

The ? of a triangle is the center of gravity of the triangular region.

The triangle balances on each median and the centroid is on each median, so the triangle balances on the centroid. As long as the weight of the cardboard is distributed evenly throughout the triangle, you can balance any triangle at its centroid. For this reason, the centroid is a very useful point of concurrency, especially in physics.

You have discovered special properties of three of the four points of concurrency—the incenter, the circumcenter, and the centroid. The incenter is the center of an inscribed circle, the circumcenter is the center of a circumscribed circle, and the centroid is the center of gravity.

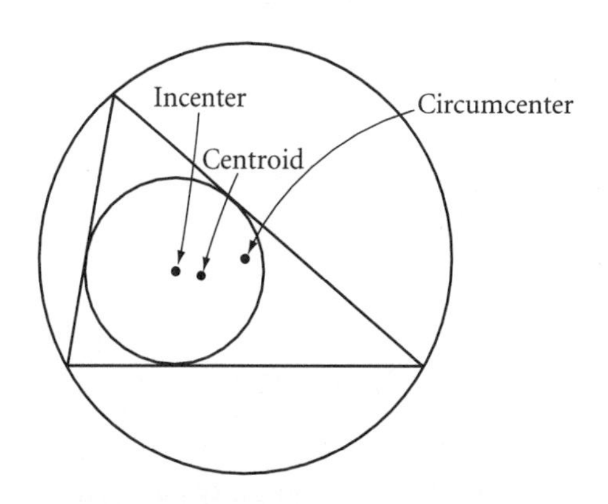

You can learn more about the orthocenter in the Project Is There More to the Orthocenter?

Science CONNECTION

In physics, the center of gravity of an object is an imaginary point where the total weight is concentrated. The center of gravity of a tennis ball, for example, would be in the hollow part, not in the actual material of the ball. The idea is useful in designing structures as complicated as bridges or as simple as furniture. Where is the center of gravity of the human body?

EXERCISES

You will need

Construction tools for Exercises **5** and **6**

Geometry software for Exercise **7**

1. Birdy McFly is designing a large triangular hang glider. She needs to locate the center of gravity for her glider. Which point does she need to locate? Birdy wishes to decorate her glider with the largest possible circle within her large triangular hang glider. Which point of concurrency does she need to locate?

In Exercises 2–4, use your new conjectures to find each length.

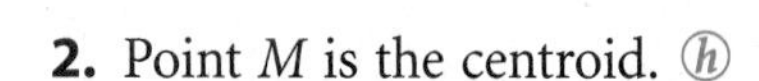

2. Point M is the centroid. ⓗ

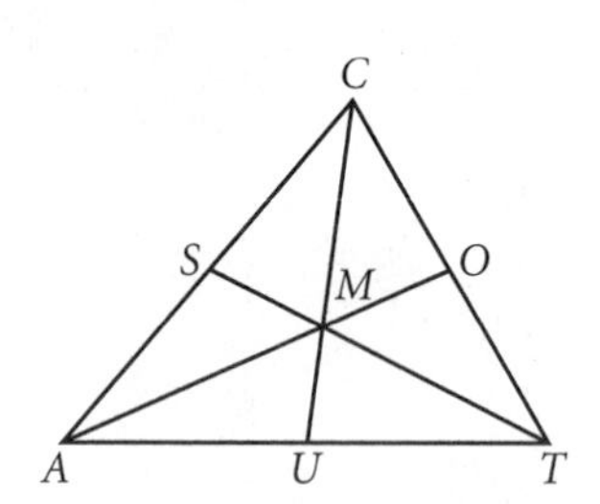

$CM = 16$

$MO = 10$

$TS = 21$

$AM = \underline{?}$

$SM = \underline{?}$

$TM = \underline{?}$

$UM = \underline{?}$

3. Point G is the centroid.

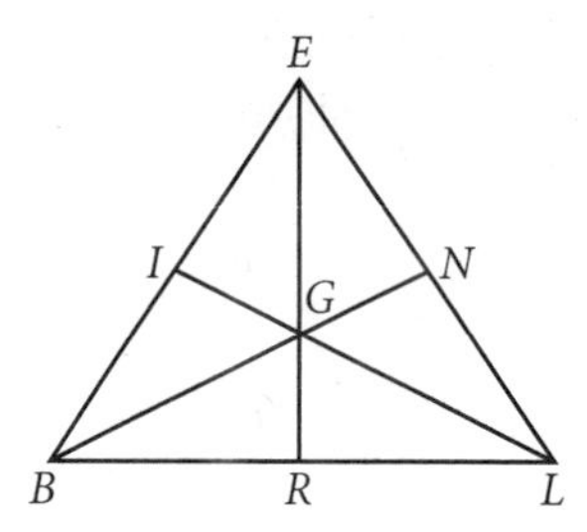

GI = GR = GN

$ER = 36$

$BG = \underline{?}$

$IG = \underline{?}$

4. Point Z is the centroid.

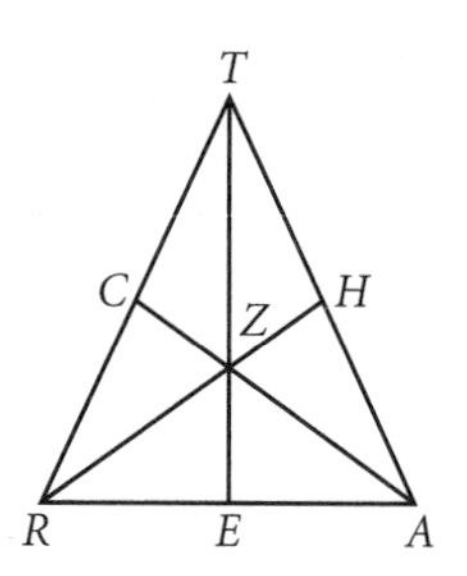

$CZ = 14$

$TZ = 30$

$RZ = AZ$

$RH = \underline{?}$

$TE = \underline{?}$

5. ***Construction*** Construct an equilateral triangle, then construct angle bisectors from two vertices, medians from two vertices, and altitudes from two vertices. What can you conclude?

6. ***Construction*** On patty paper, draw a large isosceles triangle with an acute vertex angle that measures less than 40°. Copy it onto three other pieces of patty paper. Construct the centroid on one patty paper, the incenter on a second, the circumcenter on a third, and the orthocenter on a fourth. Record the results of all four pieces of patty paper on one piece of patty paper. What do you notice about the four points of concurrency? What is the order of the four points of concurrency from the vertex to the opposite side in an acute isosceles triangle?

7. ***Technology*** Use geometry software to construct a large isosceles acute triangle. Construct the four points of concurrency. Hide all constructions except for the points of concurrency. Label them. Drag until it has an obtuse vertex angle. Now what is the order of the four points of concurrency from the vertex angle to the opposite side? When did the order change? Do the four points ever become one?

8. Where do you think the center of gravity is located on a square? A rectangle? A rhombus? In each case the center of gravity is not that difficult to find, but what about an ordinary quadrilateral? Experiment to discover a method for finding the center of gravity for a quadrilateral by geometric construction. Test your method on a large cardboard quadrilateral. ⓗ

Review

9. Sally Solar is the director of Lunar Planning for Galileo Station on the moon. She has been asked to locate the new food production facility so that it is equidistant from the three main lunar housing developments. Which point of concurrency does she need to locate?

10. Construct circle *O*. Place an arbitrary point *P* within the circle. Construct the longest chord passing through *P*. Construct the shortest chord passing through *P*. How are they related?

11. A billiard ball is hit so that it travels a distance equal to *AB* but bounces off the cushion at point *C*. Copy the figure, and sketch where the ball will rest.

12. **APPLICATION** In alkyne molecules all the bonds are single bonds except one triple bond between two carbon atoms. The first three alkynes are modeled below. The dash (–) between letters represents single bonds. The triple dash (≡) between letters represents a triple bond.

```
                          H                 H         H
                          |                 |         |
H − C ≡ C − H   H − C ≡ C − C − H   H − C − C ≡ C − C − H
                          |                 |         |
                          H                 H         H
```

Ethyne (C_2H_2)   Propyne (C_3H_4)   Butyne (C_4H_6)

Sketch the alkyne with eight carbons in the chain. What is the general rule for alkynes ($C_nH_?$)? In other words, if there are *n* carbon atoms (C), how many hydrogen atoms (H) are in the alkyne?

13. When plane figure A is rotated about the line it produces the solid figure B. What is the plane figure that produces the solid figure D?

14. Copy the diagram below. Use your Vertical Angles Conjecture and Parallel Lines Conjecture to calculate each lettered angle measure.

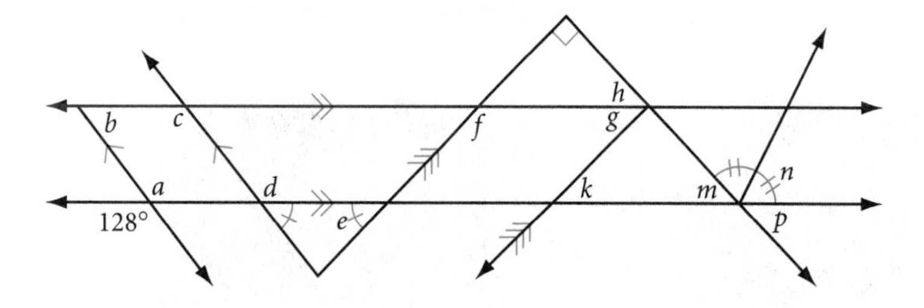

15. A brother and a sister have inherited a large triangular plot of land. The will states that the property is to be divided along the altitude from the northernmost point of the property. However, the property is covered with quicksand at the northern vertex. The will states that the heir who figures out how to draw the altitude without using the northern vertex point gets to choose his or her parcel first. How can the heirs construct the altitude? Is this a fair way to divide the land? Why or why not? ⓗ

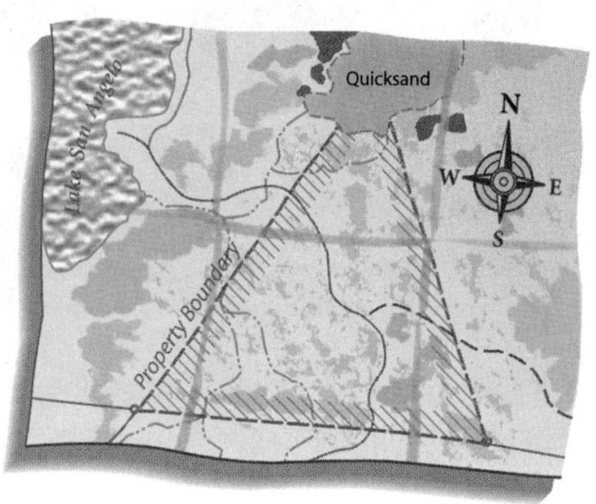

16. At the college dorm open house, each of the 20 dorm members brings two guests (usually their parents). How many greetings are possible if you do not count dorm members greeting their own guests? ⓗ

IMPROVING YOUR REASONING SKILLS

The Dealer's Dilemma

PUZZLE

In the game of bridge, the dealer deals 52 cards in a clockwise direction among four players. You are playing a game in which you are the dealer. You deal the cards, starting with the player on your left. However, in the middle of dealing you stop to answer the phone. When you return, no one can remember where the last card was dealt. (And, of course, no cards have been touched.) Without counting the number of cards in anyone's hand or the number of cards yet to be dealt, how can you rapidly finish dealing, giving each player exactly the same cards she or he would have received if you hadn't been interrupted?

Exploration

The Euler Line

In the previous lessons you discovered the four points of concurrency: circumcenter, incenter, orthocenter, and centroid. In this activity you will discover how these points relate to a special line, the Euler line.

The Euler line is named after the Swiss mathematician Leonhard Euler (1707–1783), who proved that three points of concurrency are collinear.

Activity

Three Out of Four

You will need

- patty paper

You are going to look for a relationship among the points of concurrency.

Step 1 Draw a scalene triangle and have each person in your group trace the same triangle on a separate piece of patty paper.

Step 2 Have each group member construct with patty paper a different point of the four points of concurrency for the triangle.

Step 3 Record the group's results by tracing and labeling all four points of concurrency on one of the four pieces of patty paper. What do you notice? Compare your group results with the results of other groups near you. State your discovery as a conjecture.

Euler Line Conjecture

The __?__, __?__, and __?__ are the three points of concurrency that always lie on a line.

The three special points that lie on the Euler line determine a segment called the **Euler segment.** The point of concurrency between the two endpoints of the Euler segment divides the segment into two smaller segments whose lengths have an exact ratio.

Step 4 With a compass or patty paper, compare the lengths of the two parts of the Euler segment. What is the ratio? Compare your group's results with the results of other groups and state your conjecture.

Euler Segment Conjecture

The _?_ divides the Euler segment into two parts so that the smaller part is _?_ the larger part.

Step 5 Use your conjectures to solve this problem.

$\overline{AC}$ is an Euler segment. $AC = 24$ m.
$AB = \underline{?}$ $BC = \underline{?}$

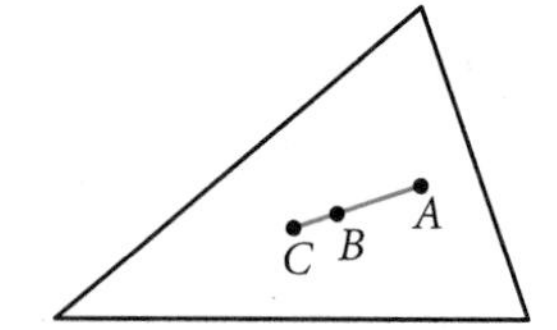

project

IS THERE MORE TO THE ORTHOCENTER?

At this point you may still wonder what's special about the orthocenter. It does lie on the Euler line. Is there anything else surprising or special about it?

Use geometry software to investigate the orthocenter. Draw a triangle *ABC* and construct its orthocenter *O*.

Drag a vertex of the triangle around, and observe the behavior of the orthocenter. Where does the orthocenter lie in an acute triangle? An obtuse triangle? A right triangle?

Drag the orthocenter. Describe how this affects the triangle.

Hide the altitudes. Draw segments from each vertex to the orthocenter, as shown, forming three triangles within the original triangle. Now find the orthocenter of each of the three triangles.

What happened? What does this mean?

Experiment dragging different points, and observe the relationships among the four orthocenters. Drag the orthocenter toward each vertex. What happens?

Write a paragraph about your findings, concluding with a conjecture about the orthocenter. Share your findings with your group members or classmates.

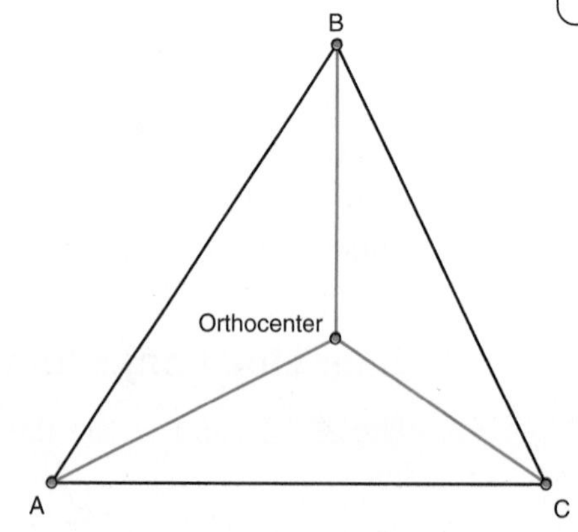

The Geometer's Sketchpad was used to create this diagram and to hide the unnecessary lines. Using Sketchpad, you can quickly construct triangles and their points of concurrency. Once you make a conjecture, you can drag to change the shape of the triangle to see whether your conjecture is true.

CHAPTER 3 REVIEW

In Chapter 1, you defined many terms that help establish the building blocks of geometry. In Chapter 2, you learned and practiced inductive reasoning skills. With the construction skills you learned in this chapter, you performed investigations that lay the foundation for geometry.

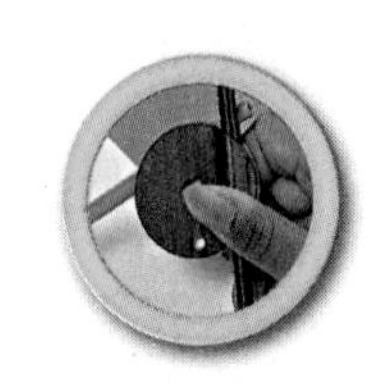

The investigation section of your notebook should be a detailed report of the mathematics you've already done. Beginning in Chapter 1 and continuing in this chapter, you summarized your work in the definition list and the conjecture list. Before you begin the review exercises, make sure your conjecture list is complete. Do you understand each conjecture? Can you draw a clear diagram that demonstrates your understanding of each definition and conjecture? Can you explain them to others? Can you use them to solve geometry problems?

EXERCISES

You will need

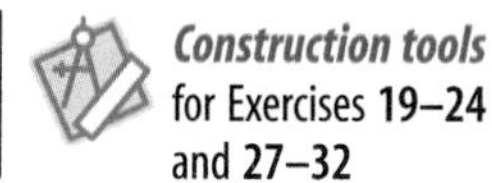

Construction tools for Exercises **19–24** and **27–32**

For Exercises 1–10, identify the statement as true or false. For each false statement, explain why it is false, or sketch a counterexample.

1. In a geometric construction, you use a protractor and a ruler.

2. A diagonal is a line segment in a polygon that connects any two vertices.

3. A trapezoid is a quadrilateral with exactly one pair of parallel sides.

4. A square is a rhombus with all angles congruent.

5. If a point is equidistant from the endpoints of a segment, then it must be the midpoint of the segment.

6. The set of all the points in the plane that are a given distance from a line segment is a pair of lines parallel to the given segment.

7. It is not possible for a trapezoid to have three congruent sides.

8. The incenter of a triangle is the point of intersection of the three angle bisectors.

9. The orthocenter of a triangle is the point of intersection of the three altitudes.

10. The incenter, the centroid, and the orthocenter are always inside the triangle.

The Principles of Perspective, Italian, ca. 1780. Victoria and Albert Museum, London, Great Britain.

For Exercises 11–18, match each geometric construction with one of the figures below.

11. Construction of a midsegment

12. Construction of an altitude

13. Construction of a centroid in a triangle

14. Construction of an incenter

15. Construction of an orthocenter in a triangle

16. Construction of a circumcenter

17. Construction of an equilateral triangle

18. Construction of an angle bisector

A.

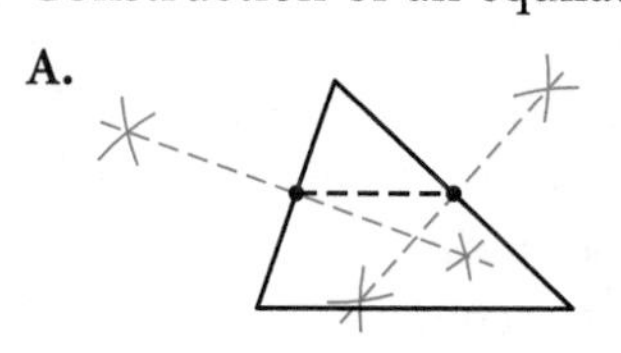

B.

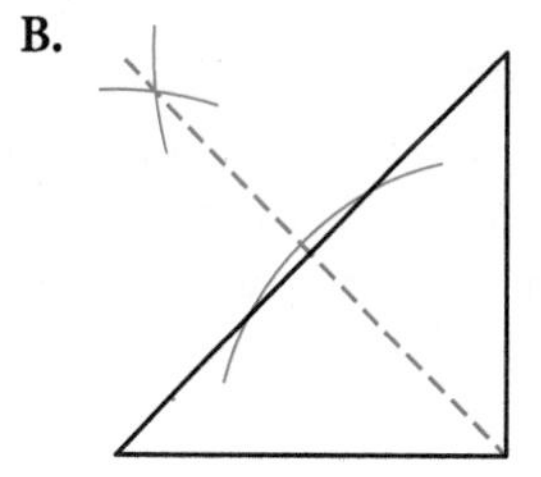

C.

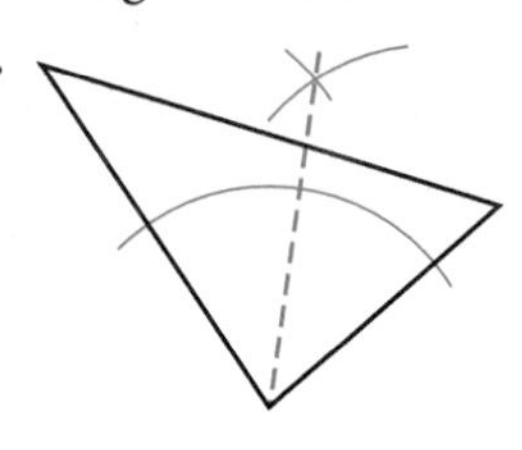

D.

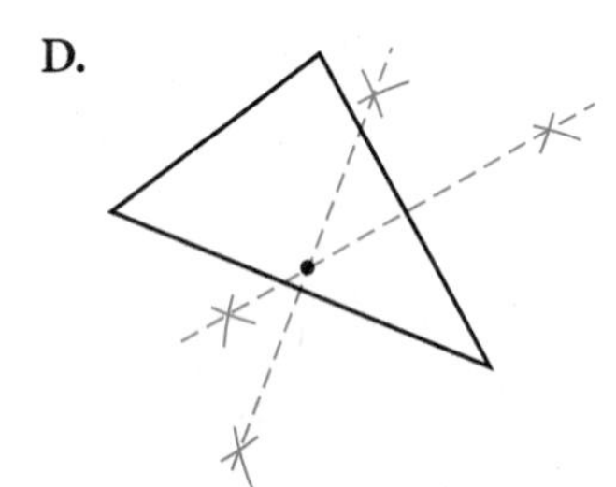

E.

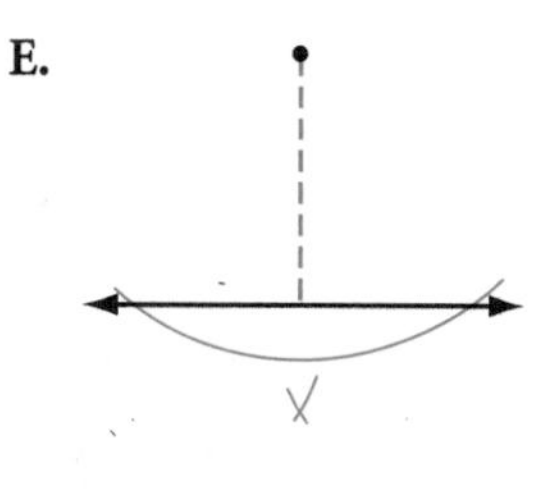

F.

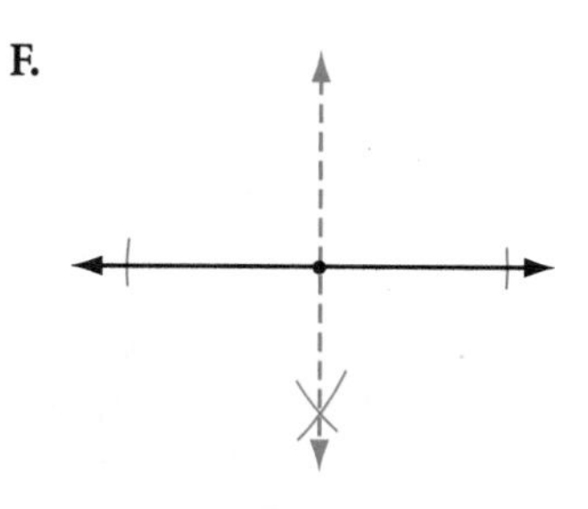

G.

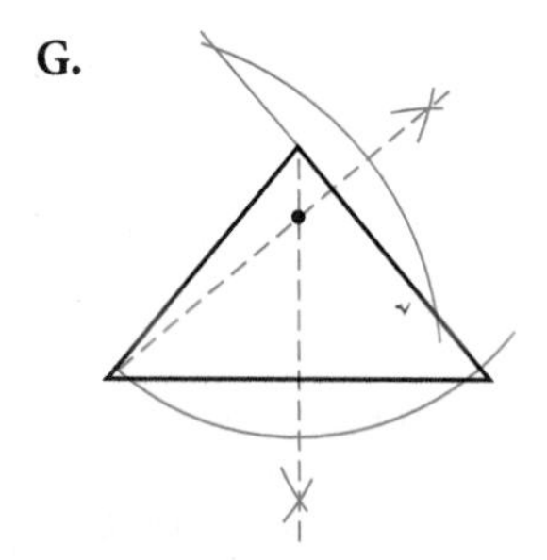

H.

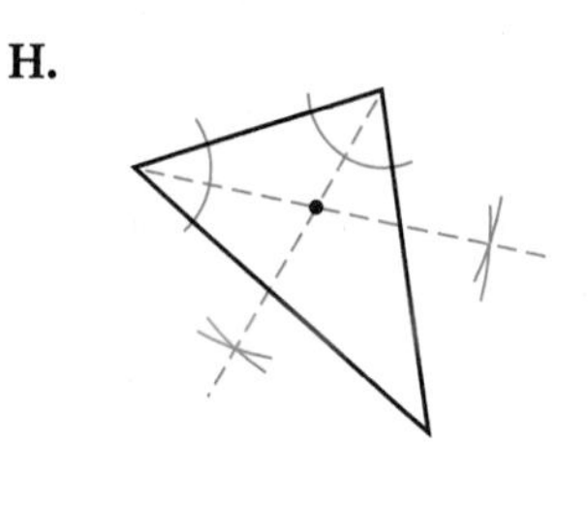

I.

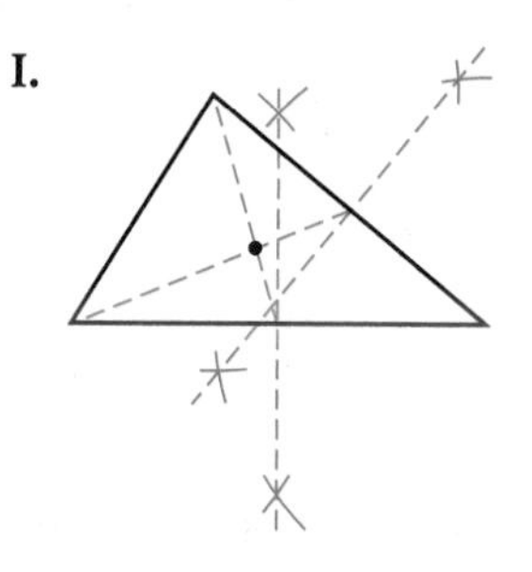

J.

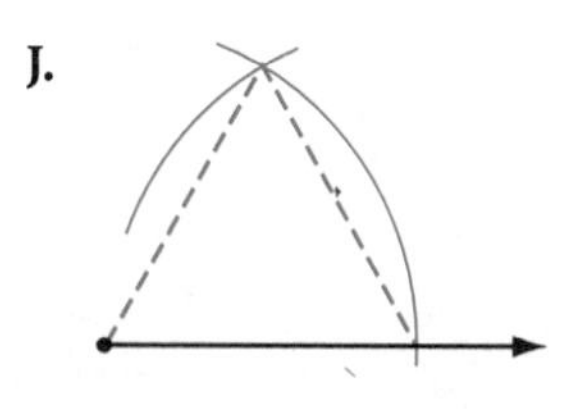

K.

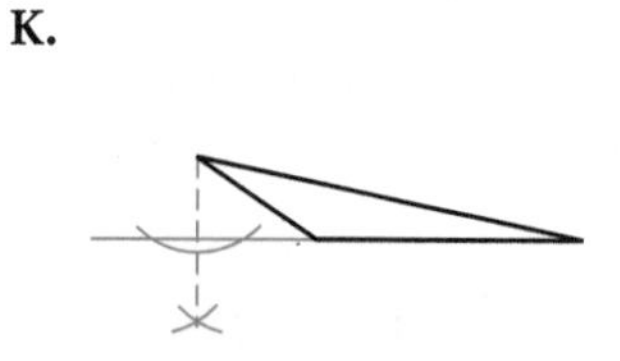

L.

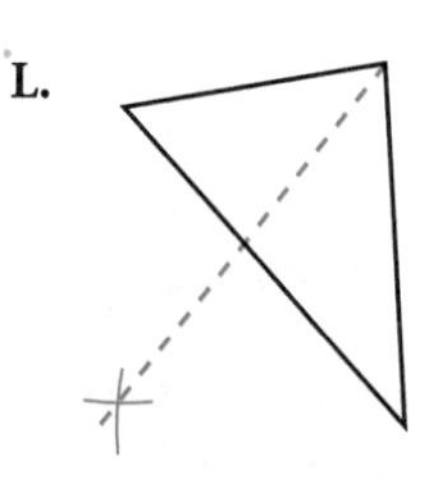

Construction For Exercises 19–24, perform a construction with compass and straightedge or with patty paper. Choose the method for each problem, but do not mix the tools in any one problem. In other words, play each construction game fairly.

19. Draw an angle and construct a duplicate of it.

20. Draw a line segment and construct its perpendicular bisector.

21. Draw a line and a point not on the line. Construct a perpendicular to the line through the point.

22. Draw an angle and bisect it.

23. Construct an angle that measures 22.5°.

24. Draw a line and a point not on the line. Construct a second line so that it passes through the point and is parallel to the first line.

25. Brad and Janet are building a home for their pet hamsters, Riff and Raff, in the shape of a triangular prism. Which point of concurrency in the triangular base do they need to locate in order to construct the largest possible circular entrance?

26. Adventurer Dakota Davis has a map that once showed the location of a large bag of gold. Unfortunately, the part of the map that showed the precise location of the gold has burned away. Dakota visits the area shown on the map anyway, hoping to find clues. To his surprise, he finds three headstones with geometric symbols on them.

The clues lead him to think that the treasure is buried at a point equidistant from the three stones. If Dakota's theory is correct, how should he go about locating the point where the bag of gold might be buried?

Construction For Exercises 27–32, use the given segments and angles to construct each figure. The lowercase letter above each segment represents the length of the segment.

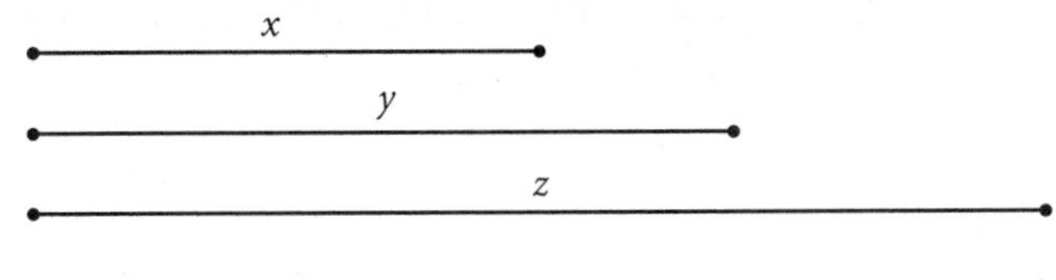

27. $\triangle ABC$ given $\angle A$, $\angle C$, and $AC = z$

28. A segment with length $2y + x - \frac{1}{2}z$ ⓗ

29. $\triangle PQR$ with $PQ = 3x$, $QR = 4x$, and $PR = 5x$

30. Isosceles triangle ABD given $\angle A$, and $AB = BD = 2y$

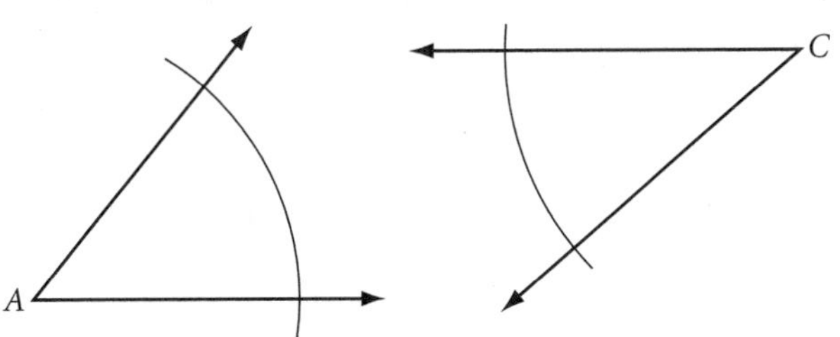

31. Quadrilateral $ABFD$ with $m\angle A = m\angle B$, $AD = BF = y$, and $AB = 4x$

32. Right triangle TRI with hypotenuse TI, $TR = x$, and $RI = y$ and a square on $\overline{TI}$, with $\overline{TI}$ as one side

MIXED REVIEW

Tell whether each symbol in Exercises 33–36 has reflectional symmetry, rotational symmetry, neither, or both. (The symbols are used in meteorology to show weather conditions.)

33. 

34. 

35. 

36. 

For Exercises 37–40, match the term with its construction.

37. Centroid

38. Circumcenter

39. Incenter

40. Orthocenter

A. 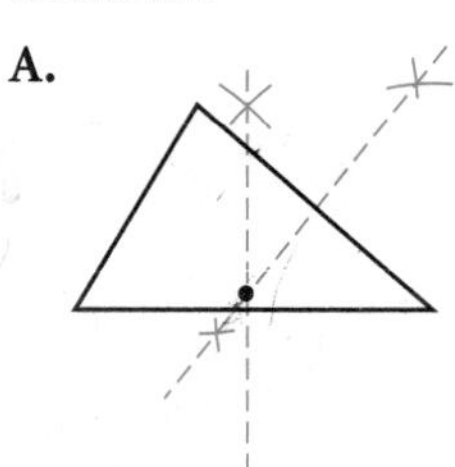

B. 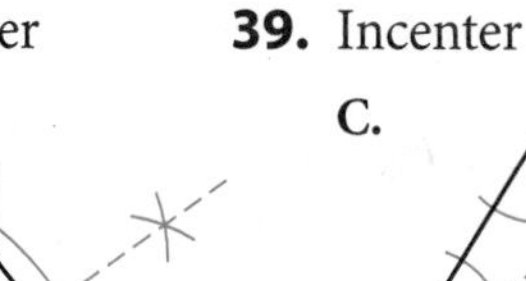

C. 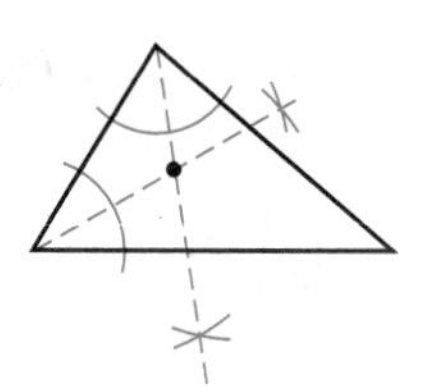

D. 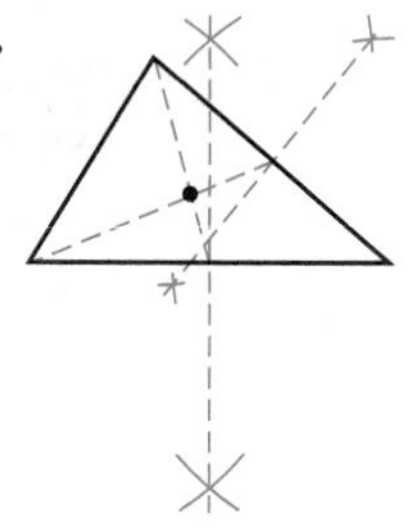

For Exercises 41–54, identify the statement as true or false. For each false statement, explain why it is false or sketch a counterexample.

41. An isosceles right triangle is a triangle with an angle measuring 90° and no two sides congruent.

42. If two parallel lines are cut by a transversal, then the alternate interior angles are congruent.

43. An altitude of a triangle must be inside the triangle.

44. The orthocenter of a triangle is the point of intersection of the three perpendicular bisectors of the sides.

45. If two lines are parallel to the same line, then they are parallel to each other.

46. If the sum of the measure of two angles is 180°, then the two angles are vertical angles.

47. Any two consecutive sides of a kite are congruent.

48. If a polygon has two pairs of parallel sides then it is a parallelogram.

49. The measure of an arc is equal to one half the measure of its central angle.

50. If $\overline{TR}$ is a median of $\triangle TIE$ and point D is the centroid, then $TD = 3DR$.

51. The shortest chord of a circle is the radius of a circle.

52. An obtuse triangle is a triangle that has one angle with measure greater than 90°.

53. Inductive reasoning is the process of showing that certain statements follow logically from accepted truths.

54. There are exactly three true statements in Exercises 41–54.

55. In the diagram, $p \parallel q$.

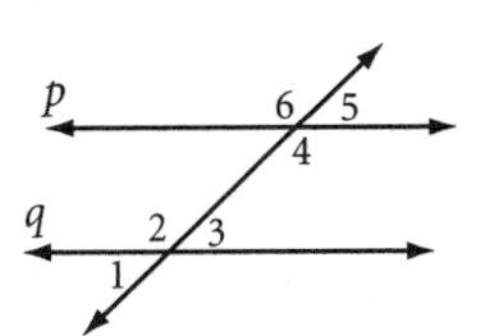

a. Name a pair of corresponding angles.

b. Name a pair of alternate exterior angles.

c. If $m\angle 3 = 42°$, what is $m\angle 6$?

In Exercises 56 and 57, use inductive reasoning to find the next number or shape in the pattern.

56. 100, 97, 91, 82, 70

57.  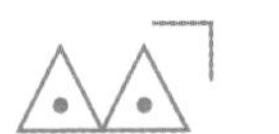 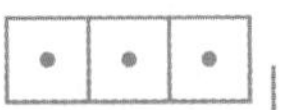 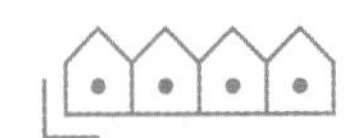

58. Consider the statement "If the month is October, then the month has 31 days."

a. Is the statement true?

b. Write the converse of this statement.

c. Is the converse true?

59. Find the point on the cushion at which a pool player should aim so that the white ball will hit the cushion and pass over point Q.

For Exercises 60 and 61, find the function rule for the sequence. Then find the 20th term.

60.

n	1	2	3	4	5	6	...	n	...	20
$f(n)$	−1	2	5	8	11	14	...		...	

61.

n	1	2	3	4	5	6	...	n	...	20
$f(n)$	0	3	8	15	24	35	...		...	

62. Calculate each lettered angle measure.

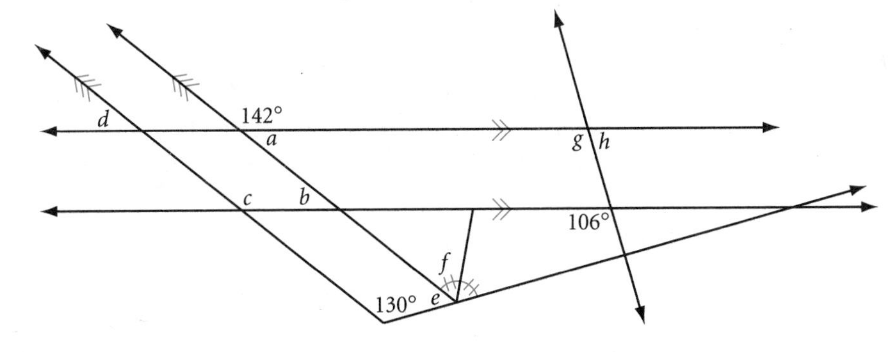

63. Draw a scalene triangle ABC. Use a straightedge and compass to construct the incenter of $\triangle ABC$.

64. What's wrong with this picture?

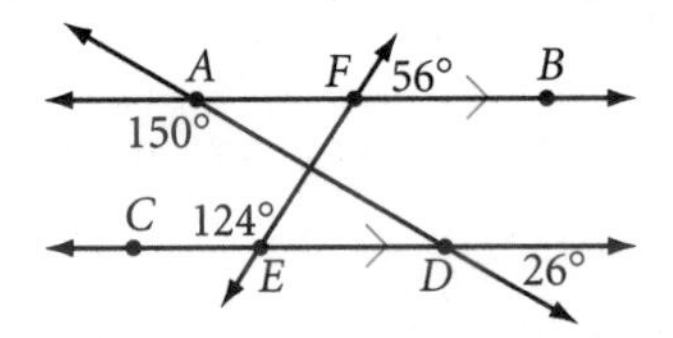

65. What is the minimum number of regions that are formed by 100 distinct lines in a plane? What is the maximum number of regions formed by 100 lines in the plane?

Assessing What You've Learned

PERFORMANCE ASSESSMENT

The subject of this chapter was the tools of geometry, so assessing what you've learned really means assessing what you can do with those tools. Can you do all the constructions you learned in this chapter? Can you show how you arrived at each conjecture? Demonstrating that you can do tasks like these is sometimes called **performance assessment.**

Look over the constructions in the Chapter Review. Practice doing any of the constructions that you're not absolutely sure of. Can you do each construction using either compass and straightedge or patty paper? Look over your conjecture list. Can you perform all the investigations that led to these conjectures?

Demonstrate at least one construction and at least one investigation for a classmate, a family member, or your teacher. Do every step from start to finish, and explain what you're doing.

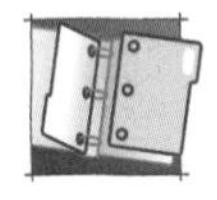

ORGANIZE YOUR NOTEBOOK

- Your notebook should have an investigation section, a definition list, and a conjecture list. Review the contents of these sections. Make sure they are complete, correct, and well organized.
- Write a one-page chapter summary from your notes.

WRITE IN YOUR JOURNAL How does the way you are learning geometry—doing constructions, looking for patterns, and making conjectures—compare to the way you've learned math in the past?

UPDATE YOUR PORTFOLIO Choose a construction problem from this chapter that you found particularly interesting and/or challenging. Describe each step, including how you figured out how to move on to the next step. Add this to your portfolio.

CHAPTER

4 Discovering and Proving Triangle Properties

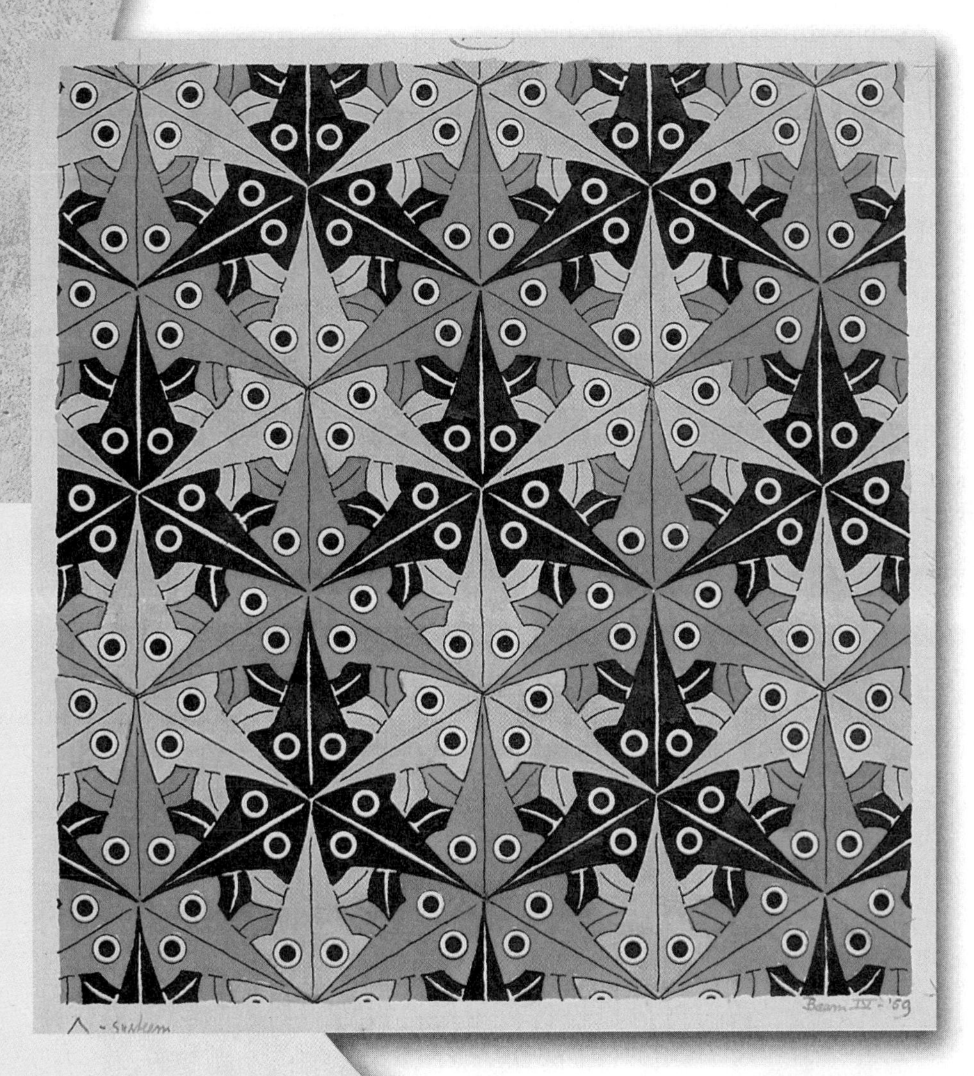

Is it possible to make a representation of recognizable figures that has no background?

M. C. ESCHER

Symmetry Drawing E103, M. C. Escher, 1959

OBJECTIVES

In this chapter you will

- learn why triangles are so useful in structures
- discover relationships between the sides and angles of triangles
- learn about the conditions that guarantee that two triangles are congruent

LESSON

4.1

Triangle Sum Conjecture

Teaching is the art of assisting discovery.

ALBERT VAN DOREN

Triangles have certain properties that make them useful in all kinds of structures, from bridges to high-rise buildings. One such property of triangles is their rigidity. If you build shelves like the first set shown at right, they will sway. But if you nail another board at the diagonal as in the second set, creating two triangles, you will have strong shelves. Rigid triangles such as these also give the bridge shown below its strength.

Steel truss bridge over the Columbia River Gorge in Oregon

Arranging congruent triangles can create parallel lines, as you can see from both the bridge and wood frame above. In this lesson you'll discover that this property is related to the sum of the angle measures in a triangle.

Architecture CONNECTION

American architect Julia Morgan (1872–1957) designed many noteworthy buildings, including Hearst Castle in central California. She often used triangular trusses made of redwood and exposed beams for strength and openness, as in this church in Berkeley, California. This building is now the Julia Morgan Center for the Arts.

Investigation
The Triangle Sum

You will need

- a protractor
- a straightedge
- scissors
- patty paper

There are an endless variety of triangles that you can draw, with different shapes and angle measures. Do their angle measures have anything in common? Start by drawing different kinds of triangles. Make sure your group has at least one acute and one obtuse triangle.

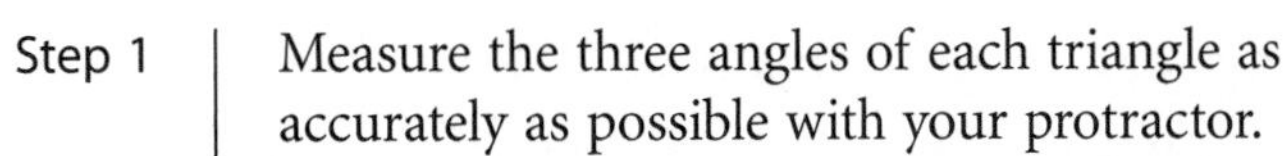

Step 1 Measure the three angles of each triangle as accurately as possible with your protractor.

Step 2 Find the sum of the measures of the three angles in each triangle. Compare results with others in your group. Does everyone get about the same result? What is it?

Step 3 Check the sum another way. Write the letters a, b, and c in the interiors of the three angles of one of the triangles, and carefully cut out the triangle.

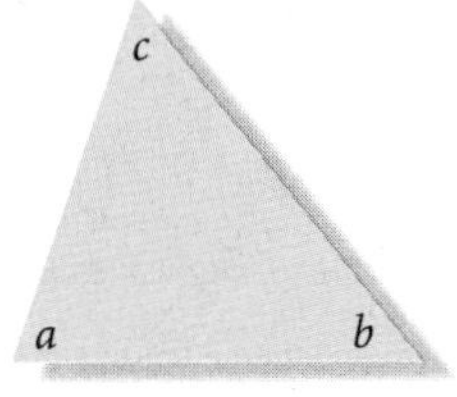

Step 4 Tear off the three angles.

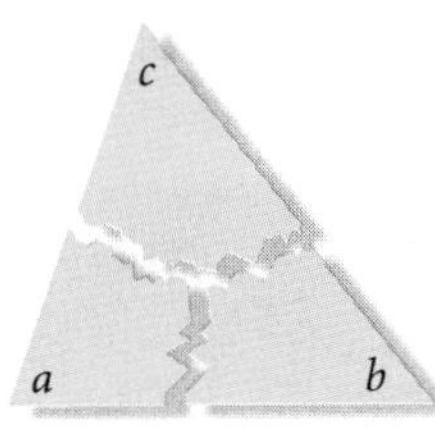

Step 5 Arrange the three angles so that their vertices meet at a point. How does this arrangement show the sum of the angle measures? Compare results with others in your group. State a conjecture.

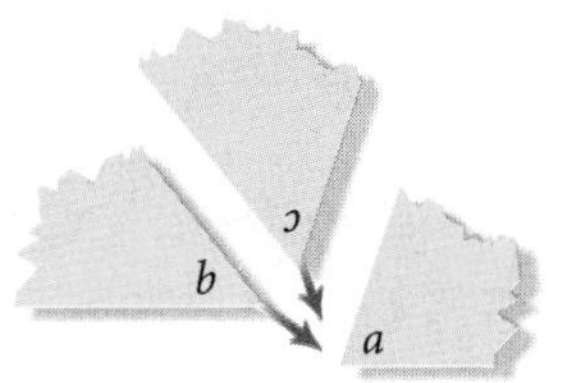

Triangle Sum Conjecture C-17

The sum of the measures of the angles in every triangle is ___?___.

Steps 1 through 5 may have convinced you that the Triangle Sum Conjecture is true, but a proof will explain *why* it is true for every triangle.

Step 6 Copy and complete the paragraph proof below to explain the connection between the Parallel Lines Conjecture and the Triangle Sum Conjecture.

Paragraph Proof: The Triangle Sum Conjecture

To prove the Triangle Sum Conjecture, you need to show that the angle measures in a triangle add up to _?_. Start by drawing any $\triangle ABC$, and $\overleftrightarrow{EC}$ parallel to side $\overline{AB}$.

$\overleftrightarrow{EC}$ is called an **auxiliary line,** because it is an extra line that helps with the proof.

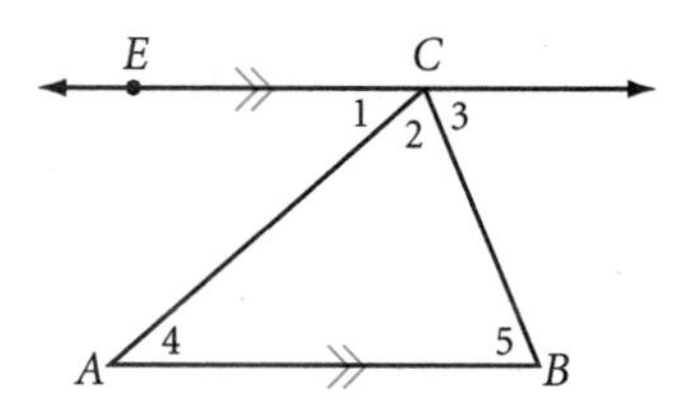

In the figure, $m\angle 1 + m\angle 2 + m\angle 3 = 180°$ if you consider $\angle 1 + \angle 2$ as one angle whose measure is $m\angle 1 + m\angle 2$, because _?_. You also know that $\overline{EC} \parallel \overline{AB}$, so $m\angle 1 = m\angle 4$ and $m\angle 3 = m\angle 5$, because _?_. So, by substituting for $m\angle 1$ and $m\angle 3$ in the first equation, you get _?_. Therefore, the measures of the angles in a triangle add up to _?_. ■

Step 7 Suppose two angles of one triangle have the same measures as two angles of another triangle. What can you conclude about the third pair of angles?

You can investigate Step 7 with patty paper. Draw a triangle on your paper. Create a second triangle on patty paper by tracing two of the angles of your original triangle, but make the side between your new angles a different length from the side between the angles you copied in the first triangle. How do the third angles in the two triangles compare?

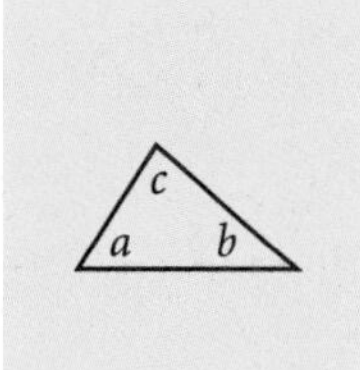

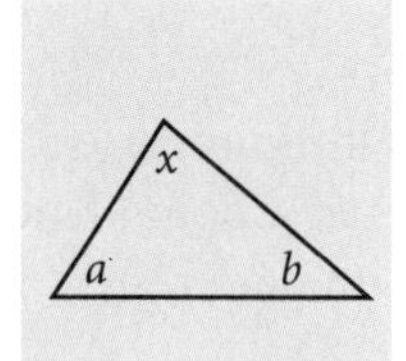

Step 8 Check your results with other students. You should be ready for your next conjecture.

Third Angle Conjecture

C-18

If two angles of one triangle are equal in measure to two angles of another triangle, then the third angle in each triangle _?_.

You can use the Triangle Sum Conjecture to show why the Third Angle Conjecture is true. You'll do this in Exercise 15.

EXAMPLE In the figure at right, is $\angle C$ congruent to $\angle E$? Write a paragraph proof explaining why.

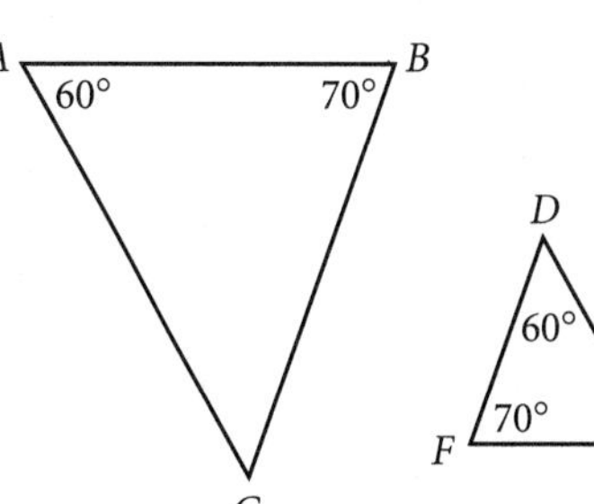

▶ Solution Yes. By the Third Angle Conjecture, because $\angle A$ and $\angle B$ are congruent to $\angle D$ and $\angle F$, then $\angle C$ must be congruent to $\angle E$.

You could also use the Triangle Sum Conjecture to find that $\angle D$ and $\angle F$ both measure 50°. Since they have the same measure, they are congruent.

EXERCISES

You will need

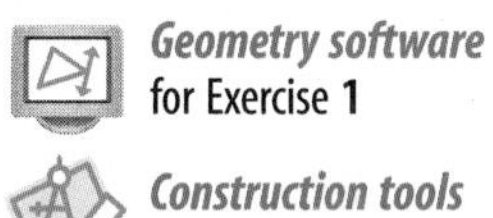

Geometry software for Exercise **1**

Construction tools for Exercises **10–13**

1. *Technology* Using geometry software, construct a triangle. Use the software to measure the three angles and calculate their sum.

a. Drag the vertices and describe your observations.

b. Repeat the process for a right triangle.

Use the Triangle Sum Conjecture to determine each lettered angle measure in Exercises 2–5. You might find it helpful to copy the diagrams so you can write on them.

2. $x = \underline{?}$

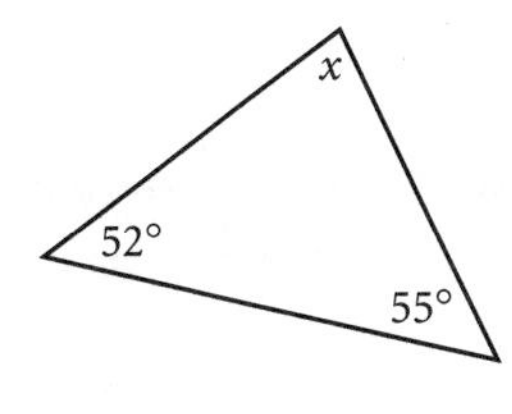

3. $v = \underline{?}$

4. $z = \underline{?}$ ⓗ

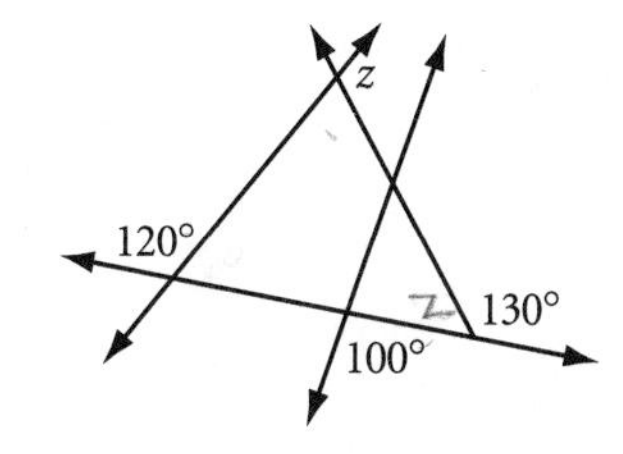

5. $w = \underline{?}$

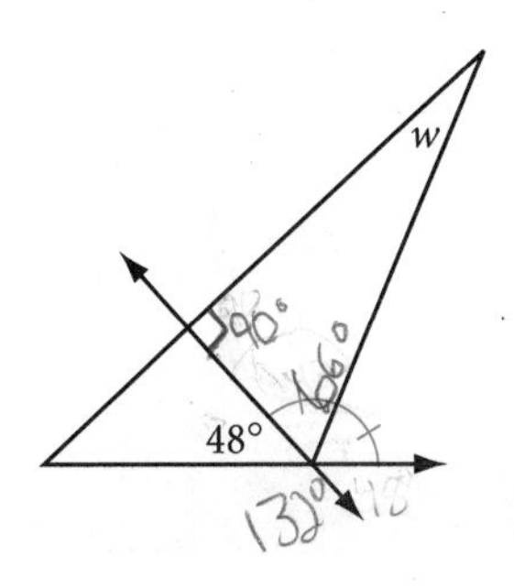

6. Find the sum of the measures of the marked angles. ⓗ

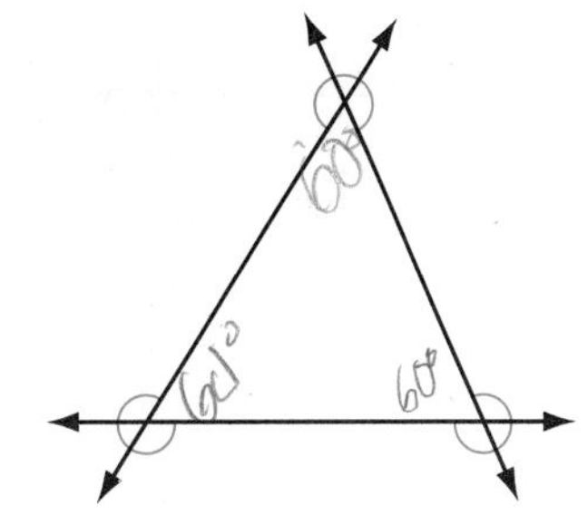

7. Find the sum of the measures of the marked angles. ⓗ

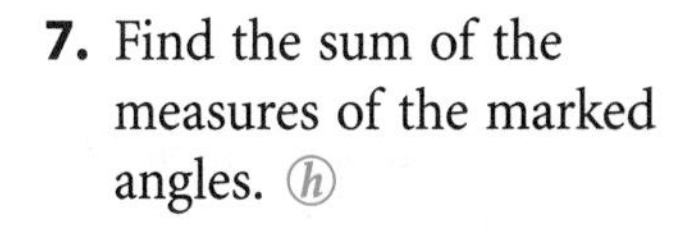

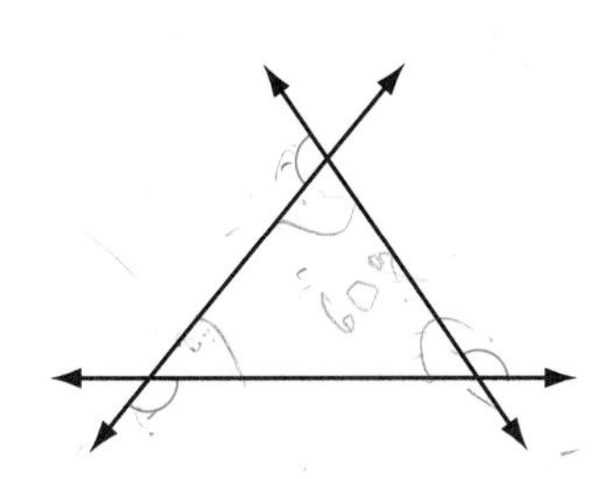

8. $a = \underline{?}$ 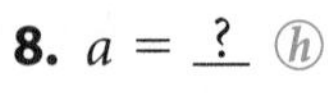
$b = \underline{?}$
$c = \underline{?}$
$d = \underline{?}$
$e = \underline{?}$

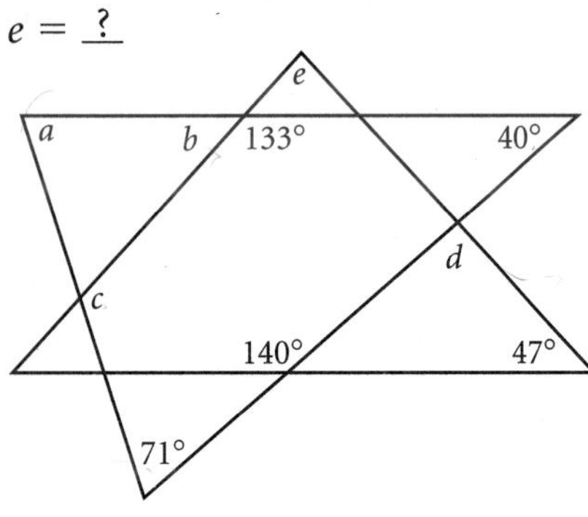

9. $m = \underline{?}$
$n = \underline{?}$
$p = \underline{?}$
$q = \underline{?}$
$r = \underline{?}$
$s = \underline{?}$
$t = \underline{?}$
$u = \underline{?}$

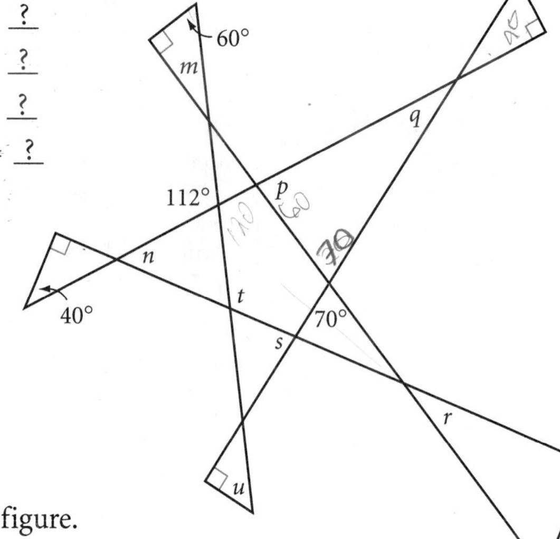

In Exercises 10–12, use what you know to construct each figure. Use only a compass and a straightedge.

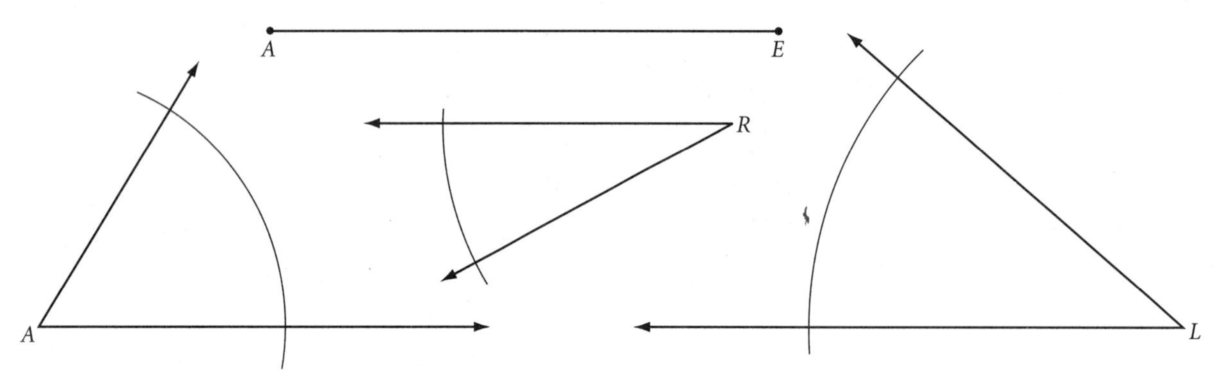

10. ***Construction*** Given $\angle A$ and $\angle R$ of $\triangle ARM$, construct $\angle M$.

11. ***Construction*** In $\triangle LEG$, $m\angle E = m\angle G$. Given $\angle L$, construct $\angle G$.

12. ***Construction*** Given $\angle A$, $\angle R$, and side $\overline{AE}$, construct $\triangle EAR$.

13. ***Construction*** Repeat Exercises 10–12 with patty-paper constructions.

14. In $\triangle MAS$ below, $\angle M$ is a right angle. Let's call the two acute angles, $\angle A$ and $\angle S$, "wrong angles." Write a paragraph proof or use algebra to show that "two wrongs make a right," at least for angles in a right triangle.

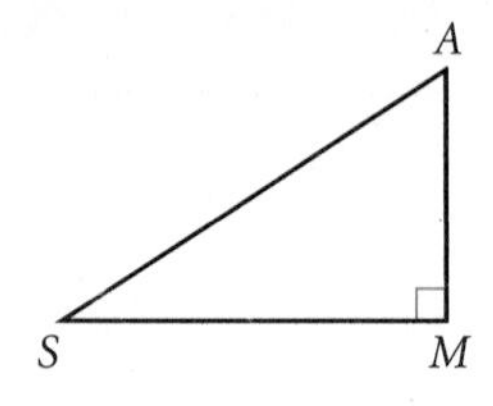

THE FAR SIDE® By GARY LARSON

"Yes, yes, I *know* that, Sidney—*everybody* knows *that*! ... But look: Four wrongs *squared*, minus two wrongs to the fourth power, divided by this formula, *do* make a right."

15. Use the Triangle Sum Conjecture and the figures at right to write a paragraph proof explaining why the Third Angle Conjecture is true. ⓗ

16. Write a paragraph proof, or use algebra, to explain why each angle of an equiangular triangle measures 60°.

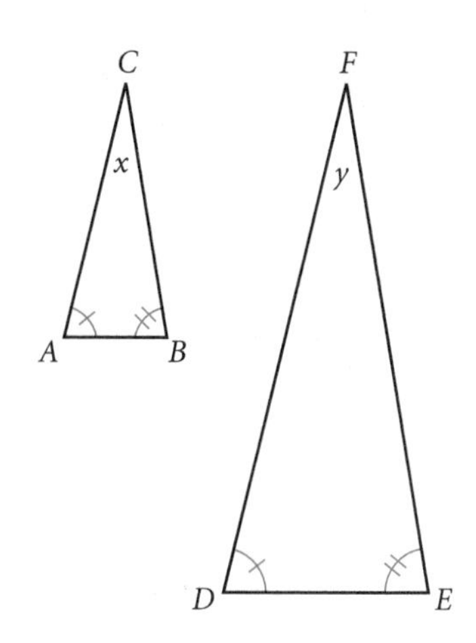

Review

In Exercises 17–21, tell whether the statement is true or false. For each false statement, explain why it is false or sketch a counterexample.

17. If two sides in one triangle are congruent to two sides in another triangle, then the two triangles are congruent.

18. If two angles in one triangle are congruent to two angles in another triangle, then the two triangles are congruent.

19. If a side and an angle in one triangle are congruent to a side and an angle in another triangle, then the two triangles are congruent.

20. If three angles in one triangle are congruent to three angles in another triangle, then the two triangles are congruent.

21. If three sides in one triangle are congruent to three sides in another triangle, then the two triangles are congruent.

22. What is the number of stories in the tallest house you can build with two 52-card decks? How many cards would it take?

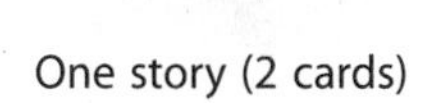

One story (2 cards)

Two stories (7 cards)

Three stories (15 cards)

IMPROVING YOUR VISUAL THINKING SKILLS

Dissecting a Hexagon I

Trace this regular hexagon twice.

1. Divide one hexagon into four congruent trapezoids.
2. Divide the other hexagon into eight congruent parts. What shape is each part?

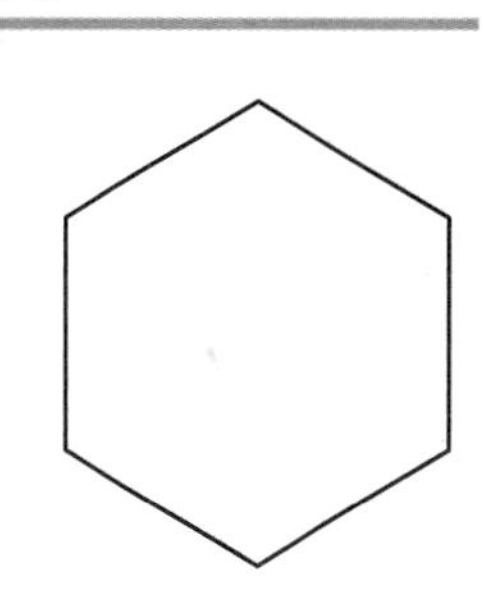

LESSON

4.2

Properties of Special Triangles

Imagination is built upon knowledge.

ELIZABETH STUART PHELPS

Recall from Chapter 1 that an isosceles triangle is a triangle with at least two congruent sides. In an isosceles triangle, the angle between the two congruent sides is called the vertex angle, and the other two angles are called the base angles. The side between the two base angles is called the base of the isosceles triangle. The other two sides are called the **legs.**

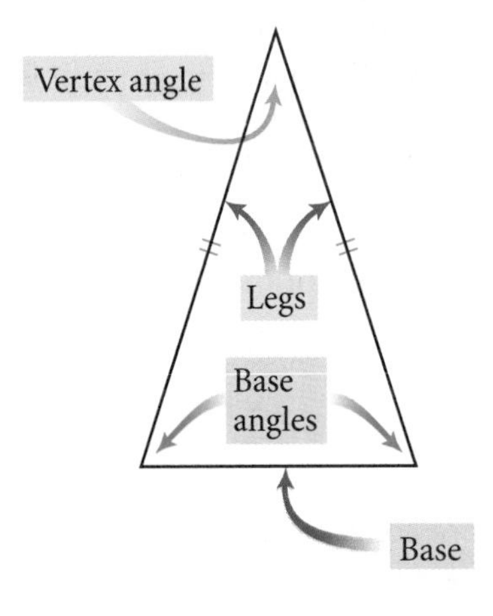

In this lesson you'll discover some properties of isosceles triangles.

The famous Transamerica Building in San Francisco contains many isosceles triangles.

The Rock and Roll Hall of Fame and Museum structure is a pyramid containing many triangles that are isosceles *and* equilateral.

Architecture CONNECTION

The Rock and Roll Hall of Fame and Museum in Cleveland, Ohio, is a dynamic structure. Its design reflects the innovative music that it honors. The front part of the museum is a large glass pyramid, divided into small triangular windows that resemble a Sierpiński tetrahedron, a three-dimensional Sierpiński triangle. The pyramid structure rests on a rectangular tower and a circular theater that looks like a performance drum. Architect I. M. Pei (b 1917) used geometric shapes to capture the resonance of rock and roll musical chords.

Investigation 1
Base Angles in an Isosceles Triangle

You will need

- patty paper
- a protractor

Let's examine the angles of an isosceles triangle. Each person in your group should draw a different angle for this investigation. Your group should have at least one acute angle and one obtuse angle.

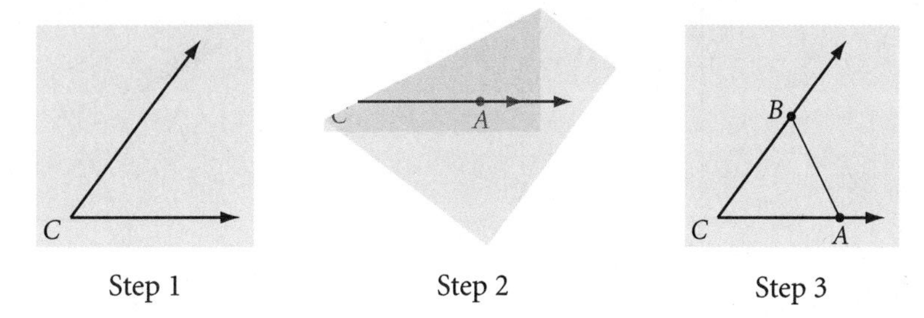

Step 1 Step 2 Step 3

Step 1 Draw an angle on patty paper. Label it $\angle C$. This angle will be the vertex angle of your isosceles triangle.

Step 2 Place a point A on one ray. Fold your patty paper so that the two rays match up. Trace point A onto the other ray.

Step 3 Label the point on the other ray point B. Draw $\overline{AB}$. You have constructed an isosceles triangle. Explain how you know it is isosceles. Name the base and the base angles.

Step 4 Use your protractor to compare the measures of the base angles. What relationship do you notice? How can you fold the paper to confirm your conclusion?

Step 5 Compare results in your group. Was the relationship you noticed the same for each isosceles triangle? State your observations as your next conjecture.

Isosceles Triangle Conjecture C-19

If a triangle is isosceles, then __?__.

Equilateral triangles have at least two congruent sides, so they fit the definition of isosceles triangles. That means any properties you discover for isosceles triangles will also apply to equilateral triangles. How does the Isosceles Triangle Conjecture apply to equilateral triangles?

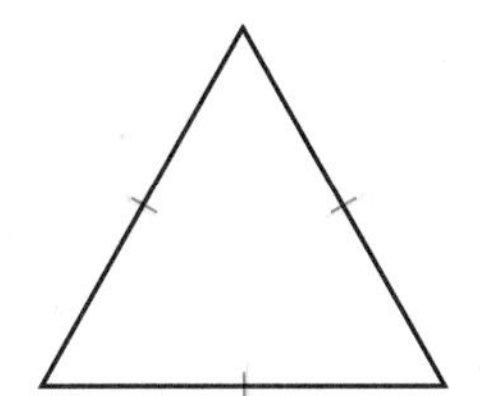

You can switch the "if" and "then" parts of the Isosceles Triangle Conjecture to obtain the converse of the conjecture. Is the converse of the Isosceles Triangle Conjecture true? Let's investigate.

Investigation 2
Is the Converse True?

You will need

- a compass
- a straightedge

Suppose a triangle has two congruent angles. Must the triangle be isosceles?

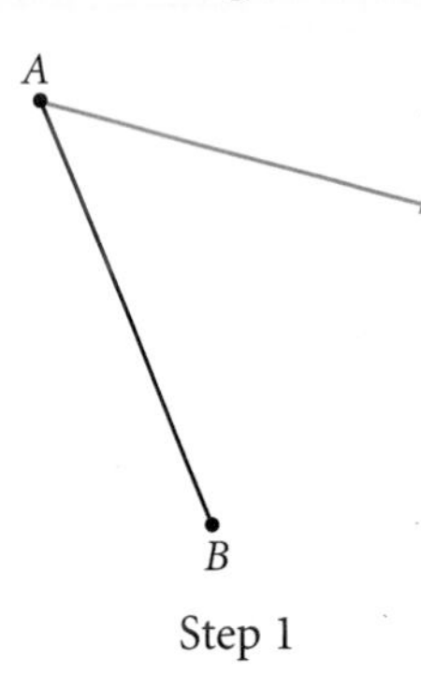

Step 1

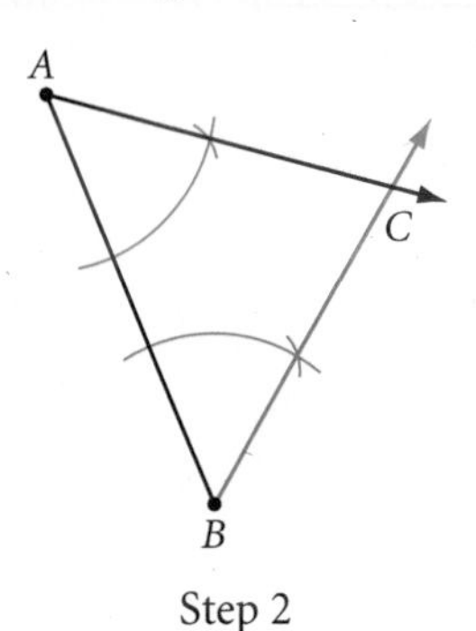

Step 2

Step 1 Draw a segment and label it $\overline{AB}$. Draw an acute angle at point A. This angle will be a base angle. (Why can't you draw an obtuse angle as a base angle?)

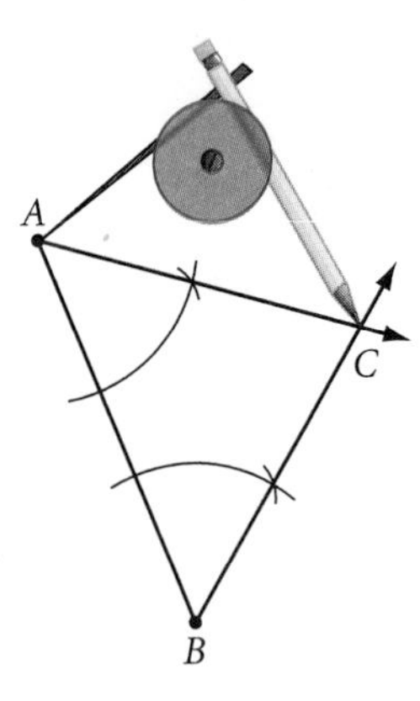

Step 2 Copy $\angle A$ at point B on the same side of $\overline{AB}$. Label the intersection of the two rays point C.

Step 3 Use your compass to compare the lengths of sides $\overline{AC}$ and $\overline{BC}$. What relationship do you notice? How can you use patty paper to confirm your conclusion?

Step 4 Compare results in your group. State your observation as your next conjecture.

Converse of the Isosceles Triangle Conjecture C-20

If a triangle has two congruent angles, then __?__.

EXERCISES

You will need

For Exercises 1–6, use your new conjectures to find the missing measures.

1. $m\angle H = $ __?__ ⓗ

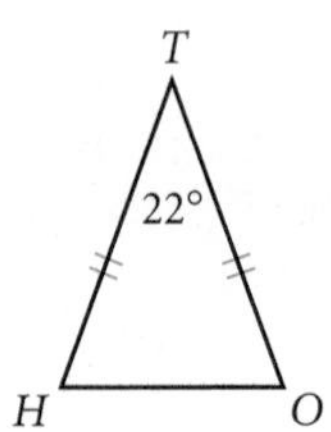

2. $m\angle G = $ __?__

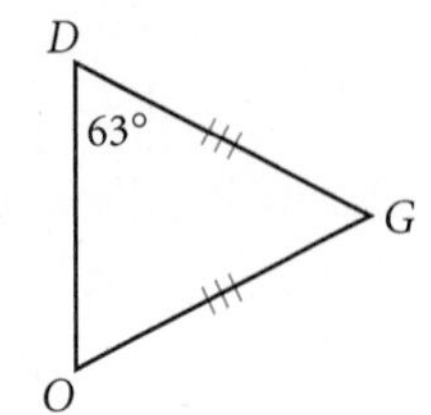

3. $m\angle OLE = $ __?__

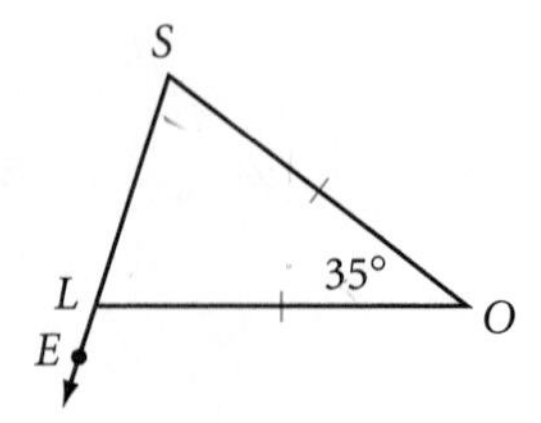

4. $m\angle R = \underline{\ ?\ }$
$RM = \underline{\ ?\ }$

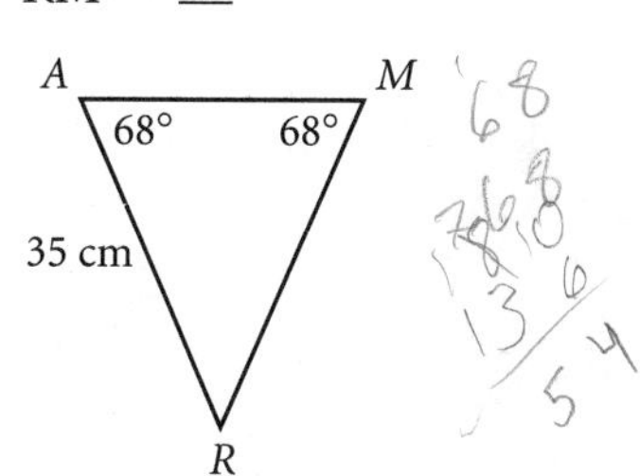

5. $m\angle Y = \underline{\ ?\ }$
$RD = \underline{\ ?\ }$

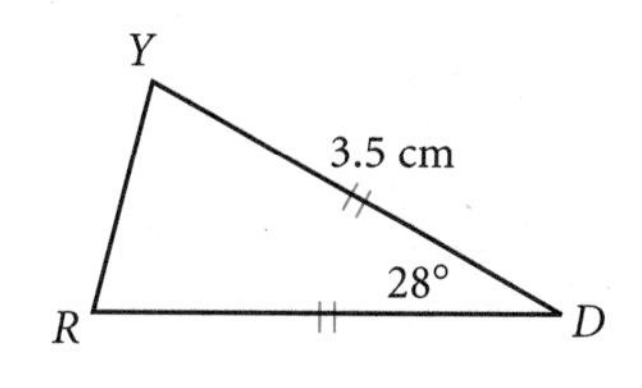

6. The perimeter of $\triangle MUD$ is 38 cm.
$m\angle D = \underline{\ ?\ }$
$MD = \underline{\ ?\ }$

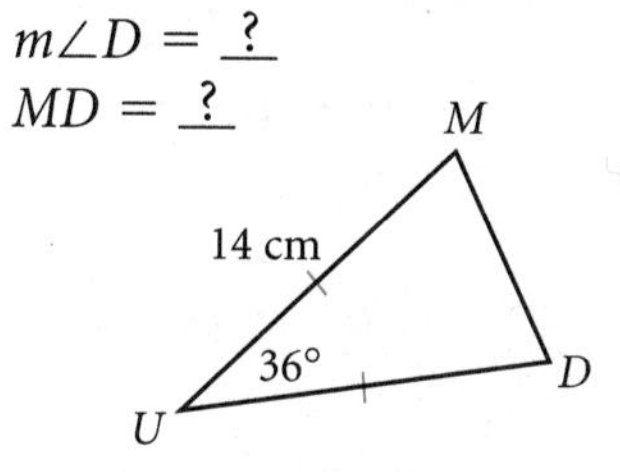

7. Copy the figure at right. Calculate the measure of each lettered angle. ⓗ

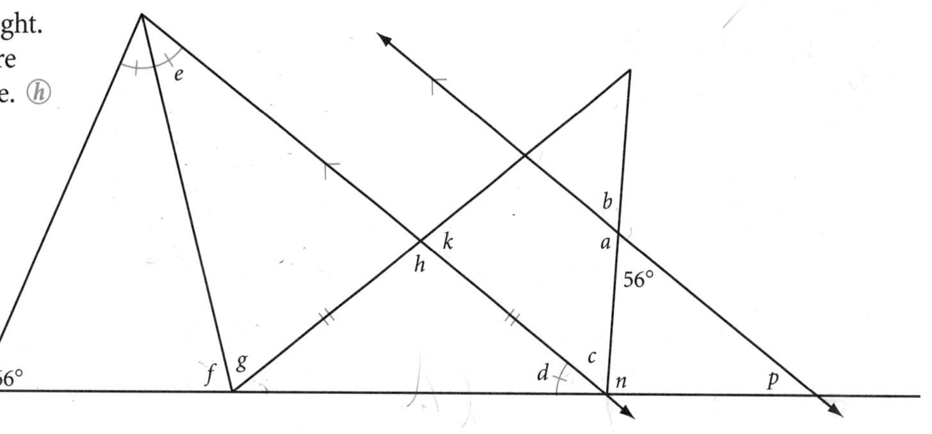

8. The Islamic design below right is based on the star decagon construction shown below left. The ten angles surrounding the center are all congruent. Find the lettered angle measures. How many triangles are not isosceles? ⓗ

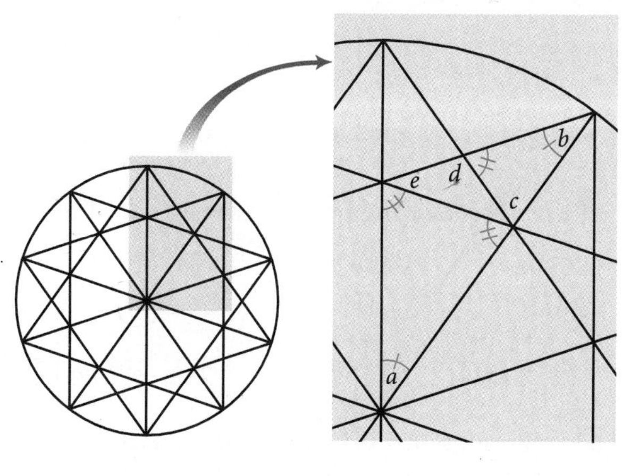

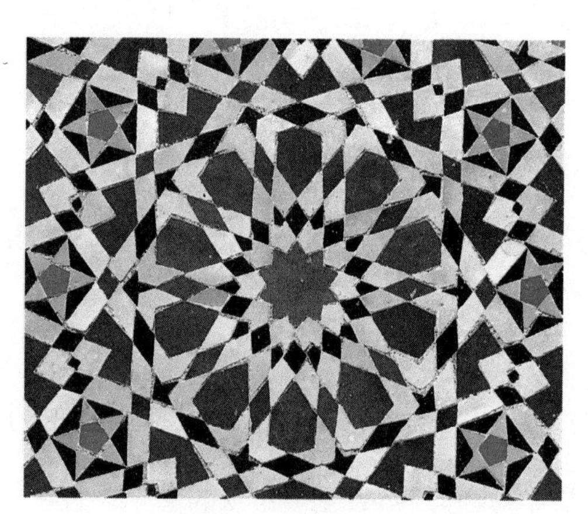

9. Study the triangles in the software constructions below. Each triangle has one vertex at the center of the circle, and two vertices on the circle.

a. Are the triangles all isosceles? Write a paragraph proof explaining why or why not.

b. If the vertex at the center of the first circle has an angle measure of 60°, find the measures of the other two angles in that triangle.

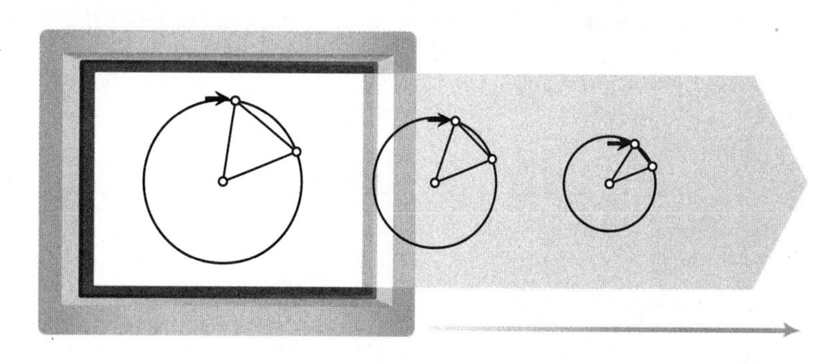

Review

In Exercises 10 and 11, complete the statement of congruence from the information given. Remember to write the statement so that corresponding parts are in order.

10. $\triangle GEA \cong \triangle$ _?_

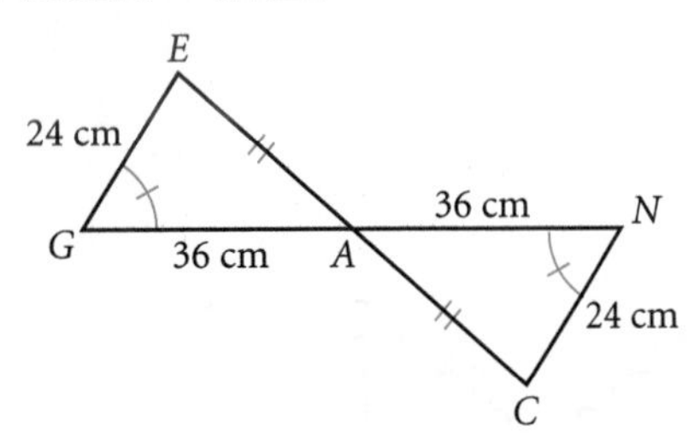

11. $\triangle JAN \cong \triangle$ _?_

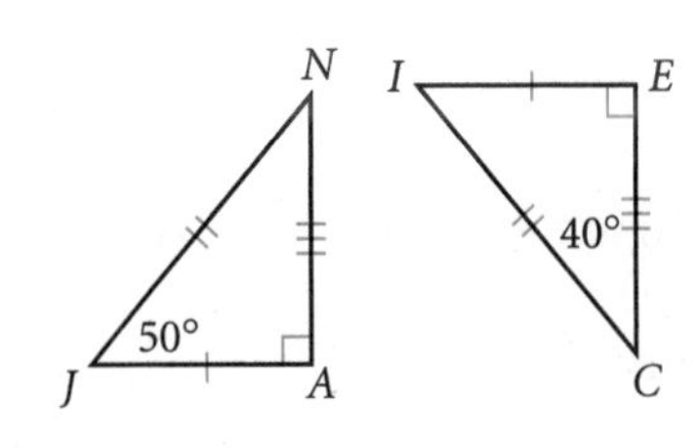

In Exercises 12 and 13, use compass and straightedge, or patty paper, to construct a triangle that is not congruent to the given triangle, but has the given parts congruent. The symbol $\ncong$ means "not congruent to."

12. ***Construction*** Construct $\triangle ABC \ncong \triangle DEF$ with $\angle A \cong \angle D$, $\angle B \cong \angle E$, and $\angle C \cong \angle F$. ⓗ

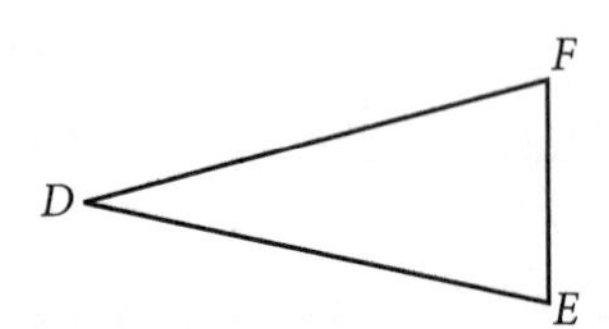

13. ***Construction*** Construct $\triangle GHK \ncong \triangle MNP$ with $\overline{HK} \cong \overline{NP}$, $\overline{GH} \cong \overline{MN}$, and $\angle G \cong \angle M$. ⓗ

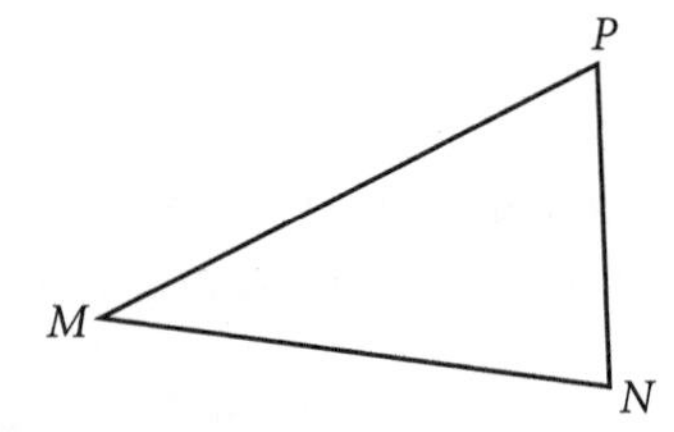

14. ***Construction*** With a straightedge and patty paper, construct an angle that measures 105°.

In Exercises 15–18, determine whether each pair of lines through the points below is parallel, perpendicular, or neither.

$A(1, 3)$ $B(6, 0)$ $C(4, 3)$ $D(1, -2)$ $E(-3, 8)$ $F(-4, 1)$ $G(-1, 6)$ $H(4, -4)$

15. $\overleftrightarrow{AB}$ and $\overleftrightarrow{CD}$ ⓗ

16. $\overleftrightarrow{FG}$ and $\overleftrightarrow{CD}$

17. $\overleftrightarrow{AD}$ and $\overleftrightarrow{CH}$

18. $\overleftrightarrow{DE}$ and $\overleftrightarrow{GH}$

19. Using the coordinate points above, is $FGCD$ a trapezoid, a parallelogram, or neither?

20. Picture the isosceles triangle below toppling side over side to the right along the line. Copy the triangle and line onto your paper, then construct the path of point P through two cycles. Where on the number line will the vertex point land?

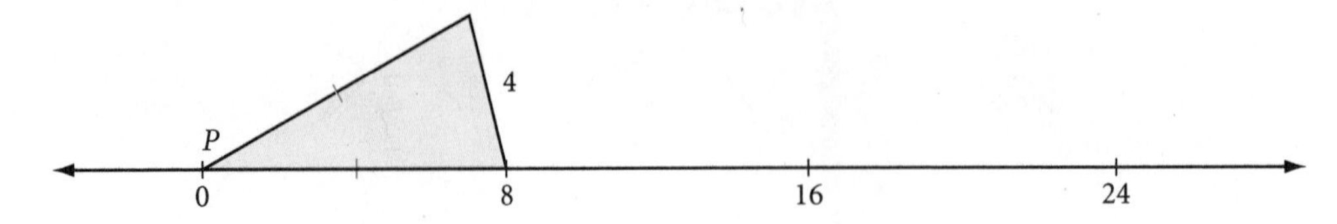

For Exercises 21 and 22, use the ordered pair rule shown to relocate each of the vertices of the given triangle. Connect the three new points to create a new triangle. Is the new triangle congruent to the original one? Describe how the new triangle has changed position from the original.

21. $(x, y) \rightarrow (x + 5, y - 3)$ ⓗ

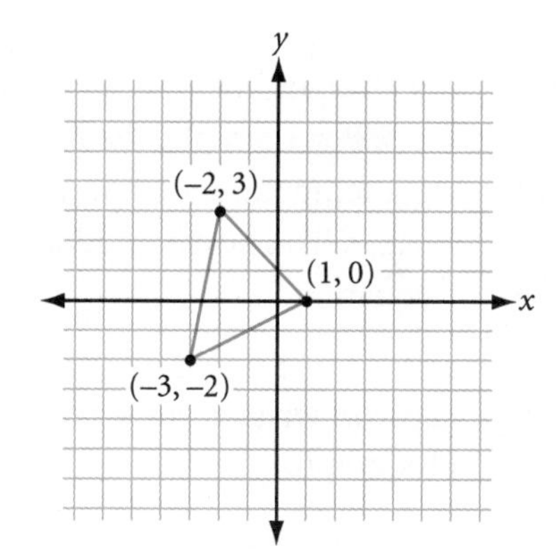

22. $(x, y) \rightarrow (x, -y)$

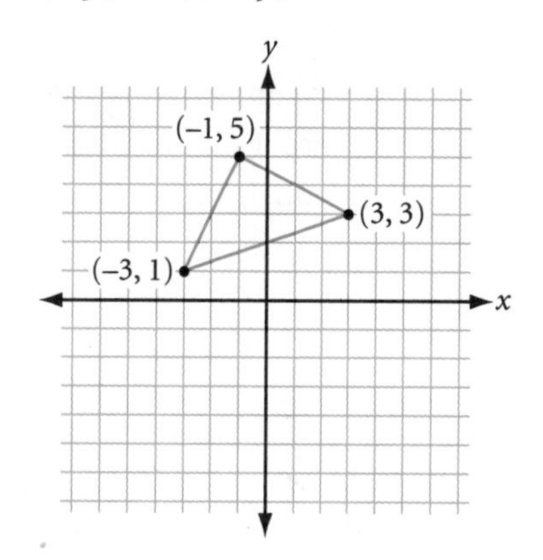

IMPROVING YOUR REASONING SKILLS

PUZZLE

Hundreds Puzzle

Fill in the blanks of each equation below. All nine digits—1 through 9—must be used, in order! You may use any combination of signs for the four basic operations (+, −, ·, ÷), parentheses, decimal points, exponents, factorial signs, and square root symbols, and you may place the digits next to each other to create two-digit or three-digit numbers.

Example: $1 + 2(3 + 4.5) + 67 + 8 + 9 = 100$

1. $1 + 2 + 3 - 4 + 5 + 6 + \underline{?} + 9 = 100$
2. $1 + 2 + 3 + 4 + 5 + \underline{?} = 100$
3. $1 + 2 + [(3)(4)(5) \div 6] + \underline{?} = 100$
4. $[(-1 - \underline{?}) \div 5] + 6 + 7 + 89 = 100$
5. $1 + 23 - 4 + \underline{?} + 9 = 100$

Writing Linear Equations

A linear equation is an equation whose graph is a straight line. Linear equations are useful in science, business, and many other areas. For example, the linear equation $f = 32 + \frac{9}{5}c$ gives the rule for converting a temperature from degrees Celsius, c, to degrees Fahrenheit, f. The numbers 32 and $\frac{9}{5}$ determine the graph of the equation.

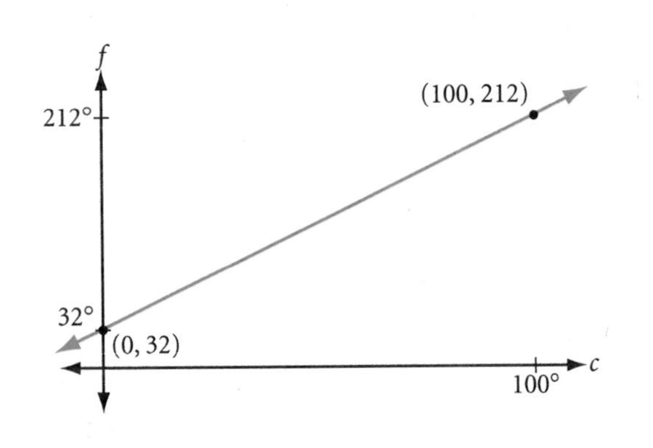

Understanding how the numbers in a linear equation determine the graph can help you write a linear equation based on information about a graph.

The y-coordinate at which a graph crosses the y-axis is called the y-intercept. The measure of steepness is called the slope. Below are the graphs of four equations. The table gives the equation, slope, and y-intercept for each graph. How do the numbers in each equation relate to the slope and y-intercept?

Equation	Slope	y-intercept
$y = 2 + 3x$	3	2
$y = 2x - 1$	2	-1
$y = -3x + 4$	-3	4
$y = -5 + \frac{3}{2}x$	$\frac{3}{2}$	-5

In each case, the slope of the line is the coefficient of x in the equation. The y-intercept is the constant that is added to, or subtracted from, the x term.

In your algebra class, you may have learned about one of these forms of a linear equation in slope-intercept form:

$y = a + bx$, where a is the y-intercept and b is the slope

$y = mx + b$, where m is the slope and b is the y-intercept

The only difference between these two forms is the order of the x term and the constant term. For example, the equation of a line with slope -3 and y-intercept 1 can be written as $y = 1 - 3x$ or $y = -3x + 1$.

Let's look at a few examples that show how you can apply what you have learned about the relationship between a linear equation and its graph.

EXAMPLE A Find the equation of $\overleftrightarrow{AB}$ from its graph.

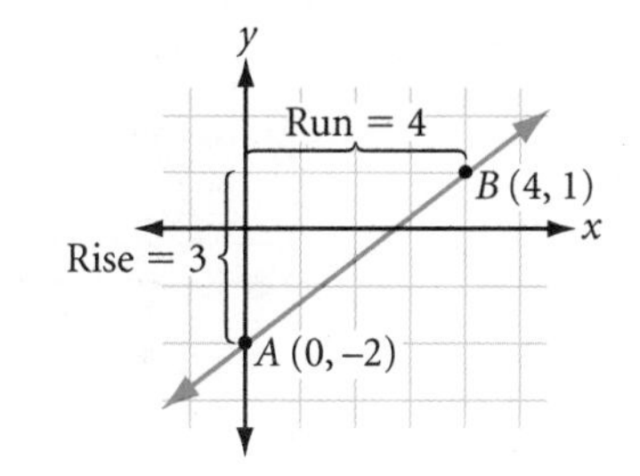

▸ Solution $\overleftrightarrow{AB}$ has y-intercept -2 and slope $\frac{3}{4}$, so the equation is

$$y = -2 + \frac{3}{4}x$$

EXAMPLE B Given points $C(4, 6)$ and $D(-2, 3)$, find the equation of $\overleftrightarrow{CD}$.

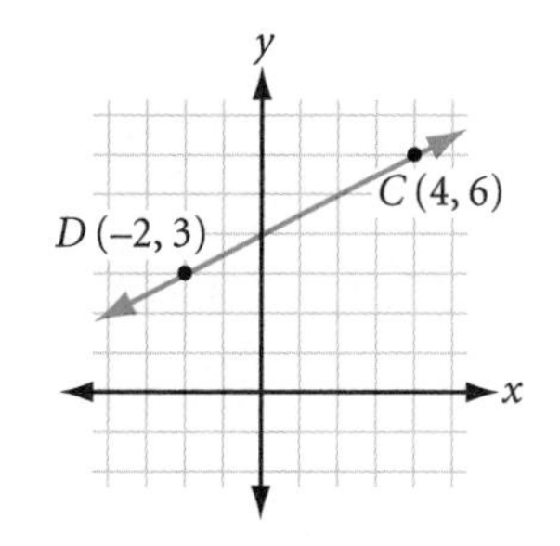

▸ Solution The slope of $\overleftrightarrow{CD}$ is $\frac{3-6}{-2-4}$, or $\frac{1}{2}$. The slope between any point (x, y) and one of the given points, say $(4, 6)$, must also be $\frac{1}{2}$.

$$\frac{y-6}{x-4} = \frac{1}{2}$$

$$y = \frac{1}{2}x + 4$$

EXAMPLE C Find the equation of the perpendicular bisector of the segment with endpoints $(2, 9)$ and $(-6, -7)$.

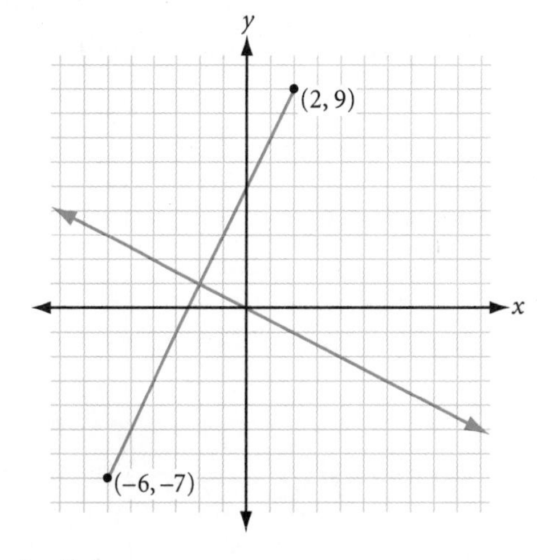

▸ Solution The perpendicular bisector passes through the midpoint. The midpoint of the segment is $\left(\frac{2+(-6)}{2}, \frac{9+(-7)}{2}\right)$, or $(-2, 1)$. The slope of the segment is $\frac{-7-9}{-6-2}$, or 2. So the slope of its perpendicular is $-\frac{1}{2}$. Write the equation of its perpendicular bisector and solve for y:

$$\frac{y-1}{x-(-2)} = -\frac{1}{2}$$

$$y = -\frac{1}{2}x$$

EXERCISES

In Exercises 1–3, graph each linear equation.

1. $y = 1 - 2x$

2. $y = \frac{4}{3}x + 4$

3. $2y - 3x = 12$

Write an equation for each line in Exercises 4 and 5.

4.

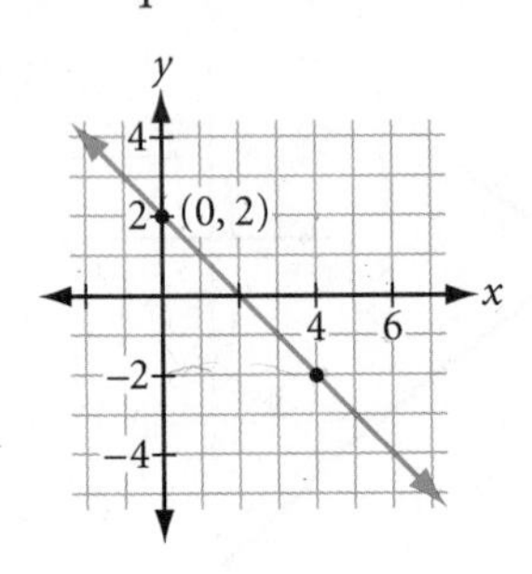

5.

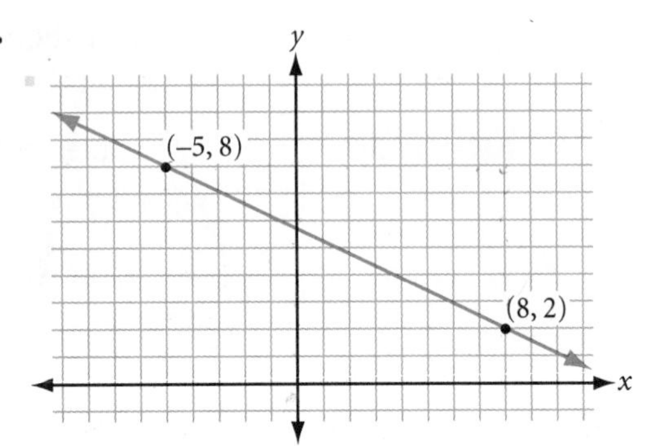

In Exercises 6–8, write an equation for the line through each pair of points.

6. (1, 2), (3, 4)

7. (1, 2), (3, −4)

8. (−1, −2), (−6, −4)

9. The math club is ordering printed T-shirts to sell for a fundraiser. The T-shirt company charges \$80 for the set-up fee and \$4 for each printed T-shirt. Using x for the number of shirts the club orders, write an equation for the total cost of the T-shirts.

10. Write an equation for the line with slope −3 that passes through the midpoint of the segment with endpoints (3, 4) and (11, 6).

11. Write an equation for the line that is perpendicular to the line $y = 4x + 5$ and that passes through the point (0, −3).

For Exercises 12–14, the coordinates of the vertices of $\triangle WHY$ are $W(0, 0)$, $H(8, 3)$, and $Y(2, 9)$.

12. Find the equation of the line containing median $\overline{WO}$.

13. Find the equation of the perpendicular bisector of side $\overline{HY}$.

14. Find the equation of the line containing altitude $\overline{HT}$.

IMPROVING YOUR REASONING SKILLS

Container Problem I

You have an unmarked 9-liter container, an unmarked 4-liter container, and an unlimited supply of water. In table, symbol, or paragraph form, describe how you might end up with exactly 3 liters in one of the containers.

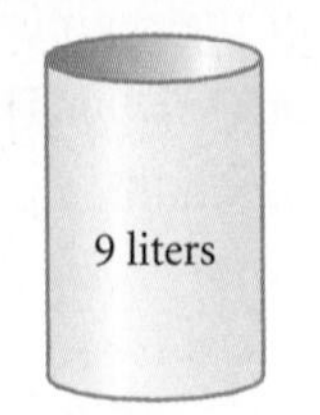

LESSON

4.3

Triangle Inequalities

Readers are plentiful, thinkers are rare.

HARRIET MARTINEAU

How long must each side of this drawbridge be so that the bridge spans the river when both sides come down?

Drawbridges over the Chicago River in Chicago, Illinois

Triangles have similar requirements. In the triangles below, the blue segments are all congruent, and the red segments are all congruent. Yet, there are a variety of triangles. Notice how the length of the yellow segment changes. Notice also how the angle measures change.

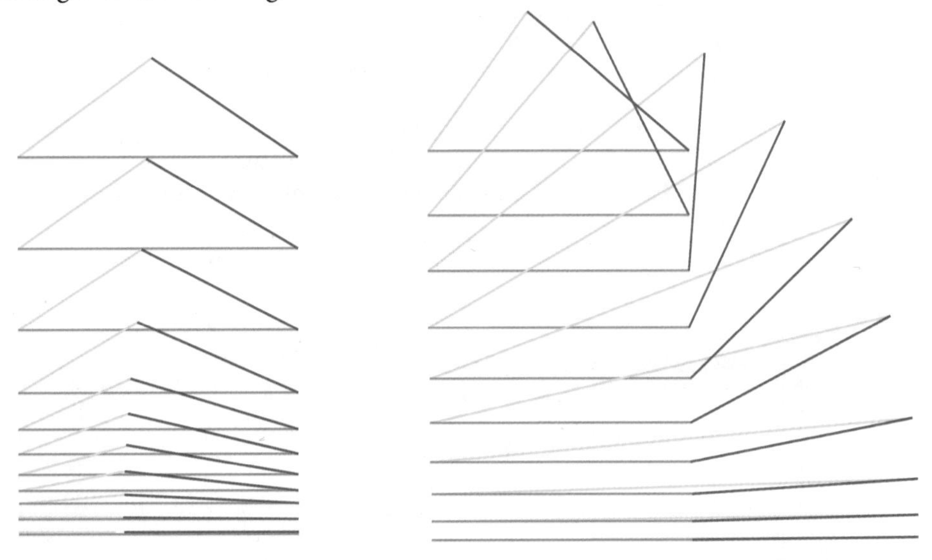

Can you form a triangle using sticks of any three lengths? How do the angle measures of a triangle relate to the lengths of its sides? In this lesson you will discover some geometric inequalities that answer these questions.

Investigation 1
What Is the Shortest Path from A to B?

You will need

- a compass
- a straightedge

Each person in your group should do each construction. Compare results when you finish.

Step 1 Construct a triangle with each set of segments as sides.

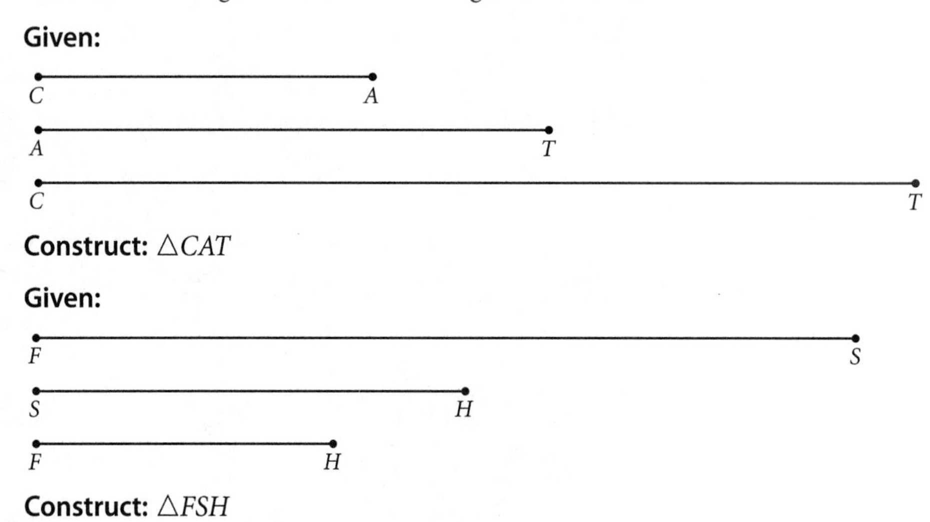

Step 2 Were you able to construct $\triangle CAT$ and $\triangle FSH$? Why or why not? Discuss your results with others. State your observations as your next conjecture.

Triangle Inequality Conjecture C-21

The sum of the lengths of any two sides of a triangle is __?__ the length of the third side.

The Triangle Inequality Conjecture relates the lengths of the three sides of a triangle. You can also think of it in another way: The shortest path between two points is along the segment connecting them. In other words, the path from A to C to B can't be shorter than the path from A to B.

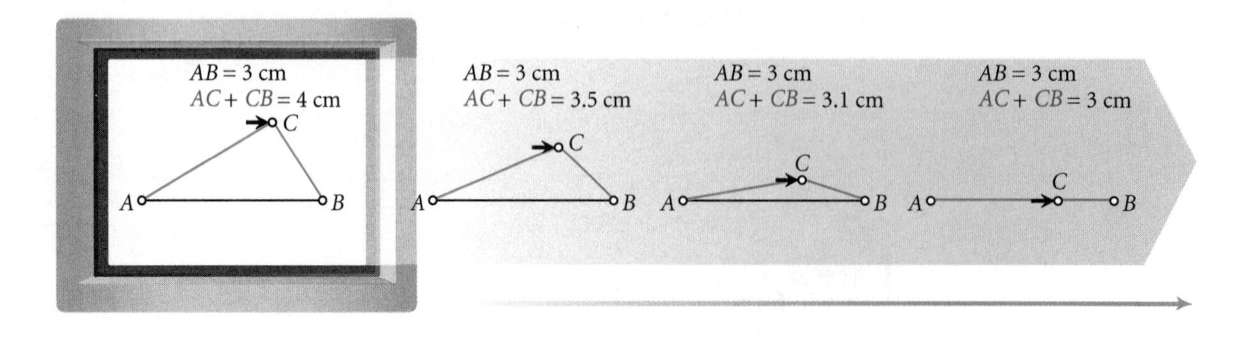

You can use geometry software to compare two different paths.

Investigation 2
Where Are the Largest and Smallest Angles?

You will need

- a ruler
- a protractor

Each person should draw a different scalene triangle for this investigation. Some group members should draw acute triangles, and some should draw obtuse triangles.

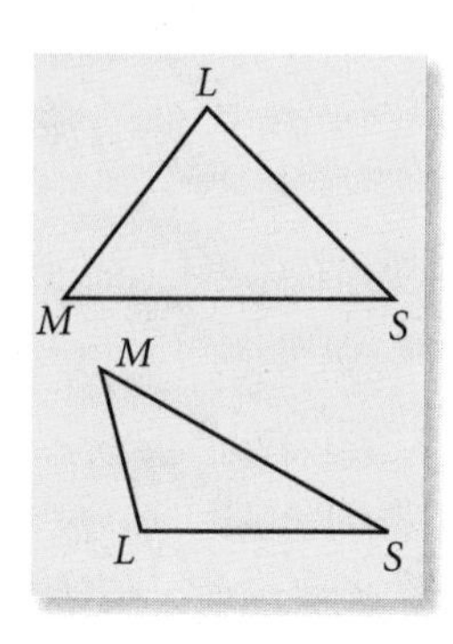

Step 1 Measure the angles in your triangle. Label the angle with greatest measure $\angle L$, the angle with second greatest measure $\angle M$, and the smallest angle $\angle S$.

Step 2 Measure the three sides. Label the longest side l, the second longest side m, and the shortest side s.

Step 3 Which side is opposite $\angle L$? $\angle M$? $\angle S$?

Discuss your results with others. Write a conjecture that states where the largest and smallest angles are in a triangle, in relation to the longest and shortest sides.

Side-Angle Inequality Conjecture C-22

In a triangle, if one side is longer than another side, then the angle opposite the longer side is _?_.

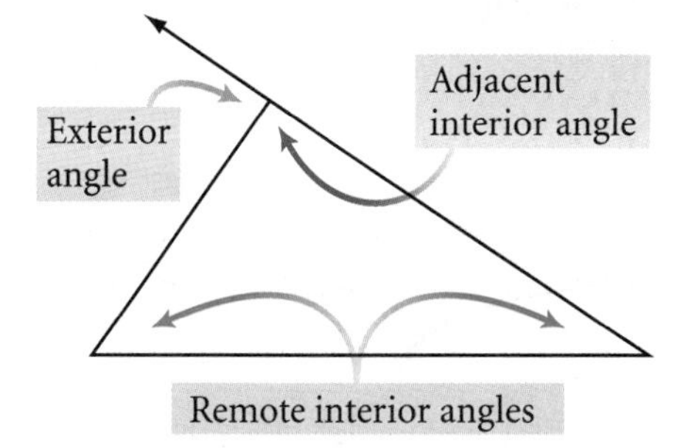

So far in this chapter, you have studied interior angles of triangles. Triangles also have exterior angles. If you extend one side of a triangle beyond its vertex, then you have constructed an **exterior angle** at that vertex.

Each exterior angle of a triangle has an **adjacent interior angle** and a pair of **remote interior angles.** The remote interior angles are the two angles in the triangle that do not share a vertex with the exterior angle.

Investigation 3
Exterior Angles of a Triangle

You will need

- a straightedge
- patty paper

Each person should draw a different scalene triangle for this investigation. Some group members should draw acute triangles, and some should draw obtuse triangles.

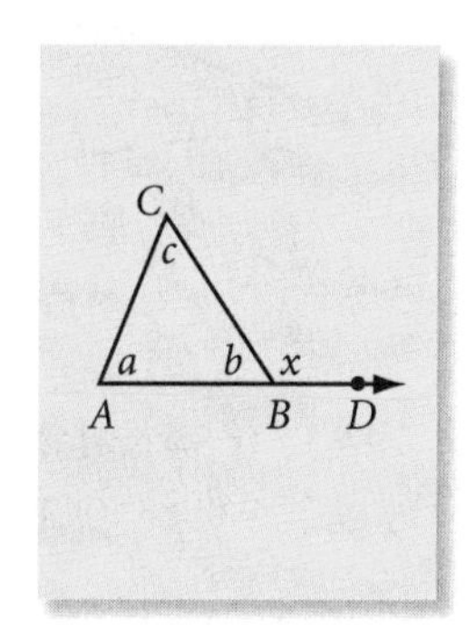

Step 1 On your paper, draw a scalene triangle, $\triangle ABC$. Extend $\overline{AB}$ beyond point B and label a point D outside the triangle on $\overrightarrow{AB}$. Label the angles as shown.

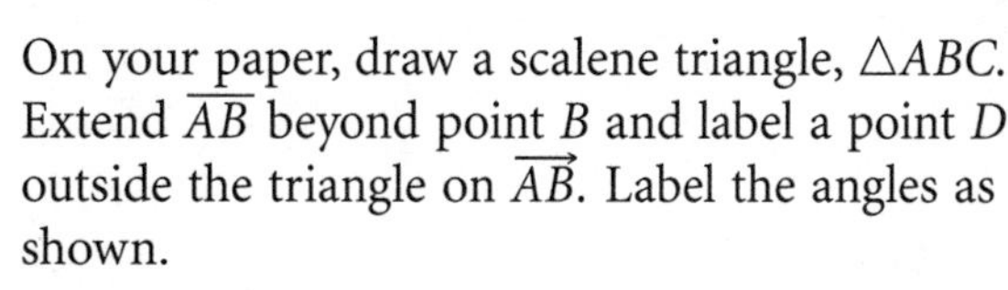

Step 2 Copy the two remote interior angles, $\angle A$ and $\angle C$, onto patty paper to show their sum.

Step 3 How does the sum of a and c compare with x? Use your patty paper from Step 2 to compare.

Step 4 Discuss your results with your group. State your observations as a conjecture.

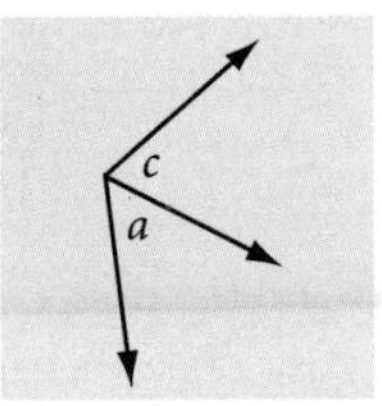

Triangle Exterior Angle Conjecture C-23

The measure of an exterior angle of a triangle $\underline{?}$.

You just discovered the Triangle Exterior Angle Conjecture by inductive reasoning. You can use the Triangle Sum Conjecture, some algebra, and deductive reasoning to show *why* the Triangle Exterior Angle Conjecture is true for all triangles.

You'll do the paragraph proof on your own in Exercise 17.

EXERCISES

In Exercises 1–4, determine whether it is possible to draw a triangle with sides of the given measures. If possible, write yes. If not possible, write no and make a sketch demonstrating why it is not possible.

1. 3 cm, 4 cm, 5 cm **2.** 4 m, 5 m, 9 m **3.** 5 ft, 6 ft, 12 ft **4.** 3.5 cm, 4.5 cm, 7 cm

In Exercises 5–10, use your new conjectures to arrange the unknown measures in order from greatest to least.

5. ⓗ **6.** **7.**

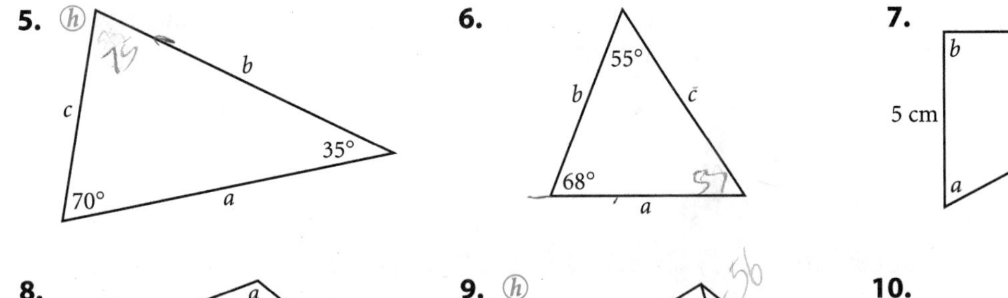

8. **9.** ⓗ **10.**

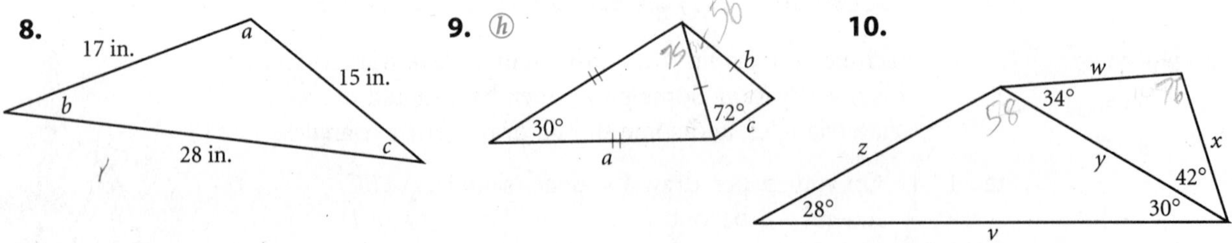

11. If 54 and 48 are the lengths of two sides of a triangle, what is the range of possible values for the length of the third side? ⓗ

12. What's wrong with this picture? Explain.

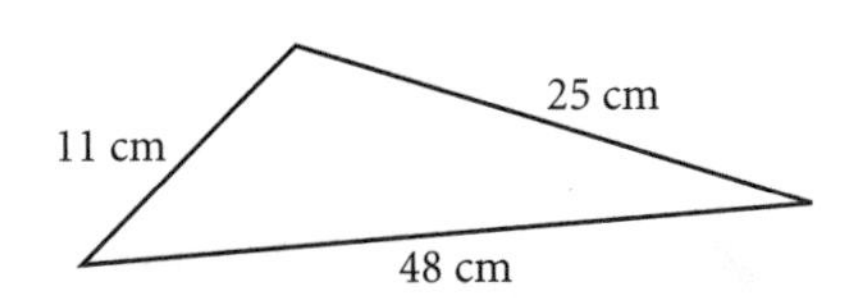

13. What's wrong with this picture? Explain. ⓗ

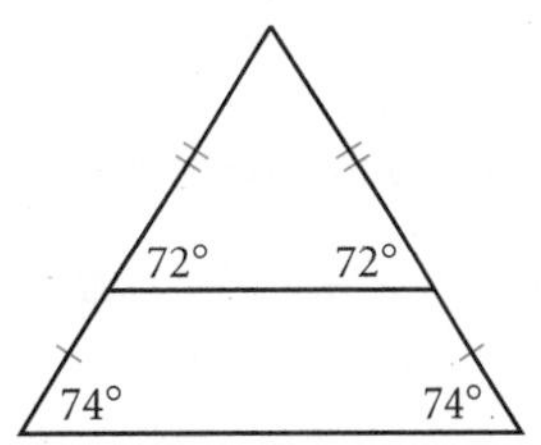

In Exercises 14–16, use one of your new conjectures to find the missing measures.

14. $t + p = \underline{?}$ ⓗ

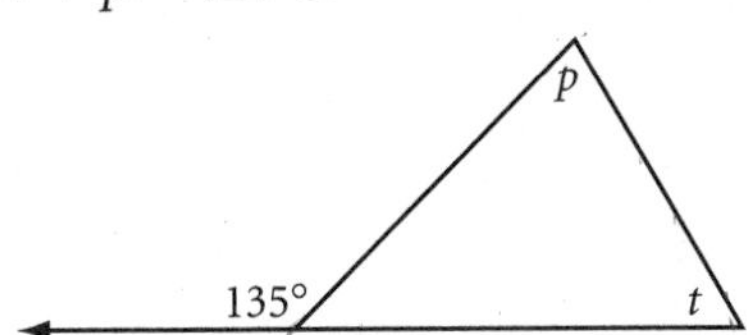

15. $r = \underline{?}$

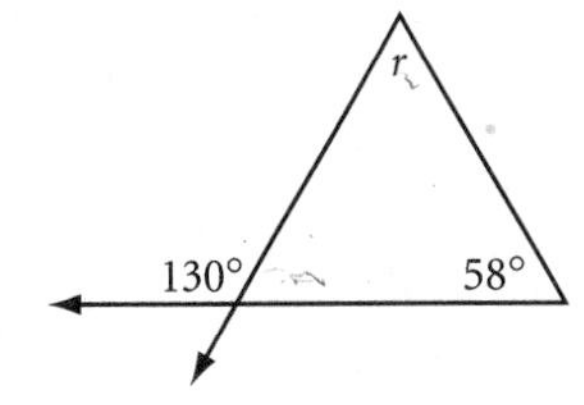

16. $x = \underline{?}$

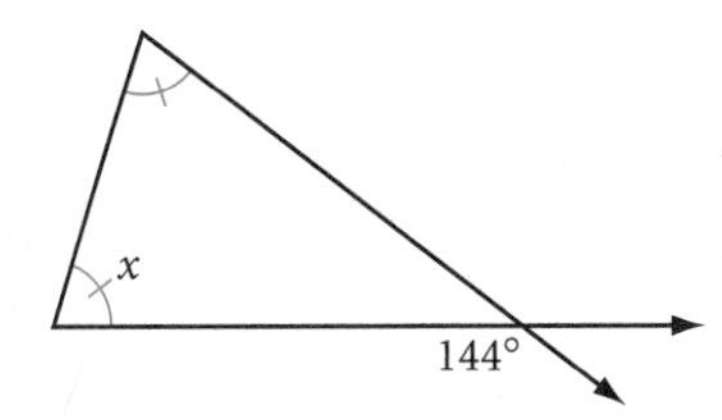

17. Use algebra and the Triangle Sum Conjecture to explain why the Triangle Exterior Angle Conjecture is true. Use the figure at right. ⓗ

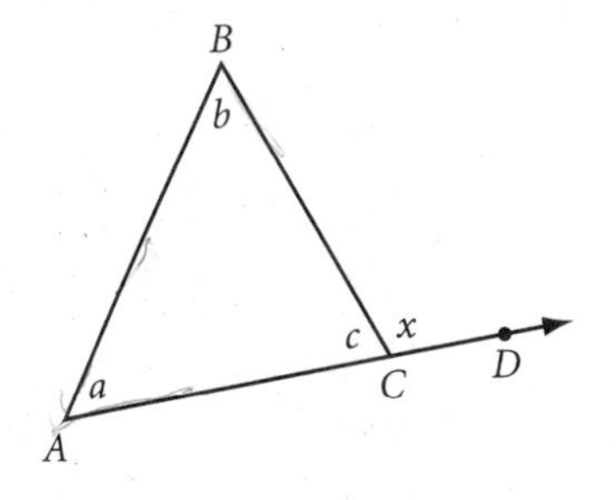

18. Read the Recreation Connection below. If you want to know the perpendicular distance from a landmark to the path of your boat, what should be the measurement of your bow angle when you begin recording?

Recreation CONNECTION

Geometry is used quite often in sailing. For example, to find the distance between the boat and a landmark on shore, sailors use a rule called *doubling the angle on the bow.* The rule says, measure the angle on the bow (the angle formed by your path and your line of sight to the landmark) at point *A*. Check your bearing until, at point *B*, the bearing is double the reading at point *A*. The distance traveled from *A* to *B* is also the distance from the landmark to your new position.

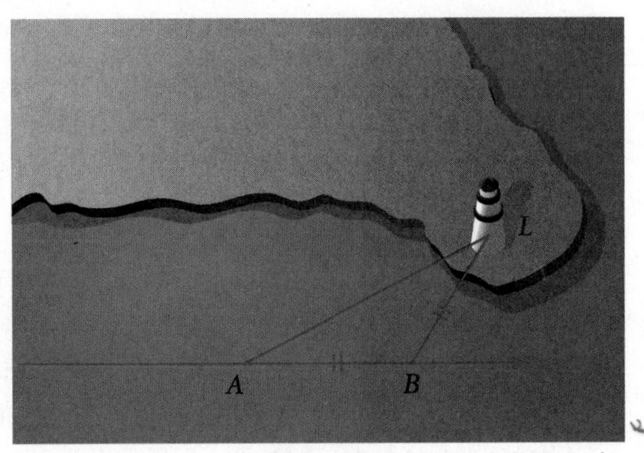

Review

In Exercises 19 and 20, calculate each lettered angle measure.

19.

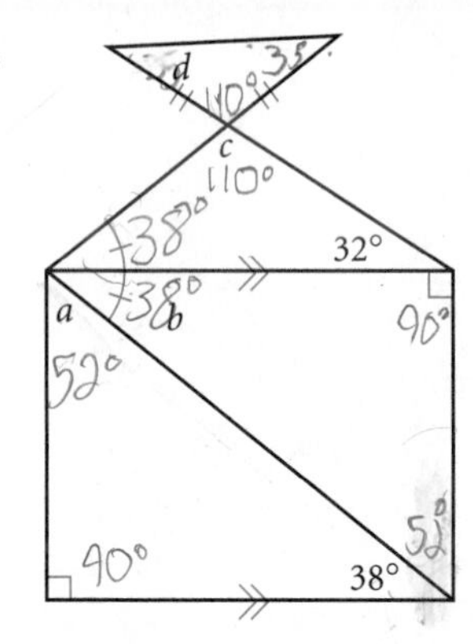

20.

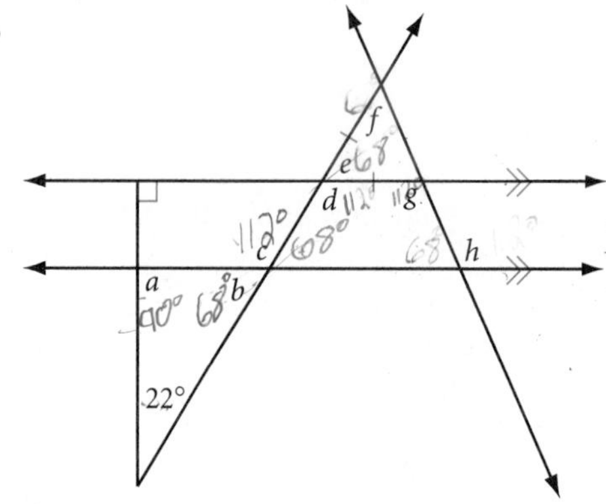

In Exercises 21–23, complete the statement of congruence.

21. $\triangle BAR \cong \triangle$ __?__ ⓗ

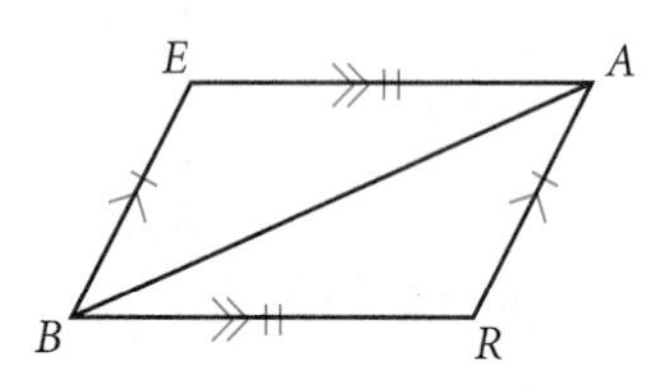

22. $\triangle FAR \cong \triangle$ __?__

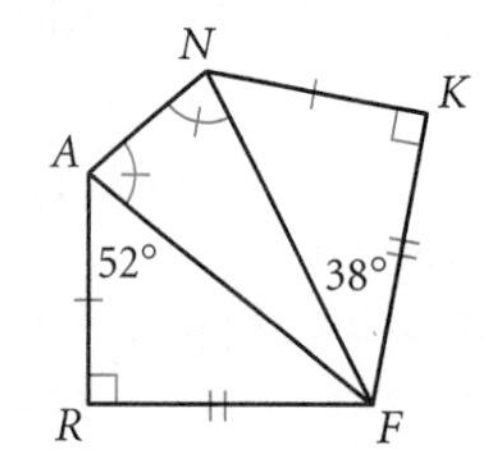

23. $\overline{HG} \cong \overline{HJ}$

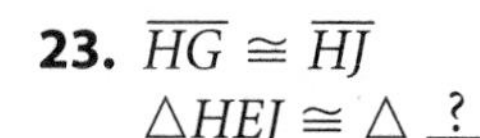

$\triangle HEJ \cong \triangle$ __?__

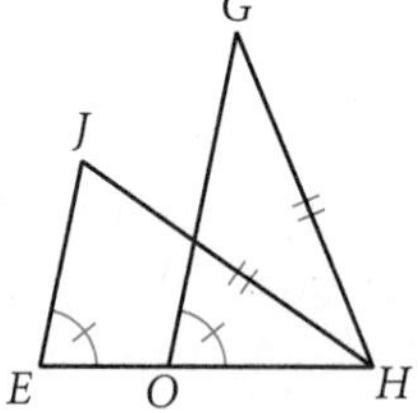

project

RANDOM TRIANGLES

Imagine you cut a 20 cm straw in two randomly selected places anywhere along its length. What is the probability that the three pieces will form a triangle? How do the locations of the cuts affect whether or not the pieces will form a triangle? Explore this situation by cutting a straw in different ways, or use geometry software to model different possibilities. Based on your informal exploration, predict the probability of the pieces forming a triangle.

Now generate a large number of randomly chosen lengths to simulate the cutting of the straw. Analyze the results and calculate the probability based on your data. How close was your prediction?

Your project should include

- Your prediction and an explanation of how you arrived at it.
- Your randomly generated data.
- An analysis of the results and your calculated probability.
- An explanation of how the location of the cuts affects the chances of a triangle being formed.

Fathom™

You can use Fathom to generate many sets of random numbers quickly. You can also set up tables to view your data, and enter formulas to calculate quantities based on your data.

LESSON

4.4 Are There Congruence Shortcuts?

The person who knows how will always have a job; the person who knows why will always be that person's boss.

ANONYMOUS

A building contractor has just assembled two massive triangular trusses to support the roof of a recreation hall. Before the crane hoists them into place, the contractor needs to verify that the two triangular trusses are identical. Must the contractor measure and compare all six parts of both triangles?

You learned from the Third Angle Conjecture that if there is a pair of angles congruent in each of two triangles, then the third angles must be congruent. But will this guarantee that the trusses are the same size? You probably need to also know something about the sides in order to be sure that two triangles are congruent. Recall from earlier exercises that *fewer* than three parts of one triangle can be congruent to corresponding parts of another triangle, without the triangles being congruent.

So let's begin looking for congruence shortcuts by comparing three parts of each triangle.

There are six different ways that the same three parts of two triangles may be congruent. They are diagrammed below. An angle that is included between two sides of a triangle is called an **included angle.** A side that is included between two angles of a triangle is called an **included side.**

Side-Side-Side (SSS)

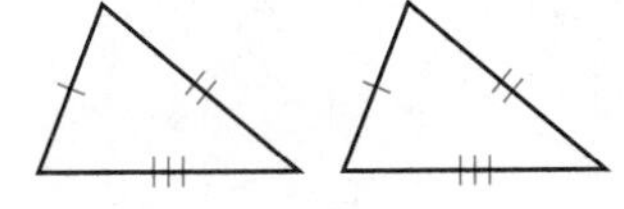

Three pairs of congruent sides

Side-Angle-Side (SAS)

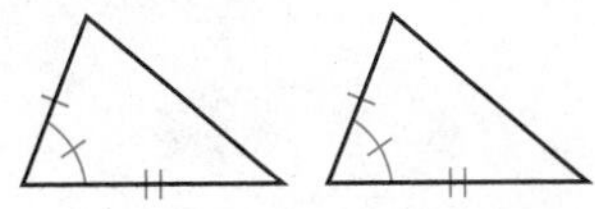

Two pairs of congruent sides and one pair of congruent angles (angles between the pairs of sides)

Angle-Side-Angle (ASA)

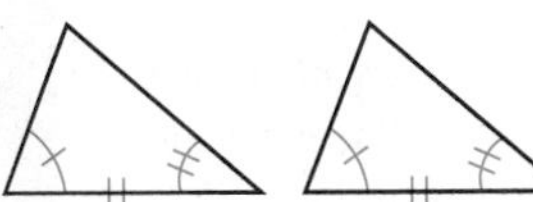

Two pairs of congruent angles and one pair of congruent sides (sides between the pairs of angles)

Side-Angle-Angle (SAA)

Two pairs of congruent angles and one pair of congruent sides (sides not between the pairs of angles)

Side-Side-Angle (SSA)

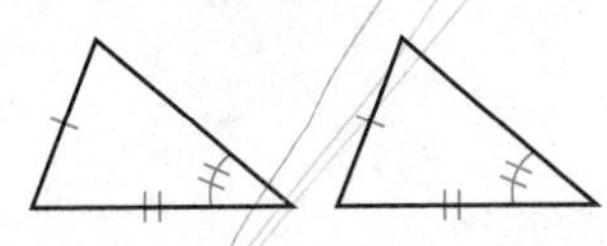

Two pairs of congruent sides and one pair of congruent angles (angles not between the pairs of sides)

Angle-Angle-Angle (AAA)

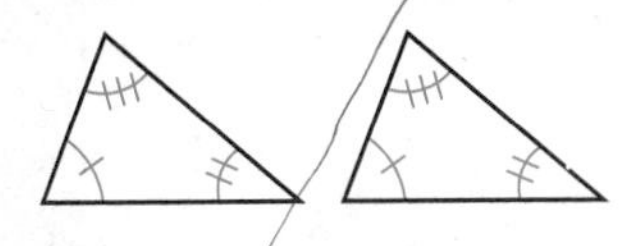

Three pairs of congruent angles

You will consider three of these cases in this lesson and three others in the next lesson. Let's begin by investigating SSS and SAS.

Investigation 1
Is SSS a Congruence Shortcut?

You will need

- a compass
- a straightedge

First you will investigate the Side-Side-Side (SSS) case. If the three sides of one triangle are congruent to the three sides of another, must the two triangles be congruent?

Step 1 Construct a triangle from the three parts shown. Be sure you match up the endpoints labeled with the same letter.

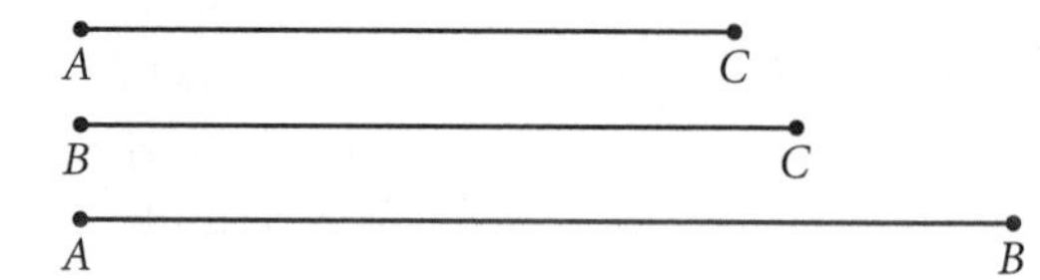

Step 2 Compare your triangle with the triangles made by others in your group. (One way to compare them is to place the triangles on top of each other and see if they coincide.) Is it possible to construct different triangles from the same three parts, or will all the triangles be congruent?

Step 3 You are now ready to complete the conjecture for the SSS case.

SSS Congruence Conjecture C-24

If the three sides of one triangle are congruent to the three sides of another triangle, then __?__.

Career CONNECTION

Congruence is very important in design and manufacturing. Modern assembly-line production relies on identical, or congruent, parts that are interchangeable. In the assembly of an automobile, for example, the same part needs to fit into each car coming down the assembly line.

Investigation 2
Is SAS a Congruence Shortcut?

You will need

- a compass
- a straightedge

Next you will consider the Side-Angle-Side (SAS) case. If two sides and the included angle of one triangle are congruent to two sides and the included angle of another, must the triangles be congruent?

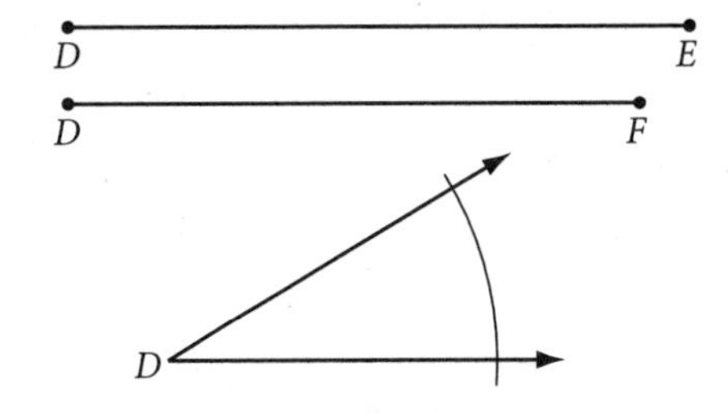

Step 1 Construct a triangle from the three parts shown. Be sure you match up the endpoints labeled with the same letter.

Step 2 Compare your triangle with the triangles made by others in your group. (One way to compare them is to place the triangles on top of each other and see if they coincide.) Is it possible to construct different triangles from the same three parts, or will all the triangles be congruent?

Step 3 You are now ready to complete the conjecture for the SAS case.

SAS Congruence Conjecture C-25

If two sides and the included angle of one triangle are congruent to two sides and the included angle of another triangle, then ?.

Next, let's look at the Side-Side-Angle (SSA) case.

EXAMPLE

If two sides and a non-included angle of one triangle are congruent to two corresponding sides and a non-included angle of another, must the triangles be congruent? In other words, can you construct only one triangle with the two sides and a non-included angle shown below?

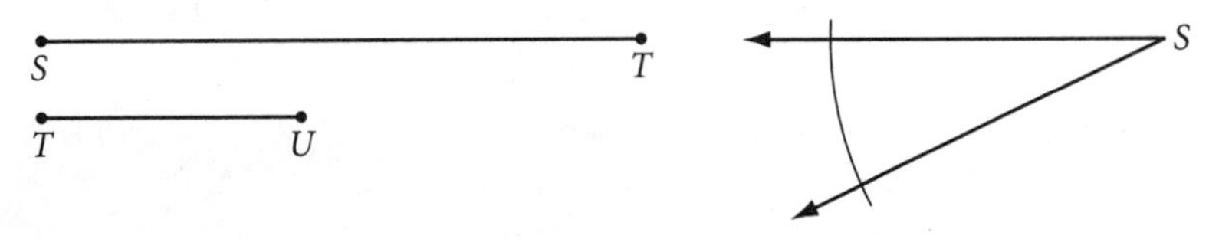

▶ Solution

Once you construct $\overline{ST}$ on a side of $\angle S$, there are two possible locations for point U on the other side of the angle.

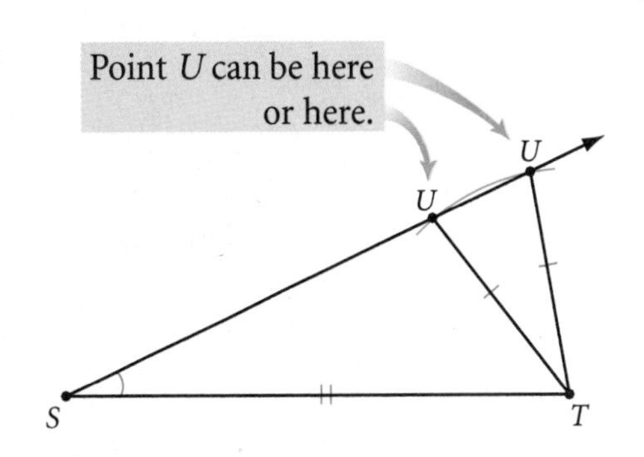

So two different triangles are possible in the SSA case, and the triangles are not necessarily congruent.

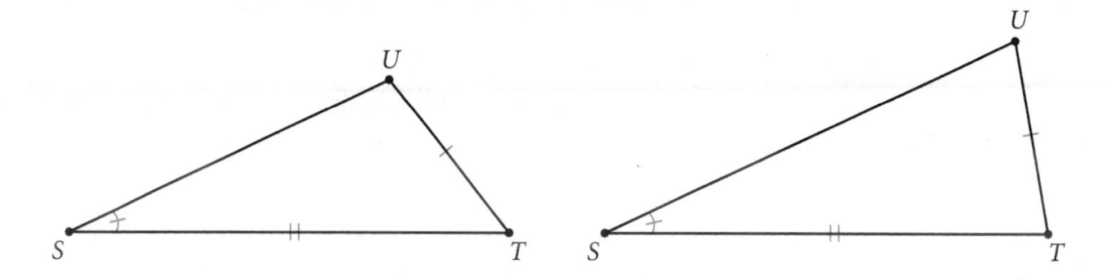

There is a counterexample for the SSA case, so it is *not* a congruence shortcut.

EXERCISES

You will need

For Exercises 1–6, decide whether the triangles are congruent, and name the congruence shortcut you used. If the triangles cannot be shown to be congruent as labeled, write "cannot be determined."

1. Which conjecture tells you $\triangle LUZ \cong \triangle IDA$? ⓗ

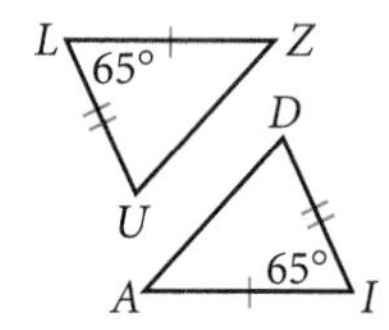

2. Which conjecture tells you $\triangle AFD \cong \triangle EFD$? ⓗ

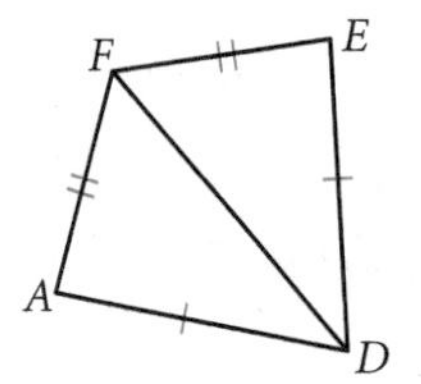

3. Which conjecture tells you $\triangle COT \cong \triangle NPA$?

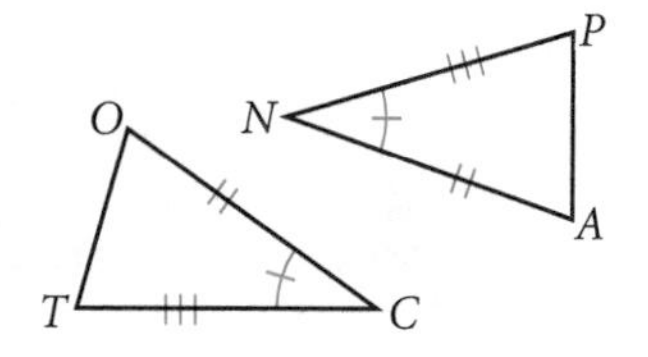

4. Which conjecture tells you $\triangle CAV \cong \triangle CEV$?

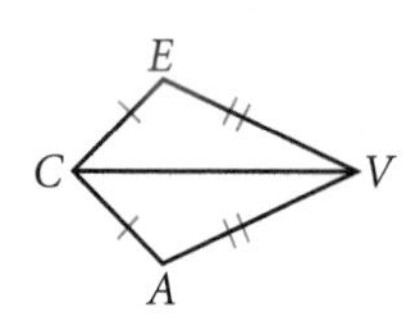

5. Which conjecture tells you $\triangle KAP \cong \triangle AKQ$?

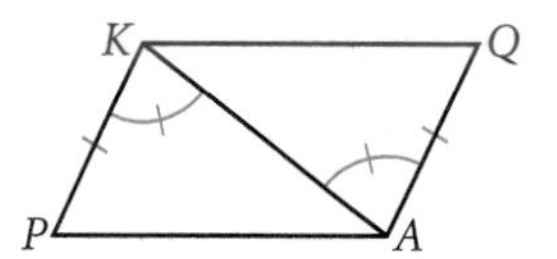

6. Y is a midpoint. Which conjecture tells you $\triangle AYB \cong \triangle RYN$?

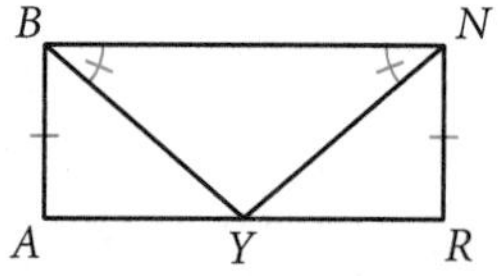

7. Explain why the boards that are nailed diagonally in the corners of this wooden gate make the gate stronger and prevent it from changing its shape under stress.

8. What's wrong with this picture?

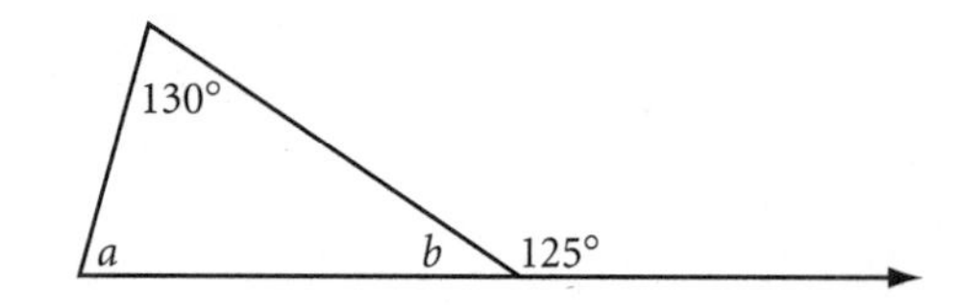

In Exercises 9–14, name a triangle congruent to the given triangle and state the congruence conjecture. If you cannot show any triangles to be congruent from the information given, write "cannot be determined" and explain why.

9. $\triangle ANT \cong \triangle$? (h)

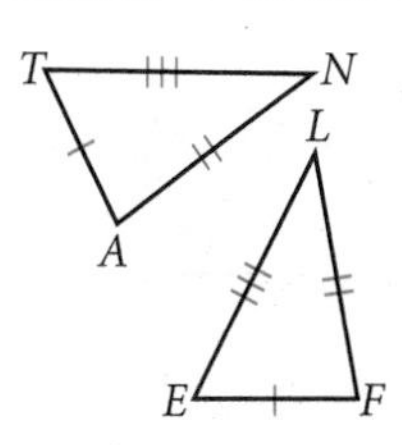

10. $\triangle RED \cong \triangle$?

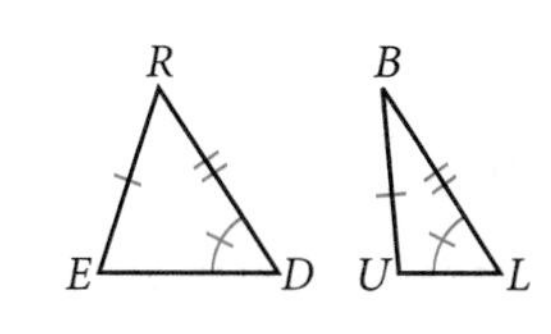

11. $\triangle WOM \cong \triangle$?

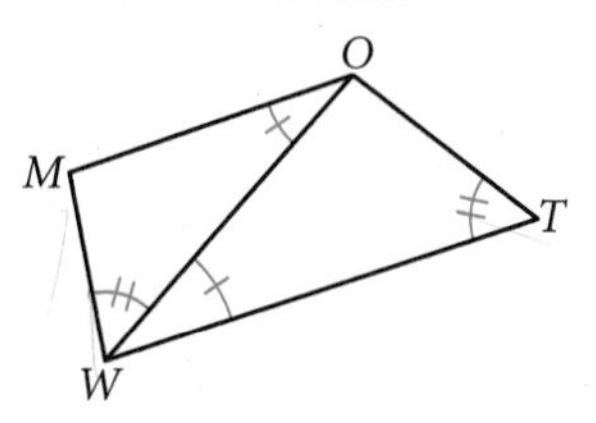

12. $\triangle MAN \cong \triangle$? (h)

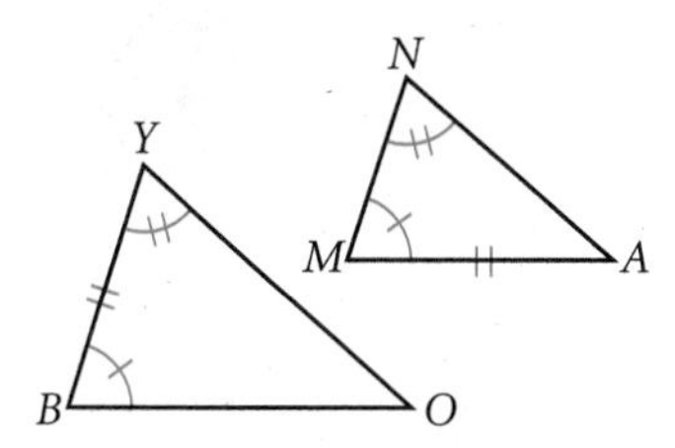

13. $\triangle SAT \cong \triangle$?

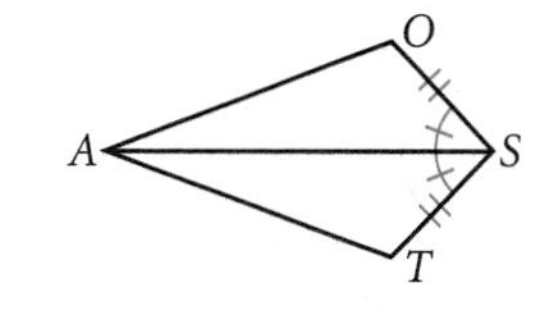

14. $\triangle GIT \cong \triangle$?

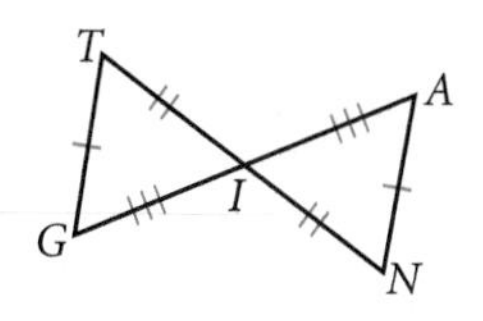

In Exercises 15 and 16, determine whether the segments or triangles in the coordinate plane are congruent and explain your reasoning.

15. $\triangle SUN \cong \triangle$? (h)

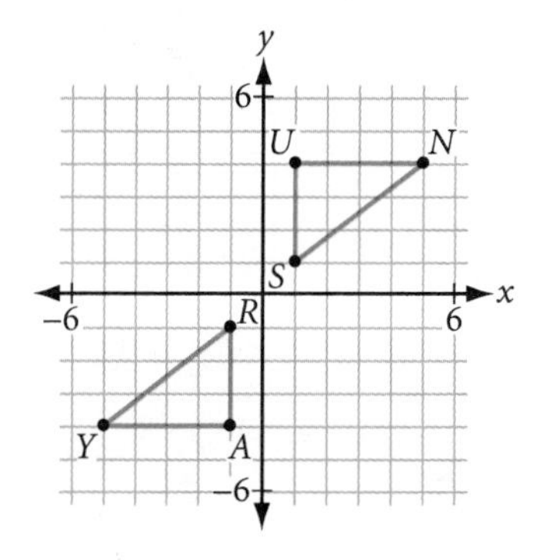

16. $\triangle DRO \cong \triangle$?

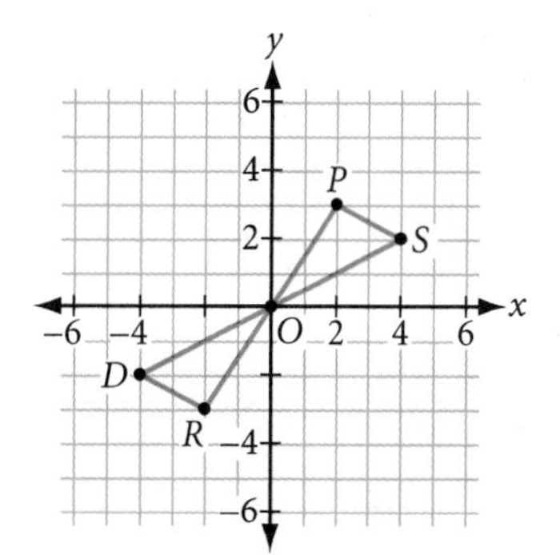

17. NASA scientists using a lunar exploration vehicle (LEV) wish to determine the distance across the deep crater shown at right. They have mapped out a path for the LEV as shown. How can the scientists use this set of measurements to calculate the approximate diameter of the crater?

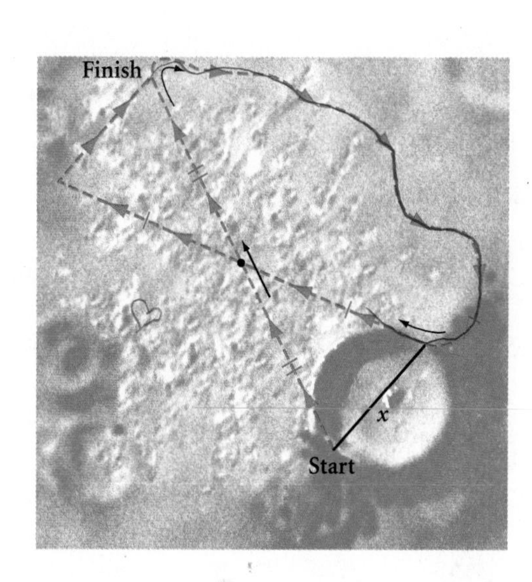

In Exercises 18 and 19, use a compass and straightedge, or patty paper, to perform these constructions.

18. ***Construction*** Draw a triangle. Use the SSS Congruence Conjecture to construct a second triangle congruent to the first.

19. ***Construction*** Draw a triangle. Use the SAS Congruence Conjecture to construct a second triangle congruent to the first.

Review

20. Copy the figure. Calculate the measure of each lettered angle.

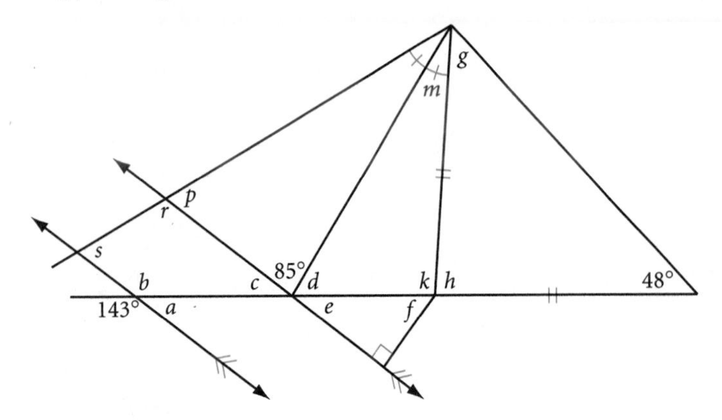

21. If two sides of a triangle measure 8 cm and 11 cm, what is the range of values for the length of the third side?

22. How many "elbow," "T," and "cross" pieces do you need to build a 20-by-20 grid? Start with the smaller grids shown below. Copy and complete the table.

Elbow: ┐
T: ┬
Cross: +

Side length	1	2	3	4	5	...	n	...	20
Elbows	4	4							
T's	0	4							
Crosses	0	1							

23. Find the point of intersection of the lines $y = \frac{2}{3}x - 1$ and $3x - 4y = 8$.

24. Isosceles right triangle ABC has vertices with coordinates $A(-8, 2)$, $B(-5, -3)$, and $C(0, 0)$. Find the coordinates of the orthocenter.

IMPROVING YOUR REASONING SKILLS

Container Problem II

You have a small cylindrical measuring glass with a maximum capacity of 250 mL. All the marks have worn off except the 150 mL and 50 mL marks. You also have a large unmarked container. It is possible to fill the large container with exactly 350 mL. How? What is the fewest number of steps required to obtain 350 mL?

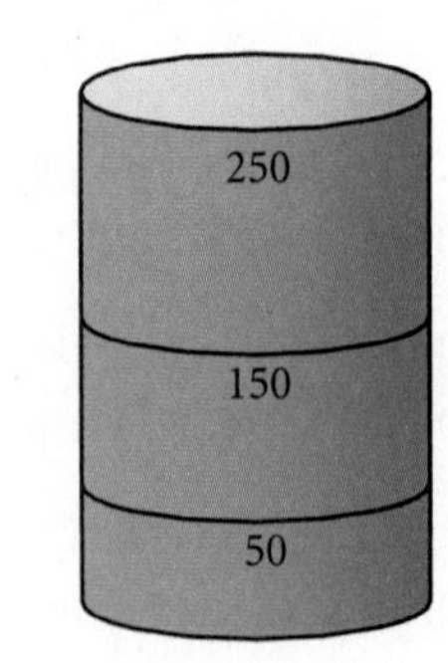

LESSON

4.5

Are There Other Congruence Shortcuts?

There is no more a math mind, than there is a history or an English mind.

GLORIA STEINEM

In the last lesson, you discovered that there are six ways that three parts of two triangles can be the same. You found that SSS and SAS both lead to the congruence of the two triangles, but that SSA does not. Is the Angle-Angle-Angle (AAA) case a congruence shortcut?

You may recall exercises that explored the AAA case. For example, these triangles have three congruent angles, but they do not have congruent sides.

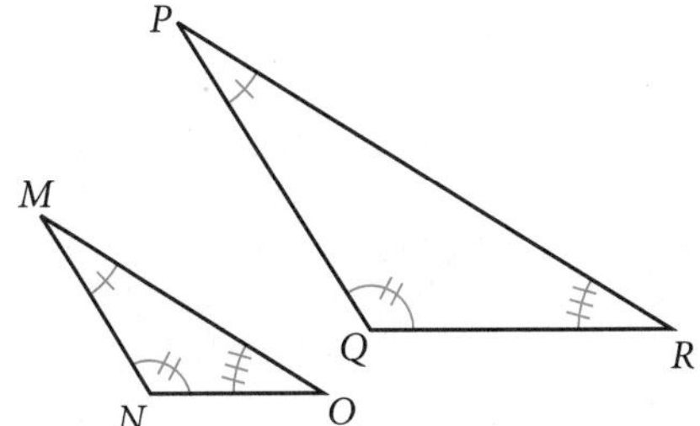

There is a counterexample for the AAA case, so it is *not* a congruence shortcut. Next, let's investigate the ASA case.

Investigation

Is ASA a Congruence Shortcut?

You will need

- a compass
- a straightedge

Consider the Angle-Side-Angle (ASA) case. If two angles and the included side of one triangle are congruent to two angles and the included side of another, must the triangles be congruent?

Step 1 Construct a triangle from the three parts shown. Be sure that the side is included between the given angles.

M T

M T

Step 2 Compare your triangle with the triangles made by others in your group. Is it possible to construct different triangles from the same three parts, or will all the triangles be congruent?

Step 3 You are now ready to complete the conjecture for the ASA case.

ASA Congruence Conjecture C-26

If two angles and the included side of one triangle are congruent to two angles and the included side of another triangle, then __?__.

Let's assume that the SSS, SAS, and ASA Congruence Conjectures are true for all pairs of triangles that have those sets of corresponding parts congruent.

The ASA case is closely related to another special case—the Side-Angle-Angle (SAA) case. You can investigate the SAA case with compass and straightedge, but you will have to use trial-and-error to accurately locate the second angle vertex because the side that is given is not the included side.

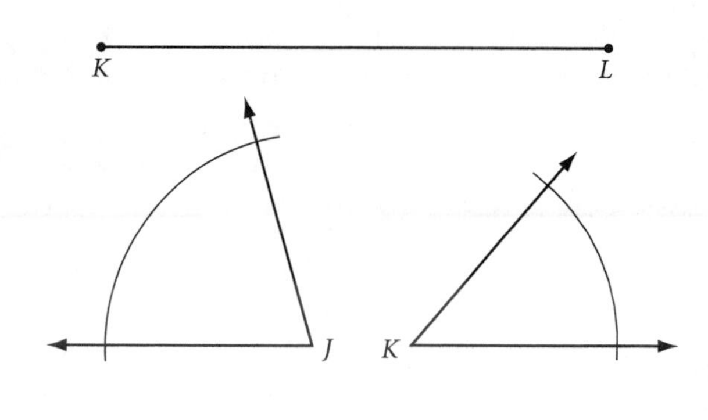

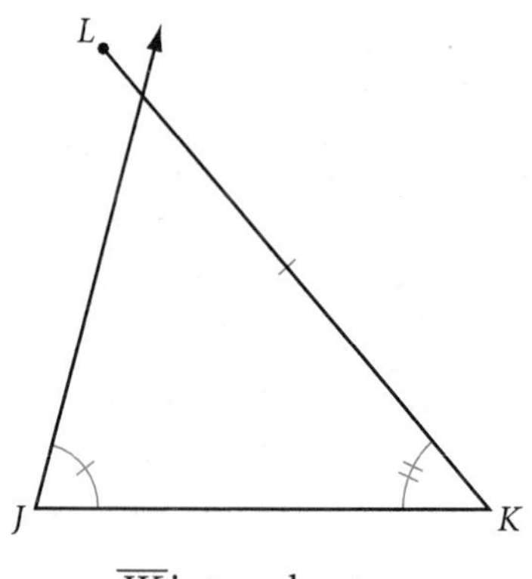

$\overline{JK}$ is too short.

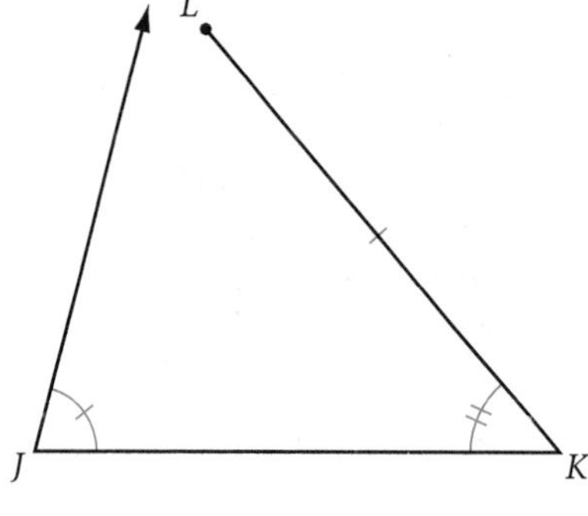

$\overline{JK}$ is too long.

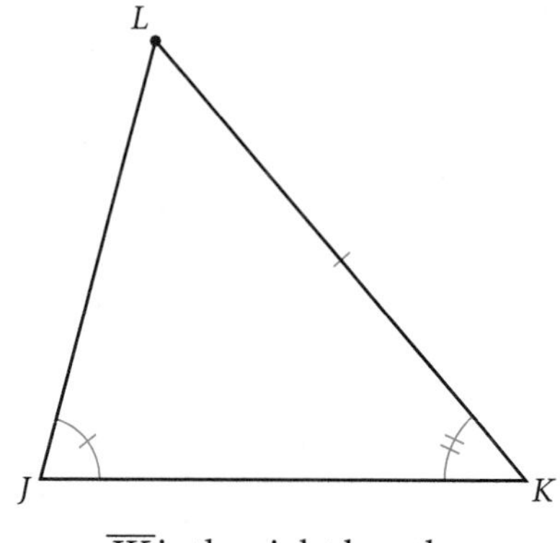

$\overline{JK}$ is the right length.

If two angles and a non-included side of one triangle are congruent to the corresponding two angles and non-included side of another, must the triangles be congruent? Let's look at it deductively.

EXAMPLE

In triangles *ABC* and *XYZ*, $\angle A \cong \angle X$, $\angle B \cong \angle Y$, and $\overline{BC} \cong \overline{YZ}$. Is $\triangle ABC \cong \triangle XYZ$? Explain your answer in a paragraph.

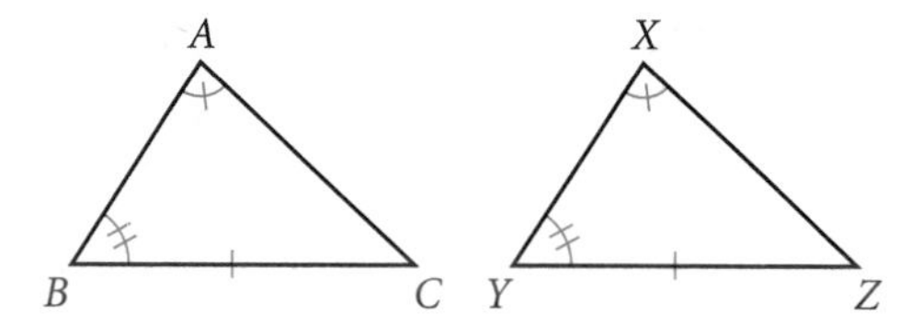

▶ Solution

Two angles in one triangle are congruent to two angles in another. The Third Angle Conjecture says that $\angle C \cong \angle Z$. The diagram now looks like this:

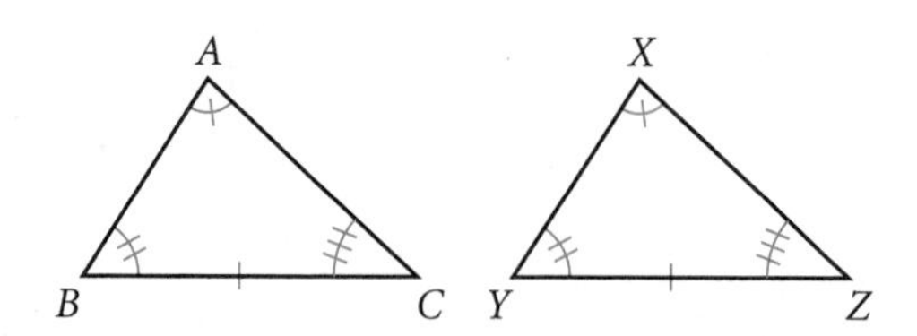

So you now have two angles and the *included* side of one triangle congruent to two angles and the included side of another. By the ASA Congruence Conjecture, $\triangle ABC \cong \triangle XYZ$.

So the SAA Congruence Conjecture follows easily from the ASA Congruence Conjecture. Complete the conjecture for the SAA case.

SAA Congruence Conjecture C-27

If two angles and a non-included side of one triangle are congruent to the corresponding angles and side of another triangle, then __?__.

Four of the six cases—SSS, SAS, ASA, and SAA—turned out to be congruence shortcuts. The diagram for each case is shown below.

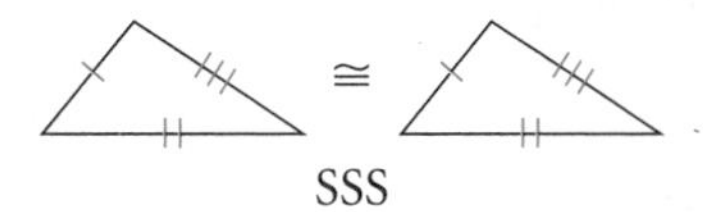

SSS

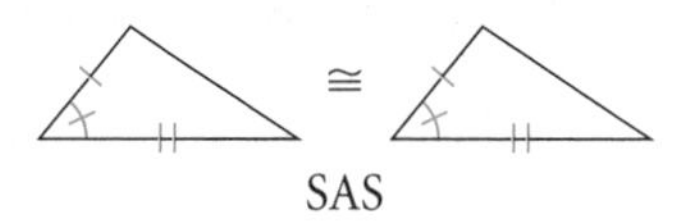

SAS

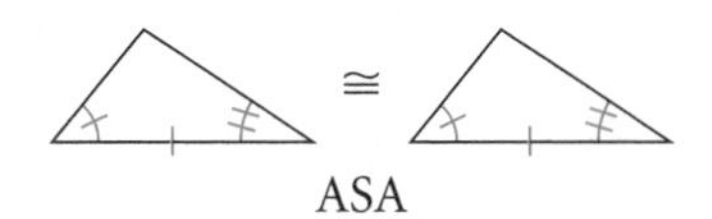

ASA

SAA

Add these diagrams, along with your congruence shortcut conjectures, to your conjecture list.

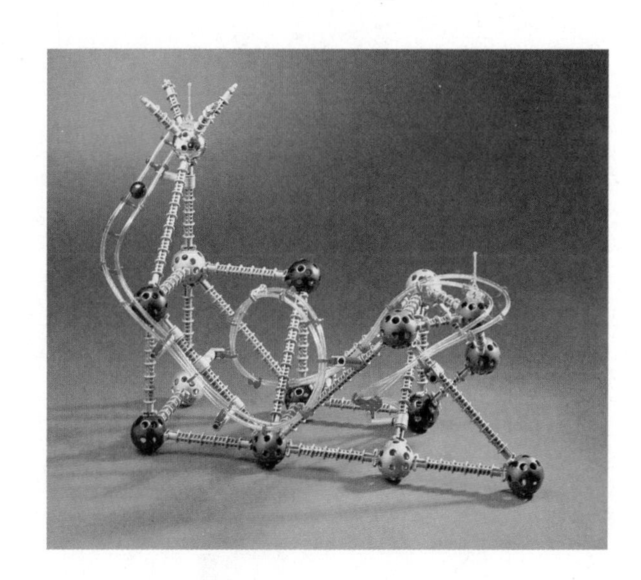

Many structures use congruent triangles for symmetry and strength. Can you tell which triangles in this toy structure are congruent?

EXERCISES

You will need

Construction tools for Exercises **17–19, 20,** and **23**

For Exercises 1–6, determine whether the triangles are congruent, and name the congruence shortcut. If the triangles cannot be shown to be congruent, write "cannot be determined."

1. $\triangle AMD \cong \triangle RMC$

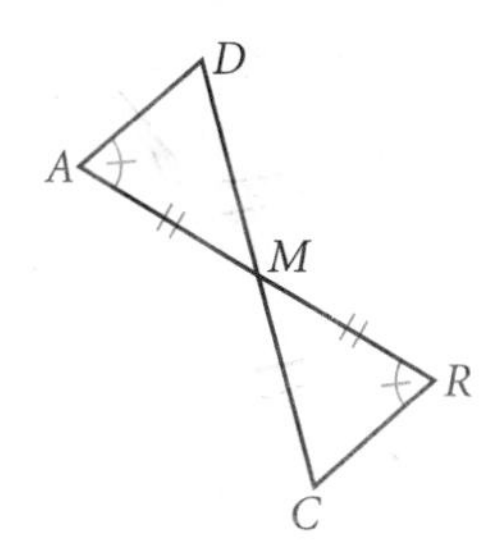

2. $\triangle BOX \cong \triangle CAR$

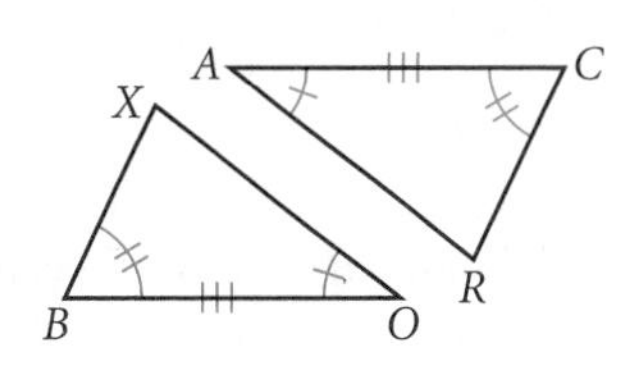

3. $\triangle GAS \cong \triangle IOL$ ⓗ

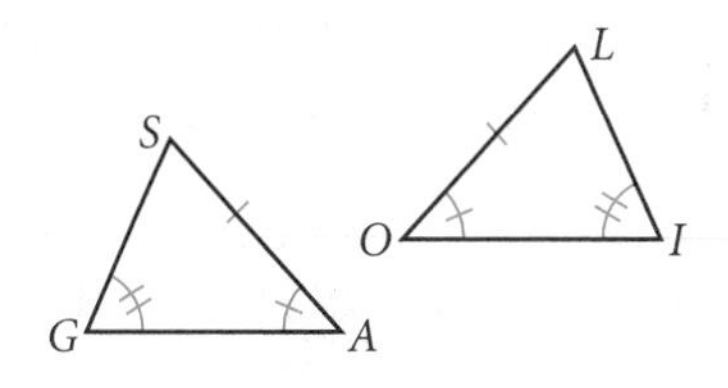

4. $\triangle HOW \cong \triangle FEW$

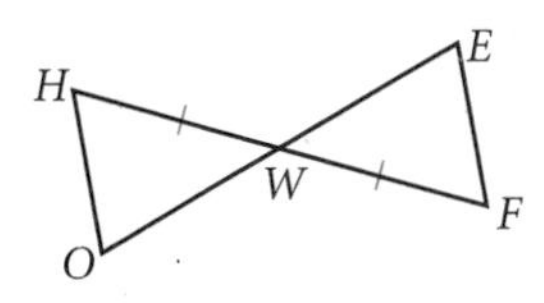

5. $\triangle FSH \cong \triangle FSI$

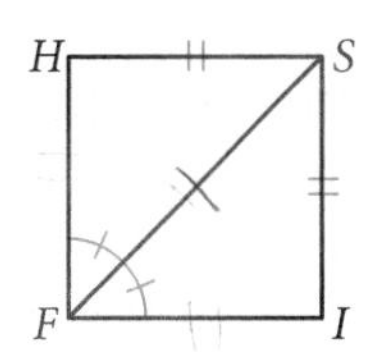

6. $\triangle ALT \cong \triangle INT$

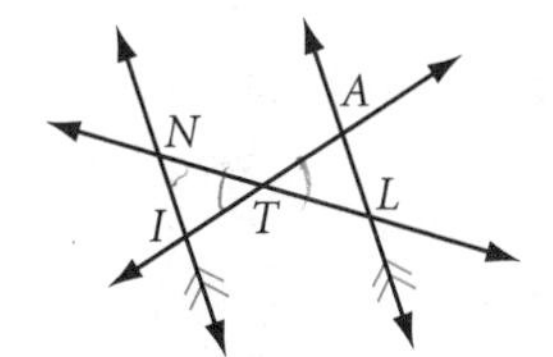

In Exercises 7–14, name a triangle congruent to the triangle given and state the congruence conjecture. If you cannot show any triangles to be congruent from the information given, write "cannot be determined" and explain why.

7. $\triangle FAD \cong \triangle$?

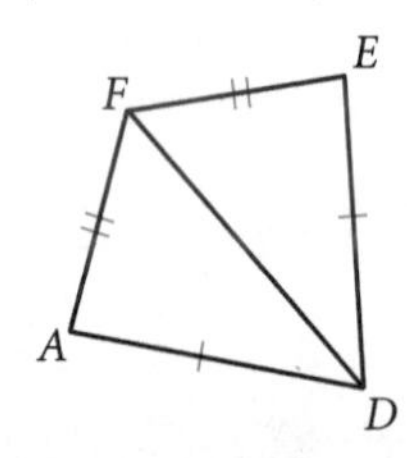

8. $\overline{OH} \parallel \overline{AT}$
$\triangle WHO \cong \triangle$?

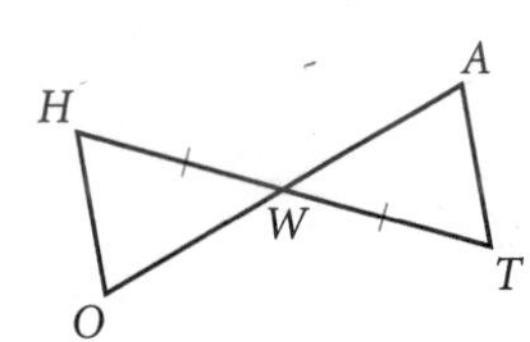

9. $\overline{AT}$ is an angle bisector.
$\triangle LAT \cong \triangle$?

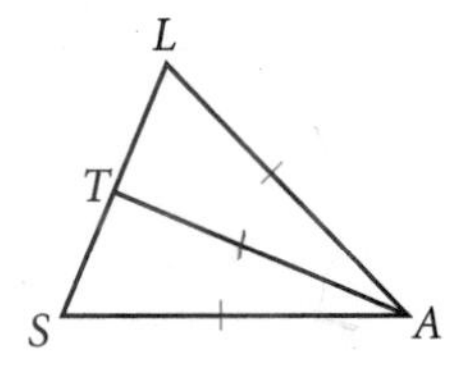

10. $PO = PR$
$\triangle POE \cong \triangle$?
$\triangle SON \cong \triangle$?

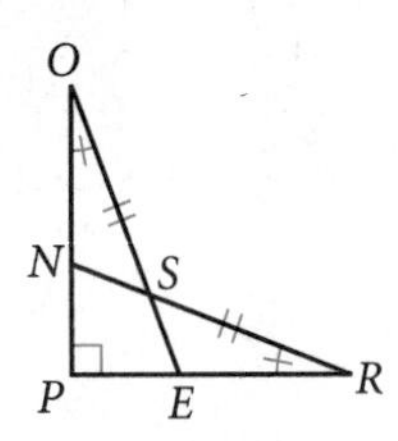

11. $\triangle$? $\cong \triangle$?

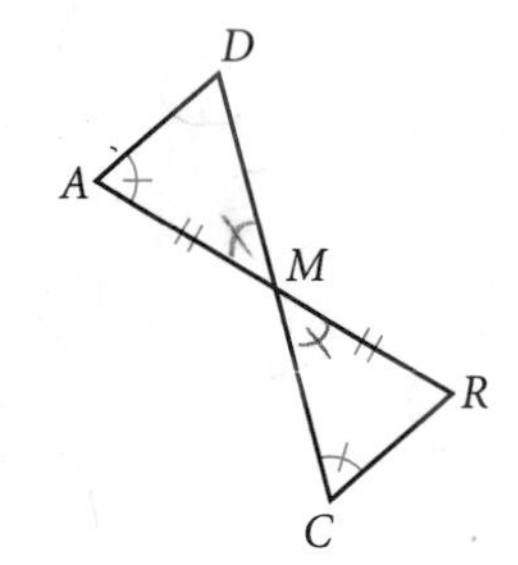

12. $\triangle RMF \cong \triangle$?

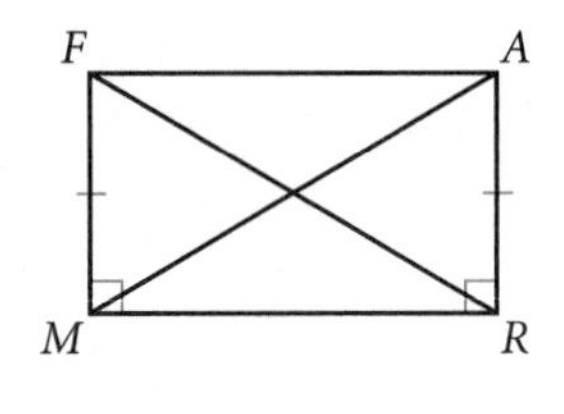

13. $\triangle BLA \cong \triangle$?

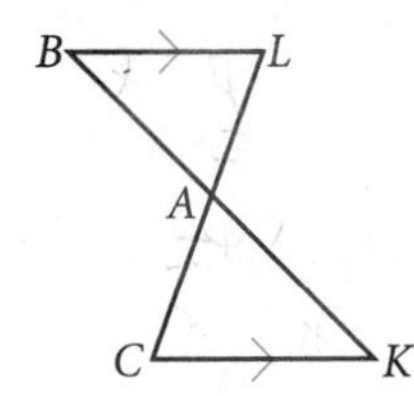

14. $\triangle LAW \cong \triangle$?

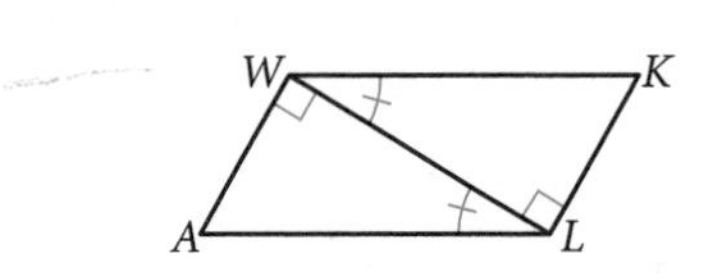

15. $\triangle SLN$ is equilateral.
Is $\triangle TIE$ equilateral? Explain.

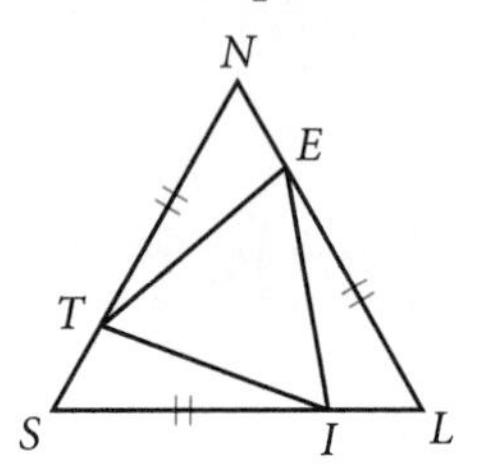

16. Use slope properties to show $\overline{AB} \perp \overline{BC}$, $\overline{CD} \perp \overline{DA}$, and $\overline{BC} \parallel \overline{DA}$.
$\triangle ABC \cong \triangle$?. Why?

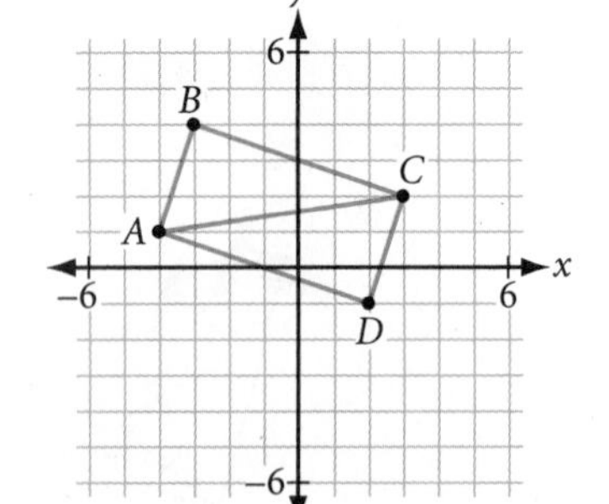

In Exercises 17–19, use a compass and a straightedge, or patty paper, to perform each construction.

17. ***Construction*** Draw a triangle. Use the ASA Congruence Conjecture to construct a second triangle congruent to the first. Write a paragraph to justify your steps.

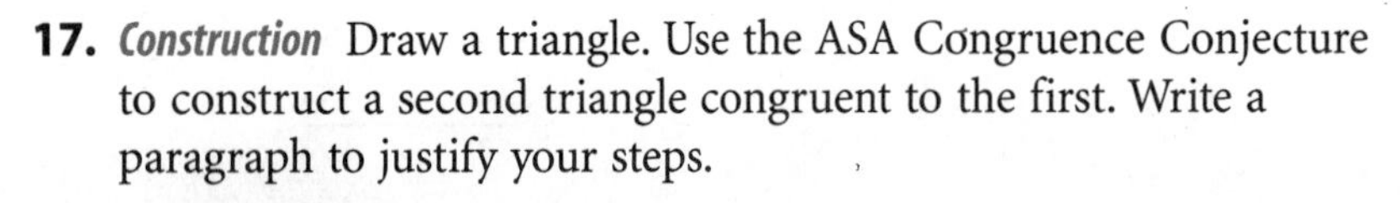

18. ***Construction*** Draw a triangle. Use the SAA Congruence Conjecture to construct a second triangle congruent to the first. Write a paragraph to justify your method.

19. ***Construction*** Construct two triangles that are not congruent, even though the three angles of one triangle are congruent to the three angles of the other. ⓗ

Review

20. *Construction* Using only a compass and a straightedge, construct an isosceles triangle with a vertex angle that measures 135°.

21. If n concurrent lines divide the plane into 250 parts then $n = \underline{?}$.

22. "If the two diagonals of a quadrilateral are perpendicular, then the quadrilateral is a rhombus." Explain why this statement is true or sketch a counterexample.

23. *Construction* Construct an isosceles right triangle with $\overline{KM}$ as one of the legs. How many noncongruent triangles can you construct? Why?

24. Sketch five lines in a plane that intersect in exactly five points. Now do this in a different way.

25. **APPLICATION** Scientists use seismograms and a method called **triangulation** to pinpoint the epicenter of an earthquake.

a. Data recorded for one quake show that the epicenter is 480 km from Eureka, California; 720 km from Elko, Nevada; and 640 km from Las Vegas, Nevada. Trace the locations of these three towns and use the scale and your construction tools to find the location of the epicenter.

b. Is it necessary to have seismogram information from three towns? Would two towns suffice? Explain.

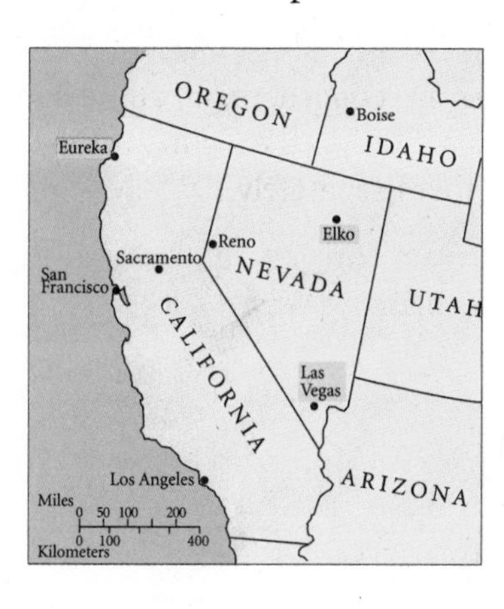

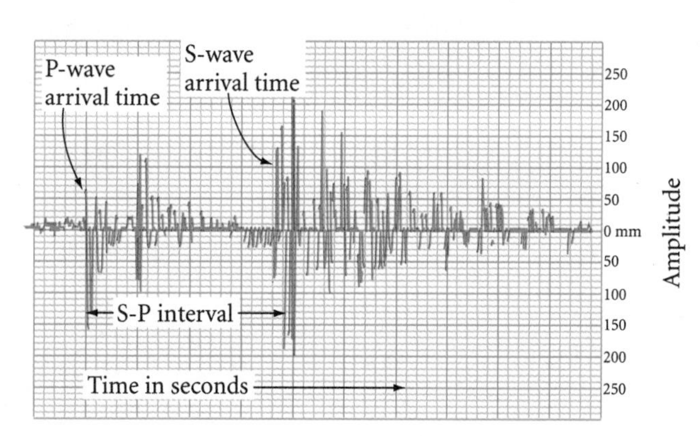

IMPROVING YOUR ALGEBRA SKILLS

Algebraic Sequences I

Find the next two terms of each algebraic sequence.

$x + 3y$, $2x + y$, $3x + 4y$, $5x + 5y$, $8x + 9y$, $13x + 14y$, $\underline{?}$, $\underline{?}$

$x + 7y$, $2x + 2y$, $4x - 3y$, $8x - 8y$, $16x - 13y$, $32x - 18y$, $\underline{?}$, $\underline{?}$

LESSON

4.6 Corresponding Parts of Congruent Triangles

The job of the younger generation is to find solutions to the solutions found by the older generation.

ANONYMOUS

In Lessons 4.4 and 4.5, you discovered four shortcuts for showing that two triangles are congruent—SSS, SAS, ASA, and SAA. The definition of congruent triangles states that if two triangles are congruent, then the ***corresponding parts of those congruent triangles are congruent***. We'll use the letters **CPCTC** to refer to the definition. Let's see how you can use congruent triangles and CPCTC.

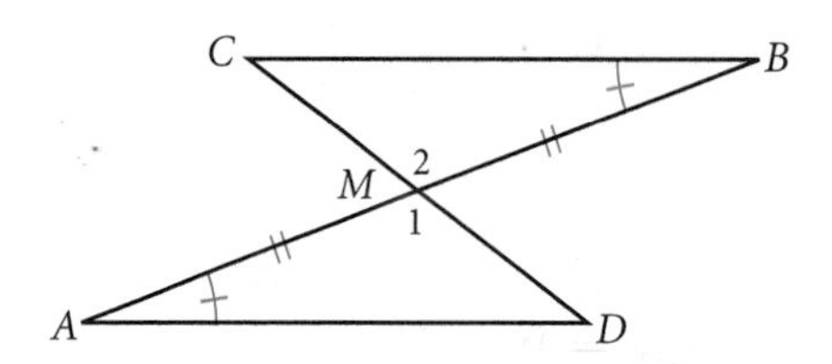

EXAMPLE A

Is $\overline{AD} \cong \overline{BC}$ in the figure above? Use a deductive argument to explain why they must be congruent.

▶ **Solution**

Here is one possible explanation: $\angle 1 \cong \angle 2$ because they are vertical angles. And it is given that $\overline{AM} \cong \overline{BM}$ and $\angle A \cong \angle B$. So, by ASA, $\triangle AMD \cong \triangle BMC$. Because the triangles are congruent, $\overline{AD} \cong \overline{BC}$ by CPCTC.

If you use a congruence shortcut to show that two triangles are congruent, then you can use CPCTC to show that any of their corresponding parts are congruent.

When you are trying to prove that triangles are congruent, it can be hard to keep track of what you know. Mark all the information on the figure. If the triangles are hard to see, use different colors or redraw them separately.

EXAMPLE B

Is $\overline{AE} \cong \overline{BD}$? Write a paragraph proof explaining why.

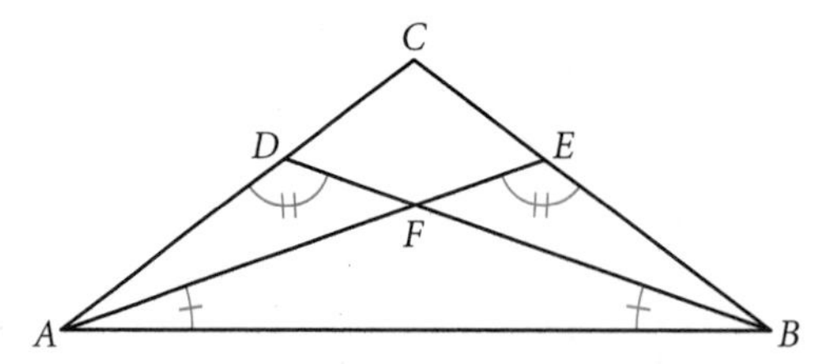

▶ **Solution**

The triangles you can use to show congruence are $\triangle ABD$ and $\triangle BAE$. You can separate or color them to see them more clearly.

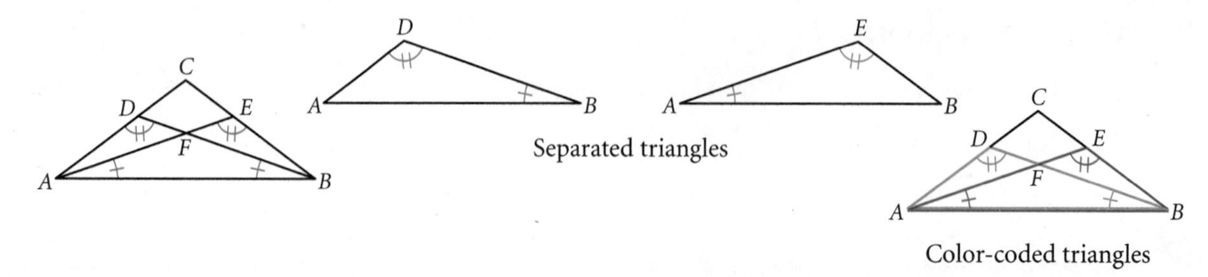

Separated triangles

Color-coded triangles

You can see that the two triangles have two pairs of congruent angles and they share a side.

Paragraph Proof: Show that $\overline{AE} \cong \overline{BD}$.

In $\triangle ABD$ and $\triangle BAE$, $\angle D \cong \angle E$ and $\angle B \cong \angle A$. Also, $\overline{AB} \cong \overline{BA}$ because they are the same segment. So $\triangle ABD \cong \triangle BAE$ by SAA. By CPCTC, $\overline{AE} \cong \overline{BD}$. ■

EXERCISES

You will need

Construction tools for Exercises **16** and **17**

For Exercises 1–9, copy the figures onto your paper and mark them with the given information. Answer the question about segment or angle congruence. If your answer is yes, write a paragraph proof explaining why. Remember to state which congruence shortcut you used. If there is not enough information to prove congruence, write "cannot be determined."

1. $\angle A \cong \angle C$, $\angle ABD \cong \angle CBD$
Is $\overline{AB} \cong \overline{CB}$? ⓗ

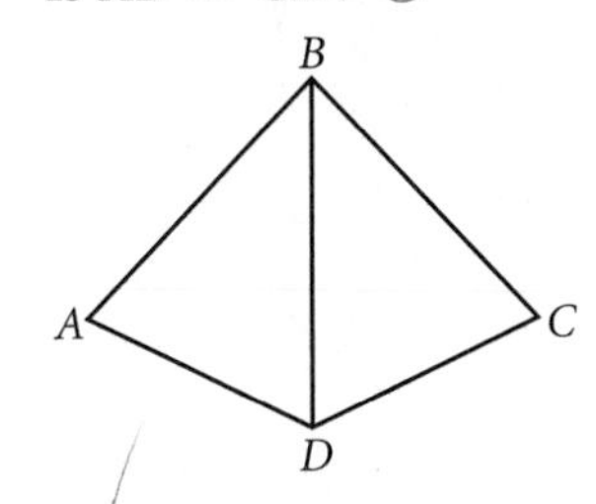

2. $\overline{CN} \cong \overline{WN}$, $\angle C \cong \angle W$
Is $\overline{RN} \cong \overline{ON}$? ⓗ

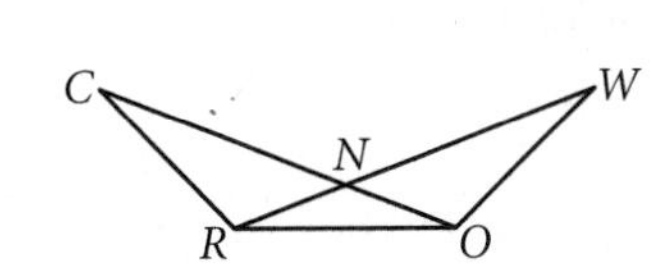

3. $\overline{CS} \cong \overline{HR}$, $\angle 1 \cong \angle 2$
Is $\overline{CR} \cong \overline{HS}$?

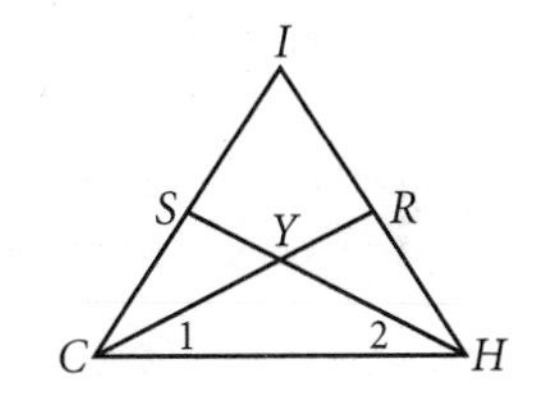

4. $\angle S \cong \angle I$, $\angle G \cong \angle A$
T is the midpoint of $\overline{SI}$.
Is $\overline{SG} \cong \overline{IA}$? ⓗ

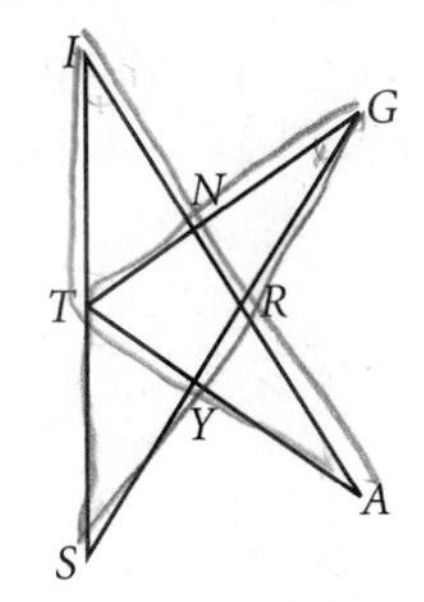

5. $\overline{FO} \cong \overline{FR}$, $\overline{UO} \cong \overline{UR}$
Is $\angle O \cong \angle R$? ⓗ

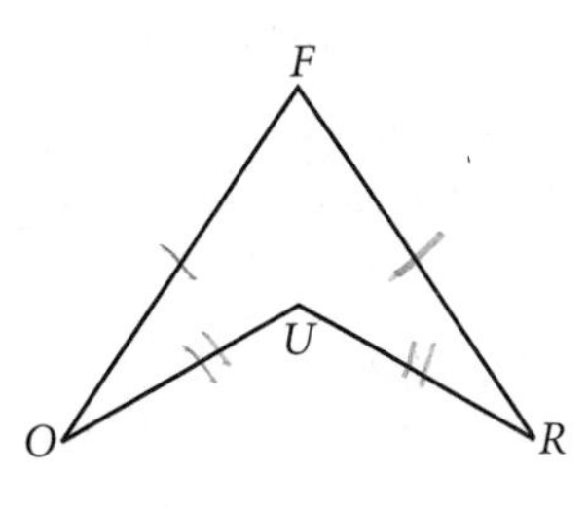

6. $\overline{MN} \cong \overline{MA}$, $\overline{ME} \cong \overline{MR}$
Is $\angle E \cong \angle R$?

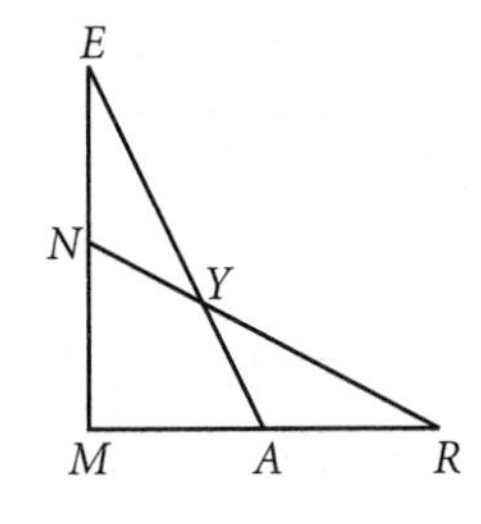

7. $\overline{BT} \cong \overline{EU}$, $\overline{BU} \cong \overline{ET}$
Is $\angle B \cong \angle E$? ⓗ

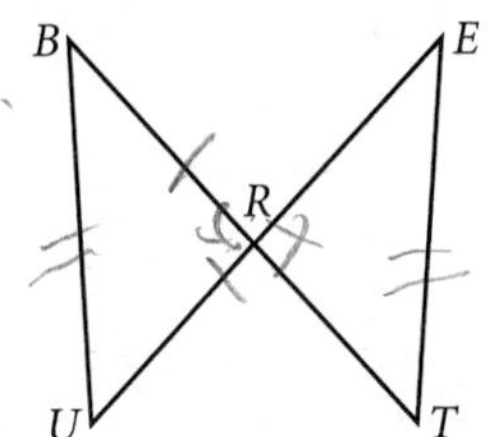

8. $HALF$ is a parallelogram.
Is $\overline{HA} \cong \overline{HF}$?

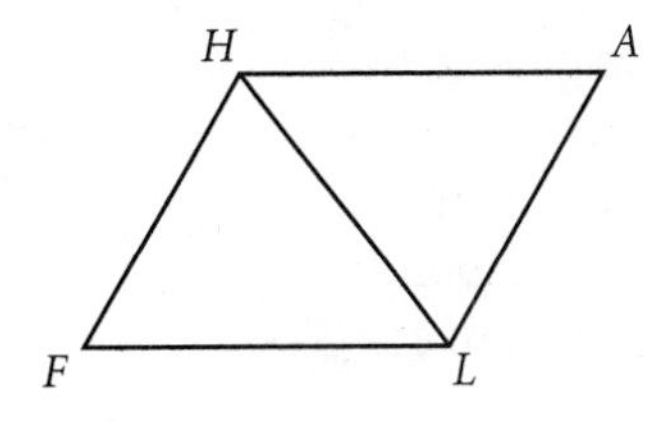

9. $\angle D \cong \angle C$, $\angle O \cong \angle A$, $\angle G \cong \angle T$. Is $\overline{TA} \cong \overline{GO}$?

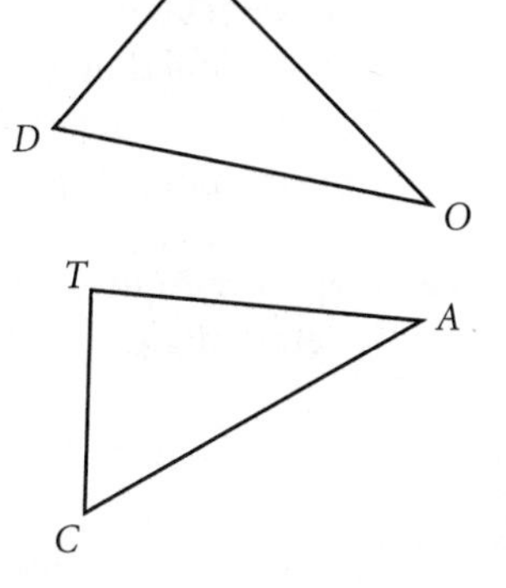

For Exercises 10 and 11, you can use the right angles and the lengths of horizontal and vertical segments shown on the grid. Answer the question about segment or angle congruence. If your answer is yes, explain why.

10. Is $\overline{FR} \cong \overline{GT}$? Why? ⓗ

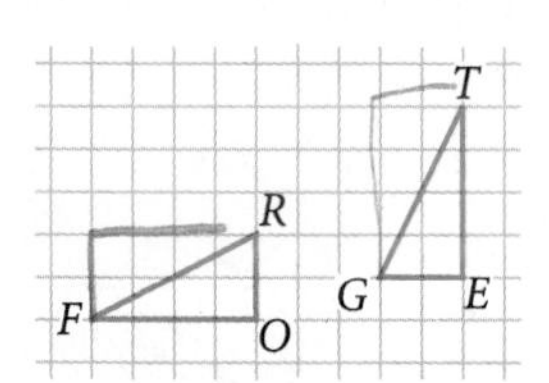

11. Is $\angle OND \cong \angle OCR$? Why?

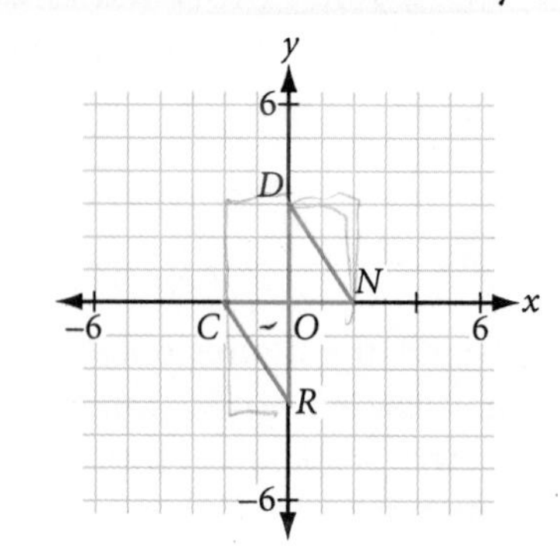

12. In Chapter 3, you used inductive reasoning to discover how to duplicate an angle using a compass and straightedge. Now you have the skills to explain *why* the construction works using deductive reasoning. The construction is shown at right. Write a paragraph proof explaining why it works.

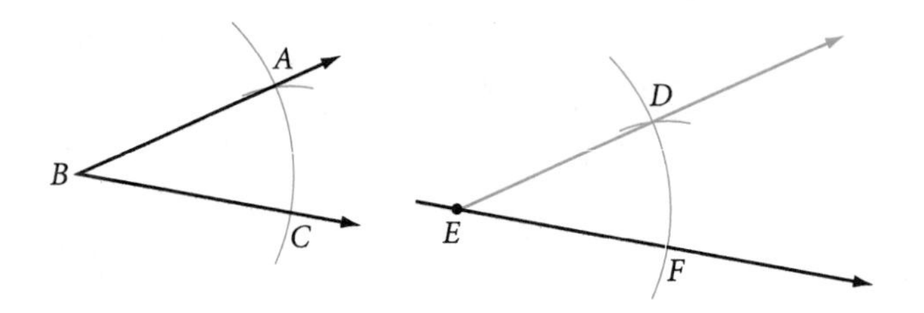

Review

In Exercises 13–15, complete each statement. If the figure does not give you enough information to show that the triangles are congruent, write "cannot be determined."

13. $\overline{AM}$ is a median.
$\triangle CAM \cong \triangle$ _?_

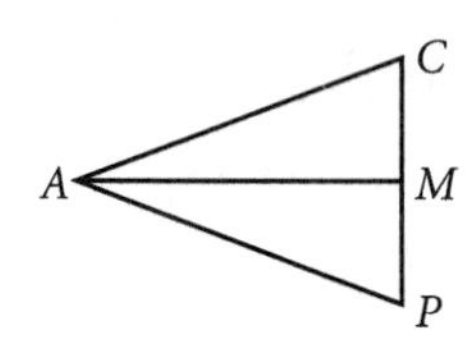

14. $\triangle HEI \cong \triangle$ _?_
Why?

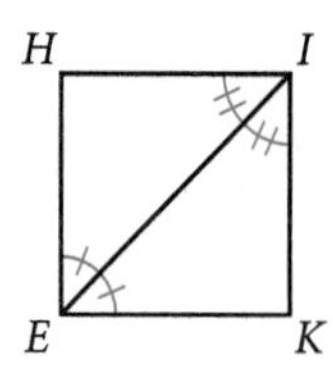

15. U is the midpoint of both $\overline{FE}$ and $\overline{LT}$. $\triangle ULF \cong \triangle$ _?_

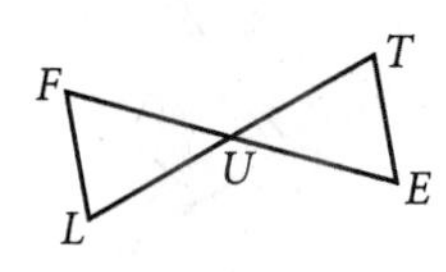

16. ***Construction*** Draw a triangle. Use the SAS Congruence Conjecture to construct a second triangle congruent to the first.

17. ***Construction*** Construct two triangles that are *not* congruent, even though two sides and a non-included angle of one triangle are congruent to two sides and a corresponding non-included angle of the other triangle. ⓗ

18. Copy the figure. Calculate the measure of each lettered angle.

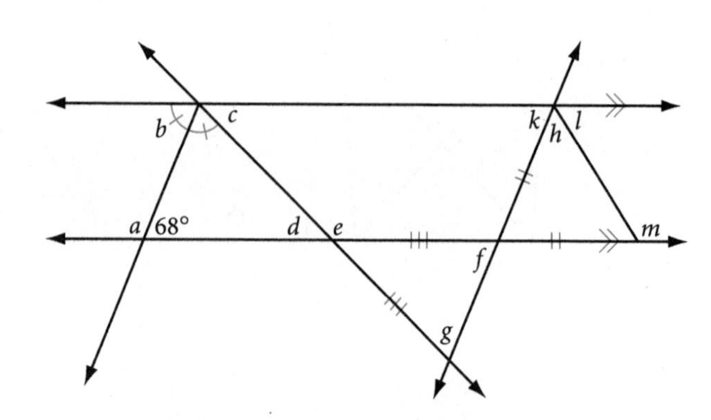

19. According to math legend, the Greek mathematician Thales (ca. 625–547 B.C.E.) could tell how far out to sea a ship was by using congruent triangles. First, he marked off a long segment in the sand. Then, from each endpoint of the segment, he drew the angle to the ship. He then remeasured the two angles on the other side of the segment away from the shore. The point where the rays of these two angles crossed located the ship. What congruence conjecture was Thales using? Explain.

20. Isosceles right triangle ABC has vertices $A(-8, 2)$, $B(-5, -3)$, and $C(0, 0)$. Find the coordinates of the circumcenter.

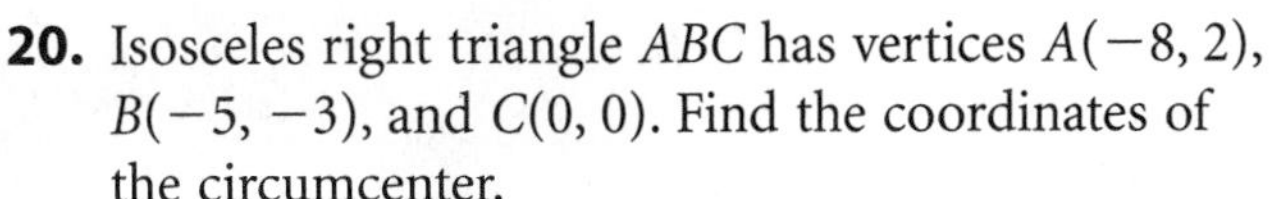

21. The SSS Congruence Conjecture explains why triangles are rigid structures though other polygons are not. By adding one "strut" (diagonal) to a quadrilateral you create a quadrilateral that consists of two triangles, and that makes it rigid. What is the minimum number of struts needed to make a pentagon rigid? A hexagon? A dodecagon? What is the minimum number of struts needed to make other polygons rigid? Complete the table and make your conjecture.

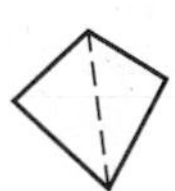
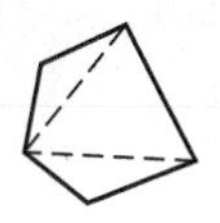
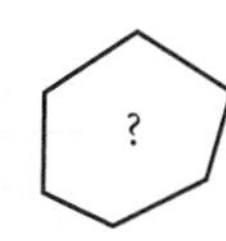

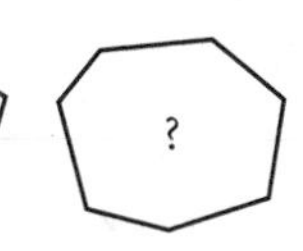

Number of sides	3	4	5	6	7	...	12	...	n
Number of struts needed to make polygon rigid						...		...	

22. Line ℓ is parallel to $\overline{AB}$. If P moves to the right along ℓ, which of the following always decreases?

a. The distance PC

b. The distance from C to $\overline{AB}$

c. The ratio $\dfrac{AB}{AP}$

d. $AC - AP$

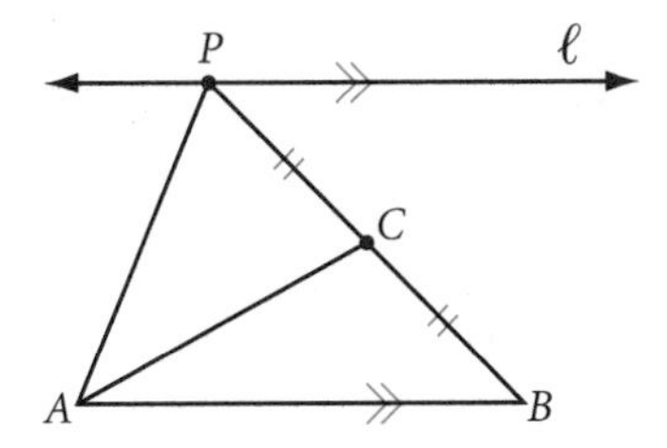

23. Find the lengths x and y. Each angle is a right angle.

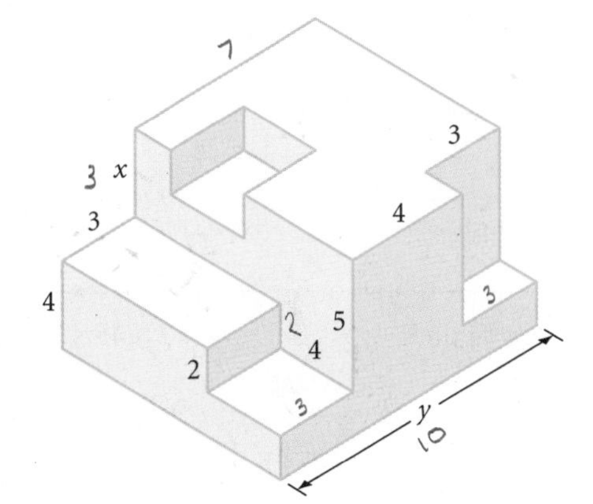

project

POLYA'S PROBLEM

George Polya (1887–1985) was a mathematician who specialized in problem-solving methods. He taught mathematics and problem solving at Stanford University for many years, and wrote the book *How to Solve It.*

He posed this problem to his students: Into how many parts will five random planes divide space?

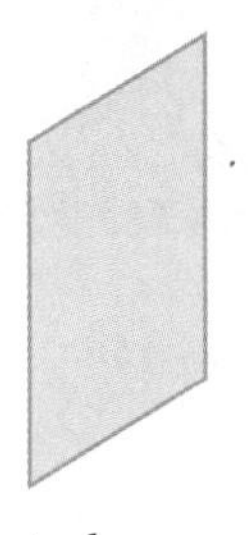

1 plane

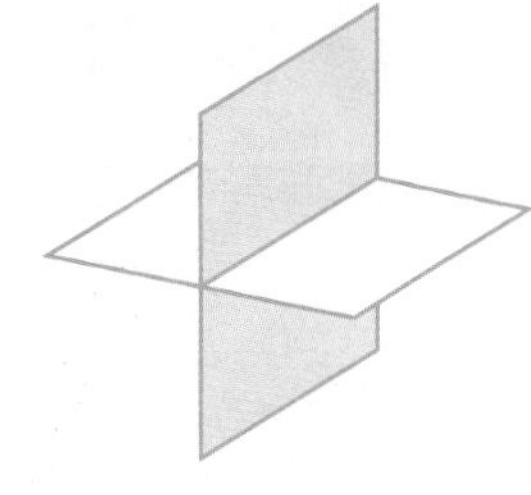

2 planes

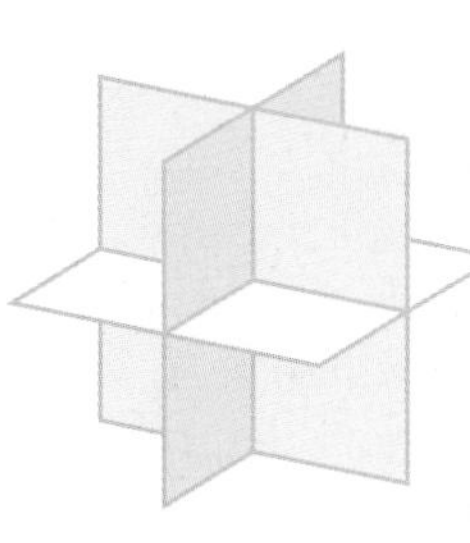

3 planes

It is difficult to visualize five random planes intersecting in space. What strategies would you use to find the answer?

Your project is to solve this problem, and to show how you know your answer is correct. Here are some of Polya's problem-solving strategies to help you.

- Understand the problem. Draw a figure or build a model. Can you restate the problem in your own words?
- Break down the problem. Have you done any simpler problems that are like this one?

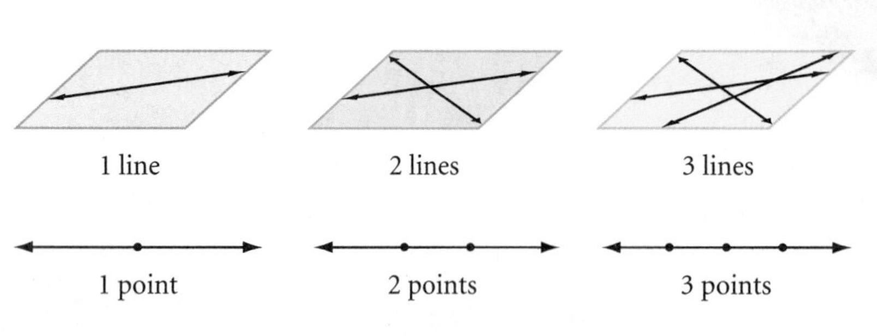

- Check your answer. Can you find the answer in a different way to show that it is correct? (The answer, by the way, is not 32!)

Your method is as important as your answer. Keep track of all the different things you try. Write down your strategies, your results, and your thinking, as well as your answer.

LESSON

4.7

Flowchart Thinking

If you can only find it, there is a reason for everything.

TRADITIONAL SAYING

You have been making many discoveries about triangles. As you try to explain why the new conjectures are true, you build upon definitions and conjectures you made before.

So far, you have written your explanations as paragraph proofs. First, we'll look at a diagram and explain why two angles must be congruent, by writing a paragraph proof, in Example A. Then we'll look at a different tool for writing proofs, and use that tool to write the same proof, in Example B.

EXAMPLE A

In the figure at right, $\overline{EC} \cong \overline{AC}$ and $\overline{ER} \cong \overline{AR}$. Is $\angle A \cong \angle E$? If so, give a logical argument to explain why they are congruent.

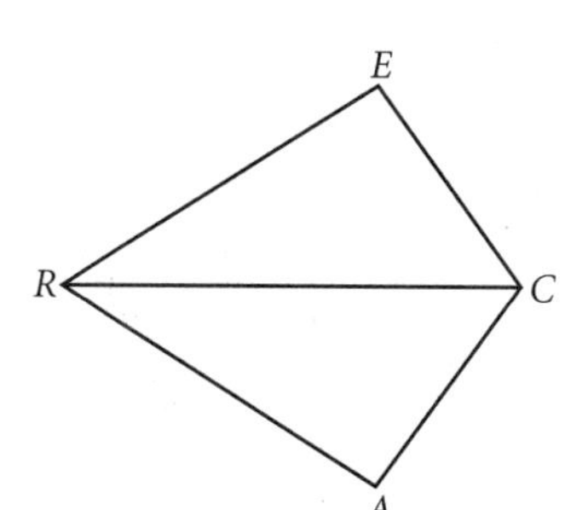

▶ Solution

First mark the given information on the figure. Then consider whether $\angle A$ is congruent to $\angle E$, and why.

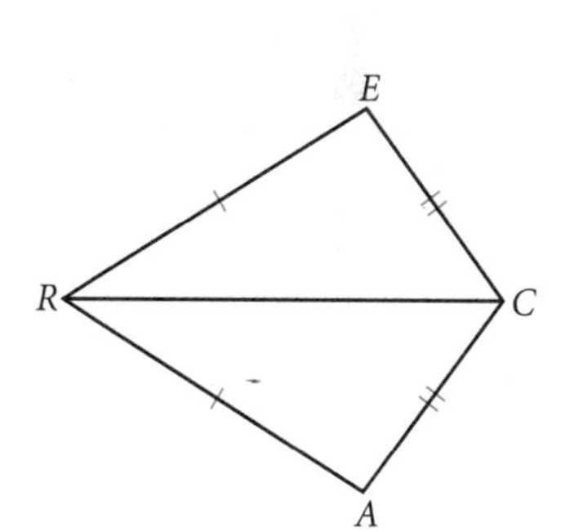

Paragraph Proof: Show that $\angle A \cong \angle E$.

$\overline{EC} \cong \overline{AC}$ and $\overline{ER} \cong \overline{AR}$ because that information is given. $\overline{RC} \cong \overline{RC}$ because it is the same segment, and any segment is congruent to itself. So, $\triangle CRE \cong \triangle CRA$ by the SSS Congruence Conjecture. If $\triangle CRE \cong \triangle CRA$, then $\angle A \cong \angle E$ by CPCTC. ■

Career CONNECTION

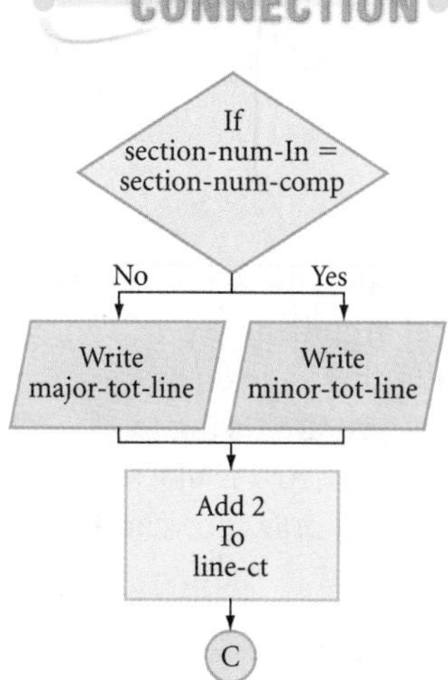

Computer programmers use programming language and detailed plans to design computer software. They often use flowcharts to plan the logic in programs.

Were you able to follow the logical steps in Example A? Sometimes a logical argument or a proof is long and complex, and a paragraph might not be the clearest way to present all the steps. In Chapter 1, you used concept maps to visualize the relationships among different kinds of polygons. A **flowchart** is a concept map that shows all the steps in a complicated procedure in proper order. Arrows connect the boxes to show how facts lead to conclusions.

Flowcharts make your logic visible so that others can follow your reasoning. To present your reasoning in flowchart form, create a **flowchart proof.** Place each statement in a box. Write the logical reason for each statement beneath its box. For example, you would write "$RC \cong RC$, because it is the same segment," as

$\overline{RC} \cong \overline{RC}$

Same segment

Here is the same logical argument that you created in Example A in flowchart proof format.

EXAMPLE B

In the figure below, $\overline{EC} \cong \overline{AC}$ and $\overline{ER} \cong \overline{AR}$. Is $\angle E \cong \angle A$? If so, write a flowchart proof to explain why.

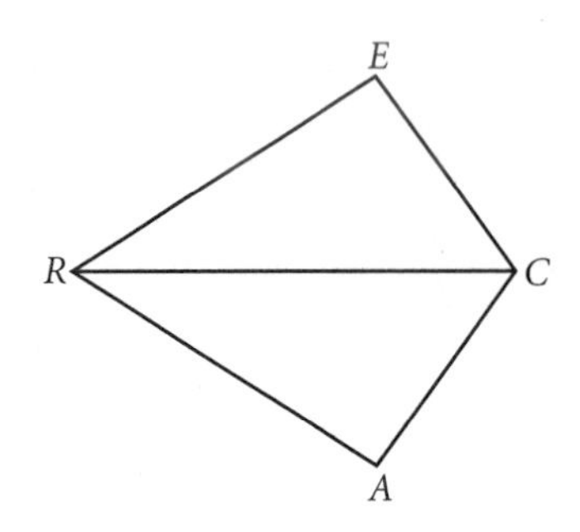

▶ Solution

First, restate the given information clearly. It helps to mark the given information on the figure. Then state what you are trying to show.

Given: $\overline{AR} \cong \overline{ER}$
$\overline{EC} \cong \overline{AC}$

Show: $\angle E \cong \angle A$

Flowchart Proof

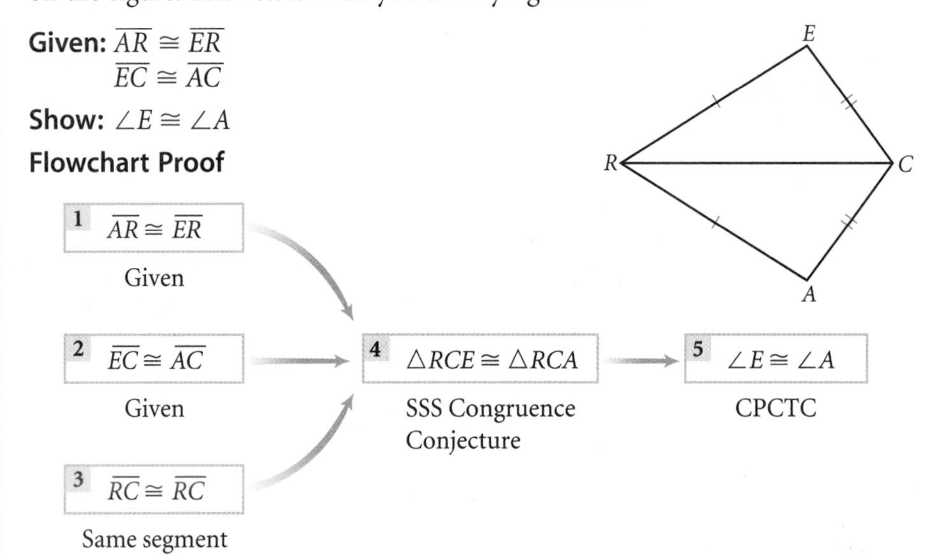

In a flowchart proof, the arrows show how the logical argument flows from the information that is given to the conclusion that you are trying to prove. Drawing an arrow is like saying "therefore." You can draw flowcharts top to bottom or left to right.

Compare the paragraph proof in Example A with the flowchart proof in Example B. What similarities and differences are there? What are the advantages of each format?

Is this contraption like a flowchart proof?

EXERCISES

1. Suppose you saw this step in a proof: Construct angle bisector CD to the midpoint of side AB in $\triangle ABC$. What's wrong with that step? Explain.

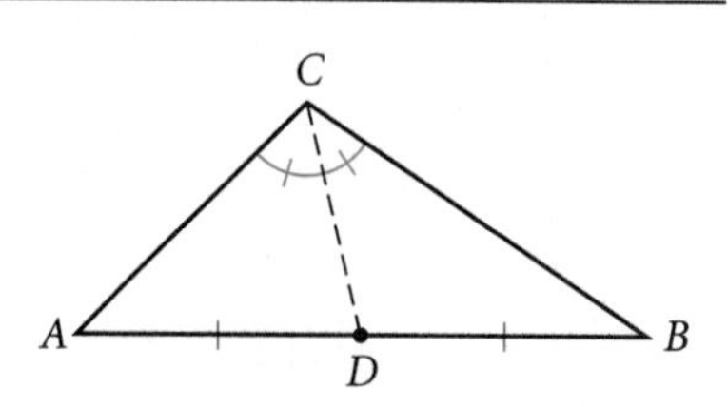

2. Copy the flowchart. Provide each missing reason or statement in the proof.

Given: $\overline{SE} \cong \overline{SU}$

$\angle E \cong \angle U$

Show: $\overline{MS} \cong \overline{OS}$

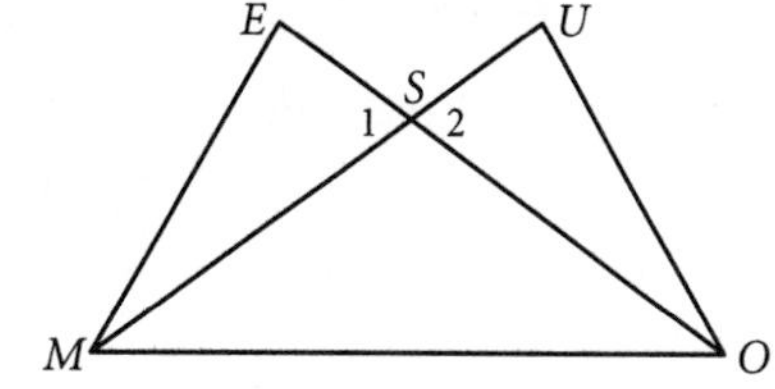

Flowchart Proof

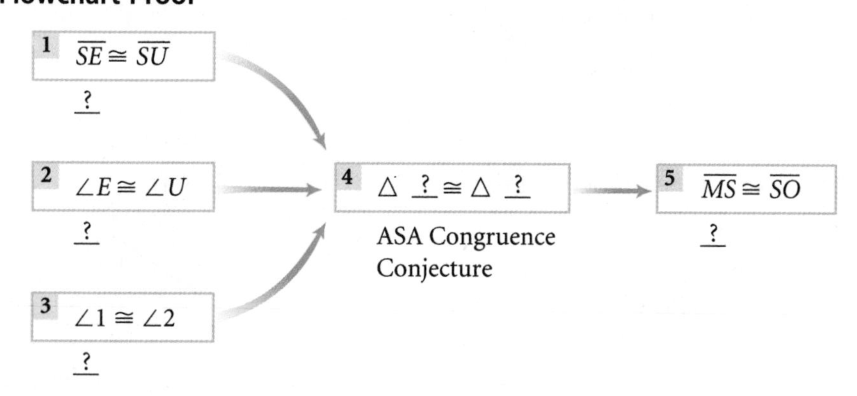

3. Copy the flowchart. Provide each missing reason or statement in the proof.

Given: I is the midpoint of $\overline{CM}$

I is the midpoint of $\overline{BL}$

Show: $\overline{CL} \cong \overline{MB}$

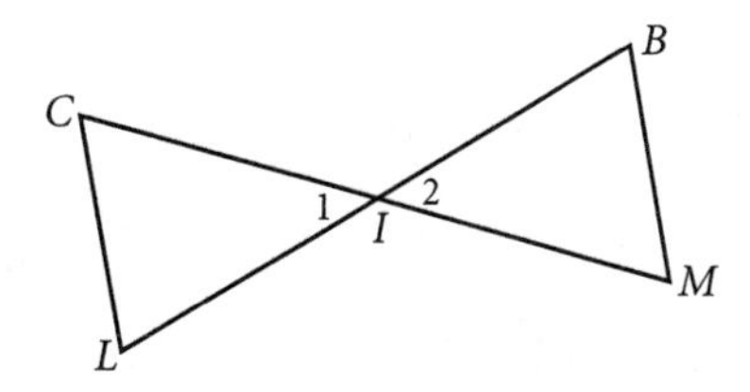

Flowchart Proof

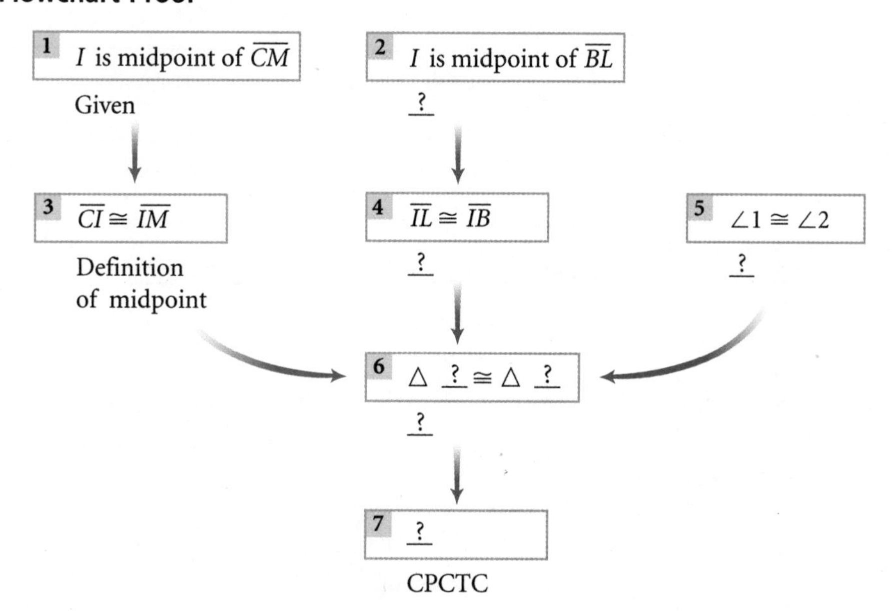

In Exercises 4–6, an auxiliary line segment has been added to the figure.

4. Complete this flowchart proof of the Isosceles Triangle Conjecture. Given that the triangle is isosceles, show that the base angles are congruent.

Given: $\triangle NEW$ is isosceles, with $\overline{WN} \cong \overline{EN}$ and median $\overline{NS}$

Show: $\angle W \cong \angle E$

Flowchart Proof

1. $\overline{NS}$ is a median — Given
2. S is a midpoint — Definition of median
3. $\overline{NS} \cong \overline{NS}$ — Same segment
4. $\overline{WS} \cong \overline{SE}$ — Definition of midpoint
5. $\overline{WN} \cong \overline{NE}$ — ?
6. $\triangle WSN \cong \triangle$? — ?
7. $\angle W \cong \angle$? — ?

5. Complete this flowchart proof of the Converse of the Isosceles Triangle Conjecture.

Given: $\triangle NEW$ with $\angle W \cong \angle E$

$\overline{NS}$ is an angle bisector

Show: $\triangle NEW$ is an isosceles triangle

Flowchart Proof

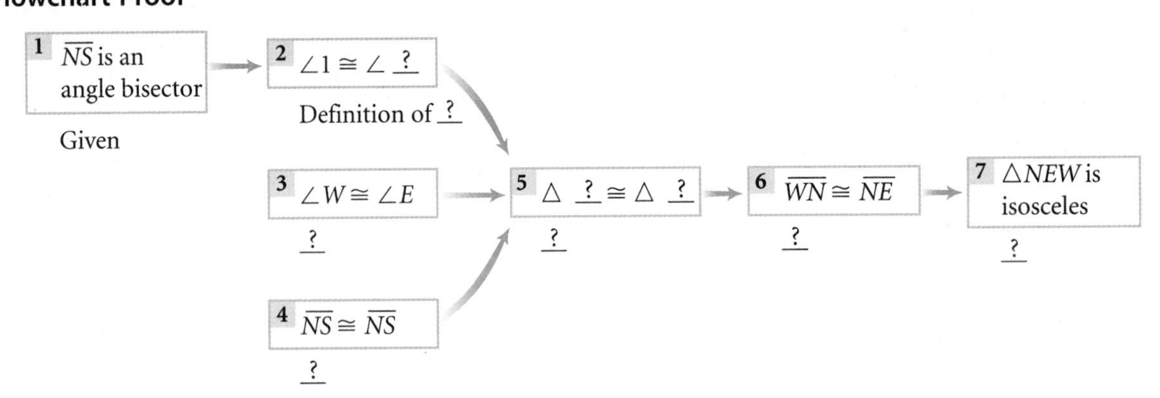

6. Complete the flowchart proof. What does this proof tell you about parallelograms?

Given: $\overline{SA} \parallel \overline{NE}$

$\overline{SE} \parallel \overline{NA}$

Show: $\overline{SA} \cong \overline{NE}$

Flowchart Proof

1. $\overline{SA} \parallel \overline{NE}$ — ?
2. $\overline{SE} \parallel \overline{NA}$ — ?
3. $\angle 3 \cong \angle 4$ — AIA Conjecture
4. ? — ?
5. $\overline{SN} \cong \overline{SN}$ — Same segment
6. $\triangle$? $\cong \triangle$? — ?
7. ? — ?

7. In Chapter 3, you learned how to construct the bisector of an angle. Now you have the skills to explain *why* the construction works, using deductive reasoning. Create a paragraph or flowchart proof to show that the construction method works. ⓗ

Given: $\angle ABC$ with $\overline{BA} \cong \overline{BC}$, $\overline{CD} \cong \overline{AD}$

Show: $\overrightarrow{BD}$ is the angle bisector of $\angle ABC$

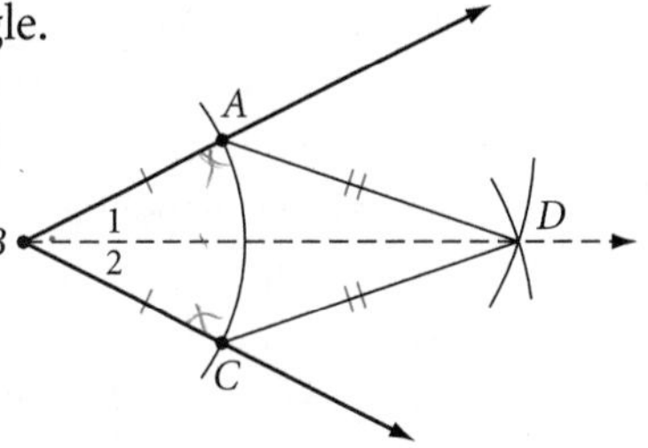

Review

8. Which segment is the shortest? Explain. ⓗ

E
30°
61°
59°
121°
29°
N
A
L

9. What's wrong with this picture? Explain.

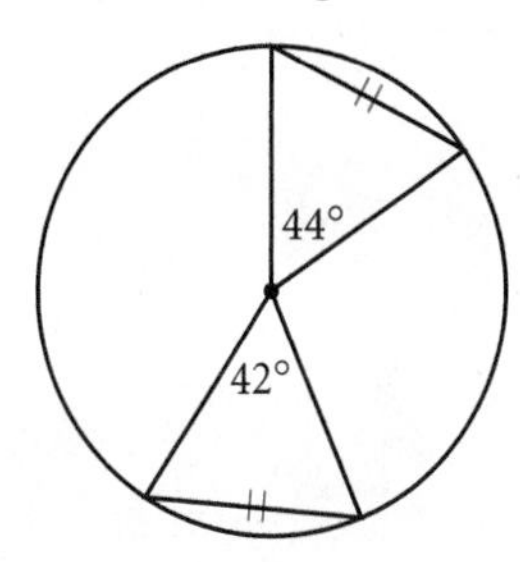

For Exercises 10–12, name the congruent triangles and explain why the triangles are congruent. If you cannot show that they are congruent, write "cannot be determined."

10. $\overline{PO} \cong \overline{PR}$
$\triangle POE \cong \triangle$ __?__
$\triangle SON \cong \triangle$ __?__ ⓗ

11. $\triangle$ __?__ $\cong \triangle$ __?__ ⓗ

12. $\overline{AC} \cong \overline{CR}$, $\overline{CK}$ is a median of $\triangle ARC$. $\triangle RCK \cong \triangle$ __?__

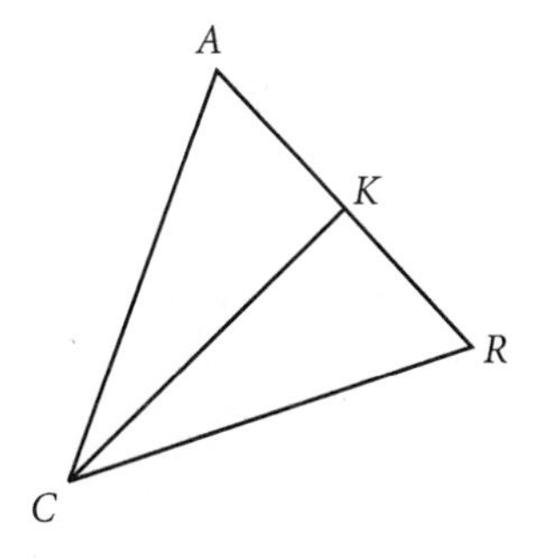

13. Copy the figure below. Calculate the measure of each lettered angle.

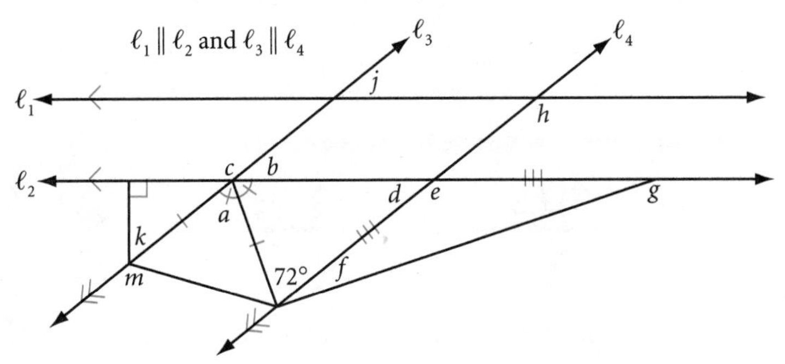

14. Which point of concurrency is equidistant from all three vertices? Explain why. Which point of concurrency is equidistant from all three sides? Explain why. ⓗ

15. Samantha is standing at the bank of a stream, wondering how wide the stream is. Remembering her geometry conjectures, she kneels down and holds her fishing pole perpendicular to the ground in front of her. She adjusts her hand on the pole so that she can see the opposite bank of the stream along her line of sight through her hand. She then turns, keeping a firm grip on the pole, and uses the same line of sight to spot a boulder on her side of the stream. She measures the distance to the boulder and concludes that this equals the distance across the stream. What triangle congruence shortcut is Samantha using? Explain.

16. What is the probability of randomly selecting one of the shortest diagonals from all the diagonals in a regular decagon?

17. Sketch the solid shown with the red and green cubes removed. ⓗ

18. Sketch the new location of rectangle $BOXY$ after it has been rotated 90° clockwise about the origin.

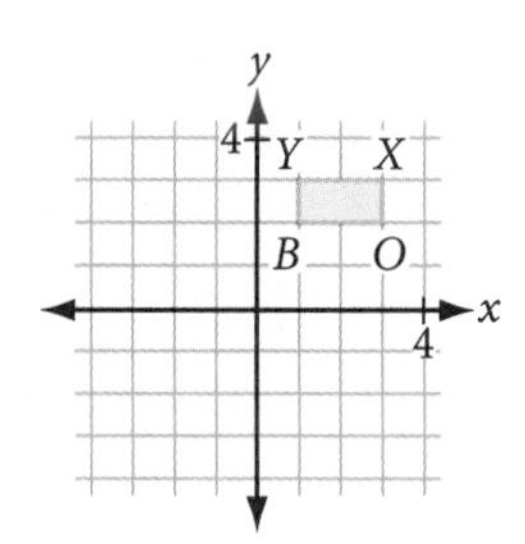

IMPROVING YOUR REASONING SKILLS

Pick a Card

Nine cards are arranged in a 3-by-3 array. Every jack borders on a king and on a queen. Every king borders on an ace. Every queen borders on a king and on an ace. (The cards border each other edge-to-edge, but not corner-to-corner.) There are at least two aces, two kings, two queens, and two jacks. Which card is in the center position of the 3-by-3 array?

LESSON

4.8

Proving Isosceles Triangle Conjectures

The right angle from which to approach any problem is the try angle.

ANONYMOUS

This boathouse is a remarkably symmetric structure with its isosceles triangle roof and the identical doors on each side. The rhombus-shaped attic window is centered on the line of symmetry of this face of the building. What might this building reveal about the special properties of the line of symmetry in an isosceles triangle?

In this lesson you will make a conjecture about a special segment in isosceles triangles. Then you will use logical reasoning to prove your conjecture is true for all isosceles triangles.

First, consider a scalene triangle. In $\triangle ARC$, $\overline{CD}$ is the altitude to the base $\overline{AR}$, $\overline{CE}$ is the angle bisector of $\angle ACR$, and $\overline{CF}$ is the median to the base $\overline{AR}$. From this example it is clear that the angle bisector, the altitude, and the median can all be different line segments. Is this true for all triangles? Can two of these ever be the same segment? Can they all be the same segment? Let's investigate.

C

A F E D R

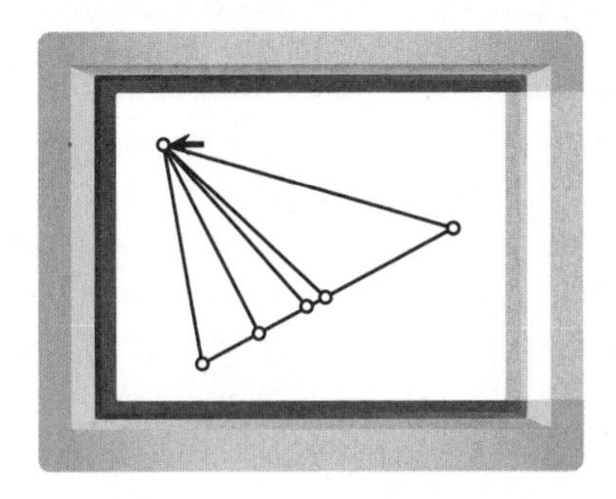

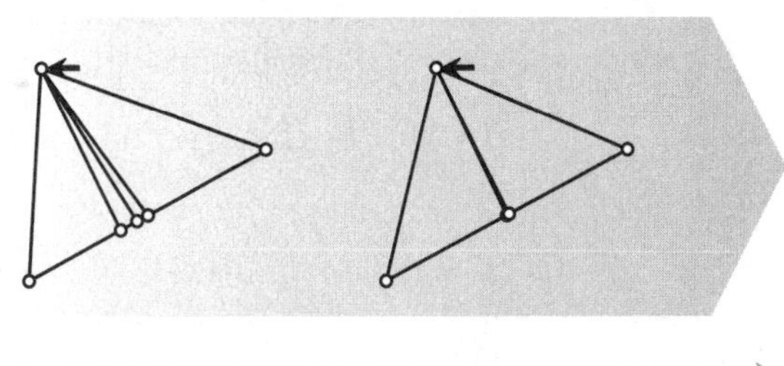

Investigation
The Symmetry Line in an Isosceles Triangle

You will need

- a compass
- a straightedge

Each person in your group should draw a different isosceles triangle for this investigation.

Step 1 Construct a large isosceles triangle on a sheet of unlined paper. Label it *ARK*, with *K* the vertex angle.

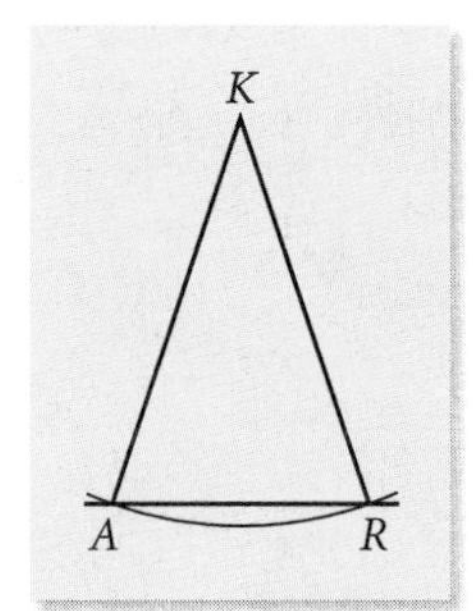

Step 2 Construct angle bisector $\overline{KD}$ with point *D* on $\overline{AR}$. Do $\triangle ADK$ and $\triangle RDK$ look congruent?

Step 3 With your compass, compare $\overline{AD}$ and $\overline{RD}$. Is *D* the midpoint of $\overline{AR}$? If *D* is the midpoint, then what type of special segment is $\overline{KD}$?

Step 4 Compare $\angle ADK$ and $\angle RDK$. Do they have equal measures? Are they supplementary? What conclusion can you make?

Step 5 Compare your conjectures with the results of other students. Now combine the two conjectures from Steps 3 and 4 into one.

Vertex Angle Bisector Conjecture C-28

In an isosceles triangle, the bisector of the vertex angle is also _?_ and _?_.

The properties you just discovered for isosceles triangles also apply to equilateral triangles. Equilateral triangles are also isosceles, although isosceles triangles are not necessarily equilateral.

You have probably noticed the following property of equilateral triangles: When you construct an equilateral triangle, each angle measures 60°. If each angle measures 60°, then all three angles are congruent. So, if a triangle is equilateral, then it is equiangular. This is called the Equilateral Triangle Conjecture.

If we agree that the Isosceles Triangle Conjecture is true, we can write the paragraph proof below.

Paragraph Proof: The Equilateral Triangle Conjecture

We need to show that if $AB = AC = BC$, then $\triangle ABC$ is equiangular. By the Isosceles Triangle Conjecture,

If $AB = AC$, then $m\angle B = m\angle C$.

If ${}_{A}\overset{C}{\triangle}_{B}$, then ${}_{A}\overset{C}{\triangle}_{B}$.

If $AB = BC$, then $m\angle A = m\angle C$.

If $_{A}\overset{C}{\triangle}_{B}$, then $_{A}\overset{C}{\triangle}_{B}$.

If $m\angle A = m\angle C$ and $m\angle B = m\angle C$, then $m\angle A = m\angle B = m\angle C$. So, $\triangle ABC$ is equiangular.

If $_{A}\overset{C}{\triangle}_{B}$ and $_{A}\overset{C}{\triangle}_{B}$, then $_{A}\overset{C}{\triangle}_{B}$. ■

The converse of the Equilateral Triangle Conjecture is called the Equiangular Triangle Conjecture, and it states: If a triangle is equiangular, then it is equilateral. Is this true? Yes, and the proof is almost identical to the proof above, except that you use the converse of the Isosceles Triangle Conjecture. So, if the Equilateral Triangle Conjecture and the Equiangular Triangle Conjecture are both true then we can combine them. Complete the conjecture below and add it to your conjecture list.

C-29

Equilateral/Equiangular Triangle Conjecture

Every equilateral triangle is _?_, and, conversely, every equiangular triangle is _?_.

The Equilateral/Equiangular Triangle Conjecture is a **biconditional** conjecture: Both the statement and its converse are true. A triangle is equilateral *if and only if* it is equiangular. One condition cannot be true unless the other is also true.

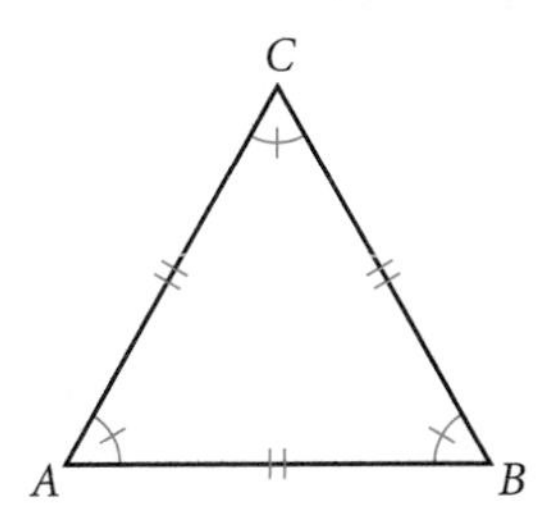

EXERCISES

You will need

Construction tools for Exercise **10**

In Exercises 1–3, $\triangle ABC$ is isosceles with $\overline{AC} \cong \overline{BC}$.

1. Perimeter $\triangle ABC = 48$ (h)
$AC = 18$
$AD = $ _?_

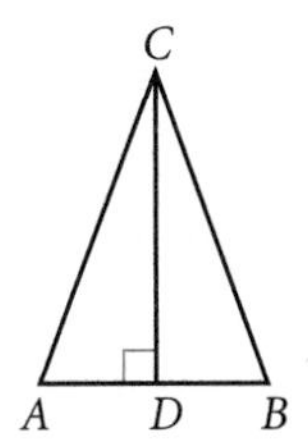

2. $m\angle ABC = 72°$
$m\angle ADC = $ _?_

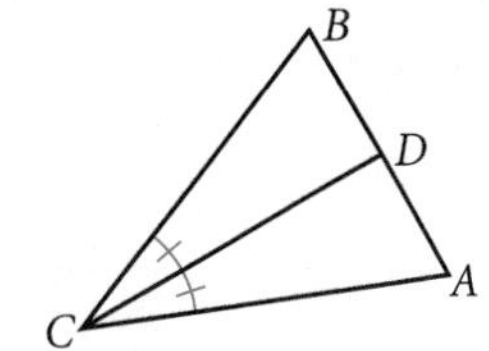

3. $m\angle CAB = 45°$
$m\angle ACD = $ _?_

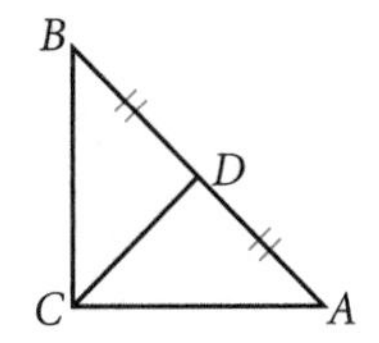

In Exercises 4–6, copy the flowchart. Supply the missing statement and reasons in the proofs of Conjectures A, B, and C shown below. These three conjectures are all part of the Vertex Angle Bisector Conjecture.

4. Complete the flowchart proof for Conjecture A.

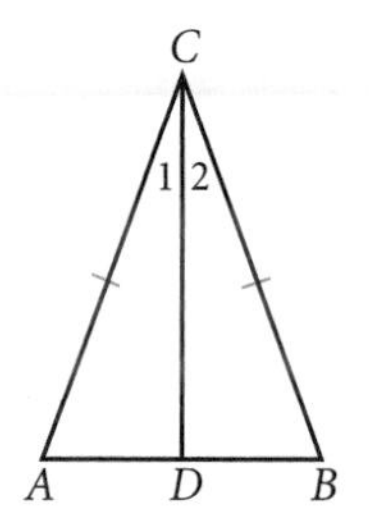

Conjecture A: The bisector of the vertex angle in an isosceles triangle divides the isosceles triangle into two congruent triangles.

Given: $\triangle ABC$ is isosceles
$\overline{AC} \cong \overline{BC}$, and $\overline{CD}$ is the bisector of $\angle C$

Show: $\triangle ADC \cong \triangle BDC$

Flowchart Proof

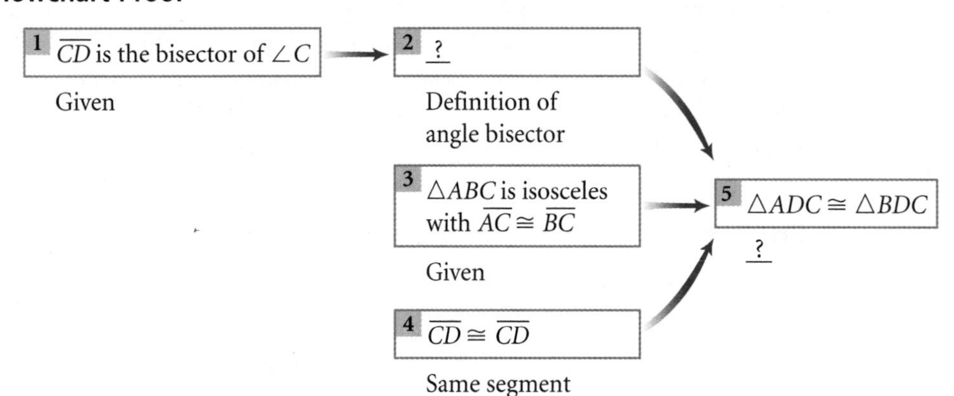

5. Complete the flowchart proof for Conjecture B.

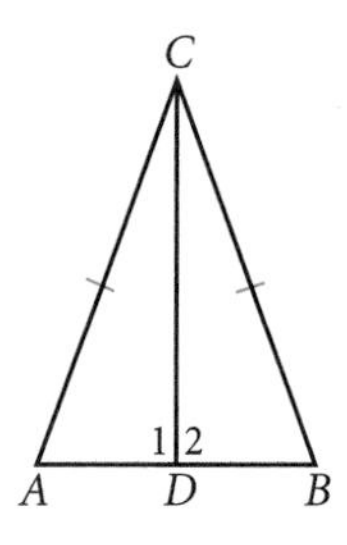

Conjecture B: The bisector of the vertex angle in an isosceles triangle is also the altitude to the base.

Given: $\triangle ABC$ is isosceles
$\overline{AC} \cong \overline{BC}$, and $\overline{CD}$ bisects $\angle C$

Show: $\overline{CD}$ is an altitude

Flowchart Proof

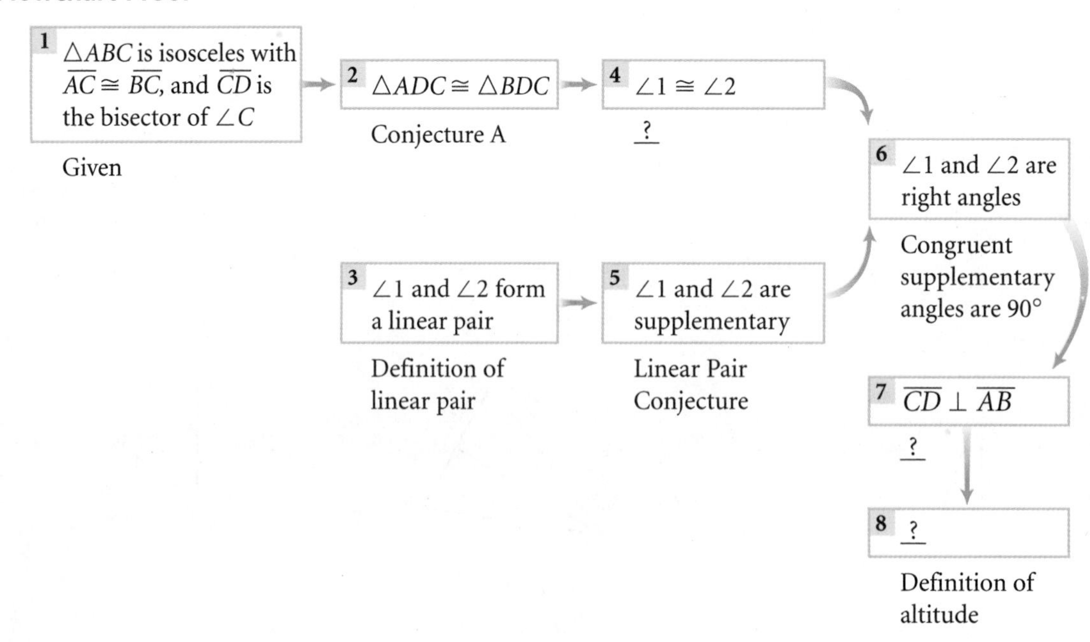

6. Create a flowchart proof for Conjecture C.

Conjecture C: The bisector of the vertex angle in an isosceles triangle is also the median to the base.

Given: $\triangle ABC$ is isosceles with $\overline{AC} \cong \overline{BC}$

$\overline{CD}$ is the bisector of $\angle C$

Show: $\overline{CD}$ is a median

7. In the figure at right, $\triangle ABC$, the plumb level is isosceles. A weight, called the plumb bob, hangs from a string attached at point C. If you place the level on a surface and the string is perpendicular to $\overline{AB}$ then the surface you are testing is level. To tell whether the string is perpendicular to $\overline{AB}$, check whether it passes through the midpoint of $\overline{AB}$. Create a flowchart proof to show that if D is the midpoint of $\overline{AB}$, then $\overline{CD}$ is perpendicular to $\overline{AB}$.

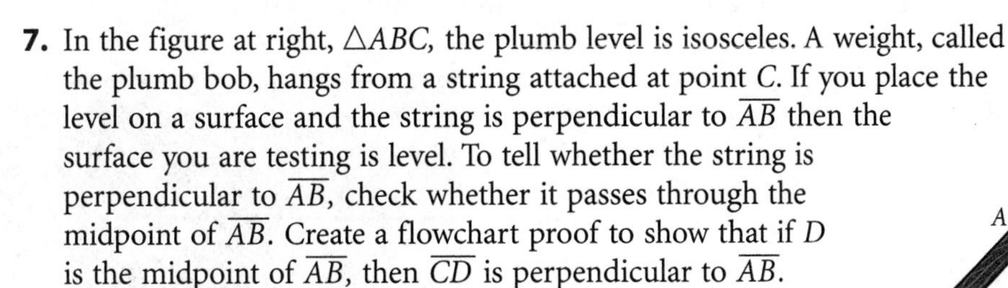

Given: $\triangle ABC$ is isosceles with $\overline{AC} \cong \overline{BC}$

D is the midpoint of $\overline{AB}$

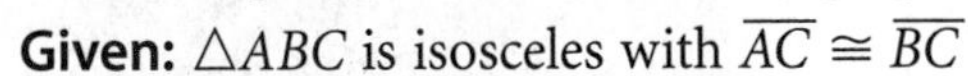

Show: $\overline{CD} \perp \overline{AB}$

History CONNECTION

Builders in ancient Egypt used a tool called a *plumb level* in building the great pyramids. With a plumb level, you can use the basic properties of isosceles triangles to determine whether a surface is level.

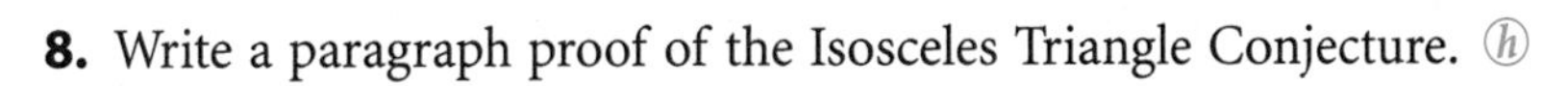

8. Write a paragraph proof of the Isosceles Triangle Conjecture. ⓗ

9. Write a paragraph proof of the Equiangular Triangle Conjecture.

10. *Construction* Use compass and straightedge to construct a 30° angle.

Review

11. Trace the figure below. Calculate the measure of each lettered angle.

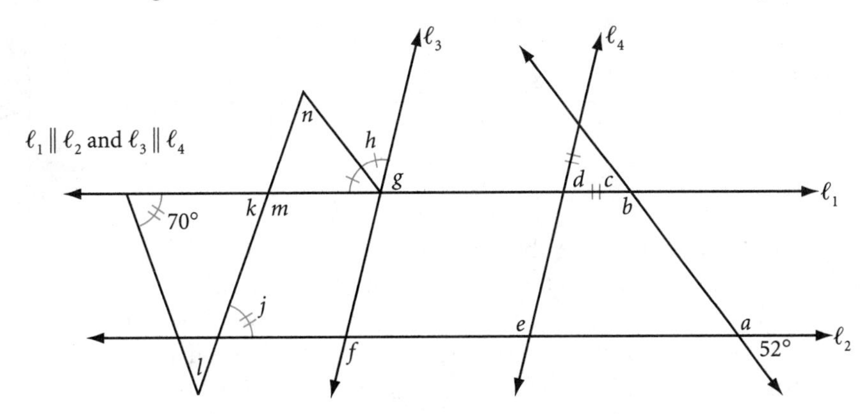

12. How many minutes after 3:00 will the hands of a clock overlap? ⓗ

13. Find the equation of the line through point C that is parallel to side $\overline{AB}$ in $\triangle ABC$. The vertices are $A(1, 3)$, $B(4, -2)$, and $C(6, 6)$. Write your answer in slope-intercept form, $y = mx + b$.

14. Sixty concurrent lines in a plane divide the plane into how many regions? ⓗ

15. If two vertices of a triangle have coordinates $A(1, 3)$ and $B(7, 3)$, find the coordinates of point C so that $\triangle ABC$ is a right triangle. Can you find any other points that would create a right triangle?

16. **APPLICATION** Hugo hears the sound of fireworks three seconds after he sees the flash. Duane hears the sound five seconds after he sees the flash. Hugo and Duane are 1.5 km apart. They know the flash was somewhere to the north. They also know that a flash can be seen almost instantly, but sound travels 340 m/sec. Do Hugo and Duane have enough information to locate the site of the fireworks? Make a sketch and label all the distances that they know or can calculate.

17. **APPLICATION** In an earlier exercise, you found the rule for the family of hydrocarbons called alkanes, or paraffins. These contain a straight chain of carbons. Alkanes can also form rings of carbon atoms. These molecules are called cycloparaffins. The first three cycloparaffins are shown below. Sketch the molecule cycloheptane. Write the general rule for cycloparaffins $(C_nH_?)$. ⓗ

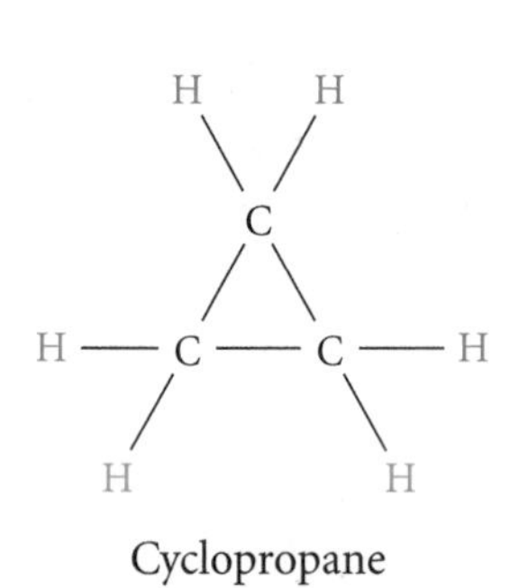

Cyclopropane

Cyclobutane

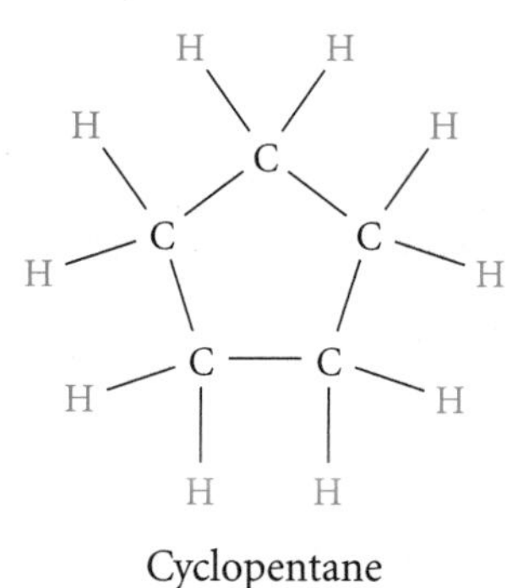

Cyclopentane

IMPROVING YOUR ALGEBRA SKILLS

Number Tricks

Try this number trick.

Double the number of the month you were born. Subtract 16 from your answer. Multiply your result by 5, then add 100 to your answer. Subtract 20 from your result, then multiply by 10. Finally, add the day of the month you were born to your answer. The number you end up with shows the month and day you were born! For example, if you were born March 15th, your answer will be 315. If you were born December 7th, your answer will be 1207.

Number tricks almost always involve algebra. Use algebra to explain why the trick works.

Exploration

Napoleon's Theorem

Portrait of Napoleon by the French painter Anne-Louis Girodet (1767–1824)

In this exploration you'll learn about a discovery attributed to French Emperor Napoleon Bonaparte (1769–1821). Napoleon was extremely interested in mathematics. This discovery, called Napoleon's Theorem, uses equilateral triangles constructed on the sides of any triangle.

Activity

Napoleon Triangles

Procedure Note

1. Construct an equilateral triangle on $\overline{AB}$.
2. Construct the centroid of the equilateral triangle.
3. Hide any medians or midpoints that you constructed for the centroid.
4. Select all three vertices, all three sides, and the centroid of the equilateral triangle.
5. Turn your new construction into a custom tool by choosing **Create New Tool** from the Custom Tools menu.

Step 1 Open a new Sketchpad sketch. Draw $\triangle ABC$.

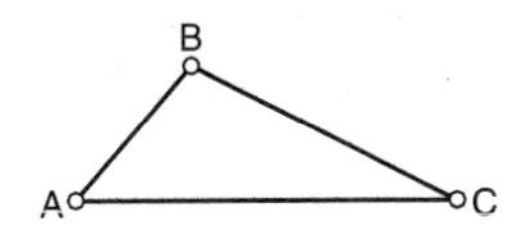

Step 2 Follow the Procedure Note to create a custom tool that constructs an equilateral triangle and its centroid given the endpoints of any segment.

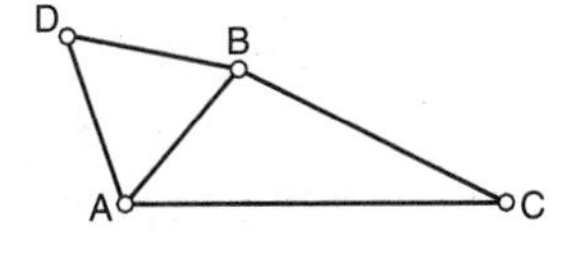

Step 3 Use your custom tool on $\overline{BC}$ and $\overline{CA}$. If an equilateral triangle falls inside your triangle, undo and try again, selecting the two endpoints in reverse order.

Step 4 Connect the centroids of the equilateral triangles. Triangle GQL is called the *outer Napoleon triangle* of $\triangle ABC$.

Drag the vertices and the sides of $\triangle ABC$ and observe what happens.

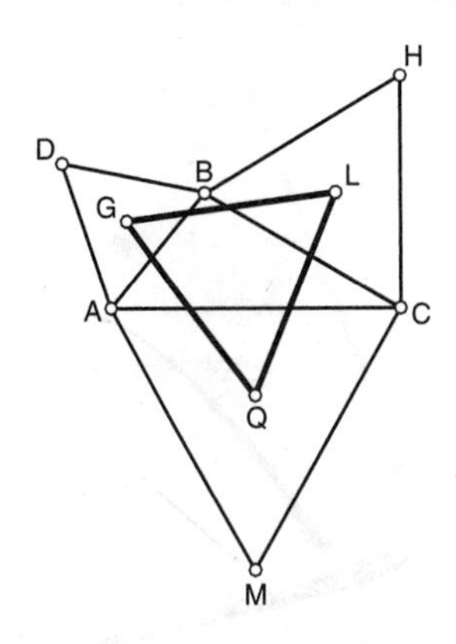

Step 5 | What can you say about the outer Napoleon triangle? Write what you think Napoleon discovered in his theorem.

Here are some extensions to this theorem for you to explore.

Step 6 | Construct segments connecting each vertex of your original triangle with the vertex of the equilateral triangle on the opposite side. What do you notice about these three segments? (This discovery was made by M. C. Escher.)

Step 7 | Construct the *inner Napoleon triangle* by reflecting each centroid across its corresponding side in the original triangle. Measure the areas of the original triangle and of the outer and inner Napoleon triangles. How do these areas compare?

LINES AND ISOSCELES TRIANGLES

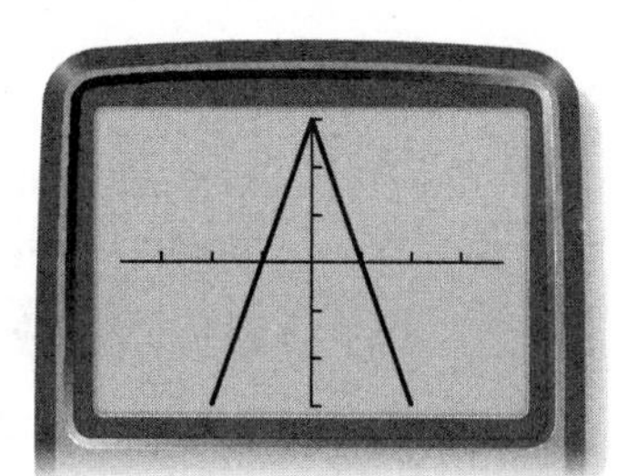

In this example, the lines $y = 3x + 3$ and $y = -3x + 3$ contain the sides of an isosceles triangle whose base is on the x-axis and whose line of symmetry is the y-axis. The window shown is $\{-4.7, 4.7, 1, -3.1, 3.1, 1\}$.

1. Find other pairs of lines that form isosceles triangles whose bases are on the x-axis and whose lines of symmetry are the y-axis.
2. Find pairs of lines that form isosceles triangles whose bases are on the y-axis and whose lines of symmetry are the x-axis.
3. A line $y = mx + b$ contains one side of an isosceles triangle whose base is on the x-axis and whose line of symmetry is the y-axis. What is the equation of the line containing the other side? Now suppose the line $y = mx + b$ contains one side of an isosceles triangle whose base is on the y-axis and whose line of symmetry is the x-axis. What is the equation of the line containing the other side?
4. Graph the lines $y = 2x - 2$, $y = \frac{1}{2}x + 1$, $y = x$, and $y = -x$. Describe the figure that the lines form. Find other sets of lines that form figures like this one.

CHAPTER 4 REVIEW

In this chapter you made many conjectures about triangles. You discovered some basic properties of isosceles and equilateral triangles. You learned different ways to show that two triangles are congruent. Do you remember them all? Triangle congruence shortcuts are an important idea in geometry. You can use them to explain why your constructions work. In later chapters, you will use your triangle conjectures to investigate properties of other polygons.

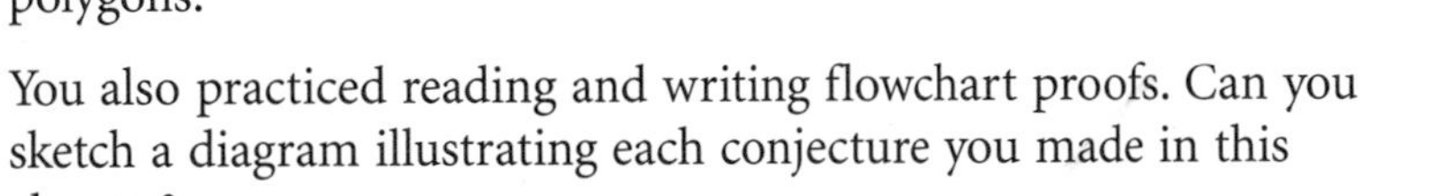

You also practiced reading and writing flowchart proofs. Can you sketch a diagram illustrating each conjecture you made in this chapter?

Check your conjecture list to make sure it is up to date. Make sure you have a clear diagram illustrating each conjecture.

EXERCISES

You will need

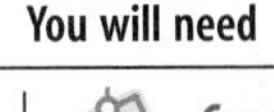

Construction tools for Exercises **33**, **34**, and **37**

1. Why are triangles so useful in structures?

2. The first conjecture of this chapter is probably the most important so far. What is it? Why do you think it is so important?

3. What special properties do isosceles triangles have?

4. What does the statement "The shortest distance between two points is the straight line between them" have to do with the Triangle Inequality Conjecture?

5. What information do you need in order to determine that two triangles are congruent? That is, what are the four congruence shortcuts?

6. Explain why SSA is not a congruence shortcut.

For Exercises 7–24, name the congruent triangles. State the conjecture or definition that supports the congruence statement. If you cannot show the triangles to be congruent from the information given, write "cannot be determined."

7. $\triangle PEA \cong \triangle$ _?_

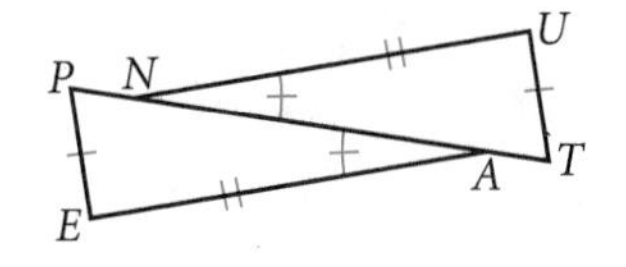

8. $\triangle TOP \cong \triangle$ _?_

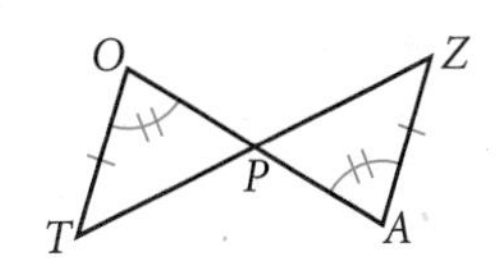

9. $\triangle MSE \cong \triangle$ _?_

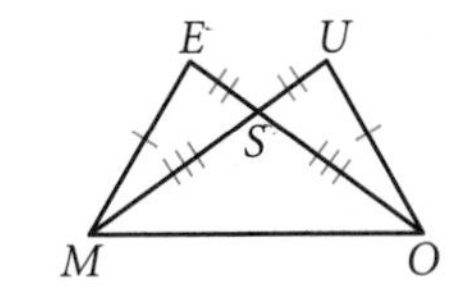

10. $\triangle TIM \cong \triangle$?

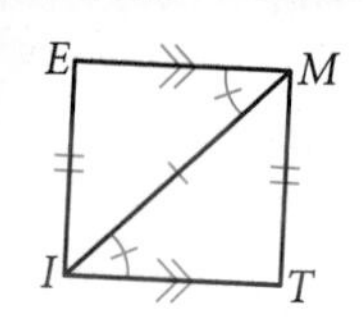

11. $\triangle TRP \cong \triangle$?

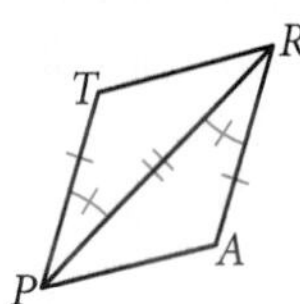

12. $\triangle CAT \cong \triangle$?

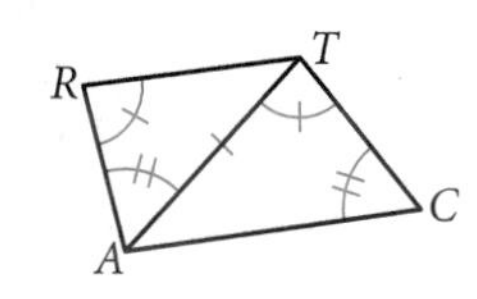

13. $\triangle CGH \cong \triangle$?

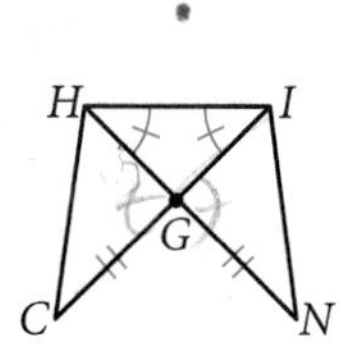

14. $\overleftrightarrow{AB} \parallel \overleftrightarrow{CD}$
$\triangle ABE \cong \triangle$?

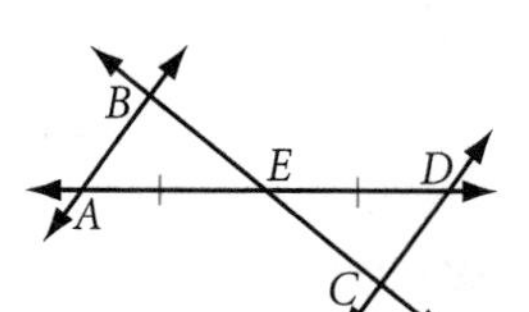

15. Polygon *CARBON* is a regular hexagon.
$\triangle ACN \cong \triangle$?

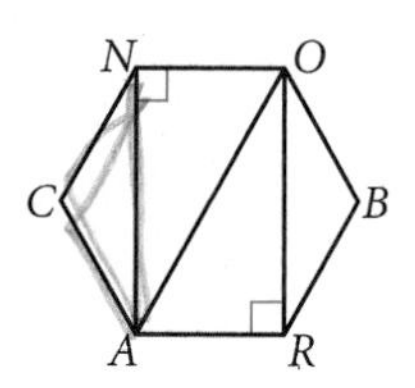

16. $\triangle$? $\cong \triangle$?, $\overline{AD} \cong$?

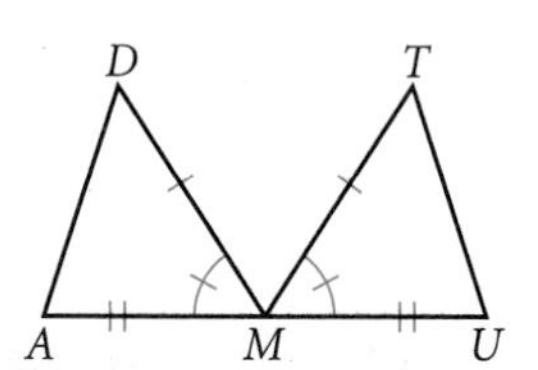

17. $\triangle$? $\cong \triangle$?

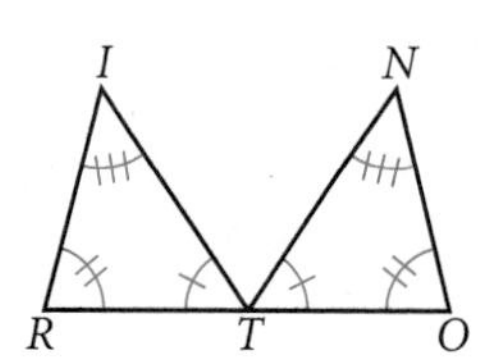

18. $\triangle$? $\cong \triangle$?

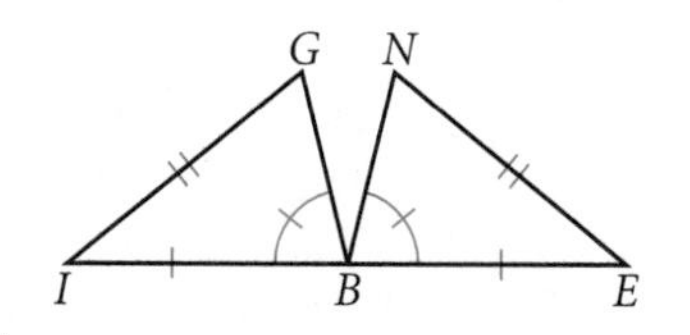

19. $\triangle$? $\cong \triangle$?, $\overline{TR} \cong$? ⓗ

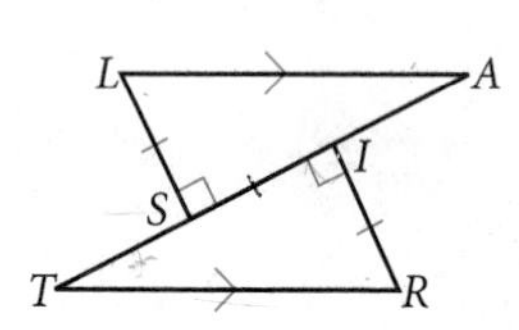

20. $\triangle$? $\cong \triangle$?, $\overline{EI} \cong$?

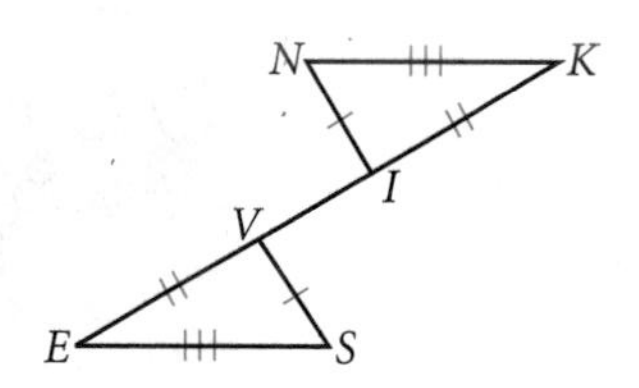

21. $\triangle$? $\cong \triangle$?
Is $\overline{WH}$ a median? ⓗ

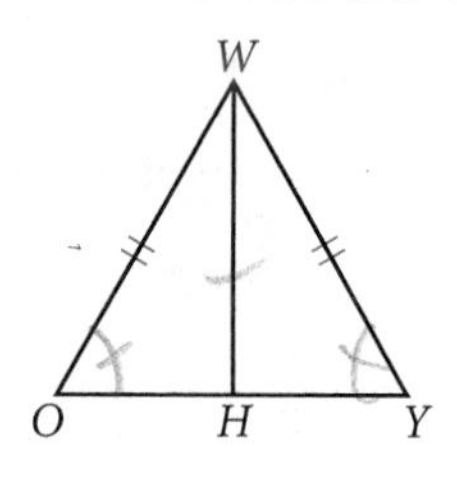

22. $\triangle$? $\cong \triangle$?
Is *NCTM* a parallelogram or a trapezoid?

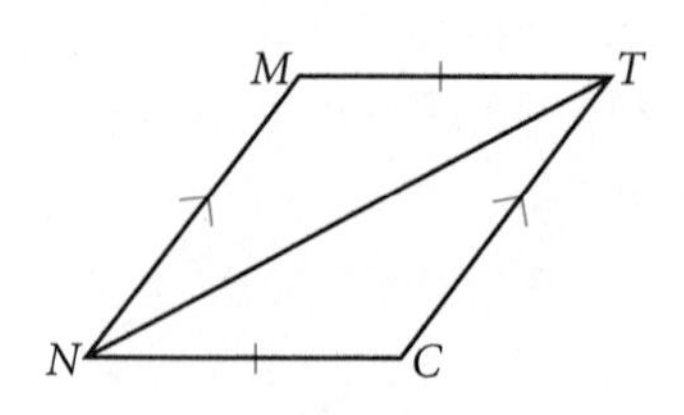

23. $\triangle$? $\cong \triangle$?
$\triangle LAI$ is isosceles with $\overline{IA} \cong \overline{LA}$. ⓗ

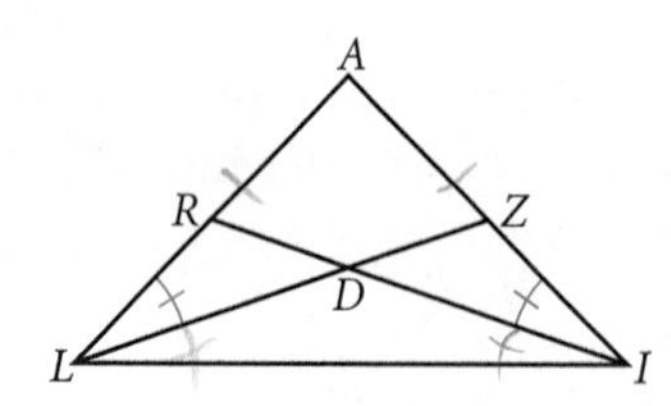

24. $\triangle$? $\cong \triangle$?
Is *STOP* a parallelogram?

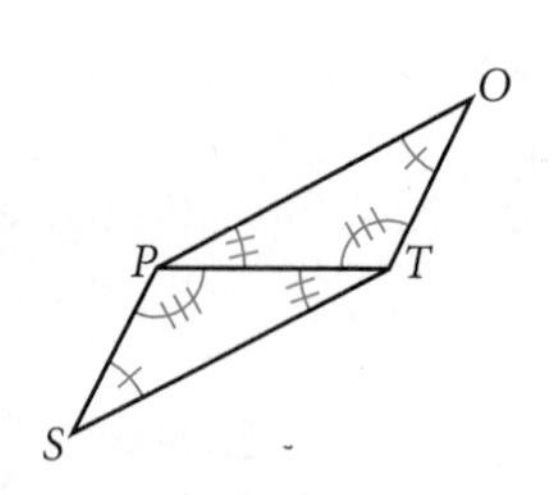

25. What's wrong with this picture? ⓗ

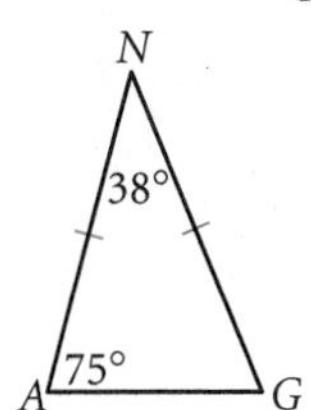

26. What's wrong with this picture?

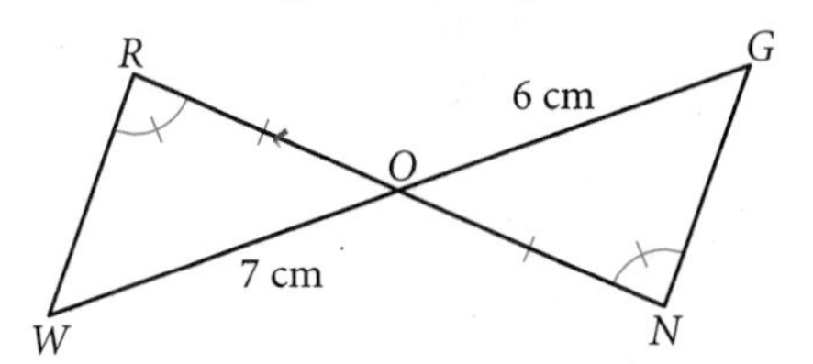

27. Quadrilateral *CAMP* has been divided into three triangles. Use the angle measures provided to determine the longest and shortest segments.

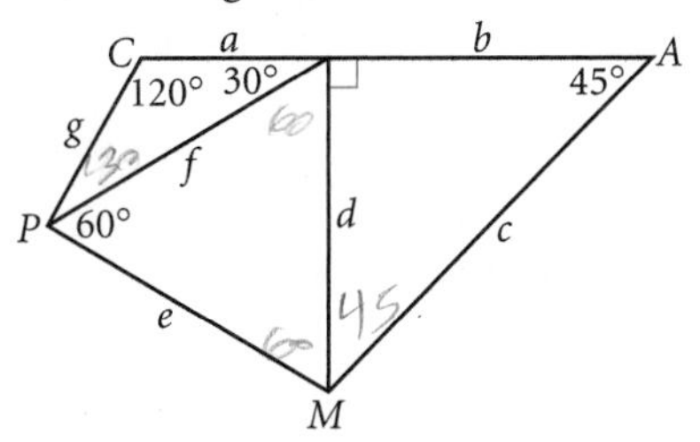

28. The measure of an angle formed by the bisectors of two angles in a triangle, as shown below, is 100°. What is angle measure *x*?

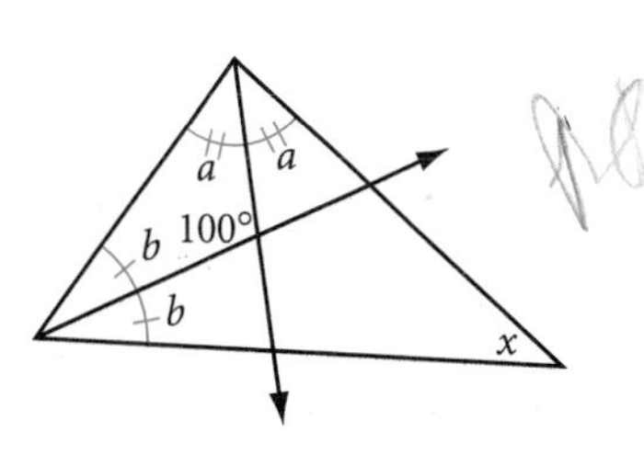

In Exercises 29 and 30, decide whether there is enough information to prove congruence. If there is, write a proof. If not, explain what is missing.

29. In the figure below, $\overline{RE} \cong \overline{AE}$, $\angle S \cong \angle T$, and $\angle ERL \cong \angle EAL$. Is $\overline{SA} \cong \overline{TR}$?

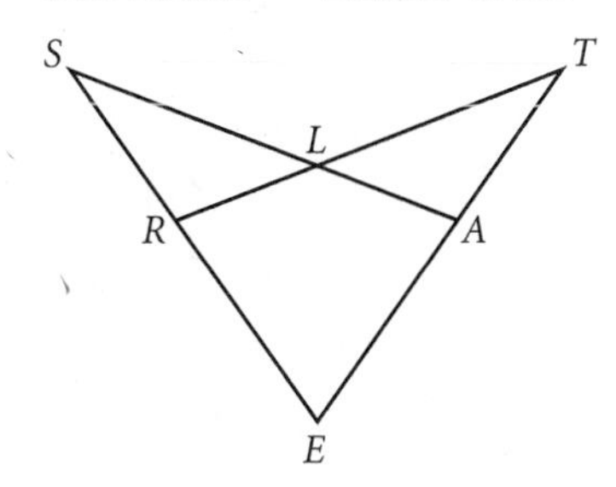

30. In the figure below, $\angle A \cong \angle M$, $\overline{AF} \perp \overline{FR}$, and $\overline{MR} \perp \overline{FR}$. Is $\triangle FRD$ isosceles?

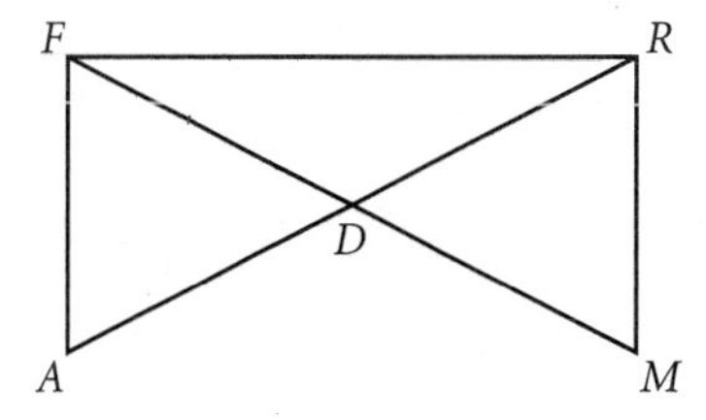

31. The measure of an angle formed by altitudes from two vertices of a triangle, as shown below, is 132°. What is angle measure *x*?

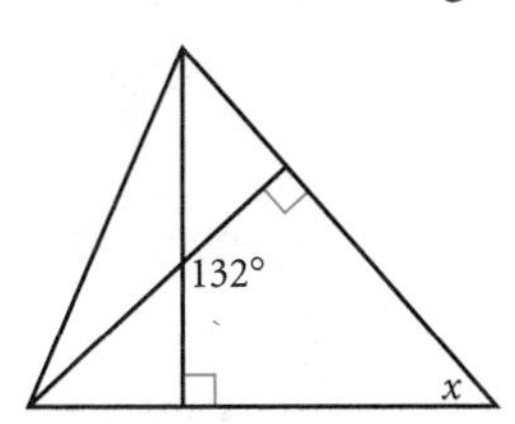

32. Connecting the legs of the chair at their midpoints as shown guarantees that the seat is parallel to the floor. Explain why. ⓗ

For Exercises 33 and 34, use the segments and the angles below. Use either patty paper or a compass and a straightedge. The lowercase letter above each segment represents the length of the segment.

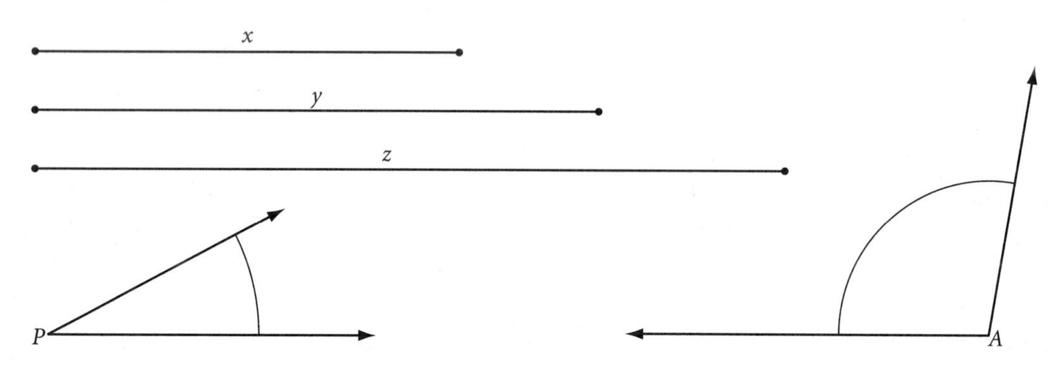

33. ***Construction*** Construct $\triangle PAL$ given $\angle P$, $\angle A$, and $AL = y$.

34. ***Construction*** Construct two triangles $\triangle PBS$ that are not congruent to each other given $\angle P$, $PB = z$, and $SB = x$.

35. In the figure at right, is $\overline{TI} \parallel \overline{RE}$? Complete the flowchart proof or explain why they are not parallel. ⓗ

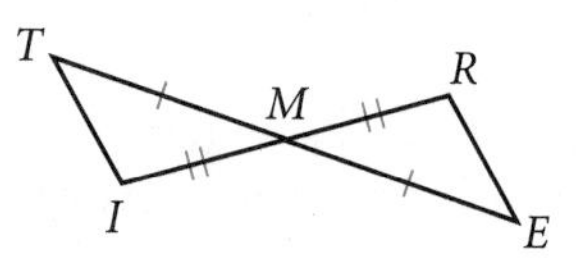

Given: M is the midpoint of both $\overline{TE}$ and $\overline{IR}$.

Show: $\overline{TI} \parallel \overline{RE}$

Flowchart Proof

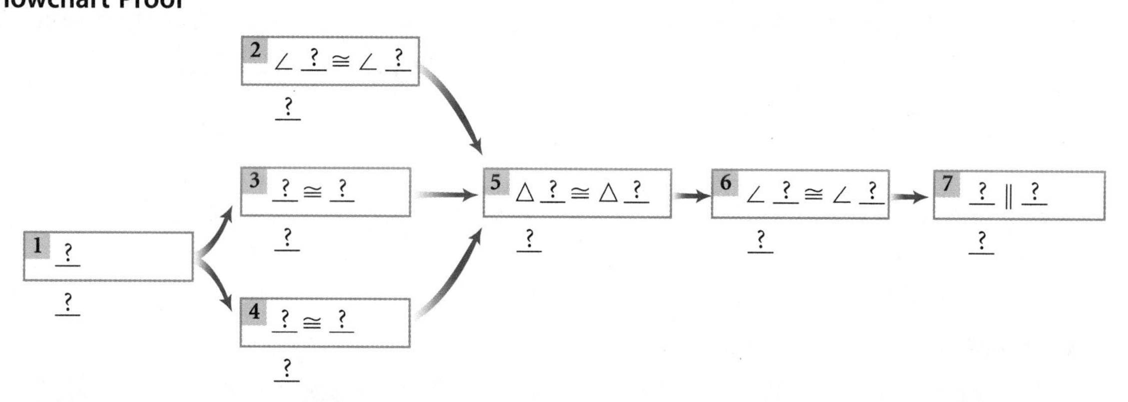

36. At the beginning of the chapter, you learned that triangles make structures more stable. Let's revisit the shelves from Lesson 4.1. Explain how the SSS congruence shortcut guarantees that the shelves on the right will retain their shape, and why the shelves on the left wobble.

37. ***Construction*** Use patty paper or compass and straightedge to construct a 75° angle. Explain your method.

TAKE ANOTHER LOOK

The section that follows, Take Another Look, gives you a chance to extend, communicate, and assess your understanding of the work you did in the investigations in this chapter. Sometimes it will lead to new, related discoveries.

1. Explore the Triangle Sum Conjecture on a sphere or a globe. Can you draw a triangle that has two or more obtuse angles? Three right angles? Write an illustrated report of your findings.

2. Investigate the Isosceles Triangle Conjecture and the Equilateral/Equiangular Triangle Conjecture on a sphere. Write an illustrated report of your findings.

3. A friend claims that if the measure of one acute angle of a triangle is half the measure of another acute angle of the triangle, then the triangle can be divided into two isosceles triangles. Try this with a computer or other tools. Describe your method and explain why it works.

4. A friend claims that if one exterior angle has twice the measure of one of the remote interior angles, then the triangle is isosceles. Use a geometry software program or other tools to investigate this claim. Describe your findings.

5. Is there a conjecture (similar to the Triangle Exterior Angle Conjecture) that you can make about exterior and remote interior angles of a convex quadrilateral? Experiment. Write about your findings.

6. Is there a conjecture you can make about inequalities among the sums of the lengths of sides and/or diagonals of a quadrilateral? Experiment. Write about your findings.

7. In Chapter 3, you discovered how to construct the perpendicular bisector of a segment. Perform this construction. Now use what you've learned about congruence shortcuts to explain why this construction method works.

8. In Chapter 3, you discovered how to construct a perpendicular through a point on a line. Perform this construction. Use a congruence shortcut to explain why the construction works.

9. Is there a conjecture similar to the SSS Congruence Conjecture that you can make about congruence between quadrilaterals? For example, is SSSS a shortcut for quadrilateral congruence? Or, if three sides and a diagonal of one quadrilateral are congruent to the corresponding three sides and diagonal of another quadrilateral, must the two quadrilaterals be congruent (SSSD)? Investigate. Write a paragraph explaining how your conjectures follow from the triangle congruence conjectures you've learned.

Assessing What You've Learned

WRITE TEST ITEMS

It's one thing to be able to do a math problem. It's another to be able to make one up. If you were writing a test for this chapter, what would it include?

Start by having a group discussion to identify the key ideas in each lesson of the chapter. Then divide the lessons among group members, and have each group member write a problem for each lesson assigned to them. Try to create a mix of problems in your group, from simple one-step exercises that require you to recall facts to more complex, multistep problems that require more thinking. An example of a simple problem might be finding a missing angle measure in a triangle. A more complex problem could be a flowchart for a logical argument, or a word problem that requires using geometry to model a real-world situation.

Share your problems with your group members and try out one another's problems. Then discuss the problems in your group: Were they representative of the content of the chapter? Were some too hard or too easy? Writing your own problems is an excellent way to assess and review what you've learned. Maybe you can even persuade your teacher to use one of your items on a real test!

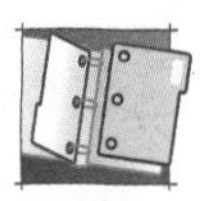

ORGANIZE YOUR NOTEBOOK Review your notebook to be sure it is complete and well organized. Write a one-page chapter summary based on your notes.

WRITE IN YOUR JOURNAL Write a paragraph or two about something you did in this class that gave you a great sense of accomplishment. What did you learn from it? What about the work makes you proud?

UPDATE YOUR PORTFOLIO Choose a piece of work from this chapter to add to your portfolio. Document the work, explaining what it is and why you chose it.

PERFORMANCE ASSESSMENT While a classmate, a friend, a family member, or your teacher observes, perform an investigation from this chapter. Explain each step, including how you arrived at the conjecture.

CHAPTER

5 Discovering and Proving Polygon Properties

The mathematicians may well nod their heads in a friendly and interested manner—I still am a tinkerer to them. And the "artistic" ones are primarily irritated. Still, maybe I'm on the right track if I experience more joy from my own little images than from the most beautiful camera in the world . . ."

Still Life and Street, M. C. Escher, 1967–1968

OBJECTIVES

In this chapter you will

- study properties of polygons
- discover relationships among their angles, sides, and diagonals
- learn about real-world applications of special polygons

LESSON

5.1

Polygon Sum Conjecture

I find that the harder I work, the more luck I seem to have.

THOMAS JEFFERSON

There are many kinds of triangles, but in Chapter 4, you discovered that the sum of their angle measures is always 180°. In this lesson you'll investigate the sum of the angle measures in quadrilaterals, pentagons, and other polygons. Then you'll look for a pattern in the sum of the angle measures in *any* polygon.

Investigation

Is There a Polygon Sum Formula?

For this investigation each person in your group should draw a different version of the same polygon. For example, if your group is investigating hexagons, try to think of different ways you could draw a hexagon.

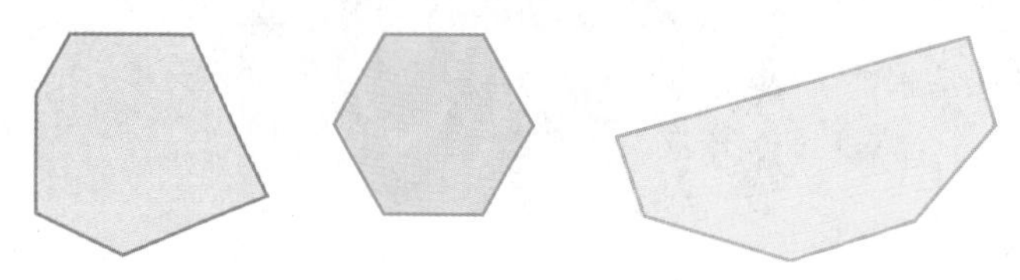

Step 1 Draw the polygon. Carefully measure all the interior angles, then find the sum.

Step 2 Share your results with your group. If you measured carefully, you should all have the same sum! If your answers aren't exactly the same, find the average.

Step 3 Copy the table below. Repeat Steps 1 and 2 with different polygons, or share results with other groups. Complete the table.

Number of sides of polygon	3	4	5	6	7	8	...	n
Sum of measures of angles	180°						...	

You can now make some conjectures.

Quadrilateral Sum Conjecture C-30

The sum of the measures of the four angles of any quadrilateral is _?_.

Pentagon Sum Conjecture C-31

The sum of the measures of the five angles of any pentagon is _?_.

If a polygon has n sides, it is called an **n-gon.**

Step 4 — Look for a pattern in the completed table. Write a general formula for the sum of the angle measures of a polygon in terms of the number of sides, n.

Polygon Sum Conjecture C-32

The sum of the measures of the n interior angles of an n-gon is _?_.

You used inductive reasoning to discover the formula. Now you can use deductive reasoning to see why the formula works.

Step 5 — Draw all the diagonals from *one* vertex of your polygon. How many triangles do the diagonals create? How does the number of triangles relate to the formula you found? How can you check that your formula is correct for a polygon with 12 sides?

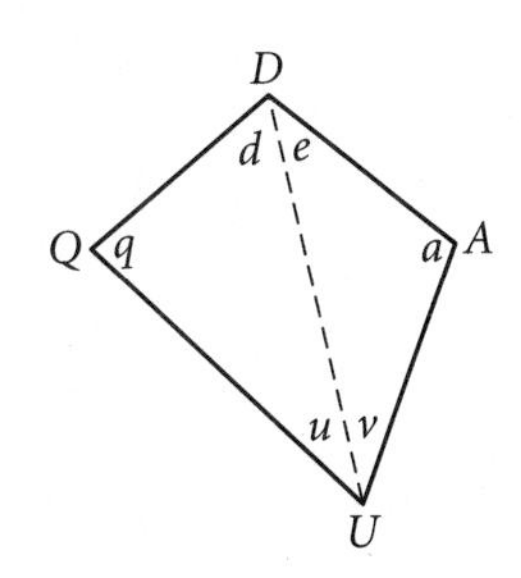

Step 6 — Write a short paragraph proof of the Quadrilateral Sum Conjecture. Use the diagram of quadrilateral *QUAD*. (Hint: Use the Triangle Sum Conjecture.)

EXERCISES

You will need

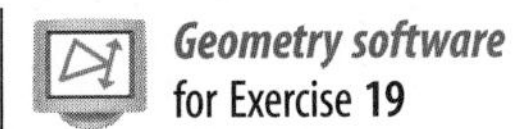

1. Use the Polygon Sum Conjecture to complete the table.

Number of sides of polygon	7	8	9	10	11	20	55	100
Sum of measures of angles								

2. What is the measure of each angle of an equiangular pentagon? An equiangular hexagon? Complete the table. ⓗ

Number of sides of equiangular polygon	5	6	7	8	9	10	12	16	100
Measures of each angle of equiangular polygon									

In Exercises 3–8, use your conjectures to calculate the measure of each lettered angle.

3. $a = \underline{?}$

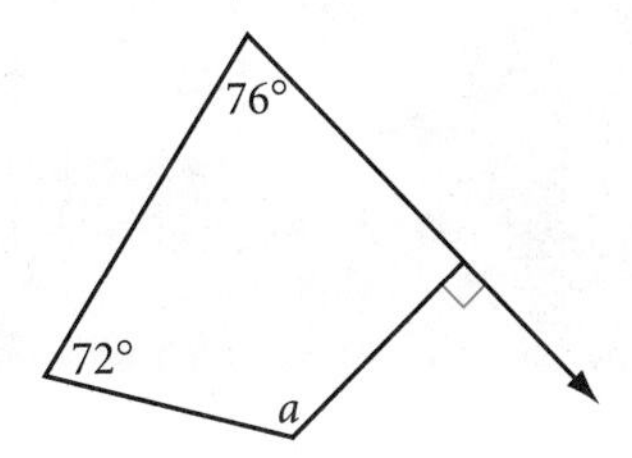

4. $b = \underline{?}$

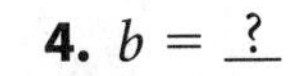

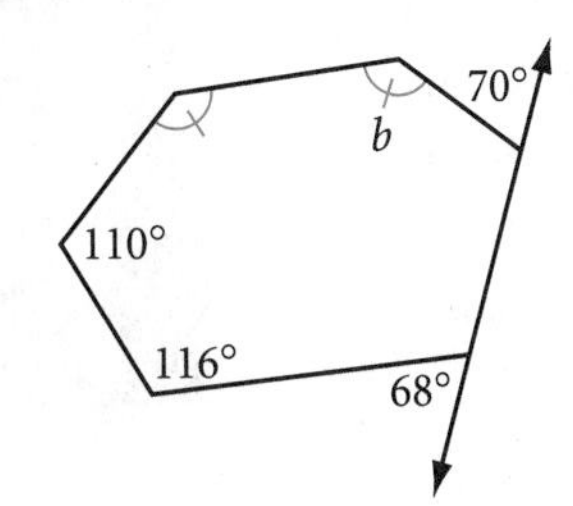

5. $e = \underline{?}$

$f = \underline{?}$

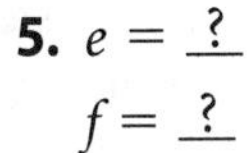

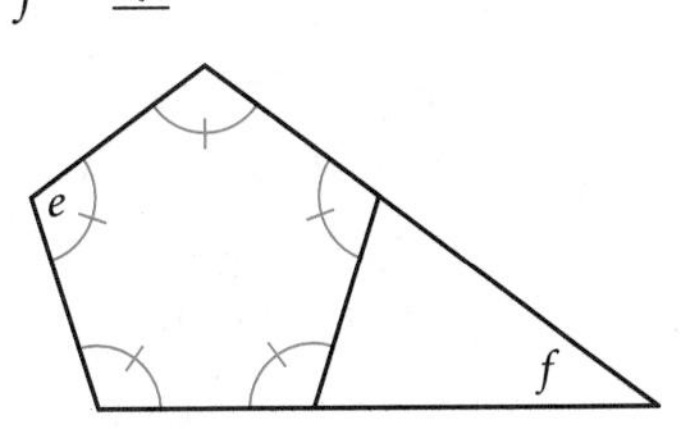

6. $c = \underline{\ ?\ }$
$d = \underline{\ ?\ }$ ⓗ

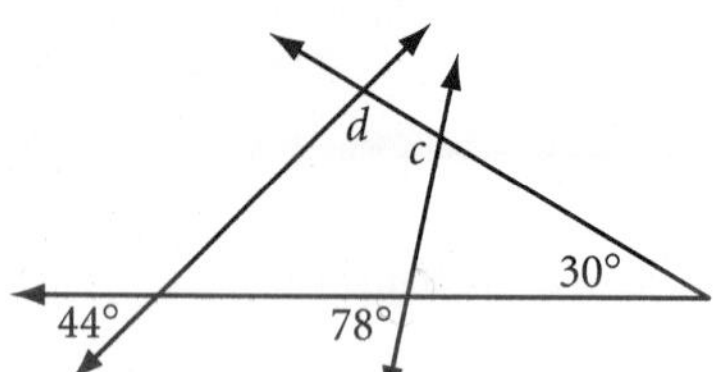

7. $g = \underline{\ ?\ }$ ⓗ
$h = \underline{\ ?\ }$

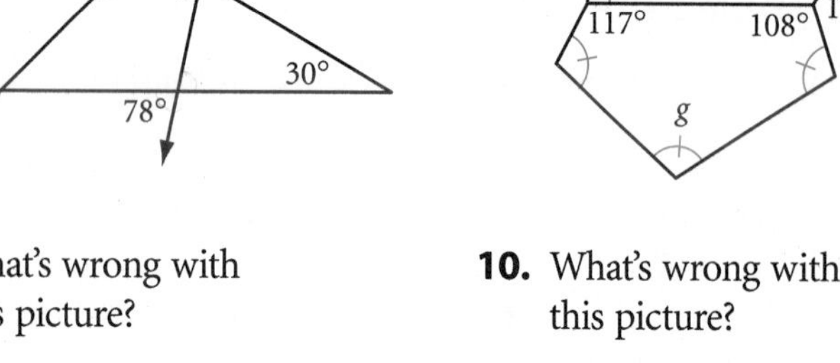

8. $j = \underline{\ ?\ }$
$k = \underline{\ ?\ }$

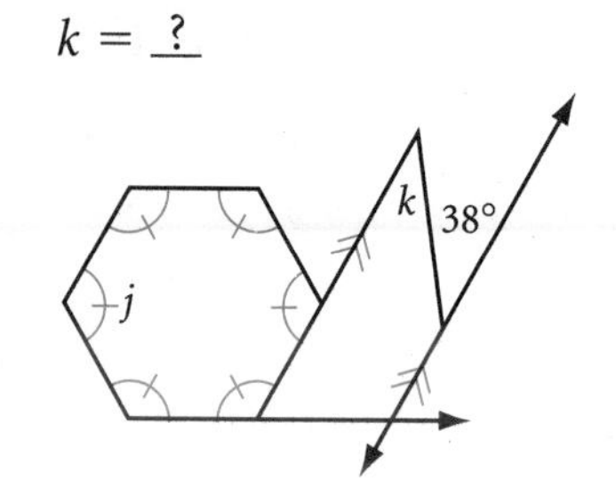

9. What's wrong with this picture?

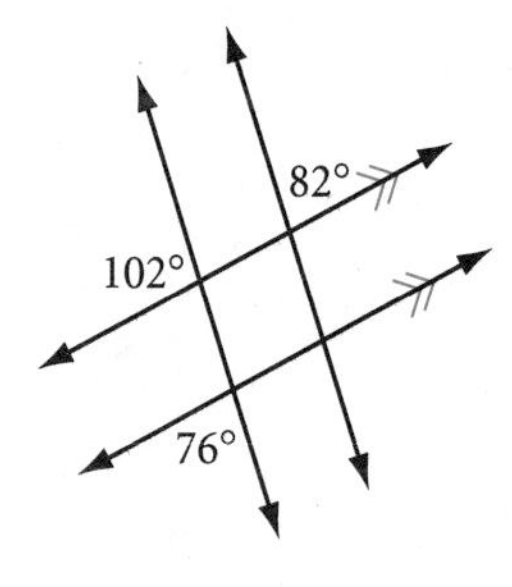

10. What's wrong with this picture?

11. Three regular polygons meet at point A. How many sides does the largest polygon have?

12. Trace the figure at right. Calculate each lettered angle measure.

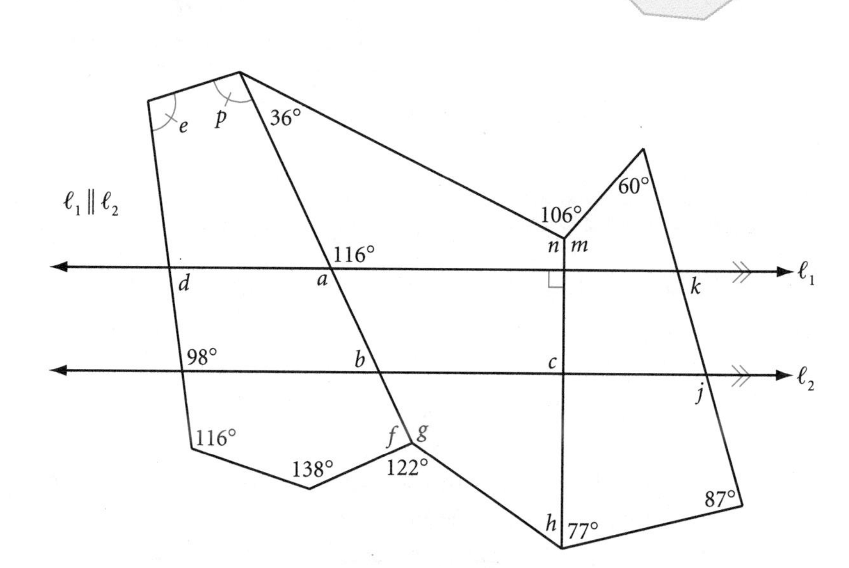

13. How many sides does a polygon have if the sum of its angle measures is 2700°? ⓗ

14. How many sides does an equiangular polygon have if each interior angle measures 156°? ⓗ

15. Archaeologist Ertha Diggs has uncovered a piece of a ceramic plate. She measures it and finds that each side has the same length and each angle has the same measure.

She conjectures that the original plate was the shape of a regular polygon. She knows that if the original plate was a regular 16-gon, it was probably a ceremonial dish from the third century. If it was a regular 18-gon, it was probably a palace dinner plate from the twelfth century.

If each angle measures 160°, from what century did the plate likely originate?

16. APPLICATION You need to build a window frame for an octagonal window like this one. To make the frame, you'll cut identical trapezoidal pieces. What are the measures of the angles of the trapezoids? Explain how you found these measures.

17. Use this diagram to prove the Pentagon Sum Conjecture. (h)

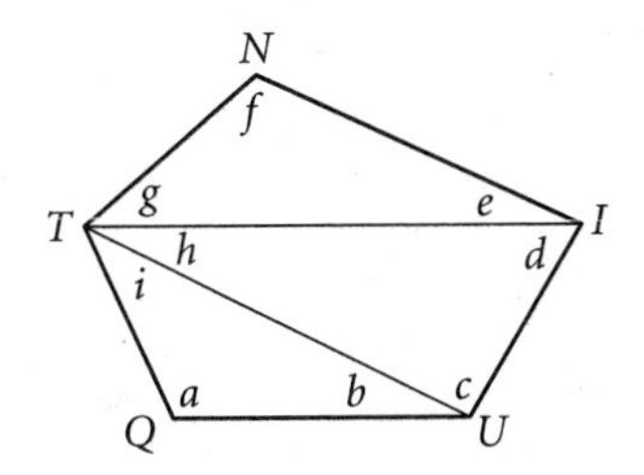

Review

18. This figure is a detail of one vertex of the tiling at the beginning of this lesson. Find the missing angle measure x.

19. *Technology* Use geometry software to construct a quadrilateral and locate the midpoints of its four sides. Construct segments connecting the midpoints of opposite sides. Construct the point of intersection of the two segments. Drag a vertex or a side so that the quadrilateral becomes concave. Observe these segments and make a conjecture.

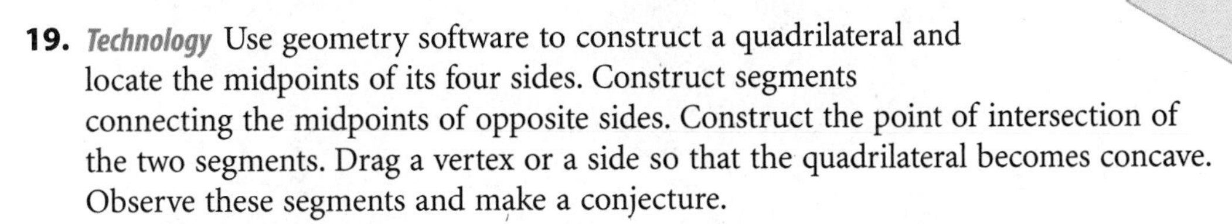

20. Write the equation of the perpendicular bisector of the segment with endpoints $(-12, 15)$ and $(4, -3)$.

21. $\triangle ABC$ has vertices $A(0, 0)$, $B(-4, -2)$, and $C(8, -8)$. What is the equation of the median to side $\overline{AB}$?

22. Line ℓ is parallel to $\overleftrightarrow{AB}$. As P moves to the right along ℓ, which of these measures will always increase?

A. The distance PA

B. The measure of $\angle APB$

C. The perimeter of $\triangle ABP$

D. The measure of $\angle ABP$

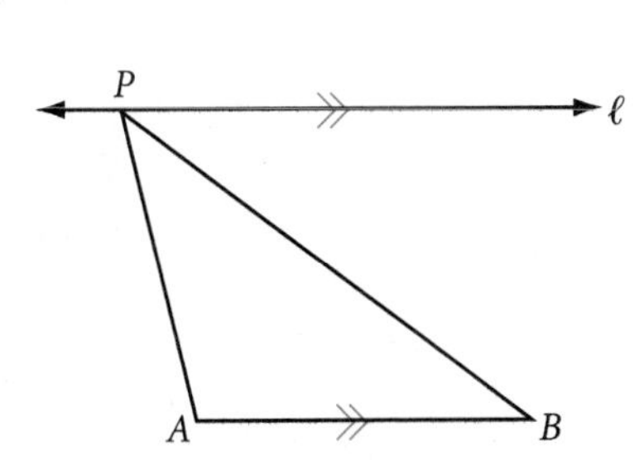

IMPROVING YOUR VISUAL THINKING SKILLS

Net Puzzle

The clear cube shown has the letters *DOT* printed on one face. When a light is shined on that face, the image of *DOT* appears on the opposite face. The image of *DOT* on the opposite face is then painted. Copy the net of the cube and sketch the painted image of the word, *DOT,* on the correct square and in the correct position.

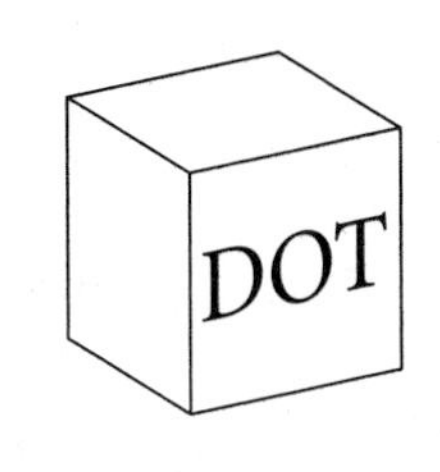

LESSON

5.2

Exterior Angles of a Polygon

If someone had told me I would be Pope someday, I would have studied harder.

POPE JOHN PAUL I

In Lesson 5.1, you discovered a formula for the sum of the measures of the *interior* angles of any polygon. In this lesson you will discover a formula for the sum of the measures of the *exterior* angles of a polygon.

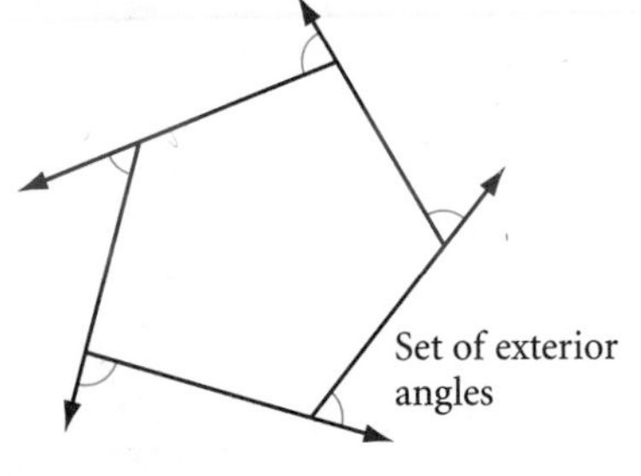

Best known for her participation in the Dada Movement, German artist Hannah Hoch (1889–1978) painted *Emerging Order* in the Cubist style. Do you see any examples of exterior angles in the painting?

Investigation

Is There an Exterior Angle Sum?

You will need

- a straightedge
- a protractor

Let's use some inductive and deductive reasoning to find the exterior angle measures in a polygon.

Each person in your group should draw the same kind of polygon for Steps 1–5.

Step 1 Draw a large polygon. Extend its sides to form a set of exterior angles.

Step 2 Measure all the *interior* angles of the polygon except one. Use the Polygon Sum Conjecture to calculate the measure of the remaining interior angle. Check your answer using your protractor.

Step 3 Use the Linear Pair Conjecture to calculate the measure of each exterior angle.

Step 4 Calculate the sum of the measures of the exterior angles. Share your results with your group members.

Step 5 Repeat Steps 1–4 with different kinds of polygons, or share results with other groups. Make a table to keep track of the number of sides and the sum of the exterior angle measures for each kind of polygon. Find a formula for the sum of the measures of a polygon's exterior angles.

Exterior Angle Sum Conjecture C-33

For any polygon, the sum of the measures of a set of exterior angles is _?_.

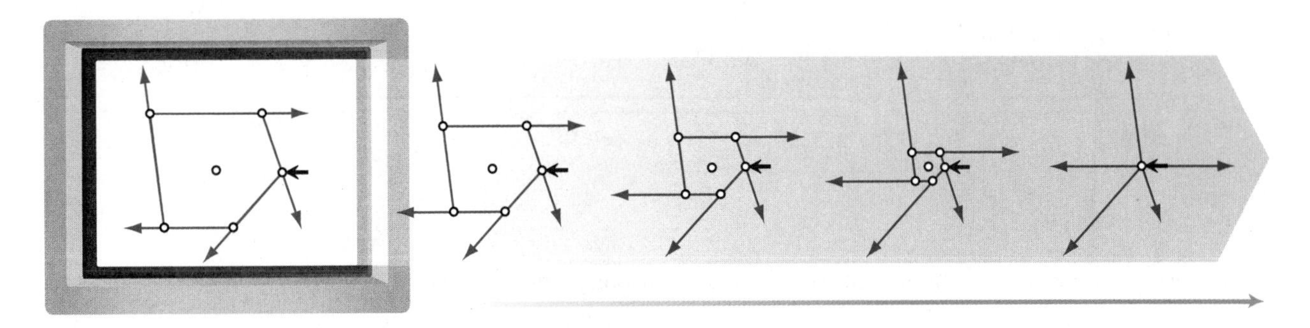

Step 6 Study the software construction above. Explain how it demonstrates the Exterior Angle Sum Conjecture.

Step 7 Using the Polygon Sum Conjecture, write a formula for the measure of each interior angle in an equiangular polygon.

Step 8 Using the Exterior Angle Sum Conjecture, write the formula for the measure of each exterior angle in an equiangular polygon.

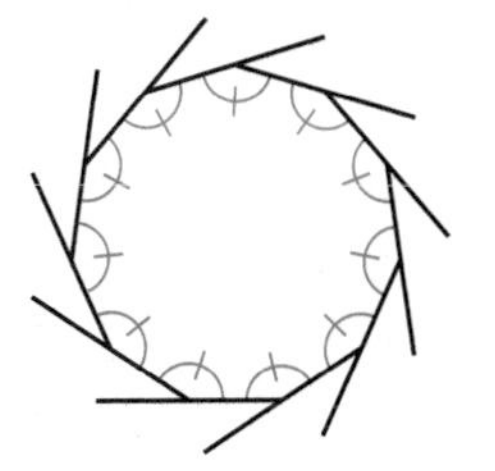

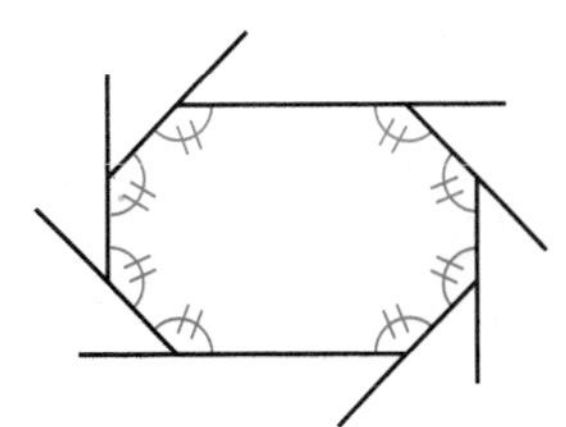

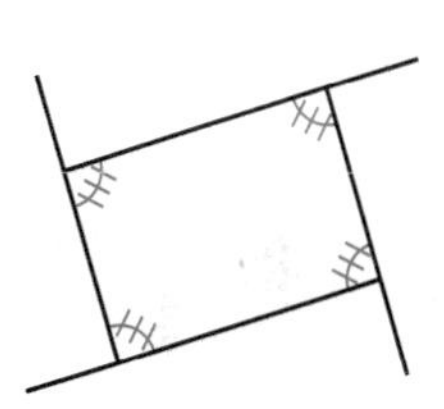

Step 9 Using your results from Step 8, you can write the formula for an interior angle a different way. How do you find the measure of an interior angle if you know the measure of its exterior angle? Complete the next conjecture.

Equiangular Polygon Conjecture C-34

You can find the measure of each interior angle of an equiangular n-gon by using either of these formulas: _?_ or _?_.

EXERCISES

1. Complete this flowchart proof of the Exterior Angle Sum Conjecture for a triangle.

Flowchart Proof

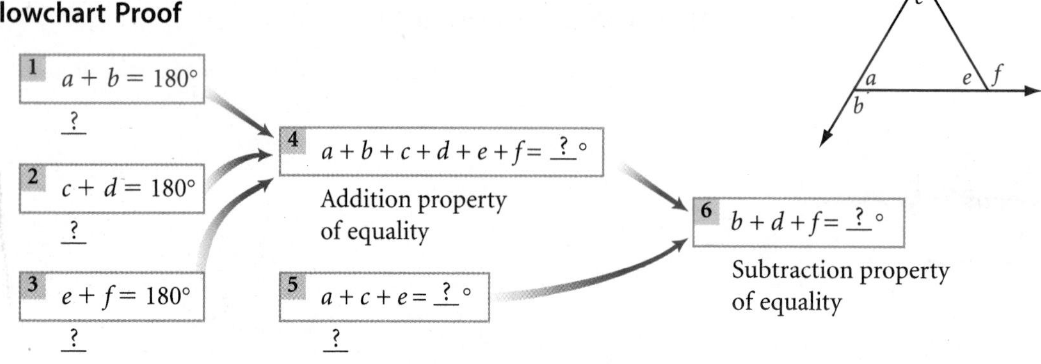

2. What is the sum of the measures of the exterior angles of a decagon?

3. What is the measure of an exterior angle of an equiangular pentagon? An equiangular hexagon?

In Exercises 4–9, use your new conjectures to calculate the measure of each lettered angle.

4.

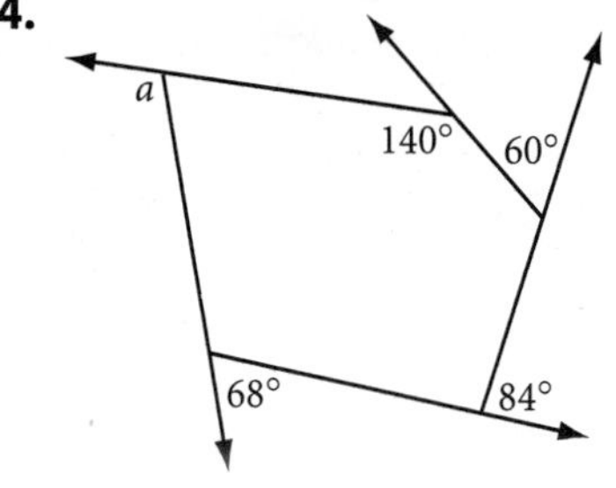

5.

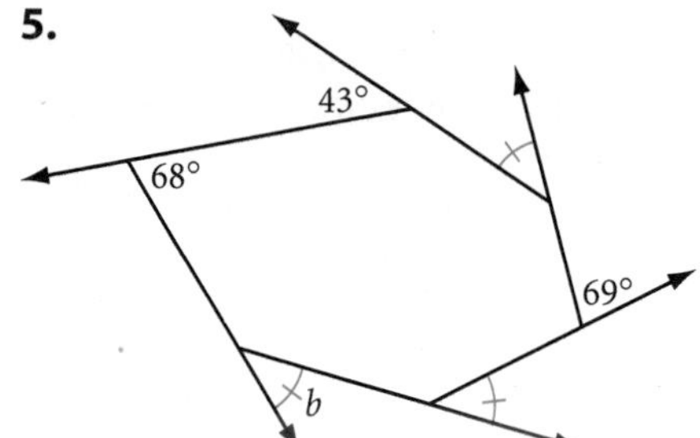

6. ⓗ

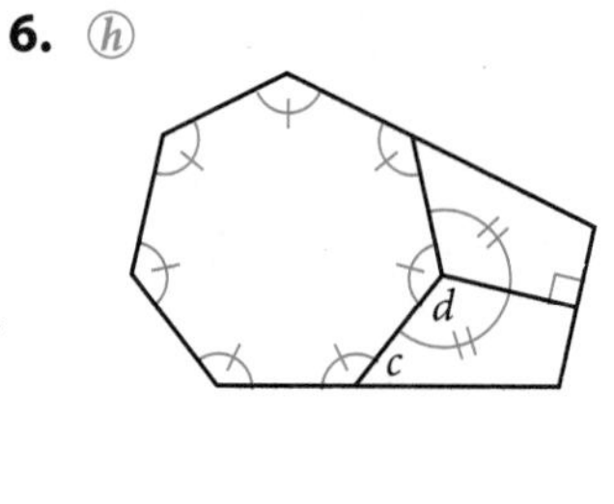

7.

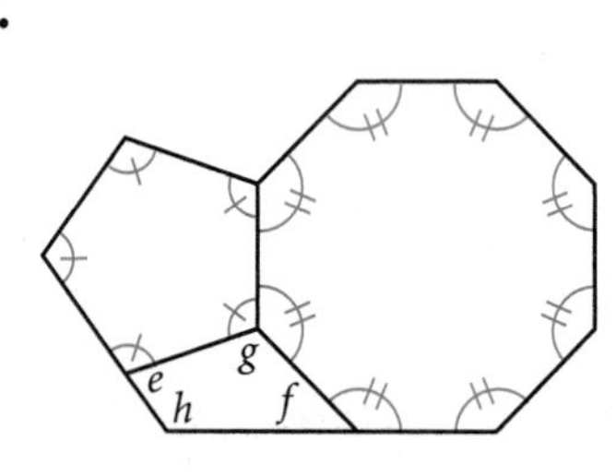

8.

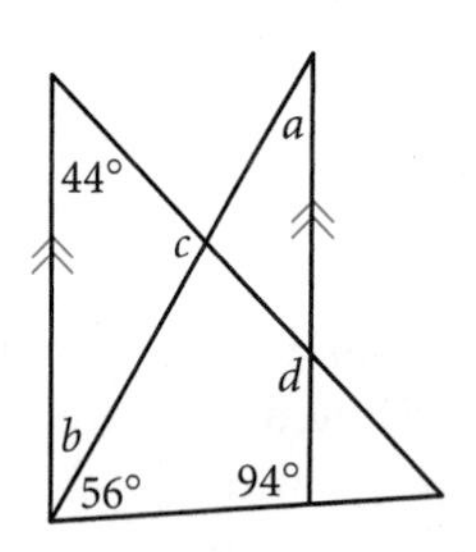

9.

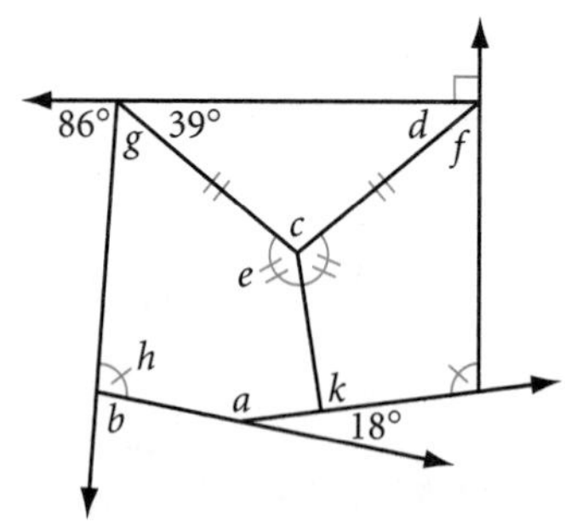

10. How many sides does a regular polygon have if each exterior angle measures 24°? ⓗ

11. How many sides does a polygon have if the sum of its interior angle measures is 7380°?

12. Is there a maximum number of obtuse exterior angles that any polygon can have? If so, what is the maximum? If not, why not? Is there a minimum number of acute interior angles that any polygon must have? If so, what is the minimum? If not, why not? ⓗ

Technology
CONNECTION

The aperture of a camera is an opening shaped like a regular polygon surrounded by thin sheets that form a set of exterior angles. These sheets move together or apart to close or open the aperture, limiting the amount of light passing through the camera's lens. How does the sequence of closing apertures shown below demonstrate the Exterior Angle Sum Conjecture? Does the number of sides make a difference in the opening and closing of the aperture?

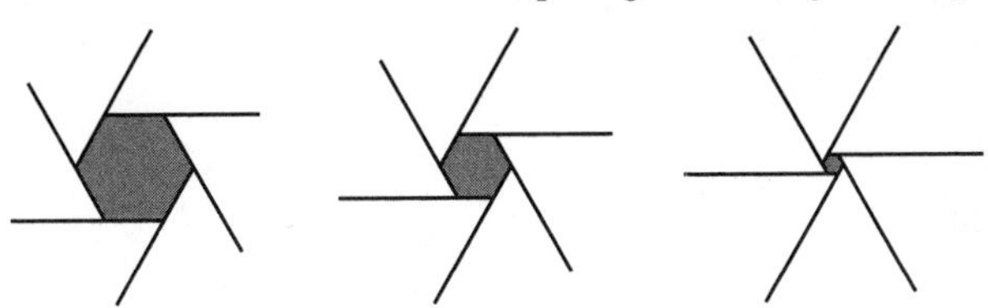

Review

13. Name the regular polygons that appear in the tiling below. Find the measures of the angles that surround point A in the tiling.

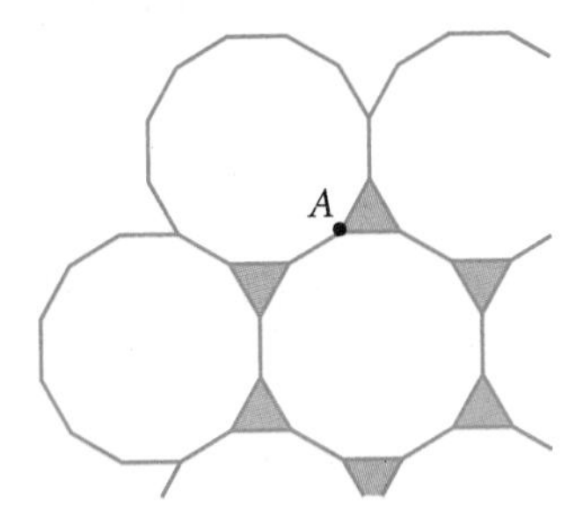

14. Name the regular polygons that appear in the tiling below. Find the measures of the angles that surround any vertex point in the tiling.

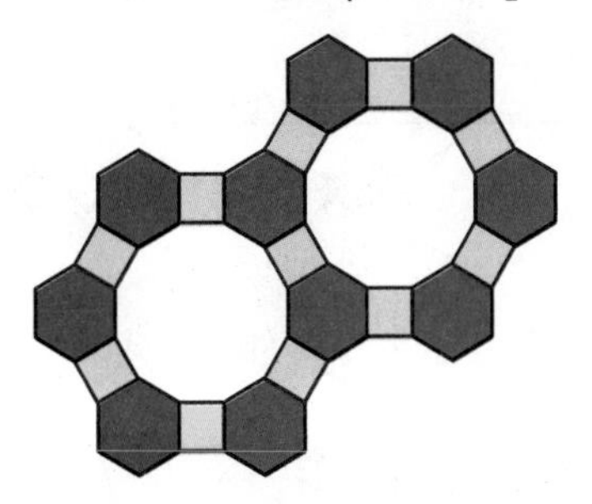

15. $\angle RAC \cong \angle DCA$, $\overline{CD} \cong \overline{AR}$, $\overline{AC} \parallel \overline{DR}$. Is $\overline{AD} \cong \overline{CR}$? Why? ⓗ

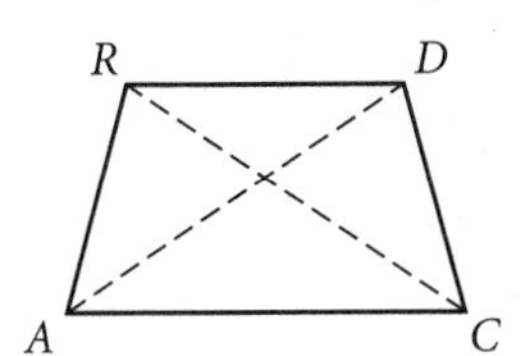

16. $\overline{DT} \cong \overline{RT}$, $\overline{DA} \cong \overline{RA}$. Is $\angle D \cong \angle R$? Why? ⓗ

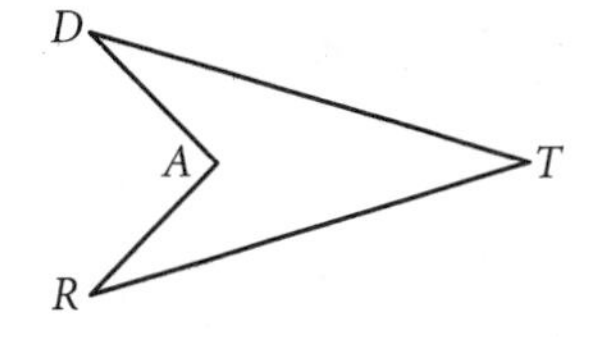

IMPROVING YOUR VISUAL THINKING SKILLS

Dissecting a Hexagon II

Make six copies of the hexagon at right by tracing it onto your paper. Then divide each hexagon into twelve identical parts in a different way.

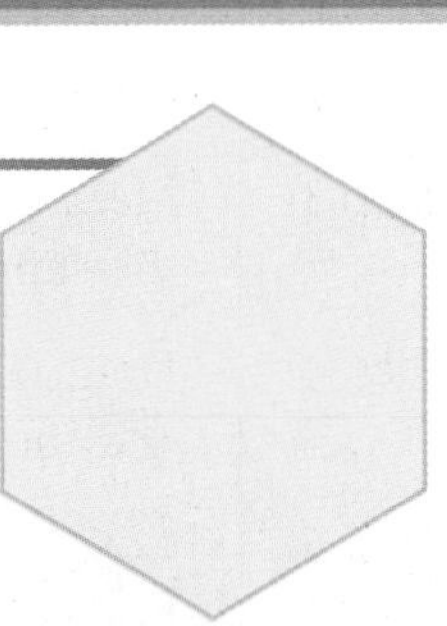

Exploration

Star Polygons

If you arrange a set of points roughly around a circle or an oval, and then you connect each point to the next with segments, you should get a convex polygon like the one at right. What do you get if you connect every second point with segments? You get a star polygon like the ones shown in the activity below.

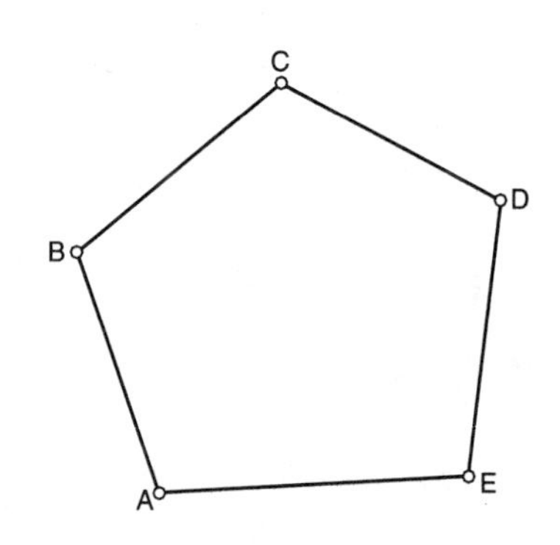

In this activity, you'll investigate the angle measure sums of star polygons.

Activity
Exploring Star Polygons

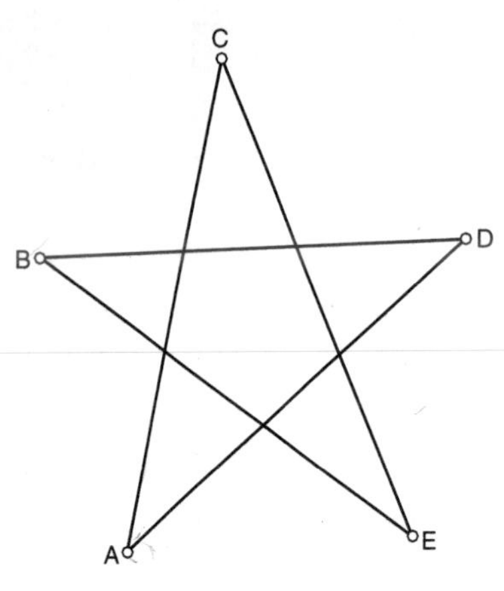

5-pointed star *ABCDE*

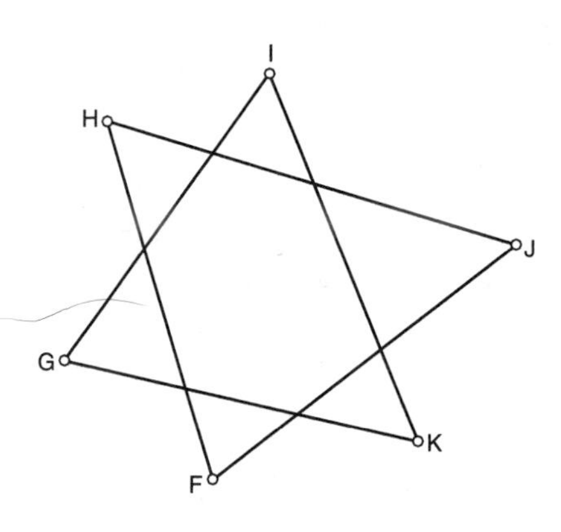

6-pointed star *FGHIJK*

Step 1 Draw five points A through E in a circular path, clockwise.

Step 2 Connect every second point with $\overline{AC}$, $\overline{CE}$, $\overline{EB}$, $\overline{BD}$, and $\overline{DA}$.

Step 3 Measure the five angles A through E at the star points. Use the calculator to find the sum of the angle measures.

Step 4 Drag each vertex of the star and observe what happens to the angle measures and the calculated sum. Does the sum change? What is the sum?

Step 5 Copy the table on page 265. Use the Polygon Sum Conjecture to complete the first column. Then enter the angle sum for the 5-pointed star.

Step 6 Repeat Steps 1–5 for a 6-pointed star. Enter the angle sum in the table. Complete the column for each n-pointed star with every second point connected.

Step 7 What happens if you connect every third point to form a star? What would be the sum of the angle measures in this star? Complete the table column for every third point.

Step 8 Use what you have learned to complete the table. What patterns do you notice? Write the rules for n-pointed stars.

Number of star points	Angle measure sums by how the star points are connected: Every point	Every 2nd point	Every 3rd point	Every 4th point	Every 5th point
5	(pentagon) 540°	(star)			
6	(hexagon) 720°				
7					

Step 9 Let's explore Step 4 a little further. Can you drag the vertices of each star polygon to make it convex? Describe the steps for turning each one into a convex polygon, and then back into a star polygon again, in the fewest steps possible.

Step 10 In Step 9, how did the sum of the angle measure change when a polygon became convex? When did it change?

This blanket by Teresa Archuleta-Sagel is titled *My Blue Vallero Heaven.* Are these star polygons? Why?

LESSON

5.3

Kite and Trapezoid Properties

Imagination is the highest kite we fly.

LAUREN BACALL

Recall that a **kite** is a quadrilateral with exactly two distinct pairs of congruent consecutive sides.

If you construct two different isosceles triangles on opposite sides of a common base and then remove the base, you have constructed a kite. In an isosceles triangle, the vertex angle is the angle between the two congruent sides. Therefore, let's call the two angles between each pair of congruent sides of a kite the **vertex angles** of the kite. Let's call the other pair the **nonvertex angles.**

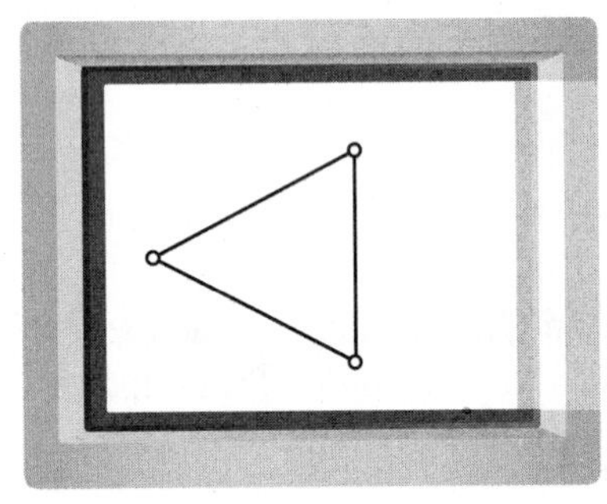

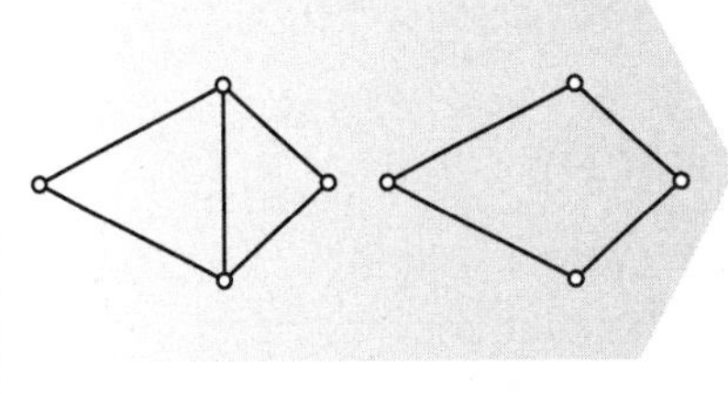

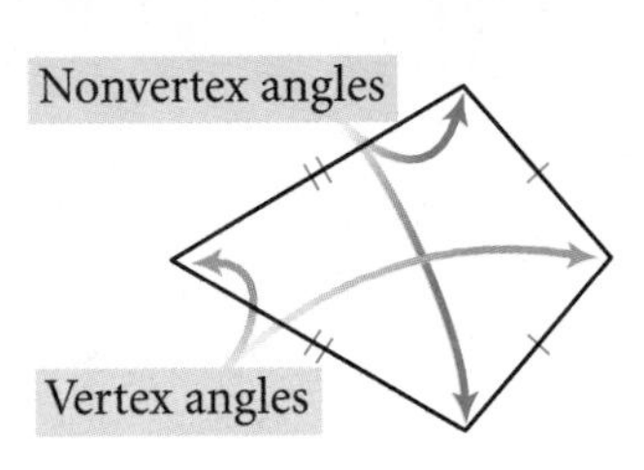

A kite also has one line of reflectional symmetry, just like an isosceles triangle. You can use this property to discover other properties of kites. Let's investigate.

Investigation 1
What Are Some Properties of Kites?

You will need

- patty paper

In this investigation you will look at angles and diagonals in a kite to see what special properties they have.

Step 1 On patty paper, draw two connected segments of different lengths, as shown. Fold through the endpoints and trace the two segments on the back of the patty paper.

Step 2 Compare the size of each pair of opposite angles in your kite by folding an angle onto the opposite angle. Are the vertex angles congruent? Are the nonvertex angles congruent? Share your observations with others near you and complete the conjecture.

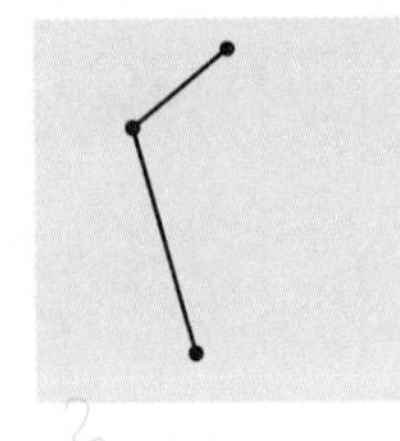

Step 1

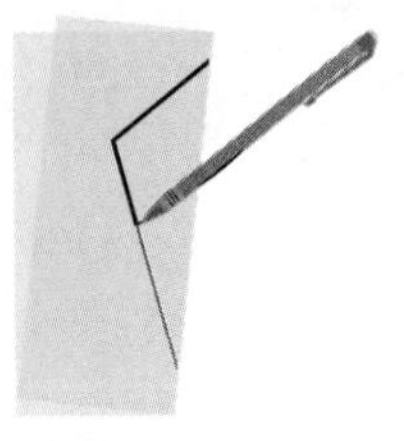

Step 2

Kite Angles Conjecture C-35

The ? angles of a kite are ?.

Step 3 Draw the diagonals. How are the diagonals related? Share your observations with others in your group and complete the conjecture.

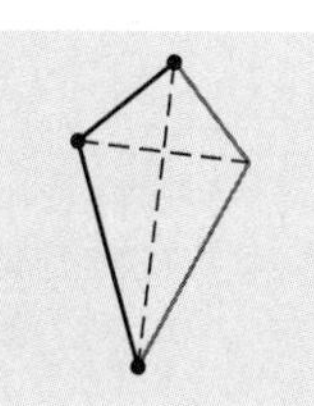

Kite Diagonals Conjecture C-36

The diagonals of a kite are ?.

What else seems to be true about the diagonals of kites?

Step 4 Compare the lengths of the segments on both diagonals. Does either diagonal bisect the other? Share your observations with others near you. Copy and complete the conjecture.

Kite Diagonal Bisector Conjecture C-37

The diagonal connecting the vertex angles of a kite is the ? of the other diagonal.

Step 5 Fold along both diagonals. Does either diagonal bisect any angles? Share your observations with others and complete the conjecture.

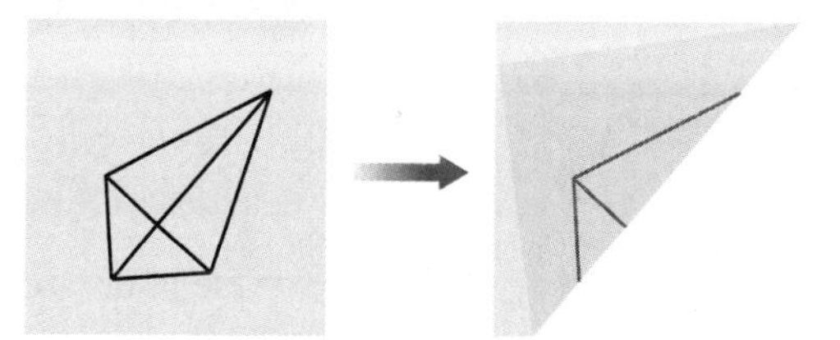

Kite Angle Bisector Conjecture C-38

The ? angles of a kite are ? by a ?.

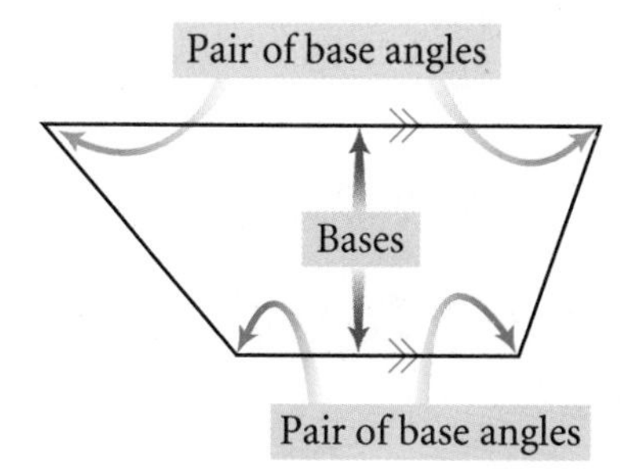

You will prove the Kite Diagonal Bisector Conjecture and the Kite Angle Bisector Conjecture as exercises after this lesson.

Let's move on to trapezoids. Recall that a **trapezoid** is a quadrilateral with exactly one pair of parallel sides.

In a trapezoid the parallel sides are called **bases.** A pair of angles that share a base as a common side are called **base angles.**

In the next investigation, you will discover some properties of trapezoids.

Science
CONNECTION

A *trapezium* is a quadrilateral with *no* two sides parallel. The words *trapezoid* and *trapezium* come from the Greek word *trapeza,* meaning table. There are bones in your wrists that anatomists call trapezoid and trapezium because of their geometric shapes.

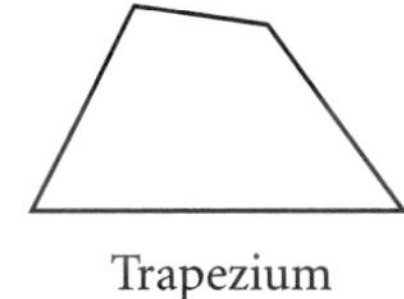
Trapezium

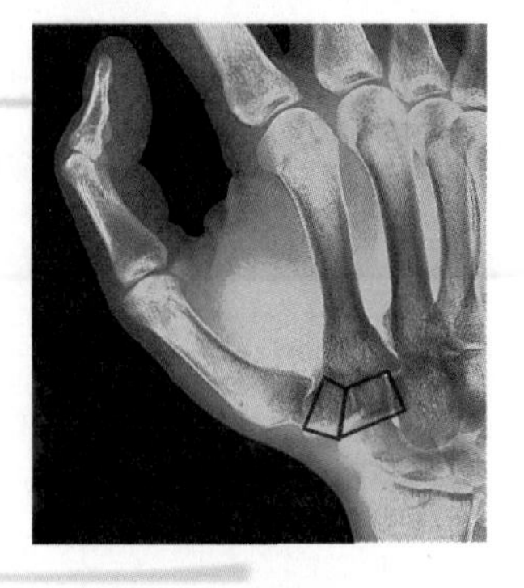

Investigation 2
What Are Some Properties of Trapezoids?

You will need

- a straightedge
- a protractor
- a compass

This is a view inside a deflating hot-air balloon. Notice the trapezoidal panels that make up the balloon.

Step 1 Use the two edges of your straightedge to draw parallel segments of unequal length. Draw two nonparallel sides connecting them to make a trapezoid.

Step 2 Use your protractor to find the sum of the measures of each pair of consecutive angles between the parallel bases. What do you notice about this sum? Share your observations with your group.

Step 3 Copy and complete the conjecture.

Trapezoid Consecutive Angles Conjecture

C-39

The consecutive angles between the bases of a trapezoid are ?.

Recall from Chapter 3 that a trapezoid whose two nonparallel sides are the same length is called an **isosceles trapezoid.** Next, you will discover a few properties of isosceles trapezoids.

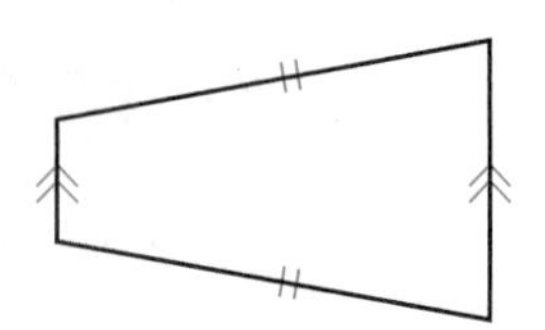

Like kites, isosceles trapezoids have one line of reflectional symmetry. Through what points does the line of symmetry pass?

Step 4 Use both edges of your straightedge to draw parallel lines. Using your compass, construct two congruent segments. Connect the four segments to make an isosceles trapezoid.

Step 5 Measure each pair of base angles. What do you notice about the pair of base angles in each trapezoid? Compare your observations with others near you.

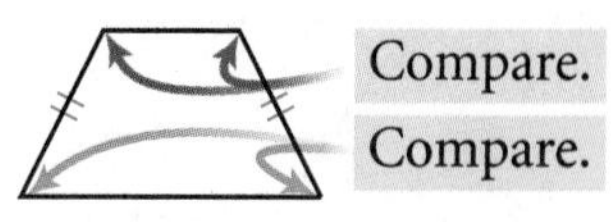

Step 6 Copy and complete the conjecture.

> **Isosceles Trapezoid Conjecture** **C-40**
>
> The base angles of an isosceles trapezoid are ?.

What other parts of an isosceles trapezoid are congruent? Let's continue.

Step 7 Draw both diagonals. Compare their lengths. Share your observations with others near you.

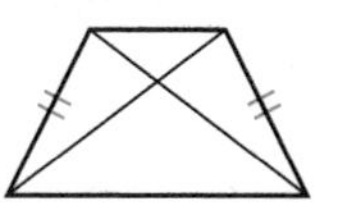

Step 8 Copy and complete the conjecture.

> **Isosceles Trapezoid Diagonals Conjecture** **C-41**
>
> The diagonals of an isosceles trapezoid are ?.

Suppose you assume that the Isosceles Trapezoid Conjecture is true. What pair of triangles and which triangle congruence conjecture would you use to explain why the Isosceles Trapezoid Diagonals Conjecture is true?

EXERCISES

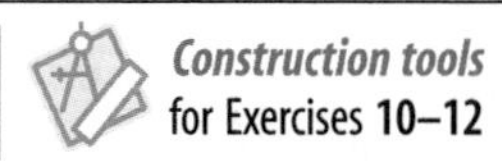

You will need

Construction tools for Exercises **10–12**

Use your new conjectures to find the missing measures.

1. Perimeter = ?

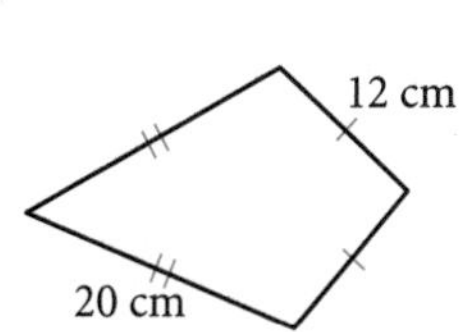

2. $x =$?
$y =$?

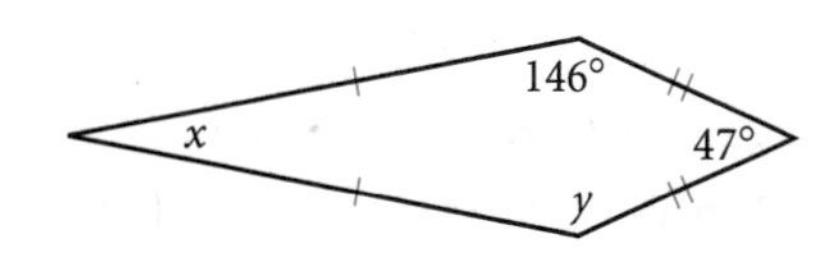

3. $x =$?
$y =$?

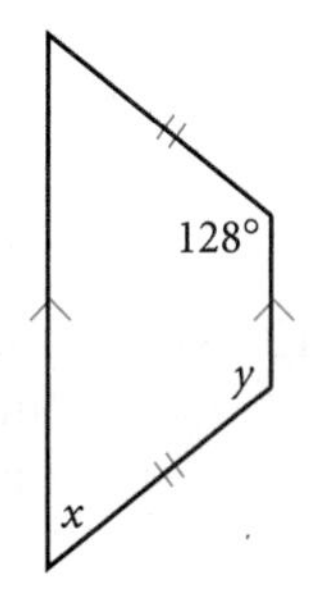

4. $x = \underline{?}$

Perimeter $= 85$ cm

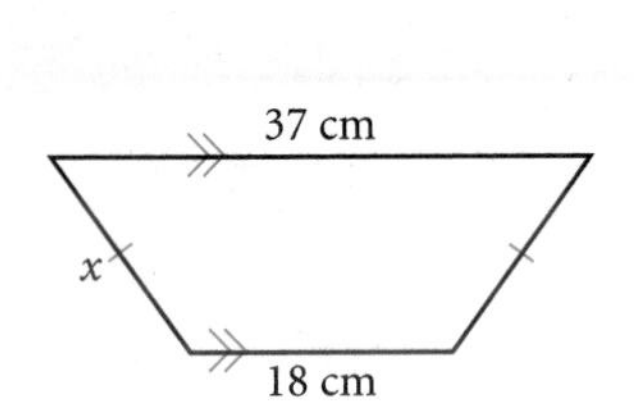

5. $x = \underline{?}$

$y = \underline{?}$

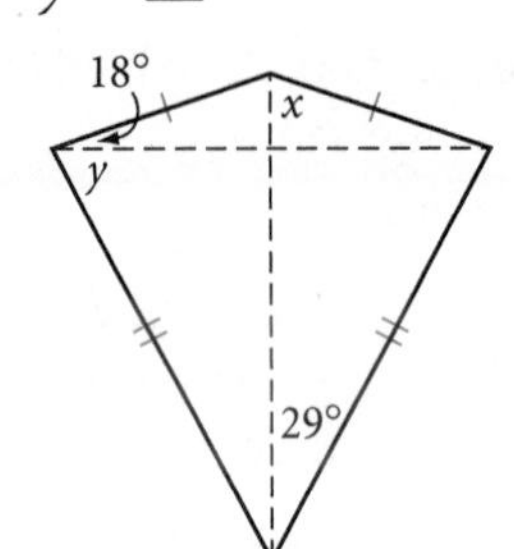

6. $x = \underline{?}$

$y = \underline{?}$

Perimeter $= 164$ cm

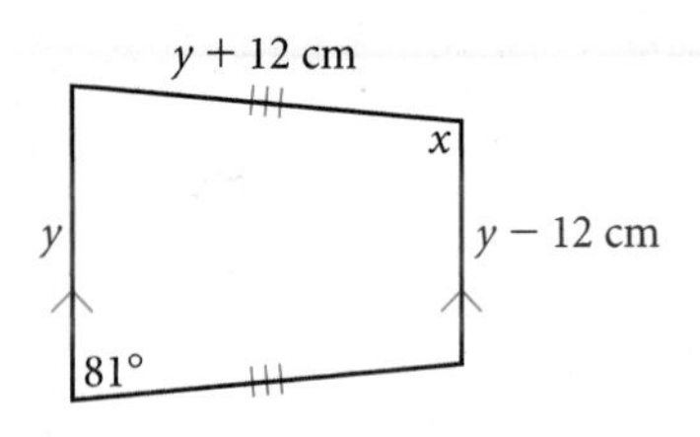

7. Sketch and label kite *KITE* with vertex angles $\angle K$ and $\angle T$ and $KI > TE$. Which angles are congruent?

8. Sketch and label trapezoid *QUIZ* with one base $\overline{QU}$. What is the other base? Name the two pairs of base angles.

9. Sketch and label isosceles trapezoid *SHOW* with one base $\overline{SH}$. What is the other base? Name the two pairs of base angles. Name the two sides of equal length.

In Exercises 10–12, use the properties of kites and trapezoids to construct each figure. You may use either patty paper or a compass and a straightedge.

10. ***Construction*** Construct kite *BENF* given sides $\overline{BE}$ and $\overline{EN}$ and diagonal $\overline{BN}$. How many different kites are possible?

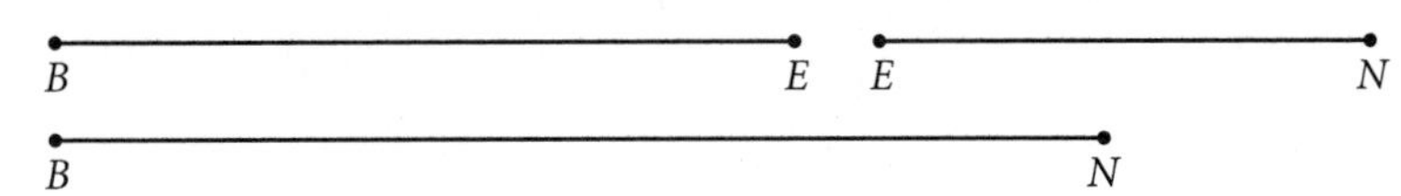

11. ***Construction*** Given $\angle W$, $\angle I$, base $\overline{WI}$, and nonparallel side $\overline{IS}$, construct trapezoid *WISH*. ⓗ

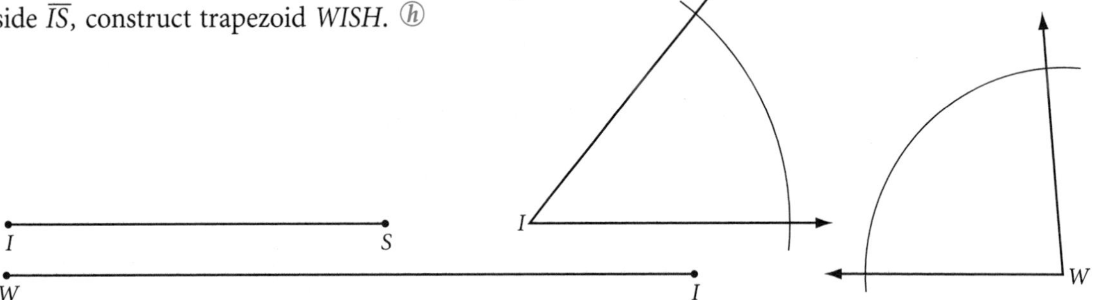

12. ***Construction*** Construct a trapezoid *BONE* with $\overline{BO} \parallel \overline{NE}$. How many different trapezoids can you construct?

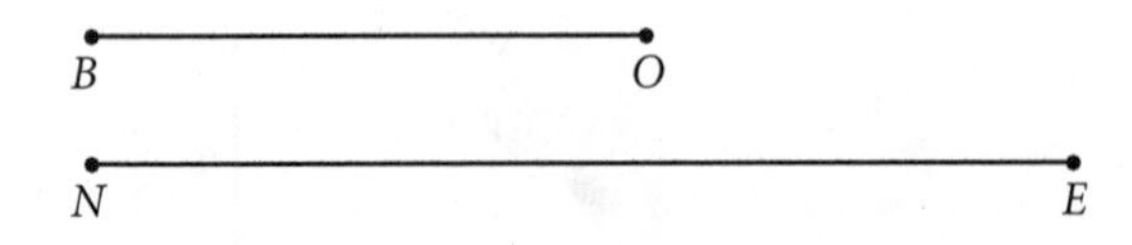

13. Write a paragraph proof or flowchart proof showing how the Kite Diagonal Bisector Conjecture logically follows from the Converse of the Perpendicular Bisector Conjecture. ⓗ

14. Copy and complete the flowchart to show how the Kite Angle Bisector Conjecture follows logically from one of the triangle congruence conjectures.

Given: Kite $BENY$ with $\overline{BE} \cong \overline{BY}$, $\overline{EN} \cong \overline{YN}$

Show: $\overline{BN}$ bisects $\angle B$

$\overline{BN}$ bisects $\angle N$

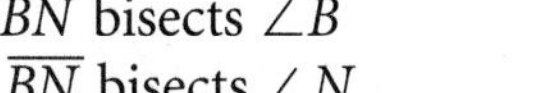

Flowchart Proof

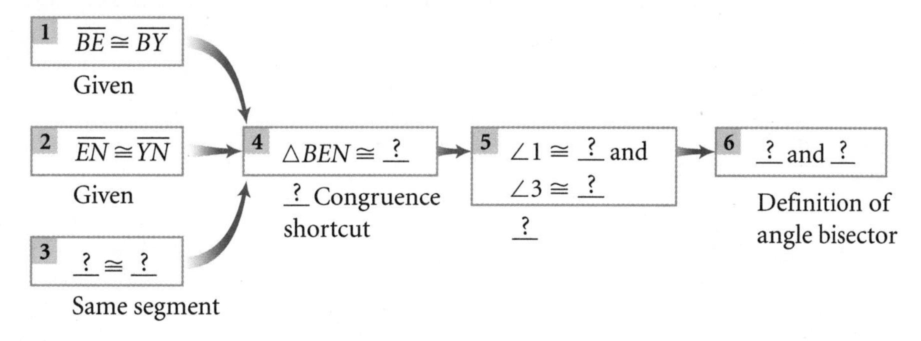

Architecture CONNECTION

The Romans used the classical arch design in bridges, aqueducts, and buildings in the early centuries of the Common Era. The classical semicircular arch is really half of a regular polygon built with wedge-shaped blocks whose faces are isosceles trapezoids. Each block supports the blocks surrounding it.

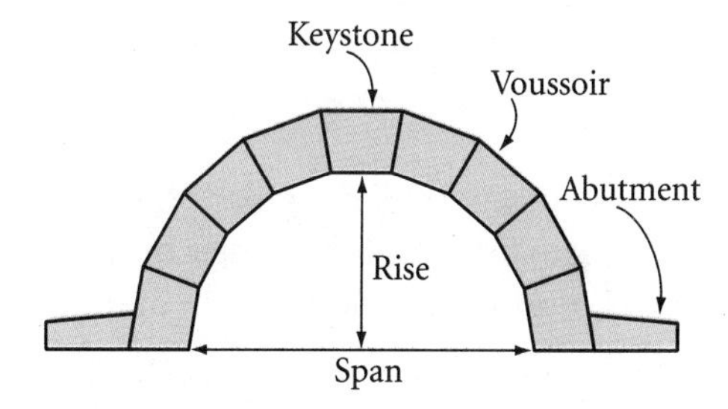

15. APPLICATION The inner edge of the arch in the diagram above right is half of a regular 18-gon. Calculate the measures of all the angles in the nine isosceles trapezoids making up the arch. Then use your geometry tools to accurately draw a nine-stone arch like the one shown.

This carton is shaped like an isosceles trapezoid block.

16. The figure below shows the path of light through a trapezoidal prism, and how an image is inverted. For the prism to work as shown, the trapezoid must be isosceles, $\angle AGF$ must be congruent to $\angle BHE$, and $\overline{GF}$ must be congruent to $\overline{EH}$. Show that if these conditions are met, then $\overline{AG}$ will be congruent to $\overline{BH}$. ⓗ

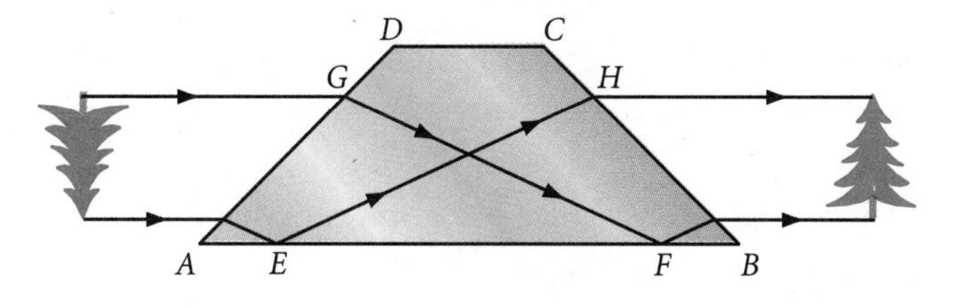

Science CONNECTION

The magnifying lenses of binoculars invert the objects you view through them, so trapezoidal prisms are used to flip the inverted images right-side-up again.

Review

17. Trace the figure below. Calculate the measure of each lettered angle.

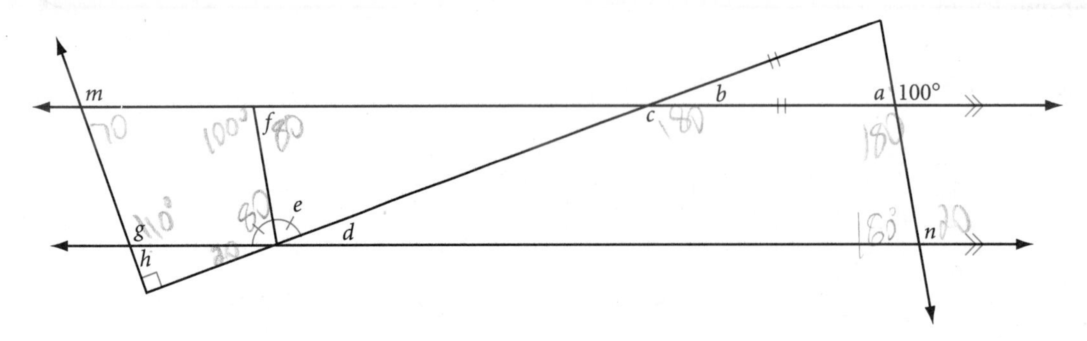

project

DRAWING REGULAR POLYGONS

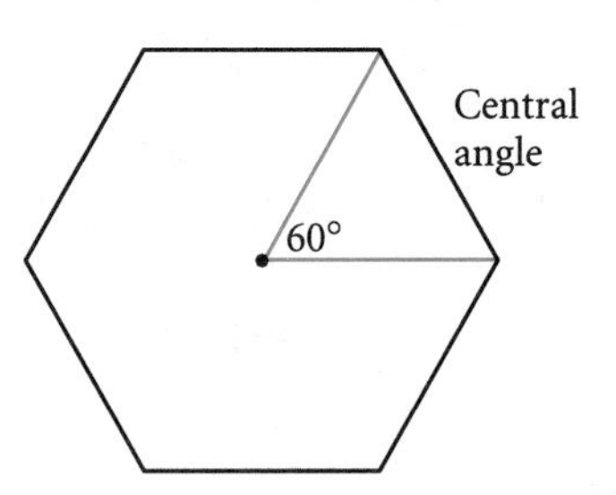

You can draw a regular polygon's central angle by extending segments from the center of the polygon to its consecutive vertices. For example, the measure of each central angle of a hexagon is 60°.

Using central angles, you can draw regular polygons on a graphing calculator. This is done with parametric equations, which give the x- and y-coordinates of a point in terms of a third variable, or parameter, t.

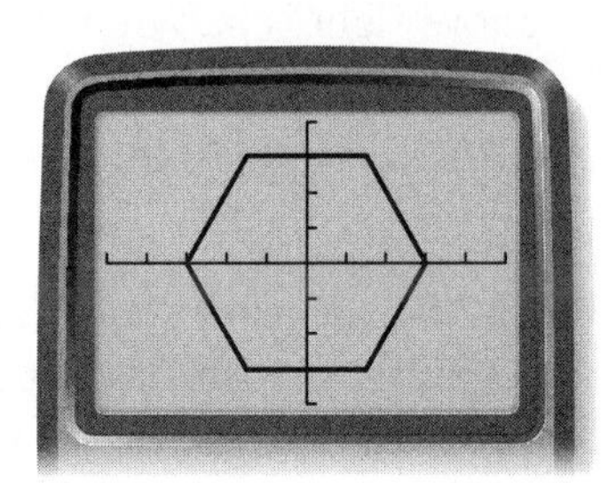

Set your calculator's mode to degrees and parametric. Set a friendly window with an x-range of -4.7 to 4.7 and a y-range of -3.1 to 3.1. Set a t-range of 0 to 360, and t-step of 60. Enter the equations $x = 3\cos t$ and $y = 3\sin t$, and graph them. You should get a hexagon.

The equations you graphed are actually the parametric equations for a circle. By using a t-step of 60 for t-values from 0 to 360, you tell the calculator to compute only six points for the circle.

Use your calculator to investigate the following. Summarize your findings.

- Choose different t-steps to draw different regular polygons, such as an equilateral triangle, a square, a regular pentagon, and so on. What is the measure of each central angle of an n-gon?
- What happens as the measure of each central angle of a regular polygon decreases?
- What happens as you draw polygons with more and more sides?
- Experiment with rotating your polygons by choosing different t-min and t-max values. For example, set a t-range of -45 to 315, then draw a square.
- Find a way to draw star polygons on your calculator. Can you explain how this works?

LESSON

5.4

Properties of Midsegments

Research is formalized curiosity. It is poking and prying with a purpose.

ZORA NEALE HURSTON

As you learned in Chapter 3, the segment connecting the midpoints of two sides of a triangle is the midsegment of a triangle. The segment connecting the midpoints of the two nonparallel sides of a trapezoid is also called the midsegment of a trapezoid.

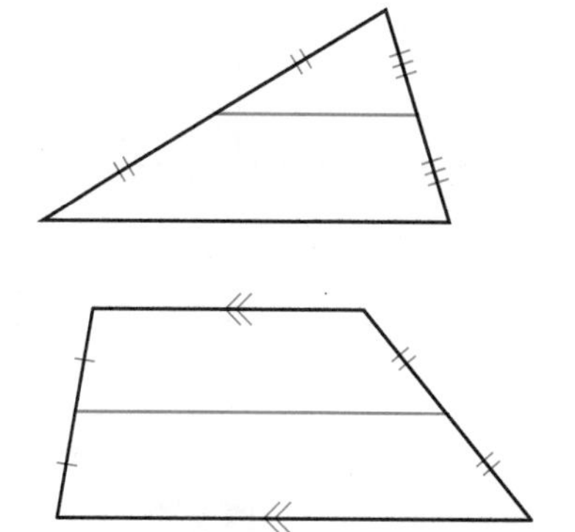

In this lesson you will discover special properties of midsegments.

Investigation 1

Triangle Midsegment Properties

You will need

- patty paper

In this investigation you will discover two properties of the midsegment of a triangle. Each person in your group can investigate a different triangle.

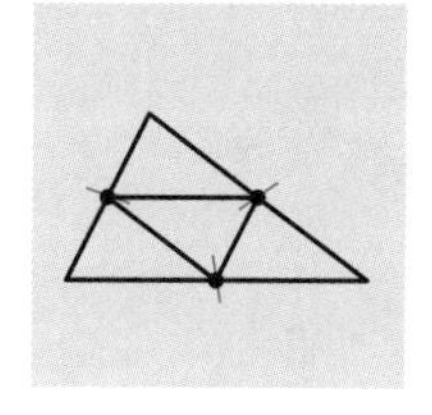

Step 1

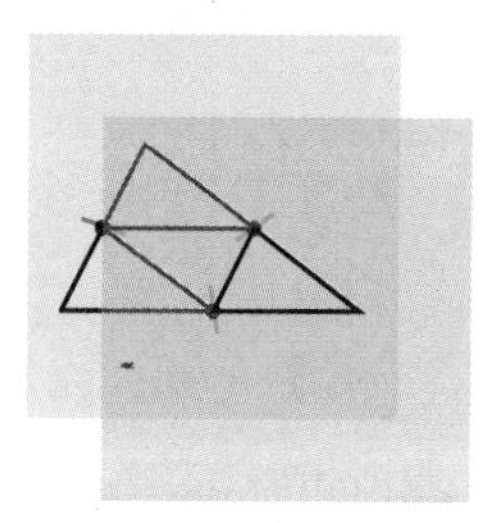

Step 2

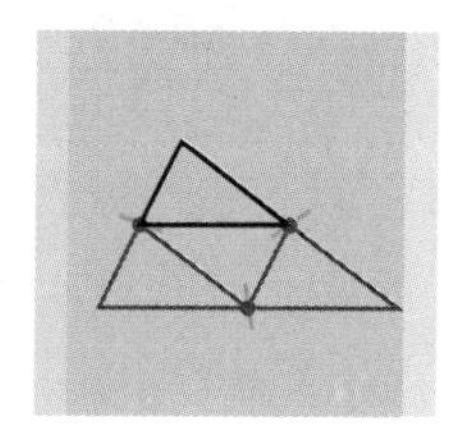

Step 3

Step 1 Draw a triangle on a piece of patty paper. Pinch the patty paper to locate midpoints of the sides. Draw the midsegments. You should now have four small triangles.

Step 2 Place a second piece of patty paper over the first and copy one of the four triangles.

Step 3 Compare all four triangles by sliding the copy of one small triangle over the other three triangles. Compare your results with the results of your group. Copy and complete the conjecture.

Three Midsegments Conjecture

C-42

The three midsegments of a triangle divide it into $\underline{?}$.

Step 4 Mark all the congruent angles in your drawing. What conclusions can you make about each midsegment and the large triangle's third side, using the Corresponding Angles Conjecture and the Alternate Interior Angles Conjecture? What do the other students in your group think?

Step 5 Compare the length of the midsegment to the large triangle's third side. How do they relate? Copy and complete the conjecture.

Triangle Midsegment Conjecture C-43

A midsegment of a triangle is _?_ to the third side and _?_ the length of _?_.

In the next investigation, you will discover two properties of the midsegment of a trapezoid.

Investigation 2
Trapezoid Midsegment Properties

You will need

- patty paper

Each person in your group can investigate a different trapezoid. Make sure you draw the two bases perfectly parallel.

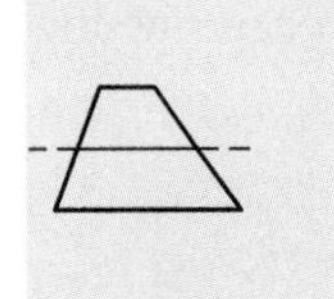

Step 1

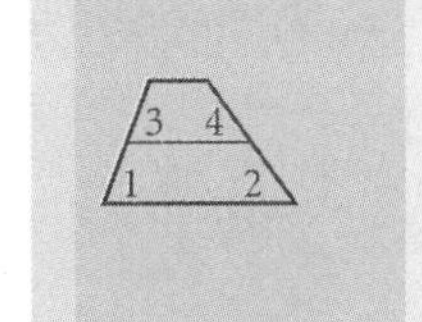

Step 2

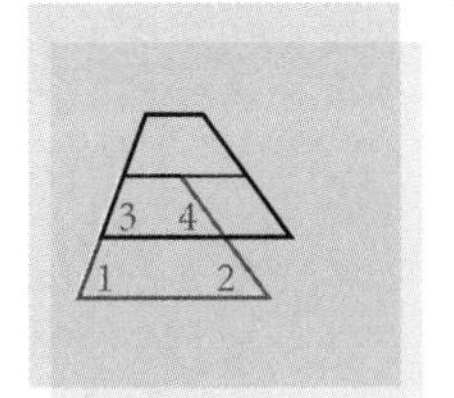

Step 3

Step 1 Draw a small trapezoid on the left side of a piece of patty paper. Pinch the paper to locate the midpoints of the nonparallel sides. Draw the midsegment.

Step 2 Label the angles as shown. Place a second piece of patty paper over the first and copy the trapezoid and its midsegment.

Step 3 Compare the trapezoid's base angles with the corresponding angles at the midsegment by sliding the copy up over the original.

Step 4 Are the corresponding angles congruent? What can you conclude about the midsegment and the bases? Compare your results with the results of other students.

The midsegment of a triangle is half the length of the third side. How does the length of the midsegment of a trapezoid compare to the lengths of the two bases? Let's investigate.

Step 5 On the original trapezoid, extend the longer base to the right by at least the length of the shorter base.

Step 6 Slide the second patty paper under the first. Show the sum of the lengths of the two bases by marking a point on the extension of the longer base.

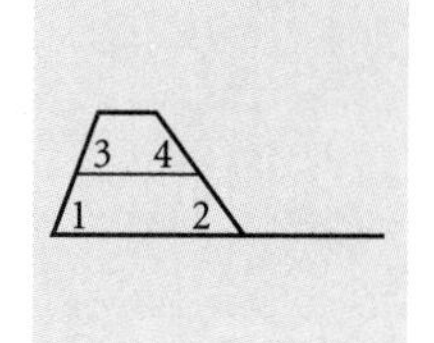

Step 5

Step 6

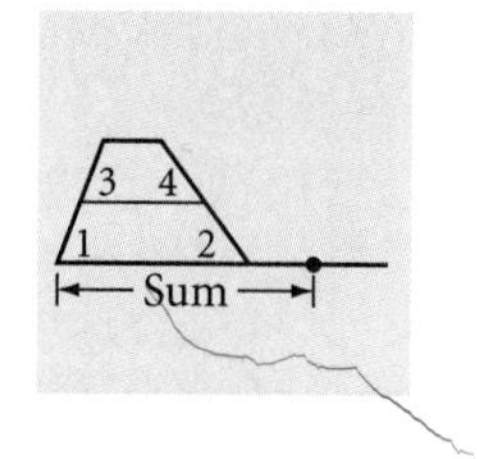

Step 7

Step 7 How many times does the midsegment fit onto the segment representing the sum of the lengths of the two bases? What do you notice about the length of the midsegment and the sum of the lengths of the two bases?

Step 8 Combine your conclusions from Steps 4 and 7 and complete this conjecture.

Trapezoid Midsegment Conjecture **C-44**

The midsegment of a trapezoid is _?_ to the bases and is equal in length to _?_.

What happens if one base of the trapezoid shrinks to a point? Then the trapezoid collapses into a triangle, the midsegment of the trapezoid becomes a midsegment of the triangle, and the Trapezoid Midsegment Conjecture becomes the Triangle Midsegment Conjecture. Do both of your midsegment conjectures work for the last figure?

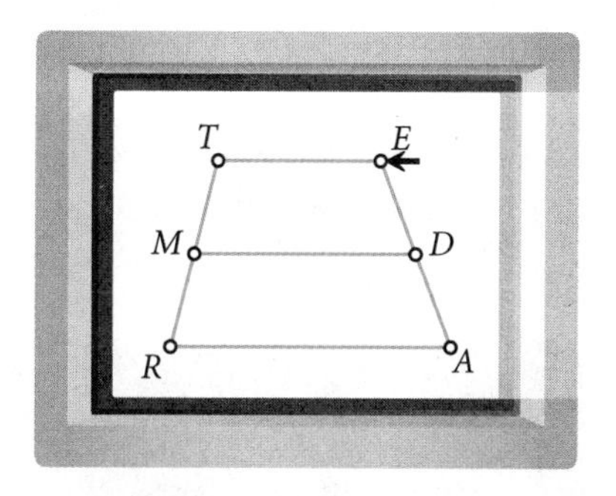

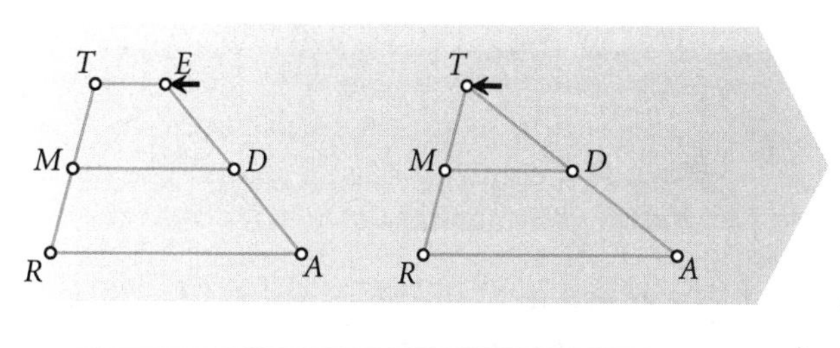

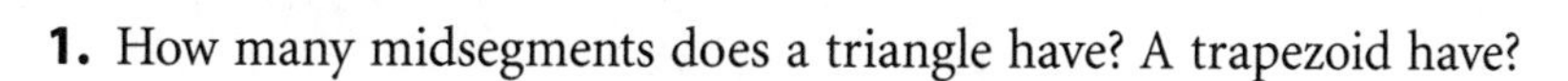

EXERCISES

You will need

1. How many midsegments does a triangle have? A trapezoid have?

2. What is the perimeter of $\triangle TOP$? ⓗ

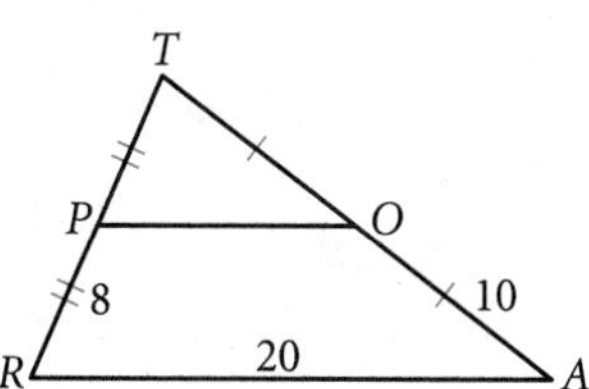

3. $x =$ _?_
$y =$ _?_

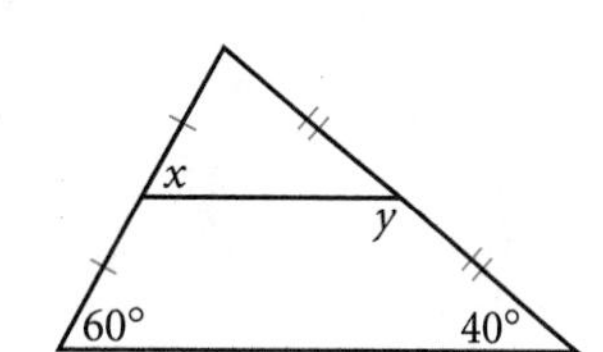

4. $z =$ _?_ ⓗ

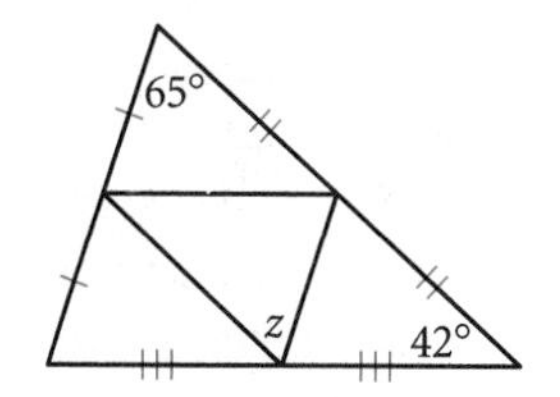

5. What is the perimeter of $\triangle TEN$?

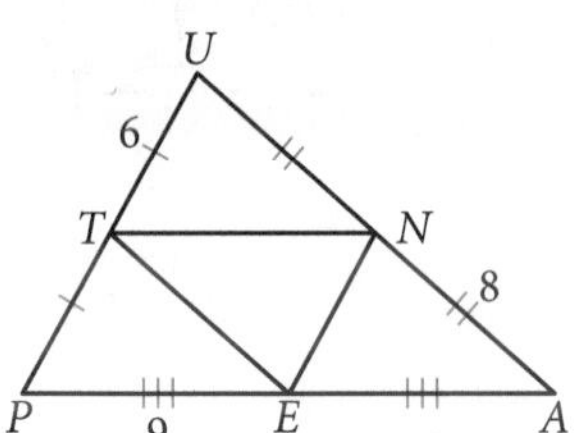
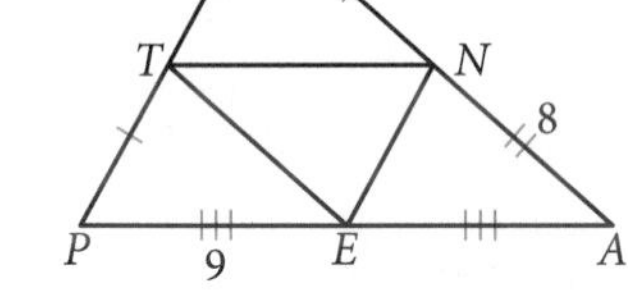

6. $m = \underline{?}$
$n = \underline{?}$
$p = \underline{?}$

7. $q = \underline{?}$

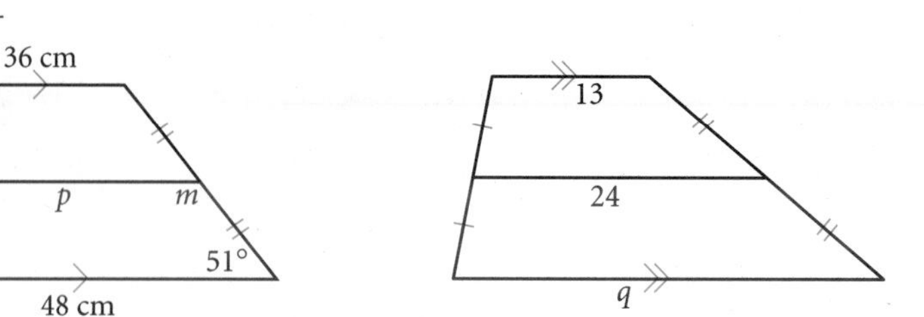

8. Copy and complete the flowchart to show that $\overline{LN} \parallel \overline{RD}$.

Given: Midsegment $\overline{LN}$ in $\triangle FOA$
Midsegment $\overline{RD}$ in $\triangle IOA$

Show: $\overline{LN} \parallel \overline{RD}$

Flowchart Proof

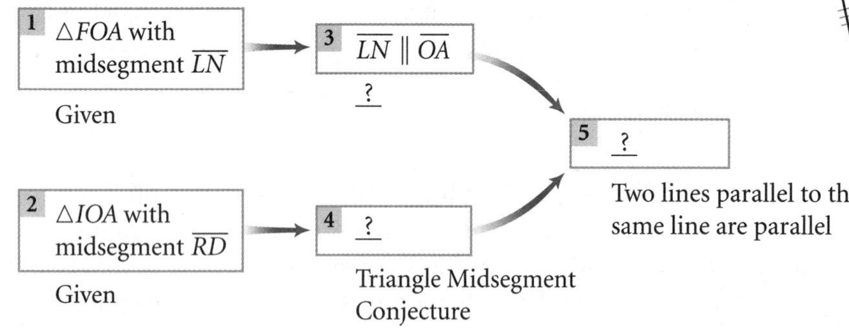
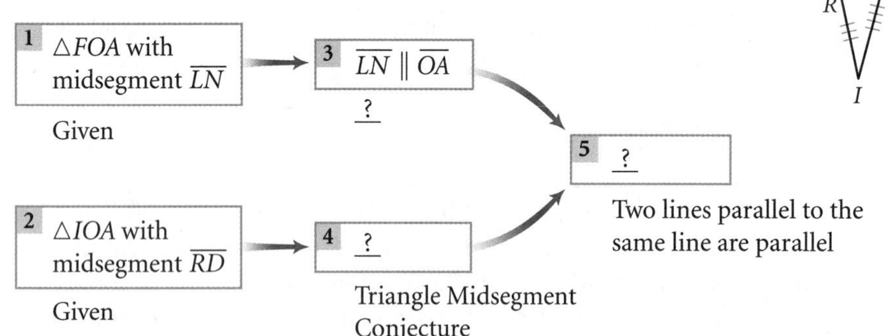

9. ***Construction*** When you connected the midpoints of the three sides of a triangle in Investigation 1, you created four congruent triangles. Draw a quadrilateral on patty paper and pinch the paper to locate the midpoints of the four sides. Connect the midpoints to form a quadrilateral. What special type of quadrilateral do you get when you connect the midpoints? Use the Triangle Midsegment Conjecture to explain your answer. ⓗ

10. Deep in a tropical rain forest, archaeologist Ertha Diggs and her assistant researchers have uncovered a square-based truncated pyramid (a square pyramid with the top part removed). The four lateral faces are isosceles trapezoids. A line of darker mortar runs along the midsegment of each lateral face. Ertha and her co-workers make some measurements and find that one of these midsegments measures 41 meters and each bottom base measures 52 meters. Now that they have this information, Ertha and her team can calculate the length of the top base without having to climb up and measure it. Can you? What is the length of the top edge? How do you know?

11. Ladie and Casey pride themselves on their estimation skills and take turns estimating distances. Casey claims that two large redwood trees visible from where they are sitting are 180 feet apart, and Ladie says they are 275 feet apart.

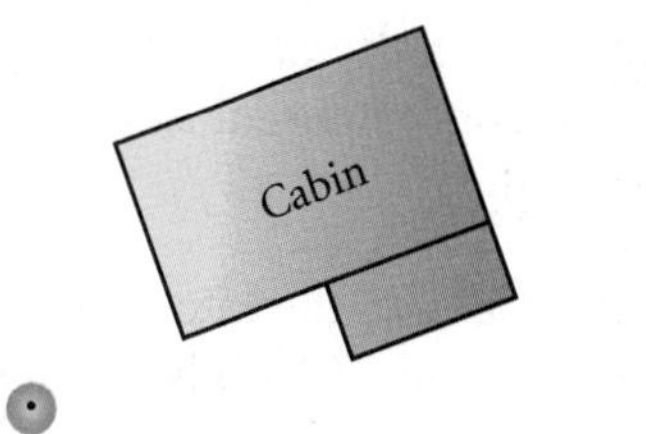

The problem is, they can't measure the distance to see whose estimate is better, because their cabin is located between the trees. All of a sudden, Ladie recalls her geometry: "Oh yeah, the Triangle Midsegment Conjecture!" She collects a tape measure, a hammer, and some wooden stakes. What is she going to do?

Review

12. The 40-by-60-by-80 cm sealed rectangular container shown at right is resting on its largest face. It is filled with a liquid to a height of 30 cm. Sketch the container resting on its smallest face. Show the height of the liquid in this new position.

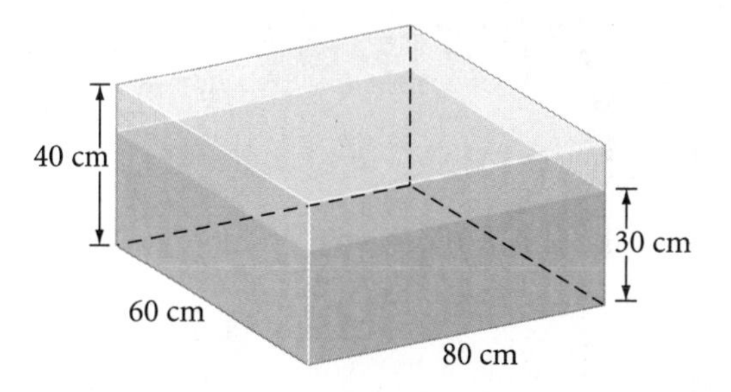

13. Write the converse of this statement: If exactly one diagonal bisects a pair of opposite angles of a quadrilateral, then the quadrilateral is a kite. Is the converse true? Is the original statement true? If either conjecture is not true, sketch a counterexample.

14. Trace the figure below. Calculate the measure of each lettered angle.

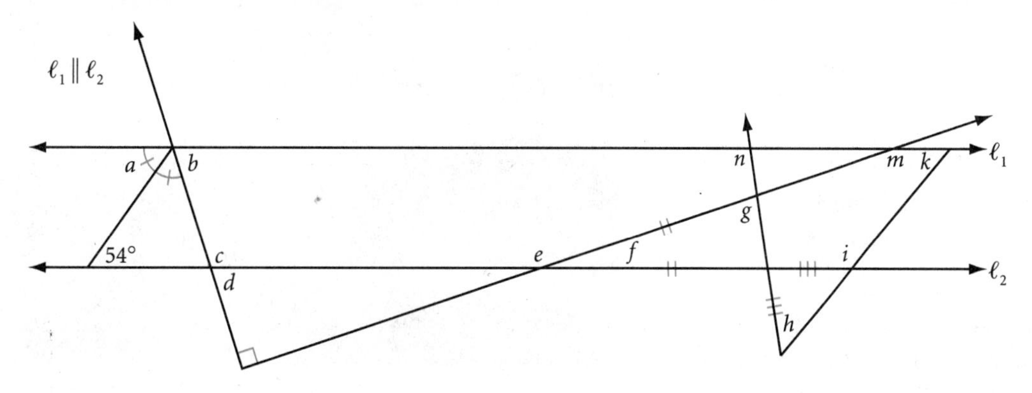

15. *CART* is an isosceles trapezoid. What are the coordinates of point *T*?

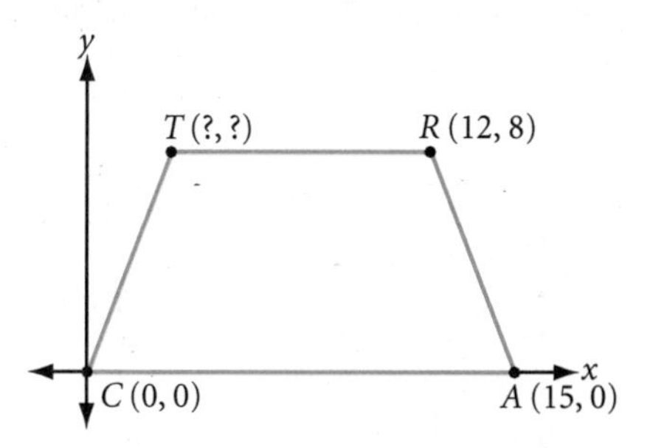

16. *HRSE* is a kite. What are the coordinates of point *R*?

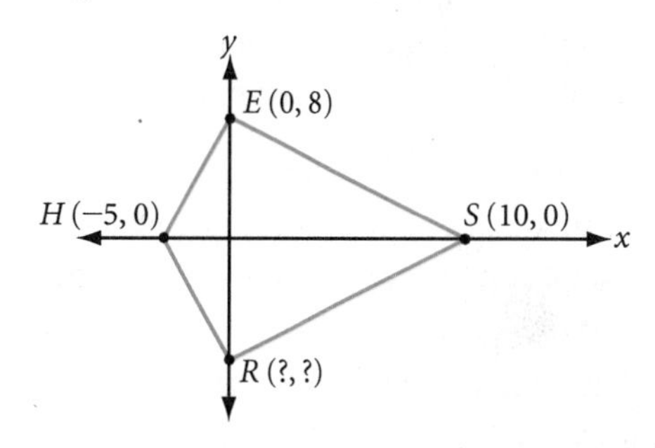

17. Find the coordinates of midpoints E and Z. Show that the slope of the line containing midsegment $\overline{EZ}$ is equal to the slope of the line containing $\overline{YT}$.

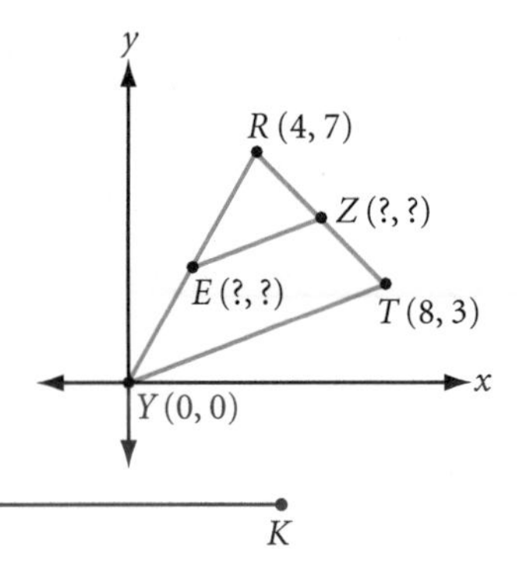

18. ***Construction*** Use the kite properties you discovered in Lesson 5.3 to construct kite $FRNK$ given diagonals $\overline{RK}$ and $\overline{FN}$ and side $\overline{NK}$. Is there only one solution?

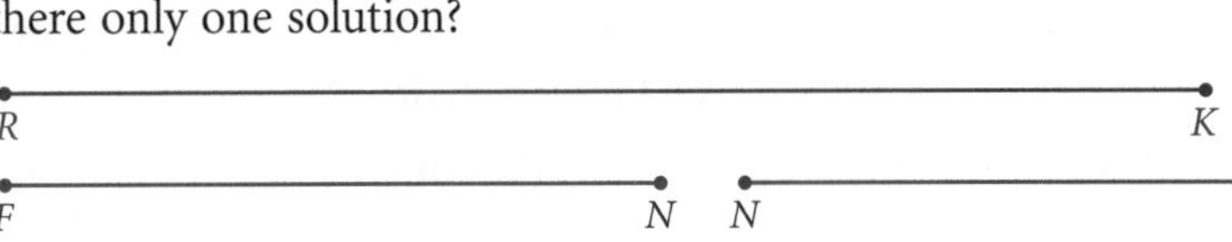

project

BUILDING AN ARCH

In this project, you'll design and build your own Roman arch.

Horseshoe Arch

Basket Arch

Tudor Arch

Lancet Arch

Arches can have a simple semicircular shape, or a pointed "broken arch" shape.

In arch construction, a wooden support holds the voussoirs in place until the keystone is placed (see arch diagram on page 271). It's said that when the Romans made an arch, they would make the architect stand under it while the wooden support was removed. That was one way to be sure architects carefully designed arches that wouldn't fall!

The arches in this Roman aqueduct, above the Gard River in France, are typical of arches you can find throughout regions that were once part of the Roman Empire. An arch can carry a lot of weight, yet it also provides an opening. The abutments on the sides of the arch keep the arch from spreading out and falling down.

What size arch would you like to build? Decide the dimensions of the opening, the thickness of the arch, and the number of voussoirs. Decide on the materials you will use. You should have your trapezoid and your materials approved by your group or your teacher before you begin construction.

Your project should include

- A scale diagram that shows the exact size and angle of the voussoirs and the keystone.
- A template for your voussoirs.
- Your arch.

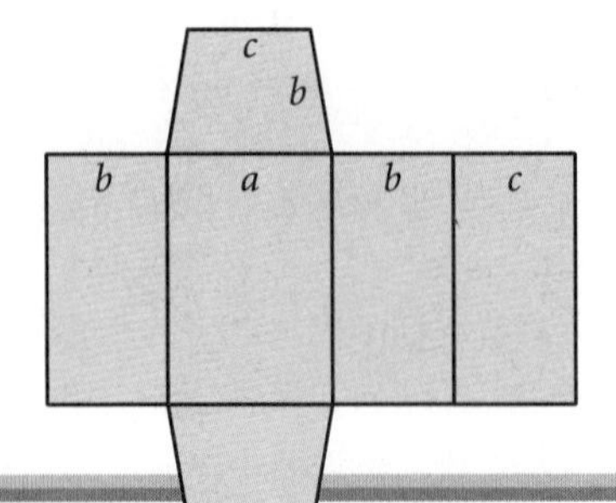

LESSON

5.5

Properties of Parallelograms

If there is an opinion, facts will be found to support it.

JUDY SPROLES

In this lesson you will discover some special properties of parallelograms. A parallelogram is a quadrilateral whose opposite sides are parallel.

Rhombuses, rectangles, and squares all fit this definition as well. Therefore, any properties you discover for parallelograms will also apply to these other shapes. However, to be sure that your conjectures will apply to *any* parallelogram, you should investigate parallelograms that don't have any other special properties, such as right angles, all congruent angles, or all congruent sides.

Investigation

Four Parallelogram Properties

You will need

- graph paper
- patty paper or a compass
- a straightedge
- a protractor

First you'll create a parallelogram.

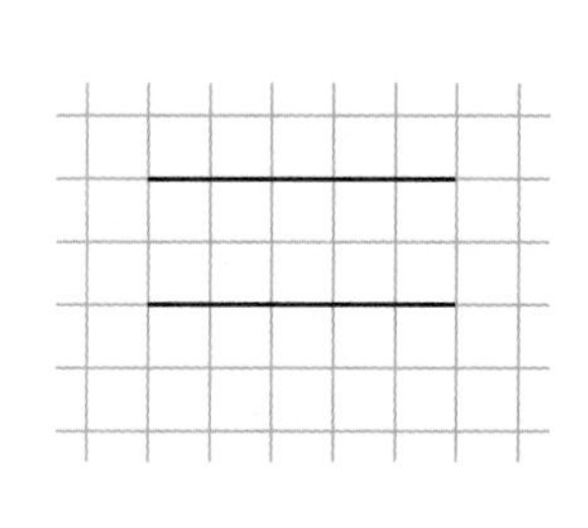

Step 1

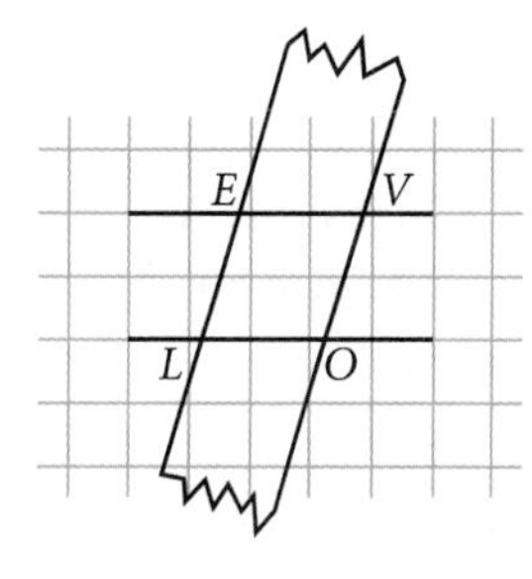

Step 2

Step 1 Using the lines on a piece of graph paper as a guide, draw a pair of parallel lines that are at least 6 cm apart. Using the parallel edges of your straightedge, make a parallelogram. Label your parallelogram *LOVE*.

Step 2 Let's look at the opposite angles. Measure the angles of parallelogram *LOVE*. Compare a pair of opposite angles using patty paper or your protractor.

Compare results with your group. Copy and complete the conjecture.

Parallelogram Opposite Angles Conjecture C-45

The opposite angles of a parallelogram are __?__.

Two angles that share a common side in a polygon are consecutive angles. In parallelogram *LOVE*, $\angle LOV$ and $\angle EVO$ are a pair of consecutive angles. The consecutive angles of a parallelogram are also related.

Step 3 Find the sum of the measures of each pair of consecutive angles in parallelogram *LOVE*.

Share your observations with your group. Copy and complete the conjecture.

Parallelogram Consecutive Angles Conjecture C-46

The consecutive angles of a parallelogram are _?_.

Step 4 Describe how to use the two conjectures you just made to find all the angles of a parallelogram with only one angle measure given.

Step 5 Next let's look at the opposite sides of a parallelogram. With your compass or patty paper, compare the lengths of the opposite sides of the parallelogram you made.

Share your results with your group. Copy and complete the conjecture.

Parallelogram Opposite Sides Conjecture C-47

The opposite sides of a parallelogram are _?_.

Step 6 Finally, let's consider the diagonals of a parallelogram. Construct the diagonals $\overline{LV}$ and $\overline{EO}$, as shown below. Label the point where the two diagonals intersect point *M*.

Step 7 Measure *LM* and *VM*. What can you conclude about point *M*? Is this conclusion also true for diagonal $\overline{EO}$? How do the diagonals relate?

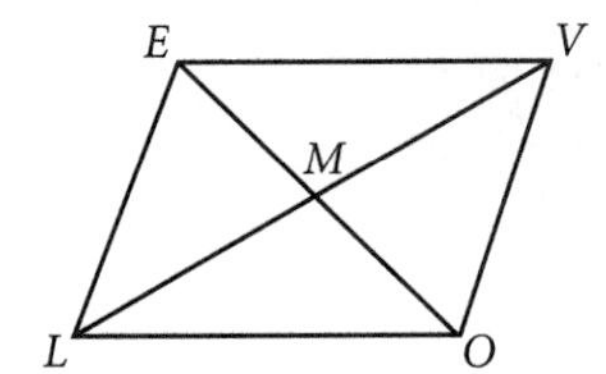

Share your results with your group. Copy and complete the conjecture.

Parallelogram Diagonals Conjecture C-48

The diagonals of a parallelogram _?_.

Parallelograms are used in vector diagrams, which have many applications in science. A **vector** is a quantity that has both magnitude and direction.

Vectors describe quantities in physics, such as velocity, acceleration, and force. You can represent a vector by drawing an arrow. The length and direction of the arrow represent the magnitude and direction of the vector. For example, a velocity vector tells you an airplane's speed and direction. The lengths of vectors in a diagram are proportional to the quantities they represent.

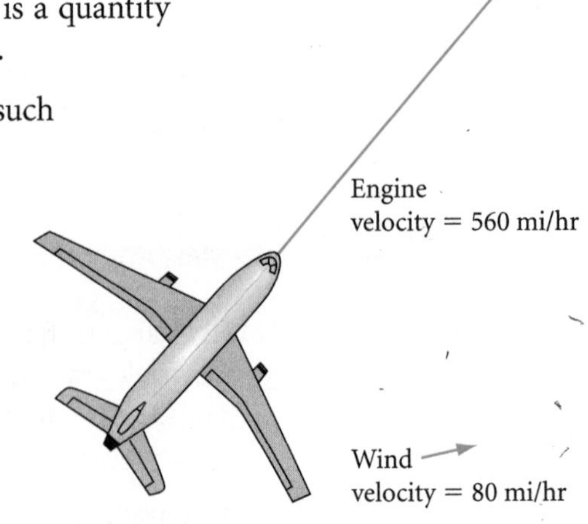

In many physics problems, you combine vector quantities acting on the same object. For example, the wind current and engine thrust vectors determine the velocity of an airplane. The **resultant vector** of these vectors is a single vector that has the same effect. It can also be called a **vector sum.** To find a resultant vector, make a parallelogram with the vectors as sides. The resultant vector is the diagonal of the parallelogram from the two vectors' tails to the opposite vertex.

In the diagram at right, the resultant vector shows that the wind will speed up the plane, and will also blow it slightly off course.

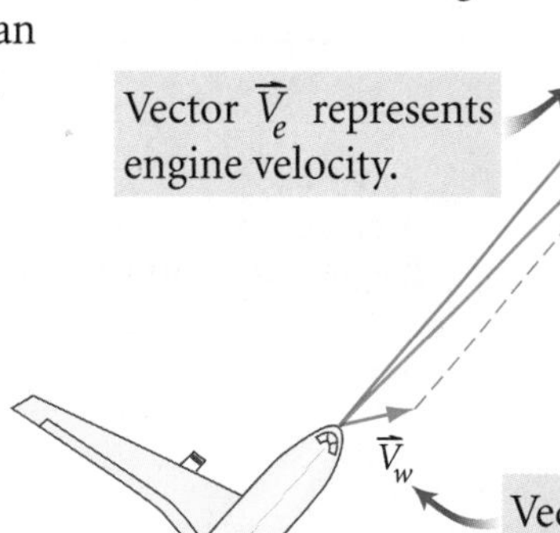

EXERCISES

You will need

Construction tools for Exercises **7** and **8**

Geometry software for Exercises **21** and **22**

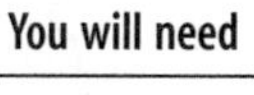

Use your new conjectures in the following exercises. In Exercises 1–6, each figure is a parallelogram.

1. $c = \underline{?}$
$d = \underline{?}$

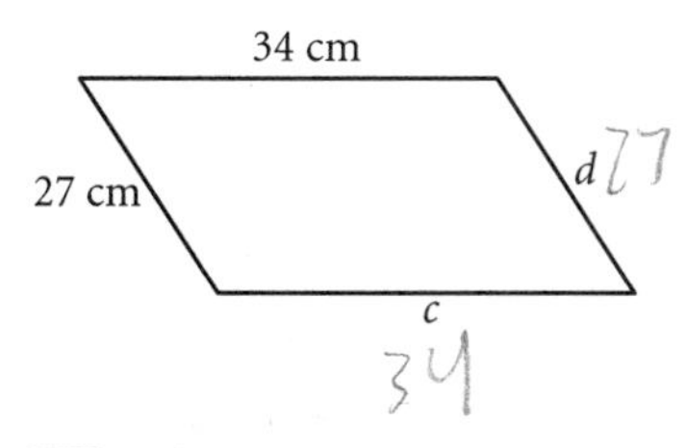

2. $a = \underline{?}$
$b = \underline{?}$

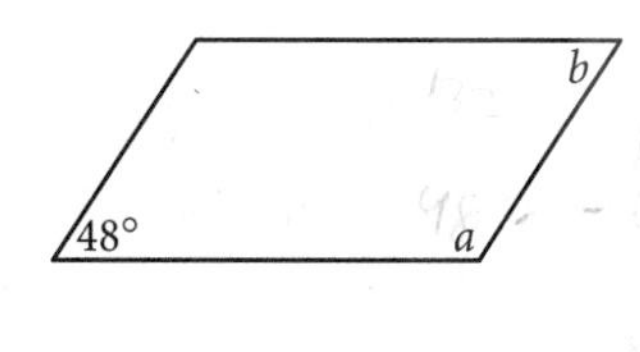

3. $g = \underline{?}$
$h = \underline{?}$

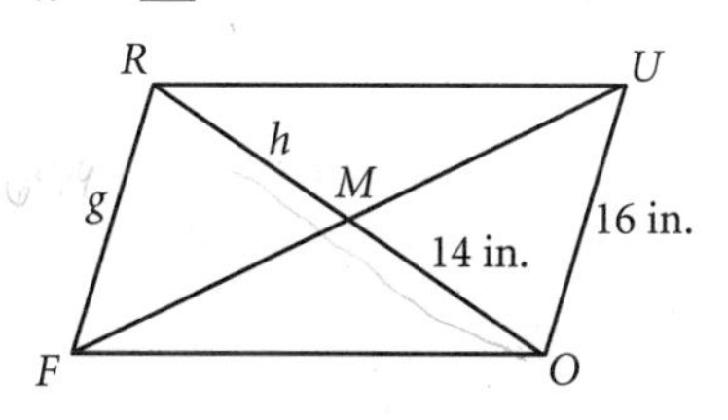

4. $VF = 36$ m
$EF = 24$ m
$EI = 42$ m
What is the perimeter of $\triangle NVI$? ⓗ

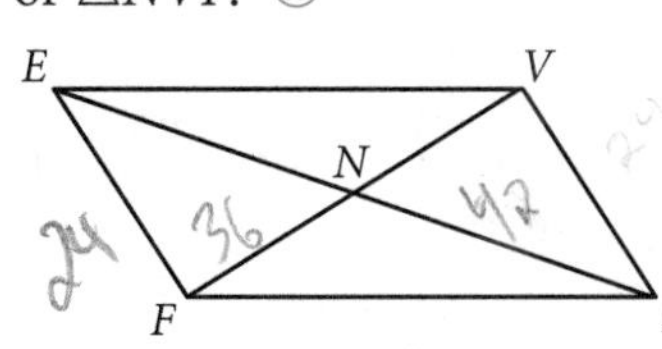

5. What is the perimeter?

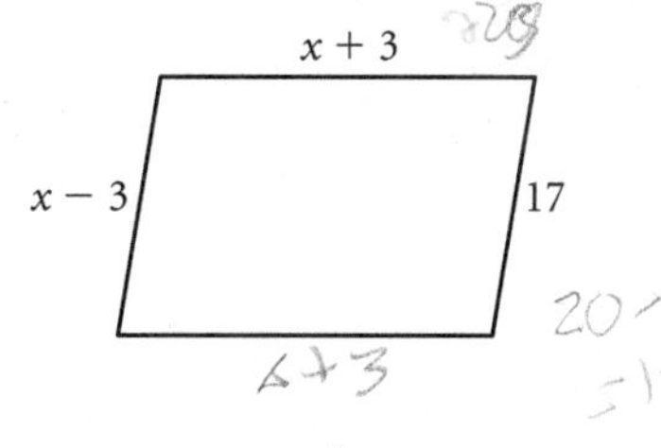

6. $e = \underline{?}$
$f = \underline{?}$

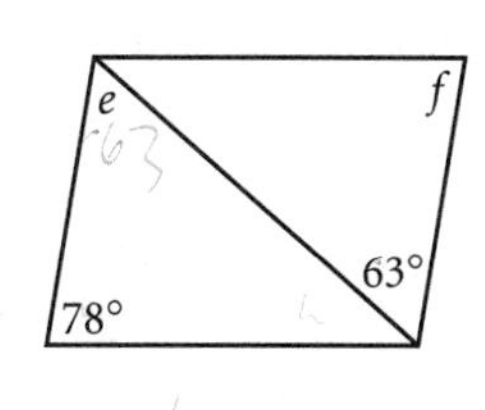

7. ***Construction*** Given side $\overline{LA}$, side $\overline{AS}$, and $\angle L$, construct parallelogram $LAST$.

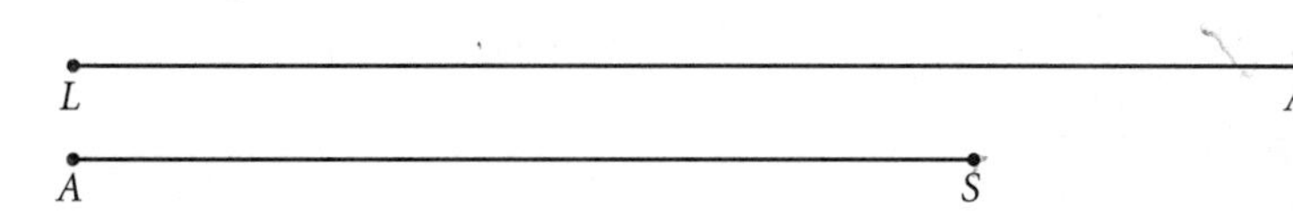

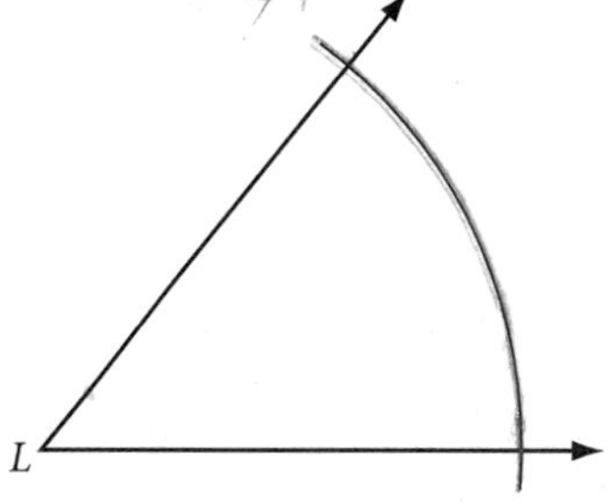

8. ***Construction*** Given side $\overline{DR}$ and diagonals $\overline{DO}$ and $\overline{PR}$, construct parallelogram *DROP*. ⓗ

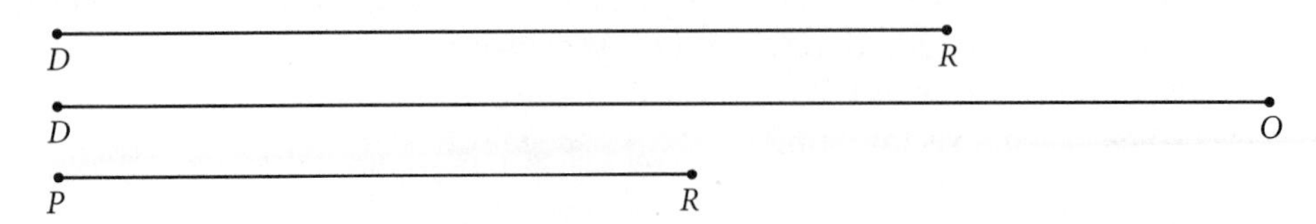

In Exercises 9 and 10, copy the vector diagram and draw the resultant vector.

9.

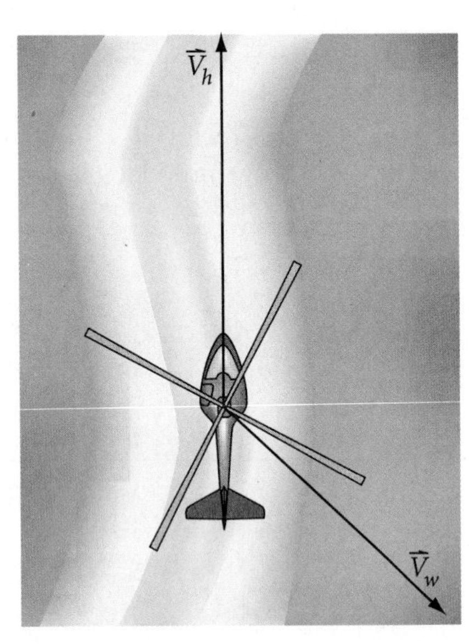

10. ⓗ

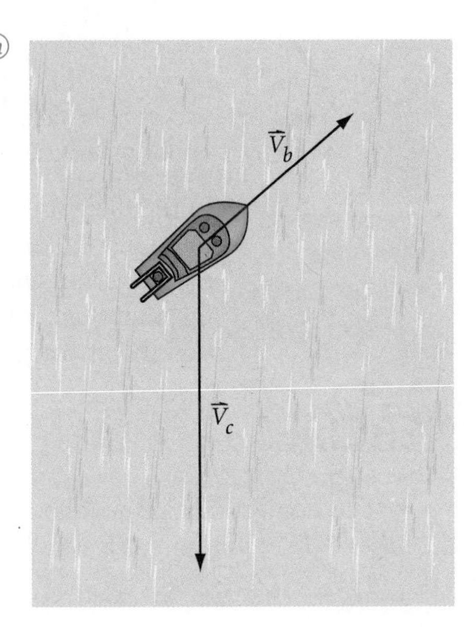

11. Find the coordinates of point *M* in parallelogram *PRAM*. ⓗ

12. Draw a quadrilateral. Make a copy of it. Draw a diagonal in the first quadrilateral. Draw the *other* diagonal in the duplicate quadrilateral. Cut each quadrilateral into two triangles along the diagonals. Arrange the four triangles into a parallelogram. Make a sketch showing how you did it.

13. Copy and complete the flowchart to show how the Parallelogram Diagonals Conjecture follows logically from other conjectures.

Given: *LEAN* is a parallelogram

Show: $\overline{EN}$ and $\overline{LA}$ bisect each other

Flowchart Proof

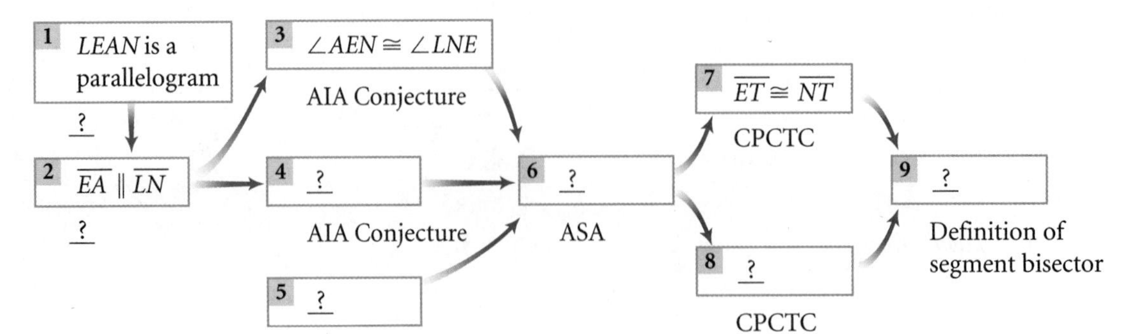

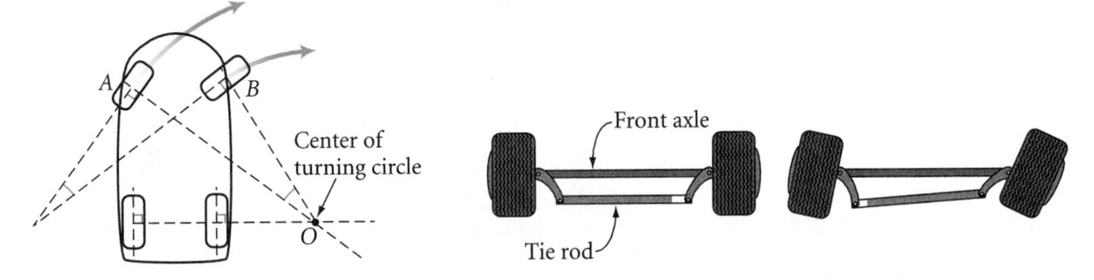

Trapezoid linkage (Top view)

Technology CONNECTION

Quadrilateral linkages are used in mechanical design, robotics, the automotive industry, and toy making. In cars, they are used to turn each front wheel the right amount for a smooth turn.

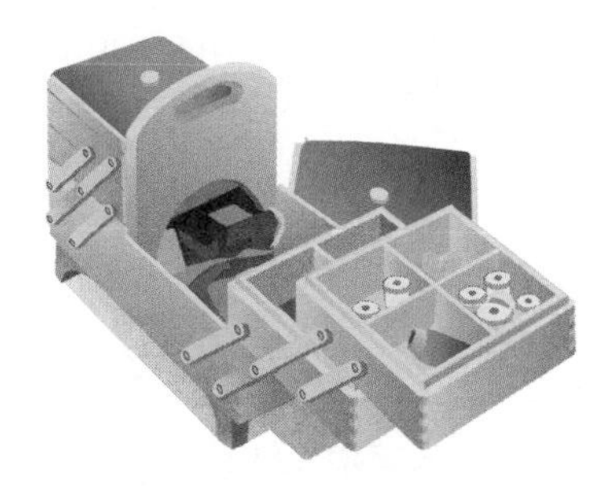

14. Study the sewing box pictured here. Sketch the box as viewed from the side, and explain why a parallelogram linkage is used.

Review

15. Find the measures of the lettered angles in this tiling of regular polygons.

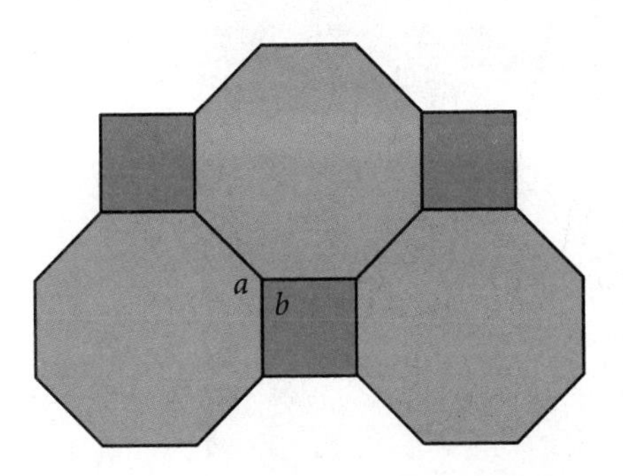

16. Trace the figure below. Calculate the measure of each lettered angle.

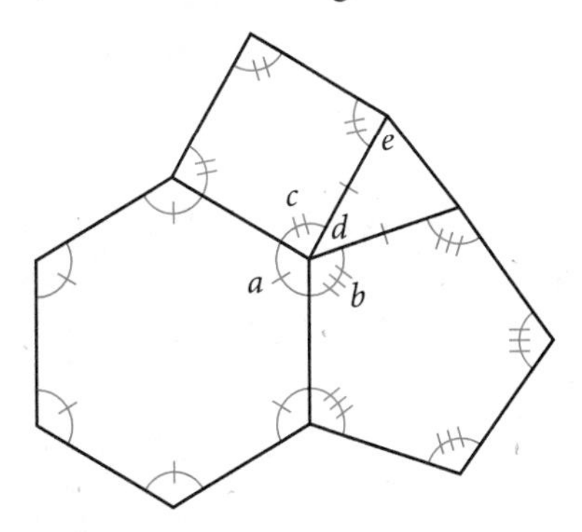

17. Find x and y. Explain.

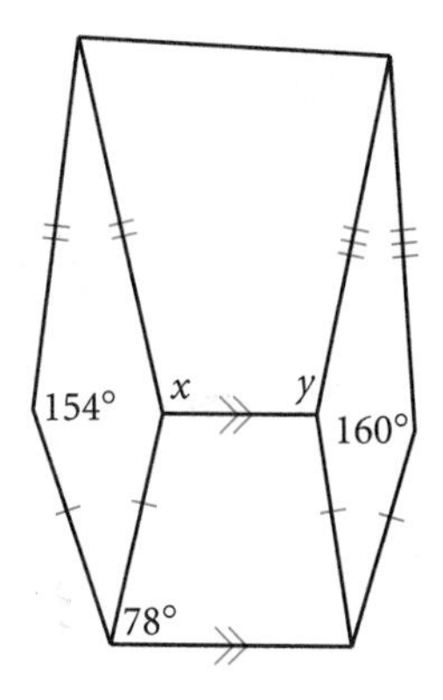

18. What is the measure of each angle in the isosceles trapezoid face of a voussoir in this 15-stone arch?

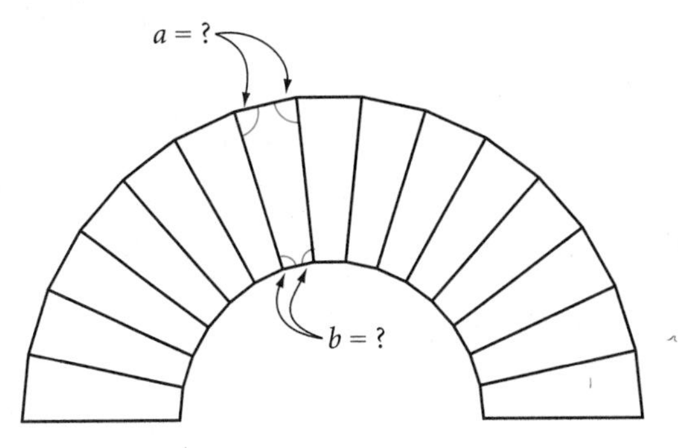

19. Is $\triangle XYW \cong \triangle WYZ$? Explain.

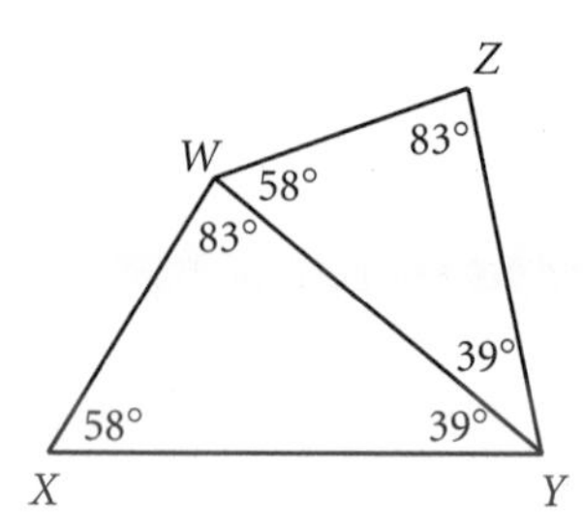

20. Sketch the section formed when this pyramid is sliced by the plane.

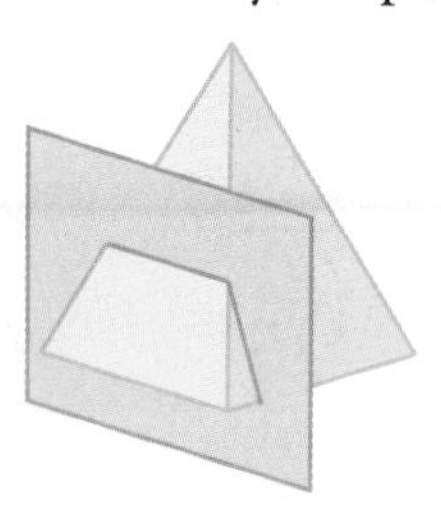

21. *Technology* Construct two segments that bisect each other. Connect their endpoints. What type of quadrilateral is this? Draw a diagram and explain why.

22. *Technology* Construct two intersecting circles. Connect the two centers and the two points of intersection to form a quadrilateral. What type of quadrilateral is this? Draw a diagram and explain why.

IMPROVING YOUR VISUAL THINKING SKILLS

PUZZLE

A Puzzle Quilt

Fourth-grade students at Public School 95, the Bronx, New York, made the puzzle quilt at right with the help of artist Paula Nadelstern. Each square has a twin made of exactly the same shaped pieces. Only the colors, chosen from traditional Amish colors, are different. For example, square A1 is the twin of square B3. Match each square with its twin.

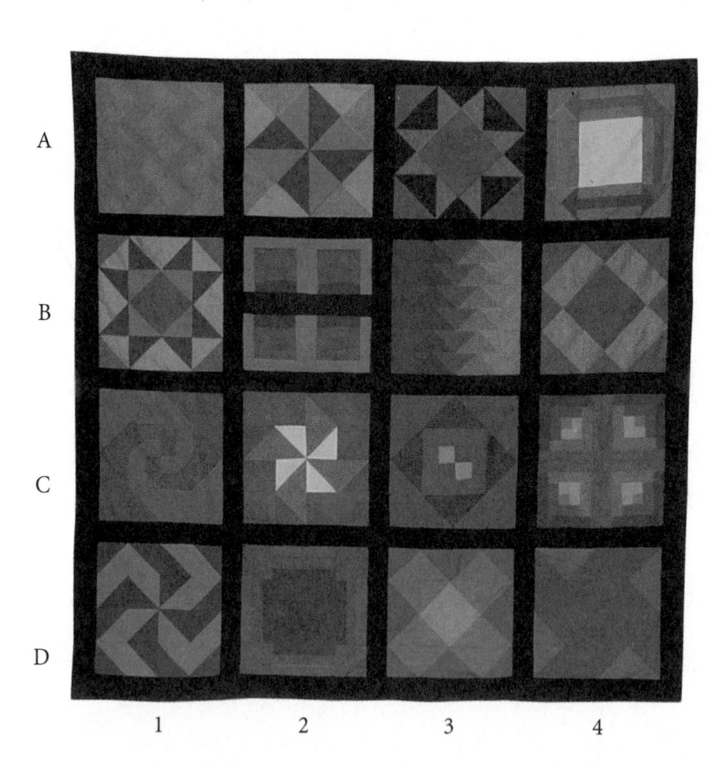

Solving Systems of Linear Equations

A *system of equations* is a set of two or more equations with the same variables. The solution of a system is the set of values that makes all the equations in the system true. For example, the system of equations below has solution $(2, -3)$. Verify this by substituting 2 for x and -3 for y in both equations.

$$\begin{cases} y = 2x - 7 \\ y = -3x + 3 \end{cases}$$

Graphically, the solution of a system is the point of intersection of the graphs of the equations.

You can estimate the solution of a system by graphing the equations. However, the point of intersection may not have convenient integer coordinates. To find the exact solution, you can use algebra. Examples A and B review how to use the *substitution* and *elimination* methods for solving systems of equations.

EXAMPLE A

Use the substitution method to solve the system $\begin{cases} 3y = 12x - 21 \\ 12x + 2y = 1 \end{cases}$.

► Solution

Start by solving the first equation for y to get $y = 4x - 7$.

Now, substitute the expression $4x - 7$ from the resulting equation for y in the second original equation.

$12x + 2y = 1$ — Second original equation.

$12x + 2(4x - 7) = 1$ — Substitute $4x - 7$ for y.

$x = \frac{3}{4}$ — Solve for x.

To find y, substitute $\frac{3}{4}$ for x in either original equation.

$3y = 12\left(\frac{3}{4}\right) - 21$ — Substitute $\frac{3}{4}$ for x in the first original equation.

$y = -4$ — Solve for y.

The solution of the system is $\left(\frac{3}{4}, -4\right)$. Verify by substituting these values for x and y in each of the original equations.

EXAMPLE B

The band sold calendars to raise money for new uniforms. Aisha sold 6 desk calendars and 10 wall calendars for a total of \$100. Ted sold 12 desk calendars and 4 wall calendars for a total of \$88. Find the price of each type of calendar by writing a system of equations and solving it using the elimination method.

▶ ***Solution***

Let d be the price of a desk calendar, and let w be the price of a wall calendar. You can write this system to represent the situation.

$$\begin{cases} 6d + 10w = 100 & \text{Aisha's sales.} \\ 12d + 4w = 88 & \text{Ted's sales.} \end{cases}$$

Solving a system by elimination involves adding or subtracting the equations to eliminate one of the variables. To solve this system, first multiply both sides of the first equation by 2.

$$\begin{cases} 6d + 10w = 100 \\ 12d + 4w = 88 \end{cases} \rightarrow \begin{cases} 12d + 20w = 200 \\ 12d + 4w = 88 \end{cases}$$

Now, subtract the second equation from the first to eliminate d.

$$\begin{aligned} 12d + 20w &= 200 \\ -(12d + 4w &= \ \ 88) \\ \hline 16w &= 112 \\ w &= 7 \end{aligned}$$

To find the value of d, substitute 7 for w in either original equation. The solution is $w = 7$ and $d = 5$, so a wall calendar costs \$7 and a desk calendar costs \$5.

EXERCISES

Solve each system of equations algebraically.

1. $\begin{cases} y = -2x + 2 \\ 6x + 2y = 3 \end{cases}$

2. $\begin{cases} x + 2y = 3 \\ 2x - y = 16 \end{cases}$

3. $\begin{cases} 5x - y = -1 \\ 15x = 2y \end{cases}$

4. $\begin{cases} -4x + 3y = 3 \\ 7x - 9y = 6 \end{cases}$

For Exercises 5 and 6 solve the systems. What happens? Graph each set of equations and use the graphs to explain your results.

5. $\begin{cases} x + 6y = 10 \\ \frac{1}{2}x + 3y = 5 \end{cases}$

6. $\begin{cases} 2x + y = 30 \\ y = -2x - 1 \end{cases}$

7. A snowboard rental company offers two different rental plans. Plan A offers \$4/hr for the rental and a \$20 lift ticket. Plan B offers \$7/hr for the rental and a free lift ticket.

a. Write the two equations that represent the costs for the two plans, using x for the number of hours. Solve for x and y.

b. Graph the two equations. What does the point of intersection represent?

c. Which is the better plan if you intend to snowboard for 5 hours? What is the most number of hours of snowboarding you can get for \$50?

8. The lines $y = 3 + \frac{2}{3}x$, $y = -\frac{1}{3}x$, and $y = -\frac{4}{3}x + 3$ intersect to form a triangle. Find the vertices of the triangle.

LESSON

5.6

Properties of Special Parallelograms

You must know a great deal about a subject to know how little is known about it.

LEO ROSTEN

The legs of the lifting platforms shown at right form rhombuses. Can you visualize how this lift would work differently if the legs formed parallelograms that weren't rhombuses?

In this lesson you will discover some properties of rhombuses, rectangles, and squares. What you discover about the diagonals of these special parallelograms will help you understand why these lifts work the way they do.

Investigation 1

What Can You Draw with the Double-Edged Straightedge?

You will need

- patty paper
- a double-edged straightedge

In this investigation you will discover the special parallelogram that you can draw using just the parallel edges of a straightedge.

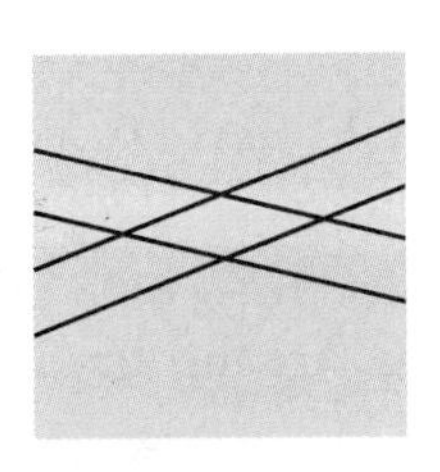

Step 1

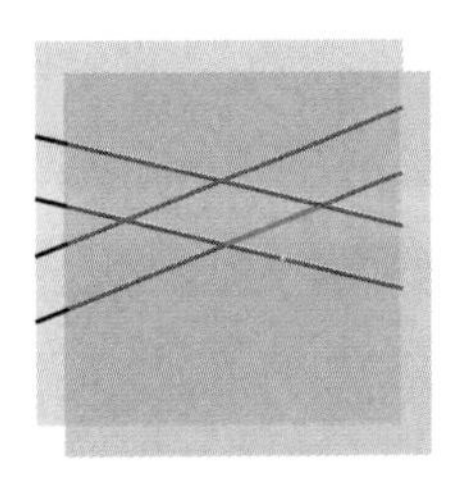

Step 2

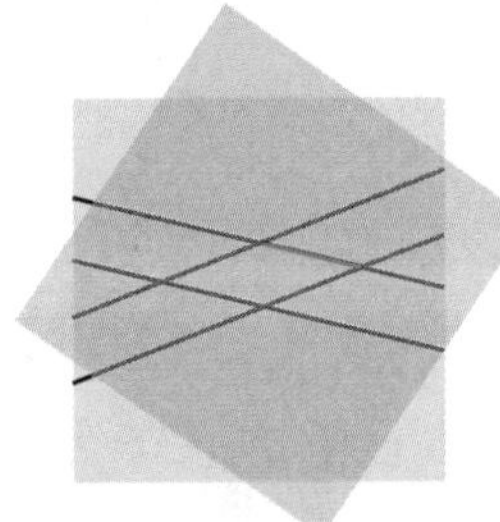

Step 3

Step 1 On a piece of patty paper, use a double-edged straightedge to draw two pairs of parallel lines that intersect each other.

Step 2 Assuming that the two edges of your straightedge are parallel, you have drawn a parallelogram. Place a second patty paper over the first and copy one of the sides of the parallelogram.

Step 3 Compare the length of the side on the second patty paper with the lengths of the other three sides of the parallelogram. How do they compare? Share your results with your group. Copy and complete the conjecture.

Double-Edged Straightedge Conjecture C-49

If two parallel lines are intersected by a second pair of parallel lines that are the same distance apart as the first pair, then the parallelogram formed is a $\underline{?}$.

In Chapter 3, you learned how to construct a rhombus using a compass and straightedge, or using patty paper. Now you know a quicker and easier way, using a double-edged straightedge. To construct a parallelogram that is *not* a rhombus, you need two double-edged staightedges of different widths.

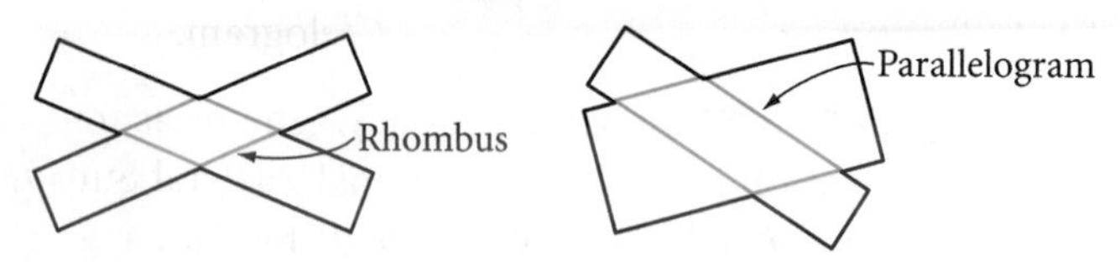

Now let's investigate some properties of rhombuses.

Investigation 2

Do Rhombus Diagonals Have Special Properties?

You will need

- patty paper

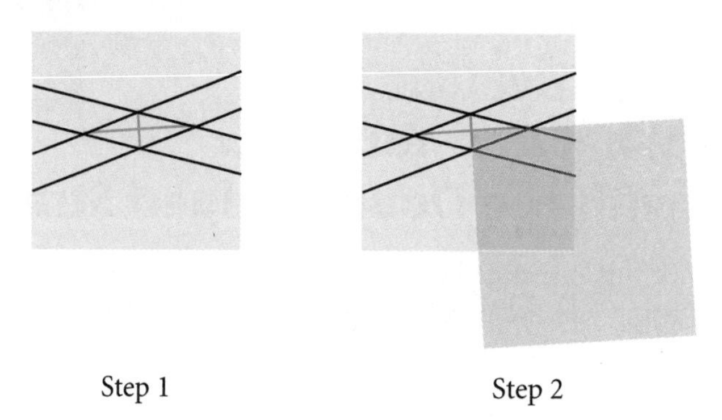

Step 1 Step 2

Step 1 Draw in both diagonals of the rhombus you created in Investigation 1.

Step 2 Place the corner of a second patty paper onto one of the angles formed by the intersection of the two diagonals. Are the diagonals perpendicular?

Compare your results with your group. Also, recall that a rhombus is a parallelogram and that the diagonals of a parallelogram bisect each other. Combine these two ideas into your next conjecture.

Rhombus Diagonals Conjecture C-50

The diagonals of a rhombus are _?_, and they _?_.

Step 3 The diagonals and the sides of the rhombus form two angles at each vertex. Fold your patty paper to compare each pair of angles. What do you observe? Compare your results with your group. Copy and complete the conjecture.

Rhombus Angles Conjecture C-51

The _?_ of a rhombus _?_ the angles of the rhombus.

So far you've made conjectures about a quadrilateral with four congruent sides. Now let's look at quadrilaterals with four congruent angles. What special properties do they have?

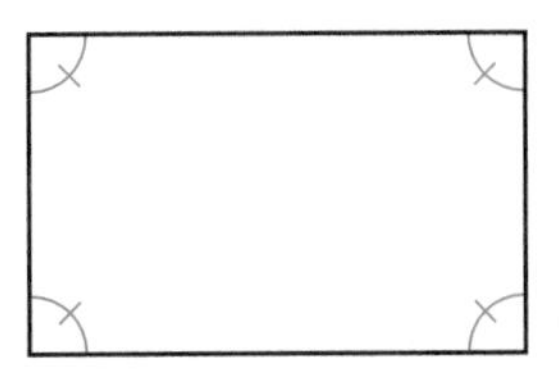

Recall the definition you created for a rectangle. A **rectangle** is an equiangular parallelogram.

Here is a thought experiment. What is the measure of each angle of a rectangle? The Quadrilateral Sum Conjecture says all four angles add up to 360°. They're congruent, so each angle must be 90°, or a right angle.

Investigation 3

Do Rectangle Diagonals Have Special Properties?

You will need

- graph paper
- a compass

Now let's look at the diagonals of rectangles.

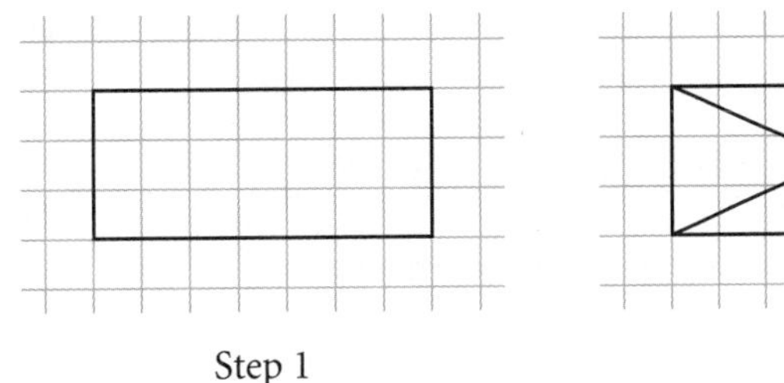

Step 1 Step 2

Step 1 Draw a large rectangle using the lines on a piece of graph paper as a guide.

Step 2 Draw in both diagonals. With your compass, compare the lengths of the two diagonals.

Compare results with your group. In addition, recall that a rectangle is also a parallelogram. So its diagonals also have the properties of a parallelogram's diagonals. Combine these ideas to complete the conjecture.

Rectangle Diagonals Conjecture C-52

The diagonals of a rectangle are __?__ and __?__.

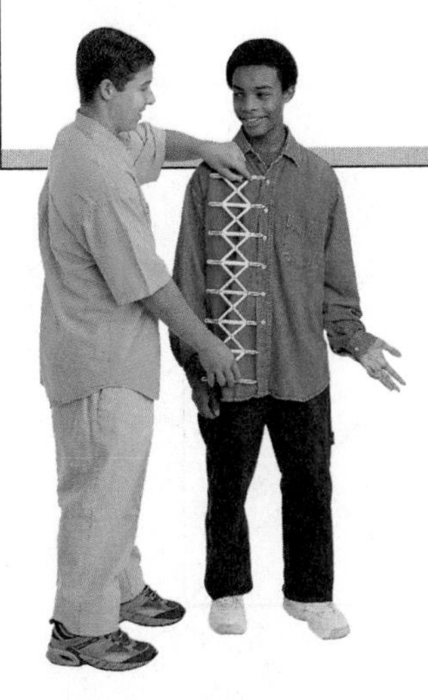

Career CONNECTION

A tailor uses a button spacer to mark the locations of the buttons. The tool opens and closes, but the tips always remain equally spaced. What quadrilateral properties make this tool work correctly?

What happens if you combine the properties of a rectangle and a rhombus? We call the shape a square, and you can think of it as a regular quadrilateral. So you can define it in two different ways.

A **square** is an equiangular rhombus.

Or

A **square** is an equilateral rectangle.

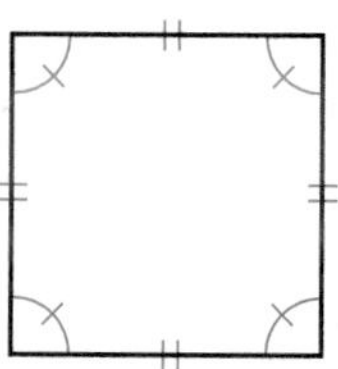

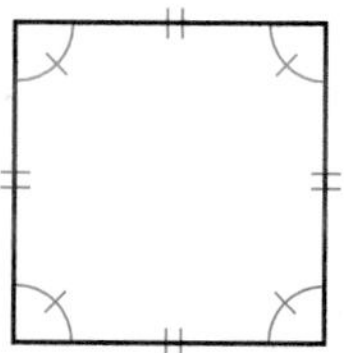

A square is a parallelogram, as well as both a rectangle and a rhombus. Use what you know about the properties of these three quadrilaterals to copy and complete this conjecture.

> **Square Diagonals Conjecture** C-53
>
> The diagonals of a square are _?_, _?_, and _?_.

EXERCISES

You will need

Construction tools for Exercises **17–19, 23, 24,** and **30**

For Exercises 1–10 identify each statement as true or false. For each false statement, sketch a counterexample or explain why it is false.

1. The diagonals of a parallelogram are congruent. ⓗ
2. The consecutive angles of a rectangle are congruent and supplementary.
3. The diagonals of a rectangle bisect each other.
4. The diagonals of a rectangle bisect the angles.
5. The diagonals of a square are perpendicular bisectors of each other.
6. Every rhombus is a square.
7. Every square is a rectangle.
8. A diagonal divides a square into two isosceles right triangles.
9. Opposite angles in a parallelogram are always congruent.
10. Consecutive angles in a parallelogram are always congruent.

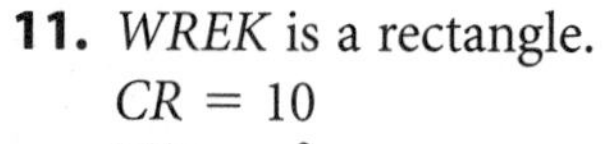

11. *WREK* is a rectangle.
$CR = 10$
$WE = \underline{?}$

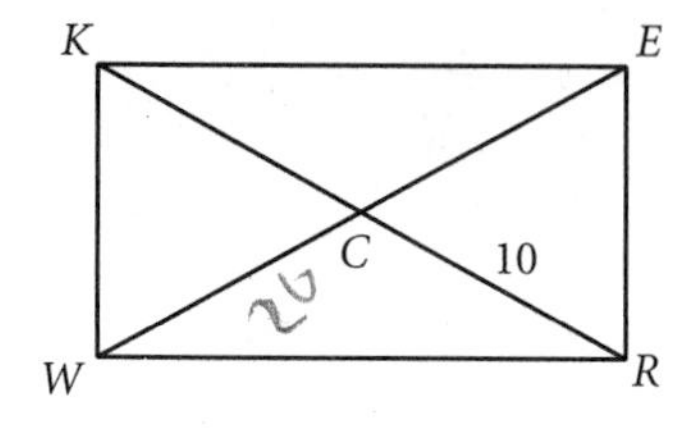

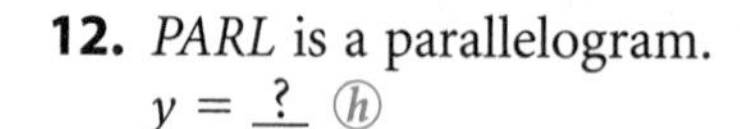

12. *PARL* is a parallelogram.
$y = \underline{?}$ ⓗ

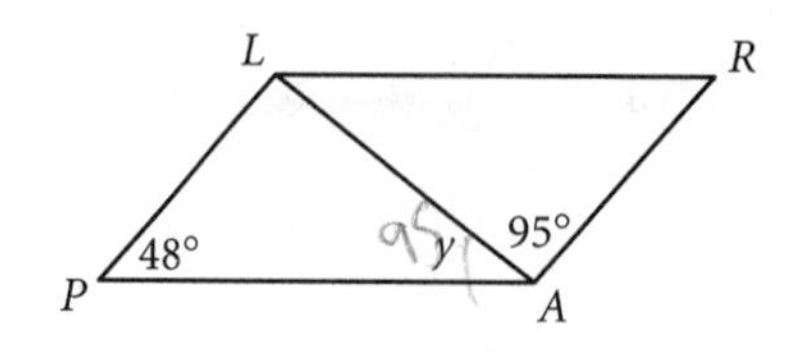

13. *SQRE* is a square.
$x = \underline{?}$
$y = \underline{?}$

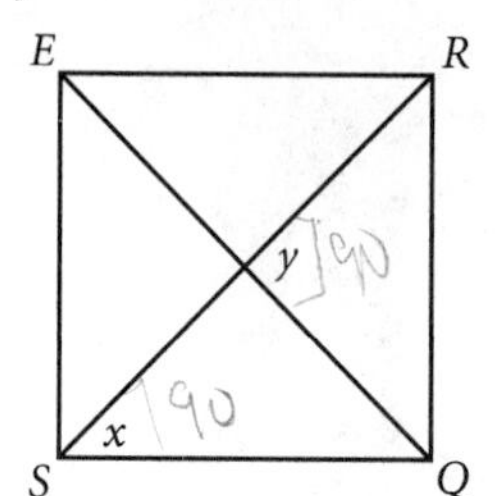

14. Is *DIAM* a rhombus? Why?

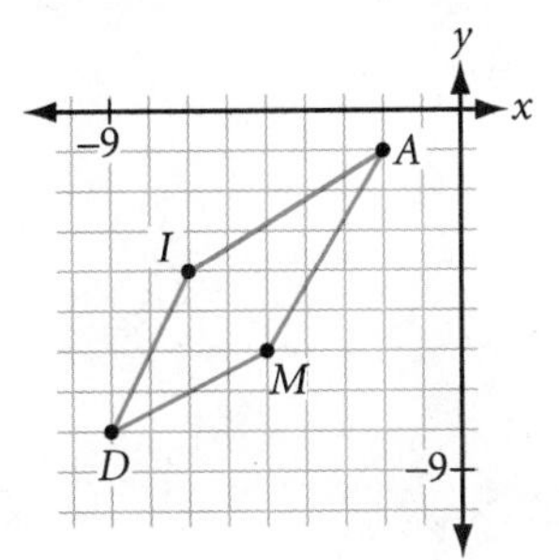

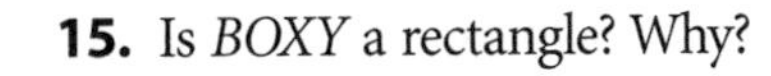
15. Is *BOXY* a rectangle? Why?

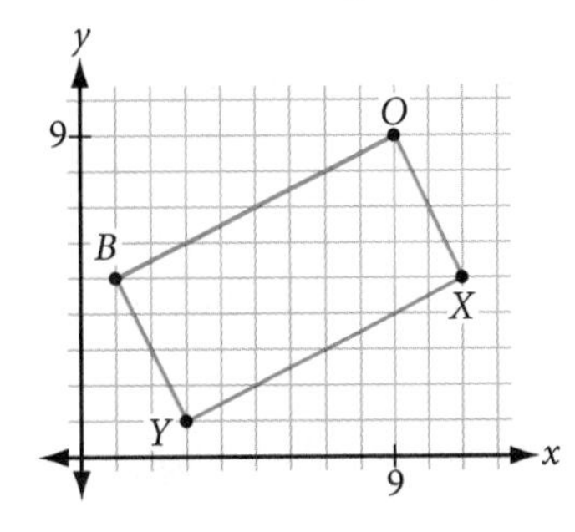

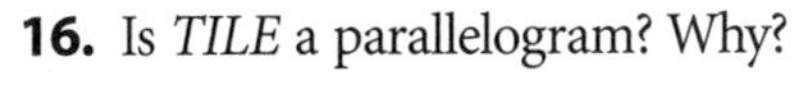
16. Is *TILE* a parallelogram? Why?

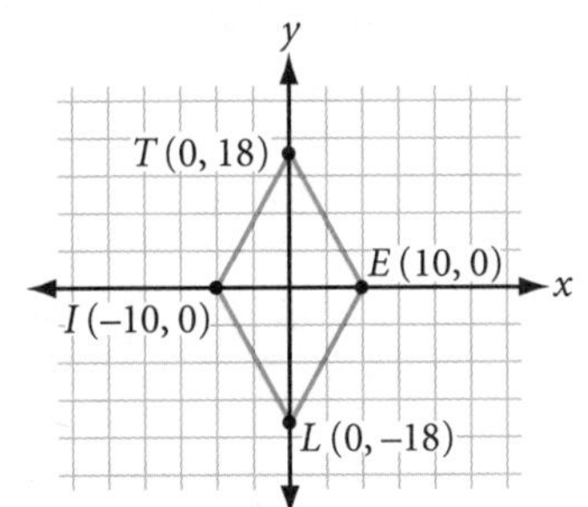

17. ***Construction*** Given the diagonal $\overline{LV}$, construct square *LOVE*. ⓗ

L ———————————— V

18. ***Construction*** Given diagonal $\overline{BK}$ and $\angle B$, construct rhombus *BAKE*. ⓗ

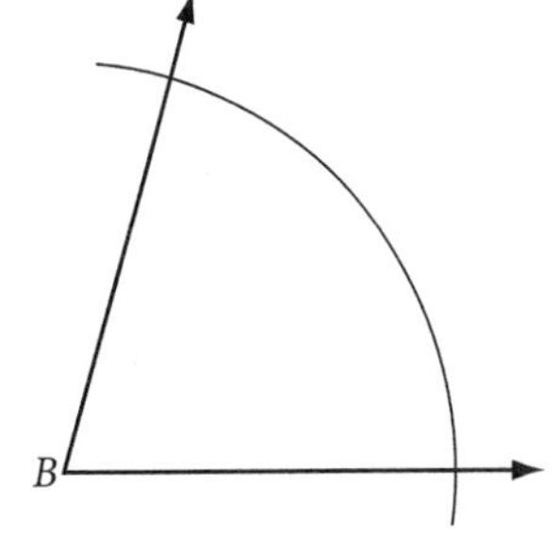

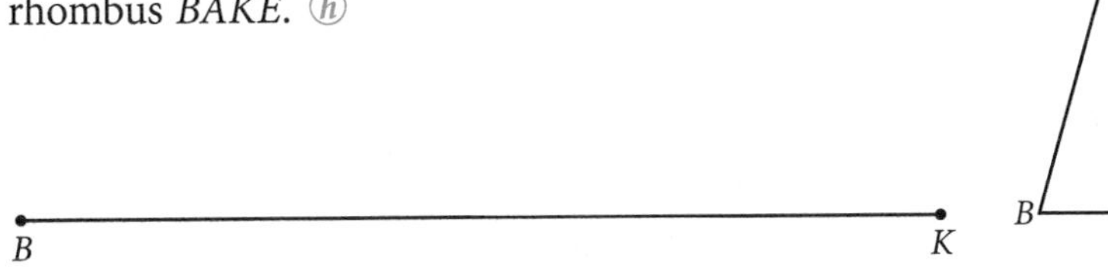

19. ***Construction*** Given side $\overline{PS}$ and diagonal $\overline{PE}$, construct rectangle *PIES*.

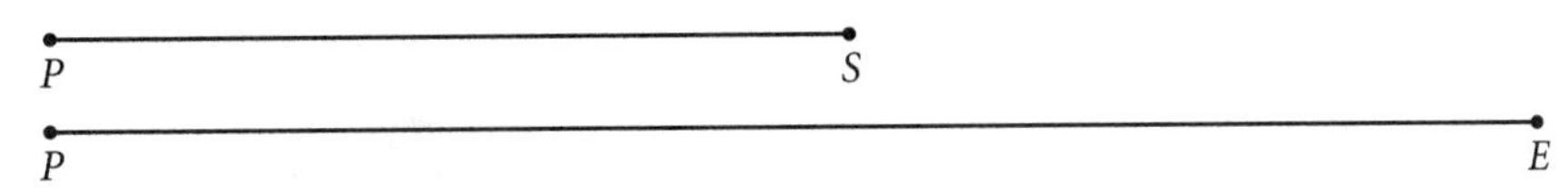

20. To make sure that a room is rectangular, builders check the two diagonals of the room. Explain what they must check, and why this works.

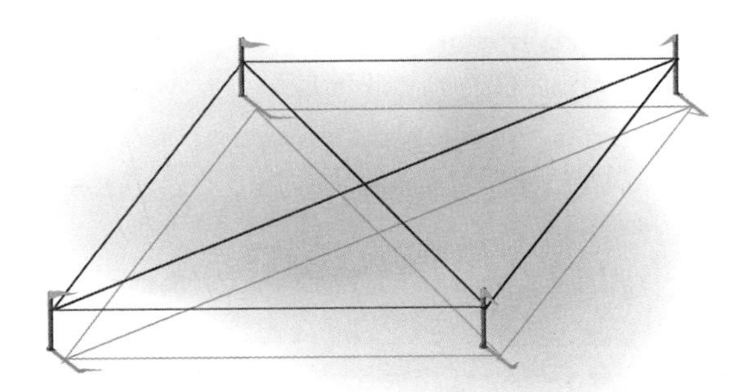

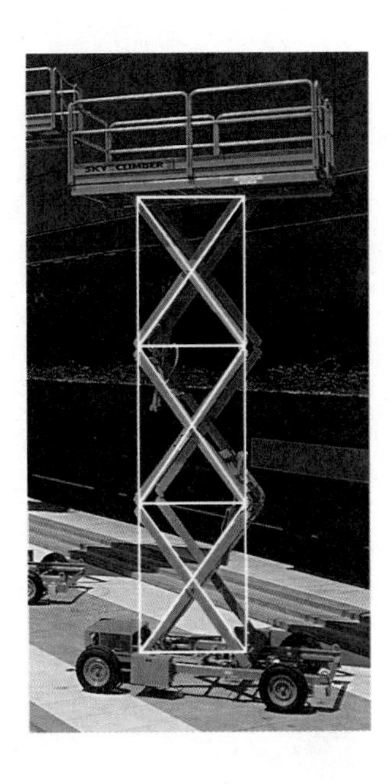

21. The platforms shown at the beginning of this lesson lift objects straight up. The platform also stays parallel to the floor. You can clearly see rhombuses in the picture, but you can also visualize the frame as the diagonals of three rectangles. Explain why the diagonals of a rectangle guarantee this vertical movement.

22. At the street intersection shown at right, one of the streets is wider than the other. Do the crosswalks form a rhombus or a parallelogram? Explain. What would have to be true about the streets if the crosswalks formed a rectangle? A square?

In Exercises 23 and 24, use only the two parallel edges of your double-edged straightedge. You may not fold the paper or use any marks on the straightedge.

23. *Construction* Draw an angle on your paper. Use your double-edged straightedge to construct the bisector of the angle. ⓗ

24. *Construction* Draw a segment on your paper. Use your double-edged straightedge to construct the perpendicular bisector of the segment. ⓗ

Review

25. Trace the figure below. Calculate the measure of each lettered angle.

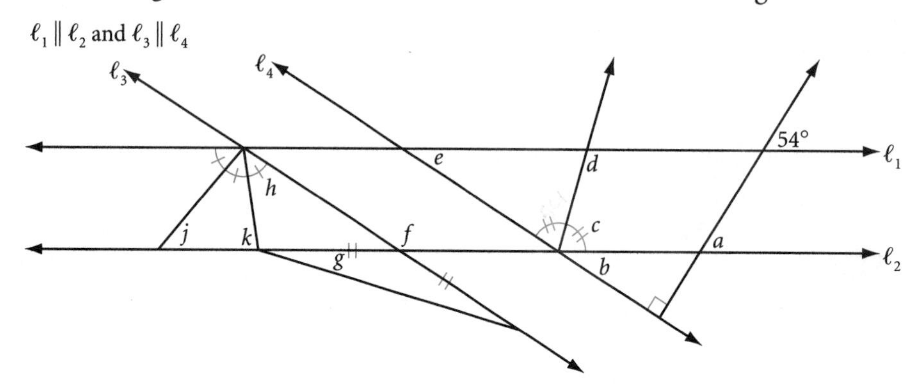

26. Complete the flowchart proof below to demonstrate logically that if a quadrilateral has four congruent sides then it is a rhombus. One possible proof for this argument has been started for you.

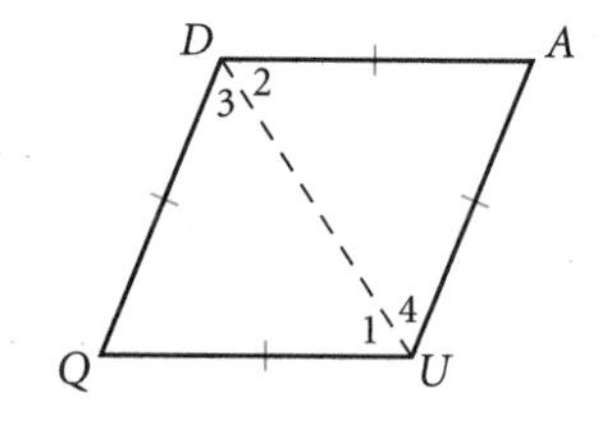

Given: Quadrilateral $QUAD$ has $\overline{QU} \cong \overline{UA} \cong \overline{AD} \cong \overline{DQ}$ with diagonal $\overline{DU}$

Show: $QUAD$ is a rhombus

Flowchart Proof

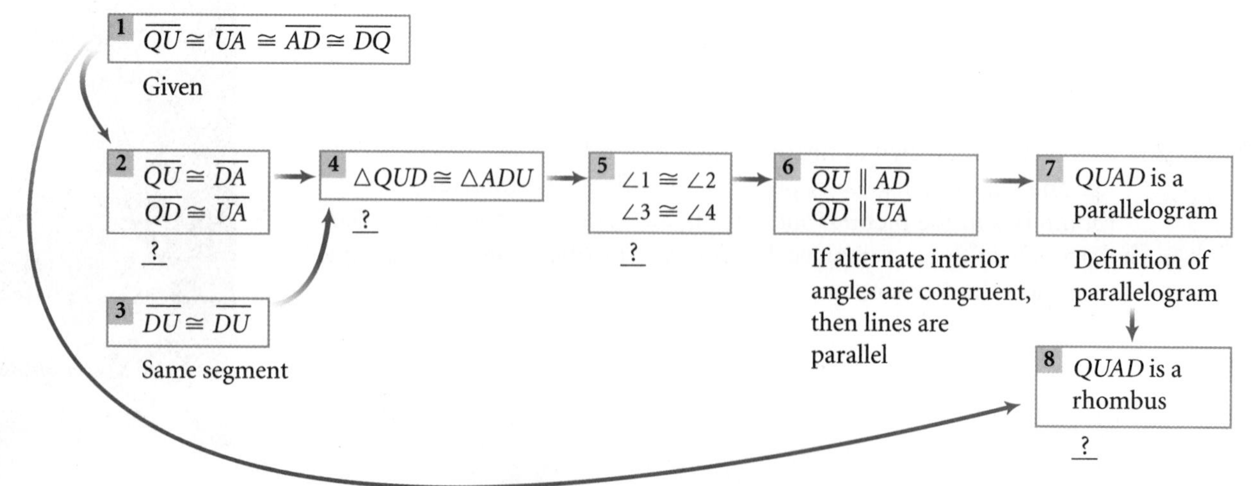

27. Find the coordinates of three more points that lie on the line passing through the points $(2, -1)$ and $(-3, 4)$.

28. Find the coordinates of the circumcenter and the orthocenter for $\triangle RGT$ with vertices $R(2, -1)$, $G(5, 2)$, and $T(-3, 4)$.

29. Draw a counterexample to show that this statement is false: If a triangle is isosceles, then its base angles are not complementary.

30. *Construction* Oran Boatwright is rowing at a 60° angle from the upstream direction as shown. Use a ruler and a protractor to draw the vector diagram. Draw the resultant vector and measure it to find his actual velocity and direction.

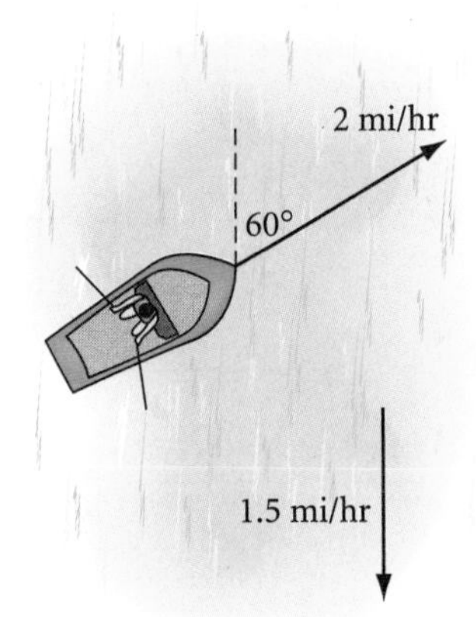

31. In Exercise 26, you proved that if the four sides of a quadrilateral are congruent, then the quadrilateral is a rhombus. So, when we defined rhombus, we did not need the added condition of it being a parallelogram. We only needed to say that it is a quadrilateral with all four sides congruent. Is this true for rectangles? Your conjecture would be, "If a quadrilateral has all four angles congruent, it must be a rectangle." Can you find a counterexample that proves it false? If you cannot, try to create a proof showing that it is true. ⓗ

IMPROVING YOUR REASONING SKILLS

PUZZLE

How Did the Farmer Get to the Other Side?

A farmer was taking her pet rabbit, a basket of prize-winning baby carrots, and her small—but hungry—rabbit-chasing dog to town. She came to a river and realized she had a problem. The little boat she found tied to the pier was big enough to carry only herself and one of the three possessions. She couldn't leave her dog on the bank with the little rabbit (the dog would frighten the poor rabbit), and she couldn't leave the rabbit alone with the carrots (the rabbit would eat all the carrots). But she still had to figure out how to cross the river safely with one possession at a time. How could she move back and forth across the river to get the three possessions safely to the other side?

LESSON

5.7

Proving Quadrilateral Properties

"For instance" is not a "proof."

JEWISH SAYING

Most of the paragraph proofs and flowchart proofs you have done so far have been set up for you to complete. Creating your own proofs requires a great deal of planning. One excellent planning strategy is "thinking backward." If you know where you are headed but are unsure where to start, start at the end of the problem and work your way back to the beginning one step at a time.

The firefighter below asks another firefighter to turn on one of the water hydrants. But which one? A mistake could mean disaster—a nozzle flying around loose under all that pressure. Which hydrant should the firefighter turn on?

Did you "think backward" to solve the puzzle? You'll find it a useful strategy as you write proofs.

To help plan a proof and visualize the flow of reasoning, you can make a flowchart. As you think backward through a proof, you draw a flowchart backward to show the steps in your thinking.

Work with a partner when you first try planning your geometry proof. Think backward to make your plan: start with the conclusion and reason back to the given. Let's look at an example.

A concave kite is sometimes called a **dart.**

EXAMPLE

Given: Dart $ADBC$ with $\overline{AC} \cong \overline{BC}$, $\overline{AD} \cong \overline{BD}$

Show: $\overline{CD}$ bisects $\angle ACB$

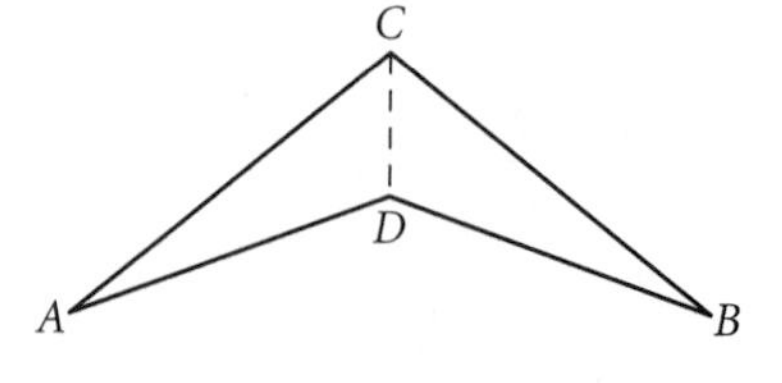

▶ Solution

Plan: Begin by drawing a diagram and marking the given information on it. Next, construct your proof by reasoning backward. Then convert this reasoning into a flowchart. Your flowchart should start with boxes containing the given information and end with what you are trying to demonstrate. The arrows indicate the flow of your logical argument. Your thinking might go something like this:

"I can show $\overline{CD}$ is the bisector of $\angle ACB$ if I can show $\angle ACD \cong \angle BCD$."

$\angle ACD \cong \angle BCD$ → $\overline{CD}$ is the bisector of $\angle ACB$

"I can show $\angle ACD \cong \angle BCD$ if they are corresponding angles in congruent triangles."

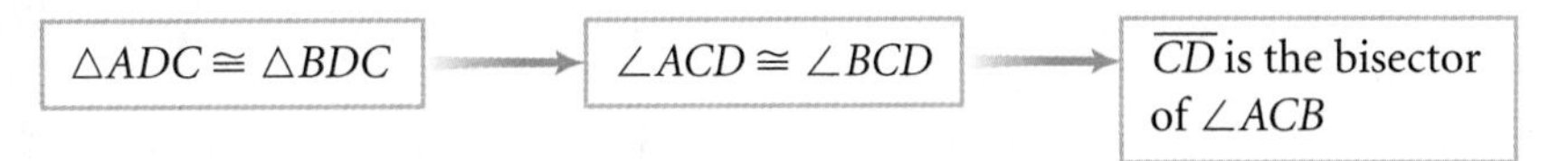

"Can I show $\triangle ADC \cong \triangle BDC$? Yes, I can, by SSS, because it is given that $\overline{AC} \cong \overline{BC}$ and $\overline{AD} \cong \overline{BD}$, and $\overline{CD} \cong \overline{CD}$ because it is the same segment in both triangles."

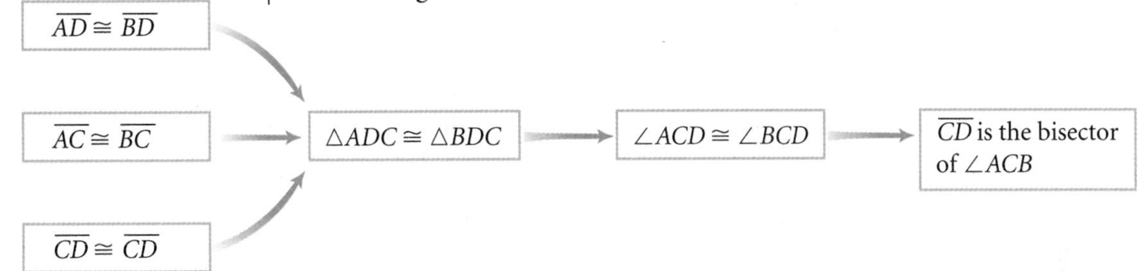

By adding the reason for each statement below each box in your flowchart, you can make the flowchart into a complete flowchart proof.

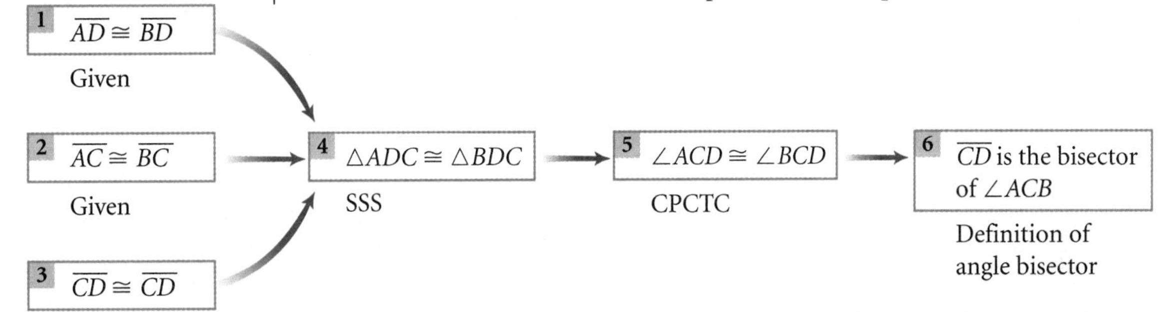

Some students prefer to write their proofs in a flowchart format, and others prefer to write out their proof as an explanation in paragraph form. By reversing the reasoning in your plan, you can make the plan into a complete paragraph proof.

"It is given that $\overline{AC} \cong \overline{BC}$ and $\overline{AD} \cong \overline{BD}$. $\overline{CD} \cong \overline{CD}$ because it is the same segment in both triangles. So, $\triangle ADC \cong \triangle BDC$ by the SSS Congruence Conjecture. So, $\angle ACD \cong \angle BCD$ by the definition of congruent triangles (CPCTC). Therefore, by the definition of angle bisectors, $\overline{CD}$ is the bisector of $\angle ACB$. Q.E.D."

The abbreviation Q.E.D. at the end of a proof stands for the Latin phrase *quod erat demonstrandum,* meaning "which was to be demonstrated." You can also think of Q.E.D. as a short way of saying "Quite Elegantly Done" at the conclusion of your proof.

In the exercises you will prove some of the special properties of quadrilaterals discovered in this chapter.

EXERCISES

1. Let's start with a puzzle. Copy the 5-by-5 puzzle grid at right. Start at square 1 and end at square 100. You can move to an adjacent square horizontally, vertically, or diagonally if you can add, subtract, multiply, or divide the number in the square you occupy by 2 or 5 to get the number in that square.

   For example, if you happen to be in square 11, you could move to square 9 by subtracting 2 or to square 55 by multiplying by 5. When you find the path from 1 to 100, show it with arrows.

   Notice that in this puzzle you may start with different moves. You could start with 1 and go to 5. From 5 you could go to 10 or 3. Or you could start with 1 and go to 2. From 2 you could go to 4. Which route should you take? ⓗ

In Exercises 2–10, each conjecture has also been stated as a "given" and a "show." Any necessary auxiliary lines have been included. Complete a flowchart proof or write a paragraph proof.

2. Prove the conjecture: The diagonal of a parallelogram divides the parallelogram into two congruent triangles.

   Given: Parallelogram $SOAK$ with diagonal $\overline{SA}$

   Show: $\triangle SOA \cong \triangle AKS$

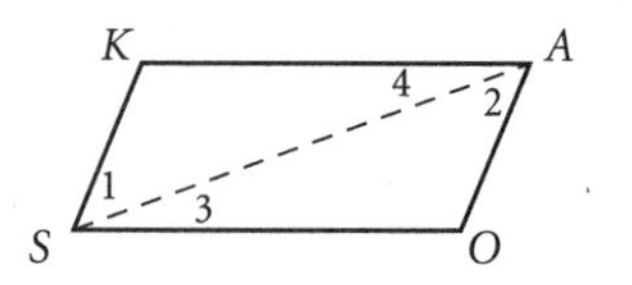

Flowchart Proof

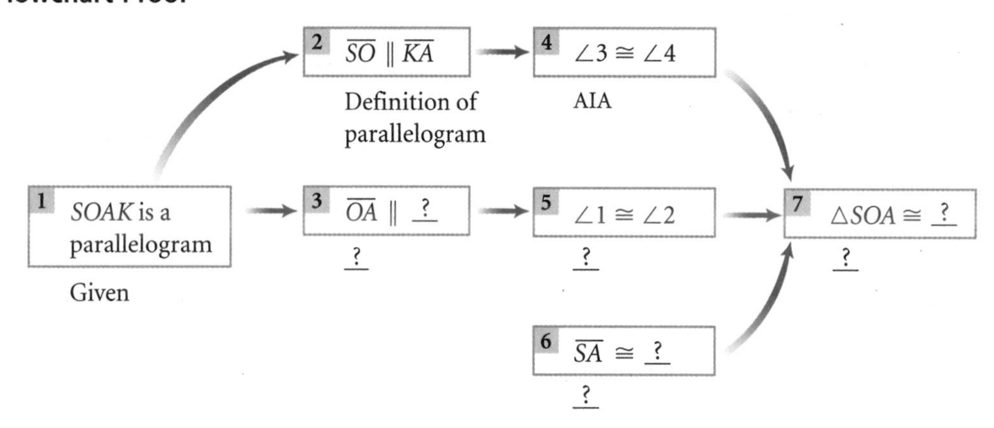

3. Prove the conjecture: The opposite angles of a parallelogram are congruent.

   Given: Parallelogram $BATH$ with diagonals $\overline{BT}$ and $\overline{HA}$

   Show: $\angle HBA \cong \angle ATH$ and $\angle BAT \cong \angle THB$

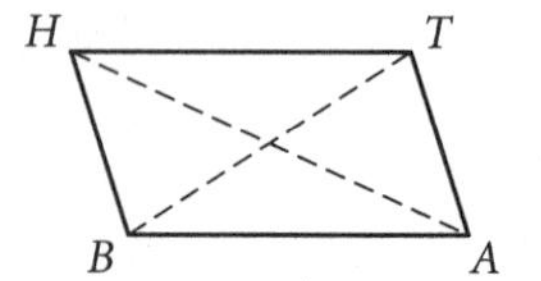

Flowchart Proof

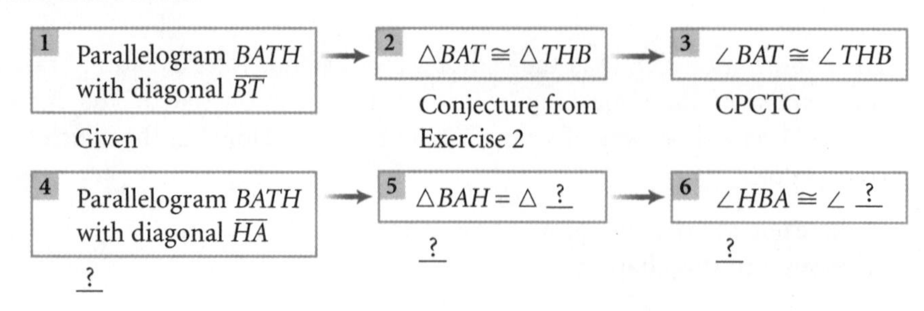

4. Prove the conjecture: If the opposite sides of a quadrilateral are congruent, then the quadrilateral is a parallelogram.

Given: Quadrilateral *WATR*, with $\overline{WA} \cong \overline{RT}$ and $\overline{WR} \cong \overline{AT}$, and diagonal $\overline{WT}$

Show: *WATR* is a parallelogram

Flowchart Proof

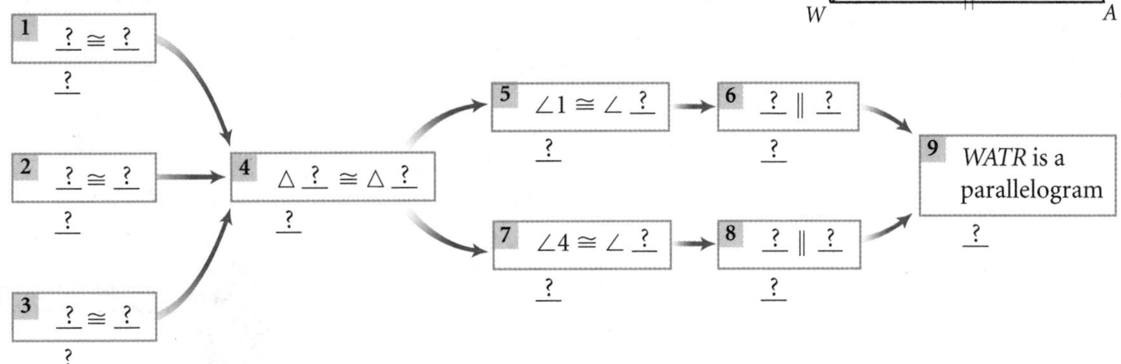

5. Write a flowchart proof to demonstrate that quadrilateral *SOAP* is a parallelogram.

Given: Quadrilateral *SOAP* with $\overline{SP} \parallel \overline{OA}$ and $\overline{SP} \cong \overline{OA}$

Show: *SOAP* is a parallelogram

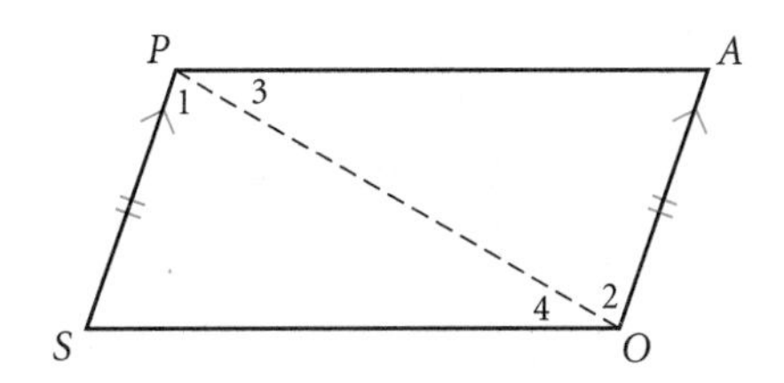

6. The results of the proof in Exercise 5 can now be stated as a proved conjecture. Complete this statement beneath your proof: "If one pair of opposite sides of a quadrilateral are both parallel and congruent, then the quadrilateral is a ___?___."

7. Prove the conjecture: The diagonals of a rectangle are congruent. ⓗ

Given: Rectangle *YOGI* with diagonals $\overline{YG}$ and $\overline{OI}$

Show: $\overline{YG} \cong \overline{OI}$

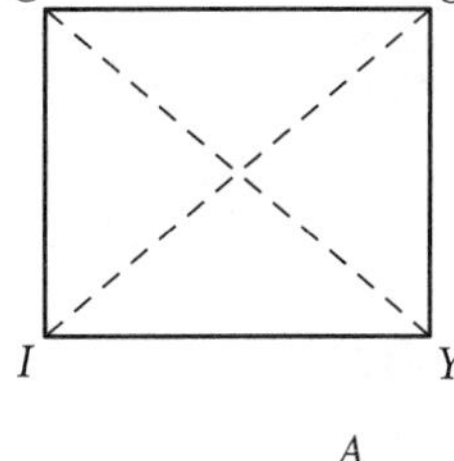

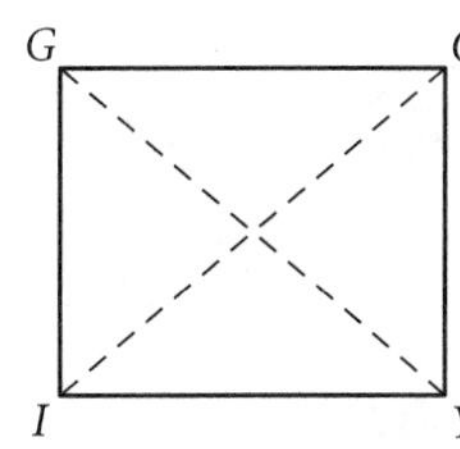

8. Prove the conjecture: If the diagonals of a parallelogram are congruent, then the parallelogram is a rectangle. ⓗ

Given: Parallelogram *BEAR*, with diagonals $\overline{BA} \cong \overline{ER}$

Show: *BEAR* is a rectangle

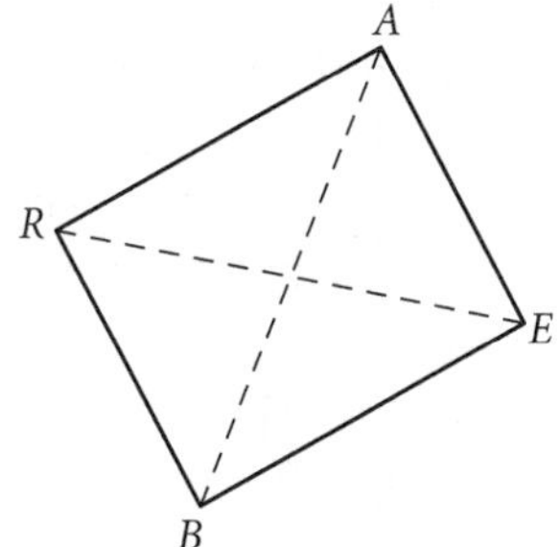

9. Prove the Isosceles Trapezoid Conjecture: The base angles of an isosceles trapezoid are congruent.

Given: Isosceles trapezoid *PART* with $\overline{PA} \parallel \overline{TR}$, $\overline{PT} \cong \overline{AR}$, and $\overline{TZ}$ constructed parallel to $\overline{RA}$

Show: $\angle TPA \cong \angle RAP$

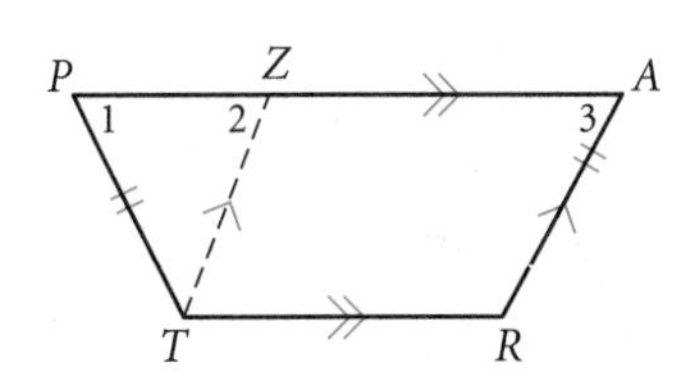

10. Prove the Isosceles Trapezoid Diagonals Conjecture: The diagonals of an isosceles trapezoid are congruent.

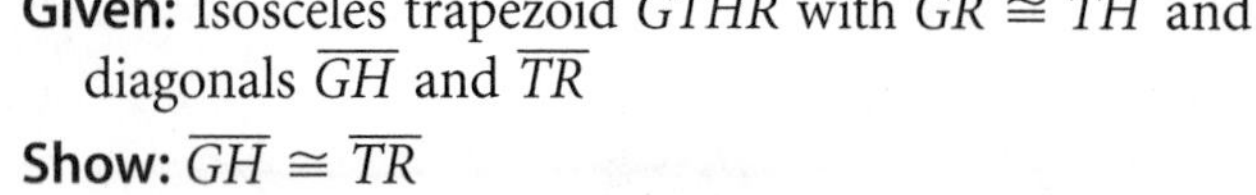

Given: Isosceles trapezoid $GTHR$ with $\overline{GR} \cong \overline{TH}$ and diagonals $\overline{GH}$ and $\overline{TR}$

Show: $\overline{GH} \cong \overline{TR}$

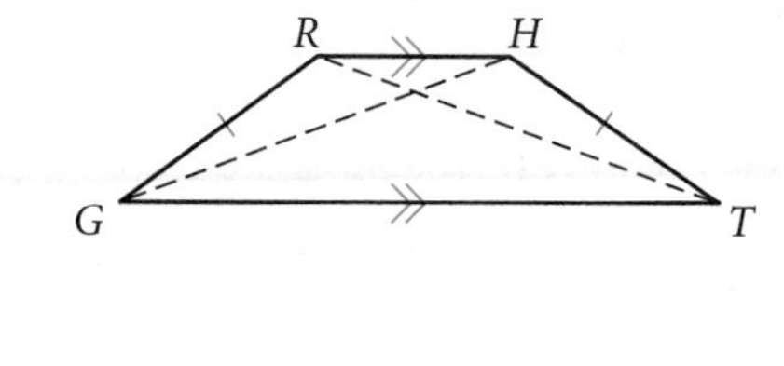

1 $\underline{?}$ — Given → 2 $\underline{?}$ — Given

3 $\overline{GT} \cong \overline{GT}$ — $\underline{?}$

4 $\underline{?}$ — Isosceles Trapezoid Conjecture

(2, 3, 4) → 5 $\triangle \underline{?} \cong \triangle \underline{?}$ — $\underline{?}$ → 6 $\underline{?}$ — $\underline{?}$

11. If an adjustable desk lamp, like the one at right, is adjusted by bending or straightening the metal arm, it will continue to shine straight down onto the desk. What property that you proved in the previous exercises explains why?

12. You have discovered that triangles are rigid but parallelograms are not. This property shows up in the making of fabric, which has warp threads and weft threads. Fabric is constructed by weaving thread at right angles, creating a grid of rectangles. What happens when you pull the fabric along the warp or weft? What happens when you pull the fabric along a diagonal (the *bias*)? ⓗ

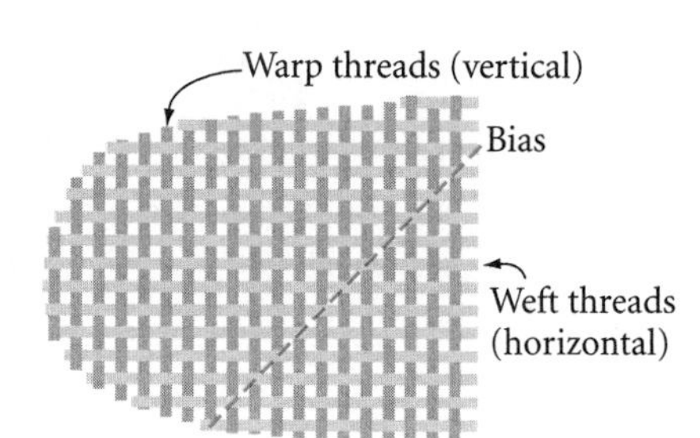

Review

13. Find the measure of the acute angles in the 4-pointed star in the Islamic tiling shown at right. The polygons are squares and regular hexagons. Find the measure of the acute angles in the 6-pointed star in the Islamic tiling on the far right. The 6-pointed star design is created by arranging six squares. Are the angles in both stars the same? ⓗ

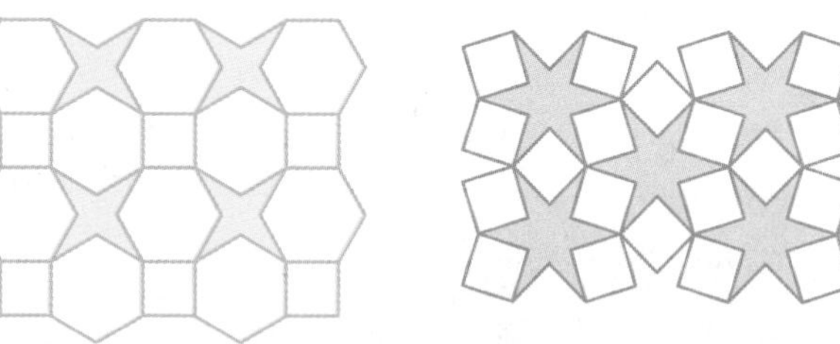

14. A contractor tacked one end of a string to each vertical edge of a window. He then handed a protractor to his apprentice and said, "Here, find out if the vertical edges are parallel." What should the apprentice do? No, he can't quit, he wants this job! Help him. ⓗ

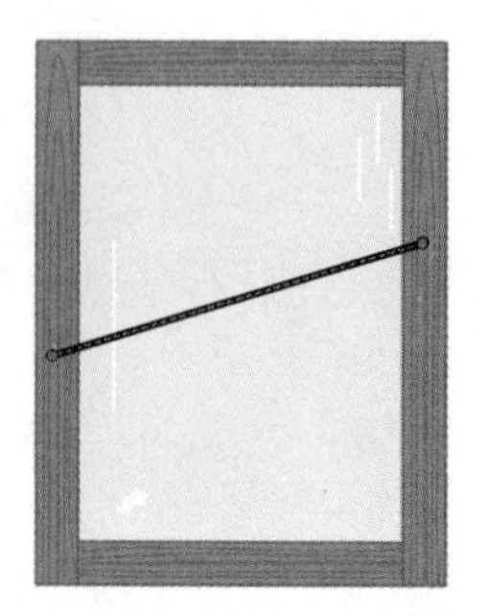

15. The last bus stops at the school some time between 4:45 and 5:00. What is the probability that you will miss the bus if you arrive at the bus stop at 4:50? ⓗ

16. The 3-by-9-by-12-inch clear plastic sealed container shown is resting on its smallest face. It is partially filled with a liquid to a height of 8 inches. Sketch the container resting on its middle-sized face. What will be the height of the liquid in the container in this position? ⓗ

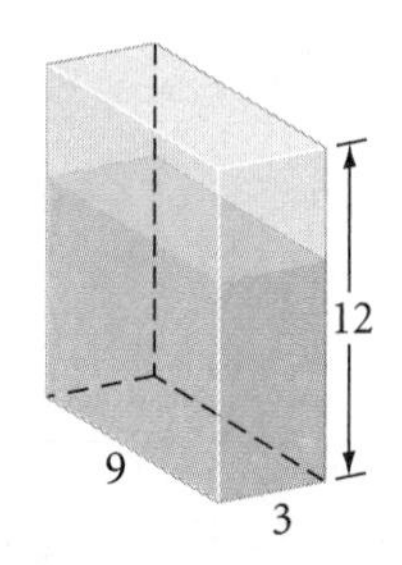

project

JAPANESE PUZZLE QUILTS

When experienced quilters first see Japanese puzzle quilts, they are often amazed (or puzzled?) because the straight rows of blocks so common to block quilts do not seem to exist. The sewing lines between apparent blocks seem jagged. At first glance, Japanese puzzle quilts look like American crazy quilts that must be handsewn and that take forever to make!

Mabry Benson designed this puzzle quilt, *Red and Blue Puzzle* (1994). Can you find any rhombic blocks that are the same? How many different types of fabric were used?

However, Japanese puzzle quilts do contain straight sewing lines. Study the Japanese puzzle quilt at right. Can you find the basic quilt block? What shape is it?

The puzzle quilt shown above is made of four different-color kites sewn into rhombuses. The

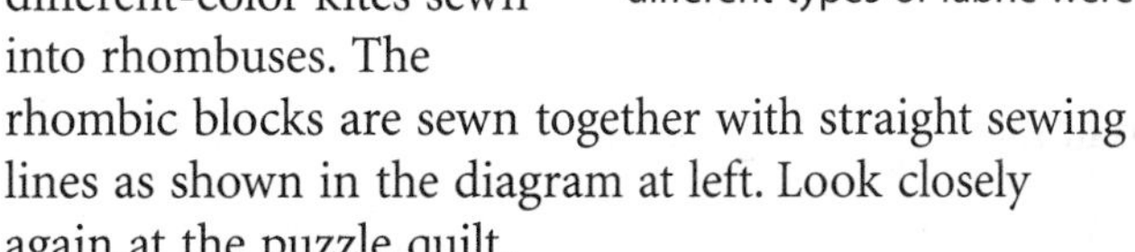

rhombic blocks are sewn together with straight sewing lines as shown in the diagram at left. Look closely again at the puzzle quilt.

Now for your project. You will need copies of the Japanese puzzle quilt grid, color pencils or markers, and color paper or fabrics.

1. To produce the zigzag effect of a Japanese puzzle quilt, you need to avoid pseudoblocks of the same color sharing an edge. How many different colors or fabrics do you need in order to make a puzzle quilt?
2. How many different types of rhombic blocks do you need for a four-color Japanese puzzle quilt? What if you want no two pseudoblocks of the same color to touch at either an edge or a vertex?
3. Can you create a four-color Japanese puzzle quilt that requires more than four different color combinations in the rhombic blocks?
4. Plan, design, and create a Japanese puzzle quilt out of paper or fabric, using the Japanese puzzle quilt technique.

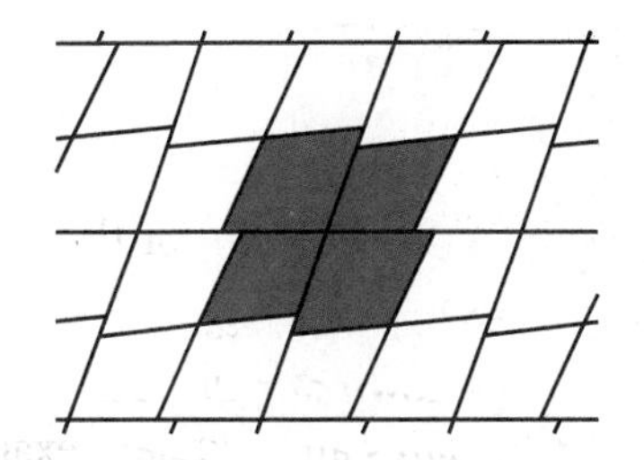

Detail of a pseudoblock

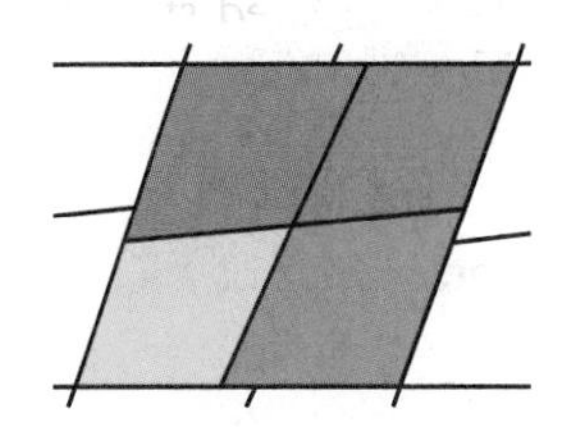

Detail of an actual block

CHAPTER 5 REVIEW

In this chapter you extended your knowledge of triangles to other polygons. You discovered the interior and exterior angle sums for all polygons. You investigated the midsegments of triangles and trapezoids and the properties of parallelograms. You learned what distinguishes various quadrilaterals and what properties apply to each class of quadrilaterals.

Along the way you practiced proving conjectures with flowcharts and paragraph proofs. Be sure you've added the new conjectures to your list. Include diagrams for clarity.

How has your knowledge of triangles helped you make discoveries about other polygons?

EXERCISES

You will need

1. How do you find the measure of one exterior angle of a regular polygon?
2. How can you find the number of sides of an equiangular polygon by measuring one of its interior angles? By measuring one of its exterior angles?
3. How do you construct a rhombus by using only a ruler or double-edged straightedge?
4. How do you bisect an angle by using only a ruler or double-edged straightedge?
5. How can you use the Rectangle Diagonals Conjecture to determine if the corners of a room are right angles?
6. How can you use the Triangle Midsegment Conjecture to find a distance between two points that you can't measure directly?

7. Find x and y.

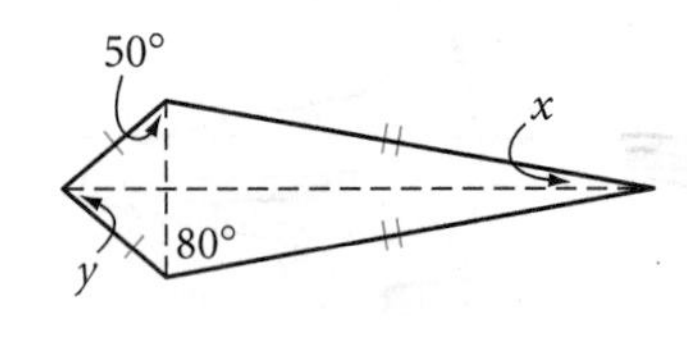

8. Perimeter = 266 cm. Find x.

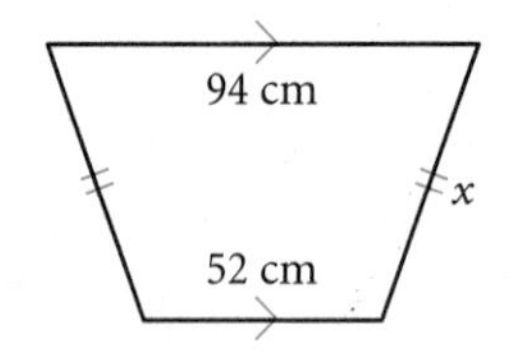

9. Find a and c.

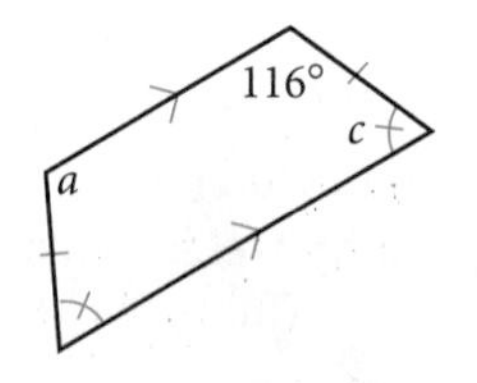

10. $\overline{MS}$ is a midsegment. Find the perimeter of $MOIS$.

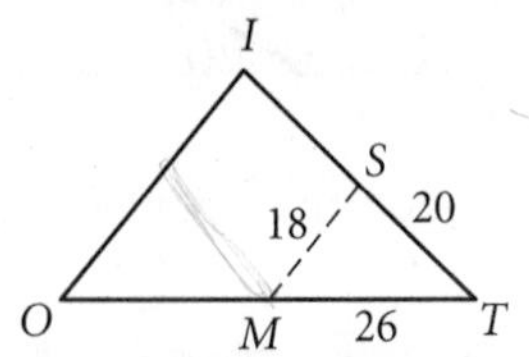

11. Find x.

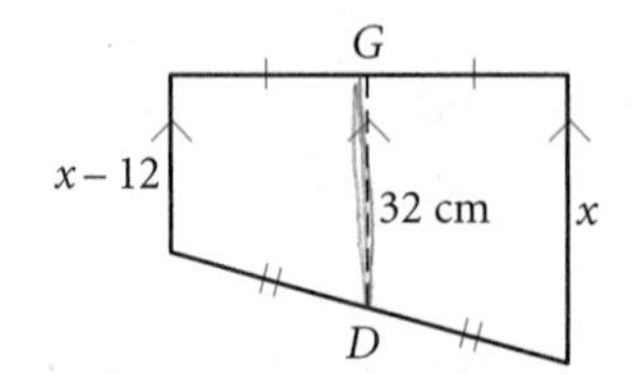

12. Find y and z.

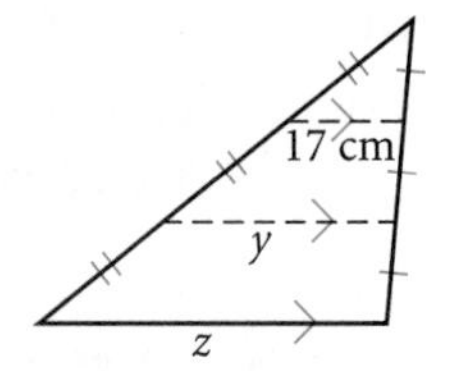

13. Copy and complete the table below by placing a yes (to mean always) or a no (to mean not always) in each empty space. Use what you know about special quadrilaterals.

	Kite	Isosceles trapezoid	Parallelogram	Rhombus	Rectangle
Opposite sides are parallel					
Opposite sides are congruent					
Opposite angles are congruent					
Diagonals bisect each other					
Diagonals are perpendicular					
Diagonals are congruent				No	
Exactly one line of symmetry	Yes				
Exactly two lines of symmetry					

14. **APPLICATION** A 2-inch-wide frame is to be built around the regular decagonal window shown. At what angles a and b should the corners of each piece be cut?

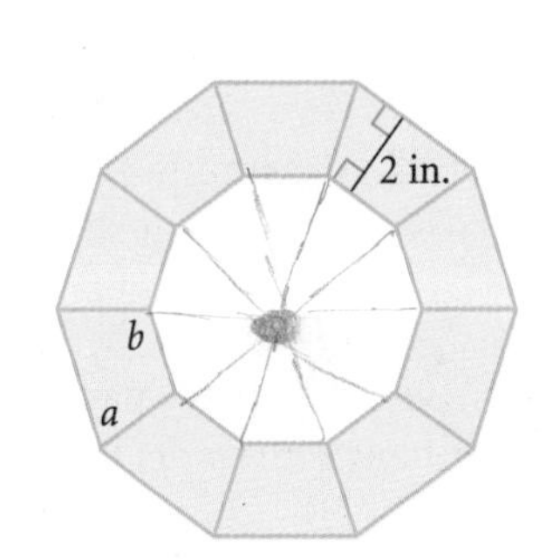

15. Find the measure of each lettered angle.

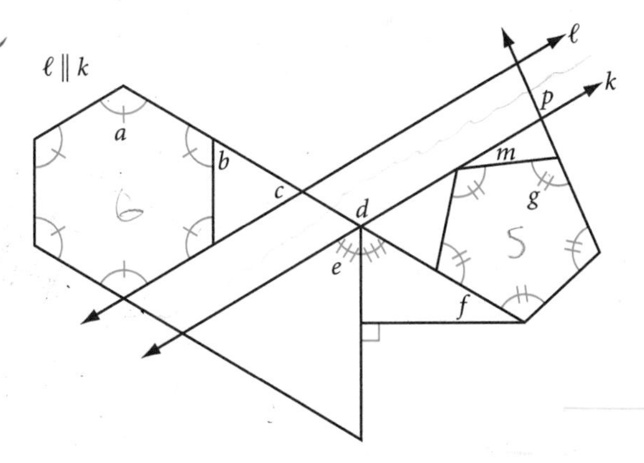

16. Archaeologist Ertha Diggs has uncovered one stone that appears to be a voussoir from a semicircular stone arch. On each isosceles trapezoidal face, the obtuse angles measure 96°. Assuming all the stones were identical, how many stones were in the original arch?

17. Kite $ABCD$ has vertices $A(-3, -2)$, $B(2, -2)$, $C(3, 1)$, and $D(0, 2)$. Find the coordinates of the point of intersection of the diagonals.

18. When you swing left to right on a swing, the seat stays parallel to the ground. Explain why.

19. ***Construction*** The tiling of congruent pentagons shown below is created from a honeycomb grid (tiling of regular hexagons). What is the measure of each lettered angle? Re-create the design with compass and straightedge.

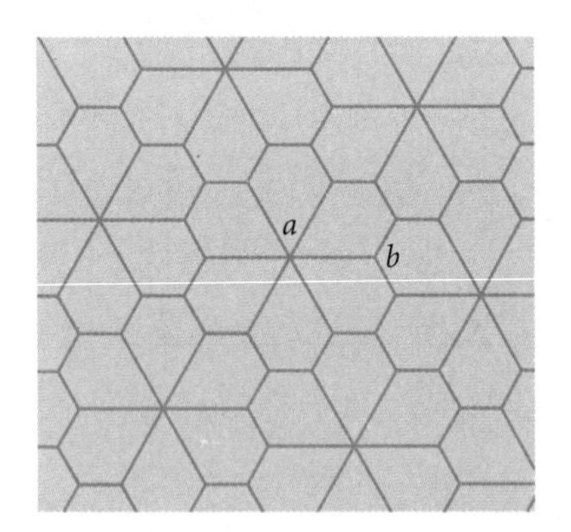

20. ***Construction*** An airplane is heading north at 900 km/hr. However, a 50 km/hr wind is blowing from the east. Use a ruler and a protractor to make a scale drawing of these vectors. Measure to find the approximate resultant velocity, both speed and direction (measured from north). ⓗ

Construction In Exercises 21–24, use the given segments and angles to construct each figure. Use either patty paper or a compass and a straightedge. The small letter above each segment represents the length of the segment.

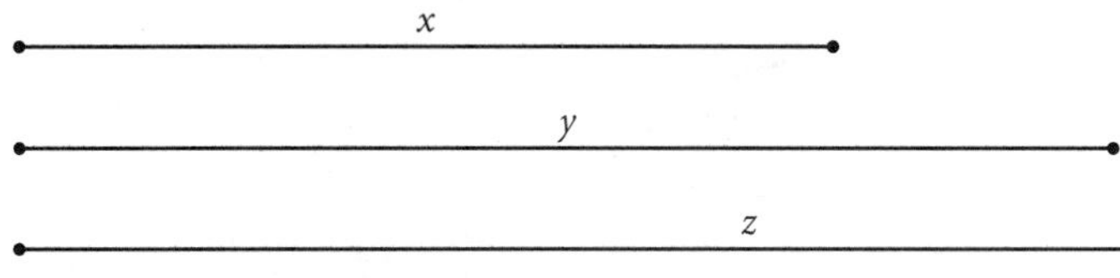

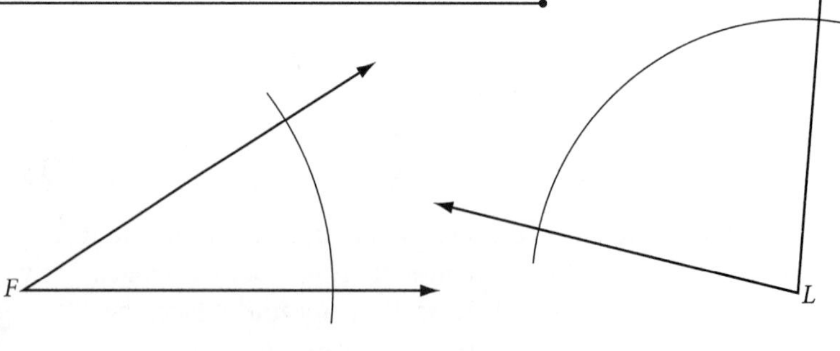

21. Construct rhombus $SQRE$ with $SR = y$ and $QE = x$.

22. Construct kite $FLYR$ given $\angle F$, $\angle L$, and $FL = x$.

23. Given bases $\overline{LP}$ with length z and $\overline{EN}$ with length y, nonparallel side $\overline{LN}$ with length x, and $\angle L$, construct trapezoid $PENL$. ⓗ

24. Given $\angle F$, $FR = x$, and $YD = z$, construct two trapezoids $FRYD$ that are not congruent to each other.

25. Three regular polygons meet at point B. Only four sides of the third polygon are visible. How many sides does this polygon have?

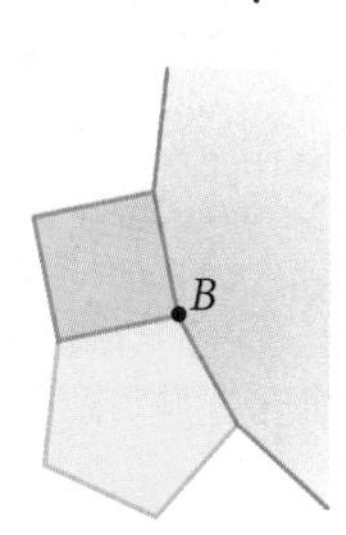

26. Find x.

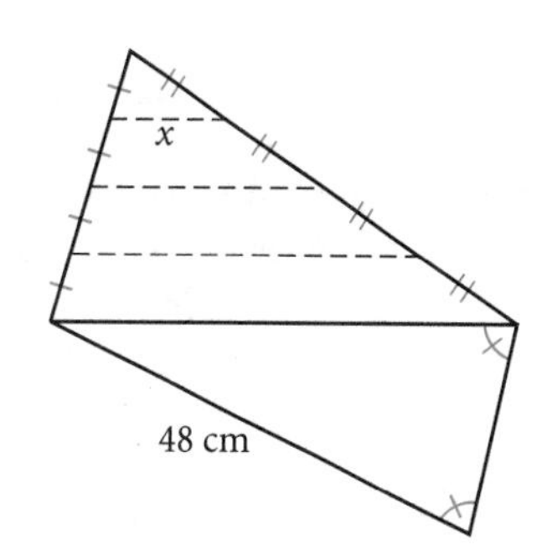

27. Prove the conjecture: The diagonals of a rhombus bisect the angles.

Given: Rhombus $DENI$, with diagonal $\overline{DN}$

Show: Diagonal $\overline{DN}$ bisects $\angle D$ and $\angle N$

Flowchart Proof

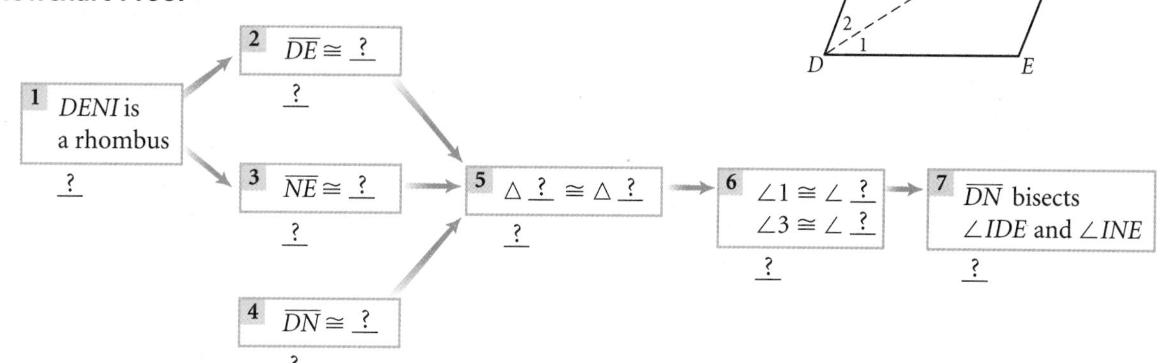

TAKE ANOTHER LOOK

1. Draw several polygons that have four or more sides. In each, draw all the diagonals from one vertex. Explain how the Polygon Sum Conjecture follows logically from the Triangle Sum Conjecture. Does the Polygon Sum Conjecture apply to concave polygons?

2. A triangle on a sphere can have three right angles. Can you find a "rectangle" with four right angles on a sphere? Investigate the Polygon Sum Conjecture on a sphere. Explain how it is related to the Triangle Sum Conjecture on a sphere. Be sure to test your conjecture on polygons with the smallest and largest possible angle measures.

The small, precise polygons in the painting, *Boy With Birds* (1953, oil on canvas), by American artist David C. Driskell (b 1931), give it a look of stained glass.

3. Draw a polygon and one set of its exterior angles. Label the exterior angles. Cut out the exterior angles and arrange them all about a point. Explain how this activity demonstrates the Exterior Angle Sum Conjecture.

4. Is the Exterior Angle Sum Conjecture also true for concave polygons? Are the kite conjectures also true for darts (concave kites)? Choose your tools and investigate.

5. Investigate exterior angle sums for polygons on a sphere. Be sure to test polygons with the smallest and largest angle measures.

Assessing What You've Learned

GIVING A PRESENTATION

Giving a presentation is a powerful way to demonstrate your understanding of a topic. Presentation skills are also among the most useful skills you can develop in preparation for almost any career. The more practice you can get in school, the better.

Choose a topic to present to your class. There are a number of things you can do to make your presentation go smoothly.

- Work with a group. Make sure your group presentation involves all group members so that it's clear everyone contributed equally.
- Choose a topic that will be interesting to your audience.
- Prepare thoroughly. Make an outline of important points you plan to cover. Prepare visual aids—like posters, models, handouts, and overhead transparencies—ahead of time. Rehearse your presentation.
- Communicate clearly. Speak up loud and clear, and show your enthusiasm about your topic.

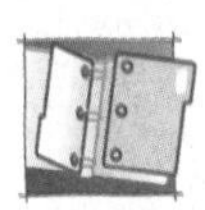

ORGANIZE YOUR NOTEBOOK Your conjecture list should be growing fast! Review your notebook to be sure it's complete and well organized. Write a one-page chapter summary.

WRITE IN YOUR JOURNAL Write an imaginary dialogue between your teacher and a parent or guardian about your performance and progress in geometry.

UPDATE YOUR PORTFOLIO Choose a piece that represents your best work from this chapter to add to your portfolio. Explain what it is and why you chose it.

PERFORMANCE ASSESSMENT While a classmate, a friend, a family member, or a teacher observes, carry out one of the investigations from this chapter. Explain what you're doing at each step, including how you arrived at the conjecture.

CHAPTER

6 Discovering and Proving Circle Properties

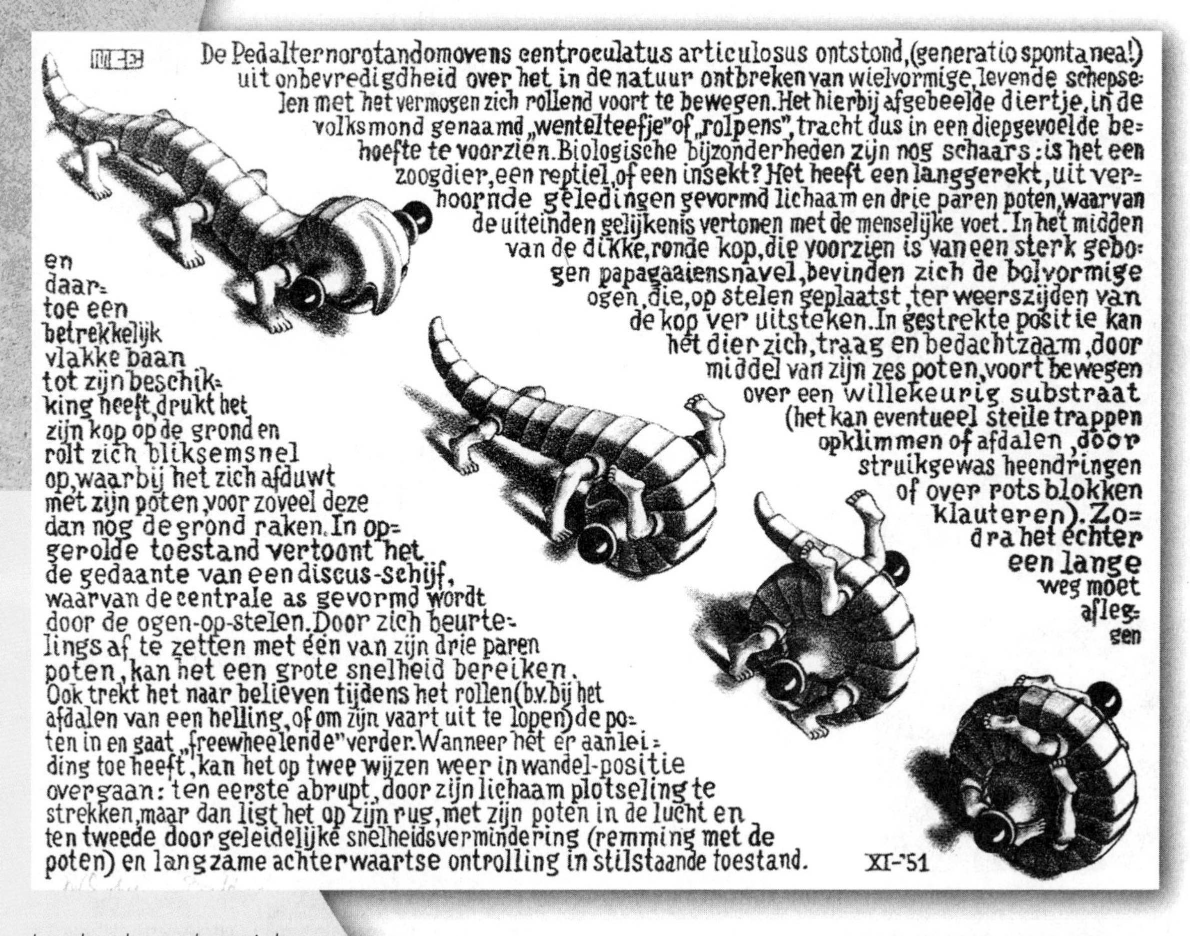

I am the only one who can judge how far I constantly remain below the quality I would like to attain.

M. C. ESCHER

Curl-Up, M. C. Escher, 1951

OBJECTIVES

In this chapter you will

- learn relationships among chords, arcs, and angles
- discover properties of tangent lines
- learn how to calculate the length of an arc
- prove circle conjectures

LESSON

6.1 Chord Properties

Let's review some basic terms before you begin discovering the properties of circles. You should be able to identify the terms below.

Match the figures at the right with the terms at the left.

Learning by experience is good, but in the case of mushrooms and toadstools, hearsay evidence is better.

ANONYMOUS

1. Congruent circles
2. Concentric circles
3. Radius
4. Chord
5. Diameter
6. Tangent
7. Minor arc
8. Major arc
9. Semicircle

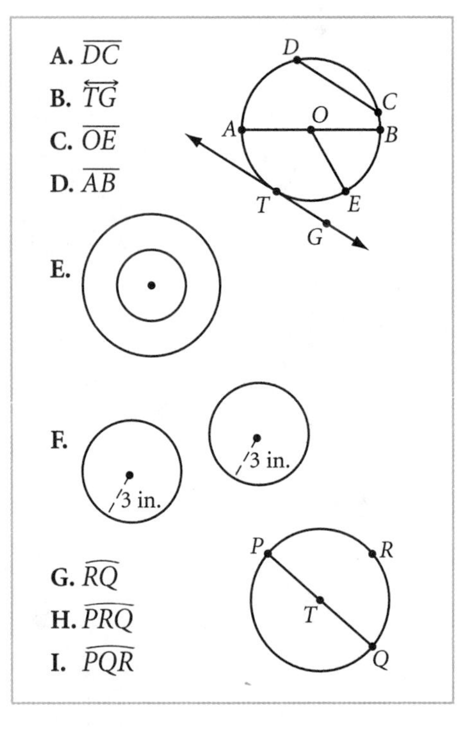

Check your answers: (1.F, 2.E, 3.C, 4.A, 5.D, 6.B, 7.G, 8.I, 9.H)

In addition to these terms, you will become familiar with two more, central angle and inscribed angle, in the next investigation.

Double Splash Evidence, part of modern California artist Gerrit Greve's *Water Series,* uses brushstrokes to produce an impression of concentric ripples in water.

Can you find parts of the water wheel that match the circle terms above?

Investigation 1
How Do We Define Angles in a Circle?

Look at the examples and non-examples for each term. Then write a definition for each. Discuss your definitions with others in your class. Agree on a common set of definitions and add them to your definition list. In your notebook, draw and label a picture to illustrate each definition.

Step 1 Define *central angle.*

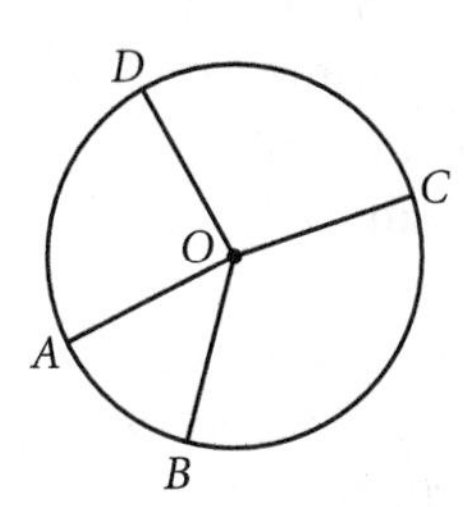

$\angle DOC$ intercepts arc $\overset{\frown}{DC}$. $\angle AOB$, $\angle BOC$, $\angle COD$, $\angle DOA$, and $\angle DOB$ are central angles of circle O.

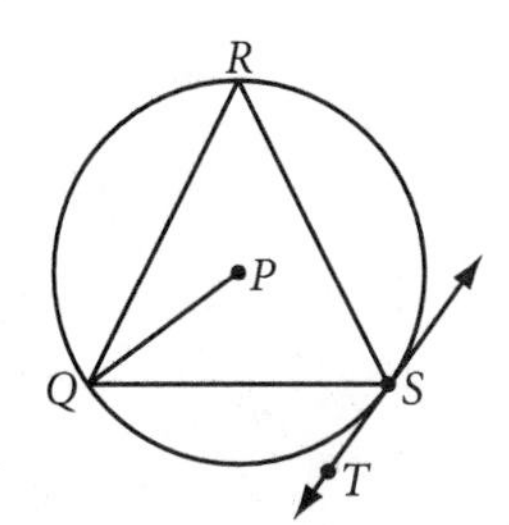

$\angle PQR$, $\angle PQS$, $\angle RST$, $\angle QST$, and $\angle QSR$ are not central angles of circle P.

Step 2 Define *inscribed angle.*

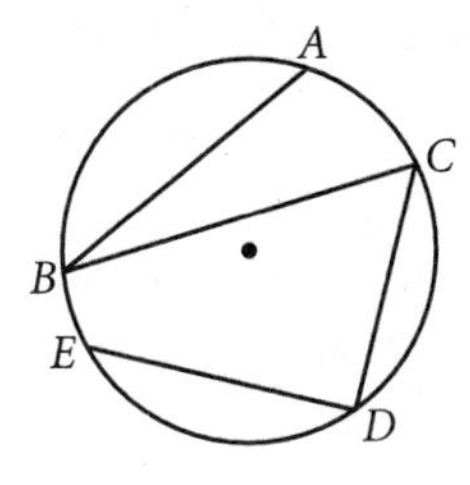

$\angle ABC$, $\angle BCD$, and $\angle CDE$ are inscribed angles. $\angle ABC$ is inscribed in $\overset{\frown}{ABC}$ and intercepts (or determines) $\overset{\frown}{AC}$.

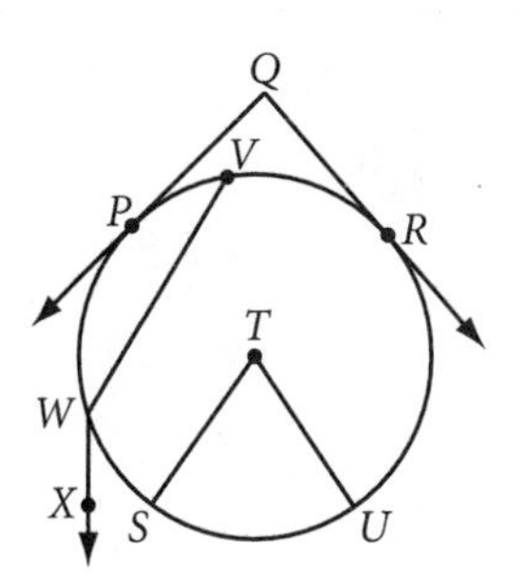

$\angle PQR$, $\angle STU$, and $\angle VWX$ are not inscribed angles.

Investigation 2
Chords and Their Central Angles

You will need

- a compass
- a straightedge
- a protractor

Next you will discover some properties of chords and central angles. You will also see a relationship between chords and arcs.

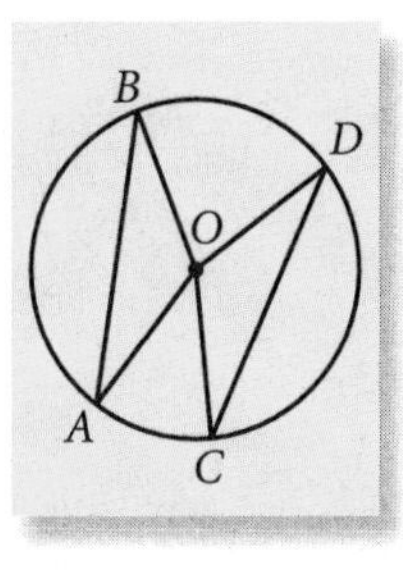

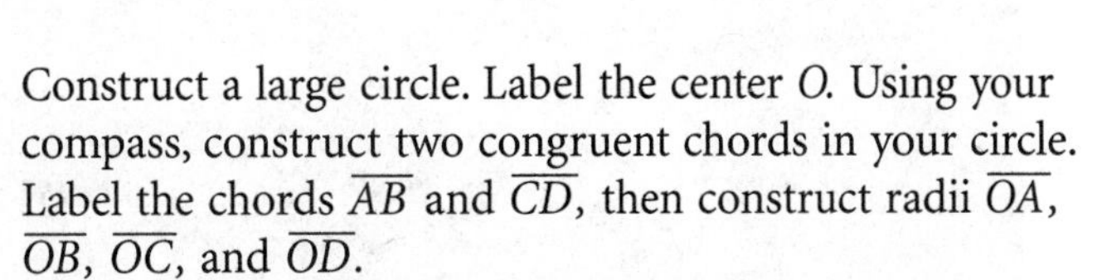

Step 1 Construct a large circle. Label the center O. Using your compass, construct two congruent chords in your circle. Label the chords $\overline{AB}$ and $\overline{CD}$, then construct radii $\overline{OA}$, $\overline{OB}$, $\overline{OC}$, and $\overline{OD}$.

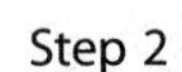

Step 2 With your protractor, measure $\angle BOA$ and $\angle COD$. How do they compare? Share your results with others in your group. Then copy and complete the conjecture.

Chord Central Angles Conjecture

C-54

If two chords in a circle are congruent, then they determine two central angles that are _?_.

Step 3 How can you fold your circle construction to check the conjecture?

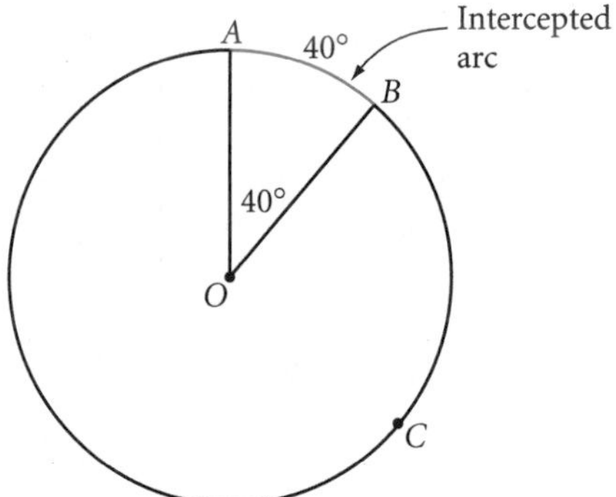

As you learned in Chapter 1, the measure of an arc is defined as the measure of its central angle. For example, the central angle, $\angle BOA$ at right, has a measure of 40°, so $m\overset{\frown}{AB} = 40°$. The measure of a major arc is 360° minus the measure of the minor arc making up the remainder of the circle. For example, the measure of major arc $\overset{\frown}{BCA}$ is $360° - 40°$, or 320°.

Your next conjecture follows from the Chord Central Angles Conjecture and the definition of arc measure.

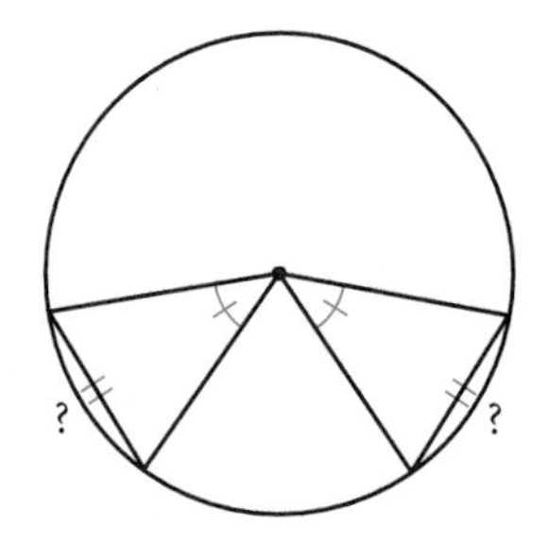

Step 4 Two congruent chords in a circle determine two central angles that are congruent. If two central angles are congruent, their intercepted arcs must be congruent. Combine these two statements to complete the conjecture.

Chord Arcs Conjecture

C-55

If two chords in a circle are congruent, then their _?_ are congruent.

"Pull the cord?! Don't I need to construct it first?"

Investigation 3
Chords and the Center of the Circle

You will need

- a compass
- a straightedge

In this investigation, you will discover relationships about a chord and the center of its circle.

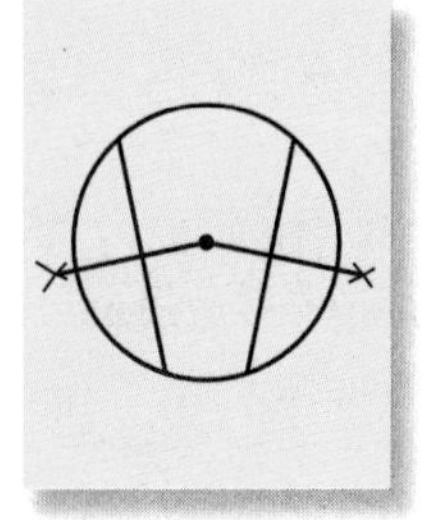

Step 1 On a sheet of paper, construct a large circle. Mark the center. Construct two nonparallel congruent chords. Then, construct the perpendiculars from the center to each chord.

Step 2 How does the perpendicular from the center of a circle to a chord divide the chord? Copy and complete the conjecture.

Perpendicular to a Chord Conjecture C-56

The perpendicular from the center of a circle to a chord is the _?_ of the chord.

Let's continue this investigation to discover a relationship between the length of congruent chords and their distances from the center of the circle.

Step 3 With your compass, compare the distances (measured along the perpendicular) from the center to the chords. Are the results the same if you change the size of the circle and the length of the chords? State your observations as your next conjecture.

Chord Distance to Center Conjecture C-57

Two congruent chords in a circle are _?_ from the center of the circle.

Investigation 4
Perpendicular Bisector of a Chord

You will need

- a compass
- a straightedge

Next you will discover a property of perpendicular bisectors of chords.

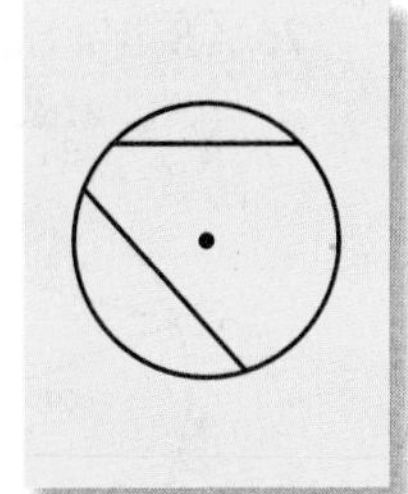

Step 1 On another sheet of paper, construct a large circle and mark the center. Construct two nonparallel chords that are not diameters. Then, construct the perpendicular bisector of each chord and extend the bisectors until they intersect.

Step 2 | What do you notice about the point of intersection? Compare your results with the results of others near you. Copy and complete the conjecture.

Perpendicular Bisector of a Chord Conjecture C-58

The perpendicular bisector of a chord _?_.

With the perpendicular bisector of a chord, you can find the center of any circle, and therefore the vertex of the central angle to any arc. All you have to do is construct the perpendicular bisectors of nonparallel chords.

EXERCISES

You will need

Construction tools for Exercises **13–15, 17,** and **21**

Solve Exercises 1–7. State which conjecture or definition you used to support your conclusion.

1. $x =$ _?_

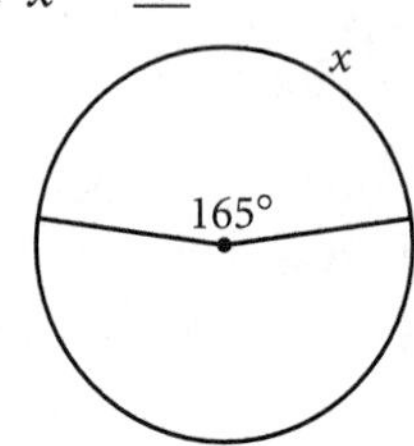

2. $z =$ _?_

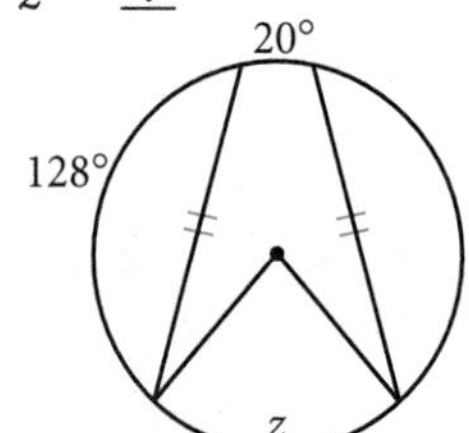

3. $w =$ _?_

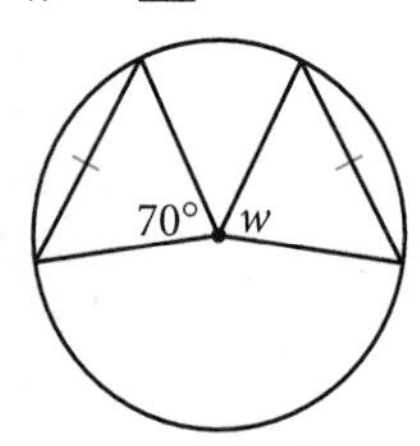

4. $y =$ _?_ ⓗ

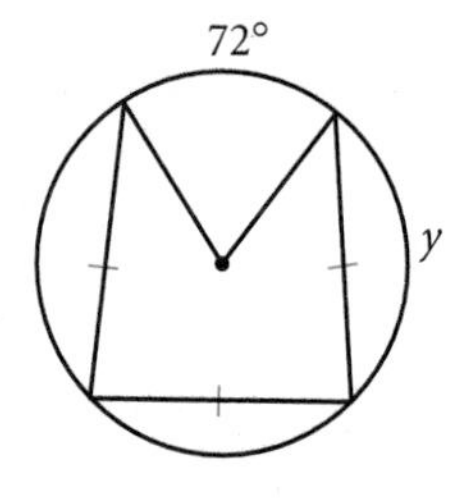

5. $AB = CD$
$PO = 8$ cm
$OQ =$ _?_

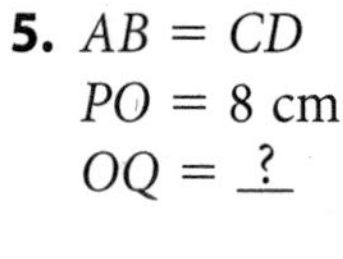

6. $AB = 6$ cm $OP = 4$ cm
$CD = 8$ cm $OQ = 3$ cm
$BD = 6$ cm
What is the perimeter of $OPBDQ$?

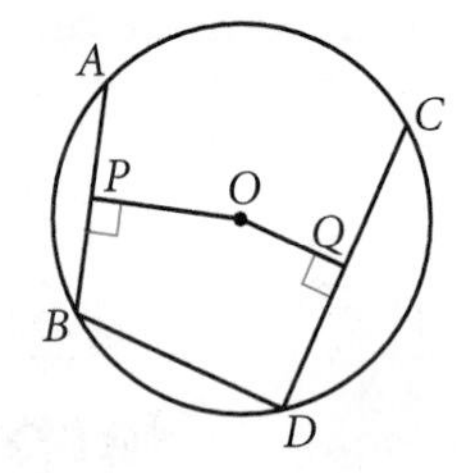

7. $\overline{AB}$ is a diameter. Find $m\widehat{AC}$ and $m\angle B$.

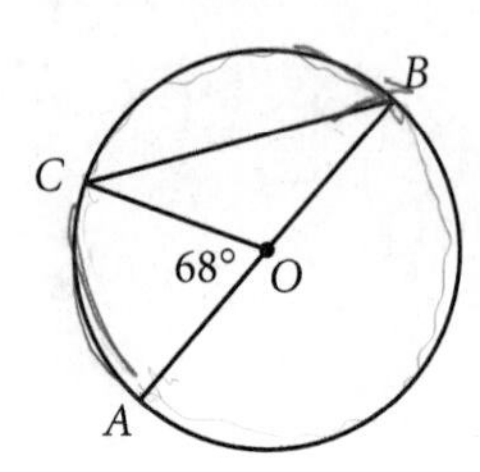

8. What's wrong with this picture?

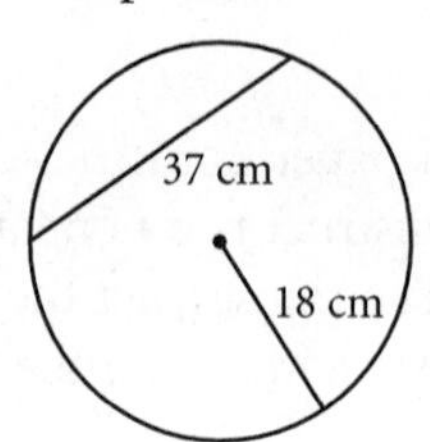

9. What's wrong with this picture?

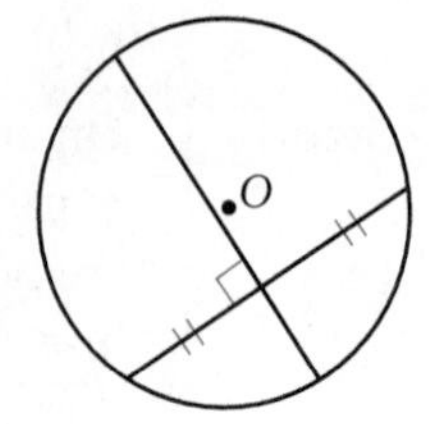

10. Draw a circle and two chords of unequal length. Which is closer to the center of the circle, the longer chord or the shorter chord? Explain.

11. Draw two circles with different radii. In each circle, draw a chord so that the chords have the same length. Draw the central angle determined by each chord. Which central angle is larger? Explain.

12. Polygon *MNOP* is a rectangle inscribed in a circle centered at the origin. Find the coordinates of points *M*, *N*, and *O*.

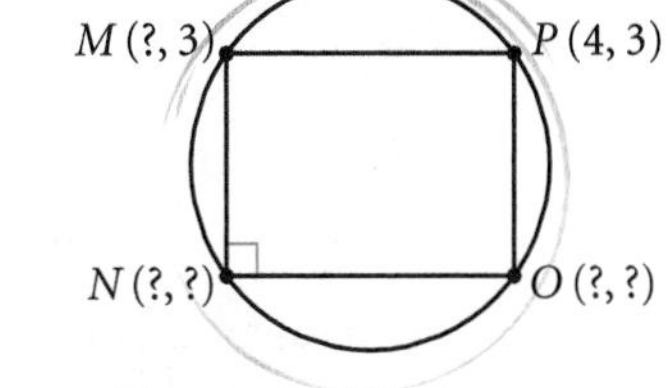

13. ***Construction*** Construct a triangle. Using the sides of the triangle as chords, construct a circle passing through all three vertices. Explain. Why does this seem familiar?

14. ***Construction*** Trace a circle onto a blank sheet of paper without using your compass. Locate the center of the circle using a compass and straightedge. Trace another circle onto patty paper and find the center by folding.

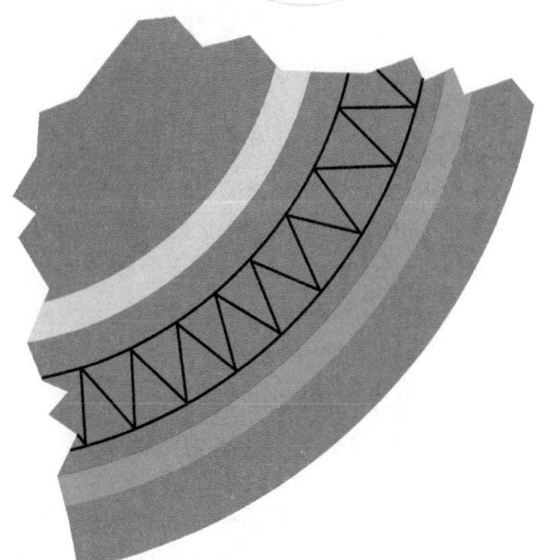

15. ***Construction*** Adventurer Dakota Davis digs up a piece of a circular ceramic plate. Suppose he believes that some ancient plates with this particular design have a diameter of 15 cm. He wants to calculate the diameter of the original plate to see if the piece he found is part of such a plate.

He has only this piece of the circular plate, shown at right, to make his calculations. Trace the outer edge of the plate onto a sheet of paper. Help him find the diameter.

16. Complete the flowchart proof shown, which proves that if two chords of a circle are congruent, then they determine two congruent central angles.

Given: Circle *O* with chords $\overline{AB} \cong \overline{CD}$

Show: $\angle AOB \cong \angle COD$

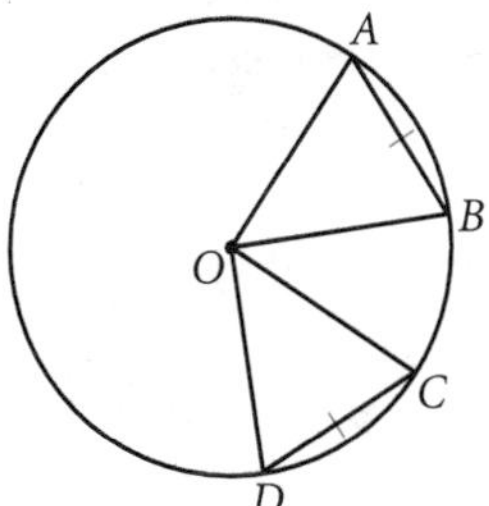

Flowchart Proof

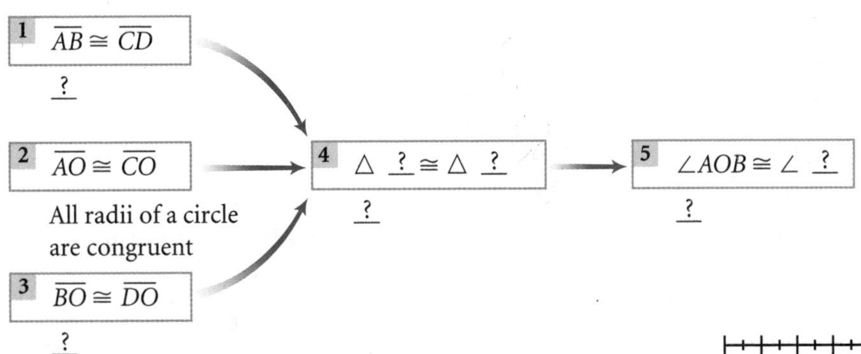

17. ***Construction*** The satellite photo at right shows only a portion of a lunar crater. How can cartographers use the photo to find its center? Trace the crater and locate its center. Using the scale shown, find its radius. To learn more about satellite photos, go to www.keymath.com/DG.

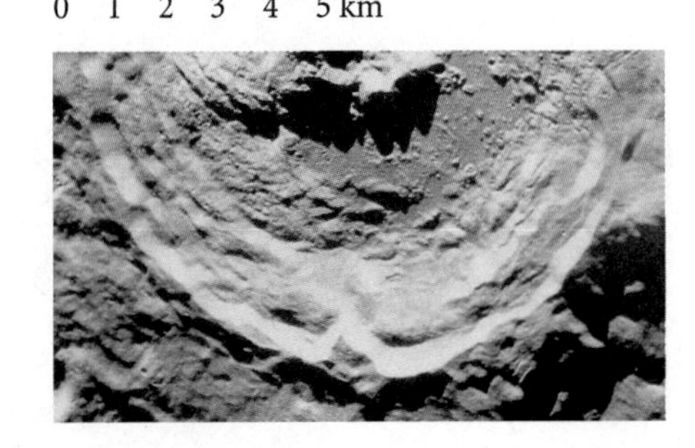

0 1 2 3 4 5 km

18. Circle O has center (0, 0) and passes through points $A(3, 4)$ and $B(4, -3)$. Find an equation to show that the perpendicular bisector of $\overline{AB}$ passes through the center of the circle. Explain your reasoning. ⓗ

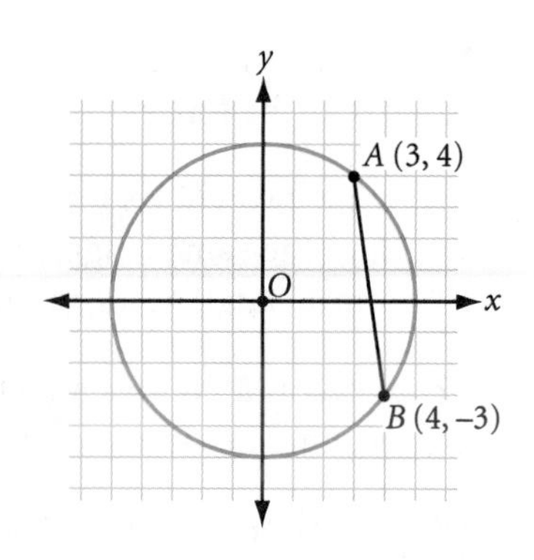

Review

19. Identify each quadrilateral from the given characteristics.

a. Diagonals are perpendicular and bisect each other.

b. Diagonals are congruent and bisect each other, but it is not a square.

c. Only one diagonal is the perpendicular bisector of the other diagonal.

d. Diagonals bisect each other.

20. A family hikes from their camp on a bearing of 15°. (A **bearing** is an angle measured clockwise from the north, so a bearing of 15° is 15° east of north.) They hike 6 km and then stop for a swim in a lake. Then they continue their hike on a new bearing of 117°. After another 9 km, they meet their friends. What is the measure of the angle between the path they took to arrive at the lake and the path they took to leave the lake? ⓗ

21. *Construction* Use a protractor and a centimeter ruler to make a careful drawing of the route the family in Exercise 20 traveled to meet their friends. Let 1 cm represent 1 km. To the nearest tenth of a kilometer, how far are they from their first camp?

22. Explain why x equals y.

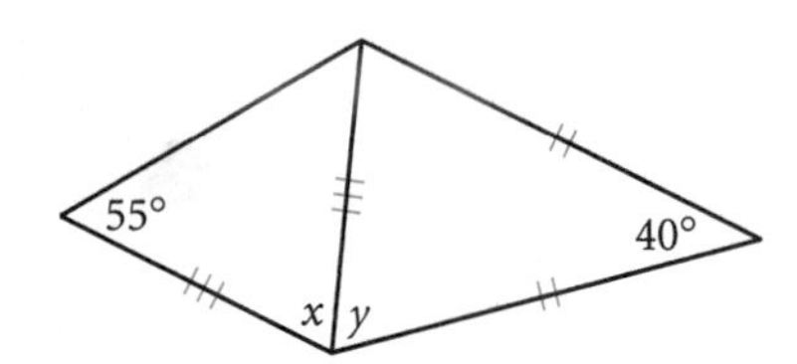

23. What is the probability of randomly selecting three points from the 3-by-3 grid below that form the vertices of a right triangle?

IMPROVING YOUR ALGEBRA SKILLS

Algebraic Sequences II

Find the next two terms of each algebraic pattern.

1. $x^6, 6x^5y, 15x^4y^2, 20x^3y^3, 15x^2y^4, \underline{?}, \underline{?}$

2. $x^7, 7x^6y, 21x^5y^2, 35x^4y^3, 35x^3y^4, 21x^2y^5, \underline{?}, \underline{?}$

3. $x^8, 8x^7y, 28x^6y^2, 56x^5y^3, 70x^4y^4, 56x^3y^5, 28x^2y^6, \underline{?}, \underline{?}$

LESSON

6.2

Tangent Properties

We are, all of us, alone
Though not uncommon
In our singularity.
Touching,
We become tangent to
Circles of common experience,
Co-incident,
Defining in collective tangency
Circles
Reciprocal in their subtle
Redefinition of us.
In tangency
We are never less alone,
But no longer
Only.

GENE MATTINGLY

Each wheel of a train theoretically touches only one point on the rail. Rails act as tangent lines to the wheels of a train. Each point where the rail and the wheel meet is a point of tangency. Why can't a train wheel touch more than one point at a time on the rail? Can a car wheel touch more than one point at a time on the road?

The rail is tangent to the wheels of the train.
The penguins' heads are tangent to each other.

Investigation 1
Going Off on a Tangent

You will need

- a compass
- a straightedge

In this investigation, you will discover the relationship between a tangent line and the radius drawn to the point of tangency.

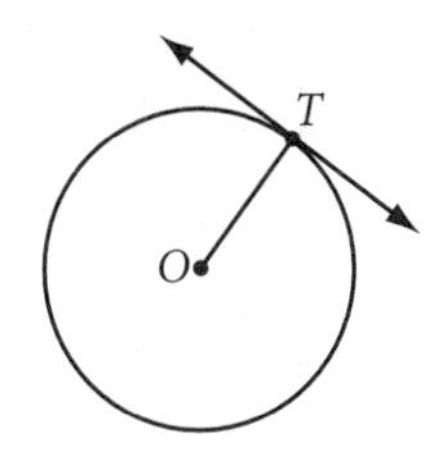

Step 1 Construct a large circle. Label the center O.

Step 2 Using your straightedge, draw a line that appears to touch the circle at only one point. Label the point T. Construct $\overline{OT}$.

Step 3 Use your protractor to measure the angles at T. What can you conclude about the radius $\overline{OT}$ and the tangent line at T?

Step 4 Share your results with your group. Then copy and complete the conjecture.

Tangent Conjecture C-59

A tangent to a circle $\underline{?}$ the radius drawn to the point of tangency.

Technology

CONNECTION

A series of rockets burning chemical fuels provide the thrust to launch satellites into orbit. Once the satellite reaches its proper orientation in space, it provides its own power for the duration of its mission, sometimes staying in space for five to ten years with the help of solar energy and a battery backup. At right is the Mir space station and the space shuttle *Atlantis* in orbit in 1995. According to the United States Space Command, there are over 8,000 objects larger than a softball circling Earth at speeds of over 18,000 miles per hour! If gravity were suddenly "turned off" somehow, these objects would travel off into space on a straight line tangent to their orbits, and not continue in a curved path.

The Tangent Conjecture has important applications related to circular motion. For example, a satellite maintains its velocity in a direction tangent to its circular orbit. This velocity vector is perpendicular to the force of gravity, which keeps the satellite in orbit.

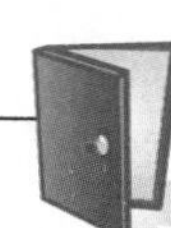

Investigation 2

Tangent Segments

You will need

- a compass
- a straightedge

In this investigation, you will discover something about the lengths of segments tangent to a circle from a point outside the circle.

Step 1 Construct a circle. Label the center E.

Step 2 Choose a point outside the circle and label it N.

Step 3 Draw two lines through point N tangent to the circle. Mark the points where these lines appear to touch the circle and label them A and G.

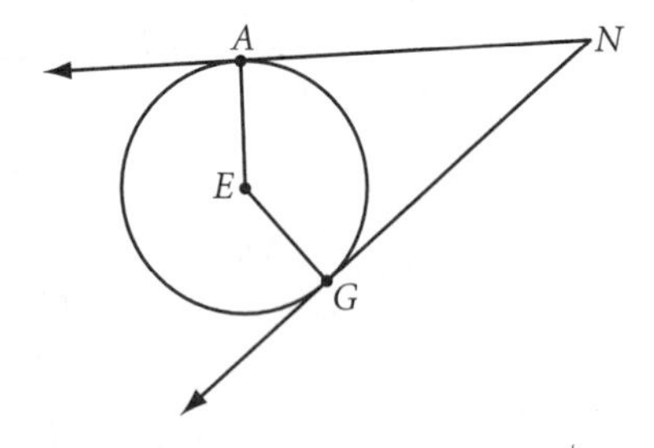

Step 4 Use your compass to compare segments NA and NG. These segments are called **tangent segments.**

Step 5 Share your results with your group. Copy and complete the conjecture.

Tangent Segments Conjecture C-60

Tangent segments to a circle from a point outside the circle are $\underline{\ ?\ }$.

Tangent circles are two circles that are tangent to the same line at the same point. They can be **internally tangent** or **externally tangent,** as shown.

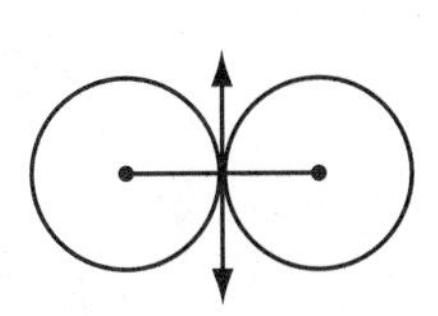

Externally tangent circles

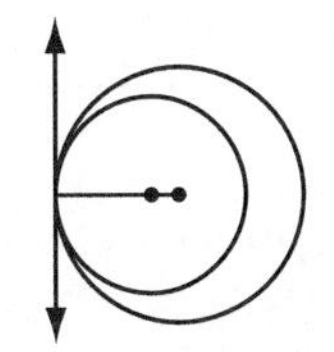

Internally tangent circles

EXERCISES

You will need

Construction tools for Exercises **8–12** and **15**

Geometry software for Exercises **13** and **26**

1. Rays m and n are tangents to circle P. $w = \underline{\ ?\ }$ ⓗ

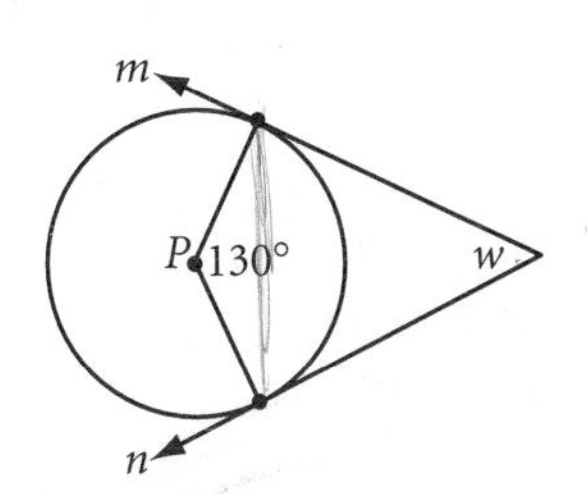

2. Rays r and s are tangent to circle Q. $x = \underline{\ ?\ }$ ⓗ

r
70°
x
Q
s

3. Ray k is tangent to circle R. $y = \underline{\ ?\ }$

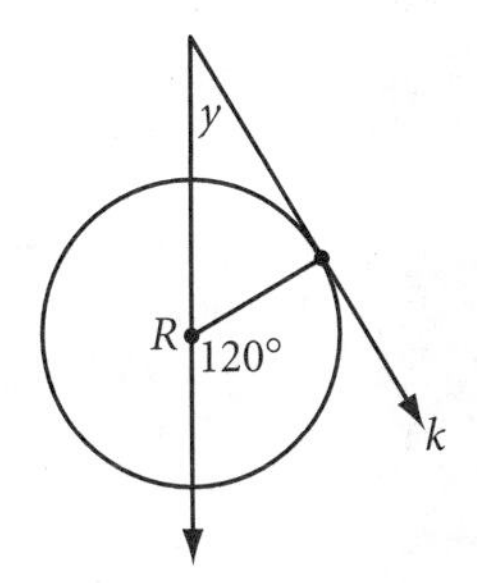

4. Line t is tangent to both circles. $z = \underline{\ ?\ }$

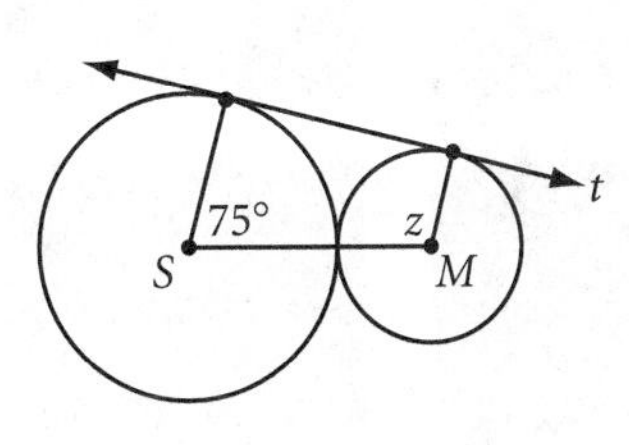

5. Quadrilateral $POST$ is circumscribed about circle Y. $OR = 13$ and $ST = 12$. What is the perimeter of $POST$? ⓗ

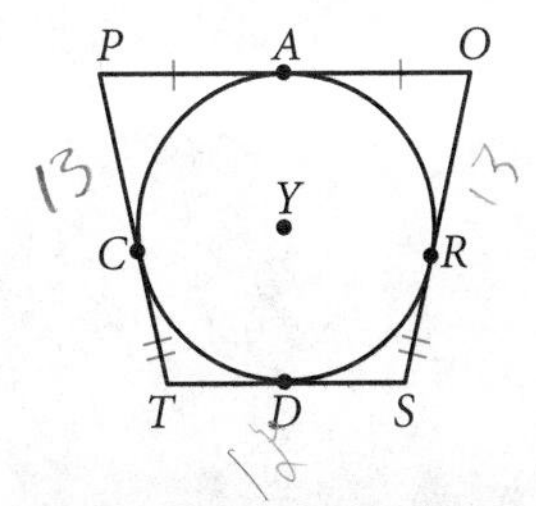

6. Pam participates in the hammer-throw event. She swings a 16 lb ball at arm's length, about eye-level. Then she releases the ball at the precise moment when the ball will travel in a straight line toward the target area. Draw an overhead view that shows the ball's circular path, her arms at the moment she releases it, and the ball's straight path toward the target area.

Pam Dukes competes in the hammer-throw event.

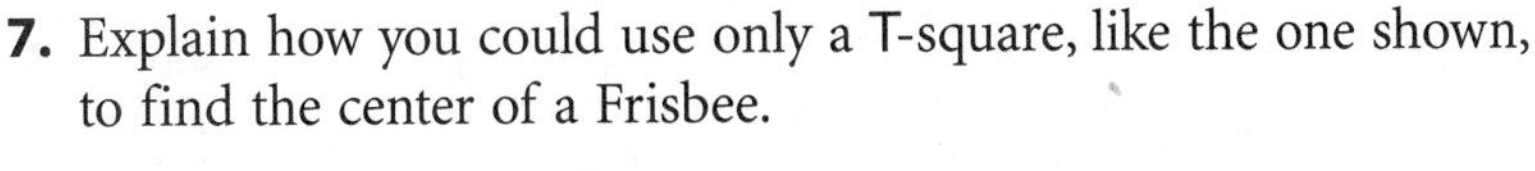

7. Explain how you could use only a T-square, like the one shown, to find the center of a Frisbee.

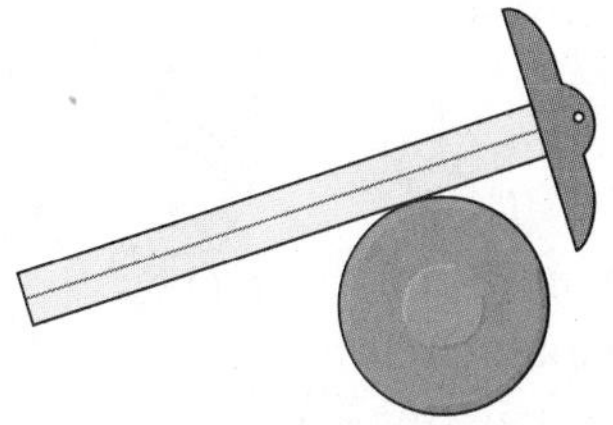

Construction For Exercises 8–12, first make a sketch of what you are trying to construct and label it. Then use the segments below, with lengths r, s, and t.

r s t

8. Construct a circle with radius r. Mark a point on the circle. Construct a tangent through this point. (h)

9. Construct a circle with radius t. Choose three points on the circle that divide it into three minor arcs and label points X, Y, and Z. Construct a triangle that is circumscribed about the circle and tangent at points X, Y, and Z.

10. Construct two congruent, externally tangent circles with radius s. Then construct a third circle that is both congruent and externally tangent to the two circles.

11. Construct two internally tangent circles with radii r and t.

12. Construct a third circle with radius s that is externally tangent to both the circles you constructed in Exercise 11.

13. *Technology* Use geometry software to construct a circle. Label three points on the circle and construct tangents through them. Drag the three points and write your observations about where the tangent lines intersect and the figures they form.

14. Find real-world examples (different from the examples shown below) of two internally tangent circles and of two externally tangent circles. Either sketch the examples or make photocopies from a book or a magazine for your notebook.

The teeth in the gears shown extend from circles that are externally tangent.

This astronomical clock in Prague, Czech Republic, has one pair of internally tangent circles. What other circle relationships can you find in the clock photo?

15. *Construction* In Taoist philosophy, all things are governed by one of two natural principles, yin and yang. Yin represents the earth, characterized by darkness, cold, or wetness. Yang represents the heavens, characterized by light, heat, or dryness. The two principles, when balanced, combine to produce the harmony of nature. The symbol for the balance of yin and yang is shown at right. Construct the yin-and-yang symbol. Start with one large circle. Then construct two circles with half the diameter that are internally tangent to the large circle and externally tangent to each other. Finally, construct small circles that are concentric to the two inside circles. Shade or color your construction. (h)

16. A satellite in geostationary orbit remains above the same point on the earth's surface even as the earth turns. If such a satellite has a 30° view of the equator of the earth, what percentage of the equator is observable from the satellite? ⓗ

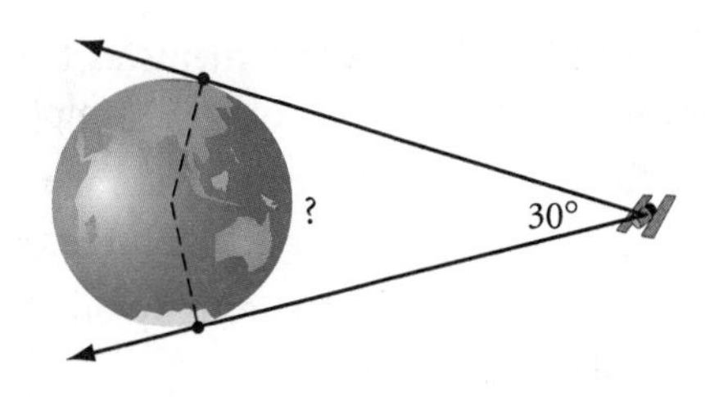

17. Circle P is centered at the origin. $\overleftrightarrow{AT}$ is tangent to circle P at $A(8, 15)$. Find the equation of $\overleftrightarrow{AT}$.

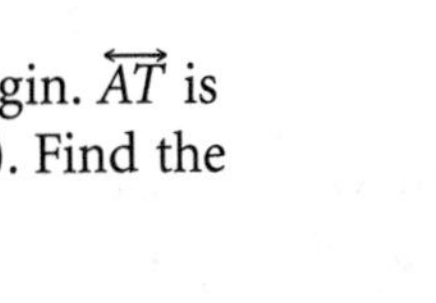

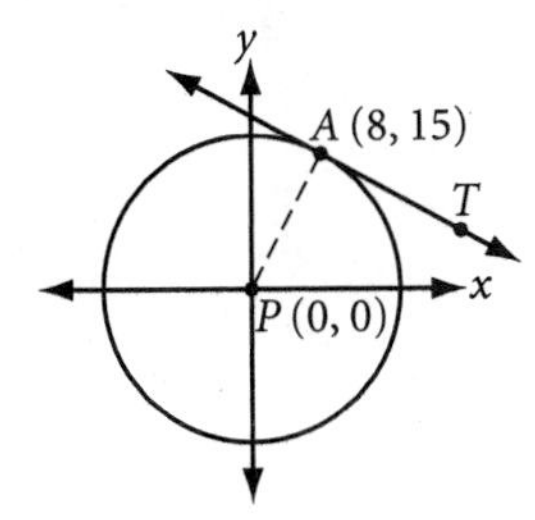

18. $\overrightarrow{PA}$ is tangent to circle Q. The line containing chord $\overline{CB}$ passes through P. Find $m\angle P$.

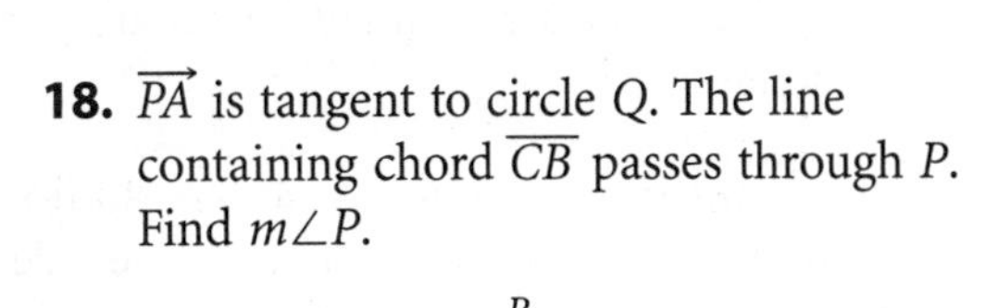

19. $\overrightarrow{TA}$ and $\overrightarrow{TB}$ are tangent to circle O. What's wrong with this picture?

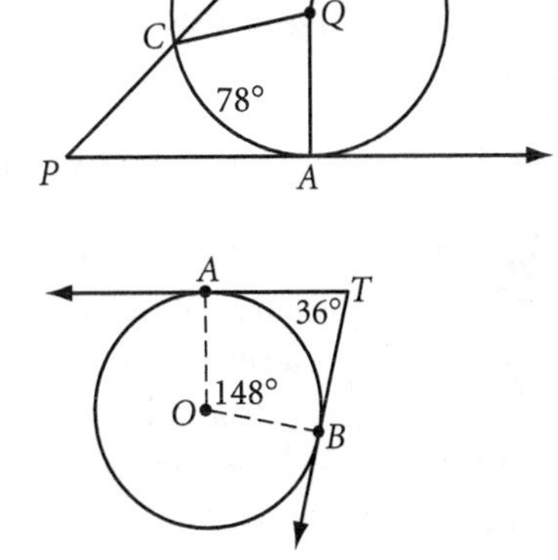

Review

20. Circle U passes through points $(3, 11)$, $(11, -1)$, and $(-14, 4)$. Find the coordinates of its center. Explain your method.

21. Complete the flowchart proof or write a paragraph proof of the Perpendicular to a Chord Conjecture: The perpendicular from the center of a circle to a chord is the bisector of the chord.

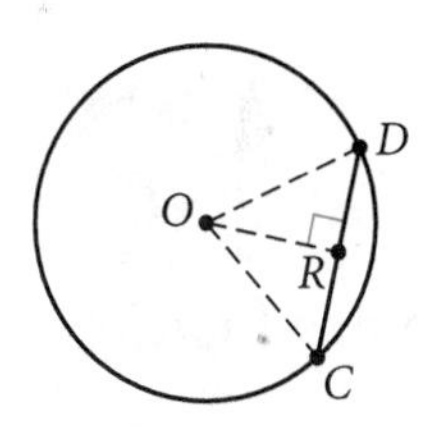

Given: Circle O with chord $\overline{CD}$, radii $\overline{OC}$ and $\overline{OD}$, and $\overline{OR} \perp \overline{CD}$

Show: $\overline{OR}$ bisects $\overline{CD}$

Flowchart Proof

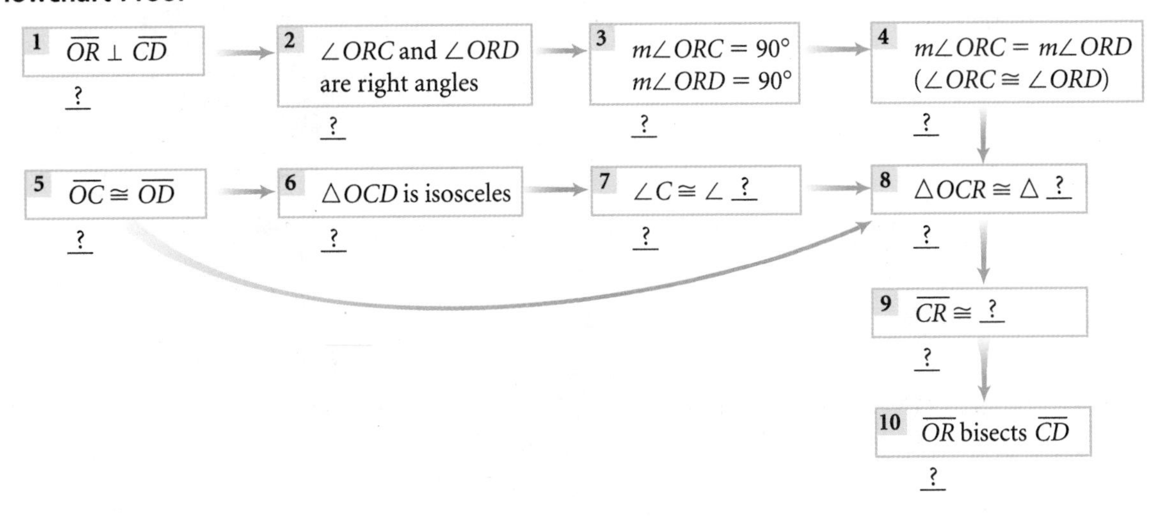

22. Identify each of these statements as true or false. If the statement is true, explain why. If it is false, give a counterexample.

a. If the diagonals of a quadrilateral are congruent, but only one is the perpendicular bisector of the other, then the quadrilateral is a kite.

b. If the quadrilateral has exactly one line of reflectional symmetry, then the quadrilateral is a kite.

c. If the diagonals of a quadrilateral are congruent and bisect each other, then it is a square.

23. Rachel and Yulia are building an art studio above their back bedroom. There will be doors on three sides leading to a small deck that surrounds the studio. They need to place an electrical junction box in the ceiling of the studio so that it is equidistant from the three light switches shown by the doors. Copy the diagram of the room and find the most efficient location for the junction box. (h)

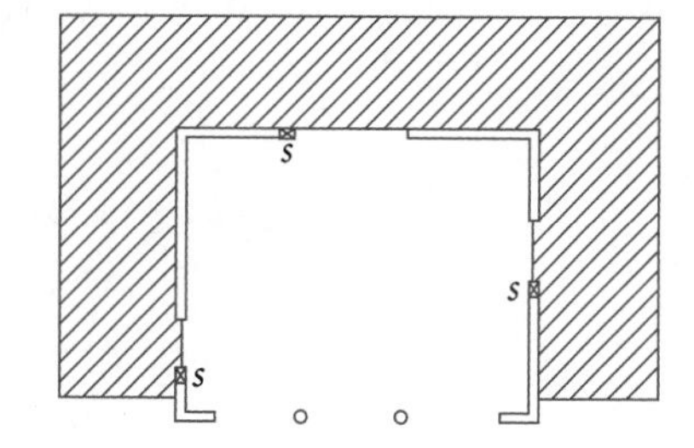

24. What will the units digit be when you evaluate 3^{23}? (h)

25. A small light-wing aircraft has made an emergency landing in a remote portion of a wildlife refuge and is sending out radio signals for help. Ranger Station Alpha receives the signal on a bearing of 38° and Station Beta receives the signal on a bearing of 312°. (Recall that a bearing is an angle measured clockwise from the north.) Stations Alpha and Beta are 8.2 miles apart, and Station Beta is on a bearing of 72° from Station Alpha. Which station is closer to the downed aircraft? Explain your reasoning. (h)

26. *Technology* Use geometry software to pick any three points. Construct an arc through all three points. (Can it be done?) How do you find the center of the circle that passes through all three points?

IMPROVING YOUR VISUAL THINKING SKILLS

Colored Cubes

Sketch the solid shown, but with the red cubes removed and the blue cube moved to cover the starred face of the green cube.

LESSON

6.3

Arcs and Angles

You will do foolish things, but do them with enthusiasm.

SIDONIE GABRIELLA COLETTE

Many arches that you see in structures are semicircular, but Chinese builders long ago discovered that arches don't have to have this shape. The Zhaozhou bridge, shown below, was completed in 605 C.E. It is the world's first stone arched bridge in the shape of a minor arc, predating other minor-arc arches by about 800 years.

In this lesson you'll discover properties of arcs and the angles associated with them.

Investigation 1
Inscribed Angle Properties

You will need

- a compass
- a straightedge
- a protractor

In this investigation, you will compare an inscribed angle and a central angle, both inscribed in the same arc. Refer to the diagram of circle *O*, with central angle *COR* and inscribed angle *CAR*.

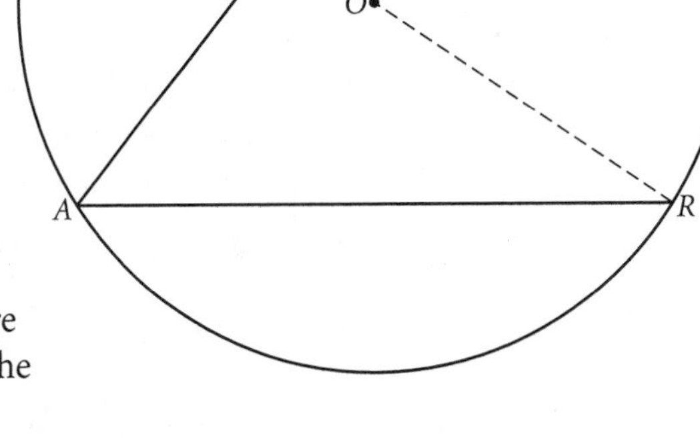

Step 1 Measure $\angle COR$ with your protractor to find $m\overset{\frown}{CR}$, the intercepted arc. Measure $\angle CAR$. How does $m\angle CAR$ compare with $m\overset{\frown}{CR}$?

Step 2 Construct a circle of your own with an inscribed angle. Draw the central angle that intercepts the same arc. What is the measure of the inscribed angle? How do the two measures compare?

Step 3 Share your results with others near you. Copy and complete the conjecture.

Inscribed Angle Conjecture C-61

The measure of an angle inscribed in a circle __?__.

Investigation 2
Inscribed Angles Intercepting the Same Arc

You will need

- a compass
- a straightedge
- a protractor

Next, let's consider two inscribed angles that intercept the same arc. In the figure at right, $\angle AQB$ and $\angle APB$ both intercept $\overset{\frown}{AB}$. Angles AQB and APB are both inscribed in $\overset{\frown}{APB}$.

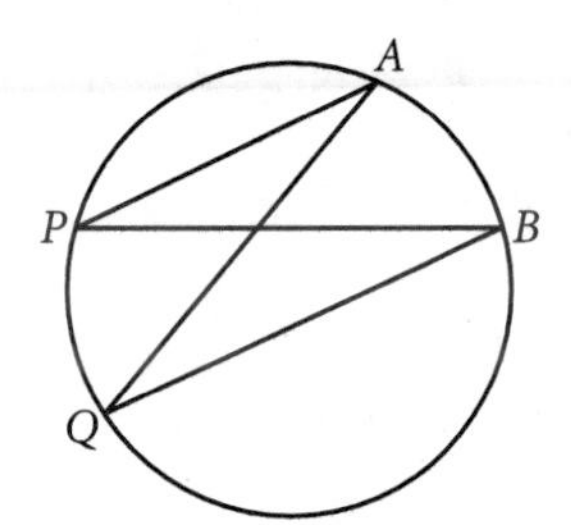

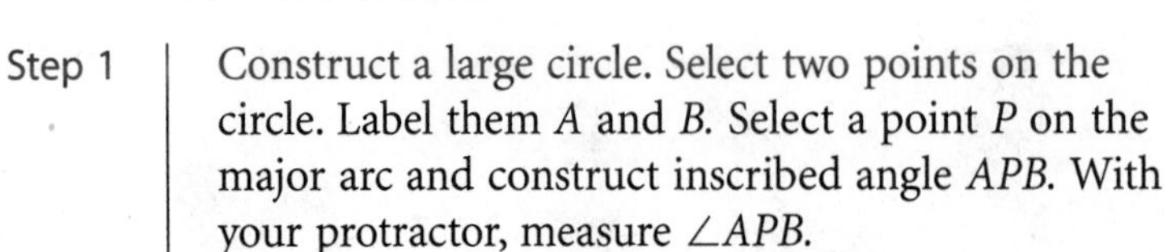

Step 1 Construct a large circle. Select two points on the circle. Label them A and B. Select a point P on the major arc and construct inscribed angle APB. With your protractor, measure $\angle APB$.

Step 2 Select another point Q on $\overset{\frown}{APB}$ and construct inscribed angle AQB. Measure $\angle AQB$.

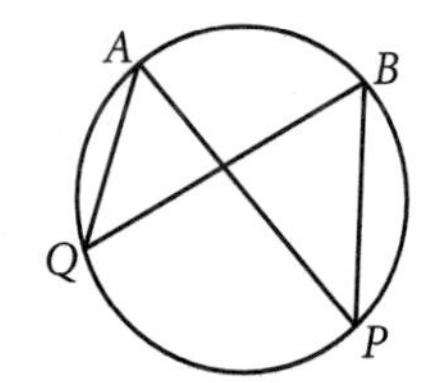

Step 3 How does $m\angle AQB$ compare with $m\angle APB$?

Step 4 Repeat Steps 1 and 2 with points P and Q selected on minor arc AB. Compare results with your group. Then copy and complete the conjecture.

Inscribed Angles Intercepting Arcs Conjecture C-62

Inscribed angles that intercept the same arc __?__.

Investigation 3
Angles Inscribed in a Semicircle

You will need

- a compass
- a straightedge
- a protractor

Next, you will investigate a property of angles inscribed in semicircles. This will lead you to a third important conjecture about inscribed angles.

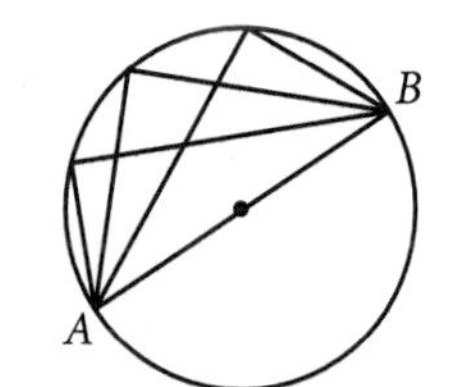

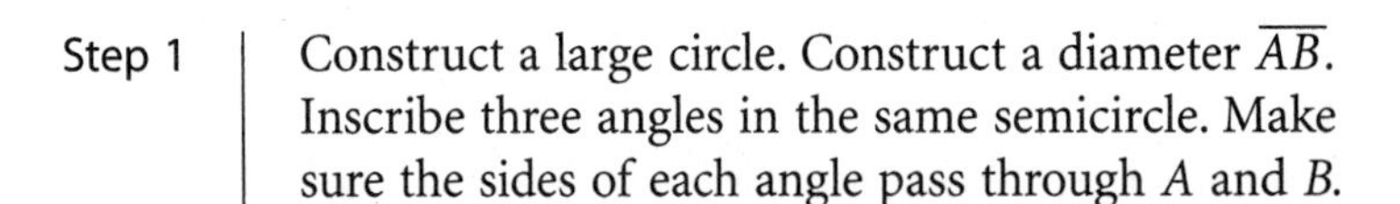

Step 1 Construct a large circle. Construct a diameter $\overline{AB}$. Inscribe three angles in the same semicircle. Make sure the sides of each angle pass through A and B.

Step 2 Measure each angle with your protractor. What do you notice? Compare your results with the results of others and make a conjecture.

Angles Inscribed in a Semicircle Conjecture C-63

Angles inscribed in a semicircle __?__.

Now you will discover a property of the angles of a quadrilateral inscribed in a circle.

Investigation 4
Cyclic Quadrilaterals

You will need

- a compass
- a straightedge
- a protractor

A quadrilateral inscribed in a circle is called a **cyclic** quadrilateral. Each of its angles is inscribed in the circle, and each of its sides is a chord of the circle.

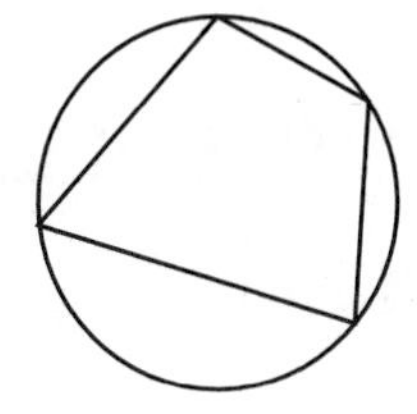

Step 1 Construct a large circle. Construct a cyclic quadrilateral by connecting four points anywhere on the circle.

Step 2 Measure each of the four inscribed angles. Write the measure in each angle. Look carefully at the sums of various angles. Share your observations with students near you. Then copy and complete the conjecture.

Cyclic Quadrilateral Conjecture C-64

The _?_ angles of a cyclic quadrilateral are _?_.

Investigation 5
Arcs by Parallel Lines

You will need

- patty paper
- a compass
- a double-edged straightedge

Next, you will investigate arcs formed by parallel lines that intersect a circle.

A line that intersects a circle in two points is called a **secant.** A secant contains a chord of the circle, and passes through the interior of a circle, while a tangent line does not.

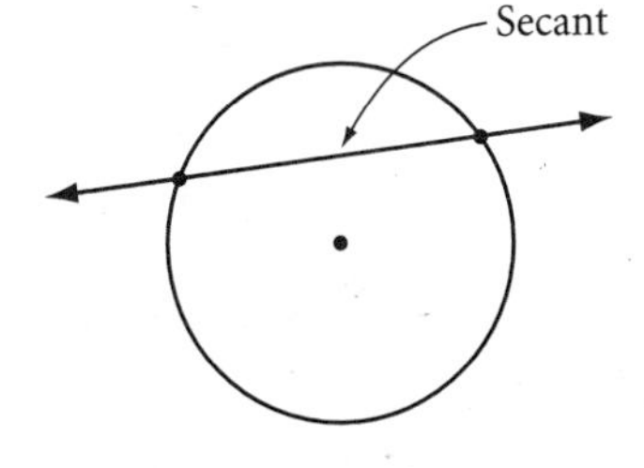

Step 1 On a piece of patty paper, construct a large circle. Lay your straightedge across the circle so that its parallel edges pass through the circle. Draw secants $\overleftrightarrow{AB}$ and $\overleftrightarrow{DC}$ along both edges of the straightedge.

Step 2 Fold your patty paper to compare $\overset{\frown}{AD}$ and $\overset{\frown}{BC}$. What can you say about $\overset{\frown}{AD}$ and $\overset{\frown}{BC}$?

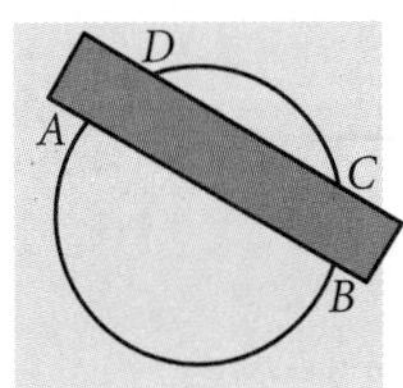

Step 3 Repeat Steps 1 and 2, using either lined paper or another object with parallel edges to construct different parallel secants. Share your results with other students. Then copy and complete the conjecture.

Parallel Lines Intercepted Arcs Conjecture C-65

Parallel lines intercept _?_ arcs on a circle.

Review these conjectures and ask yourself which quadrilaterals can be inscribed in a circle. Can any parallelogram be a cyclic quadrilateral? If two sides of a cyclic quadrilateral are parallel, then what kind of quadrilateral will it be?

EXERCISES

You will need

Geometry software for Exercises **19** and **21**

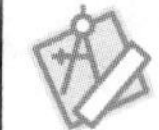

Construction tools for Exercise **24**

Use your new conjectures to solve Exercises 1–17.

1. $a = \underline{?}$

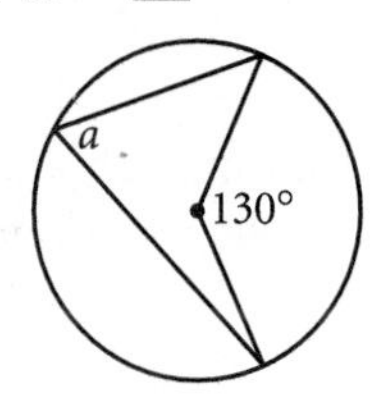

2. $b = \underline{?}$

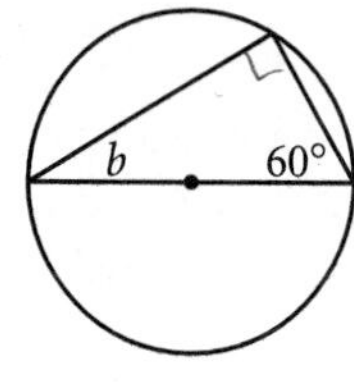

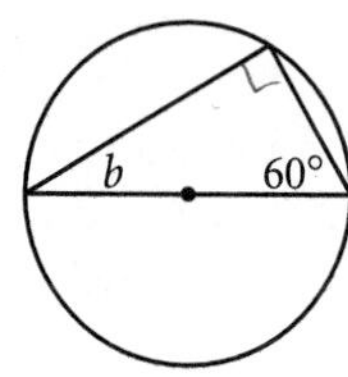

3. $c = \underline{?}$ ⓗ

95°
c
120°

4. $h = \underline{?}$ ⓗ

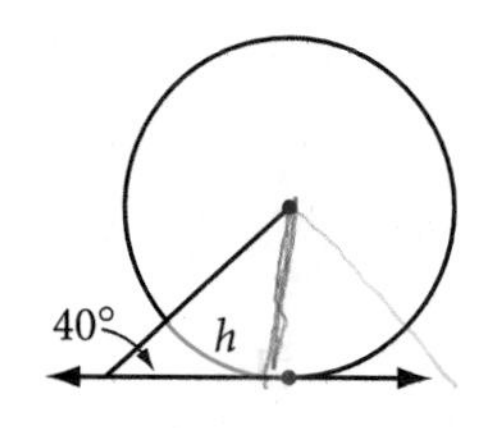

5. $d = \underline{?}$
$e = \underline{?}$

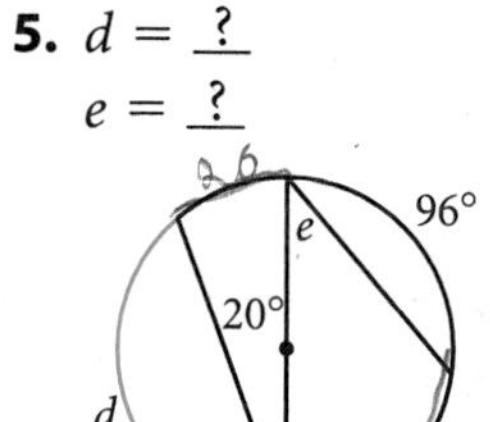

6. $f = \underline{?}$
$g = \underline{?}$

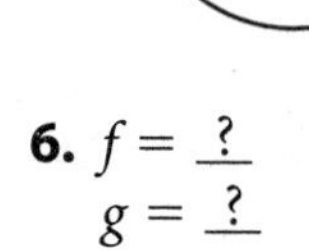

g
75°
f
110°

7. *JUST* is a rhombus.
$w = \underline{?}$

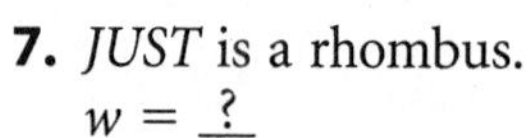
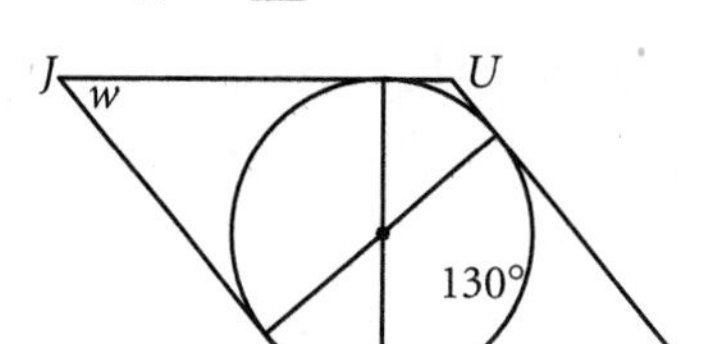

8. *CALM* is a rectangle.
$x = \underline{?}$

M
x
L
C
32°
A

9. *DOWN* is a kite.
$y = \underline{?}$

D
O
y
W
136°
N

10. $k = \underline{?}$

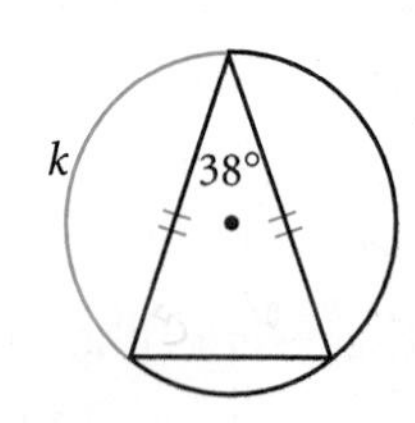

11. $r = \underline{?}$
$s = \underline{?}$

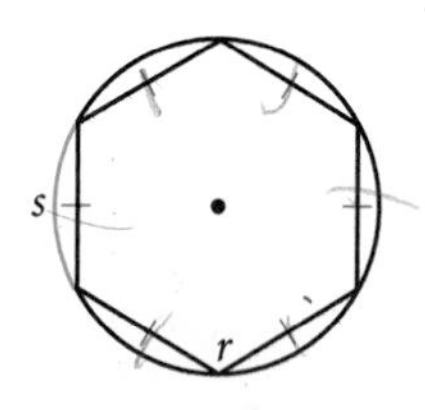

12. $m = \underline{?}$
$n = \underline{?}$

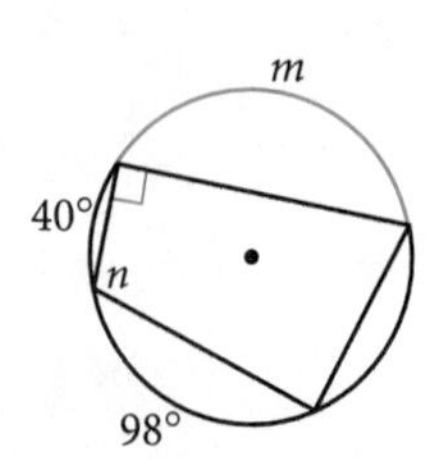

13. $\overline{AB} \parallel \overline{CD}$
$p = \underline{?}$
$q = \underline{?}$

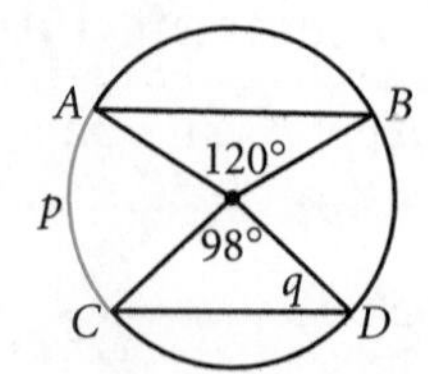

14. What is the sum of *a*, *b*, *c*, *d*, and *e*? ⓗ

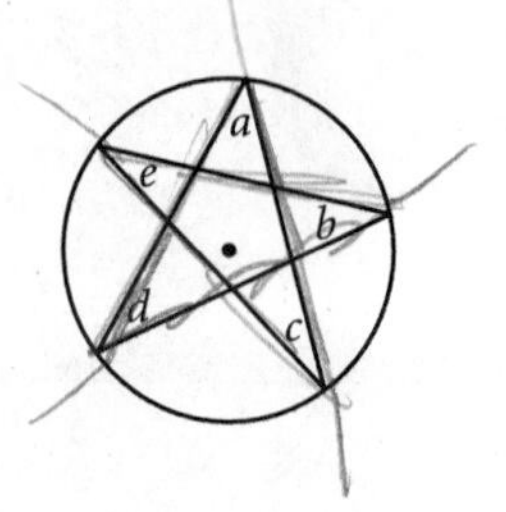

15. $y = \underline{\ ?\ }$ ⓗ

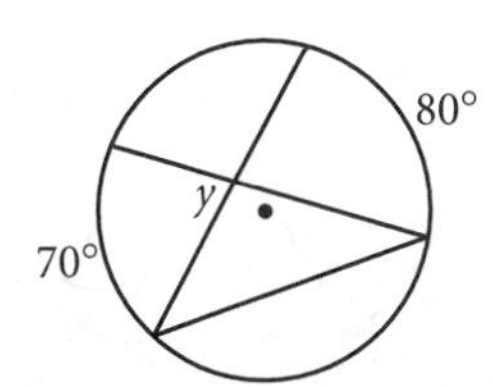

16. What's wrong with this picture?

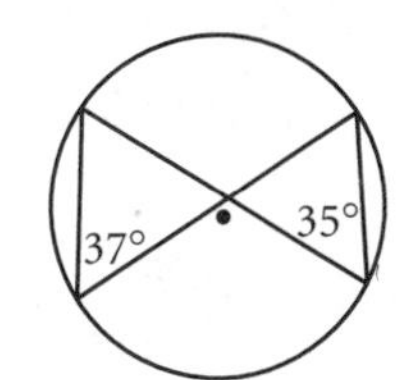

17. Explain why $\overset{\frown}{AC} \cong \overset{\frown}{CE}$.

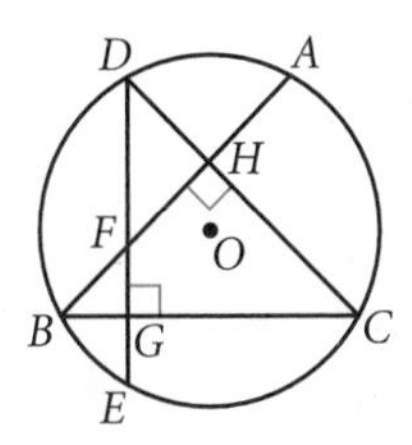

18. How can you find the center of a circle, using only the corner of a piece of paper?

19. *Technology* Chris Chisholm, a high school student in Whitmore, California, used the Angles Inscribed in a Semicircle Conjecture to discover a simpler way to find the orthocenter in a triangle. Chris constructs a circle using one of the sides of the triangle as the diameter, then he immediately finds an altitude to each of the triangle's other two sides. Use geometry software and Chris's method to find the orthocenter of a triangle. Does this method work on all kinds of triangles? ⓗ

20. APPLICATION The width of a view that can be captured in a photo depends on the camera's *picture angle.* Suppose a photographer takes a photo of your class standing in one straight row with a camera that has a 46° picture angle. Draw a line segment to represent the row. Draw a 46° angle on a piece of patty paper. Locate at least eight different points on your paper where a camera could be positioned to include all the students, filling as much of the picture as possible. What is the locus of all such camera positions? What conjecture does this activity illustrate? ⓗ

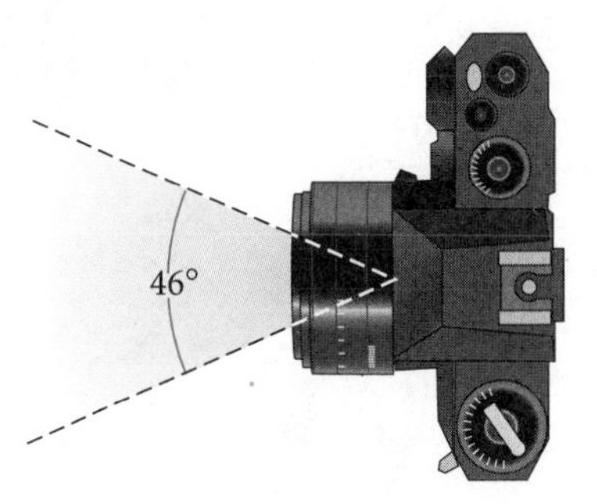

21. *Technology* Construct a circle and a diameter. Construct a point on one of the semicircles, and construct two chords from it to the endpoints of the diameter to create a right triangle. Locate the midpoint of each of the two chords. Predict, then sketch the locus of the two midpoints as the vertex of the right angle is moved around the circle. Finally, use your computer to animate the point on the circle and trace the locus of the two midpoints. What do you get?

Review

22. Find the measure of each lettered angle.

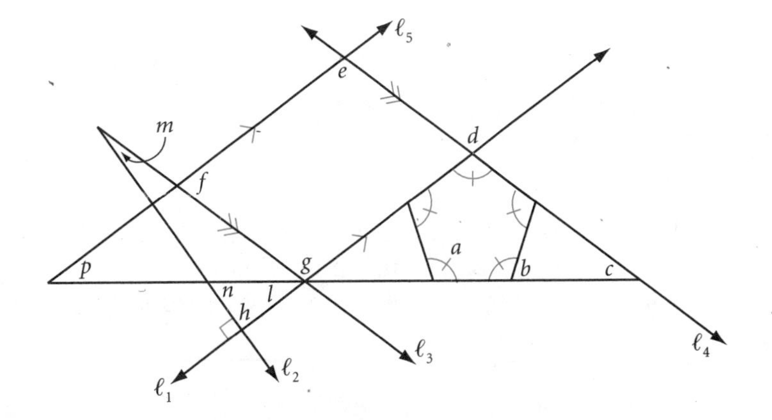

23. Use the diagram at right and the flowchart below to write a paragraph proof explaining why two congruent chords in a circle are equidistant from the center of the circle. ⓗ

Given: Circle O with $\overline{PQ} \cong \overline{RS}$ and $\overline{OT} \perp \overline{PQ}$ and $\overline{OV} \perp \overline{RS}$

Show: $\overline{OT} \cong \overline{OV}$

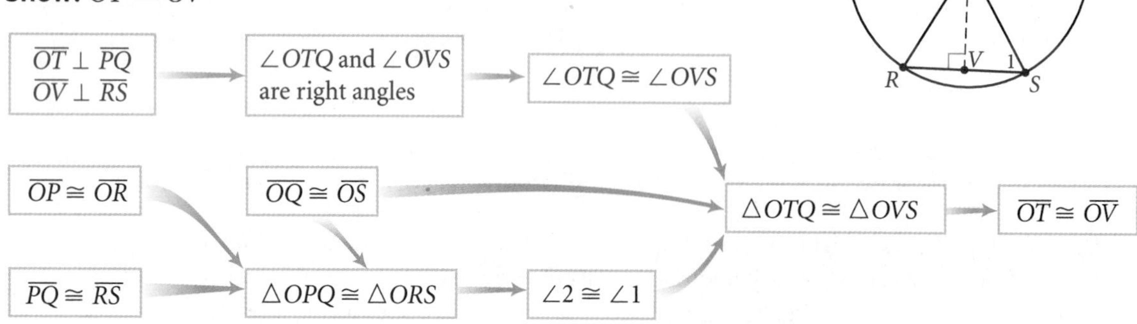

24. *Construction* Use your construction tools to re-create this design of three congruent circles all tangent to each other and internally tangent to a larger circle. ⓗ

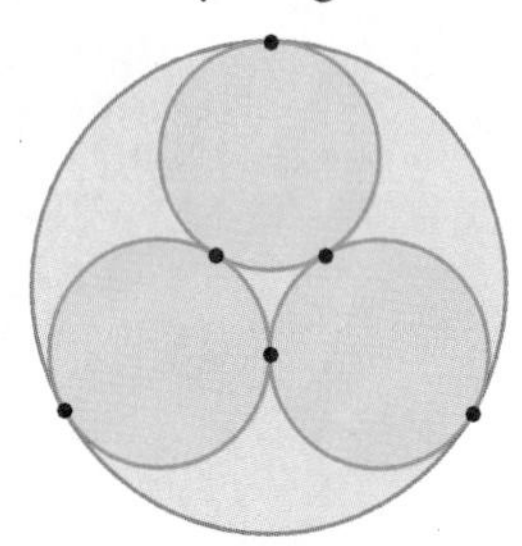

25. What's wrong with this picture?

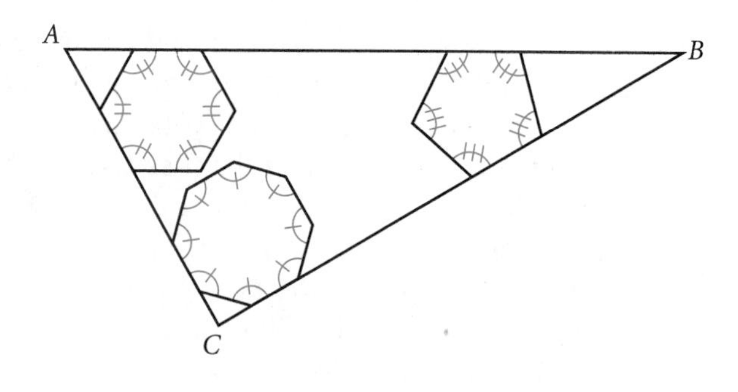

26. Consider the figure at right with line $\ell \parallel \overline{AB}$. As P moves from left to right along line ℓ, which of these lengths or distances always increases?

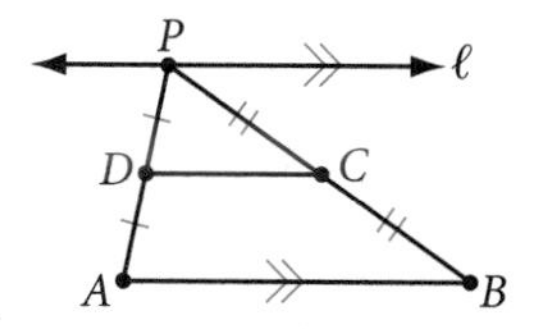

A. The distance PB

B. The distance from D to $\overleftrightarrow{AB}$

C. DC

D. Perimeter of $\triangle ABP$

E. None of the above

IMPROVING YOUR REASONING SKILLS

Think Dinosaur

If the letter in the word *dinosaur* that is three letters after the word's second vowel is also found before the sixteenth letter of the alphabet, then print the word *dinosaur* horizontally. Otherwise, print the word *dinosaur* vertically and cross out the second letter after the first vowel.

LESSON

6.4

Proving Circle Conjectures

Mistakes are a fact of life. It's the response to the error that counts.

NIKKI GIOVANNI

In Lesson 6.3, you discovered the Inscribed Angle Conjecture: The measure of an angle inscribed in a circle equals half the measure of its intercepted arc. Many other circle conjectures are logical consequences of the Inscribed Angle Conjecture. Let's start this lesson by proving it.

When you inscribe an angle in a circle, the angle will relate to the circle's center in one of the three ways described below. These three possible relationships to the center are the three cases for which you prove the Inscribed Angle Conjecture. To prove the conjecture, you must prove all three cases.

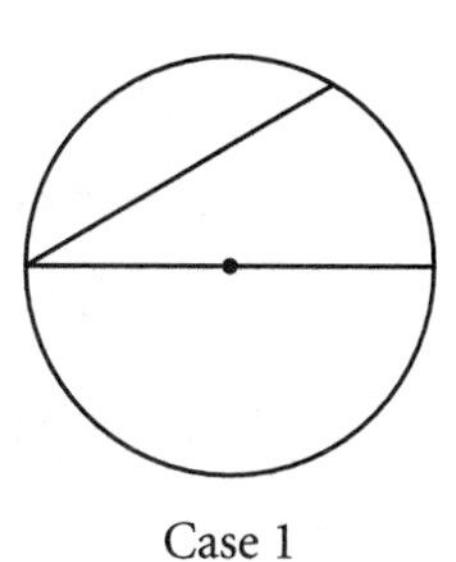

Case 1

The circle's center is on the angle.

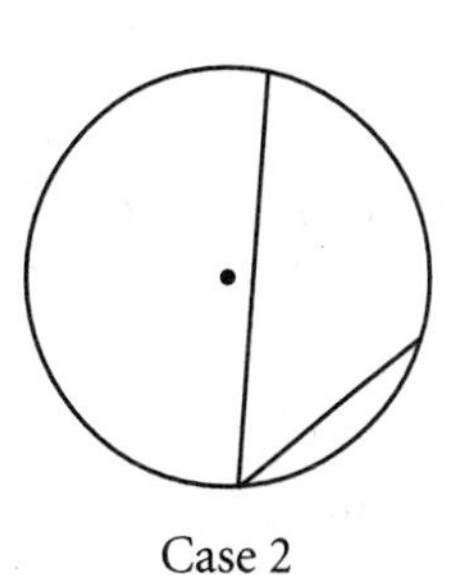

Case 2

The center is outside the angle.

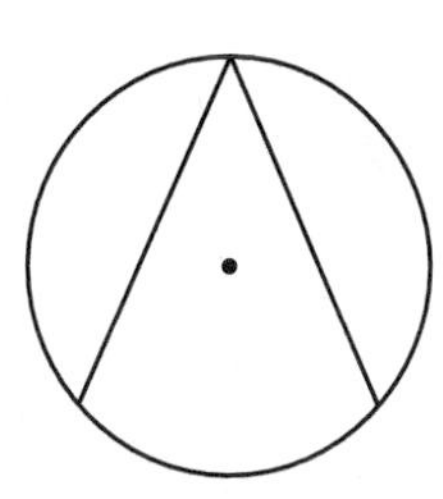

Case 3

The center is inside the angle.

You will first prove that the conjecture is true for Case 1.

Case 1 Conjecture: The measure of an inscribed angle in a circle equals half the measure of its intercepted arc when a side of the angle passes through the center of the circle.

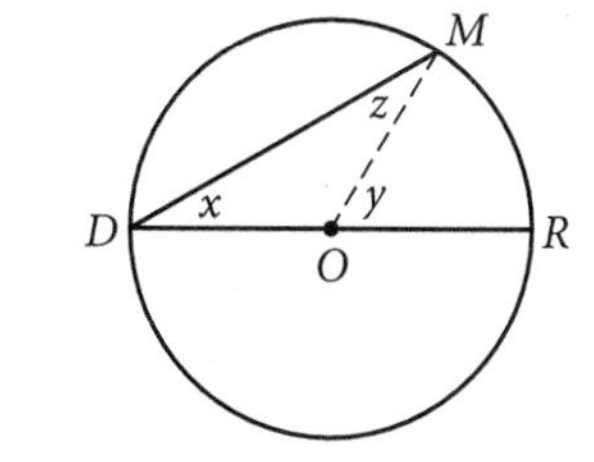

Given: Circle O with inscribed angle MDR on diameter $\overline{DR}$

Let $z = m\angle DMO$, $x = m\angle MDR$, and $y = m\angle MOR$ ($y = m\overset{\frown}{MR}$)

Show: $m\angle MDR = \frac{1}{2}m\overset{\frown}{MR}$

Work backward to formulate a plan. Ask yourself what you're trying to show and what you would need to do that.

Plan

- You need to show that $m\angle MDR = \frac{1}{2}m\overset{\frown}{MR}$. Using the variables defined in the given, this can be restated as $x = \frac{1}{2}y$.
- You want to show that $x = \frac{1}{2}y$, so you need to show that $2x = y$.
- You know that $y = x + z$ because of the Exterior Angle Conjecture.
- You also know that $x = z$ because $\triangle DOM$ is isosceles.
- So, start your flowchart proof by establishing that $\triangle DOM$ has two congruent sides.

From this plan you create a flowchart proof.

Flowchart Proof of Case 1

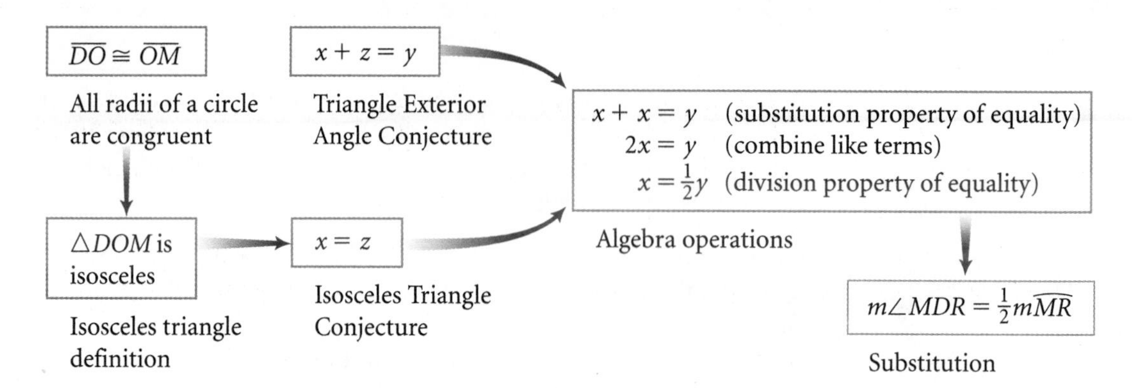

You will use Case 1 to write a paragraph proof for Case 2.

Case 2 Conjecture: The measure of an inscribed angle in a circle equals half the measure of its intercepted arc when the center of the circle is outside the angle.

Given: Circle O with inscribed angle MDK on one side of diameter $\overline{DR}$

Show: $m\angle MDK = \frac{1}{2}m\widehat{MK}$

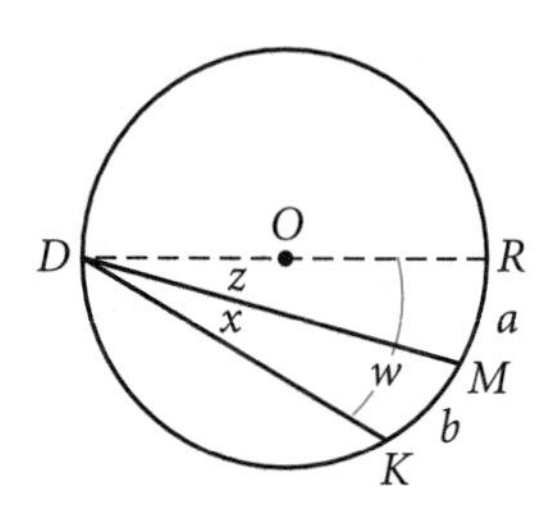

Paragraph Proof of Case 2

Let $z = m\angle MDR$, $x = m\angle MDK$, and $w = m\angle KDR$.
Then, $m\angle KDR = m\angle MDR + m\angle MDK$, so $w = x + z$.
Let $a = m\widehat{MR}$ and $b = m\widehat{MK}$. Then, $m\widehat{MR} + m\widehat{MK} = m\widehat{KR}$.
So, $m\widehat{KR} = a + b$.

Stated in terms of x and b, you wish to show that $x = \frac{b}{2}$. From Case 1, you know that $w = \frac{(a+b)}{2}$ and that $z = \frac{a}{2}$. You know that $w = x + z$, so $x = w - z$ by the subtraction property of equality. Substitute $\frac{(a+b)}{2}$ for w and $\frac{a}{2}$ for z to get $x = \frac{(a+b)}{2} - \frac{a}{2} = \frac{(a+b)-a}{2} = \frac{b}{2}$. Therefore, $m\angle MDK = \frac{1}{2}m\widehat{MK}$. ■

You will prove Case 3 in the exercises.

Science CONNECTION

Light rays from distant objects are focused on the retina of a normal eye. The rays converge short of the retina on a myopic (near-sighted) eye, causing blurry vision. With the proper lens, light rays from distant objects will focus sharply on the retina, giving the near-sighted person clear vision. When the rays are focused on the retina, what kind of angle do they form inside the eye?

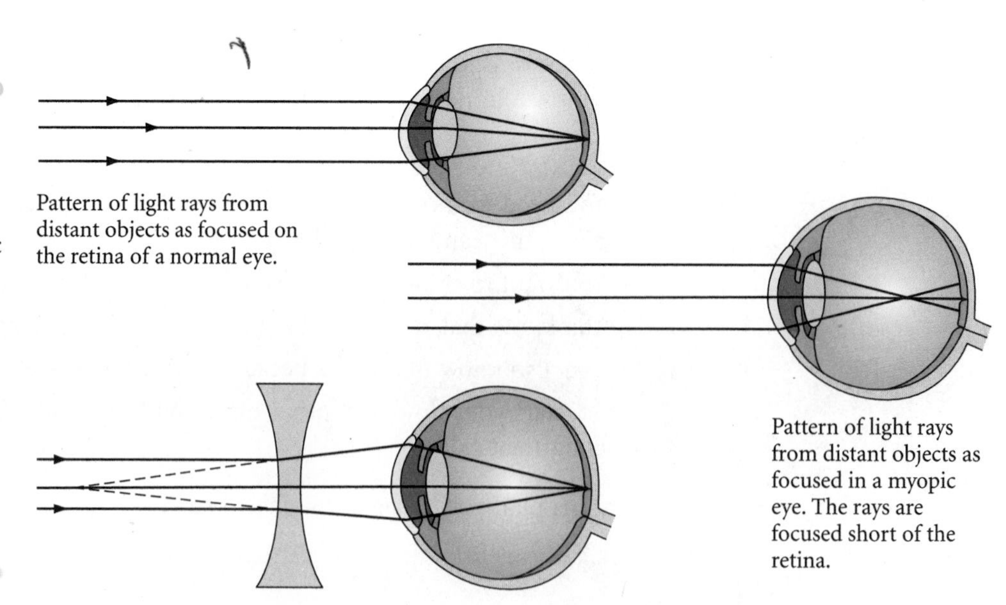

Pattern of light rays from distant objects as focused on the retina of a normal eye.

Pattern of light rays from distant objects as focused in a myopic eye. The rays are focused short of the retina.

Exercises

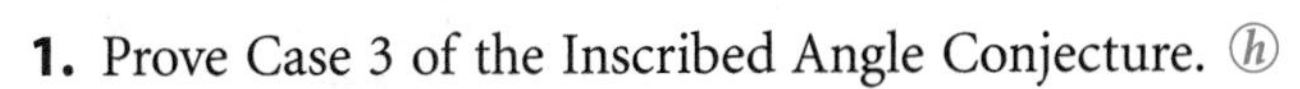

1. Prove Case 3 of the Inscribed Angle Conjecture. ⓗ

   Case 3 Conjecture: The measure of an inscribed angle in a circle equals half the measure of its intercepted arc when the center of the circle is inside the angle.

   Given: Circle O with inscribed angle MDK whose sides $\overline{DM}$ and $\overline{DK}$ lie on either side of diameter $\overline{DR}$

   Show: $m\angle MDK = \frac{1}{2}m\overarc{MK}$

In Exercises 2–5 the four conjectures are consequences of the Inscribed Angle Conjecture. Prove each conjecture by writing a paragraph proof or a flowchart proof.

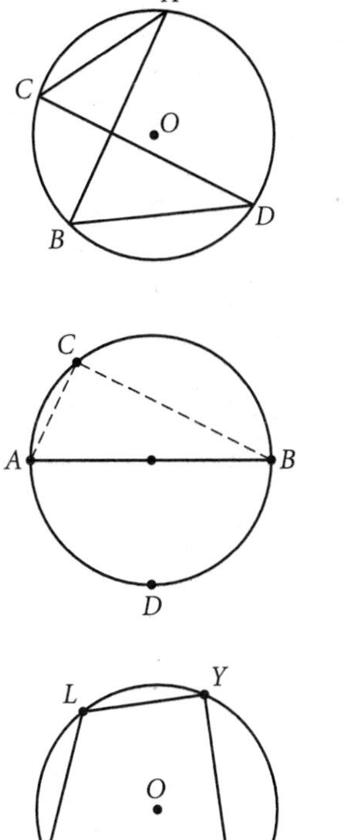

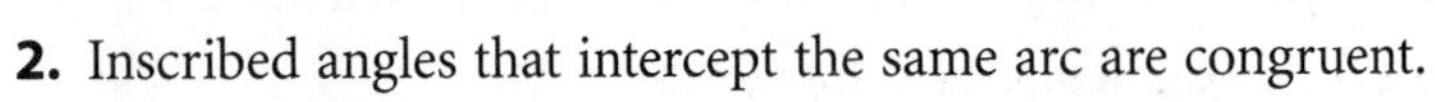

2. Inscribed angles that intercept the same arc are congruent.

   Given: Circle O with $\angle ACD$ and $\angle ABD$ inscribed in $\overarc{ACD}$

   Show: $\angle ACD \cong \angle ABD$

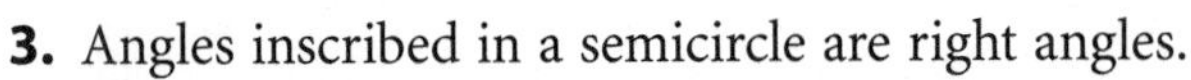

3. Angles inscribed in a semicircle are right angles.

   Given: Circle O with diameter $\overline{AB}$, and $\angle ACB$ inscribed in semicircle ACB

   Show: $\angle ACB$ is a right angle

4. The opposite angles of a cyclic quadrilateral are supplementary. ⓗ

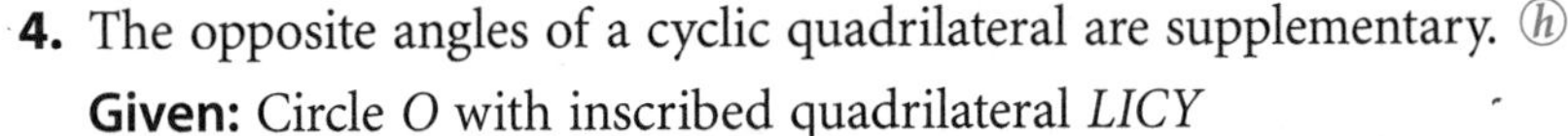

   Given: Circle O with inscribed quadrilateral $LICY$

   Show: $\angle L$ and $\angle C$ are supplementary

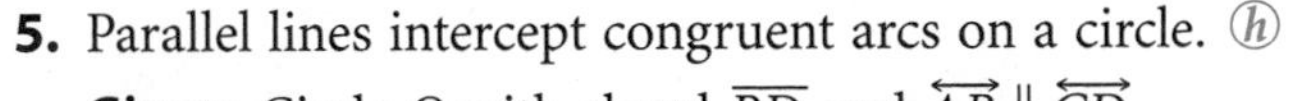

5. Parallel lines intercept congruent arcs on a circle. ⓗ

   Given: Circle O with chord $\overline{BD}$ and $\overleftrightarrow{AB} \parallel \overleftrightarrow{CD}$

   Show: $\overarc{BC} \cong \overarc{DA}$

For Exercises 6 and 7, determine whether each conjecture is true or false. If the conjecture is false, draw a counterexample. If the conjecture is true, prove it by writing either a paragraph or flowchart proof.

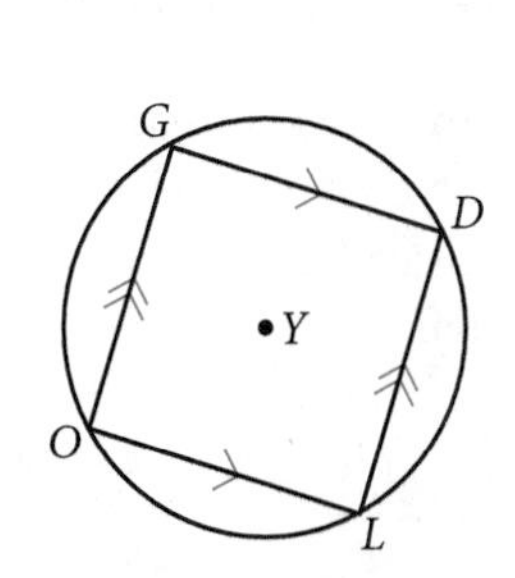

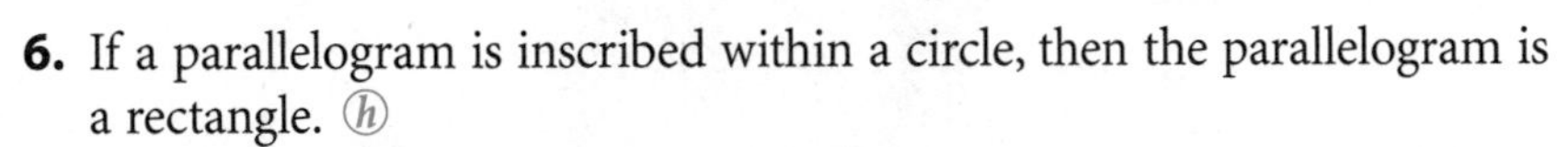

6. If a parallelogram is inscribed within a circle, then the parallelogram is a rectangle. ⓗ

   Given: Circle Y with inscribed parallelogram $GOLD$

   Show: $GOLD$ is a rectangle

7. If a trapezoid is inscribed within a circle, then the trapezoid is isosceles. (h)
Given: Circle R with inscribed trapezoid $GATE$
Show: $GATE$ is an isosceles trapezoid

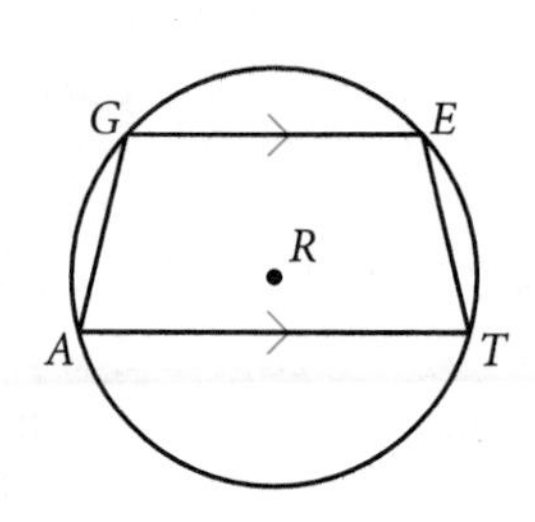

Review

8. For each of the statements below, choose the letter for the word that best fits (A stands for always, S for sometimes, and N for never). If the answer is S, give two examples, one showing how the statement is true and one showing how the statement can be false.
   a. An equilateral polygon is (A/S/N) equiangular.
   b. If a triangle is a right triangle, then the acute angles are (A/S/N) complementary.
   c. The diagonals of a kite are (A/S/N) perpendicular bisectors of each other.
   d. A regular polygon (A/S/N) has both reflectional symmetry and rotational symmetry.
   e. If a polygon has rotational symmetry, then it (A/S/N) has more than one line of reflectional symmetry.

9. Explain why m is parallel to n. (h)

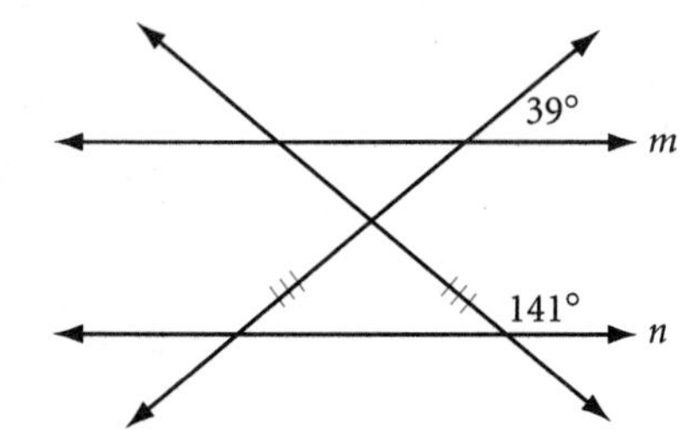

10. What is the probability of randomly selecting three collinear points from the points in the 3-by-3 grid below? (h)

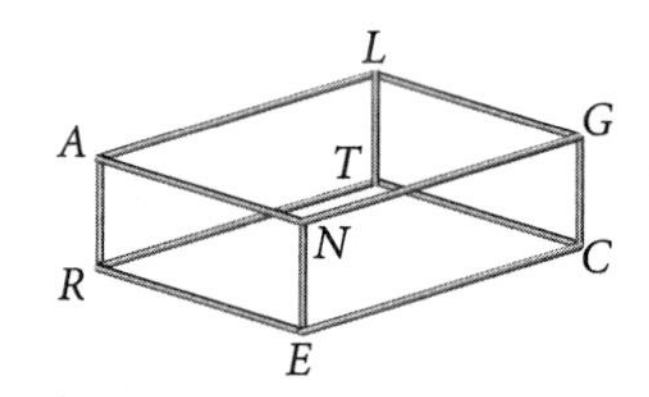

11. How many different 3-edge routes are possible from R to G along the wire frame shown? (h)

IMPROVING YOUR VISUAL THINKING SKILLS

Rolling Quarters

One of two quarters remains motionless while the other rotates around it, never slipping and always tangent to it. When the rotating quarter has completed a turn around the stationary quarter, how many turns has it made around its own center point? Try it!

Finding the Circumcenter

Suppose you know the coordinates of the vertices of a triangle. How can you find the coordinates of the circumcenter? You can graph the triangle, construct the perpendicular bisectors of the sides, and then *estimate* the coordinates of the point of concurrency. However, to find the exact coordinates, you need to use algebra. Let's look at an example.

EXAMPLE

Find the coordinates of the circumcenter of $\triangle ZAP$ with $Z(0, -4)$, $A(-4, 4)$, and $P(8, 8)$.

▶ ***Solution***

To find the coordinates of the circumcenter, you can write equations for the perpendicular bisectors of two of the sides of the triangle and then find the point where the bisectors intersect.

To review midpoint, slopes of perpendicular lines, or solving systems of equations, see the Table of Contents for those Using Your Algebra Skills topics.

To find the equation for the perpendicular bisector of $\overline{ZA}$, first find the midpoint of $\overline{ZA}$, then find its slope.

$$\text{Midpoint of } \overline{ZA} = \left(\frac{0 + (-4)}{2}, \frac{-4 + 4}{2}\right) = (-2, 0)$$

$$\text{Slope of } \overline{ZA} = \frac{4 - (-4)}{-4 - 0} = \frac{8}{-4} = -2$$

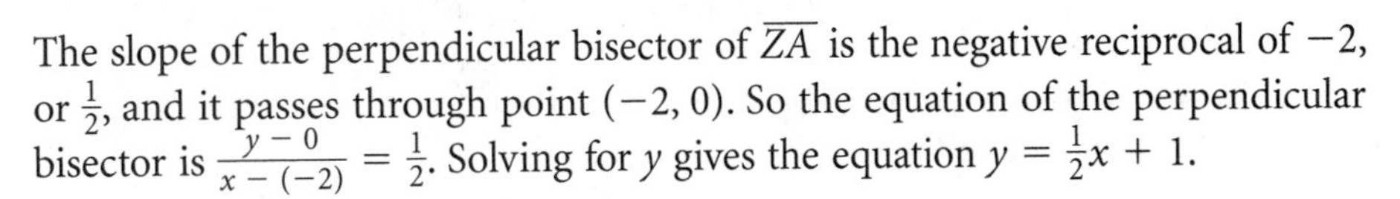

The slope of the perpendicular bisector of $\overline{ZA}$ is the negative reciprocal of -2, or $\frac{1}{2}$, and it passes through point $(-2, 0)$. So the equation of the perpendicular bisector is $\frac{y - 0}{x - (-2)} = \frac{1}{2}$. Solving for y gives the equation $y = \frac{1}{2}x + 1$.

You can use the same technique to find the equation of the perpendicular bisector of $\overline{ZP}$. The midpoint of ZP is $(4, 2)$, and the slope is $\frac{3}{2}$. So the slope of the perpendicular bisector of $\overline{ZP}$ is $-\frac{2}{3}$ and it passes through the point $(4, 2)$. The equation of the perpendicular bisector is $\frac{y - 2}{x - 4} = -\frac{2}{3}$, or $y = -\frac{2}{3}x + \frac{14}{3}$.

Since all the perpendicular bisectors intersect at the same point, you can solve these two equations to find that point. To find the point where the perpendicular bisectors intersect, solve this system by substitution.

$$\begin{cases} y = \frac{1}{2}x + 1 & \text{Perpendicular bisector of } \overline{ZA}. \\ y = -\frac{2}{3}x + \frac{14}{3} & \text{Perpendicular bisector of } \overline{ZP}. \end{cases}$$

$$y = \frac{1}{2}x + 1 \qquad \text{Original first equation.}$$

$$-\frac{2}{3}x + \frac{14}{3} = \frac{1}{2}x + 1 \qquad \text{Substitute } -\tfrac{2}{3}x + \tfrac{14}{3} \text{ (from the second equation) for } y.$$

$-4x + 28 = 3x + 6$	Multiply both sides by 6.
$22 = 7x$	Add $4x$ to both sides. Subtract 6 from both sides.
$\frac{22}{7} = x$	Divide both sides by 7.
$y = \frac{1}{2}\left(\frac{22}{7}\right) + 1$	Substitute 2 for x in the first equation.
$y = \frac{18}{7}$	Simplify.

The circumcenter is $\left(\frac{22}{7}, \frac{18}{7}\right)$. You can check this result by writing the equation for the perpendicular bisector of $\overline{AP}$ and verifying that $\left(\frac{22}{7}, \frac{18}{7}\right)$ is a point on this line.

EXERCISES

1. Triangle RES has vertices $R(0, 0)$, $E(4, -6)$, and $S(8, 4)$. Find the equation of the perpendicular bisector of $\overline{RE}$.

In Exercises 2–5, find the coordinates of the circumcenter of each triangle.

2. Triangle TRM with vertices $T(-2, 1)$, $R(4, 3)$, and $M(-4, -1)$
3. Triangle FGH with vertices $F(0, -6)$, $G(3, 6)$, and $H(12, 0)$
4. Right triangle MNO with vertices $M(-4, 0)$, $N(0, 5)$, and $O(10, -3)$
5. Isosceles triangle CDE with vertices $C(0, 6)$, $D(0, -6)$, and $E(12, 0)$
6. If a triangle is a right triangle, there is a shorter method to finding the circumcenter. What is it? Explain.
7. If a triangle is an isosceles triangle, then there is a different, perhaps shorter method to finding the circumcenter. Explain.
8. Circle P with center at $(-6, -6)$ and circle Q with center at $(11, 0)$ have a common internal tangent $\overleftrightarrow{AB}$. Find the coordinates of B if A has coordinates $(-3, -2)$.

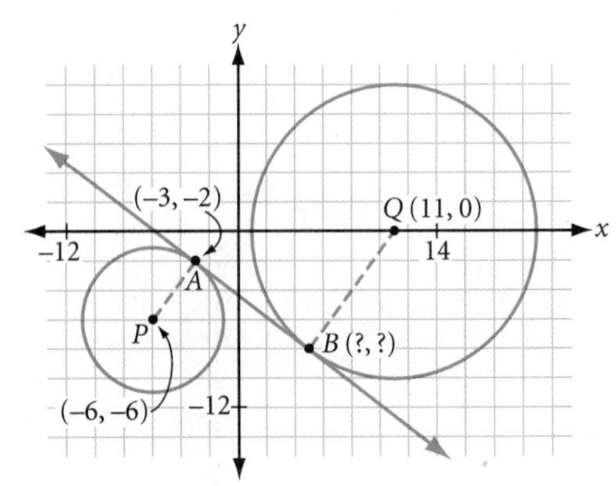

LESSON

6.5 The Circumference/ Diameter Ratio

No human investigations can ever be called true science without going through mathematical tests.

LEONARDO DA VINCI

The distance around a polygon is called the perimeter. The distance around a circle is called the **circumference.** Here is a nice visual puzzle. Which is greater, the height of a tennis-ball can or the circumference of the can? The height is approximately three tennis-ball diameters tall. The diameter of the can is approximately one tennis-ball diameter. If you have a tennis-ball can handy, try it. Wrap a string around the can to measure its circumference, then compare this measurement with the height of the can. Surprised?

If you actually compared the measurements, you discovered that the circumference of the can is greater than three diameters of the can. In this lesson you are going to discover (or perhaps rediscover) the relationship between the diameter and the circumference of every circle. Once you know this relationship, you can measure a circle's diameter and calculate its circumference.

If you measure the circumference and diameter of a circle and divide the circumference by the diameter, you get a number slightly larger than 3. The more accurate your measurements, the closer your ratio will come to a special number called π (pi), pronounced "pie," like the dessert.

History CONNECTION

In 1897, the Indiana state assembly tried to legislate the value of π. The vague language of the state's House Bill No. 246, which became known as the "Indiana Pi Bill," implies several different incorrect values for π—3.2, 3.232, 3.236, 3.24, and 4. With a unanimous vote of 67-0, the House passed the bill to the state senate, where it was postponed indefinitely.

Investigation
A Taste of Pi

You will need

- several round objects (cans, mugs, bike wheel, plates)
- a meterstick or metric measuring tape
- sewing thread or thin string

In this investigation you will find an approximate value of π by measuring circular objects and calculating the ratio of the circumference to the diameter. Let's see how close you come to the actual value of π.

Step 1 Measure the circumference of each round object by wrapping the measuring tape, or string, around its perimeter. Then measure the diameter of each object with the meterstick or tape. Record each measurement to the nearest millimeter (tenth of a centimeter).

Step 2 Make a table like the one below and record the circumference (C) and diameter (d) measurements for each round object.

Object	Circumference (C)	Diameter (d)	Ratio $\frac{C}{d}$
Can			
Mug			
Wheel			

Step 3 Calculate the ratio $\frac{C}{d}$ for each object. Record the answers in your table.

Step 4 Calculate the average of your ratios of $\frac{C}{d}$.

Compare your average with the averages of other groups. Are the $\frac{C}{d}$ ratios close? You should now be convinced that the ratio $\frac{C}{d}$ is very close to 3 for every circle. We define π as the ratio $\frac{C}{d}$. If you solve this formula for C, you get a formula for the circumference of a circle in terms of the diameter, d. The diameter is twice the radius ($d = 2r$), so you can *also* get a formula for the circumference in terms of the radius, r.

Step 5 Copy and complete the conjecture.

Circumference Conjecture C-66

If C is the circumference and d is the diameter of a circle, then there is a number π such that $C =$ _?_. If $d = 2r$ where r is the radius, then $C =$ _?_.

Mathematics CONNECTION

The number π is an irrational number—its decimal form never ends. It is also a transcendental number—the pattern of digits does not repeat. The symbol π is a letter of the Greek alphabet. Perhaps no other number has more fascinated mathematicians throughout history. Mathematicians in ancient Egypt used $\left(\frac{4}{3}\right)^4$ as their approximation of circumference to diameter. Early Chinese and Hindu mathematicians used $\sqrt{10}$. By 408 C.E., Chinese mathematicians were using $\frac{355}{113}$. Today, computers have calculated approximations of π to billions of decimal places, and there are websites devoted to π! See www.keymath.com/DG .

Accurate approximations of π have been of more interest intellectually than practically. Still, what would a carpenter say if you asked her to cut a board 3π feet long? Most calculators have a π button that gives π to eight or ten decimal places. You can use this value for most calculations, then round your answer to a specified decimal place. If your calculator doesn't have a π button, or if you don't have access to a calculator, use the value 3.14 for π. If you're asked for an exact answer instead of an approximation, state your answer in terms of π.

How do you use the Circumference Conjecture? Let's look at two examples.

EXAMPLE A If a circle has diameter 3.0 meters, what is the circumference? Use a calculator and state your answer to the nearest 0.1 meter.

▶ Solution

$C = \pi d$ Original formula.

$C = \pi(3.0)$ Substitute the value of d.

In terms of π, the answer is 3π. The circumference is about 9.4 meters.

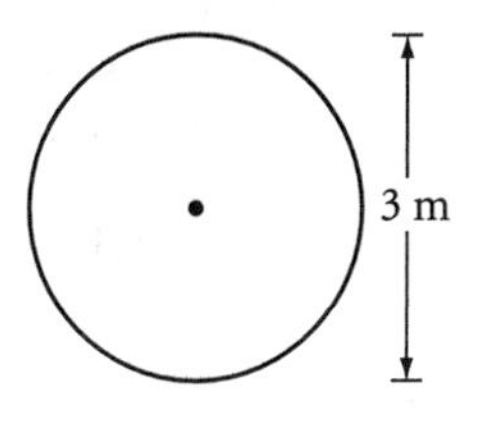

EXAMPLE B If a circle has circumference 12π meters, what is the radius?

▶ Solution

$C = 2\pi r$ Original formula.

$12\pi = 2\pi r$ Substitute the value of C.

$r = 6$ Solve.

The radius is 6 meters.

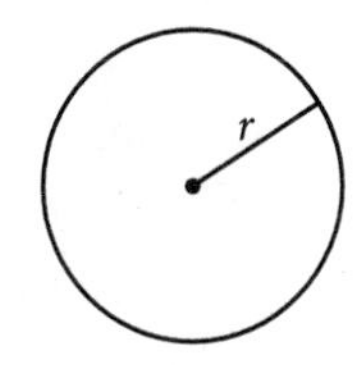

EXERCISES

You will need

A calculator for Exercises **7–10**

Use the Circumference Conjecture to solve Exercises 1–12. In Exercises 1–6, leave your answer in terms of π.

1. If $C = 5\pi$ cm, find d.

2. If $r = 5$ cm, find C.

3. If $C = 24$ m, find r.

4. If $d = 5.5$ m, find C.

5. If a circle has a diameter of 12 cm, what is its circumference?

6. If a circle has a circumference of 46π m, what is its diameter?

In Exercises 7–10, use a calculator. Round your answer to the nearest 0.1 unit. Use the symbol $\approx$ to show that your answer is an approximation.

7. If $d = 5$ cm, find C.

8. If $r = 4$ cm, find C.

9. If $C = 44$ m, find r. ⓗ

10. What's the circumference of a bicycle wheel with a 27-inch diameter?

11. If the distance from the center of a Ferris wheel to one of the seats is approximately 90 feet, what is the distance traveled by a seated person, to the nearest foot, in one revolution?

12. If a circle is inscribed in a square with a perimeter of 24 cm, what is the circumference of the circle? ⓗ

13. If a circle with a circumference of 16π inches is circumscribed about a square, what is the length of a diagonal of the square?

14. Each year a growing tree adds a new ring to its cross section. Some years the ring is thicker than others. Why do you suppose this happens?

Suppose the average thickness of growth rings in the Flintstones National Forest is 0.5 cm. About how old is "Old Fred," a famous tree in the forest, if its circumference measures 766 cm?

Science CONNECTION

Trees can live hundreds to thousands of years, and we can determine the age of one tree by counting its growth rings. A pair of rings—a light ring formed in the spring and summer and a dark one formed in the fall and early winter—represent the growth for one year. We can learn a lot about the climate of a region over a period of years by studying tree growth rings. This study is called *dendroclimatology.*

15. Pool contractor Peter Tileson needs to determine the number of 1-inch tiles to put around the edge of a pool. The pool is a rectangle with two semicircular ends as shown. How many tiles will he need?

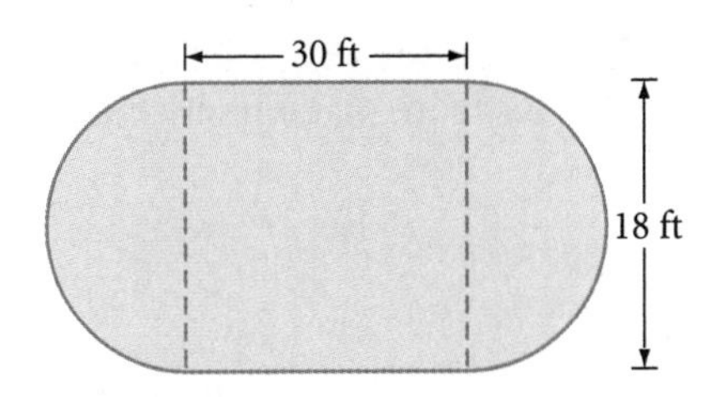

Review

16. ***Mini-Investigation*** Use these diagrams to find a relationship between $\angle AEN$ and the sum of the intercepted arc measures. Then copy and complete the conjecture.

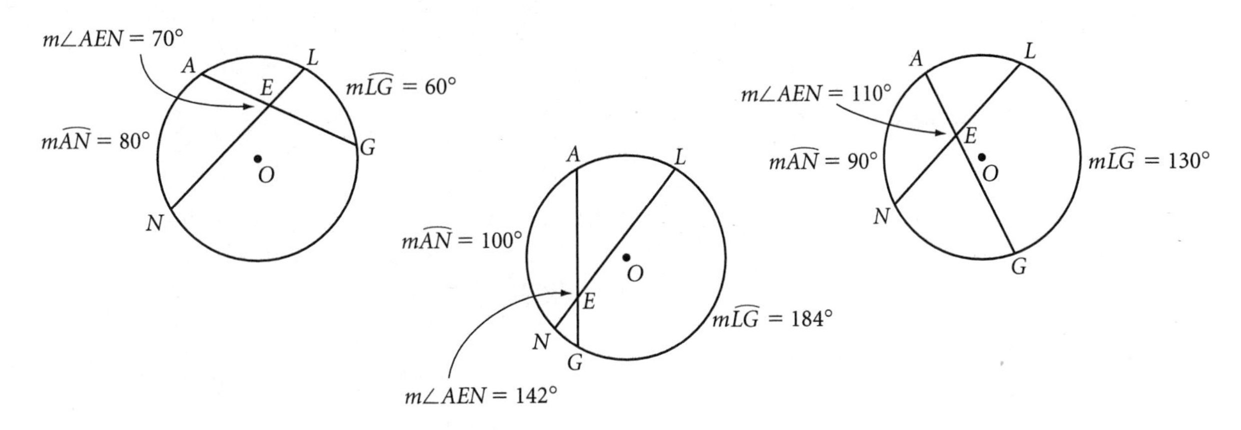

Conjecture: The measure of an angle formed by two intersecting chords is $\underline{?}$ of the two intercepted arcs.

17. Copy the diagram at right and draw $\overline{AL}$. Prove the conjecture you made in Exercise 16. ⓗ

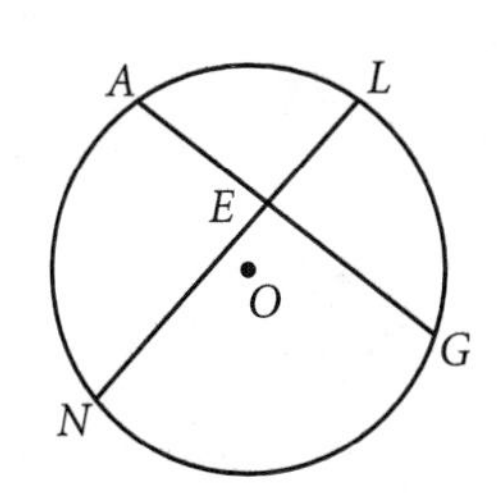

18. **Conjecture:** If two circles intersect at two points, then the segment connecting the centers is the perpendicular bisector of the common chord, the segment connecting the points of intersection.

Complete the flowchart proof of the conjecture.

Given: Circle M and circle S intersect at points A and T with radii $\overline{MA} \cong \overline{MT}$ and $\overline{SA} \cong \overline{ST}$

Show: $\overline{MS}$ is the perpendicular bisector of $\overline{AT}$

Flowchart Proof

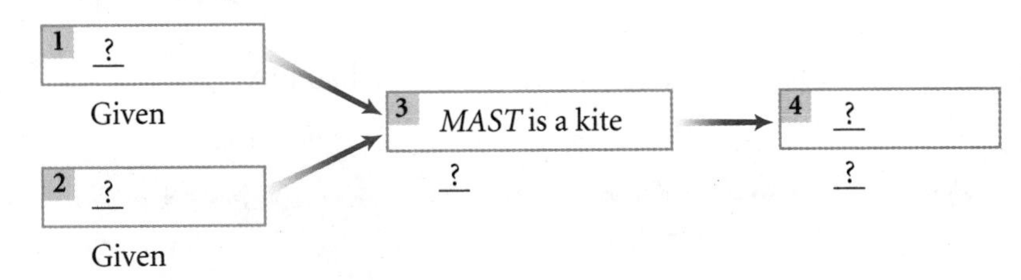

19. Trace the figure below. Calculate the measure of each lettered angle.

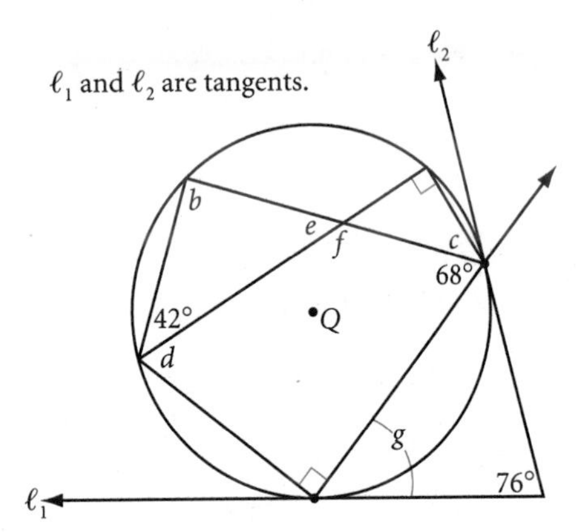

20. What is the probability that a flea strolling along the circle shown below will stop randomly on either $\overset{\frown}{AB}$ or $\overset{\frown}{CD}$? (Because you're probably not an expert in flea behavior, assume the flea will stop exactly once.)

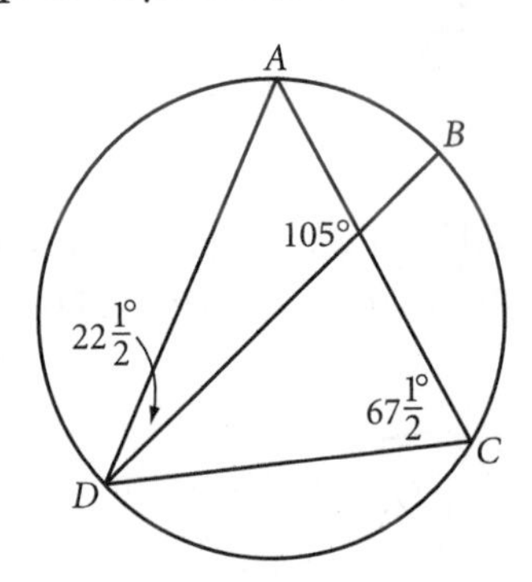

21. Explain why a and b are complementary.

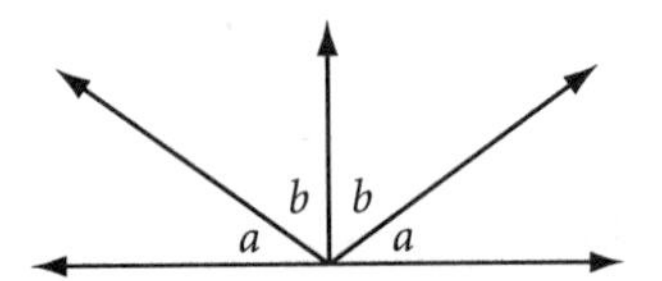

22. Assume the pattern below will continue. Write an expression for the perimeter of the tenth shape in this picture pattern.

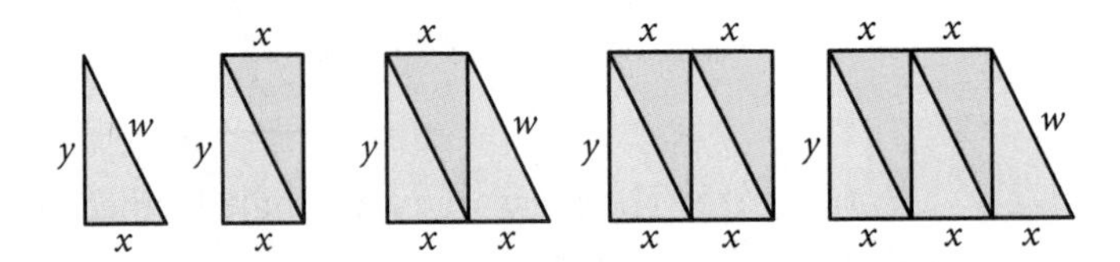

NEEDLE TOSS

If you randomly toss a needle on lined paper, what is the probability that the needle will land on one or more lines? What is the probability it will not land on any lines? The length of the needle and the distance between the lines will affect these probabilities.

Start with a toothpick of any length L as your "needle" and construct parallel lines a distance L apart. Write your predictions, then experiment. Using N as the number of times you dropped the needle, and C as the number of times the needle crossed the line, enter your results into the expression $2N/C$. As you drop the needle more and more times, the value of this expression seems to be getting close to what number?

Your project should include

- Your predictions and data.
- The calculated probabilities and your prediction of the theoretical probability.
- Any other interesting observations or conclusions.

Fathom™

Using Fathom, you can simulate many experiments. See the demonstration Buffon's Needle that comes with the software package.

LESSON 6.6

Around the World

Many application problems are related to π. Satellite orbits, the wheels of a vehicle, tree trunks, and round pizzas are just a few of the real-world examples that involve the circumference of circles. Here is a famous example from literature.

Love is like π—natural, irrational, and very important.

LISA HOFFMAN

Literature CONNECTION

In the novel *Around the World in Eighty Days* (1873), Jules Verne (1828–1905) recounts the adventures of brave Phileas Fogg and his servant Passerpartout. They begin their journey when Phileas bets his friends that he can make a trip around the world in 80 days. Phileas's precise behavior, such as monitoring the temperature of his shaving water or calculating the exact time and location of his points of travel, reflects Verne's interest in the technology boom of the late nineteenth century. His studies in geology, engineering, and astronomy aid the imaginative themes in this and his other novels, including *A Journey to the Center of the Earth* (1864) and *Twenty Thousand Leagues Under the Sea* (1870).

EXAMPLE

If the diameter of the earth is 8000 miles, find the average speed in miles per hour Phileas Fogg needs to circumnavigate the earth about the equator in 80 days.

Solution

To find the speed, you need to know the distance and the time. The distance around the equator is equal to the circumference C of a circle with a diameter of 8,000 miles.

$C = \pi d$	The equation for circumference.
$= \pi(8{,}000)$	Substitute 8,000 for d.
$\approx 25{,}133$	Round to nearest mile.

So, Phileas must travel 25,133 miles in 80 days. To find the speed v in mi/hr, you need to divide distance by time and convert days into hours.

$v = \dfrac{\text{distance}}{\text{time}}$	The formula for speed, or velocity.
$\approx \dfrac{25{,}133 \text{ mi}}{80 \text{ days}} \cdot \dfrac{1 \text{ day}}{24 \text{ hr}}$	Substitute values and convert units of time.
$\approx 13 \text{ mi/hr}$	Evaluate and round to the nearest mile per hour.

If the earth's diameter were *exactly* 8,000 miles, you could evaluate $\frac{8{,}000\pi}{80 \cdot 24}$ and get an exact answer of $\frac{25\pi}{6}$ in terms of π.

EXERCISES

You will need

A calculator for Exercises **1–8**

In Exercises 1–6, round answers to the nearest unit. You may use 3.14 as an approximate value of π. If you have a π button on your calculator, use that value and then round your final answer.

1. A satellite in a nearly circular orbit is 2000 km above Earth's surface. The radius of Earth is approximately 6400 km. If the satellite completes its orbit in 12 hours, calculate the speed of the satellite in kilometers per hour. ⓗ

2. Wilbur Wrong is flying his remote-control plane in a circle with a radius of 28 meters. His brother, Orville Wrong, clocks the plane at 16 seconds per revolution. What is the speed of the plane? Express your answer in meters per second. The brothers may be wrong, but you could be right!

3. Here is a tiring problem. The diameter of a car tire is approximately 60 cm (0.6 m). The warranty is good for 70,000 km. About how many revolutions will the tire make before the warranty is up? More than a million? A billion? (1 km = 1000 m)

4. If the front tire of this motorcycle has a diameter of 50 cm (0.5 m), how many revolutions will it make if it is pushed 1 km to the nearest gas station? In other words, how many circumferences of the circle are there in 1000 meters?

5. Goldi's Pizza Palace is known throughout the city. The small Baby Bear pizza has a 6-inch radius and sells for $9.75. The savory medium Mama Bear pizza sells for $12.00 and has an 8-inch radius. The large Papa Bear pizza is a hefty 20 inches in diameter and sells for $16.50. The edge is stuffed with cheese, and it's the best part of a Goldi's pizza. What size has the most pizza edge per dollar? What is the circumference of this pizza?

6. Felicia is a park ranger, and she gives school tours through the redwoods in a national park. Someone in every tour asks, "What is the diameter of the giant redwood tree near the park entrance?"

Felicia knows that the arm span of each student is roughly the same as his or her height. So in response, Felicia asks a few students to arrange themselves around the circular base of the tree so that by hugging the tree with arms outstretched, they can just touch fingertips to fingertips. She then asks the group to calculate the diameter of the tree.

In one group, four students with heights of 138 cm, 136 cm, 128 cm, and 126 cm were able to ring the tree. What is the approximate diameter of the redwood?

7. **APPLICATION** Zach wants a circular table so that 12 chairs, each 16 inches wide, can be placed around it with at least 8 inches between chairs. What should be the diameter of the table? Will the table fit in a 12-by-14-foot dining room? Explain.

Calvin and Hobbes by Bill Watterson

8. A 45 rpm record has a 7-inch diameter and spins at 45 revolutions per minute. A 33 rpm record has a 12-inch diameter and spins at 33 revolutions per minute. Find the difference in speeds of a point on the edge of a 33 rpm record to that of a point on the edge of a 45 rpm record, in ft/sec. ⓗ

Recreation CONNECTION

Using tinfoil records, Thomas Edison (1847–1931) invented the phonograph, the first machine to play back recorded sound, in 1877. Commonly called record players, they weren't widely reproduced until high-fidelity amplification (hi-fi) and advanced speaker systems came along in the 1930s. In 1948, records could be played at slower speeds to allow more material on the disc, creating longer-playing records (LPs). When compact discs became popular in the early 1990s, most record companies stopped making LPs. Some disc jockeys still use records instead of CDs.

Review

9. ***Mini-Investigation*** In these diagrams of circle O, find the relationship between $\angle ECA$ formed by the secants and the difference of the intercepted arc measures. Then copy and complete the conjecture.

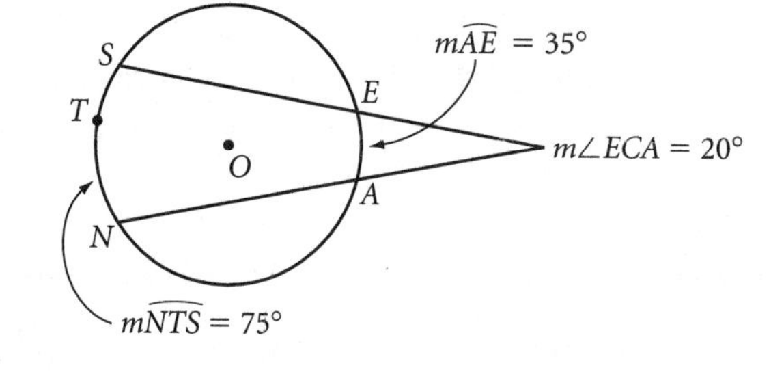

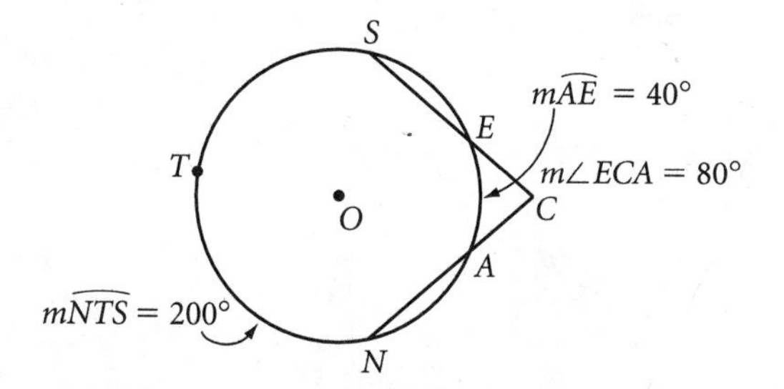

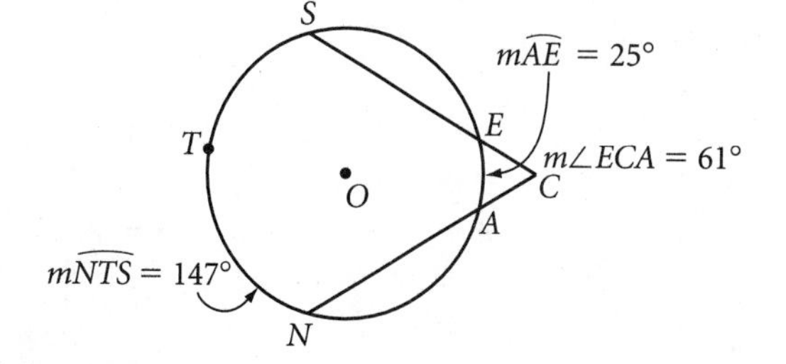

Conjecture: The measure of an angle formed by two secants through a circle is __?__.

10. Copy the diagram below. Construct $\overline{SA}$. Prove the conjecture you made in Exercise 9. ⓗ

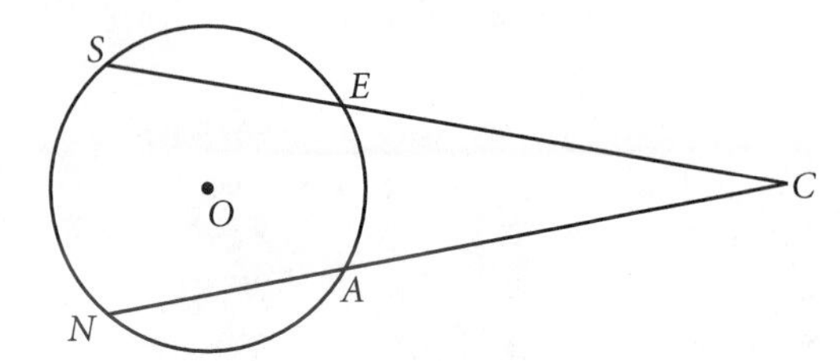

11. Explain why a equals b.

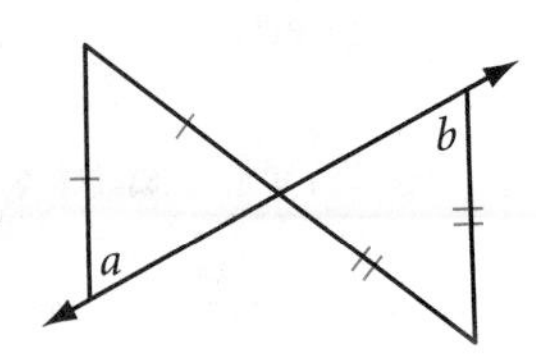

12. As P moves from A to B along the semicircle $\widehat{ATB}$, which of these measures constantly increases?

A. The perimeter of $\triangle ABP$

B. The distance from P to $\overleftrightarrow{AB}$

C. $m\angle ABP$

D. $m\angle APB$

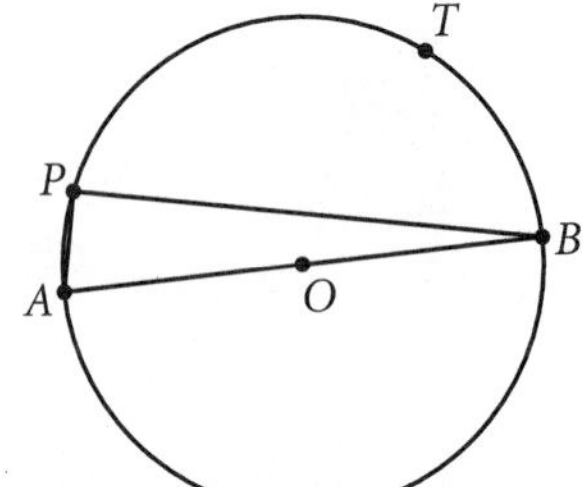
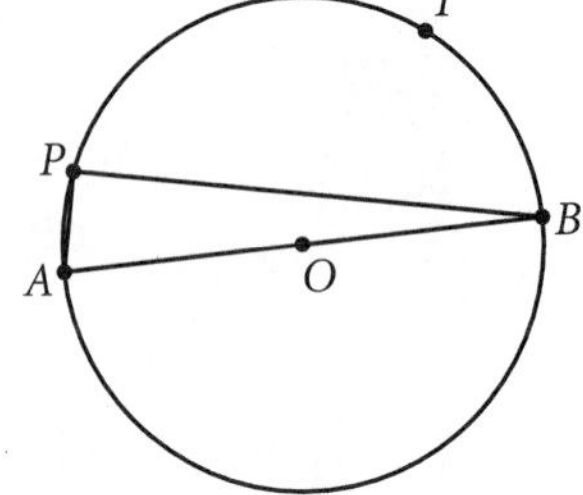

13. $a = \underline{?}$

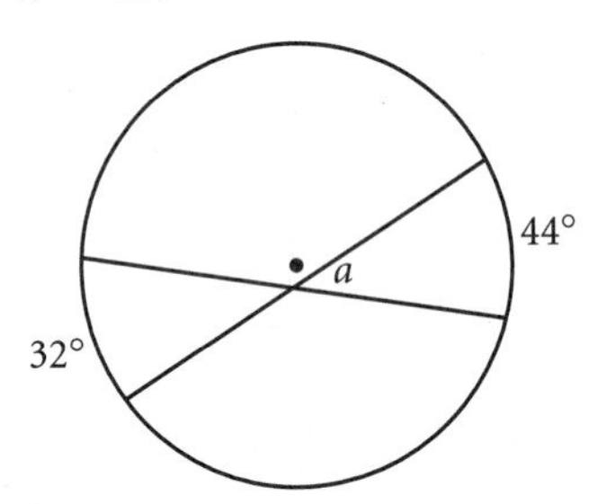

14. $b = \underline{?}$

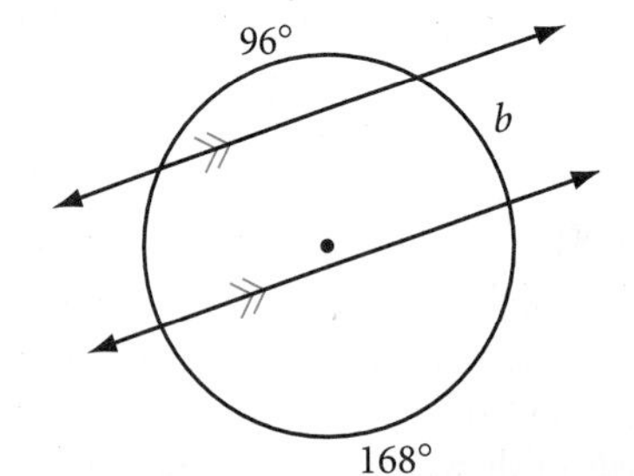

15. $d = \underline{?}$

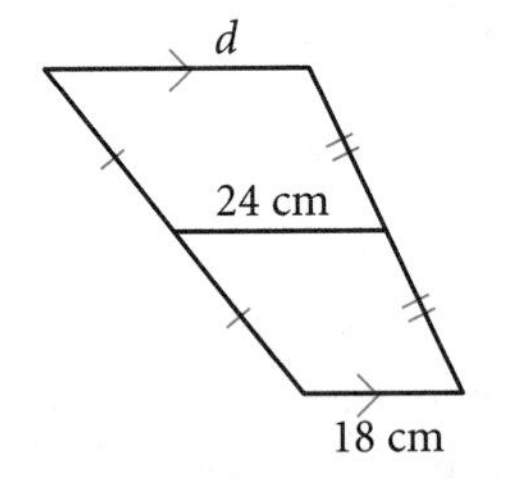

16. A helicopter has three blades each measuring about 26 feet. What is the speed in feet per second at the tips of the blades when they are moving at 400 rpm?

17. Two sides of a triangle are 24 cm and 36 cm. Write an inequality that represents the range of values for the third side. Explain your reasoning.

IMPROVING YOUR VISUAL THINKING SKILLS

PUZZLE

Picture Patterns I

Draw the next picture in each pattern. Then write the rule for the total number of squares in the nth picture of the pattern.

1. 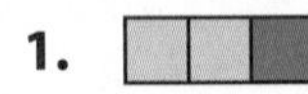 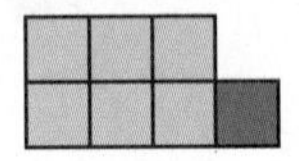 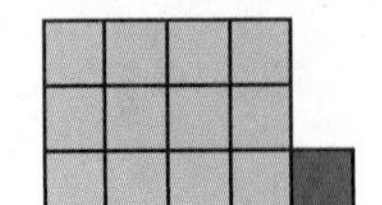

2.  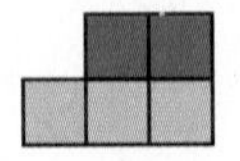 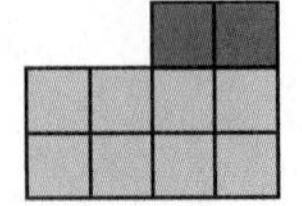 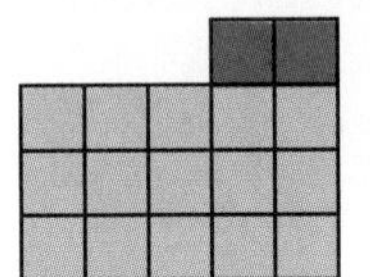

LESSON

6.7

Arc Length

Some people transform the sun into a yellow spot, others transform a yellow spot into the sun.

PABLO PICASSO

You have learned that the *measure* of an arc is equal to the measure of its central angle. On a clock, the measure of the arc from 12:00 to 1:00 is equal to the measure of the angle formed by the hour and minute hands. A circular clock is divided into 12 equal arcs, so the measure of this arc is $\frac{360°}{12}$, or 30°.

Notice that because the minute hand is longer, the tip of the minute hand must travel farther than the tip of the hour hand even though they both move 30° from 12 to 1. So the arc length is different even though the arc *measure* (the degree measure) is the same!

Let's take another look at the arc measure.

EXAMPLE A

What fraction of its circle is each arc?

a. $\widehat{AB}$ is what fraction of circle T?

b. $\widehat{CED}$ is what fraction of circle O?

c. $\widehat{EF}$ is what fraction of circle P?

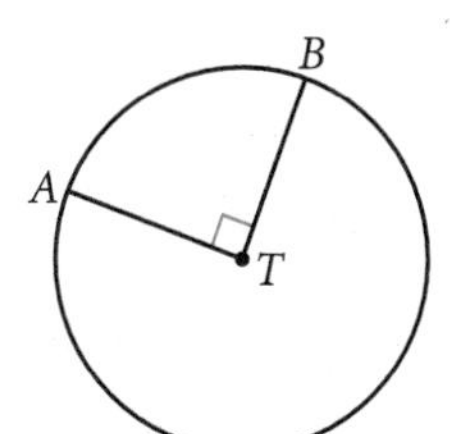

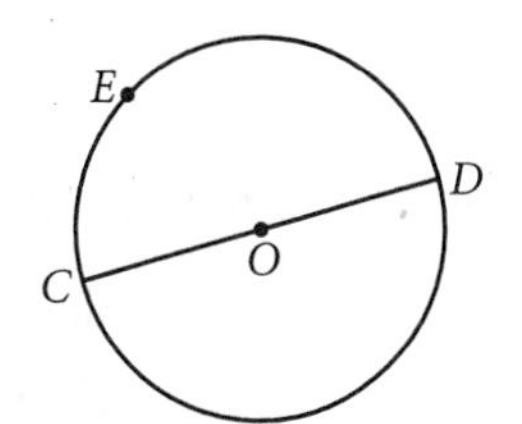

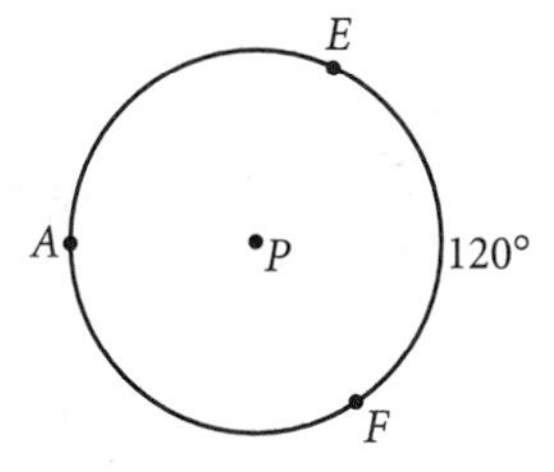

▶ Solution

In part a, you probably "just knew" that the arc is one-fourth of the circle because you have seen one-fourth of a circle so many times. Why is it one-fourth? The arc measure is 90°, a full circle measures 360°, and $\frac{90°}{360°} = \frac{1}{4}$. The arc in part b is half of the circle because $\frac{180°}{360°} = \frac{1}{2}$. In part c, you may or may not have recognized right away that the arc is one-third of the circle. The arc is one-third of the circle because $\frac{120°}{360°} = \frac{1}{3}$.

Cultural CONNECTION

This modern mosaic shows the plan of an ancient Mayan observatory. On certain days of the year light would shine through openings, indicating the seasons. This sculpture includes blocks of marble carved into arcs of concentric circles.

What do these fractions have to do with arc length? If you traveled halfway around a circle, you'd cover $\frac{1}{2}$ of its perimeter, or circumference. If you went a quarter of the way around, you'd travel $\frac{1}{4}$ of its circumference. The **length of an arc,** or arc length, is some fraction of the circumference of its circle.

The measure of an arc is calculated in units of degrees, but arc length is calculated in units of distance.

Investigation
Finding the Arcs

In this investigation you will find a method for calculating the arc length.

Step 1 For $\overset{\frown}{AB}$, $\overset{\frown}{CED}$, and $\overset{\frown}{GH}$, find what fraction of the circle each arc is.

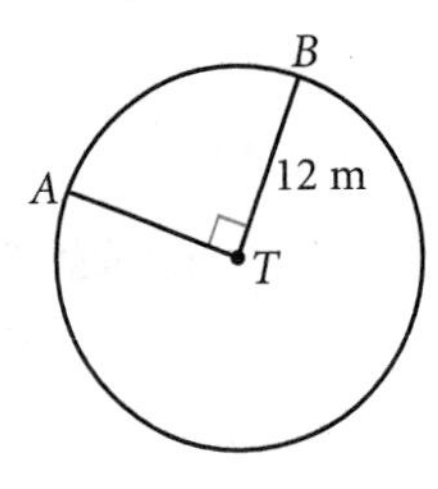

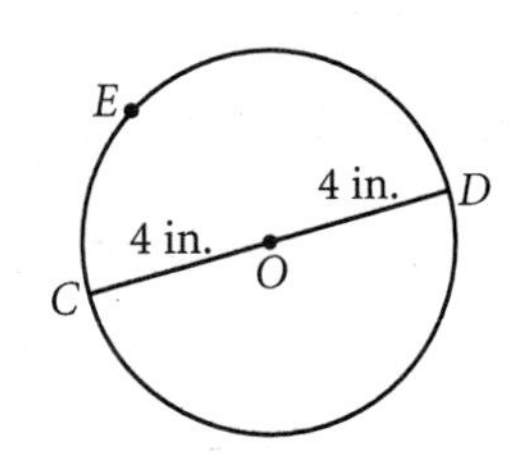

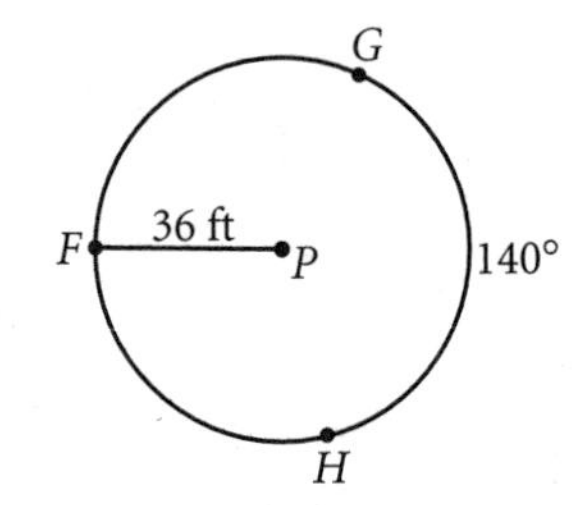

Step 2 Find the circumference of each circle.

Step 3 Combine the results of Steps 1 and 2 to find the length of each arc.

Step 4 Share your ideas for finding the length of an arc. Generalize this method for finding the length of *any* arc, and state it as a conjecture.

Arc Length Conjecture C-67

The length of an arc equals the _?_.

How do you use this new conjecture? Let's look at a few examples.

EXAMPLE B If the radius of the circle is 24 cm and $m\angle BTA = 60°$, what is the length of $\overset{\frown}{AB}$?

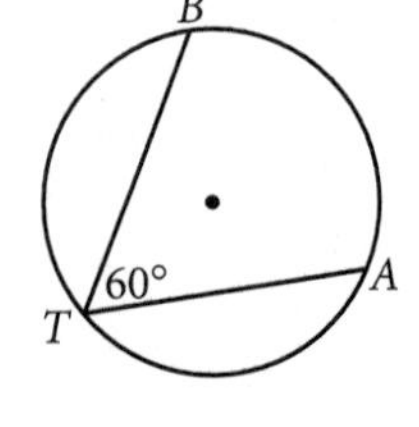

▶ Solution

$m\angle BTA = 60°$, so $m\overset{\frown}{AB} = 120°$ by the Inscribed Angle Conjecture. Then $\frac{120}{360} = \frac{1}{3}$, so the arc length is $\frac{1}{3}$ of the circumference, by the Arc Length Conjecture.

$$\text{Arc length} = \frac{1}{3}C$$

$$= \frac{1}{3}(48\pi) \quad \text{Substitute } 2\pi r \text{ for } C\text{, where } r = 24.$$

$$= 16\pi \quad \text{Simplify.}$$

The arc length is 16π cm, or approximately 50.3 cm.

EXAMPLE C If the length of $\widehat{ROT}$ is 116π meters, what is the radius of the circle?

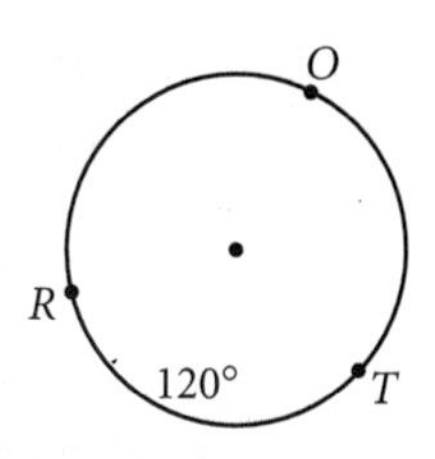

▶ Solution $m\widehat{ROT} = 240°$, so $\widehat{ROT}$ is $\frac{240}{360}$, or $\frac{2}{3}$ of the circumference.

$116\pi = \frac{2}{3}C$	Apply the Arc Length Conjecture.
$116\pi = \frac{2}{3}(2\pi r)$	Substitute $2\pi r$ for C.
$348\pi = 4\pi r$	Multiply both sides by 3.
$87 = r$	Divide both sides by 4π.

The radius is 87 m.

EXERCISES

You will need

A calculator for Exercises **9–14**

Construction tools for Exercise **16**

Geometry software for Exercise **16**

For Exercises 1–8, state your answers in terms of π.

1. Length of $\widehat{CD}$ is _?_.

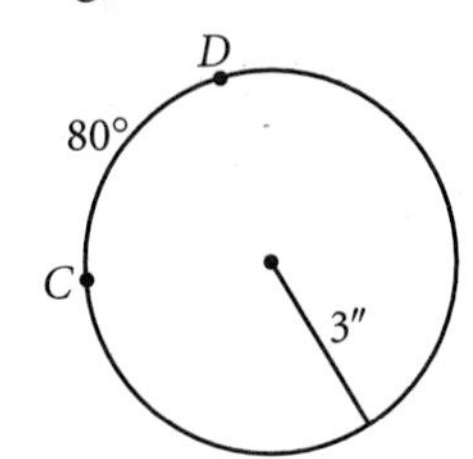

2. Length of $\widehat{EF}$ is _?_.

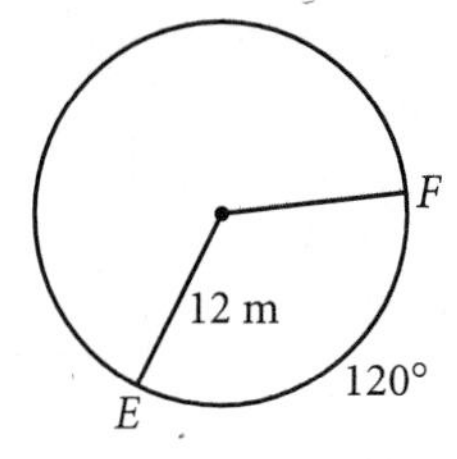

3. Length of $\widehat{BIG}$ is _?_. ⓗ

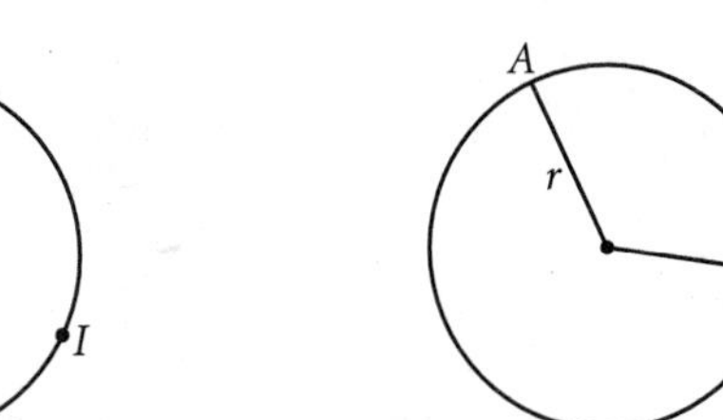

4. Length of $\widehat{AB}$ is 6π m. The radius is _?_.

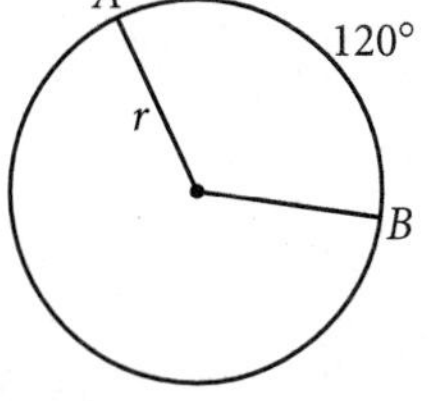

5. The radius is 18 ft. Length of $\widehat{RT}$ is _?_.

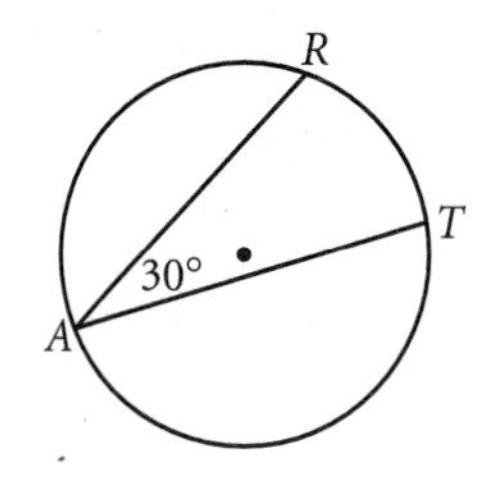

6. The radius is 9 m. Length of $\widehat{SO}$ is _?_.

7. Length of $\widehat{TV}$ is 12π in. The diameter is _?_.

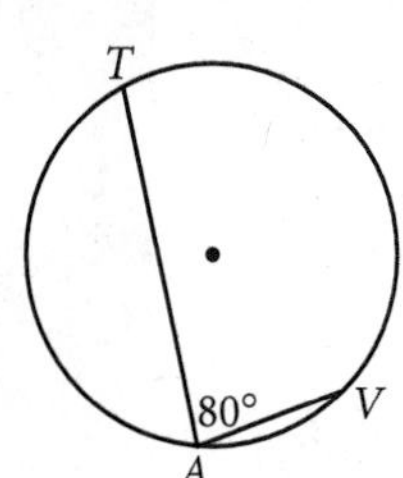

8. Length of $\widehat{AR}$ is 40π cm. $\overleftrightarrow{CA} \parallel \overleftrightarrow{RE}$. The radius is _?_. ⓗ

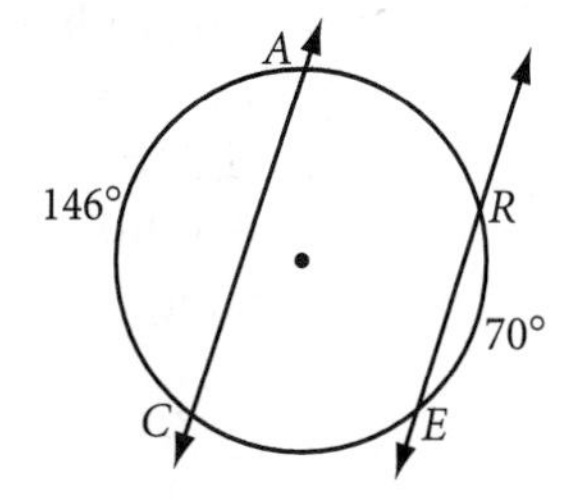

9. A go-cart racetrack has 100-meter straightaways and semicircular ends with diameters of 40 meters. Calculate the average speed in meters per minute of a go-cart if it completes 4 laps in 6 minutes. Round your answer to the nearest m/min. ⓗ

10. Astronaut Polly Hedra circles Earth every 90 minutes in a path above the equator. If the diameter of Earth is approximately 8000 miles, what distance along the equator will she pass directly over while eating a quick 15-minute lunch?

11. APPLICATION The Library of Congress reading room has desks along arcs of concentric circles. If an arc on the outermost circle with eight desks is about 12 meters long and makes up $\frac{1}{9}$ of the circle, how far are these desks from the center of the circle? How many desks would fit along an arc with the same central angle but that is half as far from the center? Explain. ⓗ

The Library of Congress, Washington, D.C.

12. A Greek mathematician who lived in the third century B.C.E., Eratosthenes, devised a clever method to calculate the circumference of Earth. He knew that the distance between Aswan (then called Syene) and Alexandria was 5000 Greek stadia (a stadium was a unit of distance at that time), or about 500 miles. At noon of the summer solstice, the Sun cast no shadow on a vertical pole in Syene, but at the same time in Alexandria a vertical pole did cast a shadow. Eratosthenes found that the angle between the vertical pole and the ray from the tip of the pole to the end of the shadow was 7.2°. From this he was able to calculate the ratio of the distance between the two cities to the circumference of Earth. Use this diagram to explain Eratosthenes' method, then use it to calculate the circumference of Earth in miles.

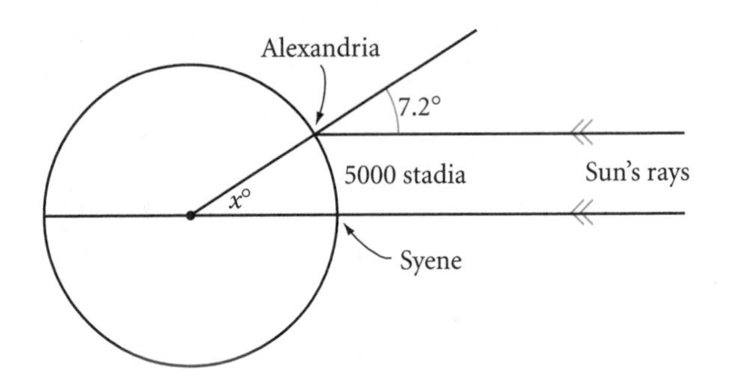

Review

13. **Angular velocity** is a measure of the rate at which an object revolves around an axis, and can be expressed in degrees per second. Suppose a carousel horse completes a revolution in 20 seconds. What is its angular velocity? Would another horse on the carousel have a different angular velocity? Why or why not?

14. **Tangential velocity** is a measure of the distance an object travels along a circular path in a given amount of time. Like speed, it can be expressed in meters per second. Suppose two carousel horses complete a revolution in 20 seconds. The horses are 8 m and 6 m from the center of the carousel, respectively. What are the tangential velocities of the two horses? Round your answers to the nearest 0.1 m/sec. Explain why the horses have equal angular velocities but different tangential velocities.

15. Calculate the measure of each lettered angle. ⓗ

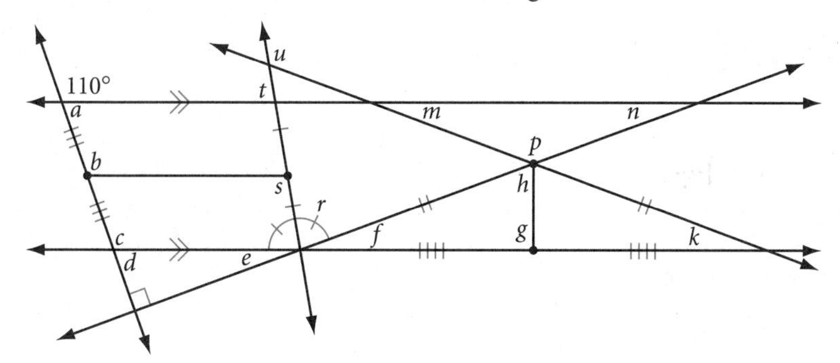

Art CONNECTION

The traceries surrounding rose windows in Gothic cathedrals were constructed with only arcs and straight lines. The photo at right shows a rose window from Reims cathedral, which was built in the thirteenth century, in Reims, a city in northeastern France. The overlaid diagram shows its constructions.

16. ***Construction*** Read the Art Connection above. Reproduce the constructions shown with your compass and straightedge or with geometry software. ⓗ

17. Find the measure of the angle formed by a clock's hands at 10:20. ⓗ

project

RACETRACK GEOMETRY

If you had to start and finish at the same line of a racetrack, which lane would you choose? The inside lane has an obvious advantage. For a race to be fair, runners in the outside lanes must be given head starts, as shown in the photo.

Design a four-lane oval track with straightaways and semicircular ends. Show start and finish lines so that an 800-meter race can be run fairly in all four lanes. The semicircular ends must have inner diameters of 50 meters. The distance of one lap in the inner lane must be 800 meters.

Your project should contain

- A detailed drawing with labeled lengths.
- An explanation of the part that radius, lane width, and straightaway length plays in the design.

Exploration

Cycloids

Imagine a bug gripping your bicycle tire as you ride down the street. What would the bug's path look like? What path does the Moon make as it rotates around Earth while Earth rotates around the Sun? Using animation in Sketchpad, you can model a rotating wheel.

Activity
Turning Wheels

In this activity, you'll investigate the path of a point on a wheel as it rolls along the ground or around another wheel. You'll start by constructing a stationary circle with a rotating spoke.

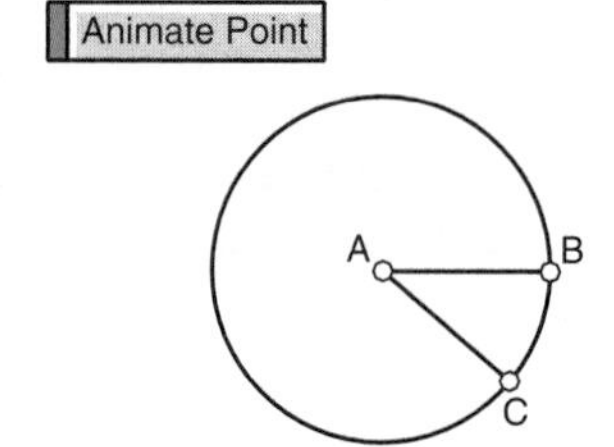

Step 1 Construct two points A and B. Select them in that order, and choose **Circle By Center+Point** from the Construct menu. Construct radius $\overline{AB}$.

Step 2 Construct a second radius, $\overline{AC}$.

Step 3 Select point C and choose **Animation** from the Action Buttons submenu of the Edit menu. Make point C move **counter-clockwise** around the circle at **fast** speed.

Step 4 Press the Animation button. Now you have a circle with one spoke that rotates in a counterclockwise direction. (Press it again to stop.)

How can you make a wheel that will roll? You can't roll your circle with the spinning spoke because if the circle moves, the spoke would have to move with it. But you can make a different circle and, as you move it, use circle A to rotate it. Here's how.

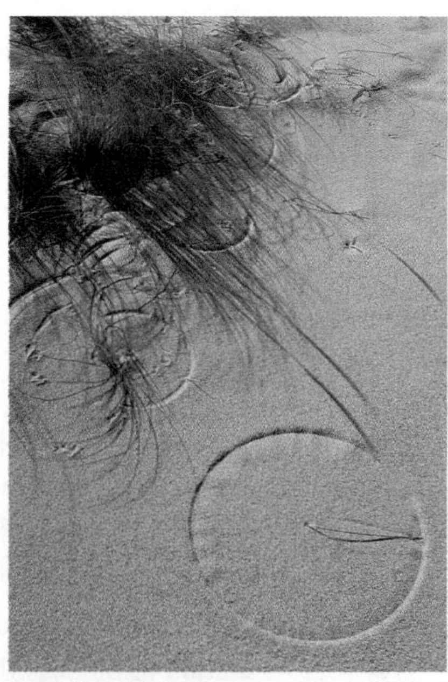
These circles in the sand were created by the wind blowing blades of grass as if they were spokes on a wheel.

Step 5 Construct a long horizontal segment DE going from right to left and a point F on the segment.

Step 6 Select points A and F, in order, and choose **Mark Vector** from the Transform menu.

Step 7 Select circle A, point C, and $\overline{AC}$, and use the Transform menu to translate by the marked vector.

Step 8 Construct a line FC' overlapping $\overline{FC'}$ and a point G anywhere on the line. Hide the line and construct $\overline{FG}$.

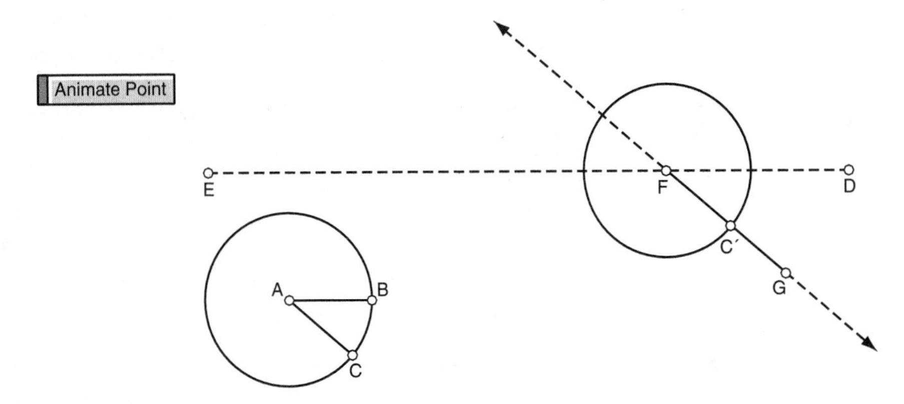

Step 9 Press the animation button. The rotating spoke $\overline{AC}$ should cause the spoke $\overline{FG}$ to spin at the same time.

Now you're ready to make that circle roll.

Step 10 Select $\overline{DE}$ and point F, and choose **Perpendicular Line** from the Construct menu.

Step 11 To construct the road, select point H, the lower point where the line intersects the circle, and $\overline{DE}$ and choose **Parallel Line** from the Construct menu. Hide the perpendicular line and point H.

Step 12 Select points F and C and create an action button that animates both point F backward along the segment at **fast** speed and point C **counterclockwise** around the circle at **fast** speed.

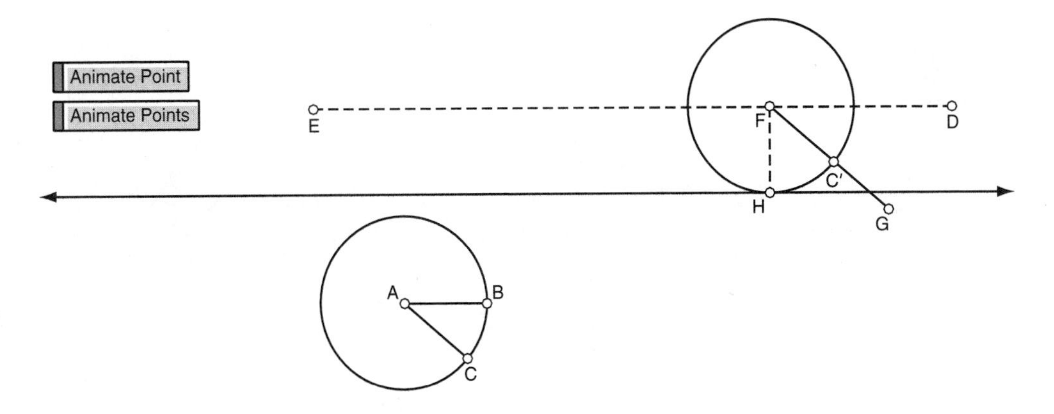

Step 13 Press this new animation button. Circle F should move to the left, rotating at the same time so that it appears to be rolling. Drag point G so that it is on the circle and choose **Trace Point** from the Display menu.

Investigate these questions.

Step 14 What does the path of a point on a rolling circle look like? On your paper, sketch the curve point G makes as the circle rolls. This curve is called a **cycloid.**

Step 15 Experiment with different cycloids made when point G is inside the circle and outside the circle. Sketch these curves onto your paper.

Step 16 A cycloid is an example of a **periodic** curve. What do you think that means? Adjust the radius AB or the length DE so that point G traces one period, or **cycle,** of the curve. Adjust these lengths so that point G traces two cycles or three cycles. How are the lengths DE and AB related to the number of cycles of the curve?

Step 17 If you apply a car's brakes on a slippery road, you'll skid, because your wheels won't turn fast enough to keep up with the car's movement down the road. If you try to accelerate on a slippery road, your wheels will spin—they'll turn faster than the car is able to go. Experiment with different speeds for points C and F by selecting the animation button and choosing **Properties** from the Edit menu. What combinations cause the wheel to spin? To skid? Explain why the wheel has good traction when the animation speeds are the same.

Step 18 Add to your sketch so that it shows a traveling bicycle or car. To make a second wheel, you can simply translate the wheel you have by a fixed distance. When you construct the vehicle on these wheels, you'll need to make sure all the parts move when the wheels do!

Step 19 The figure below shows an **epicycloid,** the path of a point on a circle that is rolling around another circle. See if you can make a construction that traces an epicycloid. (Hint: The dashed circle is important in the construction. The circle inside it is just drawn for show.)

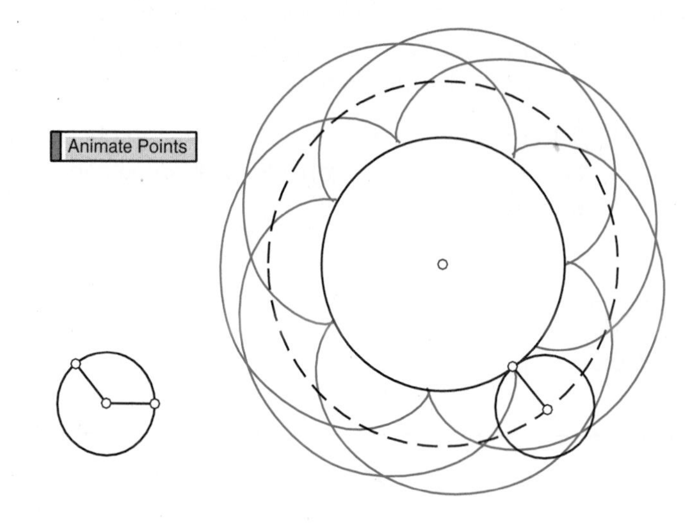

CHAPTER 6 REVIEW

In this chapter you learned some new circle vocabulary and solved real-world application problems involving circles. You discovered the relationship between a radius and a tangent line. You discovered special relationships between angles and their intercepted arcs. And you learned about the special ratio π and how to use it to calculate the circumference of a circle and the length of an arc.

You should be able to sketch these terms from memory: *chord, tangent, central angle, inscribed angle,* and *intercepted arc.* And you should be able to explain the difference between arc measure and arc length.

EXERCISES

You will need

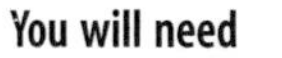

Construction tools for Exercises **21–24, 67,** and **70**

A calculator for Exercises **11, 12,** and **27–33**

1. What do you think is the most important or useful circle property you learned in this chapter? Why?
2. How can you find the center of a circle with a compass and a straightedge? With patty paper? With the right-angled corner of a carpenter's square?
3. What does the path of a satellite have to do with the Tangent Conjecture?
4. Explain the difference between the degree measure of an arc and its arc length.

Solve Exercises 5–19. If the exercise uses the "=" sign, answer in terms of π. If the exercise uses the "≈" sign, give your answer accurate to one decimal place.

5. $b = \underline{?}$ (h)

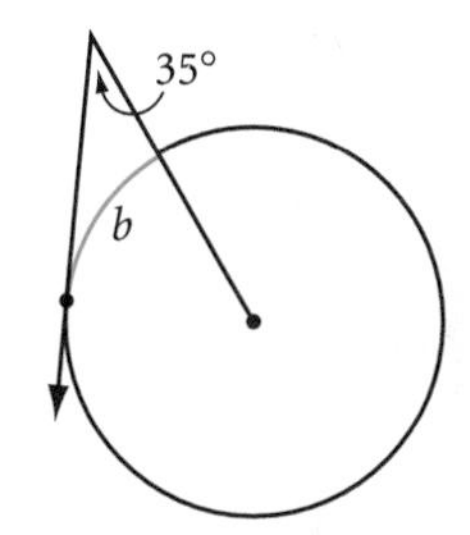

6. $a = \underline{?}$

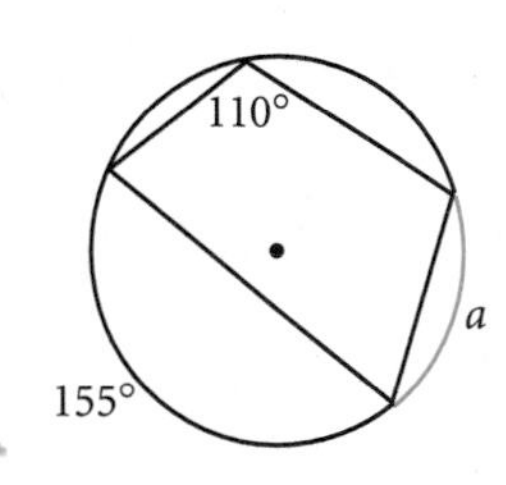

7. $c = \underline{?}$

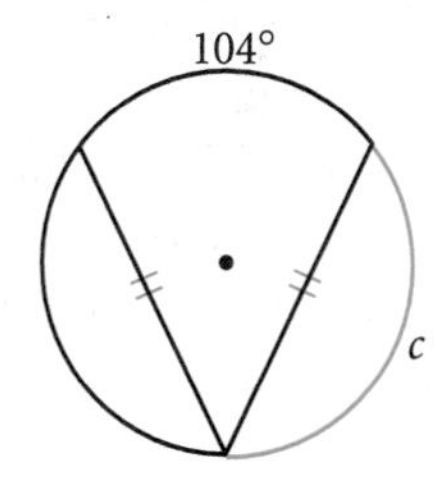

8. $e = \underline{?}$ (h)

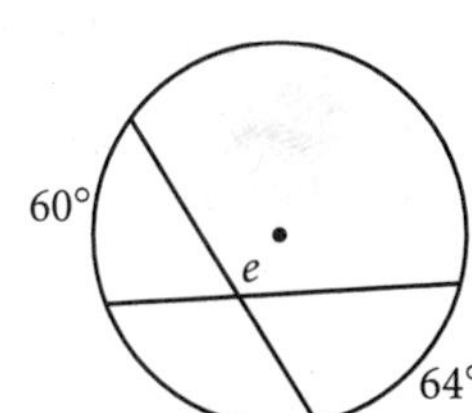

9. $d = \underline{?}$

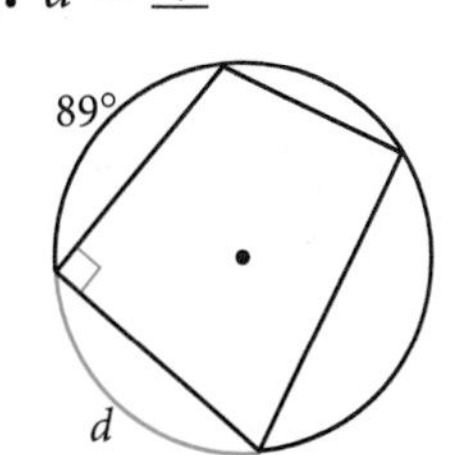

10. $f = \underline{?}$ (h)

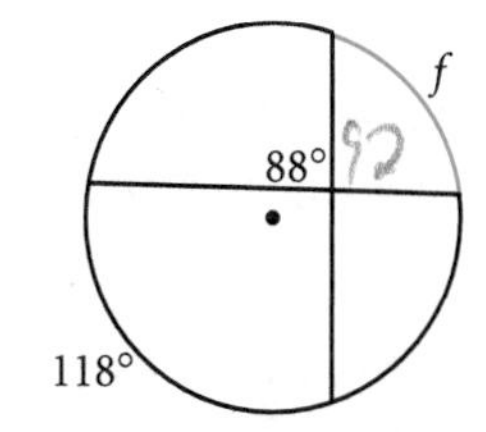

11. Circumference $\approx$ _?_

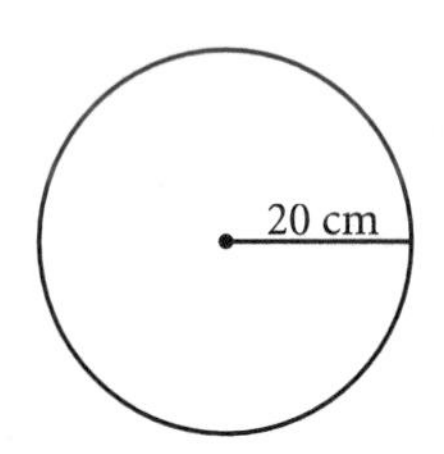

12. Circumference is 132 cm. $d \approx$ _?_ ⓗ

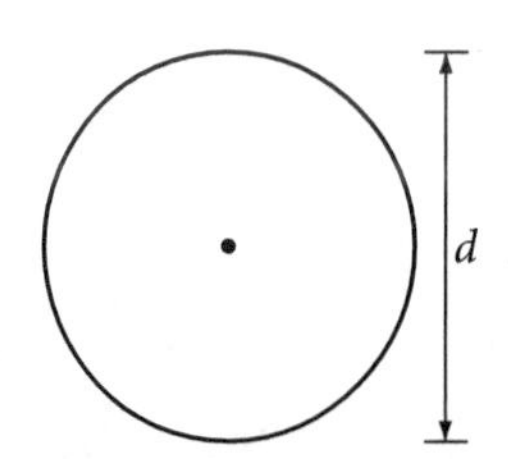

13. $r = 27$ cm. The arc length of $\overset{\frown}{AB}$ is _?_. ⓗ

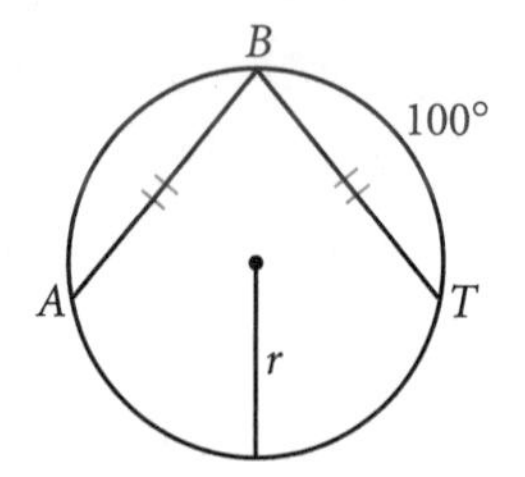

14. $r = 36$ ft. The arc length of $\overset{\frown}{CD}$ is _?_. ⓗ

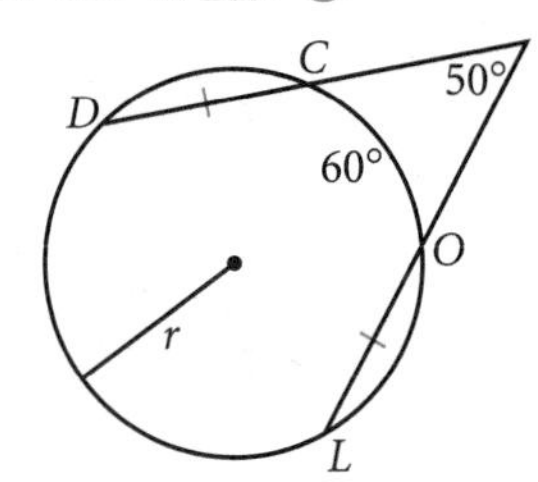

15. What's wrong with this picture?

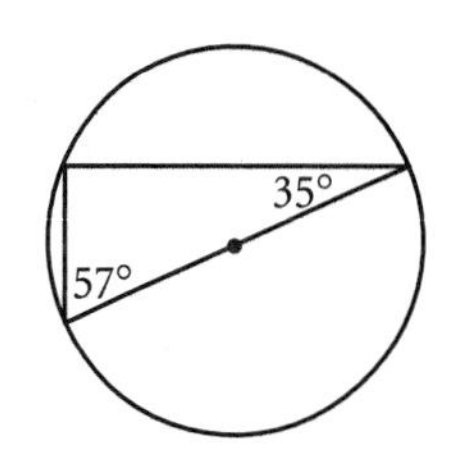

16. What's wrong with this picture?

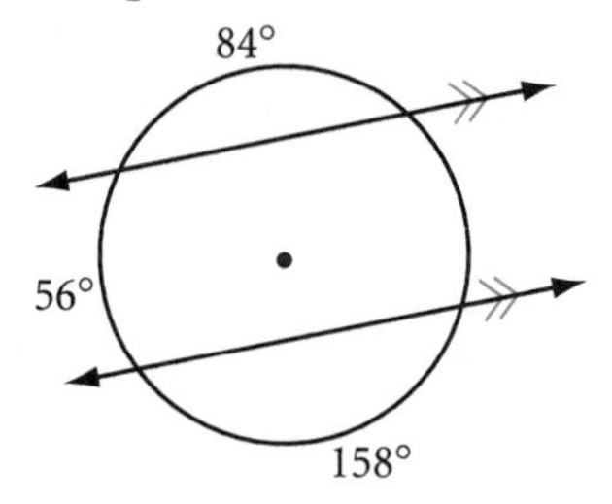

17. Explain why $\overline{KE} \parallel \overline{YL}$.

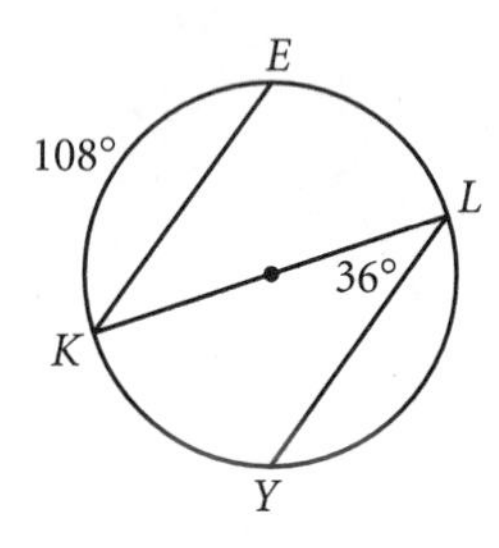

18. Explain why $\triangle JIM$ is isosceles.

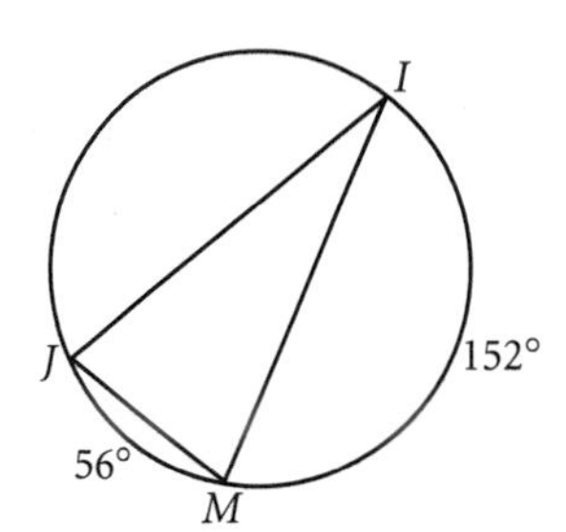

19. Explain why $\triangle KIM$ is isosceles.

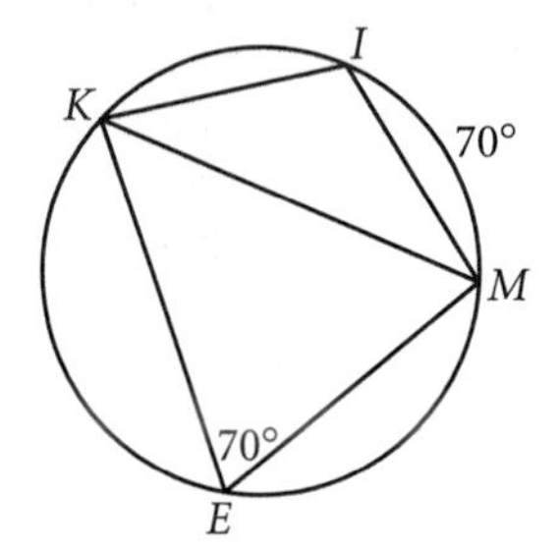

20. On her latest archaeological dig, Ertha Diggs has unearthed a portion of a cylindrical column. All she has with her is a pad of paper. How can she use it to locate the diameter of the column?

21. ***Construction*** Construct a scalene obtuse triangle. Construct the circumscribed circle.

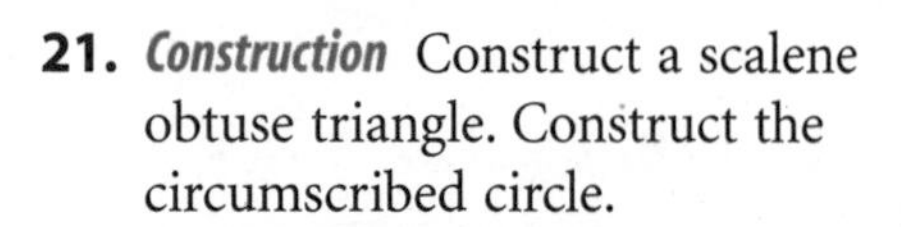

22. ***Construction*** Construct a scalene acute triangle. Construct the inscribed circle.

23. ***Construction*** Construct a rectangle. Is it possible to construct the circumscribed circle, the inscribed circle, neither, or both?

24. ***Construction*** Construct a rhombus. Is it possible to construct the circumscribed circle, the inscribed circle, neither, or both?

25. Find the equation of the line tangent to circle S centered at $(1, 1)$ if the point of tangency is $(5, 4)$.

26. Find the center of the circle passing through the points $(-7, 5)$, $(0, 6)$, and $(1, -1)$.

27. Rashid is an apprentice on a road crew for a civil engineer. He needs to find a trundle wheel similar to but larger than the one shown at right. If each rotation is to be 1 m, what should be the diameter of the trundle wheel?

28. Melanie rides the merry-go-round on her favorite horse on the outer edge, 8 meters from the center of the merry-go-round. Her sister, Melody, sits in the inner ring of horses, 3 meters in from Melanie. In 10 minutes, they go around 30 times. What is the average speed of each sister?

29. Read the Geography Connection below. Given that the polar radius of Earth is 6357 kilometers and that the equatorial radius of Earth is 6378 kilometers, use the original definition to calculate one nautical mile near a pole and one nautical mile near the equator. Show that the international nautical mile is between both values. ⓗ

One nautical mile was originally defined to be the length of one minute of arc of a great circle of Earth. (A great circle is the intersection of the sphere and a plane that cuts through its center. There are 60 minutes of arc in each degree.) But Earth is not a perfect sphere. It is wider at the great circle of the equator than it is at the great circle through the poles. So defined as one minute of arc, one nautical mile could take on a range of values. To remedy this, an international nautical mile was defined as 1.852 kilometers (about 1.15 miles).

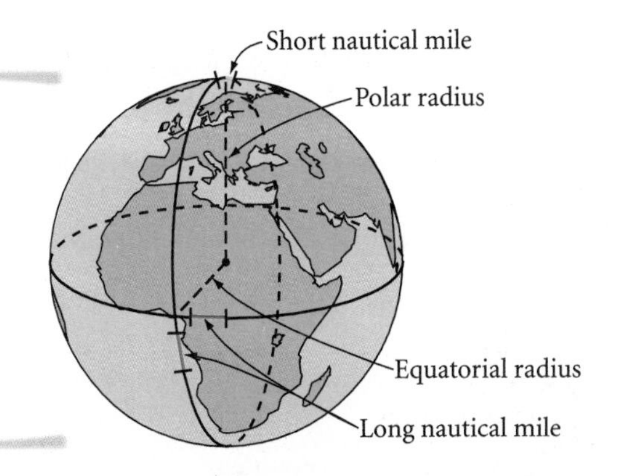

30. While talking to his friend Tara on the phone, Dmitri sees a lightning flash, and 5 seconds later he hears thunder. Two seconds after that, Tara, who lives 1 mile away, hears it. Sound travels at 1100 feet per second. Draw and label a diagram showing the possible locations of the lightning strike. ⓗ

31. King Arthur wishes to seat all his knights at a round table. He instructs Merlin to design and create an oak table large enough to seat 100 people. Each knight is to have 2 ft along the edge of the table. Help Merlin calculate the diameter of the table. ⓗ

32. If the circular moat should have been a circle of radius 10 meters instead of radius 6 meters, how much greater should the larger moat's circumference have been?

33. The part of a circle enclosed by a central angle and the arc it intercepts is called a **sector.** The sector of a circle shown below can be curled into a cone by bringing the two straight 45-cm edges together. What will be the diameter of the base of the cone?

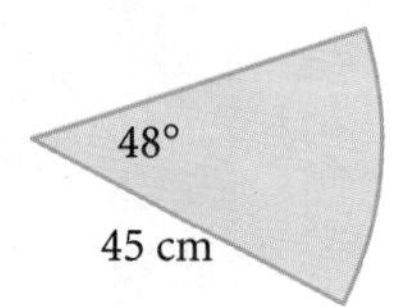

THE FAR SIDE® By GARY LARSON

Suddenly, a heated exchange took place between the king and the moat contractor.

MIXED REVIEW

In Exercises 34–56, identify the statement as true or false. For each false statement, explain why it is false or sketch a counterexample.

34. If a triangle has two angles of equal measure, then the third angle is acute.

35. If two sides of a triangle measure 45 cm and 36 cm, then the third side must be greater than 9 cm and less than 81 cm.

36. The diagonals of a parallelogram are congruent.

37. The measure of each angle of a regular dodecagon is 150°.

38. The perpendicular bisector of a chord of a circle passes through the center of the circle.

39. If $\overline{CD}$ is the midsegment of trapezoid *PLYR* with $\overline{PL}$ one of the bases, then $CD = \frac{1}{2}(PL + YR)$.

40. In $\triangle BOY$, $BO = 36$ cm, $m\angle B = 42°$, and $m\angle O = 28°$. In $\triangle GRL$, $GR = 36$ cm, $m\angle R = 28°$, and $m\angle L = 110°$. Therefore, $\triangle BOY \cong \triangle GRL$.

41. If the sum of the measures of the interior angles of a polygon is less than 1000°, then the polygon has fewer than seven sides.

42. The sum of the measures of the three angles of an obtuse triangle is greater than the sum of the measures of the three angles of an acute triangle.

43. The sum of the measures of one set of exterior angles of a polygon is always less than the sum of the measures of interior angles.

44. Both pairs of base angles of an isosceles trapezoid are supplementary.

45. If the base angles of an isosceles triangle each measure 48°, then the vertex angle has a measure of 132°.

46. Inscribed angles that intercept the same arc are supplementary.

47. The measure of an inscribed angle in a circle is equal to the measure of the arc it intercepts.

The concentric circles in the sky are actually a time exposure photograph of the movement of the stars in a night.

48. The diagonals of a rhombus bisect the angles of the rhombus.

49. The diagonals of a rectangle are perpendicular bisectors of each other.

50. If a triangle has two angles of equal measure, then the triangle is equilateral.

51. If a quadrilateral has three congruent angles, then it is a rectangle.

52. In two different circles, arcs with the same measure are congruent.

53. The ratio of the diameter to the circumference of a circle is π.

54. If the sum of the lengths of two consecutive sides of a kite is 48 cm, then the perimeter of the kite is 96 cm.

55. If the vertex angles of a kite measure 48° and 36°, then the nonvertex angles each measure 138°.

56. All but seven statements in Exercises 34–56 are false.

57. Find the measure of each lettered angle in the diagram below.

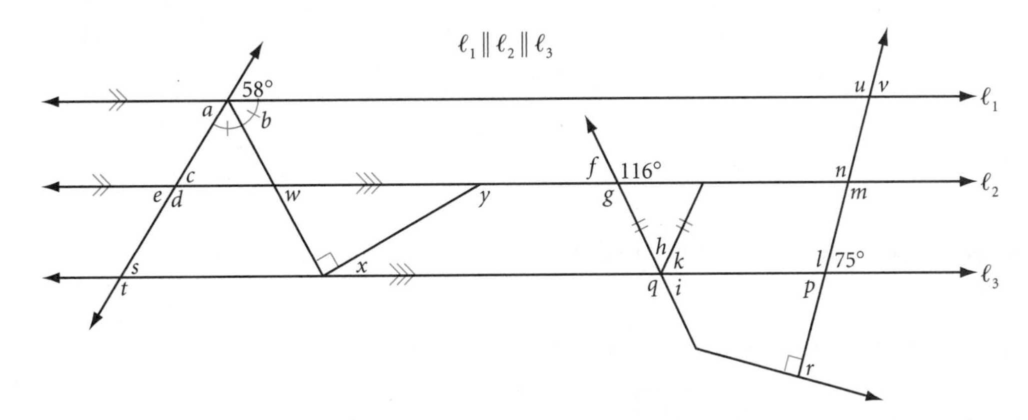

In Exercises 58–60, from the information given, determine which triangles, if any, are congruent. State the congruence conjecture that supports your congruence statement.

58. *STARY* is a regular pentagon.

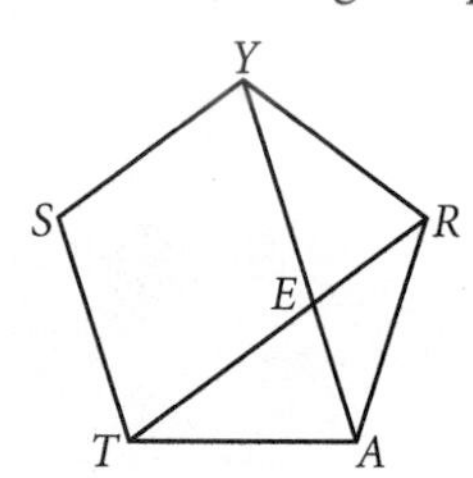

59. *FLYT* is a kite.

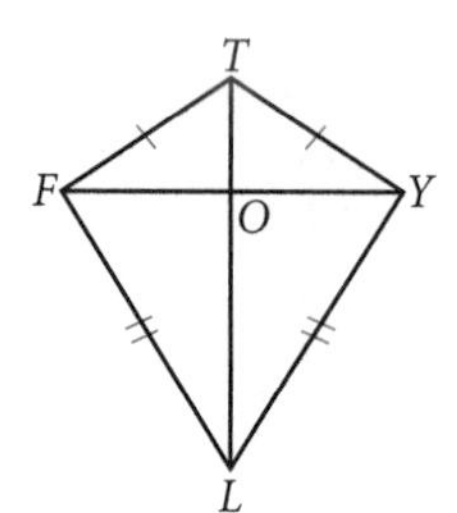

60. *PART* is an isosceles trapezoid.

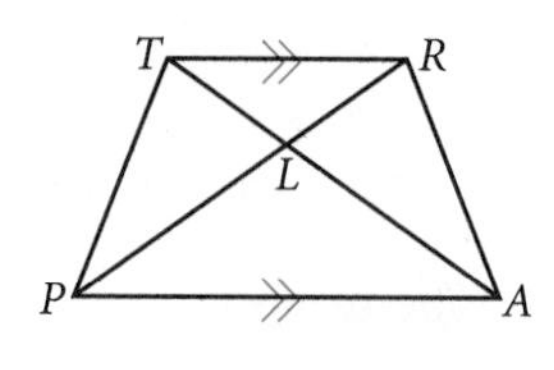

61. Adventurer Dakota Davis has uncovered a piece of triangular tile from a mosaic. A corner is broken off. Wishing to repair the mosaic, he lays the broken tile piece down on paper and traces the straight edges. With a ruler he then extends the unbroken sides until they meet. What triangle congruence shortcut guarantees that the tracing reveals the original shape?

62. Circle O has a radius of 24 inches. Find the measure and the length of $\widehat{AC}$.

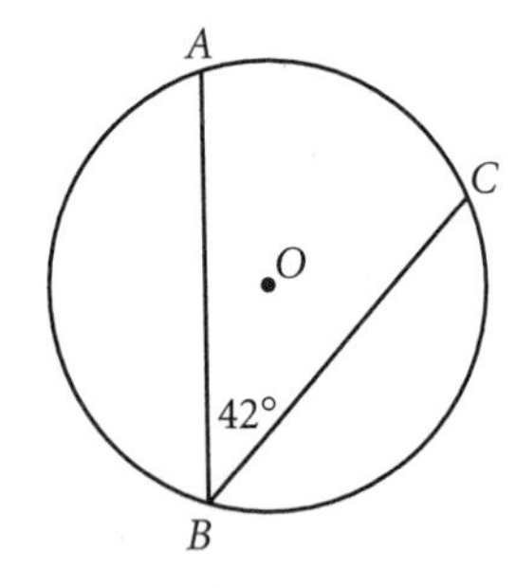

63. $\overrightarrow{EC}$ and $\overrightarrow{ED}$ are tangent to the circle, and $AB = CD$. Find the measure of each lettered angle.

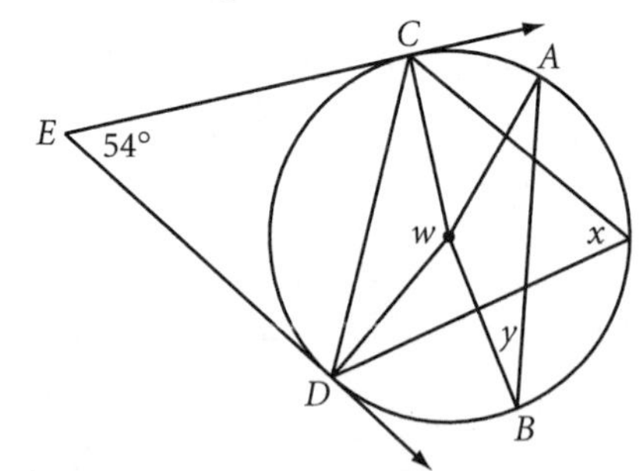

64. Use your protractor to draw and label a pair of supplementary angles that is not a linear pair.

65. Find the function rule $f(n)$ of this sequence and find the 20th term.

n	1	2	3	4	5	6	...	n	...	20
$f(n)$	5	1	−3	−7	−11	−15	...		...	

66. The design at right shows three hares joined by three ears, although each hare appears to have two ears of its own.

a. Does the design have rotational symmetry?

b. Does the design have reflectional symmetry?

Private Collection, Berkeley, California
Ceramist, Diana Hall

67. *Construction* Construct a rectangle whose length is twice its width.

68. If $AB = 15$ cm, C is the midpoint of $\overline{AB}$, D is the midpoint of $\overline{AC}$, and E is the midpoint of $\overline{DC}$, what is the length of $\overline{EB}$?

69. Draw the next shape in this pattern.

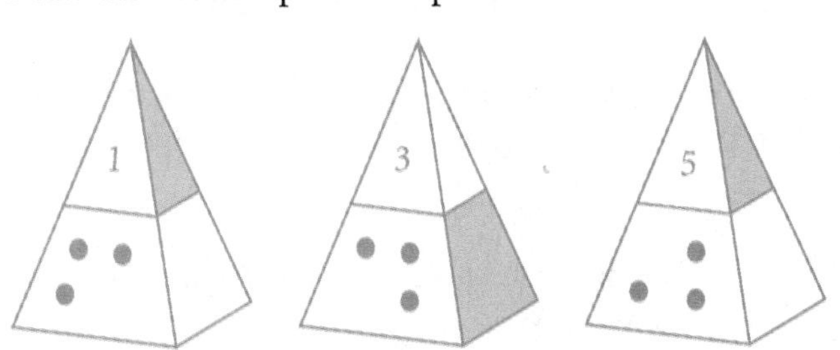

70. *Construction* Construct any triangle. Then construct its centroid.

TAKE ANOTHER LOOK

1. Show how the Tangent Segments Conjecture follows logically from the Tangent Conjecture and the converse of the Angle Bisector Conjecture.

2. Investigate the quadrilateral formed by two tangent segments to a circle and the two radii to the points of tangency. State a conjecture. Explain why your conjecture is true, based on the properties of radii and tangents.

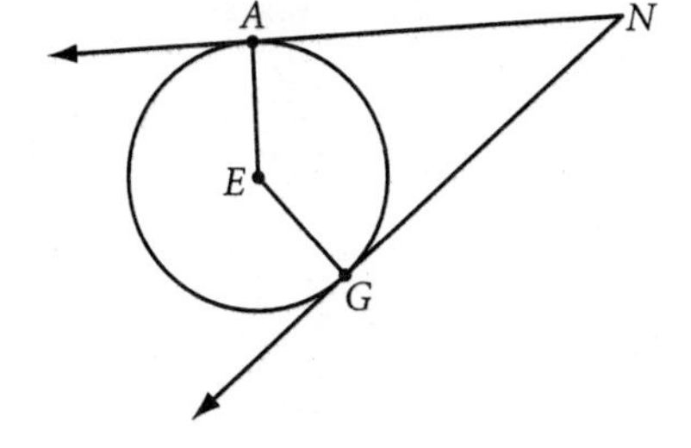

3. State the Cyclic Quadrilateral Conjecture in "if-then" form. Then state the converse of the conjecture in "if-then" form. Is the converse also true?

4. A quadrilateral that *can* be inscribed in a circle is also called a cyclic quadrilateral. Which of these quadrilaterals are always cyclic: parallelograms kites, isosceles trapezoids, rhombuses, rectangles, or squares? Which ones are never cyclic? Explain why each is or is not always cyclic.

5. Use graph paper or a graphing calculator to graph the data collected from the investigation in Lesson 6.5. Graph the diameter on the x-axis and the circumference on the y-axis. What is the slope of the best-fit line through the data points? Does this confirm the Circumference Conjecture? Explain.

Assessing What You've Learned

With the different assessment methods you've used so far, you should be getting the idea that assessment means more than a teacher giving you a grade. All the methods presented so far could be described as self-assessment techniques. Many are also good study habits. Being aware of your *own* learning and progress is the best way to stay on top of what you're doing and to achieve the best results.

WRITE IN YOUR JOURNAL

- You may be at or near the end of your school year's first semester. Look back over the first semester and write about your strengths and needs. What grade would you have given yourself for the semester? How would you justify that grade?
- Set new goals for the new semester or for the remainder of the year. Write them in your journal and compare them to goals you set at the beginning of the year. How have your goals changed? Why?

ORGANIZE YOUR NOTEBOOK Review your notebook and conjectures list to be sure they are complete and well organized. Write a one-page chapter summary.

UPDATE YOUR PORTFOLIO Choose a piece of work from this chapter to add to your portfolio. Document the work according to your teacher's instructions.

PERFORMANCE ASSESSMENT While a classmate, a friend, a family member, or a teacher observes, carry out one of the investigations or Take Another Look activities from this chapter. Explain what you're doing at each step, including how you arrived at the conjecture.

WRITE TEST ITEMS Divide the lessons from this chapter among group members and write at least one test item per lesson. Try out the test questions written by your classmates and discuss them.

GIVE A PRESENTATION Give a presentation on an investigation, exploration, Take Another Look project, or puzzle. Work with your group, or try giving a presentation on your own.

CHAPTER

7 Transformations and Tessellations

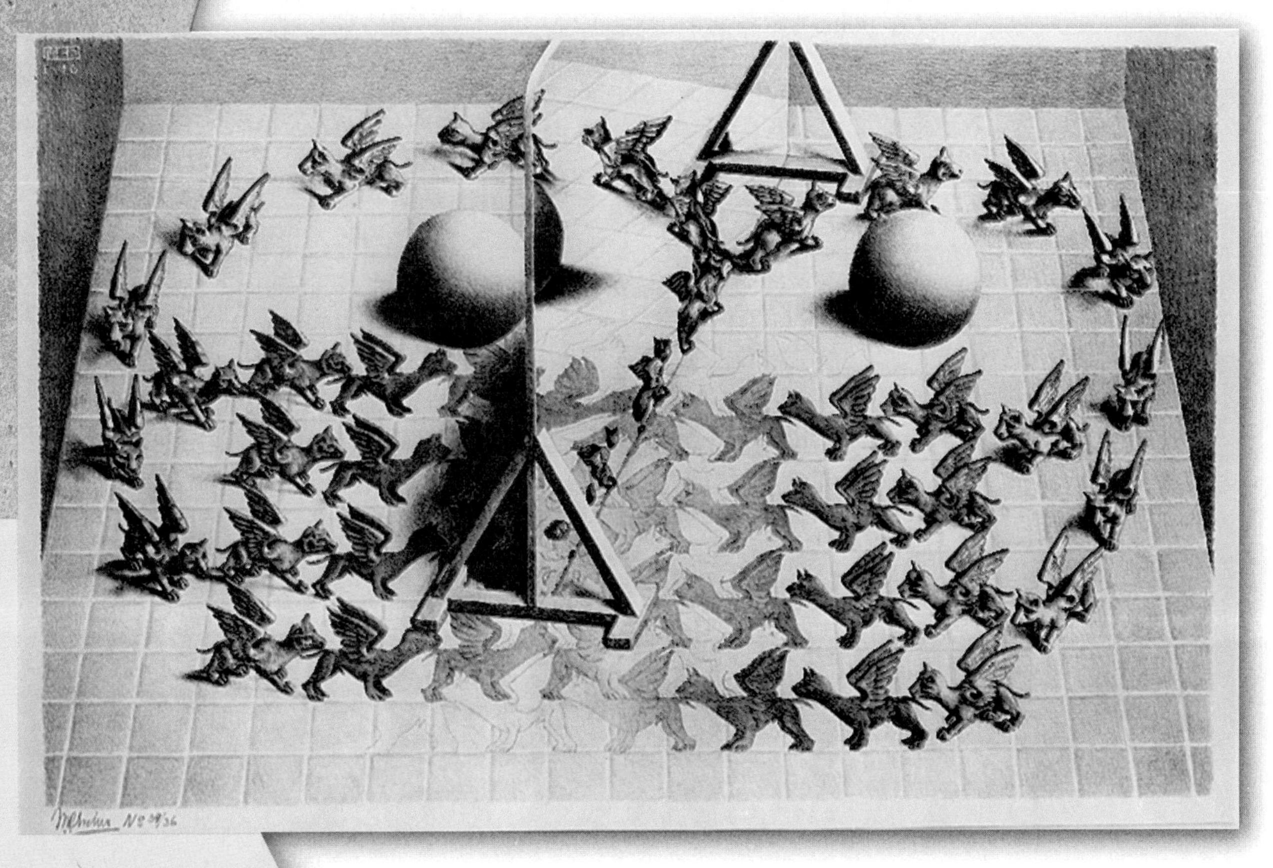

I believe that producing pictures, as I do, is almost solely a question of wanting so very much to do it well.

M. C. ESCHER

Magic Mirror, M. C. Escher, 1946

OBJECTIVES

In this chapter you will

- discover some basic properties of transformations and symmetry
- learn more about symmetry in art and nature
- create tessellations

LESSON

7.1 Transformations and Symmetry

Symmetry is one idea by which man through the ages has tried to comprehend and create order, beauty, and perfection.

HERMANN WEYL

By moving all the points of a geometric figure according to certain rules, you can create an **image** of the original figure. This process is called **transformation.** Each point on the original figure corresponds to a point on its image. The image of point A after a transformation of any type is called point A' (read "A prime"), as shown in the transformation of $\triangle ABC$ to $\triangle A'B'C'$ on the facing page.

Frieze of bowmen from the Palace of Artaxerxes II in Susa, Iran

If the image is congruent to the original figure, the process is called **rigid transformation,** or **isometry.** A transformation that does not preserve the size and shape is called **nonrigid transformation.** For example, if an image is reduced or enlarged, or if the shape changes, its transformation is nonrigid.

Each light bulb is an image of every other light bulb.

Three types of rigid transformation are translation, rotation, and reflection. You have been doing translations, rotations, and reflections in your patty-paper investigations and in exercises on the coordinate plane, using (x, y) rules.

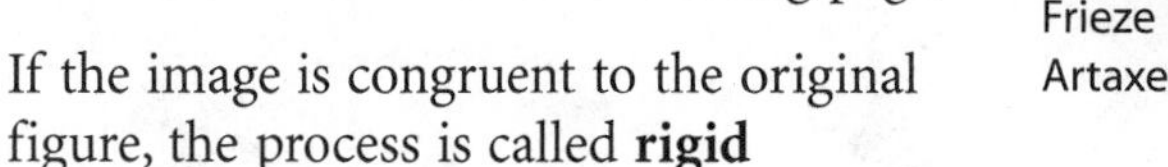

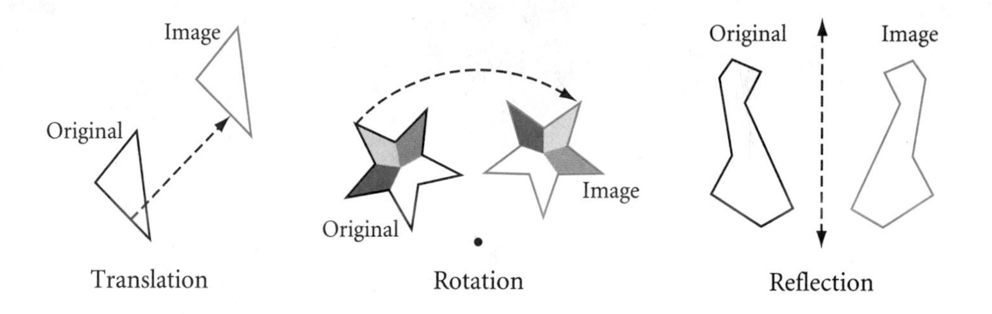

Translation is the simplest type of isometry. You can model a translation by tracing a figure onto patty paper, then sliding it along a straight path without turning it. Notice that when you slide the figure, all points move the same distance along parallel paths to form its image. That is, each point in the image is equidistant from the point that corresponds to it in the original figure. This distance, because it is the same for all points, is called the **distance** of the translation. A translation also has a particular **direction.** So you can use a **translation vector** to describe the translation.

Translation vector

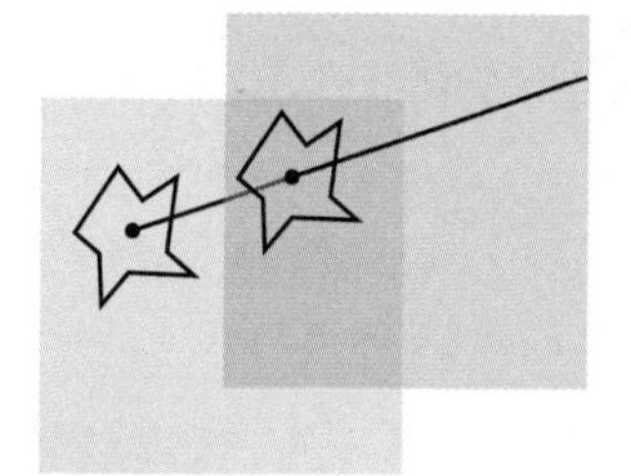

Translating with patty paper

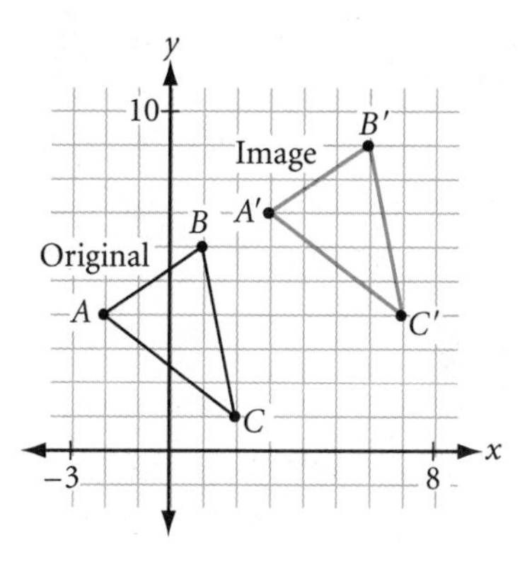

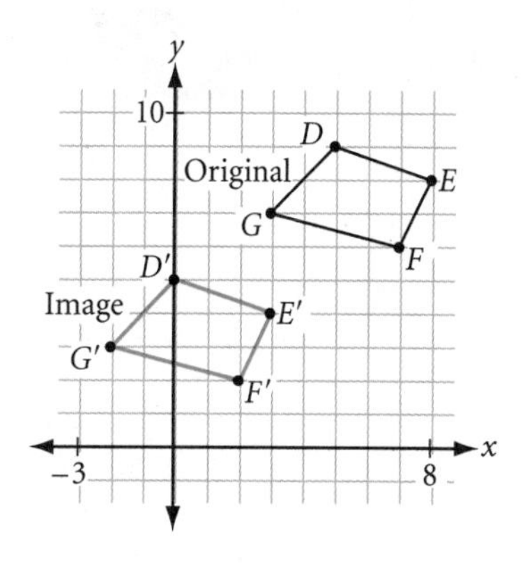

Translations on a coordinate grid

Rotation is another type of isometry. In a rotation, all the points in the original figure rotate, or turn, an identical number of degrees about a fixed center point. You can define a rotation by its center point, the number of degrees it's turned, and whether it's turned clockwise or counterclockwise. If no direction is given, assume the direction of rotation is counterclockwise.

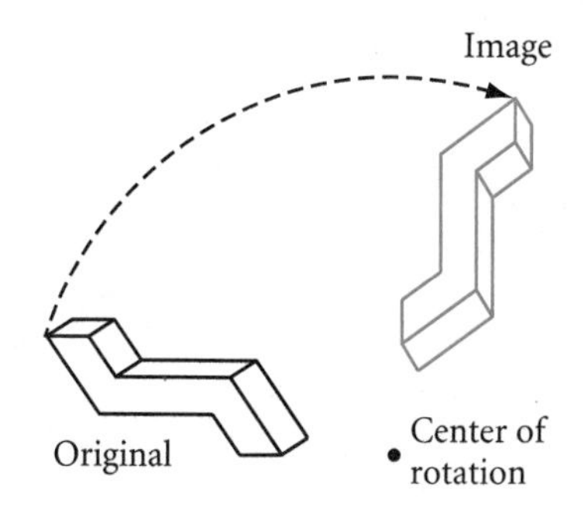

You can model a rotation by tracing over a figure, then putting your pencil point on a point on the patty paper and rotating the patty paper about the point.

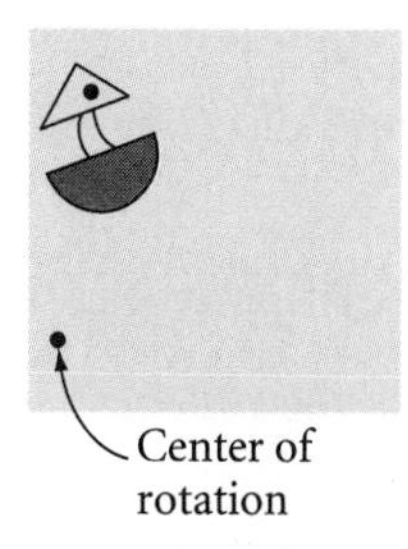

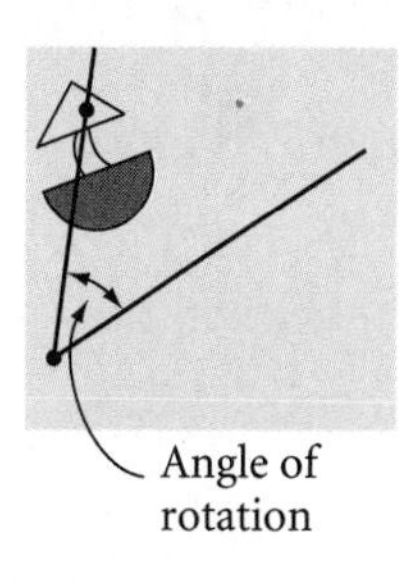

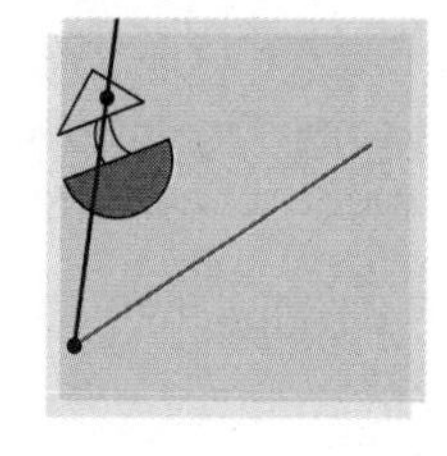

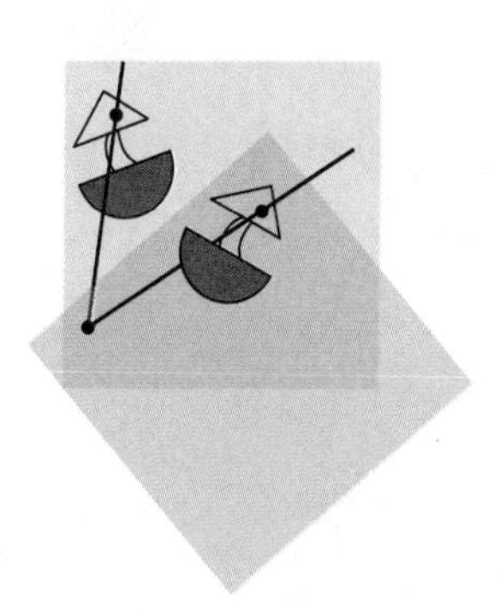

Rotating with patty paper

Reflection is a type of isometry that produces a figure's "mirror image." If you draw a figure onto a piece of paper, place the edge of a mirror perpendicular to your paper, and look at the figure in the mirror, you will see the reflected image of the figure. The line where the mirror is placed is called the **line of reflection.**

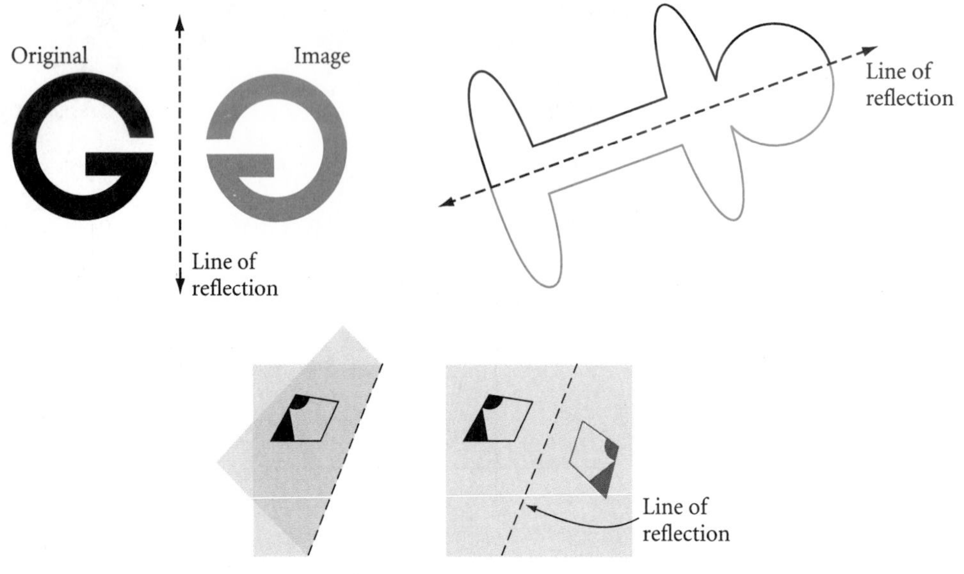

Reflecting with patty paper

History CONNECTION

Leonardo da Vinci (1452–1519, Italy) wrote his scientific discoveries backward so that others couldn't read them and punish him for his research and ideas. His knowledge and authority in almost every subject is astonishing even today. Scholars marvel at his many notebooks containing research into anatomy, mathematics, architecture, geology, meteorology, machinery, and botany, as well as his art masterpieces, like the *Mona Lisa*. Notice his plans for a helicopter in the manuscript at right!

Investigation
The Basic Property of a Reflection

You will need

- patty paper

In this investigation you'll model reflection with patty paper and discover an important property of reflections.

Step 1 Draw a polygon and a line of reflection next to it on a piece of patty paper.

Step 2 Fold your patty paper along the line of reflection and create the reflected image of your polygon by tracing it.

Step 3 Draw segments connecting each vertex with its image point. What do you notice?

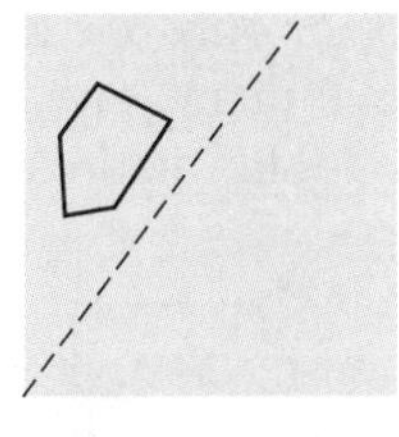

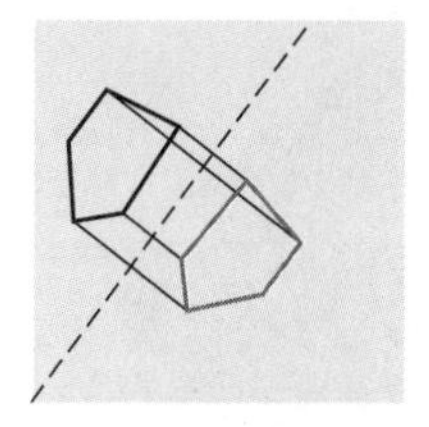

Step 1 Step 2 Step 3

Step 4 Compare your results with those of your group members. Copy and complete the following conjecture.

Reflection Line Conjecture **C-68**

The line of reflection is the ___?___ of every segment joining a point in the original figure with its image.

If a figure can be reflected over a line in such a way that the resulting image coincides with the original, then the figure has **reflectional symmetry.** The reflection line is called the **line of symmetry.** The Navajo rug shown below has two lines of symmetry.

The letter *T* has reflectional symmetry. You can test a figure for reflectional symmetry by using a mirror or by folding it.

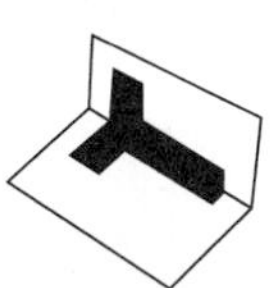

Navajo rug (two lines of symmetry)

The letter *Z* has 2-fold rotational symmetry. When it is rotated 180° and 360° about a center of rotation, the image coincides with the original figure.

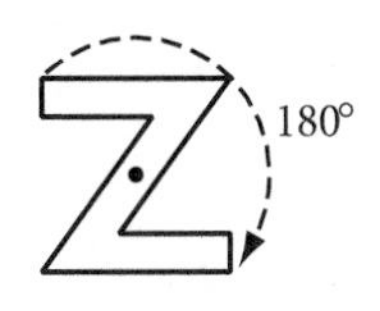

If a figure can be rotated about a point in such a way that its rotated image coincides with the original figure before turning a full 360°, then the figure has **rotational symmetry.** Of course, every image is identical to the original figure after a rotation of any multiple of 360°. However, we don't call a figure symmetric if this is the only kind of symmetry it has. You can trace a figure to test it for rotational symmetry. Place the copy exactly over the original, put your pen or pencil point on the center to hold it down, and rotate the copy. Count the number of times the copy and the original coincide until the copy is back in its original position. Two-fold rotational symmetry is also called **point symmetry.**

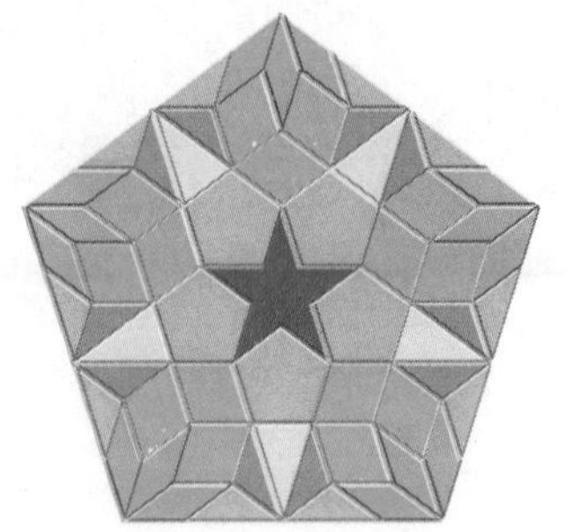

This tile pattern has both 5-fold rotational symmetry and 5-fold reflectional symmetry.

Some polygons have no symmetry, or only one kind of symmetry. Regular polygons, however, are symmetric in many ways. A square, for example, has 4-fold reflectional symmetry and 4-fold rotational symmetry.

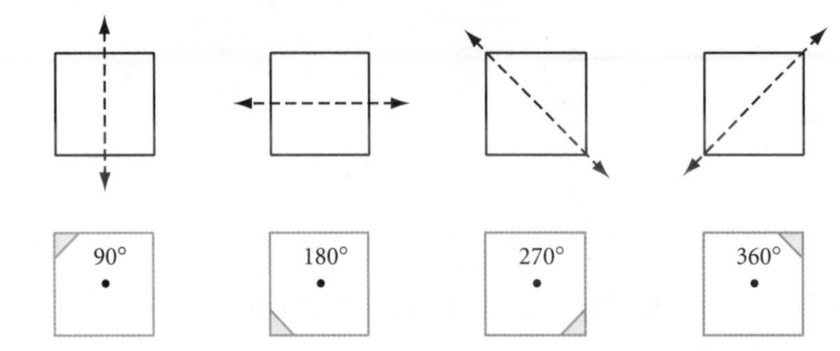

Art CONNECTION

Reflecting and rotating a letter can produce an unexpected and beautiful design. With the aid of graphics software, a designer can do this quickly and inexpensively. To see how, go to www.keymath.com/DG. This design was created by geometry student Michelle Cotter.

EXERCISES

You will need

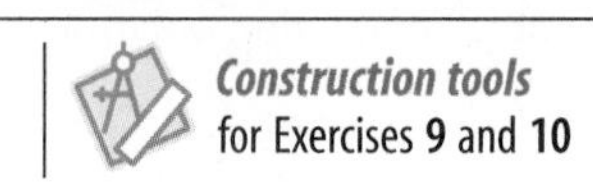

Construction tools for Exercises **9** and **10**

In Exercises 1–3, say whether the transformations are rigid or nonrigid. Explain how you know.

1.

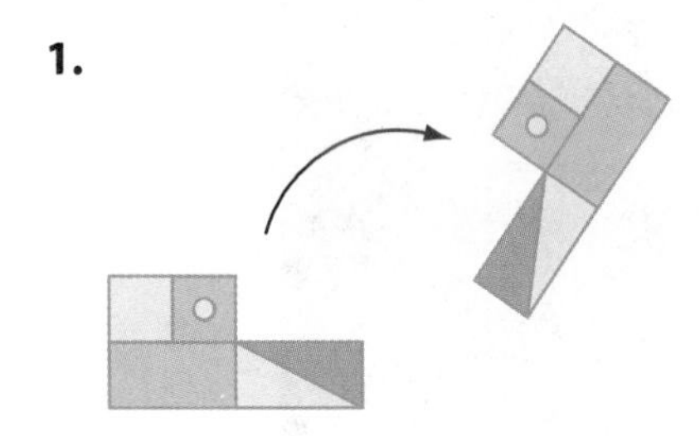

2.

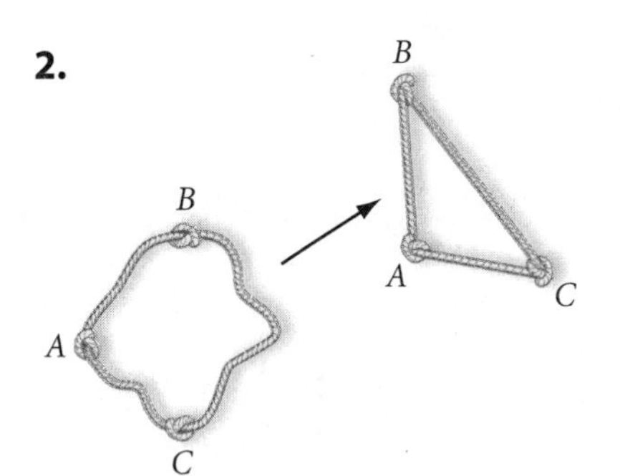

3.

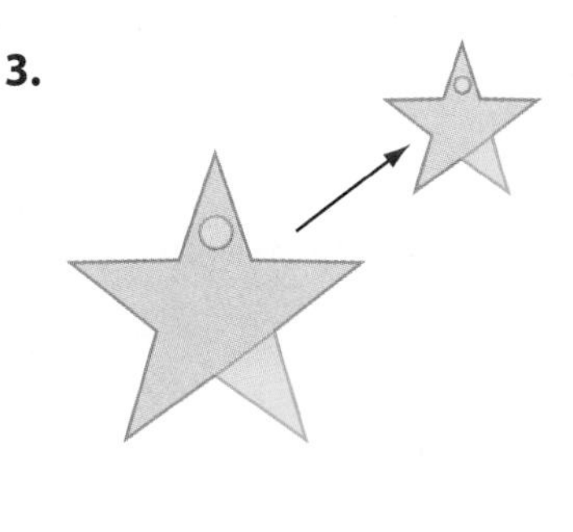

In Exercises 4–6, copy the figure onto graph or square dot paper and perform each transformation.

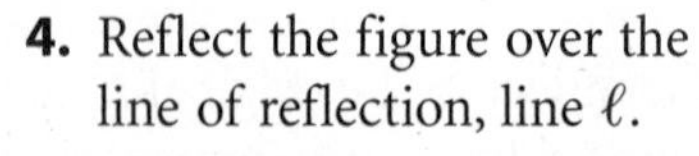

4. Reflect the figure over the line of reflection, line ℓ.

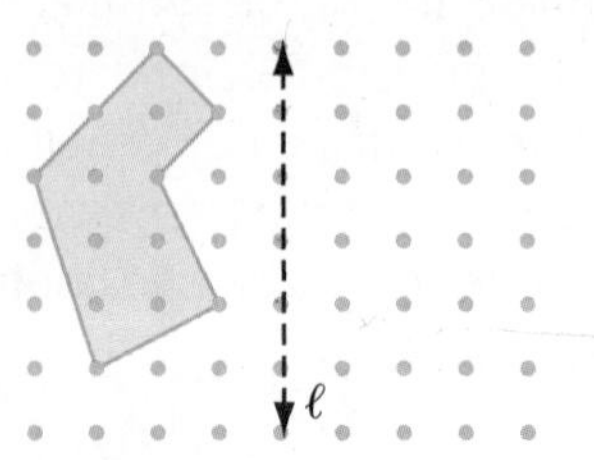

5. Rotate the figure 180° about the center of rotation, point P.

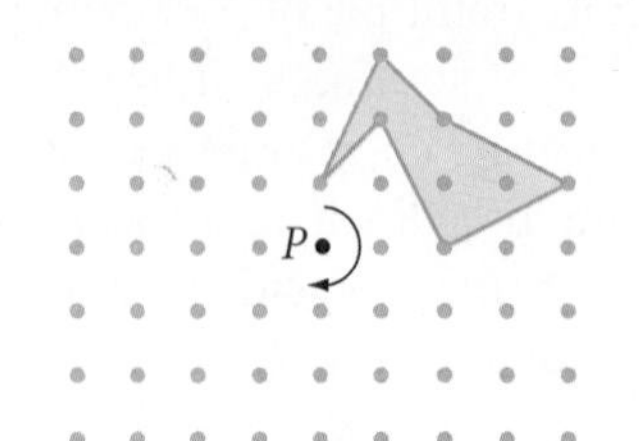

6. Translate the figure by the translation vector.

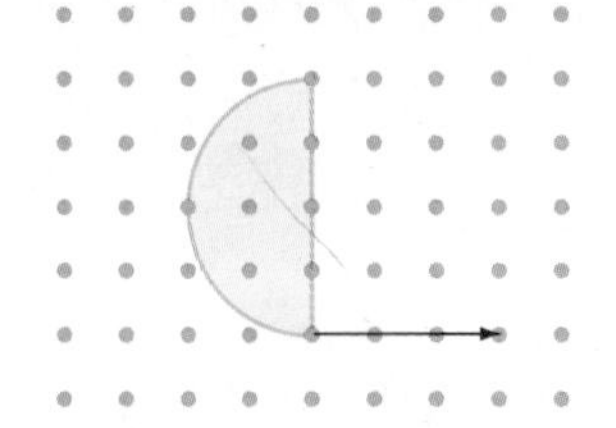

7. An ice skater gliding in one direction creates several translation transformations. Give another real-world example of translation.

8. An ice skater twirling about a point creates several rotation transformations. Give another real-world example of rotation.

In Exercises 9–11, perform each transformation. Attach your patty paper to your homework.

9. ***Construction*** Use the semicircular figure and its reflected image.

   a. Copy the figure and its reflected image onto a piece of patty paper. Locate the line of reflection. Explain your method.

   b. Copy the figure and its reflected image onto a sheet of paper. Locate the line of reflection using a compass and straightedge. Explain your method.

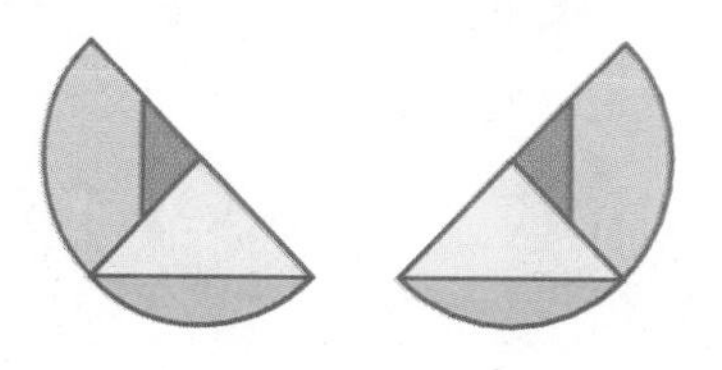

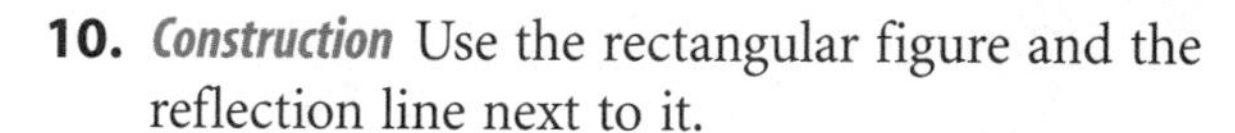

10. ***Construction*** Use the rectangular figure and the reflection line next to it.

   a. Copy the figure and the line of reflection onto a sheet of paper. Use a compass and straightedge to construct the reflected image. Explain your method.

   b. Copy the figure and the line of reflection onto a piece of patty paper. Fold the paper and trace the figure to construct the reflected image.

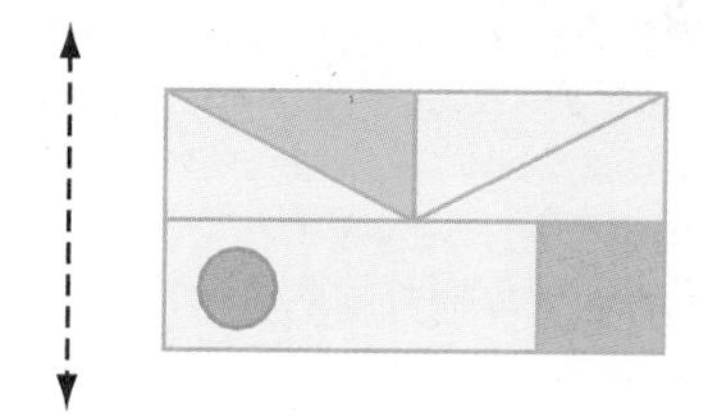

11. Trace the circular figure and the center of rotation, P. Rotate the design 90° clockwise about point P. Draw the image of the figure, as well as the dotted line.

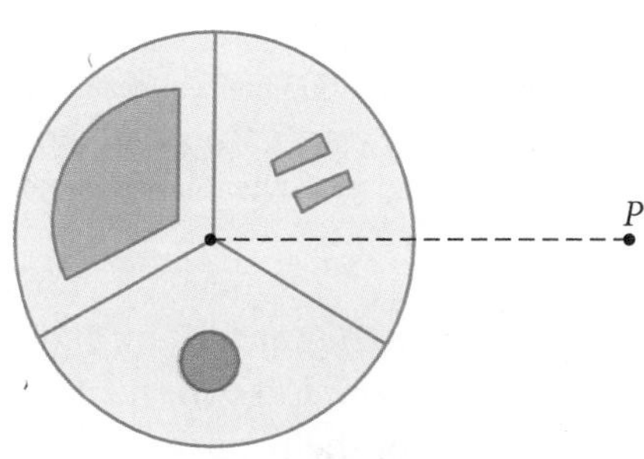

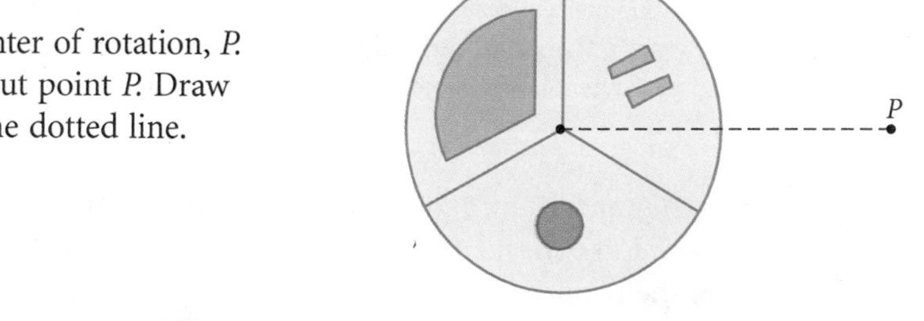

In Exercises 12–14, identify the type (or types) of symmetry in each design.

12. 

Butterfly

13. 

Hmong textile, Laos

14. 

The Temple Beth Israel, San Diego's first synagogue, built in 1889

15. All of the woven baskets from Botswana shown below have rotational symmetry and most have reflectional symmetry. Find one that has 7-fold symmetry. Find one with 9-fold symmetry. Which basket has rotational symmetry but not reflectional symmetry? What type of rotational symmetry does it have?

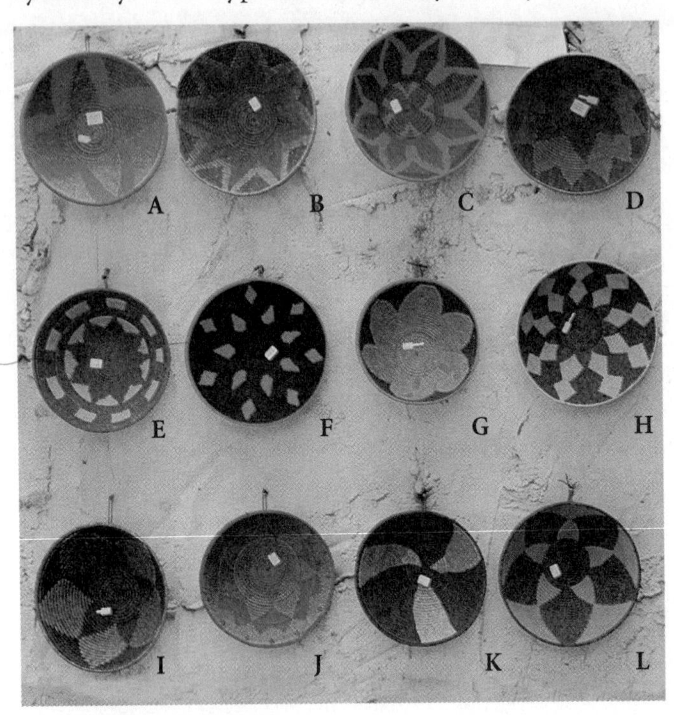

Cultural CONNECTION

For centuries, women in Botswana, a country in southern Africa, have been weaving baskets like the ones you see above, to carry and store food. Each generation passes on the tradition of weaving choice shoots from the mokola palm and decorating them in beautiful geometric patterns with natural dyes. In the past 40 years, international demand for the baskets by retailers and tourists has given economic stability to hundreds of women and their families.

16. ***Mini-Investigation*** Copy and complete the table below. If necessary, use a mirror to locate the lines of symmetry for each of the regular polygons. To find the number of rotational symmetries, you may wish to trace each regular polygon onto patty paper and rotate it. Then copy and complete the conjecture.

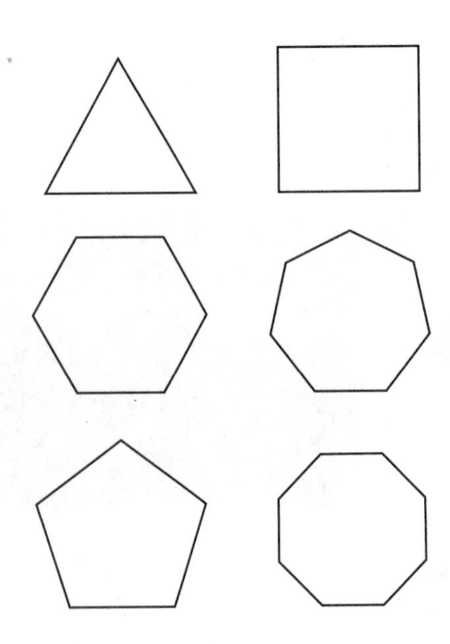

Number of sides of regular polygon	3	4	5	6	7	8	...	n
Number of reflectional symmetries		4					...	
Number of rotational symmetries ($\leq 360°$)		4					...	

A regular polygon of n sides has _?_ reflectional symmetries and _?_ rotational symmetries.

Review

In Exercises 17 and 18, sketch the next two figures.

17. ?, ?

18. 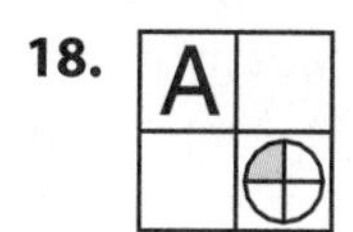

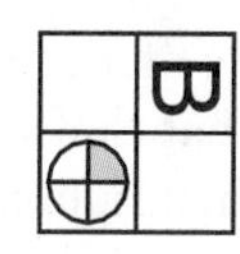

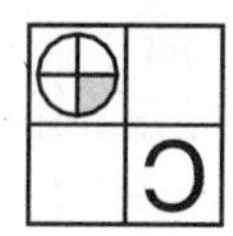

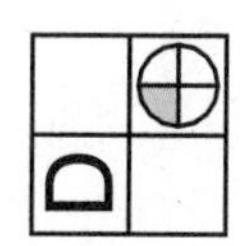

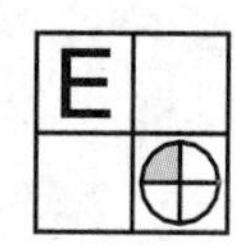

? ? ? ?
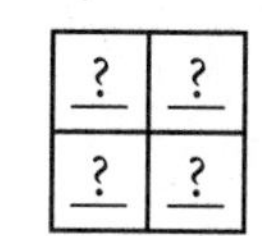

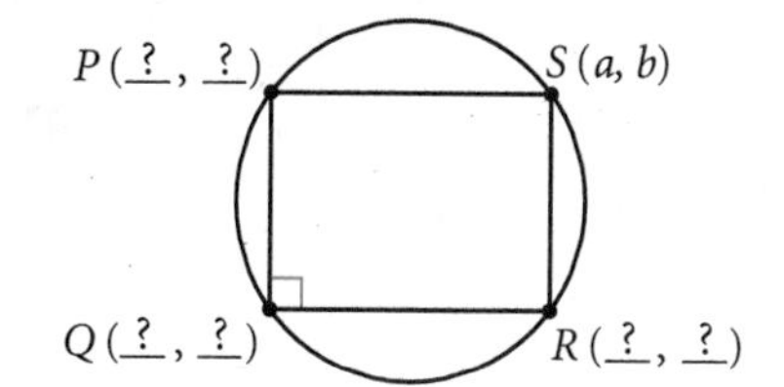

19. Polygon $PQRS$ is a rectangle inscribed in a circle centered at the origin. The slope of $\overline{PS}$ is 0. Find the coordinates of points P, Q, and R in terms of a and b.

20. Use a circular object to trace a large minor arc. Using either compass-and-straightedge construction or patty-paper construction locate a point on the arc equally distant from the arc's endpoints. Label it P.

21. If the circle pattern continues, how many total circles (shaded and unshaded) will there be in the 50th figure? How many will there be in the nth figure?

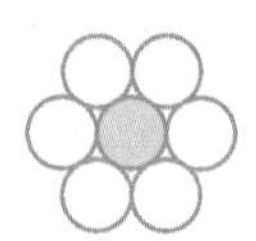
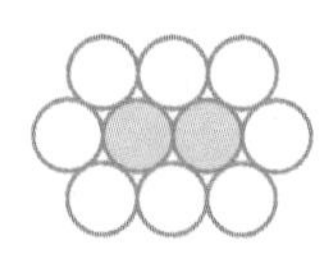
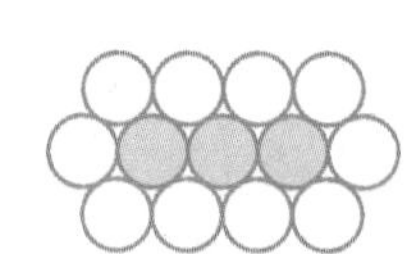

22. Quadrilateral $SHOW$ is circumscribed about circle X. $WO = 14$, $HM = 4$, $SW = 11$, and $ST = 5$. What is the perimeter of $SHOW$?

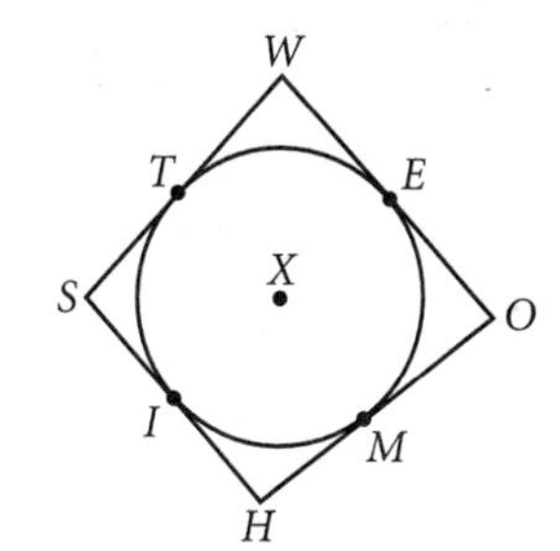

IMPROVING YOUR ALGEBRA SKILLS

The Difference of Squares

$17^2 - 16^2 = 33$	$25.5^2 - 24.5^2 = 50$	$34^2 - 33^2 = 67$
$58^2 - 57^2 = 115$	$62.1^2 - 61.1^2 = 123.2$	$76^2 - 75^2 = 151$

Can you use algebra to explain why you can just add the two base numbers to get the answer? (For example, $17 + 16 = 33$.)

LESSON

7.2

Properties of Isometries

"Why it's a looking-glass book, of course! And, if I hold it up to the glass, the words will all go the right way again."

ALICE IN *THROUGH THE LOOKING-GLASS* BY LEWIS CARROLL

In many earlier exercises, you used **ordered pair rules** to transform polygons on a coordinate plane by relocating their vertices. For any point on a figure, the ordered pair rule $(x, y) \rightarrow (x + h, y + k)$ results in a horizontal move of h units and a vertical move of k units for any numbers h and k. That is, if (x, y) is a point on the original figure, $(x + h, y + k)$ is its corresponding point on the image. Let's look at an example.

EXAMPLE A

Transform the polygon at right using the rule $(x, y) \rightarrow (x + 2, y - 3)$. Describe the type and direction of the transformation.

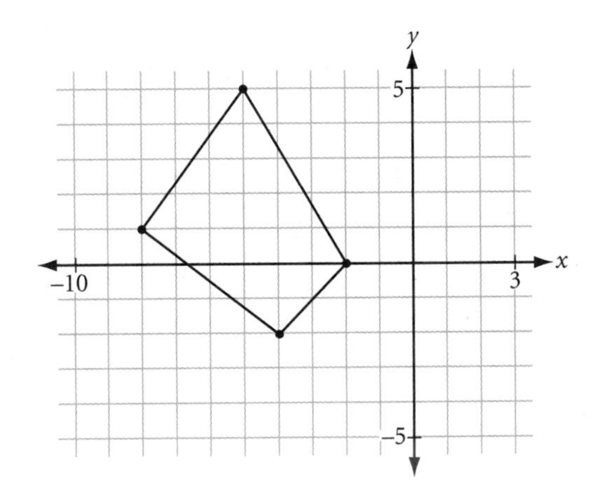

▶ Solution

Apply the rule to each ordered pair. Every point of the polygon moves right 2 units and down 3 units. This is a translation of $(2, -3)$.

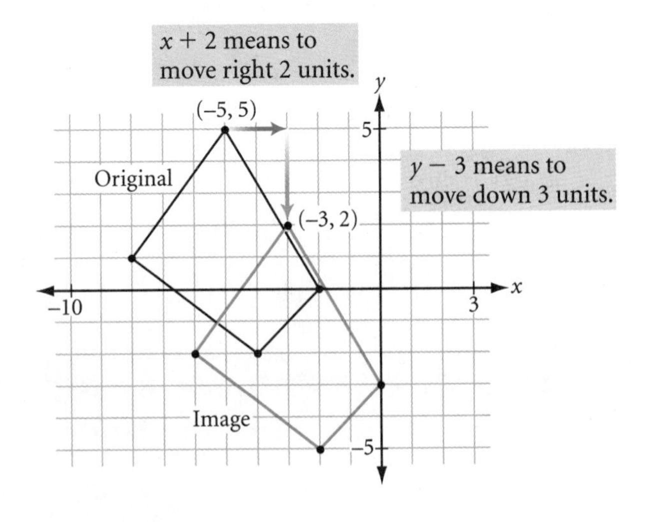

So the ordered pair rule $(x, y) \rightarrow (x + h, y + k)$ results in a translation of (h, k).

Investigation 1

Transformations on a Coordinate Plane

You will need

- graph paper
- patty paper

In this investigation you will discover (or rediscover) how four ordered pair rules transform a polygon. Each person in your group can choose a different polygon for this investigation.

Step 1 On graph paper, create and label four sets of coordinate axes. Draw the same polygon in the same position in a quadrant of each of the four graphs. Write one of these four ordered pair rules below each graph.

a. $(x, y) \rightarrow (-x, y)$

b. $(x, y) \rightarrow (x, -y)$

c. $(x, y) \rightarrow (-x, -y)$

d. $(x, y) \rightarrow (y, x)$

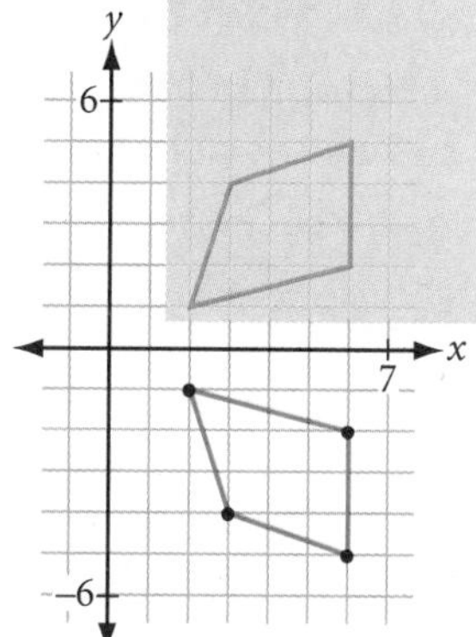

Step 2 Use the ordered pair rule you assigned to each graph to relocate the vertices of your polygon and create its image.

Step 3 Use patty paper to see if your transformation is a reflection, translation, or rotation. Compare your results with those of your group members. Complete the conjecture.

Coordinate Transformations Conjecture C-69

The ordered pair rule $(x, y) \rightarrow (-x, y)$ is a _?_ over _?_.
The ordered pair rule $(x, y) \rightarrow (x, -y)$ is a _?_ over _?_.
The ordered pair rule $(x, y) \rightarrow (-x, -y)$ is a _?_ about _?_.
The ordered pair rule $(x, y) \rightarrow (y, x)$ is a _?_ over _?_.

Let's revisit "poolroom geometry." When a ball rolls without spin into a cushion, the outgoing angle is congruent to the incoming angle. This is true because the outgoing and incoming angles are reflections of each other.

Investigation 2

Finding a Minimal Path

You will need

- patty paper
- a protractor

In Chapter 1, you used a protractor to find the path of the ball. In this investigation, you'll discover some other properties of reflections that have many applications in science and engineering. They may even help your pool game!

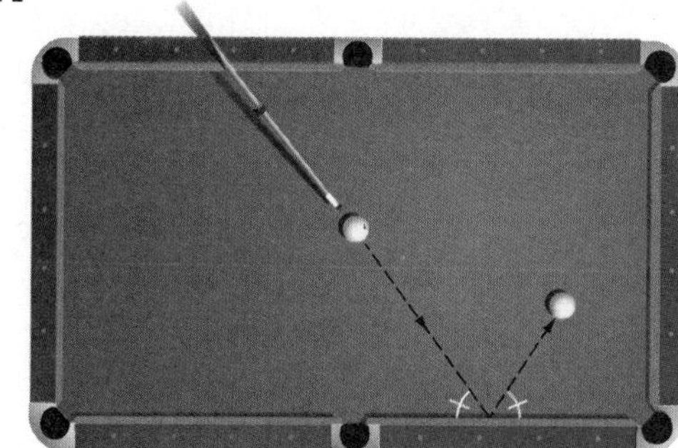

Step 1 | Draw a segment, representing a pool table cushion, near the center of a piece of patty paper. Draw two points, A and B, on one side of the segment.

Step 2 | Imagine you want to hit a ball at point A so that it bounces off the cushion and hits another ball at point B. Use your protractor to find the point C on the cushion that you should aim for.

Step 3 | Draw $\overline{AC}$ and $\overline{CB}$ to represent the ball's path.

Step 4 | Fold your patty paper to draw the reflection of point B. Label the image point B'.

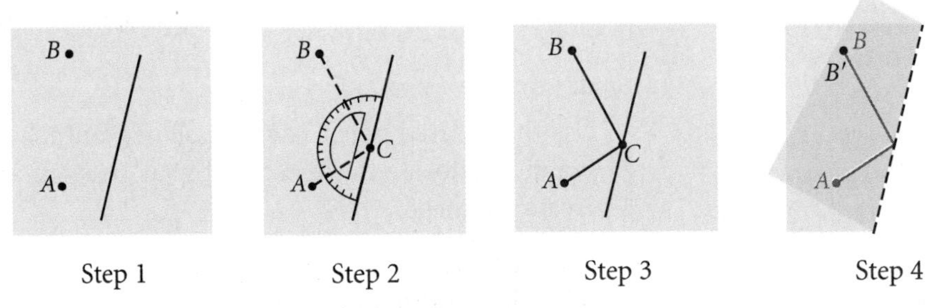

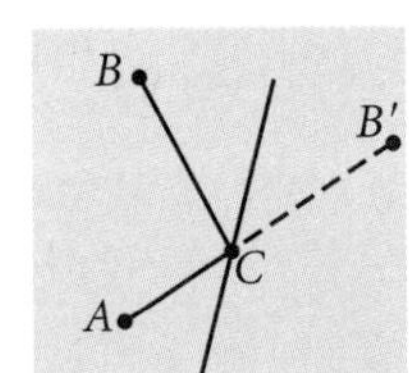

Step 5 | Unfold the paper and draw a segment from point A to point B'. What do you notice? Does point C lie on segment $\overline{AB'}$? How does the path from A to B' compare to the two-part path from A to C to B?

Step 6 | Can you draw any other path from point A to the cushion to point B that is shorter than $AC + CB$? Why or why not? The shortest path from point A to the cushion to point B is called the **minimal path.** Copy and complete the conjecture.

Minimal Path Conjecture

C-70

If points A and B are on one side of line ℓ, then the minimal path from point A to line ℓ to point B is found by __?__.

How can this discovery help your pool game? Suppose you need to hit a ball at point A into the cushion so that it will bounce off the cushion and pass through point B. To what point on the cushion should you aim? Visualize point B reflected across the cushion. Then aim directly at the reflected image.

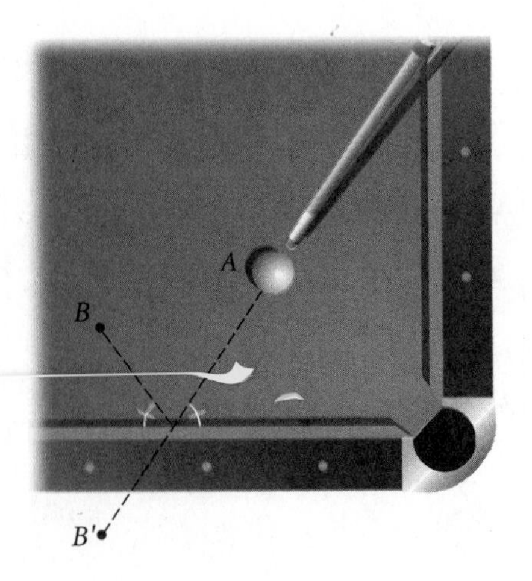

Let's look at a miniature-golf example.

EXAMPLE B

How can you hit the ball at T around the corner and into the hole at H in one shot?

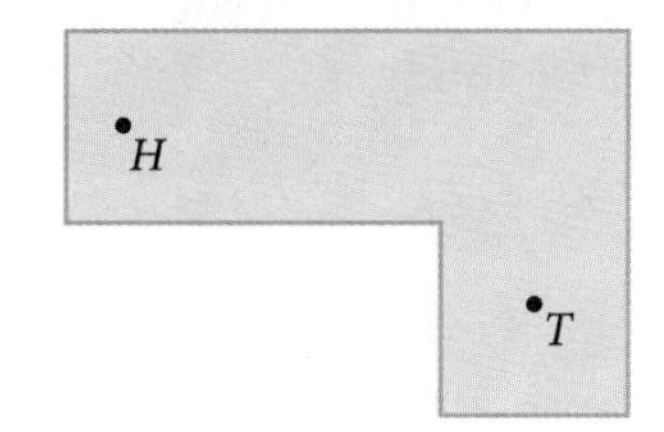

▶ Solution

First, try to get a hole-in-one with a direct shot or with just one bounce off a wall. For one bounce, decide which wall the ball should hit. Visualize the image of the hole across that wall and aim for the reflected hole. There are two possibilities.

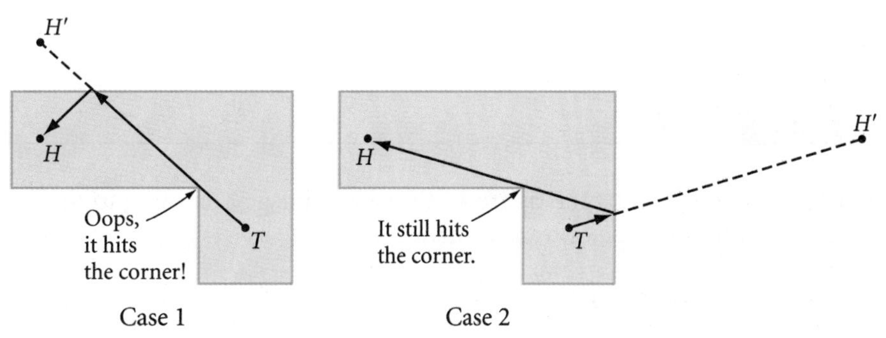

In both cases the path is blocked. It looks like you need to try two bounces. Visualize a path from the tee hitting two walls and into the hole.

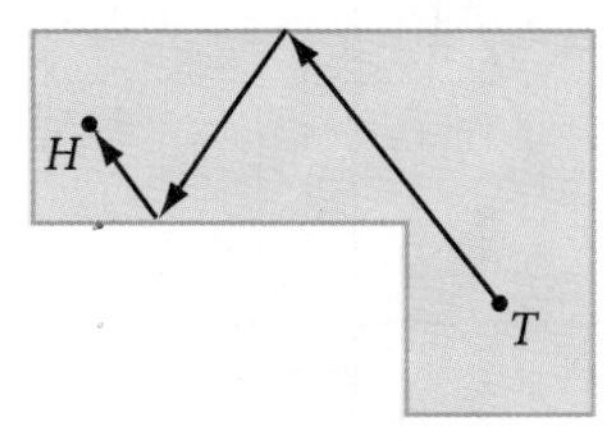

Visualize the path.

Now you can work backward. Which wall will the ball hit last? Reflect the hole across that wall creating image H'.

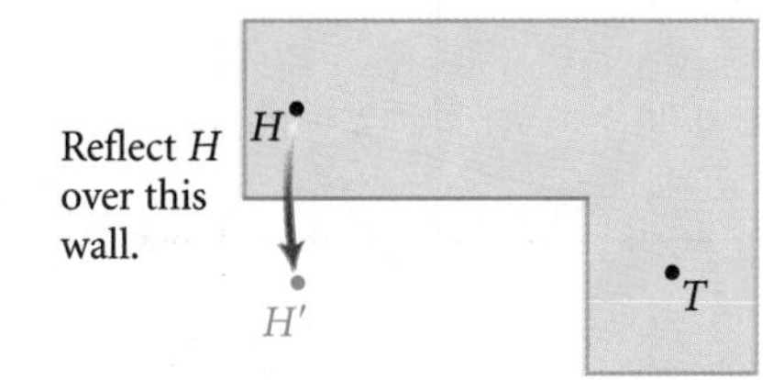

Which wall will the ball hit before it approaches the second wall? Reflect the image of the hole H' across that wall creating image H'' (read "H double prime").

Draw the path from the tee to H'', H', and H. Can you visualize other possible paths with two bounces? Three bounces? What do you suppose is the minimal path from T to H?

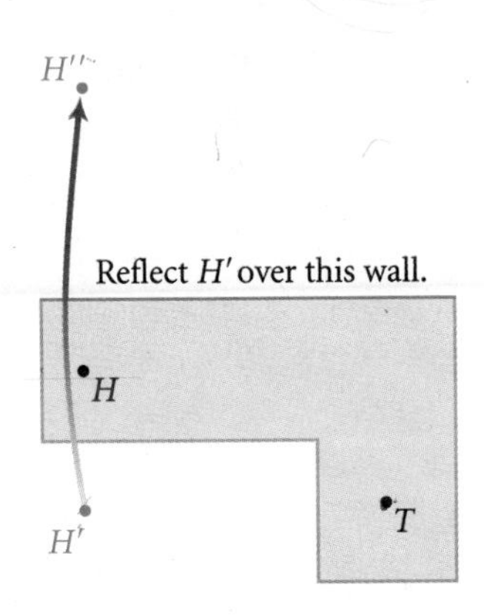

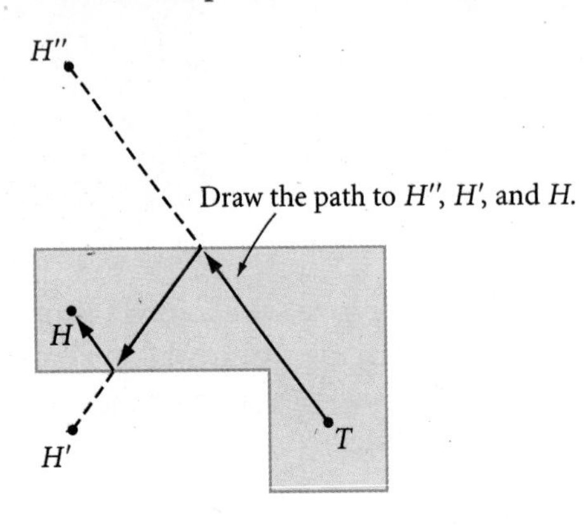

EXERCISES

You will need

In Exercises 1–5, copy the figure and draw the image according to the rule. Identify the type of transformation.

1. $(x, y) \rightarrow (x + 5, y)$

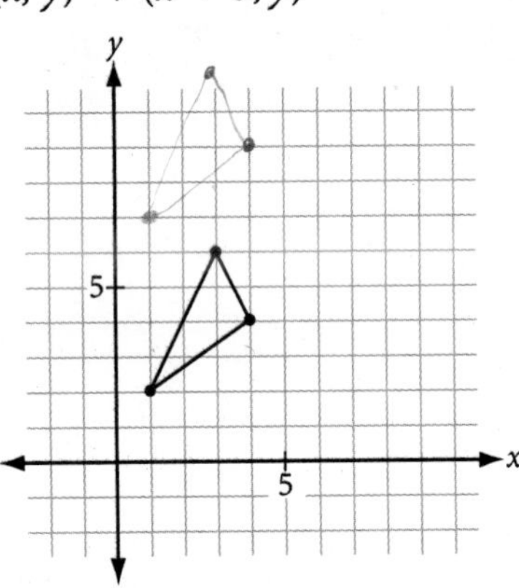

2. $(x, y) \rightarrow (x, -y)$ ⓗ

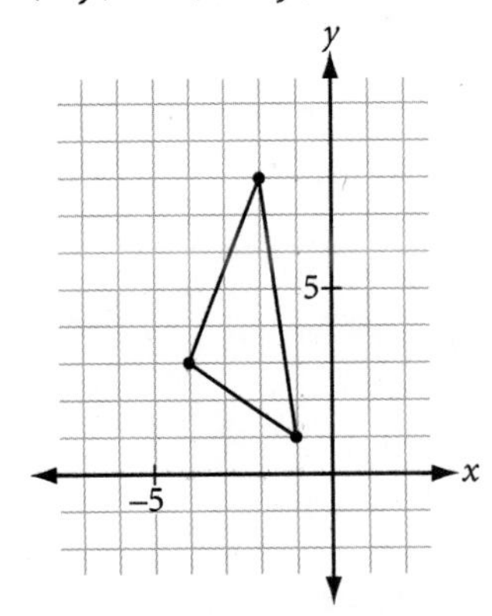

3. $(x, y) \rightarrow (y, x)$

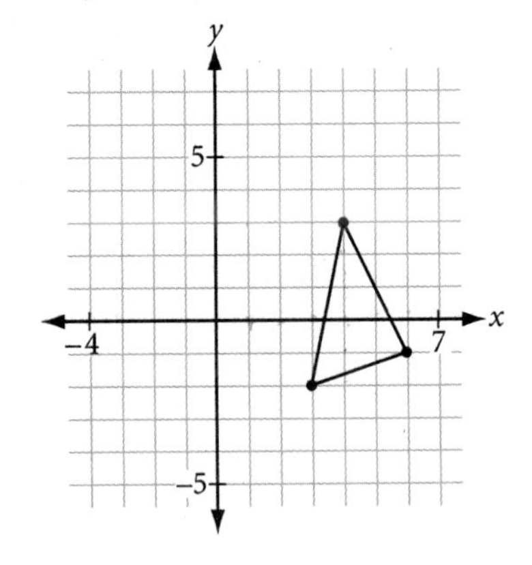

4. $(x, y) \rightarrow (8 - x, y)$

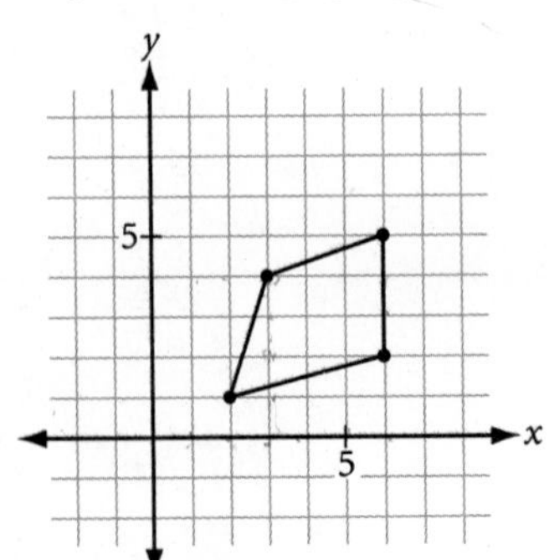

5. $(x, y) \rightarrow (-x, -y)$

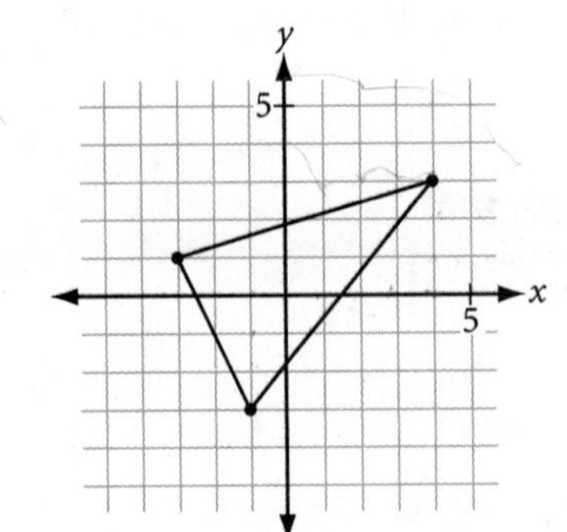

6. Look at the rules in Exercises 1–5 that produced reflections. What do these rules have in common? How about the ones that produce translations? Rotations?

In Exercises 7 and 8, complete the ordered pair rule that transforms the black triangle to its image, the red triangle.

7. $(x, y) \rightarrow (\underline{?}, \underline{?})$ ⓗ

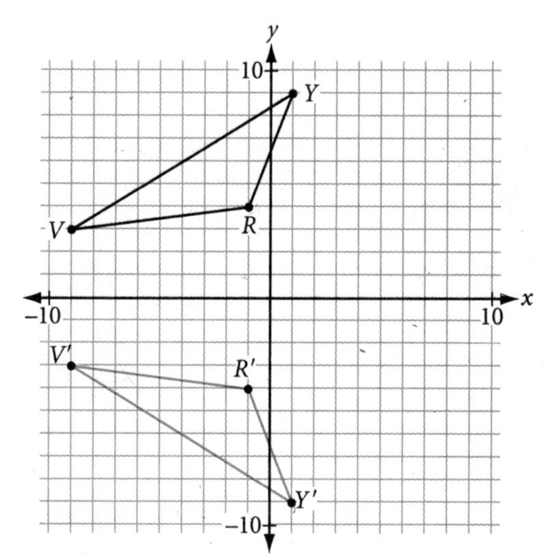

8. $(x, y) \rightarrow (\underline{?}, \underline{?})$

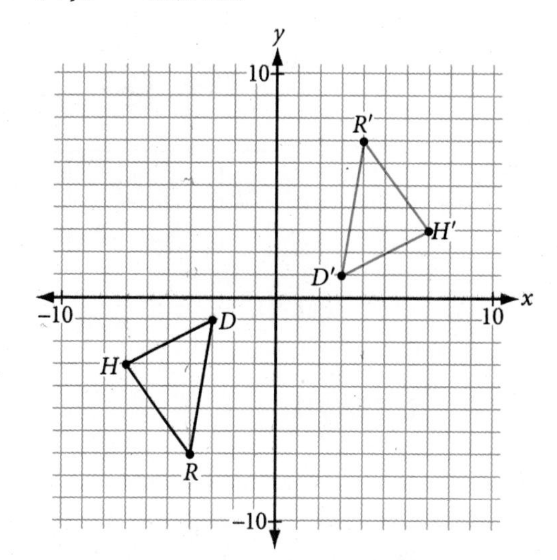

In Exercises 9–11, copy the position of each ball and hole onto patty paper and draw the path of the ball.

9. What point on the W cushion can a player aim for so that the cue ball bounces and strikes the 8-ball? What point can a player aim for on the S cushion?

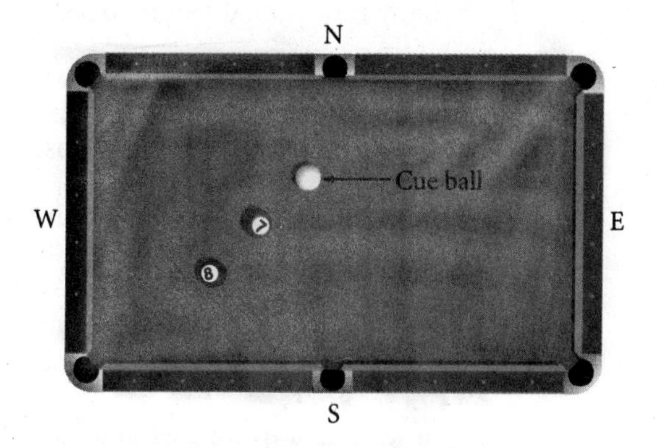

10. Starting from the tee (point T), what point on a wall should a player aim for so that the golf ball bounces off the wall and goes into the hole at H?

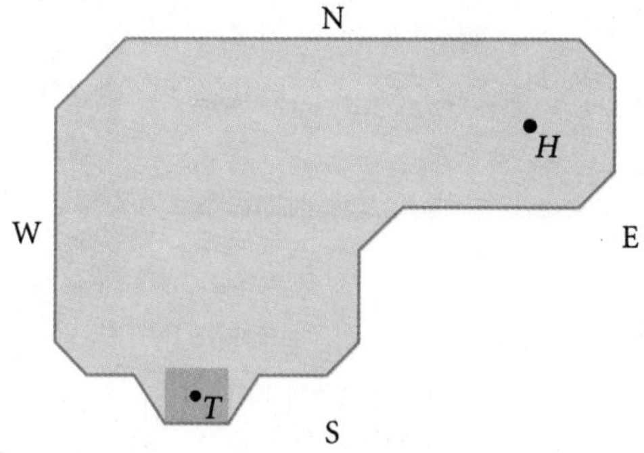

11. Starting from the tee (point T), plan a shot so that the golf ball goes into the hole at H. Show all your work.

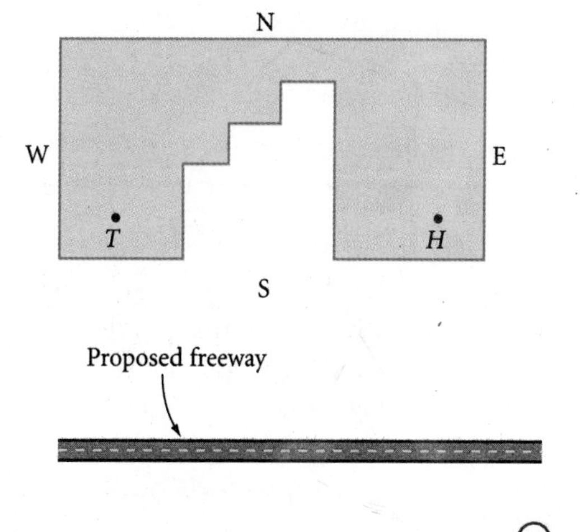

12. A new freeway is being built near the two towns of Perry and Mason. The two towns want to build roads to one junction point on the freeway. (One of the roads will be named Della Street.) Locate the junction point and draw the minimal path from Perry to the freeway to Mason. How do you know this is the shortest path?

Review

In Exercises 13 and 14, sketch the next two figures.

13. 1, 2, 3, 4, 5, ?, ?

14. 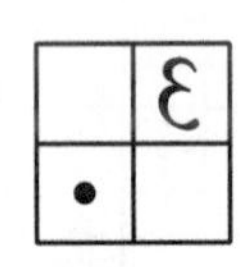 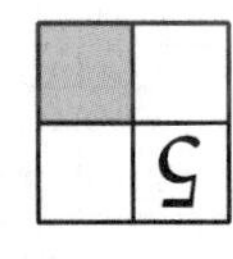 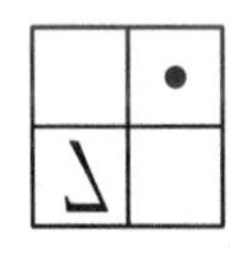  9 | ? ? ? ? | ? ? ? ?

15. The word DECODE remains unchanged when it is reflected over its horizontal line of symmetry. Find another such word with at least five letters.

←--DECODE--→

16. How many reflectional symmetries does an isosceles triangle have?

17. How many reflectional symmetries does a rhombus have?

18. Write what is actually on the T-shirt shown at right.

19. ***Construction*** Construct a kite circumscribed about a circle.

20. ***Construction*** Construct a rhombus circumscribed about a circle.

In Exercises 21 and 22, identify each statement as true or false. If true, explain why. If false, give a counterexample.

21. If two angles of a quadrilateral are right angles, then it is a rectangle.

22. If the diagonals of a quadrilateral are congruent, then it is a rectangle.

By Holland. ©1976, Punch Cartoon Library.

IMPROVING YOUR REASONING SKILLS

Chew on This for a While

If the third letter before the second consonant after the third vowel in the alphabet is in the twenty-sixth word of this puzzle, then print the fortieth word of this puzzle and then print the twenty-second letter of the alphabet after this word. Otherwise, list three uses for chewing gum.

LESSON

7.3 Compositions of Transformations

There are things which nobody would see unless I photographed them.

DIANE ARBUS

In Lesson 7.2, you reflected a point, then reflected it again to find the path of a ball. When you apply one transformation to a figure and then apply another transformation to its image, the resulting transformation is called a **composition** of transformations. Let's look at an example of a composition of two translations.

EXAMPLE

Triangle ABC with vertices $A(-1, 0)$, $B(4, 0)$, and $C(2, 6)$ is first translated by the rule $(x, y) \rightarrow (x - 6, y - 5)$, and then its image, $\triangle A'B'C'$, is translated by the rule $(x, y) \rightarrow (x + 14, y + 3)$.

a. What single translation is equivalent to the composition of these two translations?

b. What single translation brings the second image, $\triangle A''B''C''$, back to the position of the original triangle, $\triangle ABC$?

Solution

Draw $\triangle ABC$ on a set of axes and relocate its vertices using the first rule to get $\triangle A'B'C'$. Then relocate the vertices of $\triangle A'B'C'$ using the second rule to get $\triangle A''B''C''$.

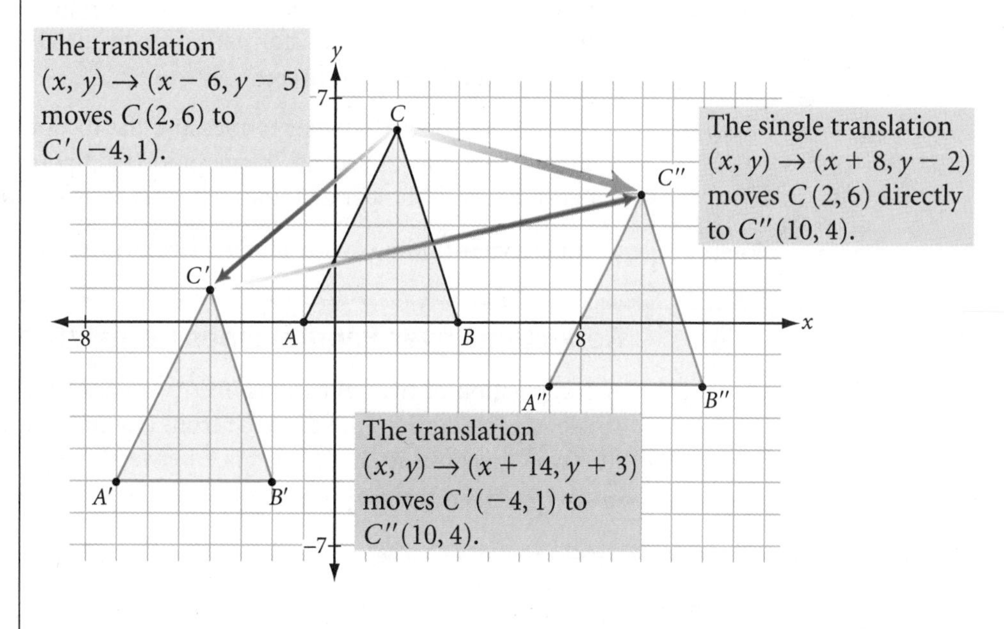

a. Each vertex is moved left 6 then right 14, and down 5 then up 3. So the equivalent single translation would be $(x, y) \rightarrow (x - 6 + 14, y - 5 + 3)$ or $(x, y) \rightarrow (x + 8, y - 2)$. You can also write this as $(8, -2)$.

b. Reversing the steps, the translation $(-8, 2)$ brings the second image, $\triangle A''B''C''$, back to $\triangle ABC$.

In the investigations you will see what happens when you compose reflections.

Investigation 1
Reflections over Two Parallel Lines

You will need

- patty paper

First consider the case of parallel lines of reflection.

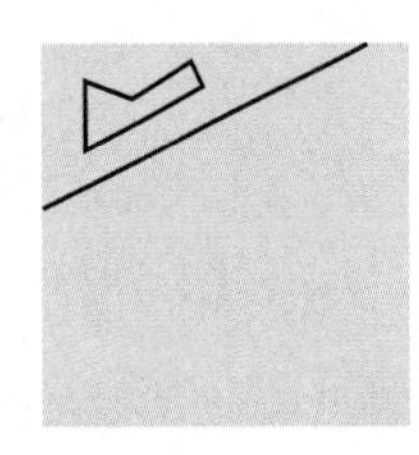
Step 1

Step 2

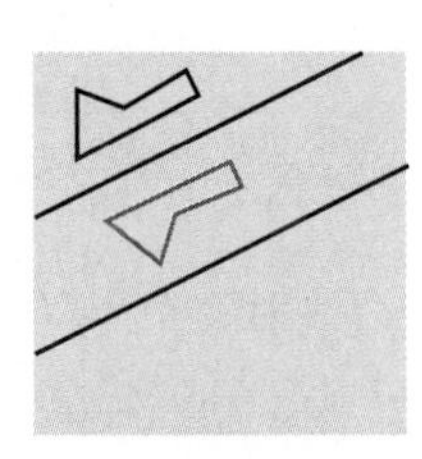
Step 3

Step 1 On a piece of patty paper, draw a figure and a line of reflection that does not intersect it.

Step 2 Fold to reflect your figure over the line of reflection and trace the image.

Step 3 On your patty paper, draw a second reflection line parallel to the first so that the image is between the two parallel reflection lines.

Step 4 Fold to reflect the image over the second line of reflection. Turn the patty paper over and trace the second image.

Step 5 How does the second image compare to the original figure? Name the single transformation that transforms the original to the second image.

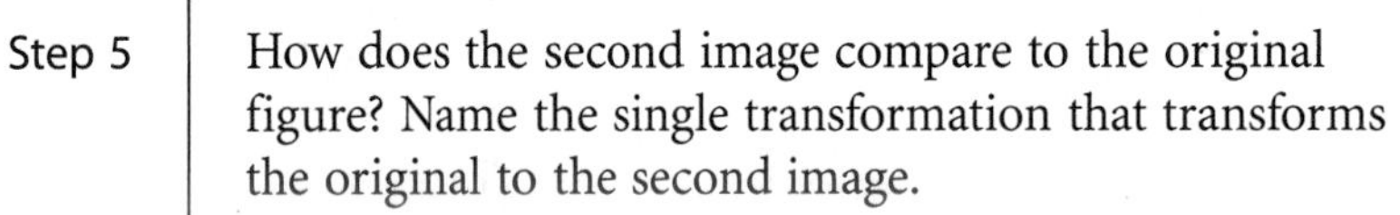

Step 6 Use a compass or patty paper to measure the distance between a point in the original figure and its second image point. Compare this distance with the distance between the parallel lines. How do they compare?

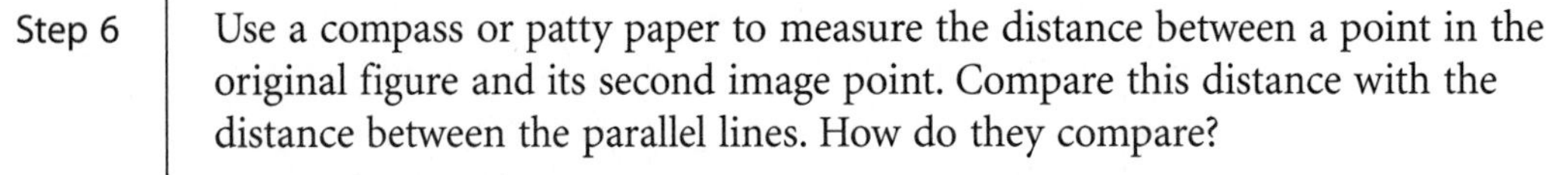

Step 7 Compare your findings with others in your group and state your conjecture.

Reflections over Parallel Lines Conjecture C-71

A composition of two reflections over two parallel lines is equivalent to a single _?_. In addition, the distance from any point to its second image under the two reflections is _?_ the distance between the parallel lines.

Is a composition of reflections always equivalent to a single reflection? If you reverse the reflections in a different order, do you still get the original figure back? Can you express a rotation as a set of reflections?

Investigation 2
Reflections over Two Intersecting Lines

You will need

- patty paper
- a protractor

Next, you will explore the case of intersecting lines of reflection.

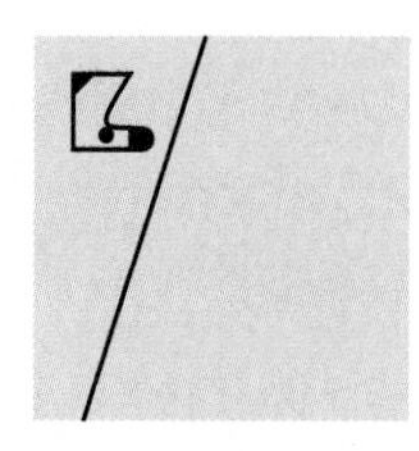
Step 1

Step 2

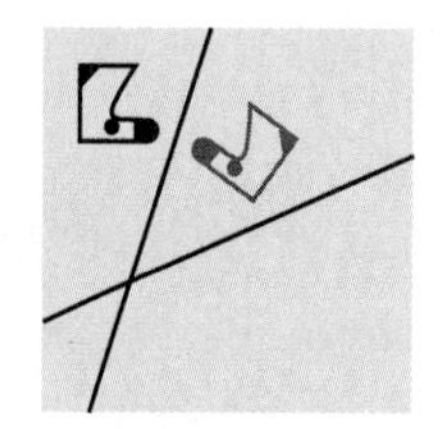
Step 3

Step 1 On a piece of patty paper, draw a figure and a reflection line that does not intersect it.

Step 2 Fold to reflect your figure over the line and trace the image.

Step 3 On your patty paper, draw a second reflection line intersecting the first so that the image is in an acute angle between the two intersecting reflection lines.

Step 4 Fold to reflect the first image over the second line and trace the second image.

Step 5

Step 6

Step 7

Step 5 Draw two rays that start at the point of intersection of the two intersecting lines and that pass through corresponding points on the original figure and its second image.

Step 6 How does the second image compare to the original figure? Name the single transformation from the original to the second image.

Step 7 With a protractor or patty paper, compare the angle created in Step 5 with the acute angle formed by the intersecting reflection lines. How do the angles compare?

Step 8 Compare findings in your group and state your next conjecture.

Reflections over Intersecting Lines Conjecture C-72

A composition of two reflections over a pair of intersecting lines is equivalent to a single ___?___. The angle of ___?___ is ___?___ the acute angle between the pair of intersecting reflection lines.

There are many other ways to combine transformations. Combining a translation with a reflection gives a special two-step transformation called a **glide reflection.** A sequence of footsteps is a common example of a glide reflection. You will explore a few other examples of glide reflection in the exercises and later in this chapter.

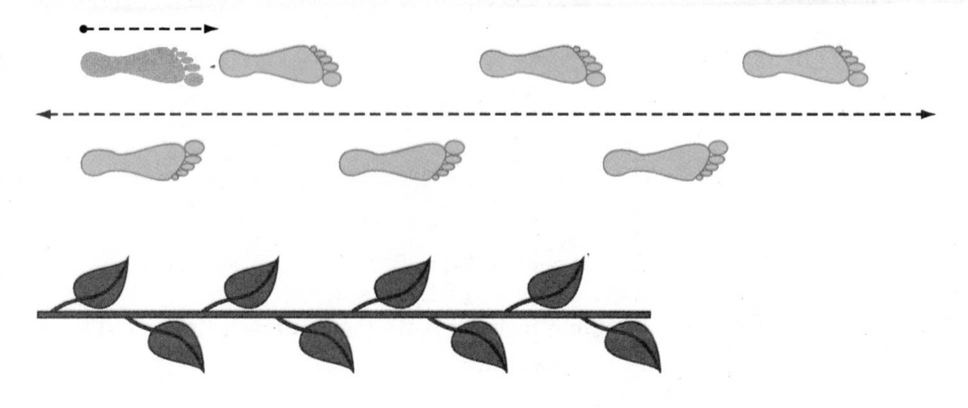

Glide-reflectional symmetry

EXERCISES

1. Name the single translation that can replace the composition of these three translations: (2, 3), then (−5, 7), then (13, 0).

2. Name the single rotation that can replace the composition of these three rotations about the same center of rotation: 45°, then 50°, then 85°. What if the centers of rotation differ? Draw a figure and try it.

3. Lines m and n are parallel and 10 cm apart.

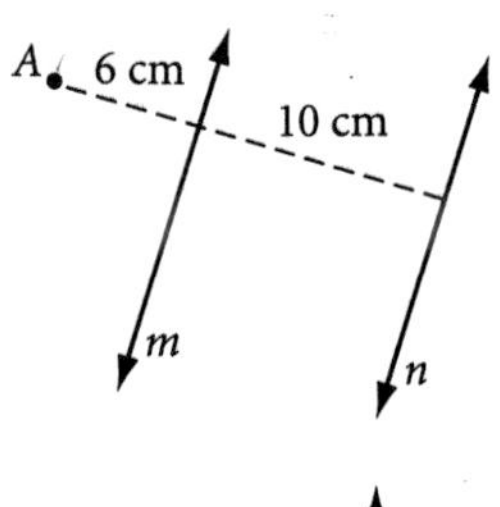

   a. Point A is 6 cm from line m and 16 cm from line n. Point A is reflected over line m, then its image, A', is reflected over line n to create a second image, point A''. How far is point A from point A''?

   b. What if A is reflected over n, and then its image is reflected over m? Find the new image and distance from A.

4. Two lines m and n intersect at point P, forming a 40° angle.

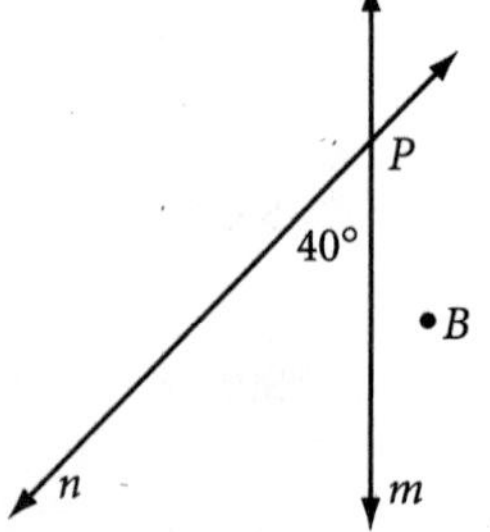

   a. You reflect point B over line m, then reflect the image of B over line n. What angle of rotation about point P rotates the second image of point B back to its original position?

   b. What if you reflect B first over n, and then reflect the image of B over m? Find the angle of rotation that rotates the second image back to the original position.

5. Copy the figure and $\angle PAL$ onto patty paper. Reflect the figure over $\overrightarrow{AP}$. Reflect the image over $\overrightarrow{AL}$. What is the equivalent rotation?

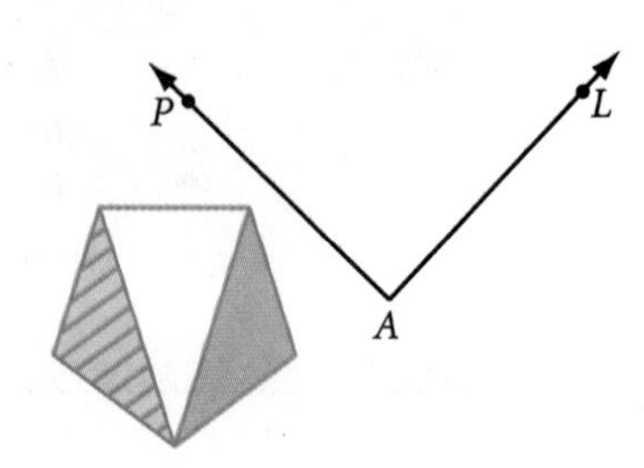

6. Copy the figure and the pair of parallel lines onto patty paper. Reflect the figure over $\overleftrightarrow{PA}$. Reflect the image over $\overleftrightarrow{RL}$. What is the length of the equivalent translation vector?

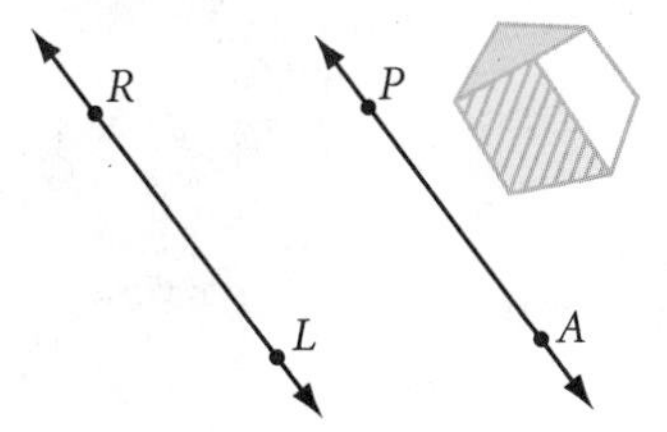

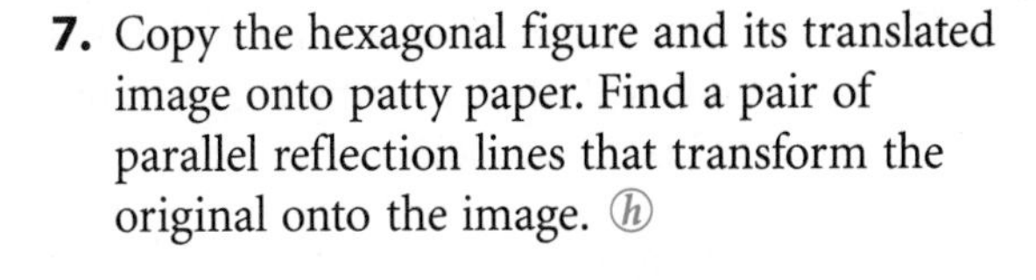

7. Copy the hexagonal figure and its translated image onto patty paper. Find a pair of parallel reflection lines that transform the original onto the image. ⓗ

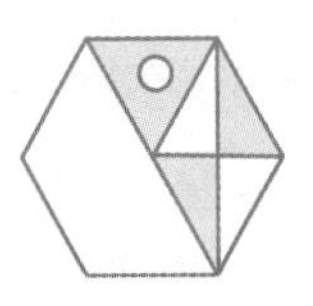

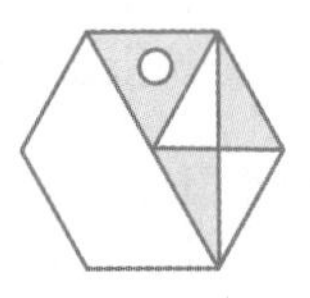

8. Copy the original figure and its translated image onto patty paper. Find a pair of intersecting reflection lines that transform the original onto the image.

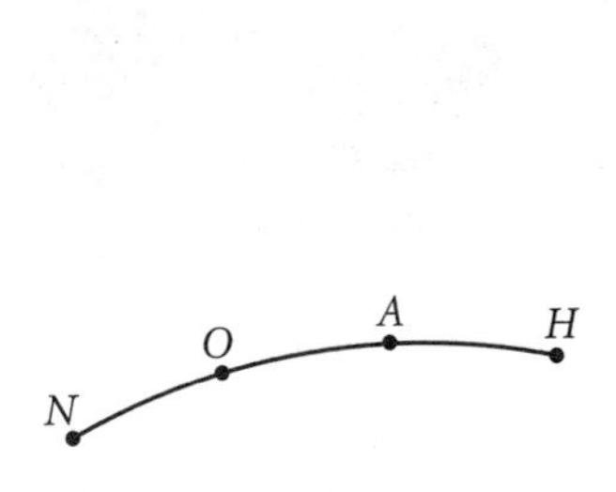

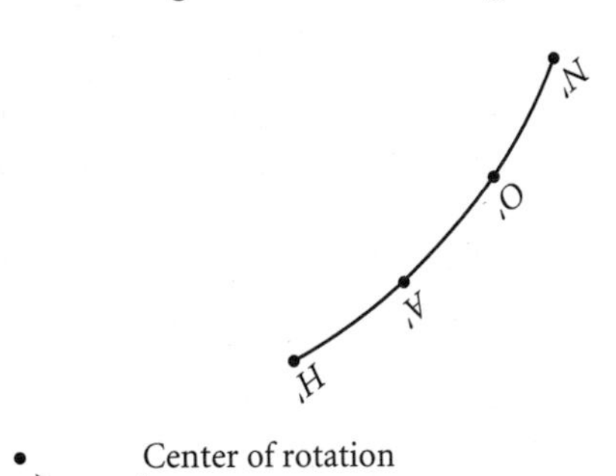

9. Copy the two figures below onto graph paper. Each figure is the glide-reflected image of the other. Continue the pattern with two more glide-reflected figures.

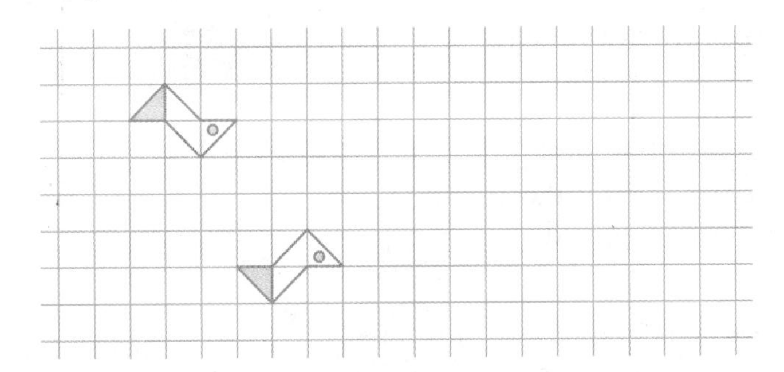

Review

In Exercises 10 and 11, sketch the next two figures.

10. AA/AA, BB/BB, CC/CC, DD/DD, EE/EE, ?, ?

11.

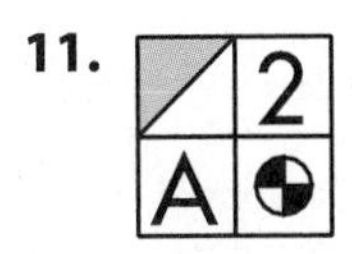

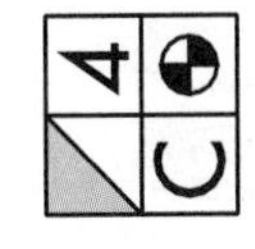

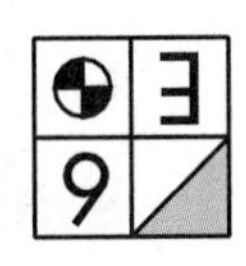

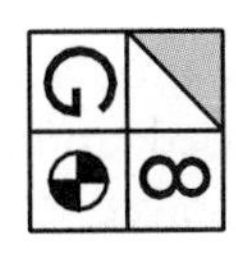

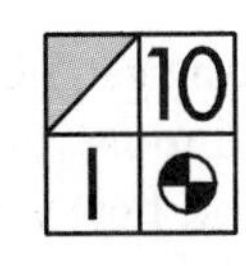

? ? / ? ?, ? ? / ? ?

12. If you draw a figure on an uninflated balloon and then blow up the balloon the figure will undergo a nonrigid transformation. Give another example of a nonrigid transformation.

13. List two objects in your home that have rotational symmetry but not reflectional symmetry. List two objects in your classroom that have reflectional symmetry but not rotational symmetry.

14. Have you noticed that some letters have both horizontal and vertical symmetries? Have you also noticed that all the letters that have both horizontal and vertical symmetries also have point symmetry? Is this a coincidence? Use what you have learned about transformations to explain why.

15. Is it possible for a triangle to have exactly one line of symmetry? Exactly two? Exactly three? Support your answers with sketches.

16. Draw two points onto a piece of paper and connect them with a curve that is point symmetric. ⓗ

project

KALEIDOSCOPES

You have probably looked through kaleidoscopes and enjoyed their beautiful designs, but do you know how they work? For a simple kaleidoscope, hinge two mirrors with tape, place a small object or photo between the mirrors and adjust them until you see four objects (the original and three images). What is the angle between the mirrors? At what angle should you hold the mirrors to see six objects? Eight objects?

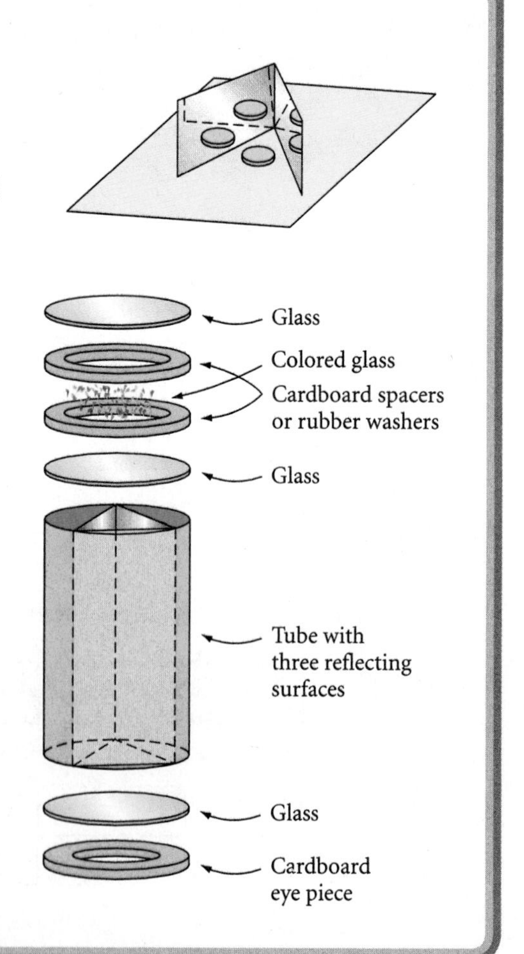

The British physicist Sir David Brewster invented the tube kaleidoscope in 1816. Some tube kaleidoscopes have colored glass or plastic pieces that tumble around in their end chambers. Some have colored liquid. Others have only a lens in the chamber—the design you see depends on where you aim it.

Design and build your own kaleidoscope using a plastic or cardboard cylinder and glass or plastic as reflecting surfaces. Try various items in the end chamber. Your project should include

- Your kaleidoscope (pass it around!).
- A report with diagrams that show the geometry properties you used, a list of the materials and tools you used, and a description of problems you had and how you solved them.

LESSON

7.4

Tessellations with Regular Polygons

I see a certain order in the universe and math is one way of making it visible.

MAY SARTON

Honeycombs are remarkably geometric structures. The hexagonal cells that bees make are ideal because they fit together perfectly without any gaps. The regular hexagon is one of many shapes that can completely cover a plane without gaps or overlaps. Mathematicians call such an arrangement of shapes a **tessellation** or a **tiling.** A tessellation that uses only one shape is called a **monohedral tiling.**

You can find tessellations in every home. Decorative floor tiles have tessellating patterns of squares. Brick walls, fireplaces, and wooden decks often display creative tessellations of rectangles. Where do you see tessellations every day?

The hexagon pattern in the honeycomb of the bee is a tessellation of regular hexagons.

Regular hexagons and equilateral triangles combine in this tiling from the 17th-century Topkapi Palace in Istanbul, Turkey.

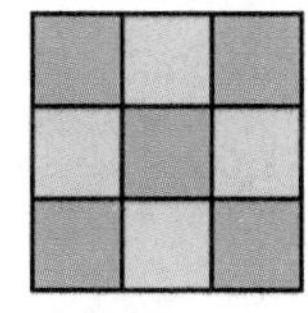

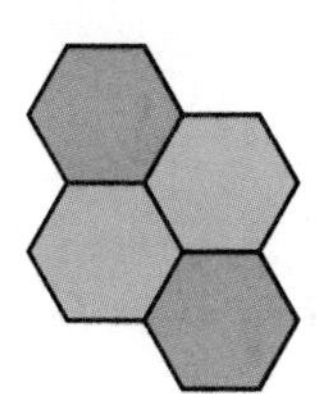

You already know that squares and regular hexagons create monohedral tessellations. Because each regular hexagon can be divided into six equilateral triangles, we can logically conclude that equilateral triangles also create monohedral tessellations. Will other regular polygons tessellate? Let's look at this question logically.

For shapes to fill the plane without gaps or overlaps, their angles, when arranged around a point, must have measures that add up to exactly 360°. If the sum is less than 360°, there will be a gap. If the sum is greater, the shapes will overlap. Six 60° angles from six equilateral triangles add up to 360°, as do four 90° angles from four squares or three 120° angles from three regular hexagons. What about regular pentagons? Each angle in a regular pentagon measures 108°, and 360 is not divisible by 108. So regular pentagons cannot be arranged around a point without overlapping or leaving a gap. What about regular heptagons?

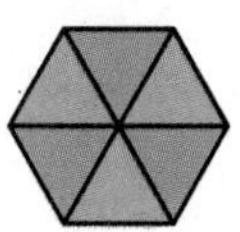

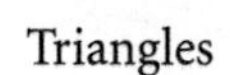

Triangles

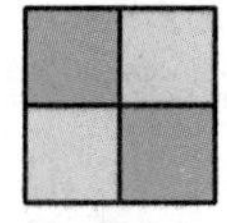

Squares

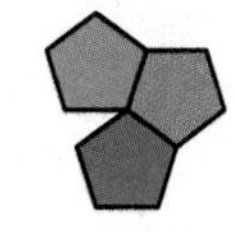

Pentagons

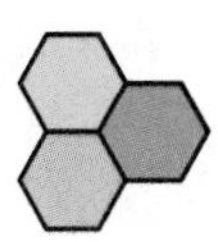

Hexagons

?

Heptagons

In any regular polygon with more than six sides, each angle has a measure greater than 120°, so no more than two angles can fit about a point without overlapping. So the only regular polygons that create monohedral tessellations are equilateral triangles, squares, and regular hexagons. A monohedral tessellation of congruent regular polygons is called a **regular tessellation.**

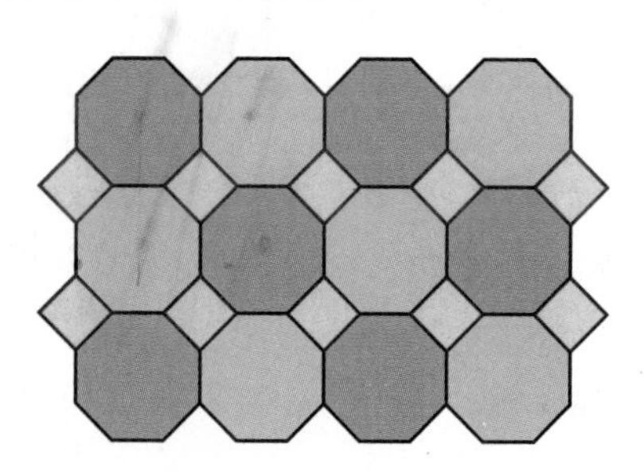

Tessellations can have more than one type of shape. You may have seen the octagon-square combination at right. In this tessellation, two regular octagons and a square meet at each vertex. Notice that you can put your pencil on any vertex and that the point is surrounded by one square and two octagons. So you can call this a 4.8.8 or a 4.8^2 tiling. The numbers give the **vertex arrangement,** or **numerical name** for the tiling.

When the same combination of regular polygons (of two or more kinds) meet in the same order at each vertex of a tessellation, it is called a **semiregular tessellation.** Below are two more examples of semiregular tessellations.

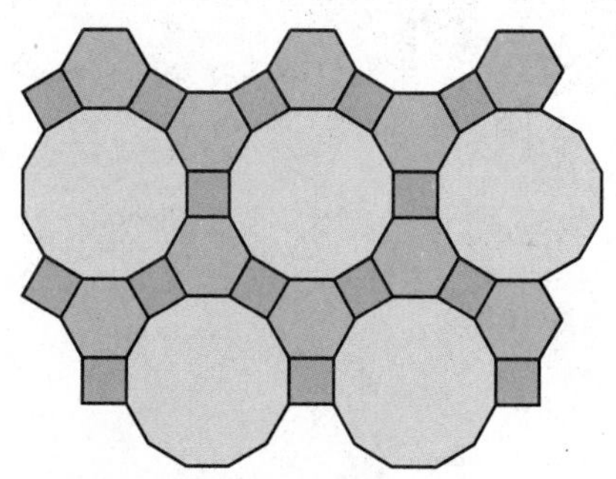

The same polygons appear in the same order at each vertex: square, hexagon, dodecagon.

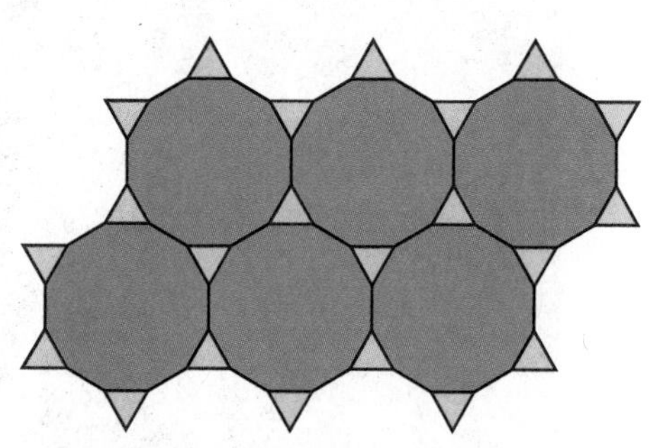

The same polygons appear in the same order at each vertex: triangle, dodecagon, dodecagon.

There are eight different semiregular tessellations. Three of them are shown above. In this investigation, you will look for the other five. To make this easier, the remaining five use only combinations of triangles, squares, or hexagons.

Investigation
The Semiregular Tessellations

You will need

- triangles, squares and hexagons from a set of pattern blocks or,
- geometry software such as GSP or,
- a set of triangles, squares and hexagons created from the set shown

Find or create a set of regular triangles, squares, and hexagons for this investigation. Then work with your group to find the remaining five of the eight semiregular tessellations. Remember, the same combination of regular polygons must meet in the same order at each vertex for the tiling to be semiregular. Also remember to check that the sum of the measures at each vertex is 360°.

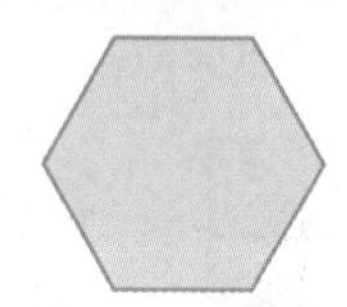

Step 1 Investigate which combinations of two kinds of regular polygons you can use to create a semiregular tessellation.

Step 2 Next, investigate which combinations of three kinds of regular polygons you can use to create a semiregular tessellation.

Step 3 Summarize your findings by sketching all eight semiregular tessellations and writing their vertex arrangements (numerical names).

The three regular tessellations and the eight semiregular tessellations you just found are called the **Archimedean tilings.** They are also called 1-uniform tilings because all the vertices in a tiling are identical.

Mathematics CONNECTION

Greek mathematician and inventor Archimedes (ca. 287–212 B.C.E.) studied the relationship between mathematics and art with tilings. He described 11 plane tilings made up of regular polygons, with each vertex being the same type. Plutarch (ca. 46–127 C.E.) wrote of Archimedes' love of geometry, "... he neglected to eat and drink and took no care of his person; that he was often carried by force to the baths, and when there he would trace geometrical figures in the ashes of the fire."

Often, different vertices in a tiling do not have the same vertex arrangement. If there are two different types of vertices, the tiling is called 2-uniform. If there are three different types of vertices, the tiling is called 3-uniform. Two examples are shown below.

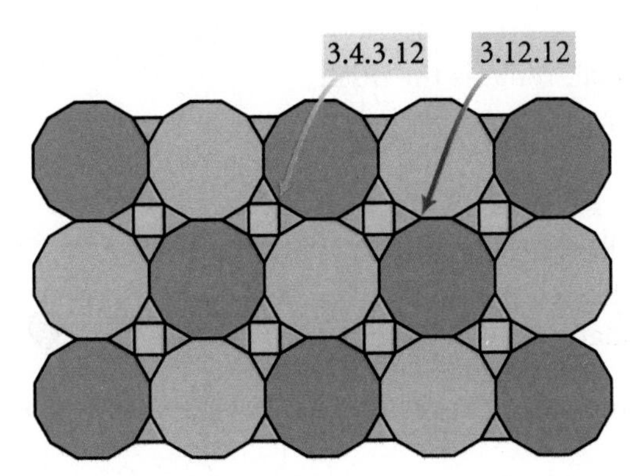

A 2-uniform tessellation: $3.4.3.12 / 3.12^2$

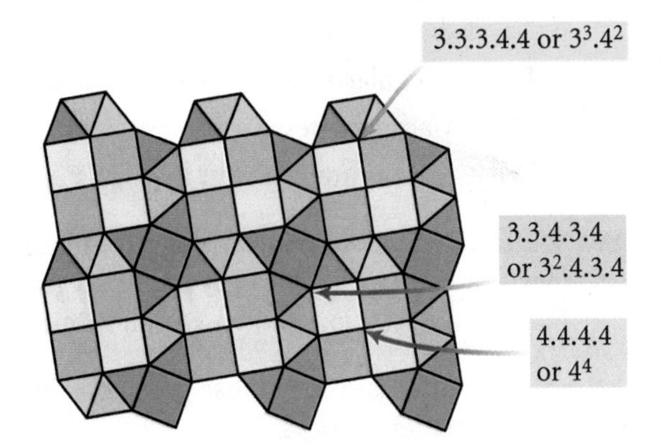

A 3-uniform tessellation: $3^3.4^2 / 3^2.4.3.4 / 4^4$

There are 20 different 2-uniform tessellations of regular polygons, and 61 different 3-uniform tilings. The number of 4-uniform tessellations of regular polygons is still an unsolved problem.

EXERCISES

You will need

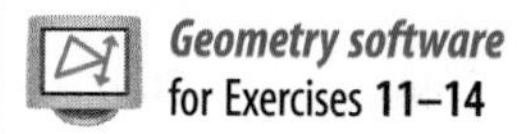

1. Sketch two objects or designs you see every day that are monohedral tessellations.

2. List two objects or designs outside your classroom that are semiregular tessellations.

In Exercises 3–5, write the vertex arrangement for each semiregular tessellation in numbers.

3. 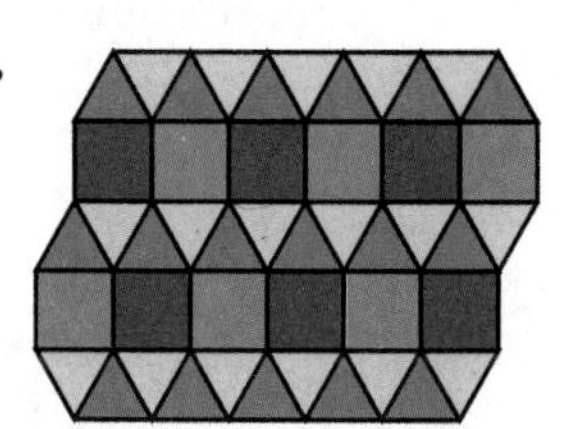

4. 

5. 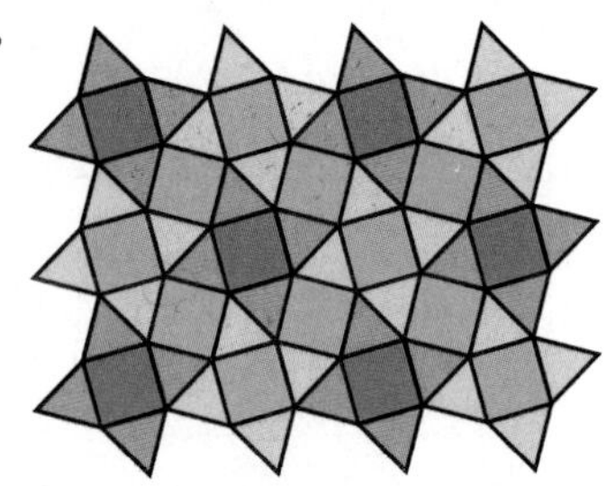

In Exercises 6–8, write the vertex arrangement for each 2-uniform tessellation in numbers.

6. 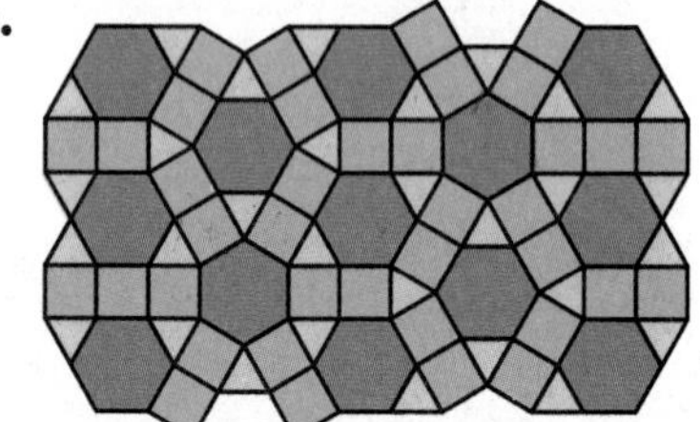

7. 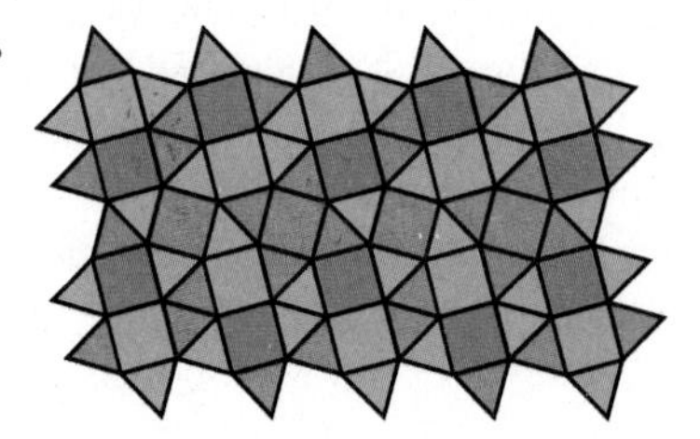

8. 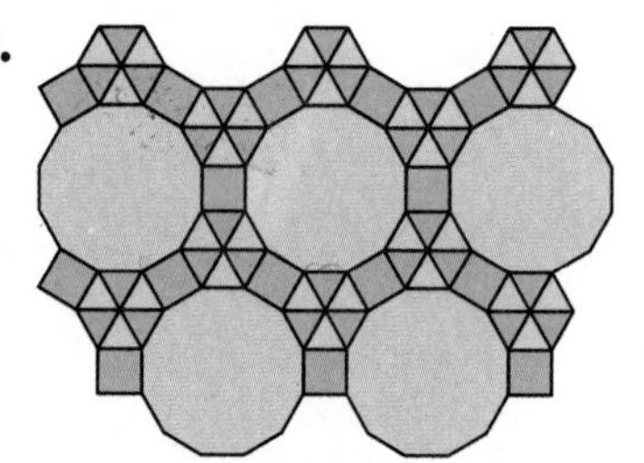

9. When you connect the center of each triangle across the common sides of the tessellating equilateral triangles at right, you get another tessellation. This new tessellation is called the **dual** of the original tessellation. Notice the dual of the equilateral triangle tessellation is the regular hexagon tessellation. Every regular tessellation of regular polygons has a dual.

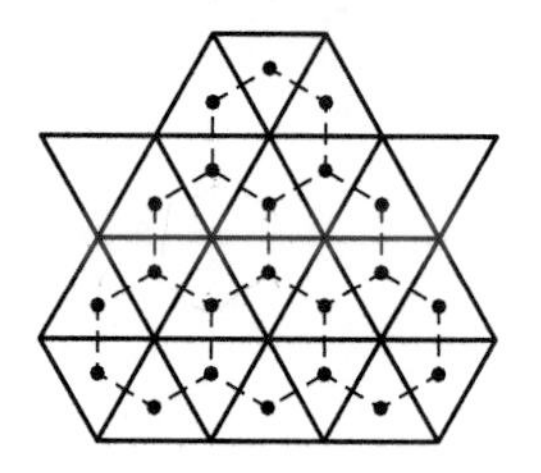

   a. Draw a regular square tessellation and make its dual. What is the dual?

   b. Draw a hexagon tessellation and make the dual of it. What is the dual?

   c. What do you notice about the duals?

10. You can make dual tessellations of semiregular tessellations, but they may not be tessellations of regular polygons. Try it. Sketch the dual of the 4.8.8 tessellation, shown at right. Describe the dual.

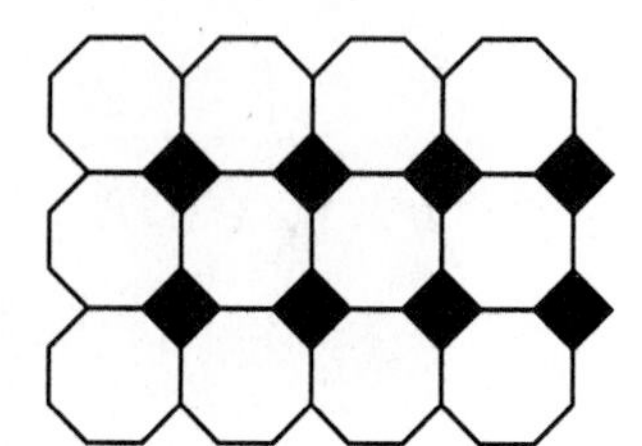

Technology In Exercises 11–14, use geometry software, templates of regular polygons, or pattern blocks.

11. Sketch and color the 3.6.3.6 tessellation. Continue it to fill an entire sheet of paper.

12. Sketch the 4.6.12 tessellation. Color it so it has reflectional symmetry but not rotational symmetry.

13. Show that two regular pentagons and a regular decagon fit about a point but that 5.5.10 does not create a semiregular tessellation. ⓗ

14. Create the tessellation 3.12.12/3.4.3.12. Draw your design onto a full sheet of paper. Color your design to highlight its symmetries.

Review

15. Design a logo with rotational symmetry for Happy Time Ice Cream Company. Or design a logo for your group or for a made-up company.

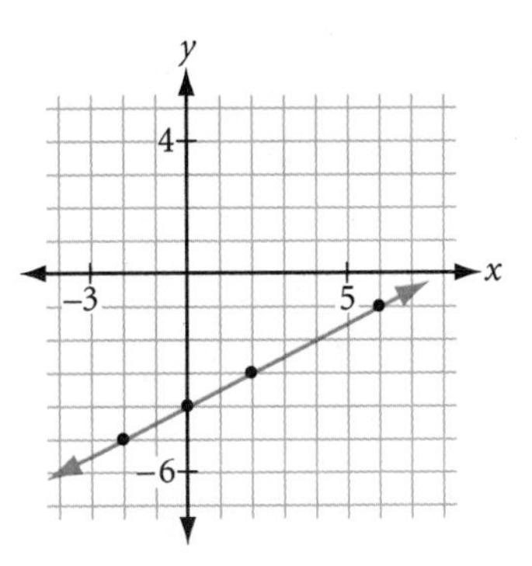

16. Reflect $y = \frac{1}{2}x - 4$ over the x-axis and find the equation of the image line.

17. Words like MOM, WOW, TOOT, and OTTO all have a vertical line of symmetry when you write them in capital letters. Find another word that has a vertical line of symmetry.

The design at left comes from *Inversions,* a book by Scott Kim. Not only does the design spell the word *mirror,* it does so with mirror symmetry!

18. Frisco Fats needs to sink the 8-ball into the NW corner pocket, but he seems trapped. Can he hit the cue ball to a point on the N cushion so that it bounces out, strikes the S cushion, and taps the 8-ball into the corner pocket? Copy the table and construct the path of the ball. ⓗ

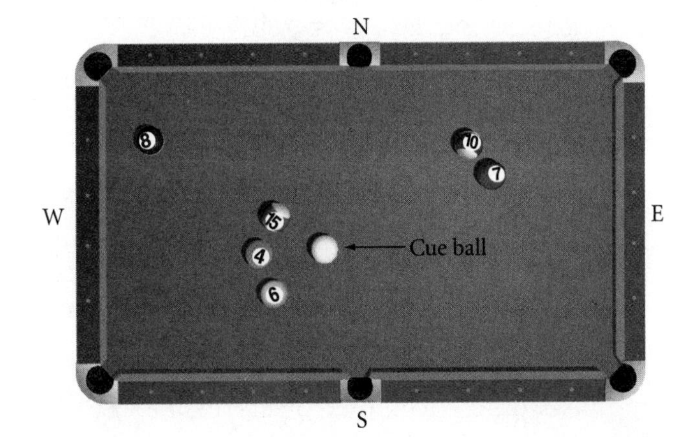

IMPROVING YOUR REASONING SKILLS

PUZZLE

Scrambled Arithmetic

In the equation $65 + 28 = 43$, all the digits are correct but they are in the wrong places! Written correctly, the equation is $23 + 45 = 68$. In each of the three equations below, the operations and the digits are correct, but some of the digits are in the wrong places. Find the correct equations.

1. $11 + 66 = 457$ **2.** $39 \cdot 11 = 75$ **3.** $\frac{78}{523} = 31$

LESSON 7.5

Tessellations with Nonregular Polygons

The most uniquely personal of all that he knows is that which he has discovered for himself.

JEROME BRUNER

In Lesson 7.4, you tessellated with regular polygons. You drew both regular and semi-regular tessellations with them. What about tessellations of nonregular polygons? For example, will a scalene triangle tessellate? Let's investigate.

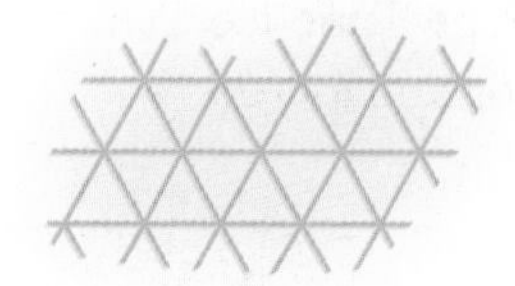

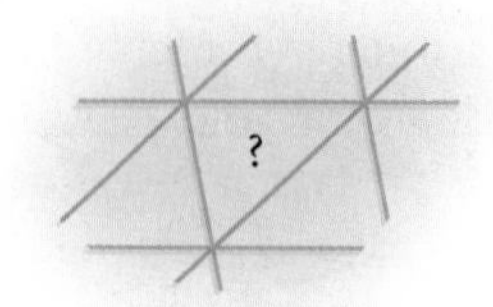

Investigation 1
Do All Triangles Tessellate?

Step 1 Make 12 congruent scalene triangles and use them to try to create a tessellation.

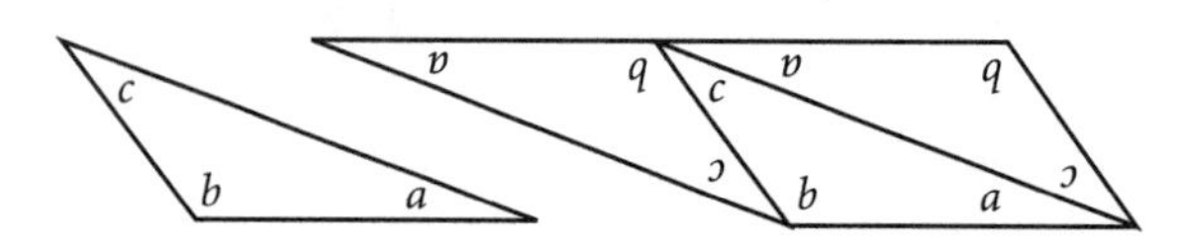

Step 2 Look at the angles about each vertex point. What do you notice?

Step 3 What is the sum of the measures of the three angles of a triangle? What is the sum of the measures of the angles that fit around each point? Compare your results with the results of others and state your next conjecture.

Procedure Note

Making Congruent Triangles

1. Stack three pieces of paper and fold them in half.
2. Draw a scalene triangle on the top half-sheet and cut it out, cutting through all six layers to get six congruent scalene triangles.
3. Use one triangle as a template and repeat. You now have 12 congruent triangles.
4. Label the corresponding angles of each triangle *a, b,* and *c,* as shown.

Tessellating Triangles Conjecture C-73

__?__ triangle will create a monohedral tessellation.

You have seen that squares and rectangles tile the plane. Can you visualize tiling with parallelograms? Will any quadrilateral tessellate? Let's investigate.

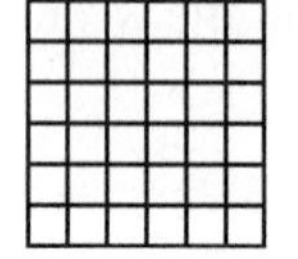

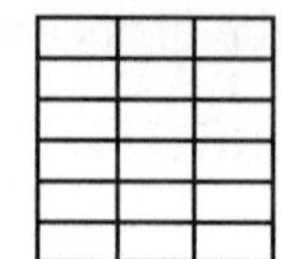

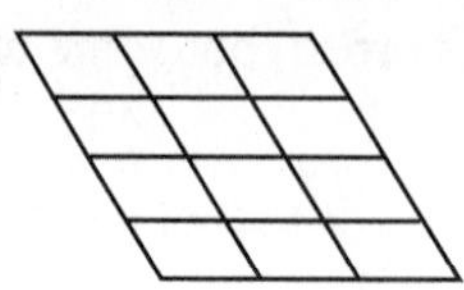

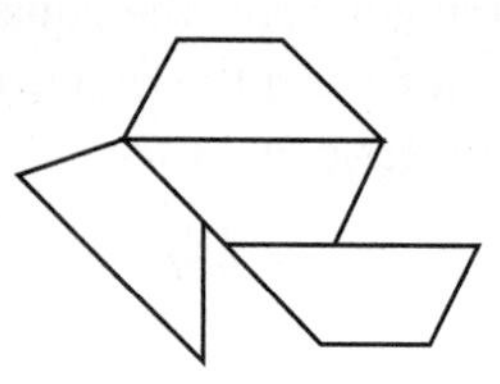

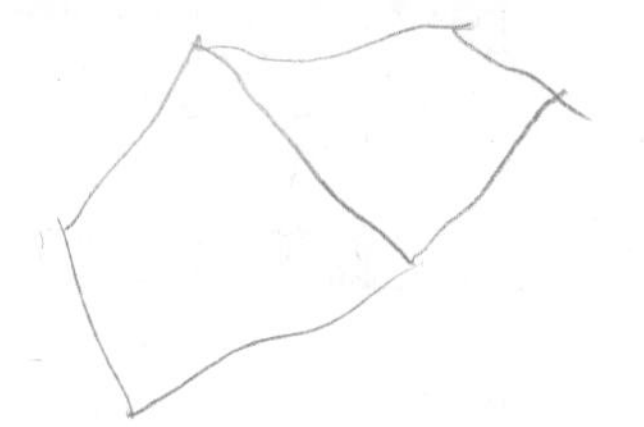

Investigation 2
Do All Quadrilaterals Tessellate?

You want to find out if *any* quadrilateral can tessellate, so you should *not* choose a special quadrilateral for this investigation.

Step 1 Cut out 12 congruent quadrilaterals. Label the corresponding angles in each quadrilateral a, b, c, and d.

Step 2 Using your 12 congruent quadrilaterals, try to create a tessellation.

Step 3 Notice the angles about each vertex point. How many times does each angle of your quadrilateral fit at each point? What is the sum of the measures of the angles of a quadrilateral? Compare your results with others. State a conjecture.

Tessellating Quadrilaterals Conjecture
C-74

? quadrilateral will create a monohedral tessellation.

A regular pentagon does not tessellate, but are there *any* pentagons that tessellate? How many?

Mathematics CONNECTION

In 1975, when Martin Gardner wrote about pentagonal tessellations in *Scientific American,* experts thought that only eight kinds of pentagons would tessellate. Soon another type was found by Richard James III. After reading about this new discovery, Marjorie Rice began her own investigations.

With no formal training in mathematics beyond high school, Marjorie Rice investigated the tessellating problem and discovered four more types of pentagons that tessellate. Mathematics professor Doris Schattschneider of Moravian College verified Rice's research and brought it to the attention of the mathematics community. Rice had indeed discovered what professional mathematicians had been unable to uncover!

In 1985, Rolf Stein, a German graduate student, discovered a fourteenth type of tessellating pentagon. Are *all* the types of convex pentagons that tessellate now known? The problem remains unsolved.

Shown at right are Marjorie Rice (left) and Dr. Doris Schattschneider.

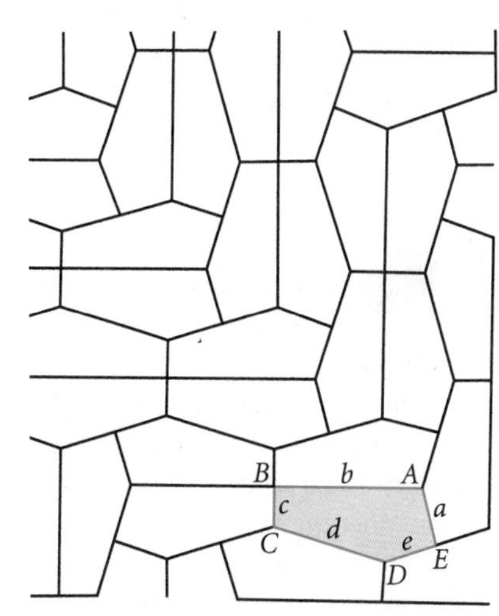

Type 13, discovered in December 1977

$B = E = 90°, 2A + D = 360°$

$2C + D = 360°$

$a = e, a + e = d$

One of the pentagonal tessellations discovered by Marjorie Rice. Capital letters represent angle measures in the shaded pentagon. Lowercase letters represent lengths of sides.

You will experiment with some pentagon tessellations in the exercises.

EXERCISES

You will need

Construction tools for Exercise 1

1. ***Construction*** The beautiful Cairo street tiling shown below uses equilateral pentagons. One pentagon is shown below left. Use a ruler and a protractor to draw the equilateral pentagon on poster board or heavy cardboard. (For an added challenge, you can try to *construct* the pentagon, as Egyptian artisans likely would have done.) Cut out the pentagon and tessellate with it. Color your design.

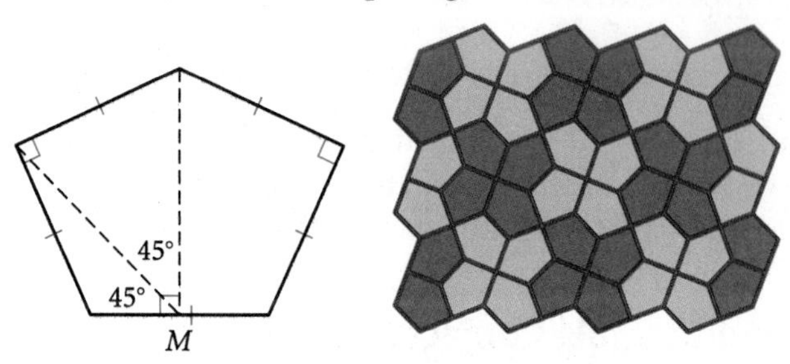

Point M is the midpoint of the base.

2. At right is Marjorie Rice's first pentagonal tiling discovery. Another way to produce a pentagonal tessellation is to make the dual of the tessellation shown in Lesson 7.4, Exercise 5. Try it. ⓗ

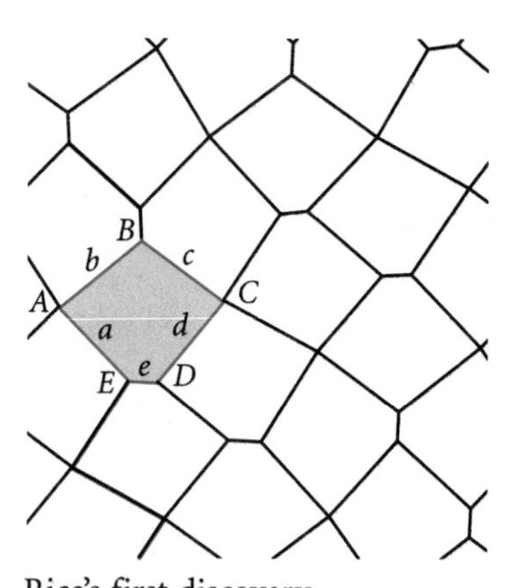

Rice's first discovery, February 1976

$2E + B = 2D + C = 360°$

$a = b = c = d$

3. A tessellation of regular hexagons can be used to create a pentagonal tessellation by dividing each hexagon as shown. Create this tessellation and color it.

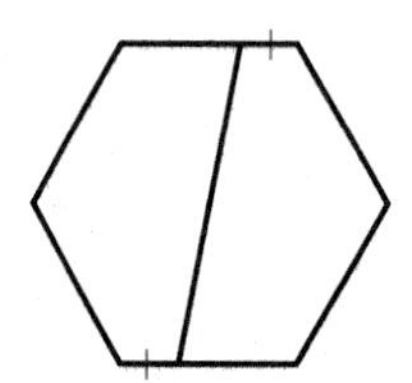

Cultural CONNECTION

Mats called *tatami* are used as a floor covering in traditional Japanese homes. *Tatami* is made from rush, a flowering plant with soft fibers, and has health benefits, such as removing carbon dioxide and regulating humidity and temperature. When arranging *tatami,* you want the seams to form T-shapes. You avoid arranging four at one vertex forming a cross because it is difficult to get a good fit within a room this way. You also want to avoid fault lines—straight seams passing all the way through a rectangular arrangement—because they make it easier for the *tatami* to slip. Room sizes are often given in *tatami* numbers (for example, a 6-mat room or an 8-mat room).

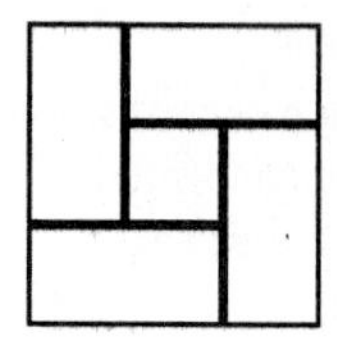

4.5-mat room

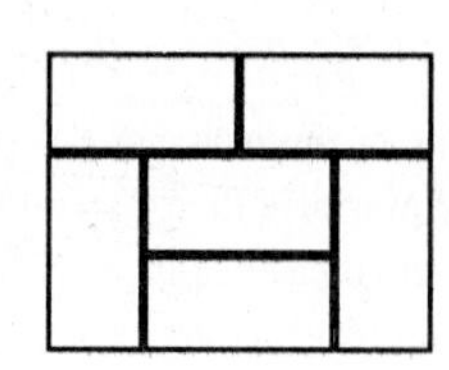

6-mat room

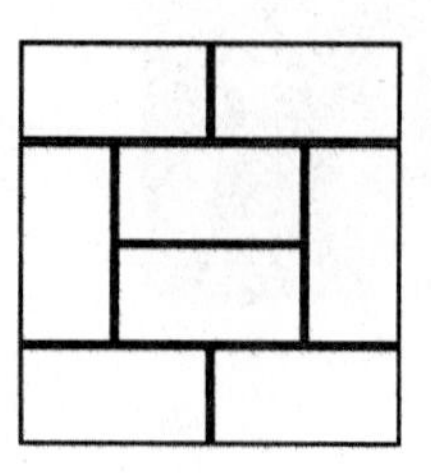

8-mat room

4. Can a concave quadrilateral like the one at right tile the plane? Try it. Create your own concave quadrilateral and try to create a tessellation with it. Decorate your drawing.

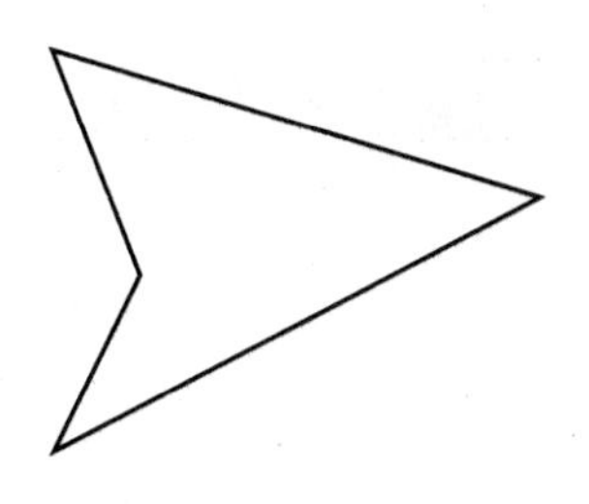

5. Write a paragraph proof explaining why you can use any triangle to create a monohedral tiling.

Review

Refer to the Cultural Connection on page 386 for Exercises 6 and 7.

6. Use graph paper to design an arrangement of *tatami* for a 10-mat room. In how many different ways can you arrange the mats so that there are no places where four mats meet at a point (no cross patterns)? Assume that the mats measure 3-by-6 feet and that each room must be at least 9 feet wide. Show all your solutions.

7. There are at least two ways to arrange a 15-mat rectangle with no fault lines. One is shown. Can you find the other?

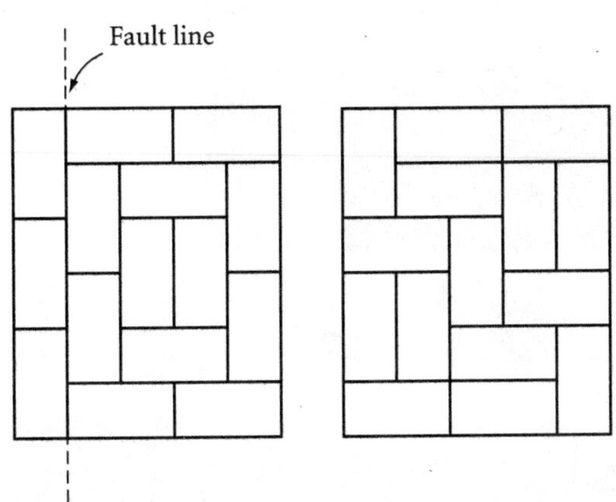

8. Reflect $y = 2x + 3$ across the y-axis and find the equation of the image line.

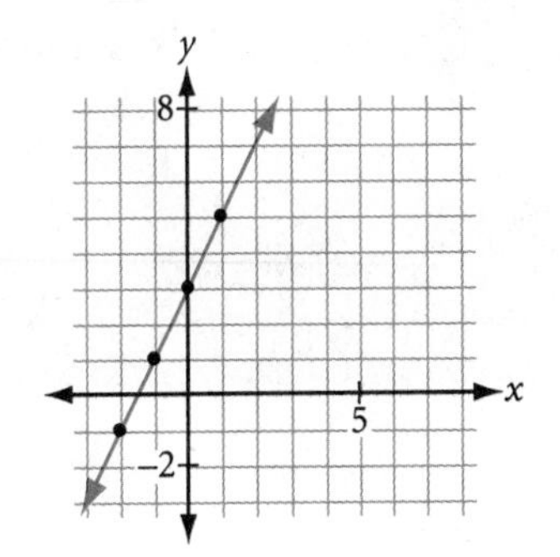

IMPROVING YOUR VISUAL THINKING SKILLS

PUZZLE

Picture Patterns II

Draw what comes next in each picture pattern.

1.

2.

project

PENROSE TILINGS

When British scientist Sir Roger Penrose of the University of Oxford is not at work on quantum mechanics or relativity theory, he's inventing mathematical games. Penrose came up with a special tiling that uses two shapes, a kite and a dart. (The *dart* is a concave kite.) The tiles must be placed so that each vertex with a dot always touches only other vertices with dots. By adding this extra requirement, Penrose's tiles make a *nonperiodic tiling*. That is, as you tessellate, the pattern does not repeat by translations.

Penrose tilings decorate the Storey Hall building in Melbourne, Australia.

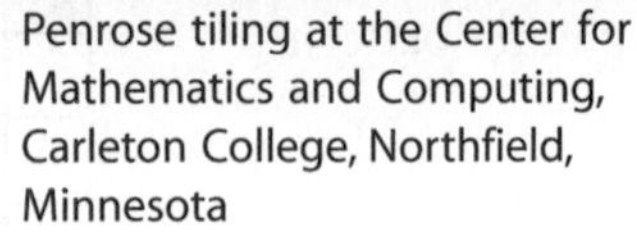

Penrose tiling at the Center for Mathematics and Computing, Carleton College, Northfield, Minnesota

Try it. Copy the two tiles shown below—the kite and the dart with their dots—onto patty paper. Use the patty-paper tracing to make two cardboard tiles. Create your own unique Penrose tiling and color it. Or, use geometry software to create and color your design.

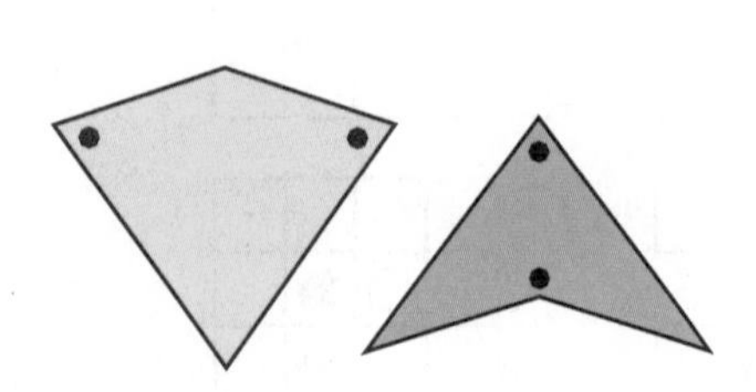

LESSON 7.6

Tessellations Using Only Translations

There are three kinds of people in this world: those who make things happen, those who watch things happen, and those who wonder what happened.

ANONYMOUS

In 1936, M. C. Escher traveled to Spain and became fascinated with the tile patterns of the Alhambra. He spent days sketching the tessellations that Islamic masters had used to decorate the walls and ceilings. Some of his sketches are shown at right. Escher wrote that the tessellations were "the richest source of inspiration" he had ever tapped.

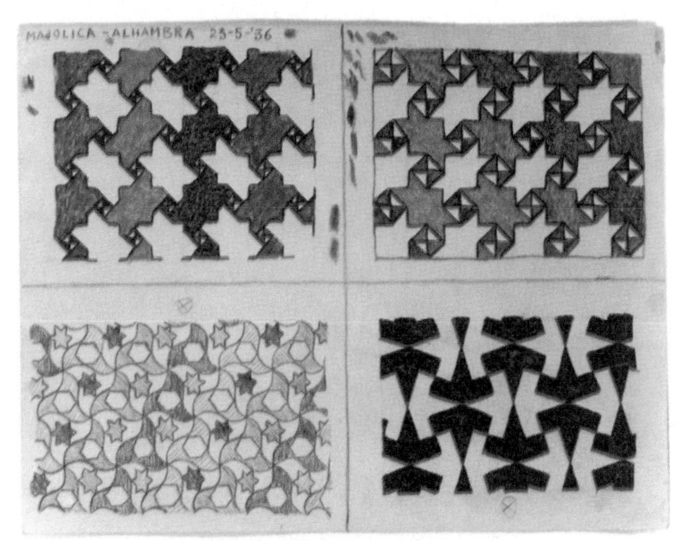

Brickwork, Alhambra, M. C. Escher

Symmetry Drawing E105, M. C. Escher, 1960

Escher spent many years learning how to use translations, rotations, and glide reflections on grids of equilateral triangles and parallelograms. But he did not limit himself to pure geometric tessellations.

The four steps below show how Escher may have created his Pegasus tessellation, shown at left. Notice how a partial outline of the Pegasus is translated from one side of a square to another to complete a single tile that fits with other tiles like itself.

You can use steps like this to create your own unique tessellation. Start with a tessellation of squares, rectangles, or parallelograms, and try translating curves on opposite sides of the tile. It may take a few tries to get a shape that looks like a person, animal, or plant. Use your imagination!

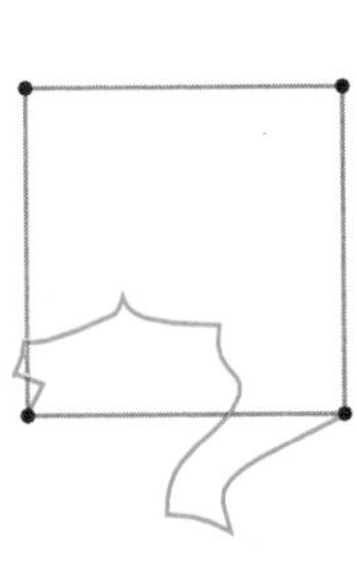

Step 1

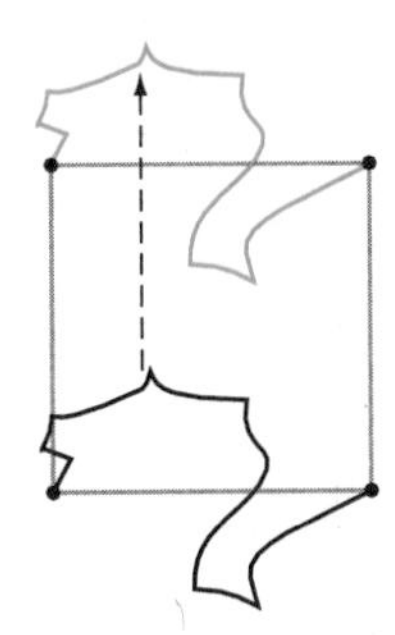

Step 2

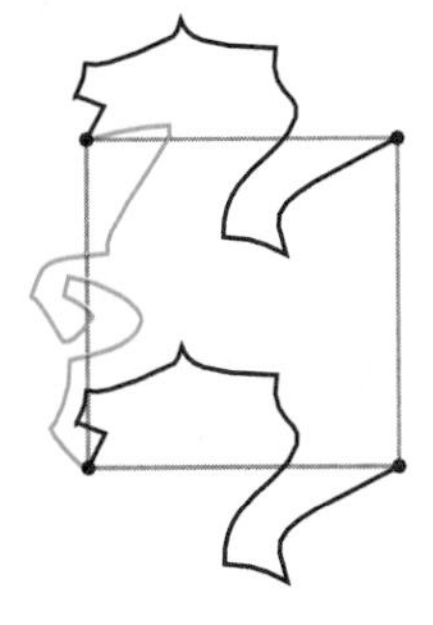

Step 3

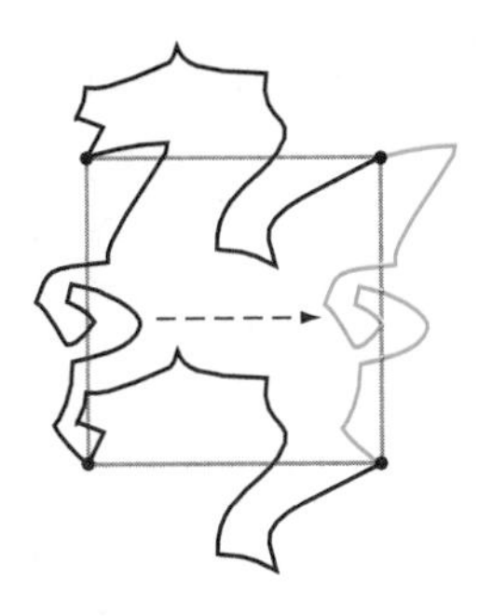

Step 4

You can also use the translation technique with regular hexagons. The only difference is that there are three sets of opposite sides on a hexagon. So you'll need to draw three sets of curves and translate them to opposite sides. The six steps below show how student Mark Purcell created his tessellation, *Monster Mix*.

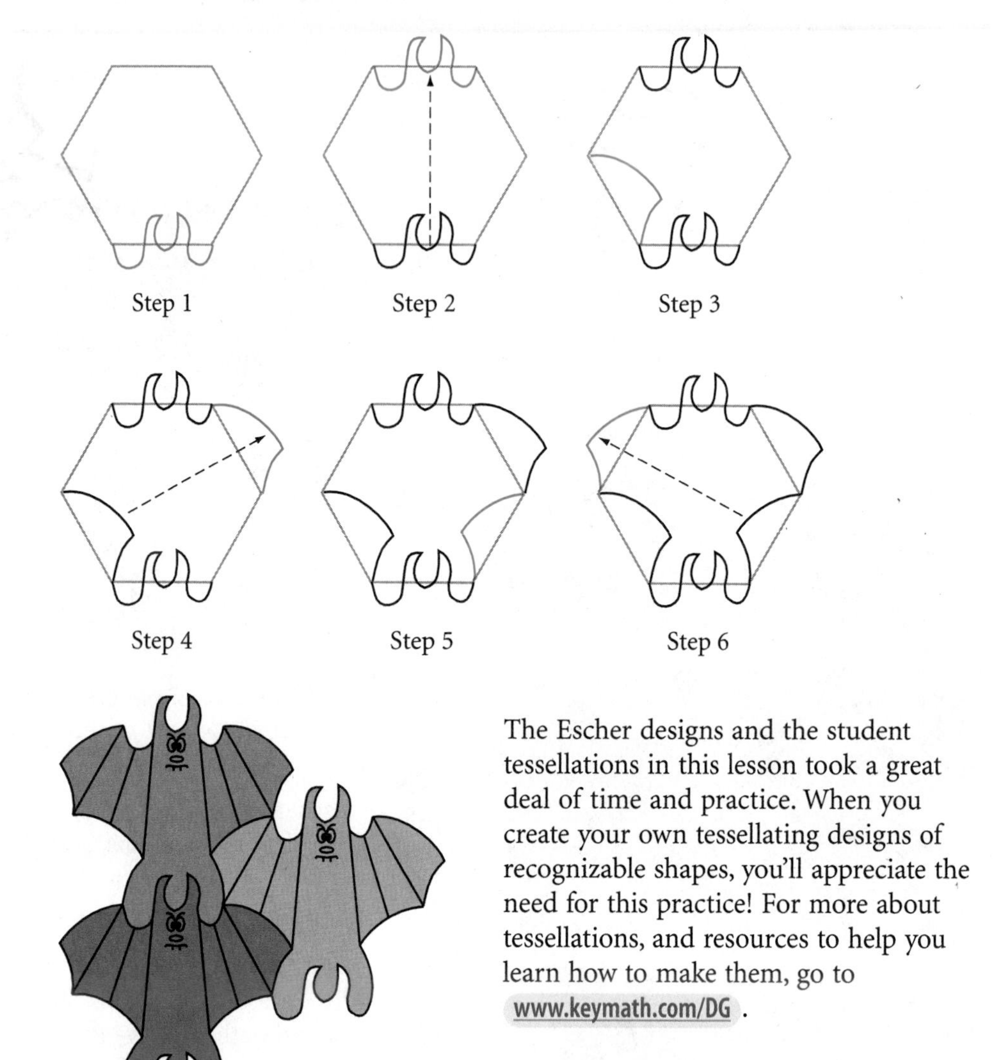

The Escher designs and the student tessellations in this lesson took a great deal of time and practice. When you create your own tessellating designs of recognizable shapes, you'll appreciate the need for this practice! For more about tessellations, and resources to help you learn how to make them, go to **www.keymath.com/DG** .

Monster Mix, Mark Purcell

EXERCISES

In Exercises 1–3, copy each tessellating shape and fill it in so that it becomes a recognizable figure.

1. 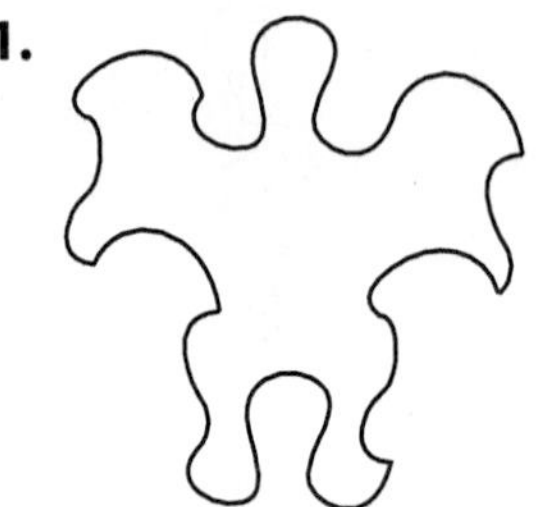

2. 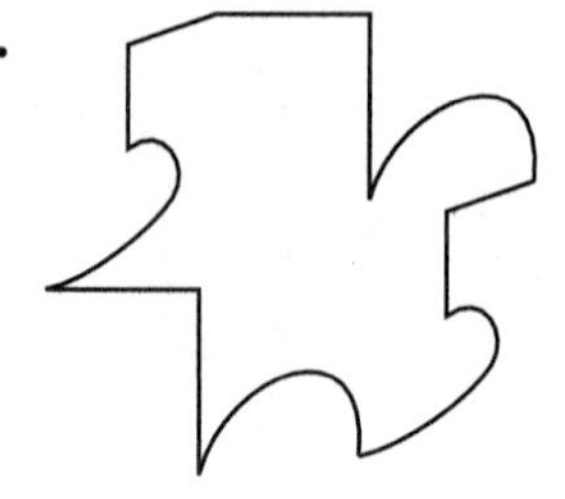

3. 

In Exercises 4–6, identify the basic tessellation grid (squares, parallelograms, or regular hexagons) that each geometry student used to create each translation tessellation.

4. **5.** **6.**

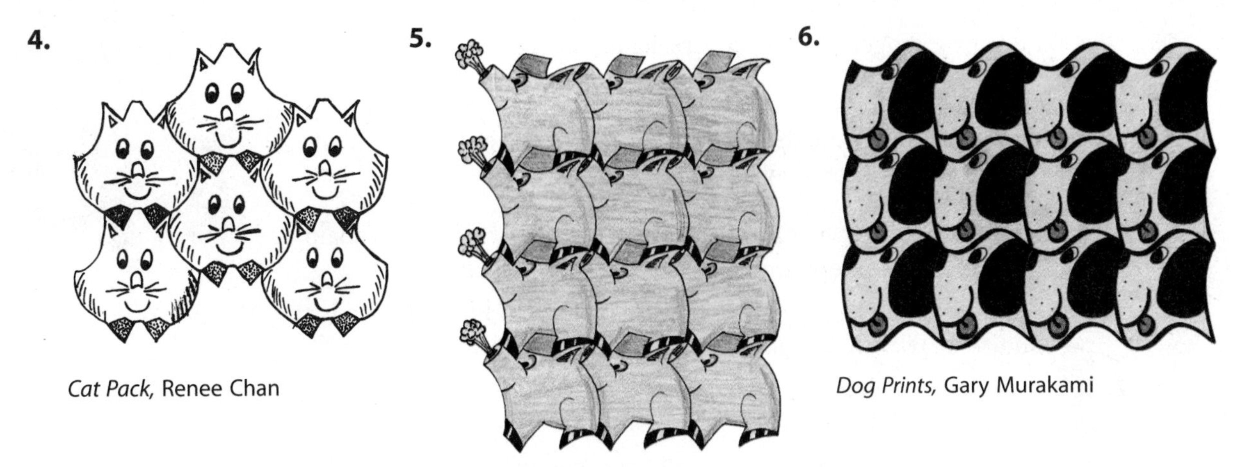

Cat Pack, Renee Chan

Snorty the Pig, Jonathan Benton

Dog Prints, Gary Murakami

In Exercises 7 and 8, copy the figure and the grid onto patty paper. Create a tessellation on the grid with the figure.

7. **8.**

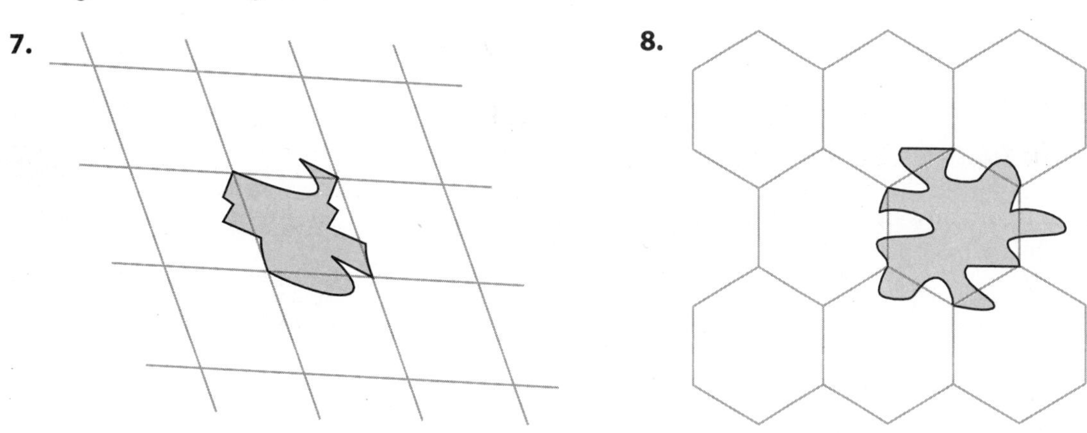

Now it's your turn. In Exercises 9 and 10, create a tessellation of recognizable shapes using the translation method you learned in this lesson. At first, you will probably end up with shapes that look like amoebas or spilled milk, but with practice and imagination, you will get recognizable images. Decorate and title your designs.

9. Use squares as the basic structure. ⓗ

10. Use regular hexagons as the basic structure.

Review

11. The route of a rancher takes him from the house at point A to the south fence, then over to the east fence, then to the corral at point B. Copy the figure at right onto patty paper and locate the points on the south and east fences that minimize the rancher's route.

12. Reflect $y = \frac{2}{3}x - 3$ over the y-axis. Write an equation for the image. How does it compare with the original equation?

13. Give the vertex arrangement for the tessellation at right.

14. A helicopter has four blades. Each blade measures about 26 feet from the center of rotation to the tip. What is the speed in feet per second at the tips of the blades when they are moving at 400 rpm?

15. Identify each of the following statements as true or false. If true, explain why. If false, give a counterexample explaining why it is false.

a. If the two diagonals of a quadrilateral are congruent but only one is the perpendicular bisector of the other, then the quadrilateral is a kite.

b. If the quadrilateral has exactly one line of reflectional symmetry, then the quadrilateral is a kite.

c. If the diagonals of a quadrilateral are congruent and bisect each other, then it is a square.

d. If a trapezoid is cyclic, then it is isosceles.

IMPROVING YOUR VISUAL THINKING SKILLS

PUZZLE

3-by-3 Inductive Reasoning Puzzle I

Sketch the figure missing in the lower right corner of this 3-by-3 pattern.

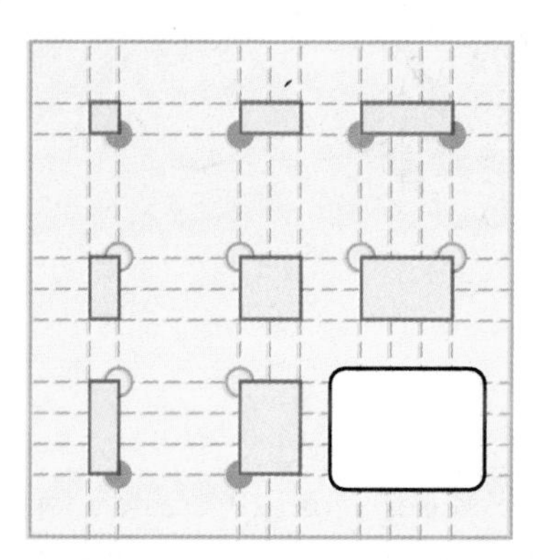

LESSON

7.7

Tessellations That Use Rotations

Einstein was once asked, "What is your phone number?" He answered, "I don't know, but I know where to find it if I need it."

ALBERT EINSTEIN

In Lesson 7.6, you created recognizable shapes by translating curves from opposite sides of a regular hexagon or square. In tessellations using only translations, all the figures face in the same direction. In this lesson you will use rotations of curves on a grid of parallelograms, equilateral triangles, or regular hexagons. The resulting tiles will fit together when you rotate them, and the designs will have rotational symmetry about points in the tiling. For example, in this Escher print, each reptile is made by rotating three different curves about three alternating vertices of a regular hexagon.

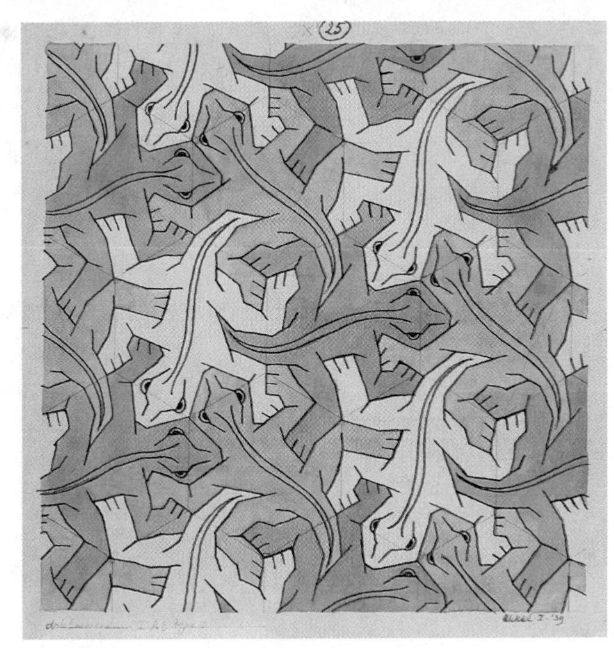

Symmetry Drawing E25, M. C. Escher, 1939

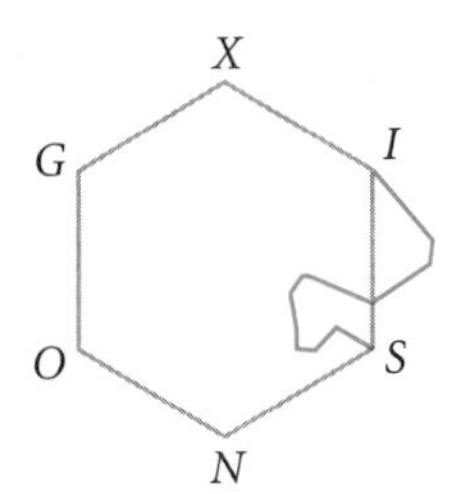

Step 1

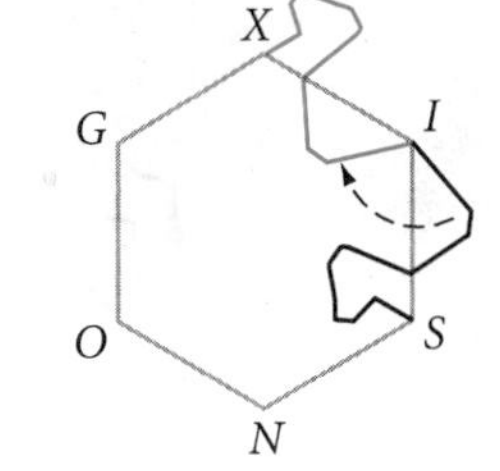

Step 2

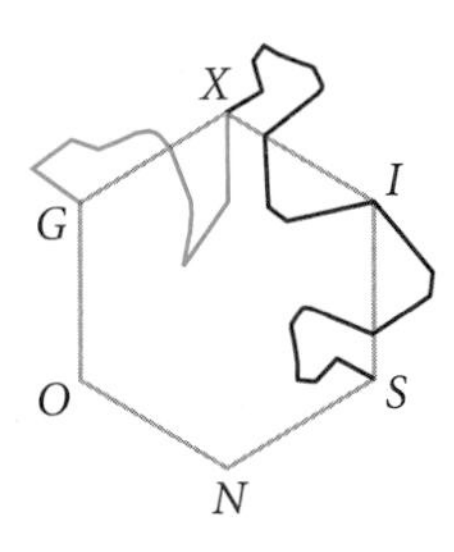

Step 3

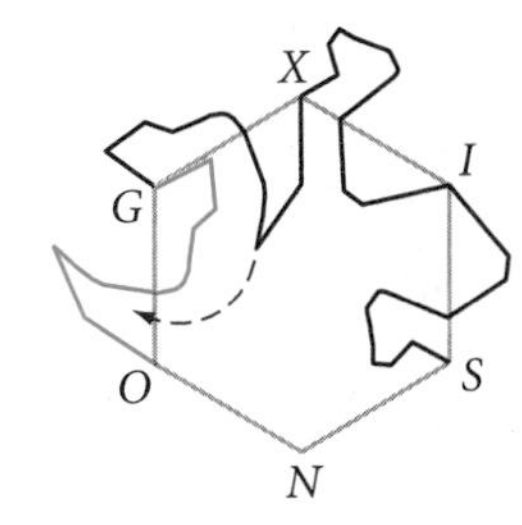

Step 4

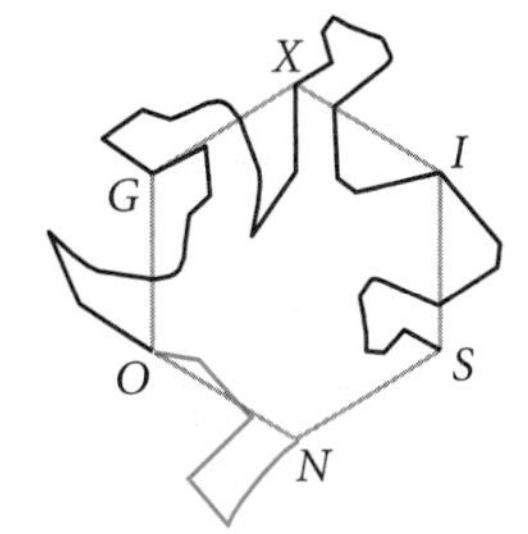

Step 5

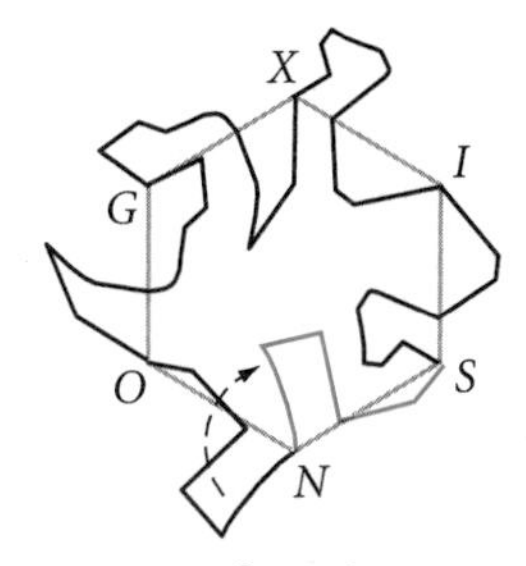

Step 6

Step 1	Connect points *S* and *I* with a curve.
Step 2	Rotate curve *SI* about point *I* so that point *S* rotates to coincide with point *X*.
Step 3	Connect points *G* and *X* with a curve.
Step 4	Rotate curve *GX* about point *G* so that point *X* rotates to coincide with point *O*.
Step 5	Create curve *NO*.
Step 6	Rotate curve *NO* about point *N* so that point *O* rotates to coincide with point *S*.

Escher worked long and hard to adjust each curve until he got what he recognized as a reptile. When you are working on your own design, keep in mind that you may have to redraw your curves a few times until something you recognize appears.

Escher used his reptile drawing in this famous lithograph. Look closely at the reptiles in the drawing. Escher loved to play with our perceptions of reality!

Reptiles, M. C. Escher, 1943

Another method used by Escher utilizes rotations on an equilateral triangle grid. Two sides of each equilateral triangle have the same curve, rotated about their common point. The third side is a curve with point symmetry. The following steps demonstrate how you might create a tessellating flying fish like that created by Escher.

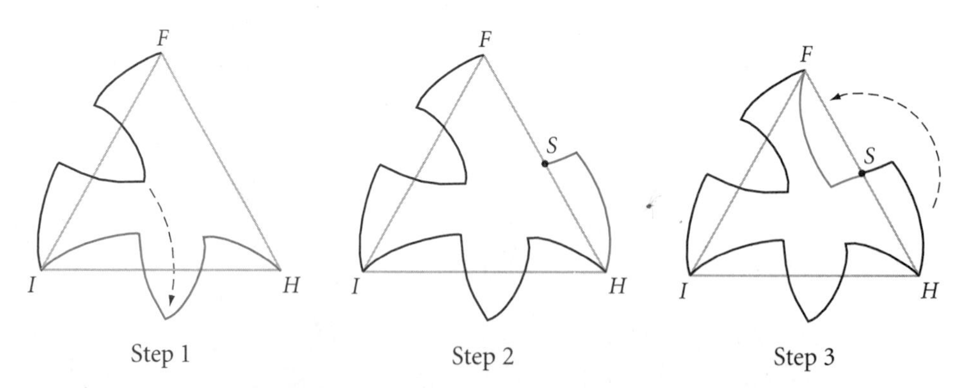

Step 1 Connect points F and I with a curve. Then rotate the curve 60° clockwise about point I so that it becomes curve IH.

Step 2 Find the midpoint S of $\overline{FH}$ and draw curve SH.

Step 3 Rotate curve SH 180° about S to produce curve FS. Together curve FS and curve SH become the point-symmetric curve FH.

With a little added detail, the design becomes a flying fish.

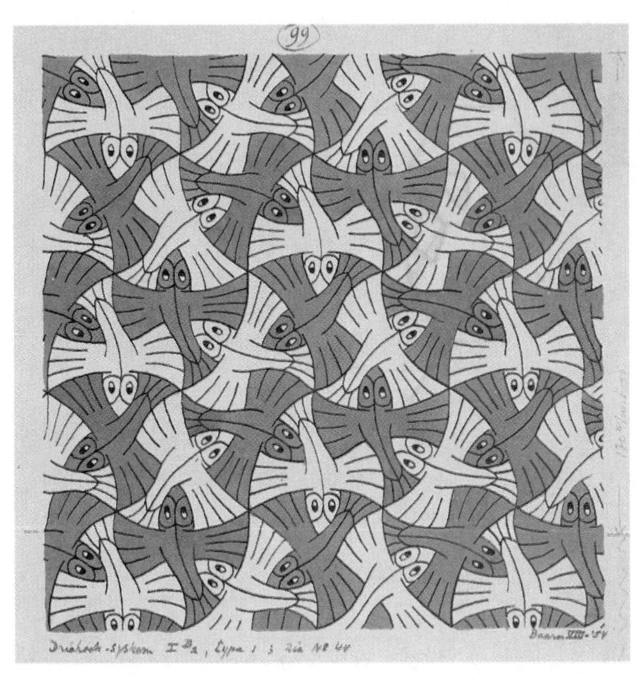

Symmetry Drawing E99,
M. C. Escher, 1954

Or, with just a slight variation in the curves, the resulting shape will appear more like a bird than a flying fish.

EXERCISES

You will need

Geometry software for Exercise **13**

In Exercises 1 and 2, identify the basic grid (equilateral triangles or regular hexagons) that each geometry student used to create the tessellation.

1.

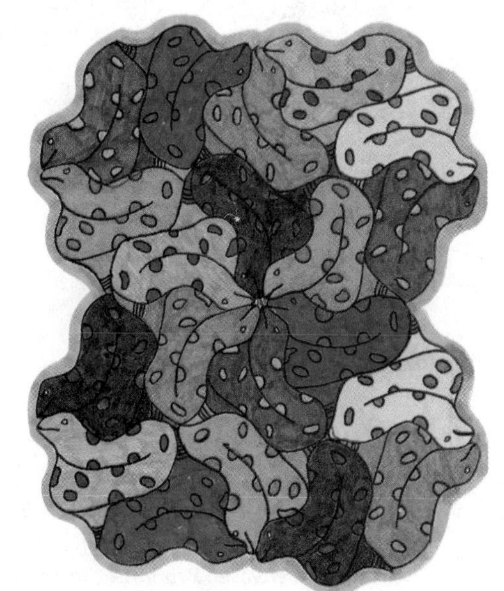

Snakes, Jack Chow

2.

Merlin, Aimee Plourdes

In Exercises 3 and 4, copy the figure and the grid onto patty paper. Show how you can use other pieces of patty paper to tessellate the figure on the grid.

3.

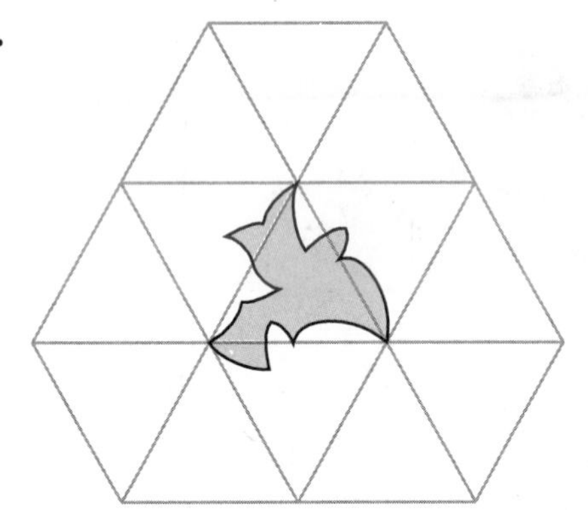

4.

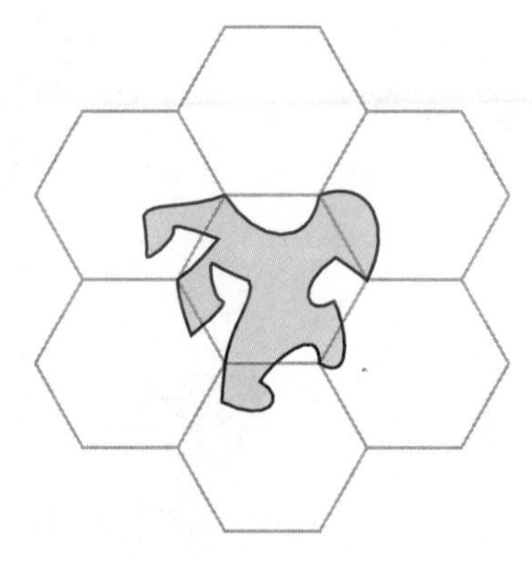

In Exercises 5 and 6, create tessellation designs by using rotations. You will need patty paper, tracing paper, or clear plastic, and grid paper or isometric dot paper.

5. Create a tessellating design of recognizable shapes by using a grid of regular hexagons. Decorate and color your art.

6. Create a tessellating design of recognizable shapes by using a grid of equilateral or isosceles triangles. Decorate and color your art.

Review

7. Study these knot designs by Rinus Roelofs. Now try creating one of your own. Select a tessellation. Make a copy and thicken the lines. Make two copies of this thick-lined tessellation. Lay one of them on top of the other and shift it slightly. Trace the one underneath onto the top copy. Erase where they overlap. Then create a knot design using what you learned in Lesson 0.5.

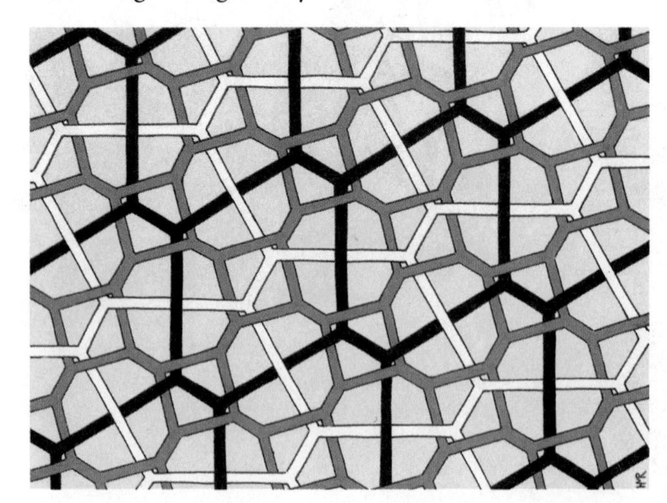

Dutch artist Rinus Roelofs (b 1954) experiments with the lines between the shapes rather than looking at the plane-filling figures. In these paintings, he has made the lines thicker and created intricate knot designs.

(Above) *Impossible Structures–III, structure 24*
(At left) *Interwoven Patterns–V, structure 17*

For Exercises 8–11, identify the statement as true or false. For each false statement, explain why it is false or sketch a counterexample.

8. If the diagonals of a quadrilateral are congruent, the quadrilateral is a parallelogram.

9. If the diagonals of a quadrilateral are congruent and bisect each other, the quadrilateral is a rectangle.

10. If the diagonals of a quadrilateral are perpendicular and bisect each other, the quadrilateral is a rhombus.

11. If the diagonals of a quadrilateral are congruent and perpendicular, the quadrilateral is a square.

12. Earth's radius is about 4000 miles. Imagine that you travel from the equator to the South Pole by a direct route along the surface. Draw a sketch of your path. How far will you travel? How long will the trip take if you travel at an average speed of 50 miles per hour?

13. *Technology* Use geometry software to construct a line and two points A and B not on the line. Reflect A and B over the line and connect the four points to form a trapezoid.

a. Is it isosceles? Why?

b. Choose a random point C inside the trapezoid and connect it to the four vertices with segments. Calculate the sum of the distances from C to the four vertices. Drag point C around. Where is the sum of the distances the greatest? The least?

IMPROVING YOUR REASONING SKILLS

PUZZLE

Logical Liars

Five students have just completed a logic contest. To confuse the school's reporter, Lois Lang, each student agreed to make one true and one false statement to her when she interviewed them. Lois was clever enough to figure out the winner. Are you? Here are the students' statements.

Frances: Kai was second. I was fourth.
Leyton: I was third. Charles was last.
Denise: Kai won. I was second.
Kai: Leyton had the best score. I came in last.
Charles: I came in second. Kai was third.

LESSON

7.8 Tessellations That Use Glide Reflections

The young do not know enough to be prudent, and therefore they attempt the impossible, and achieve it, generation after generation.

PEARL S. BUCK

In this lesson you will use glide reflections to create tessellations. In Lesson 7.6, you saw Escher's translation tessellation of the winged horse Pegasus. All the horses are facing in the same direction. In the drawings below and below left, Escher used glide reflections on a grid of glide-reflected kites to get his horsemen facing in opposite directions.

Horseman, M. C. Escher, 1946

The steps below show how you can make a tessellating design similar to Escher's *Horseman*. (The symbol indicates a glide reflection.)

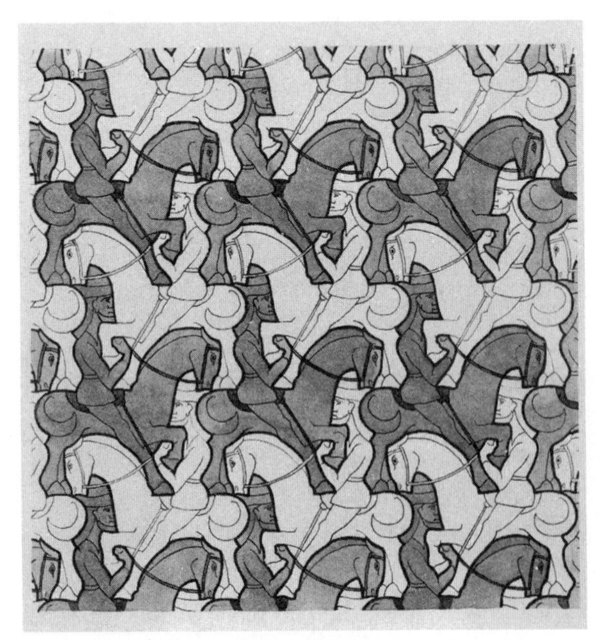

Horseman Sketch, M. C. Escher

Step 1

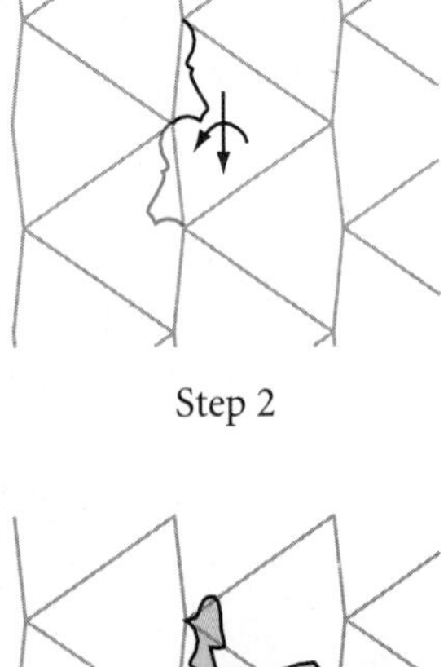

Step 2

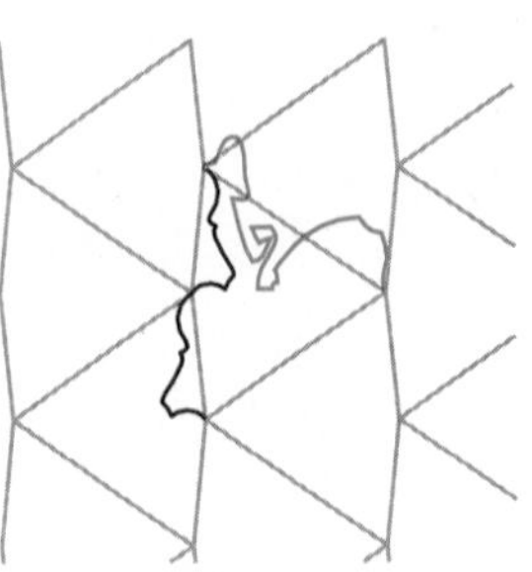

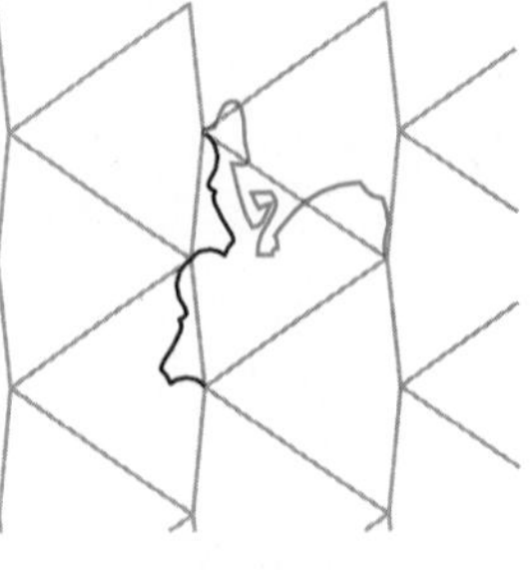

Step 3

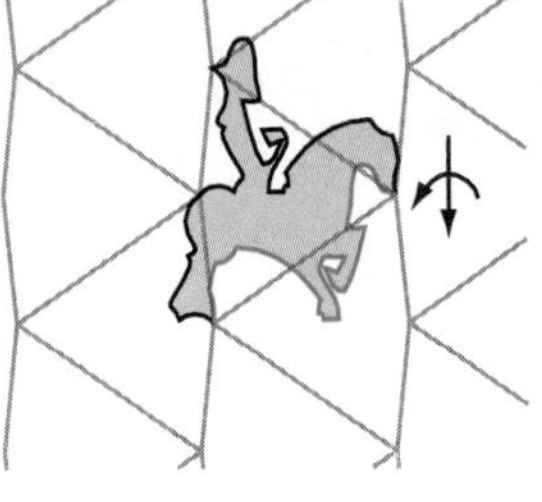

Step 4

In the tessellation of birds below left, you can see that Escher used a grid of squares. You can use the same procedure on a grid of any type of glide-reflected parallelograms. The steps below show how you might create a tessellation of birds or fishes on a parallelogram grid.

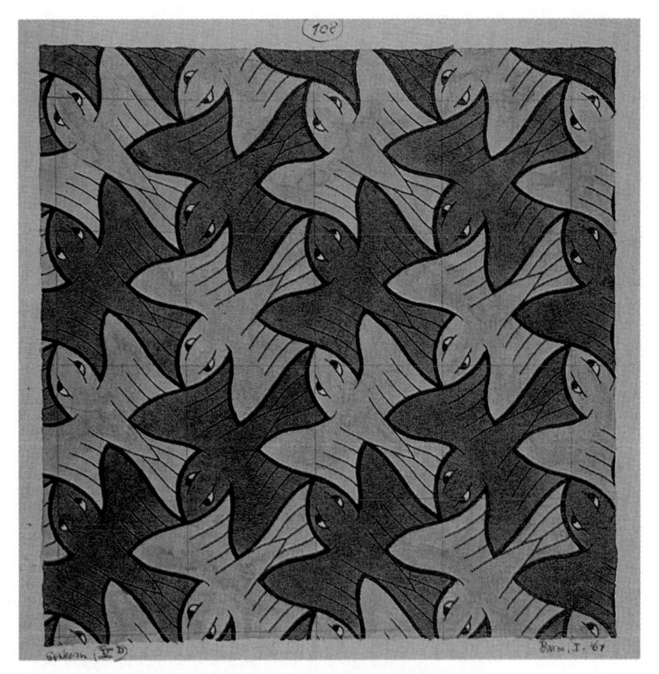

Symmetry Drawing E108, M. C. Escher, 1967

Step 1

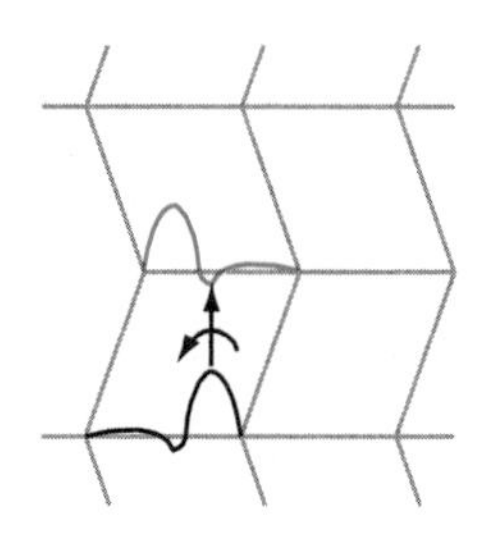

Step 2

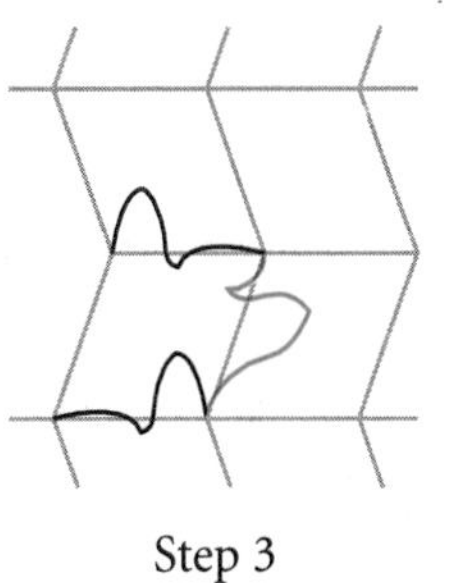

Step 3

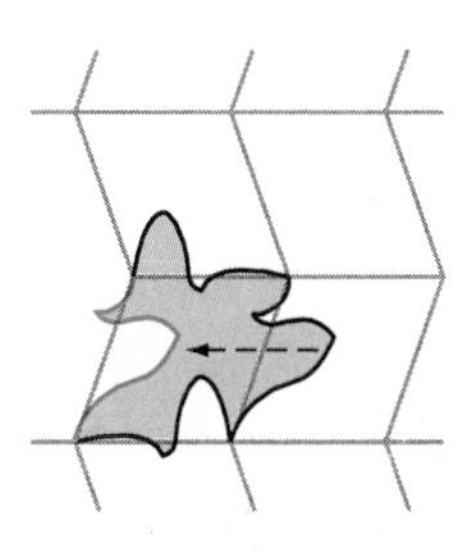

Step 4

EXERCISES

You will need

Construction tools for Exercise **8**

In Exercises 1 and 2, identify the basic tessellation grid (kites or parallelograms) that the geometry student used to create the tessellation.

1.

A Boy with a Red Scarf, Elina Uzin

2.

Glide Reflection, Alice Chan

In Exercises 3 and 4, copy the figure and the grid onto patty paper. Show how you can use other patty paper to tessellate the figure on the grid.

3.

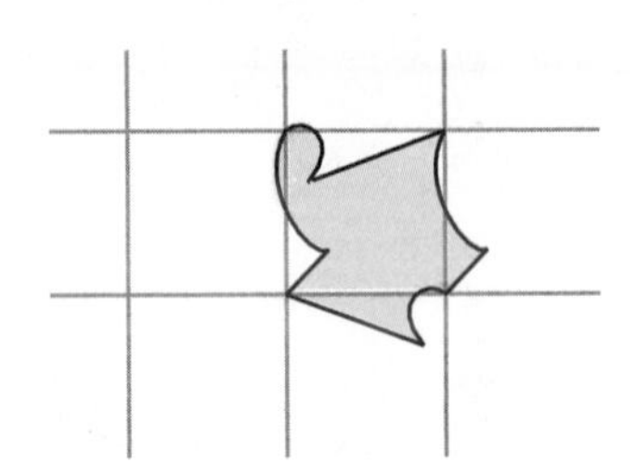

4.

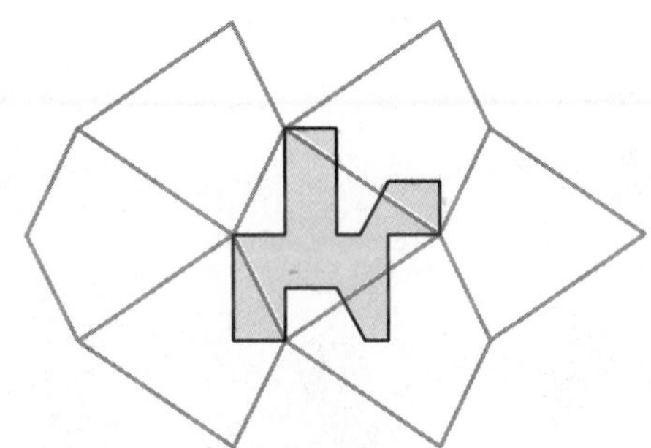

5. Create a glide-reflection tiling design of recognizable shapes by using a grid of kites. Decorate and color your art. ⓗ

6. Create a glide-reflection tiling design of recognizable shapes by using a grid of parallelograms. Decorate and color your art.

Review

7. Find the coordinates of the circumcenter and orthocenter of $\triangle FAN$ with $F(6, 0)$, $A(7, 7)$, and $N(3, 9)$.

8. ***Construction*** Construct a circle and a chord of the circle. With compass and straightedge construct a second chord parallel and congruent to the first chord.

9. Remy's friends are pulling him on a sled. One of his friends is stronger and exerts more force. The vectors in this diagram represent the forces his two friends exert on him. Copy the vectors, complete the vector parallelogram, and draw the resultant vector force on his sled.

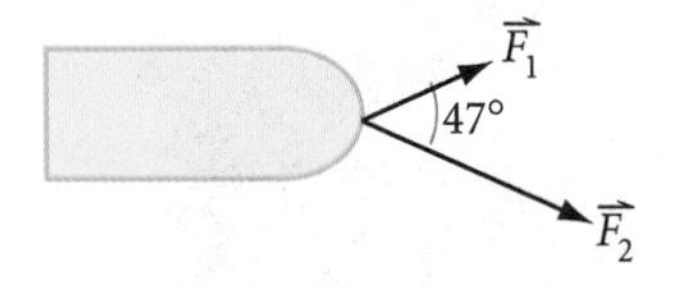

10. The green prism below right was built from the two solids below left. Copy the figure on the right onto isometric dot paper and shade in one of the two pieces to show how the complete figure was created.

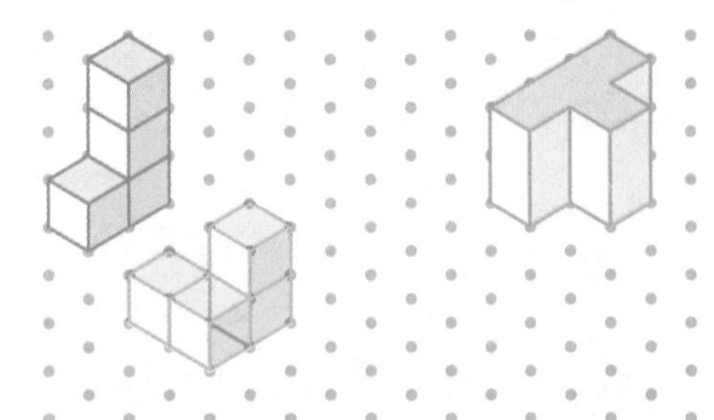

IMPROVING YOUR ALGEBRA SKILLS

Fantasy Functions

If $a \circ b = a^b$ then $3 \circ 2 = 3^2 = 9$,

and if $a \triangle b = a^2 + b^2$ then $5 \triangle 2 = 5^2 + 2^2 = 29$.

If $8 \triangle x = 17 \circ 2$, find x.

Finding the Orthocenter and Centroid

Suppose you know the coordinates of the vertices of a triangle. You have seen that you can find the coordinates of the circumcenter by writing equations for the perpendicular bisectors of two of the sides and solving the system. Similarly, you can find the coordinates of the orthocenter by finding equations for two lines containing altitudes of the triangle and solving the system.

EXAMPLE A Find the coordinates of the orthocenter of $\triangle PDQ$ with $P(0, -4)$, $D(-4, 4)$, and $Q(8, 4)$.

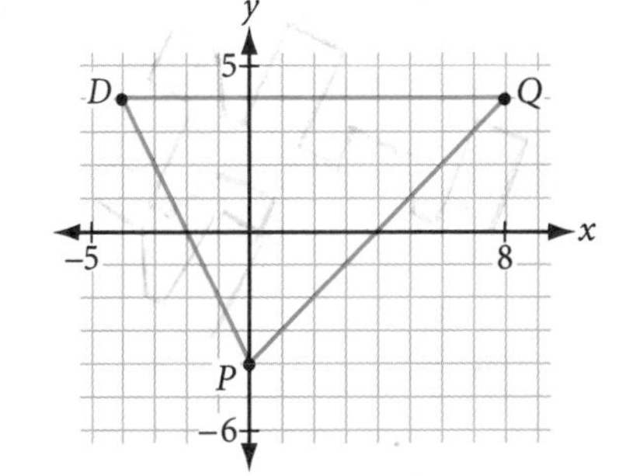

▶ **Solution** The altitude of a triangle passes through one vertex and is perpendicular to the opposite side. To find the equation for the altitude from Q to $\overline{PD}$, first you calculate the slope of $\overline{PD}$, getting -2. So the slope of the altitude to $\overline{PD}$ is $\frac{1}{2}$, the negative reciprocal of -2.

The altitude from Q to $\overline{PD}$ passes through point $Q(8, 4)$, so its equation is $\frac{y-4}{x-8} = \frac{1}{2}$. Solving for y gives $y = \frac{1}{2}x$.

Use the same technique to find the equation of the altitude from D to $\overline{PQ}$. The slope of $\overline{PQ}$ is 1. So, the slope of the altitude to $\overline{PQ}$ is -1. The altitude passes through point $D(-4, 4)$, so its equation is $\frac{y-4}{x-(-4)} = -1$, or $y = -x$.

To find the point where the altitudes intersect, use elimination to solve this system.

$$\begin{cases} y = \frac{1}{2}x \\ y = -x \end{cases}$$ Equation of the line containing the altitude from Q to $\overline{PD}$. Equation of the line containing the altitude from D to $\overline{PQ}$.

$0 = \frac{3}{2}x$ Subtract the second equation from the first equation to eliminate y.

$0 = x$ Multiply both sides by $\frac{2}{3}$.

The x-coordinate of the point of the intersection is 0. Substitute 0 for x in either original equation and you will get $y = 0$. So, the orthocenter is $(0, 0)$. You can verify your result by writing the equation of the line containing the third altitude and making sure $(0, 0)$ satisfies it.

You can also find the coordinates of the centroid of a triangle by solving a system of two lines containing medians. However, as you will see in the next example, there is a more efficient method.

EXAMPLE B

Consider $\triangle ABC$ with $A(-5, -3)$, $B(3, -5)$, and $C(-1, 2)$.

a. Find the coordinates of the centroid of $\triangle ABC$ by writing equations for two lines containing medians and finding their point of intersection.

b. Find the mean of the x-coordinates and the mean of the y-coordinates of the triangle's vertices. What do you notice?

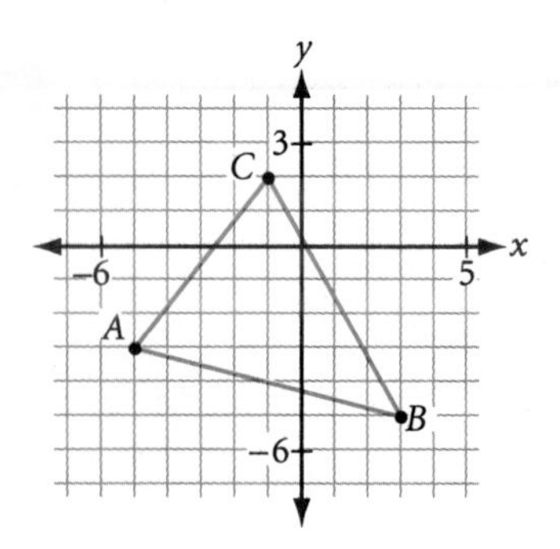

▶ **Solution**

The median of a triangle joins a vertex with the midpoint of the opposite side.

a. First, find the equation of the line containing the median from A to $\overline{BC}$. The midpoint of $\overline{BC}$ is $\left(1, -\frac{3}{2}\right)$. The slope from $A(-5, -3)$ to this midpoint is $\frac{-3-\left(\frac{-3}{2}\right)}{-5-1}$, or $\frac{1}{4}$. The equation of the line is $\frac{y-(-3)}{x-(-5)} = \frac{1}{4}$. Solving for y gives $y = \frac{1}{4}x - \frac{7}{4}$ as the equation for the median.

Next, find the equation of the line containing the median from $B(3, -5)$ to $\overline{AC}$. The midpoint of $\overline{AC}$ is $\left(-3, -\frac{1}{2}\right)$. The slope is $\frac{-3}{4}$. So you get the equation $\frac{y-(-5)}{x-3} = -\frac{3}{4}$. Solving for y gives $y = -\frac{3}{4}x - \frac{11}{4}$ as the equation for the median.

Finally, use elimination to solve this system.

$$\begin{cases} y = \frac{1}{4}x - \frac{7}{4} \\ y = -\frac{3}{4}x - \frac{11}{4} \end{cases}$$ Equation of the line containing the median from A to $\overline{BC}$. Equation of the line containing the median from B to $\overline{AC}$.

$0 = x + 1$ Subtract the second equation from the first.

$-1 = x$ Subtract 1 from both sides.

The x-coordinate of the point of intersection is -1. Use substitution to find the y-coordinate.

$y = \frac{1}{4}(-1) - \frac{7}{4}$ Substitute -1 for x in the first equation.

$y = -2$ Simplify.

The centroid is $(-1, -2)$. You can verify your result by writing the equation for the third median and making sure $(-1, -2)$ satisfies it.

b. The mean of the x-coordinates is $\frac{-5+3+(-1)}{3} = \frac{-3}{3} = -1$.

The mean of the y-coordinates is $\frac{-3+(-5)+2}{3} = \frac{-6}{3} = -2$.

Notice that these means give you the coordinates of the centroid: $(-1, -2)$.

You can generalize the findings from Example B to all triangles. The easiest way to find the coordinates of the centroid is to find the mean of the vertex coordinates.

EXERCISES

In Exercises 1 and 2, use $\triangle RES$ with vertices $R(0, 0)$, $E(4, -6)$, and $S(8, 4)$.

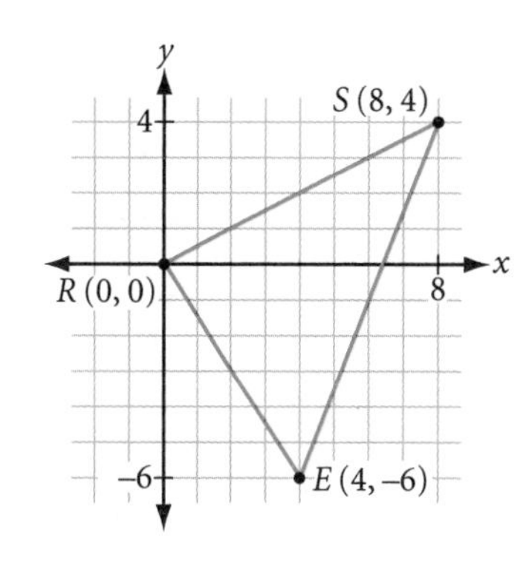

1. Find the equation of the line containing the median from R to $\overline{ES}$.

2. Find the equation of the line containing the altitude from E to $\overline{RS}$.

In Exercises 3 and 4, use algebra to find the coordinates of the centroid and the orthocenter for each triangle.

3. Right triangle *MNO*

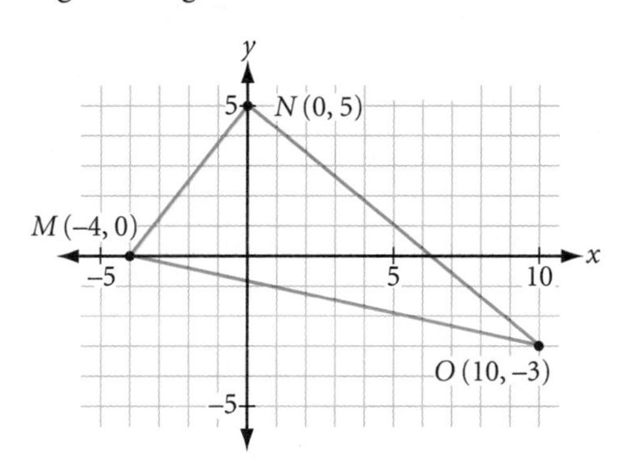

4. Isosceles triangle *CDE*

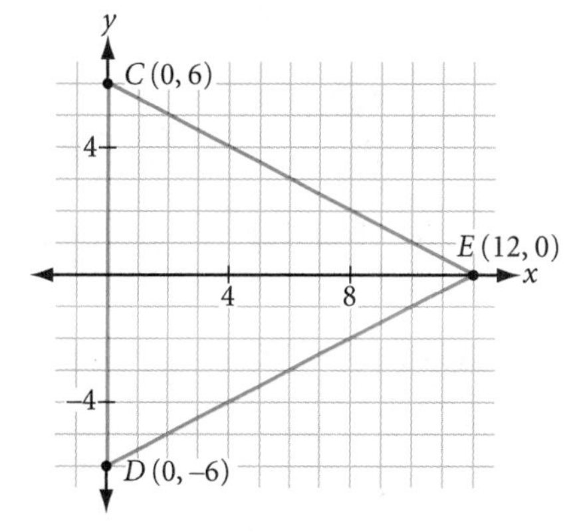

5. Find the coordinates of the centroid of the triangle formed by the x-axis, the y-axis, and the line $12x + 9y = 36$.

6. The three lines $8x + 3y = 12$, $6y - 7x = 24$, and $x + 9y + 33 = 0$ intersect to form a triangle. Find the coordinates of its centroid.

IMPROVING YOUR VISUAL THINKING SKILLS

Painted Faces I

Suppose some unit cubes are assembled into a large cube, then some of the faces of this large cube are painted. After the paint dries, the large cube is disassembled into the unit cubes and you discover that 32 of these have no paint on any of their faces. How many faces of the large cube were painted?

CHAPTER 7 REVIEW

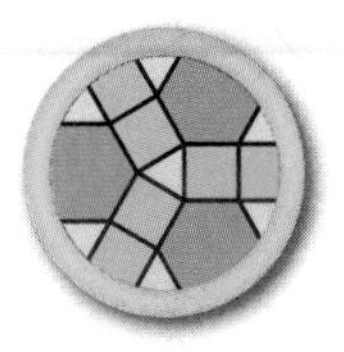

How is your memory? In this chapter you learned about rigid transformations in the plane—called isometries—and you revisited the principles of symmetry that you first learned in Chapter 0. You applied these concepts to create tessellations. Can you name the three rigid transformations? Can you describe how to compose transformations to make other transformations? How can you use reflections to improve your miniature-golf game? What types of symmetry do regular polygons have? What types of polygons will tile the plane? Review this chapter to be sure you can answer these questions.

EXERCISES

For Exercises 1–12, identify each statement as true or false. For each false statement, sketch a counterexample or explain why it is false.

1. The two transformations in which the orientation (the order of points as you move clockwise) does not change are translation and rotation.

2. The two transformations in which the image has the opposite orientation from the original are reflection and glide reflection.

3. A translation of (5, 12) followed by a translation of (−8, −6) is equivalent to a single translation of (−3, 6).

4. A rotation of 140° followed by a rotation of 260° about the same point is equivalent to a single rotation of 40° about that point.

5. A reflection across a line followed by a second reflection across a parallel line that is 12 cm from the first is equivalent to a translation of 24 cm.

6. A regular n-gon has n reflectional symmetries and n rotational symmetries.

7. The only three regular polygons that create monohedral tessellations are equilateral triangles, squares, and regular pentagons.

8. Any triangle can create a monohedral tessellation.

9. Any quadrilateral can create a monohedral tessellation.

10. No pentagon can create a monohedral tessellation.

11. No hexagon can create a monohedral tessellation.

12. There are at least three times as many true statements as false statements in Exercises 1–12.

King by Minnie Evans (1892–1987)

In Exercises 13–15, identify the type or types of symmetry, including the number of symmetries, in each design. For Exercise 15, describe how you can move candles on the menorah to make the colors symmetrical, too.

13.

Mandala,
Gary Chen, geometry student

14.

15.

16. The façade of Chartres Cathedral in France does not have reflectional symmetry. Why not? Sketch the portion of the façade that does have bilateral symmetry.

17. Find or create a logo that has reflectional symmetry. Sketch the logo and its line or lines of reflectional symmetry.

18. Find or create a logo that has rotational symmetry but not reflectional symmetry. Sketch it.

In Exercises 19 and 20, classify the tessellation and give the vertex arrangement.

19.

20.

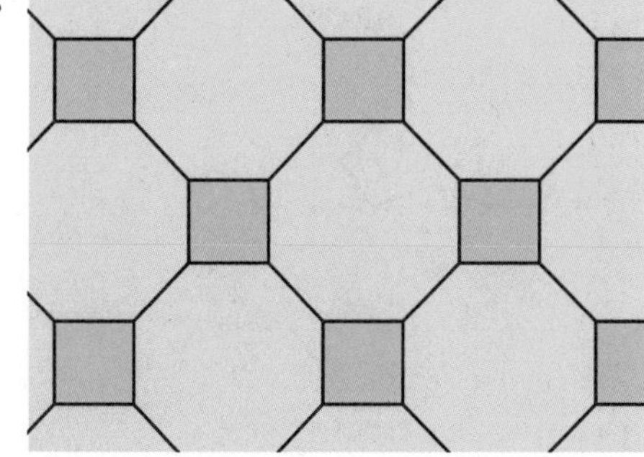

21. Experiment with a mirror to find the smallest vertical portion (y) in which you can still see your full height (x). How does y compare to x? Can you explain, with the help of a diagram and what you know about reflections, why a "full-length" mirror need not be as tall as you?

22. Miniature-golf pro Sandy Trapp wishes to impress her fans with a hole in one on the very first try. How should she hit the ball at T to achieve this feat? Explain.

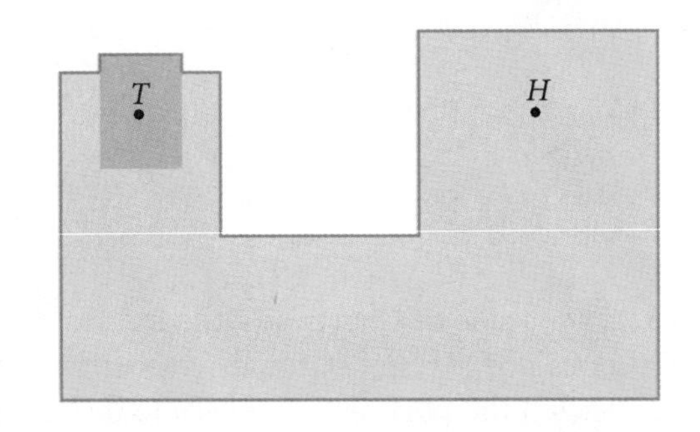

In Exercises 23–25, identify the shape of the tessellation grid and a possible method that the student used to create each tessellation.

23.

Perian Warriors, Robert Bell

24.

Doves, Serene Tam

25.

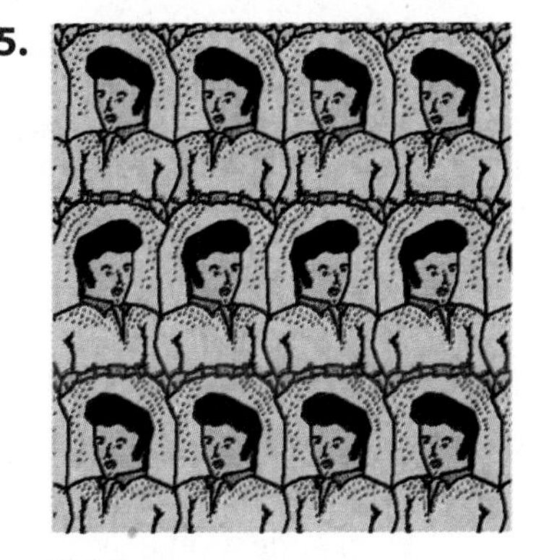

Sightings,
Peter Chua and Monica Grant

In Exercises 26 and 27, copy the figure and grid onto patty paper. Determine whether or not you can use the figure to create a tessellation on the grid. Explain your reasoning.

26.

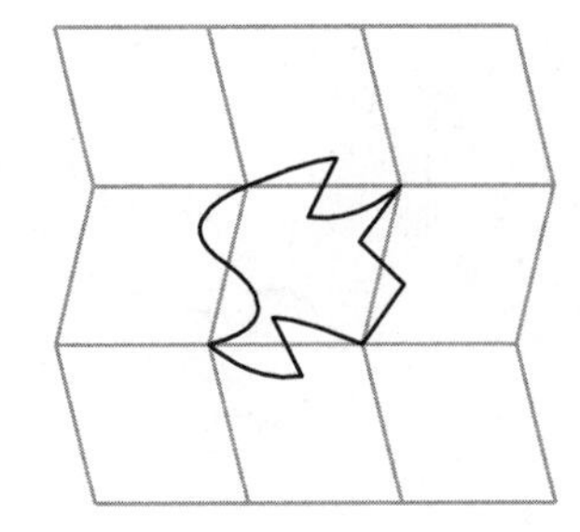

27.

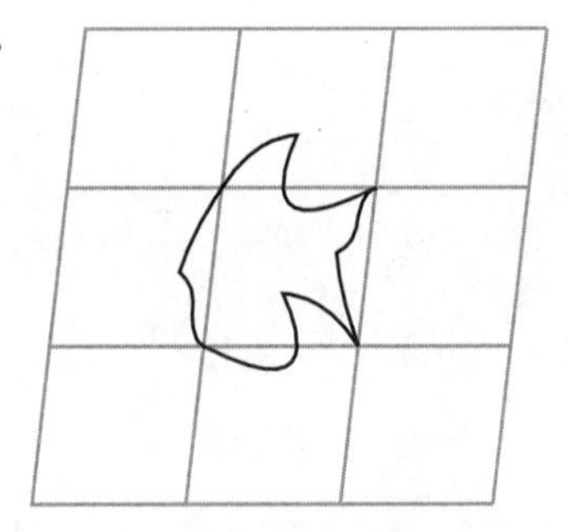

28. In his woodcut *Day and Night,* Escher gradually changes the shape of the patches of farmland into black and white birds. The birds are flying in opposite directions, so they appear to be glide reflections of each other. But notice that the tails of the white birds curve down, while the tails of the black birds curve up. So, on closer inspection, it's clear that this is not a glide-reflection tiling at all!

When two birds are taken together as one tile (a 2-motif tile), they create a translation tessellation. Use patty paper to find the 2-motif tile.

Day and Night, M. C. Escher, 1938

Career CONNECTION

Commercial tile contractors use tessellating polygons to create attractive designs for their customers. Some designs are 1-uniform or 2-uniform, and others are even more complex.

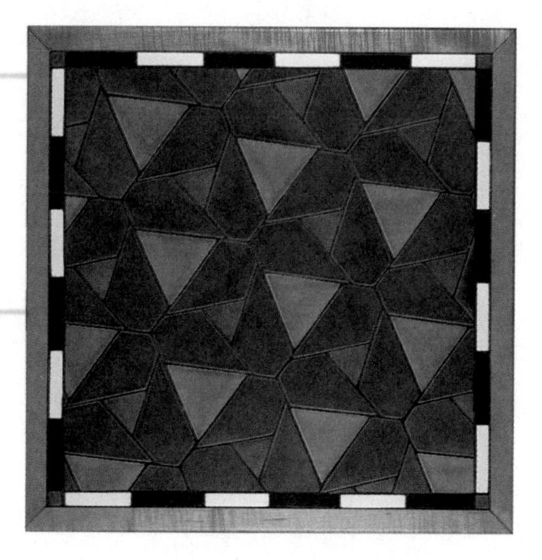

Assessing What You've Learned

Try one or more of these assessment suggestions.

UPDATE YOUR PORTFOLIO Choose one of the tessellations you did in this chapter and add it to your portfolio. Describe why you chose it and explain the transformations you used and the types of symmetry it has.

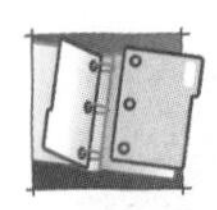

ORGANIZE YOUR NOTEBOOK Review your notebook to be sure it's complete and well organized. Are all the types of transformation and symmetry included in your definition list or conjecture list? Write a one-page chapter summary.

WRITE IN YOUR JOURNAL This chapter emphasizes applying geometry to create art. Write about connections you see between geometry and art. Does creating geometric art give you a greater appreciation for either art or geometry? Explain.

PERFORMANCE ASSESSMENT While a classmate, a friend, a family member, or a teacher observes, carry out one of the investigations from this chapter. Explain what you're doing at each step, including how you arrive at the conjecture.

GIVE A PRESENTATION Give a presentation about one of the investigations or projects you did or about one of the tessellations you created.

CHAPTER

8 Area

I could fill an entire second life with working on my prints.

M. C. ESCHER

Square Limit, M. C. Escher, 1964

OBJECTIVES

In this chapter you will

- discover area formulas for rectangles, parallelograms, triangles, trapezoids, kites, regular polygons, circles, and other shapes
- use area formulas to solve problems
- learn how to find the surface areas of prisms, pyramids, cylinders, and cones

LESSON

8.1

Areas of Rectangles and Parallelograms

A little learning is a dangerous thing—almost as dangerous as a lot of ignorance.

ANONYMOUS

People work with areas in many occupations. Carpenters calculate the areas of walls, floors, and roofs before they purchase materials for construction. Painters calculate surface areas so that they know how much paint to buy for a job. Decorators calculate the areas of floors and windows to know how much carpeting and drapery they will need. In this chapter you will discover formulas for finding the areas of the regions within triangles, parallelograms, trapezoids, kites, regular polygons, and circles.

Tile layers need to find floor area to determine how many tiles to buy.

The **area** of a plane figure is the measure of the region enclosed by the figure. You measure the area of a figure by counting the number of square units that you can arrange to fill the figure completely.

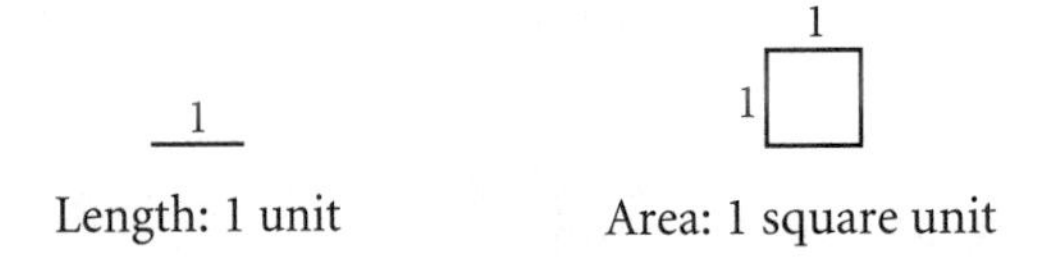

You probably already know many area formulas. Think of the investigations in this chapter as physical demonstrations of the formulas that will help you understand and remember them.

It's easy to find the area of a rectangle.

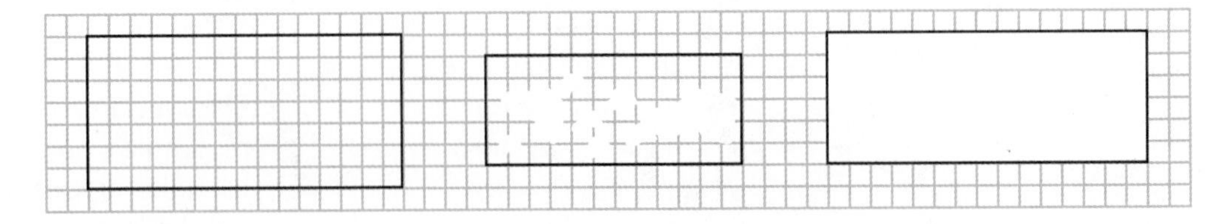

To find the area of the first rectangle, you can simply count squares. To find the areas of the other rectangles, you could draw in the lines and count the squares, but there's an easier method.

Any side of a rectangle can be called a **base.** A rectangle's **height** is the length of the side that is perpendicular to the base. For each pair of parallel bases, there is a corresponding height.

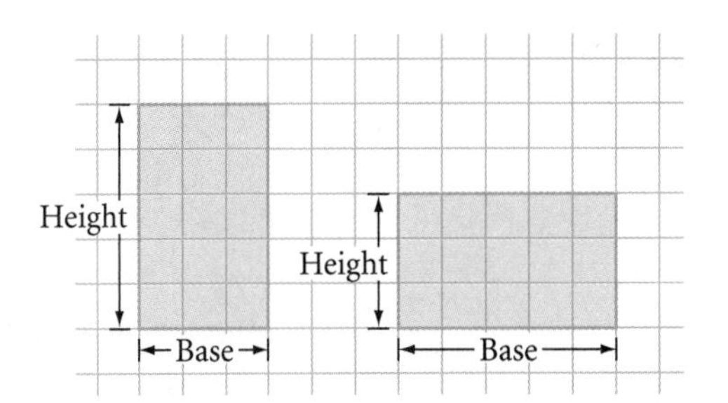

If we call the bottom side of each rectangle in the figure the base, then the length of the base is the number of squares in each row and the height is the number of rows. So you can use these terms to state a formula for the area. Add this conjecture to your list.

Rectangle Area Conjecture

C-75

The area of a rectangle is given by the formula __?__, where A is the area, b is the length of the base, and h is the height of the rectangle.

The area formula for rectangles can help you find the areas of many other shapes.

EXAMPLE A

Find the area of this shape.

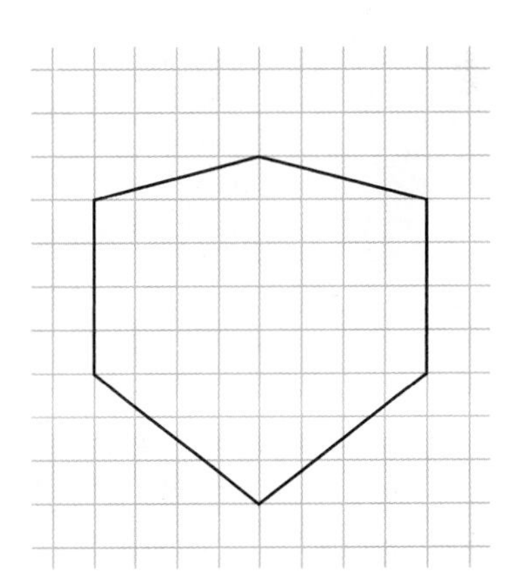

▶ Solution

The middle section is a rectangle with an area of $4 \cdot 8$, or 32 square units.

You can divide the remaining pieces into right triangles, so each piece is actually half a rectangle.

The area of the figure is $32 + 2(2) + 2(6) = 48$ square units.

There are other ways to find the area of this figure. One way is to find the area of an 8-by-8 square and subtract the areas of four right triangles.

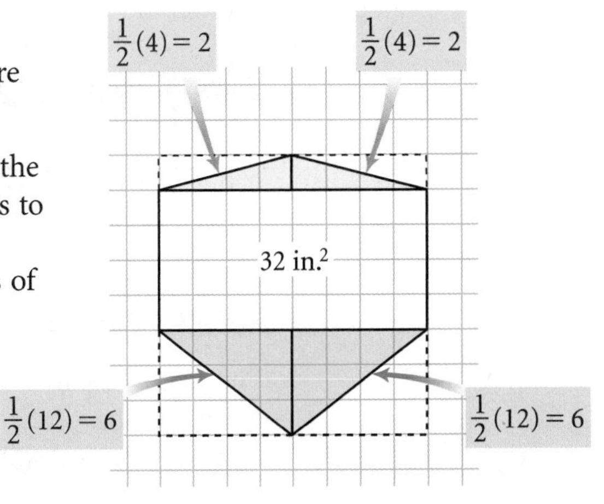

You can also use the area formula for a rectangle to find the area formula for a parallelogram.

Just as with a rectangle, any side of a parallelogram can be called a base. But the height of a parallelogram is not necessarily the length of a side. An **altitude** is any segment from one side of a parallelogram perpendicular to a line through the opposite side. The length of the altitude is the **height.**

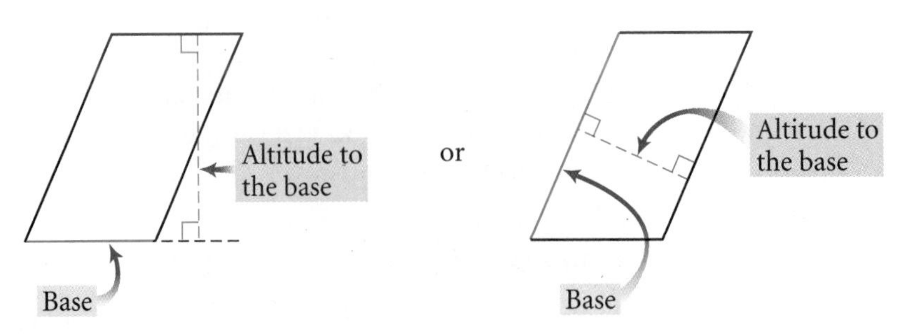

The altitude can be inside or outside the parallelogram. No matter where you draw the altitude to a base, its height should be the same, because the opposite sides are parallel.

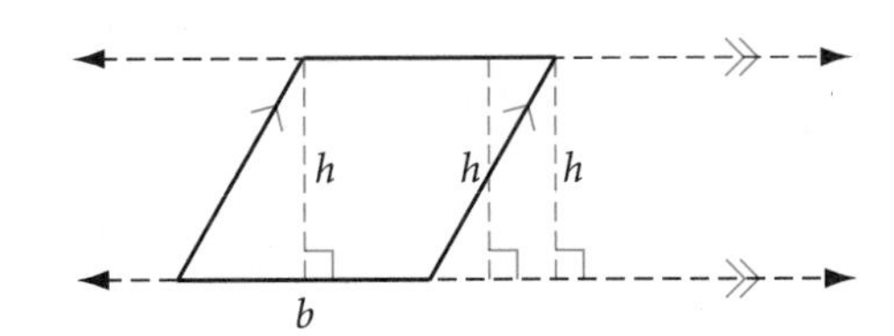

Investigation

Area Formula for Parallelograms

You will need

- heavy paper or cardboard
- a straightedge
- a compass

Using heavy paper, investigate the area of a parallelogram. Can the area be rearranged into a more familiar shape? Different members of your group should investigate different parallelograms so you can be sure your formula works for all parallelograms.

Step 1 Construct a parallelogram on a piece of heavy paper or cardboard. From the vertex of the obtuse angle adjacent to the base, draw an altitude to the side opposite the base. Label the parallelogram as shown.

Step 2 Cut out the parallelogram and then cut along the altitude. You will have two pieces—a triangle and a trapezoid. Try arranging the two pieces into other shapes without overlapping them. Is the area of each of these new shapes the same as the area of the original parallelogram? Why?

Step 3 Is one of your new shapes a rectangle? Calculate the area of this rectangle. What is the area of the original parallelogram? State your next conjecture.

Parallelogram Area Conjecture C-76

The area of a parallelogram is given by the formula __?__, where A is the area, b is the length of the base, and h is the height of the parallelogram.

If the dimensions of a figure are measured in inches, feet, or yards, the area is measured in in.^{2} (square inches), ft^2 (square feet), or yd^2 (square yards). If the dimensions are measured in centimeters or meters, the area is measured in cm^2 (square centimeters) or m^2 (square meters). Let's look at an example.

EXAMPLE B

Find the height of a parallelogram that has area 7.13 m^2 and base length 2.3 m.

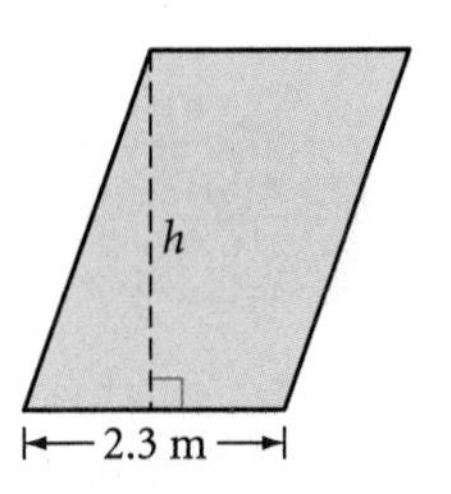

▶ Solution

$A = bh$ Write the formula.

$7.13 = (2.3)h$ Substitute the known values.

$\frac{7.13}{2.3} = h$ Solve for the height.

$h = 3.1$ Divide.

The height measures 3.1 m.

EXERCISES

In Exercises 1–6, each quadrilateral is a rectangle. A represents area and P represents perimeter. Use the appropriate unit in each answer.

1. $A = \underline{?}$

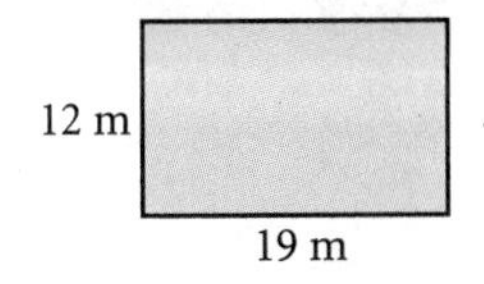

2. $A = \underline{?}$

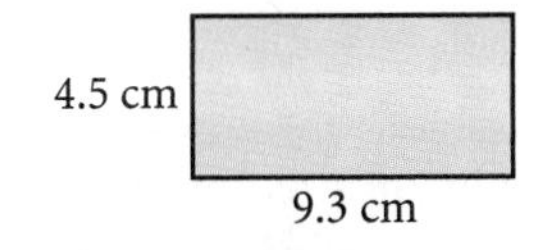

3. $A = 96$ yd^2
$b = \underline{?}$

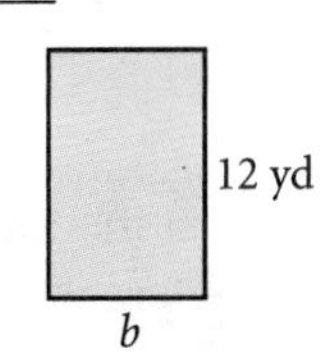

4. $A = 273$ cm^2
$h = \underline{?}$

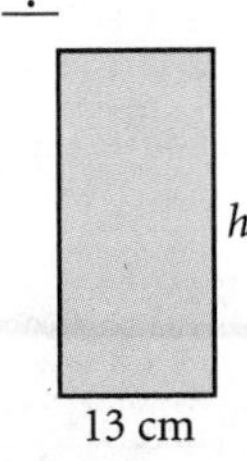

5. $P = 40$ ft
$A = \underline{?}$

7 ft

6. Shaded area $= \underline{?}$

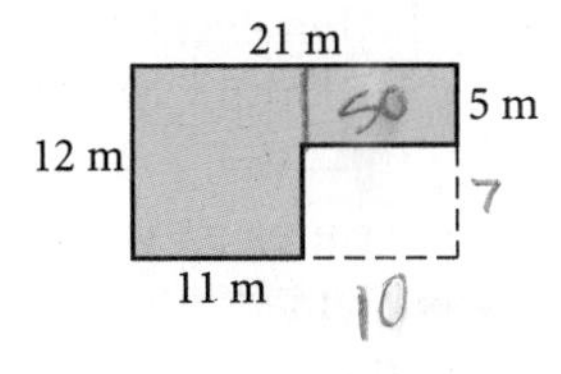

In Exercises 7–9, each quadrilateral is a parallelogram.

7. $A = \underline{?}$

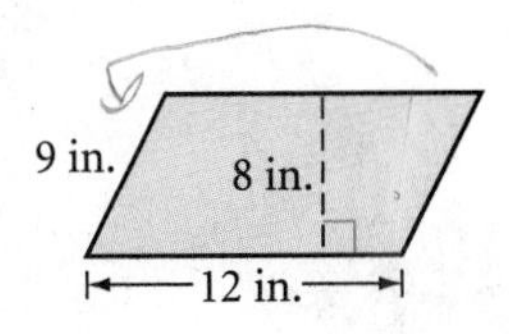

8. $A = 2508$ cm^2
$P = \underline{?}$

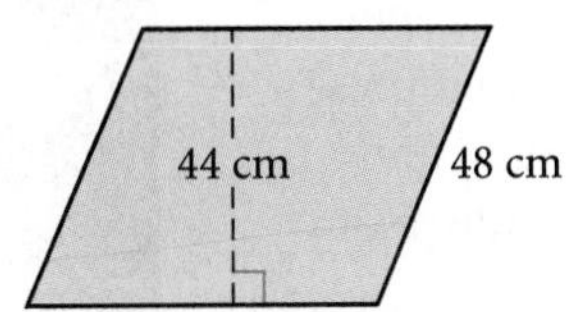

9. Find the area of the shaded region.

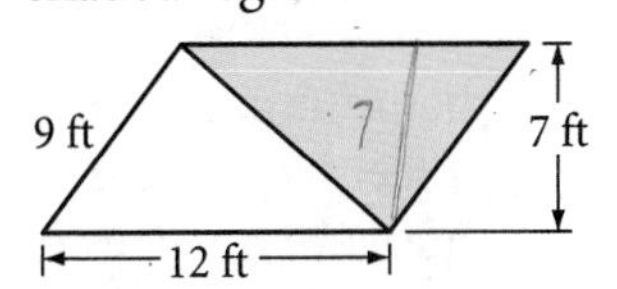

10. Sketch and label two different rectangles, each with area 48 cm^2. ⓗ

In Exercises 11 and 12, find the area of the figure and explain your method.

11.

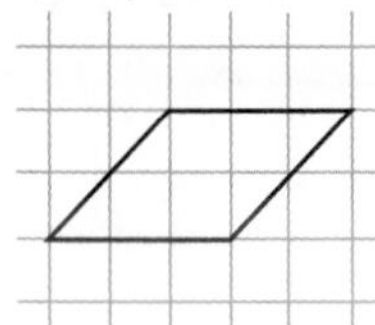

12.

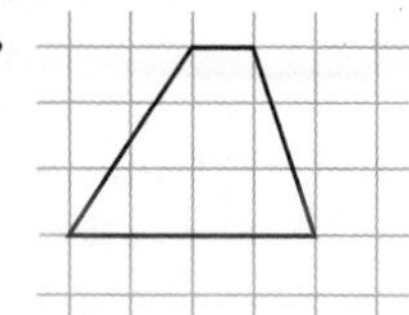

13. Sketch and label two different parallelograms, each with area 64 cm^2.

14. Draw and label a figure with area 64 cm^2 and perimeter 64 cm.

15. The photo shows a Japanese police *koban*. An arch forms part of the roof and one wall. The arch is made from rectangular panels that each measure 1 m by 0.7 m. The arch is 11 panels high and 3 panels wide. What's the total area of the arch?

Cultural CONNECTION

Koban is Japanese for "mini-station," a small police station. These stations are located in several parts of a city, and officers who work in them know the surrounding neighborhoods and people well. The presence of *kobans* in Japan helps reduce crime and provides communities with a sense of security.

16. What is the total area of the four walls of a rectangular room 4 meters long by 5.5 meters wide by 3 meters high? Ignore all doors and windows.

17. **APPLICATION** Ernesto plans to build a pen for his pet iguana. What is the area of the largest rectangular pen that he can make with 100 meters of fencing?

18. The big event at George Washington High School's May Festival each year is the Cow Drop Contest. A farmer brings his well-fed bovine to wander the football field until—well, you get the picture. Before the contest, the football field, which measures 53 yards wide by 100 yards long, is divided into square yards. School clubs and classes may purchase square yards. If one of their squares is where the first dropping lands, they win a pizza party. If the math club purchases 10 squares, what is the probability that the club wins?

19. **APPLICATION** Sarah is tiling a wall in her bathroom. It is rectangular and measures 4 feet by 7 feet. The tiles are square and measure 6 inches on each side. How many tiles does Sarah need? ⓗ

The figure at right demonstrates that $(a + b)^2 = a^2 + 2ab + b^2$. In Exercises 20 and 21, sketch and label a rectangle that demonstrates each algebraic expression.

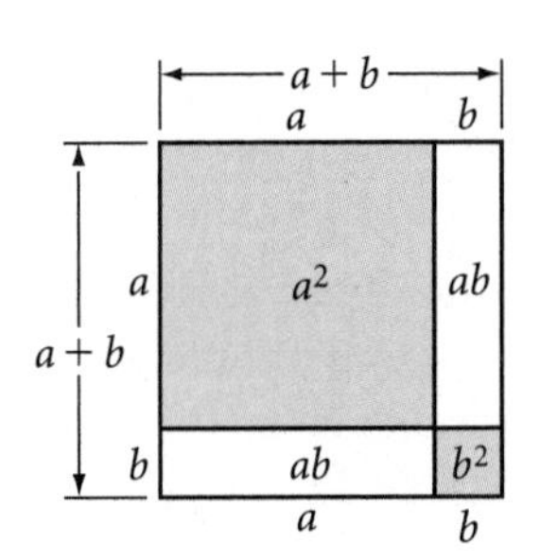

20. $(x + 3)(x + 5) = x^2 + 8x + 15$

21. $(3x + 2)(2x + 5) = 6x^2 + 19x + 10$

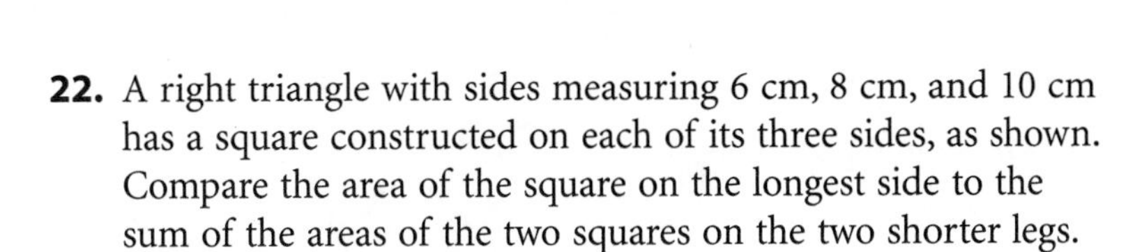

22. A right triangle with sides measuring 6 cm, 8 cm, and 10 cm has a square constructed on each of its three sides, as shown. Compare the area of the square on the longest side to the sum of the areas of the two squares on the two shorter legs.

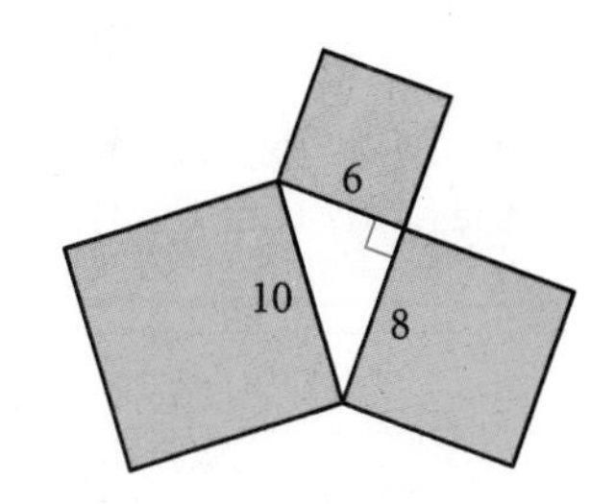

23. What is the area of the parallelogram?

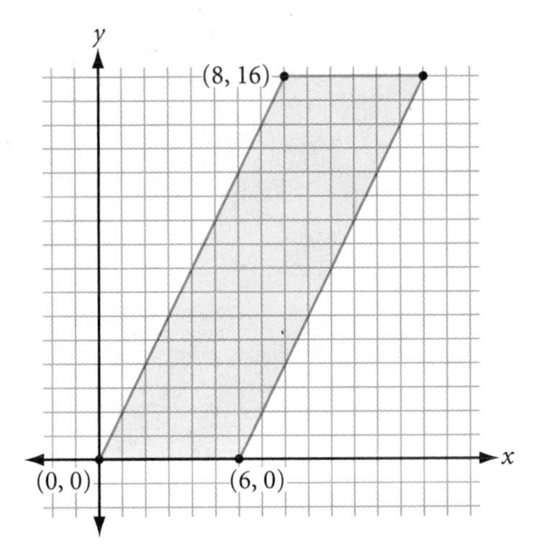

24. What is the area of the trapezoid?

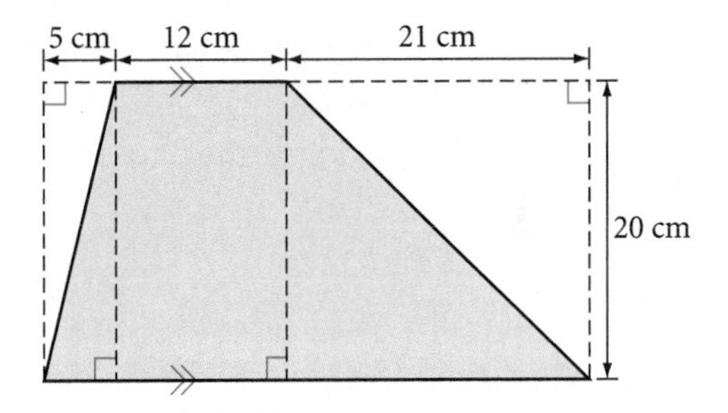

Art CONNECTION

The design at right is a quilt design called the Ohio Star. Traditional quilt block designs range from a simple nine-patch, based on 9 squares, to more complicated designs like Jacob's Ladder or Underground Railroad, which are based on 16, 25, or even 36 squares. Quiltmakers need to calculate the total area of each different type of material before they make a complete quilt.

25. **APPLICATION** The Ohio Star is a 16-square quilt design. Each block measures 12 inches by 12 inches. One block is shown above. Assume you will need an additional 20% of each fabric to allow for seams and errors.

a. Calculate the sum of the areas of all the red patches, the sum of the areas of all the blue patches, and the area of the yellow patch in a single block.

b. How many Ohio Star blocks will you need to cover an area that measures 72 inches by 84 inches, the top surface area of a king-size mattress?

c. How much fabric of each color will you need? How much fabric will you need for a 15-inch border to extend beyond the edges of the top surface of the mattress?

Review

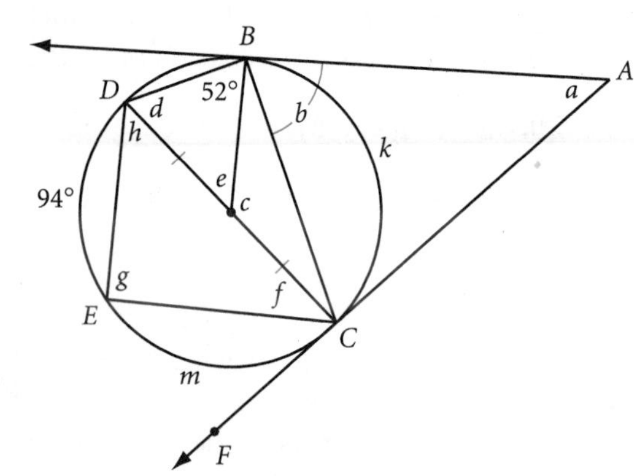

26. Copy the figure at right. Find the lettered angle measures and arc measures. $\overrightarrow{AB}$ and $\overrightarrow{AC}$ are tangents. $\overline{CD}$ is a diameter.

27. Given $\overline{AM}$ as the length of the altitude of an equilateral triangle, construct the triangle.

A •————————————• M

28. Sketch what the figure at right looks like when viewed from

a. Above the figure, looking straight down

b. In front of the figure, that is, looking straight at the red-shaded side

c. The side, looking at the blue-shaded side

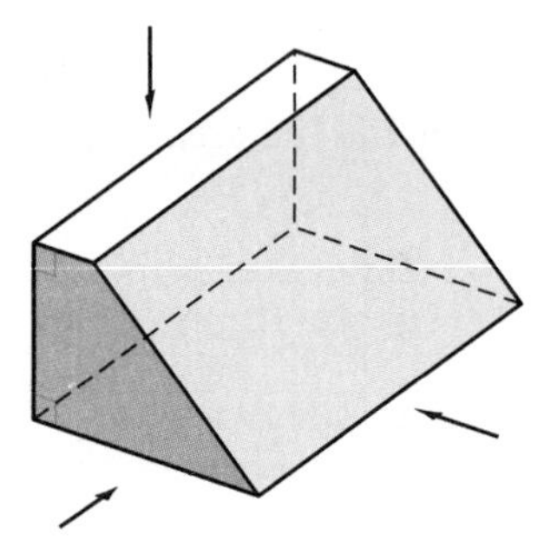

project

RANDOM RECTANGLES

What does a typical rectangle look like? A randomly generated rectangle could be long and narrow, or square-like. It could have a large perimeter, but a small area. Or it could have a large area, but a small perimeter. In this project you will randomly generate rectangles and study their characteristics using scatter plots and histograms.

Your project should include

- A description of how you created your random rectangles, including any constraints you used.
- A scatter plot of base versus height, a perimeter histogram, an area histogram, and a scatter plot of perimeter versus area.
- Any other studies or graphs you think might be interesting.
- Your predictions about the data before you made each graph.
- An explanation of why each graph looks the way it does.

Fathom™

You can use Fathom to generate random base and height values from 0 to 10. Then you can sort them by various characteristics and make a wide range of interesting graphs.

LESSON 8.2

Areas of Triangles, Trapezoids, and Kites

When you add to the truth, you subtract from it.

THE TALMUD

In Lesson 8.1, you learned the area formula for rectangles, and you used it to discover an area formula for parallelograms. In this lesson you will use those formulas to discover or demonstrate the formulas for the areas of triangles, trapezoids, and kites.

Investigation 1
Area Formula for Triangles

You will need

- heavy paper or cardboard

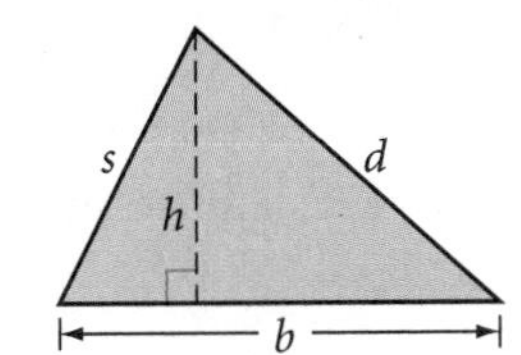

Step 1 Cut out a pair of congruent triangles. Label their corresponding parts as shown.

Step 2 Arrange the triangles to form a figure for which you already have an area formula. Calculate the area of the figure.

Step 3 What is the area of one of the triangles? Make a conjecture. Write a brief description in your notebook of how you arrived at the formula. Include an illustration.

Triangle Area Conjecture C-77

The area of a triangle is given by the formula __?__, where A is the area, b is the length of the base, and h is the height of the triangle.

Investigation 2
Area Formula for Trapezoids

You will need

- heavy paper or cardboard

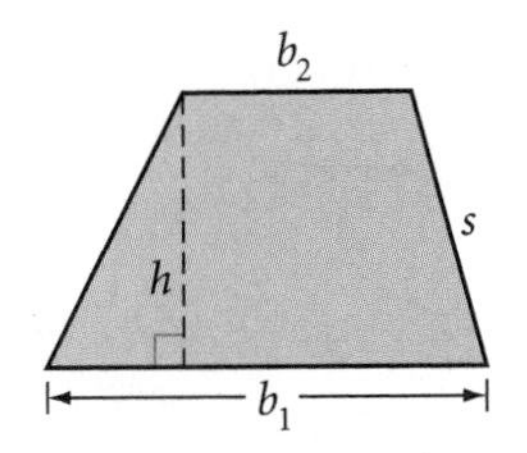

Step 1 Construct any trapezoid and an altitude perpendicular to its bases. Label the trapezoid as shown.

Step 2 Cut out the trapezoid. Make and label a copy.

Step 3 | Arrange the two trapezoids to form a figure for which you already have an area formula. What type of polygon is this? What is its area? What is the area of one trapezoid? State a conjecture.

Trapezoid Area Conjecture C-78

The area of a trapezoid is given by the formula __?__, where A is the area, b_1 and b_2 are the lengths of the two bases, and h is the height of the trapezoid.

Investigation 3
Area Formula for Kites

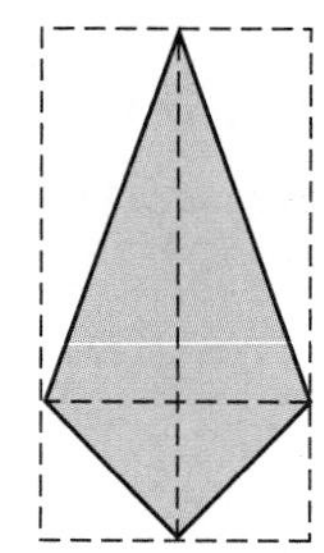

Can you rearrange a kite into shapes for which you already have the area formula? Do you recall some of the properties of a kite?

Create and carry out your own investigation to discover a formula for the area of a kite. Discuss your results with your group. State a conjecture.

Kite Area Conjecture C-79

The area of a kite is given by the formula __?__.

EXERCISES

In Exercises 1–12, use your new area conjectures to solve for the unknown measures.

1. $A =$ __?__

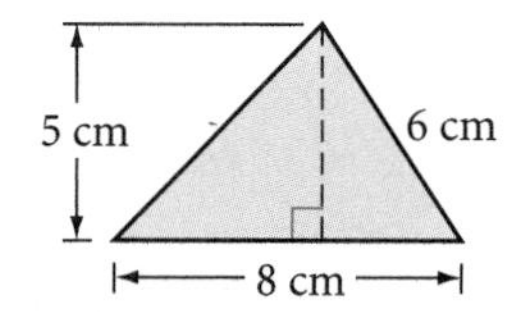

2. $A =$ __?__

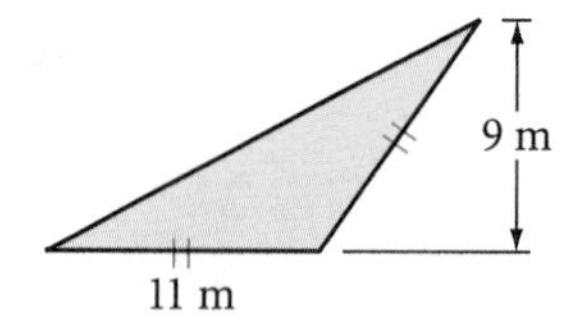

3. $A =$ __?__

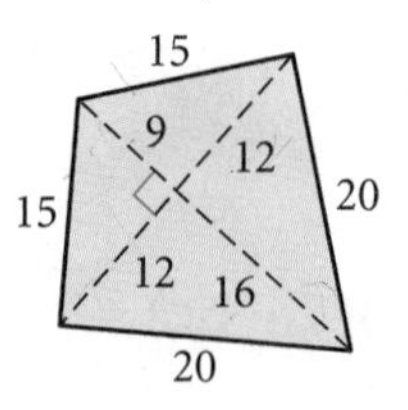

4. $A =$ __?__

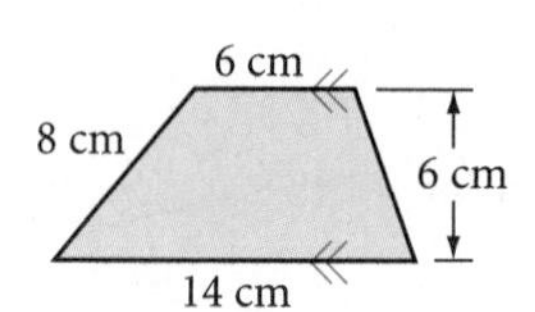

5. $A = 39\text{ cm}^2$
$h =$ __?__

13 cm
h

6. $A = 31.5\text{ ft}^2$
$b =$ __?__

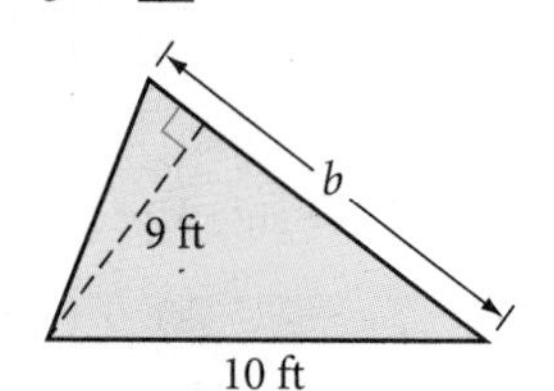

7. $A = 420\text{ ft}^2$
$LE = \underline{?}$

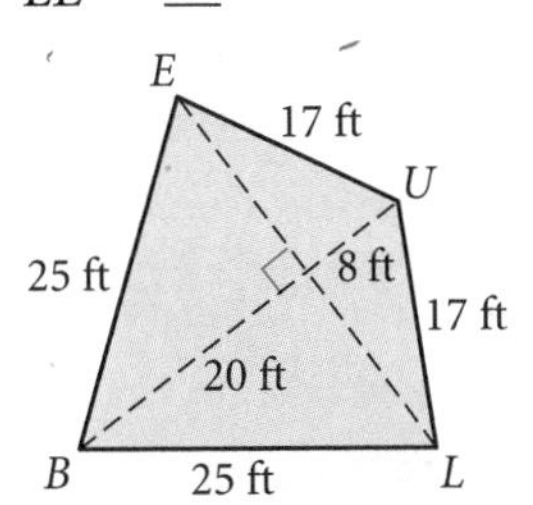

8. $A = 50\text{ cm}^2$ ⓗ
$h = \underline{?}$

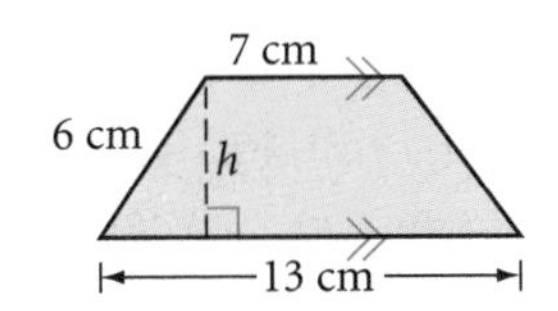

9. $A = 180\text{ m}^2$
$b = \underline{?}$

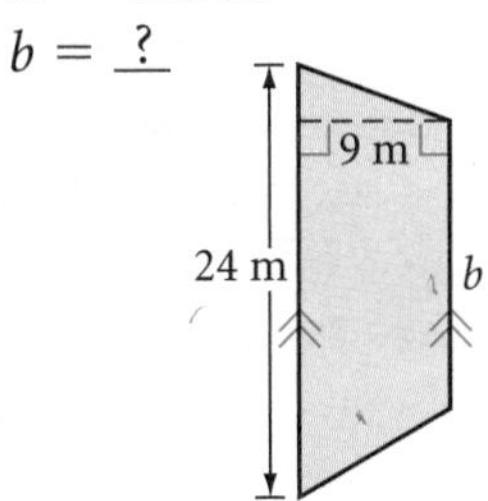

10. $A = 924\text{ cm}^2$
$P = \underline{?}$

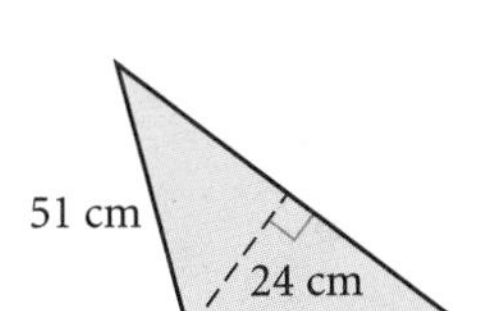

11. $A = 204\text{ cm}^2$
$P = 62\text{ cm}$
$h = \underline{?}$

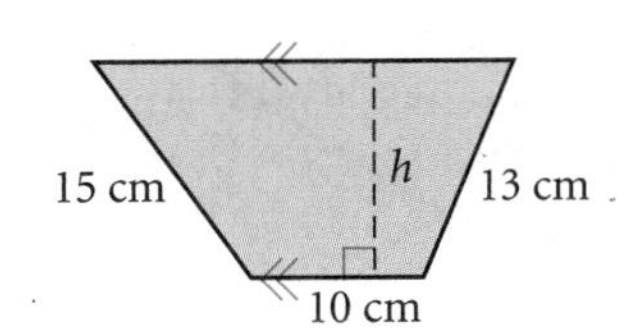

12. $x = \underline{?}$ ⓗ
$y = \underline{?}$

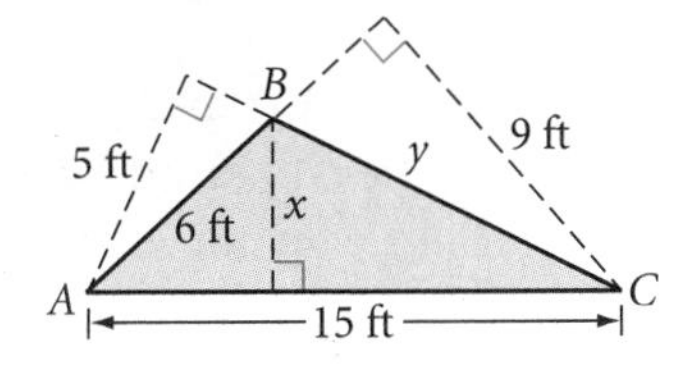

13. Sketch and label two different triangles, each with area 54 cm^2.

14. Sketch and label two different trapezoids, each with area 56 cm^2.

15. Sketch and label two different kites, each with area 1092 cm^2.

16. Sketch and label a triangle and a trapezoid with equal areas and equal heights. How does the base of the triangle compare with the two bases of the trapezoid?

17. P is a random point on side $\overline{AY}$ of rectangle $ARTY$. The shaded area is what fraction of the area of the rectangle? Why?

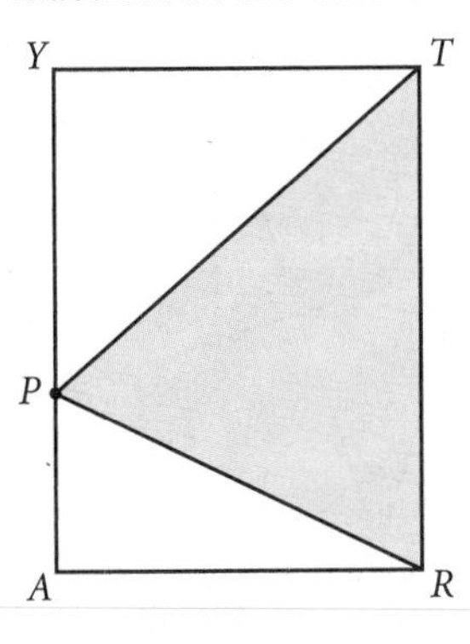

18. One playing card is placed over another, as shown. Is the top card covering half, less than half, or more than half of the bottom card? Explain.

19. **APPLICATION** Eduardo has designed this kite for a contest. He plans to cut the kite from a sheet of Mylar plastic and use balsa wood for the diagonals. He will connect all the vertices with string, and fold and glue flaps over the string.

 a. How much balsa wood and Mylar will he need?

 b. Mylar is sold in rolls 36 inches wide. What length of Mylar does Eduardo need for this kite?

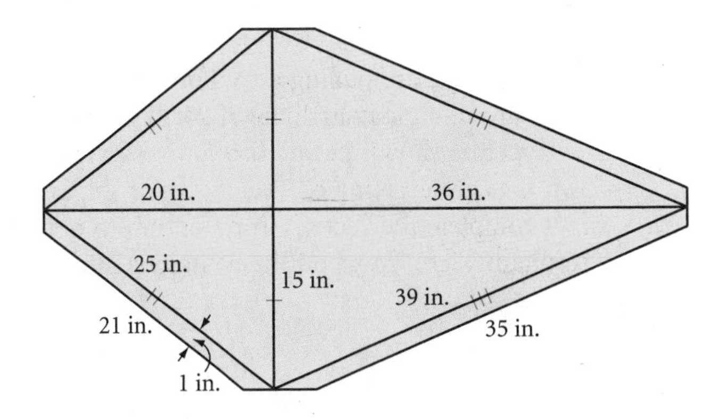

20. APPLICATION The roof on Crystal's house is formed by two congruent trapezoids and two congruent isosceles triangles, as shown. She wants to put new wood shingles on her roof. Each shingle will cover 0.25 square foot of area. (The shingles are 1 foot by 1 foot, but they overlap by 0.75 square foot.) How many shingles should Crystal buy?

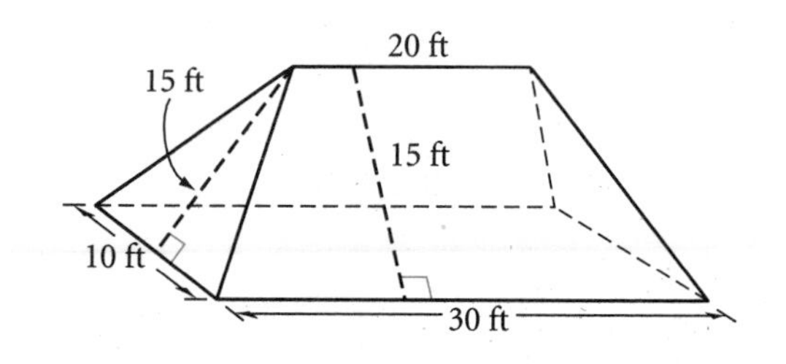

21. A trapezoid has been created by combining two congruent right triangles and an isosceles triangle, as shown. Is the isosceles triangle a right triangle? How do you know? Find the area of the trapezoid two ways: first by using the trapezoid area formula, and then by finding the sum of the areas of the three triangles.

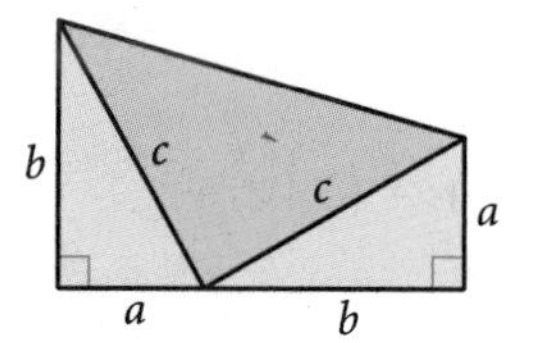

22. Divide a trapezoid into two triangles. Use algebra to derive the formula for the area of the trapezoid by expressing the area of each triangle algebraically and finding their algebraic sum. ⓗ

Review

23. $A = \underline{?}$

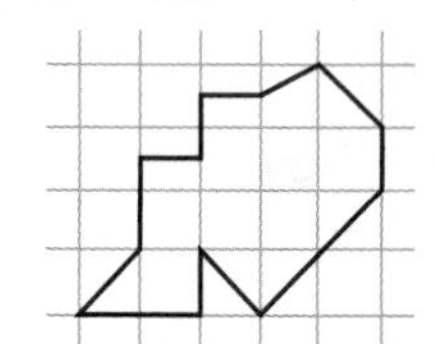

24. $A = \underline{?}$

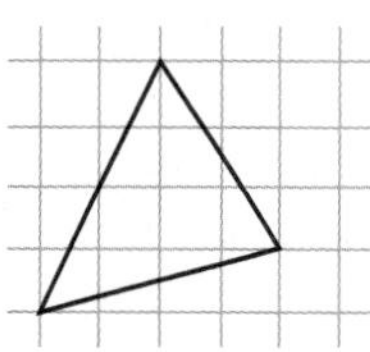

25. $A = 264\ \text{m}^2$
$P = \underline{?}$

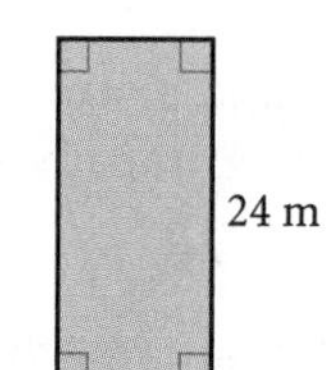

26. $P = 52$ cm
$A = \underline{?}$

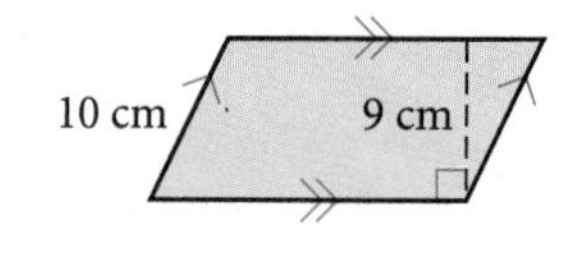

27. Trace the figure at right. Find the lettered angle measures and arc measures. $\overrightarrow{AB}$ and $\overrightarrow{AC}$ are tangents. $\overline{CD}$ is a diameter.

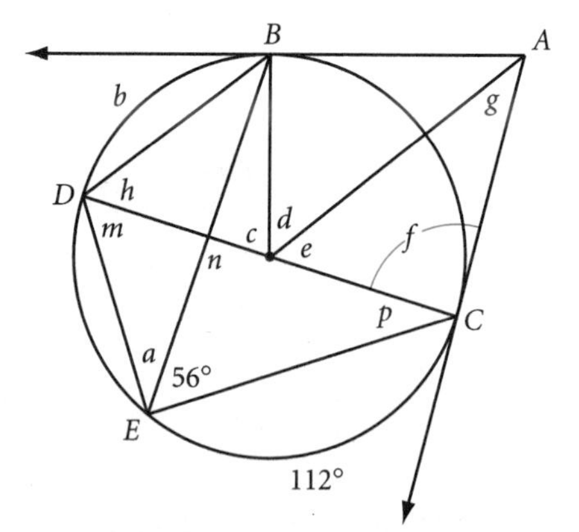

28. Two tugboats are pulling a container ship into the harbor. They are pulling at an angle of 24° between the tow lines. The vectors shown in the diagram represent the forces the two tugs are exerting on the container ship. Copy the vectors and complete the vector parallelogram to determine the resultant vector force on the container ship.

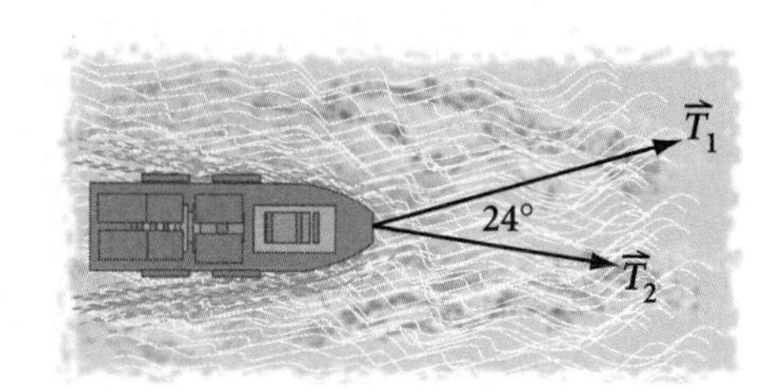

29. Two paths from C to T (traveling on the surface) are shown on the 8 cm-by-8 cm-by-4 cm prism below. M is the midpoint of edge $\overline{UA}$. Which is the shorter path from C to T: C-M-T or C-A-T? Explain. ⓗ

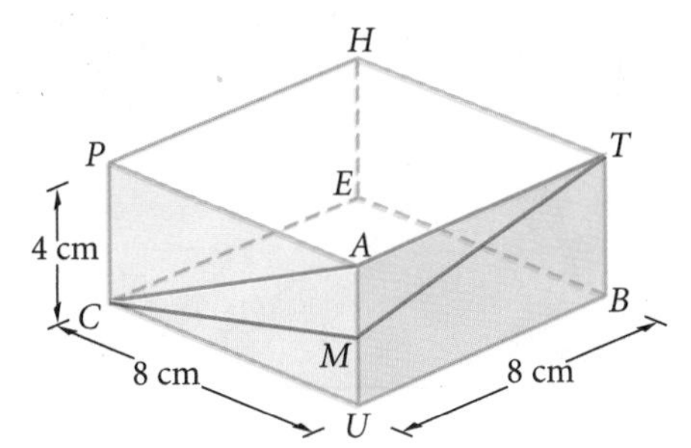

30. Give the vertex arrangement for this 2-uniform tessellation.

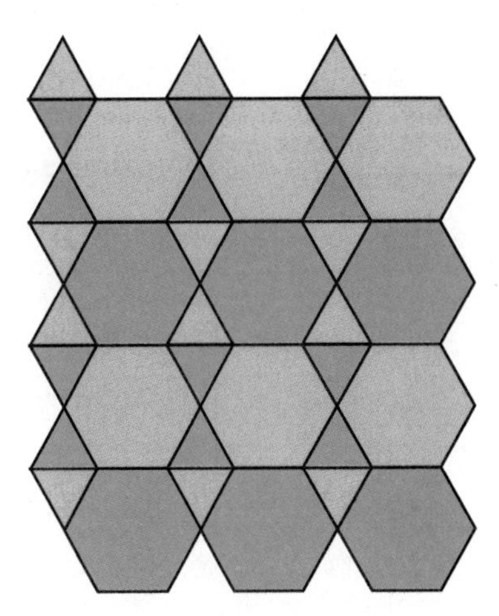

project

MAXIMIZING AREA

A farmer wants to fence in a rectangular pen using the wall of a barn for one side of the pen and the 10 meters of fencing for the remaining three sides. What dimensions will give her the maximum area for the pen?

You can use the trace feature on your calculator to find the value of x that gives the maximum area. Use your graphing calculator to investigate this problem and find the best arrangement.

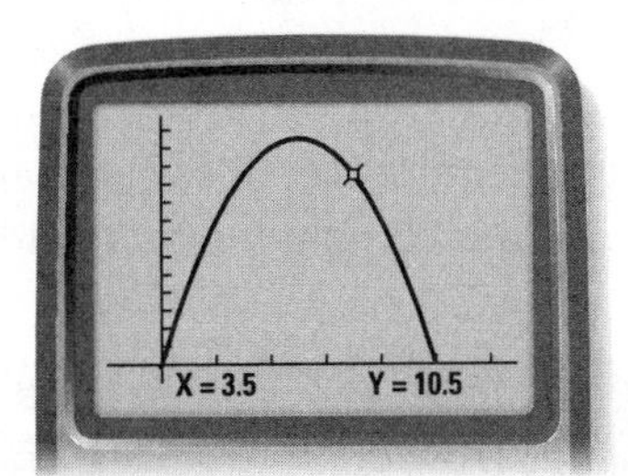

Your project should include

- An expression for the third side length, in terms of the variable x in the diagram.
- An equation and graph for the area of the pen.
- The dimensions of the best rectangular shape for the farmer's pen.

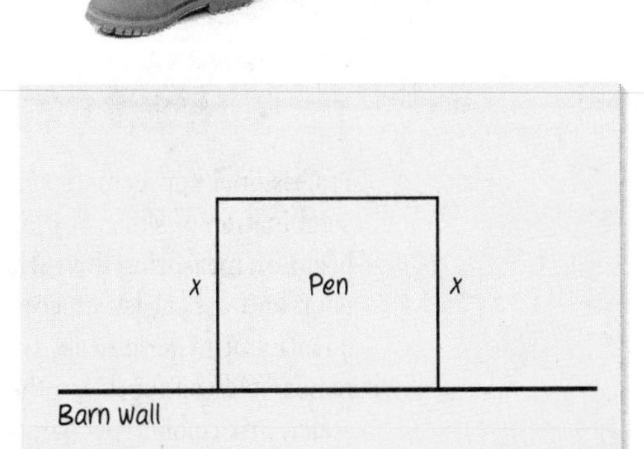

LESSON

8.3

Area Problems

Optimists look for solutions, pessimists look for excuses.

SUE SWENSON

By now, you know formulas for finding the areas of rectangles, parallelograms, triangles, trapezoids, and kites. Now let's see if you can use these area formulas to approximate the areas of irregularly shaped figures.

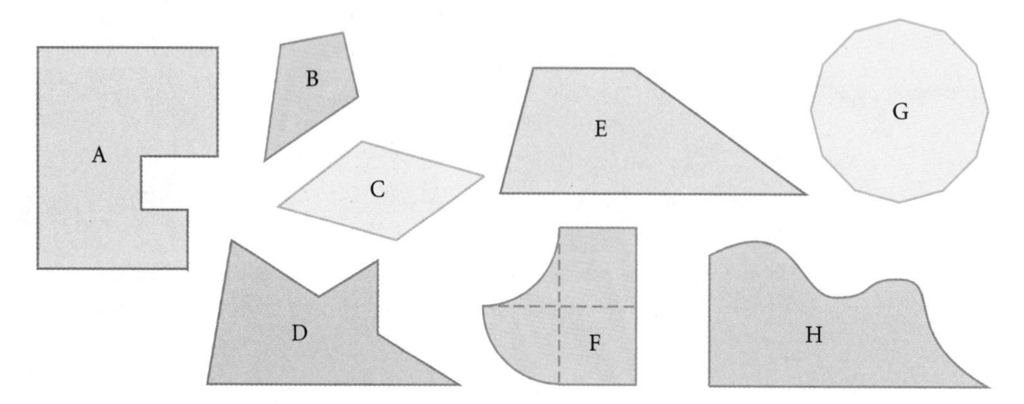

Investigation
Solving Problems with Area Formulas

You will need

- figures A–H
- centimeter rulers or meterstick

Find the area of each geometric figure your teacher provides. Before you begin to measure, discuss with your group the best strategy for each step. Discuss what units you should use. Different group members might get different results. However, your results should be close. You may average your results to arrive at one group answer. For each figure, write a sentence or two explaining how you measured the area and how accurate you think it is.

Now that you have practiced measuring and calculating area, you're ready to try some application problems. Many everyday projects require you to find the areas of flat surfaces on three-dimensional objects. You'll learn more about surface area in Lesson 8.7.

Career CONNECTION

Professional housepainters have a unique combination of skills: For large-scale jobs, they begin by measuring the surfaces that they will paint and use measurements to estimate the quantity of materials they will need. They remove old coating, clean the surface, apply sealer, mix color, apply paint, and add finishes. Painters become experienced and specialize their craft through the on-the-job training they receive during their apprenticeships.

In the exercises you will learn how to use area in buying rolls of wallpaper, gallons of paint, bundles of shingles, square yards of carpet, and square feet of tile. Keep in mind that you can't buy $12\frac{11}{16}$ gallons of paint! You must buy 13 gallons. If your calculations tell you that you need 5.25 bundles of shingles, you have to buy 6 bundles. In this type of rounding, you must always round upward.

EXERCISES

1. **APPLICATION** Tammy is estimating how much she should charge for painting 148 rooms in a new motel with one coat of base paint and one coat of finishing paint. The four walls and the ceiling of each room must be painted. Each room measures 14 ft by 16 ft by 10 ft high.
   a. Calculate the total area of all the surfaces to be painted with each coat. Ignore doors and windows.
   b. One gallon of base paint covers 500 square feet. One gallon of finishing paint covers 250 square feet. How many gallons of each will Tammy need for the job?

2. **APPLICATION** Rashad wants to wallpaper the four walls of his bedroom. The room is rectangular and measures 11 feet by 13 feet. The ceiling is 10 feet high. A roll of wallpaper at the store is 2.5 feet wide and 50 feet long. How many rolls should he buy? (Wallpaper is hung from ceiling to floor. Ignore the doors and windows.)

3. **APPLICATION** It takes 65,000 solar cells, each 1.25 in. by 2.75 in., to power the Helios Prototype, shown below. How much surface area, in square feet, must be covered with the cells? The cells on Helios are 18% efficient. Suppose they were only 12% efficient, like solar cells used in homes. How much more surface area would need to be covered to deliver the same amount of power?

Technology CONNECTION

In August 2001, the Helios Prototype, a remotely controlled, nonpolluting solar-powered aircraft, reached 96,500 feet—a record for nonrocket aircraft. Soon, the Helios will likely sustain flight long enough to enable weather monitoring and other satellite functions. For news and updates, go to www.keymath.com/DG .

For Exercises 4 and 5, refer to the floor plan at right.

Bedroom 10 ft × 9 ft
Bath 10 ft × 6 ft
Bath 7 ft × 9 ft
Hallway
Bedroom 10 ft × 8 ft
Master bedroom 13 ft × 8 ft
Entryway
Living room 13 ft × 13 ft
Kitchen 10 ft × 18 ft
Dining room 13 ft × 9 ft

4. **APPLICATION** Dareen's family is ready to have wall-to-wall carpeting installed. The carpeting they chose costs $14 per square yard, the padding $3 per square yard, and the installation $3 per square yard. What will it cost them to carpet the three bedrooms and the hallway shown? ⓗ

5. **APPLICATION** Dareen's family now wants to install 1-foot-square terra cotta tiles in the entryway and kitchen, and 4-inch-square blue tiles on each bathroom floor. The terra cotta tiles cost $5 each, and the bathroom tiles cost 45¢ each. How many of each kind will they need? What will the tiles cost?

6. **APPLICATION** Harold works at a state park. He needs to seal the redwood deck at the information center to protect the wood. He measures the deck and finds that it is a kite with diagonals 40 feet and 70 feet. Each gallon of sealant covers 400 square feet, and the sealant needs to be applied every six months. How many gallon containers should he buy to protect the deck for the next three years?

7. **APPLICATION** A landscape architect is designing three trapezoidal flowerbeds to wrap around three sides of a hexagonal flagstone patio, as shown. What is the area of the entire flowerbed? The landscape architect's fee is $100 plus $5 per square foot. What will the flowerbed cost?

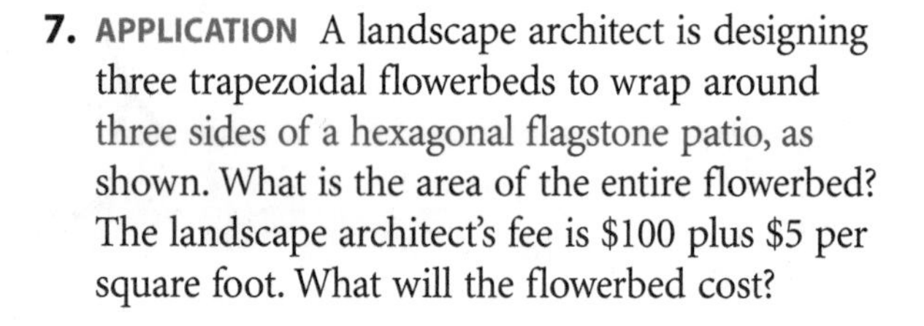

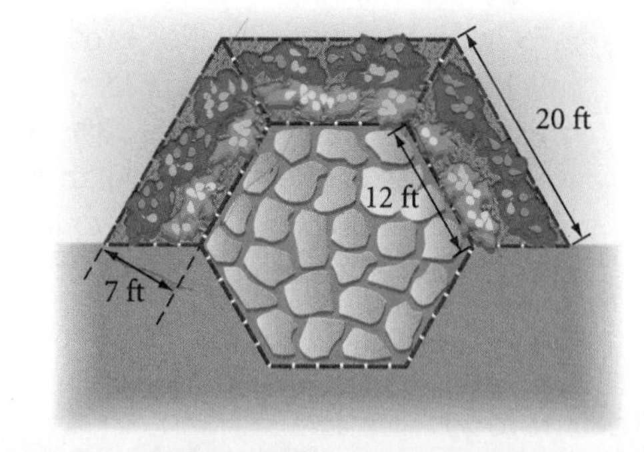

Career CONNECTION

Landscape architects have a keen eye for natural beauty. They study the grade and direction of land slopes, stability of the soil, drainage patterns, and existing structures and vegetation. They look at the various social, economic, and artistic concerns of the client. They also use science and engineering to plan environments that harmonize land features with structures, reducing the impact of urban development upon nature.

8. APPLICATION Tom and Betty are planning to paint the exterior walls of their cabin (all vertical surfaces). The paint they have selected costs \$24 per gallon and, according to the label, covers 150 to 300 square feet per gallon. Because the wood is very dry, they assume the paint will cover 150 square feet per gallon. How much will the project cost? (All measurements shown are in feet.) ⓗ

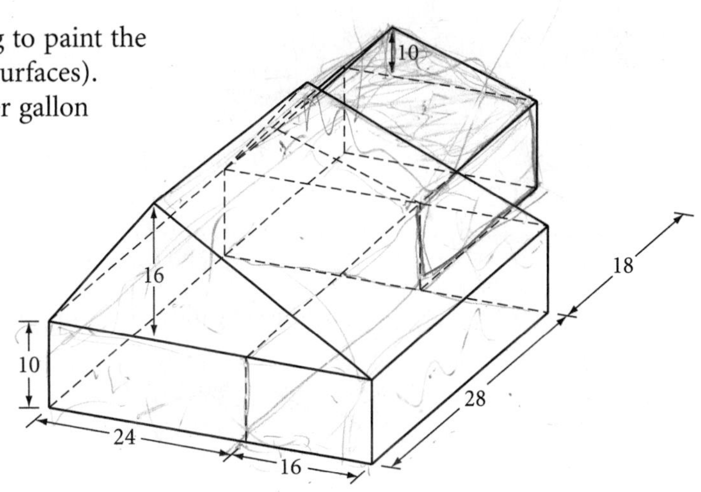

Review

9. A first-century Greek mathematician named Hero is credited with the following formula for the area of a triangle: $A = \sqrt{s(s-a)(s-b)(s-c)}$, where A is the area of the triangle, a, b, and c are the lengths of the three sides of the triangle, and s is the semiperimeter (half of the perimeter). Use Hero's formula to find the area of this triangle. Use the formula $A = \frac{1}{2}bh$ to check your answer.

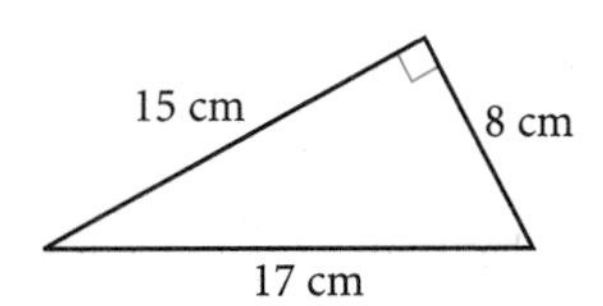

10. Explain why x must be 69° in the diagram at right.

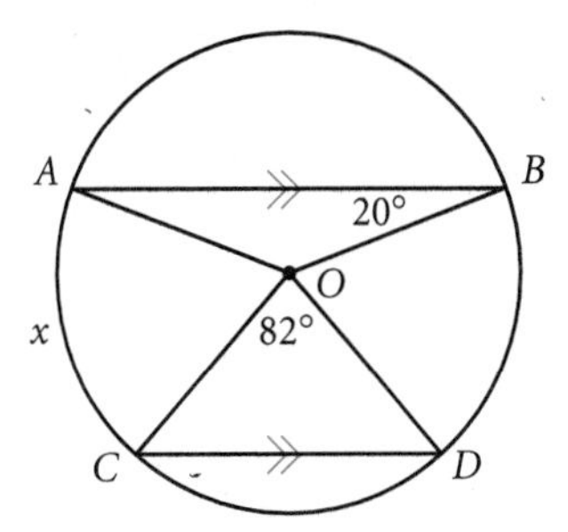

11. As P moves from left to right along ℓ, which of the following values changes?

A. The area of $\triangle ABP$

B. The area of $\triangle PDC$

C. The area of trapezoid $ABCD$

D. $m\angle A + m\angle PCD + m\angle CPD$

E. None of these

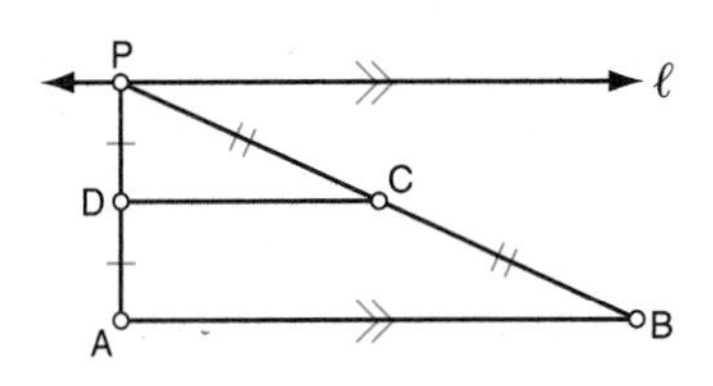

IMPROVING YOUR VISUAL THINKING SKILLS

Four-Way Split

How would you divide a triangle into four regions with equal areas? There are at least six different ways it can be done! Make six copies of a triangle and try it.

LESSON

8.4

Areas of Regular Polygons

If I had to live my life again,
I'd make the same mistakes,
only sooner.

TALLULAH BANKHEAD

You can divide a regular polygon into congruent isosceles triangles by drawing segments from the center of the polygon to each vertex. The center of the polygon is actually the center of a circumscribed circle.

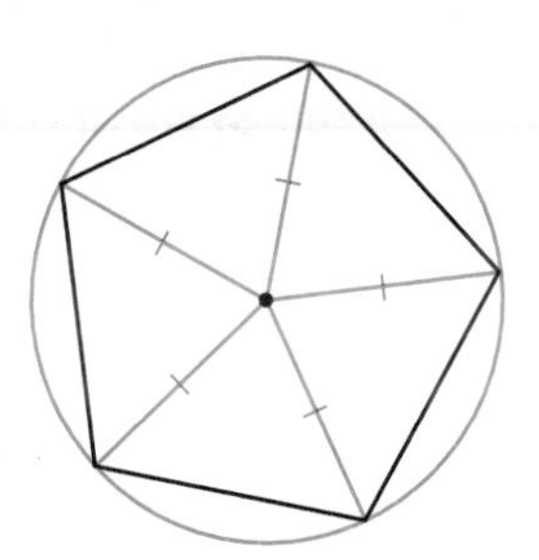

In this investigation you will divide regular polygons into triangles. Then you will write a formula for the area of any regular polygon.

Investigation
Area Formula for Regular Polygons

Consider a regular pentagon with side length s, divided into congruent isosceles triangles. Each triangle has a base s and a height a.

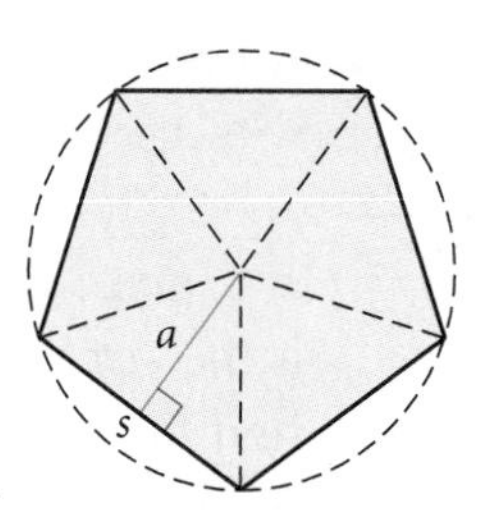

Regular pentagon

Step 1 What is the area of one isosceles triangle in terms of a and s?

Step 2 What is the area of this pentagon in terms of a and s?

Step 3 Repeat Steps 1 and 2 with other regular polygons and complete the table below.

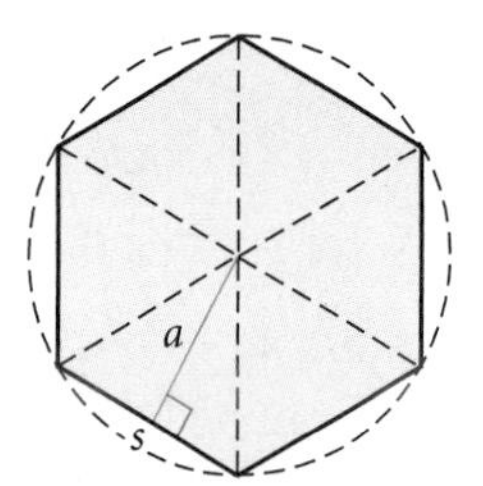

Regular hexagon

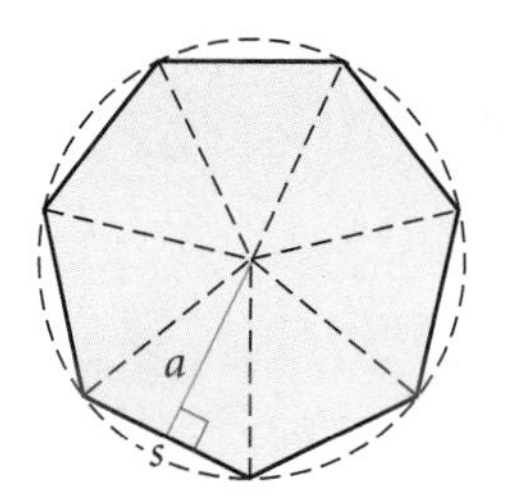

Regular heptagon

Number of sides	5	6	7	8	9	10	...	12	...	n
Area of regular polygon							...		...	

The distance a always appears in the area formula for a regular polygon, and it has a special name—apothem. An **apothem** of a regular polygon is a perpendicular segment from the center of the polygon's circumscribed circle to a side of the polygon. You may also refer to the length of the segment as the apothem.

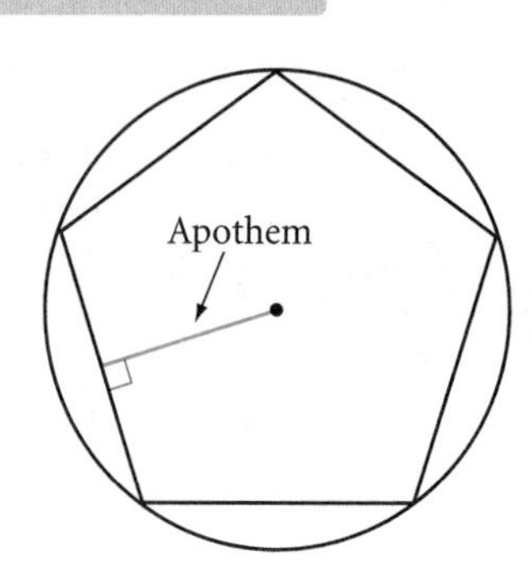

You can restate your last entry in the table as your next conjecture.

Regular Polygon Area Conjecture C-80

The area of a regular polygon is given by the formula ?, where A is the area, a is the apothem, s is the length of each side, and n is the number of sides. The length of each side times the number of sides is the perimeter, P, so $sn = P$. Thus you can also write the formula for area as $A =$? P.

EXERCISES

You will need

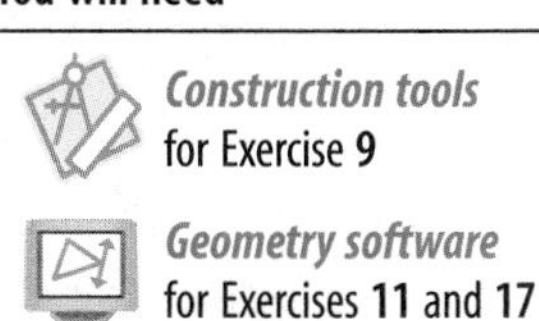

Construction tools for Exercise 9

Geometry software for Exercises 11 and 17

In Exercises 1–8, use the Regular Polygon Area Conjecture to find the unknown length accurate to the nearest unit, or the unknown area accurate to the nearest square unit. Recall that you use the symbol $\approx$ when your answer is an approximation.

1. $A \approx$?
$s = 24$ cm
$a = 24.9$ cm

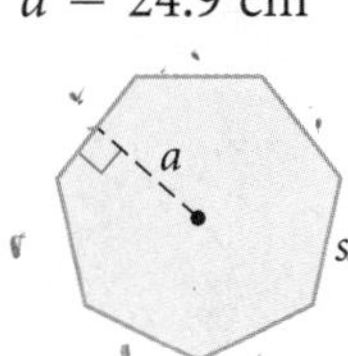

2. $a \approx$?
$s = 107.5$ cm
$A = 19{,}887.5$ cm^2

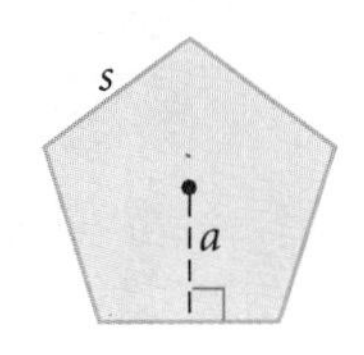

3. $P \approx$?
$a = 38.6$ cm
$A = 4940.8$ cm^2

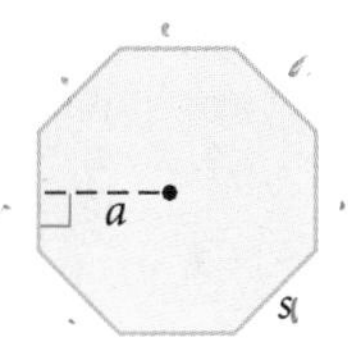

4. Regular pentagon: $a = 3$ cm and $s = 4.4$ cm, $A \approx$?

5. Regular nonagon: $a = 9.6$ cm and $A = 302.4$ cm^2, $P \approx$?

6. Regular n-gon: $a = 12$ cm and $P = 81.6$ cm, $A \approx$?

7. Find the perimeter of a regular polygon if $a = 9$ m and $A \approx 259.2$ m^2.

8. Find the length of each side of a regular n-gon if $a = 80$ feet, $n = 20$, and $A \approx 20{,}000$ square feet.

9. *Construction* Use a compass and straightedge to construct a regular hexagon with sides that measure 4 cm. Use the Regular Polygon Area Conjecture and a centimeter ruler to approximate the hexagon's area. (h)

10. Draw a regular pentagon with apothem 4 cm. Use the Regular Polygon Area Conjecture and a centimeter ruler to approximate the pentagon's area. (h)

11. *Technology* Use geometry software to construct a circle. Inscribe a pentagon that looks regular and measure its area. Now drag the vertices. How can you drag the vertices to increase the area of the inscribed pentagon? To decrease its area?

12. Find the shaded area of the regular octagon *ROADSIGN*. The apothem measures about 20 cm. Segment *GI* measures about 16.6 cm.

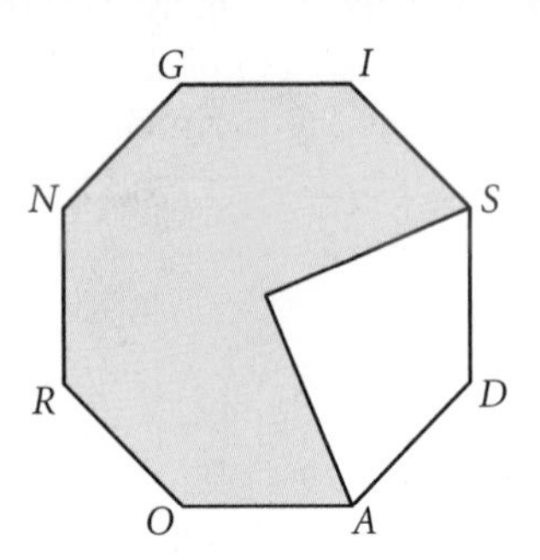

13. Find the shaded area of the regular hexagonal donut. The apothem and sides of the smaller hexagon are half as long as the apothem and sides of the large hexagon. $a \approx 6.9$ cm and $r \approx 8$ cm ⓗ

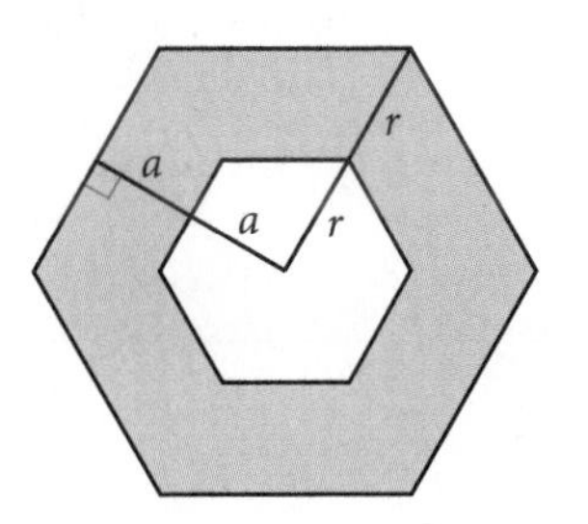

Career CONNECTION

Interior designers, unlike interior decorators, are concerned with the larger planning and technical considerations of interiors, as well as with style and color selection. They have an intuitive sense of spatial relationships. They prepare sketches, schedules, and budgets for client approval and inspect the site until the job is complete.

14. **APPLICATION** An interior designer created the kitchen plan shown. The countertop will be constructed of colored concrete. What is its total surface area? If concrete countertops 1.5 inches thick cost \$85 per square foot, what will be the total cost of this countertop?

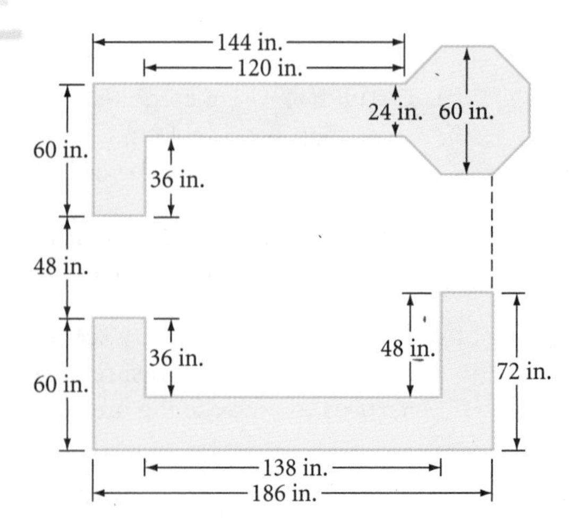

Review

In Exercises 15 and 16, graph the two lines, then find the area bounded by the x-axis, the y-axis, and both lines.

15. $y = \frac{1}{2}x + 5, y = -2x + 10$ ⓗ

16. $y = -\frac{1}{3}x + 6, y = -\frac{4}{3}x + 12$

17. *Technology* Construct a triangle and its three medians. Compare the areas of the six small triangles that the three medians formed. Make a conjecture, and support it with a convincing argument.

18. If the pattern continues, write an expression for the perimeter of the nth figure in the picture pattern.

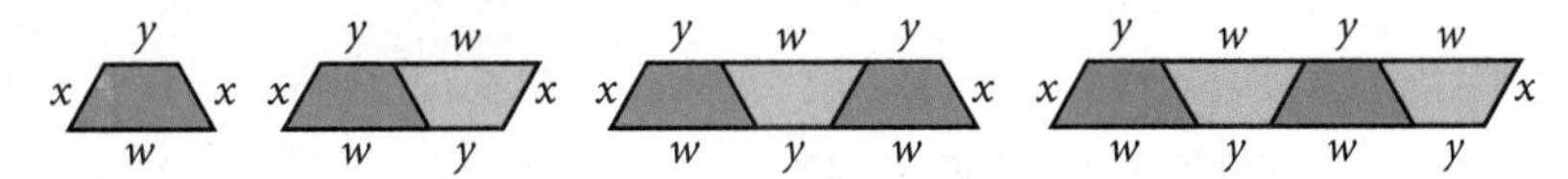

19. Identify the point of concurrency from the construction marks.

a.

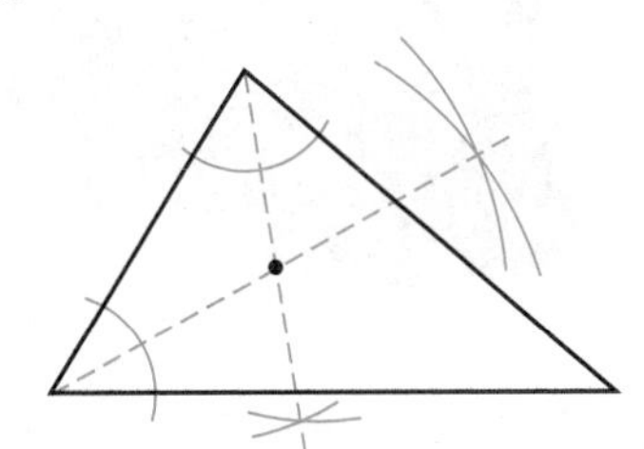

b.

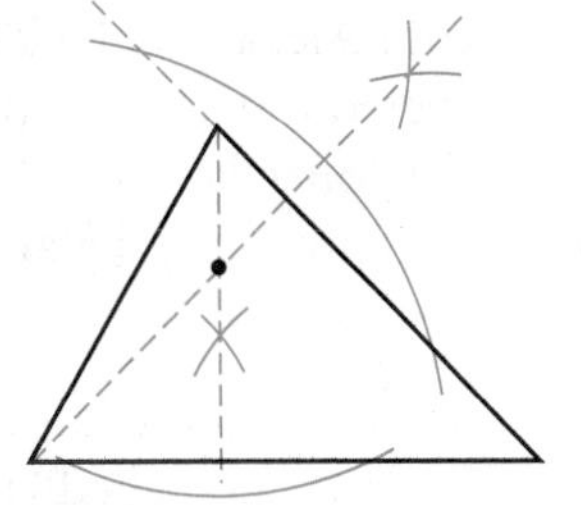

c.

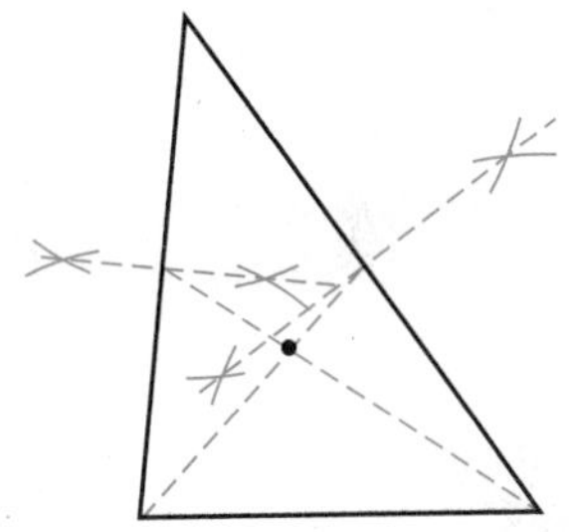

IMPROVING YOUR VISUAL THINKING SKILLS

The Squared Square Puzzle

The square shown is called a "squared square." A square 112 units on a side is divided into 21 squares. The area of square X is 50^2, or 2500, and the area of square Y is 4^2, or 16. Find the area of each of the other squares.

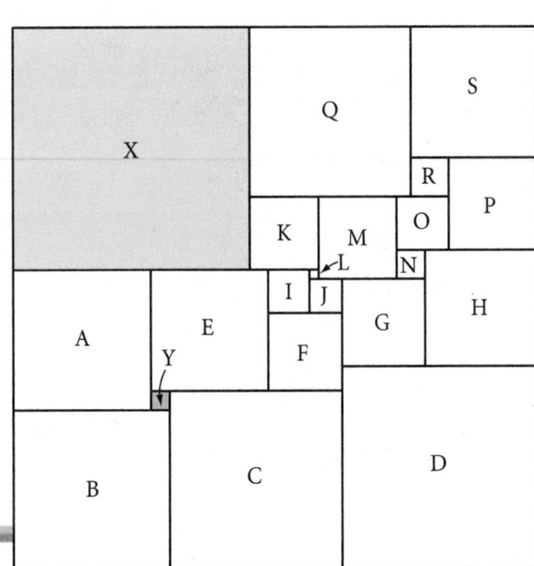

Exploration

Pick's Formula for Area

You know how to find the area of polygon regions, but how would you find the area of the dinosaur footprint at right?

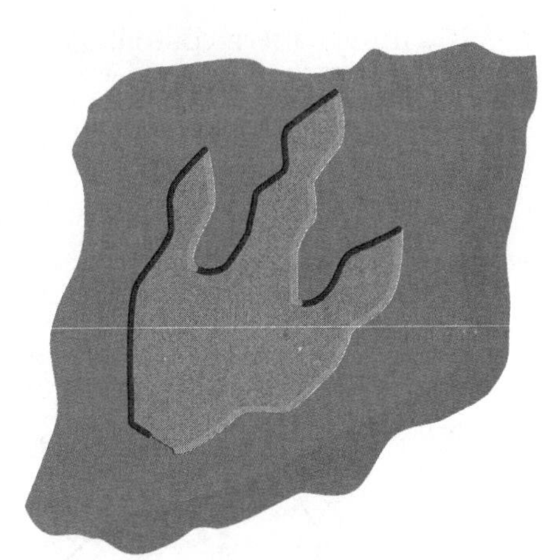

You know how to place a polygon on a grid and count squares to find the area. About a hundred years ago, Austrian mathematician Georg Alexander Pick (1859–1943) discovered a relationship, now known as Pick's formula, for finding the area of figures on a square dot grid.

Let's start by looking at polygons on a square dot grid. The dots are called lattice points. Let's count the lattice points in the interior of the polygon and those on its boundary and compare our findings to the areas of the polygon that you get by counting squares, as you did in Lesson 8.1.

Polygon A

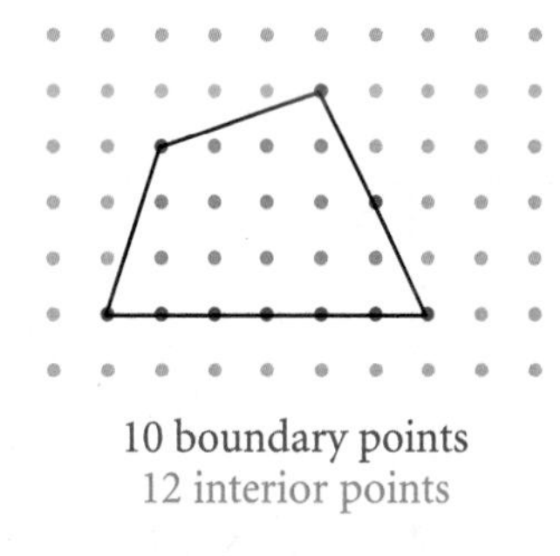

10 boundary points
12 interior points

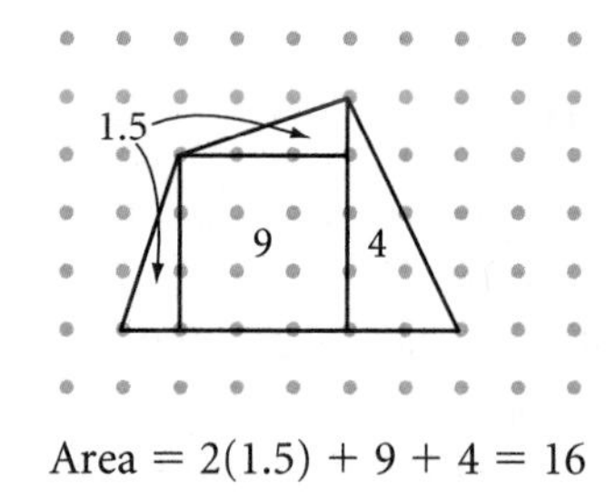

Area = 2(1.5) + 9 + 4 = 16

Polygon B

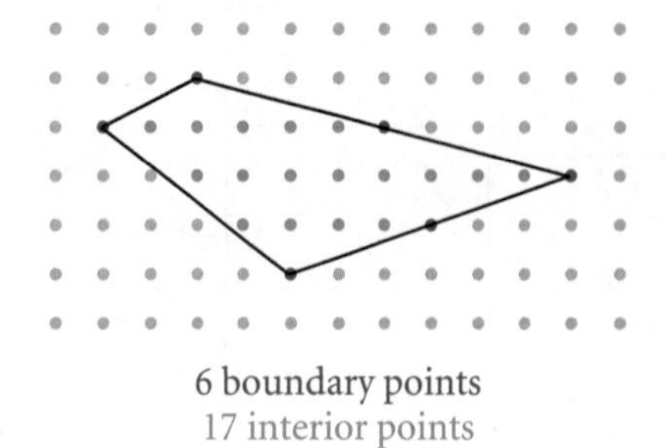

6 boundary points
17 interior points

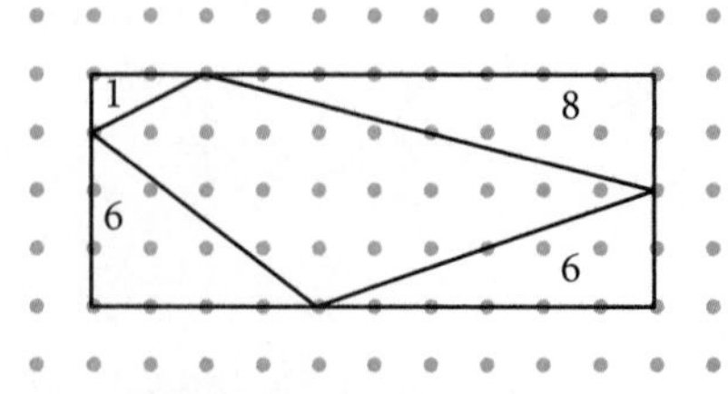

Area of rectangle = 40
Area of polygon = 40 − 1 − 8 − 2(6) = 19

How can the boundary points and interior points help us find the area? There are a lot of things to look at. It seems too difficult to find a pattern with our results. An important technique for finding patterns is to hold one variable constant and see what happens with the other variables. That's what you'll do in the activity below.

Activity

Dinosaur Footprints and Other Shapes

You will need

- square dot paper or graph paper
- geoboards (optional)

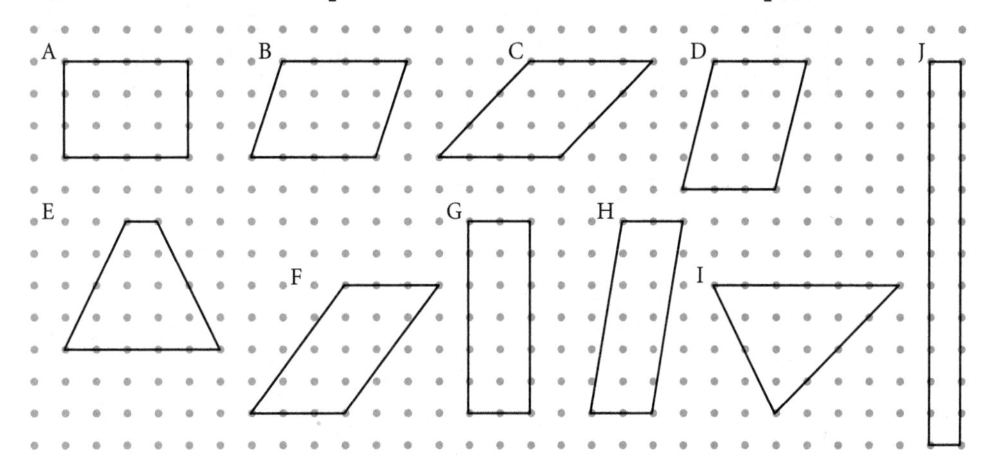

Step 1 Confirm that each polygon A through J above has area $A = 12$.

Step 2 Let b be the number of boundary points and i be the number of interior points. Create and complete a table like this one for polygons A through J.

Polygon (A = 12)	A	B	C	
Number of boundary points (b)				
Number of interior points (i)				

Step 3 Study the table for patterns. Do you see a relationship between b and i when $A = 12$? Graph the pairs (b, i) from your table and label each point with its name A through J. What do you notice? Write an equation that fits the points.

Step 4 Consider several polygons with an area of 8. Graph points (b, i) and write an equation.

Step 5 Generalize the formula you found in Steps 3 and 4. When you feel you have enough data, copy and complete the conjecture.

Pick's Formula

If A is the area of a polygon whose vertices are lattice points, b is the number of lattice points on the boundary of the polygon, and i is the number of interior lattice points, then $A = \underline{?}\,b + \underline{?}\,i + \underline{?}$.

Pick's formula is especially useful when you apply it to the areas of irregularly shaped regions. Since it relies only on lattice points, you do not need to divide the shape into rectangles or triangles.

Step 6 | Use Pick's formula to find the approximate areas of these irregular shapes.

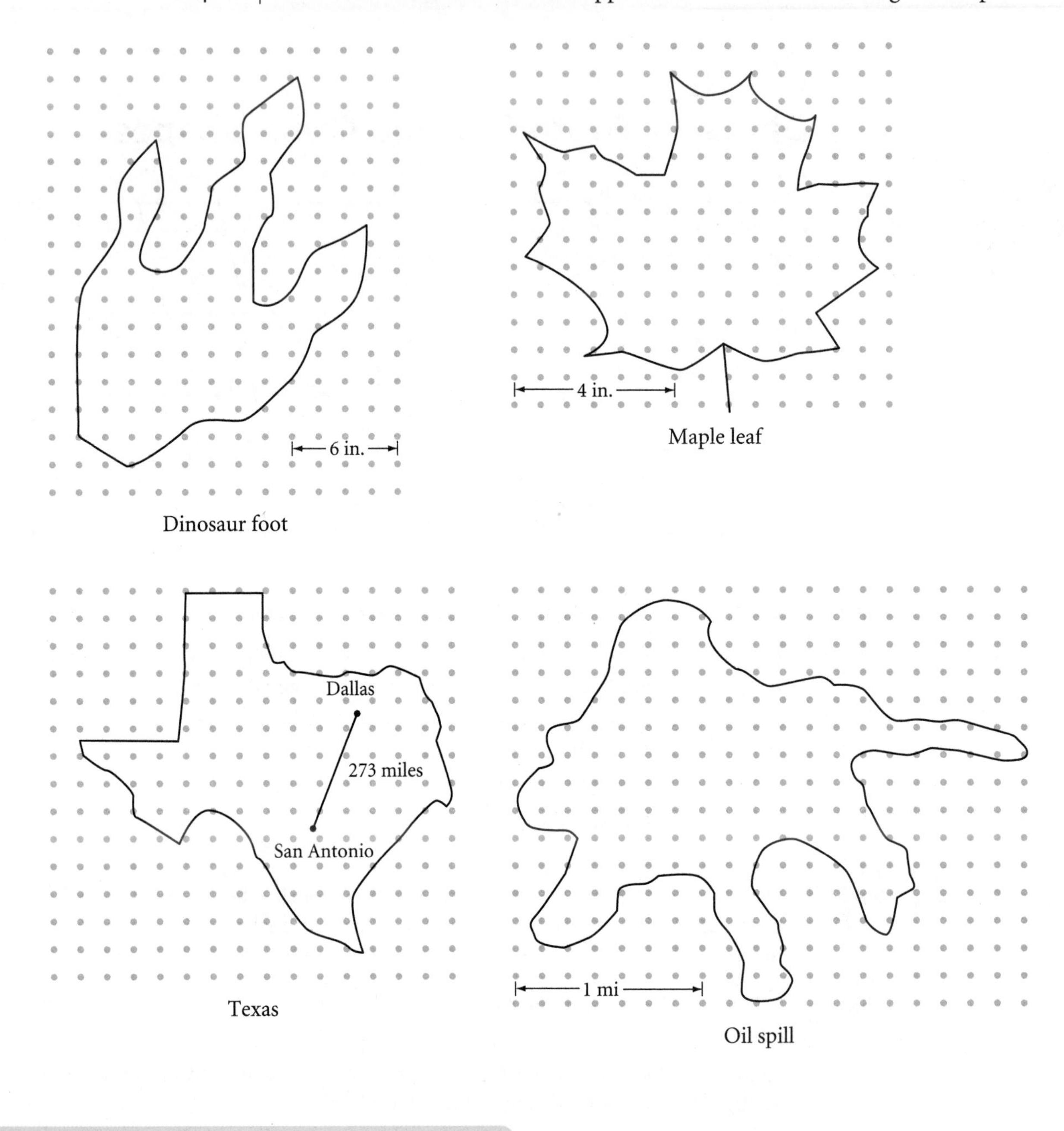

LESSON

8.5

Areas of Circles

The moon is a dream of the sun.

PAUL KLEE

So far, you have discovered the formulas for the areas of various polygons. In this lesson, you'll discover the formula for the area of a circle. Most of the shapes you have investigated in this chapter could be divided into rectangles or triangles. Can a circle be divided into rectangles or triangles? Not exactly, but in this investigation you will see an interesting way to think about the area of a circle.

Investigation

Area Formula for Circles

You will need

- a compass
- scissors

Circles do not have straight sides like polygons do. However, the area of a circle can be rearranged. Let's investigate.

Step 1 Use your compass to make a large circle. Cut out the circular region.

Step 2 Fold the circular region in half. Fold it in half a second time, then a third time and a fourth time. Unfold your circle and cut it along the folds into 16 wedges.

Step 3 Arrange the wedges in a row, alternating the tips up and down to form a shape that resembles a parallelogram.

If you cut the circle into more wedges, you could rearrange these thinner wedges to look even more like a rectangle, with fewer bumps. You would not lose or gain any area in this change, so the area of this new "rectangle," skimming off the bumps as you measure its length, would be closer to the area of the original circle.

If you could cut infinitely many wedges, you'd actually have a rectangle with smooth sides. What would its base length be? What would its height be in terms of C, the circumference of the circle?

Step 4 The radius of the original circle is r and the circumference is $2\pi r$. Give the base and the height of a rectangle made of a circle cut into infinitely many wedges. Find its area in terms of r. State your next conjecture.

Circle Area Conjecture C-81

The area of a circle is given by the formula __?__, where A is the area and r is the radius of the circle.

How do you use this new conjecture? Let's look at a few examples.

EXAMPLE A

The small apple pie has a diameter of 8 inches, and the large cherry pie has a radius of 5 inches. How much larger is the large pie?

▶ Solution

First, find each area.

Small pie

$$A = \pi r^2$$
$$= \pi(4)^2$$
$$= \pi(16)$$
$$\approx 50.2$$

Large pie

$$A = \pi r^2$$
$$= \pi(5)^2$$
$$= \pi(25)$$
$$\approx 78.5$$

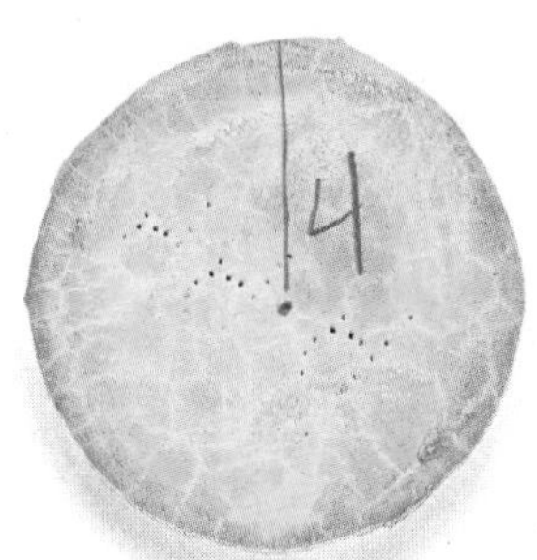

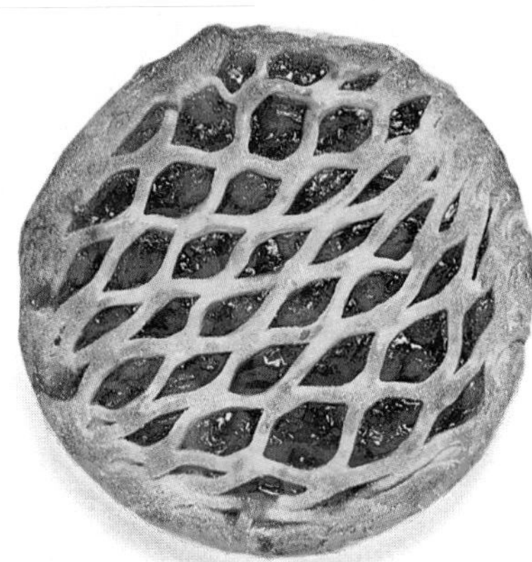

The large pie is 78.5 in^2, and the small pie is 50.2 in^2. The difference in area is about 28.3 square inches. So the large pie is more than 50% larger than the small pie, assuming they have the same thickness. Notice that we used 3.14 as an approximate value for π.

EXAMPLE B

If the area of the circle at right is 256π m^2, what is the circumference of the circle?

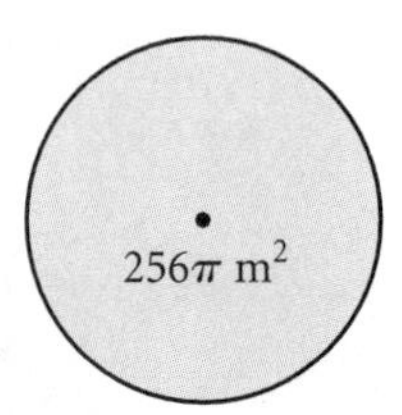

▶ Solution

Use the area to find the radius, then use the radius to find the circumference.

$$A = \pi r^2 \qquad C = 2\pi r$$
$$256\pi = \pi r^2 \qquad = 2\pi(16)$$
$$256 = r^2 \qquad = 32\pi$$
$$r = 16 \qquad \approx 100.5 \text{ m}$$

The circumference is 32π meters, or approximately 100.5 meters.

EXERCISES

You will need

Geometry software for Exercise 19

Use the Circle Area Conjecture to solve for the unknown measures in Exercises 1–8. Leave your answers in terms of π, unless the problem asks for an approximation. For approximations, use the π key on your calculator.

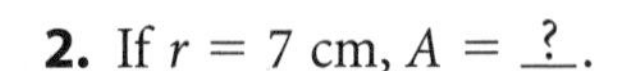

1. If $r = 3$ in., $A = \underline{?}$.

2. If $r = 7$ cm, $A = \underline{?}$.

3. If $r = 0.5$ m, $A \approx \underline{?}$.

4. If $A = 9\pi$ cm^2, then $r = \underline{?}$.

5. If $A = 3\pi$ in^2, then $r = \underline{?}$.

6. If $A = 0.785$ m^2, then $r \approx \underline{?}$.

7. If $C = 12\pi$ in., then $A = \underline{?}$.

8. If $C = 314$ m, then $A \approx \underline{?}$.

9. What is the area of the shaded region between the circle and the rectangle?

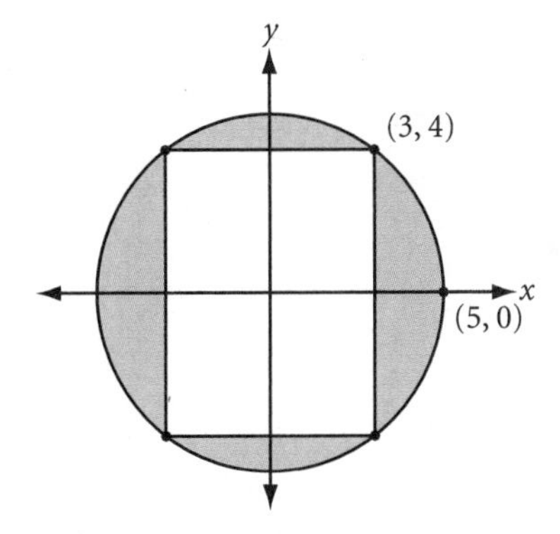

10. What is the area of the shaded region between the circle and the triangle?

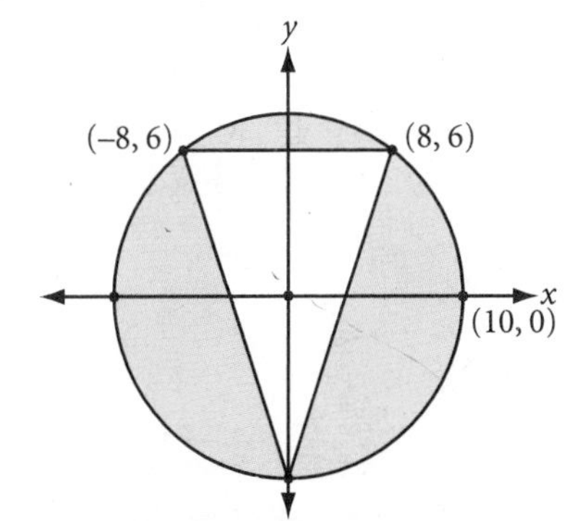

11. Sketch and label a circle with an area of 324π cm^2. Be sure to label the length of the radius.

12. **APPLICATION** The rotating sprinkler arms in the photo at right are all 16 meters long. What is the area of each circular farm? Express your answer to the nearest square meter.

13. **APPLICATION** A small college TV station can broadcast its programming to households within a radius of 60 kilometers. How many square kilometers of viewing area does the station reach? Express your answer to the nearest square kilometer.

14. Sampson's dog, Cecil, is tied to a post by a chain 7 meters long. How much play area does Cecil have? Express your answer to the nearest square meter.

15. **APPLICATION** A muscle's strength is proportional to its cross-sectional area. If the cross section of one muscle is a circular region with a radius of 3 cm, and the cross section of a second, identical type of muscle is a circular region with a radius of 6 cm, how many times stronger is the second muscle? ⓗ

Champion weight lifter Soraya Jimenez extends barbells weighing almost double her own body weight.

16. What would be a good approximation for the area of a regular 100-gon inscribed in a circle with radius r? Explain your reasoning. ⓗ

Review

17. $A = \underline{?}$

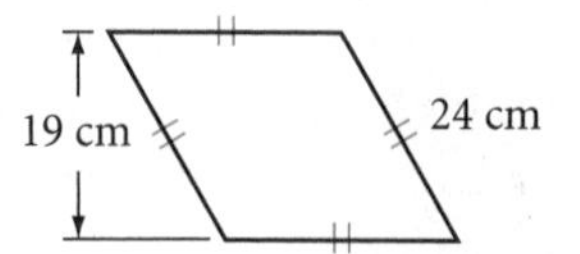

18. $A = \underline{?}$

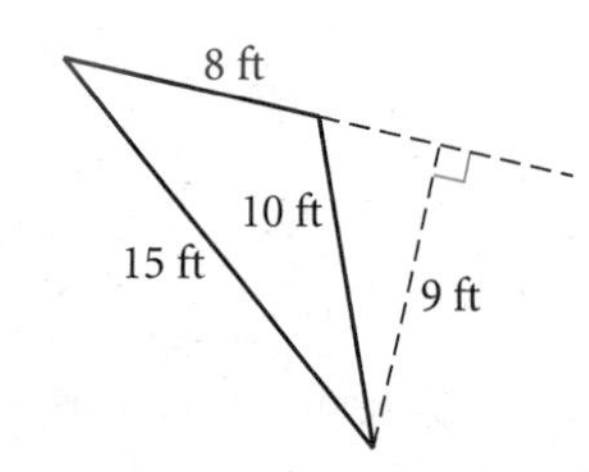

19. ***Technology*** Construct a parallelogram and a point in its interior. Construct segments from this point to each vertex, forming four triangles. Measure the area of each triangle. Move the point to find a location where all four triangles have equal area. Is there more than one such location? Explain your findings.

20. Explain why x must be 48°.

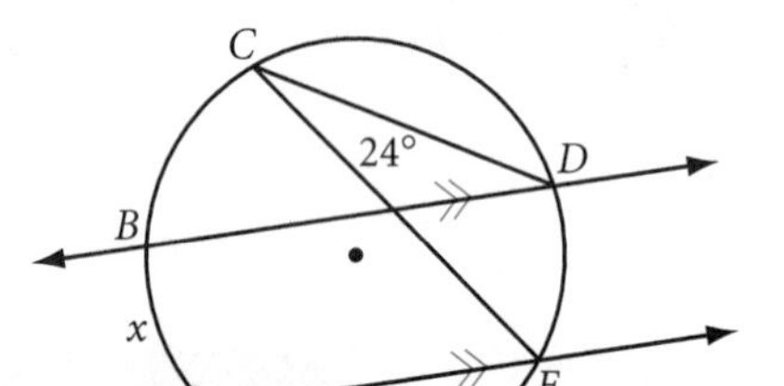

21. What's wrong with this picture?

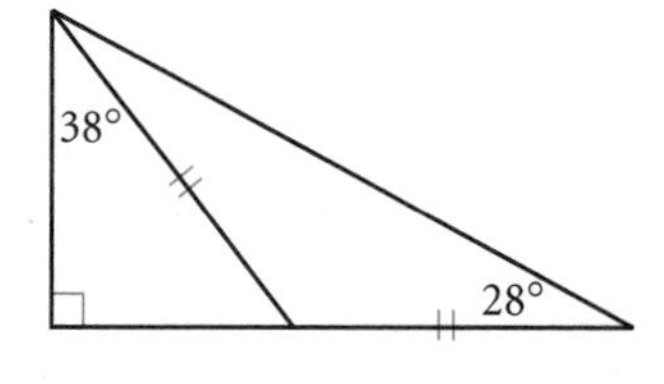

22. The 6-by-18-by-24 cm clear plastic sealed container is resting on a cylinder. It is partially filled with liquid, as shown. Sketch the container resting on its smallest face. Show the liquid level in this position.

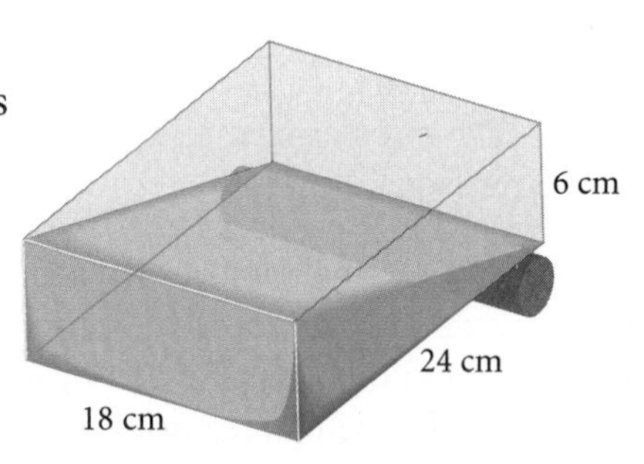

IMPROVING YOUR VISUAL THINKING SKILLS

Random Points

What is the probability of randomly selecting from the 3-by-3 grid at right three points that form the vertices of an isosceles triangle?

LESSON

8.6

Any Way You Slice It

Cut my pie into four pieces—
I don't think I could eat eight.
YOGI BERRA

In Lesson 8.5, you discovered a formula for calculating the area of a circle. With the help of your visual thinking and problem-solving skills, you can calculate the areas of different sections of a circle.

Its makers claimed this was the world's largest slice of pizza.

If you cut a slice of pizza, each slice would probably be a sector of a circle. If you could make only one straight cut with your knife, your slice would be a segment of a circle. If you don't like the crust, you'd cut out the center of the pizza; the crust shape that would remain is called an annulus.

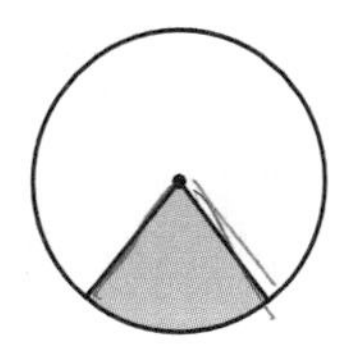

Sector of a circle

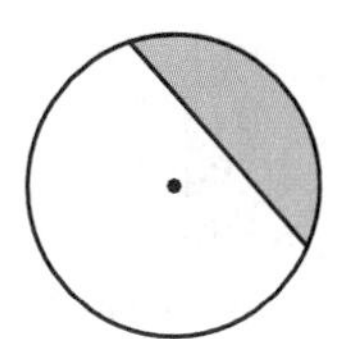

Segment of a circle

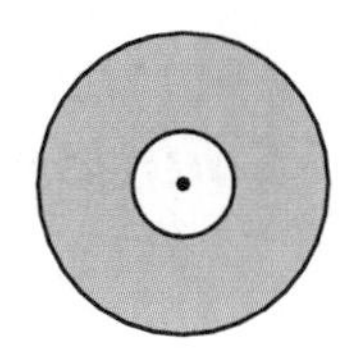

Annulus

A **sector of a circle** is the region between two radii of a circle and the included arc.

A **segment of a circle** is the region between a chord of a circle and the included arc.

An **annulus** is the region between two concentric circles.

"Picture equations" are helpful when you try to visualize the areas of these regions. The picture equations below show you how to find the area of a sector of a circle, the area of a segment of a circle, and the area of an annulus.

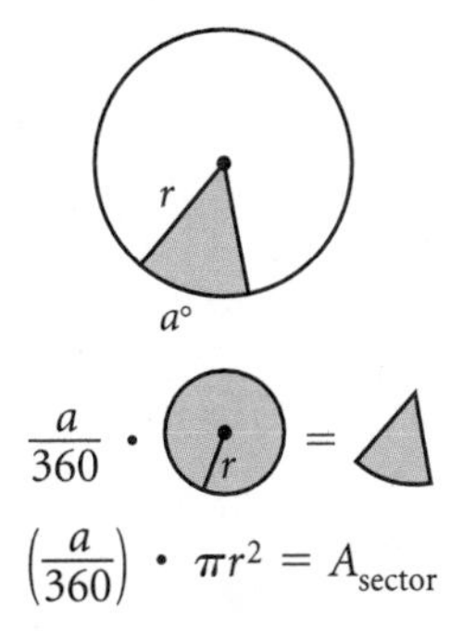

$$\left(\frac{a}{360}\right) \cdot \pi r^2 = A_{\text{sector}}$$

a° b h r r h b a°

$$\left(\frac{a}{360}\right) \cdot \pi r^2 - \frac{1}{2}bh = A_{\text{segment}}$$

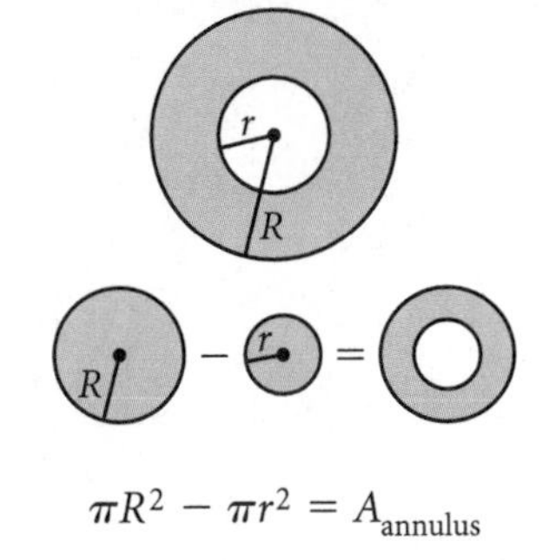

$$\pi R^2 - \pi r^2 = A_{\text{annulus}}$$

EXAMPLE A Find the area of the shaded sector.

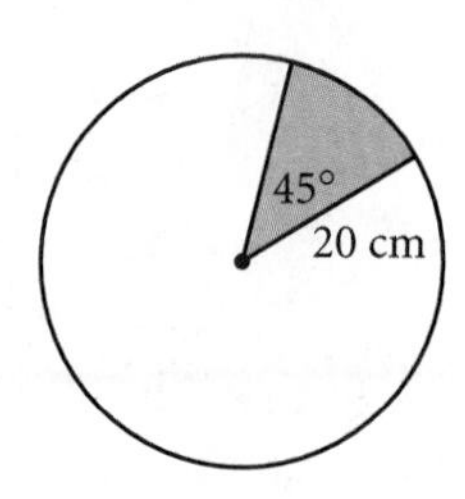

▶ Solution The sector is $\frac{45°}{360°}$, or $\frac{1}{8}$, of the circle.

$$A_{\text{sector}} = \left(\frac{a}{360}\right) \cdot \pi r^2 \qquad \text{The area formula for a sector.}$$

$$= \left(\frac{45}{360}\right) \cdot \pi(20)^2 \qquad \text{Substitute } r = 20 \text{ and } a = 45.$$

$$= \left(\frac{1}{8}\right) \cdot 400\pi \qquad \text{Reduce the fraction and square 20.}$$

$$= 50\pi \qquad \text{Multiply.}$$

The area is 50π cm^2.

EXAMPLE B Find the area of the shaded segment.

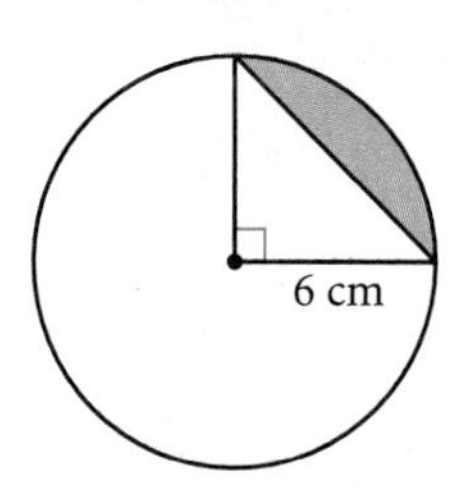

▶ Solution According to the picture equation on page 437, the area of a segment is equivalent to the area of the sector minus the area of the triangle. You can use the method in Example A to find that the area of the sector is $\left(\frac{1}{4}\right)(36\pi \text{ cm}^2)$, or 9π cm^2. The area of the triangle is $\left(\frac{1}{2}\right)(6)(6)$, or 18 cm^2. So the area of the segment is $(9\pi - 18)$ cm^2.

EXAMPLE C The shaded area is 14π cm^2, and the radius is 6 cm. Find x.

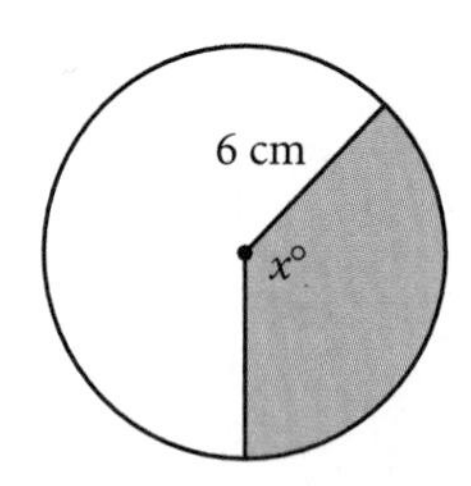

▶ Solution The sector's area is $\frac{x}{360}$ of the circle's area, which is 36π.

$$14\pi = \left(\frac{x}{360}\right)(36\pi)$$

$$\frac{(360)(14\pi)}{36\pi} = x$$

$$x = 140$$

The central angle measures 140°.

EXERCISES

You will need

Construction tools for Exercise **16**

In Exercises 1–8, find the area of the shaded region. The radius of each circle is r. If two circles are shown, r is the radius of the smaller circle and R is the radius of the larger circle.

1. $r = 6$ cm

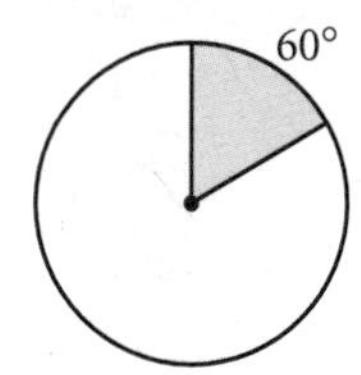

2. $r = 8$ cm

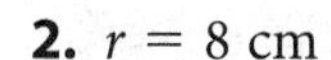

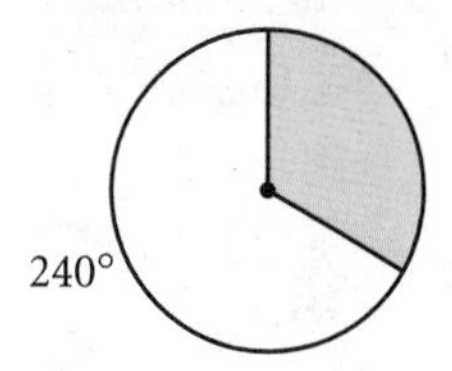

3. $r = 16$ cm

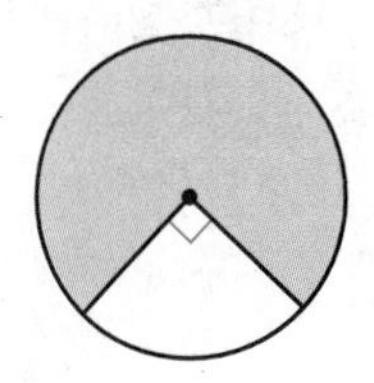

4. $r = 2$ cm

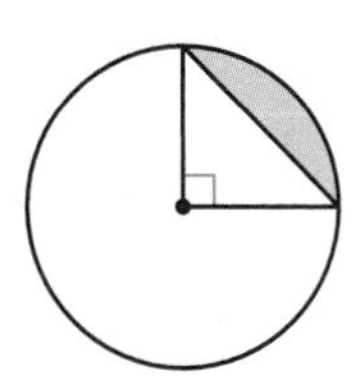

5. $r = 8$ cm

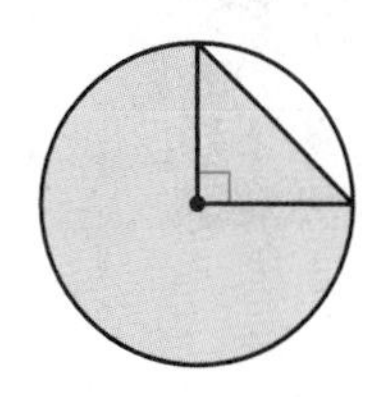

6. $R = 7$ cm
$r = 4$ cm ⓗ

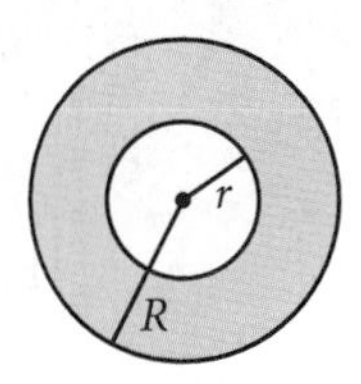

7. $r = 2$ cm

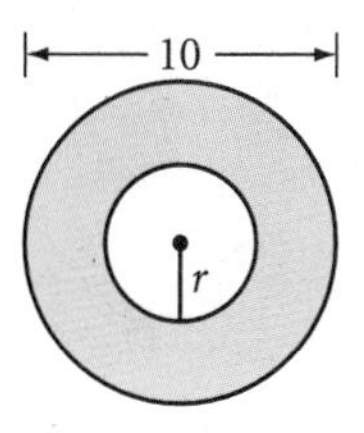

8. $R = 12$ cm
$r = 9$ cm

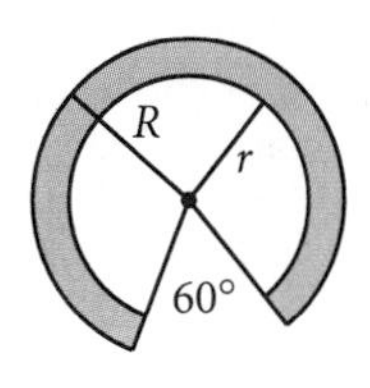

9. The shaded area is 12π cm^2. Find r.

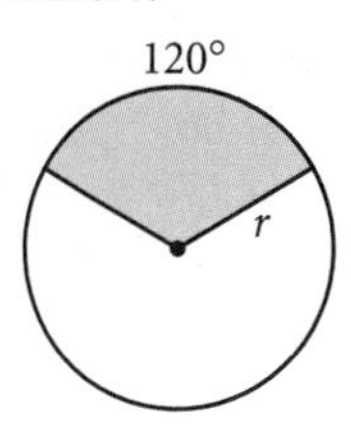

10. The shaded area is 32π cm^2. Find r.

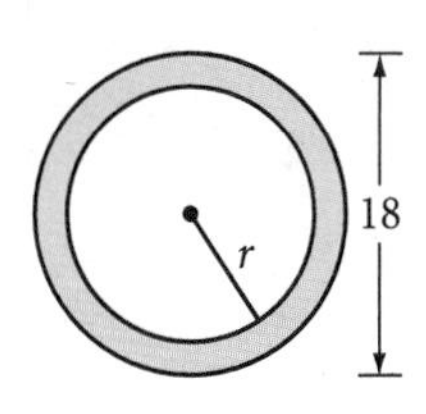

11. The shaded area is 120π cm^2, and the radius is 24 cm. Find x.

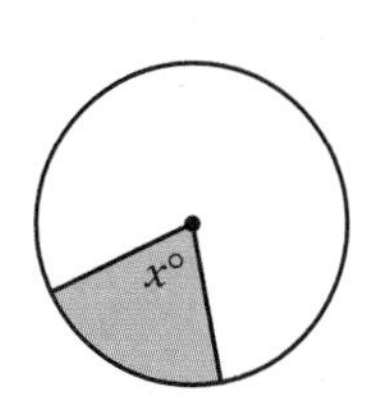

12. The shaded area is 10π cm^2. The radius of the large circle is 10 cm, and the radius of the small circle is 8 cm. Find x. ⓗ

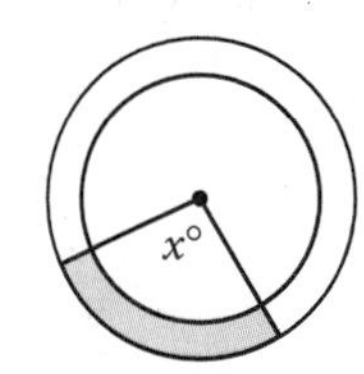

13. Suppose the pizza slice in the photo at the beginning of this lesson is a sector with a 36° arc, and the pizza has a radius of 20 ft. If one can of tomato sauce will cover 3 ft^2 of pizza, how many cans would you need to cover this slice?

14. Utopia Park has just installed a circular fountain 8 meters in diameter. The Park Committee wants to pave a 1.5-meter-wide path around the fountain. If paving costs $10 per square meter, find the cost to the nearest dollar of the paved path around the fountain.

This circular fountain at the Chateau de Villandry in Loire Valley, France, shares a center with the circular path around it. How many concentric arcs and circles do you see in the picture?

Mathematics CONNECTION

Attempts to solve the famous problem of rectifying a circle—finding a rectangle with the same area as a given circle—led to the creation of some special shapes made up of parts of circles. The diagrams below are based on some that Leonardo da Vinci sketched while attempting to solve this problem.

15. The illustrations below demonstrate how to rectify the pendulum.

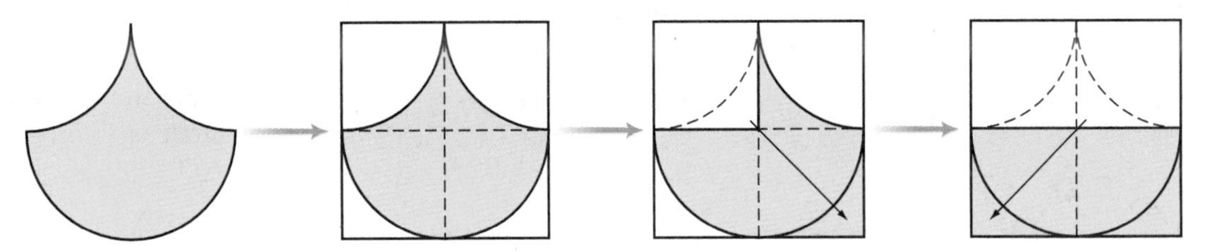

In a series of diagrams, demonstrate how to rectify each figure.

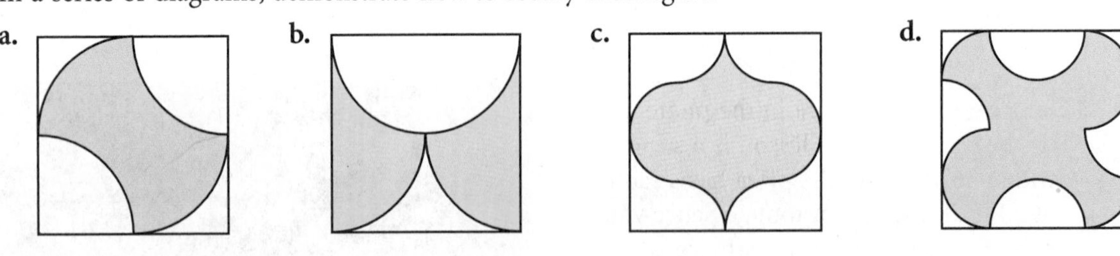

16. ***Construction*** Reverse the process you used in Exercise 15. On graph paper, draw a 12-by-6 rectangle. Use your compass to divide it into at least four parts, then rearrange the parts into a new curved figure. Draw its outline on graph paper.

Review

17. Each set of circles is externally tangent. What is the area of the shaded region in each figure? What percentage of the area of the square is the area of the circle or circles in each figure? All given measurements are in centimeters.

a. 12 × 12   b. 12 × 12   c. 12 × 12   d. 12 × 12

18. The height of a trapezoid is 15 m and the midsegment is 32 m. What is the area of the trapezoid? ⓗ

In Exercises 19–22, identify each statement as true or false. If true, explain why. If false, give a counterexample.

19. If the arc of a circle measures 90° and has an arc length of 24π cm, then the radius of the circle is 48 cm.

20. If the measure of each exterior angle of a regular polygon is 24°, then the polygon has 15 sides.

21. If the diagonals of a parallelogram bisect its angles, then the parallelogram is a square.

22. If two sides of a triangle measure 25 cm and 30 cm, then the third side must be greater than 5 cm but less than 55 cm.

IMPROVING YOUR REASONING SKILLS

Code Equations

Each code below uses the first letters of words that will make the equation true. For example, $12M$ = a Y is an abbreviation of the equation 12 Months = a Year. Find the missing words in each code.

1. $45D$ = an AA of an IRT

2. $7 = SH$

3. $90D$ = each A of a R

4. $5 = D$ in a P

Exploration

Geometric Probability II

You already know that a probability value is a number between 0 and 1 that tells you how likely something is to occur. For example, when you roll a die, three of the six rolls—namely, 2, 3, and 5—are prime numbers. Since each roll has the same chance of occurring, P(prime number) $= \frac{3}{6}$ or $\frac{1}{2}$.

In some situations, probability depends on area. For example, suppose a meteorite is headed toward Earth. If about $\frac{1}{3}$ of Earth's surface is land, the probability that the meteorite will hit land is about $\frac{1}{3}$, while the probability it will hit water is about $\frac{2}{3}$. Because Alaska has a greater area than Vermont, the probability the meteorite will land in Alaska is greater than the probability it will land in Vermont. If you knew the areas of these two states and the surface area of Earth, how could you calculate the probabilities that the meteorite would land in each state?

Activity

Where the Chips Fall

You will need

- graph paper
- a ruler
- a penny
- a dime

In this activity, you will solve several probability problems that involve area.

The Shape of Things

At each level of a computer game, you must choose one of several shapes on a coordinate grid. The computer then randomly selects a point *anywhere* on the grid. If the point is outside your shape, you move to the next level. If the point is on or inside your shape, you lose the game.

On the first level, a trapezoid, a pentagon, a square, and a triangle are displayed on a grid that goes from 0 to 12 on both axes. The table below gives the vertices of the shapes.

Shape	Vertices
Trapezoid	(1, 12), (8, 12), (7, 9), (4, 9)
Pentagon	(3, 1), (4, 4), (6, 4), (9, 2), (7, 0)
Square	(0, 6), (3, 9), (6, 6), (3, 3)
Triangle	(11, 0), (7, 4), (11, 12)

Step 1 For each shape, calculate the probability the computer will choose a point on or inside that shape. Express each probability to three decimal places.

Step 2 What is the probability the computer will choose a point that is on or inside a quadrilateral? What is the probability it will choose a point that is outside all of the shapes?

Step 3 If you choose a triangle, what is the probability you will move to the next level?

Step 4 Which shape should you choose to have the best chance of moving to the next level? Why?

Right on Target

You are playing a carnival game in which you must throw one dart at the board shown at right. The score for each region is shown on the board. If your dart lands in a Bonus section, your score is tripled. The more points you get, the better your prize will be. The radii of the circles from the inside to the outside are 4 in., 8 in., 12 in., and 16 in. Assume your aim is not very good, so the dart will hit a random spot. If you miss the board completely, you get to throw again.

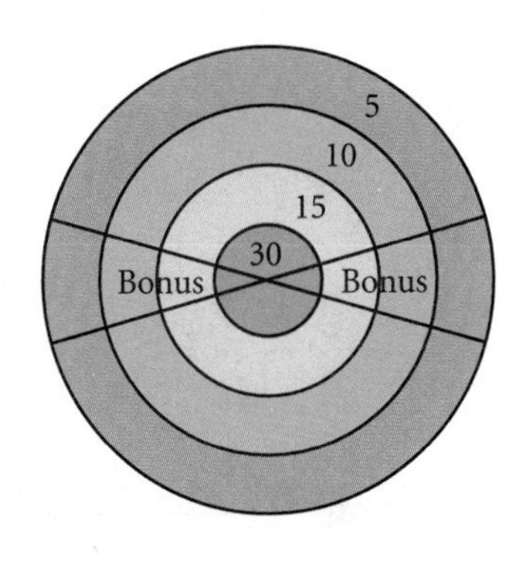

Step 5 What is the probability your dart will land in the red region? Blue region?

Step 6 Compute the probability your dart will land in a Bonus section. (The central angle measure for each Bonus section is 30°.)

Step 7 If you score 90 points, you will win the grand prize, a giant stuffed emu. What is the probability you will win the grand prize?

Step 8 If you score exactly 30 points, you win an "I ♥ Carnivals" baseball cap. What is the probability you will win the cap?

Step 9 Now imagine you have been practicing your dart game, and your aim has improved. Would your answers to Steps 5–8 change? Explain.

The Coin Toss

You own a small cafe that is popular with the mathematicians in the neighborhood. You devise a game in which the customer flips a coin onto a red-and-white checkered tablecloth with 1-inch squares. If it lands completely within a square, the customer wins, and doesn't have to pay the bill. If it lands touching or crossing the boundary of a square, the customer loses.

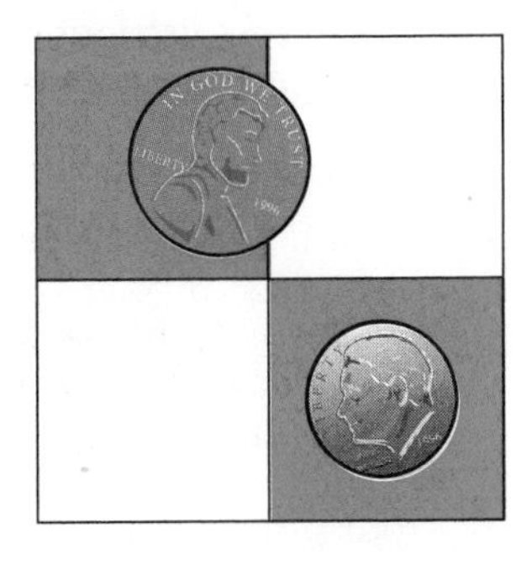

Step 10 Assuming the coin stays on the table, what is the probability of the customer winning by flipping a penny? A dime? (Hint: Where must the center of the coin land in order to win?)

Step 11 If the customer wins only if the coin falls within a red square, what is the probability of winning with a penny? A dime?

Step 12 Suppose the game is always played with a penny, and a customer wins if the penny lands completely inside any square (red or white). If your daily proceeds average $300, about how much will the game cost you per day?

On a Different Note

Two opera stars—Rigoletto and Pollione—are auditioning for a part in an upcoming production. Since the singers have similar qualifications, the director decides to have a contest to see which man can hold a note the longest. Rigoletto has been known to hold a note for any length of time from 6 to 9 minutes. Pollione has been known to hold a note for any length of time between 5 and 7 minutes.

Step 13 Draw a rectangular grid in which the bottom side represents the range of times for Rigoletto and the left side represents the range of times for Pollione. Each point in the rectangle represents one possible outcome of the contest.

Step 14 On your grid, mark all the points that represent a tie. Use your diagram to find the probability that Rigoletto will win the contest.

DIFFERENT DICE

Understanding probability can improve your chances of winning a game. If you roll a pair of standard 6-sided dice, are you more likely to roll a sum of 6 or 12? It's fairly common to roll a sum of 6, since many combinations of two dice add up to 6. But a 12 is only possible if you roll a 6 on each die.

If you rolled a pair of standard 6-sided dice over and over again, and recorded the number of times you got each sum, the histogram would look like this:

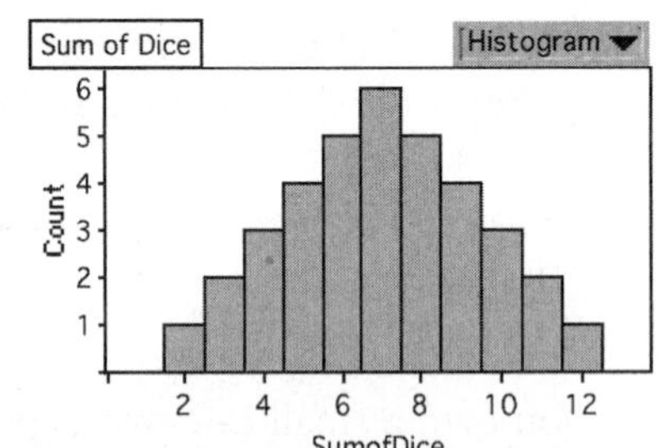

Would the distribution be different if you used different dice? What if one die had odd numbers and the other had even numbers?

What if you used 8-sided dice? What if you rolled three 6-sided dice instead of two?

Choose one of these scenarios or one that you find interesting to investigate. Make your dice and roll them 20 times. Predict what the graph will look like if you roll the dice 100 times, then check your prediction.

Your project should include

- Your dice.
- Histograms of your experimental data, and your predictions and conclusions.

Fathom™

You can use Fathom to simulate real events, such as rolling the dice that you have designed. You can obtain the results of hundreds of events very quickly, and use Fathom's graphing capabilities to make a histogram showing the distribution of different outcomes.

LESSON

8.7

Surface Area

No pessimist ever discovered the secrets of the stars, or sailed to an uncharted land, or opened a new doorway for the human spirit.

HELEN KELLER

In Lesson 8.3, you calculated the surface areas of walls and roofs. But not all building surfaces are rectangular. How would you calculate the amount of glass necessary to cover a pyramid-shaped building? Or the number of tiles needed to cover a cone-shaped roof?

In this lesson, you will learn how to find the surface areas of prisms, pyramids, cylinders, and cones. The **surface area** of each of these solids is the sum of the areas of all the faces or surfaces that enclose the solid. For prisms and pyramids, the faces include the solid's **bases** and its remaining **lateral faces.**

In a prism, the bases are two congruent polygons and the lateral faces are rectangles or parallelograms.

In a pyramid, the base can be any polygon. The lateral faces are triangles.

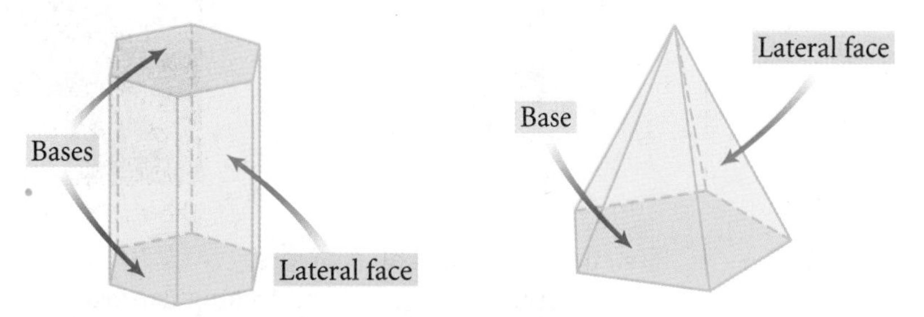

This glass pyramid was designed by I. M. Pei for the entrance of the Louvre museum in Paris, France.

A cone is part of the roof design of this Victorian house in Massachusetts.

This skyscraper in Chicago, Illinois, is an example of a prism.

This stone tower is a cylinder on top of a larger cylinder.

To find the surface areas of prisms and pyramids, follow these steps.

Steps for Finding Surface Area

1. Draw and label each face of the solid as if you had cut the solid apart along its edges and laid it flat. Label the dimensions.
2. Calculate the area of each face. If some faces are identical, you only need to find the area of one.
3. Find the total area of all the faces.

EXAMPLE A

Find the surface area of the rectangular prism.

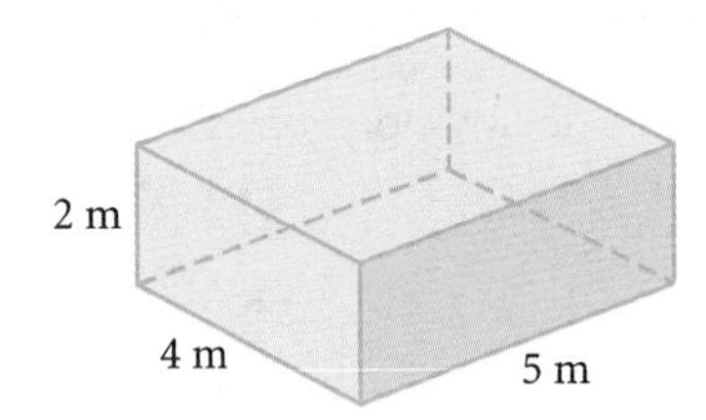

These shipping containers are rectangular prisms.

Solution

First, draw and label all six faces.
Then, find the areas of all the rectangular faces.

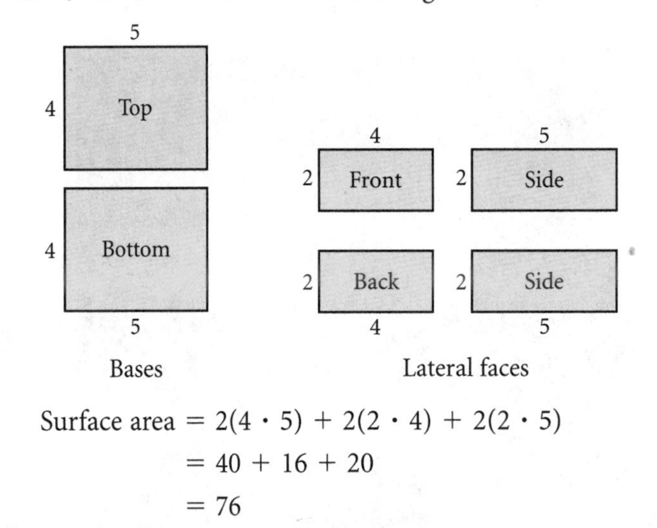

$$\begin{aligned}\text{Surface area} &= 2(4 \cdot 5) + 2(2 \cdot 4) + 2(2 \cdot 5)\\ &= 40 + 16 + 20\\ &= 76\end{aligned}$$

The surface area of the prism is 76 m^2.

EXAMPLE B

Find the surface area of the cylinder.

Solution

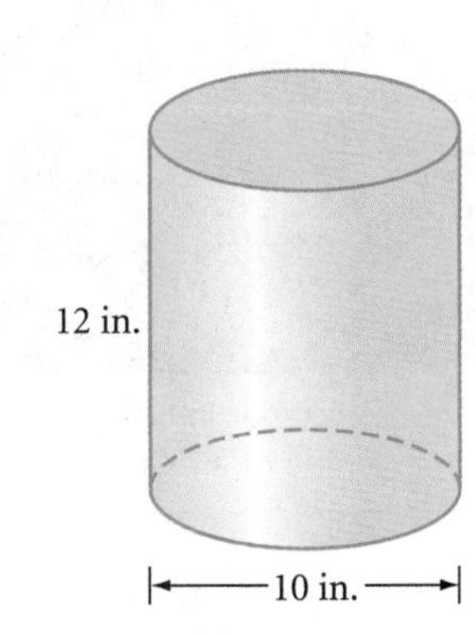

Imagine cutting apart the cylinder. The two bases are circular regions, so you need to find the areas of two circles. Think of the lateral surface as a wrapper. Slice it and lay it flat to get a rectangular region. You'll need the area of this rectangle. The height of the rectangle is the height of the cylinder. The base of the rectangle is the circumference of the circular base.

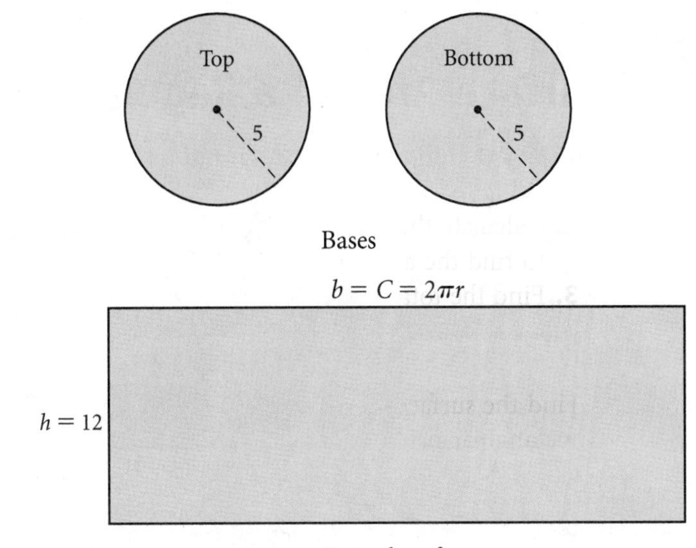

$$\begin{aligned}\text{Surface area} &= 2(\pi r^2) + (2\pi r)h \\ &= 2(\pi \cdot 5^2) + (2 \cdot \pi \cdot 5) \cdot 12 \\ &\approx 534\end{aligned}$$

The surface area of the cylinder is about 534 in^2.

This ice cream plant in Burlington, Vermont, uses cylindrical containers for its milk and cream.

These conservatories in Edmonton, Canada, are glass pyramids.

The surface area of a pyramid is the area of the base plus the areas of the triangular faces. The height of each triangular lateral face is called the **slant height.** To avoid confusing slant height with the height of the pyramid, use l rather than h for slant height.

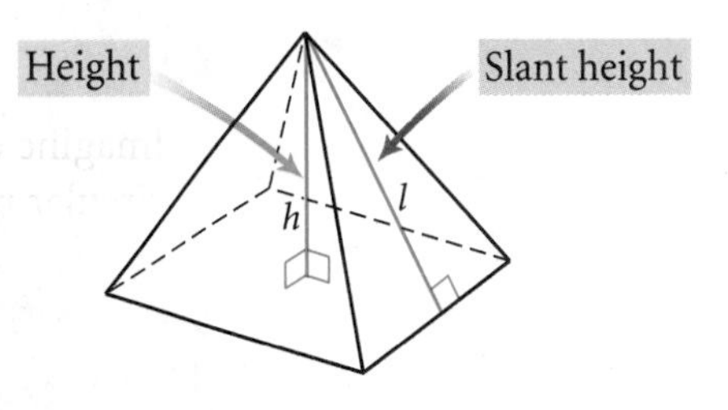

In the investigation you'll find out how to calculate the surface area of a pyramid with a regular polygon base.

Investigation 1

Surface Area of a Regular Pyramid

You can cut and unfold the surface of a regular pyramid into these shapes.

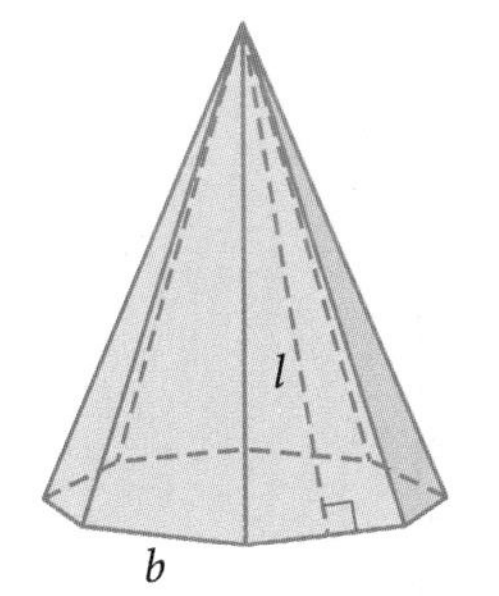

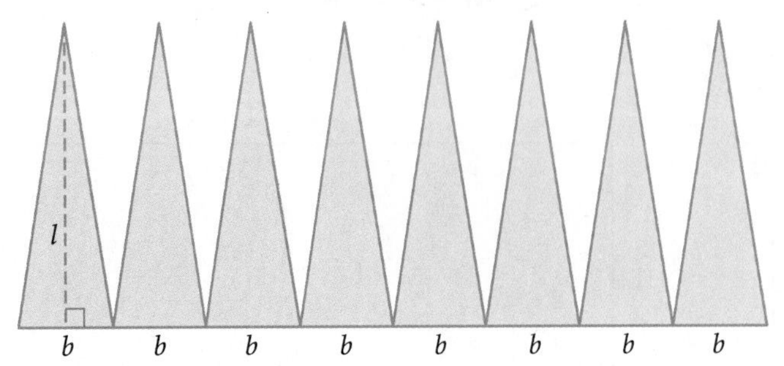

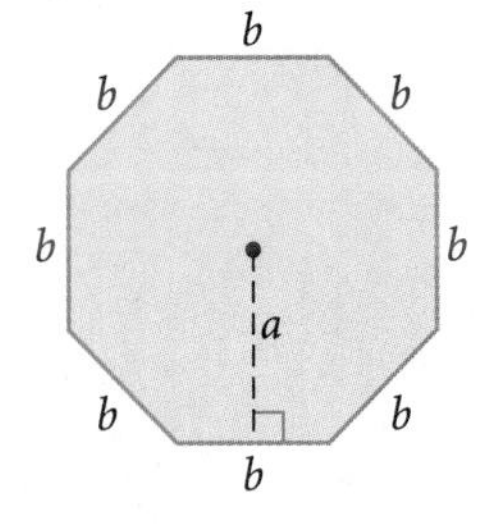

Step 1 What is the area of each lateral face?

Step 2 What is the total lateral surface area? What is the total lateral surface area for any pyramid with a regular n-gon base?

Step 3 What is the area of the base for any regular n-gon pyramid?

Step 4 Use your expressions from Steps 2 and 3 to write a formula for the surface area of a regular n-gon pyramid in terms of base length b, slant height l, and apothem a.

Step 5 Write another expression for the surface area of a regular n-gon pyramid in terms of height l, apothem a, and perimeter of the base, P.

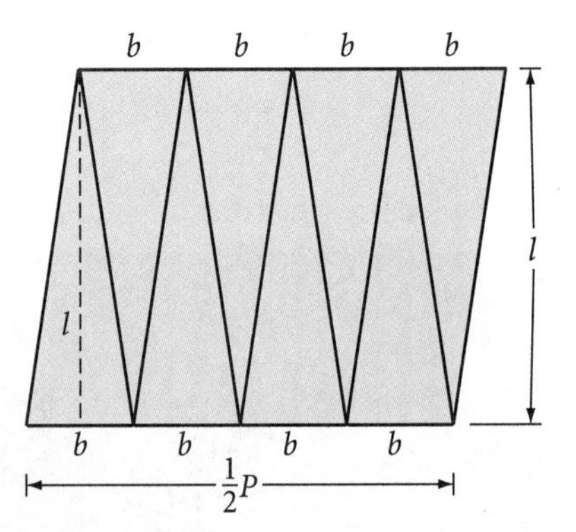

You can find the surface area of a cone using a method similar to the one you used to find the surface area of a pyramid.

Is the roof of this building in Kashan, Iran, a cone or a pyramid? What makes it hard to tell?

Investigation 2

Surface Area of a Cone

As the number of faces of a pyramid increases, it begins to look like a cone. You can think of the lateral surface as many small triangles, or as a sector of a circle.

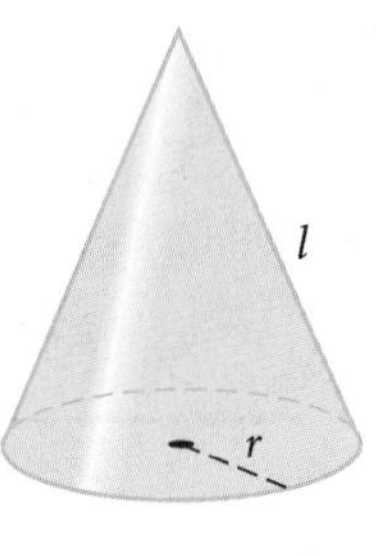

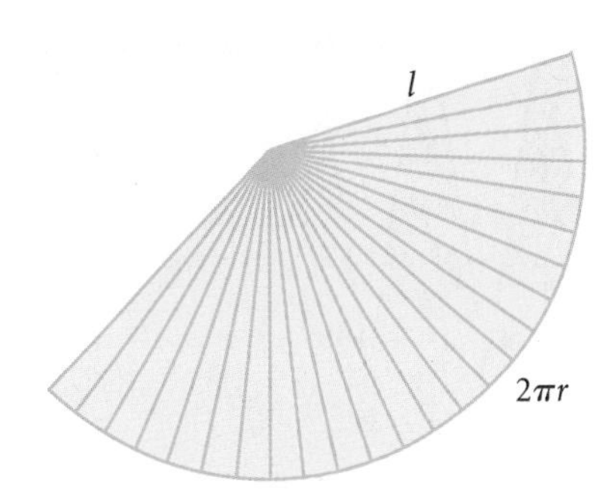

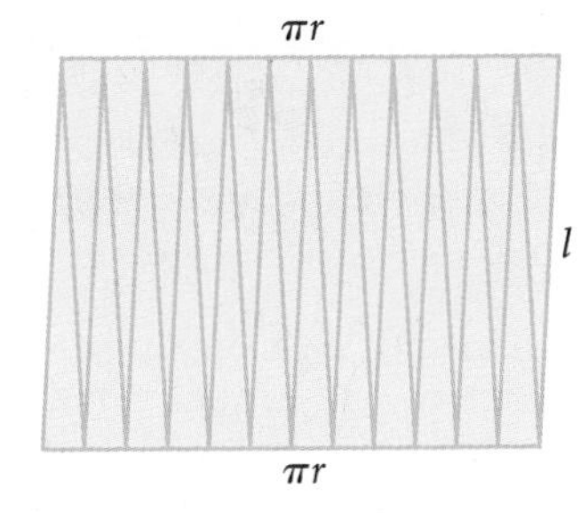

Step 1 What is the area of the base?

Step 2 What is the lateral surface area in terms of l and r? What portion is the sector of the circle? What is the area of the sector?

Step 3 Write the formula for the surface area of a cone.

This photograph of Sioux tepees was taken around 1902 in North or South Dakota. Is a tepee shaped more like a cone or a pyramid?

EXAMPLE C Find the total surface area of the cone.

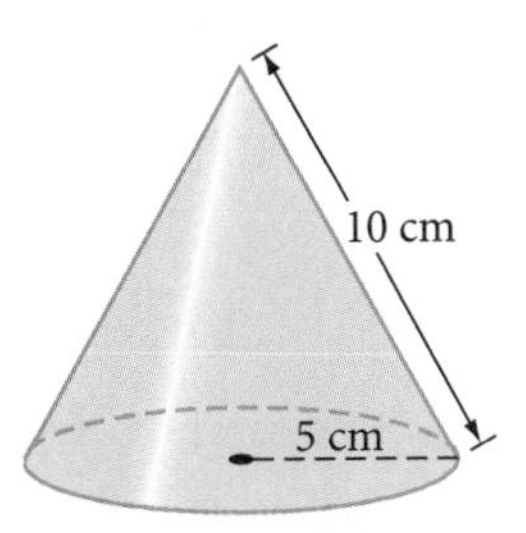

▶ Solution

$$\begin{aligned} SA &= \pi r l + \pi r^2 \\ &= (\pi)(5)(10) + \pi(5)^2 \\ &= 75\pi \\ &\approx 235.6 \end{aligned}$$

The surface area of the cone is about 236 cm^2.

EXERCISES

In Exercises 1–10, find the surface area of each solid. All quadrilaterals are rectangles, and all given measurements are in centimeters. Round your answers to the nearest 0.1 cm^2.

1.

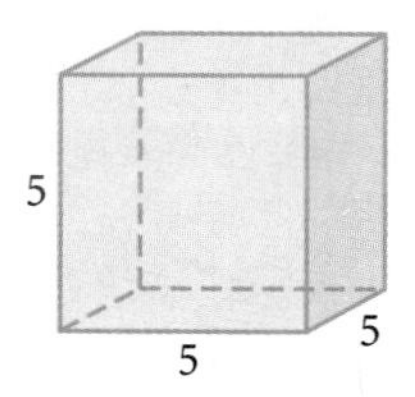

2.

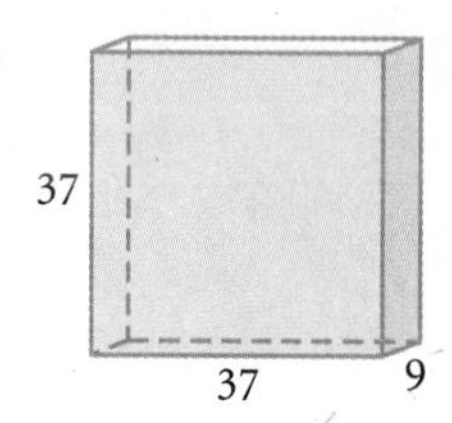

3.

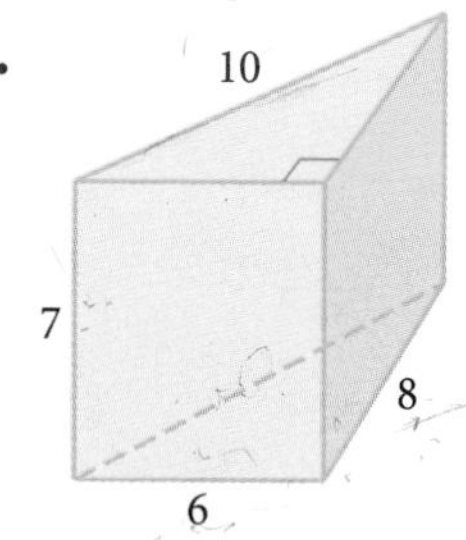

4.

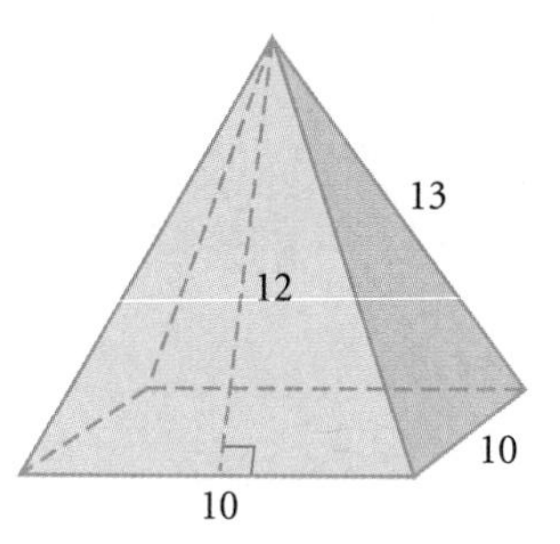

5.

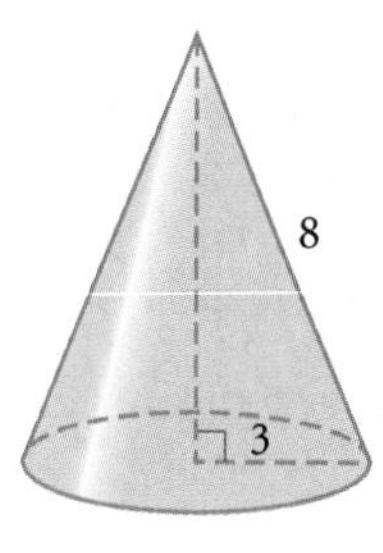

6.

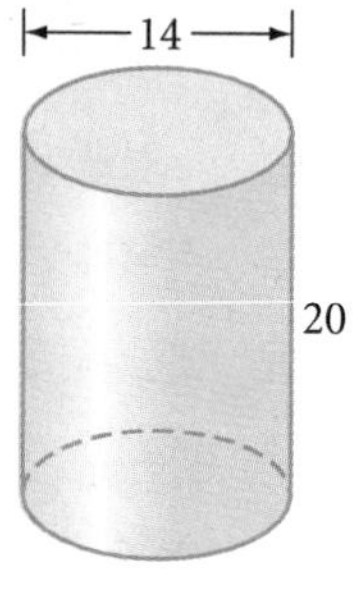

7. The base is a regular hexagon with apothem $a = 12.1$, side $s = 14$, and height $h = 7$. ⓗ

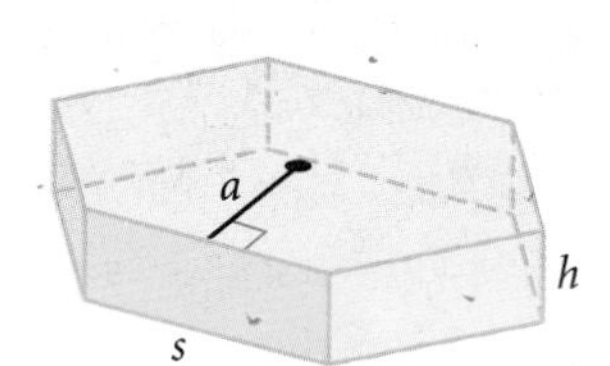

8. The base is a regular pentagon with apothem $a = 11$ and side $s = 16$. Each lateral edge $t = 17$, and the height of a face $l = 15$.

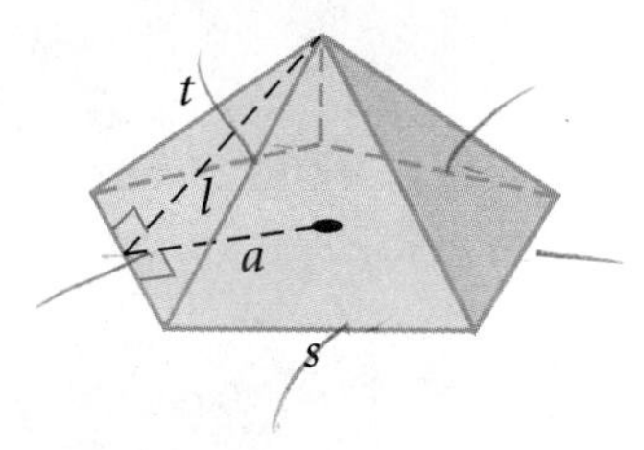

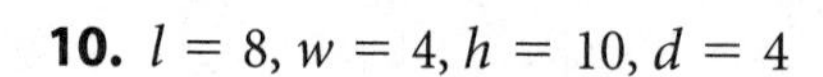

9. $D = 8$, $d = 4$, $h = 9$ ⓗ

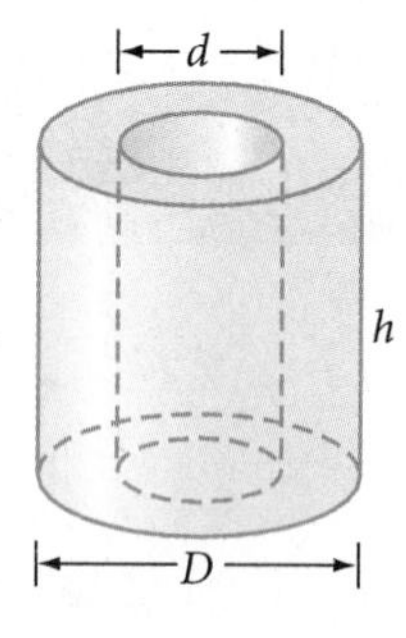

10. $l = 8$, $w = 4$, $h = 10$, $d = 4$

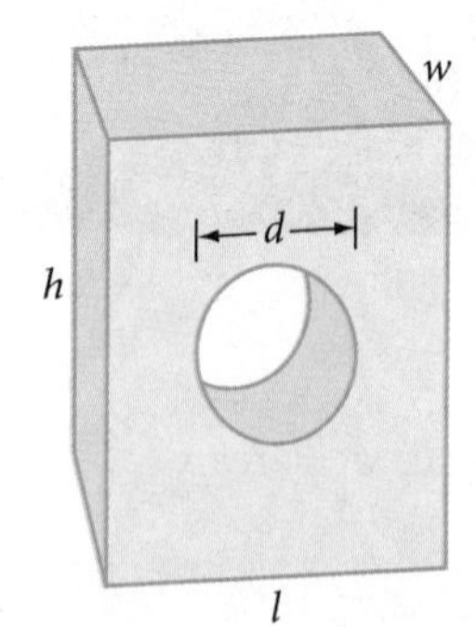

11. Explain how you would find the surface area of this obelisk.

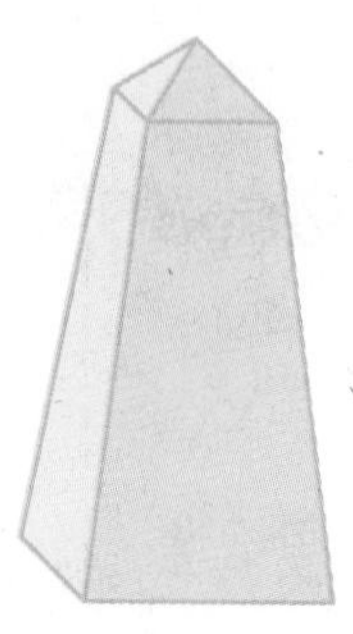

12. APPLICATION Claudette and Marie are planning to paint the exterior walls of their country farmhouse (all vertical surfaces) and to put new cedar shingles on the roof. The paint they like best costs \$25 per gallon and covers 250 square feet per gallon. The wood shingles cost \$65 per bundle, and each bundle covers 100 square feet. How much will this home improvement cost? All measurements are in feet.

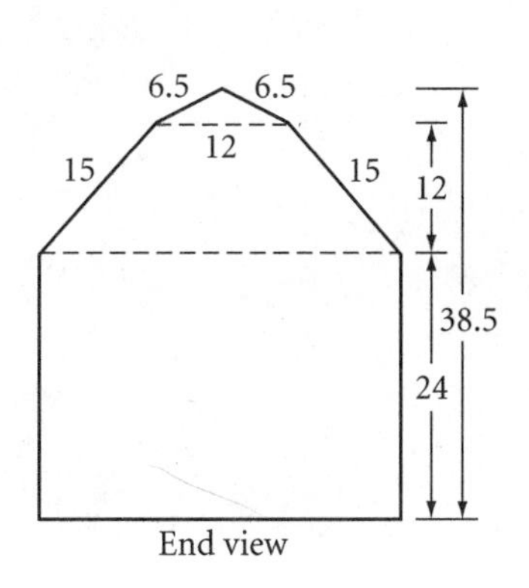

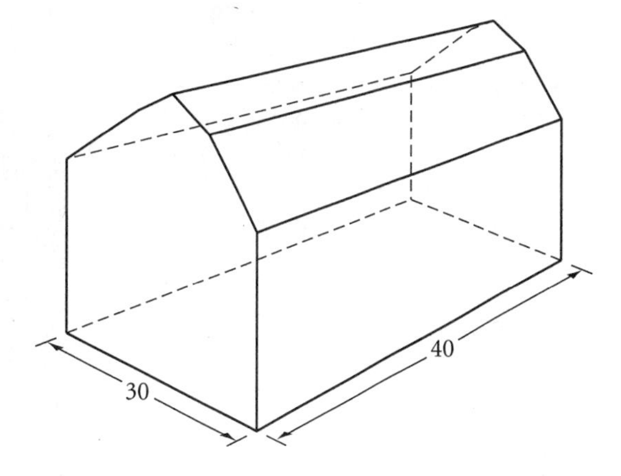

13. *Construction* The shapes of the spinning dishes in the photo are called **frustums** of cones. Think of them as cones with their tops cut off. Use your construction tools to draw pieces that you can cut out and tape together to form a frustum of a cone.

A Sri Lankan dancer balances and spins plates.

Review

14. Use patty paper, templates, or pattern blocks to create a $3^3.4^2/3^2.4.3.4/4^4$ tiling.

15. Trace the figure at right. Find the lettered angle measures and arc measures.

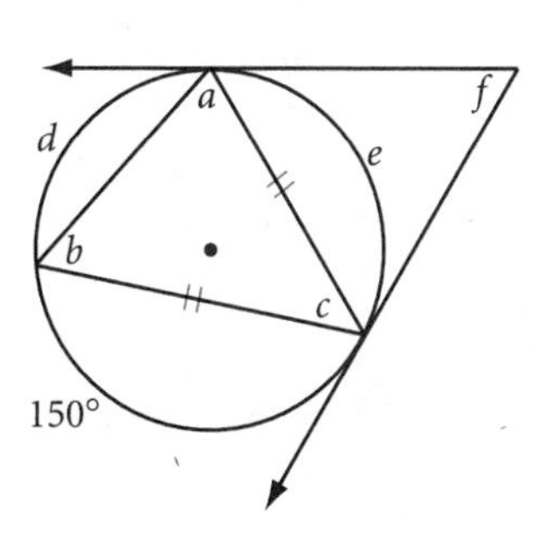

16. APPLICATION Suppose a circular ranch with a radius of 3 km was divided into 16 congruent sectors. In a one-year cycle, how long would the cattle graze in each sector? What would be the area of each sector?

History CONNECTION

In 1792, visiting Europeans presented horses and cattle to Hawaii's King Kamehameha I. Cattle ranching soon developed when Mexican *vaqueros* came to Hawaii to train Hawaiians in ranching. Today, Hawaiian cattle ranching is big business.

"Grazing geometry" is used on Hawaii's Kahua Ranch. Ranchers divide the grazing area into sectors. The cows are rotated through each sector in turn. By the time they return to the first sector, the grass has grown back and the cycle repeats.

17. Trace the figure at right. Find the lettered angle measures.

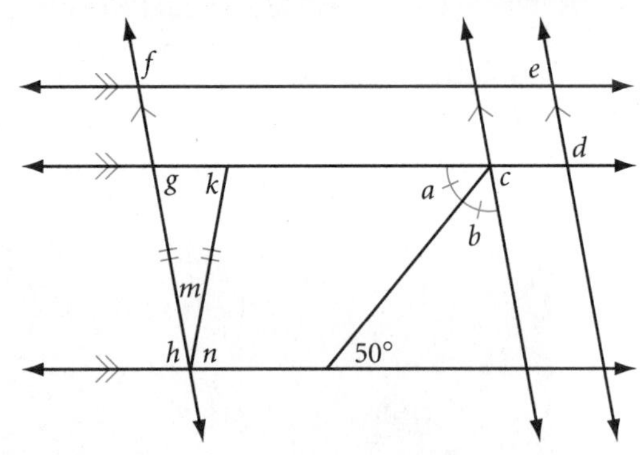

18. If the pattern of blocks continues, what will be the surface area of the 50th solid in the pattern? (Every edge of each block has length 1 unit.)

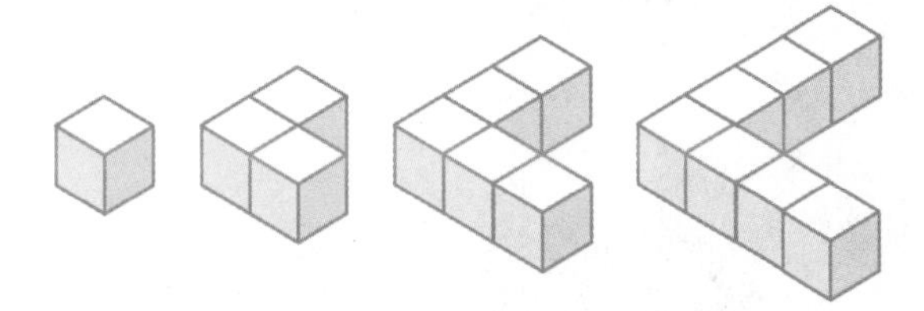

IMPROVING YOUR VISUAL THINKING SKILLS

Moving Coins

Create a triangle of coins similar to the one shown. How can you move exactly three coins so that the triangle is pointing down rather than up? When you have found a solution, use a diagram to explain it.

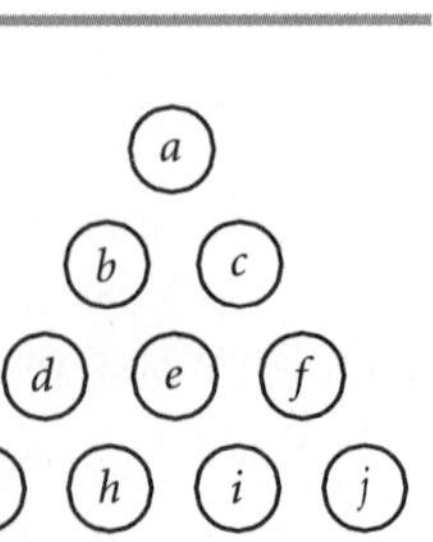

Exploration

Alternative Area Formulas

In ancient Egypt, when the yearly floods of the Nile River receded, the river often followed a different course, so the shape of farmers' fields along the banks could change from year to year. Officials then needed to measure property areas, in order to keep records and calculate taxes. Partly to keep track of land and finances, ancient Egyptians developed some of the earliest mathematics.

Historians believe that ancient Egyptian tax assessors used this formula to find the area of any quadrilateral:

$$A = \frac{1}{2}(a + c) \cdot \frac{1}{2}(b + d)$$

where *a, b, c,* and *d* are the lengths, in consecutive order, of the figure's four sides.

In this activity, you will take a closer look at this ancient Egyptian formula, and another formula called Hero's formula, named after Hero of Alexandria.

Activity
Calculating Area in Ancient Egypt

Investigate the ancient Egyptian formula for quadrilaterals.

Step 1 Construct a quadrilateral and its interior.

Step 2 Change the labels of the sides to *a, b, c,* and *d,* consecutively.

Step 3 Measure the lengths of the sides and use the Sketchpad calculator to find the area according to the ancient Egyptian formula.

Step 4 Select the polygon interior and measure its area. How does the area given by the formula compare to the actual area? Is the ancient Egyptian formula correct?

Step 5 Does the ancient Egyptian formula always favor either the tax collector or the landowner, or does it favor one in some cases and the other in other cases? Explain.

Step 6 Describe the quadrilaterals for which the formula works accurately. For what kinds of quadrilaterals is it slightly inaccurate? Very inaccurate?

Step 7 State the ancient Egyptian formula in words, using the word *mean.*

Step 8 According to Hero's formula, if s is half the perimeter of a triangle with side lengths a, b, and c, the area A is given by the formula

$$A = \sqrt{s(s-a)(s-b)(s-c)}$$

Use Sketchpad to investigate Hero's formula. Construct a triangle and its interior. Label the sides a, b, and c, and use the Sketchpad calculator to find the triangle's area according to Hero. Compare the result to the measured area of the triangle. Does Hero's formula work for all triangles?

Step 9 Devise a way of calculating the area of any quadrilateral. Use Sketchpad to test your method.

IMPROVING YOUR VISUAL THINKING SKILLS

PUZZLE

Cover the Square

Trace each diagram below onto another sheet of paper.

Cut out the four triangles in each of the two small equal squares and arrange them to exactly cover the large square.

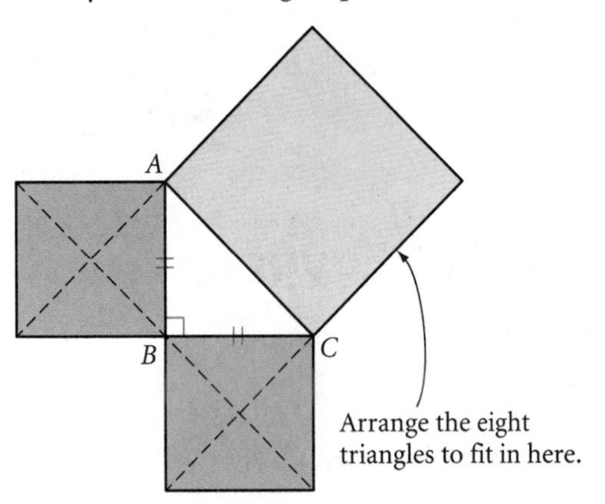

Cut out the small square and the four triangles from the square on leg $\overline{EF}$ and arrange them to exactly cover the large square.

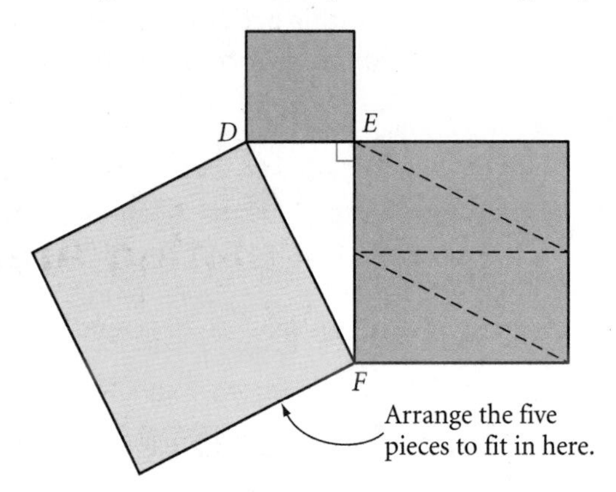

Two squares with areas x^2 and y^2 are divided into the five regions as shown. Cut out the five regions and arrange them to exactly cover a larger square with an area of z^2.

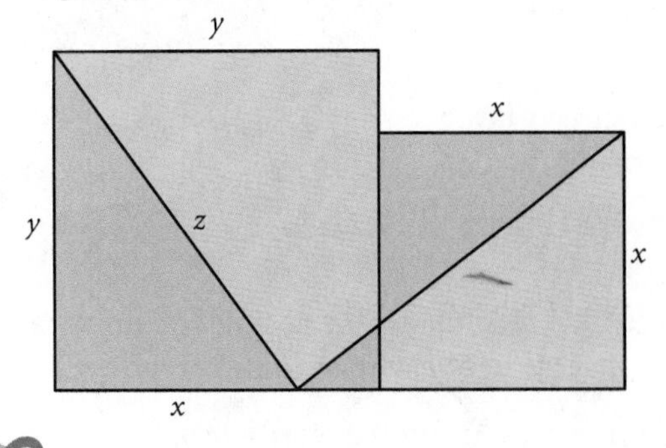

Two squares have been divided into three right triangles and two quadrilaterals. Cut out the five regions and arrange them to exactly cover a larger square.

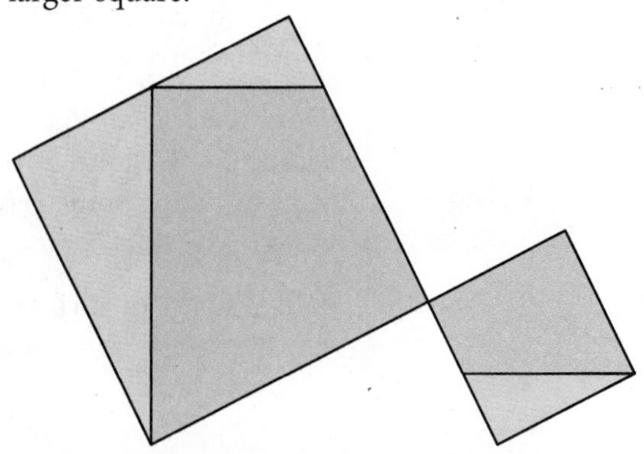

CHAPTER 8 REVIEW

You should know area formulas for rectangles, parallelograms, triangles, trapezoids, regular polygons, and circles. You should also be able to show where these formulas come from and how they're related to one another. Most importantly, you should be able to apply them to solve practical problems involving area, including the surface areas of solid figures. What occupations can you list that use area formulas?

When you use area formulas for real-world applications, you have to consider units of measurement and accuracy. Should you use inches, feet, centimeters, meters, or some other unit? If you work with a circle or a regular polygon, is your answer exact or an approximation?

EXERCISES

For Exercises 1–10, match the area formula with the shaded area.

1. $A = bh$
2. $A = 0.5bh$
3. $A = 0.5h(b_1 + b_2)$
4. $A = 0.5d_1d_2$
5. $A = 0.5aP$
6. $A = \pi r^2$
7. $A = \left(\frac{x}{360}\right)\pi r^2$
8. $A = \pi(R^2 - r^2)$
9. $SA = 2\pi rl + 2\pi r^2$
10. $LA = \pi rl$

A.  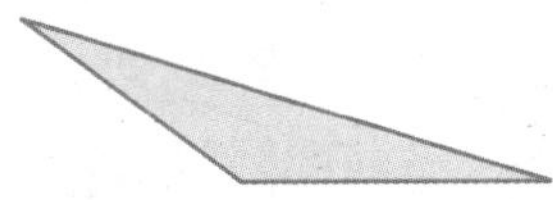

B. 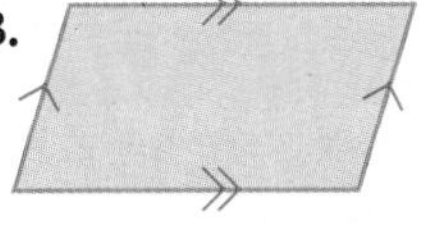

C. 

D.  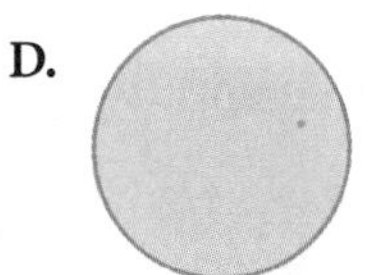

E. 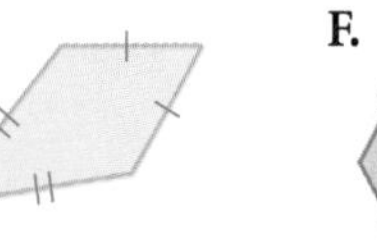

F. 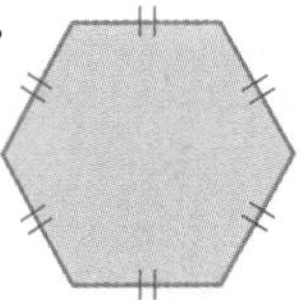

G. 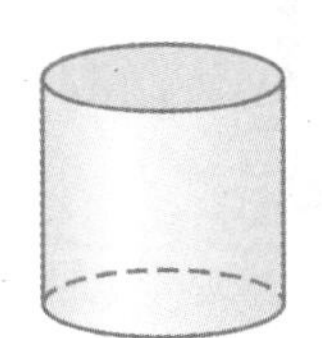

H. 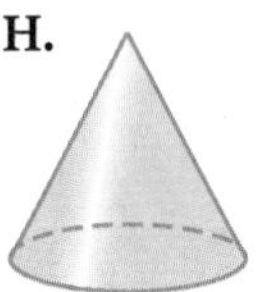

I. 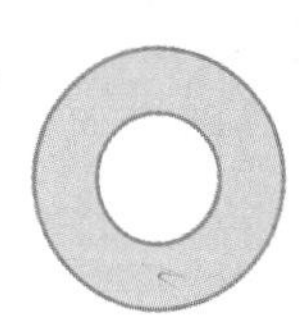

J. 

For Exercises 11–13, illustrate each term.

11. Apothem **12.** Annulus **13.** Sector of a circle

For Exercises 14–16, draw a diagram and explain in a paragraph how you derived the area formula for each figure.

14. Parallelogram **15.** Trapezoid **16.** Circle

Solve for the unknown measures in Exercises 17–25. All measurements are in centimeters.

17. $A = \underline{?}$

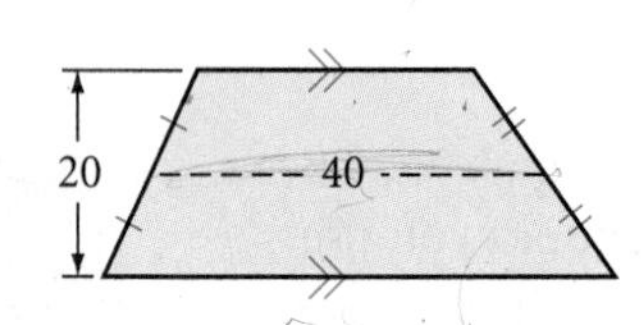

18. $A = \underline{?}$
$a = 36$
$s = 41.6$

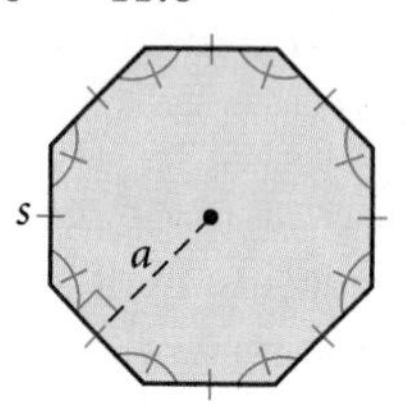

19. $A = \underline{?}$
$R = 8$
$r = 2$

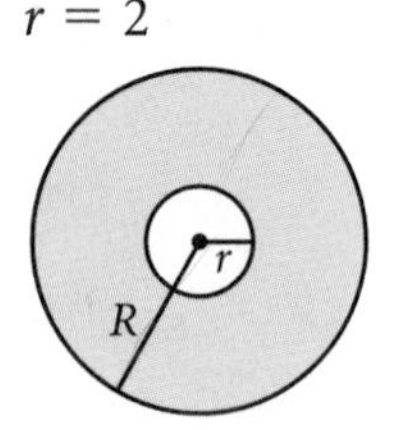

20. $A = 576\text{ cm}^2$
$h = \underline{?}$

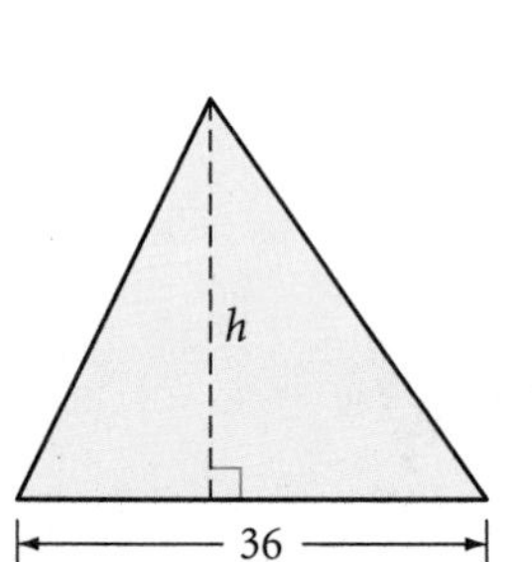

21. $A = 576\text{ cm}^2$
$d_1 = \underline{?}$

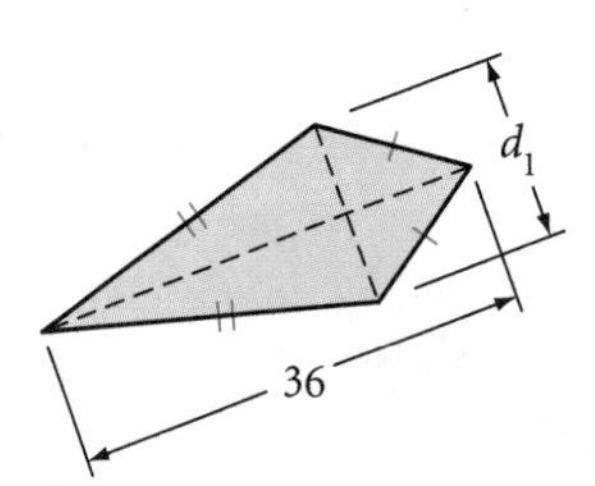

22. $A = 126\text{ cm}^2$
$a = 13\text{ cm}$
$h = 9\text{ cm}$
$b = \underline{?}$

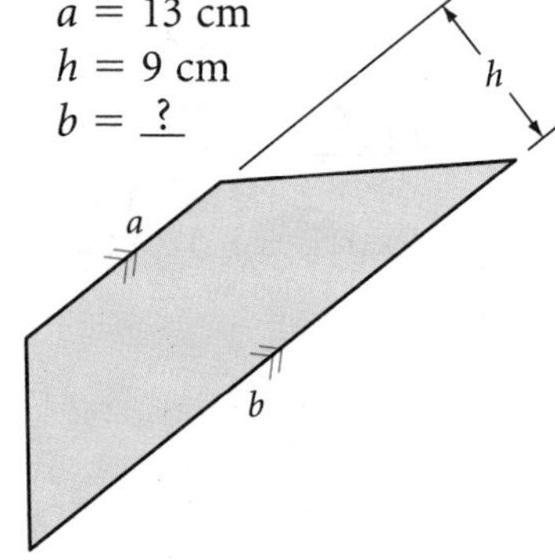

23. $C = 18\pi\text{ cm}$
$A = \underline{?}$

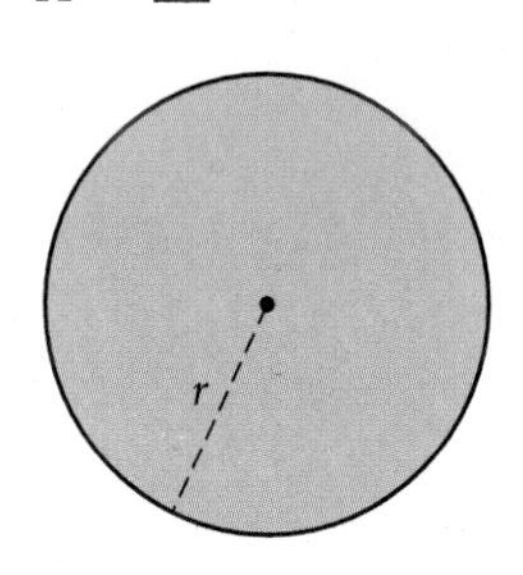

24. $A = 576\pi\text{ cm}^2$
The circumference is $\underline{?}$.

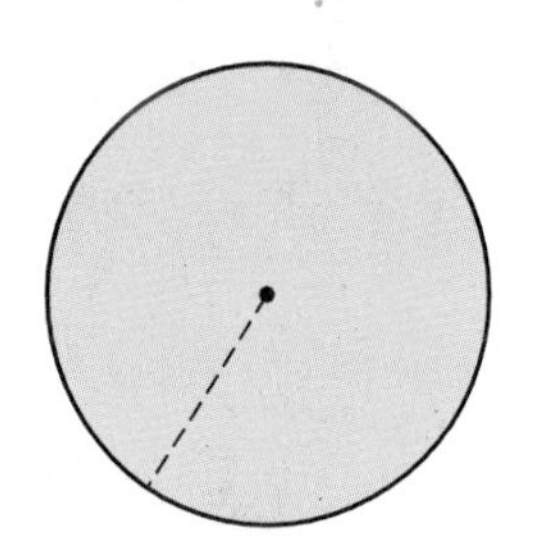

25. $A_{\text{sector}} = 16\pi\text{ cm}^2$
$m\angle FAN = \underline{?}$.

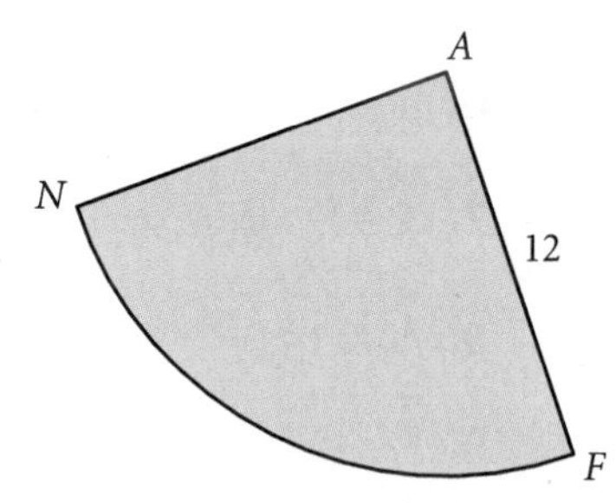

In Exercises 26–28, find the shaded area to the nearest 0.1 cm². In Exercises 27 and 28, the quadrilateral is a square and all arcs are arcs of a circle of radius 6 cm.

26.

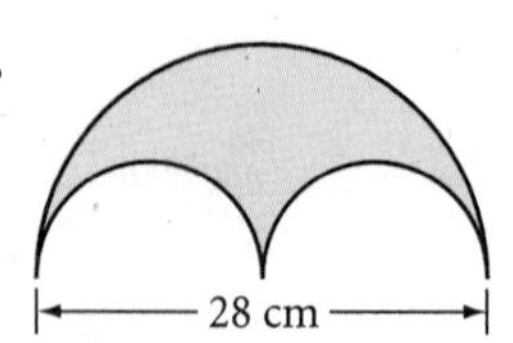

27.

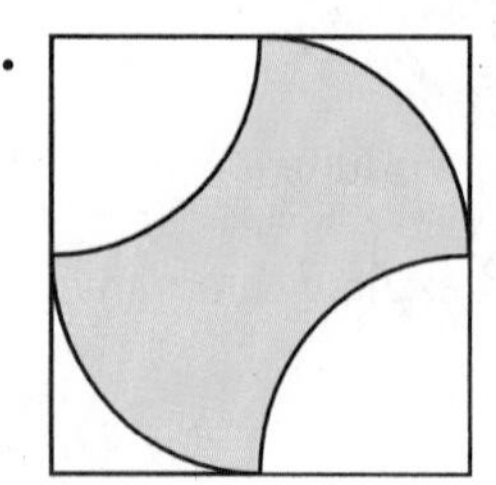

28.

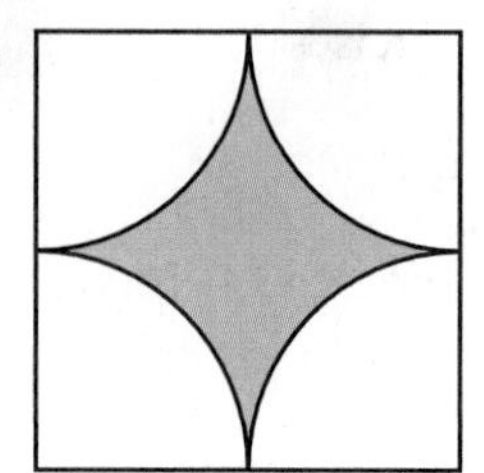

In Exercises 29–31, find the surface area of each prism or pyramid. All given measurements are in centimeters. All quadrilaterals are rectangles, unless otherwise labeled.

29.

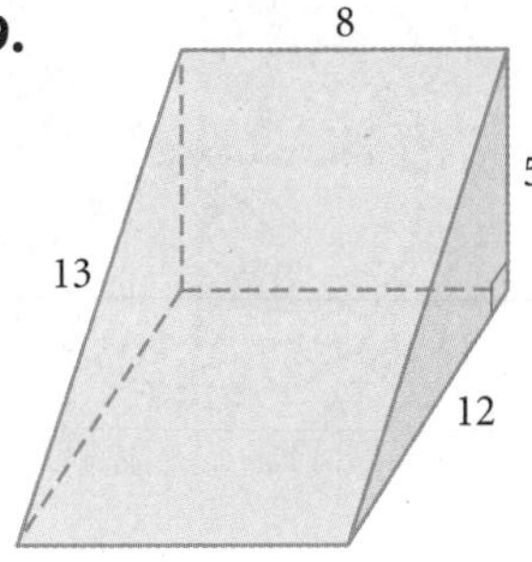

30. The base is a trapezoid.

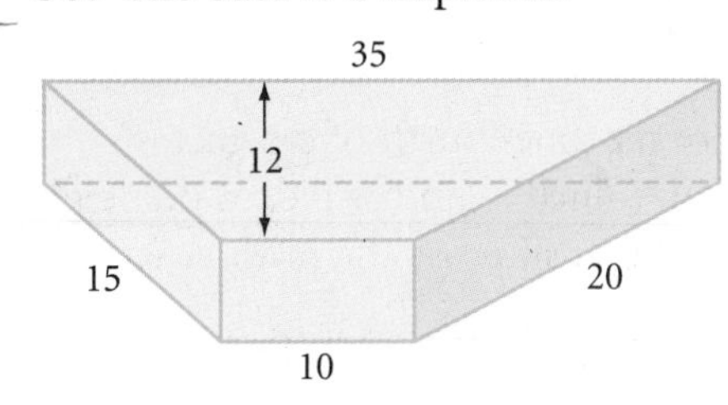

31.

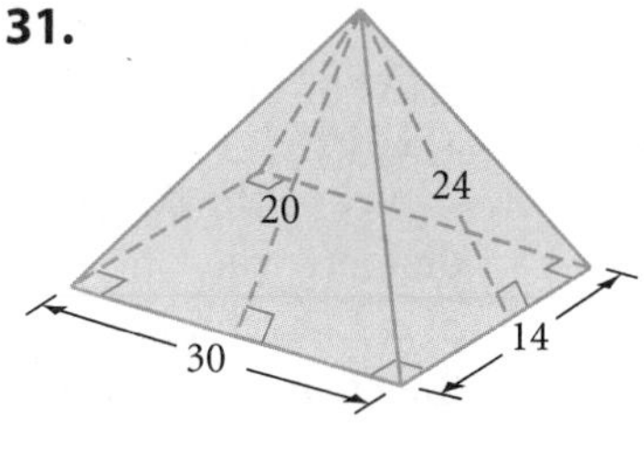

For Exercises 32 and 33, plot the vertices of each figure on graph paper, then find its area.

32. Parallelogram $ABCD$ with $A(0, 0)$, $B(14, 0)$, and $D(6, 8)$

33. Quadrilateral $FOUR$ with $F(0, 0)$, $O(4, -3)$, $U(9, 5)$, and $R(4, 15)$

34. The sum of the lengths of the two bases of a trapezoid is 22 cm, and its area is 66cm^2. What is the height of the trapezoid?

35. Find the area of a regular pentagon to the nearest tenth of a square centimeter if the apothem measures 6.9 cm and each side measures 10 cm.

36. Find three noncongruent polygons, each with an area of 24 square units, on a 6-by-6 geoboard or a 6-by-6 square dot grid.

37. Lancelot wants to make a pen for his pet, Isosceles. What is the area of the largest rectangular pen that Lancelot can make with 100 meters of fencing if he uses a straight wall of the castle for one side of the pen? ⓗ

38. If you have a hundred feet of rope to arrange into the perimeter of either a square or a circle, which shape will give you the maximum area? Explain.

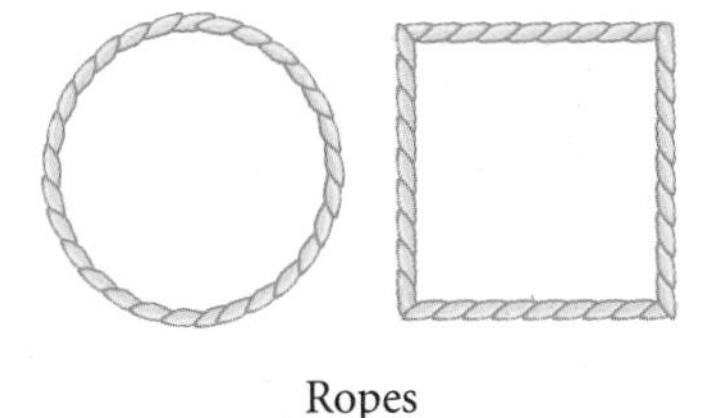

Ropes

39. Which is a better (tighter) fit: a round peg in a square hole or a square peg in a round hole? ⓗ

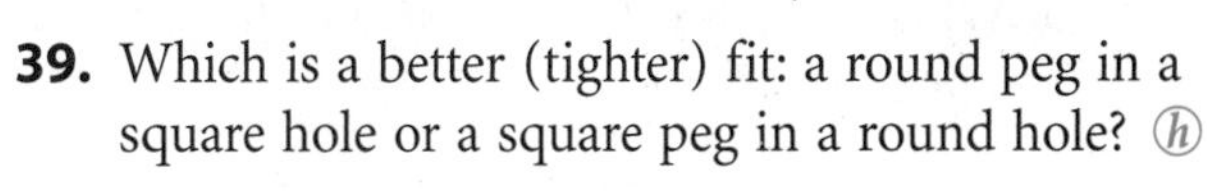

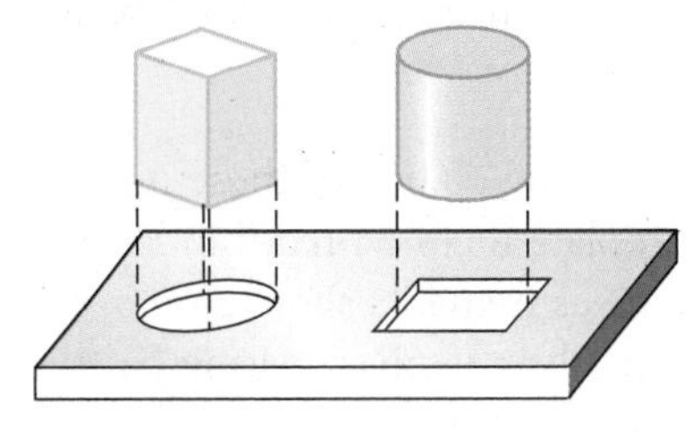

40. Al Dente's Pizzeria sells pizza by the slice, according to the sign. Which slice is the best deal (the most pizza per dollar)?

41. If you need 8 oz of dough to make a 12-inch diameter pizza, how much dough will you need to make a 16-inch pizza on a crust of the same thickness?

42. Which is the biggest slice of pie: one-fourth of a 6-inch diameter pie, one-sixth of an 8-inch diameter pie, or one-eighth of a 12-inch diameter pie? Which slice has the most crust along the curved edge?

43. The Hot-Air Balloon Club at Da Vinci High School has designed a balloon for the annual race. The panels are a regular octagon, eight squares, and sixteen isosceles trapezoids, and club members will sew them together to construct the balloon. They have built a scale model, as shown at right. The dimensions of three of the four types of panels are below, shown in feet. ⓗ

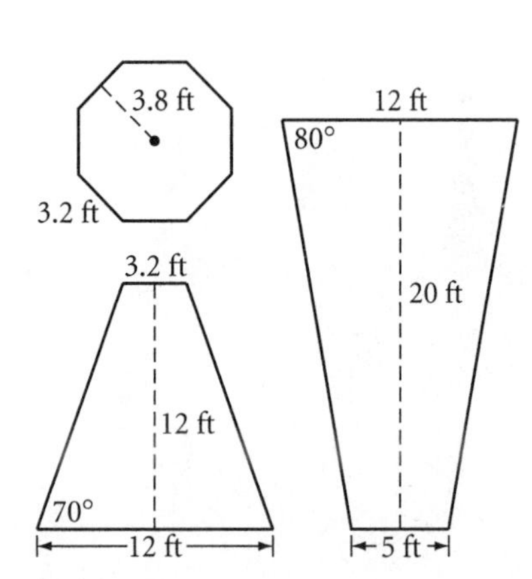

a. What will be the perimeter, to the nearest foot, of the balloon at its widest? What will be the perimeter, to the nearest foot, of the opening at the bottom of the balloon?

b. What is the total surface area of the balloon to the nearest square foot?

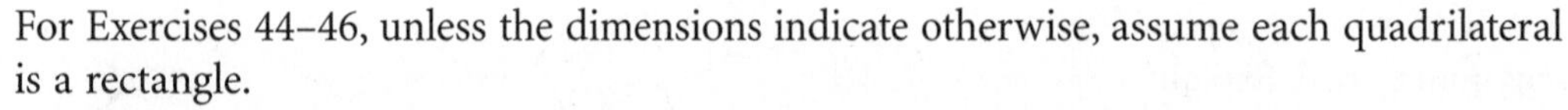

For Exercises 44–46, unless the dimensions indicate otherwise, assume each quadrilateral is a rectangle.

44. You are producing 10,000 of these metal wedges, and you must electroplate them with a thin layer of high-conducting silver. The measurements shown are in centimeters. Find the total cost for silver, if silver plating costs $1 for each 200 square centimeters.

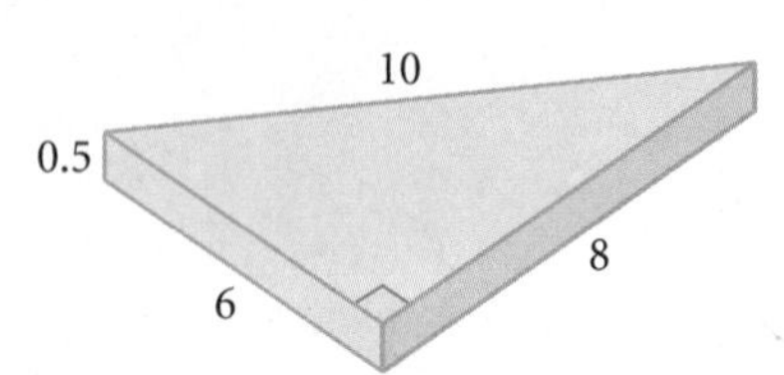

45. The measurements of a chemical storage container are shown in meters. Find the cost of painting the exterior of nine of these large cylindrical containers with sealant. The sealant costs $32 per gallon. Each gallon covers 18 square meters. Do not paint the bottom faces.

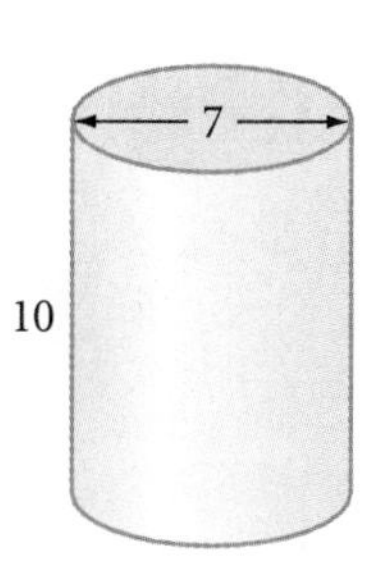

46. The measurements of a copper cone are shown in inches. Find the cost of spraying an oxidizer on 100 of these copper cones. The oxidizer costs $26 per pint. Each pint covers approximately 5000 square inches. Spray only the lateral surface.

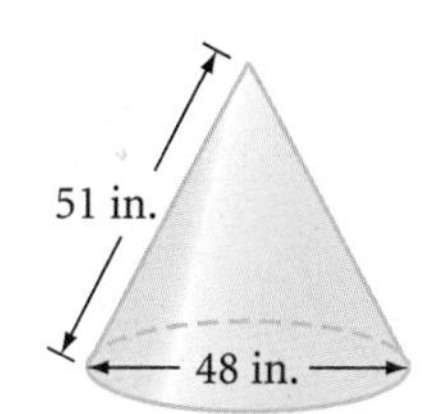

47. Hector is a very cost-conscious produce buyer. He usually buys asparagus in large bundles, each 44 cm in circumference. But today there are only small bundles that are 22 cm in circumference. Two 22 cm bundles are the same price as one 44 cm bundle. Is this a good deal or a bad deal? Why?

TAKE ANOTHER LOOK

1. Use geometry software to construct these shapes.

a. A triangle whose perimeter can vary, but whose area stays constant

b. A parallelogram whose perimeter can vary, but whose area stays constant

2. True or false? The area of a triangle is equal to half the perimeter of the triangle times the radius of the inscribed circle. Support your conclusion with a convincing argument.

3. Does the area formula for a kite hold for a dart (a concave kite)? Support your conclusion with a convincing argument.

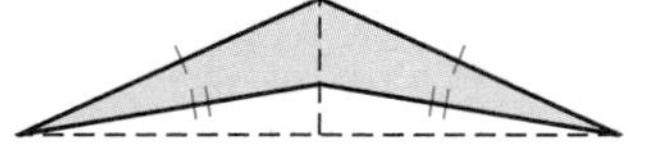

4. How can you use the Regular Polygon Area Conjecture to arrive at a formula for the area of a circle? Use a series of diagrams to help explain your reasoning.

5. Use algebra to show that the total surface area of a prism with a regular polygon base is given by the formula $SA = P(h + a)$, where h is height of the prism, a is the apothem of the base, and P is the perimeter of the base.

6. Use algebra to show that the total surface area of a cylinder is given by the formula $SA = C(h + r)$, where h is the height of the cylinder, r is the radius of the base, and C is the circumference of the base.

7. Here is a different formula for the area of a trapezoid: $A = mh$, where m is the length of the midsegment and h is the height. Does the formula work? Use algebra or a diagram to explain why or why not. Does it work for a triangle?

Assessing What You've Learned

UPDATE YOUR PORTFOLIO Choose one of the more challenging problems you did in this chapter and add it to your portfolio. Write about why you chose it, what made it challenging, what strategies you used to solve it, and what you learned from it.

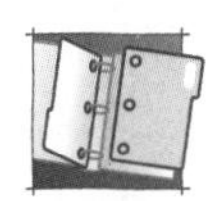

ORGANIZE YOUR NOTEBOOK Review your notebook to be sure it's complete and well organized. Be sure you have included all of this chapter's area formulas in your conjecture list. Write a one-page chapter summary.

WRITE IN YOUR JOURNAL Imagine yourself five or ten years from now, looking back on the influence this geometry class had on your life. How do you think you'll be using geometry? Will this course have influenced your academic or career goals?

PERFORMANCE ASSESSMENT While a classmate, a friend, a family member, or a teacher observes, demonstrate how to derive one or more of the area formulas. Explain each step, including how you arrive at the formula.

WRITE TEST ITEMS Work with group members to write test items for this chapter. Include simple exercises and complex application problems.

GIVE A PRESENTATION Create a poster, a model, or other visual aid, and give a presentation on how to derive one or more of the area formulas. Or present your findings from one of the Take Another Look activities.

CHAPTER

9 The Pythagorean Theorem

But serving up an action, suggesting the dynamic in the static, has become a hobby of mine . . . . The "flowing" on that motionless plane holds my attention to such a degree that my preference is to try and make it into a cycle.

M. C. ESCHER

Waterfall, M. C. Escher, 1961

OBJECTIVES

In this chapter you will

- discover the Pythagorean Theorem, one of the most important concepts in mathematics
- use the Pythagorean Theorem to calculate the distance between any two points
- use conjectures related to the Pythagorean Theorem to solve problems

LESSON

9.1 The Theorem of Pythagoras

I am not young enough to know everything.

OSCAR WILDE

In a right triangle, the side opposite the right angle is called the **hypotenuse.** The other two sides are called **legs.** In the figure at right, a and b represent the lengths of the legs, and c represents the length of the hypotenuse.

In a right triangle, the side opposite the right angle is called the **hypotenuse,** here with length c.

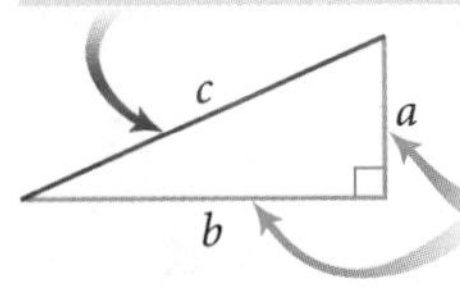

The other two sides are **legs,** here with lengths a and b.

FUNKY WINKERBEAN by Batiuk. Reprinted with special permission of North America Syndicate.

There is a special relationship between the lengths of the legs and the length of the hypotenuse. This relationship is known today as the **Pythagorean Theorem.**

Investigation
The Three Sides of a Right Triangle

You will need

- scissors
- a compass
- a straightedge
- patty paper

The puzzle in this investigation is intended to help you recall the Pythagorean Theorem. It uses a **dissection,** which means you will cut apart one or more geometric figures and make the pieces fit into another figure.

Step 1 Construct a scalene right triangle in the middle of your paper. Label the hypotenuse c and the legs a and b. Construct a square on each side of the triangle.

Step 2 To locate the center of the square on the longer leg, draw its diagonals. Label the center O.

Step 3 Through point O, construct line j perpendicular to the hypotenuse and line k perpendicular to line j. Line k is parallel to the hypotenuse. Lines j and k divide the square on the longer leg into four parts.

Step 4 Cut out the square on the shorter leg and the four parts of the square on the longer leg. Arrange them to exactly cover the square on the hypotenuse.

Step 5 | State the Pythagorean Theorem.

The Pythagorean Theorem C-82

In a right triangle, the sum of the squares of the lengths of the legs equals the square of the length of the hypotenuse. If a and b are the lengths of the legs, and c is the length of the hypotenuse, then $\underline{?}$.

History CONNECTION

Pythagoras of Samos (ca. 569–475 B.C.E.), depicted in this statue, is often described as "the first pure mathematician." Samos was a principal commercial center of Greece and is located on the island of Samos in the Aegean Sea. The ancient town of Samos now lies in ruins, as shown in the photo at right.

Mysteriously, none of Pythagoras's writings still exist, and we know very little about his life. He founded a mathematical society in Croton, in what is now Italy, whose members discovered irrational numbers and the five regular solids. They proved what is now called the Pythagorean Theorem, although it was discovered and used 1000 years earlier by the Chinese and Babylonians. Some math historians believe that the ancient Egyptians also used a special case of this property to construct right angles.

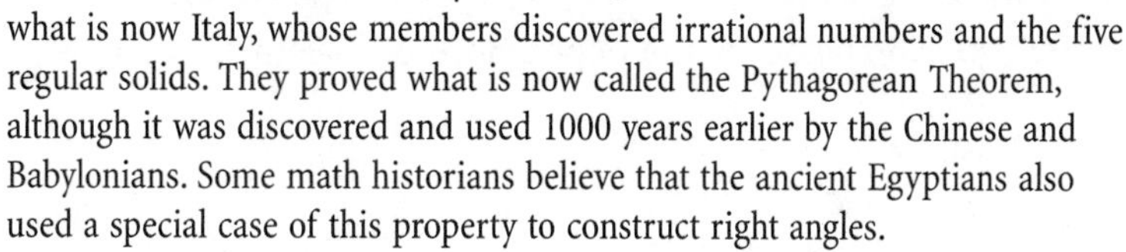

A **theorem** is a conjecture that has been proved. Demonstrations like the one in the investigation are the first step toward proving the Pythagorean Theorem.

Believe it or not, there are more than 200 proofs of the Pythagorean Theorem. Elisha Scott Loomis's *Pythagorean Proposition*, first published in 1927, contains original proofs by Pythagoras, Euclid, and even Leonardo da Vinci and U. S. President James Garfield. One well-known proof of the Pythagorean Theorem is included below. You will complete another proof as an exercise.

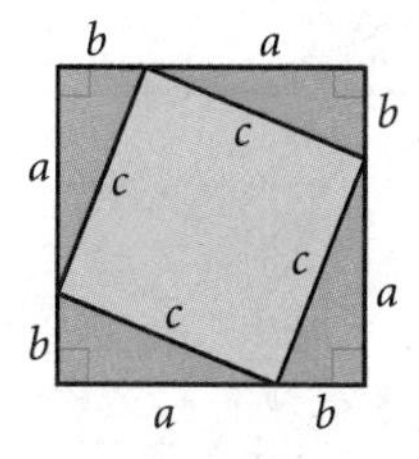

Paragraph Proof: The Pythagorean Theorem

You need to show that $a^2 + b^2$ equals c^2 for the right triangles in the figure at left. The area of the entire square is $(a + b)^2$ or $a^2 + 2ab + b^2$. The area of any triangle is $\left(\frac{1}{2}\right)ab$, so the sum of the areas of the four triangles is $2ab$. The area of the quadrilateral in the center is $\left(a^2 + 2ab + b^2\right) - 2ab$, or $a^2 + b^2$.

If the quadrilateral in the center is a square then its area also equals c^2. You now need to show that it is a square. You know that all the sides have length c, but you also need to show that the angles are right angles. The two acute angles in the right triangle, along with any angle of the quadrilateral, add up to 180°. The acute angles in a right triangle add up to 90°. Therefore the quadrilateral angle measures 90° and the quadrilateral is a square. If it is a square with side length c, then its area is c^2. So, $a^2 + b^2 = c^2$, which proves the Pythagorean Theorem. ■

The Pythagorean Theorem works for right triangles, but does it work for all triangles? A quick check demonstrates that it doesn't hold for other triangles.

Acute triangle

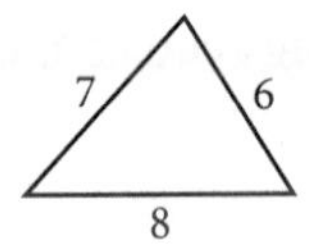

$6^2 + 7^2 > 8^2$

Obtuse triangle

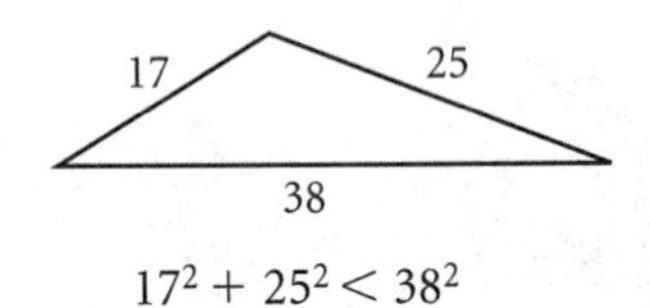

$17^2 + 25^2 < 38^2$

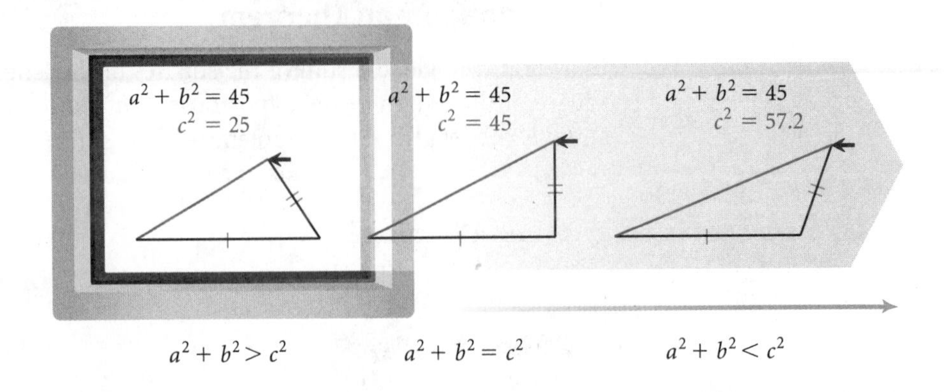

Let's look at a few examples to see how you can use the Pythagorean Theorem to find the distance between two points.

EXAMPLE A

How high up on the wall will a 20-foot ladder touch if the foot of the ladder is placed 5 feet from the wall?

20 ft h

5 ft

▶ Solution

The ladder is the hypotenuse of a right triangle, so $a^2 + b^2 = c^2$.

$(5)^2 + (h)^2 = (20)^2$	Substitute.
$25 + h^2 = 400$	Multiply.
$h^2 = 375$	Subtract 25 from both sides.
$h = \sqrt{375} \approx 19.4$	Take the square root of each side.

The top of the ladder will touch the wall about 19.4 feet up from the ground.

Notice that the exact answer in Example A is $\sqrt{375}$. However, this is a practical application, so you need to calculate the approximate answer.

EXAMPLE B

Find the area of the rectangular rug if the width is 12 feet and the diagonal measures 20 feet.

▶ Solution

Use the Pythagorean Theorem to find the length.

$$a^2 + b^2 = c^2$$
$$(12)^2 + (L)^2 = (20)^2$$
$$144 + L^2 = 400$$
$$L^2 = 256$$
$$L = \sqrt{256}$$
$$L = 16$$

The length is 16 feet. The area of the rectangle is $12 \cdot 16$, or 192 square feet.

EXERCISES

In Exercises 1–11, find each missing length. All measurements are in centimeters. Give approximate answers accurate to the nearest tenth of a centimeter.

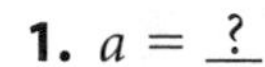

1. $a = \underline{?}$

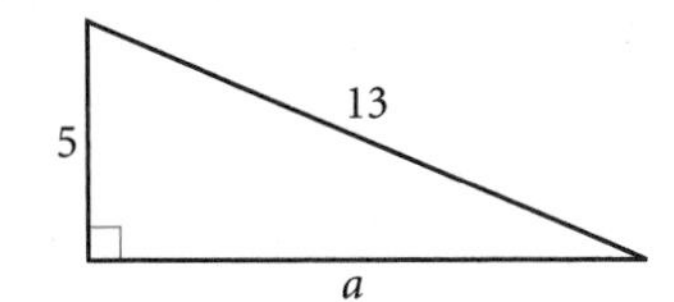

2. $c \approx \underline{?}$

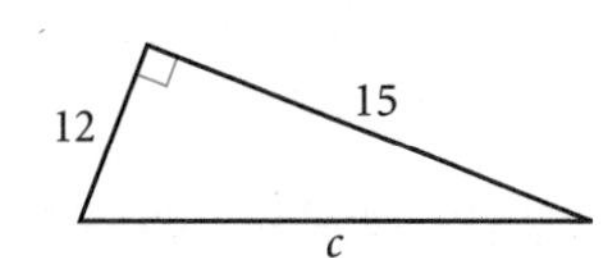

3. $a \approx \underline{?}$

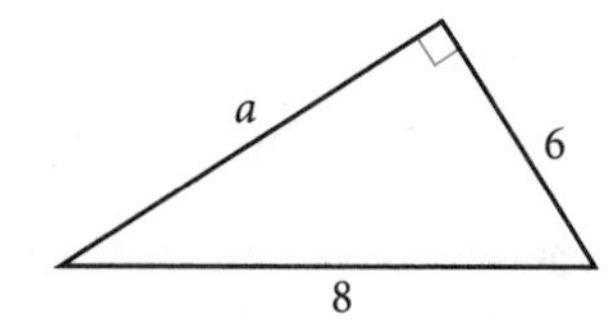

4. $d = \underline{?}$

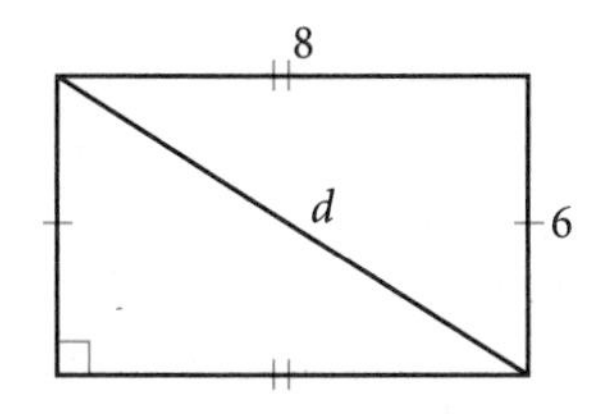

5. $s = \underline{?}$

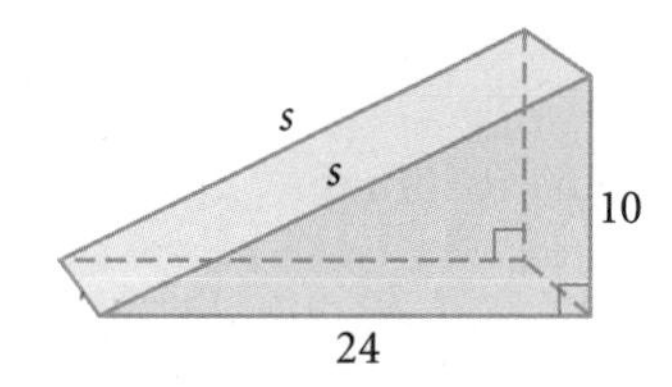

6. $c \approx \underline{?}$ ⓗ

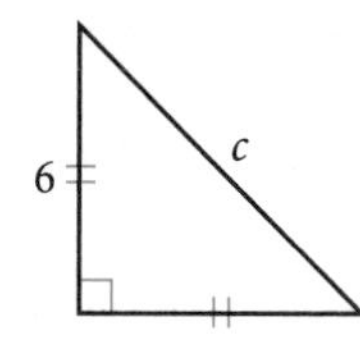

7. $b = \underline{?}$

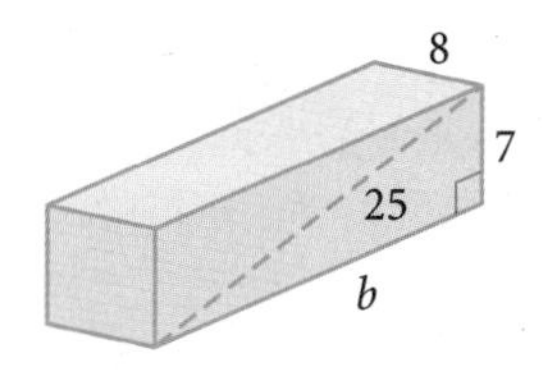

8. $x = \underline{?}$

9. The base is a circle.
$x = \underline{?}$

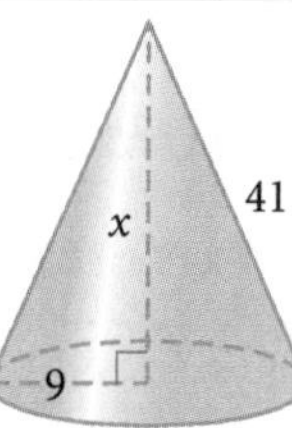

10. $s \approx \underline{?}$

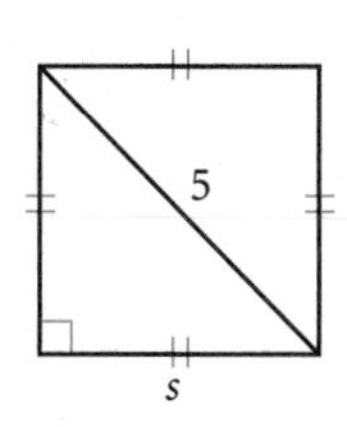

11. $r = \underline{?}$ ⓗ

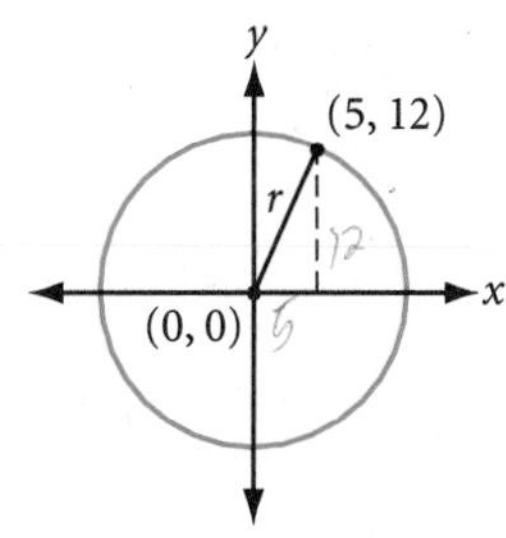

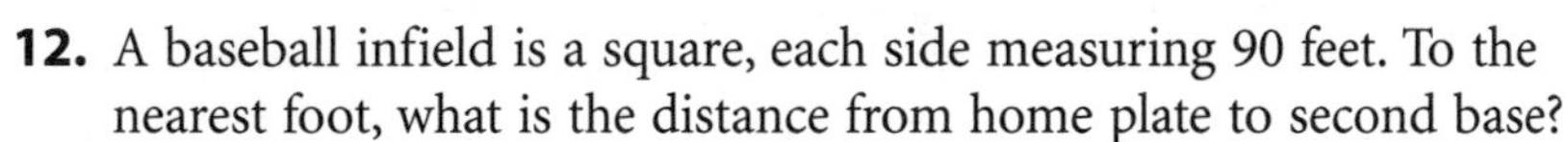

12. A baseball infield is a square, each side measuring 90 feet. To the nearest foot, what is the distance from home plate to second base?

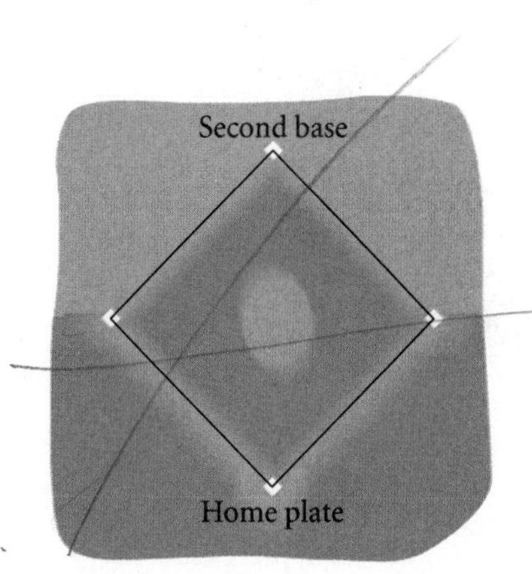

13. The diagonal of a square measures 32 meters. What is the area of the square? ⓗ

14. What is the length of the diagonal of a square whose area is 64 cm^2?

15. The lengths of the three sides of a right triangle are consecutive integers. Find them. ⓗ

16. A rectangular garden 6 meters wide has a diagonal measuring 10 meters. Find the perimeter of the garden.

17. One very famous proof of the Pythagorean Theorem is by the Hindu mathematician Bhaskara. It is often called the "Behold" proof because, as the story goes, Bhaskara drew the diagram at right and offered no verbal argument other than to exclaim, "Behold." Use algebra to fill in the steps, explaining why this diagram proves the Pythagorean Theorem. ⓗ

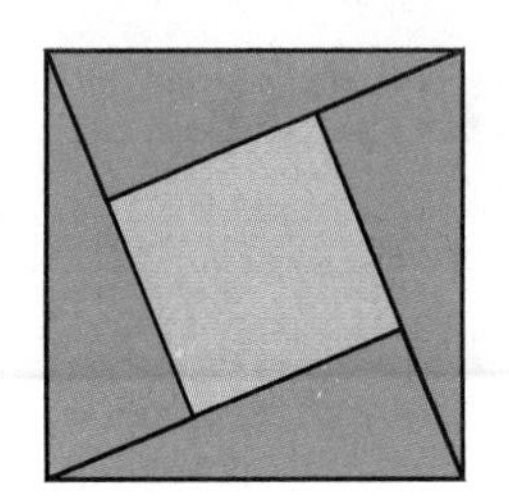

History CONNECTION

Bhaskara (1114–1185, India) was one of the first mathematicians to gain a thorough understanding of number systems and how to solve equations, several centuries before European mathematicians. He wrote six books on mathematics and astronomy, and led the astronomical observatory at Ujjain.

18. Is $\triangle ABC \cong \triangle XYZ$? Explain your reasoning.

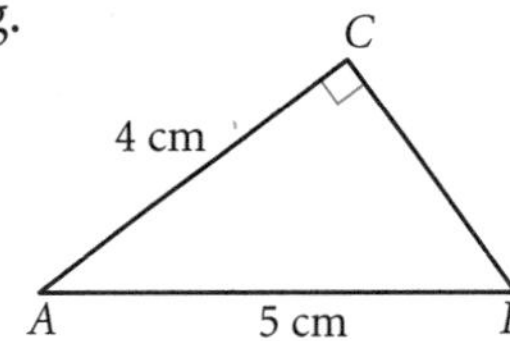

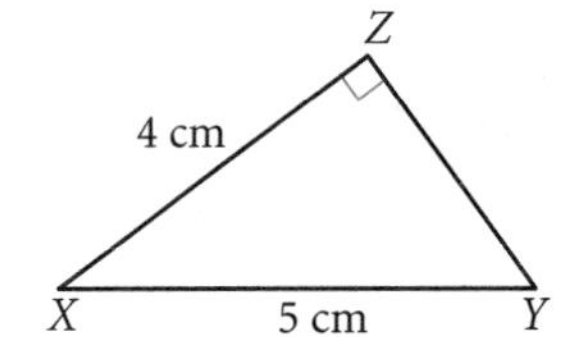

Review

19. The two quadrilaterals are squares. Find x.

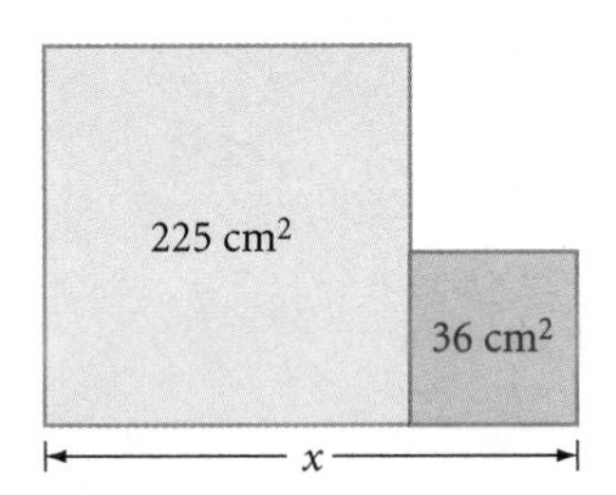

20. Give the vertex arrangement for the 2-uniform tessellation.

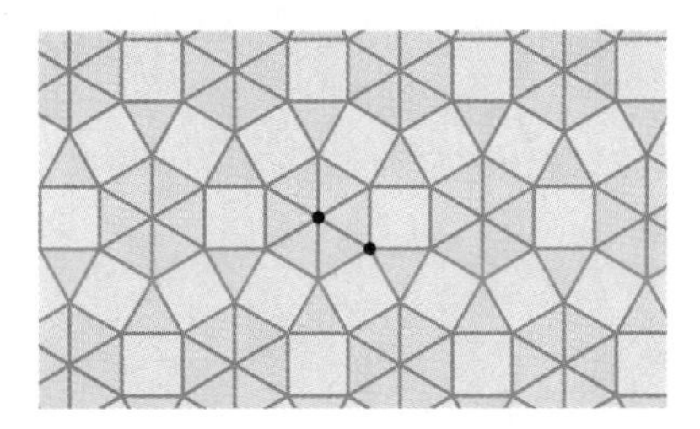

21. Explain why $m + n = 120°$.

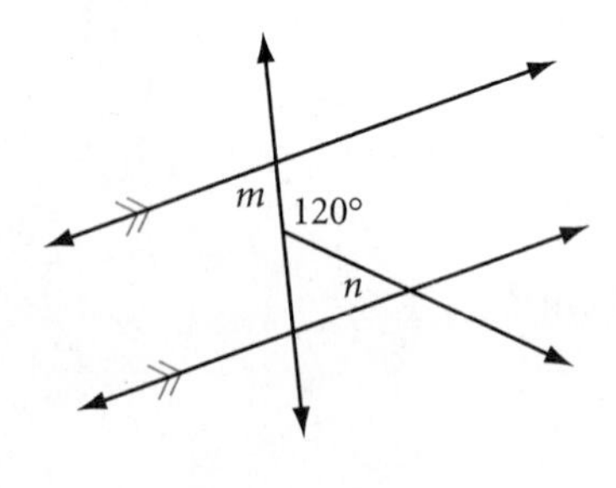

22. Calculate each lettered angle, measure, or arc. $\overline{EF}$ is a diameter; ℓ_1 and ℓ_2 are tangents.

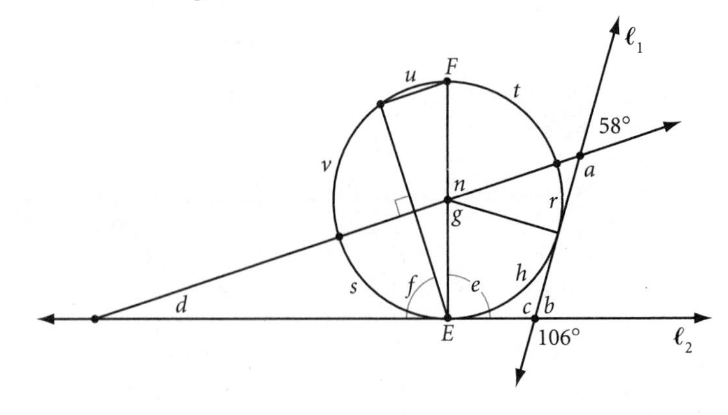

project

CREATING A GEOMETRY FLIP BOOK

Have you ever fanned the pages of a flip book and watched the pictures seem to move? Each page shows a picture slightly different from the previous one. Flip books are basic to animation technique. For more information about flip books, see www.keymath.com/DG .

These five frames start off the photo series titled *The Horse in Motion,* by photographer, innovator, and motion picture pioneer Eadweard Muybridge (1830–1904).

Here are two dissections that you can animate to demonstrate the Pythagorean Theorem. (You used another dissection in the Investigation The Three Sides of a Right Triangle.)

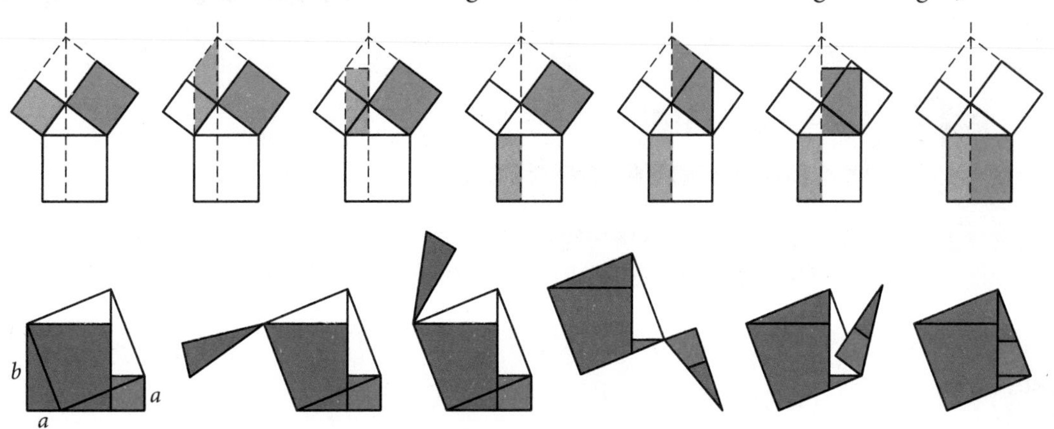

You could also animate these drawings to demonstrate area formulas.

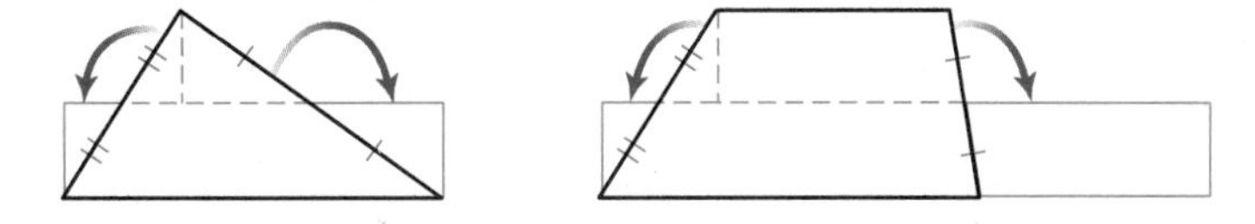

Choose one of the animations mentioned above and create a flip book that demonstrates it. Be ready to explain how your flip book demonstrates the formula you chose.

Here are some practical tips.

- ▶ Draw your figures in the same position on each page so they don't jump around when the pages are flipped. Use graph paper or tracing paper to help.
- ▶ The smaller the change from picture to picture, and the more pictures there are, the smoother the motion will be.
- ▶ Label each picture so that it's clear how the process works.

LESSON

9.2

The Converse of the Pythagorean Theorem

Any time you see someone more successful than you are, they are doing something you aren't.

MALCOLM X

In Lesson 9.1, you saw that if a triangle is a right triangle, then the square of the length of its hypotenuse is equal to the sum of the squares of the lengths of the two legs. What about the converse? If x, y, and z are the lengths of the three sides of a triangle and they satisfy the Pythagorean equation, $a^2 + b^2 = c^2$, must the triangle be a right triangle? Let's find out.

Investigation

Is the Converse True?

You will need

- string
- a ruler
- paper clips
- a piece of patty paper

Three positive integers that work in the Pythagorean equation are called **Pythagorean triples.** For example, 8-15-17 is a Pythagorean triple because $8^2 + 15^2 = 17^2$. Here are nine sets of Pythagorean triples.

3-4-5	5-12-13	7-24-25	8-15-17
6-8-10	10-24-26		16-30-34
9-12-15			
12-16-20			

Step 1 Select one set of Pythagorean triples from the list above. Mark off four points, A, B, C, and D, on a string to create three consecutive lengths from your set of triples.

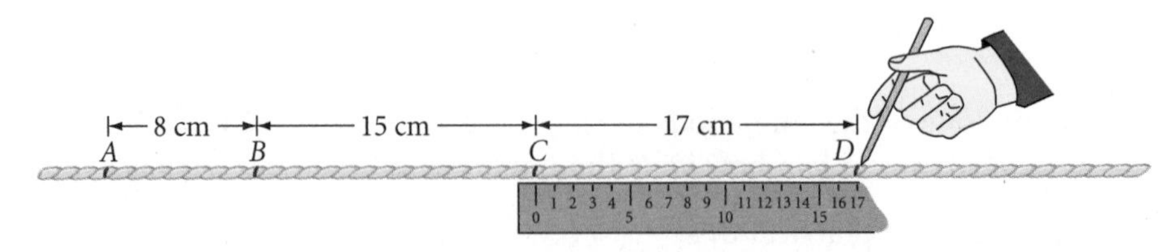

Step 2 Loop three paper clips onto the string. Tie the ends together so that points A and D meet.

Step 3 Three group members should each pull a paper clip at point A, B, or C to stretch the string tight.

Step 4 With your paper, check the largest angle. What type of triangle is formed?

Step 5 Select another set of triples from the list. Repeat Steps 1–4 with your new lengths.

Step 6 Compare results in your group. State your results as your next conjecture.

Converse of the Pythagorean Theorem C-83

If the lengths of the three sides of a triangle satisfy the Pythagorean equation, then the triangle _?_.

This ancient Babylonian tablet, called Plimpton 322, dates sometime between 1900 and 1600 B.C.E. It suggests several advanced Pythagorean triples, such as 1679-2400-2929.

History CONNECTION

Some historians believe Egyptian "rope stretchers" used the Converse of the Pythagorean Theorem to help reestablish land boundaries after the yearly flooding of the Nile and to help construct the pyramids. Some ancient tombs show workers carrying ropes tied with equally spaced knots. For example, 13 equally spaced knots would divide the rope into 12 equal lengths. If one person held knots 1 and 13 together, and two others held the rope at knots 4 and 8 and stretched it tight, they could have created a 3-4-5 right triangle.

The proof of the Converse of the Pythagorean Theorem is very interesting because it is one of the few instances where the original theorem is used to prove the converse. Let's take a look. One proof is started for you below. You will finish it as an exercise.

Proof: Converse of the Pythagorean Theorem

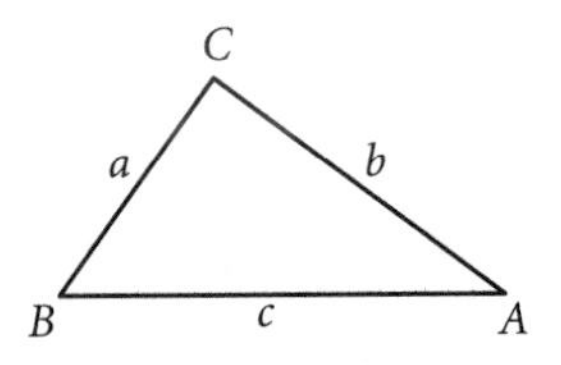

Conjecture: If the lengths of the three sides of a triangle work in the Pythagorean equation, then the triangle is a right triangle.

Given: a, b, c are the lengths of the sides of $\triangle ABC$ and $a^2 + b^2 = c^2$

Show: $\triangle ABC$ is a right triangle

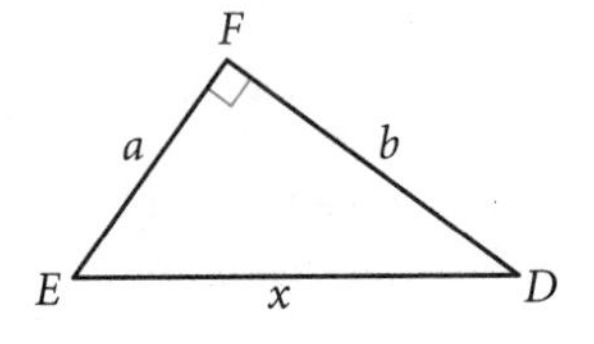

Plan: Begin by constructing a second triangle, right triangle DEF (with $\angle F$ a right angle), with legs of lengths a and b and hypotenuse of length x. The plan is to show that $x = c$, so that the triangles are congruent. Then show that $\angle C$ and $\angle F$ are congruent. Once you show that $\angle C$ is a right angle, then $\triangle ABC$ is a right triangle and the proof is complete.

EXERCISES

In Exercises 1–6, use the Converse of the Pythagorean Theorem to determine whether each triangle is a right triangle.

1.

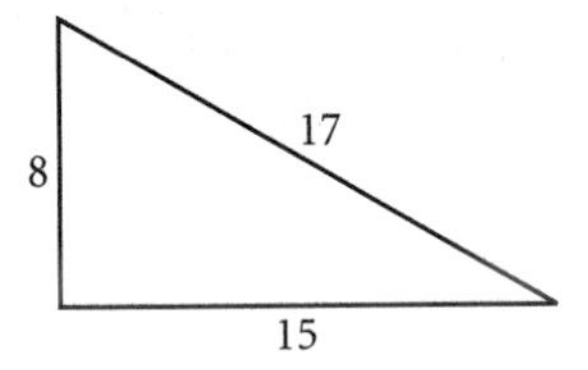

2.

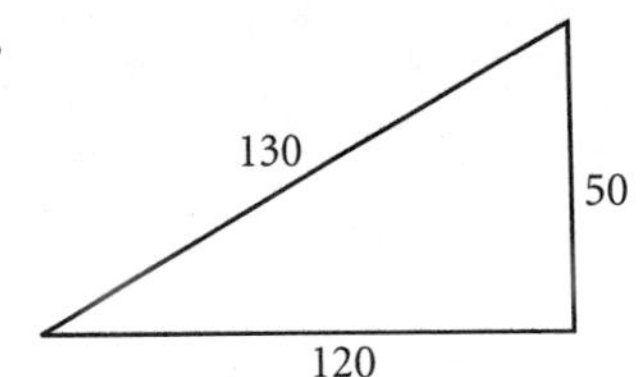

3.

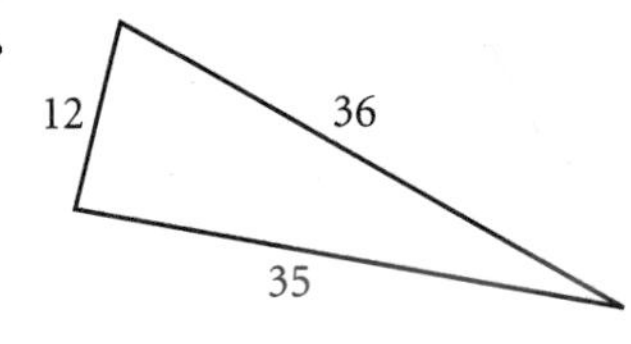

4.

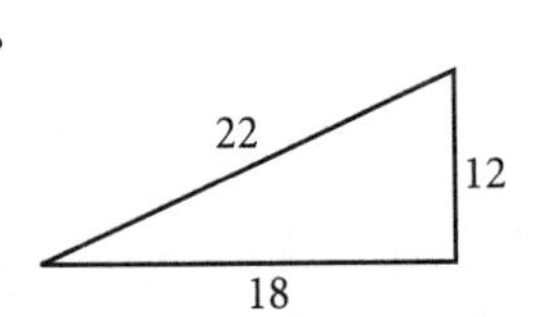

5.

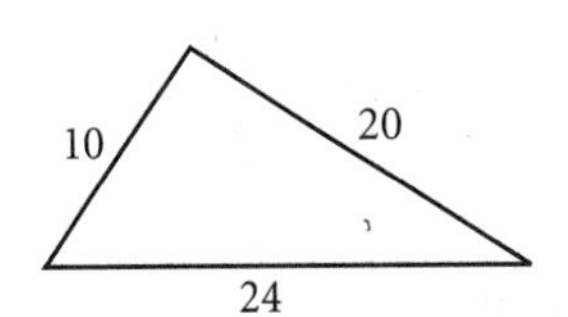

6. ⓗ

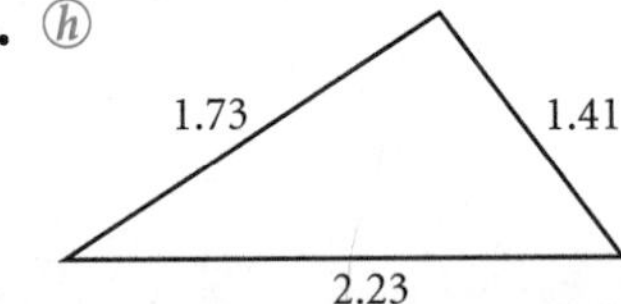

In Exercises 7 and 8, use the Converse of the Pythagorean Theorem to solve each problem.

7. Is a triangle with sides measuring 9 feet, 12 feet, and 18 feet a right triangle?

8. A window frame that seems rectangular has height 408 cm, length 306 cm, and one diagonal with length 525 cm. Is the window frame really rectangular? Explain.

In Exercises 9–11, find y.

9. Both quadrilaterals are squares.

15 cm

y

25 cm²

10. Ⓗ

y

$(-7, y)$

25

x

11.

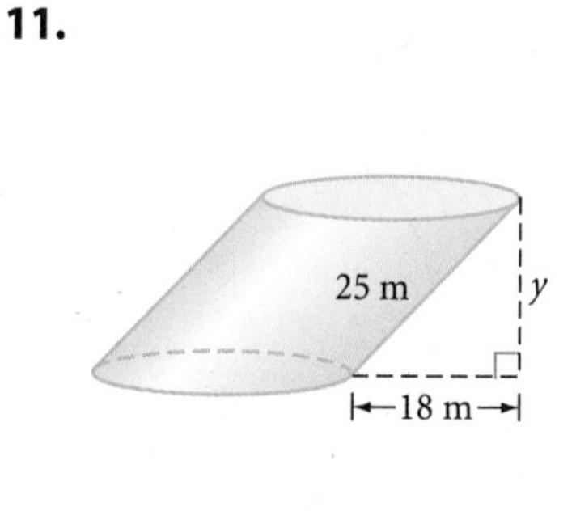

12. The lengths of the three sides of a right triangle are consecutive even integers. Find them. Ⓗ

13. Find the area of a right triangle with hypotenuse length 17 cm and one leg length 15 cm.

14. How high on a building will a 15-foot ladder touch if the foot of the ladder is 5 feet from the building?

15. The congruent sides of an isosceles triangle measure 6 cm, and the base measures 8 cm. Find the area.

16. Find the amount of fencing in linear feet needed for the perimeter of a rectangular lot with a diagonal length 39 m and a side length 36 m.

17. A rectangular piece of cardboard fits snugly on a diagonal in this box.

a. What is the area of the cardboard rectangle?

b. What is the length of the diagonal of the cardboard rectangle?

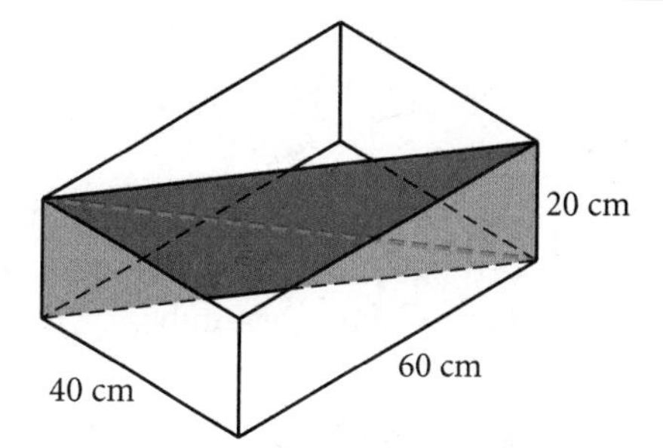

18. Look back at the start of the proof of the Converse of the Pythagorean Theorem. Copy the conjecture, the given, the show, the plan, and the two diagrams. Use the plan to complete the proof.

19. What's wrong with this picture?

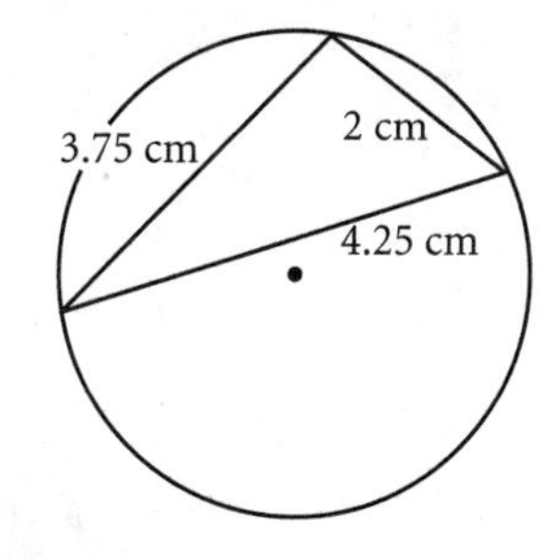

20. Explain why $\triangle ABC$ is a right triangle.

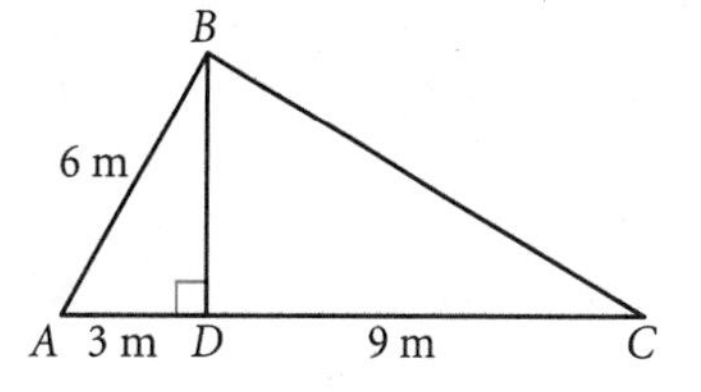

Review

21. Identify the point of concurrency from the construction marks.

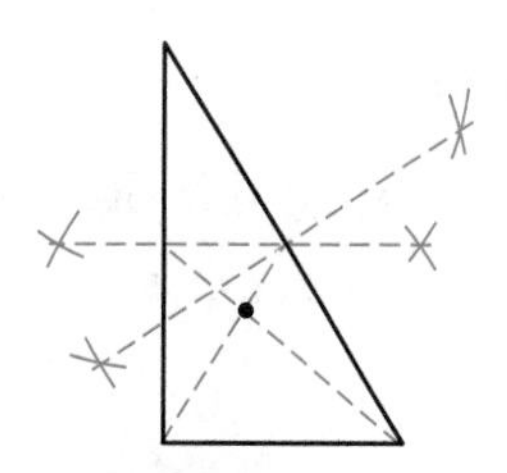

22. Line CF is tangent to circle D at C. The arc measure of $\overset{\frown}{CE}$ is a. Explain why $x = \left(\frac{1}{2}\right)a$. ⓗ

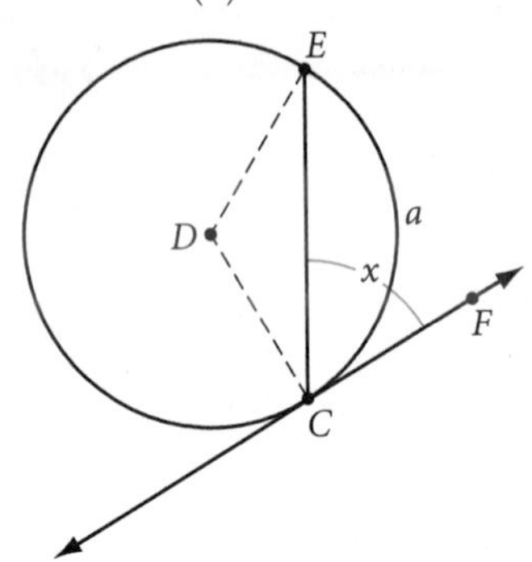

23. What is the probability of randomly selecting three points that form an isosceles triangle from the 10 points in this isometric grid?

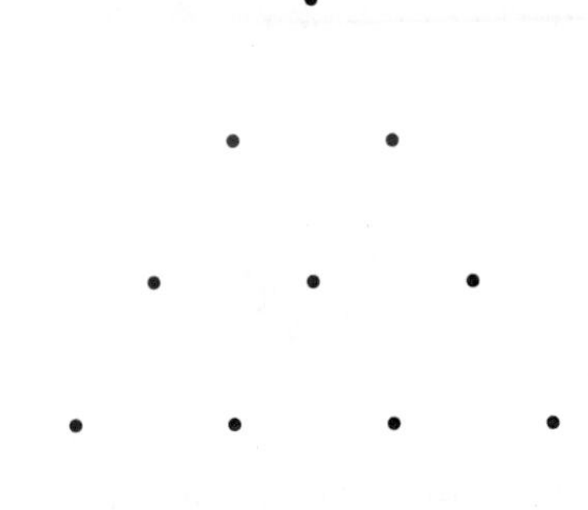

24. If the pattern of blocks continues, what will be the surface area of the 50th solid in the pattern?

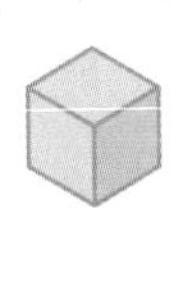
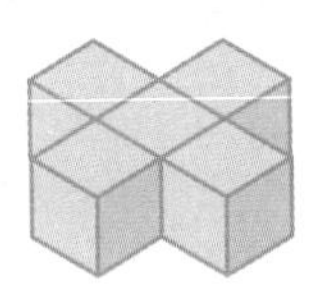
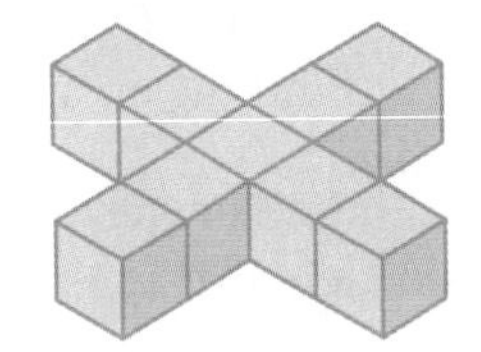
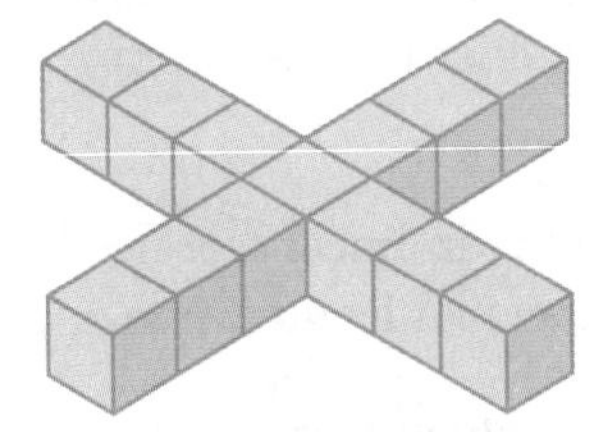

25. Sketch the solid shown, but with the two blue cubes removed and the red cube moved to cover the visible face of the green cube.

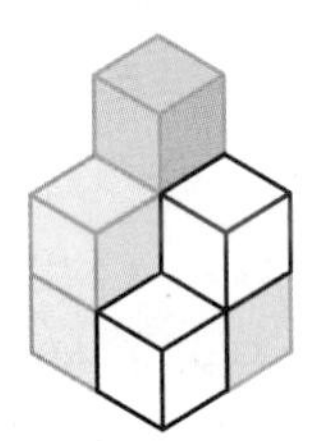

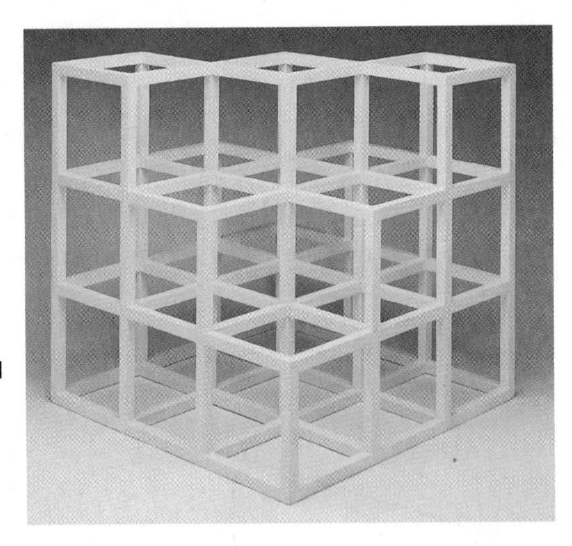

The outlines of stacked cubes create a visual impact in this untitled module unit sculpture by conceptual artist Sol Lewitt.

IMPROVING YOUR ALGEBRA SKILLS

Algebraic Sequences III

Find the next three terms of this algebraic sequence.

$x^9, 9x^8y, 36x^7y^2, 84x^6y^3, 126x^5y^4, 126x^4y^5, 84x^3y^6, \underline{?}, \underline{?}, \underline{?}$

Radical Expressions

When you work with the Pythagorean Theorem, you often get radical expressions, such as $\sqrt{50}$. Until now you may have left these expressions as radicals, or you may have found a decimal approximation using a calculator. Some radical expressions can be simplified. To simplify a square root means to take the square root of any perfect-square factors of the number under the radical sign. Let's look at an example.

EXAMPLE A | Simplify $\sqrt{50}$.

▶ ***Solution*** | One way to simplify a square root is to look for perfect-square factors.

The largest perfect-square factor of 50 is 25.

$$\sqrt{50} = \sqrt{25 \cdot 2} = \sqrt{25} \cdot \sqrt{2} = 5\sqrt{2}$$

25 is a perfect square, so you can take its square root.

Another approach is to factor the number as far as possible with prime factors.

Write 50 as a set of prime factors.

Look for any square factors (factors that appear twice).

$$\sqrt{50} = \sqrt{5 \cdot 5 \cdot 2} = \sqrt{5^2 \cdot 2} = \sqrt{5^2} \cdot \sqrt{2} = 5\sqrt{2}$$

Squaring and taking the square root are inverse operations—they undo each other. So, $\sqrt{5^2}$ equals 5.

You might argue that $5\sqrt{2}$ doesn't look any simpler than $\sqrt{50}$. However, in the days before calculators with square root buttons, mathematicians used paper-and-pencil algorithms to find approximate values of square roots. Working with the smallest possible number under the radical made the algorithms easier to use.

Giving an exact answer to a problem involving a square root is important in a number of situations. Some patterns are easier to discover with simplified square roots than with decimal approximations. Standardized tests often express answers in simplified form. And when you multiply radical expressions, you often have to simplify the answer.

EXAMPLE B | Multiply $3\sqrt{6}$ by $5\sqrt{2}$.

▶ **Solution** | To multiply radical expressions, associate and multiply the quantities outside the radical sign, and associate and multiply the quantities inside the radical sign.

$$(3\sqrt{6})(5\sqrt{2}) = 3 \cdot 5 \cdot \sqrt{6 \cdot 2} = 15 \cdot \sqrt{12} = 15 \cdot \sqrt{4 \cdot 3} = 15 \cdot 2\sqrt{3} = 30\sqrt{3}$$

EXERCISES

In Exercises 1–5, express each product in its simplest form.

1. $(\sqrt{3})(\sqrt{2})$ **2.** $(\sqrt{5})^2$ **3.** $(3\sqrt{6})(2\sqrt{3})$ **4.** $(7\sqrt{3})^2$ **5.** $(2\sqrt{2})^2$

In Exercises 6–20, express each square root in its simplest form.

6. $\sqrt{18}$ **7.** $\sqrt{40}$ **8.** $\sqrt{75}$ **9.** $\sqrt{85}$ **10.** $\sqrt{96}$

11. $\sqrt{576}$ **12.** $\sqrt{720}$ **13.** $\sqrt{722}$ **14.** $\sqrt{784}$ **15.** $\sqrt{828}$

16. $\sqrt{2952}$ **17.** $\sqrt{5248}$ **18.** $\sqrt{8200}$ **19.** $\sqrt{11808}$ **20.** $\sqrt{16072}$

21. What is the next term in the pattern? $\sqrt{2952}, \sqrt{5248}, \sqrt{8200}, \sqrt{11808}, \sqrt{16072}, \ldots$

IMPROVING YOUR VISUAL THINKING SKILLS

PUZZLE

Folding Cubes II

Each cube has designs on three faces. When unfolded, which figure at right could it become?

1.

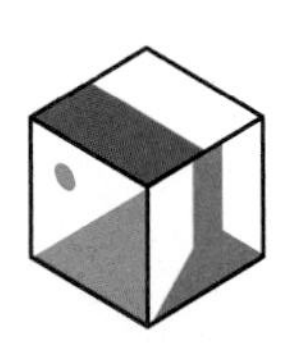

A.

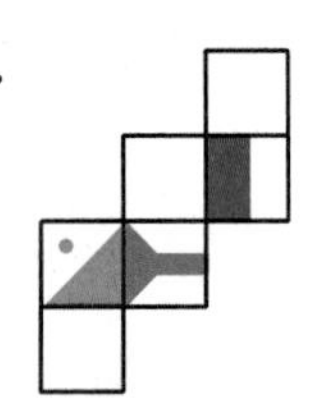

B.

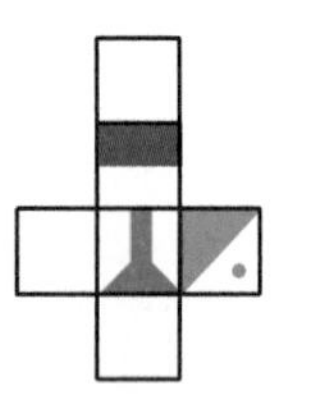

C.

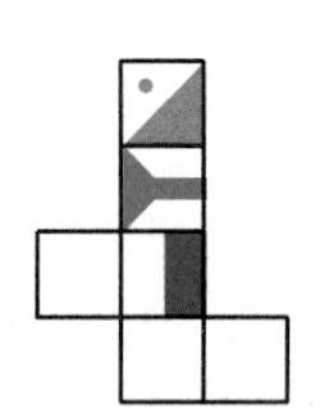

D.

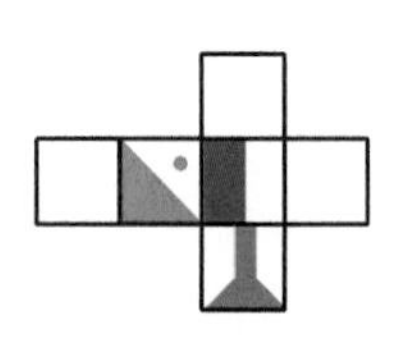

2.

A.

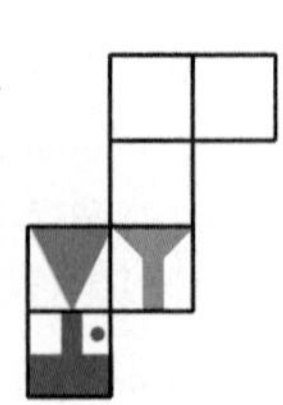

B.

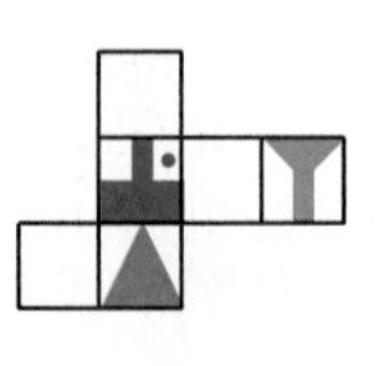

C.

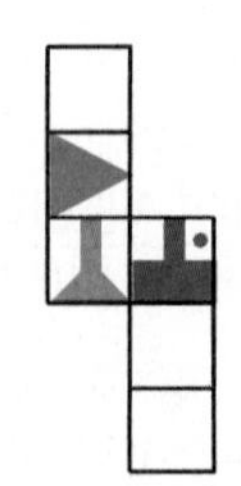

D.

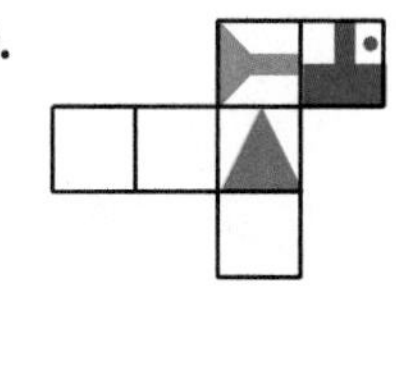

LESSON

9.3

Two Special Right Triangles

In an isosceles triangle, the sum of the square roots of the two equal sides is equal to the square root of the third side.

THE SCARECROW IN THE 1939 FILM *THE WIZARD OF OZ*

In this lesson you will use the Pythagorean Theorem to discover some relationships between the sides of two special right triangles.

One of these special triangles is an isosceles right triangle, also called a 45°-45°-90° triangle. Each isosceles right triangle is half a square, so they show up often in mathematics and engineering. In the next investigation, you will look for a shortcut for finding the length of an unknown side in a 45°-45°-90° triangle.

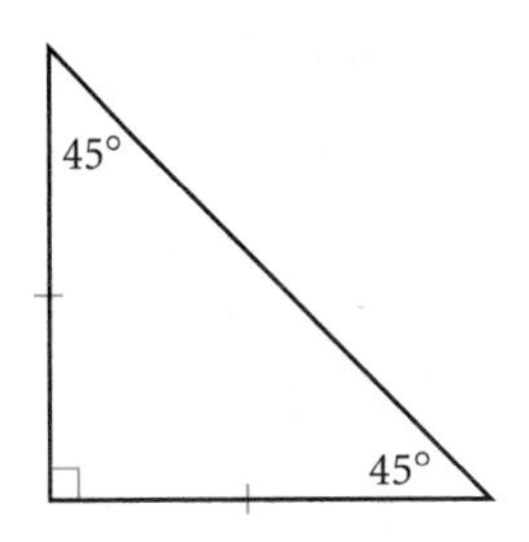

An isosceles right triangle

Investigation 1
Isosceles Right Triangles

Step 1 Sketch an isosceles right triangle. Label the legs l and the hypotenuse h.

Step 2 Pick any integer for l, the length of the legs. Use the Pythagorean Theorem to find h. Simplify the square root.

Step 3 Repeat Step 2 with several different values for l. Share results with your group. Do you see any pattern in the relationship between l and h?

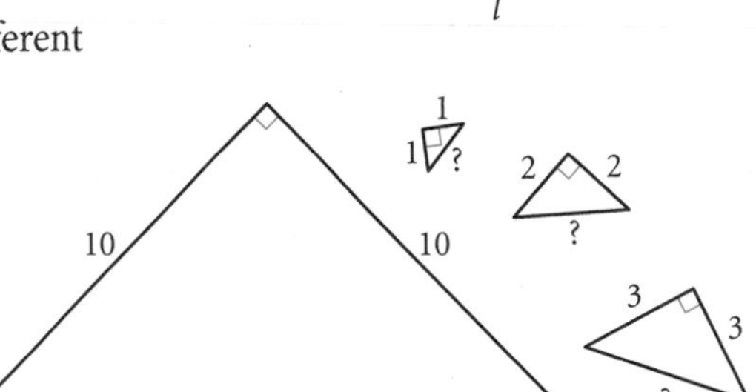

Step 4 State your next conjecture in terms of length l.

Isosceles Right Triangle Conjecture C-84

In an isosceles right triangle, if the legs have length l, then the hypotenuse has length __?__.

You can also demonstrate this property on a geoboard or graph paper, as shown at right.

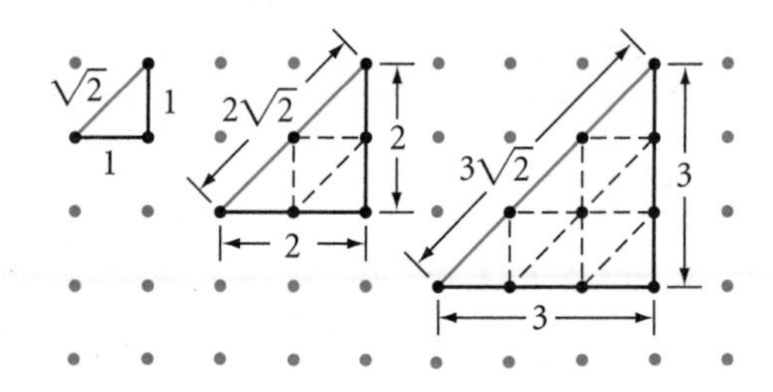

The other special right triangle is a 30°-60°-90° triangle. If you fold an equilateral triangle along one of its altitudes, the triangles you get are 30°-60°-90° triangles. A 30°-60°-90° triangle is half an equilateral triangle, so it also shows up often in mathematics and engineering. Let's see if there is a shortcut for finding the lengths of its sides.

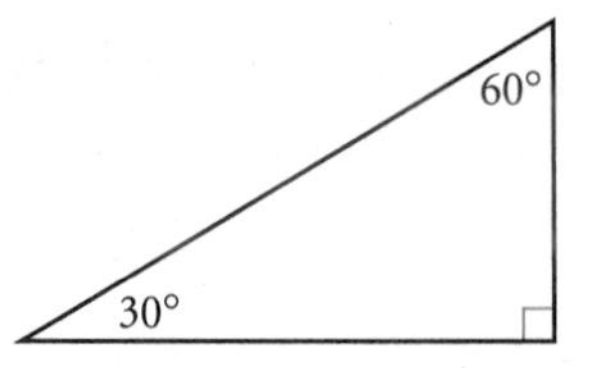

A 30°-60°-90° triangle

Investigation 2
30°-60°-90° Triangles

Let's start by using a little deductive thinking to find the relationships in 30°-60°-90° triangles. Triangle ABC is equilateral, and $\overline{CD}$ is an altitude.

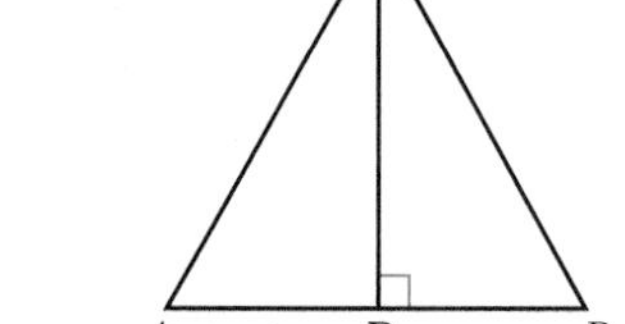

Step 1 What are $m\angle A$ and $m\angle B$? What are $m\angle ACD$ and $m\angle BCD$? What are $m\angle ADC$ and $m\angle BDC$?

Step 2 Is $\triangle ADC \cong \triangle BDC$? Why?

Step 3 Is $\overline{AD} \cong \overline{BD}$? Why? How do AC and AD compare? In a 30°-60°-90° triangle, will this relationship between the hypotenuse and the shorter leg always hold true? Explain.

Step 4 Sketch a 30°-60°-90° triangle. Choose any integer for the length of the shorter leg. Use the relationship from Step 3 and the Pythagorean Theorem to find the length of the other leg. Simplify the square root.

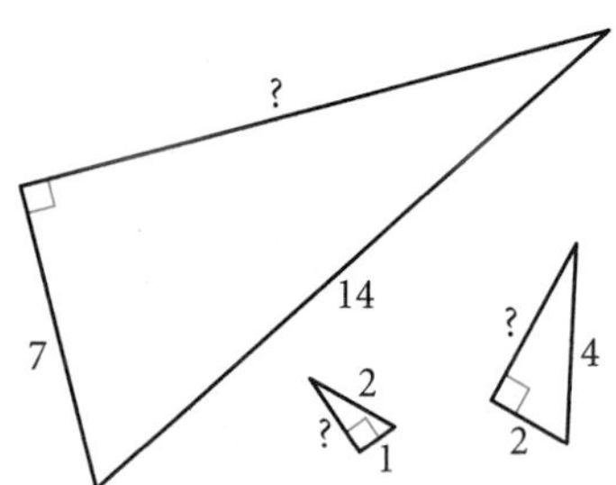

Step 5 Repeat Step 4 with several different values for the length of the shorter leg. Share results with your group. What is the relationship between the lengths of the two legs? You should notice a pattern in your answers.

Step 6 State your next conjecture in terms of the length of the shorter leg, a.

30°-60°-90° Triangle Conjecture **C-85**

In a 30°-60°-90° triangle, if the shorter leg has length a, then the longer leg has length _?_ and the hypotenuse has length _?_.

You can use algebra to verify that the conjecture will hold true for any 30°-60°-90° triangle.

Proof: 30°-60°-90° Triangle Conjecture

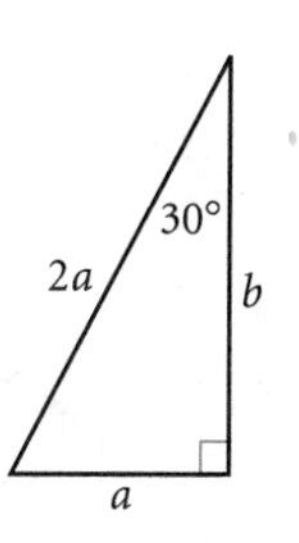

$(2a)^2 = a^2 + b^2$	Start with the Pythagorean Theorem.
$4a^2 = a^2 + b^2$	Square $2a$.
$3a^2 = b^2$	Subtract a^2 from both sides.
$a\sqrt{3} = b$	Take the square root of both sides.

Although you investigated only integer values, the proof shows that any number, even a non-integer, can be used for a. You can also demonstrate this property for integer values on isometric dot paper.

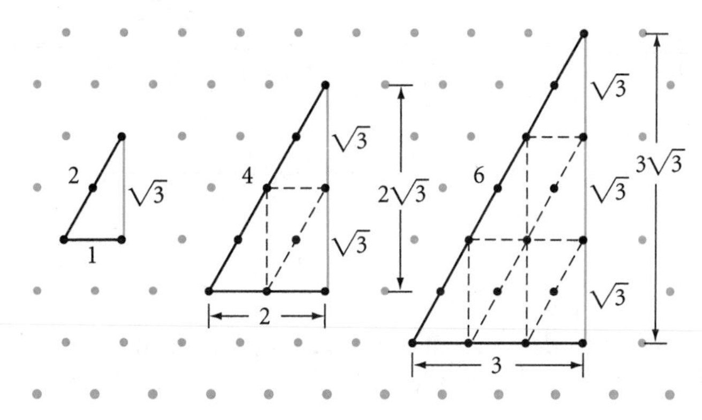

EXERCISES

You will need

Construction tools for Exercises **19** and **20**

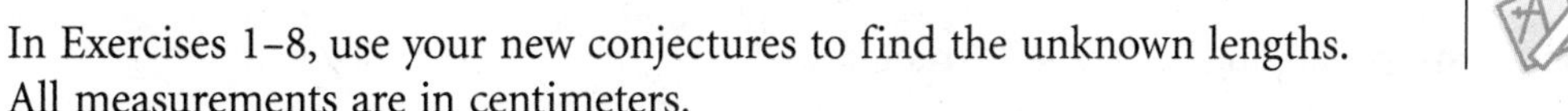

In Exercises 1–8, use your new conjectures to find the unknown lengths. All measurements are in centimeters.

1. $a = \underline{?}$

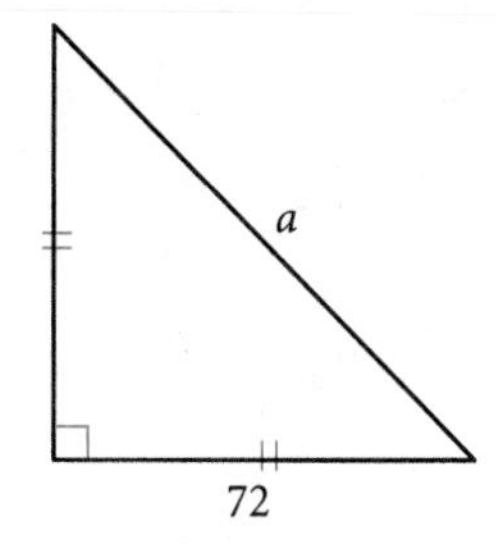

2. $b = \underline{?}$ (h)

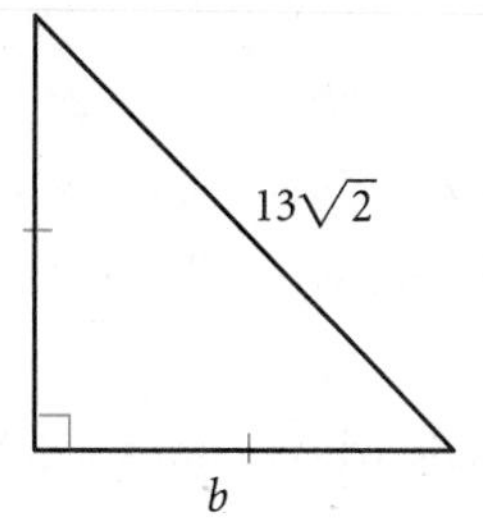

3. $a = \underline{?}$, $b = \underline{?}$

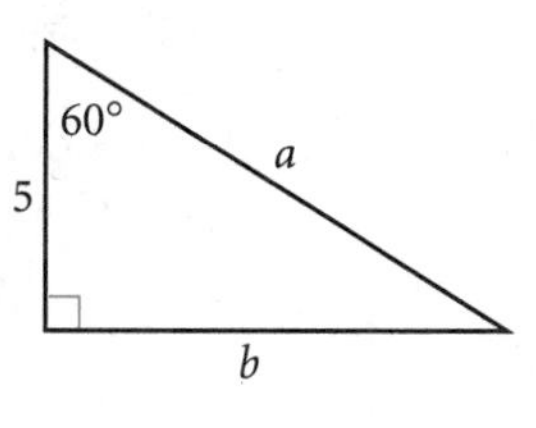

4. $c = \underline{?}$, $d = \underline{?}$ (h)

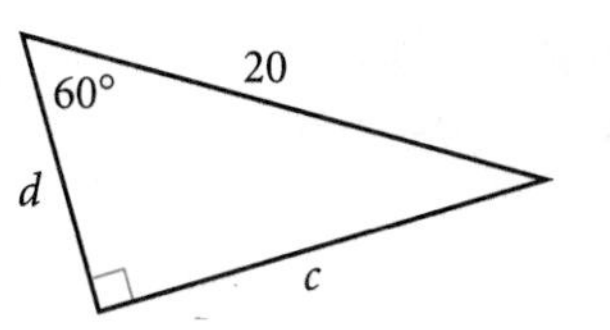

5. $e = \underline{?}$, $f = \underline{?}$

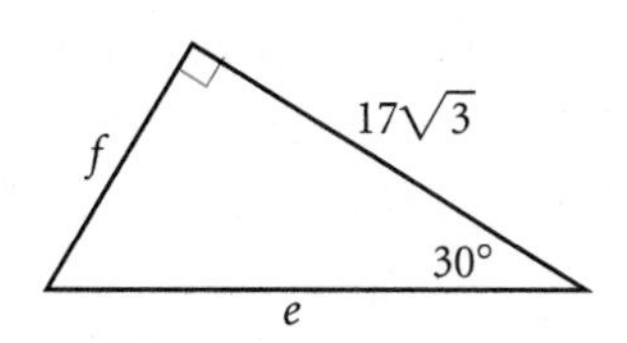

6. What is the perimeter of square $SQRE$?

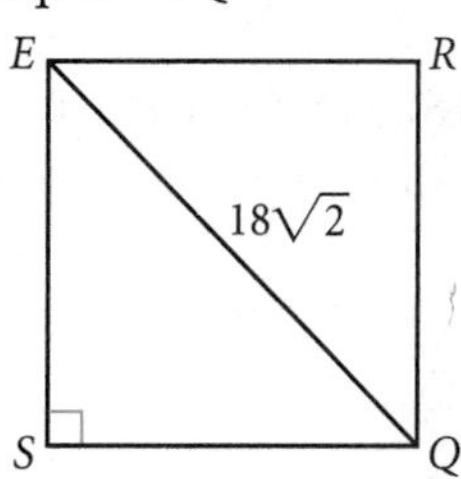

7. The solid is a cube.
$d = \underline{?}$ ⓗ

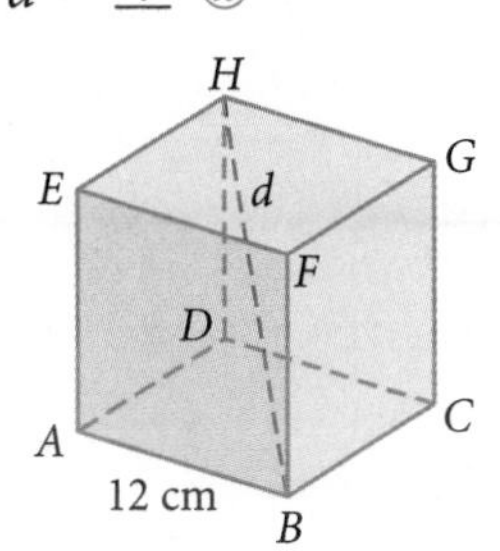

8. $g = \underline{?}$, $h = \underline{?}$

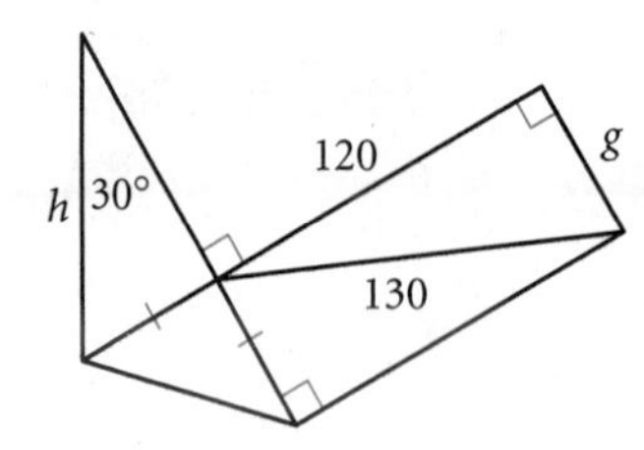

9. What is the area of the triangle? ⓗ

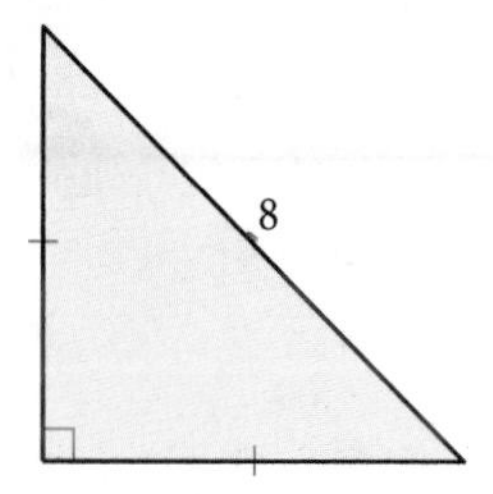

10. Find the coordinates of *P*.

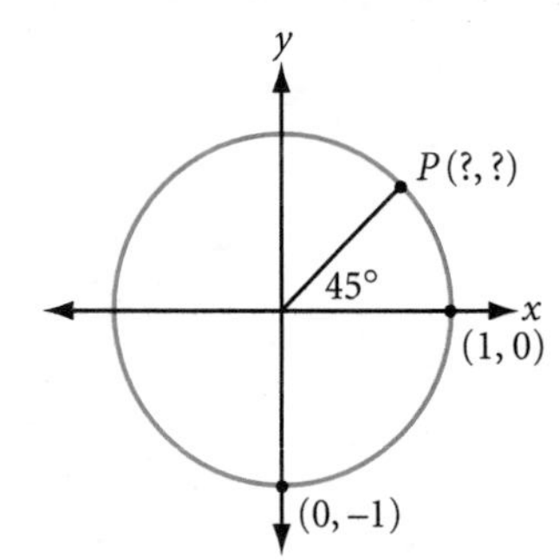

11. What's wrong with this picture?

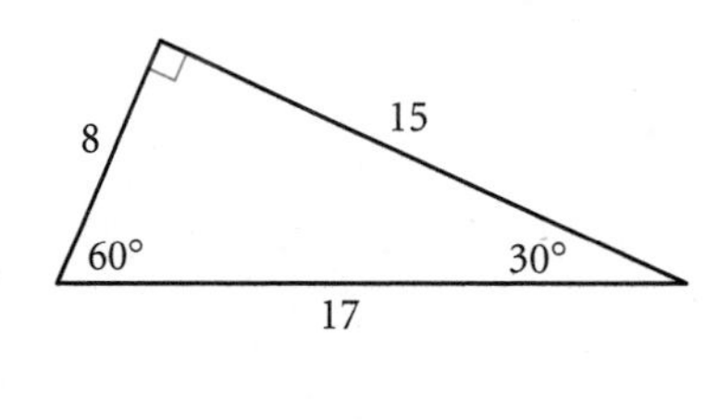

12. Sketch and label a figure to demonstrate that $\sqrt{27}$ is equivalent to $3\sqrt{3}$. (Use isometric dot paper to aid your sketch.) ⓗ

13. Sketch and label a figure to demonstrate that $\sqrt{32}$ is equivalent to $4\sqrt{2}$. (Use square dot paper or graph paper.)

14. In equilateral triangle *ABC*, $\overline{AE}$, $\overline{BF}$, and $\overline{CD}$ are all angle bisectors, medians, and altitudes simultaneously. These three segments divide the equilateral triangle into six overlapping 30°-60°-90° triangles and six smaller, non-overlapping 30°-60°-90° triangles.

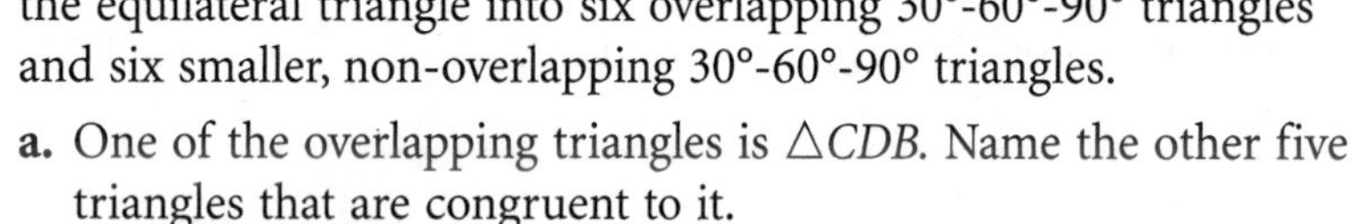

a. One of the overlapping triangles is $\triangle CDB$. Name the other five triangles that are congruent to it.

b. One of the non-overlapping triangles is $\triangle MDA$. Name the other five triangles congruent to it.

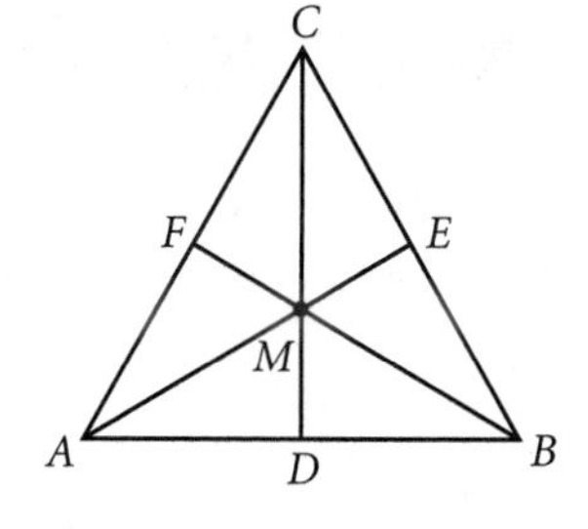

15. Use algebra and deductive reasoning to show that the Isosceles Right Triangle Conjecture holds true for any isosceles right triangle. Use the figure at right.

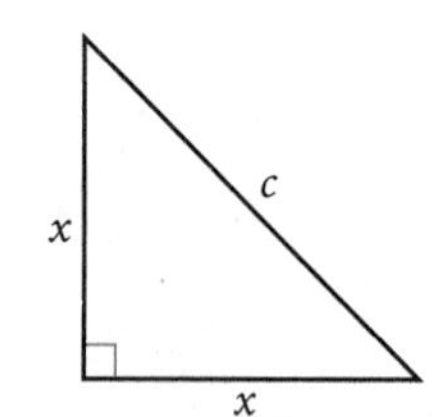

16. Find the area of an equilateral triangle whose sides measure 26 meters. ⓗ

17. An equilateral triangle has an altitude that measures 26 meters. Find the area of the triangle to the nearest square meter.

18. Sketch the largest 45°-45°-90° triangle that fits in a 30°-60°-90° triangle. What is the ratio of the area of the 30°-60°-90° triangle to the area of the 45°-45°-90° triangle?

Review

Construction In Exercises 19 and 20, choose either patty paper or a compass and straightedge and perform the constructions.

19. Given the segment with length a below, construct segments with lengths $a\sqrt{2}$, $a\sqrt{3}$, and $a\sqrt{5}$. ⓗ

a

20. ***Mini-Investigation*** Draw a right triangle with sides of lengths 6 cm, 8 cm, and 10 cm. Locate the midpoint of each side. Construct a semicircle on each side with the midpoints of the sides as centers. Find the area of each semicircle. What relationship do you notice among the three areas?

21. The *Jiuzhang suanshu* is an ancient Chinese mathematics text of 246 problems. Some solutions use the *gou gu*, the Chinese name for what we call the Pythagorean Theorem. The *gou gu* reads $(gou)^2 + (gu)^2 = (xian)^2$. Here is a *gou gu* problem translated from the ninth chapter of *Jiuzhang*.

A rope hangs from the top of a pole with three *chih* of it lying on the ground. When it is tightly stretched so that its end just touches the ground, it is eight *chih* from the base of the pole. How long is the rope?

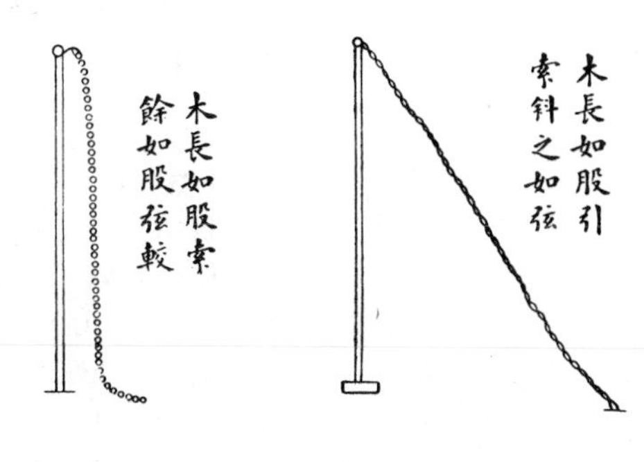

22. Explain why $m\angle 1 + m\angle 2 = 90°$. ⓗ

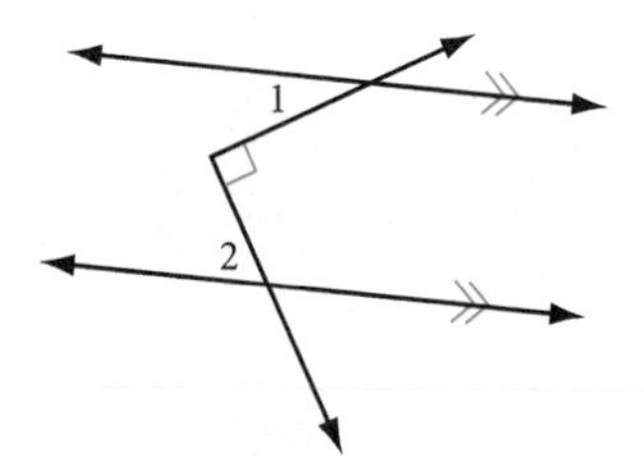

23. The lateral surface area of the cone below is unwrapped into a sector. What is the angle at the vertex of the sector?

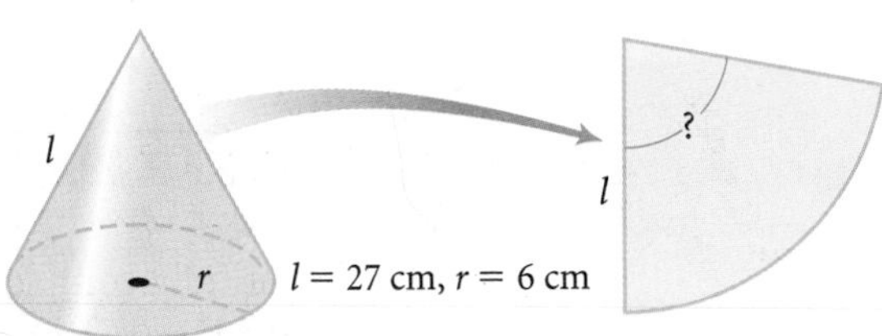

IMPROVING YOUR VISUAL THINKING SKILLS

PUZZLE

Mudville Monsters

The 11 starting members of the Mudville Monsters football team and their coach, Osgood Gipper, have been invited to compete in the Smallville Punt, Pass, and Kick Competition. To get there, they must cross the deep Smallville River. The only way across is with a small boat owned by two very small Smallville football players. The boat holds just one Monster visitor or the two Smallville players. The Smallville players agree to help the Mudville players across if the visitors agree to pay $5 each time the boat crosses the river. If the Monsters have a total of $100 among them, do they have enough money to get all players and the coach to the other side of the river?

Exploration

A Pythagorean Fractal

If you wanted to draw a picture to state the Pythagorean Theorem without words, you'd probably draw a right triangle with squares on each of the three sides. This is the way you first explored the Pythagorean Theorem in Lesson 9.1.

Another picture of the theorem is even simpler: a right triangle divided into two right triangles. Here, a right triangle with hypotenuse c is divided into two smaller triangles, the smaller with hypotenuse a and the larger with hypotenuse b. Clearly, their areas add up to the area of the whole triangle. What's surprising is that all three triangles have the same angle measures. Why? Though different in size, the three triangles all have the same shape. Figures that have the same shape but not necessarily the same size are called **similar figures.** You'll use these similar triangles to prove the Pythagorean Theorem in a later chapter.

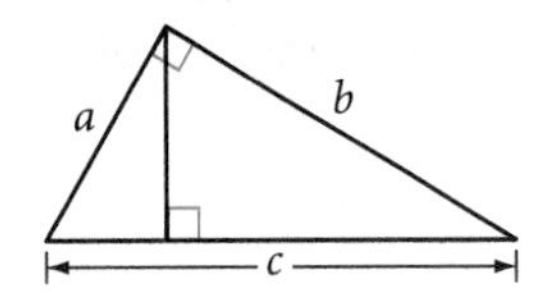

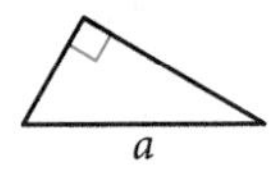

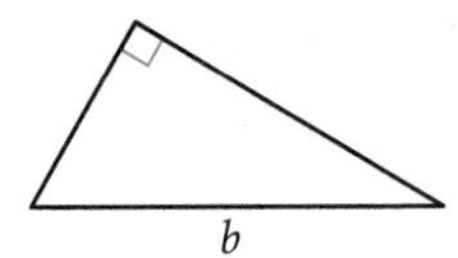

A beautifully complex fractal combines both of these pictorial representations of the Pythagorean Theorem. The fractal starts with a right triangle with squares on each side. Then similar triangles are built onto the squares. Then squares are built onto the new triangles, and so on. In this exploration, you'll create this fractal.

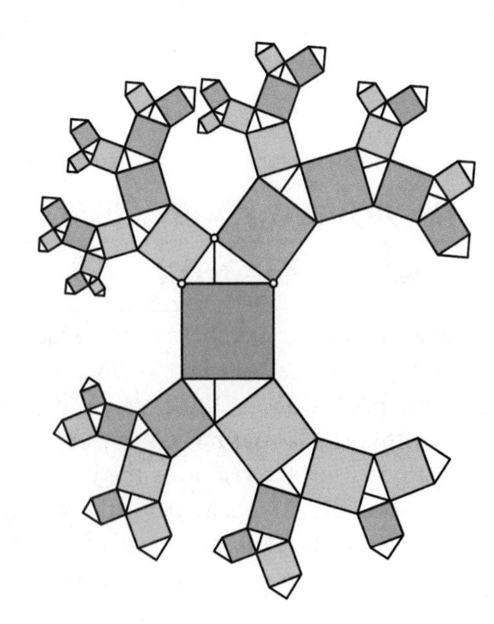

Activity

The Right Triangle Fractal

You will need

- the worksheet The Right Triangle Fractal (optional)

The Geometer's Sketchpad software uses custom tools to save the steps of repeated constructions. They are very helpful for fractals like this one.

Procedure Note

1. Use the diameter of a circle and an inscribed angle to make a triangle that always remains a right triangle.
2. It is important to construct the altitude to the hypotenuse in each triangle in order to divide it into similar triangles.
3. Create custom tools to make squares and similar triangles repeatedly.

Step 1 Use The Geometer's Sketchpad to create the fractal on page 480. Follow the Procedure Note.

Notice that each square has two congruent triangles on two opposite sides. Use a reflection to guarantee that the triangles are congruent.

After you successfully make the Pythagorean fractal, you're ready to investigate its fascinating patterns.

Step 2 First, try dragging a vertex of the original triangle.

Step 3 Does the Pythagorean Theorem still apply to the branches of this figure? That is, does the sum of the areas of the branches on the legs equal the area of the branch on the hypotenuse? See if you can answer without actually measuring all the areas.

Step 4 Consider your original sketch to be a single right triangle with a square built on each side. Call this sketch Stage 0 of your fractal. Explore these questions.

a. At Stage 1, you add three triangles and six squares to your construction. On a piece of paper, draw a rough sketch of Stage 1. How much area do you add to this fractal between Stage 0 and Stage 1? (Don't measure any areas to answer this.)

b. Draw a rough sketch of Stage 2. How much area do you add between Stage 1 and Stage 2?

c. How much area is added at any new stage?

d. A true fractal exists only after an infinite number of stages. If you could build a true fractal based on the construction in this activity, what would be its total area?

Step 5 Give the same color and shade to sets of squares that are congruent. What do you notice about these sets of squares other than their equal area? Describe any patterns you find in sets of congruent squares.

Step 6 Describe any other patterns you can find in the Pythagorean fractal.

LESSON

9.4 Story Problems

You may be disappointed if you fail, but you are doomed if you don't try.

BEVERLY SILLS

You have learned that drawing a diagram will help you to solve difficult problems. By now you know to look for many special relationships in your diagrams, such as congruent polygons, parallel lines, and right triangles.

FUNKY WINKERBEAN by Batiuk. Reprinted with special permission of North America Syndicate.

EXAMPLE What is the longest stick that will fit inside a 24-by-30-by-18-inch box?

Solution Draw a diagram.

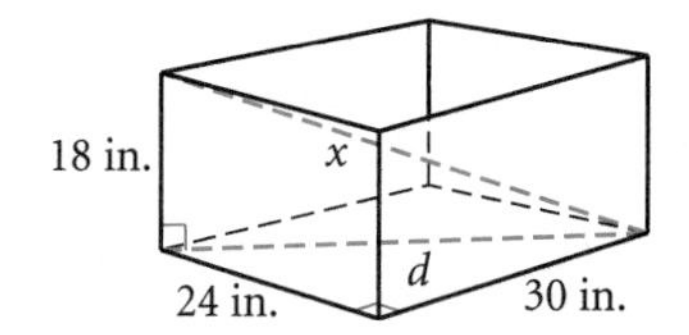

You can lay a stick with length d diagonally at the bottom of the box. But you can position an even longer stick with length x along the diagonal of the box, as shown. How long is this stick?

Both d and x are the hypotenuses of right triangles, but finding d^2 will help you find x.

$$30^2 + 24^2 = d^2$$
$$900 + 576 = d^2$$
$$d^2 = 1476$$

$$d^2 + 18^2 = x^2$$
$$1476 + 18^2 = x^2$$
$$1476 + 324 = x^2$$
$$1800 = x^2$$
$$x \approx 42.4$$

The longest possible stick is about 42.4 in.

EXERCISES

1. A giant California redwood tree 36 meters tall cracked in a violent storm and fell as if hinged. The tip of the once beautiful tree hit the ground 24 meters from the base. Researcher Red Woods wishes to investigate the crack. How many meters up from the base of the tree does he have to climb? ⓗ

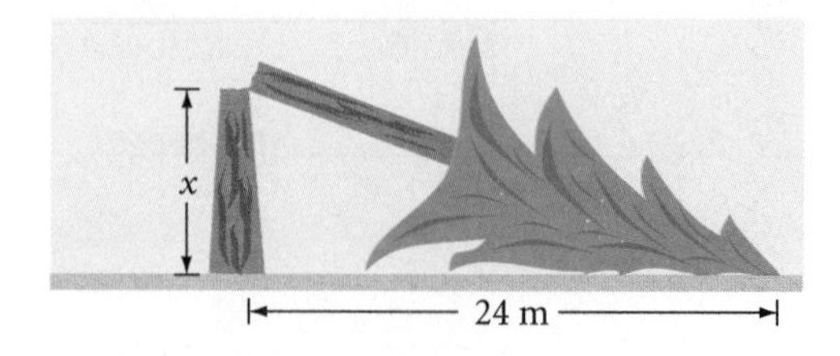

2. Amir's sister is away at college, and he wants to mail her a 34 in. baseball bat. The packing service sells only one kind of box, which measures 24 in. by 2 in. by 18 in. Will the box be big enough?

3. Meteorologist Paul Windward and geologist Rhaina Stone are rushing to a paleontology conference in Pecos Gulch. Paul lifts off in his balloon at noon from Lost Wages, heading east for Pecos Gulch Conference Center. With the wind blowing west to east, he averages a land speed of 30 km/hr. This will allow him to arrive in 4 hours, just as the conference begins. Meanwhile, Rhaina is 160 km north of Lost Wages. At the moment of Paul's lift off, Rhaina hops into an off-roading vehicle and heads directly for the conference center. At what average speed must she travel to arrive at the same time Paul does? ⓗ

Career
CONNECTION

Meteorologists study the weather and the atmosphere. They also look at air quality, oceanic influence on weather, changes in climate over time, and even other planetary climates. They make forecasts using satellite photographs, weather balloons, contour maps, and mathematics to calculate wind speed or the arrival of a storm.

4. A 25-foot ladder is placed against a building. The bottom of the ladder is 7 feet from the building. If the top of the ladder slips down 4 feet, how many feet will the bottom slide out? (It is not 4 feet.) ⓗ

5. The front and back walls of an A-frame cabin are isosceles triangles, each with a base measuring 10 m and legs measuring 13 m. The entire front wall is made of glass 1 cm thick that cost $120/m^2$. What did the glass for the front wall cost? ⓗ

6. A regular hexagonal prism fits perfectly inside a cylindrical box with diameter 6 cm and height 10 cm. What is the surface area of the prism? What is the surface area of the cylinder? ⓗ

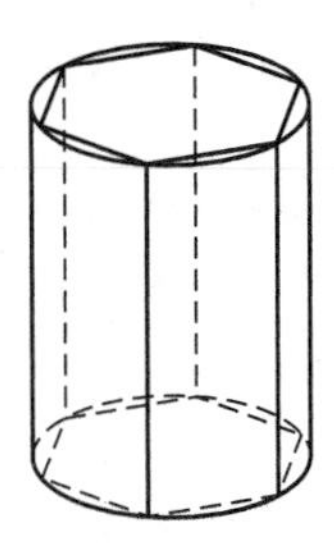

7. Find the perimeter of an equilateral triangle whose median measures 6 cm.

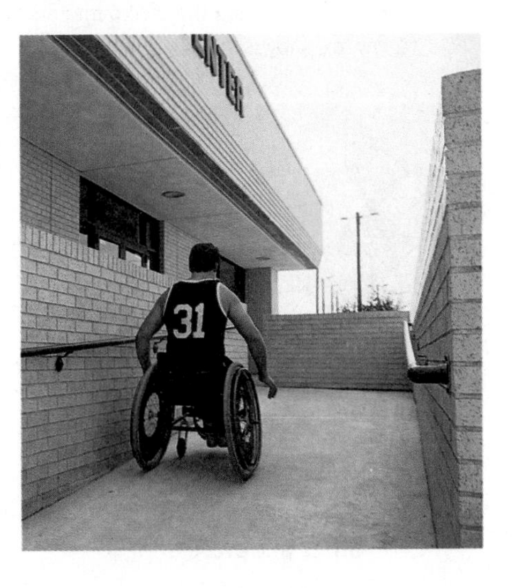

8. **APPLICATION** According to the Americans with Disabilities Act, the slope of a wheelchair ramp must be no greater than $\frac{1}{12}$. What is the length of ramp needed to gain a height of 4 feet? Read the Science Connection on the top of page 484 and then figure out how much force is required to go up the ramp if a person and a wheelchair together weigh 200 pounds.

Science
CONNECTION

It takes less effort to roll objects up an *inclined plane,* or ramp, than to lift them straight up. *Work* is a measure of force applied over distance, and you calculate it as a product of force and distance. For example, a force of 100 pounds is required to hold up a 100-pound object. The work required to lift it 2 feet is 200 foot-pounds. But if you use a 4-foot-long ramp to roll it up, you'll do the 200 foot-pounds of work over a 4-foot distance. So you need to apply only 50 pounds of force at any given moment.

For Exercises 9 and 10, refer to the above Science Connection about inclined planes.

9. Compare what it would take to lift an object these three different ways.

a. How much work, in foot-pounds, is necessary to lift 80 pounds straight up 2 feet?

b. If a ramp 4 feet long is used to raise the 80 pounds up 2 feet, how much force, in pounds, will it take?

c. If a ramp 8 feet long is used to raise the 80 pounds up 2 feet, how much force, in pounds, will it take?

10. If you can exert only 70 pounds of force and you need to lift a 160-pound steel drum up 2 feet, what is the minimum length of ramp you should set up?

Review

Recreation
CONNECTION

This set of enameled porcelain *qi qiao* bowls can be arranged to form a 37-by-37 cm square (as shown) or other shapes, or used separately. Each bowl is 10 cm deep. Dishes of this type are usually used to serve candies, nuts, dried fruits, and other snacks on special occasions.

The *qi qiao,* or tangram puzzle, originated in China and consists of seven pieces—five isosceles right triangles, a square, and a parallelogram. The puzzle involves rearranging the pieces into a square, or hundreds of other shapes (a few are shown below).

Private collection, Berkeley, California. Photo by Cheryl Fenton.

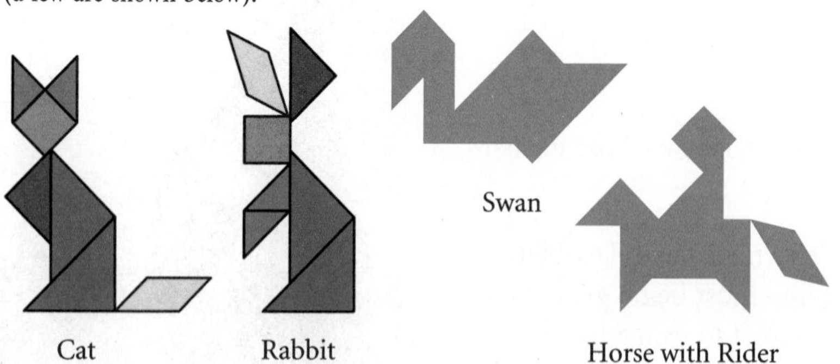

11. If the area of the red square piece is 4 cm^2, what are the dimensions of the other six pieces?

12. Make a set of your own seven tangram pieces and create the Cat, Rabbit, Swan, and Horse with Rider as shown on page 484.

13. Find the radius of circle Q.

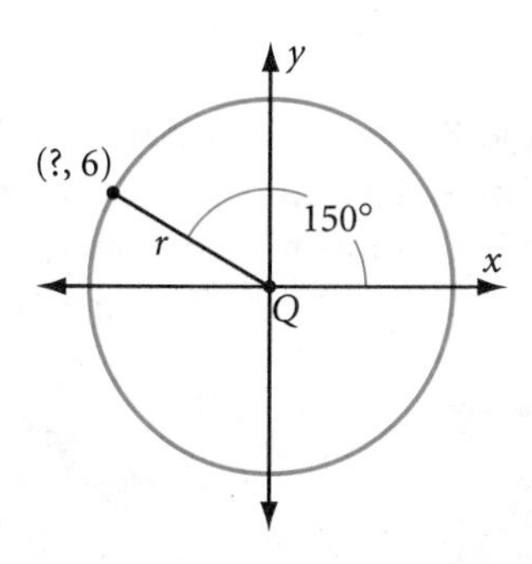

14. Find the length of $\overline{AC}$.

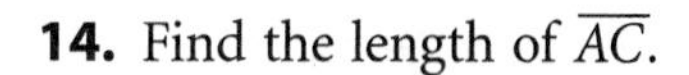

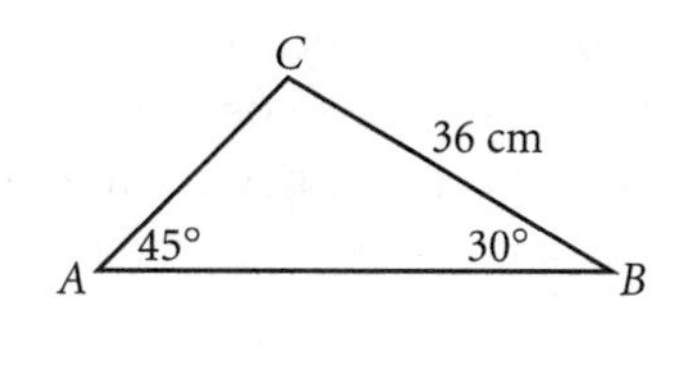

15. The two rays are tangent to the circle. What's wrong with this picture?

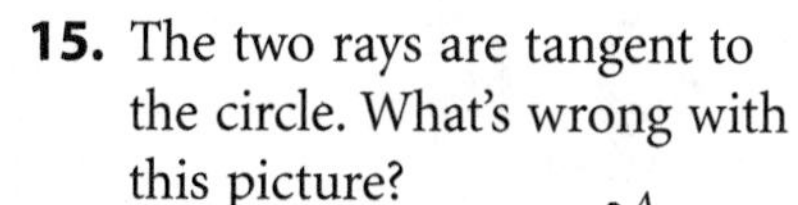

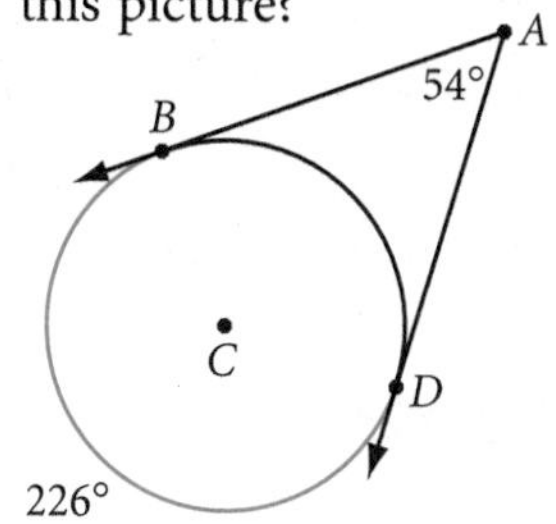

16. In the figure below, point A' is the image of point A after a reflection over $\overrightarrow{OT}$. What are the coordinates of A'? ⓗ

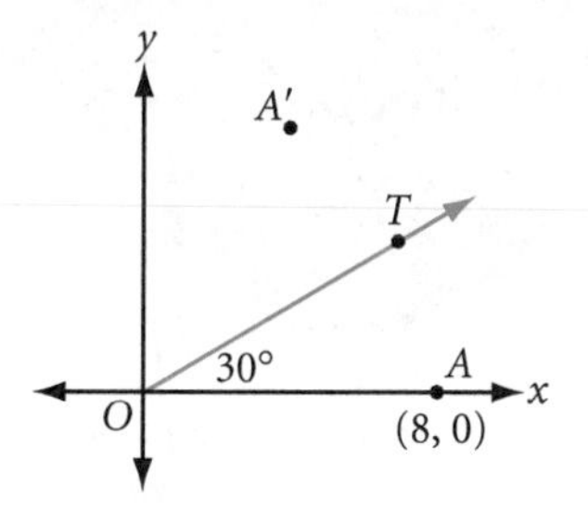

17. Which congruence shortcut can you use to show that $\triangle ABP \cong \triangle DCP$?

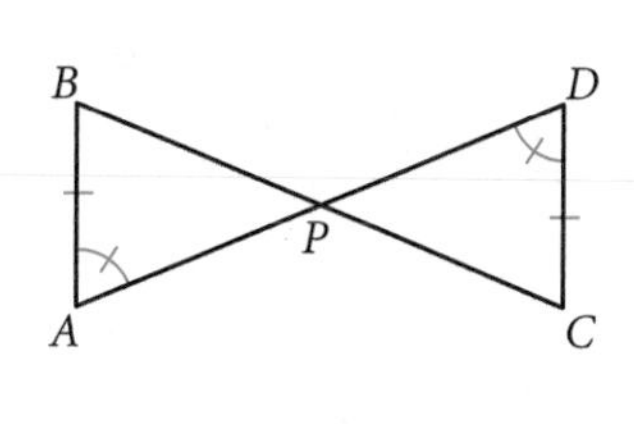

18. Identify the point of concurrency in $\triangle QUO$ from the construction marks.

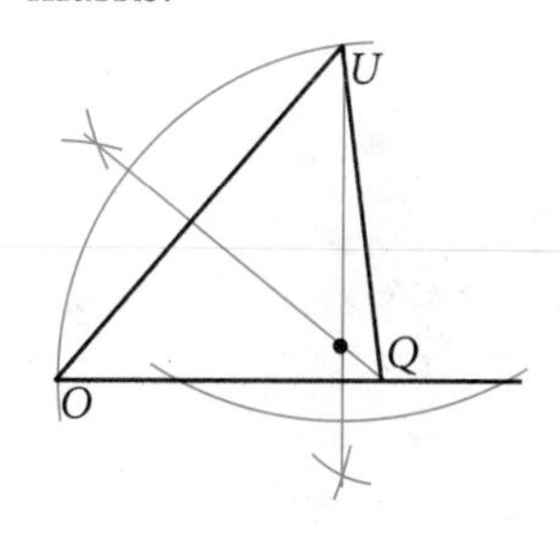

19. In parallelogram $QUID$, $m\angle Q = 2x + 5°$ and $m\angle I = 4x - 55°$. What is $m\angle U$?

20. In $\triangle PRO$, $m\angle P = 70°$ and $m\angle R = 45°$. Which side of the triangle is the shortest?

IMPROVING YOUR VISUAL THINKING SKILLS

PUZZLE

Fold, Punch, and Snip

A square sheet of paper is folded vertically, a hole is punched out of the center, and then one of the corners is snipped off. When the paper is unfolded it will look like the figure at right.

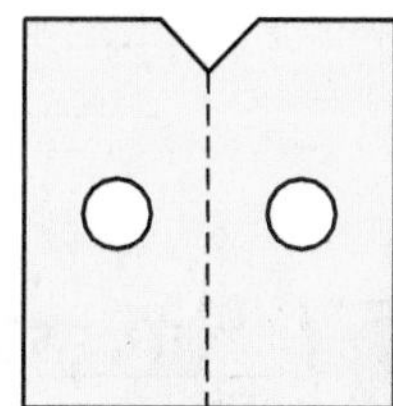

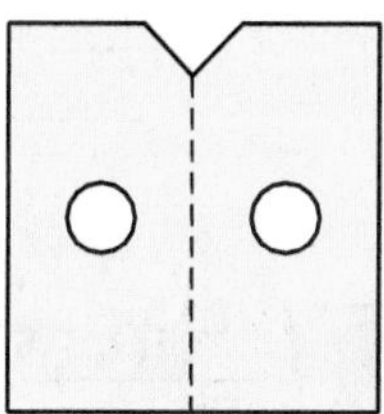

Sketch what a square sheet of paper will look like when it is unfolded after the following sequence of folds, punches, and snips.

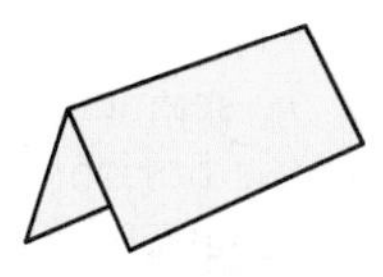

Fold once.

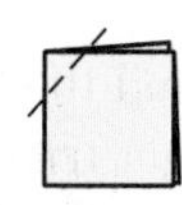

Fold twice.

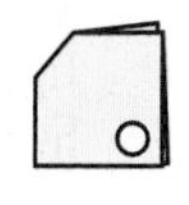

Snip double-fold corner.

Punch opposite corner.

LESSON 9.5

Distance in Coordinate Geometry

We talk too much; we should talk less and draw more.

JOHANN WOLFGANG VON GOETHE

Viki is standing on the corner of Seventh Street and 8th Avenue, and her brother Scott is on the corner of Second Street and 3rd Avenue. To find her shortest sidewalk route to Scott, Viki can simply count blocks. But if Viki wants to know her diagonal distance to Scott, she would need the Pythagorean Theorem to measure across blocks.

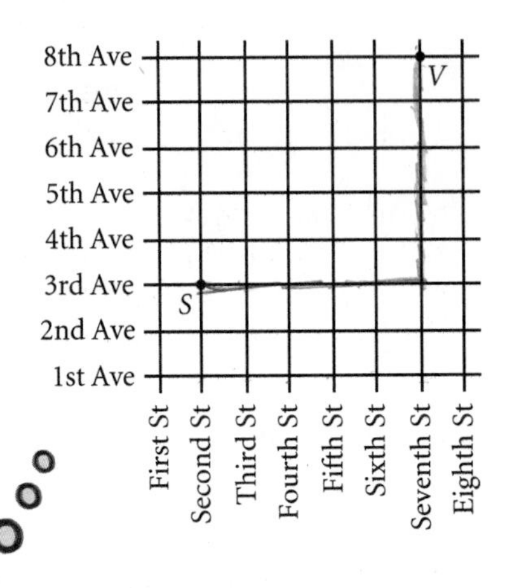

You can think of a coordinate plane as a grid of streets with two sets of parallel lines running perpendicular to each other. Every segment in the plane that is not in the x- or y-direction is the hypotenuse of a right triangle whose legs are in the x- and y-directions. So you can use the Pythagorean Theorem to find the distance between any two points on a coordinate plane.

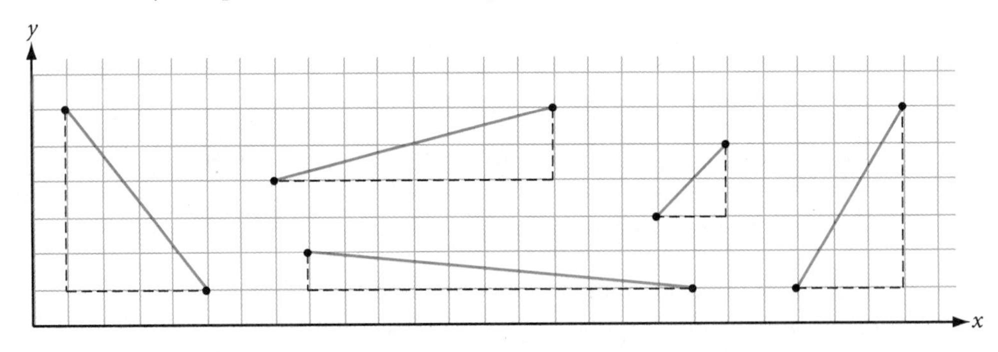

Investigation 1
The Distance Formula

You will need

- graph paper

In Steps 1 and 2, find the length of each segment by using the segment as the hypotenuse of a right triangle. Simply count the squares on the horizontal and vertical legs, then use the Pythagorean Theorem to find the length of the hypotenuse.

Step 1 | Copy graphs a–d from the next page onto your own graph paper. Use each segment as the hypotenuse of a right triangle. Draw the legs along the grid lines. Find the length of each segment.

a.

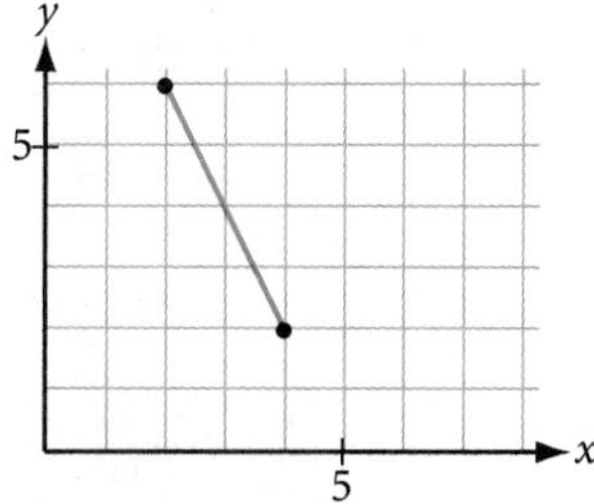

b.

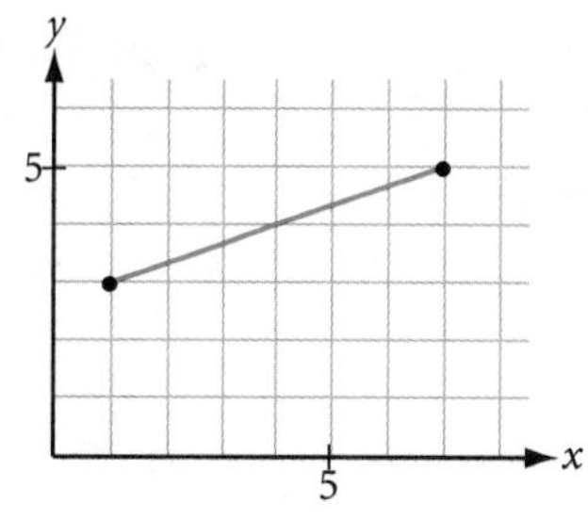

c.

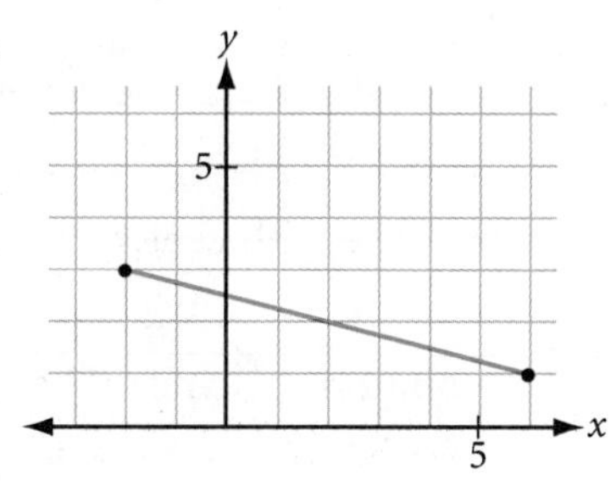

d.

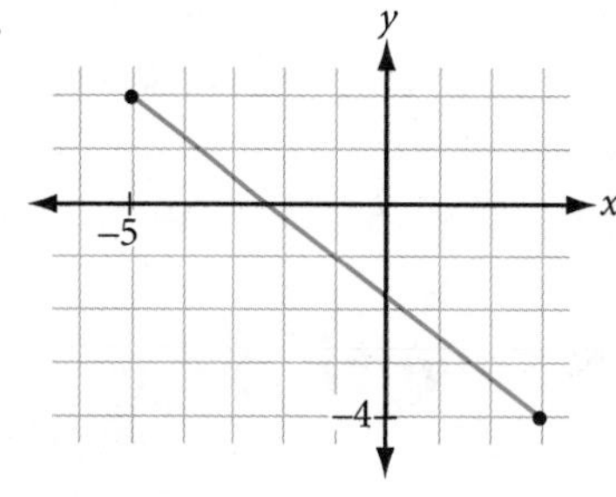

Step 2 Graph each pair of points, then find the distances between them.

a. $(-1, -2), (11, -7)$ **b.** $(-9, -6), (3, 10)$

What if the points are so far apart that it's not practical to plot them? For example, what is the distance between the points $A(15, 34)$ and $B(42, 70)$? A formula that uses the coordinates of the given points would be helpful. To find this formula, you first need to find the lengths of the legs in terms of the x- and y-coordinates. From your work with slope triangles, you know how to calculate horizontal and vertical distances.

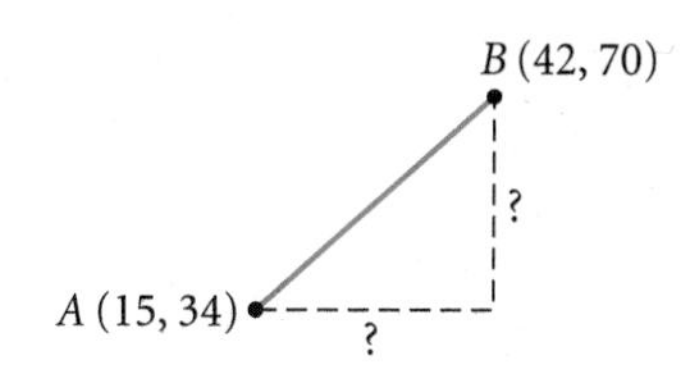

Step 3 Write an expression for the length of the horizontal leg using the x-coordinates.

Step 4 Write a similar expression for the length of the vertical leg using the y-coordinates.

Step 5 Use your expressions from Steps 3 and 4, and the Pythagorean Theorem, to find the distance between points $A(15, 34)$ and $B(42, 70)$.

Step 6 Generalize what you have learned about the distance between two points in a coordinate plane. Copy and complete the conjecture below.

Distance Formula C-86

The distance between points $A(x_1, y_1)$ and $B(x_2, y_2)$ is given by

$$(AB)^2 = (\underline{?})^2 + (\underline{?})^2 \text{ or } AB = \sqrt{(\underline{?})^2 + (\underline{?})^2}$$

Let's look at an example to see how you can apply the distance formula.

EXAMPLE A | Find the distance between $A(8, 15)$ and $B(-7, 23)$.

▶ Solution

$(AB)^2 = (x_2 - x_1)^2 + (y_2 - y_1)^2$	The distance formula.
$= (-7 - 8)^2 + (23 - 15)^2$	Substitute 8 for x_1, 15 for y_1, −7 for x_2, and 23 for y_2.
$= (-15)^2 + (8)^2$	Subtract.
$(AB)^2 = 289$	Square −15 and 8 and add.
$AB = 17$	Take the square root of both sides.

The distance formula is also used to write the equation of a circle.

EXAMPLE B | Write an equation for the circle with center (5, 4) and radius 7 units.

▶ Solution

Let (x, y) represent any point on the circle. The distance from (x, y) to the circle's center, (5, 4), is 7. Substitute this information in the distance formula.

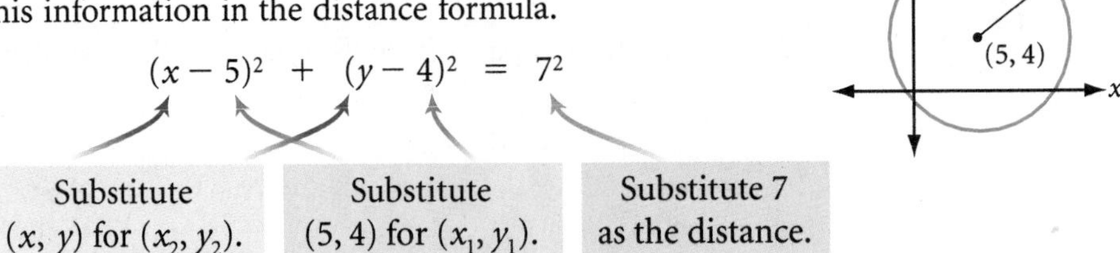

So, the equation in standard form is $(x - 5)^2 + (y - 4)^2 = 7^2$.

Investigation 2
The Equation of a Circle

You will need

- graph paper

Find equations for a few more circles and then generalize the equation for any circle with radius r and center (h, k).

Step 1 Given its center and radius, graph each circle on graph paper.

a. Center = $(1, -2)$, $r = 8$ **b.** Center = $(0, 2)$, $r = 6$

c. Center = $(-3, -4)$, $r = 10$

Step 2 Select any point on each circle; label it (x, y). Use the distance formula to write an equation expressing the distance between the center of each circle and (x, y).

Step 3 Copy and complete the conjecture for the equation of a circle.

Equation of a Circle

C-87

The equation of a circle with radius r and center (h, k) is

$$(x - \underline{?})^2 + (y - \underline{?})^2 = (\underline{?})^2$$

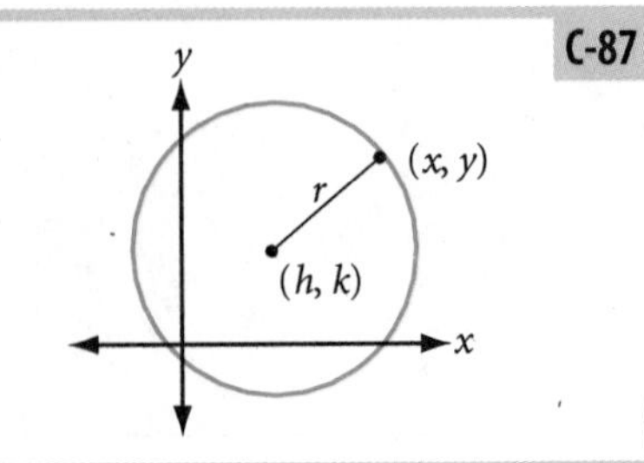

Let's look at an example that uses the equation of a circle in reverse.

EXAMPLE C | Find the center and radius of the circle $(x + 2)^2 + (y - 5)^2 = 36$.

▶ Solution | Rewrite the equation of the circle in the standard form.

$$(x - h)^2 + (y - k)^2 = r^2$$
$$(x - (-2))^2 + (y - 5)^2 = 6^2$$

Identify the values of h, k, and r. The center is $(-2, 5)$ and the radius is 6.

EXERCISES

In Exercises 1–3, find the distance between each pair of points.

1. (10, 20), (13, 16)

2. (15, 37), (42, 73)

3. (−19, −16), (−3, 14)

4. Look back at the diagram of Viki's and Scott's locations on page 486. Assume each block is approximately 50 meters long. What is the shortest distance from Viki to Scott to the nearest meter?

5. Find the perimeter of $\triangle ABC$ with vertices $A(2, 4)$, $B(8, 12)$, and $C(24, 0)$.

6. Determine whether $\triangle DEF$ with vertices $D(6, -6)$, $E(39, -12)$, and $F(24, 18)$ is scalene, isosceles, or equilateral.

For Exercises 7 and 8, find the equation of the circle.

7. Center = (0, 0), $r = 4$

8. Center = (2, 0), $r = 5$

For Exercises 9 and 10, find the radius and center of the circle.

9. $(x - 2)^2 + (y + 5)^2 = 6^2$

10. $x^2 + (y - 1)^2 = 81$

11. The center of a circle is (3, −1). One point on the circle is (6, 2). Find the equation of the circle. ⓗ

12. ***Mini-Investigation*** How would you find the distance between two points in a three-dimensional coordinate system? Investigate and make a conjecture. ⓗ

a. What is the distance from the origin (0, 0, 0) to (2, −1, 3)?

b. What is the distance between $P(1, 2, 3)$ and $Q(5, 6, 15)$?

c. Complete this conjecture:
If $A(x_1, y_1, z_1)$ and $B(x_2, y_2, z_2)$ are two points in a three-dimensional coordinate system, then the distance AB is $\sqrt{(\underline{?})^2 + (\underline{?})^2 + (\underline{?})^2}$.

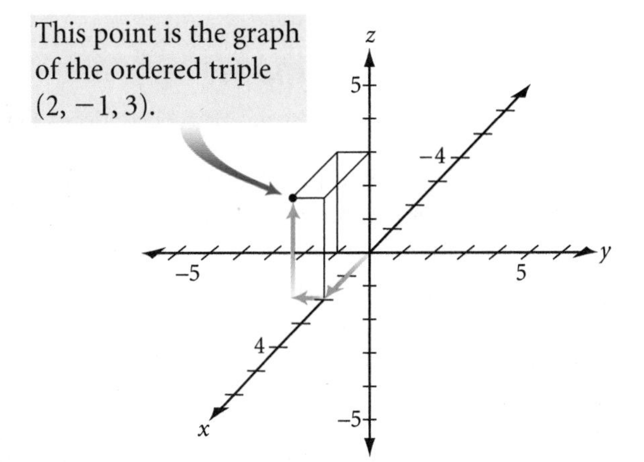

Review

13. Find the coordinates of A.

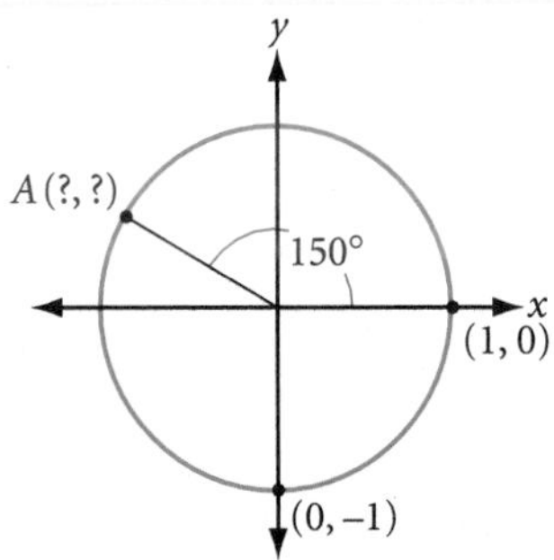

14. $k = \underline{\ ?\ }$, $m = \underline{\ ?\ }$ ⓗ

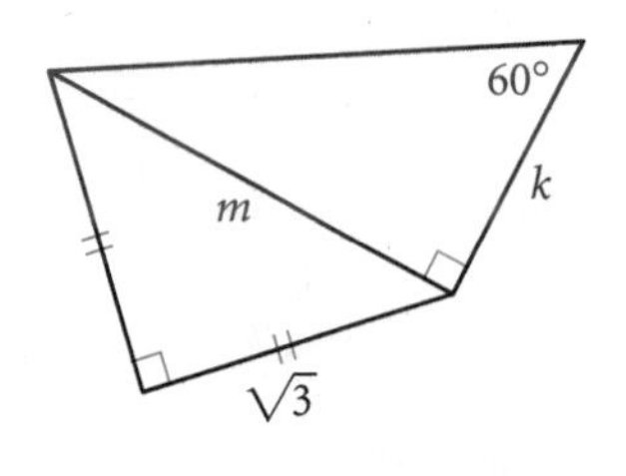

15. The large triangle is equilateral. Find x and y.

16. Antonio is a biologist studying life in a pond. He needs to know how deep the water is. He notices a water lily sticking straight up from the water, whose blossom is 8 cm above the water's surface. Antonio pulls the lily to one side, keeping the stem straight, until the blossom touches the water at a spot 40 cm from where the stem first broke the water's surface. How is Antonio able to calculate the depth of the water? What is the depth? ⓗ

17. $C'U'R'T'$ is the image of $CURT$ under a rotation transformation. Copy the polygon and its image onto patty paper. Find the center of rotation and the measure of the angle of rotation. Explain your method.

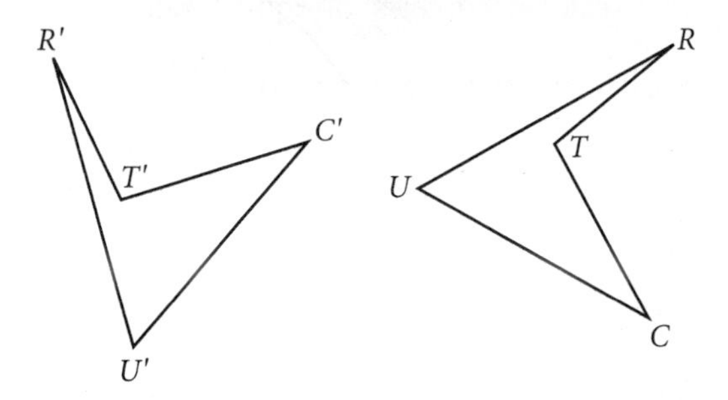

IMPROVING YOUR VISUAL THINKING SKILLS

PUZZLE

The Spider and the Fly

(attributed to the British puzzlist Henry E. Dudeney, 1857–1930)

In a rectangular room, measuring 30 by 12 by 12 feet, a spider is at point A on the middle of one of the end walls, 1 foot from the ceiling. A fly is at point B on the center of the opposite wall, 1 foot from the floor. What is the shortest distance that the spider must crawl to reach the fly, which remains stationary? The spider never drops or uses its web, but crawls fairly.

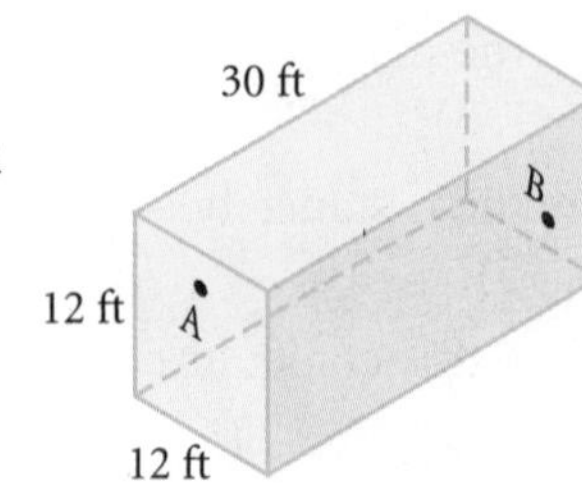

Exploration

Ladder Climb

Suppose a house painter rests a 20-foot ladder against a building, then decides the ladder needs to rest 1 foot higher against the building. Will moving the ladder 1 foot toward the building do the job? If it needs to be 2 feet lower, will moving the ladder 2 feet away from the building do the trick? Let's investigate.

Activity
Climbing the Wall

You will need

- a graphing calculator

Sketch a ladder leaning against a vertical wall, with the foot of the ladder resting on horizontal ground. Label the sketch using y for height reached by the ladder and x for the distance from the base of the wall to the foot of the ladder.

Step 1 Write an equation relating x, y, and the length of the ladder and solve it for y. You now have a function for the height reached by the ladder in terms of the distance from the wall to the foot of the ladder. Enter this equation into your calculator.

Step 2 Before you graph the equation, think about the settings you'll want for the graph window. What are the greatest and least values possible for x and y? Enter reasonable settings, then graph the equation.

Step 3 Describe the shape of the graph.

Step 4 Trace along the graph, starting at $x = 0$. Record values (rounded to the nearest 0.1 unit) for the height reached by the ladder when $x = 3, 6, 9$, and 12. If you move the foot of the ladder away from the wall 3 feet at a time, will each move result in the same change in the height reached by the ladder? Explain.

Step 5 Find the value for x that gives a y-value approximately equal to x. How is this value related to the length of the ladder? Sketch the ladder in this position. What angle does the ladder make with the ground?

Step 6 Should you lean a ladder against a wall in such a way that x is greater than y? Explain. How does your graph support your explanation?

LESSON

9.6 Circles and the Pythagorean Theorem

You must do things you think you cannot do.

ELEANOR ROOSEVELT

In Chapter 6, you discovered a number of properties that involved right angles in and around circles. In this lesson you will use the conjectures you made, along with the Pythagorean Theorem, to solve some challenging problems. Let's review two of the most useful conjectures.

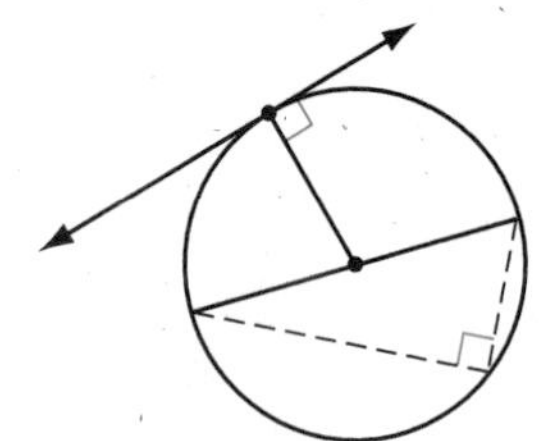

Tangent Conjecture: A tangent to a circle is perpendicular to the radius drawn to the point of tangency.

Angles Inscribed in a Semicircle Conjecture: Angles inscribed in a semicircle are right angles.

Here are two examples that use these conjectures along with the Pythagorean Theorem.

EXAMPLE A

$\overrightarrow{TA}$ is tangent to circle N at A. $TA = 12\sqrt{3}$ cm. Find the area of the shaded region.

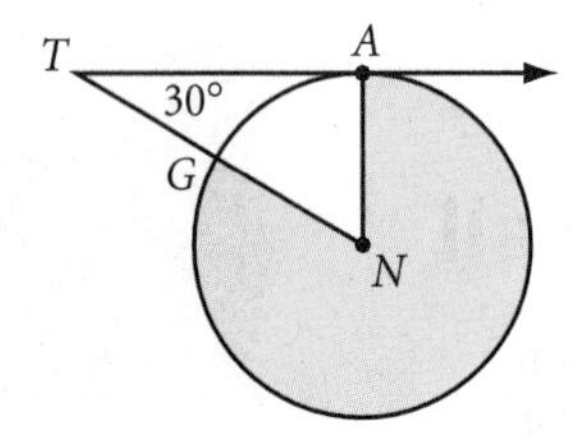

▶ Solution

$\overrightarrow{TA}$ is tangent at A, so $\angle TAN$ is a right angle and $\triangle TAN$ is a 30°-60°-90° triangle. The longer leg is $12\sqrt{3}$ cm, so the shorter leg (also the radius of the circle) is 12 cm. The area of the entire circle is 144π cm^2. The area of the shaded region is $\frac{360-60}{360}$, or $\frac{5}{6}$, of the area of the circle. Therefore the shaded area is $\frac{5}{6}(144\pi)$, or 120π cm^2.

EXAMPLE B

$AB = 6$ cm and $BC = 8$ cm. Find the area of the circle.

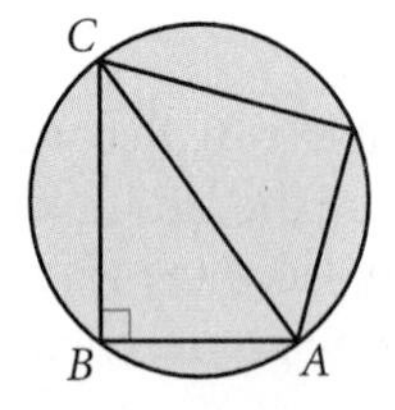

▶ Solution

Inscribed angle ABC is a right angle, so $\overparen{ABC}$ is a semicircle and $\overline{AC}$ is a diameter. By the Pythagorean Theorem, if $AB = 6$ cm and $BC = 8$ cm, then $AC = 10$ cm. Therefore the radius of the circle is 5 cm and the area of the circle is 25π cm^2.

Exercises

In Exercises 1–4, find the area of the shaded region in each figure. Assume lines that appear tangent are tangent at the labeled points.

1. $OD = 24$ cm (h) **2.** $HT = 8\sqrt{3}$ cm **3.** $HA = 8\sqrt{3}$ cm (h) **4.** $HO = 8\sqrt{3}$ cm

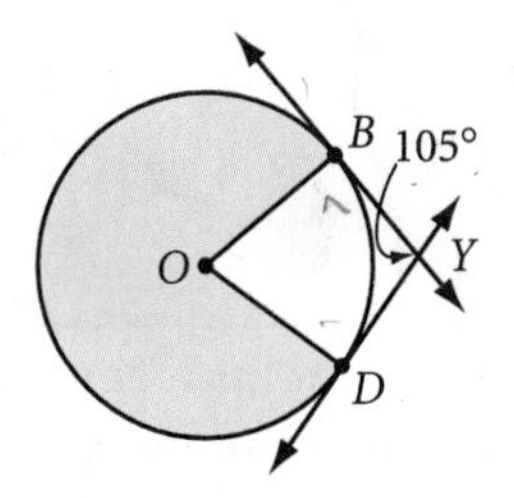

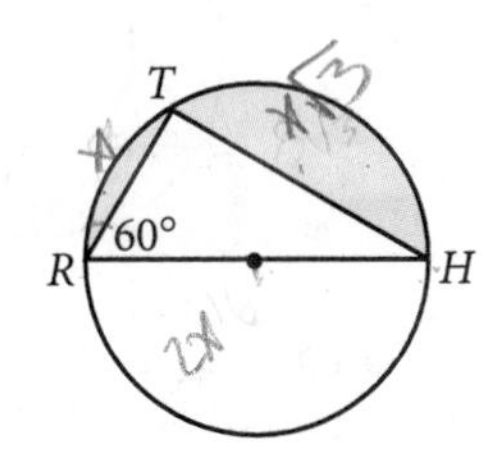

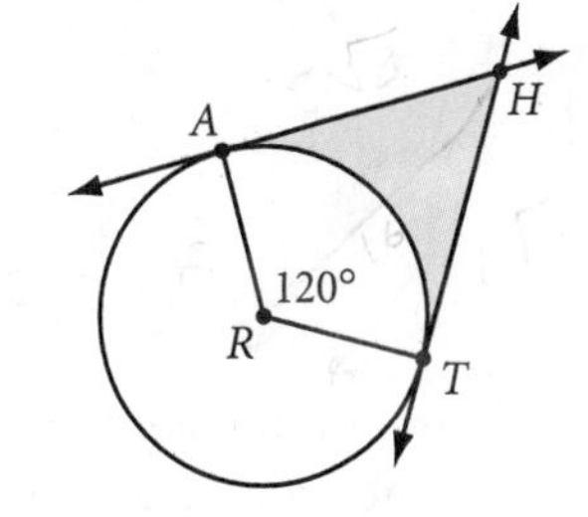

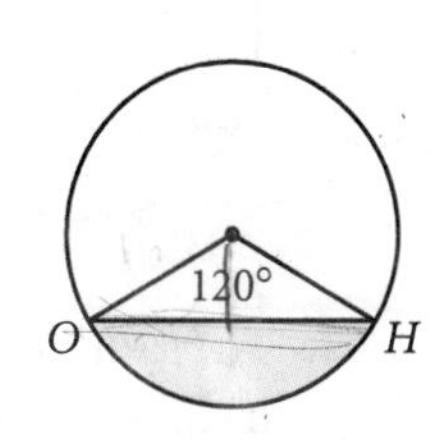

5. A 3-meter-wide circular track is shown at right. The radius of the inner circle is 12 meters. What is the longest straight path that stays on the track? (In other words, find AB.) (h)

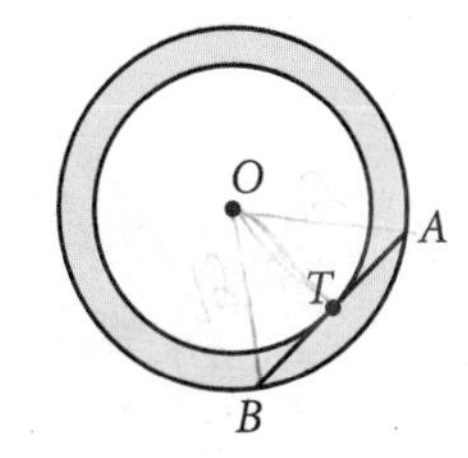

6. An annulus has a 36 cm chord of the outer circle that is also tangent to the inner concentric circle. Find the area of the annulus.

7. In her latest expedition, Ertha Diggs has uncovered a portion of circular, terra-cotta pipe that she believes is part of an early water drainage system. To find the diameter of the original pipe, she lays a meterstick across the portion and measures the length of the chord at 48 cm. The depth of the portion from the midpoint of the chord is 6 cm. What was the pipe's original diameter?

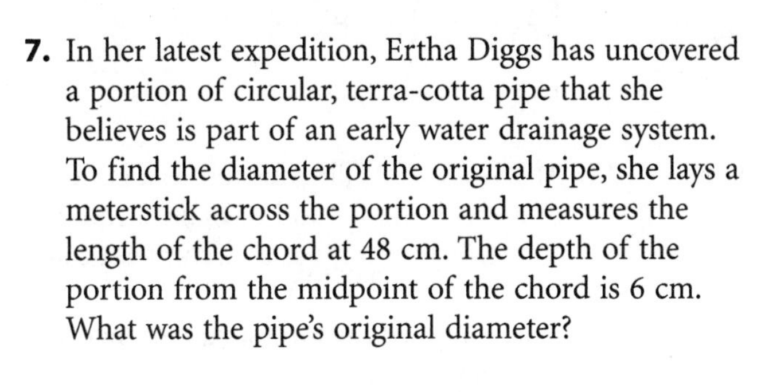

8. APPLICATION A machinery belt needs to be replaced. The belt runs around two wheels, crossing between them so that the larger wheel turns the smaller wheel in the opposite direction. The diameter of the larger wheel is 36 cm, and the diameter of the smaller is 24 cm. The distance between the centers of the two wheels is 60 cm. The belt crosses 24 cm from the center of the smaller wheel. What is the length of the belt? (h)

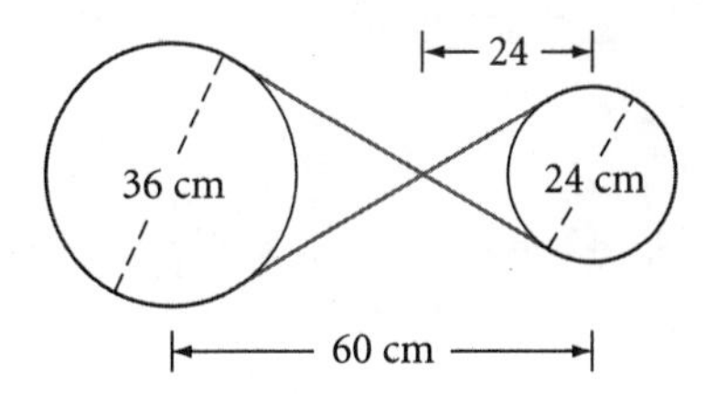

9. A circle of radius 6 has chord $\overline{AB}$ of length 6. If point C is selected randomly on the circle, what is the probability that $\triangle ABC$ is obtuse?

In Exercises 10 and 11, each triangle is equilateral. Find the area of the inscribed circle and the area of the circumscribed circle. How many times greater is the area of the circumscribed circle than the area of the inscribed circle?

10. $AB = 6$ cm ⓗ

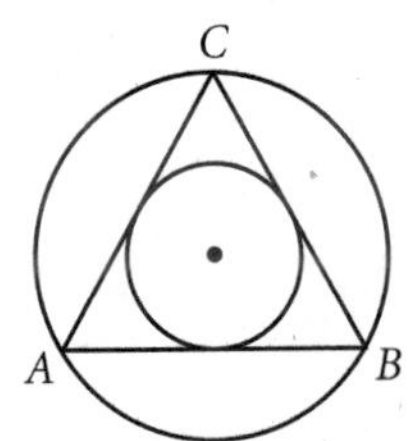

11. $DE = 2\sqrt{3}$ cm

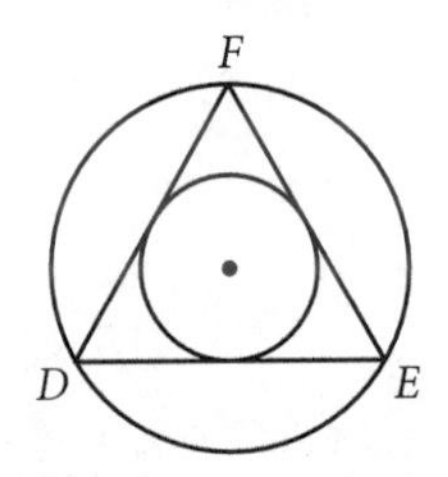

12. The Gothic arch is based on the equilateral triangle. If the base of the arch measures 80 cm, what is the area of the shaded region?

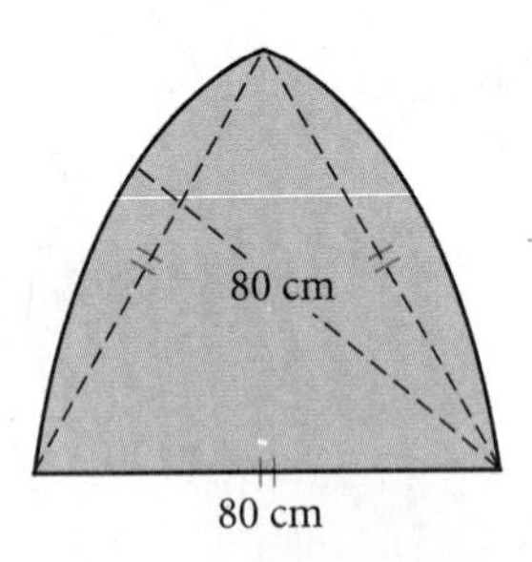

13. Each of three circles of radius 6 cm is tangent to the other two, and they are inscribed in a rectangle, as shown. What is the height of the rectangle?

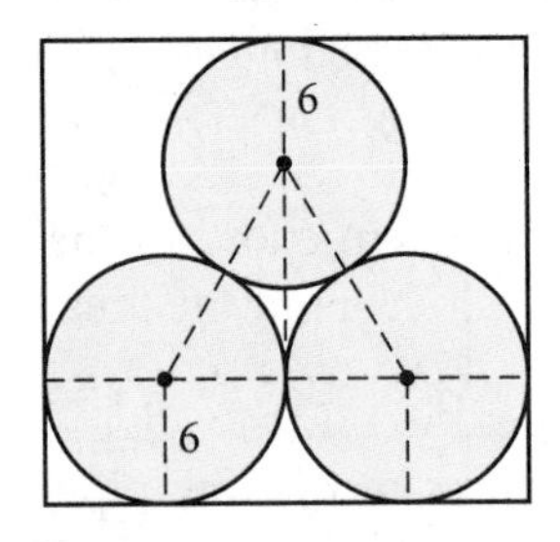

14. Sector ARC has a radius of 9 cm and an angle that measures 80°. When sector ARC is cut out and $\overline{AR}$ and $\overline{RC}$ are taped together, they form a cone. The length of $\overset{\frown}{AC}$ becomes the circumference of the base of the cone. What is the height of the cone? ⓗ

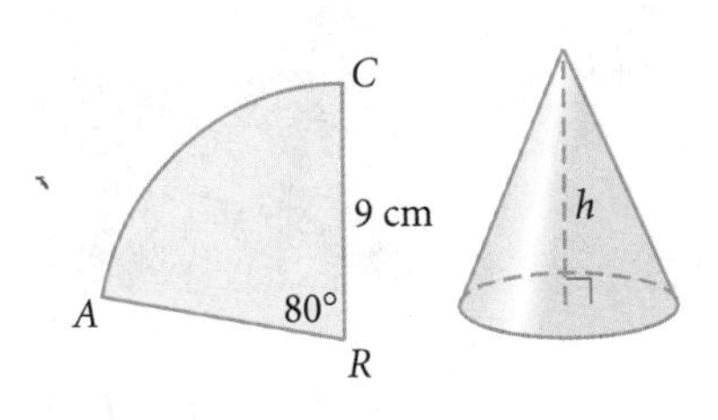

15. **APPLICATION** Will plans to use a circular cross section of wood to make a square table. The cross section has a circumference of 336 cm. To the nearest centimeter, what is the side length of the largest square that he can cut from it?

16. Find the coordinates of point M.

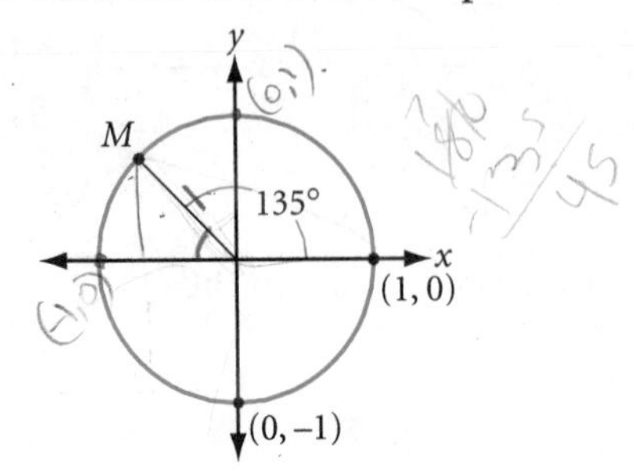

17. Find the coordinates of point K.

y
210°
x
(1, 0)
K
(0, –1)

Review

18. Find the equation of a circle with center (3, 3) and radius 6.

19. Find the radius and center of a circle with the equation $x^2 + y^2 - 2x + 1 = 100$. ⓗ

20. *Construction* Construct a circle and a chord in a circle. With compass and straightedge, construct a second chord parallel and congruent to the first chord. Explain your method.

21. Explain why the opposite sides of a regular hexagon are parallel.

22. Find the rule for this number pattern:

$$1 \cdot 3 - 3 = 4 \cdot 0$$
$$2 \cdot 4 - 3 = 5 \cdot 1$$
$$3 \cdot 5 - 3 = 6 \cdot 2$$
$$4 \cdot 6 - 3 = 7 \cdot 3$$
$$5 \cdot 7 - 3 = 8 \cdot 4$$
$$\vdots$$
$$n \cdot (\underline{?}) - (\underline{?}) = (\underline{?}) \cdot (\underline{?})$$

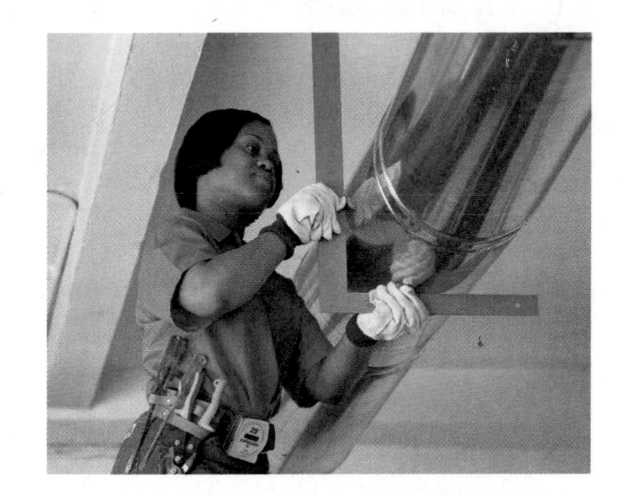

23. APPLICATION Felice wants to determine the diameter of a large heating duct. She places a carpenter's square up to the surface of the cylinder, and the length of each tangent segment is 10 inches.

a. What is the diameter? Explain your reasoning.

b. Describe another way she can find the diameter of the duct.

IMPROVING YOUR REASONING SKILLS

Reasonable 'rithmetic I

Each letter in these problems represents a different digit.

1. What is the value of *B*?

	3	7	2
	3	8	4
+	9	B	4
C	7	C	A

2. What is the value of *J*?

	E	F	6
	×	D	7
D	D	F	D
J	E	D	
H	G	E	D

CHAPTER 9 REVIEW

If 50 years from now you've forgotten everything else you learned in geometry, you'll probably still remember the Pythagorean Theorem. (Though let's hope you don't really forget everything else!) That's because it has practical applications in the mathematics and science that you encounter throughout your education.

It's one thing to remember the equation $a^2 + b^2 = c^2$. It's another to know what it means and to be able to apply it. Review your work from this chapter to be sure you understand how to use special triangle shortcuts and how to find the distance between two points in a coordinate plane.

EXERCISES

You will need

Construction tools for Exercise **30**

For Exercises 1–4, measurements are given in centimeters.

1. $x = \underline{?}$

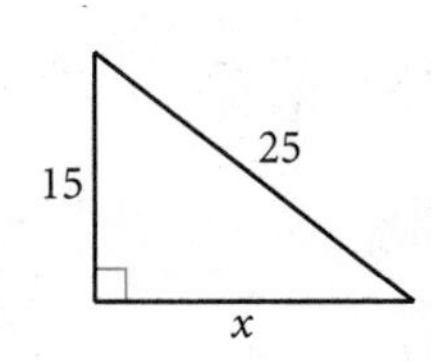

2. $AB = \underline{?}$

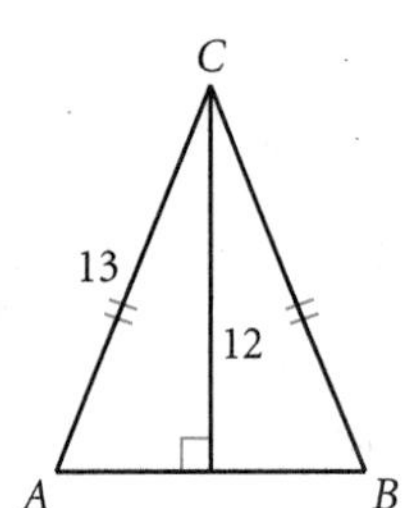

3. Is $\triangle ABC$ an acute, obtuse, or right triangle?

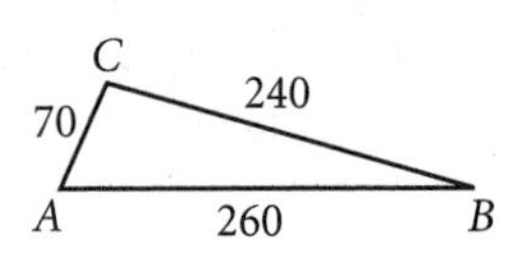

4. The solid is a rectangular prism. $AB = \underline{?}$

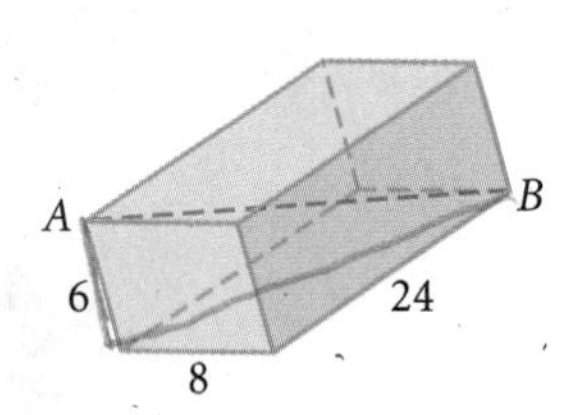

5. Find the coordinates of point U.

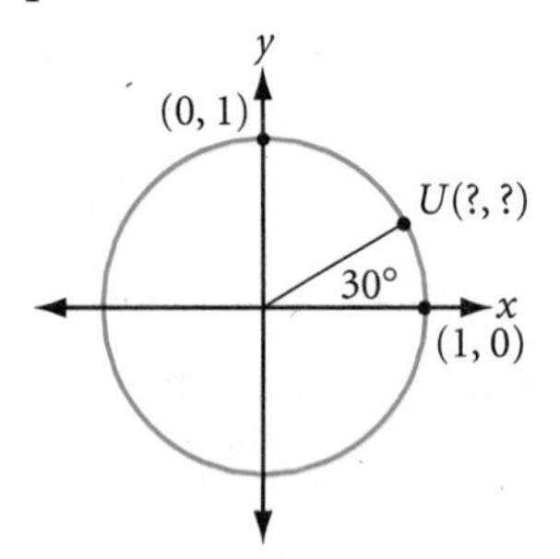

6. Find the coordinates of point V.

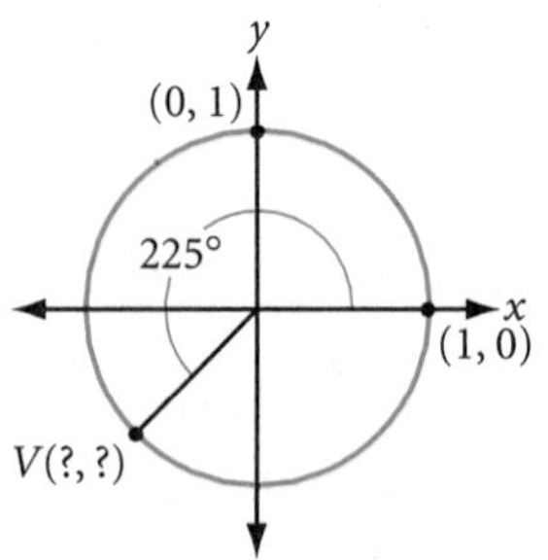

7. What is the area of the triangle?

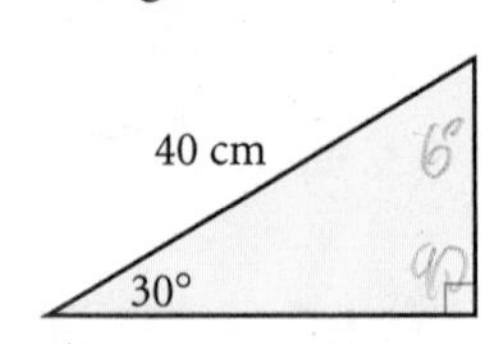

8. The area of this square is 144 cm^2. Find d.

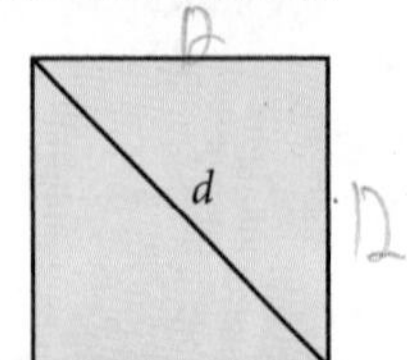

9. What is the area of trapezoid $ABCD$?

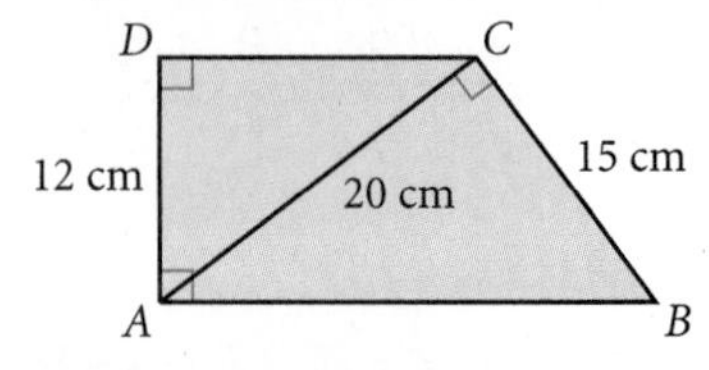

10. The arc is a semicircle. What is the area of the shaded region? ⓗ

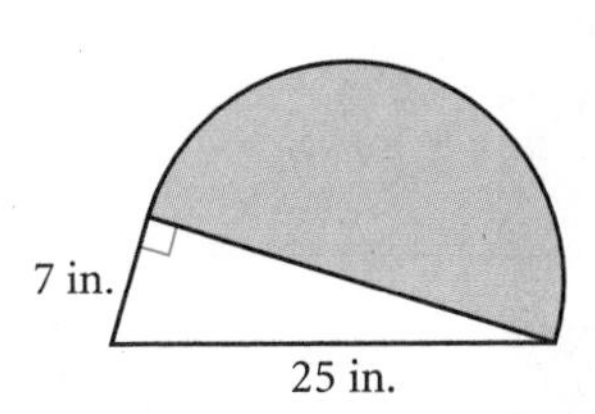

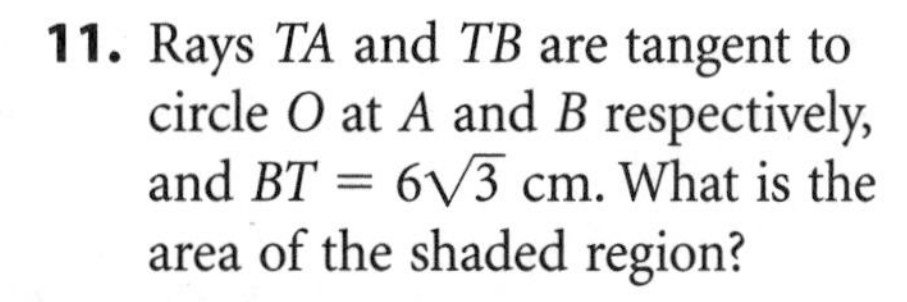

11. Rays TA and TB are tangent to circle O at A and B respectively, and $BT = 6\sqrt{3}$ cm. What is the area of the shaded region?

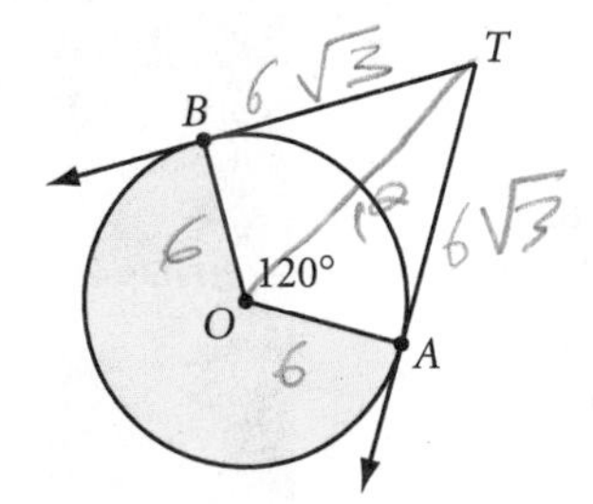

12. The quadrilateral is a square, and $QE = 2\sqrt{2}$ cm. What is the area of the shaded region? ⓗ

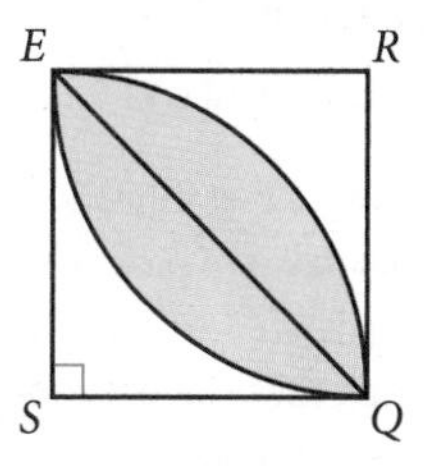

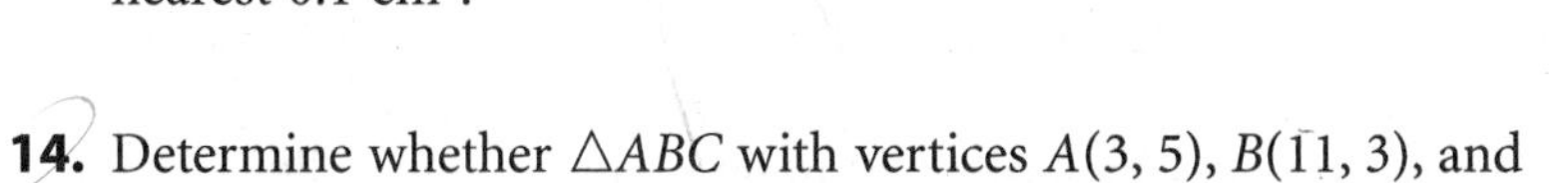

13. The area of circle Q is 350 cm². Find the area of square $ABCD$ to the nearest 0.1 cm².

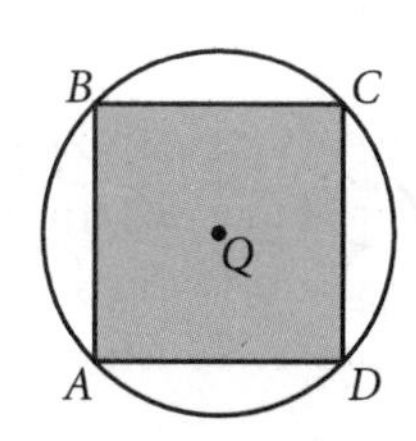

14. Determine whether $\triangle ABC$ with vertices $A(3, 5)$, $B(11, 3)$, and $C(8, 8)$ is an equilateral, isosceles, or isosceles right triangle.

15. Sagebrush Sally leaves camp on her dirt bike traveling east at 60 km/hr with a full tank of gas. After 2 hours, she stops and does a little prospecting—with no luck. So she heads north for 2 hours at 45 km/hr. She stops again, and this time hits pay dirt. Sally knows that she can travel at most 350 km on one tank of gas. Does she have enough fuel to get back to camp? If not, how close can she get? ⓗ

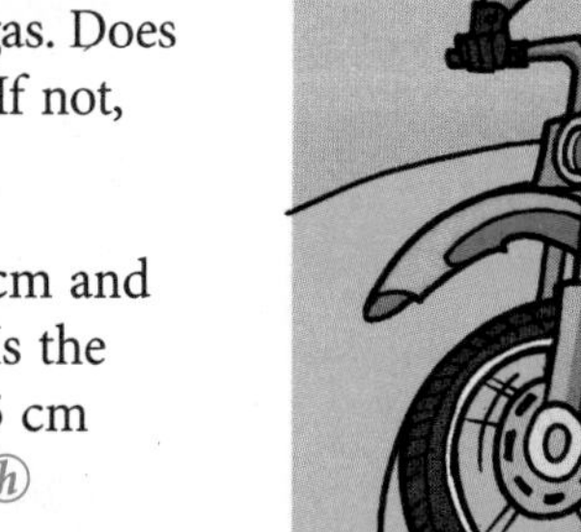

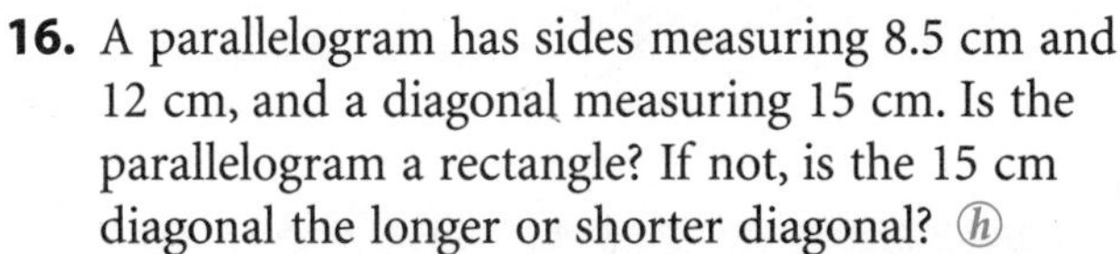

16. A parallelogram has sides measuring 8.5 cm and 12 cm, and a diagonal measuring 15 cm. Is the parallelogram a rectangle? If not, is the 15 cm diagonal the longer or shorter diagonal? ⓗ

17. After an argument, Peter and Paul walk away from each other on separate paths at a right angle to each other. Peter is walking 2 km/hr, and Paul is walking 3 km/hr. After 20 min, Paul sits down to think. After 30 min, Peter stops. Both decide to apologize. How far apart are they? How long will it take them to reach each other if they both start running straight toward each other at 5 km/hr?

18. Flora is away at camp and wants to mail her flute back home. The flute is 24 inches long. Will it fit diagonally within a box whose inside dimensions are 12 by 16 by 14 inches?

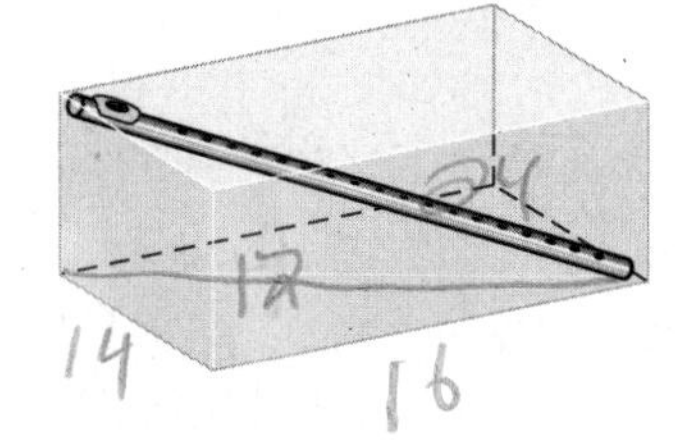

19. To the nearest foot, find the original height of a fallen flagpole that cracked and fell as if hinged, forming an angle of 45 degrees with the ground. The tip of the pole hit the ground 12 feet from its base.

20. You are standing 12 feet from a cylindrical corn-syrup storage tank. The distance from you to a point of tangency on the tank is 35 feet. What is the radius of the tank?

r
35 feet
r
12 feet

Technology CONNECTION

Radio and TV stations broadcast from high towers. Their signals are picked up by radios and TVs in homes within a certain radius. Because Earth is spherical, these signals don't get picked up beyond the point of tangency.

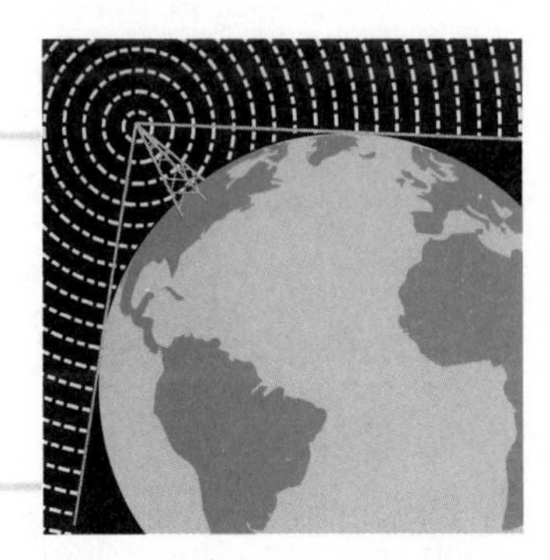

21. APPLICATION Read the Technology Connection above. What is the maximum broadcasting radius from a radio tower 1800 feet tall (approximately 0.34 mile)? The radius of Earth is approximately 3960 miles, and you can assume the ground around the tower is nearly flat. Round your answer to the nearest 10 miles.

22. A diver hooked to a 25-meter line is searching for the remains of a Spanish galleon in the Caribbean Sea. The sea is 20 meters deep and the bottom is flat. What is the area of circular region that the diver can explore?

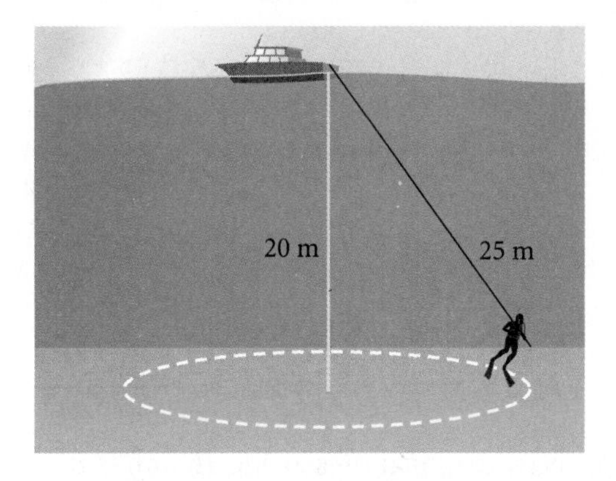

23. What are the lengths of the two legs of a 30°-60°-90° triangle if the length of the hypotenuse is $12\sqrt{3}$?

24. Find the side length of an equilateral triangle with an area of $36\sqrt{3}$ m^2.

25. Find the perimeter of an equilateral triangle with a height of $7\sqrt{3}$.

26. Al baked brownies for himself and his two sisters. He divided the square pan of brownies into three parts. He measured three 30° angles at one of the corners so that two pieces formed right triangles and the middle piece formed a kite. Did he divide the pan of brownies equally? Draw a sketch and explain your reasoning.

27. A circle has a central angle AOB that measures 80°. If point C is selected randomly on the circle, what is the probability that $\triangle ABC$ is obtuse?

28. One of the sketches below shows the greatest area that you can enclose in a right-angled corner with a rope of length *s*. Which one? Explain your reasoning.

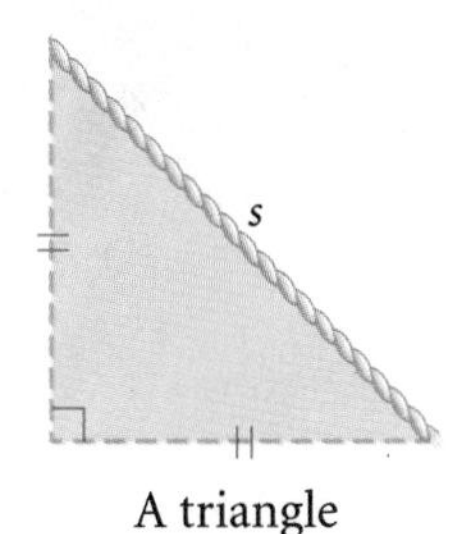

A triangle

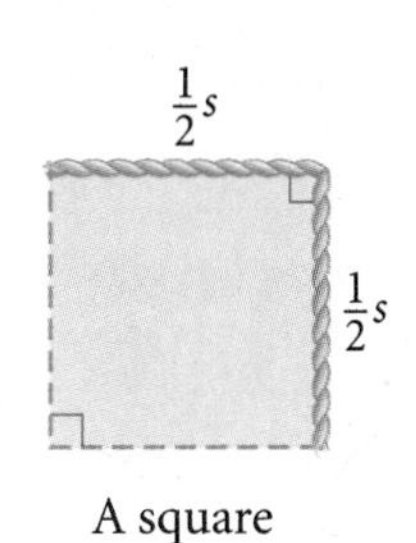

A square

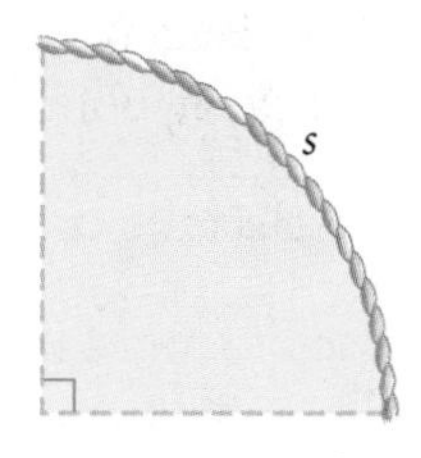

A quarter-circle

29. A wire is attached to a block of wood at point *A*. The wire is pulled over a pulley as shown. How far will the block move if the wire is pulled 1.4 meters in the direction of the arrow?

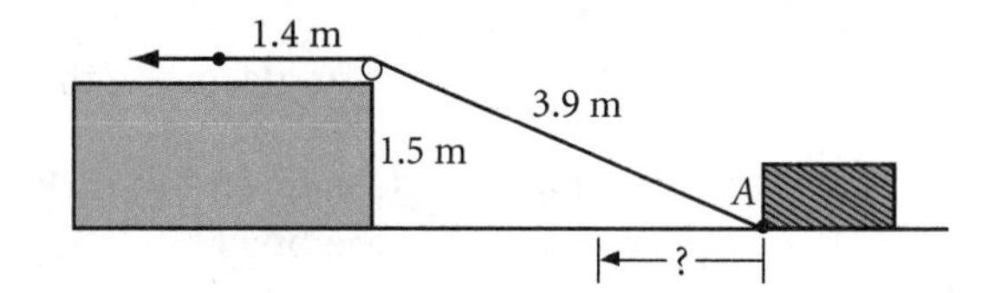

MIXED REVIEW

30. ***Construction*** Construct an isosceles triangle that has a base length equal to half the length of one leg.

31. In a regular octagon inscribed in a circle, how many diagonals pass through the center of the circle? In a regular nonagon? a regular 20-gon? What is the general rule?

32. A bug clings to a point two inches from the center of a spinning fan blade. The blade spins around once per second. How fast does the bug travel in inches per second?

In Exercises 33–40, identify the statement as true or false. For each false statement, explain why it is false or sketch a counterexample.

33. The area of a rectangle and the area of a parallelogram are both given by the formula $A = bh$, where A is the area, b is the length of the base, and h is the height.

34. When a figure is reflected over a line, the line of reflection is perpendicular to every segment joining a point on the original figure with its image.

35. In an isosceles right triangle, if the legs have length x, then the hypotenuse has length $x\sqrt{3}$.

36. The area of a kite or a rhombus can be found by using the formula $A = (0.5)d_1d_2$, where A is the area and d_1 and d_2 are the lengths of the diagonals.

37. If the coordinates of points A and B are (x_1, y_1) and (x_2, y_2), respectively, then $AB = \sqrt{(x_1 - y_1)^2 + (x_2 - y_2)^2}$.

38. A glide reflection is a combination of a translation and a rotation.

39. Equilateral triangles, squares, and regular octagons can be used to create monohedral tessellations.

40. In a 30°-60°-90° triangle, if the shorter leg has length x, then the longer leg has length $x\sqrt{3}$ and the hypotenuse has length $2x$.

In Exercises 41–46, select the correct answer.

41. The hypotenuse of a right triangle is always _?_.

A. opposite the smallest angle and is the shortest side.

B. opposite the largest angle and is the shortest side.

C. opposite the smallest angle and is the longest side.

D. opposite the largest angle and is the longest side.

42. The area of a triangle is given by the formula _?_, where A is the area, b is the length of the base, and h is the height.

A. $A = bh$

B. $A = \frac{1}{2}bh$

C. $A = 2bh$

D. $A = b^2h$

43. If the lengths of the three sides of a triangle satisfy the Pythagorean equation, then the triangle must be a(n) _?_ triangle.

A. right

B. acute

C. obtuse

D. scalene

44. The ordered pair rule $(x, y) \rightarrow (y, x)$ is a _?_.

A. reflection over the x-axis

B. reflection over the y-axis

C. reflection over the line $y = x$

D. rotation 90° about the origin

45. The composition of two reflections over two intersecting lines is equivalent to _?_.

A. a single reflection

B. a translation

C. a rotation

D. no transformation

46. The total surface area of a cone is equal to _?_, where r is the radius of the circular base and l is the slant height.

A. $\pi r^2 + 2\pi r$

B. πrl

C. $\pi rl + 2\pi r$

D. $\pi rl + \pi r^2$

47. Create a flowchart proof to show that the diagonal of a rectangle divides the rectangle into two congruent triangles.

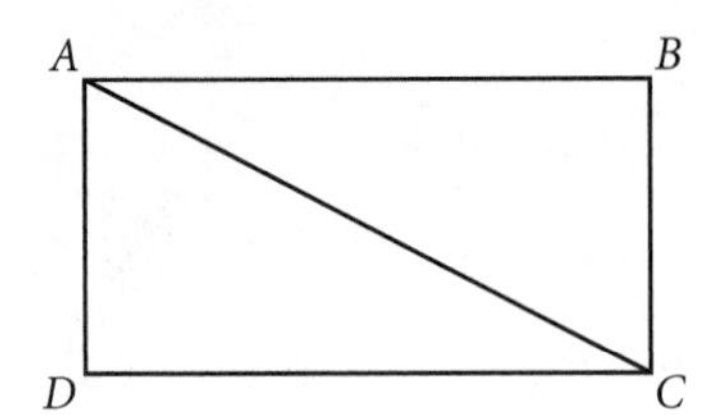

48. Copy the ball positions onto patty paper.

a. At what point on the S cushion should a player aim so that the cue ball bounces off and strikes the 8-ball? Mark the point with the letter *A*.

b. At what point on the W cushion should a player aim so that the cue ball bounces off and strikes the 8-ball? Mark the point with the letter *B*.

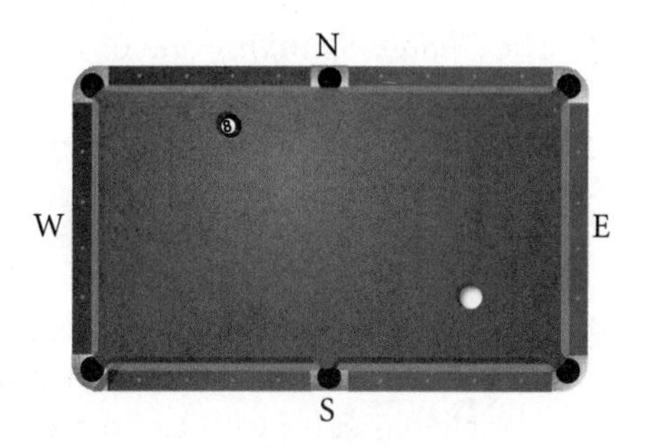

49. Find the area and the perimeter of the trapezoid.

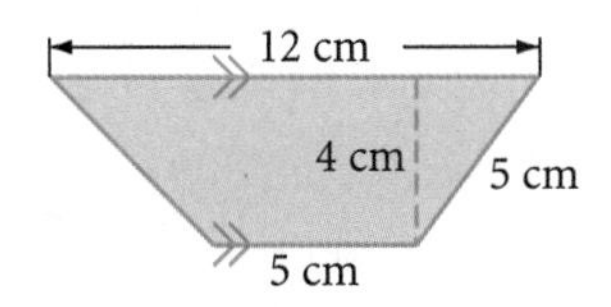

50. Find the area of the shaded region.

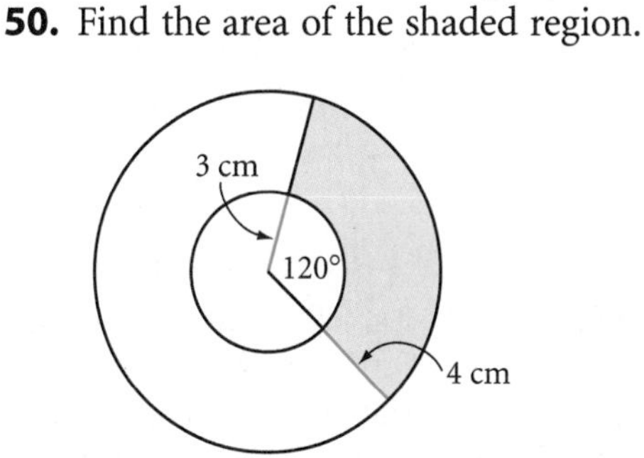

51. An Olympic swimming pool has length 50 meters and width 25 meters. What is the diagonal distance across the pool?

52. The side length of a regular pentagon is 6 cm, and the apothem measures about 4.1 cm. What is the area of the pentagon?

53. The box below has dimensions 25 cm, 36 cm, and *x* cm. The diagonal shown has length 65 cm. Find the value of *x*.

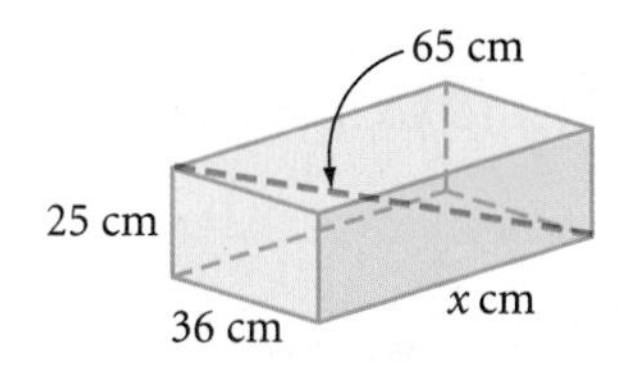

54. The cylindrical container below has an open top. Find the surface area of the container (inside and out) to the nearest square foot.

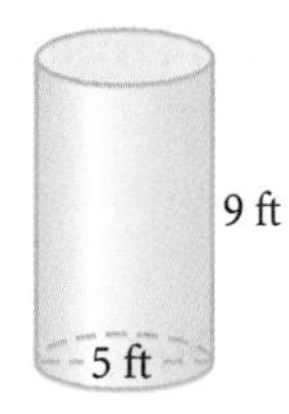

TAKE ANOTHER LOOK

1. Use geometry software to demonstrate the Pythagorean Theorem. Does your demonstration still work if you use a shape other than a square—for example, an equilateral triangle or a semicircle?

2. Find Elisha Scott Loomis's *Pythagorean Proposition* and demonstrate one of the proofs of the Pythagorean Theorem from the book.

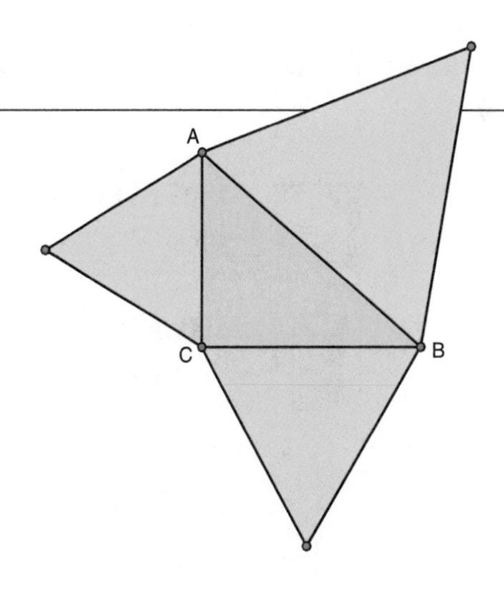

3. The *Zhoubi Suanjing*, one of the oldest sources of Chinese mathematics and astronomy, contains the diagram at right demonstrating the Pythagorean Theorem (called *gou gu* in China). Find out how the Chinese used and proved the *gou gu*, and present your findings.

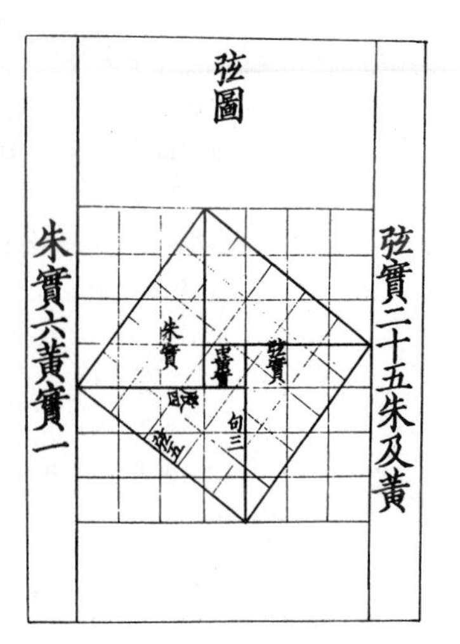

4. Use the SSS Congruence Conjecture to verify the converse of the 30°-60°-90° Triangle Conjecture. That is, show that if a triangle has sides with lengths x, $x\sqrt{3}$, and $2x$, then it is a 30°-60°-90° triangle.

5. Starting with an isosceles right triangle, use geometry software or a compass and straightedge to start a right triangle like the one shown. Continue constructing right triangles on the hypotenuse of the previous triangle at least five more times. Calculate the length of each hypotenuse and leave them in radical form.

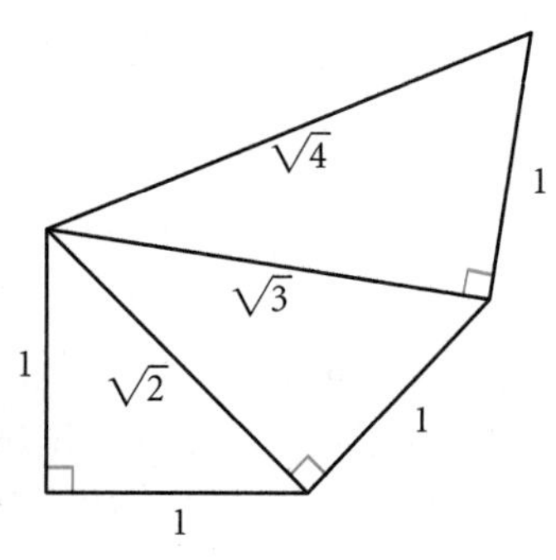

Assessing What You've Learned

UPDATE YOUR PORTFOLIO Choose a challenging project, Take Another Look activity, or exercise you did in this chapter and add it to your portfolio. Explain the strategies you used.

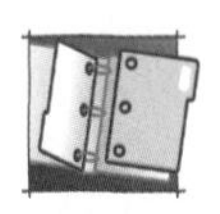

ORGANIZE YOUR NOTEBOOK Review your notebook and your conjecture list to be sure they are complete. Write a one-page chapter summary.

WRITE IN YOUR JOURNAL Why do you think the Pythagorean Theorem is considered one of the most important theorems in mathematics?

WRITE TEST ITEMS Work with group members to write test items for this chapter. Try to demonstrate more than one way to solve each problem.

GIVE A PRESENTATION Create a visual aid and give a presentation about the Pythagorean Theorem.

CHAPTER

10 Volume

Perhaps all I pursue is astonishment and so I try to awaken only astonishment in my viewers. Sometimes "beauty" is a nasty business.

M. C. ESCHER

Verblifa tin, M. C. Escher, 1963

OBJECTIVES

In this chapter you will

- explore and define many three-dimensional solids
- discover formulas for finding the volumes of prisms, pyramids, cylinders, cones, and spheres
- learn how density is related to volume
- derive a formula for the surface area of a sphere

LESSON

10.1

The Geometry of Solids

Everything in nature adheres to the cone, the cylinder, and the cube.

PAUL CEZANNE

Most of the geometric figures you have worked with so far have been flat plane figures with two dimensions—base and height. In this chapter you will work with solid figures with three dimensions—length, width, and height. Most real-world solids, like rocks and plants, are very irregular, but many others are geometric. Some real-world geometric solids occur in nature: viruses, oranges, crystals, the earth itself. Others are human-made: books, buildings, baseballs, soup cans, ice cream cones.

Still Life With a Basket (1888–1890) by French post-impressionist painter Paul Cézanne (1839–1906) uses geometric solids to portray everyday objects.

This amethyst crystal is an irregular solid, but parts of it have familiar shapes.

Science CONNECTION

Three-dimensional geometry plays an important role in the structure of molecules. For example, when carbon atoms are arranged in a very rigid network, they form diamonds, one of the earth's hardest materials. But when carbon atoms are arranged in planes of hexagonal rings, they form graphite, a soft material used in pencil lead.

Carbon atoms can also bond into very large molecules. Named fullerenes, after U.S. engineer Buckminster Fuller (1895–1983), these carbon molecules have the same symmetry as a soccer ball, as shown at left. They are popularly called buckyballs.

The geometry of diamonds

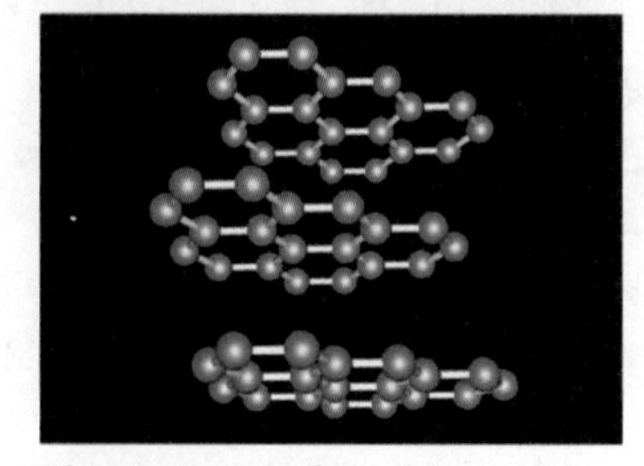

The geometry of graphite

A solid formed by polygons that enclose a single region of space is called a **polyhedron.** The flat polygonal surfaces of a polyhedron are called its **faces.** Although a face of a polyhedron includes the polygon and its interior region, we identify the face by naming the polygon that encloses it. A segment where two faces intersect is called an **edge.** The point of intersection of three or more edges is called a **vertex** of the polyhedron.

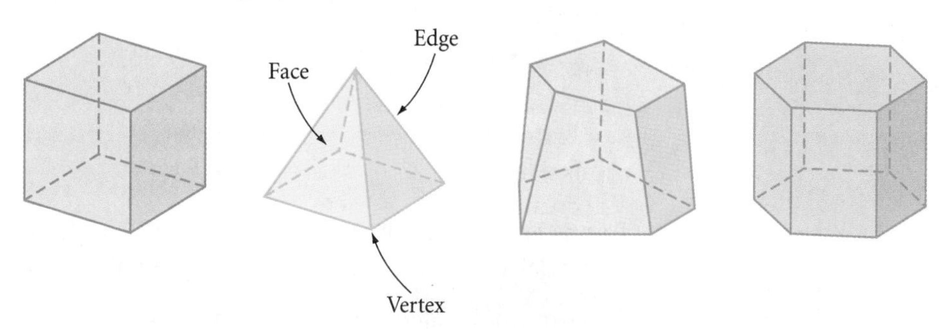

Just as a polygon is classified by its number of sides, a polyhedron is classified by its number of faces. The prefixes for polyhedrons are the same as they are for polygons with one exception: A polyhedron with four faces is called a **tetrahedron.** Here are some examples of polyhedrons.

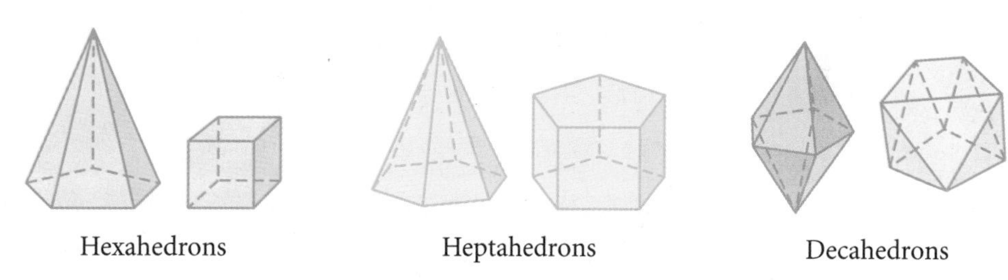

If each face of a polyhedron is enclosed by a regular polygon, and each face is congruent to the other faces, and the faces meet at each vertex in exactly the same way, then the polyhedron is called a **regular polyhedron.** The regular polyhedron shown at right is called a regular dodecahedron because it has 12 faces.

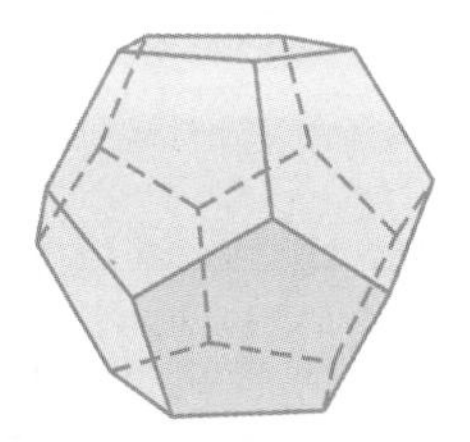
Regular dodecahedron

The Ramat Polin housing complex in Jerusalem, Israel, has many polyhedral shapes.

A **prism** is a special type of polyhedron, with two faces called **bases,** that are congruent, parallel polygons. The other faces of the polyhedron, called **lateral faces,** are parallelograms that connect the corresponding sides of the bases.

The lateral faces meet to form the **lateral edges.** Each solid shown below is a prism.

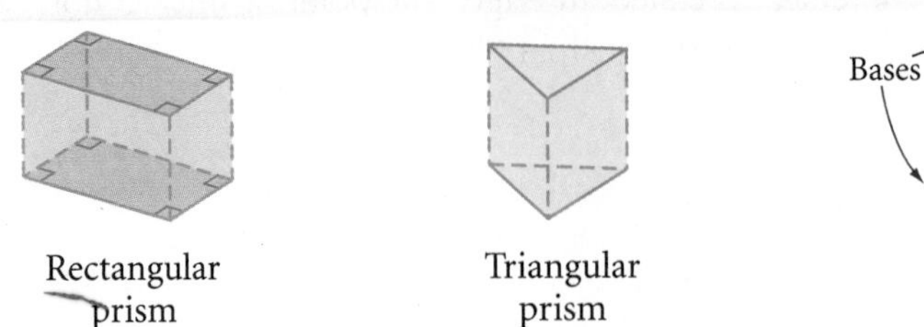

Rectangular prism | Triangular prism | Hexagonal prism

Prisms are classified by their bases. For example, a prism with triangular bases is a triangular prism, and a prism with hexagonal bases is a hexagonal prism.

A prism whose lateral faces are rectangles is called a **right prism.** Its lateral edges are perpendicular to its bases. A prism that is not a right prism is called an **oblique prism.** The **altitude** of a prism is any perpendicular segment from one base to the plane of the other base. The length of an altitude is the **height** of the prism.

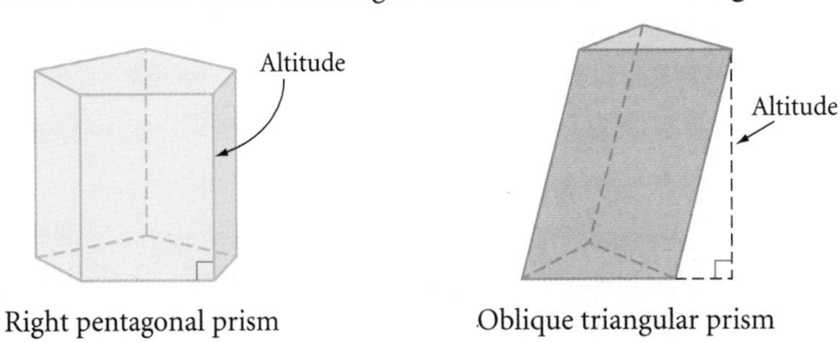

Right pentagonal prism | Oblique triangular prism

A **pyramid** is another special type of polyhedron. Pyramids have only one base. Like a prism, the other faces are called the lateral faces, and they meet to form the lateral edges. The common vertex of the lateral faces is the vertex of the pyramid.

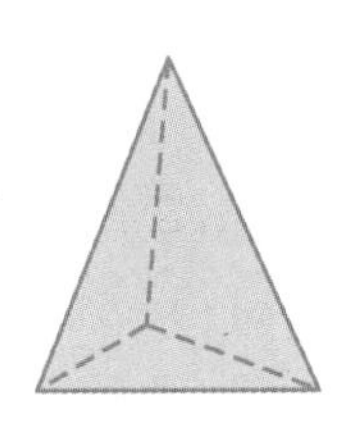

Triangular pyramid

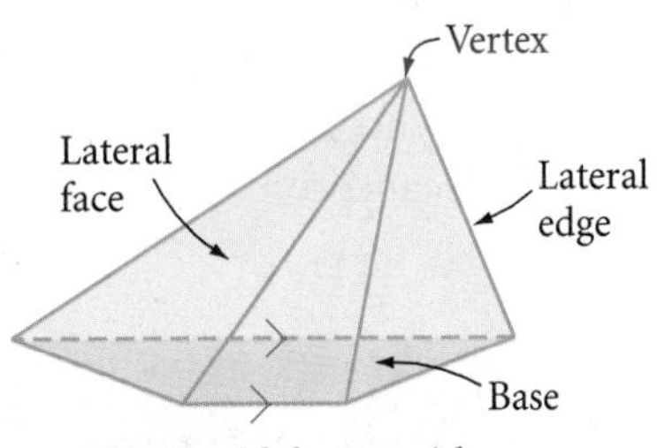

Trapezoidal pyramid

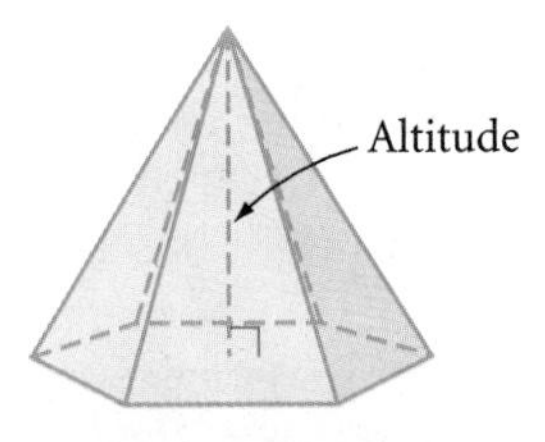

Hexagonal pyramid

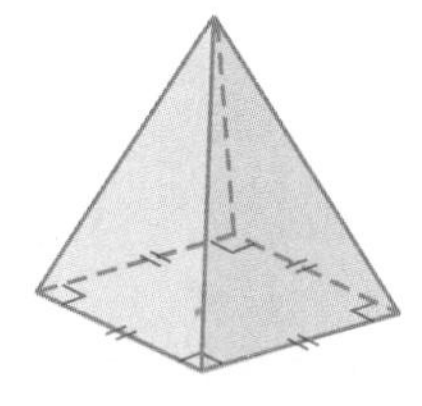

Square pyramid

Like prisms, pyramids are also classified by their bases. The pyramids of Egypt are square pyramids because they have square bases.

The altitude of the pyramid is the perpendicular segment from its vertex to the plane of its base. The length of the altitude is the height of the pyramid.

Polyhedrons are geometric solids with flat surfaces. There are also geometric solids that have curved surfaces. One that all sports fans know well is the ball, or sphere—you can think of it as a three-dimensional circle. An orange is one example of a sphere found in nature. What are some others?

A **sphere** is the set of all points in space at a given distance from a given point.

The given distance is called the **radius** of the sphere, and the given point is the **center** of the sphere. A **hemisphere** is half a sphere and its circular base. The circle that encloses the base of a hemisphere is called a **great circle** of the sphere. Every plane that passes through the center of a sphere determines a great circle. All the longitude lines on a globe of Earth are great circles. The equator is the only latitude line that is a great circle.

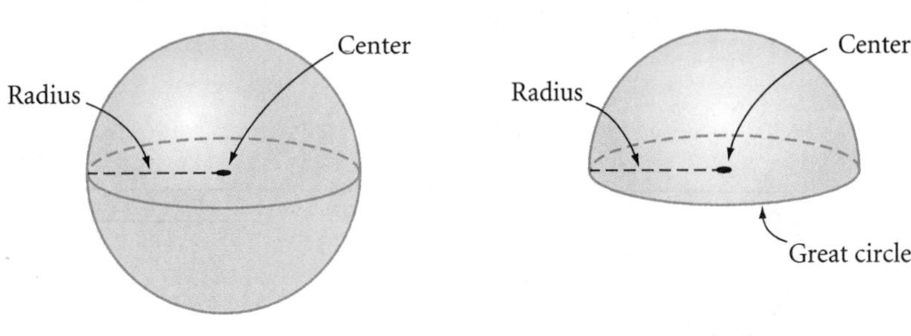

100 Cans (1962 oil on canvas), by pop art artist Andy Warhol (1925–1987), repeatedly uses the cylindrical shape of a soup can to make an artistic statement with a popular image.

Another solid with a curved surface is a **cylinder.** Soup cans, compact discs (CDs), and plumbing pipes are shaped like cylinders. Like a prism, a cylinder has two bases that are both parallel and congruent. Instead of polygons, however, the bases of cylinders are circles and their interiors. The segment connecting the centers of the bases is called the **axis** of the cylinder. The **radius** of the cylinder is the radius of a base.

If the axis of a cylinder is perpendicular to the bases, then the cylinder is a **right cylinder.** A cylinder that is not a right cylinder is an **oblique cylinder.**

The altitude of a cylinder is any perpendicular segment from the plane of one base to the plane of the other. The height of a cylinder is the length of an altitude.

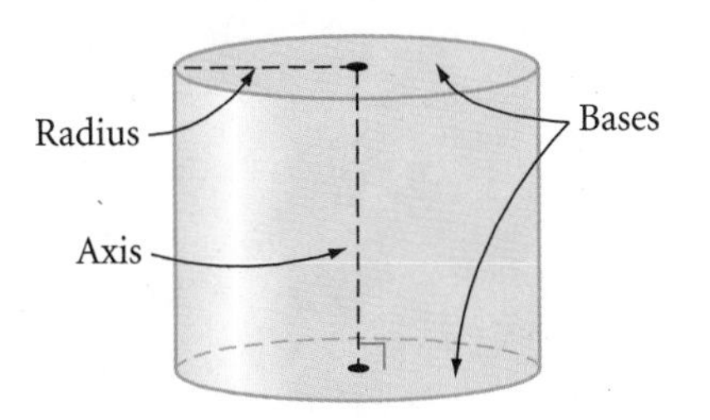

Right cylinder

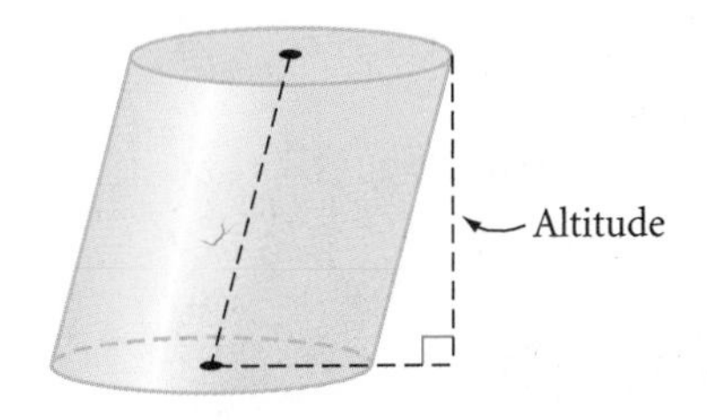

Oblique cylinder

A third type of solid with a curved surface is a **cone.** Funnels and ice cream cones are shaped like cones. Like a pyramid, a cone has a base and a vertex.

The base of a cone is a circle and its interior. The radius of a cone is the radius of the base. The vertex of a cone is the point that is the greatest perpendicular distance from the base. The altitude of a cone is the perpendicular segment from the vertex to the plane of the base. The length of the altitude is the height of a cone. If the line segment connecting the vertex of a cone with the center of its base is perpendicular to the base, then it is a **right cone.**

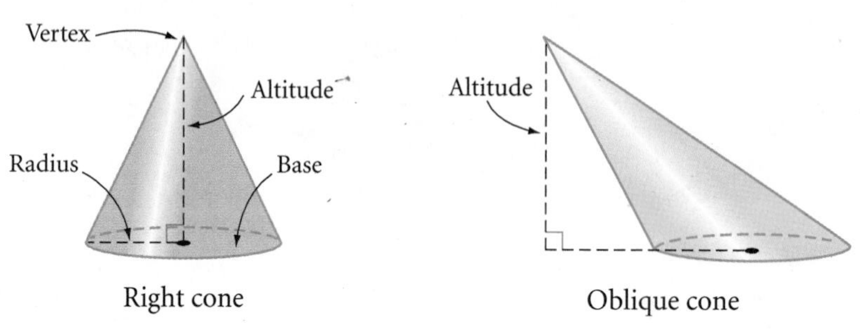

EXERCISES

1. Complete this definition:
   A pyramid is a _?_ with one _?_ face (called the base) and whose other faces (lateral faces) are _?_ formed by segments connecting the vertices of the base to a common point (the vertex) not on the base.

For Exercises 2–9, refer to the figures below. All measurements are in centimeters.

2. Name the bases of the prism.
3. Name all the lateral faces of the prism.
4. Name all the lateral edges of the prism.
5. What is the height of the prism?
6. Name the base of the pyramid.
7. Name the vertex of the pyramid.
8. Name all the lateral edges of the pyramid.
9. What is the height of the pyramid?

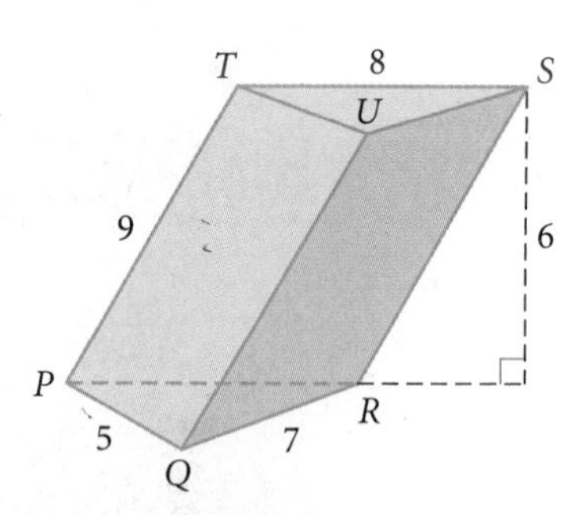

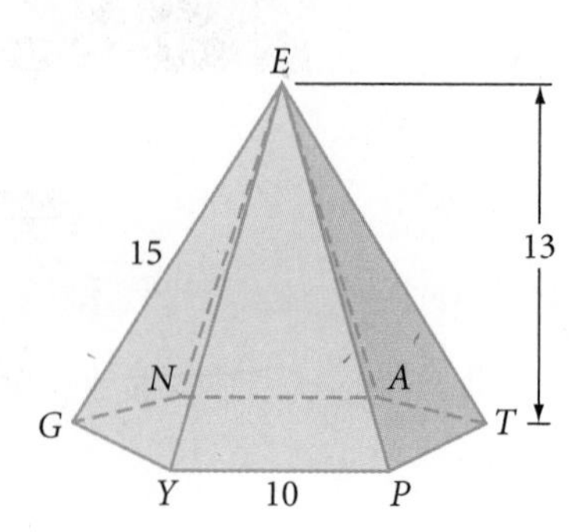

For Exercises 10–22, match each real object with a geometry term. You may use a geometry term more than once or not at all.

10. Die

11. Tomb of Egyptian rulers

12. Holder for a scoop of ice cream

13. Wedge or doorstop

14. Box of breakfast cereal

15. Plastic bowl with lid

16. Ingot of silver

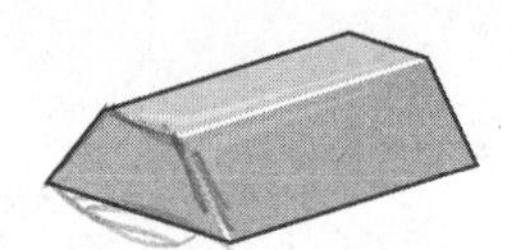

17. Honeycomb

18. Stop sign

19. Moon

20. Can of tuna fish

21. Book

22. Pup tent

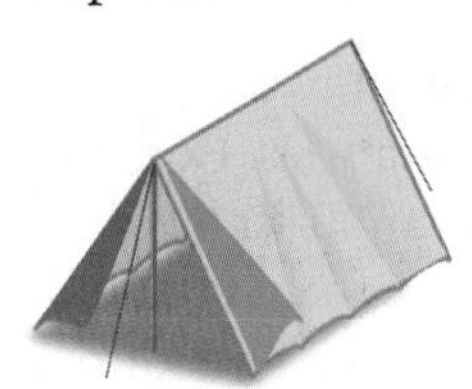

A. Cylinder
B. Cone
C. Square prism
D. Square pyramid
E. Sphere
F. Triangular pyramid
G. Octagonal prism
H. Triangular prism
I. Trapezoidal prism
J. Rectangular prism
K. Heptagonal pyramid
L. Hexagonal prism
M. Hemisphere

For Exercises 23–26, draw and label each solid. Use dashed lines to show the hidden edges.

23. A triangular pyramid whose base is an equilateral triangular region (use the proper marks to show that the base is equilateral)

24. A hexahedron with two trapezoidal faces

25. A cylinder with a height that is twice the diameter of the base (use x and $2x$ to indicate the height and the diameter)

26. A right cone with a height that is half the diameter of the base

For Exercises 27–35, identify each statement as true or false. Sketch a counterexample for each false statement or explain why it is false.

27. A lateral face of a pyramid is always a triangular region.

28. A lateral edge of a pyramid is always perpendicular to the base.

29. Every slice of a prism cut parallel to the bases is congruent to the bases. ⓗ

30. When the lateral surface of a right cylinder is unwrapped and laid flat, it is a rectangle.

31. When the lateral surface of a right circular cone is unwrapped and laid flat, it is a triangle. ⓗ

32. Every section of a cylinder, parallel to the base, is congruent to the base.

33. The length of a segment from the vertex of a cone to the circular base is the height of the cone.

34. The length of the axis of a right cylinder is the height of the cylinder.

35. All slices of a sphere passing through the sphere's center are congruent.

36. Write a paragraph describing the visual tricks that Belgian artist René Magritte (1898–1967) plays in his painting at right.

The Promenades of Euclid (1935 oil on canvas), René Magritte

37. An **antiprism** is a polyhedron with two congruent bases and lateral faces that are triangles. Complete the tables below for prisms and antiprisms. Describe any relationships you see between the number of lateral faces, total faces, edges, and vertices of related prisms and antiprisms.

	Triangular prism	Rectangular prism	Pentagonal prism	Hexagonal prism		*n*-gonal prism
Lateral faces	3				. . .	
Total faces		6			. . .	
Edges				18	. . .	
Vertices			10		. . .	

	Triangular antiprism	Rectangular antiprism	Pentagonal antiprism	Hexagonal antiprism		*n*-gonal antiprism
Lateral faces	6				. . .	
Total faces		10			. . .	
Edges				24	. . .	
Vertices			10		. . .	

Review

For Exercises 38 and 39, how many cubes measuring 1 cm on each edge will fit into the container?

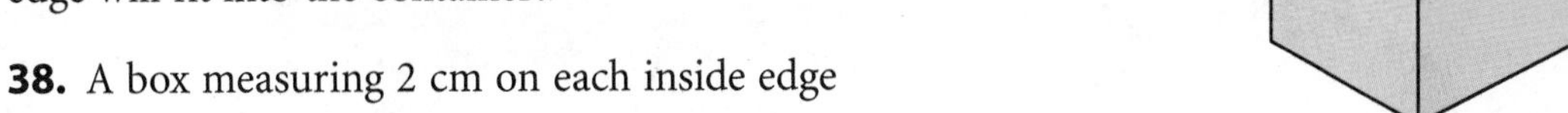

38. A box measuring 2 cm on each inside edge

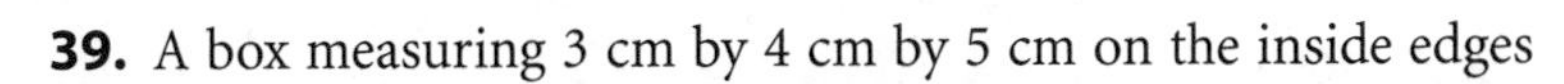

39. A box measuring 3 cm by 4 cm by 5 cm on the inside edges

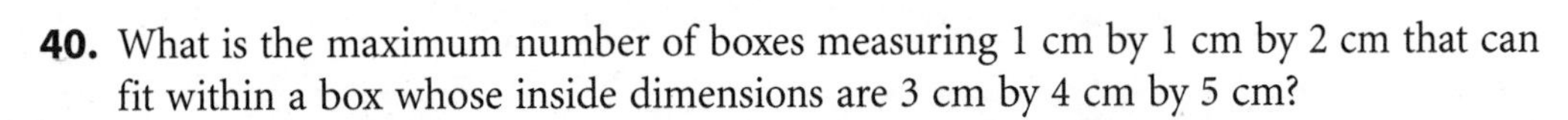

40. What is the maximum number of boxes measuring 1 cm by 1 cm by 2 cm that can fit within a box whose inside dimensions are 3 cm by 4 cm by 5 cm?

41. For each net, decide whether it folds to make a box. If it does, copy the net and mark each pair of opposite faces with the same symbol.

a. 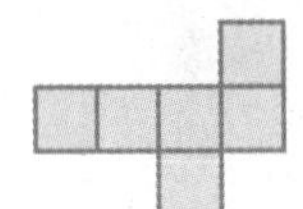

b. 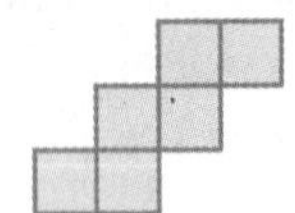

c. 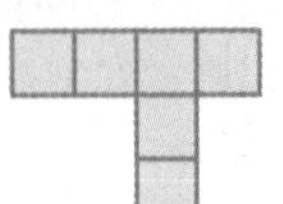

d. 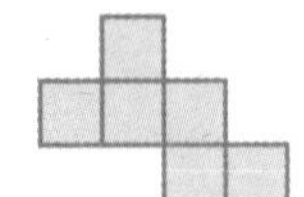

IMPROVING YOUR VISUAL THINKING SKILLS

PUZZLE

Piet Hein's Puzzle

In 1936, while listening to a lecture on quantum physics, the Danish mathematician Piet Hein (1905–1996) devised the following visual thinking puzzle:

> What are all the possible nonconvex solids that can be created by joining four or fewer cubes face-to-face?

A nonconvex polyhedron is a solid that has at least one diagonal that is exterior to the solid. For example, four cubes in a row, joined face-to-face, form a convex polyhedron. But four cubes joined face-to-face into an L-shape form a nonconvex polyhedron.

Use isometric dot paper to sketch the nonconvex solids that solve Piet Hein's puzzle. There are seven solids.

Exploration

Euler's Formula for Polyhedrons

In this activity you will discover a relationship among the vertices, edges, and faces of a polyhedron. This relationship is called Euler's Formula for Polyhedrons, named after Leonhard Euler. Let's first build some of the polyhedrons you learned about in Lesson 10.1.

Toothpick Polyhedrons

You will need

- toothpicks
- modeling clay, gumdrops, or dried peas

First, you'll model polyhedrons using toothpicks as edges and using small balls of clay, gumdrops, or dried peas as connectors.

Step 1 Build and save the polyhedrons shown in parts a–d below and described in parts e–i on the top of page 513. You may have to cut or break some sticks. Share the tasks among the group.

a.

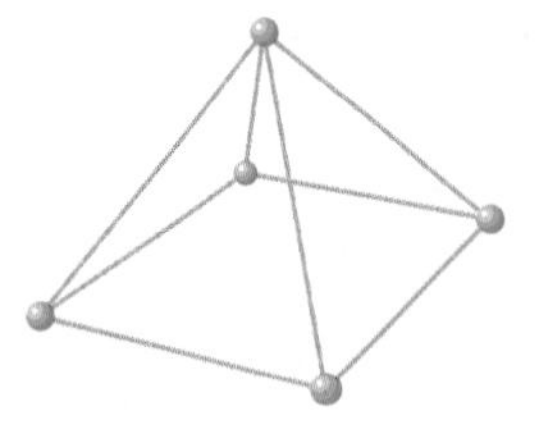

b.

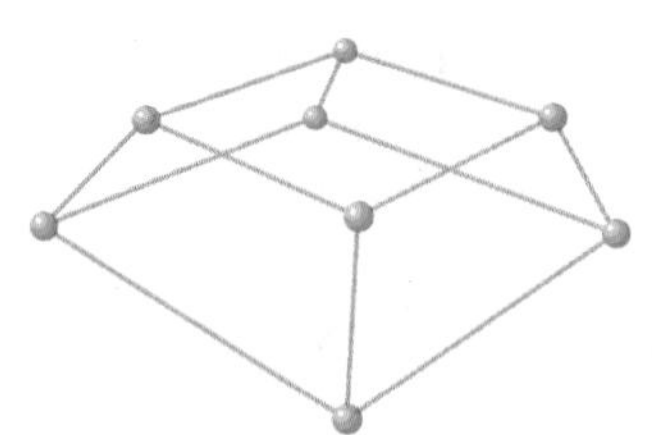

c.

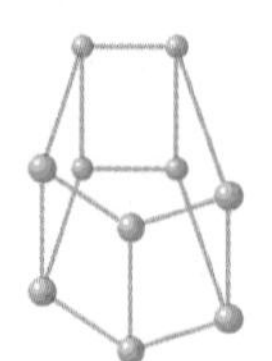

d.

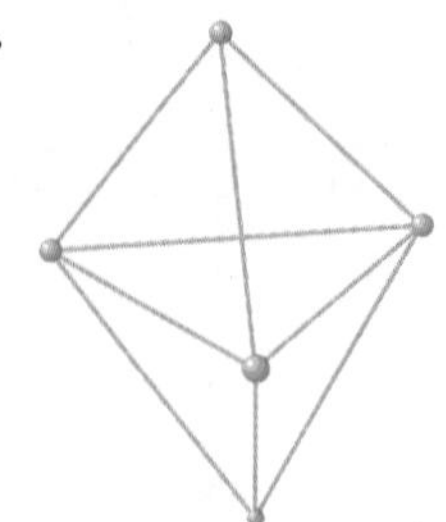

e. Build a tetrahedron.

f. Build an octahedron.

g. Build a nonahedron.

h. Build at least two different-shaped decahedrons.

i. Build at least two different-shaped dodecahedrons.

Step 2 Classify all the different polyhedrons your class built as prisms, pyramids, regular polyhedrons, or just polyhedrons.

Next, you'll look for a relationship among the vertices, faces, and edges of the polyhedrons.

Step 3 Count the number of vertices (V), edges (E), and faces (F) of each polyhedron model. Copy and complete a chart like this one.

Polyhedron	Vertices (V)	Faces (F)	Edges (E)
⋮	⋮	⋮	⋮

Step 4 Look for patterns in the table. By adding, subtracting, or multiplying V, F, and E (or a combination of two or three of these operations), you can discover a formula that is commonly known as Euler's Formula for Polyhedrons.

Step 5 Now that you have discovered the formula relating the number of vertices, edges, and faces of a polyhedron, use it to answer each of these questions.

a. Which polyhedron has 4 vertices and 6 edges? Can you build another polyhedron with a different number of faces that also has 4 vertices and 6 edges?

b. Which polyhedron has 6 vertices and 12 edges? Can you build another polyhedron with a different number of faces that also has 6 vertices and 12 edges?

c. If a solid has 8 faces and 12 vertices, how many edges will it have?

d. If a solid has 7 faces and 12 edges, how many vertices will it have?

e. If a solid has 6 faces, what are all the possible combinations of vertices and edges it can have?

LESSON

10.2

Volume of Prisms and Cylinders

How much deeper would oceans be if sponges didn't live there?

STEVEN WRIGHT

In real life you encounter many volume problems. For example, when you shop for groceries, it's a good idea to compare the volumes and the prices of different items to find the best buy. When you fill a car's gas tank or when you fit last night's leftovers into a freezer dish, you fill the volume of an empty container.

Many occupations also require familiarity with volume. An engineer must calculate the volume and the weight of sections of a bridge to avoid too much stress on any one section. Chemists, biologists, physicists, and geologists must all make careful volume measurements in their research. Carpenters, plumbers, and painters also know and use volume relationships. A chef must measure the correct volume of each ingredient in a cake to ensure a tasty success.

American artist Wayne Thiebaud (b 1920) painted *Bakery Counter* in 1962.

Volume is the measure of the amount of space contained in a solid. You use cubic units to measure volume: cubic inches (in.^3), cubic feet (ft^3), cubic yards (yd^3), cubic centimeters (cm^3), cubic meters (m^3), and so on. The volume of an object is the number of unit cubes that completely fill the space within the object.

1

Length: 1 unit

1 1 1

Volume: 1 cubic unit

Volume: 20 cubic units

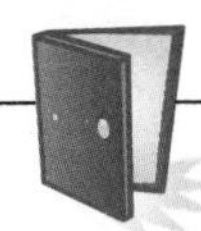

Investigation

The Volume Formula for Prisms and Cylinders

Step 1 | Find the volume of each right rectangular prism below in cubic centimeters. That is, how many 1 cm-by-1 cm-by-1 cm cubes will fit into each solid? Within your group, discuss different strategies for finding each volume. How could you find the volume of any right rectangular prism?

a.

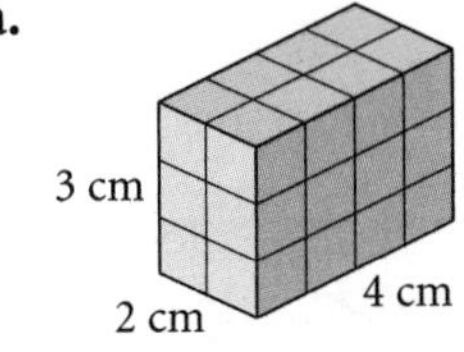

b.

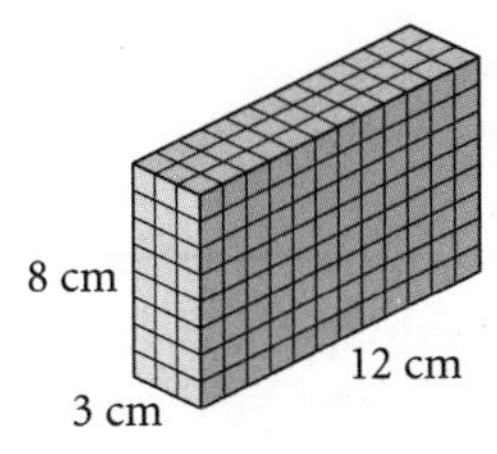

c.

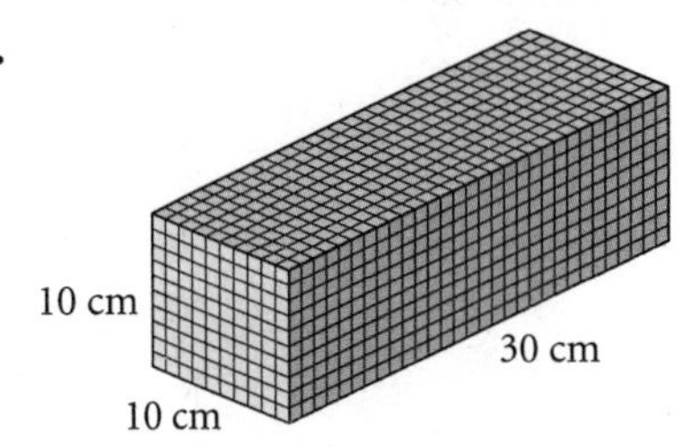

Notice that the number of cubes resting on the base equals the number of square units in the area of the base. The number of layers of cubes equals the number of units in the height of the prism. So you can multiply the area of the base by the height of the prism to calculate the volume.

Step 2 | Complete the conjecture.

Conjecture A

C-88a

If B is the area of the base of a right rectangular prism and H is the height of the solid, then the formula for the volume is $V = \underline{?}$.

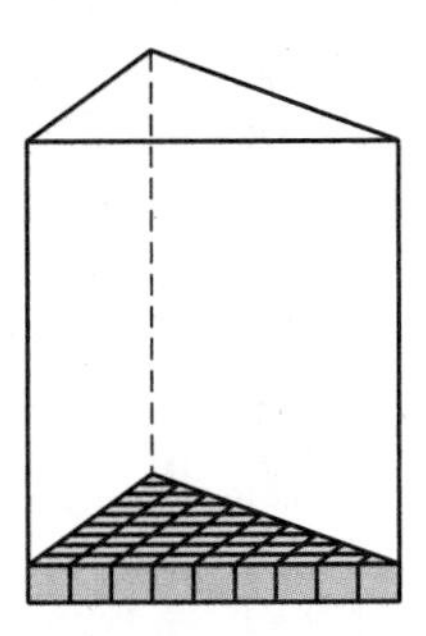

In Chapter 8, you discovered that you can reshape parallelograms, triangles, trapezoids, and circles into rectangles to find their area. You can use the same method to find the areas of bases that have these shapes. Then you can multiply the area of the base by the height of the prism to find its volume. For example, to find the volume of a right triangular prism, find the area of the triangular base (the number of cubes resting on the base) and multiply it by the height (the number of layers of cubes).

So you can extend Conjecture A (the volume of right rectangular prisms) to all right prisms and right cylinders.

Step 3 | Complete the conjecture.

Conjecture B

C-88b

If B is the area of the base of a right prism (or cylinder) and H is the height of the solid, then the formula for the volume is $V = \underline{?}$.

What about the volume of an oblique prism or cylinder? You can approximate the shape of this oblique rectangular prism with a staggered stack of three reams of 8.5-by-11-inch paper. If you nudge the individual pieces of paper into a slanted stack, then your approximation can be even better.

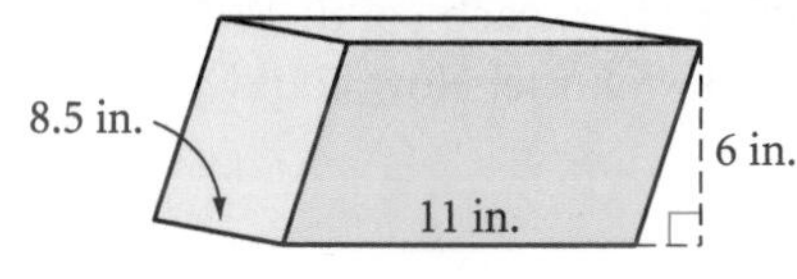

Oblique rectangular prism

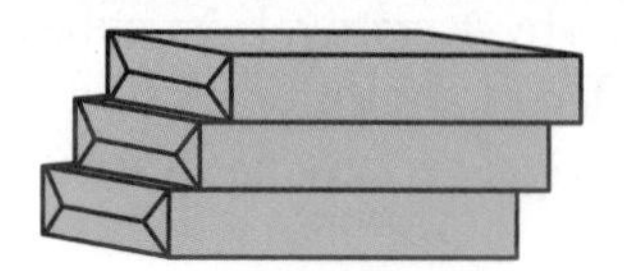
Stacked reams of 8.5-by-11-inch paper

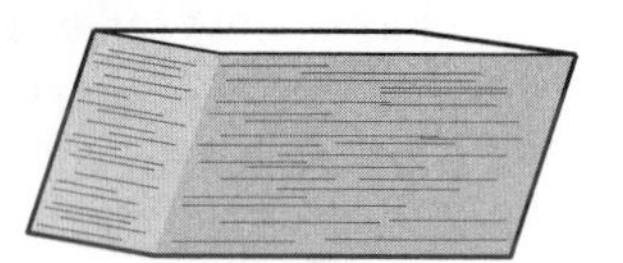
Stacked sheets of paper

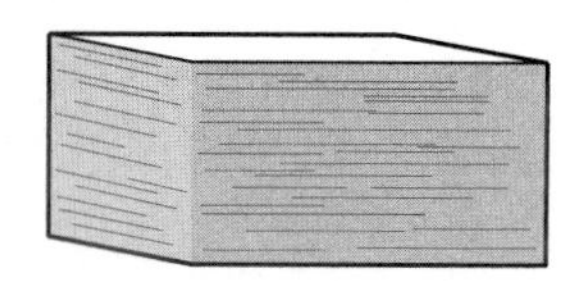
Sheets of paper stacked straight

Rearranging the paper into a right rectangular prism changes the shape, but certainly the volume of paper hasn't changed. The area of the base, 8.5 by 11 inches, didn't change and the height, 6 inches, didn't change, either.

In the same way, you can use coffee filters, coins, candies, or chemistry filter papers to show that an oblique cylinder has the same volume as a right cylinder with the same base and height.

Step 4 | Use the stacking model to extend Conjecture B (the volume of right prisms and cylinders) to oblique prisms and cylinders. Complete the conjecture.

Conjecture C C-88c

The volume of an oblique prism (or cylinder) is the same as the volume of a right prism (or cylinder) that has the same _?_ and the same _?_.

Finally, you can combine Conjectures A, B, and C into one conjecture for finding the volume of any prism or cylinder, whether it's right or oblique.

Step 5 | Copy and complete the conjecture.

Prism-Cylinder Volume Conjecture C-88

The volume of a prism or a cylinder is the _?_ multiplied by the _?_.

If you successfully completed the investigation, you saw that the same volume formula applies to all prisms and cylinders, regardless of the shapes of their bases. To calculate the volume of a prism or cylinder, first calculate the area of the base using the formula appropriate to its shape. Then multiply the area of the base by the height of the solid. In oblique prisms and cylinders, the lateral edges are no longer at right angles to the bases, so you do *not* use the length of the lateral edge as the height.

EXAMPLE A

Find the volume of a right trapezoidal prism that has a height of 15 cm. The two bases of the trapezoid measure 4 cm and 8 cm, and its height is 5 cm.

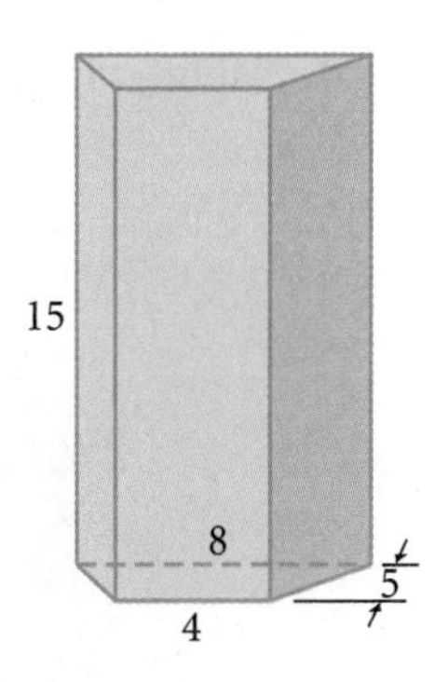

▶ Solution

Use $B = \frac{1}{2}h(b_1 + b_2)$ for the area of the trapezoidal base.

$V = BH$	The formula for volume of a prism, where B is the area of its base and H is its height.
$= \frac{1}{2}(5)(4 + 8) \cdot (15)$	Substitute $\frac{1}{2}(5)(4 + 8)$ for B, applying the formula for area of a trapezoid. Substitute 15 for H, the height of the prism.
$= (30)(15)$	Simplify.
$= 450$	

The volume is 450 cm^3.

EXAMPLE B

Find the volume of an oblique cylinder that has a base with a radius of 6 inches and a height of 7 inches.

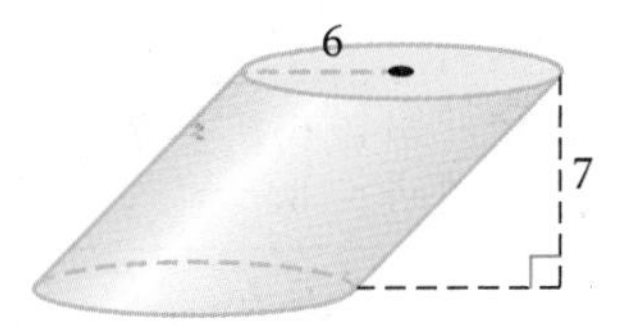

▶ Solution

Use $B = \pi r^2$ for the area of the circular base.

$V = BH$	The formula for volume of a cylinder.
$= (\pi \cdot 6^2)(7)$	Substitute $(\pi \cdot 6^2)$ for B, applying the formula for area of a circle.
$= 36\pi(7)$	Simplify.
$= 252\pi \approx 791.68$	Use the π key on your calculator to get an approximate answer.

The volume is 252π in.^{3}, or about 791.68 in.^{3}.

EXERCISES

Find the volume of each solid in Exercises 1–6. All measurements are in centimeters. Round approximate answers to two decimal places.

1. Oblique rectangular prism

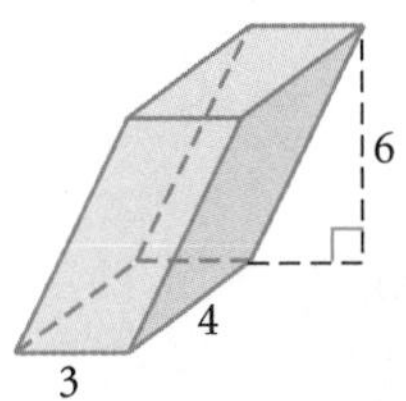

2. Right triangular prism ⓗ

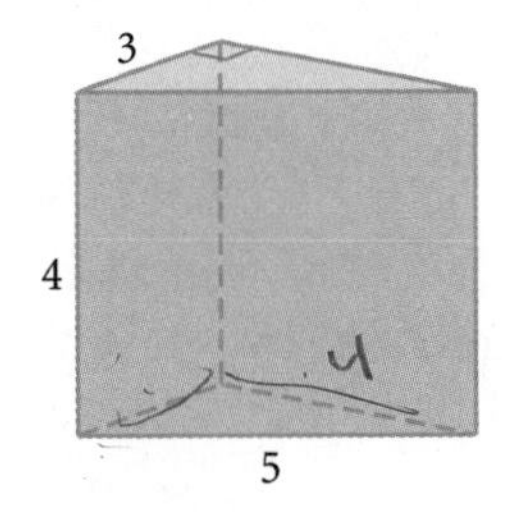

3. Right trapezoidal prism

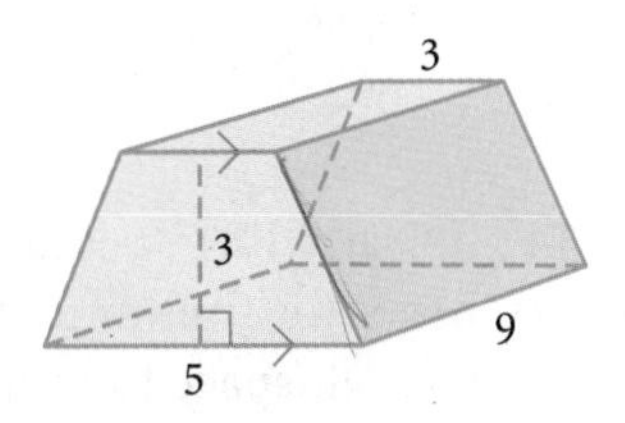

4. Right cylinder

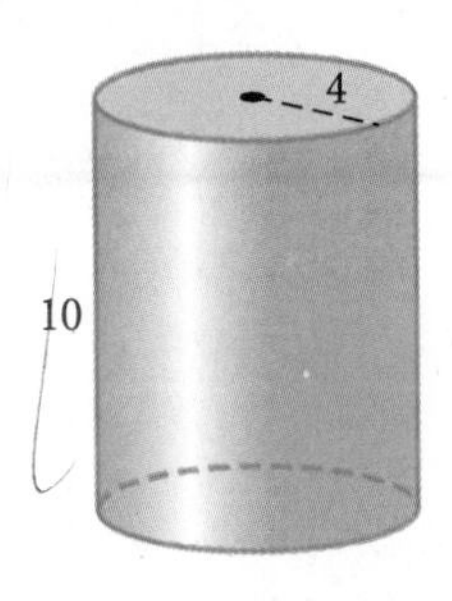

5. Right semicircular cylinder ⓗ

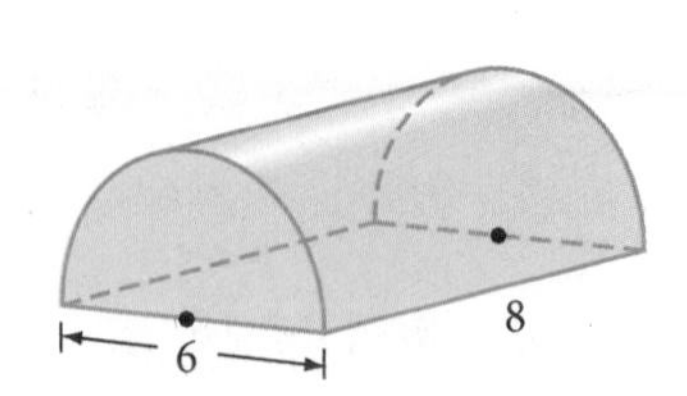

6. Right cylinder with a 90° slice removed ⓗ

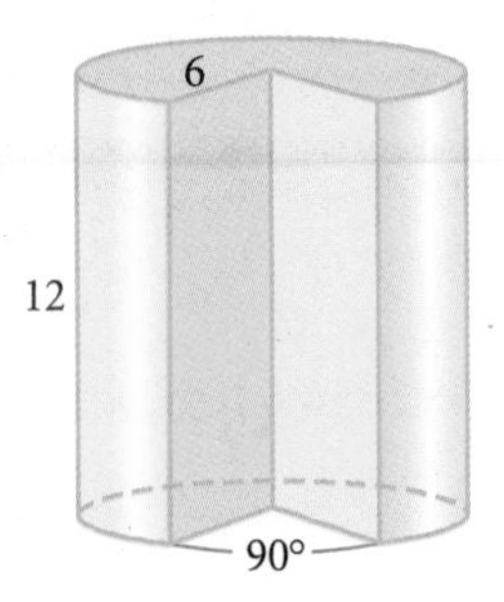

7. Use the information about the base and height of each solid to find the volume. All measurements are given in centimeters.

Information about base of solid	Height of solid	Right triangular prism	Right rectangular prism	Right trapezoidal prism	Right cylinder
$b = 6, b_2 = 7,$ $h = 8, r = 3$	$H = 20$	**a.** $V =$ ⓗ	**d.** $V =$	**g.** $V =$	**j.** $V =$
$b = 9, b_2 = 12,$ $h = 12, r = 6$	$H = 20$	**b.** $V =$	**e.** $V =$	**h.** $V =$	**k.** $V =$
$b = 8, b_2 = 19,$ $h = 18, r = 8$	$H = 23$	**c.** $V =$	**f.** $V =$	**i.** $V =$	**l.** $V =$

For Exercises 8–9, sketch and label each solid described, then find the volume.

8. An oblique trapezoidal prism. The trapezoidal base has a height of 4 in. and bases that measure 8 in. and 12 in. The height of the prism is 24 in.

9. A right circular cylinder with a height of T. The radius of the base is $\sqrt{Q}$. ⓗ

10. Sketch and label two different rectangular prisms each with a volume of 288 cm^3.

In Exercises 11–13, express the volume of each solid with the help of algebra.

11. Right rectangular prism

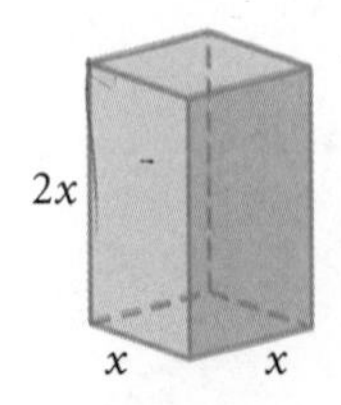

12. Oblique cylinder

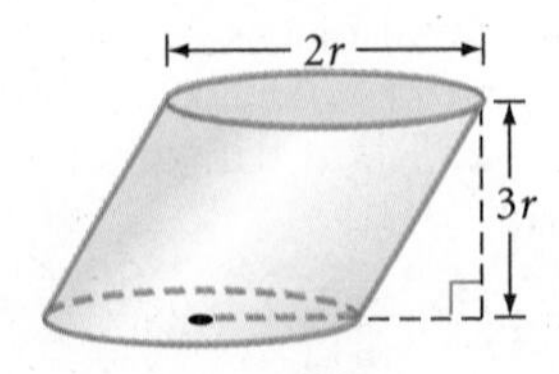

13. Right rectangular prism with a rectangular hole

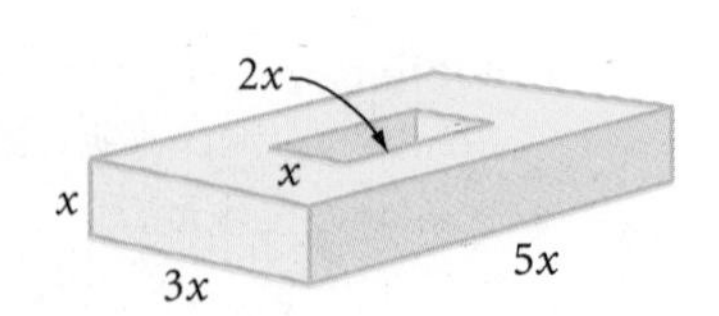

14. **APPLICATION** A cord of firewood is 128 cubic feet. Margaretta has three storage boxes for firewood that each measure 2 feet by 3 feet by 4 feet. Does she have enough space to order a full cord of firewood? A half cord? A quarter cord? Explain.

Career CONNECTION

In construction and landscaping, sand, rocks, gravel, and fill dirt are often sold by the "yard," which actually means a cubic yard.

15. **APPLICATION** A contractor needs to build up a ramp as shown at right from the street to the front of a garage door. How many cubic yards of fill will she need?

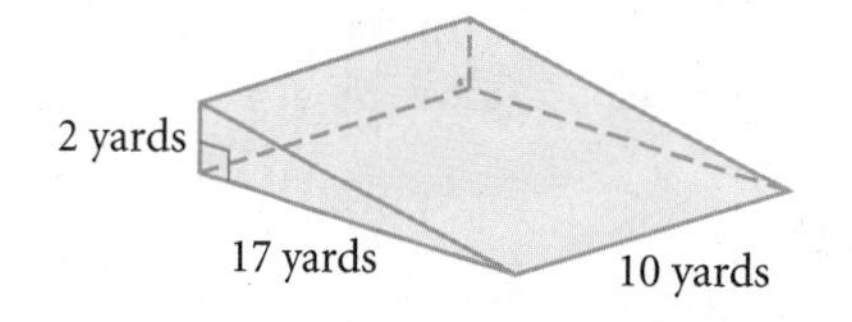

16. If an average rectangular block of limestone used to build the Great Pyramid of Khufu at Giza is approximately 2.5 feet by 3 feet by 4 feet and limestone weighs approximately 170 pounds per cubic foot, what is the weight of one of the nearly 2,300,000 limestone blocks used to build the pyramid?

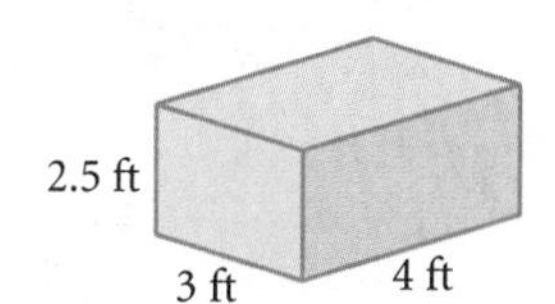

The Great Pyramid of Khufu at Giza, Egypt, was built around 2500 B.C.E.

17. Although the Exxon *Valdez* oil spill (11 million gallons of oil) is one of the most notorious oil spills, it was small compared to the 250 million gallons of crude oil that were spilled during the 1991 Persian Gulf War. A gallon occupies 0.13368 cubic foot. How many rectangular swimming pools, each 20 feet by 30 feet by 5 feet, could be filled with 250 million gallons of crude oil?

18. When folded, a 12-by-12-foot section of the AIDS Memorial Quilt requires about 1 cubic foot of storage. In 1996, the quilt consisted of 32,000 3-by-6-foot panels. What was the quilt's volume in 1996? If the storage facility had a floor area of 1,500 square feet, how high did the quilt panels need to be stacked?

The NAMES Project AIDS Memorial Quilt memorializes persons all around the world who have died of AIDS. In 1996, the 32,000 panels represented less than 10% of the AIDS deaths in the United States alone, yet the quilt could cover about 19 football fields.

Review

For Exercises 19 and 20, draw and label each solid. Use dashed lines to show the hidden edges.

19. An octahedron with all triangular faces and another octahedron with at least one nontriangular face

20. A cylinder with both radius and height r, a cone with both radius and height r resting flush on one base of the cylinder, and a hemisphere with radius r resting flush on the other base of the cylinder

For Exercises 21 and 22, identify each statement as true or false. Sketch a counterexample for each false statement or explain why it is false.

21. A prism always has an even number of vertices.

22. A section of a cube is either a square or a rectangle.

23. The tower below is an unusual shape. It's neither a cylinder nor a cone. Sketch a two-dimensional figure and an axis such that if you spin your figure about the axis, it will create a solid of revolution shaped like the tower.

The Sugong Tower Mosque in Turpan, China

24. Do research to find a photo or drawing of a chemical model of a crystal. Sketch it. What type of polyhedral structure does it exhibit? You will find helpful Internet links at www.keymath.com/DG .

Science CONNECTION

Ice is a well-known crystal structure. If ice were denser than water, it would sink to the bottom of the ocean, away from heat sources. Eventually the oceans would fill from the bottom up with ice, and we would have an ice planet. What a cold thought!

25. Six points are equally spaced around a circular track with a 20 m radius. Ben runs around the track from one point, past the second, to the third. Al runs straight from the first point to the second, and then straight to the third. How much farther does Ben run than Al?

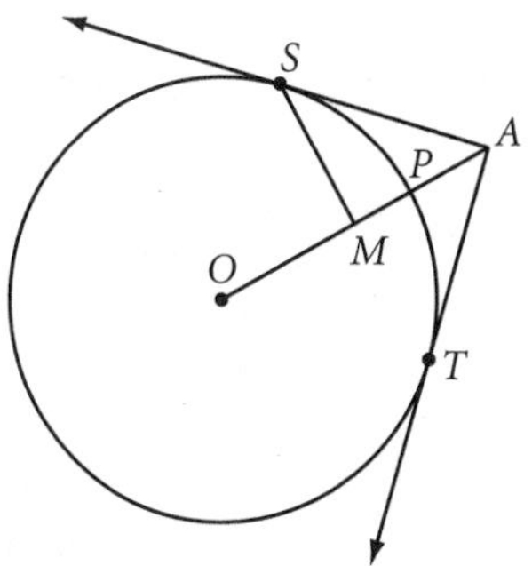

26. $\overrightarrow{AS}$ and $\overrightarrow{AT}$ are tangent to circle O at S and T, respectively. $m\angle SMO = 90°$, $m\angle SAT = 90°$, $SM = 6$. Find the exact value of PA. ⓗ

project

THE SOMA CUBE

If you solved Piet Hein's puzzle at the end of the previous lesson, you now have sketches of the seven nonconvex polyhedrons that can be assembled using four or fewer cubes. These seven polyhedrons consist of a total of 27 cubes: 6 sets of 4 cubes and 1 set of 3 cubes. You are ready for the next rather amazing discovery. These pieces can be arranged to form a 3-by-3-by-3 cube. The puzzle of how to put them together in a perfect cube is known as the Soma Cube puzzle.

Use cubes (wood, plastic, or sugar cubes) to build one set of the seven pieces of the Soma Cube. Use glue, tape, or putty to connect the cubes.

Solve the Soma Cube puzzle. Put the pieces together to make a 3-by-3-by-3 cube.

Now build these other shapes. How do you build the sofa? The tunnel? The castle? The winners' podium?

Sofa

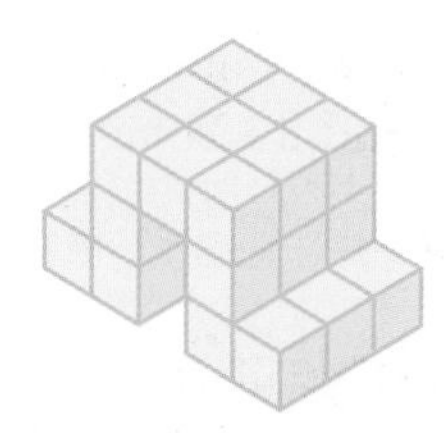

Tunnel

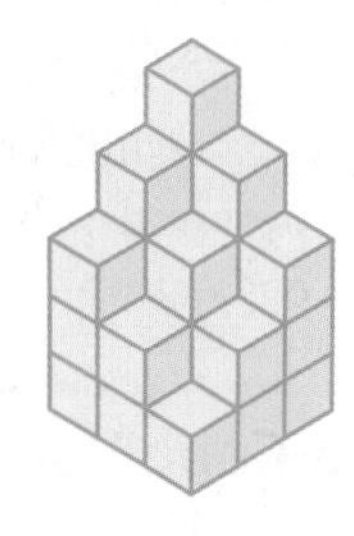

Castle

The Winners' Podium

Create a shape of your own that uses all the pieces. Go to www.keymath.com/DG to learn more about the Soma Cube.

LESSON

10.3

Volume of Pyramids and Cones

If I had influence with the good fairy . . . I should ask that her gift to each child in the world be a sense of wonder so indestructible that it would last throughout life.

RACHEL CARSON

There is a simple relationship between the volumes of prisms and pyramids with congruent bases and the same height, and between cylinders and cones with congruent bases and the same height. You'll discover this relationship in the investigation.

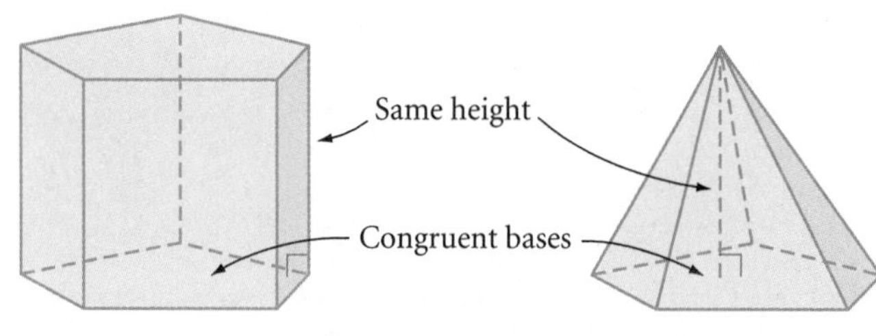

Investigation
The Volume Formula for Pyramids and Cones

You will need
- container pairs of prisms and pyramids
- container pairs of cylinders and cones
- sand, rice, birdseed, or water

Step 1 Choose a prism and a pyramid that have congruent bases and the same height.

Step 2 Fill the pyramid, then pour the contents into the prism. About what fraction of the prism is filled by the volume of one pyramid?

Step 3 Check your answer by repeating Step 2 until the prism is filled.

Step 4 Choose a cone and a cylinder that have congruent bases and the same height and repeat Steps 2 and 3.

Step 5 Compare your results with the results of others. Did you get similar results with both your pyramid-prism pair and the cone-cylinder pair? You should be ready to make a conjecture.

Pyramid-Cone Volume Conjecture C-89

If B is the area of the base of a pyramid or a cone and H is the height of the solid, then the formula for the volume is $V = \underline{?}$.

If you successfully completed the investigation, you probably noticed that the volume formula is the same for all pyramids and cones, regardless of the type of base they have. To calculate the volume of a pyramid or cone, first find the area of its base. Then find the product of that area and the height of the solid, and multiply by the fraction you discovered in the investigation.

EXAMPLE A

Find the volume of a regular hexagonal pyramid with a height of 8 cm. Each side of its base is 6 cm.

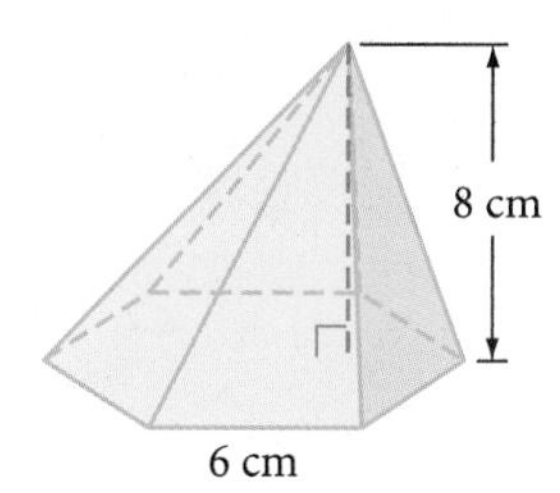

▶ Solution

First find the area of the base. To find the area of a regular hexagon, you need the apothem. By the 30°-60°-90° Triangle Conjecture, the apothem is $3\sqrt{3}$ cm.

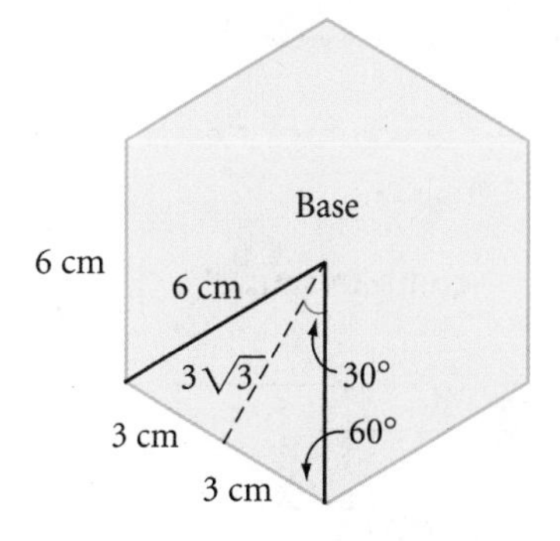

$B = \left(\frac{1}{2}\right)ap$ — The area of a regular polygon is one-half the apothem times the perimeter.

$B = \left(\frac{1}{2}\right)(3\sqrt{3})(36)$ — Substitute $3\sqrt{3}$ for a and 36 for p.

$B = 54\sqrt{3}$ — Multiply.

The base has an area of $54\sqrt{3}$ cm^2. Now find the volume of the pyramid.

$V = \left(\frac{1}{3}\right)BH$ — The volume of a pyramid is one-third the area of the base times the height.

$V = \left(\frac{1}{3}\right)(54\sqrt{3})(8)$ — Substitute $54\sqrt{3}$ for B and 8 for H.

$V = 144\sqrt{3}$ — Multiply.

The volume is $144\sqrt{3}$ cm^3 or approximately 249.4 cm^3.

EXAMPLE B

A cone has a base radius of 3 in. and a volume of 24π in.^{3}. Find the height.

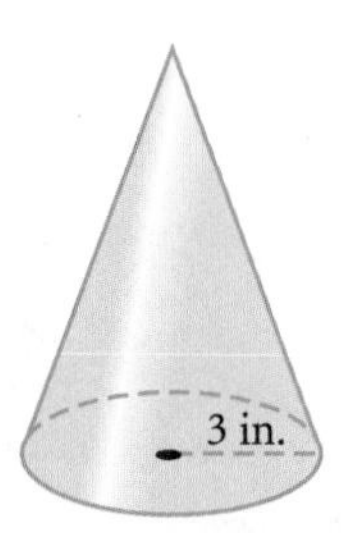

▶ ***Solution***

Start with the volume formula and solve for H.

$V = \frac{1}{3}BH$	Volume formula for pyramids and cones.
$V = \frac{1}{3}(\pi r^2)(H)$	The base of a cone is a circle.
$24\pi = \frac{1}{3}(\pi \cdot 3^2)(H)$	Substitute 24π for the volume and 3 for the radius.
$24\pi = 3\pi H$	Square the 3 and multiply by $\frac{1}{3}$.
$8 = H$	Solve for H.

The height of the cone is 8 in.

EXERCISES

Find the volume of each solid named in Exercises 1–6. All measurements are in centimeters.

1. Square pyramid

2. Cone

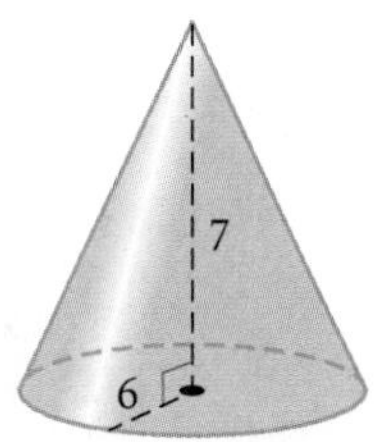

3. Trapezoidal pyramid ⓗ

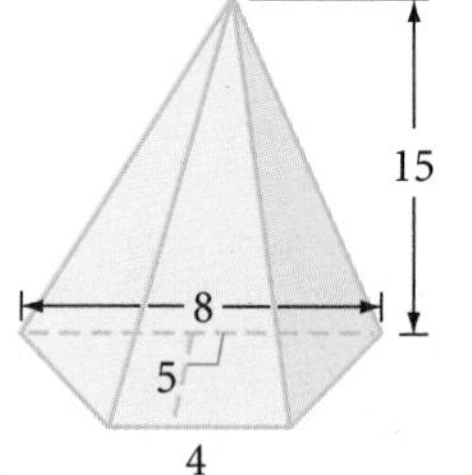

4. Triangular pyramid

5. Semicircular cone

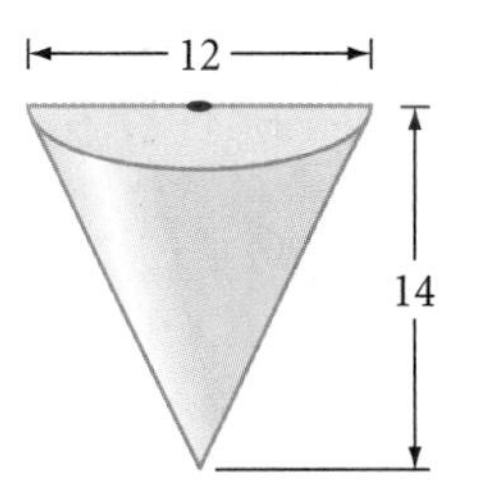

6. Cylinder with cone removed ⓗ

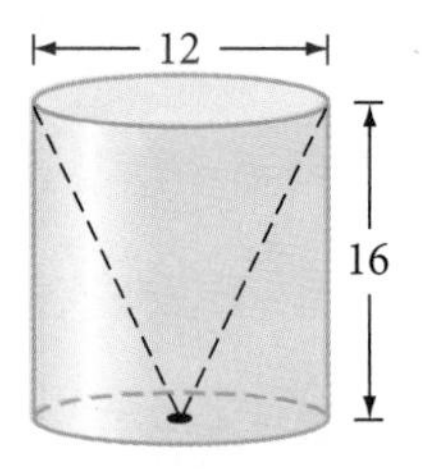

In Exercises 7–9, express the total volume of each solid with the help of algebra. In Exercise 9, what percentage of the volume is filled with the liquid? All measurements are in centimeters.

7. Square pyramid

8. Cone

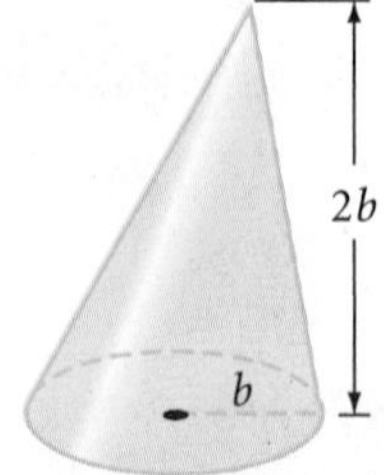

9. Cone

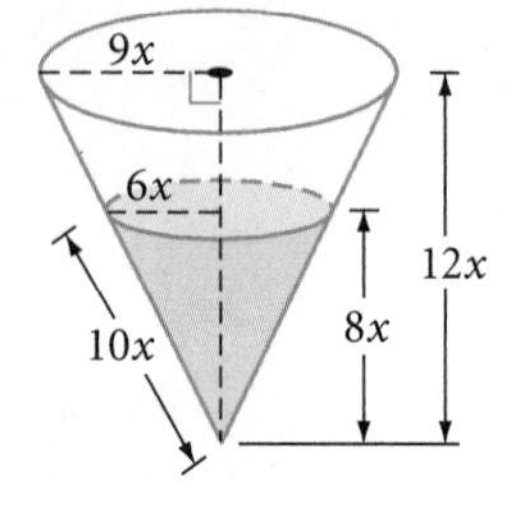

10. Use the information about the base and height of each solid to find the volume. All measurements are given in centimeters.

Information about base of solid	Height of solid	Triangular pyramid	Rectangular pyramid	Trapezoidal pyramid	Cone
$b = 6, b_2 = 7,$ $h = 6, r = 3$	$H = 20$	**a.** $V =$ ⓗ	**d.** $V =$	**g.** $V =$	**j.** $V =$
$b = 9, b_2 = 22,$ $h = 8, r = 6$	$H = 20$	**b.** $V =$	**e.** $V =$	**h.** $V =$	**k.** $V =$
$b = 13, b_2 = 29,$ $h = 17, r = 8$	$H = 24$	**c.** $V =$	**f.** $V =$	**i.** $V =$	**l.** $V =$

For Exercises 11 and 12, sketch and label the solids described.

11. Sketch and label a square pyramid with height H feet and each side of the base M feet. The altitude meets the square base at the intersection of the two diagonals. Find the volume in terms of H and M.

12. Sketch two different circular cones each with a volume of 2304π cm^3.

13. Mount Fuji, the active volcano in Honshu, Japan, is 3776 m high and has a slope of approximately 30°. Mount Etna, in Sicily, is 3350 m high and approximately 50 km across the base. If you assume they both can be approximated by cones, which volcano is larger?

Mount Fuji is Japan's highest mountain. Legend claims that an earthquake created it.

14. Bretislav has designed a crystal glass sculpture. Part of the piece is in the shape of a large regular pentagonal pyramid, shown at right. The apothem of the base measures 27.5 cm. How much will this part weigh if the glass he plans to use weighs 2.85 grams per cubic centimeter?

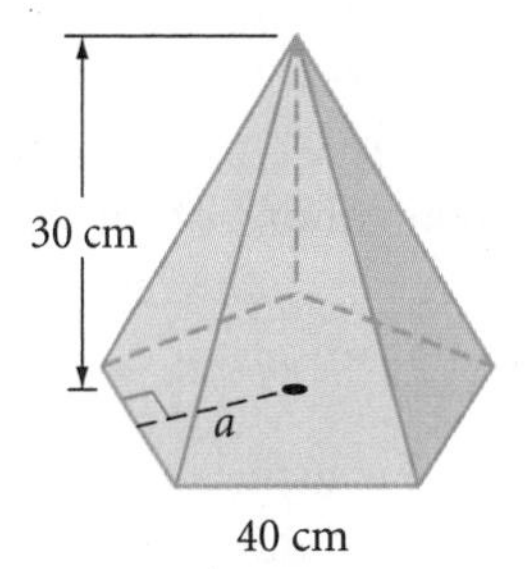

15. Jamala has designed a container that she claims will hold 50 in.^{3}. The net is shown at right. Check her calculations. What is the volume of the solid formed by this net? ⓗ

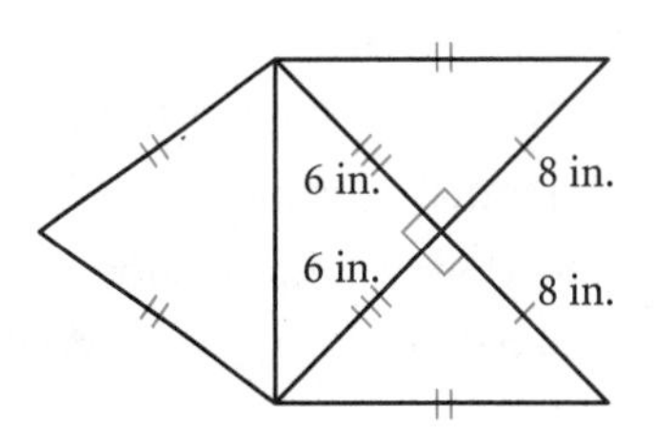

16. Find the volume of the solid formed by rotating the shaded figure about the x-axis.

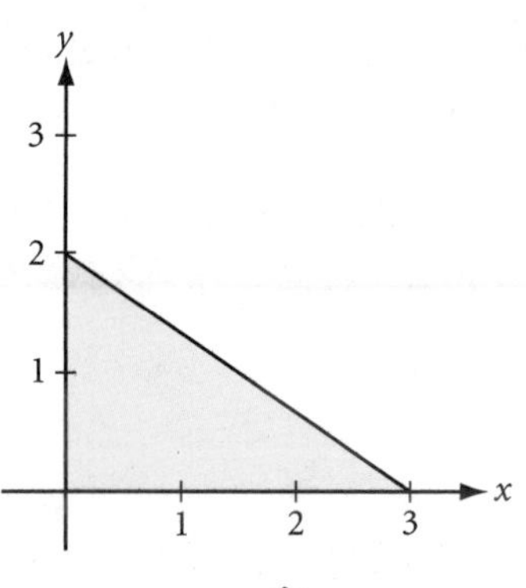

Review

17. Find the volume of the liquid in this right rectangular prism. All measurements are given in centimeters.

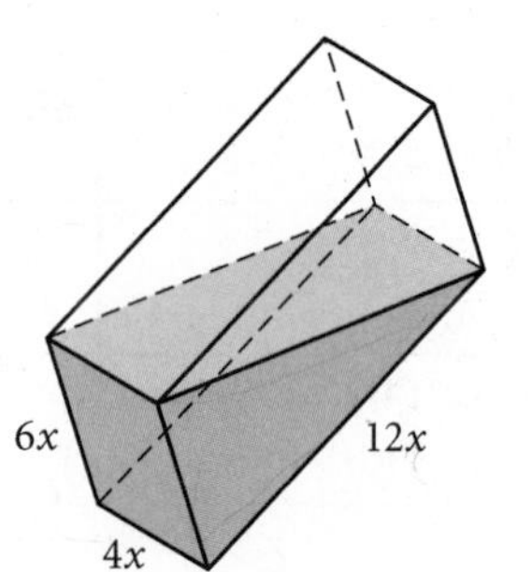

18. **APPLICATION** A swimming pool is in the shape of this prism. A cubic foot of water is about 7.5 gallons. How many gallons of water can the pool hold? If a pump is able to pump water into the pool at a rate of 15 gallons per minute, how long will it take to fill the pool? ⓗ

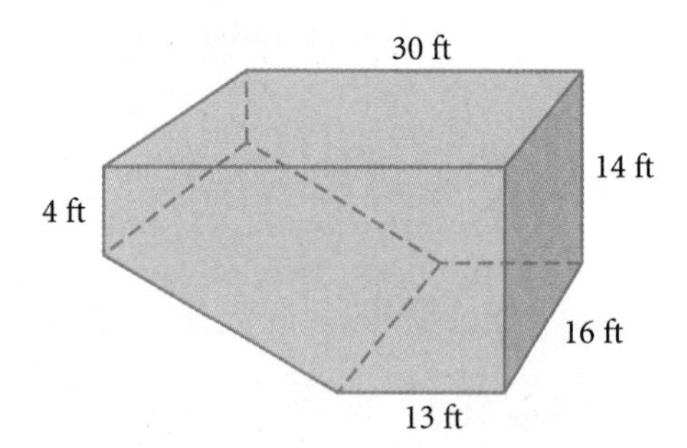

19. **APPLICATION** A landscape architect is building a stone retaining wall as sketched at right. How many cubic feet of stone will she need?

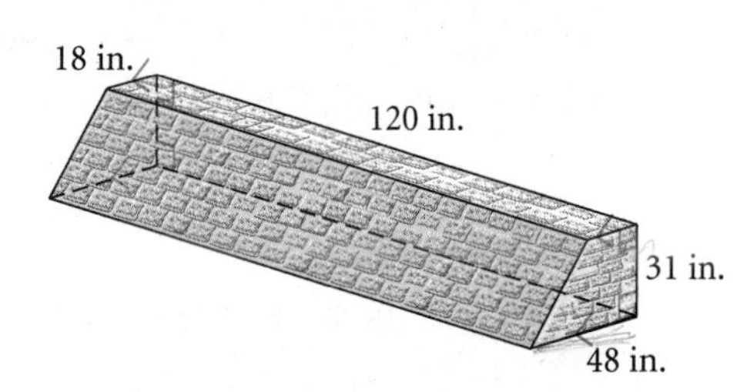

20. As bad as tanker oil spills are, they are only about 12% of the 3.5 million tons of oil that enters the oceans each year. The rest comes from routine tanker operations, sewage treatment plants' runoff, natural sources, and offshore oil rigs. One month's maintenance and routine operation of a single supertanker produces up to 17,000 gallons of oil sludge that gets into the ocean! If a cylindrical barrel is about 1.6 feet in diameter and 2.8 feet tall, how many barrels are needed to hold 17,000 gallons of oil sludge? Recall that a cubic foot of water is about 7.5 gallons.

21. Find the surface area of each of the following polyhedrons. (See the shapes on page 528.) Give *exact* answers.

a. A regular tetrahedron with an edge of 4 cm

b. A regular hexahedron with an edge of 4 cm

c. A regular icosahedron with an edge of 4 cm

d. The dodecahedron shown at right, made of four congruent rectangles and eight congruent triangles

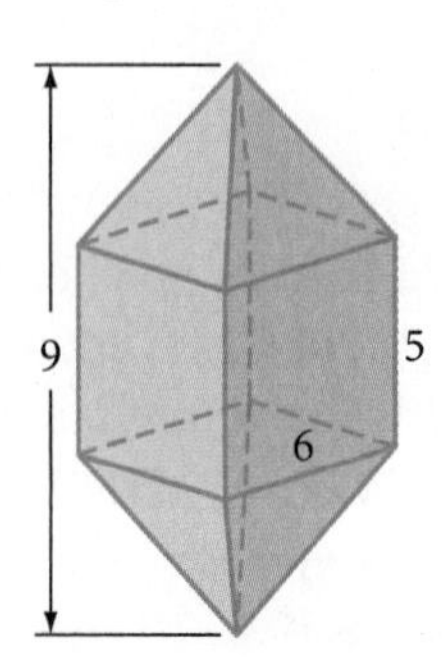

22. Given the triangle at right, reflect D over $\overline{AC}$ to D'. Then reflect D over $\overline{BC}$ to D''. Explain why D', C, D'' are collinear.

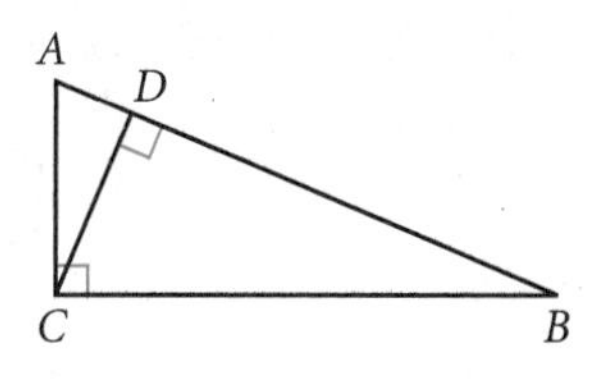

23. In each diagram, $WXYZ$ is a parallelogram. Find the coordinates of Y.

a.

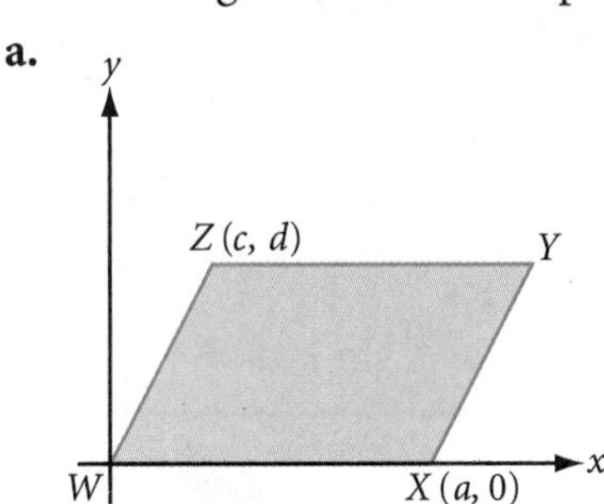

b.

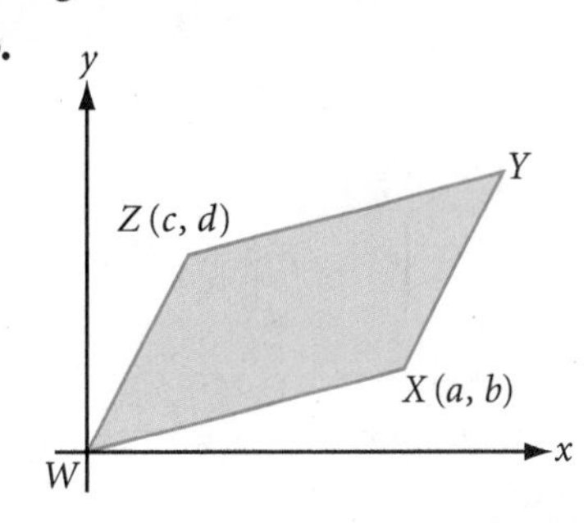

c.

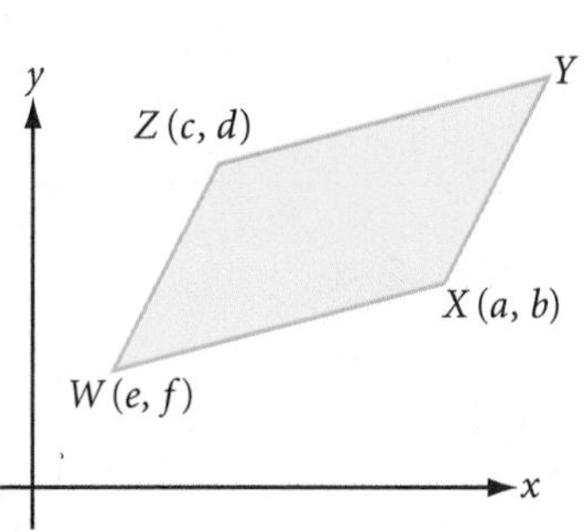

project

THE WORLD'S LARGEST PYRAMID

The pyramid at Cholula, Mexico, shown at right, was built between the second and eighth centuries C.E. Like most pyramids of the Americas, it has a flat top. In fact, it is really two flat-topped pyramids.

Some people claim it is the world's largest pyramid—even larger than the Great Pyramid of Khufu at Giza (shown on page 519) erected around 2500 B.C.E. Is it? Which of the two has the greater volume?

Your project should include

- Volume calculations for both pyramids.
- Scale models of both pyramids.

This church in Cholula appears to be on a hill, but it is actually built on top of an ancient pyramid!

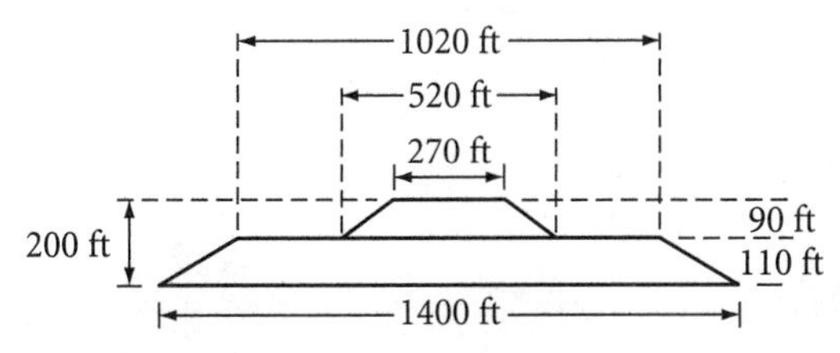

Side view of the two pyramids at Cholula

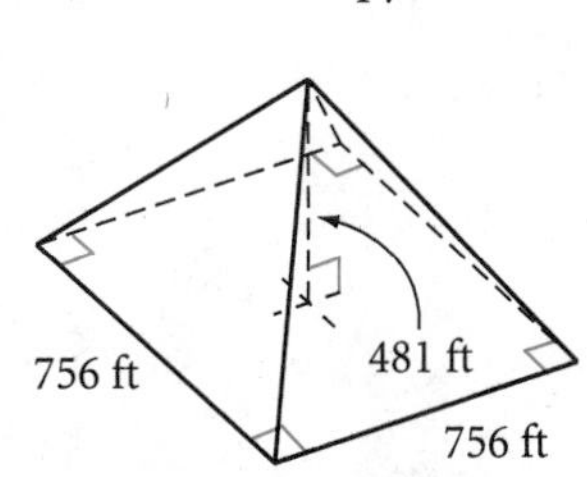

Dimensions of the Great Pyramid at Giza

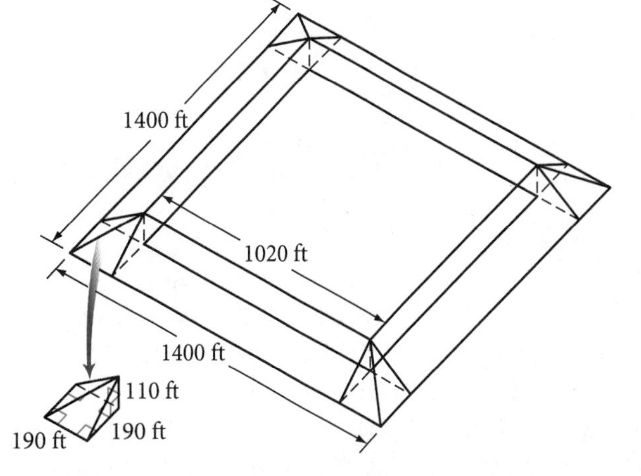

Dissected view of bottom pyramid at Cholula

Exploration

The Five Platonic Solids

Regular polyhedrons have intrigued mathematicians for thousands of years. Greek philosophers saw the principles of mathematics and science as the guiding forces of the universe. Plato (429–347 B.C.E.) reasoned that because all objects are three-dimensional, their smallest parts must be in the shape of regular polyhedrons. There are only five regular polyhedrons, and they are commonly called the **Platonic solids.**

Plato

Plato assigned each regular solid to one of the five "atoms": the tetrahedron to fire, the icosahedron to water, the octahedron to air, the cube or hexahedron to earth, and the dodecahedron to the cosmos.

Fire

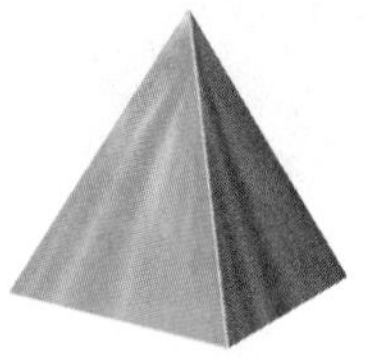

Regular tetrahedron (4 faces)

Water

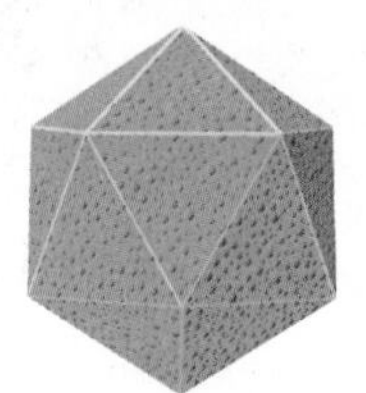

Regular icosahedron (20 faces)

Air

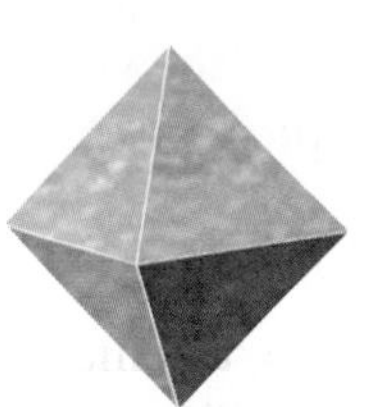

Regular octahedron (8 faces)

Earth

Regular hexahedron (6 faces)

Cosmos

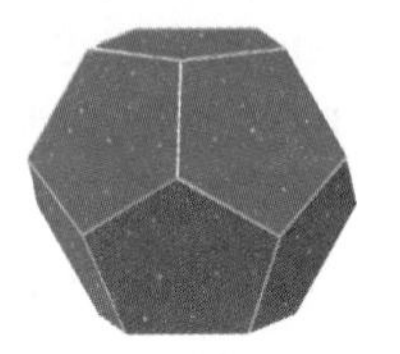

Regular dodecahedron (12 faces)

Activity
Modeling the Platonic Solids

You will need

- poster board or cardboard
- a compass and straightedge
- scissors
- glue, paste, or cellophane tape
- colored pens or pencils for decorating the solids (*optional*)

What would each of the five Platonic solids look like when unfolded? There is more than one way to unfold each polyhedron. Recall that a flat figure that you can fold into a polyhedron is called its net.

Step 1 One face is missing in the net at right. Complete the net to show what the regular tetrahedron would look like if it were cut open along the three lateral edges and unfolded into one piece.

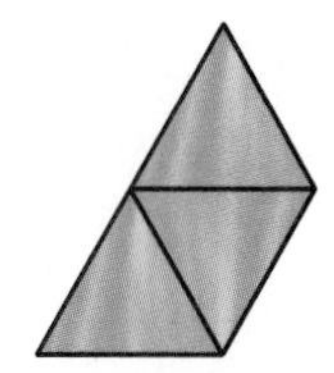

Step 2 Two faces are missing in the net at right. Complete the net to show what the regular hexahedron would look like if it were cut open along the lateral edges and three top edges, then unfolded.

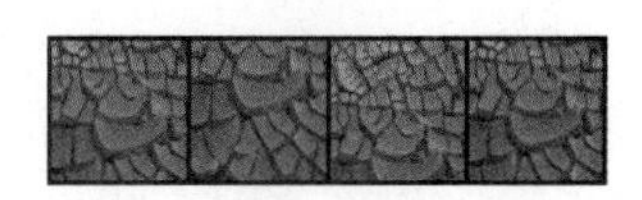

Step 3 Here is one possible net for the regular icosahedron. When the net is folded together, the five top triangles meet at one top point. Which edge—*a, b,* or *c*—does the edge labeled *x* line up with?

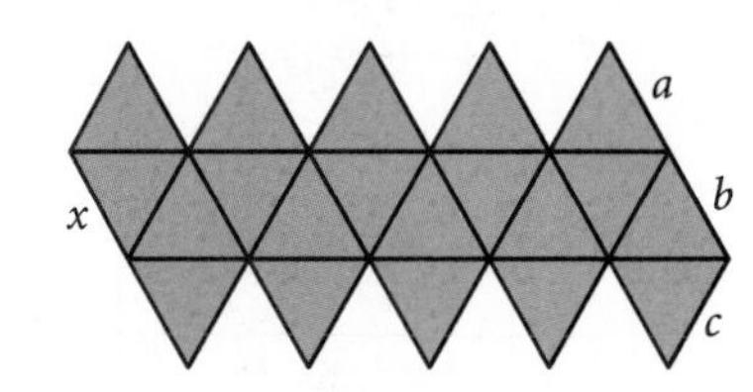

Step 4 The regular octahedron is similar to the icosahedron but has only eight equilateral triangles. Complete the octahedron net at right. Two faces are missing.

Step 5 The regular dodecahedron has 12 regular pentagons as faces. If you cut the dodecahedron into two equal parts, they would look like two flowers, each having five pentagon-shaped petals around a center pentagon. Complete the net for half a dodecahedron.

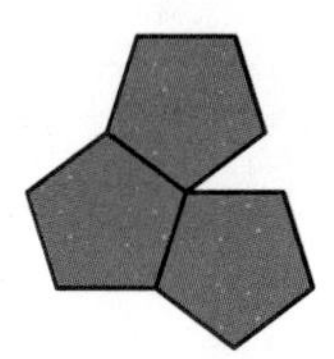

Now you know what the nets of the five Platonic solids could look like. Let's use the nets to construct and assemble models of the five Platonic solids. See the Procedure Note for some tips.

Procedure Note

1. To save time, build solids with the same shape faces at the same time.
2. Construct a regular polygon template for each shape face.
3. Erase the unnecessary line segments.
4. Leave tabs on some edges for gluing.

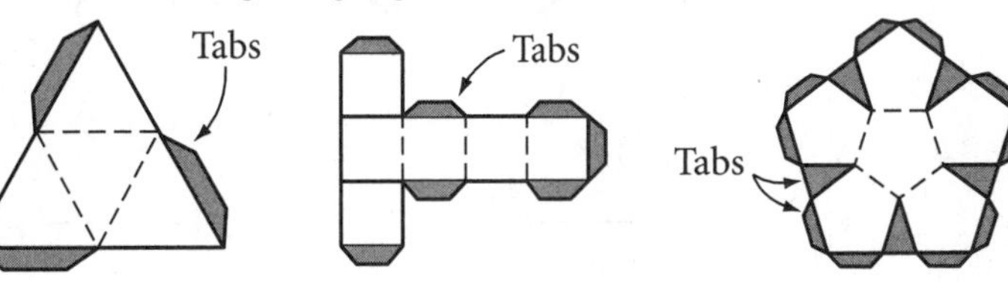

5. Decorate each solid before you cut it out.
6. Score on both sides of the net by running a pen or compass point over the fold lines.

Step 6 Construct the nets for icosahedron, octahedron, and tetrahedron with equilateral triangles.

Step 7	Construct a net for the hexahedron, or cube, with squares.
Step 8	You will construct a net for the dodecahedron with regular pentagons. To construct a regular pentagon, follow Steps A–F below. Construct a circle. Construct two perpendicular diameters. Find M, the midpoint of $\overline{OA}$. Swing an arc with radius BM intersecting $\overline{OC}$ at point D.

Step A

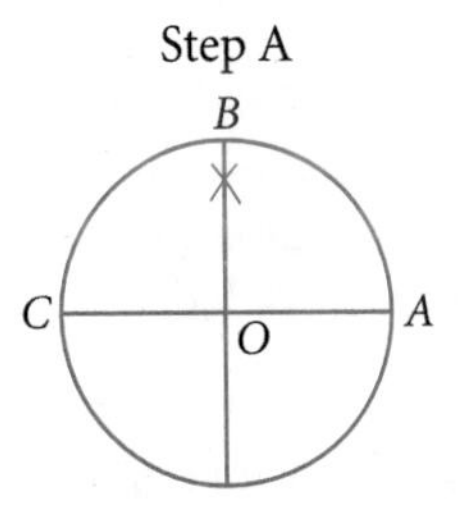

Step B

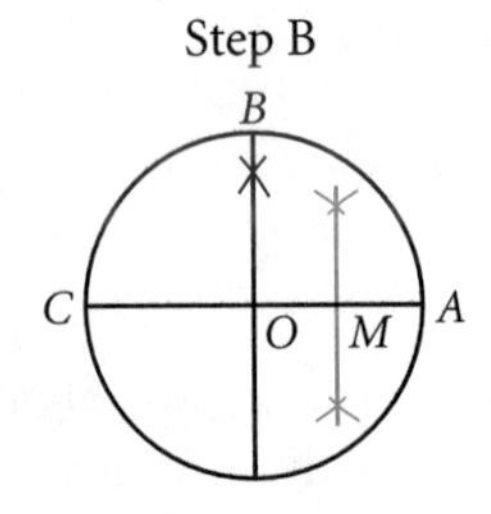

Step C

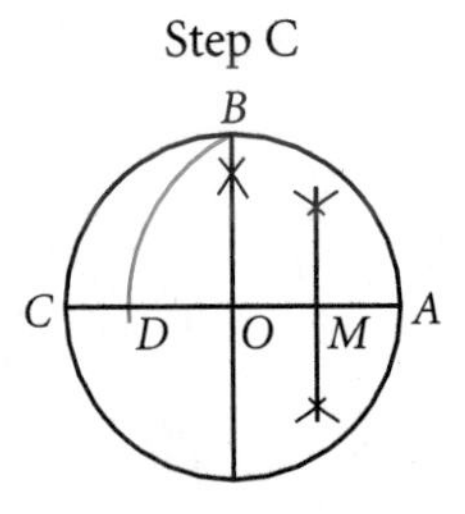

BD is the length of each side of the pentagon. Starting at point B, mark off BD on the circumference five times. Connect the points to form a pentagon.

Step D

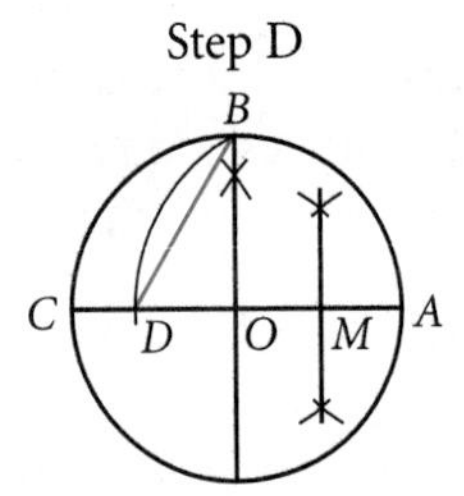

Step E

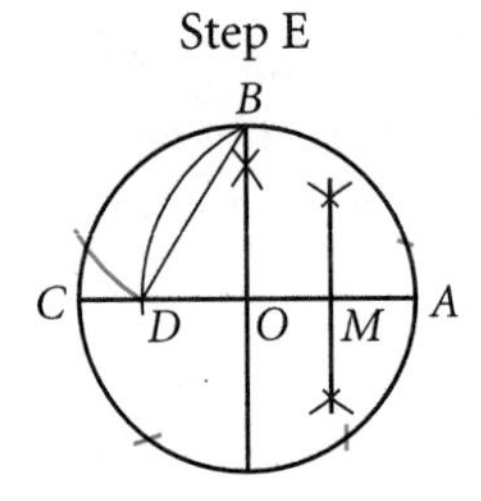

Step F

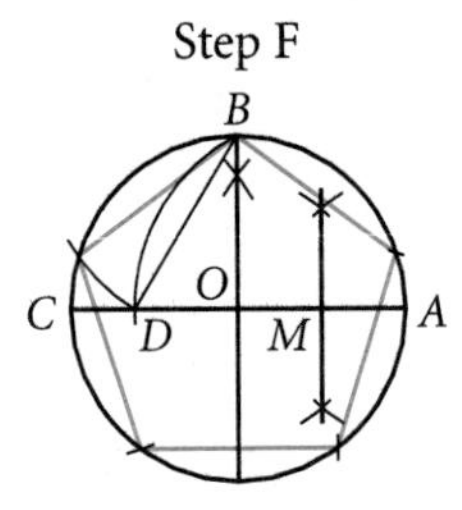

Step 9	Follow Steps A–D below to construct half the net for the dodecahedron. Construct a large regular pentagon. Lightly draw all the diagonals. The smaller regular pentagon will be one of the 12 faces of the dodecahedron.

Step A

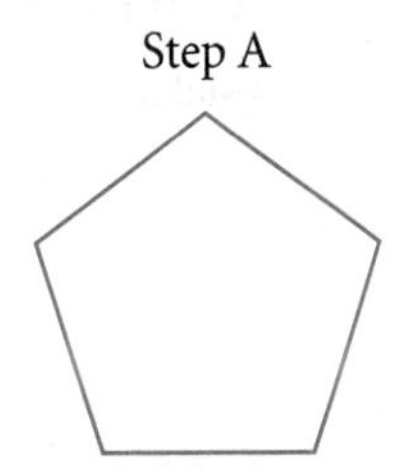

Step B

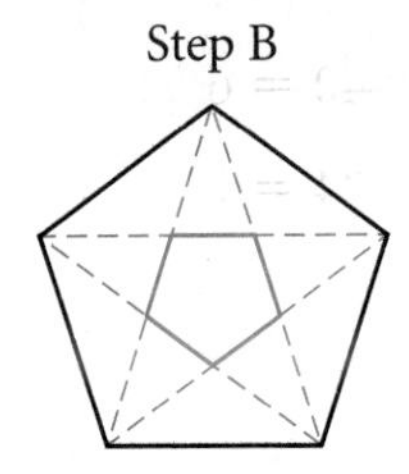

Draw the diagonals of the central pentagon and extend them to the sides of the larger pentagon. Find the five pentagons that encircle the central pentagon.

Step C

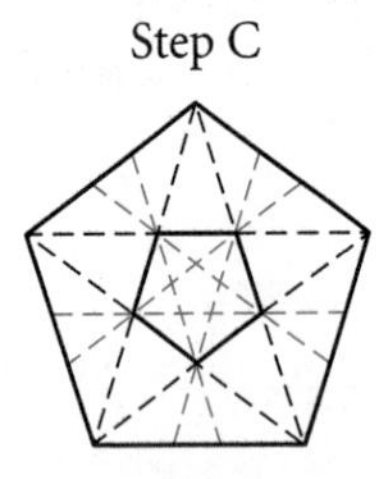

Step D

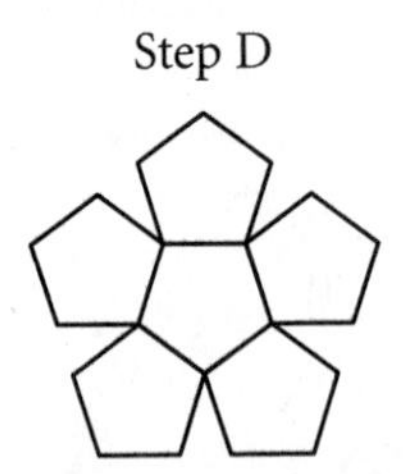

Step 10	Can you explain from looking at the nets why there are only five Platonic solids?

LESSON

10.4

Volume Problems

If you have made mistakes . . . there is always another chance for you . . . for this thing we call "failure" is not the falling down, but the staying down.

MARY PICKFORD

Volume has applications in science, medicine, engineering, and construction. For example, a chemist needs to accurately measure the volume of reactive substances. A doctor may need to calculate the volume of a cancerous tumor based on a body scan. Engineers and construction personnel need to determine the volume of building supplies such as concrete or asphalt. The volume of the rooms in a completed building will ultimately determine the size of mechanical devices such as air conditioning units.

Sometimes, if you know the volume of a solid, you can calculate an unknown length of a base or the solid's height. Here are two examples.

EXAMPLE A

The volume of this right triangular prism is 1440 cm^3. Find the height of the prism.

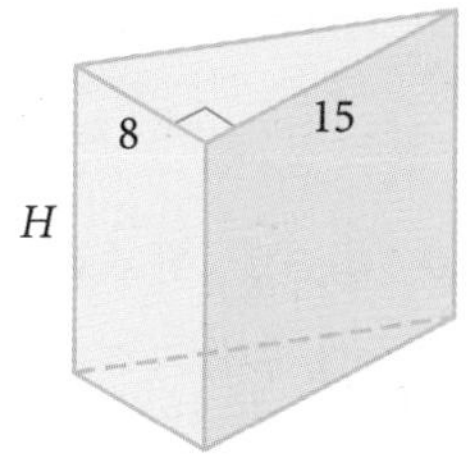

▶ Solution

$V = BH$	Volume formula for prisms and cylinders.
$V = \frac{1}{2}(bh)H$	The base of the prism is a triangle.
$1440 = \frac{1}{2}(8)(15)H$	Substitute 1440 for the volume, 8 for the base of the triangle, and 15 for the height of the triangle.
$1440 = 60H$	Multiply.
$24 = H$	Solve for H.

The height of the prism is 24 cm.

EXAMPLE B

The volume of this sector of a right cylinder is 2814 m^3. Find the radius of the base of the cylinder to the nearest m.

▶ Solution

The volume is the area of the base times the height. To find the area of the sector, you first find what fraction the sector is of the whole circle: $\frac{40}{360} = \frac{1}{9}$.

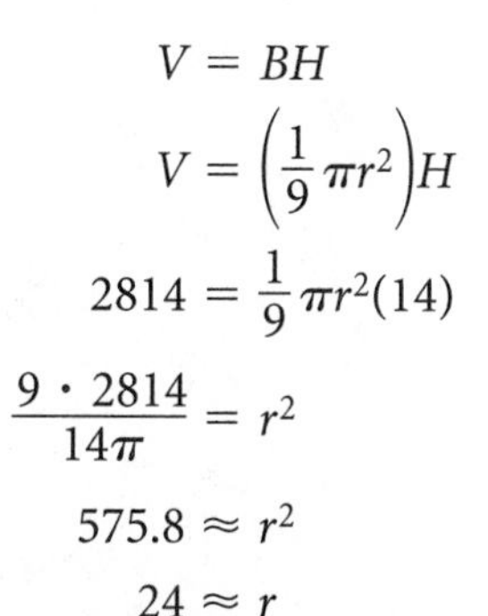

$$V = BH$$

$$V = \left(\frac{1}{9}\pi r^2\right)H$$

$$2814 = \frac{1}{9}\pi r^2(14)$$

$$\frac{9 \cdot 2814}{14\pi} = r^2$$

$$575.8 \approx r^2$$

$$24 \approx r$$

The radius is about 24 m.

EXERCISES

You will need

Construction tools for Exercise **18**

1. If you cut a 1-inch square out of each corner of an 8.5-by-11-inch piece of paper and fold it into a box without a lid, what is the volume of the container?

2. The prism at right has equilateral triangle bases with side lengths of 4 cm. The height of the prism is 8 cm. Find the volume.

3. A triangular pyramid has a volume of 180 cm^3 and a height of 12 cm. Find the length of a side of the triangular base if the triangle's height from that side is 6 cm.

4. A trapezoidal pyramid has a volume of 3168 cm^3, and its height is 36 cm. The lengths of the two bases of the trapezoidal base are 20 cm and 28 cm. What is the height of the trapezoidal base? ⓗ

5. The volume of a cylinder is 628 cm^3. Find the radius of the base if the cylinder has a height of 8 cm. Round your answer to the nearest 0.1 cm.

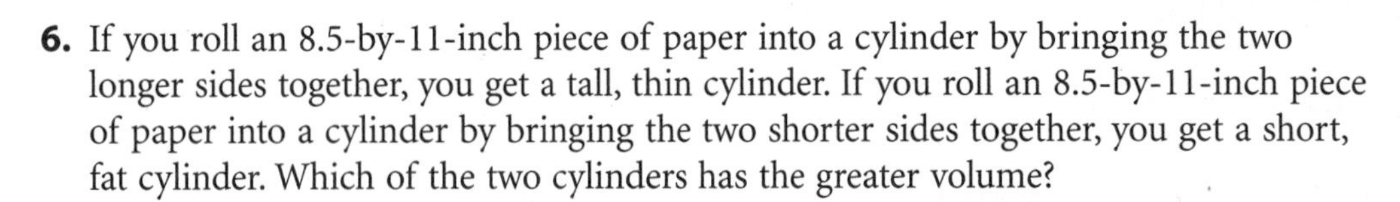

6. If you roll an 8.5-by-11-inch piece of paper into a cylinder by bringing the two longer sides together, you get a tall, thin cylinder. If you roll an 8.5-by-11-inch piece of paper into a cylinder by bringing the two shorter sides together, you get a short, fat cylinder. Which of the two cylinders has the greater volume?

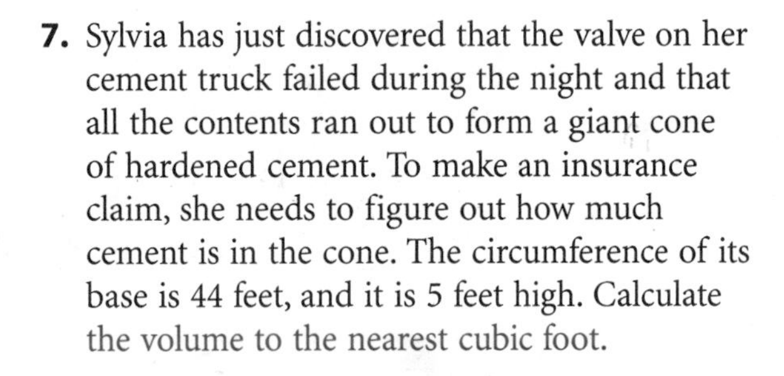

7. Sylvia has just discovered that the valve on her cement truck failed during the night and that all the contents ran out to form a giant cone of hardened cement. To make an insurance claim, she needs to figure out how much cement is in the cone. The circumference of its base is 44 feet, and it is 5 feet high. Calculate the volume to the nearest cubic foot.

8. A sealed rectangular container 6 cm by 12 cm by 15 cm is sitting on its smallest face. It is filled with water up to 5 cm from the top. How many centimeters from the bottom will the water level reach if the container is placed on its largest face?

9. To test his assistant, noted adventurer Dakota Davis states that the volume of the regular hexagonal ring at right is equal to the volume of the regular hexagonal hole in its center. The assistant must confirm or refute this, using dimensions shown in the figure. What should he say to Dakota? ⓗ

Use this information to solve Exercises 10–12: Water weighs about 63 pounds per cubic foot, and a cubic foot of water is about 7.5 gallons.

10. APPLICATION A king-size waterbed mattress measures 5.5 feet by 6.5 feet by 8 inches deep. To the nearest pound, how much does the water in this waterbed weigh? ⓗ

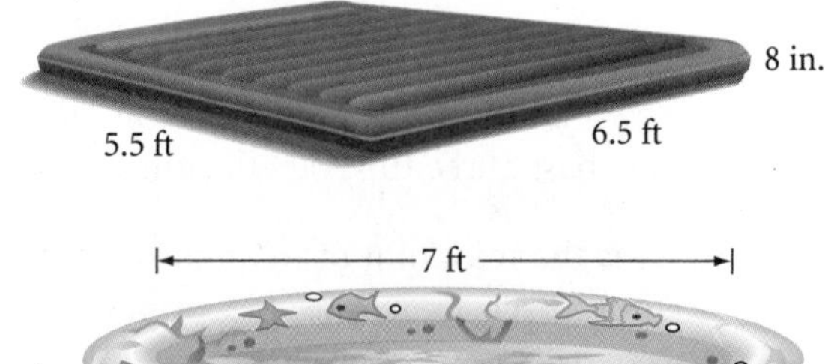

11. A child's wading pool has a diameter of 7 feet and is 8 inches deep. How many gallons of water can the pool hold? Round your answer to the nearest 0.1 gallon.

12. Madeleine's hot tub has the shape of a regular hexagonal prism. The chart on the hot-tub heater tells how long it takes to warm different amounts of water by 10°F. Help Madeleine determine how long it will take to raise the water temperature from 93°F to 103°F.

Minutes to Raise Temperature 10°F

Gallons	350	400	450	500	550	600	650	700
Minutes	9	10	11	12	14	15	16	18

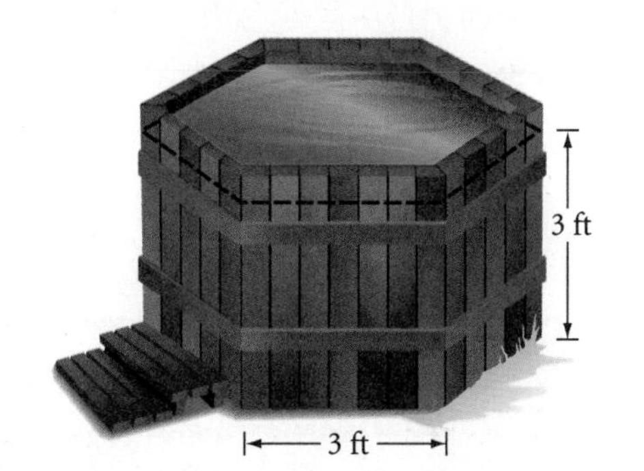

13. A standard juice box holds 8 fluid ounces. A fluid ounce of liquid occupies 1.8 in.^{3}. Design a cylindrical can that will hold about the same volume as one juice box. What are some possible dimensions of the can?

14. The photo at right shows an ice tray that is designed for a person who has the use of only one hand—each piece of ice will rotate out of the tray when pushed with one finger. Suppose the tray has a length of 12 inches and a height of 1 inch. Approximate the volume of water the tray holds if it is filled to the top. (Ignore the thickness of the plastic.)

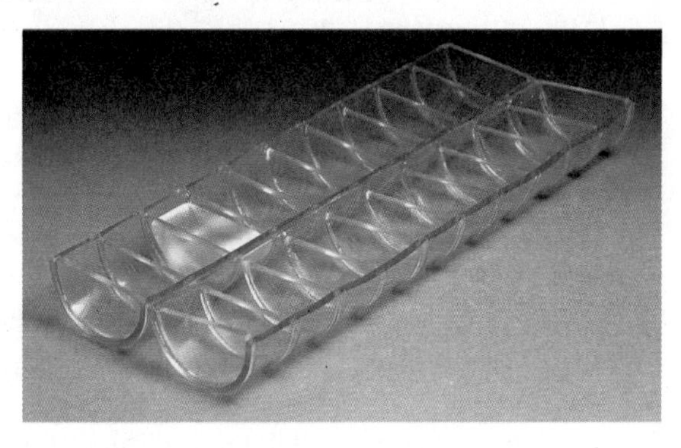

Review

15. Find the height of this right square pyramid. Give your answer to the nearest 0.1 cm.

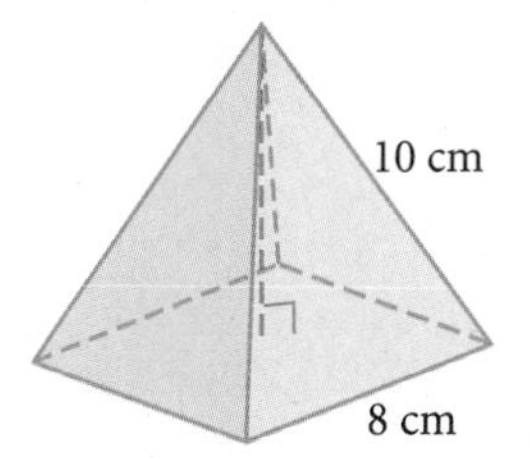

16. $\overleftrightarrow{EC}$ is tangent at C. $\overleftrightarrow{ED}$ is tangent at D. Find x.

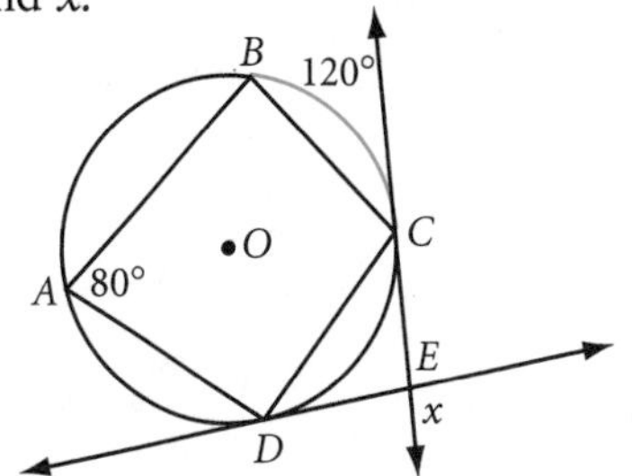

17. In the figure at right, *ABCE* is a parallelogram and *BCDE* is a rectangle. Write a paragraph proof showing that $\triangle ABD$ is isosceles.

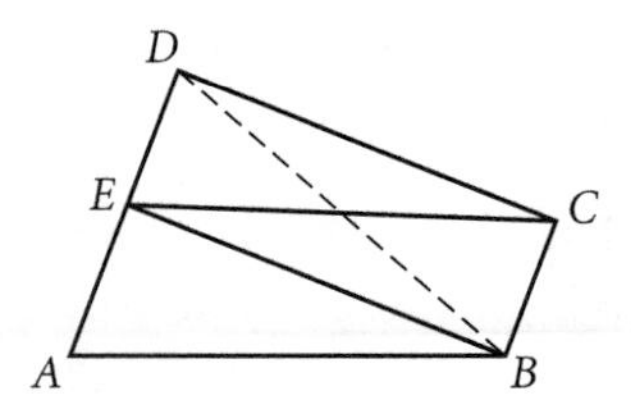

18. ***Construction*** Use your compass and straightedge to construct an isosceles trapezoid with a base angle of 45° and the length of one base three times the length of the other base.

19. *M* is the midpoint of $\overline{AC}$ and $\overline{BD}$. For each statement, select always (A), sometimes (S), or never (N).

a. *ABCD* is a parallelogram.

b. *ABCD* is a rhombus.

c. *ABCD* is a kite.

d. $\triangle AMD \cong \triangle AMB$

e. $\angle DAM \cong \angle BCM$

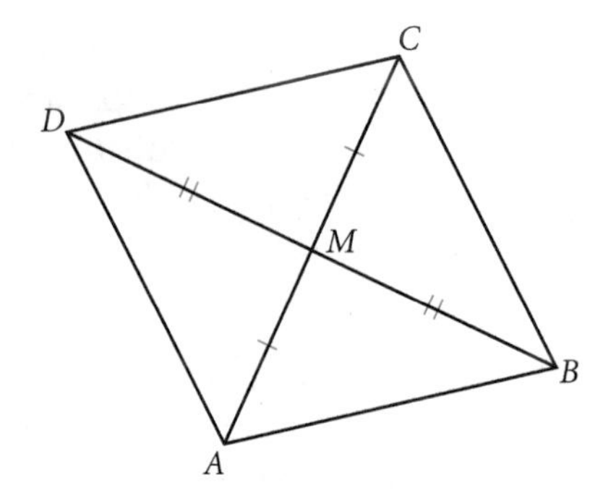

IMPROVING YOUR REASONING SKILLS

Bert's Magic Hexagram

Bert is the queen's favorite jester. He entertains himself with puzzles. Bert is creating a magic hexagram on the front of a grid of 19 hexagons. When Bert's magic hexagram (like its cousin the magic square) is completed, it will have the same sum in every straight hexagonal row, column, or diagonal (whether it is three, four, or five hexagons long). For example, $B + 12 + 10$ is the same sum as $B + 2 + 5 + 6 + 9$, which is the same sum as $C + 8 + 6 + 11$. Bert planned to use just the first 19 positive integers (his age in years), but he only had time to place the first 12 integers before he was interrupted. Your job is to complete Bert's magic hexagram. What are the values for *A, B, C, D, E, F,* and *G*?

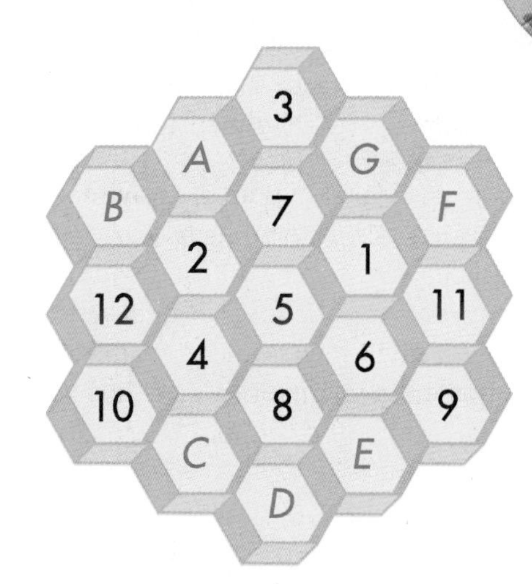

LESSON

10.5

Displacement and Density

Eureka! I have found it!

ARCHIMEDES

What happens if you step into a bathtub that is filled to the brim? If you add a scoop of ice cream to a glass filled with root beer? In each case, you'll have a mess! The volume of the liquid that overflows in each case equals the volume of the solid below the liquid level. This volume is called an object's **displacement.**

EXAMPLE A

Mary Jo wants to find the volume of an irregularly shaped rock. She puts some water into a rectangular prism with a base that measures 10 cm by 15 cm. When the rock is put into the container, Mary Jo notices that the water level rises 2 cm because the rock displaces its volume of water. This new "slice" of water has a volume of (2)(10)(15), or 300 cm^3. So the volume of the rock is 300 cm^3.

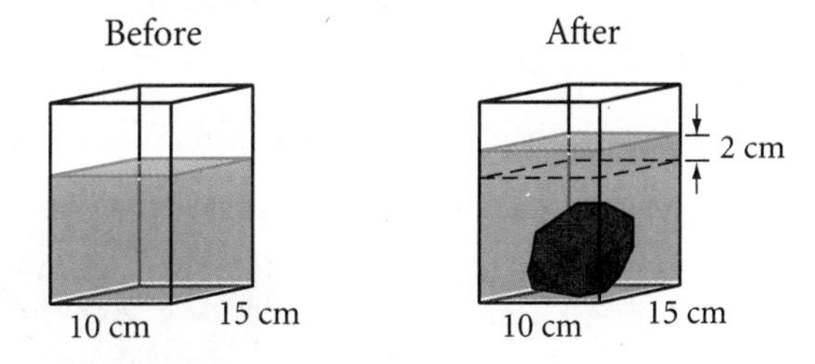

An important property of a material is its density. **Density** is the mass of matter in a given volume. You can find the mass of an object by weighing it. You calculate density by dividing the mass by the volume,

$$\text{density} = \frac{\text{mass}}{\text{volume}}$$

EXAMPLE B

A clump of metal weighing 351.4 g is dropped into a cylindrical container, causing the water level to rise 1.1 cm. The radius of the base of the container is 3.0 cm. What is the density of the metal? Given the table, and assuming the metal is pure, what is the metal?

Metal	Density	Metal	Density
Aluminum	2.81 g/cm^3	Nickel	8.89 g/cm^3
Copper	8.97 g/cm^3	Platinum	21.40 g/cm^3
Gold	19.30 g/cm^3	Potassium	0.86 g/cm^3
Lead	11.30 g/cm^3	Silver	10.50 g/cm^3
Lithium	0.54 g/cm^3	Sodium	0.97 g/cm^3

▶ Solution

First, find the volume of displaced water. Then, divide the weight by the volume to get the density of the metal.

$$\begin{aligned} \text{Volume} &= \pi(3.0)^2(1.1) \\ &= (\pi)(9)(1.1) \\ &\approx 31.1 \end{aligned} \qquad \begin{aligned} \text{Density} &\approx \frac{351.4}{31.1} \\ &\approx 11.3 \end{aligned}$$

The density is 11.3 g/cm^3. Therefore the metal is lead.

History CONNECTION

Archimedes solved the problem of how to tell if a crown was made of genuine gold by weighing the crown under water. Legend has it that the insight came to him while he was bathing. Thrilled by his discovery, Archimedes ran through the streets shouting "Eureka!" wearing just what he'd been wearing in the bathtub.

EXERCISES

1. When you put a rock into a container of water, it raises the water level 3 cm. If the container is a rectangular prism whose base measures 15 cm by 15 cm, what is the volume of the rock?

2. You drop a solid glass ball into a cylinder with a radius of 6 cm, raising the water level 1 cm. What is the volume of the glass ball?

3. A fish tank 10 by 14 by 12 inches high is the home of a large goldfish named Columbia. She is taken out when her owner cleans the tank, and the water level in the tank drops $\frac{1}{3}$ inch. What is Columbia's volume?

For Exercises 4–9, refer to the table on page 535.

4. How much does a solid block of aluminum weigh if its dimensions are 4 cm by 8 cm by 20 cm?

5. Which weighs more: a solid cylinder of gold with a height of 5 cm and a diameter of 6 cm or a solid cone of platinum with a height of 21 cm and a diameter of 8 cm?

6. Chemist Dean Dalton is given a clump of metal and is told that it is sodium. He finds that the metal weighs 145.5 g. He places it into a nonreactive liquid in a square prism whose base measures 10 cm on each edge. If the metal is indeed sodium, how high should the liquid level rise? ⓗ

7. A square-prism container with a base 5 cm by 5 cm is partially filled with water. You drop a clump of metal that weighs 525 g into the container, and the water level rises 2 cm. What is the density of the metal? Assuming the metal is pure, what is the metal?

8. When ice floats in water, one-eighth of its volume floats above the water level and seven-eighths floats beneath the water level. A block of ice placed into an ice chest causes the water in the chest to rise 4 cm. The right rectangular chest measures 35 cm by 50 cm by 30 cm high. What is the volume of the block of ice? ⓗ

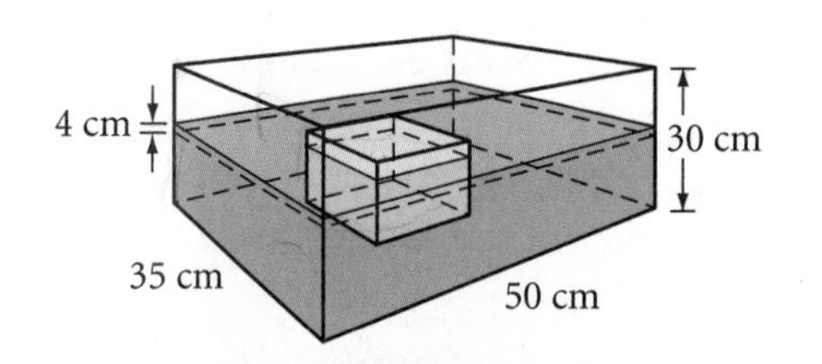

Science CONNECTION

Buoyancy is the tendency of an object to float in either a liquid or a gas. For an object to float on the surface of water, it must sink enough to displace the volume of water equal to its weight.

9. Sherlock Holmes rushes home to his chemistry lab, takes a mysterious medallion from his case, and weighs it. "It weighs 3088 grams. Now, let's check its volume." He pours water into a graduated glass container with a 10-by-10 cm square base, and records the water level, which is 53.0 cm. He places the medallion into the container and reads the new water level, 54.6 cm. He enjoys a few minutes of mental calculation, then turns to Dr. Watson. "This confirms my theory. Quick, Watson! Off to the train station."

"Holmes, you amaze me. Is it gold?" questions the good doctor.

"If it has a density of 19.3 grams per cubic centimeter, it is gold," smiles Mr. Holmes. "If it is gold, then Colonel Banderson is who he says he is. If it is a fake, then so is the Colonel."

"Well?" Watson queries.

Holmes smiles and says, "It's elementary, my dear Watson. Elementary geometry, that is."

What is the volume of the medallion? Is it gold? Is Colonel Banderson who he says he is?

Review

10. What is the volume of the slice removed from this right cylinder? Give your answer to the nearest cm^3.

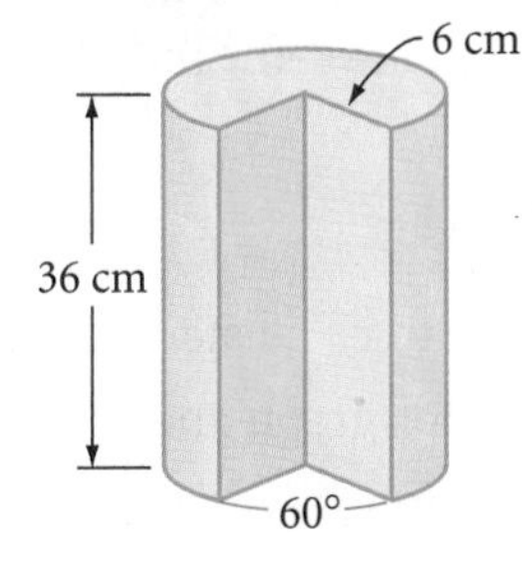

11. APPLICATION Ofelia has brought home a new aquarium shaped like the regular hexagonal prism shown at right. She isn't sure her desk is strong enough to hold it. The aquarium, without water, weighs 48 pounds. How much will it weigh when it is filled? (Water weighs 63 pounds per cubic foot.) If a small fish needs about 180 cubic inches of water to swim around in, about how many small fish can this aquarium house?

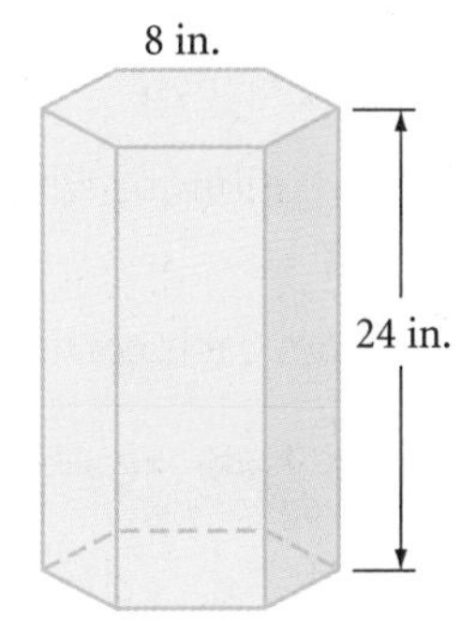

12. $\triangle ABC$ is equilateral. M is the centroid. $AB = 6$ Find the area of $\triangle CEA$.

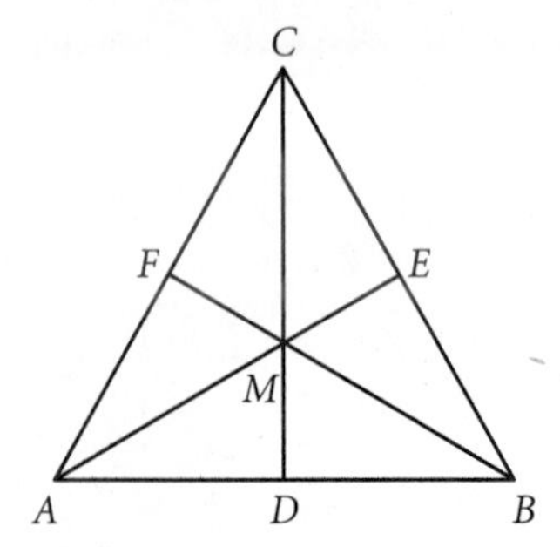

13. Give a paragraph or flowchart proof explaining why M is the midpoint of $\overline{PQ}$.

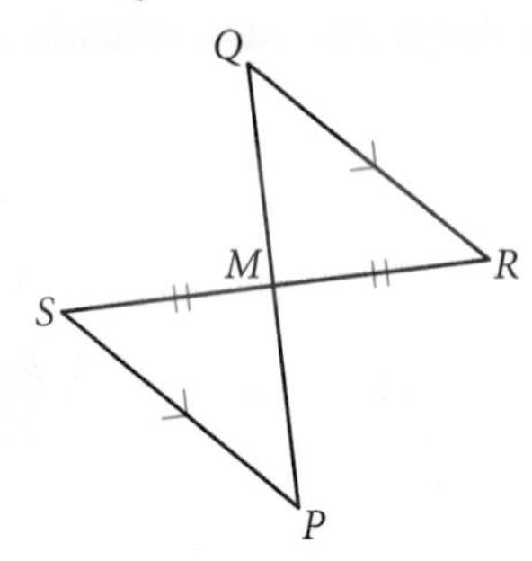

14. The three polygons are regular polygons. How many sides does the red polygon have?

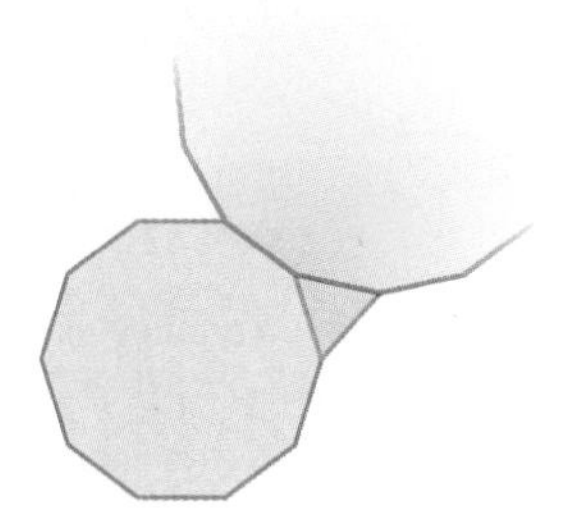

15. A circle passes through the three points (4, 7), (6, 3), and (1, −2).

a. Find the center. **b.** Find the equation.

16. A secret rule matches the following numbers:

$2 \rightarrow 4, 3 \rightarrow 7, 4 \rightarrow 10, 5 \rightarrow 13$

Find $20 \rightarrow$ _?_, and $n \rightarrow$ _?_.

project

MAXIMIZING VOLUME

Suppose you have a 10-inch-square sheet of metal and you want to make a small box by cutting out squares from the corners of the sheet and folding up the sides. What size corners should you cut out to get the biggest box possible?

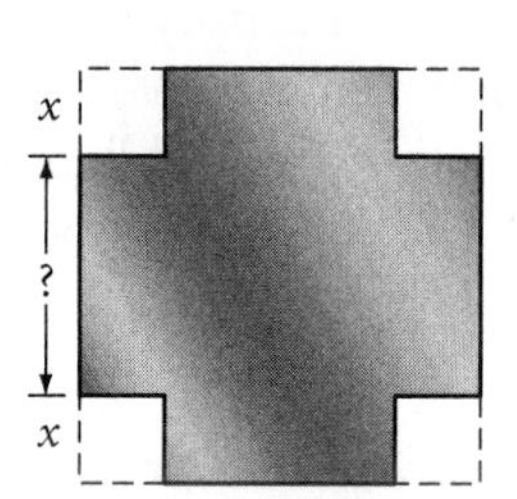

To answer this question, consider the length of the corner cut x and write an equation for the volume of the box, y, in terms of x. Graph your equation using reasonable window values. You should see your graph touch the x-axis in at least two places and reach a maximum somewhere in between. Study these points carefully to find their significance.

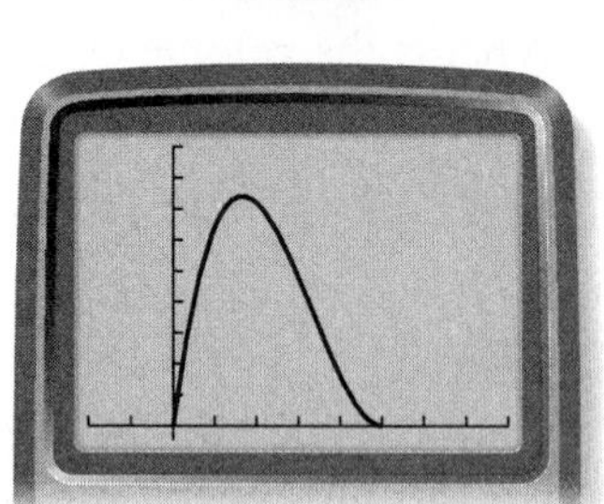

Your project should include

- The equation you used to calculate volume in terms of x and y.
- A sketch of the calculator graph and the graphing window you used.
- An explanation of important points, for example, when the graph touches the x-axis.
- A solution for what size corners make the biggest volume.

Lastly, generalize your findings. For example, what fraction of the side length could you cut from each corner of a 12-inch-square sheet to make a box of maximum volume?

Exploration

Orthographic Drawing

If you have ever put together a toy from detailed instructions, or built a birdhouse from a kit, or seen blueprints for a building under construction, you have seen isometric drawings.

Isometric means "having equal measure," so the edges of a cube drawn isometrically all have the same length. In contrast, recall that when you drew a cube in two-point perspective, you needed to use edges of different lengths to get a natural look.

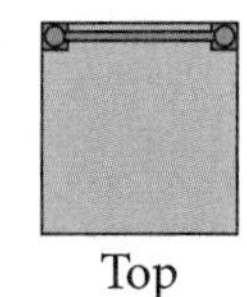

Top

When you buy a product from a catalog or off the Internet, you want to see it from several angles. The top, front, and right side views are given in an **orthographic drawing.** Ortho means "straight," and the views of an orthographic drawing show the faces of a solid as though you are looking at them "head-on."

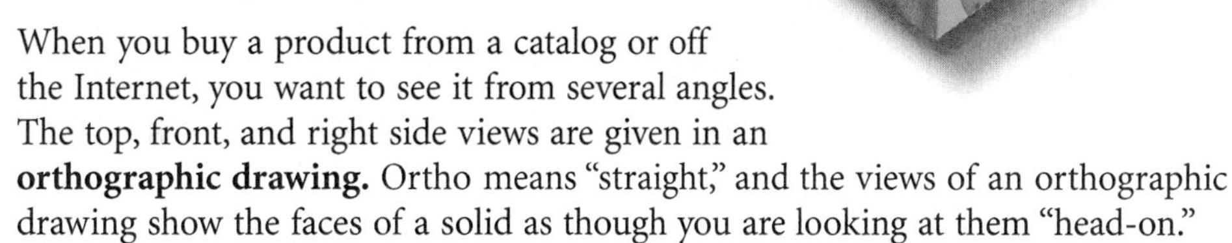

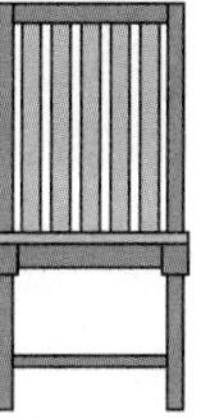

Front

An isometric drawing

A two-point perspective drawing

An orthographic drawing

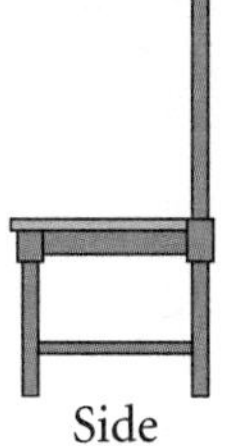

Side

Isometric

Career CONNECTION

Architects create blueprints for their designs, as shown at right. Architectural drawing plans use orthographic techniques to describe the proposed design from several angles. These front and side views are called building elevations.

EXAMPLE A

Make an orthographic drawing of the solid shown in the isometric drawing at right.

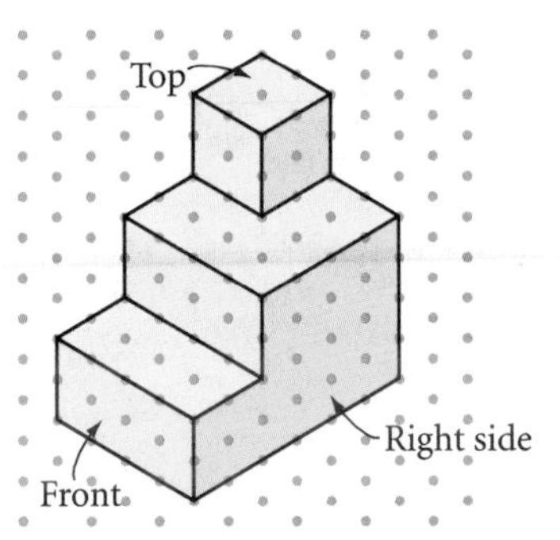

▶ ***Solution***

Visualize how the building would look from the top, the front, and the right side. Draw an edge wherever there is a change of depth. The top and front views must have the same width, and the front and right side views must have the same height.

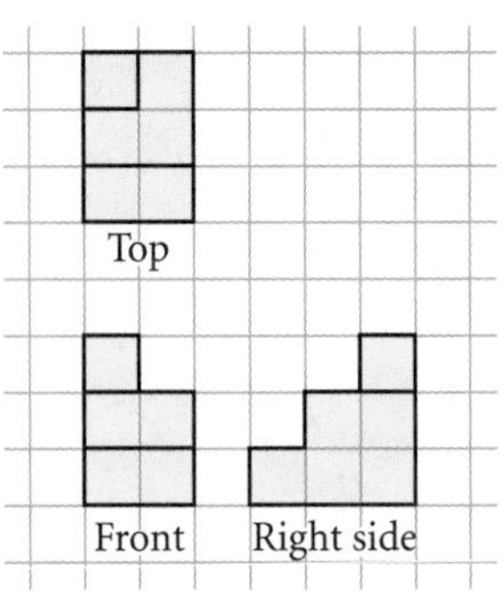

EXAMPLE B

Draw the isometric view of the object shown here as an orthographic drawing. The dashed lines mean that there is an invisible edge.

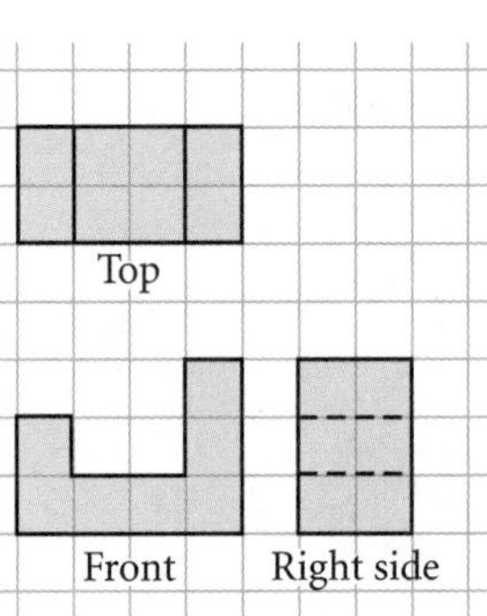

▶ ***Solution***

Find the vertices of the front face and make the shape. Use the width of the side and top views to extend parallel lines. Complete the back edges. You can shade parallel planes to show depth.

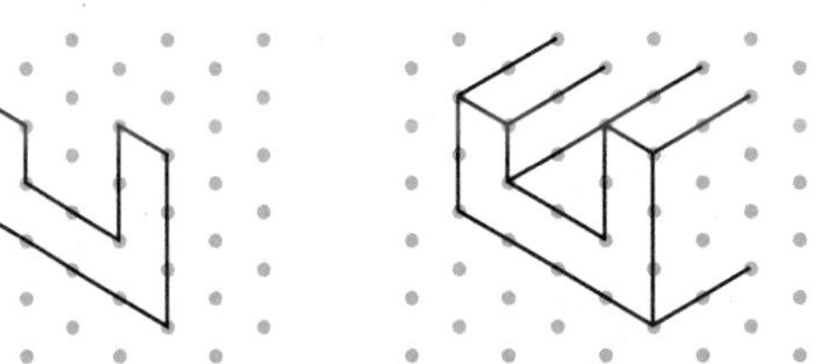

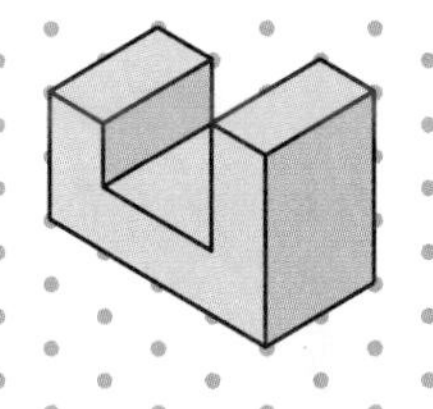

British pop artist David Hockney (b 1937) titled this photographic collage *Sunday Morning Mayflower Hotel, N.Y., Nov-28, 1982*. Each photo shows the view from a different angle, just as an orthographic drawing shows multiple angles at once.

Activity

Isometric and Orthographic Drawings

You will need

- isometric dot paper
- graph paper
- 12 cubes

In this investigation you'll build block models and draw their isometric and orthographic views.

Step 1 Practice drawing a cube on isometric dot paper. What is the shape of each visible face? Are they congruent? What should the orthographic views of a cube look like?

Step 2 Stack three cubes to make a two-step "staircase." Turn the structure so that you look at it the way you would walk up stairs. Call that view the front. Next, identify the top and right sides. How many planes are visible from each view? Make an isometric drawing of the staircase on dot paper and the three orthographic views on graph paper.

Step 3 Build solids A–D from their orthographic views, then draw their isometric views.

A

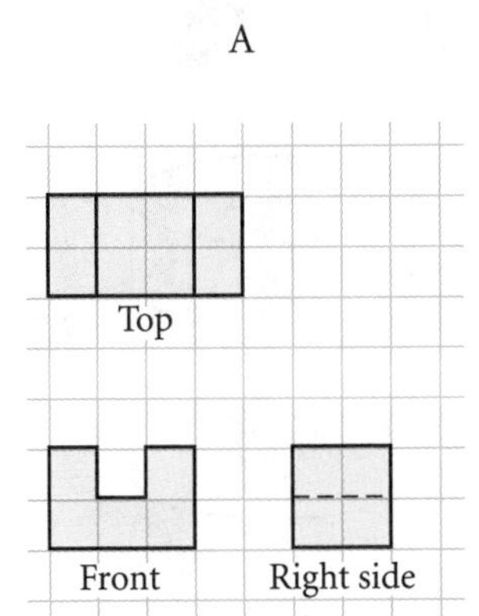

B

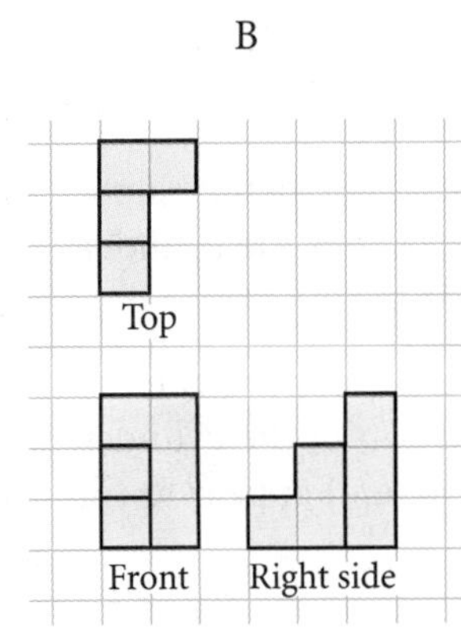

C

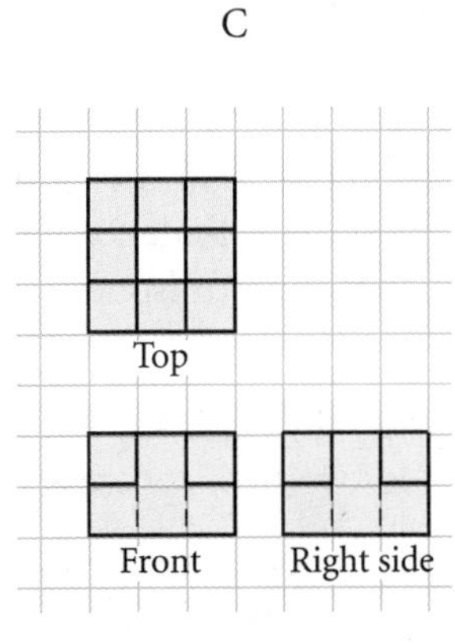

D

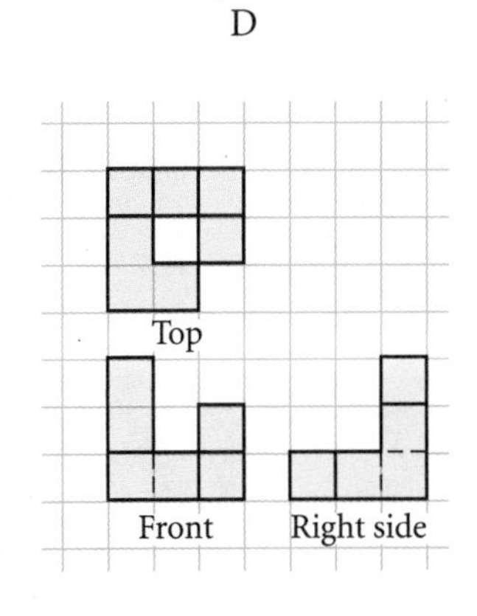

Step 4 Make your own original 8- to 12-cube structure and agree on the orthographic views that represent it. Then, trade places with another group and draw the orthographic views of their structure.

Step 5 Make orthographic views for solids E and F, and sketch the isometric views of solids G and H.

E

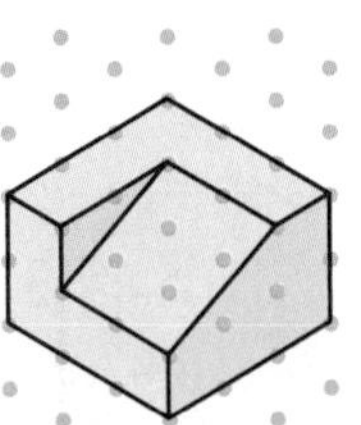

F

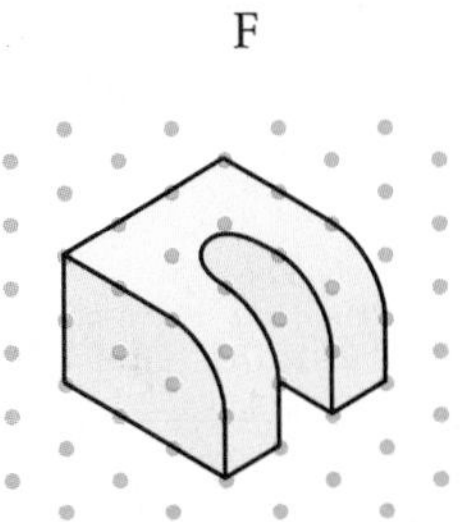

G

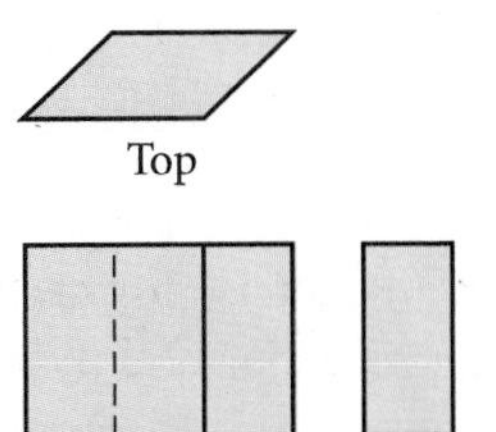

H

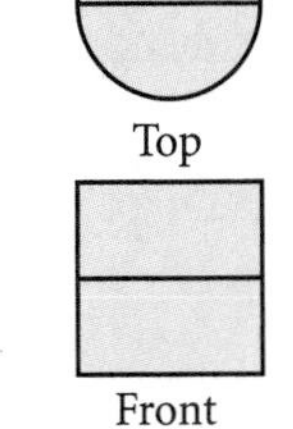
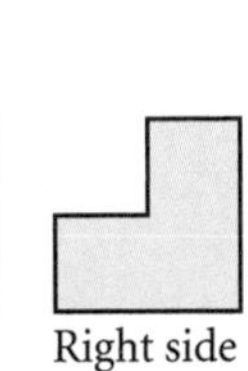

LESSON

10.6 Volume of a Sphere

Satisfaction lies in the effort, not in the attainment. Full effort is full victory.

MOHANDAS K. GANDHI

In this lesson you will develop a formula for the volume of a sphere. In the investigation you'll compare the volume of a right cylinder to the volume of a hemisphere.

Investigation
The Formula for the Volume of a Sphere

You will need

- cylinder and hemisphere with the same radius
- sand, rice, birdseed, or water

This investigation demonstrates the relationship between the volume of a hemisphere with radius r and the volume of a right cylinder with base radius r and height $2r$—that is, the smallest cylinder that encloses a given sphere.

Step 1 Fill the hemisphere.

Step 2 Carefully pour the contents of the hemisphere into the cylinder. What fraction of the cylinder does the hemisphere appear to fill?

Step 3 Fill the hemisphere again and pour the contents into the cylinder. What fraction of the cylinder do two hemispheres (one sphere) appear to fill?

Step 4 If the radius of the cylinder is r and its height is $2r$, then what is the volume of the cylinder in terms of r?

Step 5 The volume of the sphere is the fraction of the cylinder's volume that was filled by two hemispheres. What is the formula for the volume of a sphere? State it as your conjecture.

Sphere Volume Conjecture C-90

The volume of a sphere with radius r is given by the formula __?__.

EXAMPLE A As an exercise for her art class, Mona has cast a plaster cube, 12 cm on each side. Her assignment is to carve the largest possible sphere from the cube. What percentage of the plaster will be carved away?

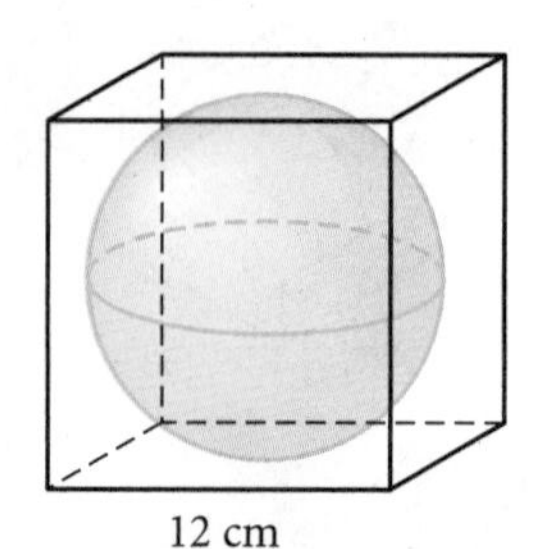

▸ ***Solution***

The largest possible sphere will have a diameter of 12 cm, so its radius is 6 cm. Applying the formula for volume of a sphere, you get $V = \left(\frac{4}{3}\right)\pi r^3 = \left(\frac{4}{3}\right)\pi \cdot 6^3 = \left(\frac{4}{3}\right)\pi \cdot 216 = 288\pi$ or about 905 cm^3. The volume of the plaster cube is 12^3 or 1728 cm^3. You subtract the volume of the sphere from the volume of the cube to get the amount carved away, which is about 823 cm^3. Therefore the percentage carved away is $\frac{823}{1728} \approx 48\%$.

EXAMPLE B

Find the volume of plastic (to the nearest cubic inch) needed for this hollow toy component. The outer-hemisphere diameter is 5.0 in. and the inner-hemisphere diameter is 4.0 in.

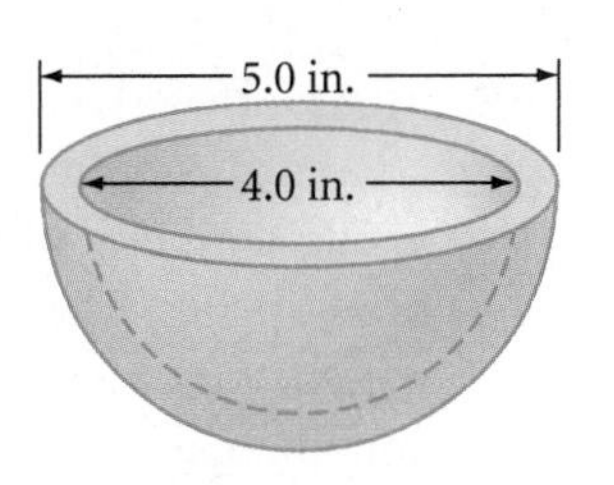

▸ ***Solution***

The formula for volume of a sphere is $V = \left(\frac{4}{3}\right)\pi r^3$, so the volume of a hemisphere is half of that, $V = \left(\frac{2}{3}\right)\pi r^3$. A radius is half a diameter.

Outer Hemisphere	**Inner Hemisphere**
$V_o = \left(\frac{2}{3}\right)\pi r^3$	$V_i = \left(\frac{2}{3}\right)\pi r^3$
$= \left(\frac{2}{3}\right)\pi(2.5)^3$	$= \left(\frac{2}{3}\right)\pi(2)^3$
$= \left(\frac{2}{3}\right)\pi(15.625)$	$= \left(\frac{2}{3}\right)\pi(8)$
≈ 33	≈ 17

Subtracting the volume of the inner hemisphere from the volume of the outer one, 16 in.^{3} of plastic are needed.

EXERCISES

You will need

Construction tools for Exercise **21**

In Exercises 1–6, find the volume of each solid. All measurements are in centimeters.

1.

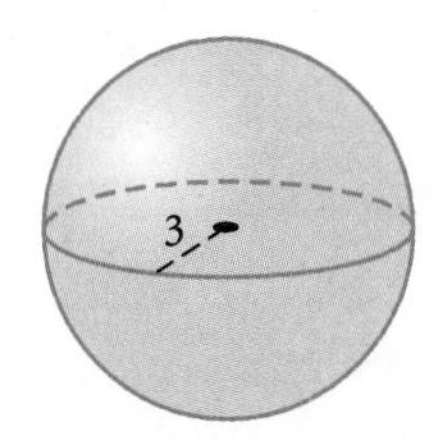

2.

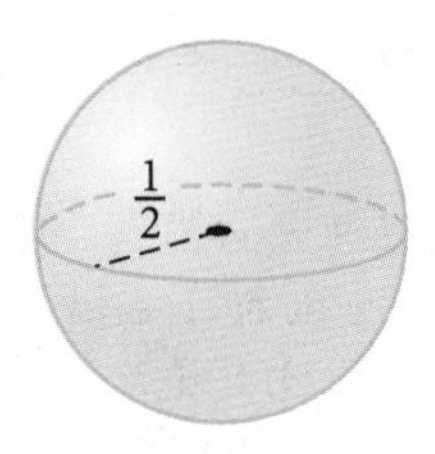

3.

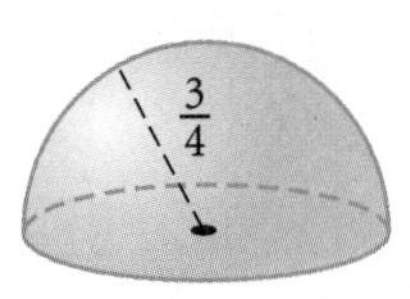

4. ⓗ

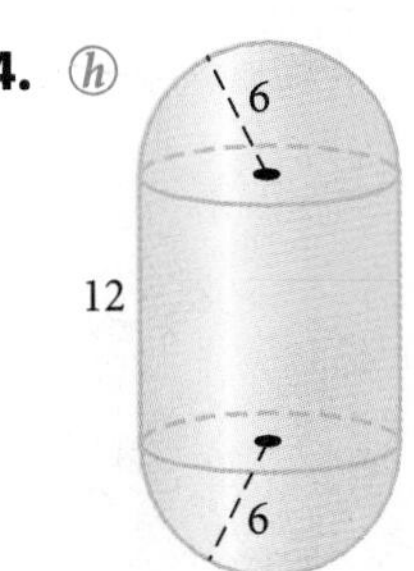

5.

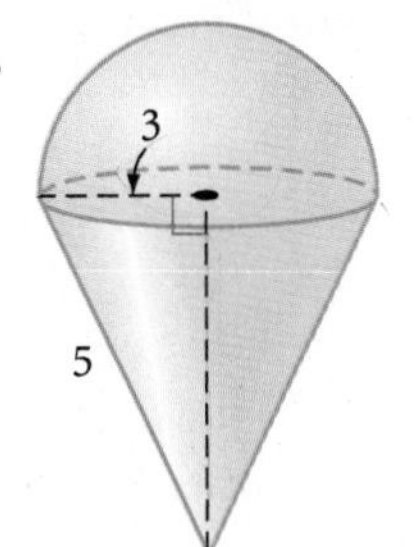

6. ⓗ

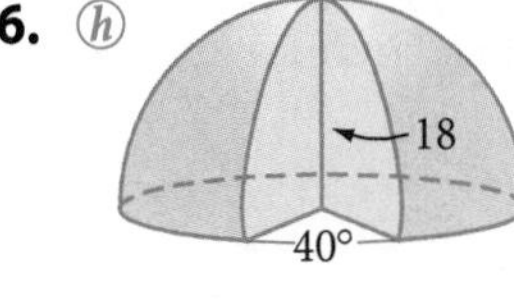

7. What is the volume of the largest hemisphere that you could carve out of a wooden block whose edges measure 3 m by 7 m by 7 m?

8. A sphere of ice cream is placed onto your ice cream cone. Both have a diameter of 8 cm. The height of your cone is 12 cm. If you push the ice cream into the cone, will all of it fit?

9. **APPLICATION** Lickety Split ice cream comes in a cylindrical container with an inside diameter of 6 inches and a height of 10 inches. The company claims to give the customer 25 scoops of ice cream per container, each scoop being a sphere with a 3-inch diameter. How many scoops will each container really hold?

10. Find the volume of a spherical shell with an outer diameter of 8 meters and an inner diameter of 6 meters.

11. Which is greater, the volume of a hemisphere with radius 2 cm or the total volume of two cones with radius 2 cm and height 2 cm?

12. A sphere has a volume of 972π in.^{3}. Find its radius. ⓗ

13. A hemisphere has a volume of 18π cm^3. Find its radius.

14. The base of a hemisphere has an area of 256π cm^2. Find its volume.

15. If the diameter of a student's brain is about 6 inches, and you assume its shape is approximately a hemisphere, then what is the volume of the student's brain?

THE FAR SIDE® By GARY LARSON

"Mr. Osborne, may I be excused? My brain is full."

16. A cylindrical glass 10 cm tall and 8 cm in diameter is filled to 1 cm from the top with water. If a golf ball 4 cm in diameter is placed into the glass, will the water overflow?

17. **APPLICATION** This underground gasoline storage tank is a right cylinder with a hemisphere at each end. How many gallons of gasoline will the tank hold? (1 gallon = 0.13368 cubic foot) If the service station fills twenty 15-gallon tanks from the storage tank per day, how many days will it take to empty the storage tank?

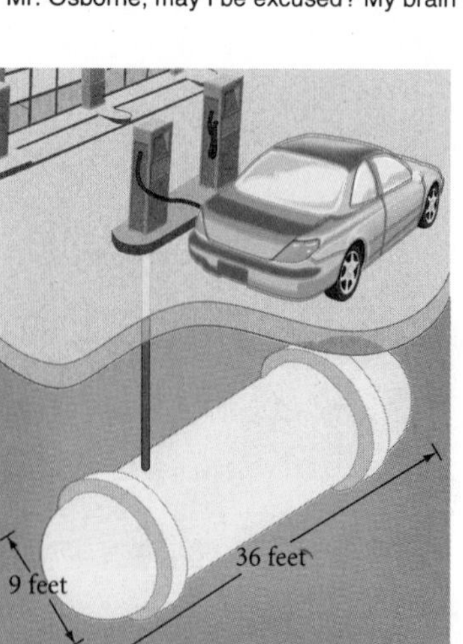

Review

18. Inspector Lestrade has sent a small piece of metal to the crime lab. The lab technician finds that its mass is 54.3 g. It appears to be lithium, sodium, or potassium, all highly reactive with water. Then the technician places the metal into a graduated glass cylinder of radius 4 cm that contains a nonreactive liquid. The metal causes the level of the liquid to rise 2.0 cm. Which metal is it? (Refer to the table on page 535.)

19. City law requires that any one-story commercial building supply a parking area equal in size to the floor area of the building. A-Round Architects has designed a cylindrical building with a 150-foot diameter. They plan to ring the building with parking. How far from the building should the parking lot extend? Round your answer to the nearest foot.

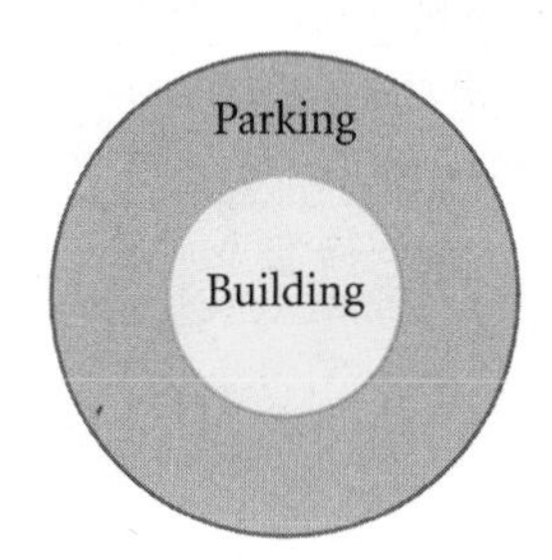

20. Plot A, B, C, and D onto graph paper.

A is $(3, -5)$.

C is the reflection of A over the x-axis.

B is the rotation of C 180° around the origin.

D is a transformation of A by the rule $(x, y) \rightarrow (x + 6, y + 10)$.

What kind of quadrilateral is $ABCD$? Give reasons for your answer.

21. *Construction* Use your geometry tools to construct an inscribed and circumscribed circle for an equilateral triangle.

22. Find w, x, and y.

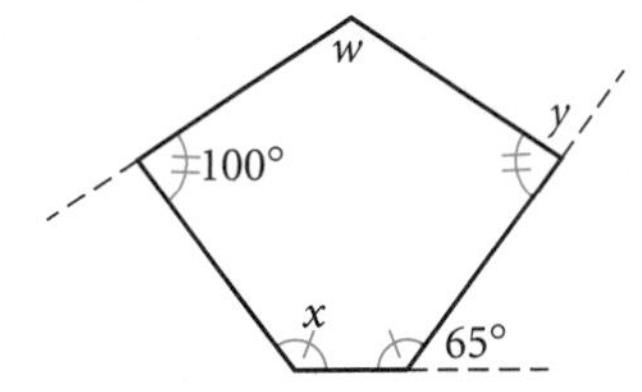

IMPROVING YOUR VISUAL THINKING SKILLS

Patchwork Cubes

The large cube at right is built from 13 double cubes like the one shown plus one single cube. What color must the single cube be, and where must it be positioned?

LESSON

10.7

Surface Area of a Sphere

Sometimes it's better to talk about difficult subjects lying down; the change in posture sort of tilts the world so you can get a different angle on things.

MARY WILLIS WALKER

Earth is so large that it is reasonable to use area formulas for plane figures—rectangles, triangles, and circles—to find the areas of most small land regions. But, to find Earth's entire surface area, you need a formula for the surface area of a sphere. Now that you know how to find the volume of a sphere, you can use that knowledge to arrive at the formula for the surface area of a sphere.

Earth rises over the Moon's horizon. From there, the Moon's surface seems flat.

Investigation
The Formula for the Surface Area of a Sphere

In this investigation you'll visualize a sphere's surface covered by tiny shapes that are nearly flat. So the surface area, S, of the sphere is the sum of the areas of all the "nearly polygons." If you imagine radii connecting each of the vertices of the "nearly polygons" to the center of the sphere, you are mentally dividing the volume of the sphere into many "nearly pyramids." Each of the "nearly polygons" is a base for a pyramid, and the radius, r, of the sphere is the height of the pyramid. So the volume, V, of the sphere is the sum of the volumes of all the pyramids. Now get ready for some algebra.

A horsefly's eyes resemble spheres covered by "nearly polygons."

Step 1 Divide the surface of the sphere into 1000 "nearly polygons" with areas $B_1, B_2, B_3, \ldots, B_{1000}$. Then you can write the surface area, S, of the sphere as the sum of the 1000 B's:

$$S = B_1 + B_2 + B_3 + \ldots + B_{1000}$$

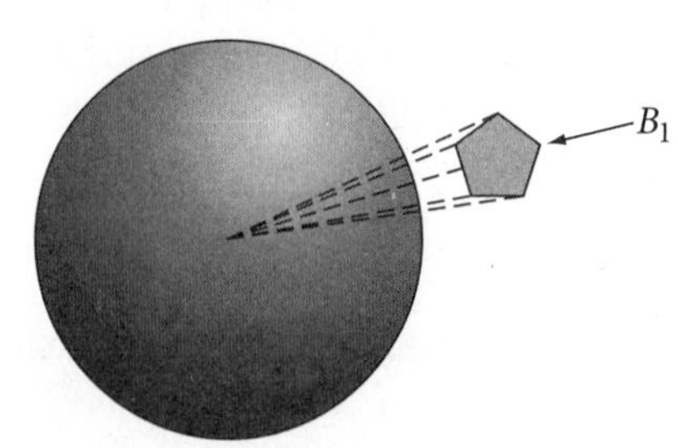

Step 2 The volume of the pyramid with base B_1 is $\frac{1}{3}(B_1)(r)$, so the total volume of the sphere, V, is the sum of the volumes of the 1000 pyramids:

$$V = \frac{1}{3}(B_1)(r) + \frac{1}{3}(B_2)(r) + \ldots + \frac{1}{3}(B_{1000})(r)$$

What common expression can you factor from each of the terms on the right side? Rewrite the last equation showing the results of your factoring.

Step 3 But the volume of the sphere is $V = \frac{4}{3}\pi r^3$. Rewrite your equation from Step 2 by substituting $\frac{4}{3}\pi r^3$ for V and substituting for S the sum of the areas of all the "nearly polygons."

Step 4 Solve the equation from Step 3 for the surface area, S. You now have a formula for finding the surface area of a sphere in terms of its radius. State this as your next conjecture and add it to your conjecture list.

Sphere Surface Area Conjecture C-91

The surface area, S, of a sphere with radius r is given by the formula $\underline{?}$.

EXAMPLE Find the surface area of a sphere whose volume is $12{,}348\pi$ m³.

▶ ***Solution*** First, use the volume formula for a sphere to find its radius. Then, use the radius to find the surface area.

Radius Calculation

$$V = \left(\frac{4}{3}\right)\pi r^3$$
$$12{,}348\pi = \left(\frac{4}{3}\right)\pi r^3$$
$$\left(\frac{3}{4}\right)12{,}348 = r^3$$
$$9261 = r^3$$
$$r = 21$$

Surface Area Calculation

$$S = 4\pi r^2$$
$$= 4\pi(21)^2$$
$$= 4\pi(441)$$
$$S = 1764\pi \approx 5541.8$$

The radius is 21 m, and the surface area is 1764π m², or about 5541.8 m².

EXERCISES

You will need

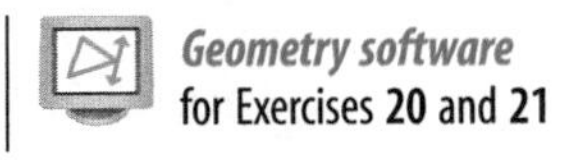

For Exercises 1–3, find the volume and total surface area of each solid. All measurements are in centimeters.

1. (h)

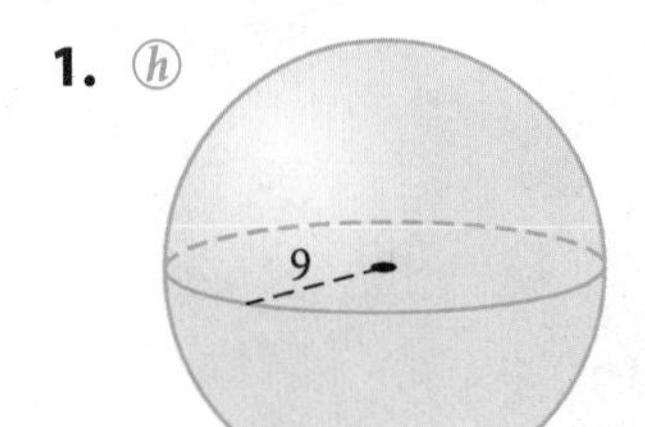

2.

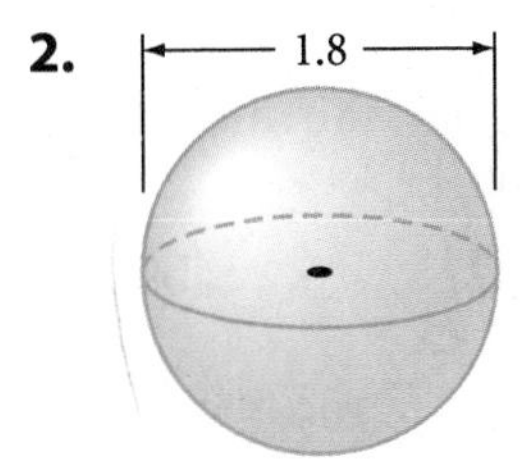

3. (h)

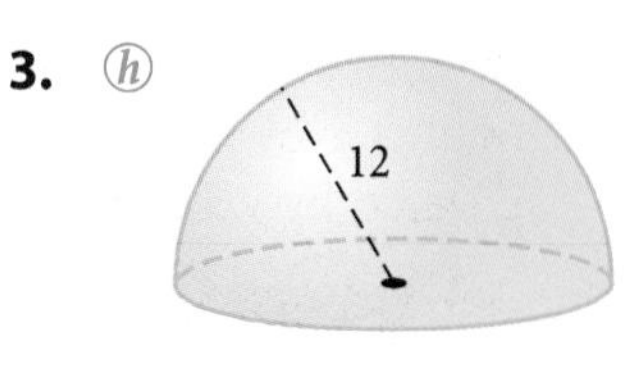

4. The shaded circle at right has area 40π cm^2. Find the surface area of the sphere. ⓗ

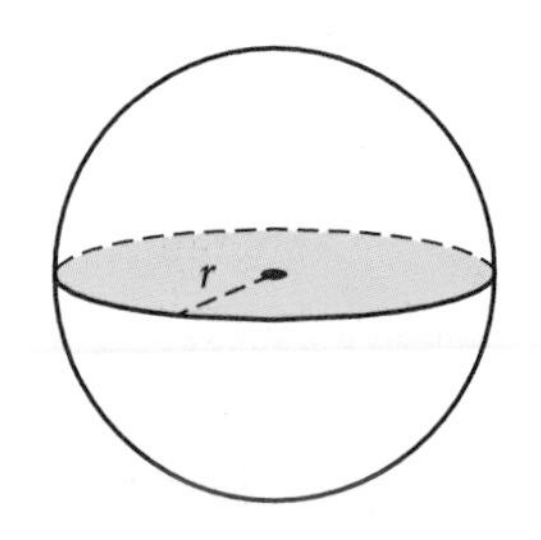

5. Find the volume of a sphere whose surface area is 64π cm^2.

6. Find the surface area of a sphere whose volume is 288π cm^3.

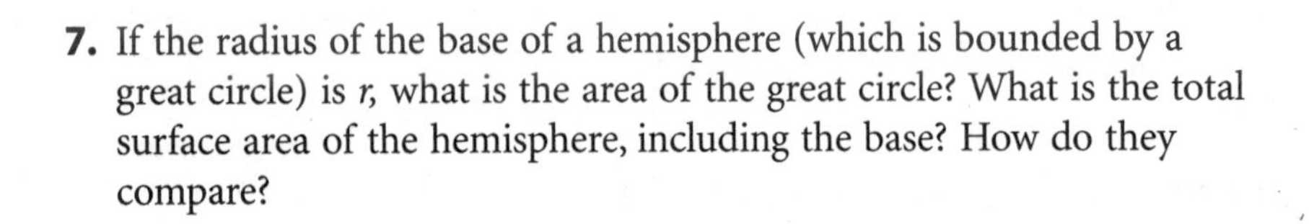

7. If the radius of the base of a hemisphere (which is bounded by a great circle) is r, what is the area of the great circle? What is the total surface area of the hemisphere, including the base? How do they compare?

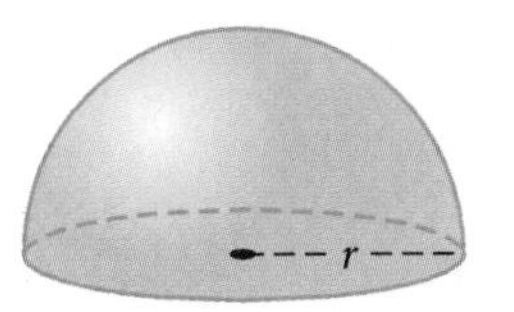

8. If Jose used 4 gallons of wood sealant to cover the hemispherical ceiling of his vacation home, how many gallons of wood sealant are needed to cover the floor?

9. Assume a Kickapoo wigwam is a semicylinder with a half-hemisphere on each end. The diameter of the semicylinder and each of the half-hemispheres is 3.6 meters. The total length is 7.6 meters. What is the volume of the wigwam and the surface area of its roof?

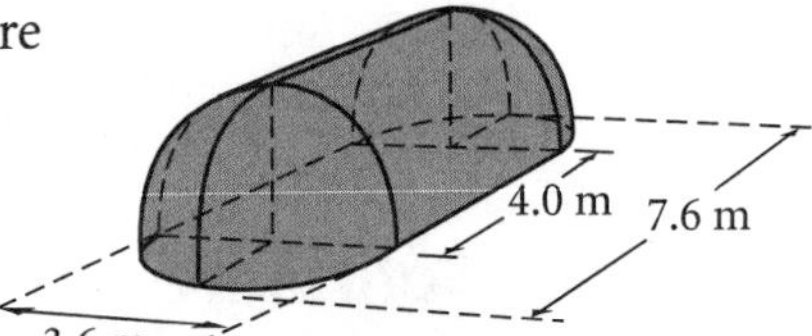

Cultural CONNECTION

A wigwam was a domed structure that Native American woodland tribes, such as the Kickapoo, Iroquois, and Cherokee, used for shelter and warmth in the winter. They designed each wigwam with an oval floor pattern, set tree saplings vertically into the ground around the oval, bent the tips of the saplings into an arch, and tied all the pieces together to support the framework. They then wove more branches horizontally around the building and added mats over the entire dwelling, except for the doorway and smoke hole.

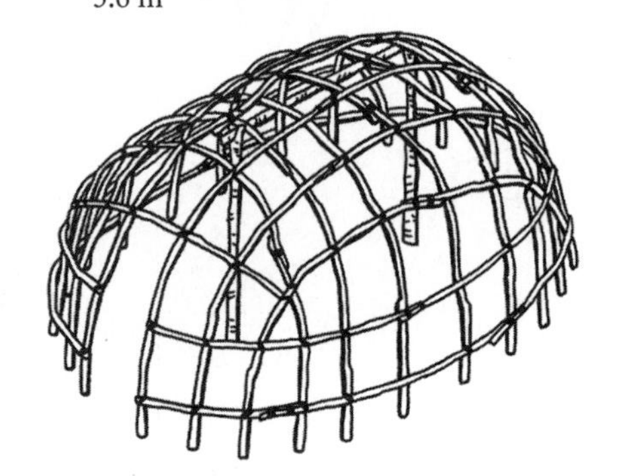

10. **APPLICATION** A farmer must periodically resurface the interior (wall, floor, and ceiling) of his silo to protect it from the acid created by the silage. The height of the silo to the top of the hemispherical dome is 50 ft, and the diameter is 18 ft.

    a. What is the approximate surface area that needs to be treated?

    b. If 1 gallon of resurfacing compound covers about 250 ft^2, how many gallons are needed?

    c. There is 0.8 bushel per ft^3. Calculate the number of bushels of grain this silo will hold.

11. About 70% of Earth's surface is covered by water. If the diameter of Earth is about 12,750 km, find the area not covered by water to the nearest 100,000 km^2.

History
CONNECTION

From the early 13th century to the late 17th century, the Medici family of Florence, Italy, were successful merchants and generous patrons of the arts. The Medici family crest, shown here, features six spheres—five red spheres and one that resembles the earth. The use of three gold spheres to advertise a pawnshop could have been inspired by the Medici crest.

12. A sculptor has designed a statue that features six hemispheres (inspired by the Medici crest) and three spheres (inspired by the pawnshop logo). He wants to use gold electroplating on the six hemispheres (diameter 6 cm) and the three spheres (diameter 8 cm), which will cost about 14¢/cm^2. (The bases of the hemispheres will *not* be electroplated.) Will he be able to stay under his $150 budget? If not, what diameter spheres should he make to stay under budget?

13. Earth has a thin outer layer called the *crust,* which averages about 24 km thick. Earth's diameter is about 12,750 km. What percentage of the volume of Earth is the crust?

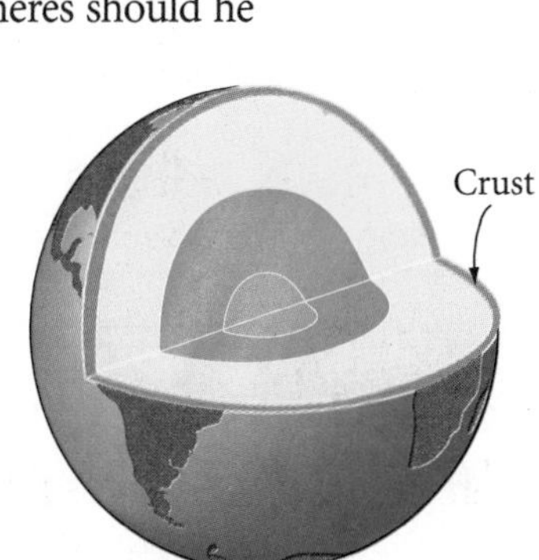

Review

14. A piece of wood placed in a cylindrical container causes the container's water level to rise 3 cm. This type of wood floats half out of the water, and the radius of the container is 5 cm. What is the volume of the piece of wood?

15. Find the ratio of the area of the circle inscribed in an equilateral triangle to the area of the circumscribed circle.

16. Find the ratio of the area of the circle inscribed in a square to the area of the circumscribed circle.

17. Find the ratio of the area of the circle inscribed in a regular hexagon to the area of the circumscribed circle.

18. Make a conjecture as to what happens to the ratio in Exercises 15–17 as the number of sides of the regular polygon increases. Make sketches to support your conjecture.

19. Use inductive reasoning to complete each table.

a.

n	1	2	3	4	5	6	...	n	...	200
$f(n)$	−2	1	4	7			...		...	

b.

n	1	2	3	4	5	6	...	n	...	200
$f(n)$	0	$\frac{1}{3}$	$\frac{1}{2}$	$\frac{3}{5}$			...		...	

20. ***Technology*** Use geometry software to construct a segment $\overline{AB}$, and its midpoint C. Trace C and B, and drag B around to sketch a shape. Compare the shapes they trace.

21. ***Technology*** Use geometry software to construct a circle. Choose a point A on the circle and a point B not on the circle, and construct the perpendicular bisector of $\overline{AB}$. Trace the perpendicular bisector as you animate A around the circle. Describe the locus of points traced.

IMPROVING YOUR REASONING SKILLS

PUZZLE

Reasonable 'rithmetic II

Each letter in these problems represents a different digit.

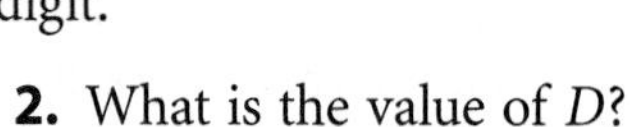

1. What is the value of C?

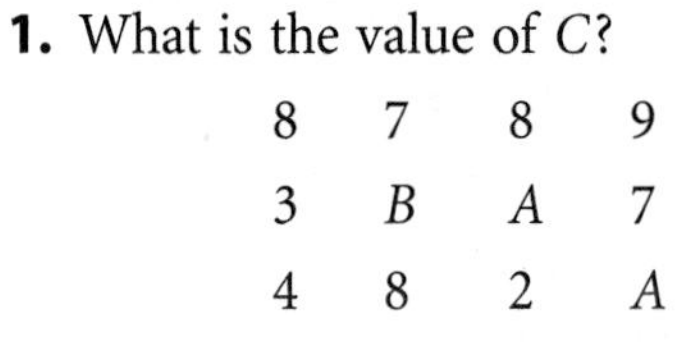

$$\begin{array}{rrrrr} & 8 & 7 & 8 & 9 \\ & 3 & B & A & 7 \\ & 4 & 8 & 2 & A \\ + & 7 & A & B & 5 \\ \hline 2 & C & 2 & 8 & 7 \end{array}$$

2. What is the value of D?

$$\begin{array}{rrrrr} & D & E & F & F \\ - & E & 2 & F & 6 \\ \hline & 1 & 9 & 9 & 7 \end{array}$$

3. What is the value of K?

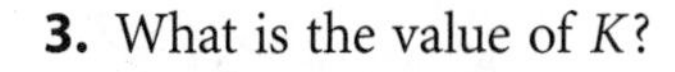

$$\begin{array}{r} GJ \\ 7\overline{)HGK} \\ \underline{21} \\ HK \\ \underline{HK} \end{array}$$

4. What is the value of N?

$$\begin{array}{r} 52 \\ LQ\overline{)NM2} \\ \underline{NP} \\ M2 \\ \underline{M2} \end{array}$$

Exploration

Sherlock Holmes and Forms of Valid Reasoning

"That's logical!" You've probably heard that expression many times. What do we mean when we say someone is thinking logically? One dictionary defines *logical* as "capable of reasoning or using reason in an orderly fashion that brings out fundamental points."

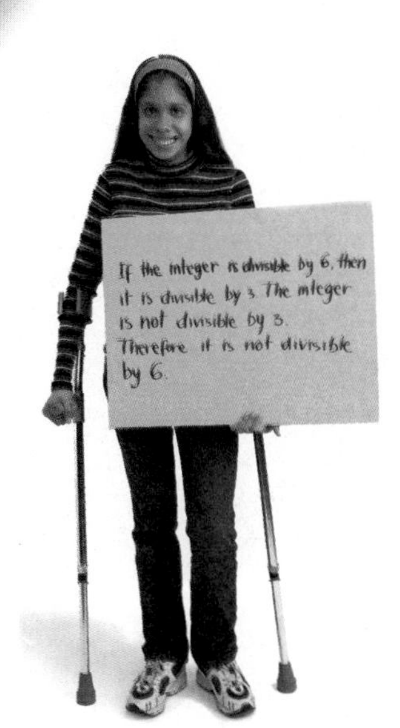

"Prove it!" That's another expression you've probably heard many times. It is an expression that is used by someone concerned with logical thinking. In daily life, proving something often means you can present some facts to support a point.

In Chapter 2 you learned that in geometry—as in daily life—a conclusion is valid when you present rules and facts to support it. You have often used given information and previously proven conjectures to prove new conjectures in paragraph proofs and flowchart proofs.

When you apply deductive reasoning, you are "being logical" like detective Sherlock Holmes. The statements you take as true are called premises, and the statements that follow from them are conclusions.

When you translate a deductive argument into symbolic form, you use capital letters to stand for simple statements. When you write "If *P* then *Q*," you are writing a **conditional statement.** Here are two examples.

English argument	Symbolic translation
If Watson has chalk between his fingers, then he has been playing billiards. Watson has chalk between his fingers. Therefore, Watson has been playing billiards.	*P*: Watson has chalk between his fingers. *Q*: Watson has been playing billiards. If *P* then *Q*. *P* ∴ Q
If triangle *ABC* is isosceles, then the base angles are congruent. Triangle *ABC* is isosceles. Therefore, its base angles are congruent.	*P*: Triangle *ABC* is isosceles. *Q*: Triangle *ABC*'s base angles are congruent. If *P* then *Q*. *P* ∴ Q

The symbol $\therefore$ means "therefore." So you can read the last two lines "P, $\therefore$ Q" as "P, therefore Q" or "P is true, so Q is true."

Both of these examples illustrate one of the well-accepted forms of valid reasoning. According to ***Modus Ponens*** (MP), if you accept "If P then Q" as true and you accept P as true, then you must logically accept Q as true.

In geometry—as in daily life—we often encounter "not" in a statement. "Not P" is the **negation** of statement P. If P is the statement "It is raining," then "not P," symbolized $\sim P$, is the statement "It is not raining" or "It is not the case that it is raining." To remove negation from a statement, you remove the not. The negation of the statement "It is not raining" is "It is raining." You can also negate a "not" by adding yet another "not." So you can also negate the statement "It is not raining" by saying "It is not the case that it is not raining." This property is called **double negation.**

According to ***Modus Tollens*** (MT), if you accept "If P then Q" as true and you accept $\sim Q$ as true, then you must logically accept $\sim P$ as true. Here are two examples.

English argument	Symbolic translation
If Watson wished to invest money with Thurston, then he would have had his checkbook with him. Watson did not have his checkbook with him. Therefore Watson did not wish to invest money with Thurston.	P: Watson wished to invest money with Thurston. Q: Watson had his checkbook with him. If P then Q. $\sim Q$ $\therefore \sim P$
If $\overline{AC}$ is the longest side in $\triangle ABC$, then $\angle B$ is the largest angle in $\triangle ABC$. $\angle B$ is not the largest angle in $\triangle ABC$. Therefore $\overline{AC}$ is not the longest side in $\triangle ABC$.	P: $\overline{AC}$ is the longest side in $\triangle ABC$. Q: $\angle B$ is the largest angle in $\triangle ABC$. If P then Q. $\sim Q$ $\therefore \sim P$

Activity

It's Elementary!

In this activity you'll apply what you have learned about *Modus Ponens* (MP) and *Modus Tollens* (MT). You'll also get practice using the symbols of logic such as P and $\sim P$ as statements and $\therefore$ for "so" or "therefore." To shorten your work even further you can symbolize the conditional "If P then Q" as $P \rightarrow Q$. Then *Modus Ponens* and *Modus Tollens* written symbolically look like this:

Modus Ponens	*Modus Tollens*
$P \rightarrow Q$	$R \rightarrow S$
P	$\sim S$
$\therefore Q$	$\therefore \sim R$

Step 1 Use logic symbols to translate parts a–e. Tell whether *Modus Ponens* or *Modus Tollens* is used to make the reasoning valid.

a. If Watson was playing billiards, then he was playing with Thurston. Watson was playing billiards. Therefore Watson was playing with Thurston.

b. Every cheerleader at Washington High School is in the 11th grade. Mark is a cheerleader at Washington High School. Therefore, Mark is in the 11th grade.

c. If Carolyn studies, then she does well on tests. Carolyn did not do well on her tests, so she must not have studied.

d. If $\overline{ED}$ is a midsegment in $\triangle ABC$, then $\overline{ED}$ is parallel to a side of $\triangle ABC$. $\overline{ED}$ is a midsegment in $\triangle ABC$. Therefore $\overline{ED}$ is parallel to a side of $\triangle ABC$.

e. If $\overline{ED}$ is a midsegment in $\triangle ABC$, then $\overline{ED}$ is parallel to a side of $\triangle ABC$. $\overline{ED}$ is not parallel to a side of $\triangle ABC$. Therefore $\overline{ED}$ is a not a midsegment in $\triangle ABC$.

Step 2 Use logic symbols to translate parts a–d. If the two premises fit the valid reasoning pattern of *Modus Ponens* or *Modus Tollens*, state the conclusion symbolically and translate it into English. Tell whether *Modus Ponens* or *Modus Tollens* is used to make the reasoning valid. Otherwise write “no valid conclusion.”

a. If Aurora passes her Spanish test, then she will graduate. Aurora passes the test.

b. The diagonals of *ABCD* are not congruent. If *ABCD* is a rectangle, then its diagonals are congruent.

c. If yesterday was Thursday, then there is no school tomorrow. There is no school tomorrow.

d. If you don’t use Shining Smile toothpaste, then you won’t be successful. You do not use Shining Smile toothpaste.

e. If squiggles are flitz, then ruggles are bodrum. Ruggles are not bodrum.

Step 3 Identify each symbolic argument as *Modus Ponens* or *Modus Tollens*. If the argument is not valid, write “no valid conclusion.”

a. $P \rightarrow S$
P
$\therefore S$

b. $\sim T \rightarrow P$
$\sim T$
$\therefore P$

c. $R \rightarrow \sim Q$
Q
$\therefore \sim R$

d. $Q \rightarrow S$
S
$\therefore \sim Q$

e. $Q \rightarrow P$
$\sim Q$
$\therefore \sim P$

f. $\sim R \rightarrow S$
$\sim S$
$\therefore R$

g. $\sim P \rightarrow (R \rightarrow Q)$
$\sim P$
$\therefore (R \rightarrow Q)$

h. $(T \rightarrow \sim P) \rightarrow Q$
$\sim Q$
$\therefore \sim(T \rightarrow \sim P)$

i. $P \rightarrow (\sim R \rightarrow P)$
$(\sim R \rightarrow P)$
$\therefore P$

CHAPTER 10 REVIEW

In this chapter you discovered a number of formulas for finding volumes. It's as important to remember how you discovered these formulas as it is to remember the formulas themselves. For example, if you recall pouring the contents of a cone into a cylinder with the same base and height, you may recall that the volume of the cone is one-third the volume of the cylinder. Making connections will help, too. Recall that prisms and cylinders share the same volume formula because their shapes—two congruent bases connected by lateral faces—are alike.

You should also be able to find the surface area of a sphere. The formula for the surface area of a sphere was intentionally not included in Chapter 8, where you first learned about surface area. Look back at the investigations in Lesson 8.7 and explain why the surface area formula requires that you know volume.

As you have seen, volume formulas can be applied to many practical problems. Volume also has many extensions such as calculating displacement and density.

EXERCISES

1. How are a prism and a cylinder alike?
2. What does a cone have in common with a pyramid?

For Exercises 3–8, find the volume of each solid. Each quadrilateral is a rectangle. All solids are right (not oblique). All measurements are in centimeters.

3.

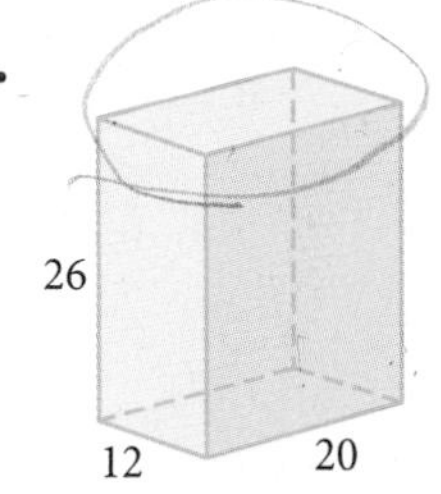

4.

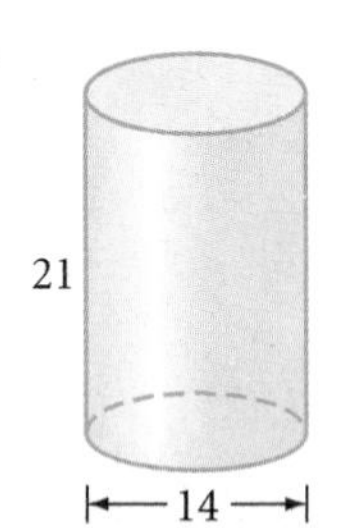

5. ⓗ

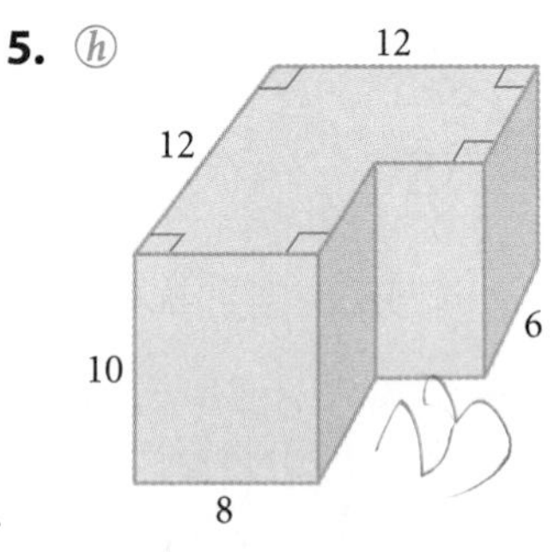

6.

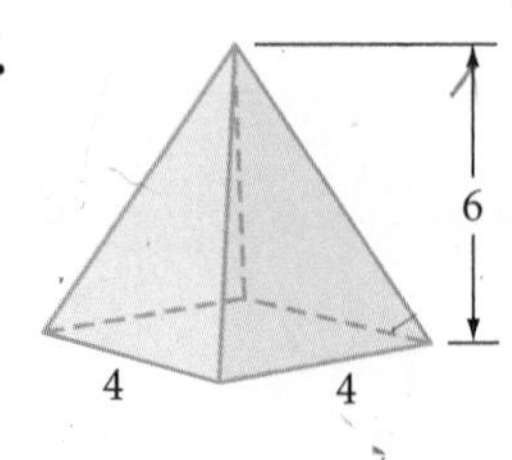

7.

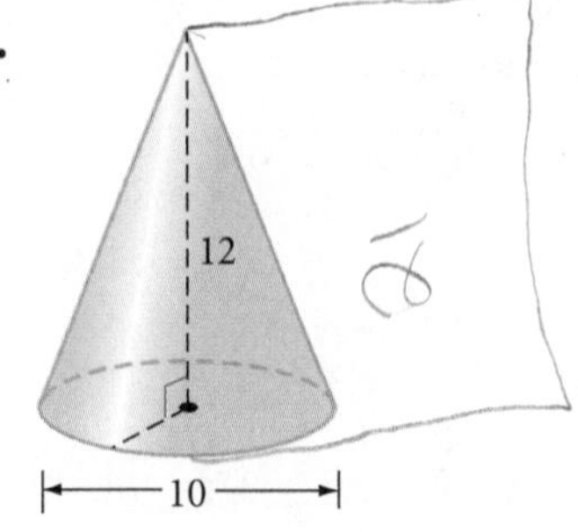

8.

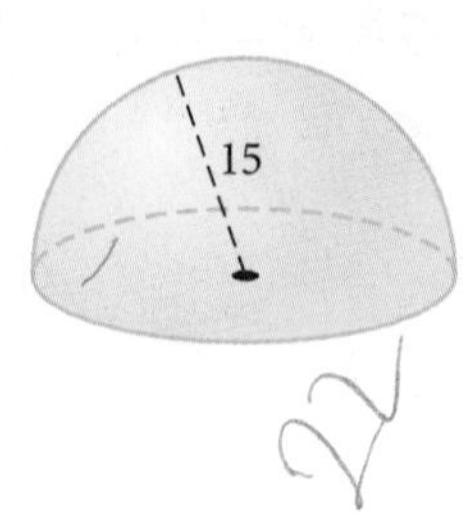

For Exercises 9–12, calculate each unknown length given the volume of the solid. All measurements are in centimeters.

9. Find H. $V = 768\ \text{cm}^3$

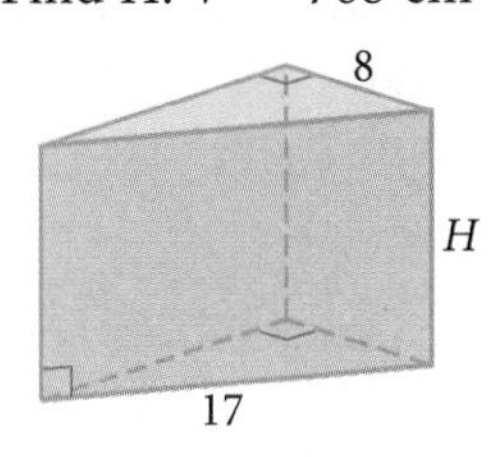

10. Find h. $V = 896\ \text{cm}^3$

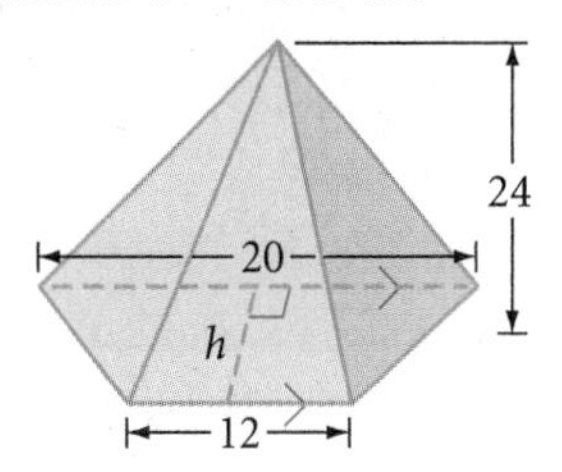

11. Find r. $V = 1728\pi\ \text{cm}^3$

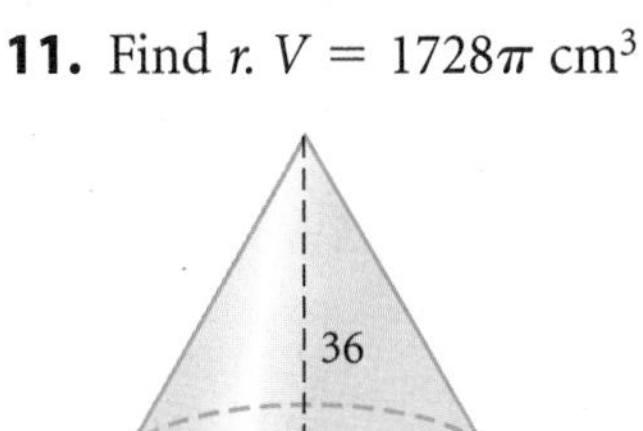

12. Find r. $V = 256\pi\ \text{cm}^3$ ⓗ

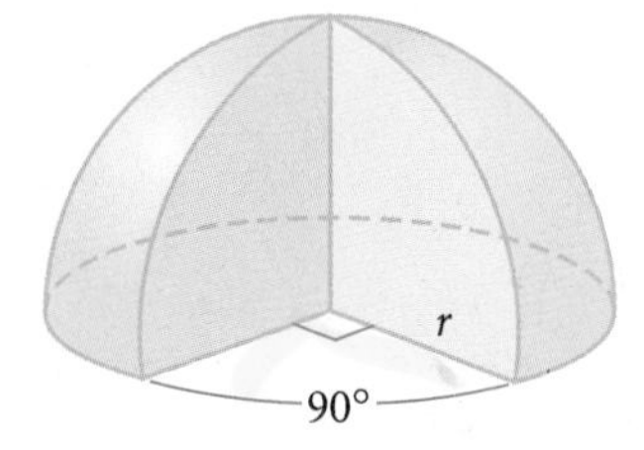

13. Find the volume of a rectangular prism whose dimensions are twice those of another rectangular prism that has a volume of $120\ \text{cm}^3$.

14. Find the height of a cone with a volume of 138π cubic meters and a base area of 46π square meters.

15. Find the volume of a regular hexagonal prism that has a cylinder drilled from its center. Each side of the hexagonal base measures 8 cm. The height of the prism is 16 cm. The cylinder has a radius of 6 cm. Express your answer to the nearest cubic centimeter.

16. Two rectangular prisms have equal heights but unequal bases. Each dimension of the smaller solid's base is half each dimension of the larger solid's base. The volume of the larger solid is how many times as great as the volume of the smaller solid?

17. The "extra large" popcorn container is a right rectangular prism with dimensions 3 in. by 3 in. by 6 in. The "jumbo" is a cone with height 12 in. and diameter 8 in. The "colossal" is a right cylinder with diameter 10 in. and height 10 in.

a. Find the volume of all three containers.

b. Approximately how many times as great is the volume of the "colossal" than the "extra large"?

18. Two solid cylinders are made of the same material. Cylinder A is six times as tall as cylinder B, but the diameter of cylinder B is four times the diameter of cylinder A. Which cylinder weighs more? How many times as much?

19. **APPLICATION** Rosa Avila is a plumbing contractor. She needs to deliver 200 lengths of steel pipe to a construction site. Each cylindrical steel pipe is 160 cm long, has an outer diameter of 6 cm, and has an inner diameter of 5 cm. Rosa needs to know if her quarter-tonne truck can handle the weight of the pipes. To the nearest kilogram, what is the weight of these 200 pipes? How many loads will Rosa have to transport to deliver the 200 lengths of steel pipe? (Steel has a density of about 7.7 g/cm^3. One tonne equals 1000 kg.) ⓗ

20. A ball is placed snugly into the smallest possible box that will completely contain the ball. What percentage of the box is filled by the ball?

21. **APPLICATION** The blueprint for a cement slab floor is shown at right. How many cubic yards of cement are needed for ten identical floors that are each 4 inches thick?

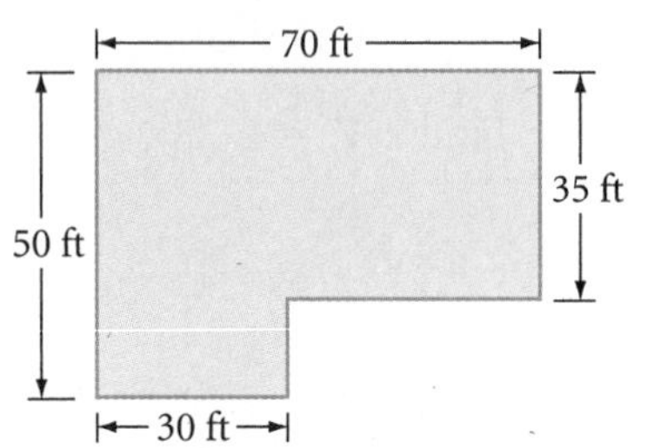

22. A prep chef has just made two dozen meatballs. Each meatball has a 2-inch diameter. Right now, before the meatballs are added, the sauce is 2 inches from the top of the 14-inch-diameter pot. Will the sauce spill over when the chef adds the meatballs to the pot?

23. To solve a crime, Betty Holmes, who claims to be Sherlock's distant cousin, and her friend Professor Hilton Gardens must determine the density of a metal art deco statue that weighs 5560 g. She places it into a graduated glass prism filled with water and finds that the level rises 4 cm. Each edge of the glass prism's regular hexagonal base measures 5 cm. Professor Gardens calculates the statue's volume, then its density. Next, Betty Holmes checks the density table (see page 535) to determine if the statue is platinum. If so, it is the missing piece from her client's collection and Inspector Clouseau is the thief. If not, then the Baron is guilty of fraud. What is the statue made of?

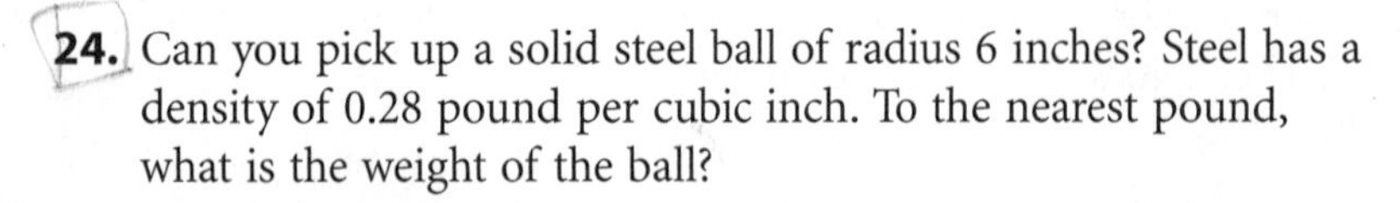

24. Can you pick up a solid steel ball of radius 6 inches? Steel has a density of 0.28 pound per cubic inch. To the nearest pound, what is the weight of the ball?

25. To the nearest pound, what is the weight of a hollow steel ball with an outer diameter of 14 inches and a thickness of 2 inches?

26. A hollow steel ball has a diameter of 14 inches and weighs 327.36 pounds. Find the thickness of the ball. ⓗ

27. A water barrel that is 1 m in diameter and 1.5 m long is partially filled. By tapping on its sides, you estimate that the water is 0.25 m deep at the deepest point. What is the volume of the water in cubic meters? ⓗ

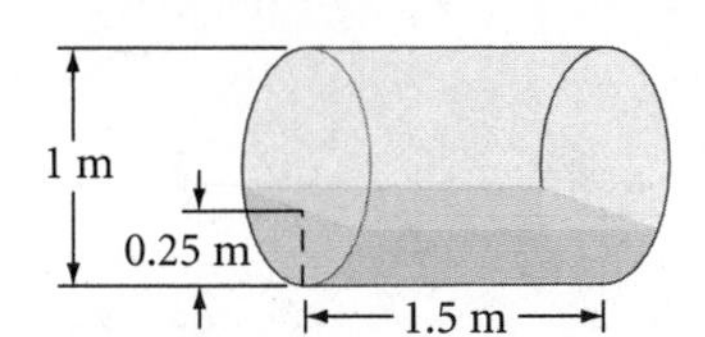

28. Find the volume of the solid formed by rotating the shaded figure about the y-axis.

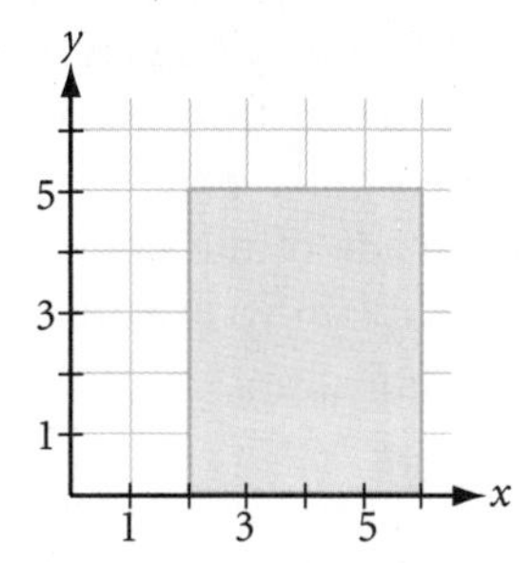

TAKE ANOTHER LOOK

1. You may be familiar with the area model of the expression $(a + b)^2$, shown below. Draw or build a volume model of the expression $(a + b)^3$. How many distinct pieces does your model have? What's the volume of each type of piece? Use your model to write the expression $(a + b)^3$ in expanded form.

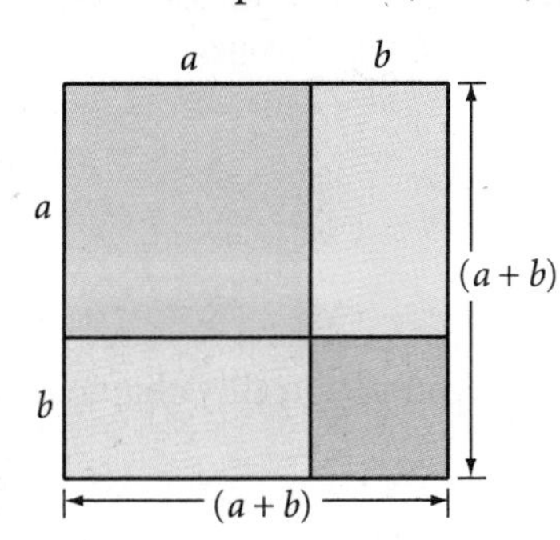

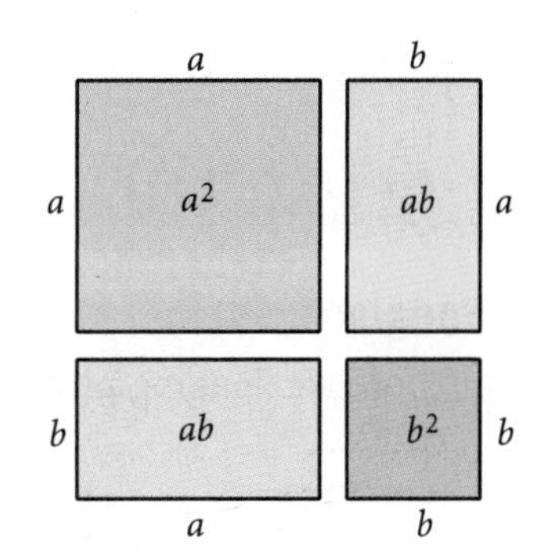

2. Use algebra to show that if you double all three dimensions of a prism, a cylinder, a pyramid, or a cone, the volume is increased eightfold but the surface area is increased only four times.

3. Any sector of a circle can be rolled into a cone. Find a way to calculate the volume of a cone given the radius and central angle of the sector.

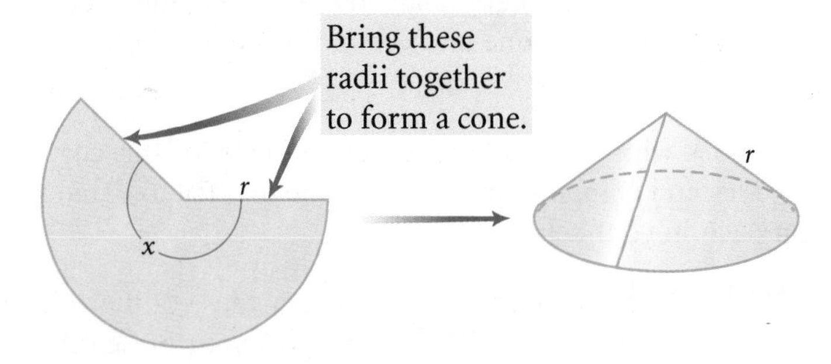

4. Build a model of three pyramids with equal volumes that you can assemble into a prism.

5. Derive the Sphere Volume Conjecture by using a pair of hollow shapes different from those you used in the Investigation The Formula for the Volume of a Sphere. Or use two solids made of the same material and compare weights. Explain what you did and how it demonstrates the conjecture.

6. Spaceship Earth, located at the Epcot center in Orlando, Florida, is made of polygonal regions arranged in little pyramids. The building appears spherical, but the surface is not smooth. If a perfectly smooth sphere had the same volume as Spaceship Earth, would it have the same surface area? If not, which would be greater, the surface area of the smooth sphere or of the bumpy sphere? Explain.

Spaceship Earth, the Epcot center's signature structure, opened in 1982 in Walt Disney World. It features a ride that chronicles the history of communication technology.

Assessing What You've Learned

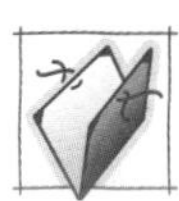

UPDATE YOUR PORTFOLIO Choose a project, a Take Another Look activity, or one of the more challenging problems or puzzles you did in this chapter to add to your portfolio.

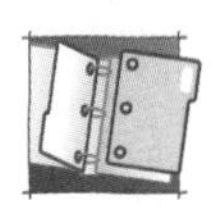

WRITE IN YOUR JOURNAL Describe your own problem-solving approach. Are there certain steps you follow when you solve a challenging problem? What are some of your most successful problem-solving strategies?

ORGANIZE YOUR NOTEBOOK Review your notebook to be sure it's complete and well organized. Be sure you have all the conjectures in your conjecture list. Write a one-page summary of Chapter 10.

PERFORMANCE ASSESSMENT While a classmate, friend, family member, or teacher observes, demonstrate how to derive one or more of the volume formulas. Explain what you're doing at each step.

WRITE TEST ITEMS Work with classmates to write test items for this chapter. Include simple exercises and complex application problems. Try to demonstrate more than one approach in your solutions.

GIVE A PRESENTATION Give a presentation about one or more of the volume conjectures. Use posters, models, or visual aids to support your presentation.

CHAPTER

11 Similarity

Nobody can draw a line that is not a boundary line, every line separates a unity into a multiplicity. In addition, every closed contour no matter what its shape, pure circle or whimsical splash accidental in form, evokes the sensation of "inside" and "outside," followed quickly by the suggestion of "nearby" and "far off," of object and background.

M. C. ESCHER

Path of Life I, M. C. Escher, 1958

OBJECTIVES

In this chapter you will

- review ratio and proportion
- define similar polygons and solids
- discover shortcuts for similar triangles
- learn about area and volume relationships in similar polygons and solids
- use the definition of similarity to solve problems

Proportion and Reasoning

Working with similar geometric figures involves ratios and proportions. You may be a little rusty with these topics, so let's review.

A **ratio** is an expression that compares two quantities by division. You can write the ratio of quantity a to quantity b in these three ways:

$\frac{a}{b}$ a to b $a{:}b$

In this book you will write ratios in fraction form. As with fractions, you can multiply or divide both parts of a ratio by the same number to get an equivalent ratio.

A **proportion** is a statement of equality between two ratios. The equality $\frac{6}{18} = \frac{1}{3}$ is an example of a proportion. Proportions are useful for solving problems involving comparisons.

EXAMPLE A

In a photograph, Dan is 2.5 inches tall and his sister Emma is 1.5 inches tall. Dan's actual height is 70 inches. What is Emma's actual height?

▶ Solution

The ratio of Dan's height to Emma's height is the same in real life as it is in the photo. Let x represent Emma's height and set up a proportion.

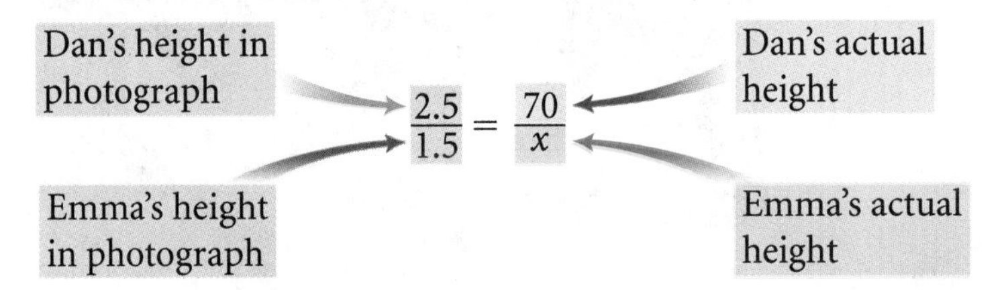

Find Emma's height by solving for x.

$\frac{2.5}{1.5} = \frac{70}{x}$ Original proportion.

$\frac{2.5}{1.5}x = 70$ Multiply both sides by x.

$2.5x = 105$ Multiply both sides by 1.5.

$x = 42$ Divide both sides by 2.5.

Emma is 42 inches tall.

There are other proportions you could have used to solve the problem in Example A. For instance, the ratio of Dan's actual height to his height in the photo is equal to the ratio of Emma's actual height to her height in the photo. So you could have found Emma's height by solving $\frac{70}{2.5} = \frac{x}{1.5}$. What other correct proportion could you use?

Some proportions require more algebra to solve.

EXAMPLE B Solve $\frac{306}{24} = \frac{x + 50}{20}$.

▸ Solution

$\frac{306}{24} = \frac{x + 50}{20}$	Original proportion.
$20 \cdot \frac{306}{24} = x + 50$	Multiply both sides by 20.
$255 = x + 50$	Multiply and divide on the left side.
$205 = x$	Subtract 50 from both sides.

EXERCISES

1. Look at the rectangle at right. Find the ratio of the shaded area to the area of the whole figure. Find the ratio of the shaded area to the unshaded area.

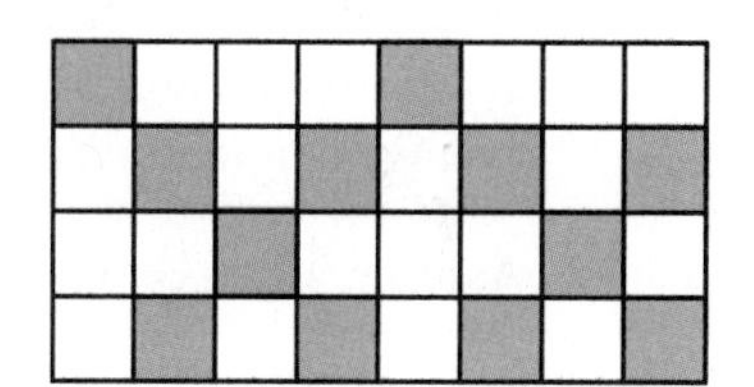

2. Use the figure below to find these ratios: $\frac{AC}{CD}$, $\frac{CD}{BD}$, and $\frac{BD}{BC}$.

3 cm 5 cm 8 cm
A C D B

3. Consider these triangles.

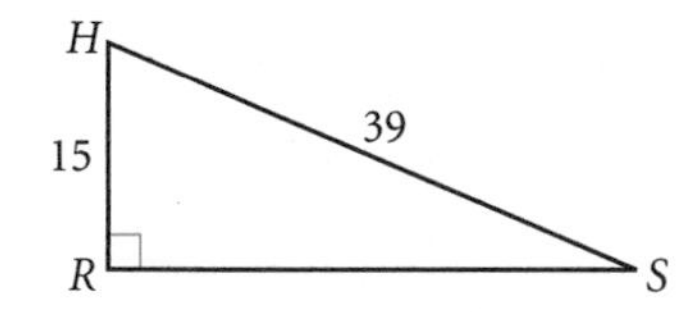

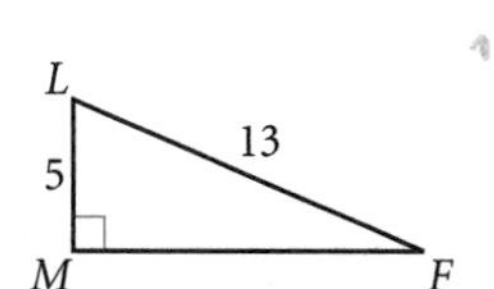

   a. Find the ratio of the perimeter of $\triangle RSH$ to the perimeter of $\triangle MFL$.

   b. Find the ratio of the area of $\triangle RSH$ to the area of $\triangle MFL$.

In Exercises 4–12, solve the proportion.

4. $\frac{7}{21} = \frac{a}{18}$

5. $\frac{10}{b} = \frac{15}{24}$

6. $\frac{20}{13} = \frac{60}{c}$

7. $\frac{4}{5} = \frac{x}{7}$

8. $\frac{2}{y} = \frac{y}{32}$

9. $\frac{14}{10} = \frac{x + 9}{15}$

10. $\frac{10}{10 + z} = \frac{35}{56}$

11. $\frac{d}{5} = \frac{d + 3}{20}$

12. $\frac{y}{y + 2} = \frac{15}{21}$

13. Solve this proportion for x. Assume $c \neq 0$ and $z \neq 0$.

$$\frac{x}{c} = \frac{b}{z}$$

In Exercises 14–17, use a proportion to solve the problem.

14. APPLICATION A car travels 106 miles on 4 gallons of gas. How far can it go on a full tank of 12 gallons?

15. APPLICATION Ernie is a baseball pitcher. He gave up 34 runs in 152 innings last season. What is Ernie's earned run average—the number of runs he would give up in 9 innings? Give your answer accurate to two decimal places.

16. APPLICATION The floor plan of a house is drawn to the scale of $\frac{1}{4}$ in. = 1 ft. The master bedroom measures 3 in. by $3\frac{3}{4}$ in. on the blueprints. What is the actual size of the room?

17. Altor and Zenor are ambassadors from Titan, the largest moon of Saturn. The sum of the lengths of any Titan's antennae is a direct measure of that Titan's age. Altor has antennae with lengths 8 cm, 10 cm, 13 cm, 16 cm, 14 cm, and 12 cm. Zenor is 130 years old, and her seven antennae have an average length of 17 cm. How old is Altor?

18. Assume $\frac{AB}{XY} = \frac{BC}{YZ}$. Find AB and BC.

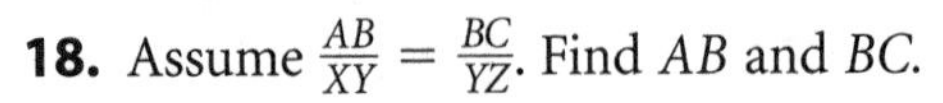

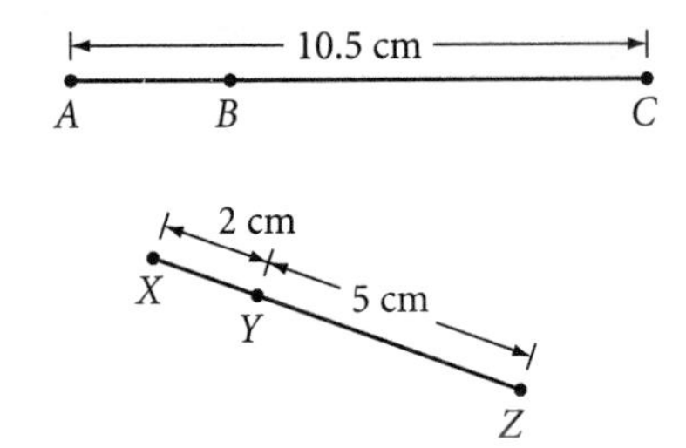

IMPROVING YOUR ALGEBRA SKILLS

Algebraic Magic Squares II

In this algebraic magic square, the sum of the entries in every row, column, and diagonal is the same. Find the value of x.

$8 - x$	15	14	$11 - x$
12	$x - 1$	x	9
8	$x + 3$	$x + 4$	5
$2x - 1$	3	2	$2x + 2$

LESSON

11.1

Similar Polygons

He that lets
the small things bind him
Leaves the great
undone behind him.
PIET HEIN

You know that figures that have the same shape and size are congruent figures. Figures that have the same shape but not necessarily the same size are **similar figures.** To say that two figures have the same shape but not necessarily the same size is not, however, a precise definition of similarity.

Is your reflection in a fun-house mirror similar to a regular photograph of you? The images have a lot of features in common, but they are not mathematically similar. In mathematics, you can think of similar shapes as enlargements or reductions of each other with no irregular distortions.

Are all rectangles similar? They have common characteristics, but they are not all similar. That is, you could not enlarge or reduce a given rectangle to fit perfectly over every other rectangle. What about other geometric figures: squares, circles, triangles?

The uneven surface of a fun-house mirror creates a distorted image of you. Your true proportions look different in your reflection.

Rectangles A and B are not similar. You could not enlarge or reduce one to fit perfectly over the other.

A

B

Art CONNECTION

Movie scenes are scaled down to small images on strips of film. Then they are scaled up to fit a large screen. So the film image and the projected image are similar. If the distance between the projector and the screen is decreased by half, each dimension of the screen image is cut in half.

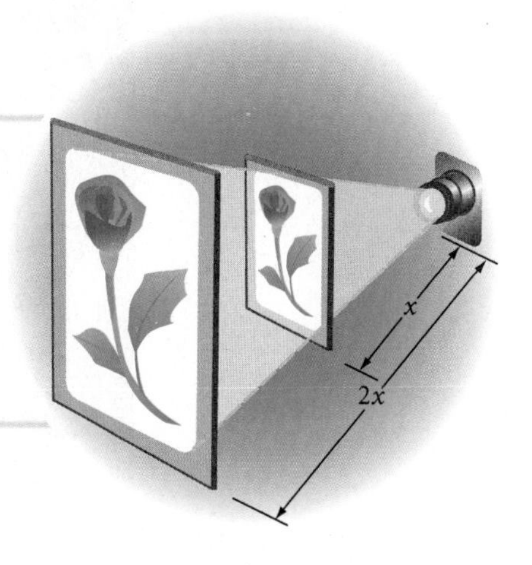

Investigation 1
What Makes Polygons Similar?

You will need

- patty paper
- a ruler

Let's explore what makes polygons similar. Hexagon *PQRSTU* is an enlargement of hexagon *ABCDEF*—they are similar.

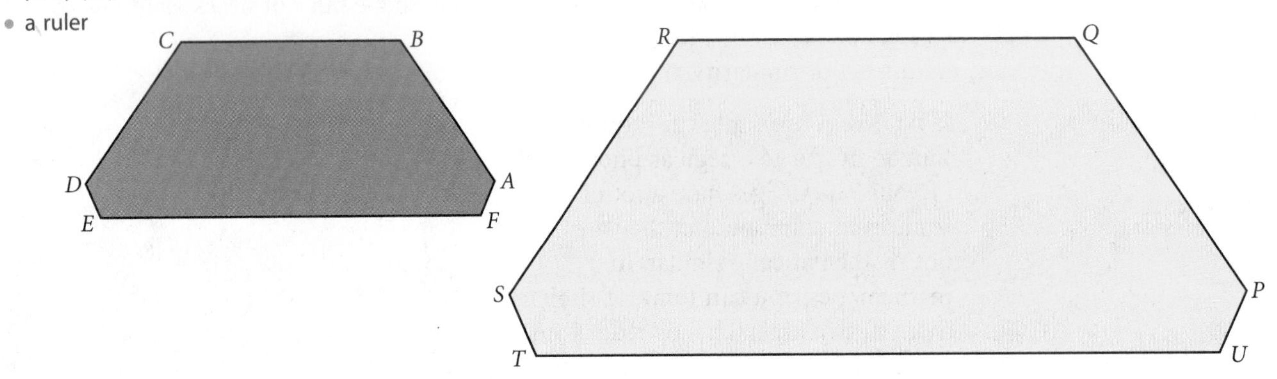

Step 1 Use patty paper to compare all corresponding angles. How do the corresponding angles compare?

Step 2 Measure the corresponding segments in both hexagons.

Step 3 Find the ratios of the lengths of corresponding sides. How do the ratios of corresponding sides compare?

Similar objects are often used to create unique buildings. The giant basket shown here is actually an office building for a basket manufacturer. The giant donut advertises a donut shop in Los Angeles, California.

From the investigation, you should be able to state a mathematical definition of similar polygons. Two polygons are **similar polygons** if and only if the corresponding angles are congruent and the corresponding sides are proportional. Similarity is the state of being similar.

The statement $CORN \sim PEAS$ says that quadrilateral *CORN* is similar to quadrilateral *PEAS.* Just as in statements of congruence, the order of the letters tells you which segments and which angles in the two polygons correspond.

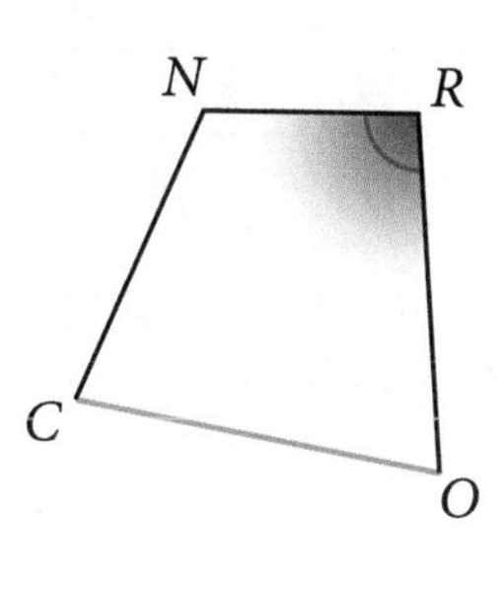

Corresponding angles are congruent:

$\angle C \cong \angle P$ $\quad \angle R \cong \angle A$

$\angle O \cong \angle E$ $\quad \angle N \cong \angle S$

Corresponding segments are proportional:

$$\frac{CO}{PE} = \frac{OR}{EA} = \frac{RN}{AS} = \frac{NC}{SP}$$

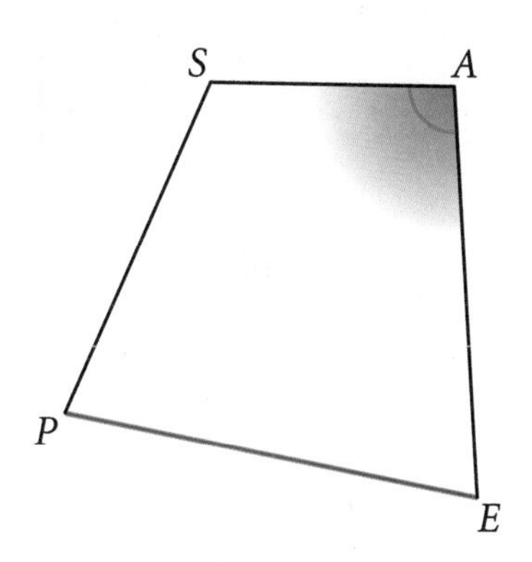

Do you need both conditions—congruent angles and proportional sides—to guarantee that the two polygons are similar? For example, if you know only that the corresponding angles of two polygons are congruent, can you conclude that the polygons have to be similar? Or, if corresponding sides of two polygons are proportional, are the polygons necessarily similar? These counterexamples show that both answers are no.

In the figures below, corresponding angles of square *SQUE* and rectangle *RCTL* are congruent, but their corresponding sides are not proportional.

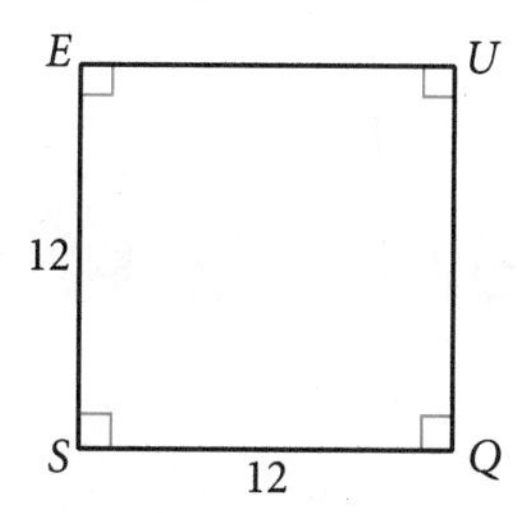

$$\frac{12}{10} \neq \frac{12}{18}$$

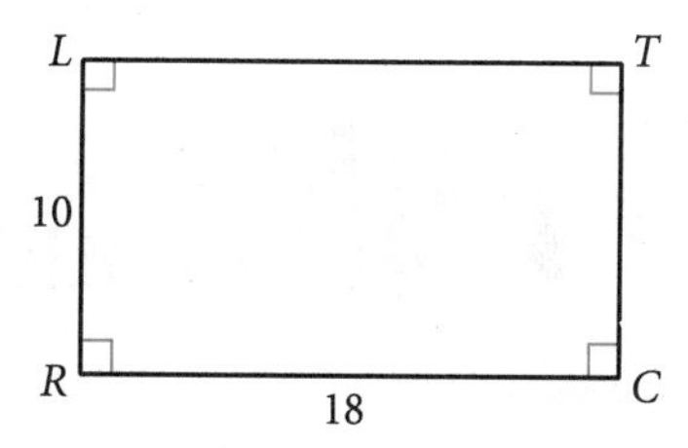

In the figures below, corresponding sides of square *SQUE* and rhombus *RHOM* are proportional, but their corresponding angles are not congruent.

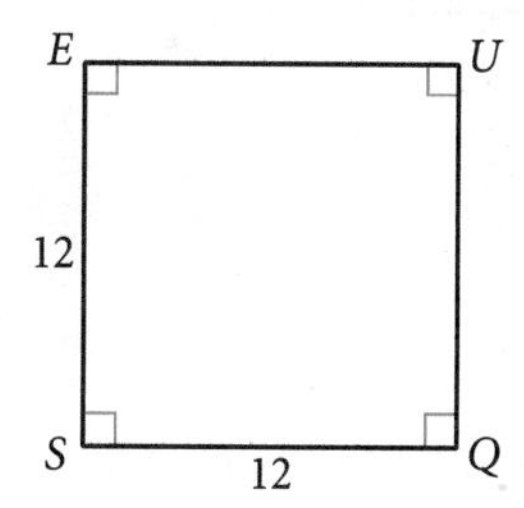

$$\frac{12}{18} = \frac{12}{18}$$

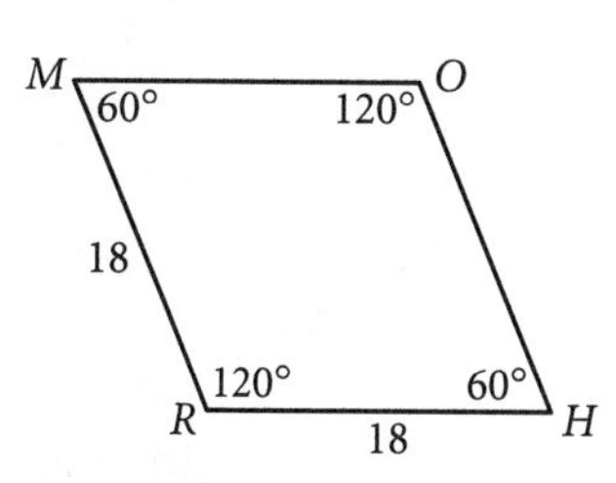

Clearly, neither pair of polygons is similar. You cannot conclude that two polygons are similar given only the fact that their corresponding angles are congruent or given only the fact that their corresponding sides are proportional.

You can use the definition of similar polygons to find missing measures in similar polygons.

EXAMPLE

SMAL ~ *BIGE*
Find x and y.

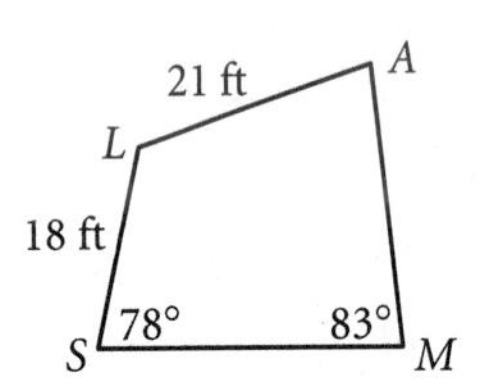

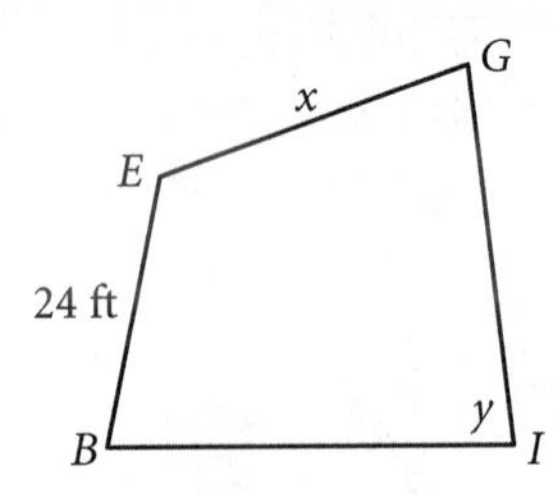

▶ **Solution**

The quadrilaterals are similar, so you can use a proportion to find x.

$$\frac{18}{24} = \frac{21}{x}$$ A proportion of corresponding sides.

$$18x = (24)(21)$$ Multiply both sides by $24x$ and reduce.

$$x = 28$$ Divide both sides by 18.

The measure of the side labeled x is 28 ft.

In similar polygons, corresponding angles are congruent, so $\angle M \cong \angle I$. The measure of the angle labeled y is therefore 83°.

Earlier in this book you worked with translations, rotations, and reflections. These rigid transformations preserve both size and shape—the images are congruent to the original figures. One type of nonrigid transformation is called a **dilation.** Let's look at an image after a dilation transformation.

Investigation 2
Dilations on a Coordinate Plane

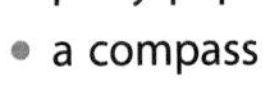

- graph paper
- a straightedge
- patty paper
- a compass

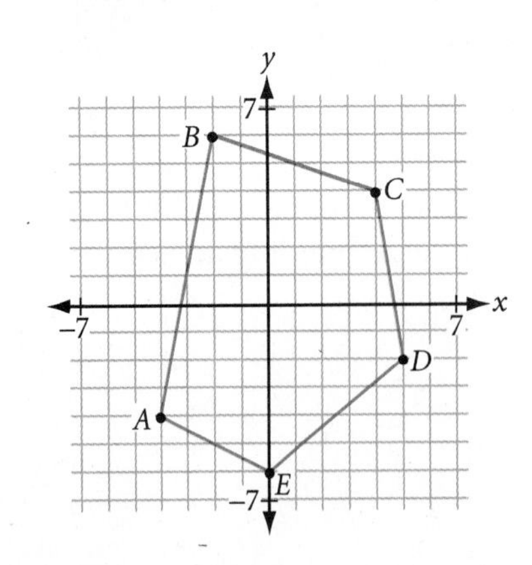

Step 1 To dilate a pentagon on a coordinate plane, first copy this pentagon onto your graph paper.

Step 2 Have each member of your group multiply the coordinates of the vertices by one of these numbers: $\frac{1}{2}$, $\frac{3}{4}$, 2, or 3. Each of these factors is called a **scale factor.**

Step 3 Locate these new coordinates on your graph paper and draw the new pentagon.

Step 4 | Copy the original pentagon onto patty paper. Compare the corresponding angles of the two pentagons. What do you notice?

Step 5 | Compare the corresponding sides with a compass or with patty paper. The length of each side of the new pentagon is how many times as long as the length of the corresponding side of the original pentagon?

Step 6 | Compare results with your group. You should be ready to state a conjecture.

Dilation Similarity Conjecture **C-92**

If one polygon is the image of another polygon under a dilation, then $\underline{?}$.

History CONNECTION

Similarity plays an important role in human history. For example, accurate maps of regions of China have been found dating back to the second century B.C.E. Neolithic cave paintings 8000–6000 B.C.E. contain small-scale drawings of the animals people hunted. Giant geoglyphs made by the Nazca people of Peru (110 B.C.E.–800 C.E.) are some of the largest scale drawings ever made.

In order for a map to be accurate, cartographers need to use similarity to reduce the earth's attributes to a smaller scale. This sixteenth-century French map, a plan of Constantinople, included a mariner's chart of America, Europe, Africa, and Asia.

This cave art is part of a grouping of over 15,000 drawings in Tassili N'Ajjier National Park of the Algerian Sahara. Interestingly, these drawings depict animals and landscapes that are absent from the region today, such as these elephants or vast lakes.

This monkey is a geoglyph found in the Pampa region of Peru in 1920. Called the Nazca Lines, the figure measures over 400 feet long and can only be clearly seen from the air.

EXERCISES

You will need

Construction tools for Exercises **22** and **25**

For Exercises 1 and 2, match the similar figures.

1. 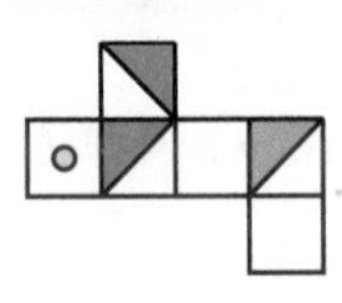

A. 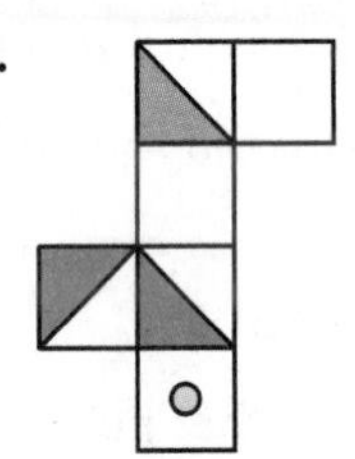

B. 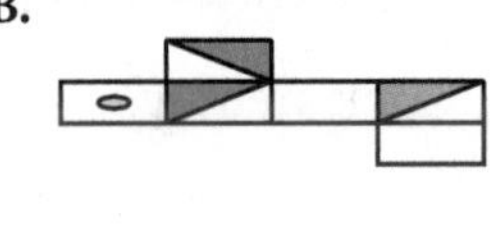

C. 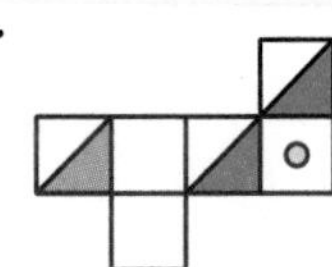

2. 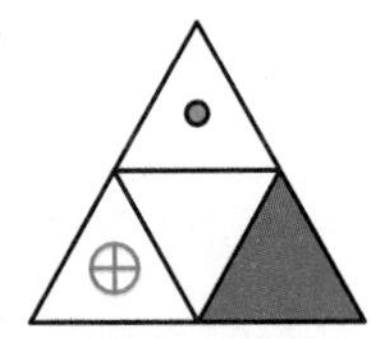

A. 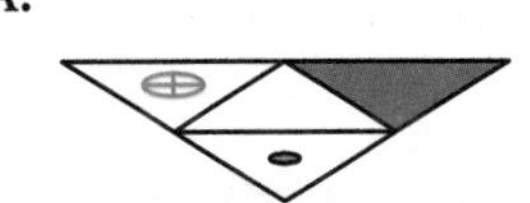

B. 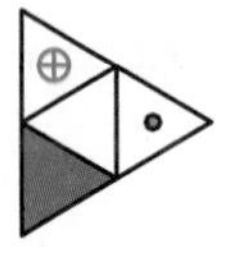

C. 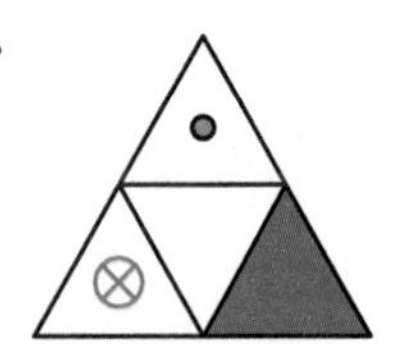

For Exercises 3–5, sketch on graph paper a similar, but not congruent, figure.

3. 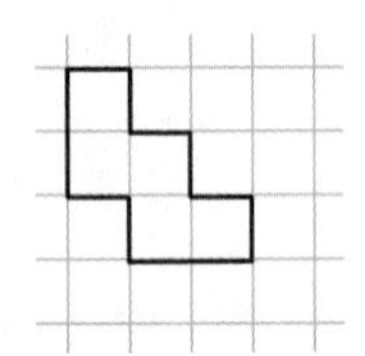

4. 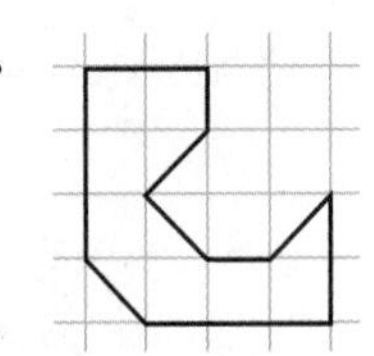

5. 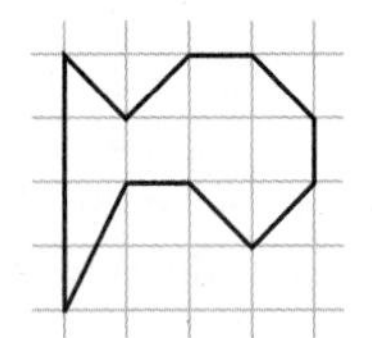

6. Complete the statement: If Figure A is similar to Figure B and Figure B is similar to Figure C, then _?_. Draw and label figures to illustrate the statement.

For Exercises 7–14, use the definition of similar polygons. All measurements are in centimeters.

7. $THINK \sim LARGE$
Find AL, RA, RG, and KN.

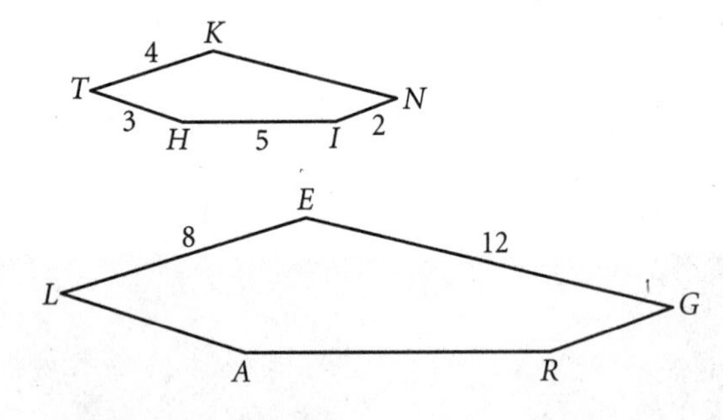

8. Are these polygons similar? Explain why or why not. ⓗ

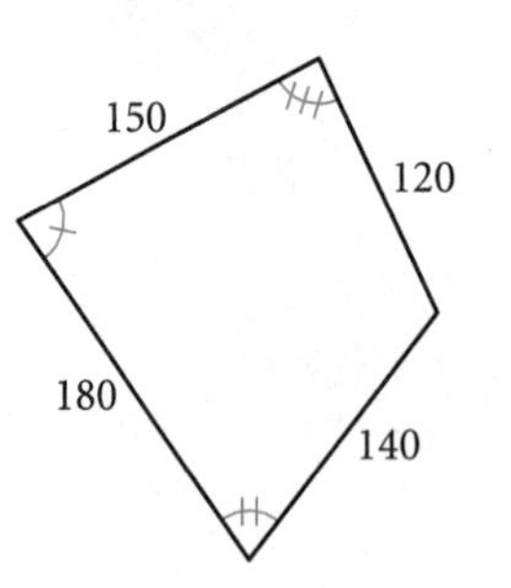

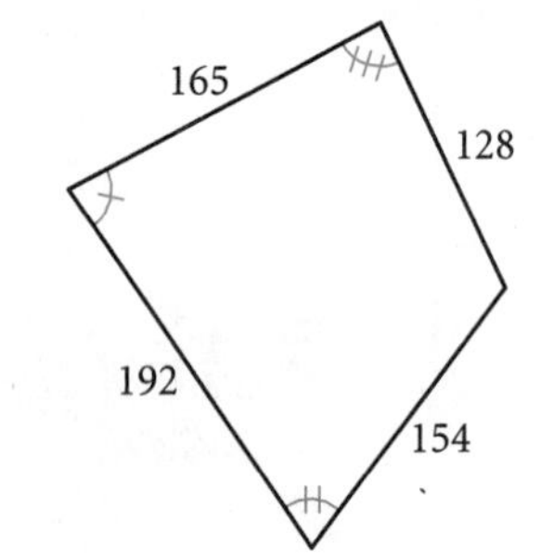

9. $SPIDER \sim HNYCMB$
Find NY, YC, CM, and MB.

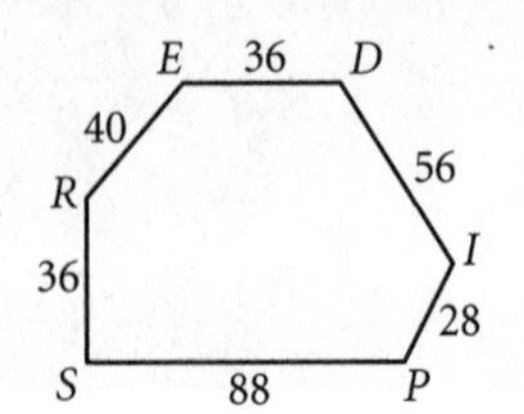

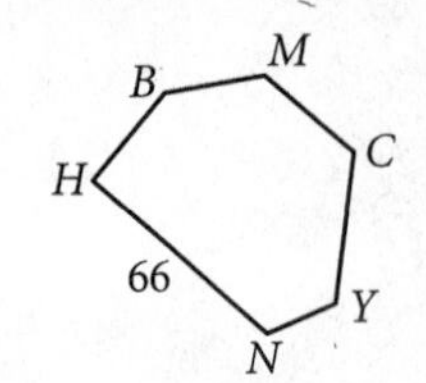

10. Are these polygons similar? Explain why or why not.

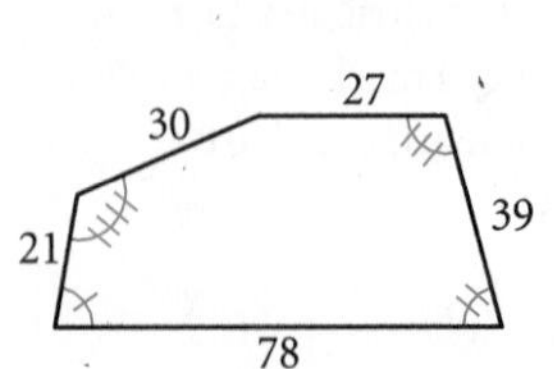

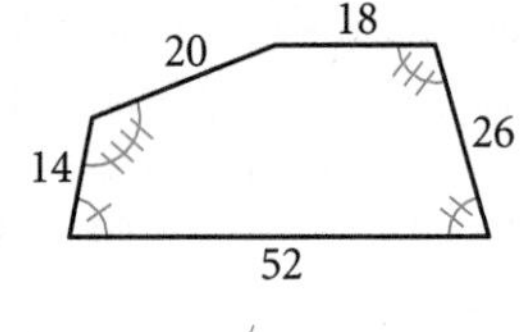

11. $\triangle ACE \sim \triangle IKS$
Find x and y.

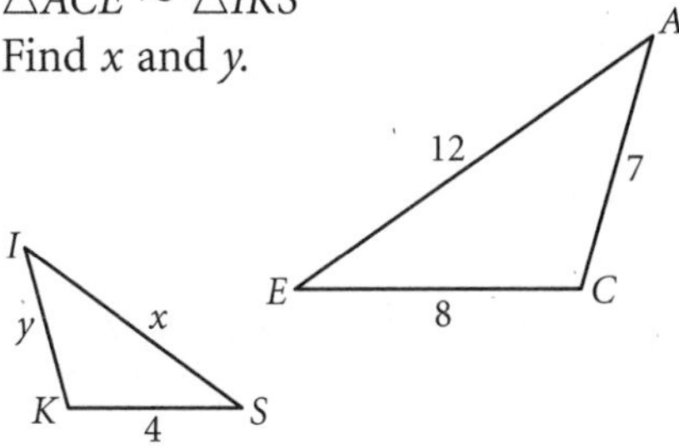

12. $\triangle RAM \sim \triangle XAE$
Find z.

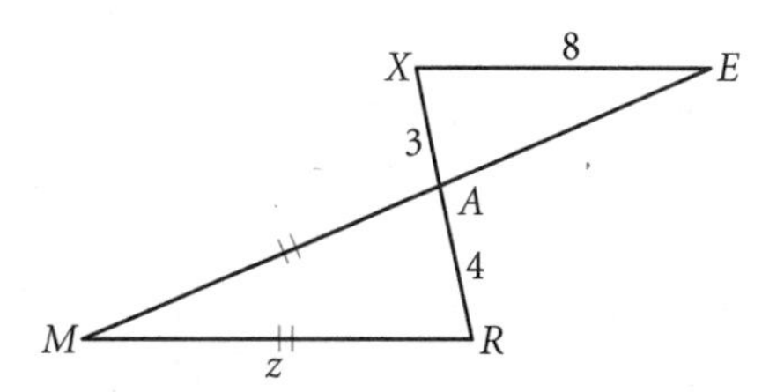

13. $\overline{DE} \parallel \overline{BC}$
Are the corresponding angles congruent in $\triangle AED$ and $\triangle ABC$? Are the corresponding sides proportional? Is $\triangle AED \sim \triangle ABC$? ⓗ

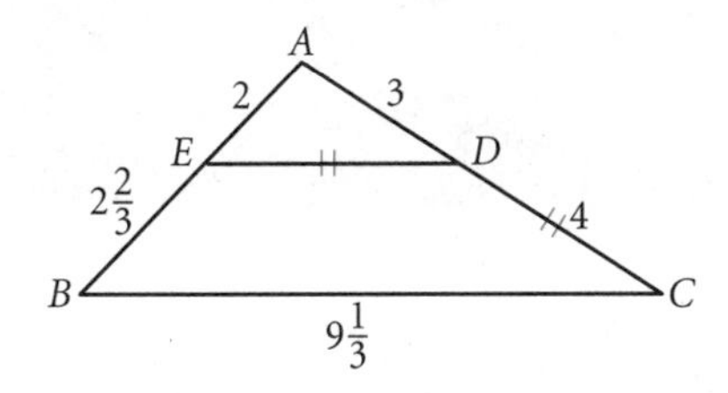

14. $\triangle ABC \sim \triangle DBA$
Find m and n.

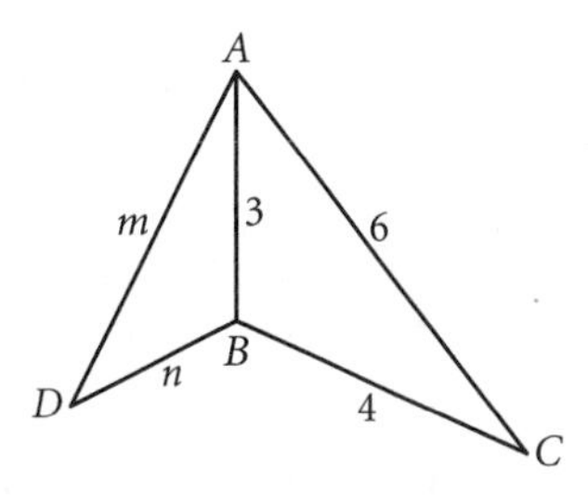

15. Copy $\triangle ROY$ onto your graph paper. Sketch its dilation with a scale factor of 3. What is the ratio of the perimeter of the smaller triangle to the perimeter of the larger triangle?

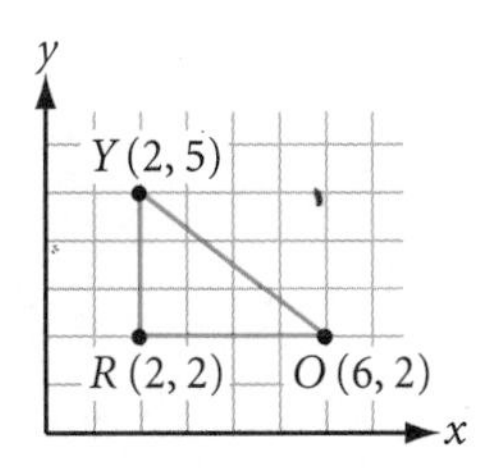

16. Copy this quadrilateral onto your graph paper. Draw a similar quadrilateral with each side half the length of its corresponding side in the original quadrilateral.

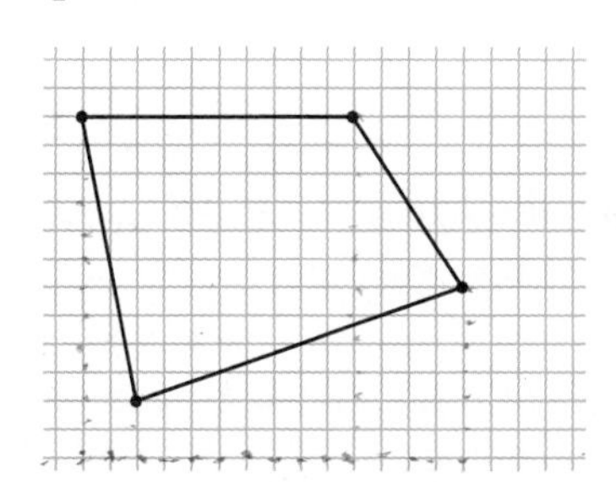

17. **APPLICATION** The photo at right shows the Crazy Horse Memorial and a scale model of the complete monument's design. The head of the Crazy Horse Memorial, from the chin to the top of the forehead, is 87.5 ft high. When the arms are carved, how long will each be? Use the photo and explain how you got your answer.

The Crazy Horse Memorial is located in South Dakota. Started in 1948, it will be the world's largest sculpture when complete. You can learn more about this monument using the links at www.keymath.com/DG.

Review

For Exercises 18–20, use algebra to answer each proportion question. Assume that a, b, c, and d are all nonzero values.

18. If $\frac{15}{a} = \frac{20}{a + 12}$, then $a = \underline{\ ?\ }$.

19. If $\frac{a}{b} = \frac{c}{d}$, then $ad = \underline{\ ?\ }$.

20. If $\frac{a}{b} = \frac{c}{d}$, then $\frac{b}{a} = \underline{\ ?\ }$.

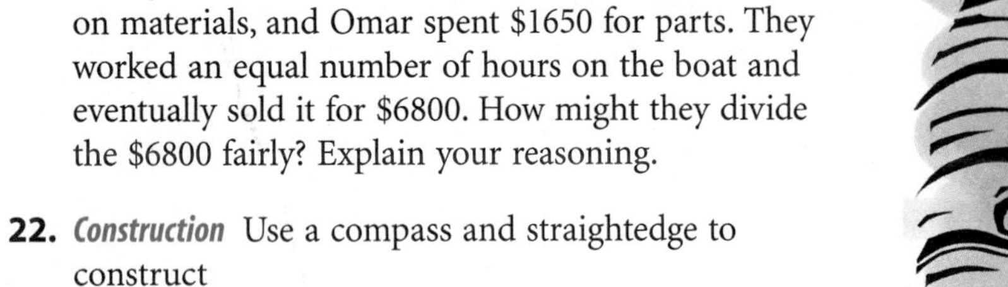

21. APPLICATION Jade and Omar each put in $1000 to buy an old boat to fix up. Later Jade spent $825 on materials, and Omar spent $1650 for parts. They worked an equal number of hours on the boat and eventually sold it for $6800. How might they divide the $6800 fairly? Explain your reasoning.

22. *Construction* Use a compass and straightedge to construct

a. A rhombus with a 60° angle.

b. A second rhombus of different size with a 60° angle.

23. A cubic foot of liquid is about 7.5 gallons. How many gallons of liquid are in this tank?

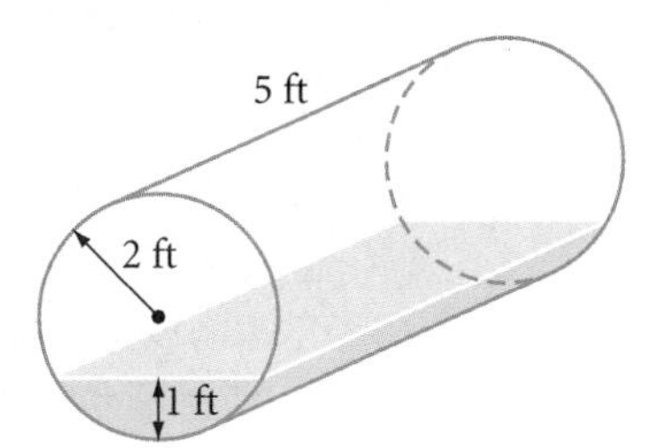

24. Triangle PQR has side lengths 18 cm, 24 cm, and 30 cm. Is $\triangle PQR$ a right triangle? Explain why or why not.

25. *Construction* Use the triangular figure at right and its rotated image. ⓗ

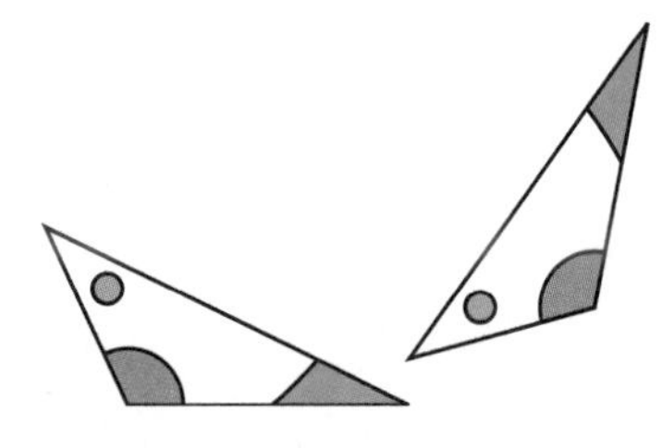

a. Copy the figure and its image onto a piece of patty paper. Locate the center of rotation. Explain your method.

b. Copy the figure and its image onto a sheet of paper. Locate the center of rotation using a compass and straightedge. Explain your method.

Career CONNECTION

Similarity plays an important part in the design of cars, trucks, and airplanes, which is done with small-scale drawings and models.

This model airplane is about to be tested in a wind tunnel.

project

MAKING A MURAL

A mural is a large work of art that usually fills an entire wall. This project gives you a chance to use similarity and make your own mural.

One way to create a mural from a small picture is to draw a grid of squares lightly over the small picture. Then divide the mural surface into a similar but larger grid. Proceeding square by square, draw the lines and curves of the small grid in each corresponding square of the mural grid. Complete the mural by coloring or painting the regions and erasing the grid lines.

You project should include

- An original drawing, a cartoon, or a photograph divided into a grid of squares.
- A finished mural drawn on a large sheet of paper.

You can learn more about the art of mural making through the links at www.keymath.com/DG .

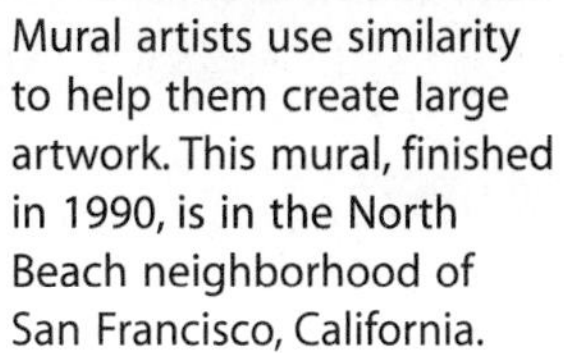

Mural artists use similarity to help them create large artwork. This mural, finished in 1990, is in the North Beach neighborhood of San Francisco, California.

LESSON

11.2

Similar Triangles

Life is change. Growth is optional. Choose wisely.

KAREN KAISER CLARK

In Lesson 11.1, you concluded that you must know about both the angles and the sides of two quadrilaterals in order to make a valid conclusion about their similarity.

However, triangles are unique. Recall from Chapter 4 that you found four shortcuts for triangle congruence: SSS, SAS, ASA, and SAA. Are there shortcuts for triangle similarity as well? Let's first look for shortcuts using only angles.

The figures below illustrate that you cannot conclude that two triangles are similar given that only one set of corresponding angles are congruent.

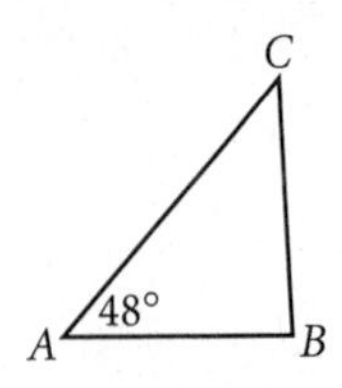

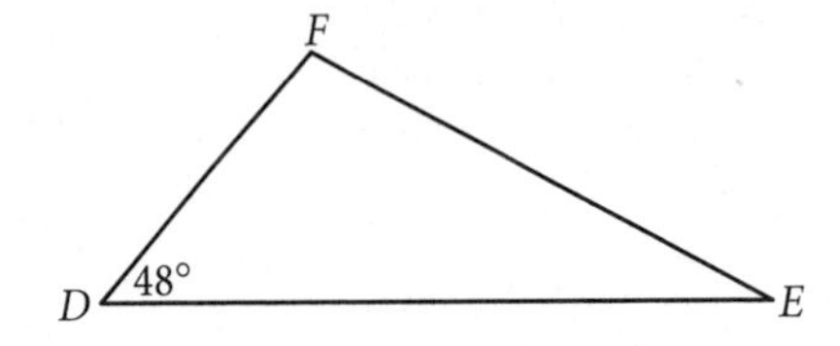

$\angle A \cong \angle D$, but $\triangle ABC$ is not similar to $\triangle DEF$.

How about two sets of congruent angles?

Investigation 1

Is AA a Similarity Shortcut?

You will need

- a compass
- a ruler

If two angles of one triangle are congruent to two angles of another triangle, must the two triangles be similar?

Step 1 Draw any triangle *ABC*.

Step 2 Construct a second triangle, *DEF*, with $\angle D \cong \angle A$ and $\angle E \cong \angle B$. What will be true about $\angle C$ and $\angle F$? Why?

Step 3 Carefully measure the lengths of the sides of both triangles. Compare the ratios of the corresponding sides. Is $\frac{AB}{DE} \approx \frac{AC}{DF} \approx \frac{BC}{EF}$?

Step 4 Compare your results with the results of others near you. You should be ready to state a conjecture.

AA Similarity Conjecture C-93

If _?_ angles of one triangle are congruent to _?_ angles of another triangle, then _?_.

As you may have guessed from Step 2 of the investigation, there is no need to investigate the AAA Similarity Conjecture. Thanks to the Third Angle Conjecture, the AA Similarity Conjecture is all you need.

Now let's look for shortcuts for similarity that use only sides. The figures below illustrate that you cannot conclude that two triangles are similar given that two sets of corresponding sides are proportional.

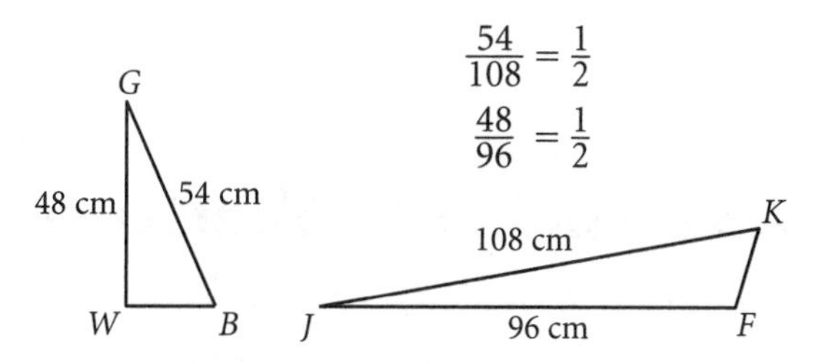

$\frac{GB}{JK} = \frac{GW}{JF}$, but $\triangle GWB$ is not similar to $\triangle JFK$.

How about all three sets of corresponding sides?

Investigation 2
Is SSS a Similarity Shortcut?

You will need

- a compass
- a straightedge
- a protractor

If three sides of one triangle are proportional to the three sides of another triangle, must the two triangles be similar?

Draw any triangle *ABC*. Then construct a second triangle, *DEF*, whose side lengths are a multiple of the original triangle. (Your second triangle can be larger or smaller.)

Compare the corresponding angles of the two triangles. Compare your results with the results of others near you and state a conjecture.

SSS Similarity Conjecture C-94

If the three sides of one triangle are proportional to the three sides of another triangle, then the two triangles are __?__.

Many dollhouses and other toys are scale models of real objects.

In Investigations 1 and 2, you discovered two shortcuts for triangle similarity: AA and SSS. But if AA is a shortcut, then so are ASA, SAA, and AAA. That leaves SAS and SSA as possible shortcuts to consider.

Investigation 3
Is SAS a Similarity Shortcut?

You will need

- a compass
- a protractor
- a ruler

Is SAS a shortcut for similarity? Try to construct two different triangles that are not similar but have two pairs of sides proportional and the pair of included angles equal in measure.

Compare the measures of corresponding sides and corresponding angles. Share your results with others near you and state a conjecture.

SAS Similarity Conjecture C-95

If two sides of one triangle are proportional to two sides of another triangle and __?__, then the __?__.

One question remains: Is SSA a shortcut for similarity? Recall from Chapter 4 that SSA did not work for congruence because you could create two different triangles. Those two different triangles were neither congruent nor similar. So, no, SSA is not a shortcut for similarity.

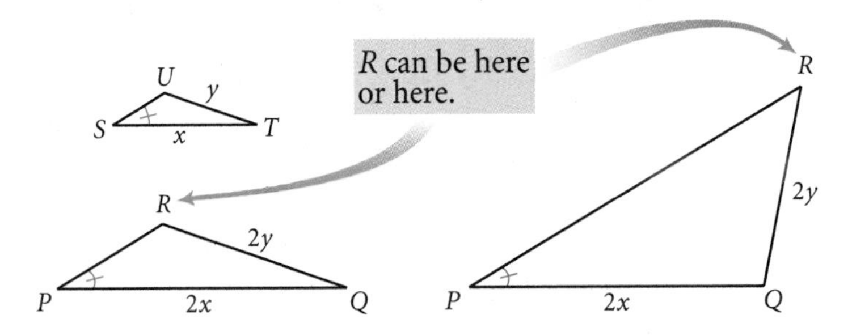

Exercises

For Exercises 1–14, use your new conjectures. All measurements are in centimeters.

1. $g = \underline{?}$

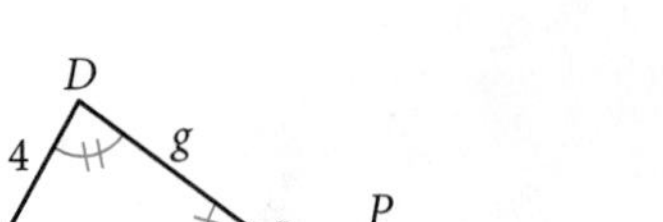

2. $h = \underline{?}$, $k = \underline{?}$

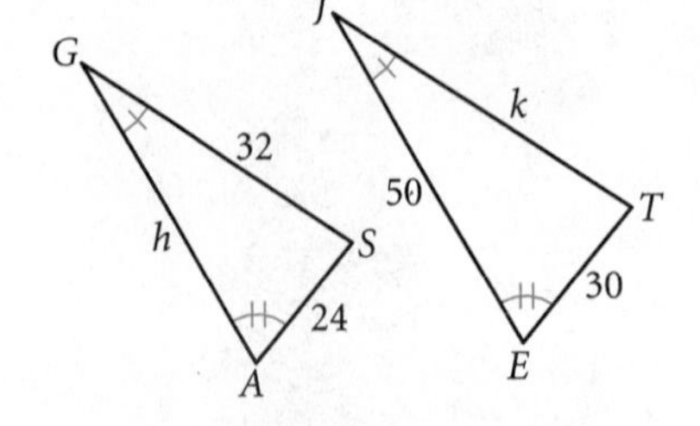

3. $m = \underline{?}$ ⓗ

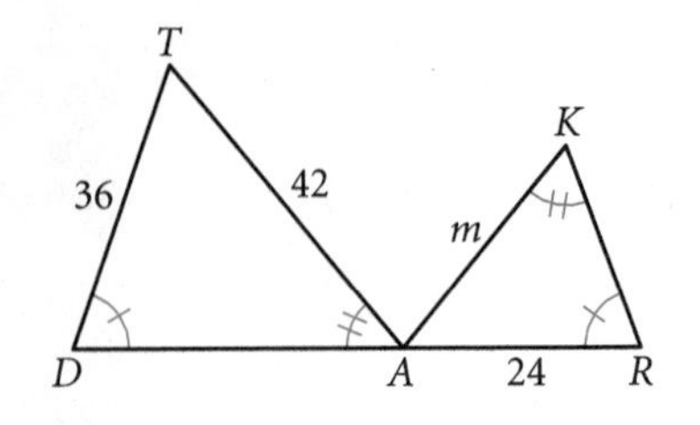

4. $n = \underline{?}$, $s = \underline{?}$

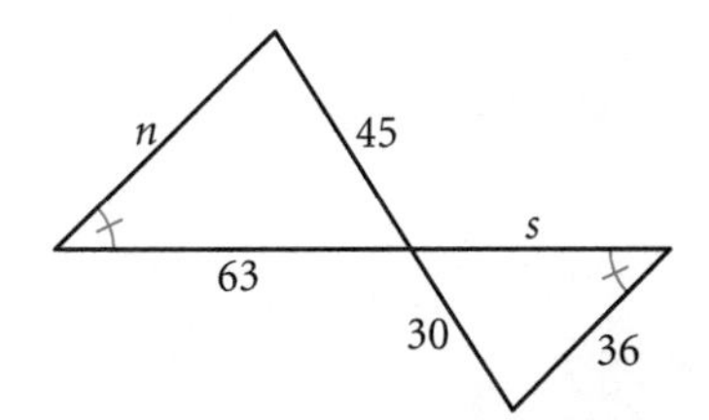

5. Is $\triangle AUL \sim \triangle MST$? Explain why or why not.

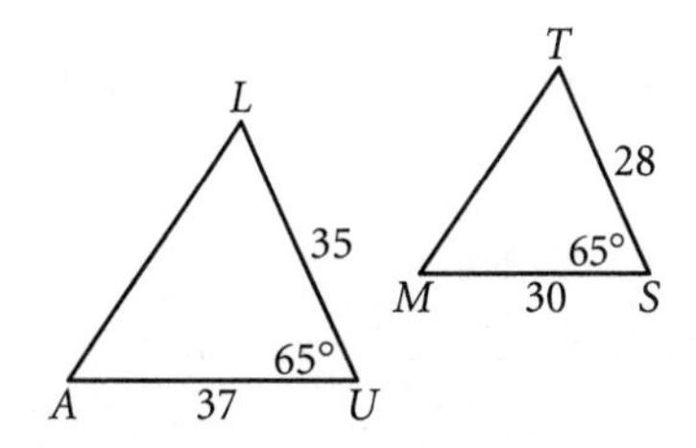

6. Is $\triangle MOY \sim \triangle NOT$? Explain why or why not.

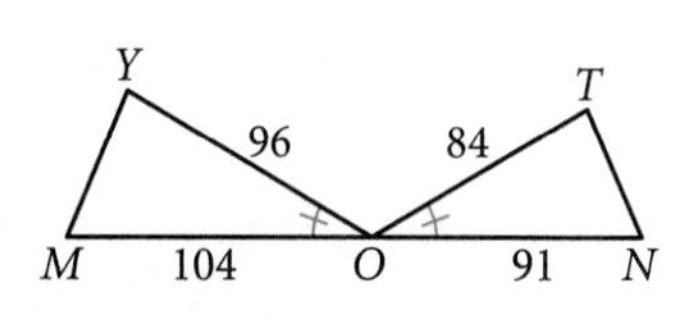

7. Is $\triangle PHY \sim \triangle YHT$? Is $\triangle PTY$ a right triangle? Explain why or why not.

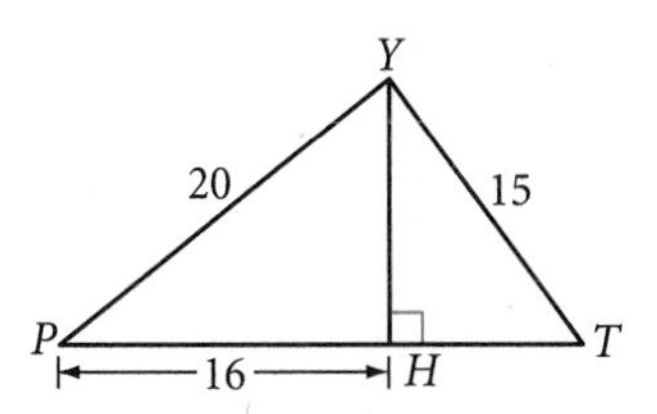

8. Why is $\triangle TMR \sim \triangle THM \sim \triangle MHR$? Find x, y, and h. ⓗ

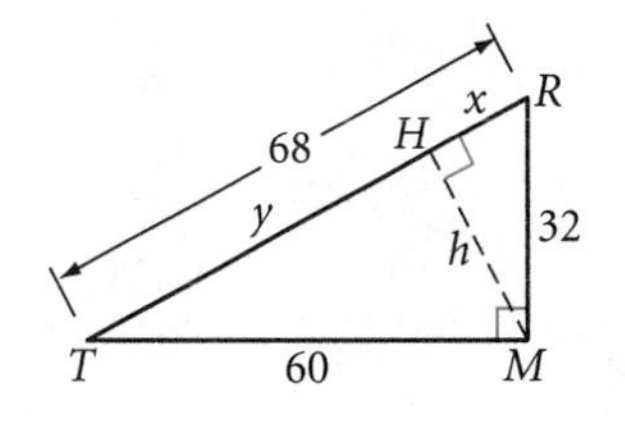

9. $\overline{TA} \parallel \overline{UR}$
Is $\angle QTA \cong \angle TUR$?
Is $\angle QAT \cong \angle ARU$?
Why is $\triangle QTA \sim \triangle QUR$?
$e = \underline{?}$

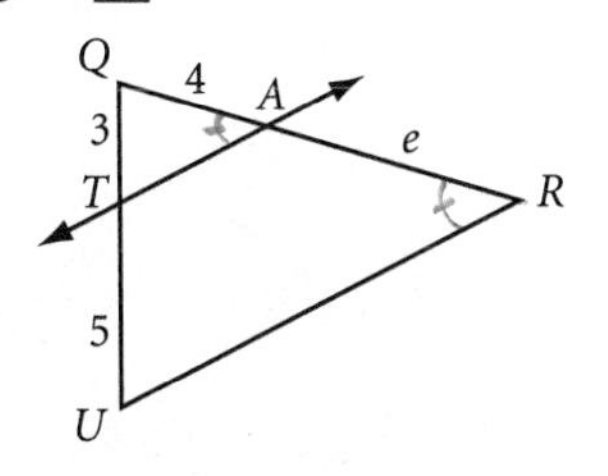

10. $\overline{OR} \parallel \overline{UE} \parallel \overline{NT}$
$f = \underline{?}$, $g = \underline{?}$

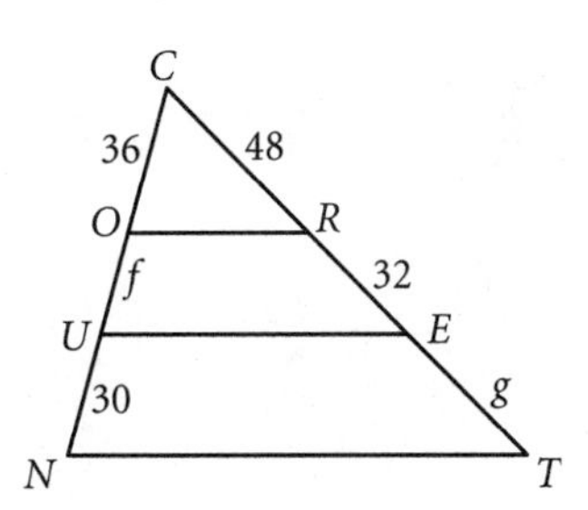

11. Is $\angle THU \cong \angle GDU$? Is $\angle HTU \cong \angle DGU$?
$p = \underline{?}$, $q = \underline{?}$ ⓗ

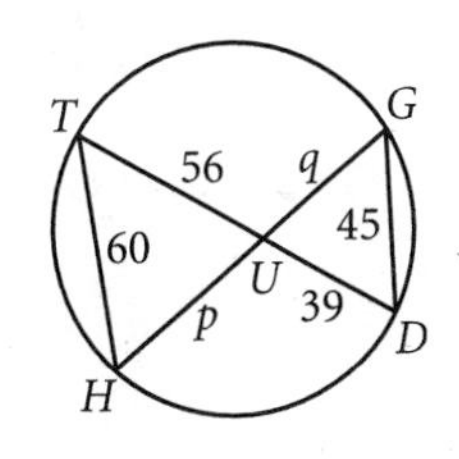

12. Why is $\triangle SUN \sim \triangle TAN$?
$r = \underline{?}$, $s = \underline{?}$

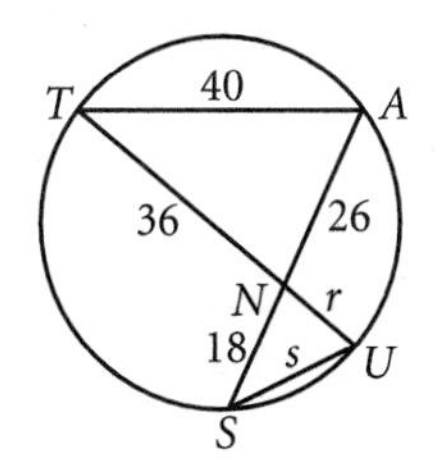

13. *FROG* is a trapezoid.
Is $\angle RGO \cong \angle FRG$?
Is $\angle GOF \cong \angle RFO$?
Why is $\triangle GOS \sim \triangle RFS$?
$t = \underline{?}$, $s = \underline{?}$

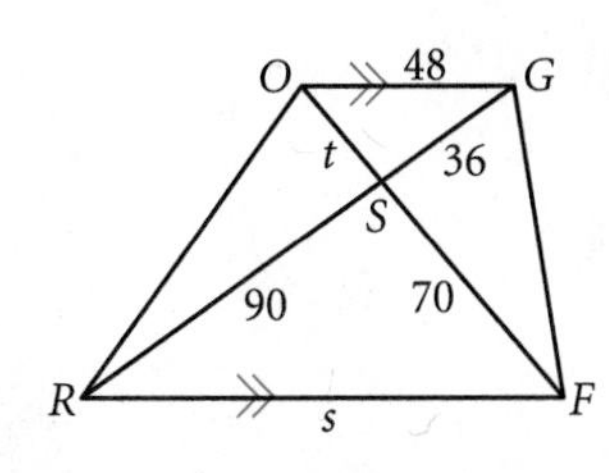

14. *TOAD* is a trapezoid.
$w = \underline{?}$, $x = \underline{?}$

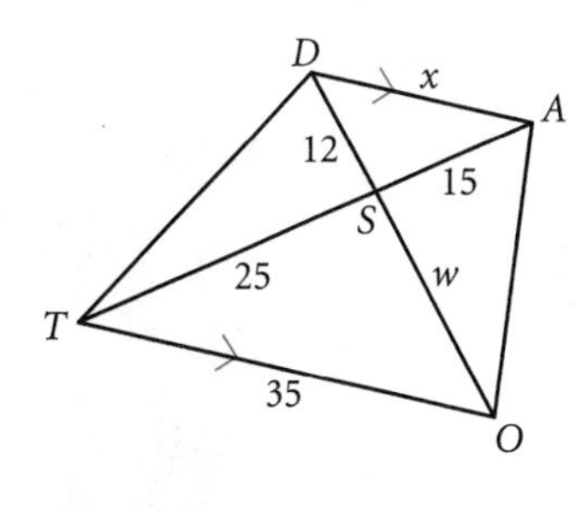

15. Find x and y. ⓗ

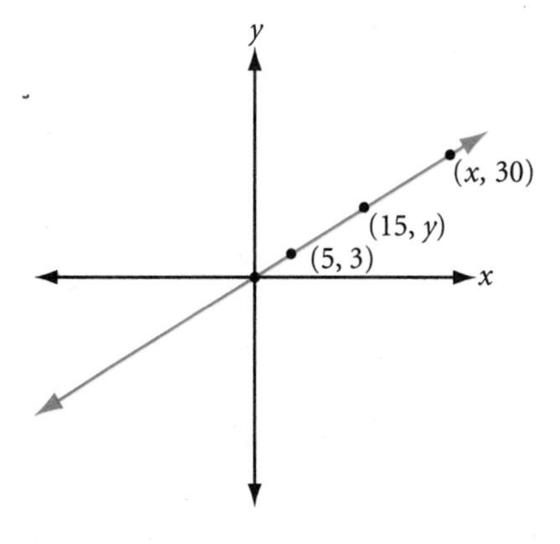

Review

16. In the figure below right, find the radius, r, of one of the small circles in terms of the radius, R, of the large circle. ⓗ

This Tibetan mandala is a complex design with a square inscribed within a circle and tangent circles inscribed within the corners of a larger circumscribed square.

17. **APPLICATION** Phoung volunteers at an SPCA that always houses 8 dogs. She notices that she uses seven 35-pound bags of dry dog food every two months. A new, larger SPCA facility that houses 20 dogs will open soon. Help Phoung estimate the amount of dry dog food that the facility should order every three months. Explain your reasoning.

18. **APPLICATION** Ramon and Sabina are oceanography students studying the habitat of a Hawaiian fish called Humuhumunukunukuapuaʻa. They are going to use the capture-recapture method to determine the fish population. They first capture and tag 84 fish, which they release back into the ocean. After one week, Ramon and Sabina catch another 64. Only 12 have tags. Can you estimate the population of Humuhumunukunukuapuaʻa?

19. Points $A(-9, 5)$, $B(4, 13)$, and $C(1, -7)$ are connected to form a triangle. Find the area of $\triangle ABC$.

20. Use the ordered pair rule, $(x, y) \rightarrow \left(\frac{1}{2}x, \frac{1}{2}y\right)$, to relocate the coordinates of the vertices of parallelogram $ABCD$. Call the new parallelogram $A'B'C'D'$. Is $A'B'C'D'$ similar to $ABCD$? If they are similar, what is the ratio of the perimeter of $ABCD$ to the perimeter of $A'B'C'D'$? What is the ratio of their areas?

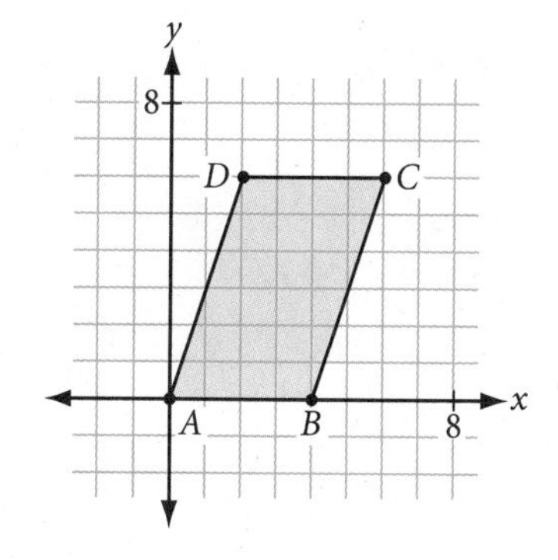

21. The photo below shows a fragment from an ancient statue of the Roman Emperor Constantine. Use this photo to estimate how tall the entire statue was. List the measurements you need to make. List any assumptions you need to make. Explain your reasoning.

History CONNECTION

The Emperor Constantine the Great (Roman Emperor 306–337 C.E.) adopted Christianity as the official religion of the Roman Empire. The Roman Catholic Church regards him as Saint Constantine, and the city of Constantinople was named for him. The colossal statue of Constantine was built between 315 and 330 C.E., and broke when sculptors tried to add the extra weight of a beard to its face. The pieces of the statue remain close to its original location in Rome, Italy.

IMPROVING YOUR VISUAL THINKING SKILLS

Build a Two-Piece Puzzle

Construct two copies of Figure A, shown at right. Here's how to construct the figure.

- Construct a regular hexagon.
- Construct an equilateral triangle on two alternating edges, as shown.
- Construct a square on the edge between the two equilateral triangles, as shown.

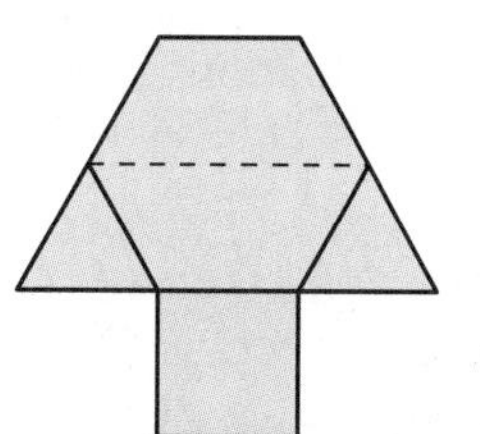

Figure A

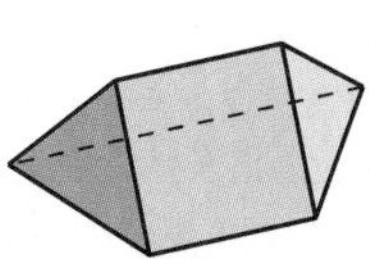

Figure B

Cut out each copy and fold them into two identical solids, as shown in Figure B. Tape the edges. Now arrange your two solids to form a regular tetrahedron.

Exploration

Constructing a Dilation Design

In Lesson 11.1, you saw how to dilate a polygon on a coordinate plane. You can also use a simple construction to dilate any polygon. Draw rays from any point P through the vertices of the polygon. Use a compass to measure the distance from point P to one of the vertices. Then mark this distance two more times along the ray; that will give you a scale factor of 3. Repeat this process for each of the other vertices using the same scale factor. When you connect the image of each vertex, you will get a similar polygon. Try it yourself. How would you create a similar polygon with a scale factor of 2? Of $\frac{1}{2}$?

Now take a closer look at *Path of Life I,* the M. C. Escher woodcut that begins this chapter. Notice that dilations transform the black fishlike creatures, shrinking them again and again as they approach the picture's center. The same is true for the white fish. (The black-and-white fish around the outside border are congruent to one another, but they're not similar to the other fish.) Also notice that rotations repeat the dilations in eight sectors. With Sketchpad, you can make a similar design.

Activity

Dilation Creations

Step 1 Construct a circle with center point A and point B on the circle. Construct $\overline{AB}$.

Step 2 Use the Transform menu to mark point A as center, then rotate point B by an angle of 45°. Your new point is B'. Construct $\overrightarrow{AB'}$.

Step 3 Construct a larger circle with center point A and point C on $\overrightarrow{AB'}$. Hide $\overrightarrow{AB'}$ and construct $\overline{AC}$.

Step 4 Rotate $\overline{AC}$, point B', and point C by an angle of 45°. You now have $\overline{AC'}$, point B'', and point C'.

Step 5 Construct $\overline{C'D}$ and $\overline{DC}$, where D is any point in the region between the circles and between $\overline{B''C'}$ and $\overline{B'C}$.

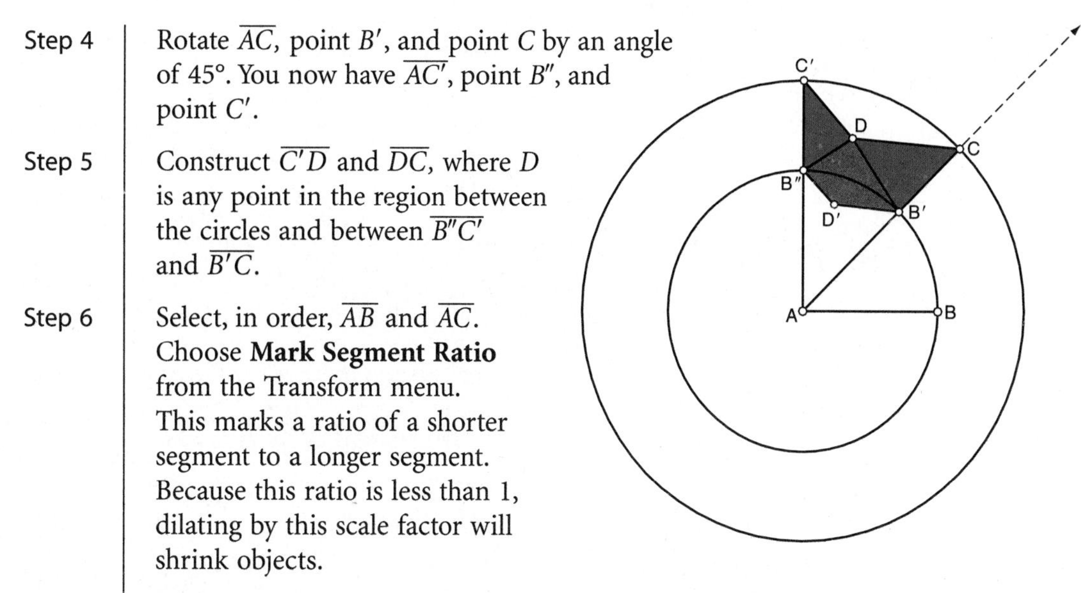

Step 6 Select, in order, $\overline{AB}$ and $\overline{AC}$. Choose **Mark Segment Ratio** from the Transform menu. This marks a ratio of a shorter segment to a longer segment. Because this ratio is less than 1, dilating by this scale factor will shrink objects.

Step 7 Select $\overline{C'D}$, $\overline{DC}$, and point D. Choose **Dilate** from the Transform menu and dilate by the marked ratio. The dilated images are $\overline{B''D'}$, $\overline{D'B'}$, and D'.

Step 8 Construct three polygon interiors—two triangles and a quadrilateral.

Step 9 Select the three polygon interiors and dilate them by the marked ratio. Repeat this process two or three times.

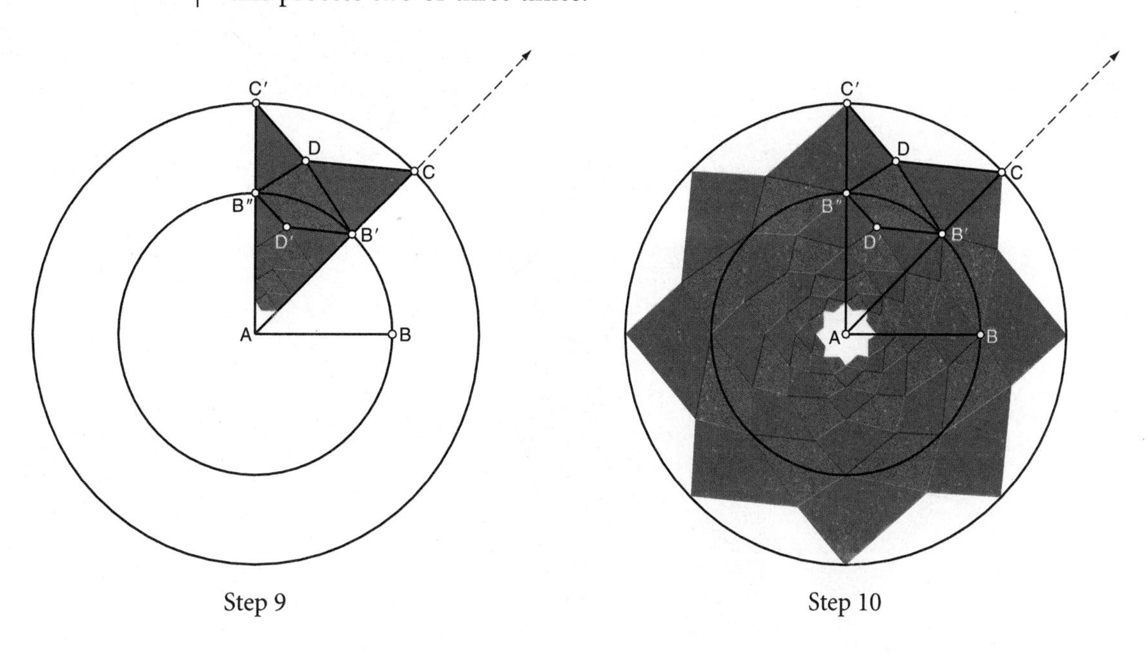

Step 9   Step 10

Step 10 Select all the polygon interiors in the sector and rotate them by an angle of 45°. Repeat the rotation by 45° until you've gone all the way around the circle.

Now you have a design that has the same basic mathematical properties as Escher's *Path of Life I.*

Step 11 Experiment with changing the design by moving different points. Answer these questions.

a. What locations of point D result in both rotational and reflectional symmetry?

b. Drag point C away from point A. What does this do to the numerical dilation ratio? What effect does that have on the geometric figure?

c. Drag point C toward point A. What happens when the circle defined by point C becomes smaller than the circle defined by point B? Why?

Experiment with other dilation-rotation designs of your own. Try different angles of rotation or different polygons. Here are some examples.

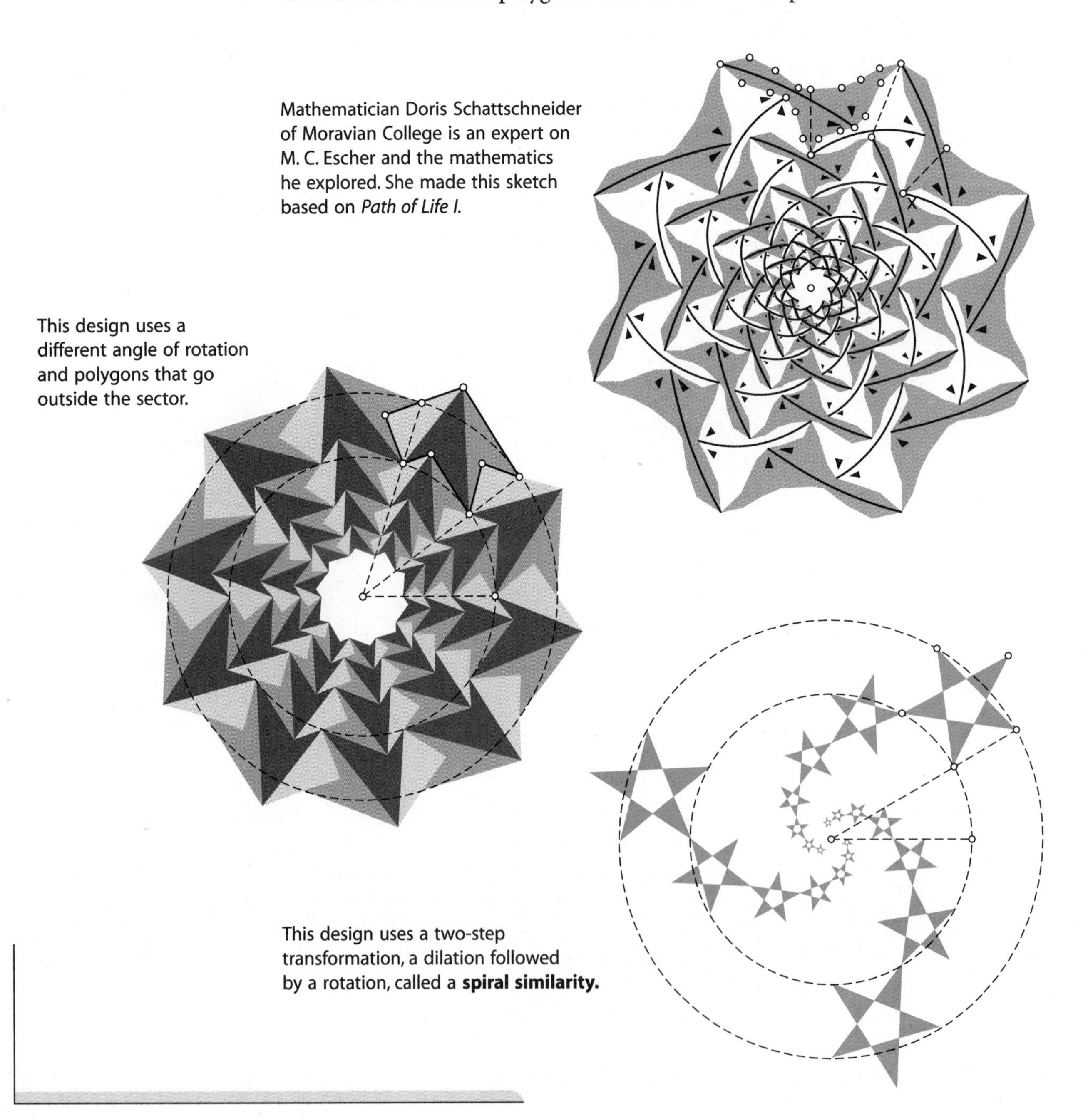

Mathematician Doris Schattschneider of Moravian College is an expert on M. C. Escher and the mathematics he explored. She made this sketch based on *Path of Life I*.

This design uses a different angle of rotation and polygons that go outside the sector.

This design uses a two-step transformation, a dilation followed by a rotation, called a **spiral similarity.**

LESSON

11.3

Indirect Measurement with Similar Triangles

Never be afraid to sit awhile and think.

LORRAINE HANSBERRY

You can use similar triangles to calculate the height of tall objects that you can't reach. This is called **indirect measurement.** One method uses mirrors. Try it in the next investigation.

Investigation
Mirror, Mirror

You will need

- metersticks
- masking tape or a soluble pen
- a mirror

Choose a tall object with a height that would be difficult to measure directly, such as a football goalpost, a basketball hoop, a flagpole, or the height of your classroom.

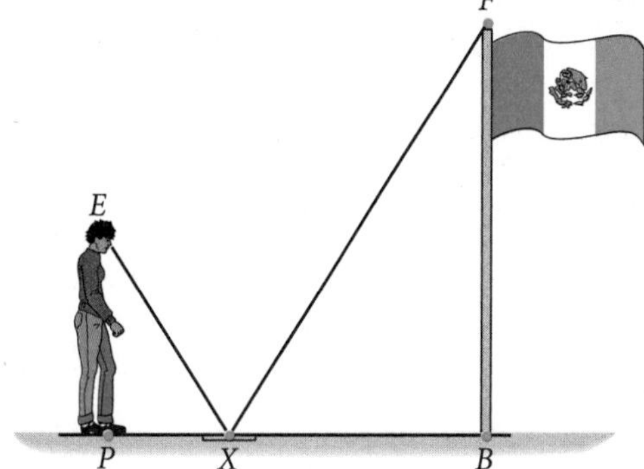

Step 1 Mark crosshairs on your mirror. Use tape or a soluble pen. Call the intersection point X. Place the mirror on the ground several meters from your object.

Step 2 An observer should move to a point P in line with the object and the mirror in order to see the reflection of an identifiable point F at the top of the object at point X on the mirror. Make a sketch of your setup, like this one.

Step 3 Measure the distance PX and the distance from X to a point B at the base of the object directly below F. Measure the distance from P to the observer's eye level, E.

Step 4 Think of $\overline{FX}$ as a light ray that bounces back to the observer's eye along $\overline{XE}$. Why is $\angle B \cong \angle P$? Name two similar triangles. Tell why they are similar.

Step 5 Set up a proportion using corresponding sides of similar triangles. Use it to calculate FB, the approximate height of the tall object.

Step 6 Write a summary of what you and your group did in this investigation. Discuss possible causes for error.

Another method of indirect measurement uses shadows.

EXAMPLE

A person 5 feet 3 inches tall casts a 6-foot shadow. At the same time of day, a lamppost casts an 18-foot shadow. What is the height of the lamppost?

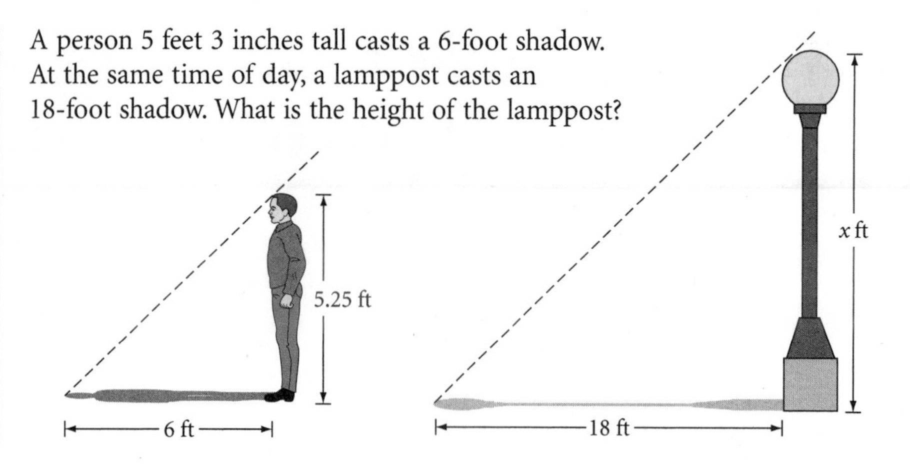

▶ Solution

The light rays that create the shadows hit the ground at congruent angles. Assuming both the person and the lamppost are perpendicular to the ground, you have similar triangles by the AA Similarity Conjecture. Solve a proportion that relates corresponding lengths.

$$\frac{5.25}{6} = \frac{x}{18}$$

$$18 \cdot \frac{5.25}{6} = x$$

$$15.75 = x$$

The height of the lamppost is 15 feet 9 inches.

EXERCISES

You will need

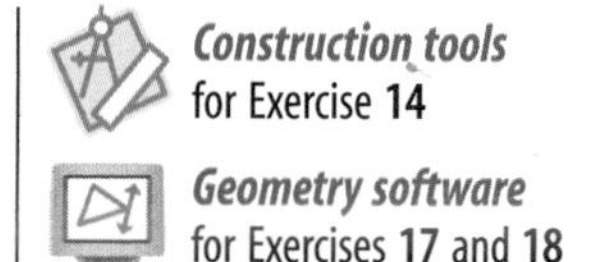
Construction tools for Exercise **14**

Geometry software for Exercises **17** and **18**

1. A flagpole 4 meters tall casts a 6-meter shadow. At the same time of day, a nearby building casts a 24-meter shadow. How tall is the building?

2. Five-foot-tall Melody casts an 84-inch shadow. How tall is her friend if, at the same time of day, his shadow is 1 foot shorter than hers?

3. A 10 m rope from the top of a flagpole reaches to the end of the flagpole's 6 m shadow. How tall is the nearby football goalpost if, at the same moment, it has a shadow of 4 m? ⓗ

4. Private eye Samantha Diamond places a mirror on the ground between herself and an apartment building and stands so that when she looks into the mirror, she sees into a window. The mirror's crosshairs are 1.22 meters from her feet and 7.32 meters from the base of the building. Sam's eye is 1.82 meters above the ground. How high is the window?

5. **APPLICATION** Juanita, who is 1.82 meters tall, wants to find the height of a tree in her backyard. From the tree's base, she walks 12.20 meters along the tree's shadow to a position where the end of her shadow exactly overlaps the end of the tree's shadow. She is now 6.10 meters from the end of the shadows. How tall is the tree?

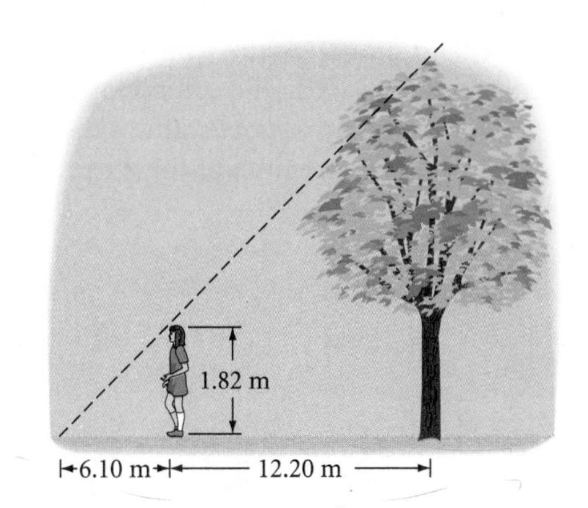

6. While vacationing in Egypt, the Greek mathematician Thales calculated the height of the Great Pyramid. According to legend, Thales placed a pole at the tip of the pyramid's shadow and used similar triangles to calculate its height. This involved some estimating since he was unable to measure the distance from directly beneath the height of the pyramid to the tip of the shadow. From the diagram, explain his method. Calculate the height of the pyramid from the information given in the diagram.

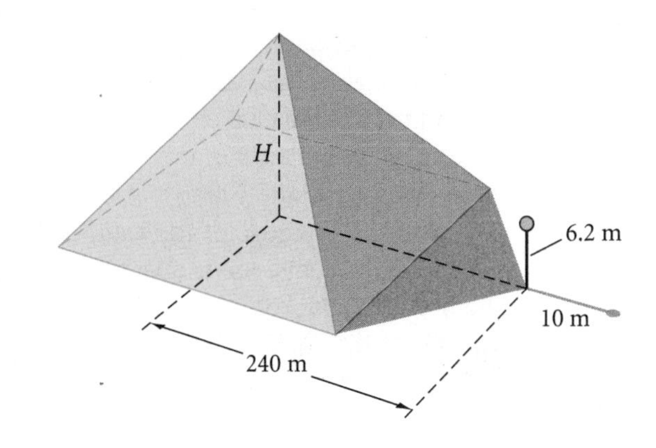

7. Calculate the distance across this river, PR, by sighting a pole, at point P, on the opposite bank. Points R and O are collinear with point P. Point C is chosen so that $\overline{OC} \perp \overline{PO}$. Lastly, point E is chosen so that P, E, and C are collinear and that $\overline{RE} \perp \overline{PO}$. Also explain why $\triangle PRE \sim \triangle POC$. ⓗ

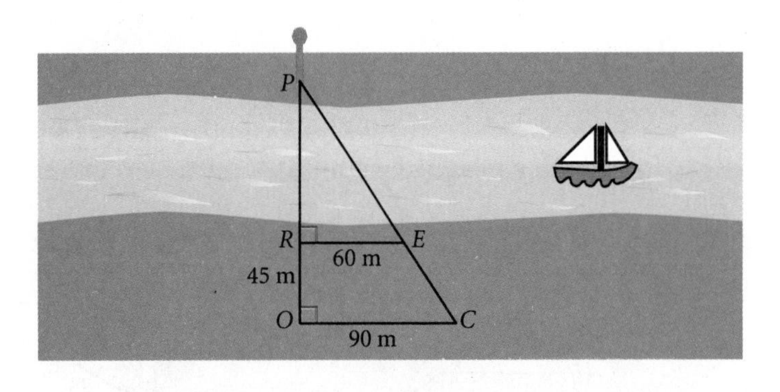

8. A pinhole camera is a simple device. Place unexposed film at one end of a shoe box, and make a pinhole at the opposite end. When light comes through the pinhole, an inverted image is produced on the film. Suppose you take a picture of a painting that is 30 cm wide by 45 cm high with a pinhole box camera that is 20 cm deep. How far from the painting should the pinhole be to make an image that is 2 cm wide by 3 cm high? Sketch a diagram of this situation. ⓗ

9. **APPLICATION** A guy wire attached to a high tower needs to be replaced. The contractor does not know the height of the tower or the length of the wire. Find a method to measure the length of the wire indirectly. ⓗ

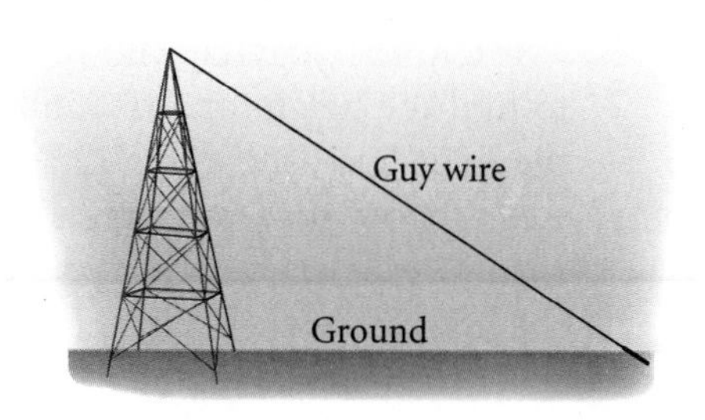

10. Kristin has developed a new method for indirectly measuring the height of her classroom. Her method uses string and a ruler. She tacks a piece of string to the base of the wall and walks back from the wall holding the other end of the string to her eye with her right hand. She holds a 12-inch ruler parallel to the wall in her left hand and adjusts her distance to the wall until the bottom of the ruler is in line with the bottom edge of the wall and the top of the ruler is in line with the top edge of the wall. Now with two measurements, she is able to calculate the height of the room. Explain her method. If the distance from her eye to the bottom of the ruler is 23 inches and the distance from her eye to the bottom of the wall is 276 inches, calculate the height of the room.

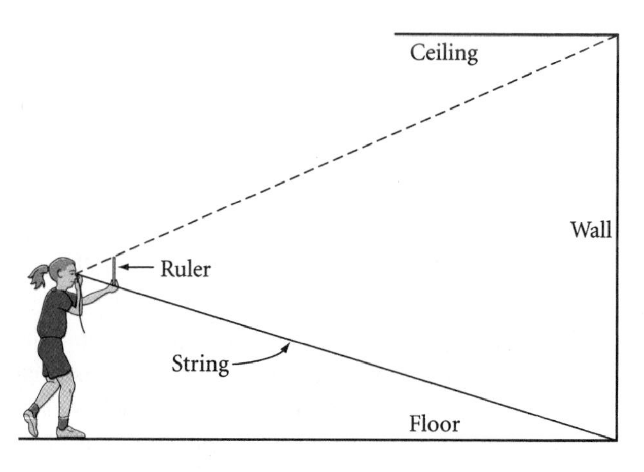

Review

For Exercises 11–13, first identify similar triangles and explain why they are similar. Then find the missing lengths.

11. Find x. ⓗ

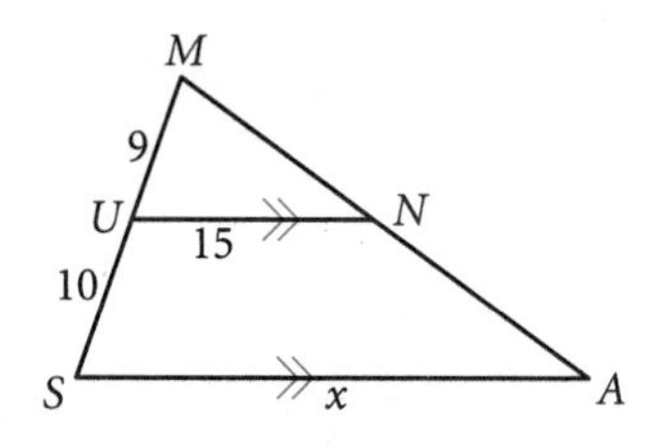

12. Find y.

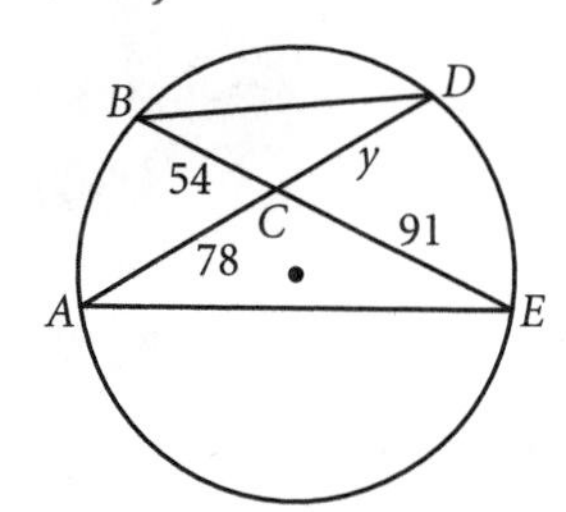

13. Find x, y, and h. ⓗ

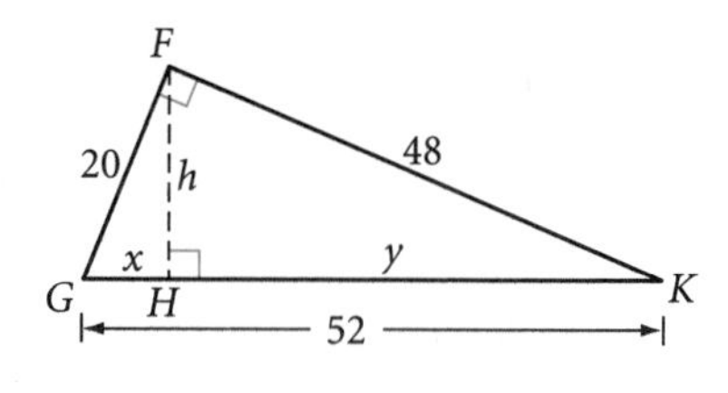

14. ***Construction*** Draw an obtuse triangle.

a. Use a compass and straightedge to construct two altitudes.

b. Use a ruler to measure both altitudes and their corresponding bases.

c. Calculate the area using both altitude-base pairs. Compare your results.

15. Find the radius of the circle.

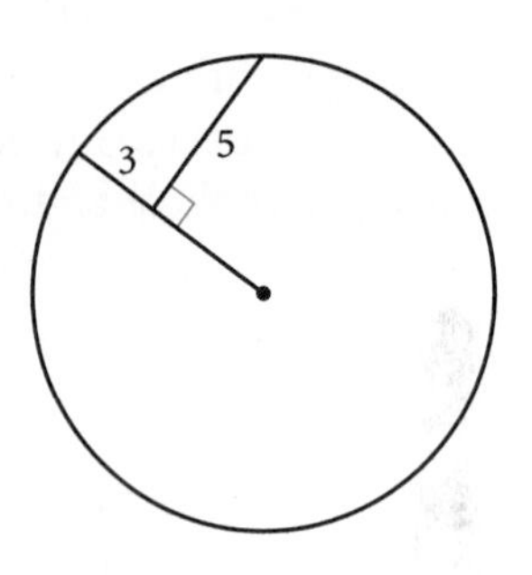

16. Give the vertex arrangement of each tessellation.

a.

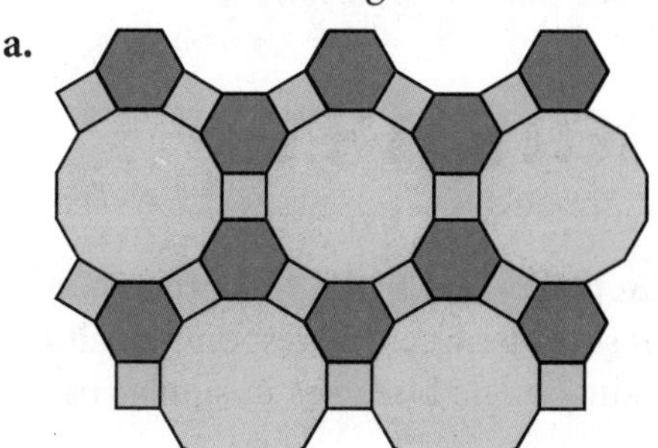

b.

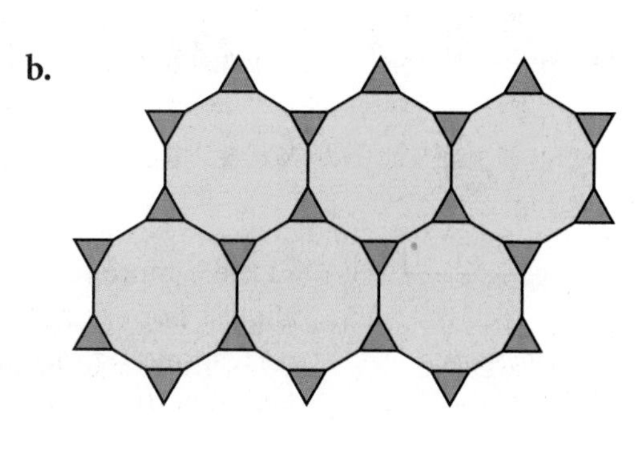

17. ***Technology*** On a segment AB, point X is called the **golden cut** if $\frac{AB}{AX} = \frac{AX}{XB}$, where $AX > XB$. The **golden ratio** is the value of $\frac{AB}{AX}$ and $\frac{AX}{XB}$ when they are equal. Use geometry software to explore the location of the golden cut on any segment AB. What is the value of the golden ratio? Find a way to construct the golden cut. ⓗ

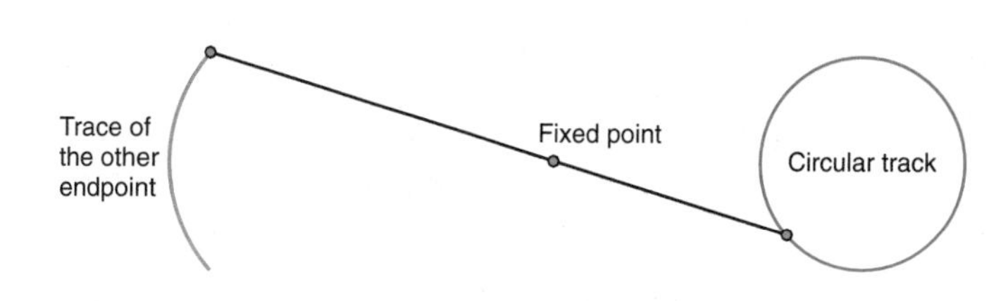

18. ***Technology*** Imagine that a rod of a given length is attached at one end to a circular track and passes through a fixed pivot point. As one endpoint moves around the circular track, the other endpoint traces a curve.

a. Predict what type of curve will be traced.

b. Model this situation with geometry software. Describe the curve that is traced.

c. Experiment with changing the size of the circular track, the length of the rod, or the location of the pivot point. Describe your results.

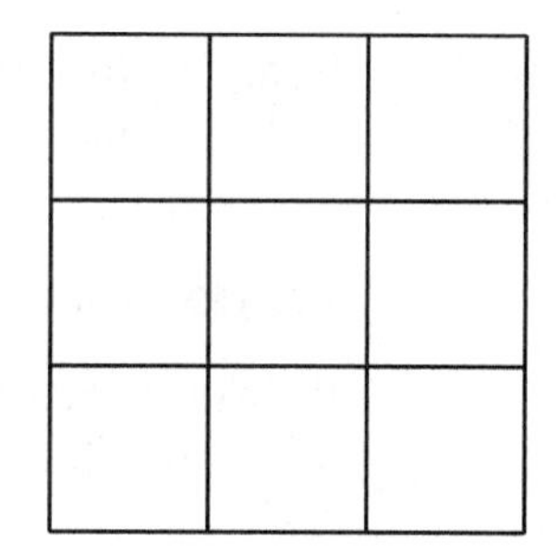

IMPROVING YOUR VISUAL THINKING SKILLS

TIC-TAC-NO!

Is it possible to shade six of the nine squares of a 3-by-3 grid so that no three of the shaded squares are in a straight line (row, column, or diagonal)?

LESSON

11.4 Corresponding Parts of Similar Triangles

Big doesn't necessarily mean better. Sunflowers aren't better than violets.

EDNA FERBER

Is there more to similar triangles than just proportional sides and congruent angles? For example, are there relationships between corresponding altitudes, corresponding medians, or corresponding angle bisectors in similar triangles? Let's investigate.

Investigation 1
Corresponding Parts

You will need

- a compass
- a straightedge

Use unlined paper for this investigation.

Step 1 Draw any triangle and construct a triangle of a different size similar to it. State the scale factor you used.

Step 2 Construct a pair of corresponding altitudes and use your compass to compare their lengths. How do they compare? How does the comparison relate to the scale factor you used?

Step 3 Construct a pair of corresponding medians. How do their lengths compare?

Step 4 Construct a pair of corresponding angle bisectors. How do their lengths compare?

Step 5 Compare your results with the results of others near you. You should be ready to make a conjecture.

Proportional Parts Conjecture C-96

If two triangles are similar, then the corresponding _?_, _?_, and _?_ are _?_ to the corresponding sides.

The discovery you made in the investigation probably seems very intuitive. Let's see how you can prove one part of your conjecture. You will prove the other two parts in the exercises.

EXAMPLE Prove that corresponding medians of similar triangles are proportional to corresponding sides.

▸ Solution

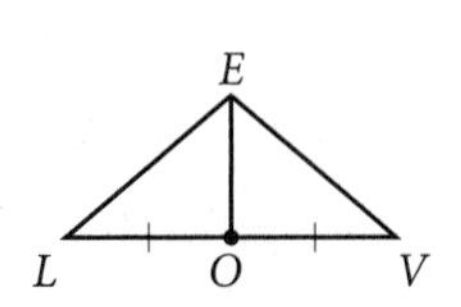

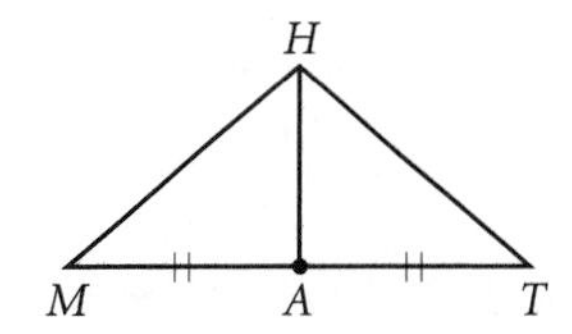

Consider similar triangles $\triangle LVE$ and $\triangle MTH$ with corresponding medians $\overline{EO}$ and $\overline{HA}$. You need to show that the corresponding medians are proportional to corresponding sides, for example $\frac{EO}{HA} = \frac{EL}{HM}$. If you show that $\triangle LOE \sim \triangle MAH$ then you can show that $\frac{EO}{HA} = \frac{EL}{HM}$.

If you accept the SAS Similarity Conjecture as true, then you can show that $\triangle LOE \sim \triangle MAH$. You already know that $\angle L \cong \angle M$. Use algebra to show that $\frac{EL}{HM} = \frac{LO}{MA}$.

$\frac{EL}{HM} = \frac{LV}{MT}$ — Corresponding sides of similar triangles $\triangle LVE$ and $\triangle MTH$ are proportional.

$\frac{EL}{HM} = \frac{LO + OV}{MA + AT}$ — $LV = LO + OV$ and $MT = MA + AT$. Substitute.

$\frac{EL}{HM} = \frac{LO + LO}{MA + MA}$ — Since $\overline{EO}$ and $\overline{HA}$ are medians, O and A are midpoints. Since O and A are midpoints, $OV = LO$ and $AT = MA$. Substitute.

$\frac{EL}{HM} = \frac{2LO}{2MA}$ — Add.

$\frac{EL}{HM} = \frac{LO}{MA}$ — Reduce.

So, $\triangle LOE \sim \triangle MAH$ by the SAS Similarity Conjecture. Therefore you can also set up the proportion $\frac{EO}{HA} = \frac{EL}{HM}$, which shows that the corresponding medians are proportional to corresponding sides.

Recall when you first saw an angle bisector in a triangle. You may have thought that the bisector of an angle in a triangle divides the opposite side into two equal parts as well. A counterexample shows that this is not necessarily true. In $\triangle ROE$, $\overline{RT}$ bisects $\angle R$, but point T does not bisect $\overline{OE}$.

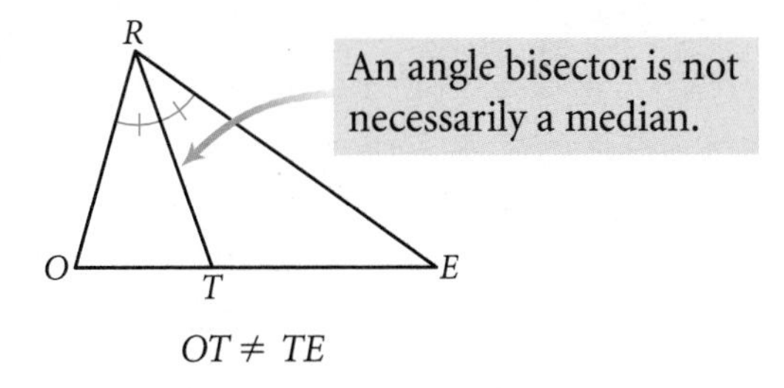

The angle bisector does, however, divide the opposite side in a particular way.

Investigation 2
Opposite Side Ratios

You will need

- a compass
- a ruler

In this investigation you'll discover that there is a proportional relationship involving angle bisectors.

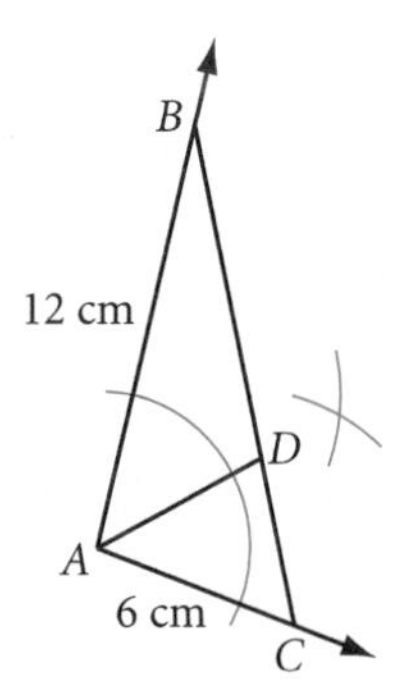

Step 1 Draw any angle. Label it A.

Step 2 On one ray, locate point C so that AC is 6 cm. Use the same compass setting and locate point B on the other ray so that AB is 12 cm. Draw $\overline{BC}$ to form $\triangle ABC$.

Step 3 Construct the bisector of $\angle A$. Locate point D where the bisector intersects side $\overline{BC}$.

Step 4 Measure and compare CD and BD.

Step 5 Calculate and compare the ratios $\frac{CA}{BA}$ and $\frac{CD}{BD}$.

Step 6 Repeat Steps 1–5 with $AC = 10$ cm and $AB = 15$ cm.

Step 7 Compare your results with the results of others near you. State a conjecture.

Angle Bisector/Opposite Side Conjecture

C-97

A bisector of an angle in a triangle divides the opposite side into two segments whose lengths are in the same ratio as _?_.

EXERCISES

You will need

Construction tools for Exercises **16** and **23**

For Exercises 1–13, use your new conjectures. All measurements are in centimeters.

1. $\triangle ICE \sim \triangle AGE$
$h =$ _?_

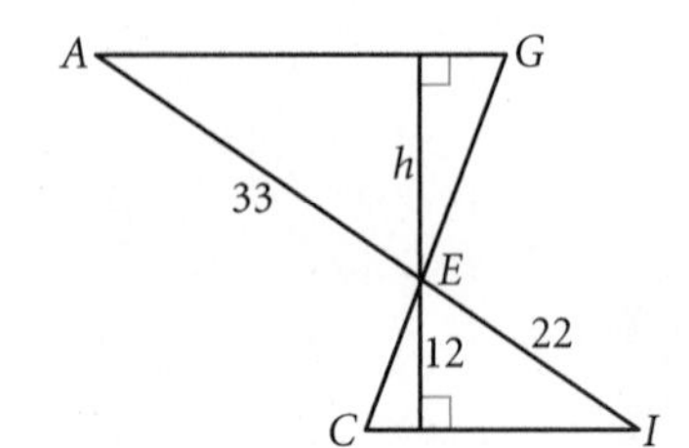

2. $\triangle SKI \sim \triangle JMP$
$x =$ _?_

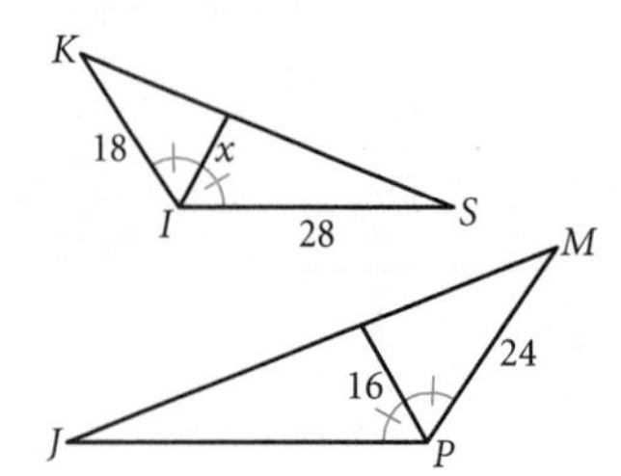

3. $\triangle PIE \sim \triangle SIC$
Point S is the midpoint of PI.
$CL =$ _?_, $CS =$ _?_ ⓗ

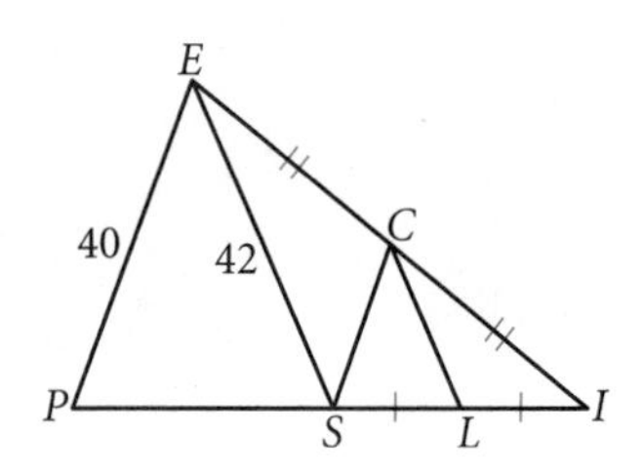

4. $\triangle CAP \sim \triangle DAY$
$FD = \underline{?}$

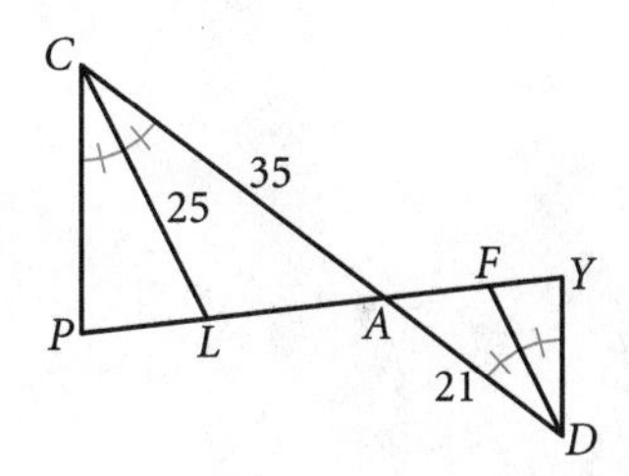

5. $\triangle HAT \sim \triangle CLD$
$x = \underline{?}$

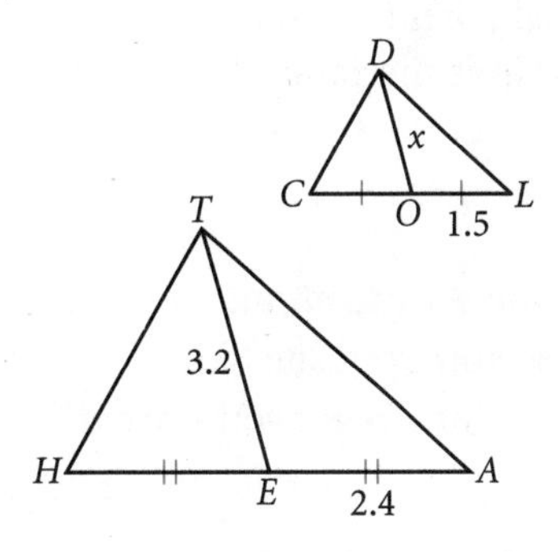

6. $\triangle ARM \sim \triangle LEG$
Area of $\triangle ARM = \underline{?}$
Area of $\triangle LEG = \underline{?}$

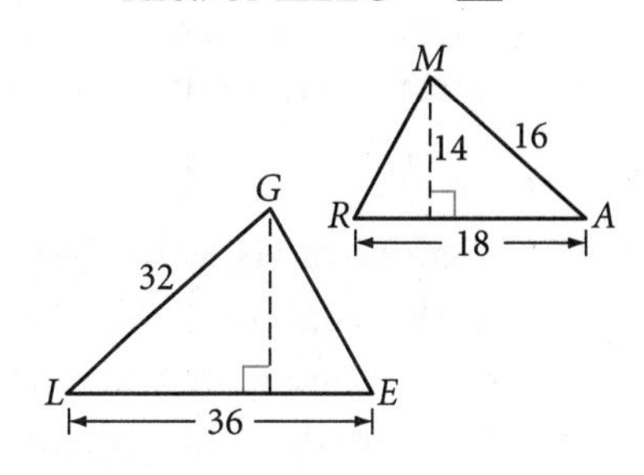

7. $v = \underline{?}$

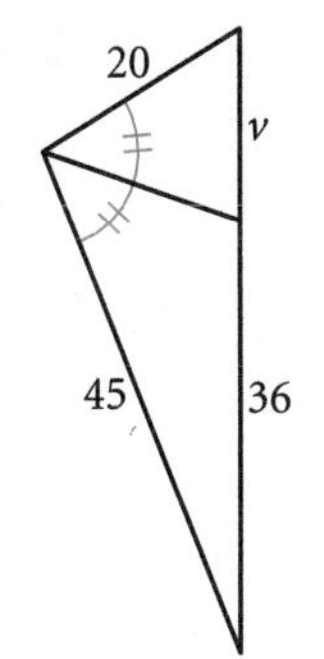

8. $y = \underline{?}$

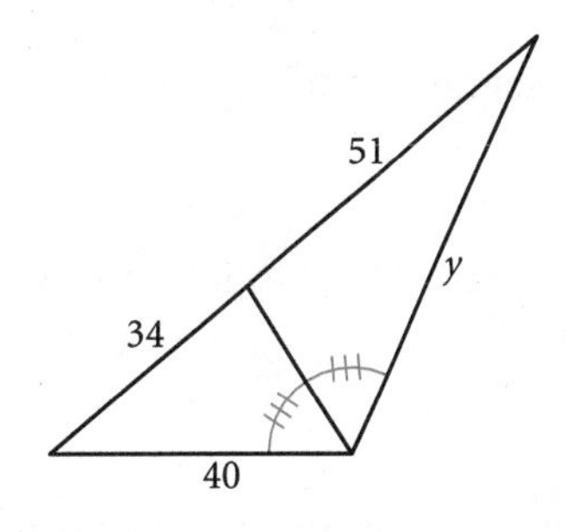

9. $x = \underline{?}$ ⓗ

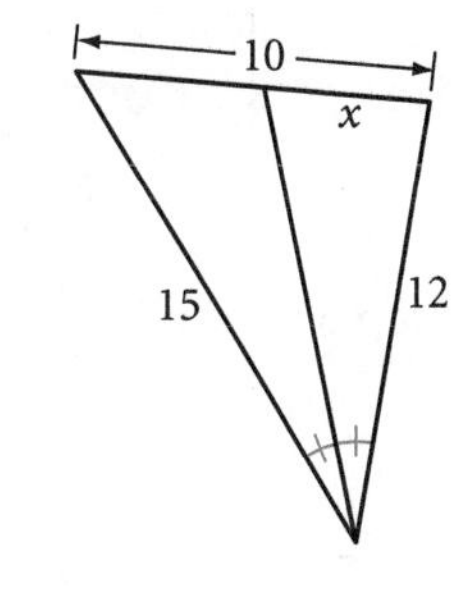

10. $\frac{a}{b} = \underline{?}$, $\frac{a}{p} = \underline{?}$

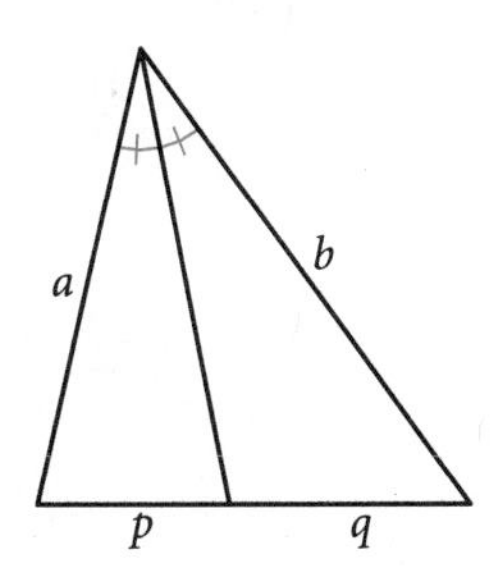

11. $k = \underline{?}$

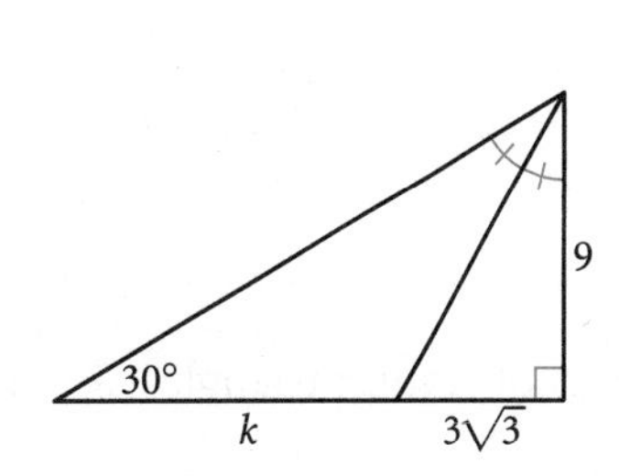

12. $x = \underline{?}$ ⓗ

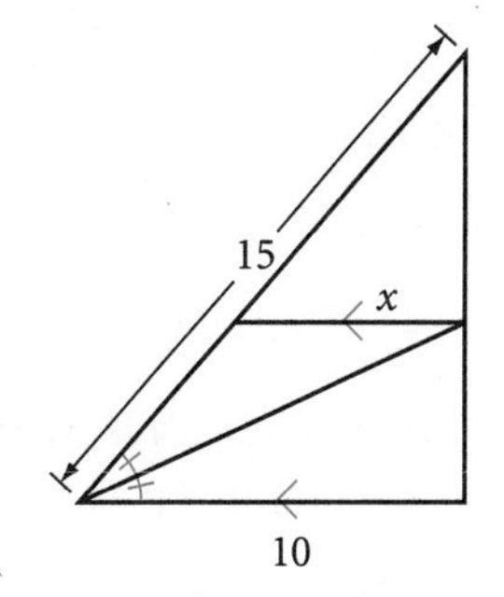

13. $x = \underline{?}$, $y = \underline{?}$

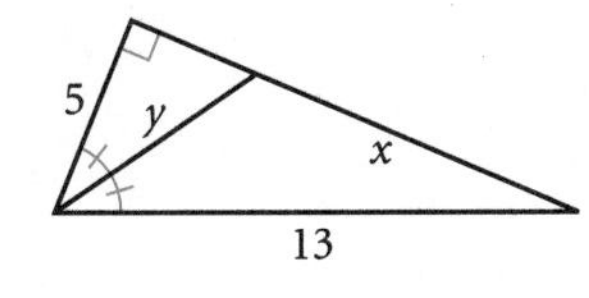

14. Triangle PQR is the image of $\triangle ABC$ under a dilation. Find the coordinates of B and R. Find the ratio $\frac{k}{h}$.

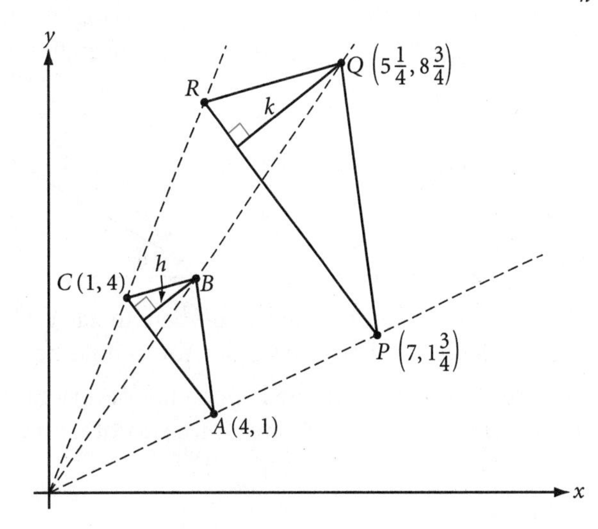

15. Aunt Florence has willed to her two nephews a plot of land in the shape of an isosceles right triangle. The land is to be divided into two unequal parts by bisecting one of the two congruent angles. What is the ratio of the greater area to the lesser area?

16. ***Construction*** How would you divide a segment into lengths with a ratio of $\frac{2}{3}$? The Angle Bisector/Opposite Side Conjecture gives you a way to do this. To get you started, here are the first four steps. ⓗ

Step 1 Construct any segment AB.

A ——— B

Step 2 Construct a second segment. Call its length x.

x

Step 3 Construct two more segments with lengths $2x$ and $3x$.

$2x$

$3x$

Step 4 Construct a triangle with lengths $2x$, $3x$, and AB.

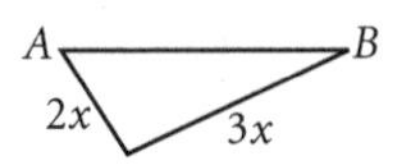

You're on your own from here!

17. Prove that corresponding angle bisectors of similar triangles are proportional to corresponding sides. ⓗ

18. Prove that corresponding altitudes of similar triangles are proportional to corresponding sides.

19. ***Mini-Investigation*** This investigation is in two parts. You will need to complete the conjecture in part a before moving on to part b.

a. The altitude to the hypotenuse has been constructed in each right triangle below. This construction creates two smaller right triangles within each original right triangle. Calculate the measures of the acute triangles in each diagram.

i. $a = \underline{?}$, $b = \underline{?}$, $c = \underline{?}$ **ii.** $a = \underline{?}$, $b = \underline{?}$, $c = \underline{?}$ **iii.** $a = \underline{?}$, $b = \underline{?}$, $c = \underline{?}$

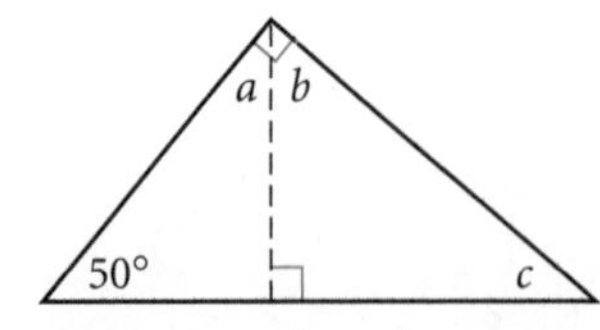

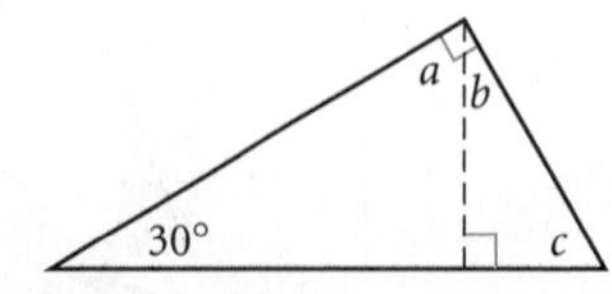

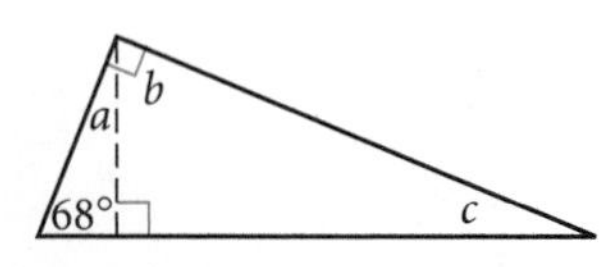

How do the smaller right triangles compare in each diagram? How do they compare to the original right triangle? You should be ready to state a conjecture.

Conjecture: The altitude to the hypotenuse of a right triangle divides the triangle into two right triangles that are $\underline{?}$ to each other and to the original $\underline{?}$.

b. Complete each proportion for these right triangles.

i. $\frac{h}{r} = \frac{s}{?}$

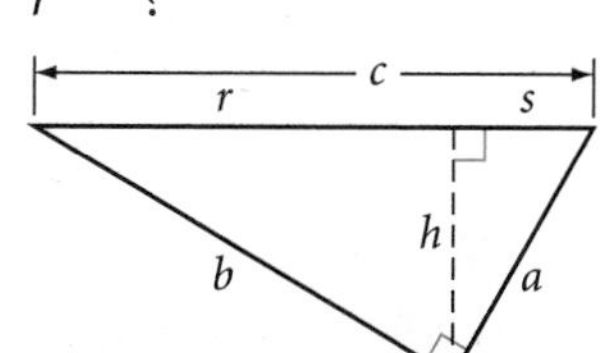

ii. $\frac{y}{h} = \frac{?}{x}$

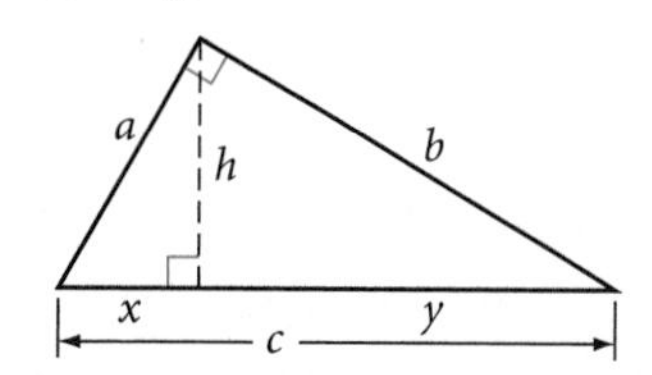

iii. $\frac{n}{h} = \frac{?}{?}$

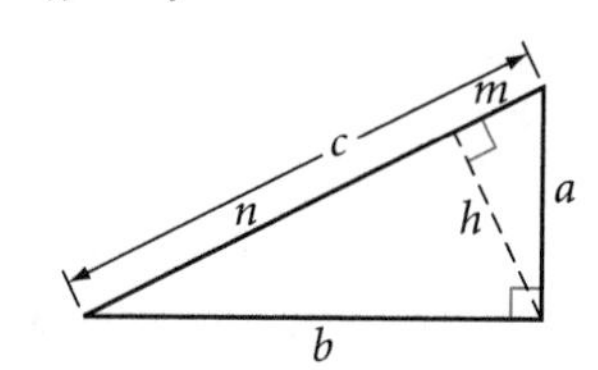

Review the proportions you wrote. How are they alike? You should be ready to state a conjecture.

Conjecture: The altitude (length h) to the hypotenuse of a right triangle divides the hypotenuse into two segments (lengths p and q), such that $\frac{p}{?} = \frac{?}{q}$.

Add these conjectures to your notebook.

Review

20. Use algebra to show that if $\frac{a}{b} = \frac{c}{d}$, then $\frac{a+b}{b} = \frac{c+d}{d}$. ⓗ

21. A rectangle is divided into four rectangles, each similar to the original rectangle. What is the ratio of short side to long side in the rectangles?

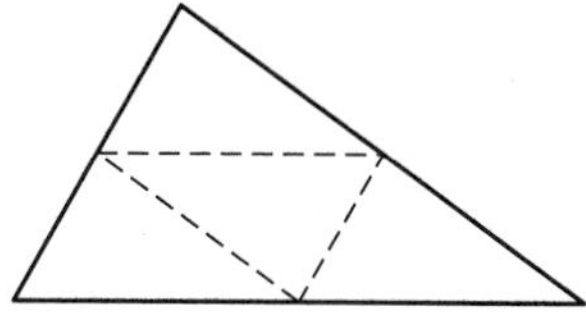

22. In Chapter 5, you discovered that when you construct the three midsegments in a triangle, they divide the triangle into four congruent triangles. Are the four triangles similar to the original? Explain why.

23. ***Construction*** Draw any triangle ABC. Select any point X on $\overline{AB}$. Construct a line through X parallel to $\overline{AC}$ that intersects $\overline{BC}$ in point Y. Find a proportion that relates AX, XB, BY, and YC.

24. A rectangle has sides a and b. For what values of a and b is another rectangle with sides $2a$ and $\frac{b}{2}$

a. Congruent to the original?

b. Equal in perimeter to the original?

c. Equal in area to the original?

d. Similar but not congruent to the original?

25. Find the volume of this truncated cone.

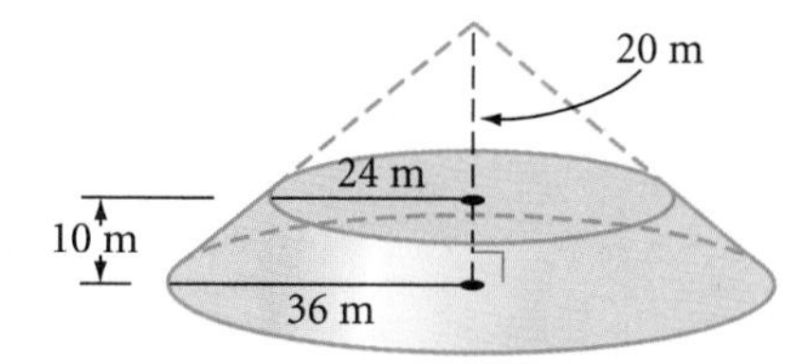

26. The large circles are tangent to the square and tangent to each other. The smaller circle is tangent to each larger circle. Find the radius of the smaller circle in terms of s, the length of each side of the square. ⓗ

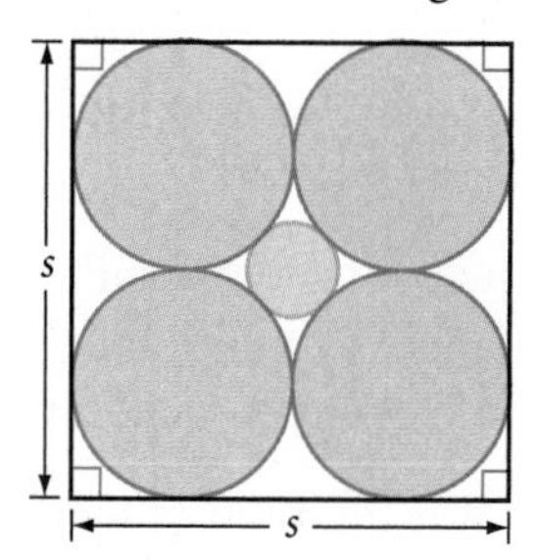

LESSON

11.5

Proportions with Area and Volume

It is easy to show that a hare could not be as large as a hippopotamus, or a whale as small as a herring. For every type of animal there is a most convenient size, and a large change in size inevitably carries with it a change of form.

J. B. S. HALDANE

You can use similarity to find the surface areas and volumes of objects that are geometrically similar. Suppose an artist wishes to gold-plate a sculpture. If it costs \$250 to gold-plate a model that is half as long in each dimension, how much will it cost for the full-size sculpture? Not \$500, but \$1000! If the model weighs 40 pounds, how much will the full-size sculpture weigh? Not 80 pounds, but 320 pounds! In this lesson you will discover why these answers may not be what you expected.

You might recognize this giant golden man as the Academy Awards statuette. Also called the Oscar, smaller versions of the figure are handed out annually for excellence in the motion picture industry. If this statuette were real gold, it would be very expensive and incredibly heavy.

Investigation 1
Area Ratios

You will need

- graph paper

In this investigation you will find the relationship between areas of similar figures.

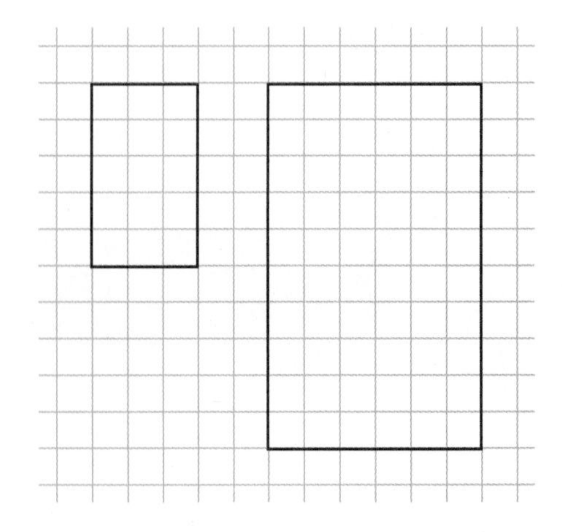

Step 1 Draw a rectangle on graph paper. Calculate its area.

Step 2 Draw a rectangle similar to your first rectangle by multiplying its sides by a scale factor. Calculate this area.

Step 3 What is the ratio of side lengths (larger to smaller) for your two rectangles? What is the ratio of their areas (larger to smaller)?

Step 4 How many copies of the smaller rectangle would you need to fill the larger rectangle? Draw lines in your larger rectangle to show how you would place the copies to fill the area.

Step 5 Compare your results with the results of others near you.

Step 6 Repeat Steps 1–5 using triangles instead of rectangles.

Step 7 Discuss whether or not your findings would apply to any pair of similar polygons. Would they apply to similar circles or other curved figures? You should be ready to state a conjecture.

Proportional Areas Conjecture

C-98

If corresponding sides of two similar polygons or the radii of two circles compare in the ratio $\frac{m}{n}$, then their areas compare in the ratio __?__.

Similar solids are solids that have the same shape but not necessarily the same size. All cubes are similar, but not all prisms are similar. All spheres are similar, but not all cylinders are similar. Two polyhedrons are similar if all their corresponding faces are similar and the lengths of their corresponding edges are proportional. Two right cylinders (or right cones) are similar if their radii and heights are proportional.

EXAMPLE A Are these right rectangular prisms similar?

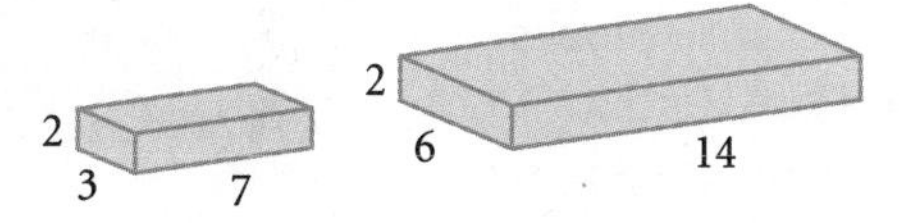

▶ ***Solution*** The two prisms are not similar because the corresponding edges are not proportional.

$$\frac{2}{2} \neq \frac{3}{6} = \frac{7}{14}$$

EXAMPLE B Are these right circular cones similar?

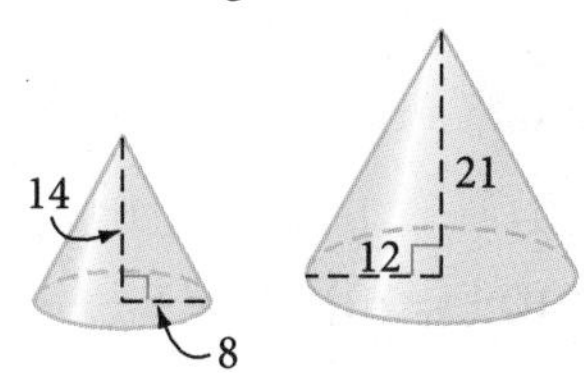

▶ ***Solution*** The two cones are similar because the radii and heights are proportional.

$$\frac{8}{12} = \frac{14}{21}$$

Investigation 2
Volume Ratios

You will need

- interlocking cubes

How does the ratio of lengths of corresponding edges of similar solids compare with the ratio of their volumes? Let's find out.

Step 1 Use blocks to build the "snake." Calculate its volume.

Step 2 Build a similar snake by multiplying every dimension by a scale factor of 3. Calculate this volume.

Step 3 What is the ratio of side lengths (larger to smaller) for your two snakes? What is the ratio of volumes (larger to smaller)?

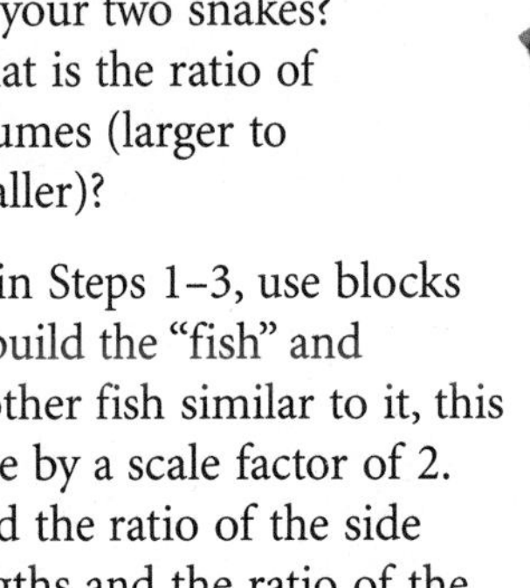

Step 4 As in Steps 1–3, use blocks to build the "fish" and another fish similar to it, this time by a scale factor of 2. Find the ratio of the side lengths and the ratio of the volumes.

Step 5 How do your results compare with the results in Investigation 1? Discuss how you would calculate the volume of a snake increased by a scale factor of 5. Discuss how you would calculate the volume of a fish increased by a scale factor of 4.

Step 6 You should be ready to state a conjecture.

Proportional Volumes Conjecture C-99

If corresponding edges (or radii, or heights) of two similar solids compare in the ratio $\frac{m}{n}$, then their volumes compare in the ratio $\underline{\ ?\ }$.

EXERCISES

1. $\triangle CAT \sim \triangle MSE$
Area of $\triangle CAT = 72\text{ cm}^2$
Area of $\triangle MSE = \underline{\ ?\ }$ (h)

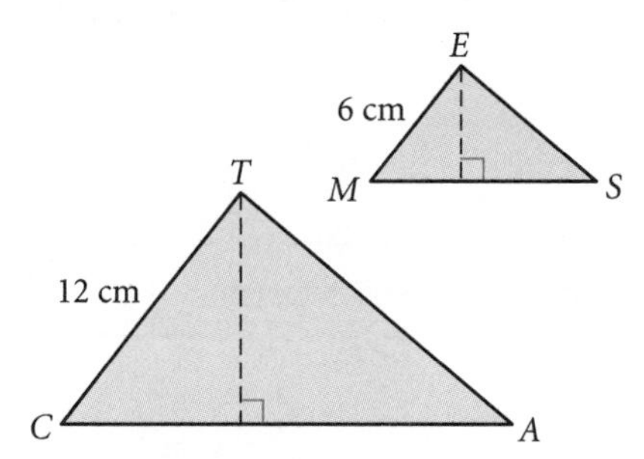

2. $RECT \sim ANGL$
$\dfrac{\text{Area of } RECT}{\text{Area of } ANGL} = \dfrac{9}{16}$
$TR = \underline{\ ?\ }$

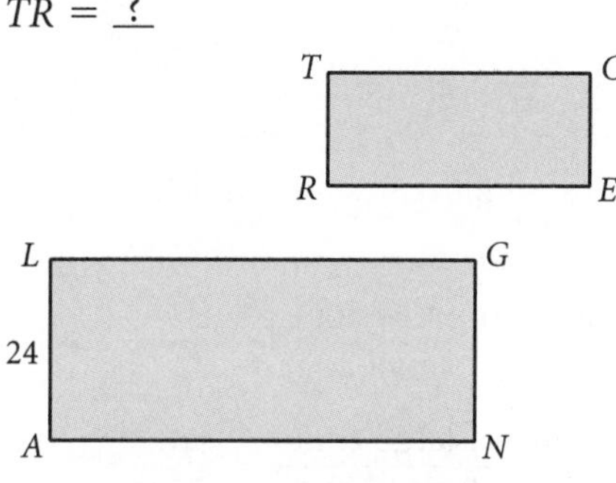

3. $TRAP \sim ZOID$ (h)
$\dfrac{\text{Area of } ZOID}{\text{Area of } TRAP} = \dfrac{16}{25}$
$a = \underline{\ ?\ }$, $b = \underline{\ ?\ }$

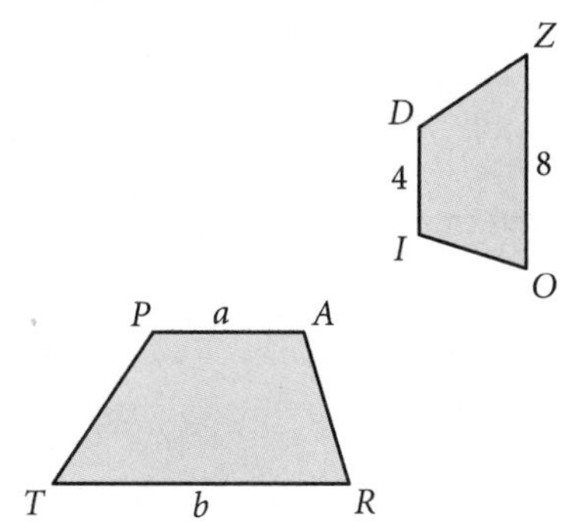

4. semicircle $R \sim$ semicircle S
$\dfrac{r}{s} = \dfrac{3}{5}$
Area of semicircle $S = 75\pi\text{ cm}^2$
Area of semicircle $R = \underline{\ ?\ }$

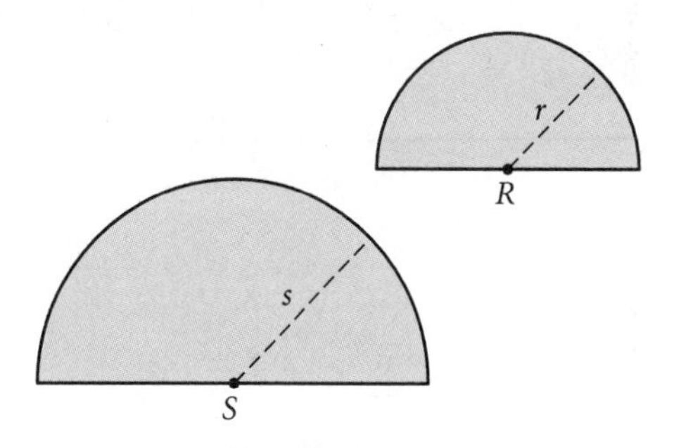

5. The ratio of the lengths of corresponding diagonals of two similar kites is $\frac{1}{7}$. What is the ratio of their areas?

6. The ratio of the areas of two similar trapezoids is $\frac{1}{9}$. What is the ratio of the lengths of their altitudes?

7. The ratio of the lengths of the edges of two cubes is $\frac{m}{n}$. What is the ratio of their surface areas? (h)

8. The celestial sphere shown at right has radius 9 inches. The planet in the sphere's center has radius 3 inches. What is the ratio of the volume of the planet to the volume of the celestial sphere? What is the ratio of the surface area of the planet to the surface area of the celestial sphere?

9. **APPLICATION** Annie works in a magazine's advertising department. A client has requested that his 5 cm-by-12 cm ad be enlarged: "Double the length and double the width, then send me the bill." The original ad cost \$1500. How much should Annie charge for the larger ad? Explain your reasoning.

10. The pentagonal pyramids are similar.

$\frac{h}{H} = \frac{4}{7}$

Volume of large pyramid = $\underline{?}$

Volume of small pyramid = 320 cm^3

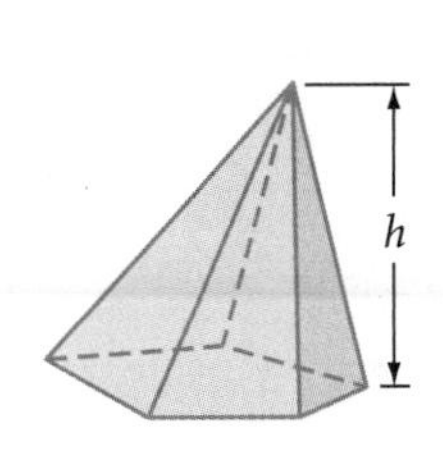

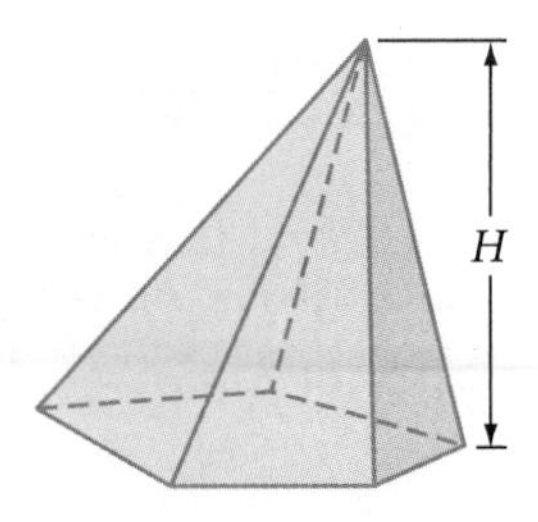

11. These right cones are similar.

$H = \underline{?}$, $h = \underline{?}$

Volume of large cone = $\underline{?}$

Volume of small cone = $\underline{?}$

$\frac{\text{Volume of large cone}}{\text{Volume of small cone}} = \underline{?}$

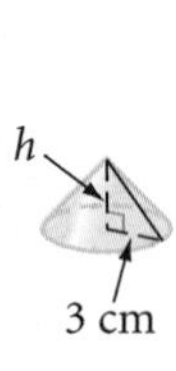

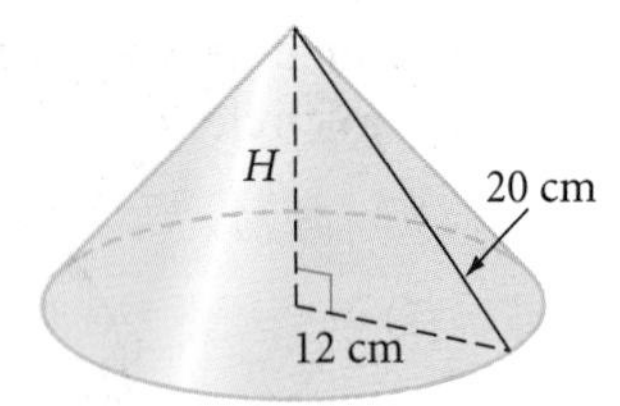

12. These right trapezoidal prisms are similar.

Volume of small prism = 324 cm^3

$\frac{\text{Area of base of small prism}}{\text{Area of base of large prism}} = \frac{9}{25}$

$\frac{h}{H} = \underline{?}$

$\frac{\text{Volume of large prism}}{\text{Volume of small prism}} = \underline{?}$

Volume of large prism = $\underline{?}$ ⓗ

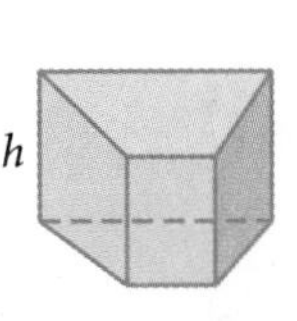

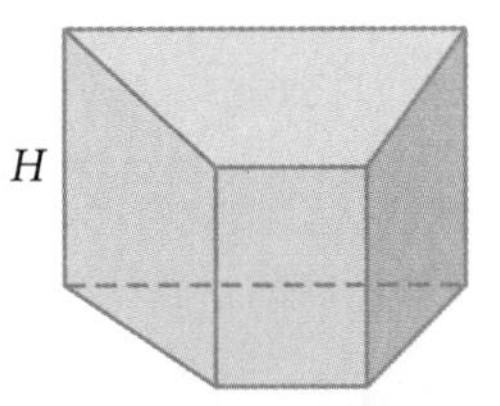

13. These right cylinders are similar.

Volume of large cylinder = 4608π ft^3

Volume of small cylinder = $\underline{?}$

$\frac{\text{Volume of large cylinder}}{\text{Volume of small cylinder}} = \underline{?}$

$H = \underline{?}$

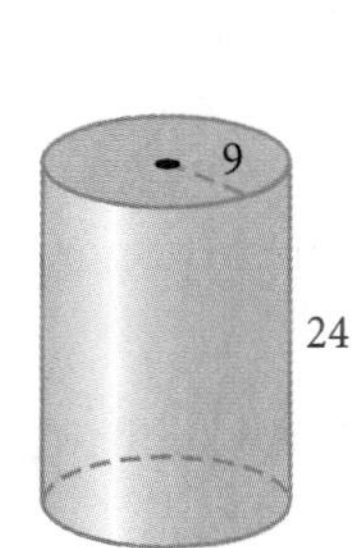

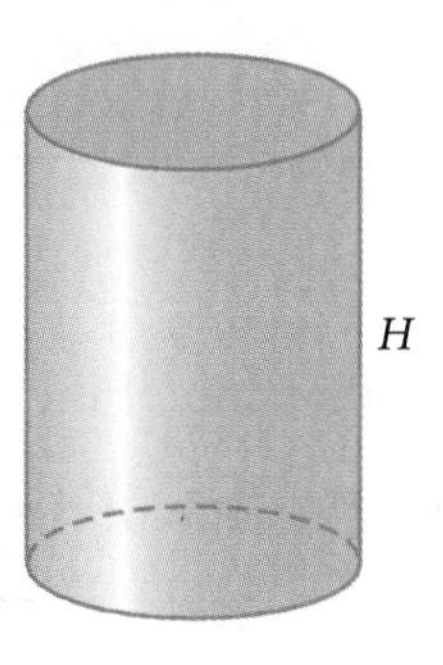

14. The ratio of the lengths of corresponding edges of two similar triangular prisms is $\frac{5}{3}$. What is the ratio of their volumes?

15. The ratio of the volumes of two similar pentagonal prisms is $\frac{8}{125}$. What is the ratio of their heights?

16. The ratio of the weights of two spherical steel balls is $\frac{8}{27}$. What is the ratio of their diameters?

17. **APPLICATION** The energy (and cost) needed to operate an air conditioner is proportional to the volume of the space that is being cooled. It costs ZAP Electronics about \$125 per day to run an air conditioner in their small rectangular warehouse. The company's large warehouse, a few blocks away, is 2.5 times as long, wide, and high as the small warehouse. Estimate the daily cost of cooling the large warehouse with the same model of air conditioner. ⓗ

18. APPLICATION A sculptor creates a small bronze statue that weighs 38 lb. She plans to make a version that will be four times as large in each dimension. How much will this larger statue weigh if it is also bronze?

This bronze sculpture by Camille Claudel (1864–1943) is titled *La Petite Chatelaine.* Claudel was a notable French artist and student of Auguste Rodin, whose famous sculptures include *The Thinker.*

19. A tabloid magazine at a supermarket checkout exclaims, "Scientists Breed 4-Foot Tall Chicken." A photo shows a giant chicken that supposedly weighs 74 pounds and will solve the world's hunger problem. What do you think about this headline? Assuming an average chicken stands 14 inches tall and weighs 7 pounds, would a 4-foot chicken weigh 74 pounds? Is it possible for a chicken to be 4 feet tall? Explain your reasoning.

20. The African goliath frog shown in this photo is the largest known frog—about 0.3 m long and 3.2 kg in weight. The Brazilian gold frog is one of the smallest known frogs—about 9.8 mm long. Approximate the weight of a gold frog. What assumptions do you need to make? Explain your reasoning.

Review

21. Make four copies of the trapezoid at right. Arrange them into a similar but larger trapezoid. Sketch the final trapezoid and show how the smaller trapezoids fit inside it.

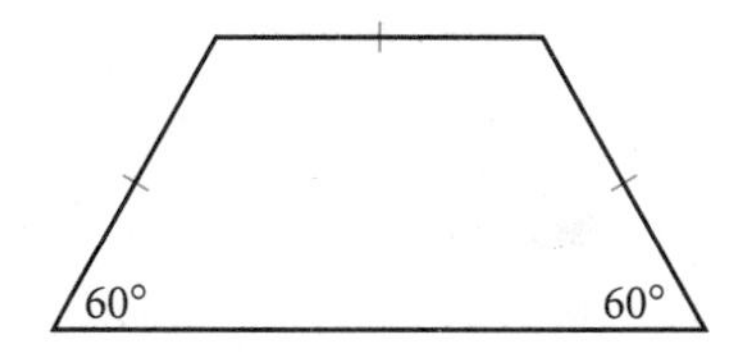

22. Sara rents her goat, Munchie, as a lawn mower. Munchie is tied to a stake with a 10 m rope. Sara wants to find an efficient pattern for Munchie's stake positions so that all grass in a field 42 m by 42 m is mowed but overlap is minimized. Make a sketch showing all the stake positions needed.

23. $x = \underline{?}$, $y = \underline{?}$

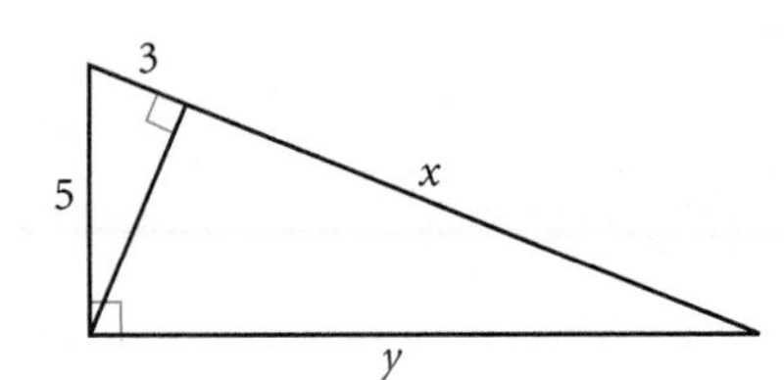

24. $XY = \underline{?}$

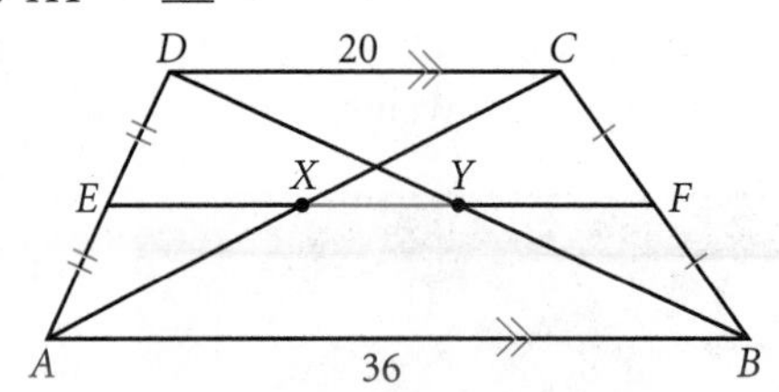

25. Find the area of a regular decagon with an apothem 5.7 cm and a perimeter 37 cm.

26. Find the area of a triangle whose sides measure 13 feet, 13 feet, and 10 feet. ⓗ

27. True or false? Every cross section of a pyramid has the same shape as, but a different size from the base. ⓗ

28. What's wrong with this picture?

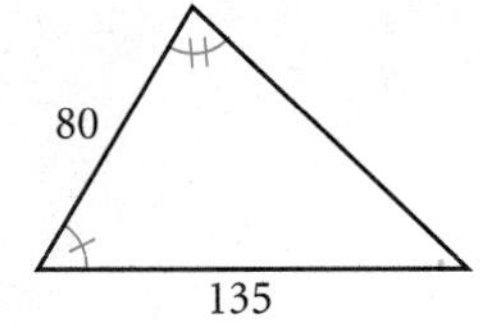

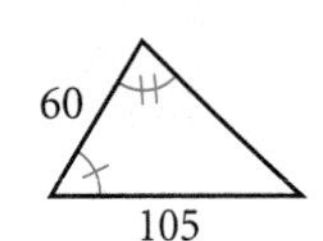

project

IN SEARCH OF THE PERFECT RECTANGLE

A square is a perfectly symmetric quadrilateral. Yet, people rarely make books, posters, or magazines that are square. Instead, most people seem to prefer rectangles. In fact, some people believe that a particular type of rectangle is more appealing because its proportions fit the golden ratio. You can learn more about the historical importance of the golden ratio by doing research with the links at www.keymath.com/DG .

Do people have a tendency to choose a particular length/width ratio when they design or build common objects? Find at least ten different rectangular objects in your classroom or home: books, postcards, desks, doors, and other everyday items. Measure the longer side and the shorter side of each one. Predict what a graph of your data will look like.

Now graph your data. Is there a pattern? Find the line of best fit. How well does it fit the data? What is the range of length/width ratios? What ratio do points on the line of best fit represent?

Your project should include

- A table and graph of your data.
- Your predictions and your analysis.
- A paragraph explaining your opinion about whether or not people have a tendency to choose a particular type of rectangle and why.

Fathom™

You can use Fathom to graph your data and find the line of best fit. Choose different types of graphs to get other insights into what your data mean.

Exploration

Why Elephants Have Big Ears

The relationship between surface area and volume is of critical importance to all living things. It explains why elephants have big ears, why hippos and rhinos have short, thick legs and must spend a lot of time in water, and why movie monsters like King Kong and Godzilla can't exist.

Activity

Convenient Sizes

Body Temperature

Every living thing processes food for energy. This energy creates heat that radiates from its surface.

Step 1 Imagine two similar animals, one with dimensions three times as large as those of the other.

Step 2 How would the surface areas of these two animals compare? How much more heat could the larger animal radiate through its surface?

Step 3 How would the volumes of these two animals compare? If the animals' bodies produce energy in proportion to their volumes, how many times as much heat would the larger animal produce?

Step 4 Review your answers from Steps 2 and 3. How many times as much heat must each square centimeter of the larger animal radiate? Would this be good or bad?

Step 5 Use what you have concluded to answer these questions. Consider size, surface area, and volume.

a. Why do large objects cool more slowly than similar small objects?

b. Why is a beached whale more likely than a beached dolphin to experience overheating?

c. Why are larger mammals found closer to the poles than the equator?

d. If a woman and a small child fall into a cold lake, why is the child in greater danger of hypothermia?

e. When the weather is cold, iguanas hardly move. When it warms up, they become active. If a small iguana and a large iguana are sunning themselves in the morning sun, which one will become active first? Why? Which iguana will remain active longer after sunset? Why?

f. Why do elephants have big ears?

Bone and Muscle Strength

The strength of a bone or a muscle is proportional to its cross-sectional area.

Step 6 Imagine a 7-foot-tall basketball player and a 42-inch-tall child. How would the dimensions of the bones of the basketball player compare to the corresponding bones of the child?

Step 7 How would the cross-sectional areas of their corresponding bones compare?

Step 8 How would their weights compare?

Step 9 How many times as much weight would each cross-sectional square inch of bone have to support in the basketball player? What are some factors that may explain why basketball players' bones don't usually break?

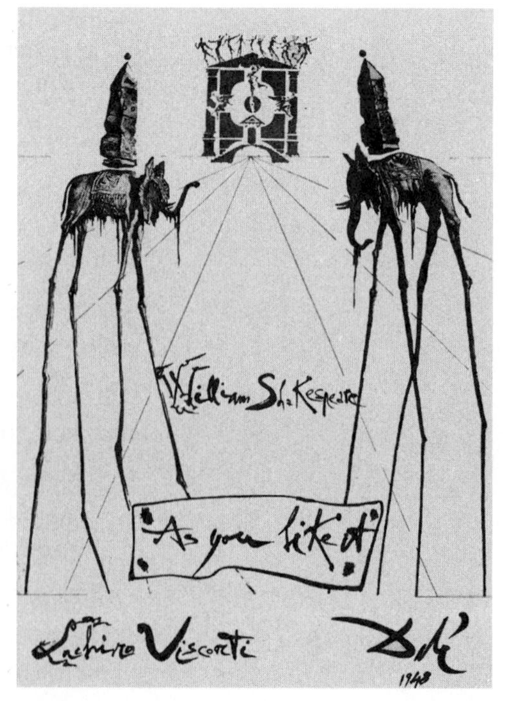

The Spanish artist Salvador Dalí (1904–1989) designed this cover for a 1948 program of the ballet *As You Like It.* What is the effect created by the elephants on long spindly legs? Could these animals really exist? Explain.

Step 10 Use what you have concluded to answer these questions. Consider size, surface area, and volume.

a. Why are the largest living mammals, the whales, confined to the sea?

b. Why do hippos and rhinos have short, thick legs?

c. Why are champion weight lifters seldom able to lift more than twice their weight?

d. Thoroughbred racehorses are fast runners but break their legs easily, while draft horses are slow moving and rarely break their legs. Why is this?

e. Assume that a male gorilla can weigh as much as 450 pounds and can reach about 6 feet tall. King Kong is 30 feet tall. Could King Kong really exist?

f. Professional basketball players are not typically similar in shape to professional football players. Discuss the advantages and disadvantages of each body type in each sport.

Similarity is a theme in *King Kong* (1933), a movie about a giant gorilla similar in shape to a real gorilla. Willis O'Brien (1886–1962), an innovator of stop-motion animation, designed the models of Kong that brought the gorilla to life for moviegoers everywhere. Here, special effects master Ray Harryhausen holds a model that was used to create this captivating scene.

Gravity and Air Resistance

Objects in a vacuum fall at the same rate. However, an object falling through air is slowed by air resistance. Air resistance is proportional to the surface area of the falling object.

Step 11 Imagine a rat that is 8 times the length, width, and height of a similar mouse. Both animals fall from a cliff.

Step 12 How would the volumes of the two animals compare?

Step 13 How would the air resistance against the two animals compare?

Step 14 The mass (related to weight) of an object is a factor in the force of the impact of the object with the ground. But air resistance on an object slows its fall, counteracting some of the force of impact. Assume that the weights of the rat and mouse are proportional to their volumes. Compare the force of the rat's impact with the ground to the force of the mouse's impact. Which is more likely to survive the fall?

Step 15 Use what you have concluded to answer this question. Consider size, surface area, and volume.

An ant can fall 100 times its height and live. This is not true for a human. Why?

IMPROVING YOUR VISUAL THINKING SKILLS

Painted Faces II

Small cubes are assembled to form a larger cube, and then some of the faces of this larger cube are painted. After the paint dries, the larger cube is taken apart. Exactly 60 of the small cubes have no paint on any of their faces. What were the dimensions of the larger cube? How many of its faces were painted?

LESSON 11.6

Proportional Segments Between Parallel Lines

Mistakes are the portals of discovery.

JAMES JOYCE

In the figure below, $\overleftrightarrow{MT} \parallel \overline{LU}$. Is $\triangle LUV$ similar to $\triangle MTV$? Yes, it is. A short paragraph proof can support this observation.

Given: $\triangle LUV$ with $\overleftrightarrow{MT} \parallel \overline{LU}$

Show: $\triangle LUV \sim \triangle MTV$

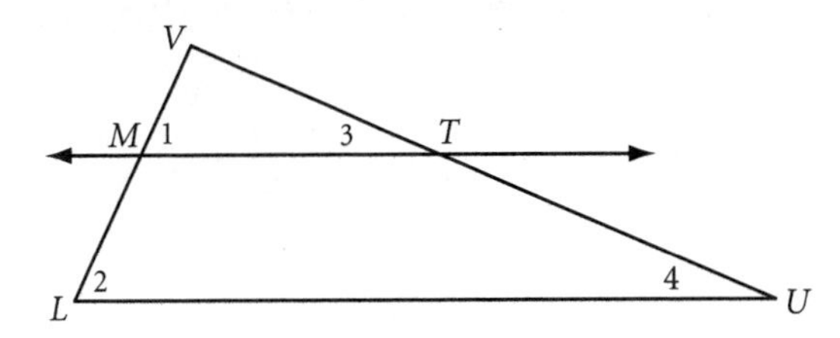

Paragraph Proof

First assume that the Corresponding Angles Conjecture and the AA Similarity Conjecture are true.

If $\overleftrightarrow{MT} \parallel \overline{LU}$, then $\angle 1 \cong \angle 2$ and $\angle 3 \cong \angle 4$ by the Corresponding Angles Conjecture.

If $\angle 1 \cong \angle 2$ and $\angle 3 \cong \angle 4$, then $\triangle LUV \sim \triangle MTV$ by the AA Similarity Conjecture. ■

Let's see how you can use this observation to solve problems.

EXAMPLE A

$\overline{EO} \parallel \overline{LN}$

$y = \underline{?}$

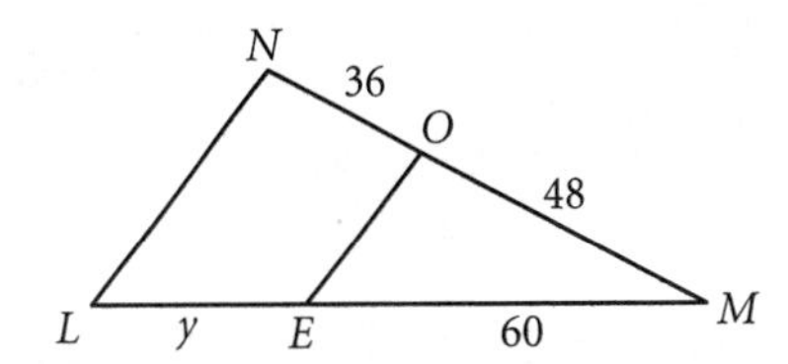

▶ Solution

Use the fact that $\triangle EMO \sim \triangle LMN$ to write a proportion with the lengths of corresponding sides.

$\frac{MO}{MN} = \frac{ME}{ML}$ — Corresponding sides of similar triangles are proportional.

$\frac{48}{48 + 36} = \frac{60}{60 + y}$ — Substitute lengths given in the figure.

$\frac{4}{7} = \frac{60}{60 + y}$ — Reduce the left side of the equation.

$240 + 4y = 420$ — Multiply both sides by $7(60 + y)$, reduce, and distribute.

$4y = 180$ — Subtract 240 from both sides.

$y = 45$ — Divide by 4.

Look back at the figure in Example A. Notice that the ratio $\frac{LE}{EM}$ is the same as the ratio $\frac{NO}{OM}$. So there are more relationships in the figure than the ones we find in similar triangles. Let's investigate.

Investigation 1
Parallels and Proportionality

You will need
- a ruler
- a protractor

In this investigation, we'll look at the ratios of segments that have been cut by parallel lines.

Step 1 In each figure below, find x. Then find numerical values for the ratios.

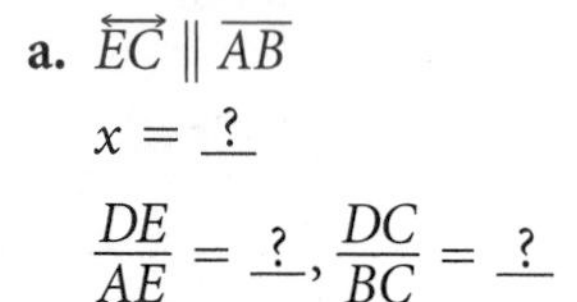

a. $\overleftrightarrow{EC} \parallel \overline{AB}$

$x = \underline{\ ?\ }$

$\frac{DE}{AE} = \underline{\ ?\ }, \frac{DC}{BC} = \underline{\ ?\ }$

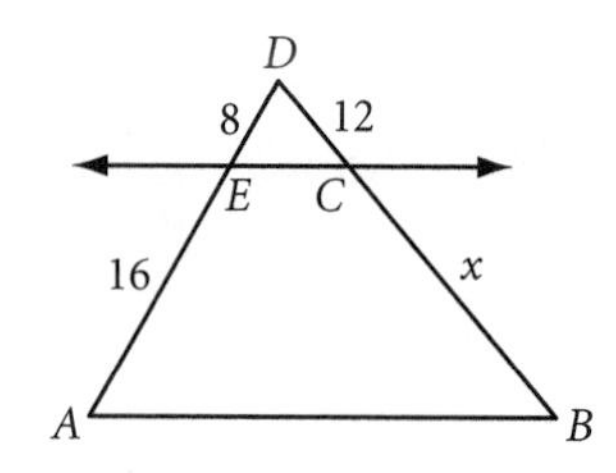

b. $\overleftrightarrow{KH} \parallel \overline{FG}$

$x = \underline{\ ?\ }$

$\frac{JK}{KF} = \underline{\ ?\ }, \frac{JH}{HG} = \underline{\ ?\ }$

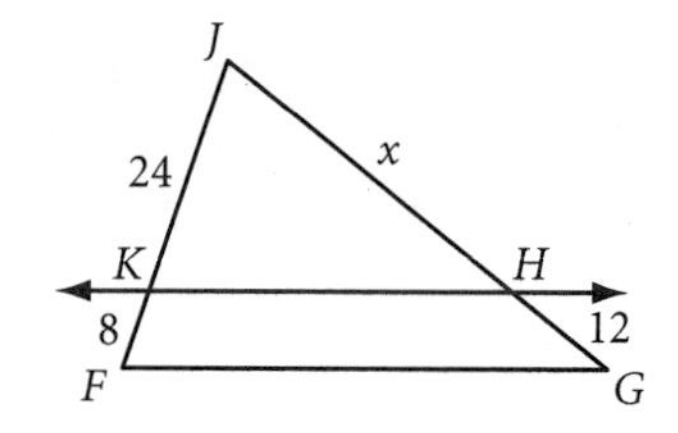

c. $\overleftrightarrow{QN} \parallel \overline{LM}$

$x = \underline{\ ?\ }$

$\frac{PQ}{QL} = \underline{\ ?\ }, \frac{PN}{MN} = \underline{\ ?\ }$

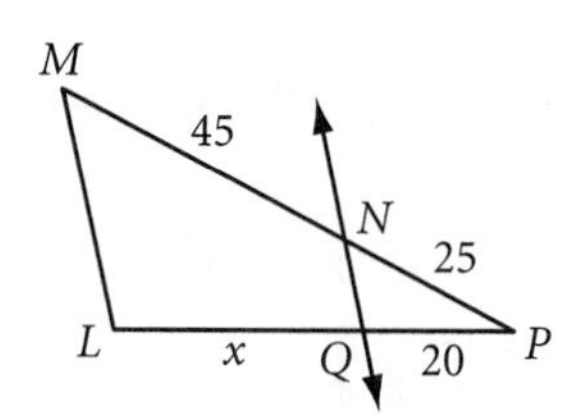

Step 2 What do you notice about the ratios of the lengths of the segments that have been cut by the parallel lines?

Is the converse true? That is, if a line divides two sides of a triangle proportionally, is it parallel to the third side? Let's see.

Step 3 Draw an acute angle, P.

Step 4 Beginning at point P, use your ruler to mark off lengths of 8 cm and 10 cm on one ray. Label the points A and B.

Step 5 Mark off lengths of 12 cm and 15 cm on the other ray. Label the points C and D. Notice that $\frac{8}{10} = \frac{12}{15}$.

Step 6 Draw $\overline{AC}$ and $\overline{BD}$.

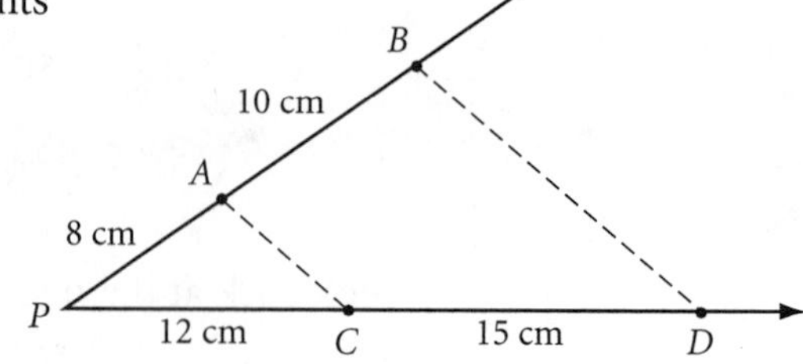

Step 7 With a protractor, measure $\angle PAC$ and $\angle PBD$. Are $\overline{AC}$ and $\overline{BD}$ parallel?

Step 8 Repeat Steps 3–7, but this time use your ruler to create your own lengths such that $\frac{PA}{AB} = \frac{PC}{CD}$.

Step 9 Compare your results with the results of others near you.

You should be ready to combine your observations from Steps 2 and 9 into one conjecture.

Parallel/Proportionality Conjecture **C-100**

If a line parallel to one side of a triangle passes through the other two sides, then it divides the other two sides _?_. Conversely, if a line cuts two sides of a triangle proportionally, then it is _?_ to the third side.

If you assume that the AA Similarity Conjecture is true, you can use algebra to prove the Parallel/Proportionality Conjecture. Here's the first part.

EXAMPLE B

Given: $\triangle ABC$ with $\overleftrightarrow{XY} \parallel \overline{BC}$

Show: $\frac{a}{c} = \frac{b}{d}$

(Assume that the lengths a, b, c, and d are all nonzero.)

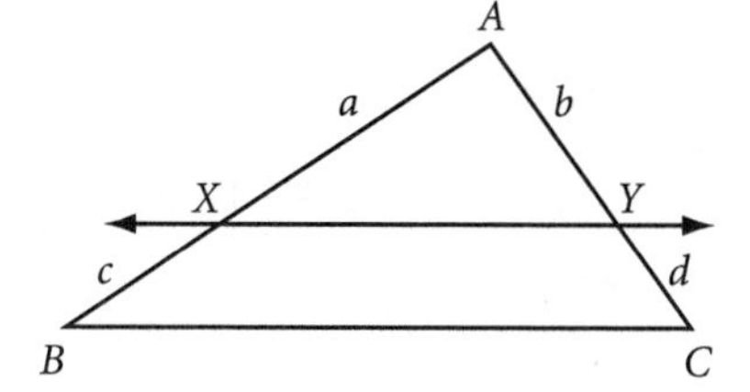

▸ *Solution*

First, you know that $\triangle AXY \sim \triangle ABC$ (see the proof on page 603). Use a proportion of corresponding sides.

$\frac{a}{a+c} = \frac{b}{b+d}$	Lengths of corresponding sides of similar triangles are proportional.
$\frac{a(a+c)(b+d)}{(a+c)} = \frac{b(a+c)(b+d)}{(b+d)}$	Multiply both sides by $(a + c)(b + d)$.
$a(b+d) = b(a+c)$	Reduce.
$ab + ad = ba + bc$	Apply the distributive property.
$ab + ad = ab + bc$	Commute *ba* to *ab*.
$ad = bc$	Subtract *ab* from both sides.
$\frac{ad}{cd} = \frac{bc}{cd}$	We want *c* and *d* in the denominator, so divide both sides by *cd*.
$\frac{a}{c} = \frac{b}{d}$	Reduce.

You'll prove the converse of the Parallel/Proportionality Conjecture in Exercise 18.

Can the Parallel/Proportionality Conjecture help you divide segments into several proportional parts? Let's investigate.

Investigation 2
Extended Parallel/Proportionality

Step 1 Use the Parallel/Proportionality Conjecture to find each missing length. Are the ratios equal?

a. $\overline{FT} \parallel \overline{LA} \parallel \overline{GR}$

$x = \underline{?}$, $y = \underline{?}$

Is $\frac{FL}{LG} = \frac{TA}{AR}$?

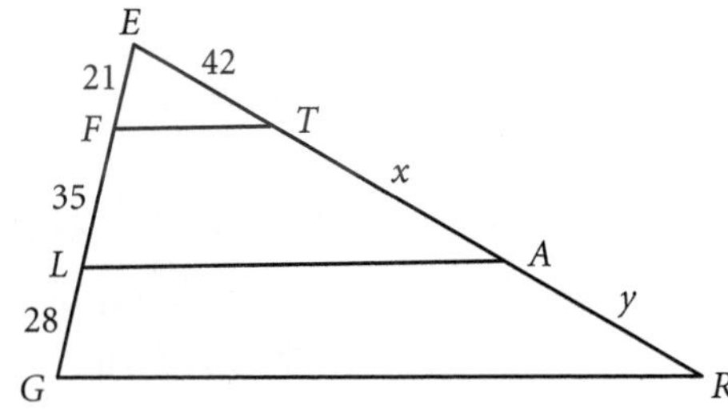

b. $\overline{ZE} \parallel \overline{OP} \parallel \overline{IA} \parallel \overline{DR}$

$a = \underline{?}$, $b = \underline{?}$, $c = \underline{?}$

Is $\frac{DI}{IO} = \frac{RA}{AP}$? Is $\frac{IO}{OZ} = \frac{AP}{PE}$?

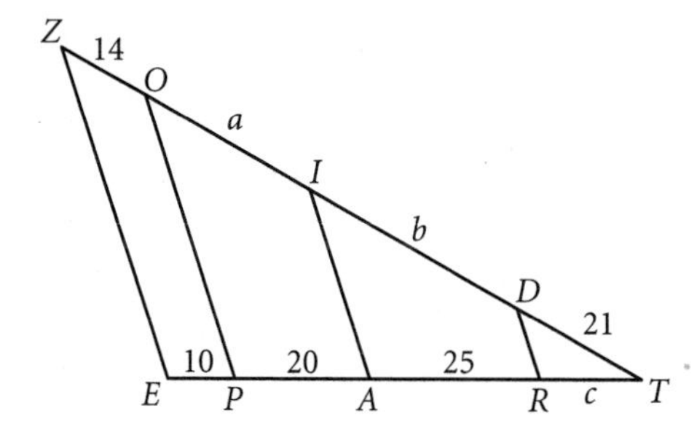

Step 2 Compare your results with the results of others near you. Complete the conjecture below.

Extended Parallel/Proportionality Conjecture C-101

If two or more lines pass through two sides of a triangle parallel to the third side, then they divide the two sides $\underline{?}$.

Exploring the converse of this conjecture has been left for you as a Take Another Look activity.

You already know how to use a perpendicular bisector to divide a segment into two, four, or eight equal parts. Now you can use your new conjecture to divide a segment into *any* number of equal parts.

EXAMPLE C Divide any segment AB into three congruent parts using only a compass and straightedge.

▶ ***Solution*** Draw segment AB. From one endpoint of $\overline{AB}$, draw any ray to form an angle. On the ray, mark off three congruent segments with your compass. Connect the third compass mark to the other endpoint of $\overline{AB}$ to form a triangle.

Finally, through the other two compass marks on the ray, construct lines parallel to the third side of the triangle. The two parallel lines divide $\overline{AB}$ into three equal parts.

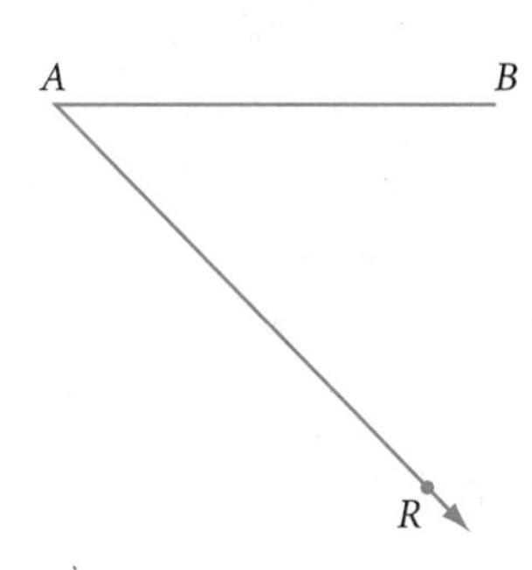

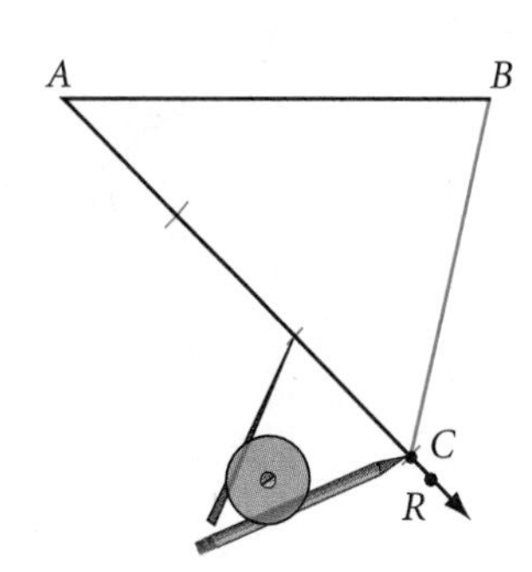

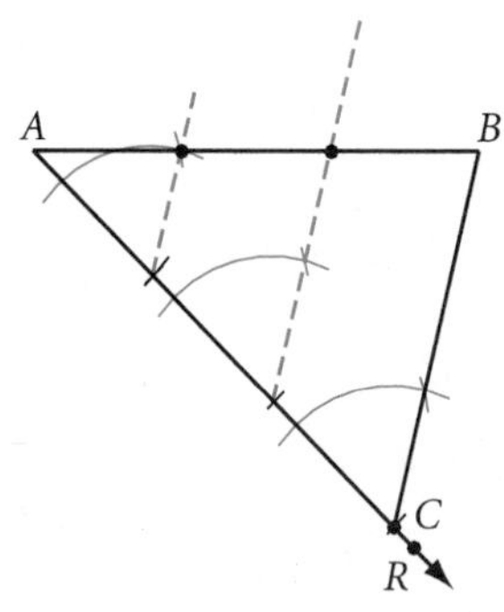

EXERCISES

You will need

Construction tools for Exercises **13** and **14**

Geometry software for Exercise **26**

For Exercises 1–12, all measurements are in centimeters.

1. $\ell \parallel \overline{WE}$
$a = \underline{?}$ ⓗ

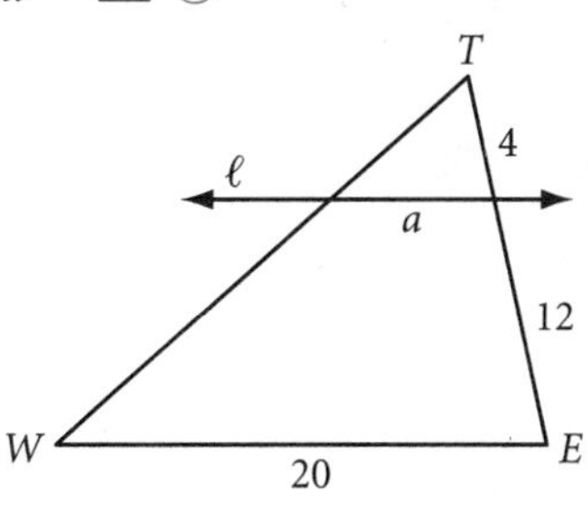

2. $m \parallel \overline{DR}$
$b = \underline{?}$

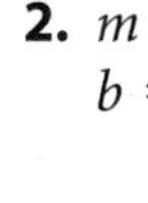
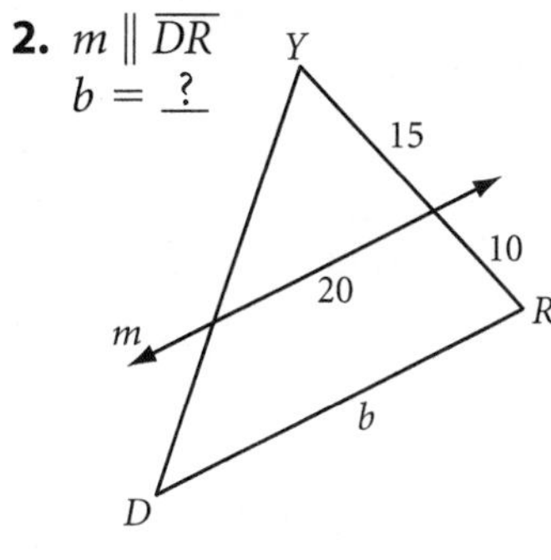

3. $n \parallel \overline{SN}$
$c = \underline{?}$ ⓗ

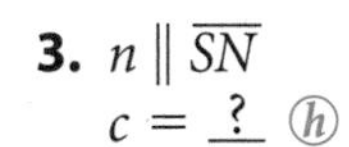
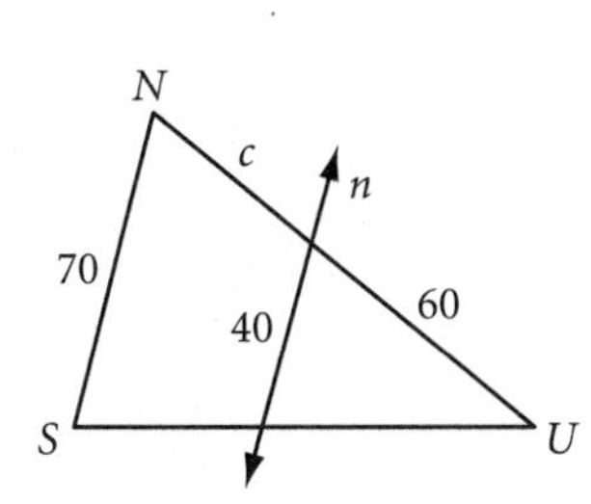

4. $\ell \parallel \overline{RA}$
$d = \underline{?}$ ⓗ

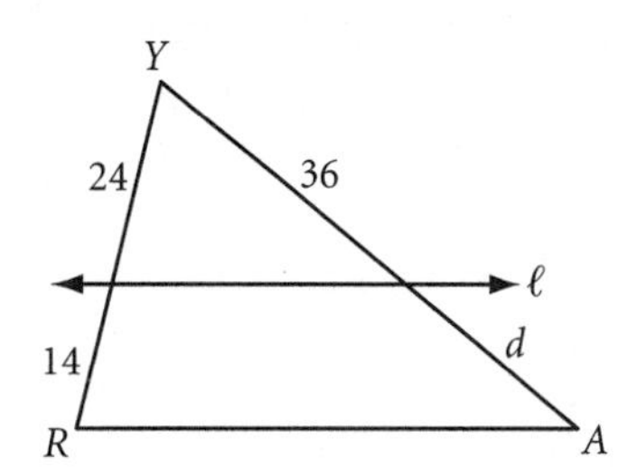

5. $m \parallel \overline{BA}$
$e = \underline{?}$

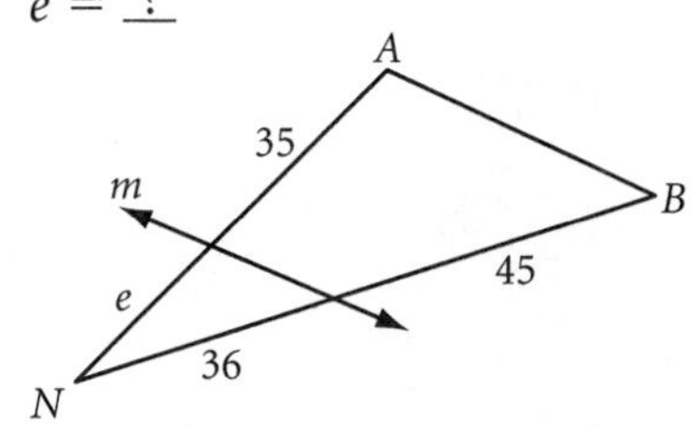

6. Is $r \parallel \overline{AN}$? ⓗ

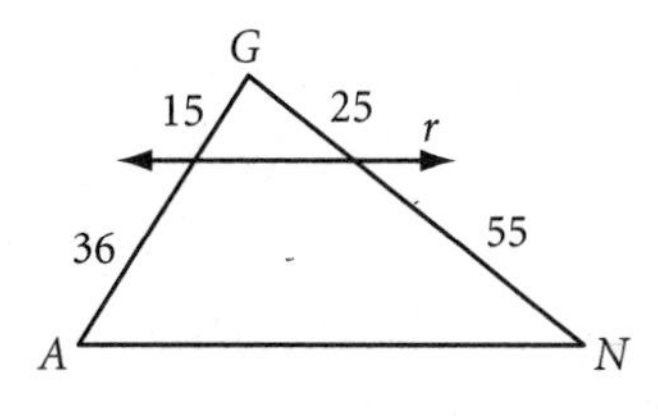

Similarity is used to create integrated circuits. Electrical engineers use large-scale maps of extremely small silicon chips. This engineer is making a scale drawing of a computer chip.

7. Is $m \parallel \overline{FL}$?

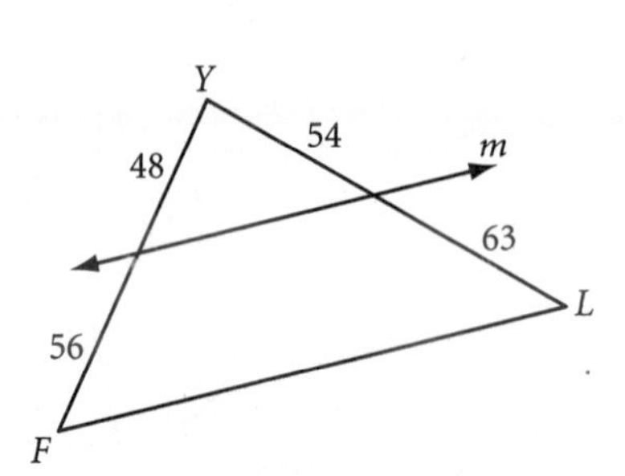

8. $r \parallel s \parallel \overline{OU}$
$m = \underline{\ ?\ }$, $n = \underline{\ ?\ }$

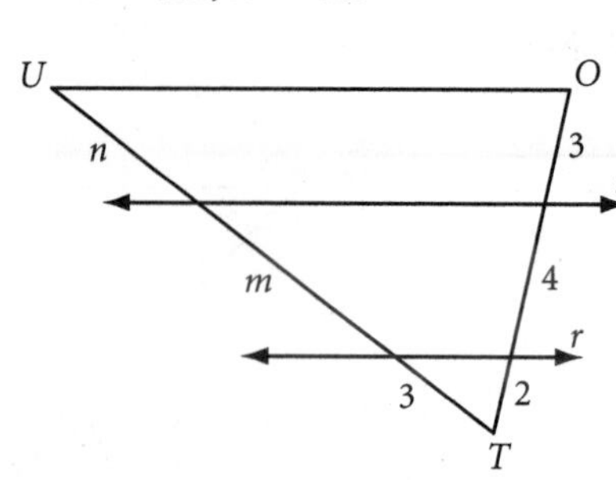

9. $\overline{MR} \parallel p \parallel q$
$w = \underline{\ ?\ }$, $x = \underline{\ ?\ }$

10. Is $m \parallel \overline{EA}$?
Is $n \parallel \overline{EA}$?
Is $m \parallel n$?

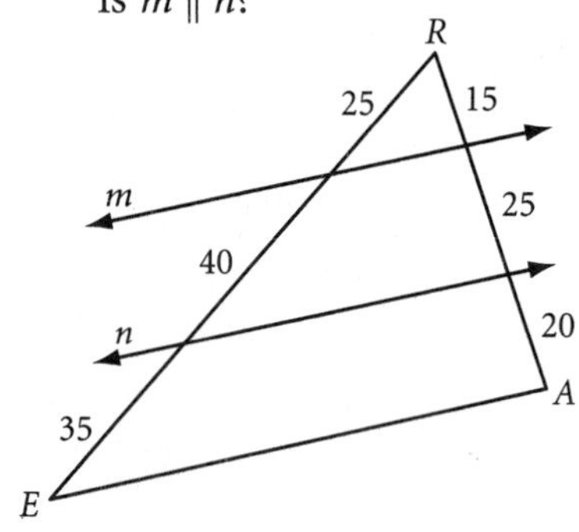

11. Is $\overline{XY} \parallel \overline{GO}$?
Is $\overline{XY} \parallel \overline{FR}$?
Is *FROG* a trapezoid?

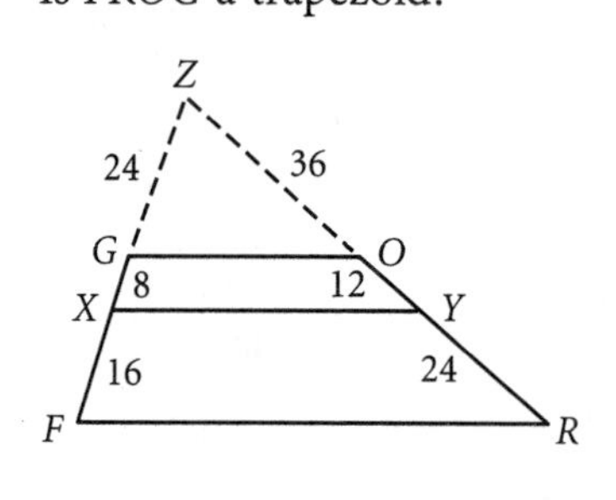

12. $a = \underline{\ ?\ }$, $b = \underline{\ ?\ }$ ⓗ

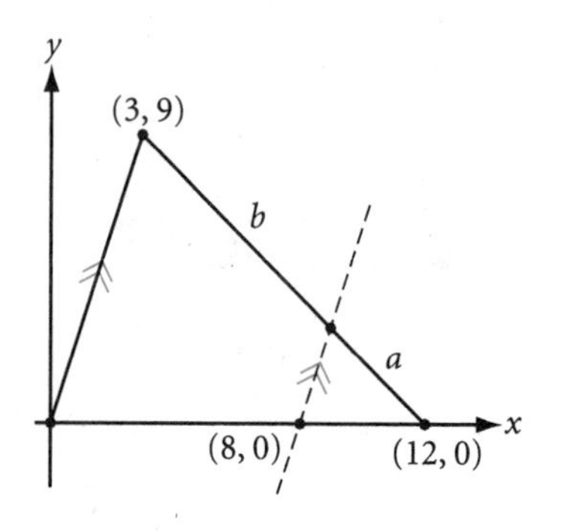

13. ***Construction*** Draw segment *EF*. Use compass and straightedge to divide it into five equal parts.

14. ***Construction*** Draw segment *IJ*. Construct a regular hexagon with *IJ* as the perimeter.

15. You can use a sheet of lined paper to divide a segment into equal parts. Draw a segment on a piece of patty paper, and divide it into five equal parts by placing it over lined paper. What conjecture explains why this works?

16. The drafting tool shown at right is called a sector compass. You position a given segment between the 100-marks. What points on the compass should you connect to construct a segment that is three-fourths (or 75%) of *BC*? Explain why this works.

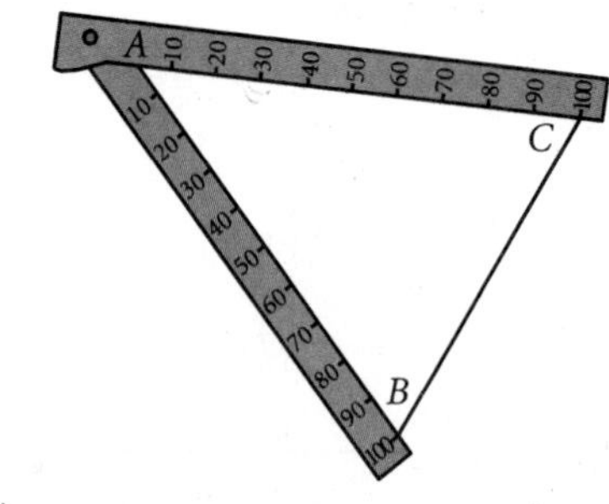

17. This truncated cone was formed by cutting off the top of a cone with a slice parallel to the base of the cone. What is the volume of the truncated cone? ⓗ

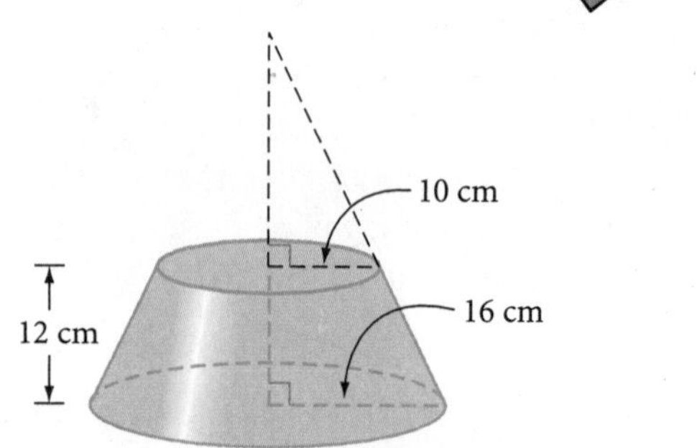

18. Assume that the Corresponding Angles Conjecture and the AA Similarity Conjecture are true. Write a proof to show that if a line cuts two sides of a triangle proportionally, then it is parallel to the third side.

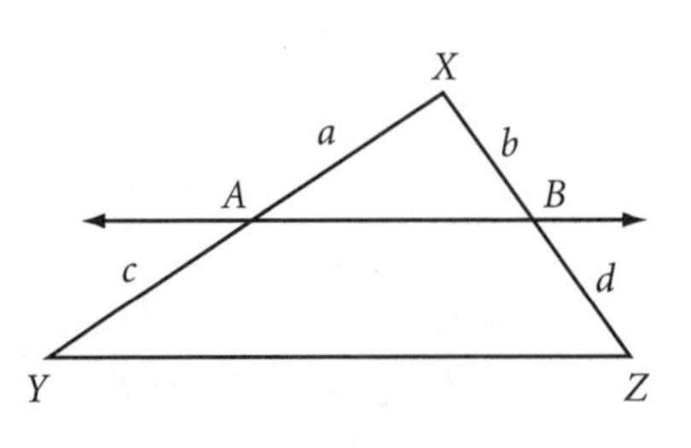

Given: $\frac{a}{c} = \frac{b}{d}$ (Assume $c \neq 0$ and $d \neq 0$.)

Show: $\overleftrightarrow{AB} \parallel \overline{YZ}$

19. Another drafting tool used to construct segments is a pair of proportional dividers, shown at right. Two styluses of equal length are connected by a screw. The tool is adjusted for different proportions by moving the screw. Where should the screw be positioned so that AB is three-fourths of CD?

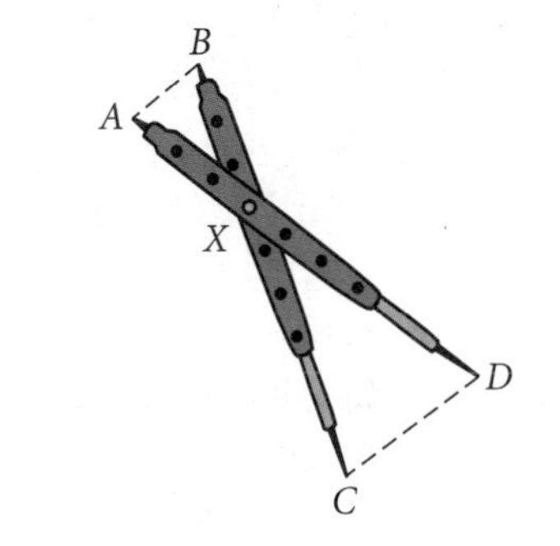

The Extended Parallel/Proportionality Conjecture can be extended even further. That is, you don't necessarily need a triangle. If three or more parallel lines intercept two other lines (transversals) in the same plane, they do so proportionally. For Exercises 20 and 21 use this extension.

20. Find x and y.

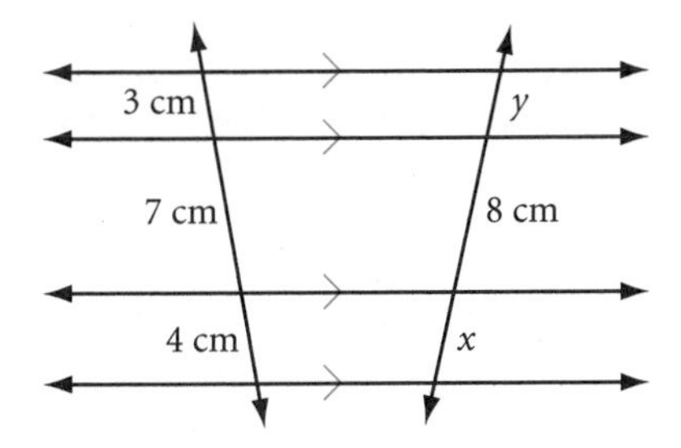

21. A real estate developer has parceled land between a river and River Road as shown. The land has been divided by segments perpendicular to the road. What is the "river frontage" (lengths x, y, and z) for each of the three lots?

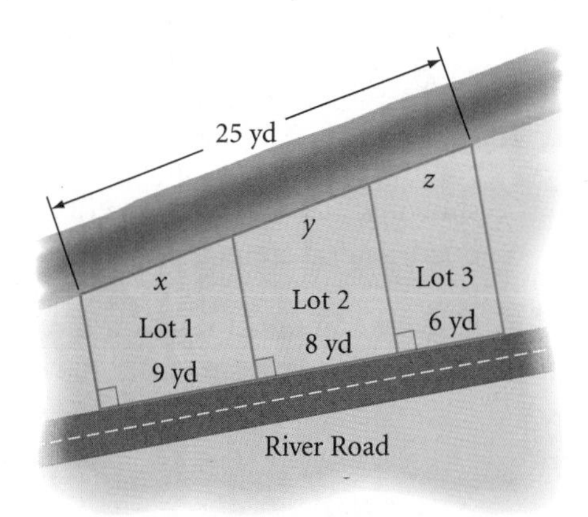

Review

22. The ratio of the surface areas of two cubes is $\frac{49}{81}$. What is the ratio of their volumes?

23. Romunda's original recipe for her special "cannonball" cookies makes 36 spheres with 4 cm diameters. She reasons that she can make 36 cannonballs with 8 cm diameters by doubling the amount of dough. Is she correct? If not, how many 8 cm diameter cannonballs can she make by doubling the recipe?

24. Find the surface area of a cube with edge x. Find the surface area of a cube with edge $2x$. Find the surface area of a cube with edge $3x$.

25. A circle of radius r has a chord of length r. Find the length of the minor arc.

26. *Technology* In Lesson 11.3, Exercise 17, you learned about the golden cut and the golden ratio. A **golden rectangle** is a rectangle in which the ratio of the length to the width is the golden ratio. That is, a golden rectangle's length, l, and width, w, satisfy the proportion

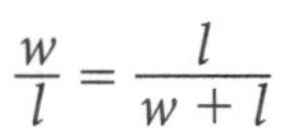

$$\frac{w}{l} = \frac{l}{w + l}$$

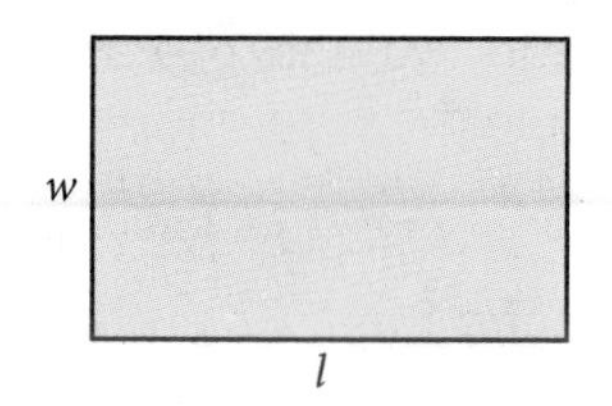

A golden rectangle

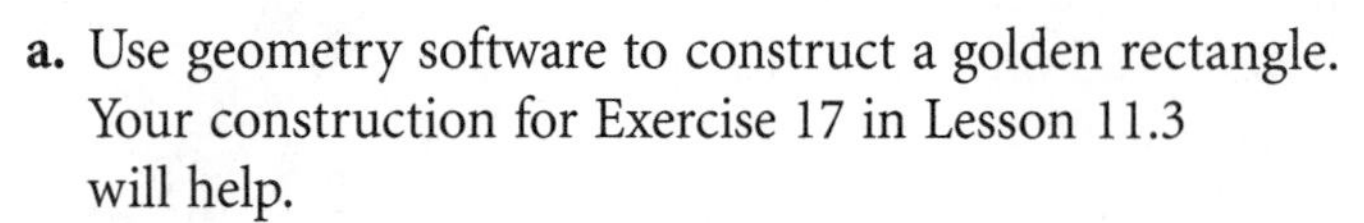

a. Use geometry software to construct a golden rectangle. Your construction for Exercise 17 in Lesson 11.3 will help.

b. When a square is cut off one end of a golden rectangle, the remaining rectangle is a smaller, similar golden rectangle. If you continue this process over and over again, and then connect opposite vertices of the squares with quarter-circles, you create a curve called the golden spiral. Use geometry software to construct a **golden spiral.** The first three quarter-circles are shown below.

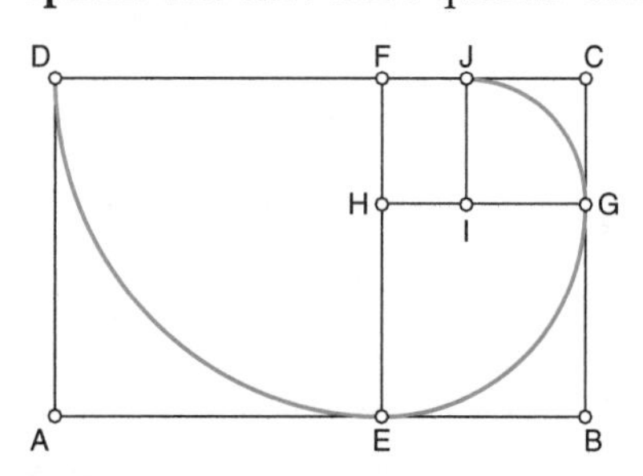

ABCD is a golden rectangle.

EBCF is a golden rectangle.

HGCF is a golden rectangle.

IJFH is a golden rectangle.

The curve from *D* to *E* to *G* to *J* is the beginning of a golden spiral.

Some researchers believe Greek architects used golden rectangles to design the Parthenon. You can learn more about the historical importance of golden rectangles using the links at www.keymath.com/DG .

27. A circle is inscribed in a quadrilateral. Write a proof showing that the two sums of the opposite sides of the quadrilateral are equal.

28. Copy the figure at right onto your own paper. Divide it into four figures similar to the original figure.

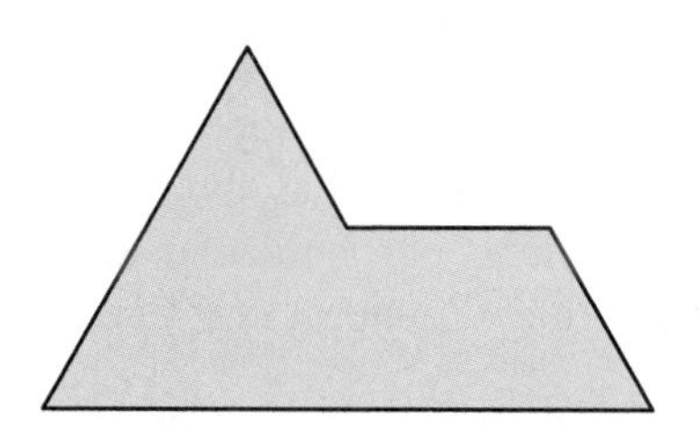

IMPROVING YOUR VISUAL THINKING SKILLS

Connecting Cubes

The two objects shown at right can be placed together to form each of the shapes below except one. Which one?

A.

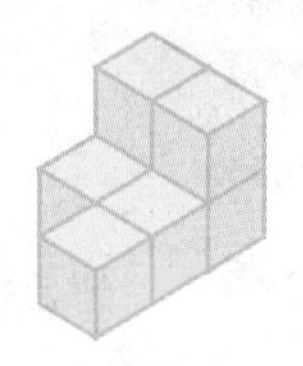

B.

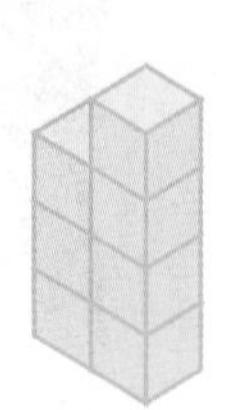

C.

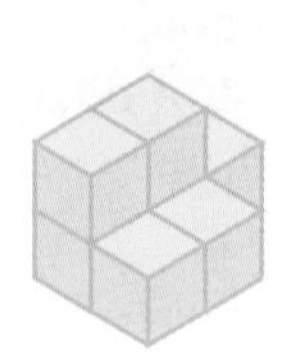

D.

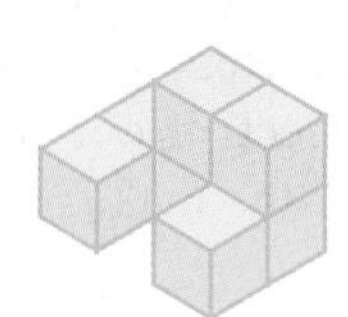

Exploration

Two More Forms of Valid Reasoning

In the Chapter 10 Exploration Sherlock Holmes and Valid Forms of Reasoning, you learned about *Modus Ponens* and *Modus Tollens.* A third valid form of reasoning is called the Law of Syllogism.

According to the **Law of Syllogism** (LS), if you accept "If P then Q" as true and if you accept "If Q then R" as true, then you must logically accept "If P then R" as true.

Here is an example of the law of syllogism.

English statement	Symbolic translation
If I eat pizza after midnight, then I will have nightmares. If I have nightmares, then I will get very little sleep. Therefore, if I eat pizza after midnight, then I will get very little sleep.	P: I eat pizza after midnight. Q: I will have nightmares R: I will get very little sleep. $P \to Q$ $Q \to R$ $\therefore P \to R$

To work on the next law, you need some new statement forms. Every conditional statement has three other conditionals associated with it. To get the converse of a statement, you switch the "if" and "then" parts. To get the **inverse,** you negate both parts. To get the **contrapositive,** you reverse and negate the two parts. These new forms may be true or false.

Statement	If two angles are vertical angles, then they are congruent.	$P \to Q$	true
Converse	If two angles are congruent, then they are vertical angles.	$Q \to P$	false
Inverse	If two angles are not vertical angles, then they are not congruent.	$\sim P \to \sim Q$	false
Contrapositive	If two angles are not congruent, then they are not vertical angles.	$\sim Q \to \sim P$	true

Notice that the original conditional statement and its contrapositive have the same truth value. This leads to a fourth form of valid reasoning. The **Law of Contrapositive** (LC) says that if a conditional statement is true, then its contrapositive is also true. Conversely, if the contrapositive is true, then the original conditional statement must also be true. This also means that if a conditional statement is false, so is its contrapositive.

Often, a logical argument contains multiple steps, applying the same rule more than once, or applying more than one rule. Here is an example.

English statement	Symbolic translation
If the consecutive sides of a parallelogram are congruent, then it is a rhombus. If a parallelogram is a rhombus, then its diagonals are perpendicular bisectors of each other. The diagonals are not perpendicular bisectors of each other. Therefore the consecutive sides of the parallelogram are not congruent.	P: The consecutive sides of a parallelogram are congruent. Q: The parallelogram is a rhombus. R: The diagonals are perpendicular bisectors of each other. $P \rightarrow Q$ $Q \rightarrow R$ $\sim R$ $\therefore \sim P$

You can show that this argument is valid in three logical steps.

Step 1	$P \rightarrow Q$ $Q \rightarrow R$ $\therefore P \rightarrow R$	by the Law of Syllogism
Step 2	$P \rightarrow R$ $\therefore \sim R \rightarrow \sim P$	by the Law of Contrapositive
Step 3	$\sim R \rightarrow \sim P$ $\sim R$ $\therefore \sim P$	by *Modus Ponens*

Literature CONNECTION

Lewis Carroll was the pseudonym of the English novelist and mathematician Charles Lutwidge Dodgson (1832–1898). He is often associated with his famous children's book *Alice's Adventures in Wonderland.*

In 1886 he published *The Game of Logic,* which used a game board and counters to solve logic problems. In 1896 he published *Symbolic Logic, Part I,* which was an elementary book intended to teach symbolic logic. Here is one of the silly problems from *Symbolic Logic.* What conclusion follows from these premises?

> Babies are illogical.
> Nobody is despised who can manage a crocodile.
> Illogical persons are despised.

Lewis Carroll enjoyed incorporating mathematics and logic into all of his books. Here is a quote from *Through the Looking Glass.* Is Tweedledee using valid reasoning?

> "Contrariwise," said Tweedledee, "if it was so, it might be; and if it were so, it would be, but as it isn't, it ain't. That's logic."

So far, you have learned four basic forms of valid reasoning.

Now let's apply them in symbolic proofs.

Four Forms of Valid Reasoning

$P \rightarrow Q$	$P \rightarrow Q$	$P \rightarrow Q$	$P \rightarrow Q$
P	$\sim Q$	$Q \rightarrow R$	$\therefore \sim Q \rightarrow \sim P$
$\therefore Q$	$\therefore \sim P$	$\therefore P \rightarrow R$	
by MP	by MT	by LS	by LC

Activity

Symbolic Proofs

Step 1 Determine whether or not each logical argument is valid. If it is valid, state what reasoning form or forms it follows. If it is not valid, write "no valid conclusion."

a. $P \rightarrow \sim Q$
Q
$\therefore \sim P$

b. $\sim S \rightarrow P$
$R \rightarrow \sim S$
$\therefore R \rightarrow P$

c. $\sim Q \rightarrow \sim R$
$\sim Q$
$\therefore \sim R$

d. $R \rightarrow P$
$T \rightarrow \sim P$
$\therefore R \rightarrow T$

e. $\sim P \rightarrow \sim R$
R
$\therefore P$

f. $P \rightarrow Q$
$\sim R \rightarrow \sim Q$
$\therefore P \rightarrow R$

Step 2 Translate parts a–c into symbols, and give the reasoning form(s) or state that the conclusion is not valid.

a. If I study all night, then I will miss my late-night talk show. If Jeannine comes over to study, then I study all night. Jeannine comes over to study. Therefore I will miss my late-night talk show.

b. If I don't earn money, then I can't buy a computer. If I don't get a job, then I don't earn money. I have a job. Therefore I can buy a computer.

c. If $\overline{EF}$ is not parallel to side $\overline{AB}$ in trapezoid $ABCD$, then $\overline{EF}$ is not a midsegment of trapezoid $ABCD$. If $\overline{EF}$ is parallel to side $\overline{AB}$, then $ABFE$ is a trapezoid. $\overline{EF}$ is a midsegment of trapezoid $ABCD$. Therefore $ABFE$ is a trapezoid.

Step 3 Show how you can use *Modus Ponens* and the Law of Contrapositive to make the same logical conclusions as *Modus Tollens.*

CHAPTER 11 REVIEW

Similarity, like area, volume, and the Pythagorean Theorem, has many applications. Any scale drawing or model, anything that is reduced or enlarged, is governed by the properties of similar figures. So engineers, visual artists, and film-makers all use similarity. It is also useful in indirect measurement. Do you recall the two indirect measurement methods you learned in this chapter? The ratios of area and volume in similar figures are also related to the ratios of their dimensions. But recall that as the dimensions increase, the area increases by a squared factor and volume increases by a cubed factor.

EXERCISES

You will need

Construction tools for Exercise **9**

For Exercises 1–4, solve each proportion.

1. $\frac{x}{15} = \frac{8}{5}$ **2.** $\frac{4}{11} = \frac{24}{x}$ **3.** $\frac{4}{x} = \frac{x}{9}$ **4.** $\frac{x}{x+3} = \frac{34}{40}$

In Exercises 5 and 6, measurements are in centimeters.

5. $ABCDE \sim FGHIJ$
$w = \underline{?}, x = \underline{?}, y = \underline{?}, z = \underline{?}$

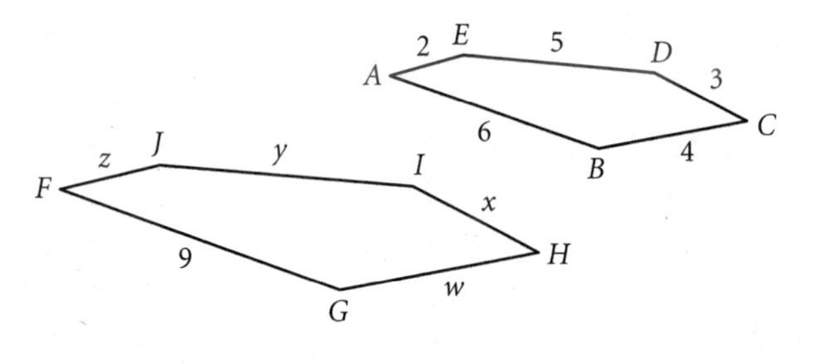

6. $\triangle ABC \sim \triangle DBA$
$x = \underline{?}, y = \underline{?}$

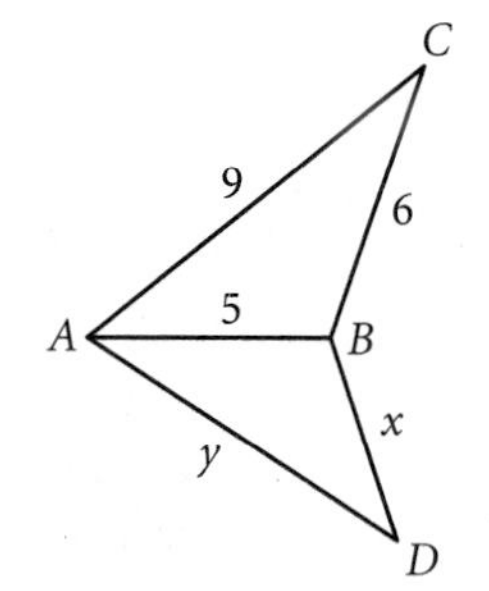

7. APPLICATION David is 5 ft 8 in. tall and wants to find the height of an oak tree in his front yard. He walks along the shadow of the tree until his head is in a position where the end of his shadow exactly overlaps the end of the tree's shadow. He is now 11 ft 3 in. from the foot of the tree and 8 ft 6 in. from the end of the shadows. How tall is the oak tree?

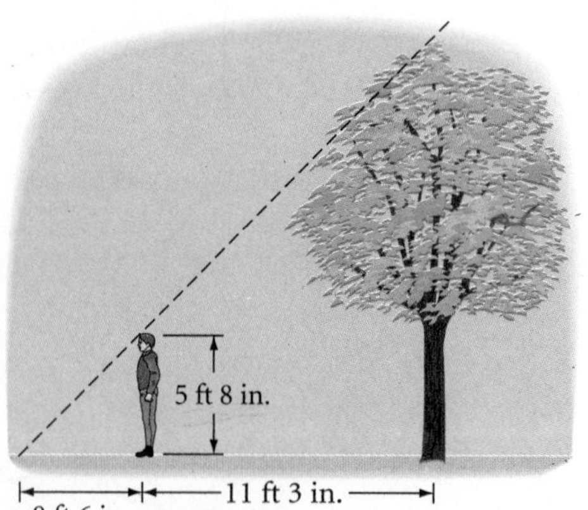

8. A certain magnifying glass when held 6 in. from an object creates an image that is 10 times the size of the object being viewed. What is the measure of a 20° angle under this magnifying glass?

9. ***Construction*** Construct $\overline{KL}$. Then find a point P that divides $\overline{KL}$ into two segments that have a ratio $\frac{3}{4}$. ⓗ

10. Patsy does a juggling act. She sits on a stool that sits on top of a rotating ball that spins at the top of a 20-meter pole. The diameter of the ball is 4 meters, and Patsy's eye is approximately 2 meters above the ball. Seats for the show are arranged on the floor in a circle so that each spectator can see Patsy's eyes. Find the radius of the circle of seats to the nearest meter. ⓗ

11. Charlie builds a rectangular box home for his pet python and uses 1 gallon of paint to cover its surface. Lucy also builds a box for Charlie's pet, but with dimensions twice as great. How many gallons of paint will Lucy need to paint her box? How many times as much volume does her box have?

12. Suppose you had a real clothespin similar to the sculpture at right and made of the same material. What measurements would you make to calculate the weight of the sculpture? Explain your reasoning.

13. The ratio of the perimeters of two similar parallelograms is $\frac{3}{7}$. What is the ratio of their areas?

14. The ratio of the areas of two circles is $\frac{25}{16}$. What is the ratio of their radii?

15. **APPLICATION** The Jones family paid \$150 to a painting contractor to stain their 12-by-15-foot deck. The Smiths have a similar deck that measures 16 ft by 20 ft. What price should the Smith family expect to pay to have their deck stained?

This sculpture, called *Clothespin* (1976), was created by Swedish-American sculptor Claes Oldenburg (b 1929). His art reflects how everyday objects can be intriguing.

16. The dimensions of the smaller cylinder are two-thirds of the dimensions of the larger cylinder. The volume of the larger cylinder is 2160π cm^3. Find the volume of the smaller cylinder.

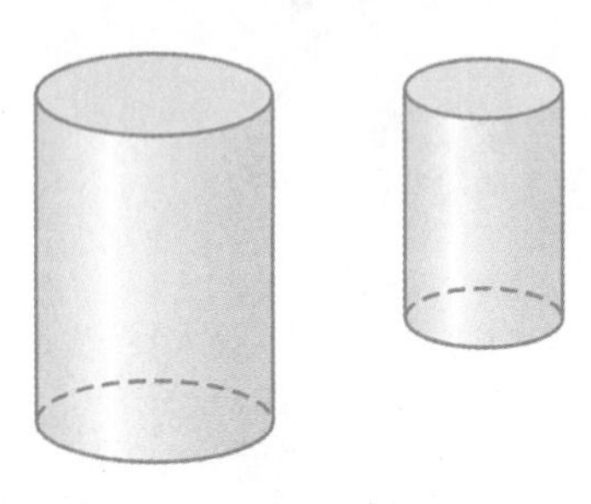

17. $\mathfrak{z} \parallel \ell \parallel \mathfrak{g} \parallel \mathfrak{h}$

$w = \underline{?}, x = \underline{?}, y = \underline{?}, z = \underline{?}$

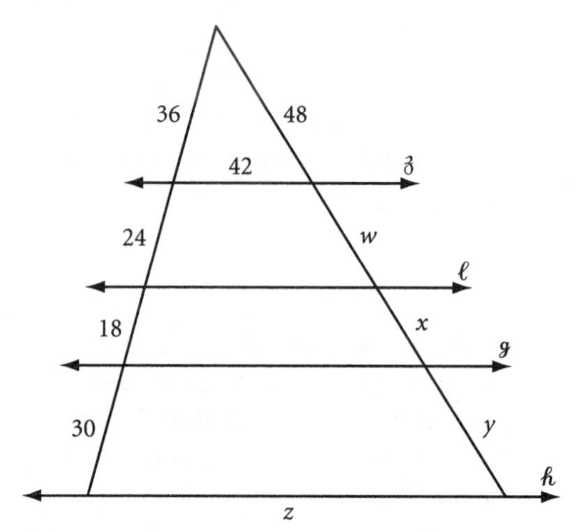

18. Below is a 58-foot statue of Bahubali, in Sravanabelagola, India. Every 12 years, worshipers of the Jain religion bathe the statue with coconut milk. Suppose the milk of one coconut is just enough to cover the surface of the similar 2-foot statuette shown at right. How many coconuts would be required to cover the surface of the full-size statue?

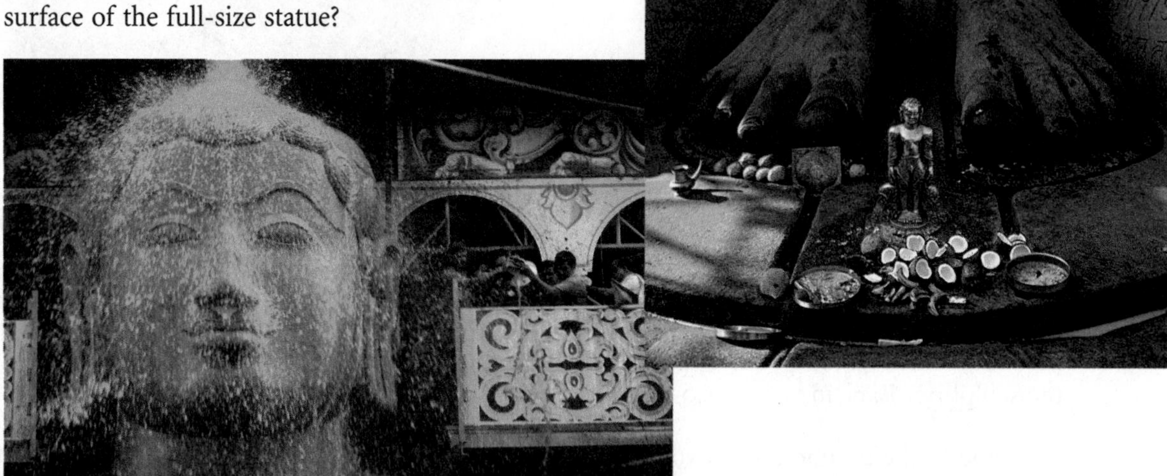

This 58-foot statue is carved from a single stone.

19. Greek mathematician Archimedes liked the design at right so much that he wanted it on his tombstone. ⓗ

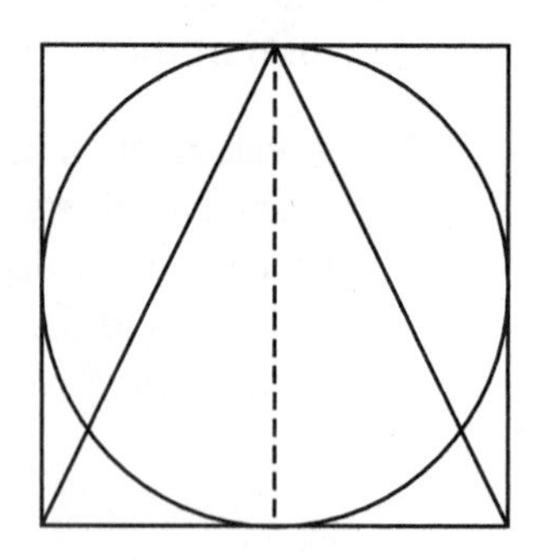

a. Calculate the ratio of the area of the square, the area of the circle, and the area of the isosceles triangle. Copy and complete this statement of proportionality.

Area of square to Area of circle to Area of triangle is $\underline{?}$ to $\underline{?}$ to $\underline{?}$.

b. When each of the figures is revolved about the vertical line of symmetry, it generates a solid of revolution—a cylinder, a sphere, and a cone. Calculate their volumes. Copy and complete this statement of proportionality.

Volume of cylinder to Volume of sphere to Volume of cone is __?__ to __?__ to __?__.

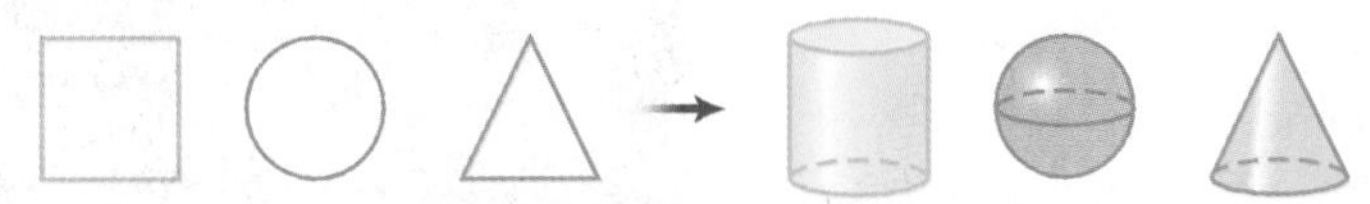

c. What is so special about this design?

20. Many fanciful stories are about people who accidentally shrink to a fraction of their original height. If a person shrank to one-twentieth his original height, how would that change the amount of food he'd require, or the amount of material needed to clothe him, or the time he'd need to get to different places? Explain.

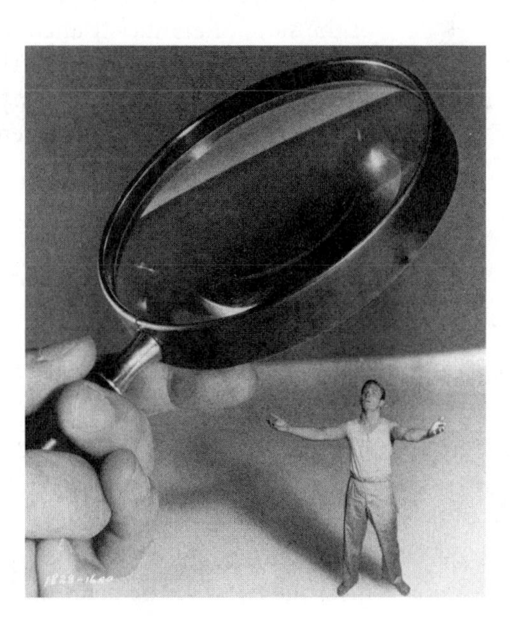

This scene is from the 1957 science fiction movie *The Incredible Shrinking Man.*

21. Would 15 pounds of 1-inch ice cubes melt faster than a 15-pound block of ice? Explain.

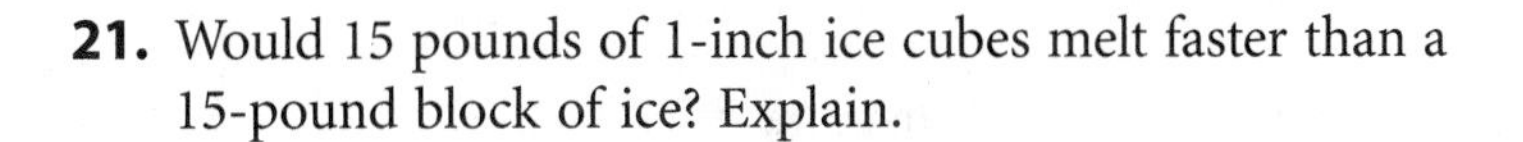

TAKE ANOTHER LOOK

1. You've learned that an ordered pair rule such as $(x, y) \rightarrow (x + b, y + c)$ is a translation. You discovered in this chapter that an ordered pair rule such as $(x, y) \rightarrow (kx, ky)$ is a dilation in the coordinate plane, centered at the origin. What transformation is described by the rule $(x, y) \rightarrow (kx + b, ky + c)$? Investigate.

2. In Lesson 11.1, you dilated figures in the coordinate plane, using the origin as the center of dilation. What happens if a different point in the plane is the center of dilation? Copy the polygon at right onto graph paper. Draw the polygon's image under a dilation with a scale factor of 2 and with point A as the center of dilation. Draw another image using a scale factor of $\frac{2}{3}$. Explain how you found the image points. How does dilating about point A differ from dilating about the origin?

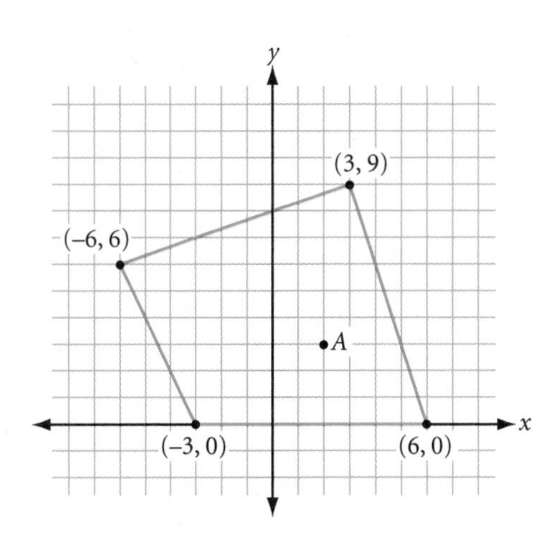

3. True or false? The angle bisector of one of the nonvertex angles of a kite will divide the diagonal connecting the vertex angles into two segments whose lengths are in the same ratio as two unequal sides of the kite. If true, explain why. If false, show a counterexample that proves it false.

4. A total eclipse of the Sun can occur because the ratio of the Moon's diameter to its distance from Earth is about the same as the ratio of the Sun's diameter to its distance to Earth. Draw a diagram and use similar triangles to explain why it works.

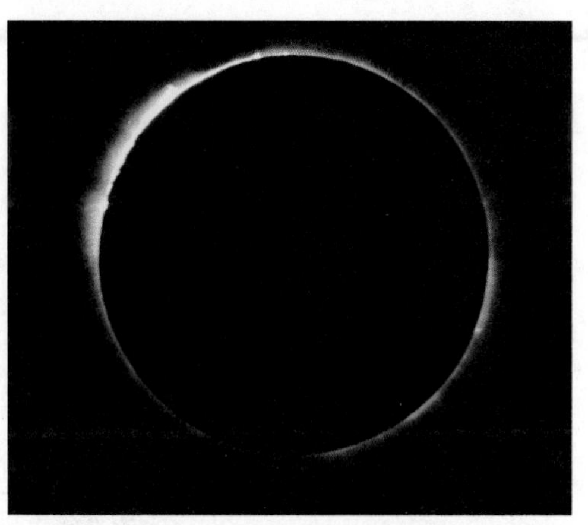

A solar eclipse

5. It is possible for the three angles and two of the sides of one triangle to be congruent to the three angles and two of the sides of another triangle, and yet the two triangles won't be congruent. Two such triangles are shown below. Use geometry software or patty paper to find another pair of similar (but not congruent) triangles in which five parts of one are congruent to five parts of the other.

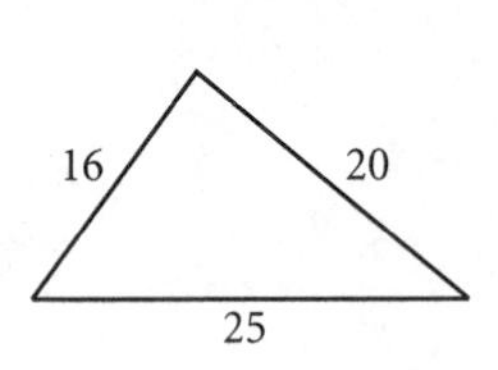

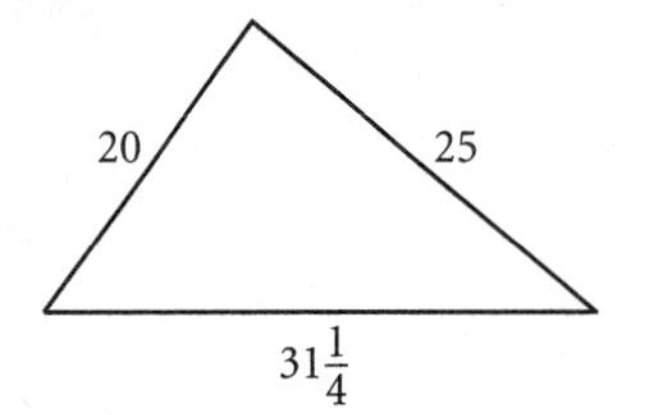

Explain why these sets of side lengths work. Use algebra to explain your reasoning.

6. Is the converse of the Extended Parallel Proportionality Conjecture true? That is, if two lines intersect two sides of a triangle, dividing the two sides proportionally, must the two lines be parallel to the third side? Prove that it is true or find a counterexample showing that it is not true.

7. If the three sides of one triangle are each parallel to one of the three sides of another triangle, what might be true about the two triangles? Use geometry software to investigate. Make a conjecture and explain why you think your conjecture is true.

Assessing What You've Learned

UPDATE YOUR PORTFOLIO If you did the Project Making a Mural, add your mural to your portfolio.

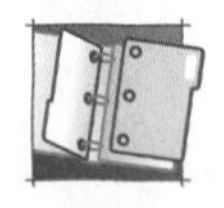

ORGANIZE YOUR NOTEBOOK Review your notebook to be sure it's complete and well organized. Be sure you have each definition and the conjecture. Write a one-page summary of Chapter 11.

GIVE A PRESENTATION Give a presentation about one or more of the similarity conjectures. You could even explain how an overhead projector produces similar figures!

CHAPTER

12 Trigonometry

I wish I'd learn to draw a little better! What exertion and determination it takes to try and do it well. . . . It is really just a question of carrying on doggedly, with continuous and, if possible, pitiless self-criticism.

M. C. ESCHER

Belvedere, M. C. Escher, 1958

OBJECTIVES

In this chapter you will

- learn about the branch of mathematics called trigonometry
- define three important ratios between the sides of a right triangle
- use trigonometry to solve problems involving right triangles
- discover how trigonometry extends beyond right triangles

LESSON

12.1

Trigonometric Ratios

Research is what I am doing when I don't know what I'm doing.

WERNHER VON BRAUN

Trigonometry is the study of the relationships between the sides and the angles of triangles. In this lesson you will discover some of these relationships for right triangles.

Science CONNECTION

Trigonometry has origins in astronomy. The Greek astronomer Claudius Ptolemy (100–170 C.E.) used tables of chord ratios in his book known as *Almagest.* These chord ratios and their related angles were used to describe the motion of planets in what were thought to be circular orbits. This woodcut shows Ptolemy using astronomy tools.

When studying right triangles, early mathematicians discovered that whenever the ratio of the shorter leg's length to the longer leg's length was close to a specific fraction, the angle opposite the shorter leg was close to a specific measure. They found this (and its converse) to be true for all similar right triangles. For example, in every right triangle in which the ratio of the shorter leg's length to the longer leg's length is $\frac{3}{5}$, the angle opposite the shorter leg is approximately 31°.

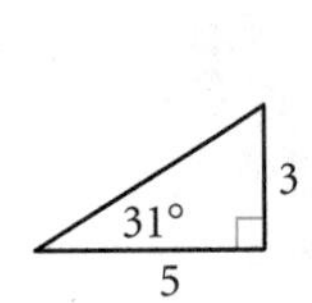

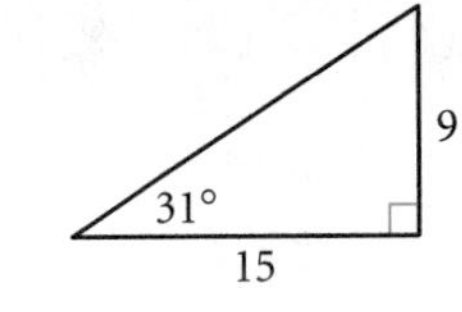

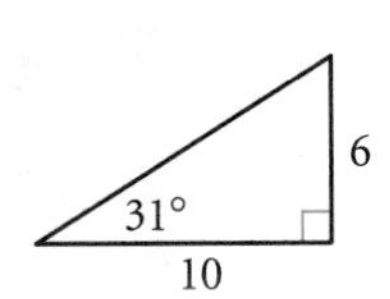

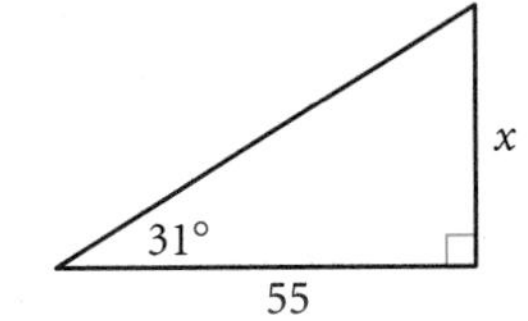

What is a good approximation for x?

What early mathematicians discovered is supported by what you know about similar triangles. If two right triangles each have an acute angle of the same measure, then the triangles are similar by the AA Similarity Conjecture. And if the triangles are similar, then corresponding sides are proportional. For example, in the similar right triangles shown below, these proportions are true:

$$\frac{BC}{AB} = \frac{EF}{DE} = \frac{HI}{GH} = \frac{KL}{JK}$$

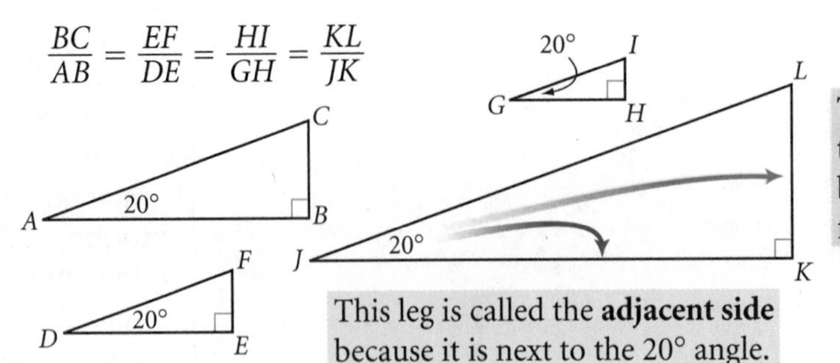

This leg is called the **opposite side** because it is across from the 20° angle.

This leg is called the **adjacent side** because it is next to the 20° angle.

The ratio of the length of the opposite side to the length of the adjacent side in a right triangle came to be called the **tangent** of the angle.

In Chapter 11, you used mirrors and shadows to measure heights indirectly. Trigonometry gives you another indirect measuring method.

EXAMPLE A

At a distance of 36 meters from a tree, the angle from the ground to the top of the tree is 31°. Find the height of the tree.

▶ Solution

As you saw in the right triangles on page 620, the ratio of the length of the side opposite a 31° angle divided by the length of the side adjacent to a 31° angle is approximately $\frac{3}{5}$, or 0.6. You can set up a proportion using this tangent ratio.

$$\frac{HT}{HA} \approx \tan 31°$$ The definition of tangent.

$$\frac{HT}{HA} \approx 0.6$$ The tangent of 31° is approximately 0.6.

$$\frac{HT}{36} \approx 0.6$$ Substitute 36 for HA.

$$HT \approx (36)(0.6)$$ Multiply both sides by 36 and reduce the left side.

$$HT \approx 22$$ Multiply.

The height of the tree is approximately 22 meters.

Deg.	Sin	Cos	Tan
12.0	0.2079	0.9781	0.2126
.1	.2096	.9778	.2144
.2	.2113	.9774	.2162
.3	.2130	.9770	.2180
.4	.2147	.9767	.2199
.5	.2164	.9763	.2217
.6	.2181	.9759	.2235
.7	.2198	.9755	.2254
.8	.2215	.9751	.2272
.9	.2233	.9748	.2290
13.0	0.2250	0.9744	0.2309
.1	.2267	.9740	.2327
.2	.2284	.9736	.2345
.3	.2300	.9732	.2364
.4	.2317	.9728	.2382
.5	.2334	.9724	.2401
.6	.2351	.9720	.2419
.7	.2368	.9715	.2438
.8	.2385	.9711	.2456
.9	.2402	.9707	.2475
14.0	0.2419	0.9703	0.2493
.1	.2436	.9699	.2512
.2	.2453	.9694	.2530

In order to solve problems like Example A, early mathematicians made tables that related ratios of side lengths to angle measures. They named six possible ratios. You will work with these three: **sine, cosine,** and **tangent,** abbreviated sin, cos, and tan. **Sine** is the ratio of the length of the opposite side to the length of the hypotenuse. **Cosine** is the ratio of the length of the adjacent side to the length of the hypotenuse.

This excerpt from a trigonometric table shows sine, cosine, and tangent ratios for angles measuring from 12.0° to 14.2°.

Trigonometric Ratios

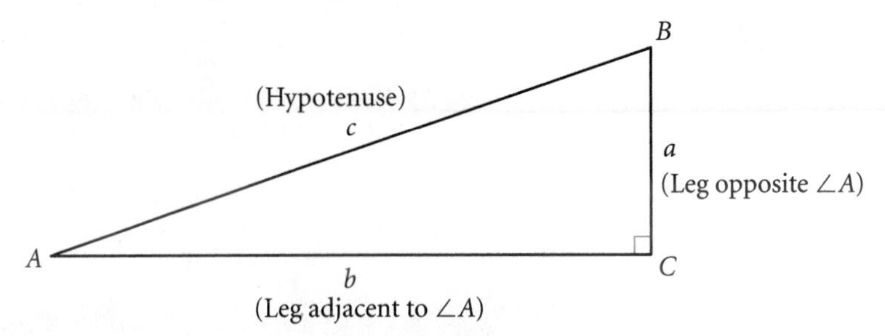

For an acute angle A in any right triangle ABC:

$$\text{sine of } \angle A = \frac{\text{length of leg opposite } \angle A}{\text{length of hypotenuse}} \quad \text{or} \quad \sin A = \frac{a}{c}$$

$$\text{cosine of } \angle A = \frac{\text{length of leg adjacent to } \angle A}{\text{length of hypotenuse}} \quad \text{or} \quad \cos A = \frac{b}{c}$$

$$\text{tangent of } \angle A = \frac{\text{length of leg opposite } \angle A}{\text{length of leg adjacent to } \angle A} \quad \text{or} \quad \tan A = \frac{a}{b}$$

Investigation
Trigonometric Tables

You will need

- a protractor
- a ruler

In this investigation you will make a small table of trigonometric ratios for angles measuring 20° and 70°.

Step 1 Use your protractor to make a large right triangle ABC with $m\angle A = 20°$, $m\angle B = 90°$, and $m\angle C = 70°$.

Step 2 Measure AB, AC, and BC to the nearest millimeter.

Step 3 Use your side lengths and the definitions of sine, cosine, and tangent to complete a table like this. Round your calculations to the nearest thousandth.

$m\angle A$	$\sin A$	$\cos A$	$\tan A$
20°			

$m\angle C$	$\sin C$	$\cos C$	$\tan C$
70°			

Step 4 Share your results with your group. Calculate the average of each ratio within your group. Create a new table with your group's average values.

Step 5 Discuss your results. What observations can you make about the trigonometric ratios you found? What is the relationship between the values for 20° and the values for 70°? Explain why you think these relationships exist.

Go to **www.keymath.com/DG** to find complete tables of trigonometric ratios.

Today, trigonometric tables have been replaced by calculators that have sin, cos, and tan keys.

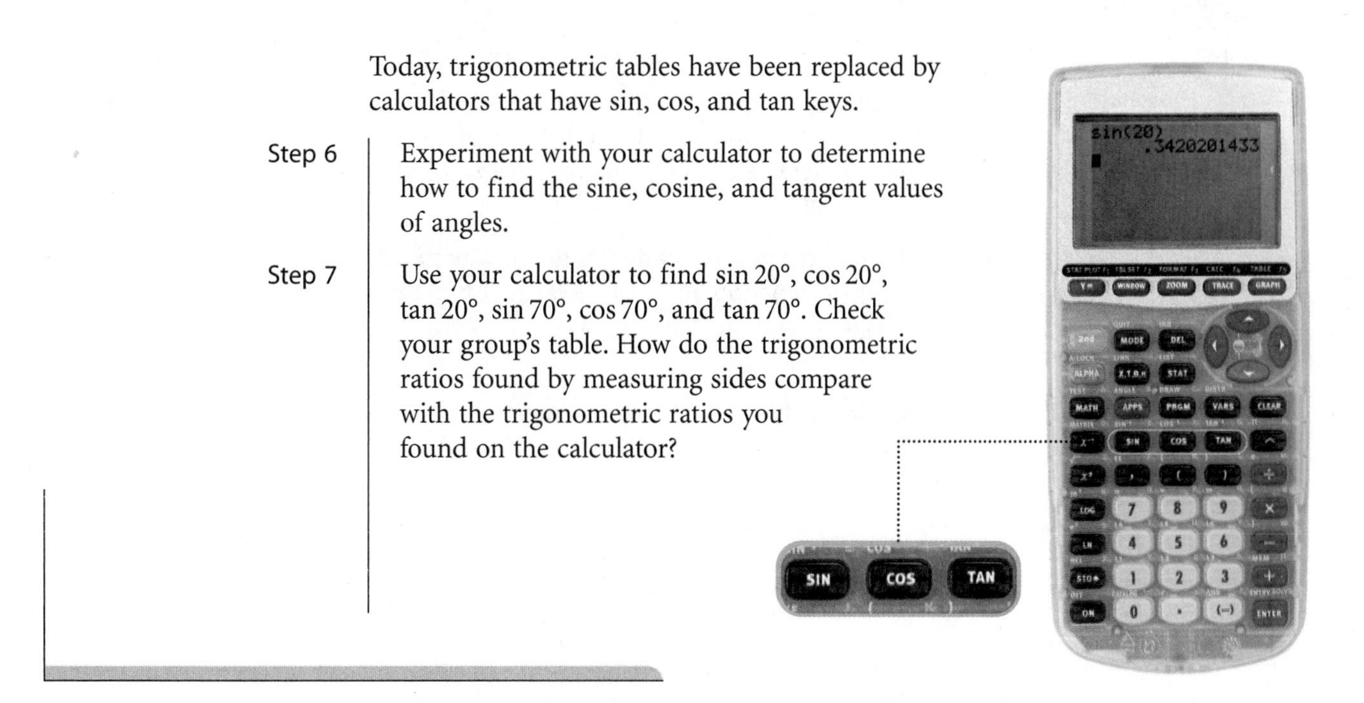

Step 6 Experiment with your calculator to determine how to find the sine, cosine, and tangent values of angles.

Step 7 Use your calculator to find $\sin 20°$, $\cos 20°$, $\tan 20°$, $\sin 70°$, $\cos 70°$, and $\tan 70°$. Check your group's table. How do the trigonometric ratios found by measuring sides compare with the trigonometric ratios you found on the calculator?

Using a table of trigonometric ratios, or using a calculator, you can find the approximate lengths of the sides of a right triangle given the measures of any acute angle and any side.

EXAMPLE B Find the length of the hypotenuse of a right triangle if an acute angle measures 20° and the side opposite the angle measures 410 feet.

▶ Solution

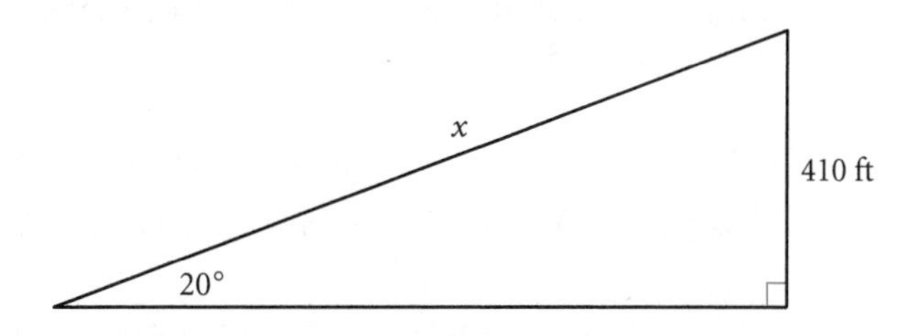

Sketch a diagram. The trigonometric ratio that relates the lengths of the opposite side and the hypotenuse is the sine ratio.

$$\sin 20° = \frac{410}{x}$$ Substitute 20° for the measure of $\angle A$ and substitute 410 for the length of the opposite side. The length of the hypotenuse is unknown, so use x.

$$x(\sin 20°) = 410$$ Multiply both sides by x and reduce the right side.

$$x = \frac{410}{\sin 20°}$$ Divide both sides by $\sin 20°$ and reduce the left side.

From your table in the investigation, or from a calculator, you know that $\sin 20°$ is approximately 0.342.

$$x \approx \frac{410}{0.342}$$ Sin 20° is approximately 0.342.

$$x \approx 1199$$ Divide.

The length of the hypotenuse is approximately 1199 feet.

With the help of a calculator, it is also possible to determine the size of either acute angle in a right triangle if you know the length of any two sides of that triangle. For instance, if you know the ratio of the legs in a right triangle, you can find the measure of one acute angle by using the **inverse tangent,** or $\tan^{-1}$, function. Let's look at an example.

The inverse tangent of x is defined as the measure of the acute angle whose tangent is x. The tangent function and inverse tangent function undo each other. That is, $\tan^{-1}(\tan A) = A$ and $\tan(\tan^{-1} x) = x$.

EXAMPLE C

A right triangle has legs of length 8 inches and 15 inches. Find the measure of the angle opposite the 8-inch leg.

▶ ***Solution***

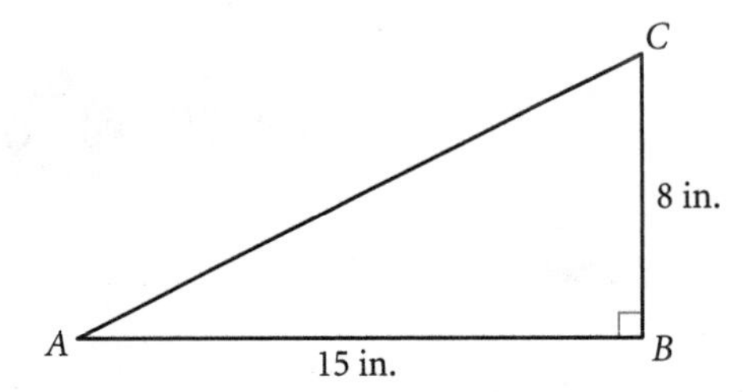

Sketch a diagram. In this sketch the angle opposite the 8-inch side is $\angle A$. The trigonometric ratio that relates the lengths of the opposite side and the adjacent side is the tangent ratio.

$$\tan A = \frac{8}{15}$$ Substitute 8 for the length of the opposite side and substitute 15 for the length of the adjacent side.

To find the angle that has an approximate tangent value of $\frac{8}{15}$, you can use a calculator to find the inverse tangent of $\frac{8}{15}$, or $\tan^{-1}\left(\frac{8}{15}\right)$.

$$A \approx \tan^{-1}\left(\frac{8}{15}\right)$$ Take the inverse tangent of both sides.

$$A \approx 28$$ Use your calculator to evaluate $\tan^{-1}\left(\frac{8}{15}\right)$.

The measure of the angle opposite the 8-inch side is approximately 28°.

You can also use inverse sine, or $\sin^{-1}$, and inverse cosine, or $\cos^{-1}$, to find angle measures.

EXERCISES

You will need

A calculator for Exercises **1–6** and **10–22**

For Exercises 1–3, use a calculator to find each trigonometric ratio accurate to four decimal places.

1. $\sin 37°$ **2.** $\cos 29°$ **3.** $\tan 8°$

For Exercises 4–6, solve for x. Express each answer accurate to two decimal places.

4. $\sin 40° = \frac{x}{18}$ **5.** $\cos 52° = \frac{19}{x}$ **6.** $\tan 29° = \frac{x}{112}$

For Exercises 7–9, find each trigonometric ratio.

7. $\sin A = \underline{?}$
$\cos A = \underline{?}$
$\tan A = \underline{?}$ ⓗ

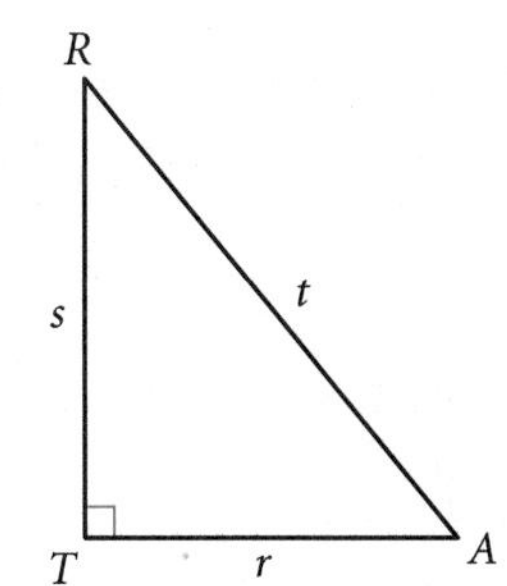

8. $\sin \theta = \underline{?}$
$\cos \theta = \underline{?}$
$\tan \theta = \underline{?}$

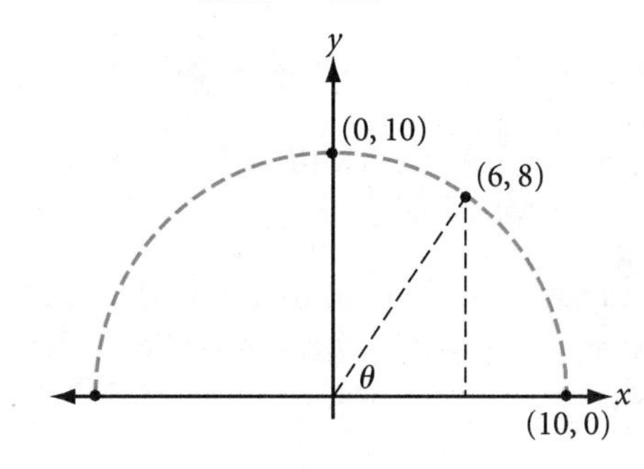

9. $\sin A = \underline{?}$ $\sin B = \underline{?}$
$\cos A = \underline{?}$ $\cos B = \underline{?}$
$\tan A = \underline{?}$ $\tan B = \underline{?}$

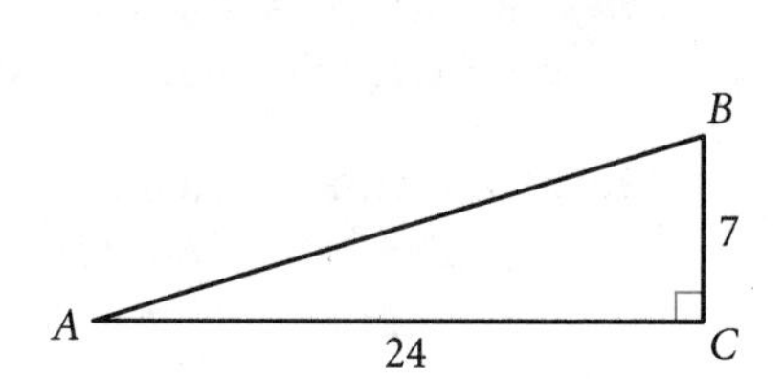

For Exercises 10–13, find the measure of each angle accurate to the nearest degree.

10. $\sin A = 0.5$ ⓗ

11. $\cos B = 0.6$

12. $\tan C = 0.5773$

13. $\tan x = \frac{48}{106}$

For Exercises 14–20, find the values of *a*–*g* accurate to the nearest whole unit.

14. ⓗ

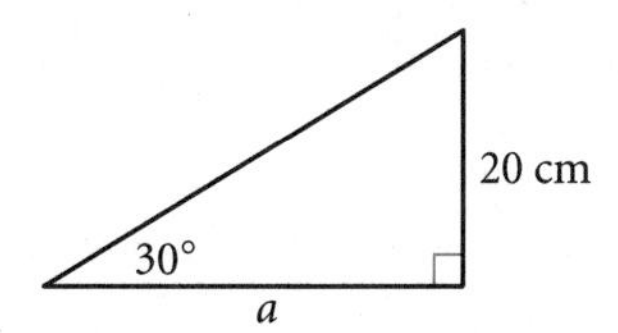

15.

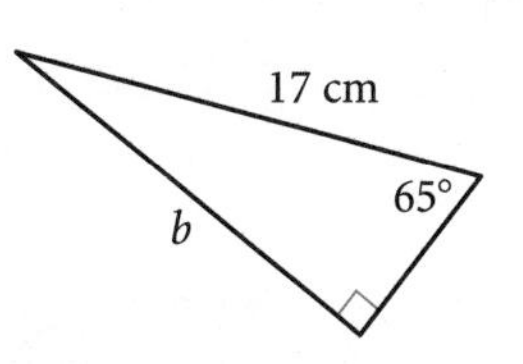

16.

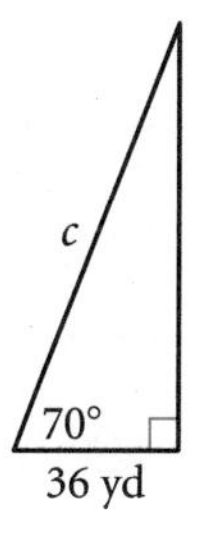

17.

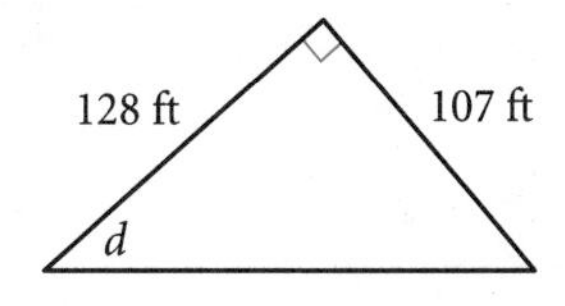

18.

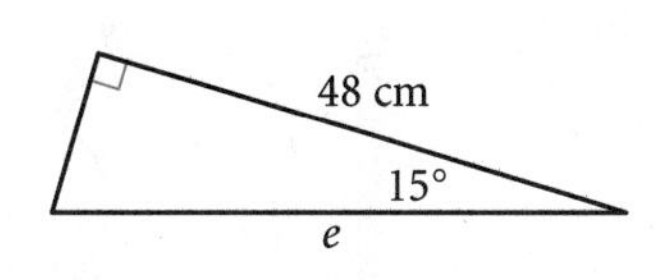

19.

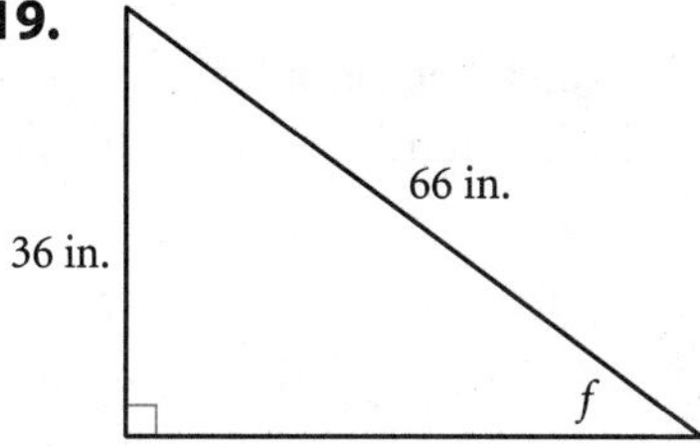

20.

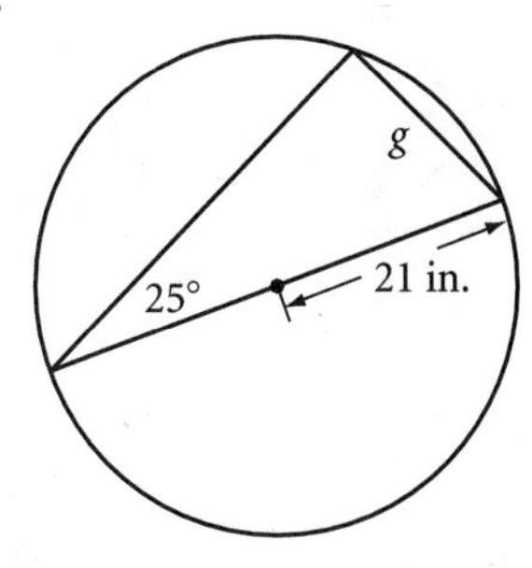

21. Find the perimeter of this quadrilateral. ⓗ

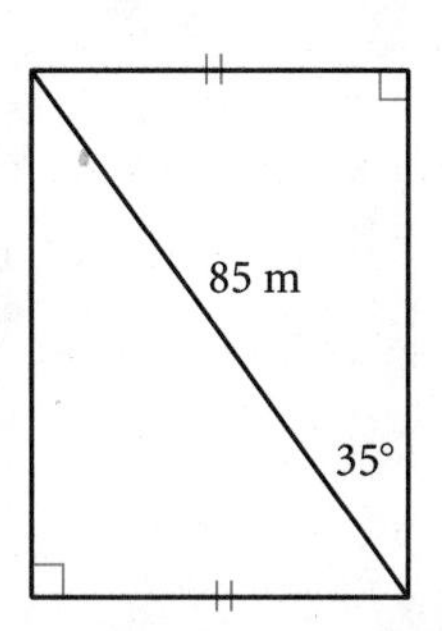

22. Find *x*.

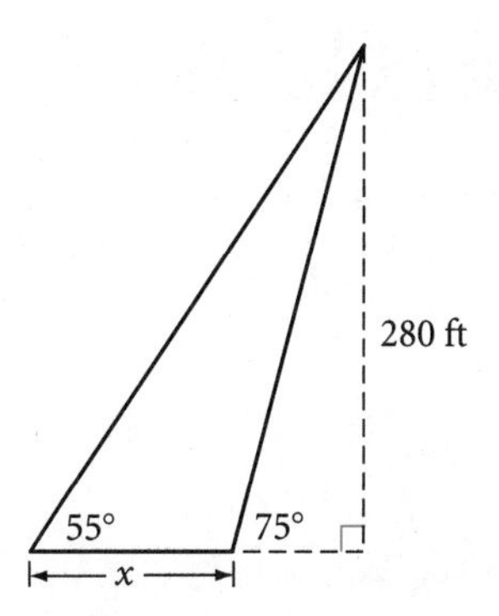

Review

For Exercises 23 and 24, solve for x.

23. $\frac{x}{3} = \frac{17}{8}$

24. $\frac{5}{x} = \frac{25}{11}$

25. APPLICATION Which is the better buy? A pizza with a 16-inch diameter for \$12.50, or a pizza with a 20-inch diameter for \$20.00?

26. APPLICATION Which is the better buy? Ice cream in a cylindrical container with a base diameter of 6 inches and a height of 8 inches for \$3.98, or ice cream in a box (square prism) with a base edge of 6 inches and a height of 8 inches for \$4.98?

27. A diameter of a circle is cut at right angles by a chord into a 12 cm segment and a 4 cm segment. How long is the chord? ⓗ

28. Find the volume and surface area of this sphere.

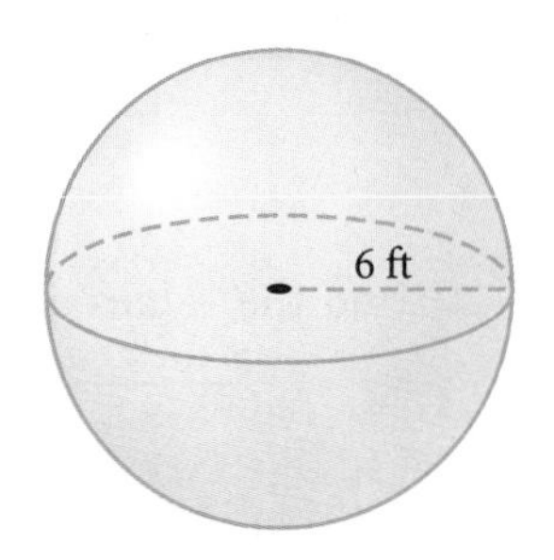

IMPROVING YOUR VISUAL THINKING SKILLS

3-by-3 Inductive Reasoning Puzzle II

Sketch the figure missing in the lower right corner of this pattern.

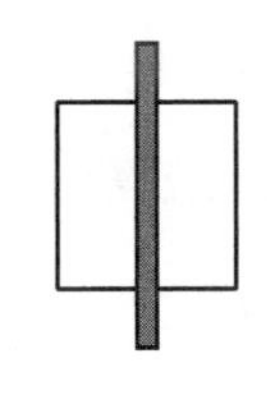
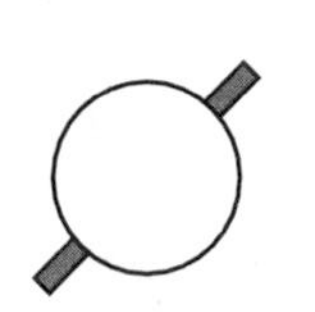
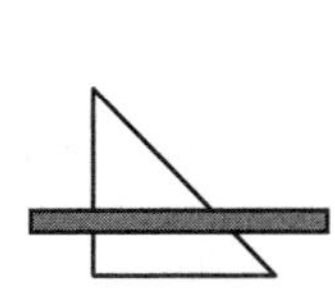
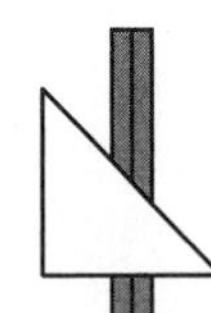

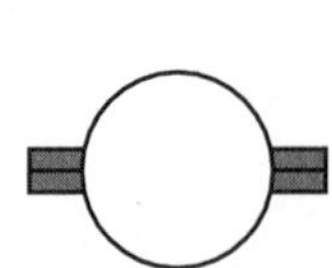
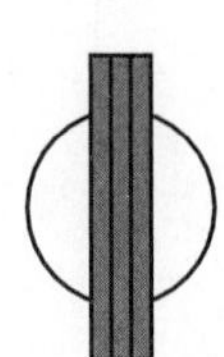
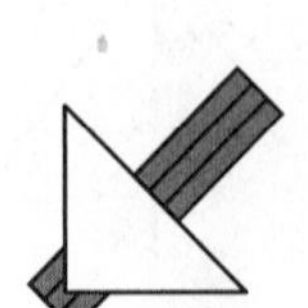

LESSON

12.2

Problem Solving with Right Triangles

What science can there be more noble, more excellent, more useful . . . than mathematics?

BENJAMIN FRANKLIN

Right triangle trigonometry is often used indirectly to find the height of a tall object. To solve a problem of this type, measure the angle from the horizontal to your line of sight when you look at the top or bottom of the object.

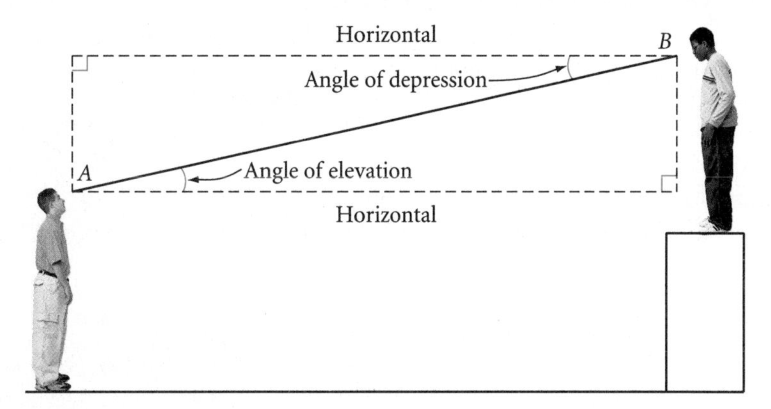

If you look up, you measure the **angle of elevation.** If you look down, you measure the **angle of depression.**

Here's an example.

EXAMPLE

The angle of elevation from a sailboat to the top of a 121-foot lighthouse on the shore measures 16°. To the nearest foot, how far is the sailboat from shore?

▸ Solution

The height of the lighthouse is opposite the 16° angle. The unknown distance is the adjacent side. Set up a tangent ratio.

$$\tan 16^\circ = \frac{121}{d}$$

$$d(\tan 16^\circ) = 121$$

$$d = \frac{121}{\tan 16^\circ}$$

$$d \approx 422$$

The sailboat is approximately 422 feet from shore.

EXERCISES

You will need

A calculator for Exercises **1–19**

1. According to a Chinese legend from the Han dynasty (206 B.C.E.–220 C.E.), General Han Xin flew a kite over the palace of his enemy to determine the distance between his troops and the palace. If the general let out 800 meters of string and the kite was flying at a 35° angle of elevation, how far away was the palace from General Han Xin's position?

2. Benny is flying a kite directly over his friend, Frank, who is 125 meters away. When he holds the kite string down to the ground, the string makes a 39° angle with the level ground. How high is Benny's kite?

3. **APPLICATION** The angle of elevation from a ship to the top of a 42-meter lighthouse on the shore measures 33°. How far is the ship from the shore? (Assume the horizontal line of sight meets the bottom of the lighthouse.)

4. **APPLICATION** A salvage ship's sonar locates wreckage at a 12° angle of depression. A diver is lowered 40 meters to the ocean floor. How far does the diver need to walk along the ocean floor to the wreckage?

5. **APPLICATION** A meteorologist shines a spotlight vertically onto the bottom of a cloud formation. He then places an angle-measuring device 65 meters from the spotlight and measures a 74° angle of elevation from the ground to the spot of light on the clouds. How high are the clouds?

The distance from the ground to a cloud formation is called the cloud *ceiling*.

6. **APPLICATION** Meteorologist Wendy Stevens uses a theodolite (an angle-measuring device) on a 1-meter-tall tripod to find the height of a weather balloon. She views the balloon at a 44° angle of elevation. A radio signal from the balloon tells her that it is 1400 meters from her theodolite.
   a. How high is the balloon? ⓗ
   b. How far is she from the point directly below the balloon?
   c. If Wendy's theodolite were on the ground rather than on a tripod, would your answers change? Explain your reasoning.

Science CONNECTION

Weather balloons carry into the atmosphere what is called a *radiosonde,* an instrument with sensors that detect information about wind direction, temperature, air pressure, and humidity. Twice a day across the world, this upper-air data is transmitted by radio waves to a receiving station. Meteorologists use the information to forecast the weather.

7. **APPLICATION** A ship's officer sees a lighthouse at a 42° angle to the path of the ship. After the ship travels 1800 m, the lighthouse is at a 90° angle to the ship's path. What is the distance between the ship and the lighthouse at this second sighting? ⓗ

When there are no visible landmarks, sailors at sea depend on the location of stars or the Sun for navigation. For example, in the Northern Hemisphere, Polaris (the North Star), stays approximately at the same angle above the horizon for a given latitude. If Polaris appears higher overhead or closer to the horizon, sailors can tell whether their course is taking them north or south.

This painting by Winslow Homer (1836–1910) is titled *Breezing Up* (1876).

For Exercises 8–16, find each length or angle measure accurate to the nearest whole unit.

8. $a \approx$?

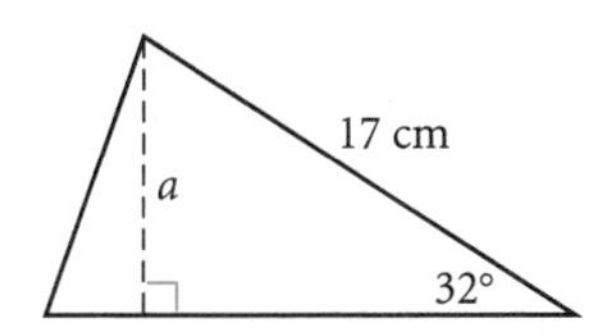

9. $x \approx$?

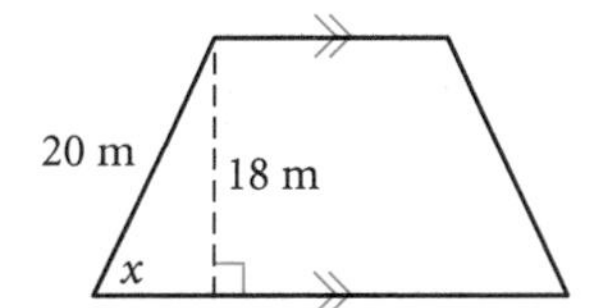

10. $r \approx$?

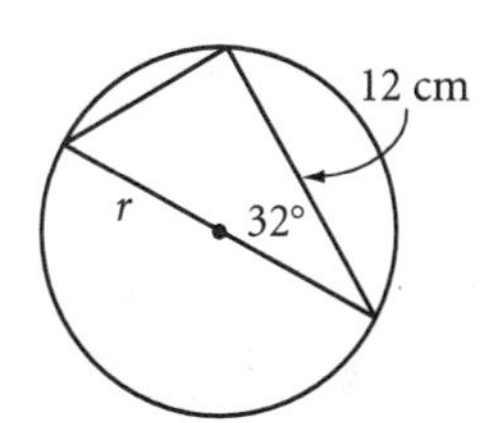

11. $e \approx$?

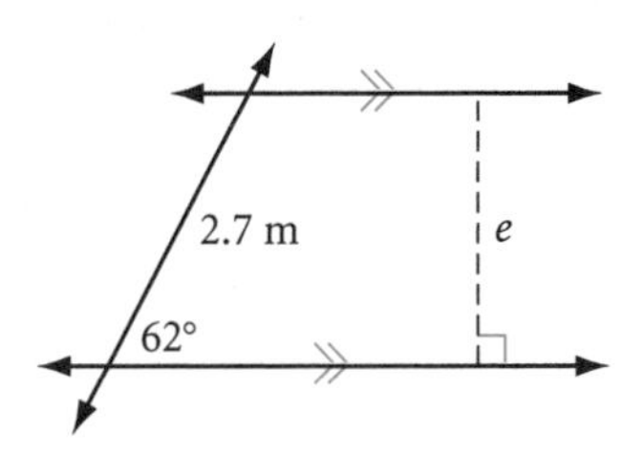

12. $d_1 \approx$? ⓗ

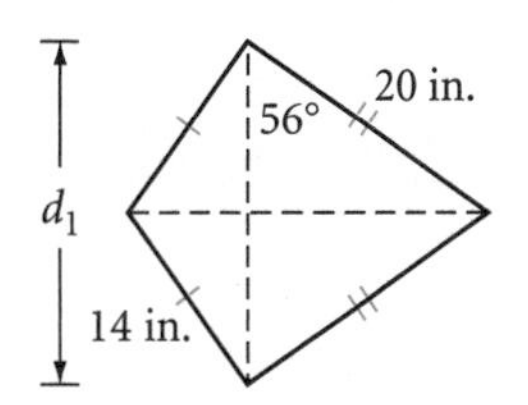

13. $f \approx$?

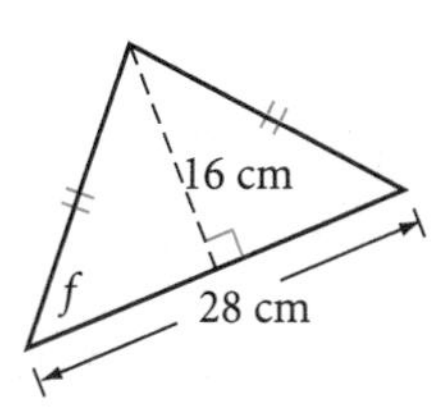

14. $\theta \approx$?

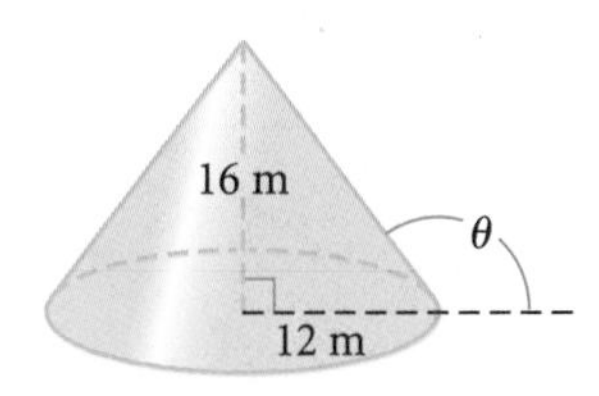

15. $\beta \approx$? ⓗ

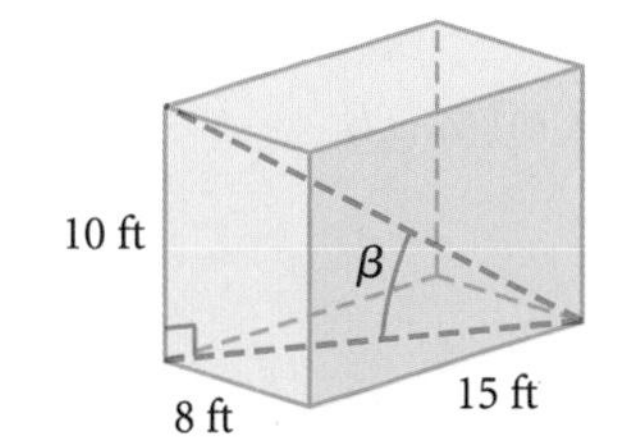

16. $h \approx$?

Review

For Exercises 17–19, find the measure of each angle to the nearest degree.

17. $\sin D = 0.7071$ **18.** $\tan E = 1.7321$ **19.** $\cos F = 0.5$

Technology CONNECTION

The earliest known navigation tool was used by the Polynesians, yet it didn't measure angles. Early Polynesians carried several different-length hooks made from split bamboo and shells. A navigator held a hook at arm's length, positioned the bottom of the hook on the horizon, and sighted the North Star through the top of the hook. The length of the hook indicated the navigator's approximate latitude. Can you use trigonometry to explain how this method works?

20. Solve for x.

a. $4.7 = \frac{x}{3.2}$

b. $8 = \frac{16.4}{x}$

c. $0.3736 = \frac{x}{14}$

d. $0.9455 = \frac{2.5}{x}$

21. Find x and y.

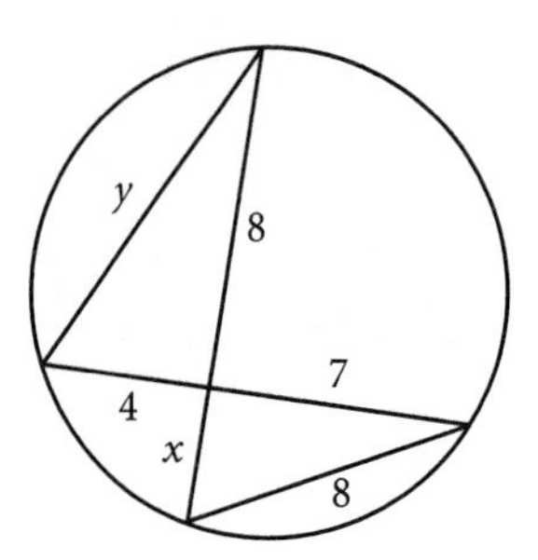

22. A 3-by-5-by-6 cm block of wood is dropped into a cylindrical container of water with radius 5 cm. The level of the water rises 0.8 cm. Does the block sink or float? Explain how you know.

23. Scalene triangle ABC has altitudes $\overline{AX}$, $\overline{BY}$, and $\overline{CZ}$. If $AB > BC > AC$, write an inequality that relates the heights.

24. In the diagram at right, $\overrightarrow{PT}$ and $\overrightarrow{PS}$ are tangent to circle O at points T and S, respectively. As point P moves to the right along $\overleftrightarrow{AB}$, describe what happens to each of these measures or ratios.

a. $m\angle TPS$ b. OD

c. $m\angle ATB$ d. Area of $\triangle ATB$

e. $\frac{AP}{BP}$ f. $\frac{AD}{BD}$

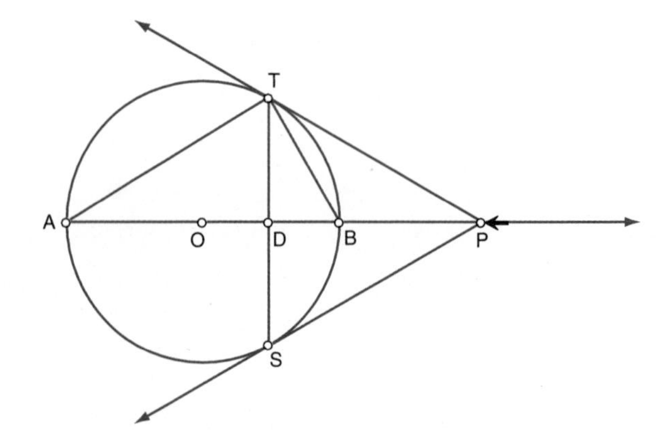

25. Points S and Q, shown at right, are consecutive vertices of square $SQRE$. Find coordinates for the other two vertices, R and E. There are two possible answers. Try to find both.

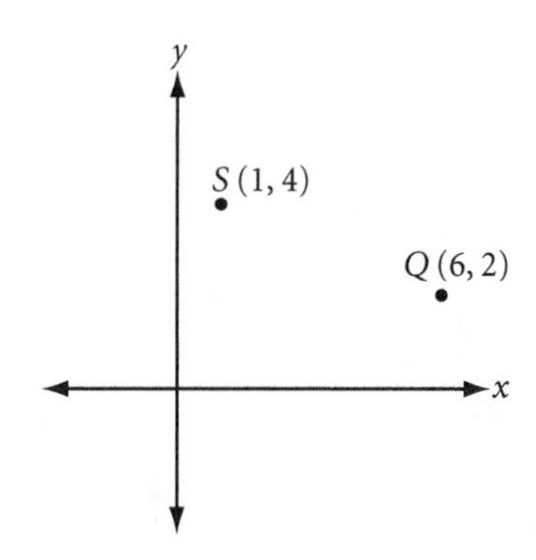

project

LIGHT FOR ALL SEASONS

You have seen that roof design is a practical application of slope—steep roofs shed snow and rain. But have you thought about the overhang of a roof?

In a hot climate, a deep overhang shelters windows from the sun.

In a cold climate, a narrow overhang lets in more light and warmth.

What roof design is common for homes in your area? What factors would an architect consider in the design of a roof relative to the position, size, and orientation of the windows? Do some research and build a shoebox model of the roof design you select.

What design is best for your area will depend on your latitude, because that determines the angle of the sun's light in different seasons. Research the astronomy of solar angles, then use trigonometry and a movable light source to illustrate the effects on your model.

Your project should include

- Research notes on seasonal solar angles.
- A narrative explanation, with mathematical support, for your choice of roof design, roof overhang, and window placement.
- Detailed, labeled drawings showing the range of light admitted from season to season, at a given time of day.
- A model with a movable light source.

Exploration

Indirect Measurement

In Chapter 11, you used shadows, mirrors, and similar triangles to measure the height of tall objects that you couldn't measure directly. Right triangle trigonometry gives you yet another method of indirect measurement.

In this exploration, you will use two or three different methods of indirect measurement. Then you will compare your results from each method.

Activity

Using a Clinometer

You will need

- a measuring tape or metersticks
- a clinometer (use the Making a Clinometer worksheet or make one of your own design)
- a mirror

In this activity, you will use a **clinometer**—a protractor-like tool used to measure angles. You probably will want to make your clinometer in advance, based on one of the designs below. Practice using it before starting the activity.

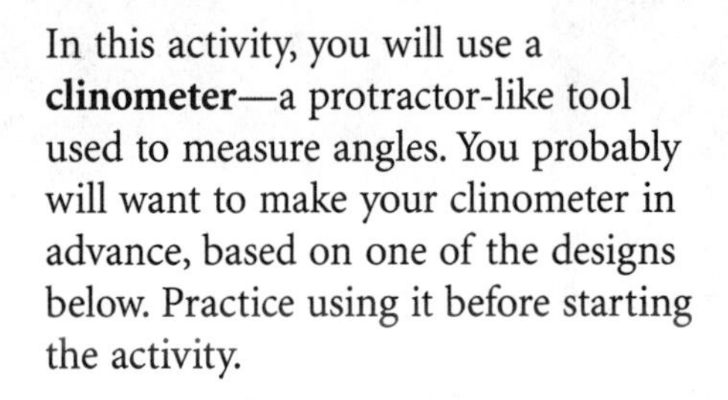

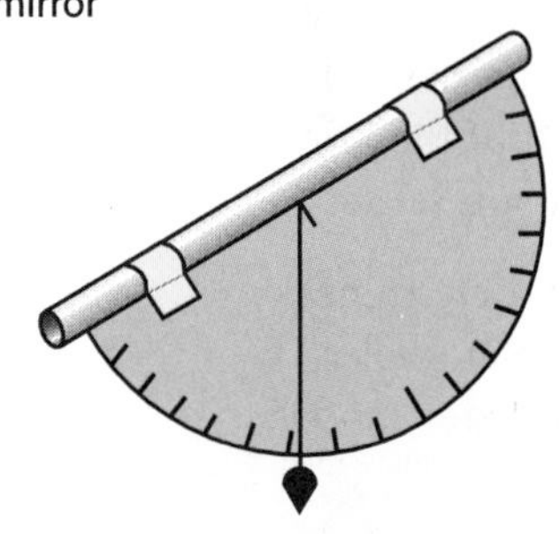

Clinometer 1

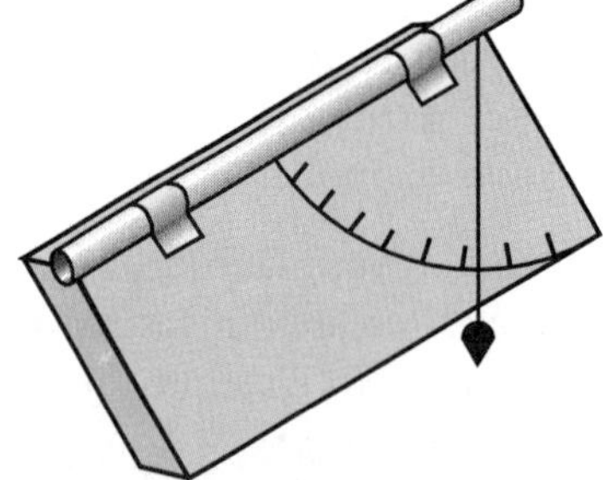

Clinometer 2

Step 1 Locate a tall object that would be difficult to measure directly. Start a table like this one.

Name of object	Viewing angle	Height of observer's eye	Distance from observer to object	Calculated height of object

Step 2 Use your clinometer to measure the viewing angle from the horizontal to the top of the object.

Step 3 Measure the observer's eye height. Measure the distance from the observer to the base of the object.

Step 4 Calculate the approximate height of the object.

U.S. Forest Service Ranger Al Sousi uses a clinometer to measure the angle of a mountain slope. In snowy conditions, a slope steeper than 35° can be a high avalanche hazard.

Step 5 Use either the shadow method or the mirror method or both to measure the height of the same object. How do your results compare? If you got different results, explain what part of each process could contribute to the differences.

Step 6 Repeat Steps 1–5 for another tall object. If you measure the height of the same object as another group, compare your results when you finish.

IMPROVING YOUR VISUAL THINKING SKILLS

Puzzle Shapes

Make five of these shapes and assemble them to form a square. Does it take three, four, or five of the shapes to make a square?

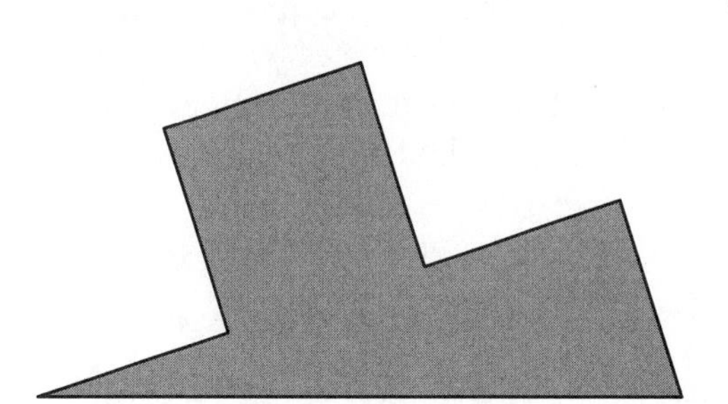

LESSON

12.3 The Law of Sines

To think and to be fully alive are the same.

HANNAH ARENDT

So far you have used trigonometry only to solve problems with right triangles. But you can use trigonometry with any triangle. For example, if you know the measures of two angles and one side of a triangle, you can find the other two sides with a trigonometric property called the **Law of Sines.** The Law of Sines is related to the area of a triangle. Let's first see how trigonometry can help you find area.

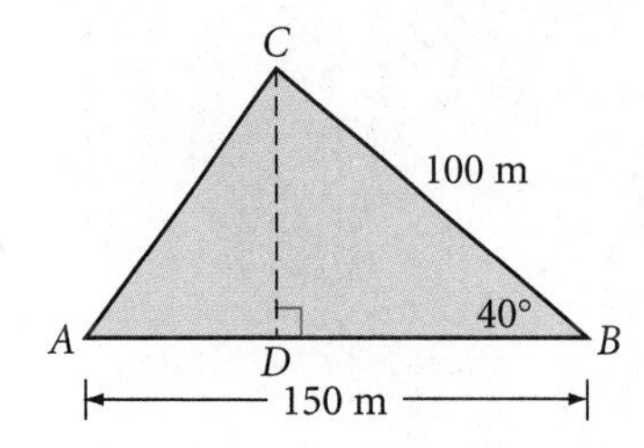

EXAMPLE A Find the area of $\triangle ABC$.

Solution Consider $\overline{AB}$ as the base and use trigonometry to find the height, CD.

$\sin 40° = \frac{CD}{100}$ — In $\triangle BCD$, CD is the length of the opposite side and 100 is the length of the hypotenuse.

$(100)(\sin 40°) = CD$ — Multiply both sides by 100 and reduce the right side.

Now find the area.

$A = 0.5bh$ — Area formula for a triangle.

$A = (0.5)(AB)(CD)$ — Substitute AB for the length of the base and CD for the height.

$A = (0.5)(150)[(100)(\sin 40°)]$ — Substitute 150 for AB and substitute the expression $(100)(\sin 40°)$ for CD.

$A \approx 4821$ — Evaluate.

The area is approximately 4821 m^2.

In the next investigation, you will find a general formula for the area of a triangle given the lengths of two sides and the measure of the included angle.

Investigation 1
Area of a Triangle

Step 1 Find the area of each triangle. Use Example A as a guide.

a.

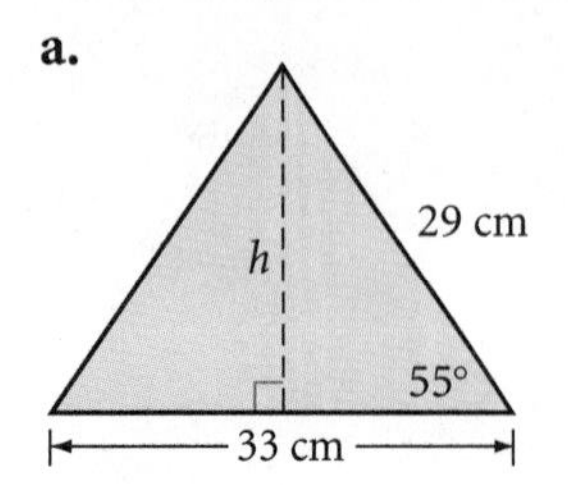

b.

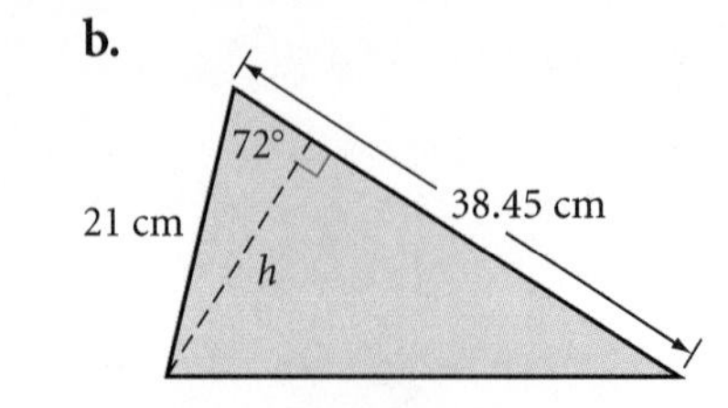

c.

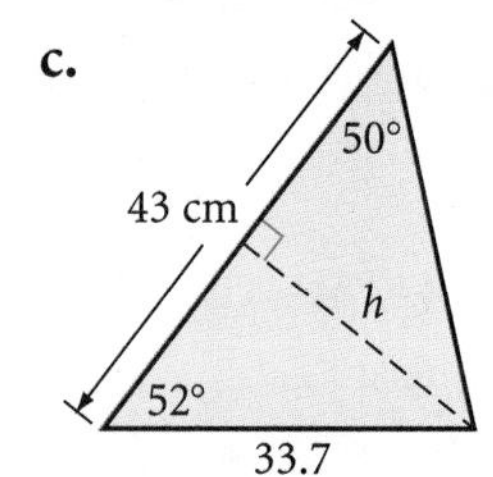

Step 2 Generalize Step 1 to find the area of this triangle in terms of a, b, and $\angle C$. State your general formula as your next conjecture.

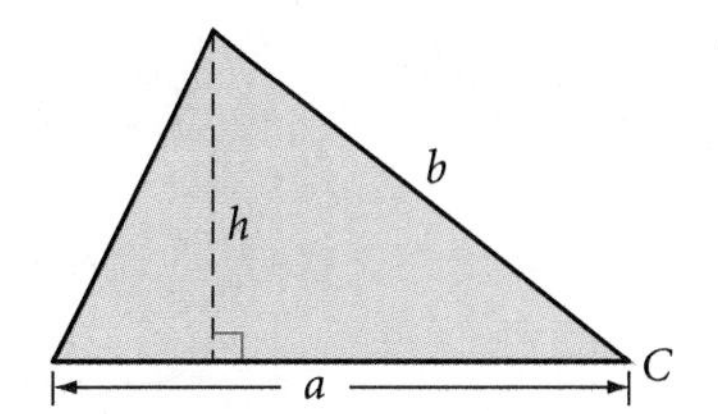

SAS Triangle Area Conjecture

C-102

The area of a triangle is given by the formula $A = \underline{\ ?\ }$, where a and b are the lengths of two sides and C is the angle between them.

Now use what you've learned about finding the area of a triangle to derive the property called the Law of Sines.

Investigation 2
The Law of Sines

Consider $\triangle ABC$ with height h.

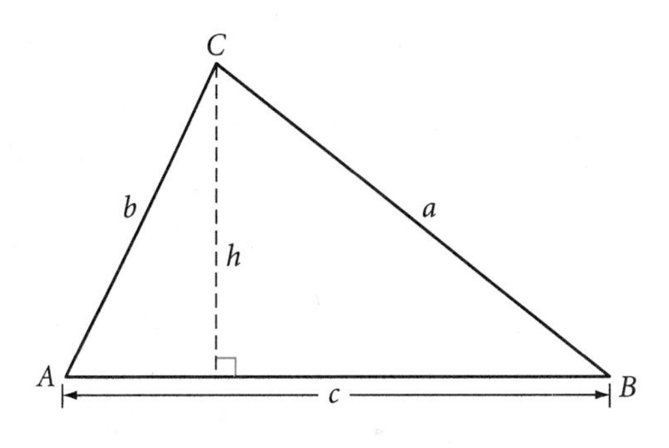

Step 1 Find h in terms of a and the sine of an angle.

Step 2 Find h in terms of b and the sine of an angle.

Step 3 Use algebra to show

$$\frac{\sin A}{a} = \frac{\sin B}{b}$$

Now consider the same $\triangle ABC$ using a different height, k.

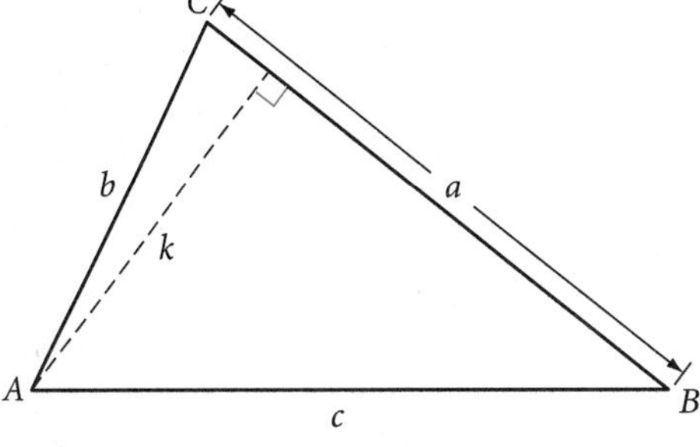

Step 4 Find k in terms of c and the sine of an angle.

Step 5 Find k in terms of b and the sine of an angle.

Step 6 Use algebra to show

$$\frac{\sin B}{b} = \frac{\sin C}{c}$$

Step 7 Combine Steps 3 and 6. Complete this conjecture.

Law of Sines

C-103

For a triangle with angles A, B, and C and sides of lengths a, b, and c (a opposite A, b opposite B, and c opposite C),

$$\frac{\sin A}{?} = \frac{?}{b} = \frac{?}{?}$$

Did you notice that you used deductive reasoning rather than inductive reasoning to discover the Law of Sines?

You can use the Law of Sines to find the lengths of a triangle's sides when you know one side's length and two angles' measures.

EXAMPLE B

Find the length of side $\overline{AC}$ in $\triangle ABC$.

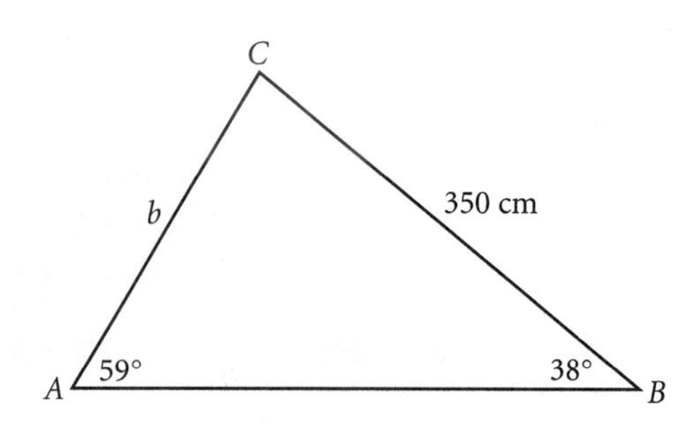

▶ Solution

Start with the Law of Sines, and solve for b.

$\frac{\sin A}{a} = \frac{\sin B}{b}$	The Law of Sines.
$b \sin A = a \sin B$	Multiply both sides by ab and reduce.
$b = \frac{a \sin B}{\sin A}$	Divide both sides by $\sin A$ and reduce the left side.
$b = \frac{(350)(\sin 38°)}{\sin 59°}$	Substitute 350 for a, 38° for B, and 59° for A.
$b \approx 251$	Multiply and divide.

The length of side $\overline{AC}$ is approximately 251 cm.

You can also use the Law of Sines to find the measure of a missing angle, but only if you know whether the angle is acute or obtuse. Recall from Chapter 4 that SSA failed as a congruence shortcut. For example, if you know in $\triangle ABC$ that $BC = 160$ cm, $AC = 260$ cm, and $m\angle A = 36°$, you would not be able to find $m\angle B$. There are two possible measures for $\angle B$, one acute and one obtuse.

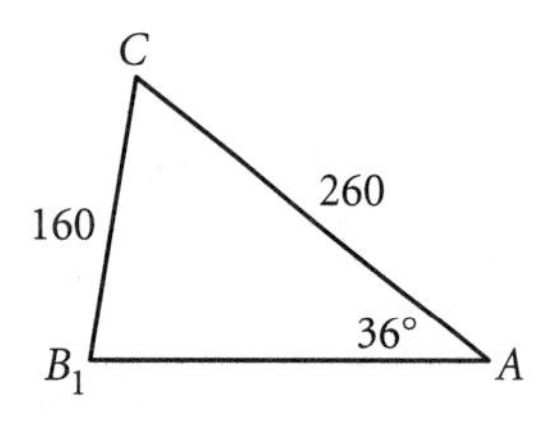

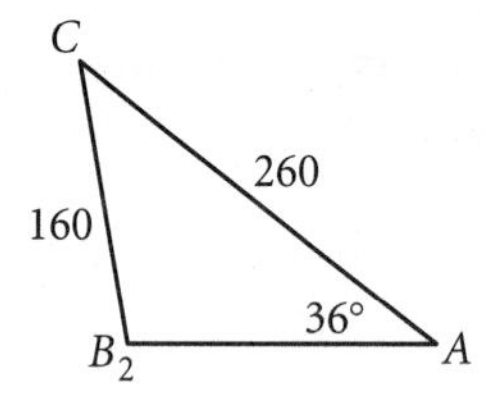

Because you've defined trigonometric ratios only for acute angles, you'll be asked to find only acute angle measures.

EXAMPLE C

Find the measure of acute angle B in $\triangle ABC$.

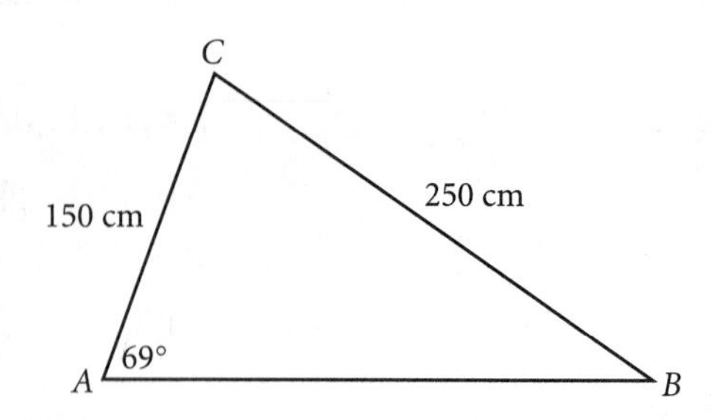

▶ ***Solution***

Start with the Law of Sines, and solve for B.

$$\frac{\sin A}{a} = \frac{\sin B}{b}$$ The Law of Sines.

$$\sin B = \frac{b \sin A}{a}$$ Solve for $\sin B$.

$$\sin B = \frac{(150)(\sin 69°)}{250}$$ Substitute known values.

$$B = \sin^{-1}\left[\frac{(150)(\sin 69°)}{250}\right]$$ Take the inverse sine of both sides.

$$B \approx 34$$ Use your calculator to evaluate.

The measure of $\angle B$ is approximately 34°.

EXERCISES

You will need

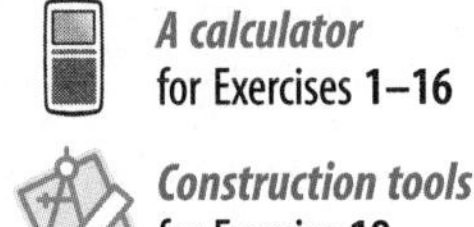

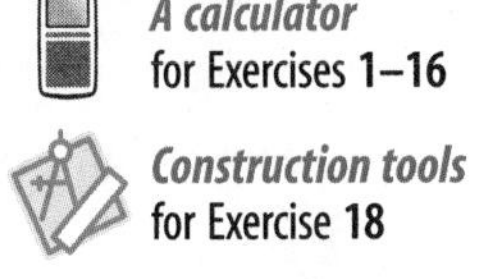

In Exercises 1–4, find the area of each polygon to the nearest square centimeter.

1.

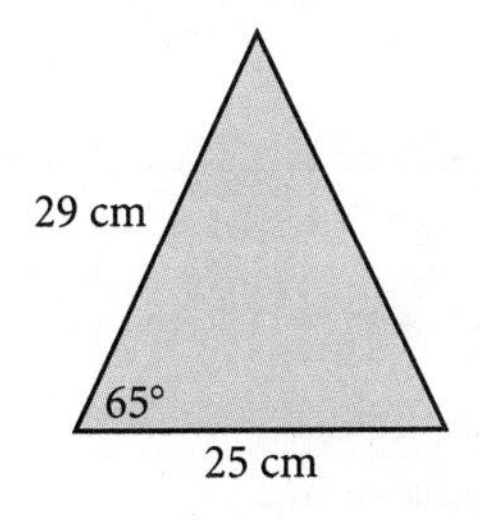

2.

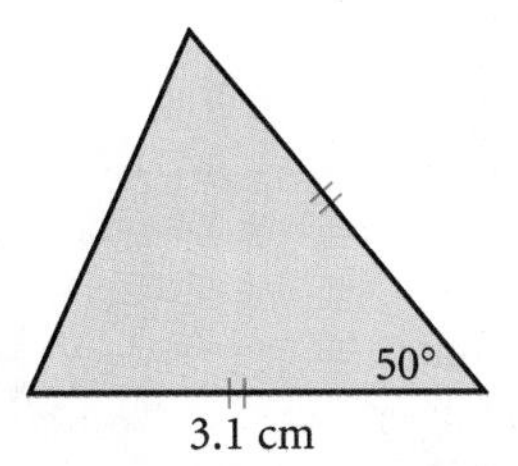

3. ⓗ

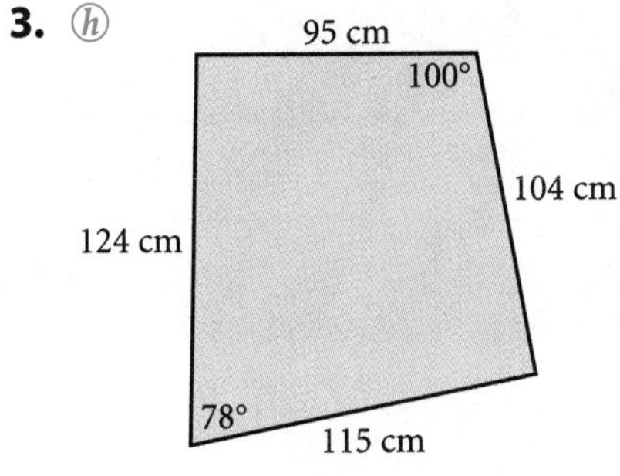

4. ⓗ

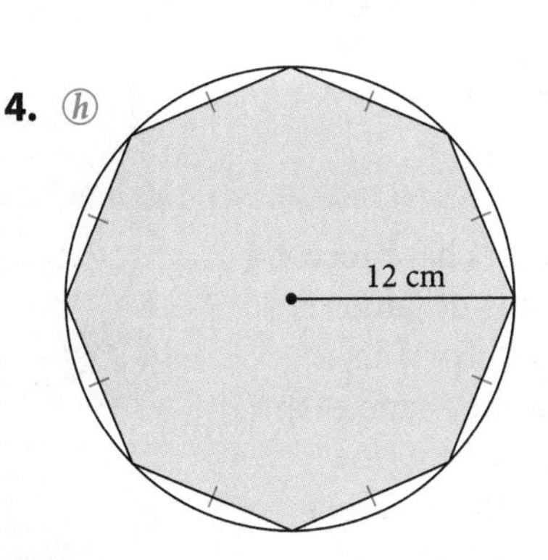

In Exercises 5–7, find each length to the nearest centimeter.

5. $w \approx$ __?__ ⓗ

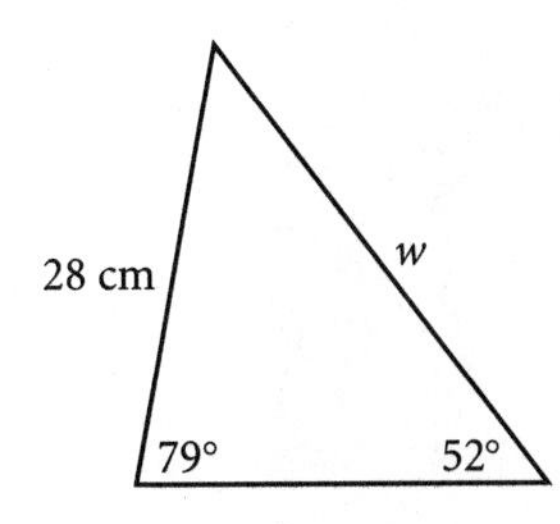

6. $x \approx$ __?__

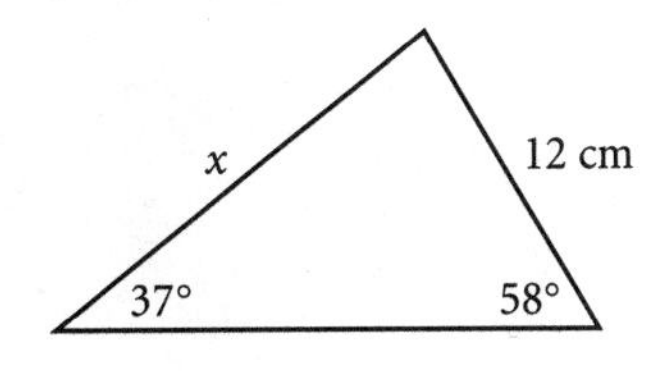

7. $y \approx$ __?__

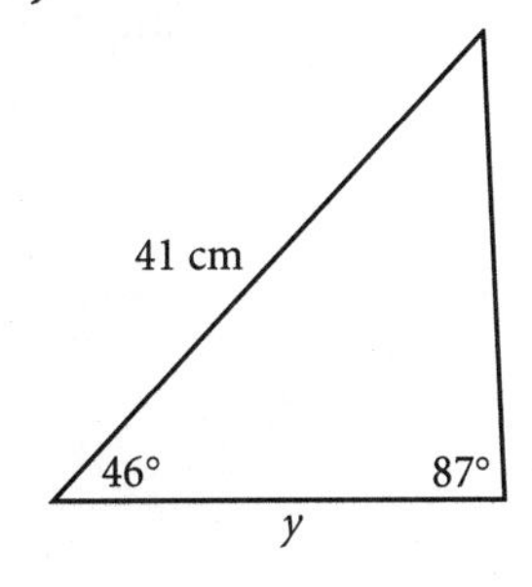

For Exercises 8–10, each triangle is an acute triangle. Find each angle measure to the nearest degree.

8. $m\angle A \approx$ _?_

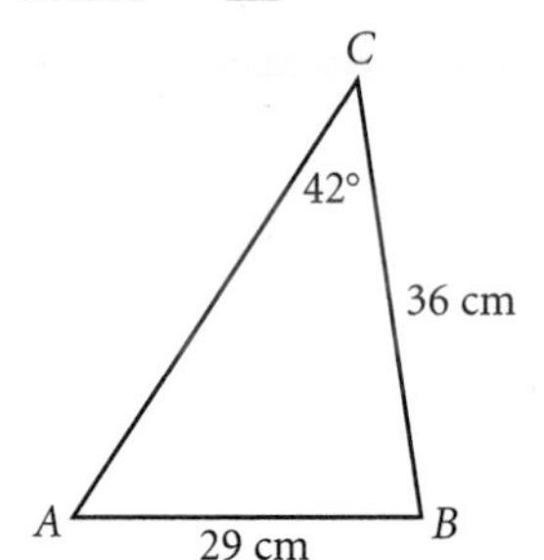

9. $m\angle B \approx$ _?_

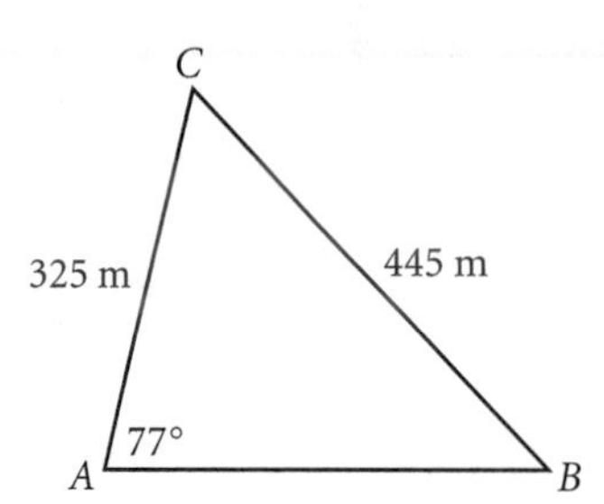

10. $m\angle C \approx$ _?_

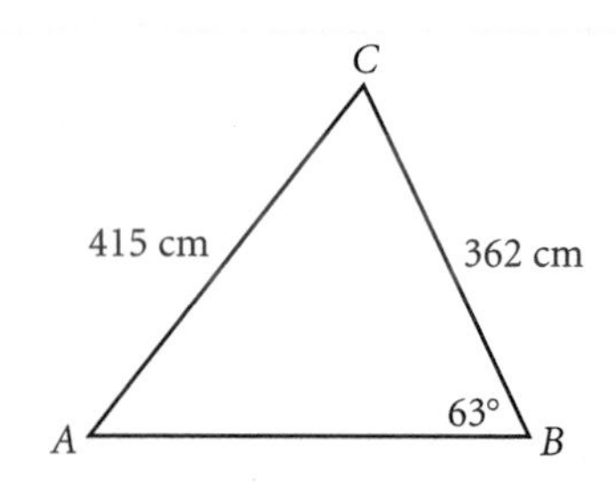

11. Alphonse (point A) is over a 2500-meter landing strip in a hot-air balloon. At one end of the strip, Beatrice (point B) sees Alphonse with an angle of elevation measuring 39°. At the other end of the strip, Collette (point C) sees Alphonse with an angle of elevation measuring 62°.

a. What is the distance between Alphonse and Beatrice?

b. What is the distance between Alphonse and Collette?

c. How high up is Alphonse?

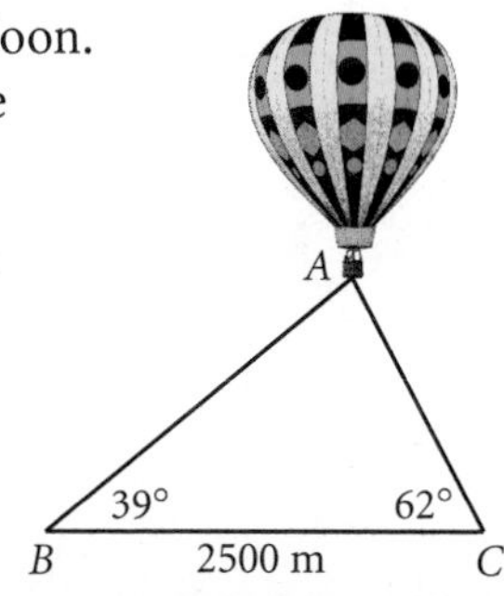

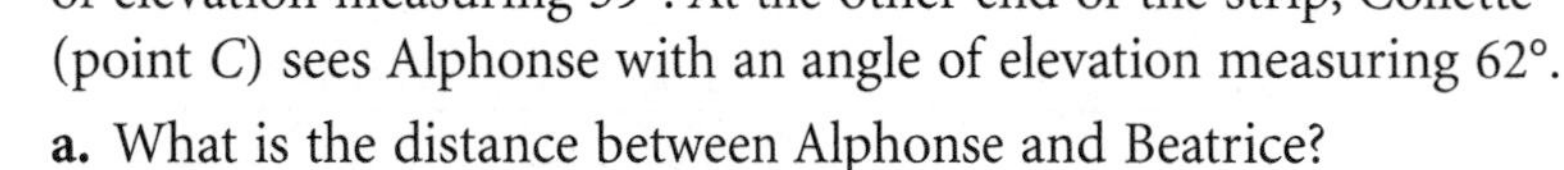

History CONNECTION

For over 200 years, people believed that the entire site of James Fort was washed into the James River. Archaeologists have recently uncovered over 250 feet of the fort's wall, as well as hundreds of thousands of artifacts dating to the early 1600s.

12. **APPLICATION** Archaeologists have recently started uncovering remains of James Fort (also known as Jamestown Fort) in Virginia. The fort was in the shape of an isosceles triangle. Unfortunately, one corner has disappeared into the James River. If the remaining complete wall measures 300 feet and the remaining corners measure 46.5° and 87°, how long were the two incomplete walls? What was the approximate area of the original fort?

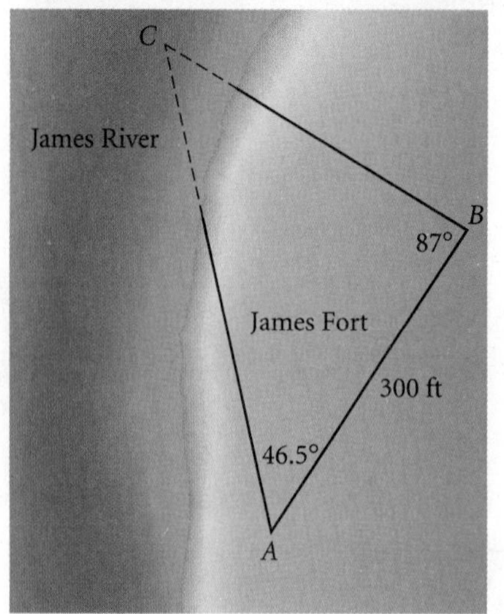

13. A tree grows vertically on a hillside. The hill is at a 16° angle to the horizontal. The tree casts an 18-meter shadow up the hill when the angle of elevation of the sun measures 68°. How tall is the tree? ⓗ

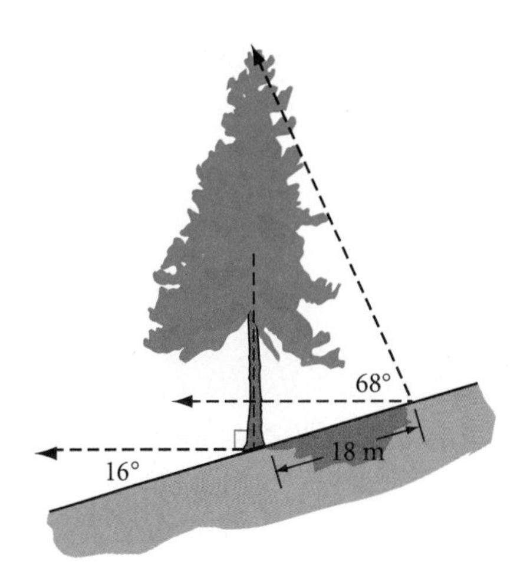

Review

14. Read the History Connection below. Each step of El Castillo is 30 cm deep by 26 cm high. How tall is the pyramid, not counting the platform at the top? What is the angle of ascent?

History CONNECTION

One of the most impressive Mayan pyramids is El Castillo in Chichén Itzá, Mexico. Built in approximately 800 C.E., it has 91 steps on each of its four sides, or 364 steps in all. The top platform adds a level, so the pyramid has 365 levels to represent the number of days in the Mayan year.

15. According to legend, Galileo (1564–1642, Italy) used the Leaning Tower of Pisa to conduct his experiments in gravity. Assume that when he dropped objects from the top of the 55-meter tower (this is the measured length, not the height, of the tower), they landed 4.8 meters from the tower's base. What was the angle that the tower was leaning from the vertical?

16. Find the volume of this cone.

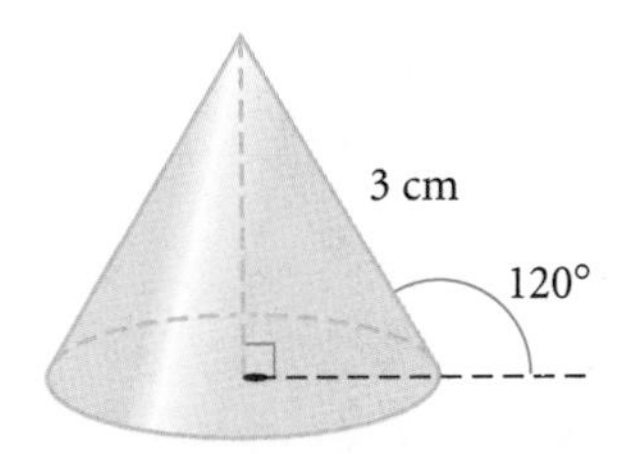

17. Use the circle diagram at right and write a paragraph proof to show that $\triangle ABE$ is isosceles.

18. *Construction* Put two points on patty paper. Assume these points are opposite vertices of a square. Find the two missing vertices.

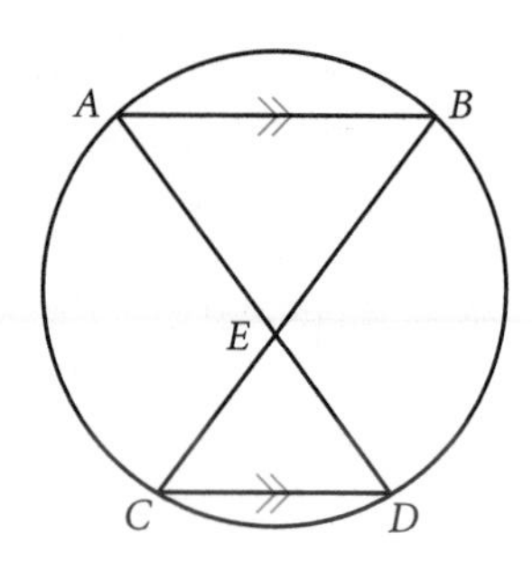

19. Find AC, AE, and AF. All measurements are in centimeters.

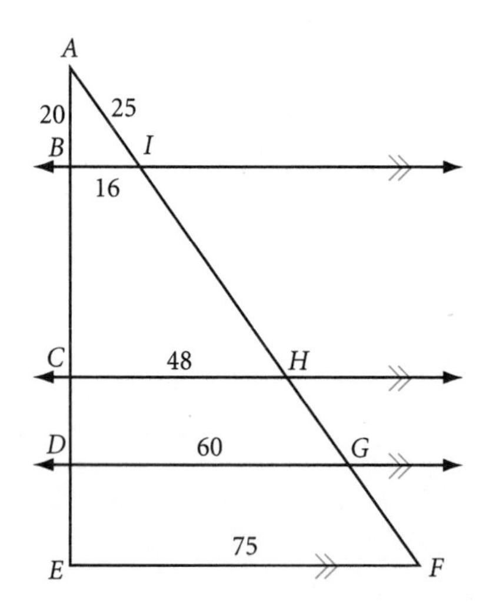

20. Both boxes are right rectangular prisms. In which is the diagonal rod longer?

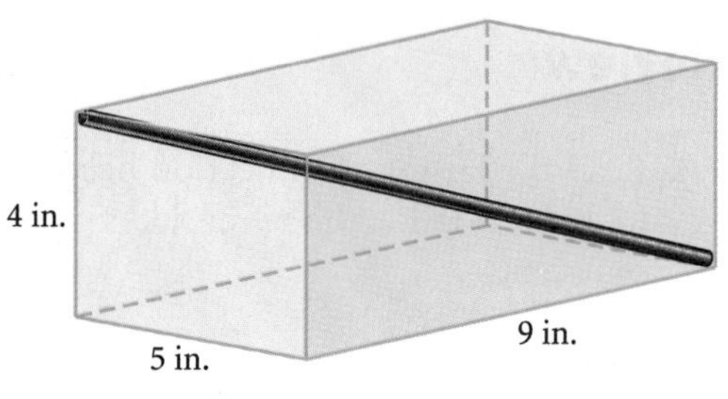

Box 1

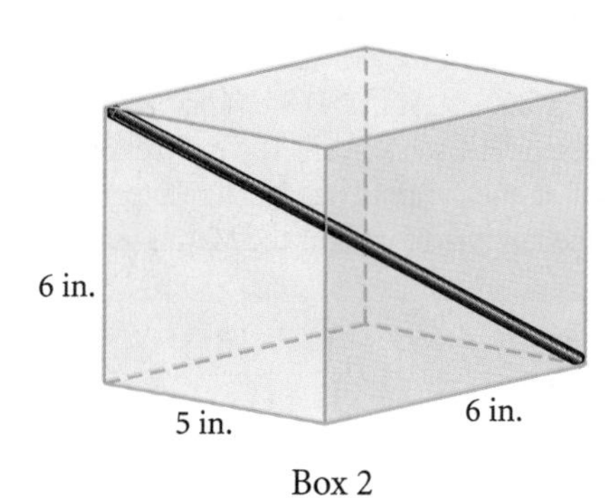

Box 2

IMPROVING YOUR VISUAL THINKING SKILLS

Rope Tricks

Each rope will be cut 50 times as shown. For each rope, how many pieces will result?

1.

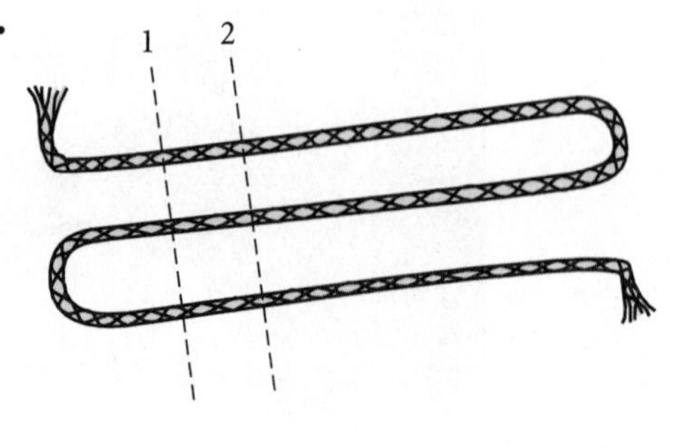

2.

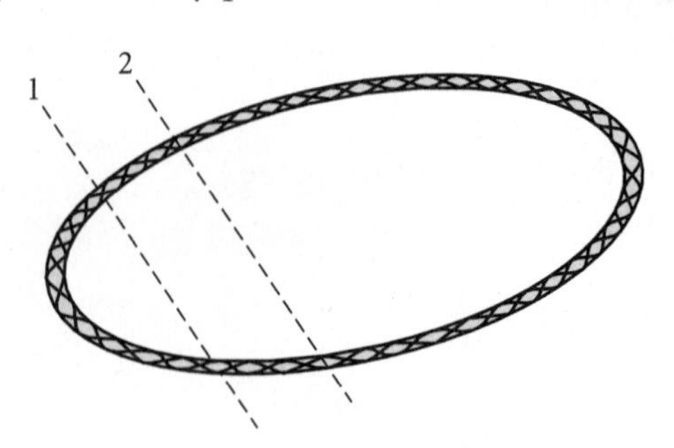

LESSON

12.4

The Law of Cosines

A ship in a port is safe, but that is not what ships are built for.

JOHN A. SHEDD

You've solved a variety of problems with the Pythagorean Theorem. It is perhaps your most important geometry conjecture. In Chapter 9, you found that the distance formula was really just the Pythagorean Theorem. You even used the Pythagorean Theorem to derive the equation of a circle.

You can also derive trigonometry relationships from the Pythagorean Theorem. These are called Pythagorean identities. Complete the steps below to derive one of the Pythagorean identities.

Investigation
A Pythagorean Identity

Step 1 Pick any measure of $\angle A$ and find

$$(\sin A)^2 + (\cos A)^2 = \underline{\ ?\ }$$

Step 2 Repeat Step 1 for several different measures of $\angle A$. When you are ready, make a tentative conjecture.

Let's see if you can derive your conjecture. Use this triangle for Steps 3–6.

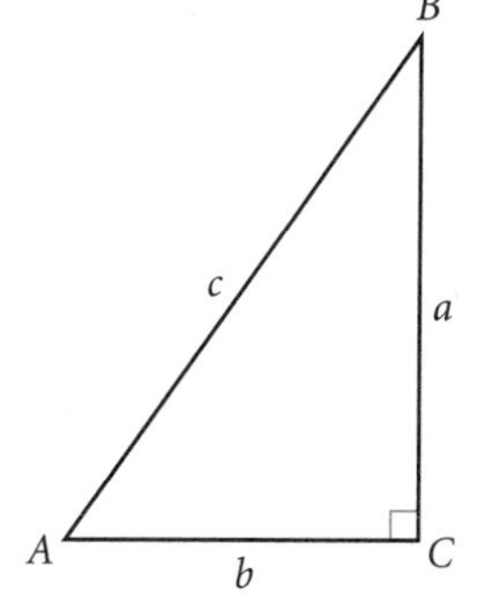

Step 3 Find ratios for $\sin A$ and $\cos A$.

Step 4 Substitute your results from Step 3 into this equation.

$$(\sin A)^2 + (\cos A)^2 = \left(\frac{?}{?}\right)^2 + \left(\frac{?}{?}\right)^2$$

Step 5 Add the two fractions on the right side of your equation.

Step 6 Triangle ABC is a right triangle. How can you use the Pythagorean Theorem to further simplify your equation?

Step 7 Does your result in Step 6 support your conjecture in Step 2? You should now be ready to state the Pythagorean identity.

Pythagorean Identity C-104

For any angle A, $\underline{\ ?\ }$.

The Pythagorean Theorem is very powerful, but its use is still limited to right triangles. Recall from Chapter 9 that the Pythagorean Theorem does not work for acute triangles or obtuse triangles. You might ask, "What happens to the Pythagorean equation for acute triangles or obtuse triangles?"

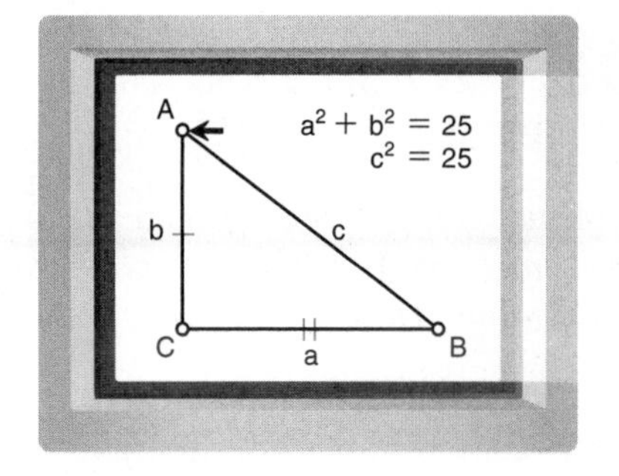

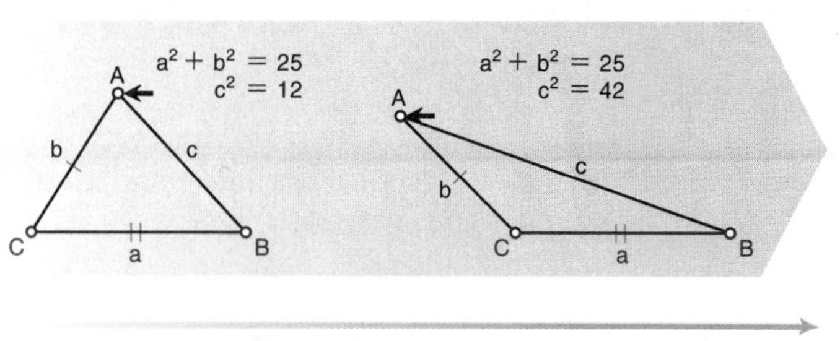

In this right triangle
$c^2 = a^2 + b^2$

In this acute triangle
$c^2 < a^2 + b^2$

In this obtuse triangle
$c^2 > a^2 + b^2$

If the legs of a right triangle are brought closer together so that the right angle becomes an acute angle, you'll find that $c^2 < a^2 + b^2$. In order to make this inequality into an equality, you would have to subtract something from $a^2 + b^2$.

$$c^2 = a^2 + b^2 - \textit{something}$$

If the legs are widened to form an obtuse angle, you'll find that $c^2 > a^2 + b^2$. Here, you'd have to add something to make an equality.

$$c^2 = a^2 + b^2 + \textit{something}$$

Mathematicians found that the "something" was $2ab \cos C$. The Pythagorean Theorem generalizes to all triangles with a trigonometric property called the **Law of Cosines.** The steps used to derive the Law of Cosines are left for you as a Take Another Look activity.

Law of Cosines

C-105

For any triangle with sides of lengths a, b, and c, and with C the angle opposite the side with length c,

$$c^2 = a^2 + b^2 - 2ab \cos C$$

You can use the Law of Cosines when you are given three side lengths or two side lengths and the angle measure between them (SSS or SAS). Again, you'll be asked to work only with acute angles.

EXAMPLE A

Find the length of side $\overline{CT}$ in acute triangle CRT.

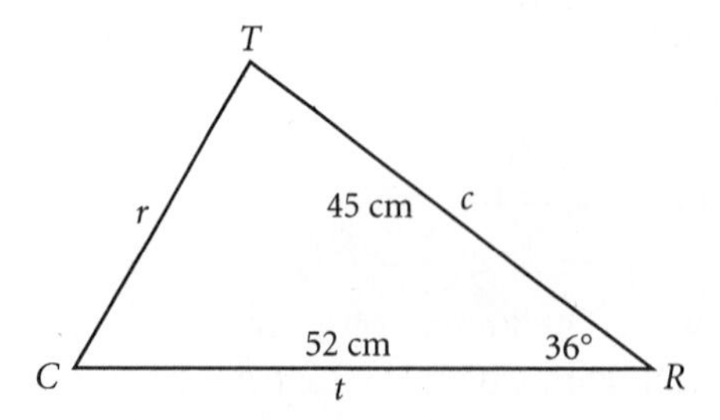

▸ Solution

To find r, use the Law of Cosines:

$c^2 = a^2 + b^2 - 2ab \cos C$ The Law of Cosines.

Using the variables in this problem, the Law of Cosines becomes

$$r^2 = c^2 + t^2 - 2ct \cos R$$ Substitute r for c, c for a, t for b, and R for C.

$$r^2 = 45^2 + 52^2 - 2(45)(52)(\cos 36°)$$ Substitute 45 for c, 52 for t, and 36° for R.

$$r = \sqrt{45^2 + 52^2 - 2(45)(52)(\cos 36°)}$$ Take the positive square root of both sides.

$$r \approx 31$$ Evaluate.

The length of side $\overline{CT}$ is about 31 cm.

EXAMPLE B

Find the measure of $\angle Q$ in acute triangle QED.

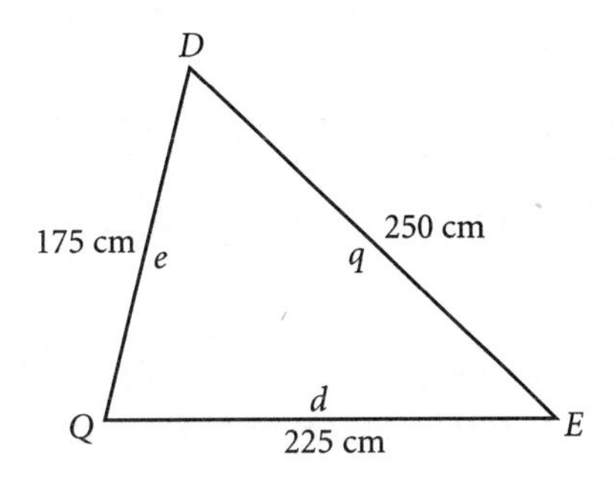

▶ Solution

Use the Law of Cosines and solve for Q.

$$q^2 = e^2 + d^2 - 2ed \cos Q$$ The Law of Cosines with respect to $\angle Q$.

$$\cos Q = \frac{q^2 - e^2 - d^2}{-2ed}$$ Solve for cos Q.

$$\cos Q = \frac{250^2 - 175^2 - 225^2}{-2(175)(225)}$$ Substitute known values.

$$Q = \cos^{-1}\left(\frac{250^2 - 175^2 - 225^2}{-2(175)(225)}\right)$$ Take the inverse cosine of both sides.

$$Q \approx 76$$ Evaluate.

The measure of $\angle Q$ is about 76°.

EXERCISES

You will need

A calculator for Exercises **1–14**

Construction tools for Exercises **20 and 21**

Geometry software for Exercise **22**

In Exercises 1–3, find each length to the nearest centimeter.

1. $w \approx$? ⓗ

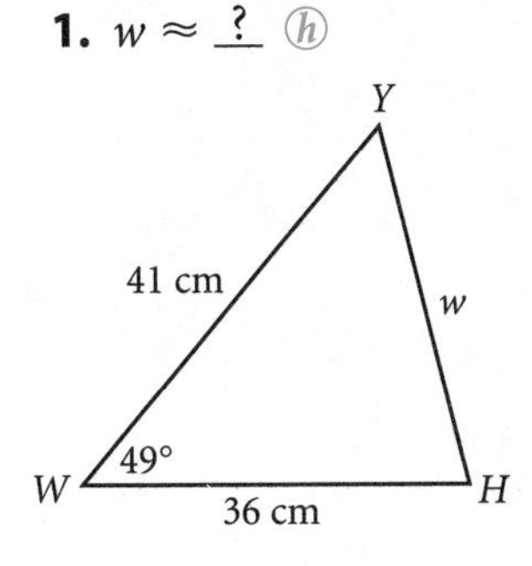

2. $y \approx$?

E
y
32 cm
78°
S
42 cm
Y

3. $x \approx$?

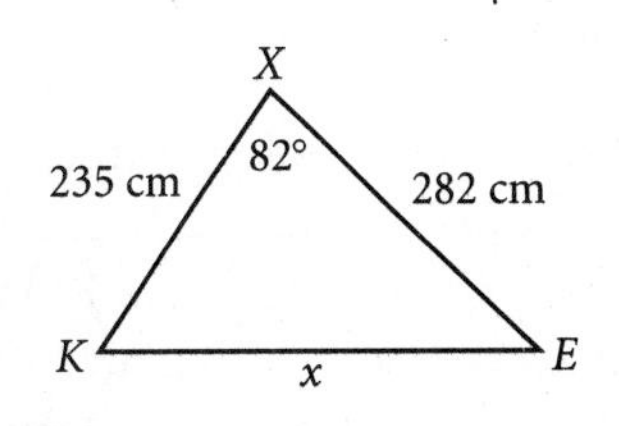

In Exercises 4–6, each triangle is an acute triangle. Find each angle measure to the nearest degree.

4. $m\angle A \approx$? ⓗ

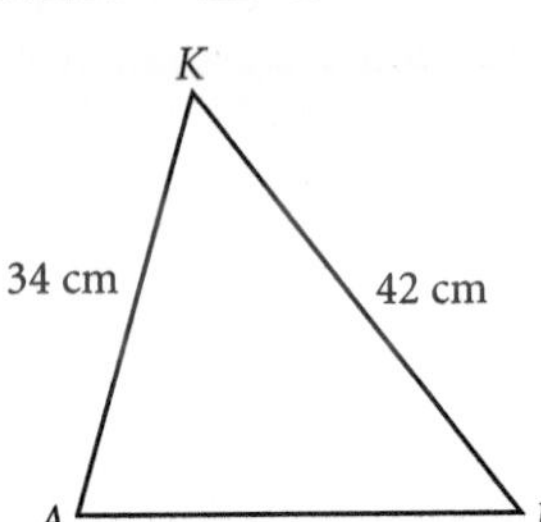

5. $m\angle B \approx$?

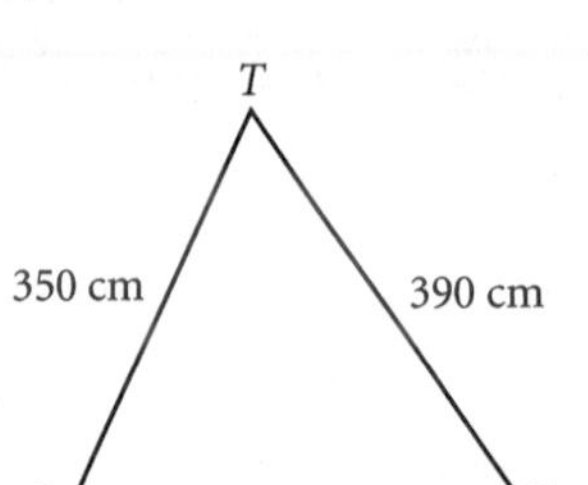

6. $m\angle C \approx$?

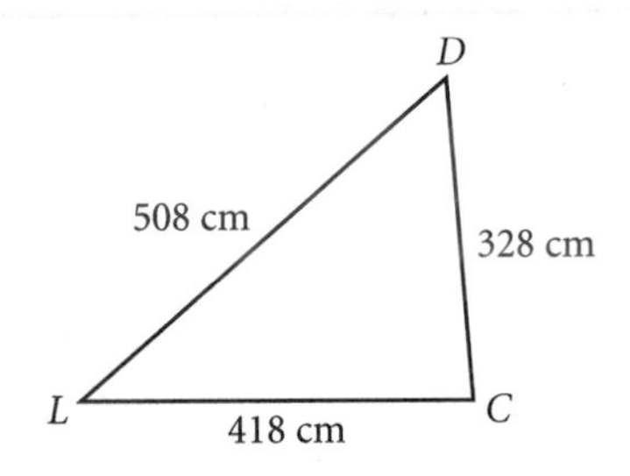

7. Two 24-centimeter radii of a circle form a central angle measuring 126°. What is the length of the chord connecting the two radii?

8. Find the measure of the smallest angle in an acute triangle whose side lengths are 4 m, 7 m, and 8 m. ⓗ

9. Two sides of a parallelogram measure 15 cm and 20 cm, and one of the diagonals measures 19 cm. What are the measures of the angles of the parallelogram to the nearest degree?

10. **APPLICATION** Captain Malloy is flying a passenger jet. He is heading east at 720 km/hr when he sees an electrical storm straight ahead. He turns the jet 20° to the north to avoid the storm and continues in this direction for 1 hr. Then he makes a second turn, back toward his original flight path. Eighty minutes after his second turn, he makes a third turn and is back on course. By avoiding the storm, how much time did Captain Malloy lose from his original flight plan? ⓗ

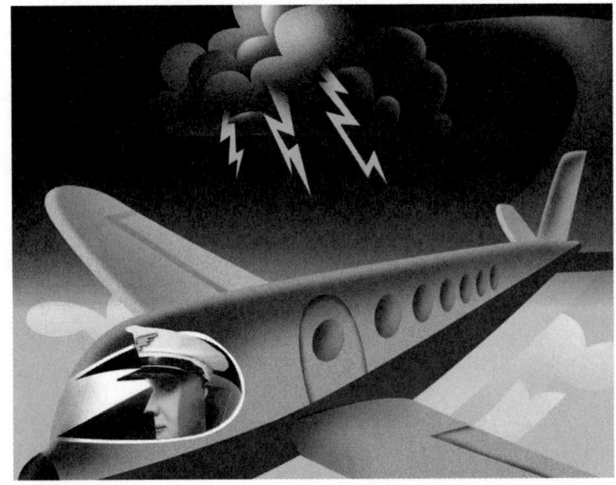

Review

11. **APPLICATION** A cargo company loads truck trailers into ship cargo containers. The trucks drive up a ramp to a horizontal loading platform 30 ft off the ground, but they have difficulty driving up a ramp at an angle steeper than 20°. What is the minimum length that the ramp needs to be?

12. **APPLICATION** An archaeologist uncovers the remains of a square-based Egyptian pyramid. The base is intact and measures 130 meters on each side. The top of the pyramid has eroded away, but what remains of each face of the pyramid forms a 65° angle with the ground. What was the original height of the pyramid? ⓗ

13. APPLICATION A lighthouse 55 meters above sea level spots a distress signal from a sailboat. The angle of depression to the sailboat measures 21°. How far away is the sailboat from the base of the lighthouse?

14. A painting company has a general safety rule to place ladders at an angle measuring between 55° and 75° from the level ground. Regina places the foot of her 25 ft ladder 6 ft from the base of a wall. What is the angle of the ladder? Is the ladder placed safely? If not, how far from the base of the wall should she place the ladder?

15. Show that $\frac{\sin A}{\cos A} = \tan A$.

16. *TRAP* is an isosceles trapezoid. ⓗ

a. Find *PR* in terms of *x*.

b. Write a paragraph proof to show that $m\angle TPR = 90°$.

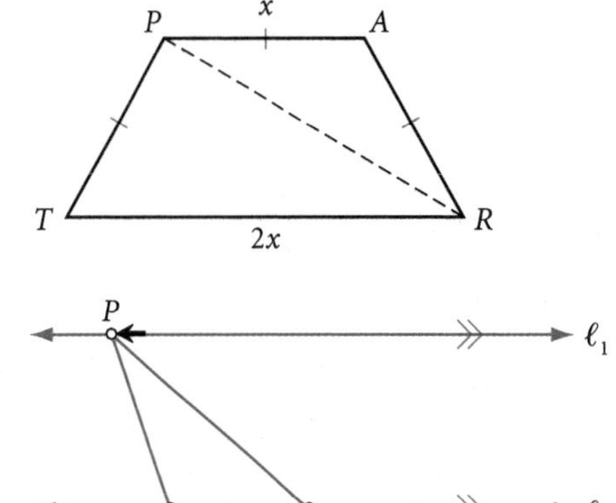

17. As *P* moves to the right on line ℓ_1, describe what happens to

a. $m\angle PAB$

b. $m\angle APB$

18. Which of these figures, the cone or the square pyramid, has the greater

a. Base perimeter?

b. Volume?

c. Surface area?

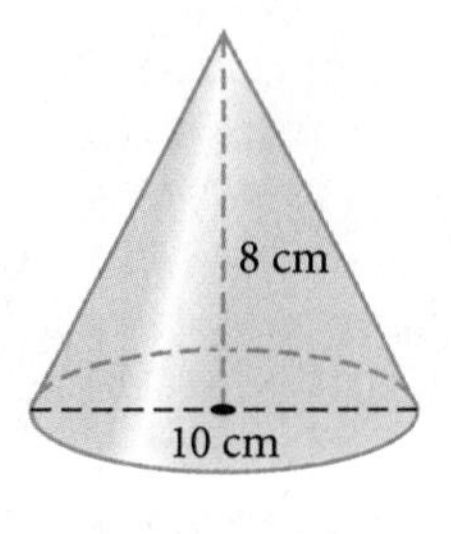

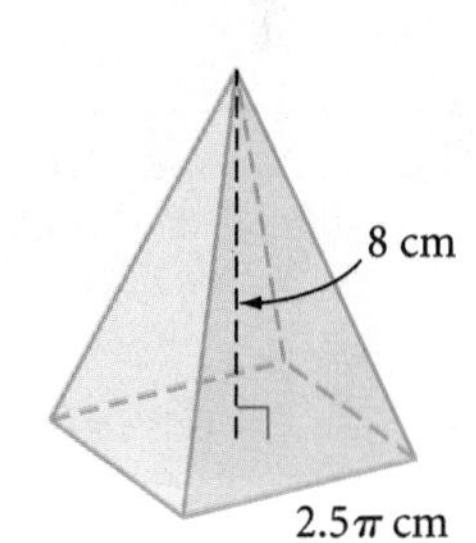

19. What single transformation is equivalent to the composition of each pair of functions? Write a rule for each.

a. A reflection over the line $x = -2$ followed by a reflection over the line $x = 3$

b. A reflection over the *x*-axis followed by a reflection over the *y*-axis

20. *Construction* Construct two rectangles that are not similar.

21. *Construction* Construct two isosceles trapezoids that are similar.

22. *Technology* Use geometry software to construct two circles. Connect the circles with a segment and construct the midpoint of the segment. Animate the endpoints of the segment around the circles and trace the midpoint of the segment. What shape does the midpoint of the segment trace? Try adjusting the relative size of the radii of the circles; try changing the distance between the centers of the circles; try starting the endpoints of the segment in different positions; or try animating the endpoints of the segment in different directions. Explain how these changes affect the shape traced by the midpoint of the segment.

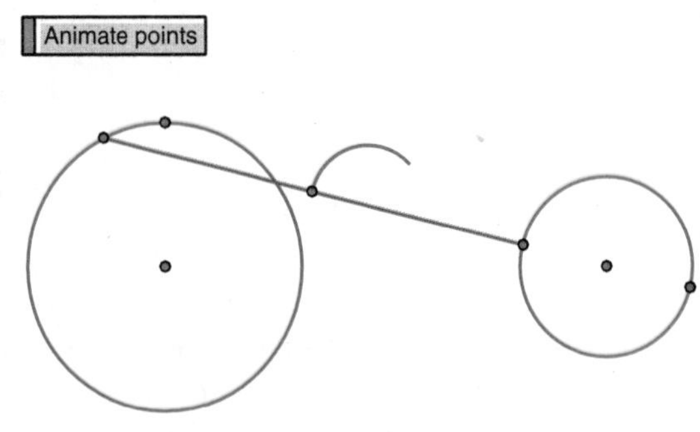

project

JAPANESE TEMPLE TABLETS

For centuries it has been customary in Japan to hang colorful wooden tablets in Shinto shrines to honor the gods of this native religion. During Japan's historical period of isolation (1639–1854), this tradition continued with a mathematical twist. Merchants, farmers, and others who were dedicated to mathematical learning made tablets containing mathematical problems, called *sangaku,* to inspire and challenge visitors. See if you can answer this *sangaku* problem.

These circles are tangent to each other and to the line. How are the radii of the three circles related?

r_1 r_2 r_3

Research other *sangaku* problems, then design your own tablet. Your project should include

- ▶ Your solution to the problem above.
- ▶ Some problems you found during your research and your sources.
- ▶ Your own decorated *sangaku* tablet with its solution on the back.

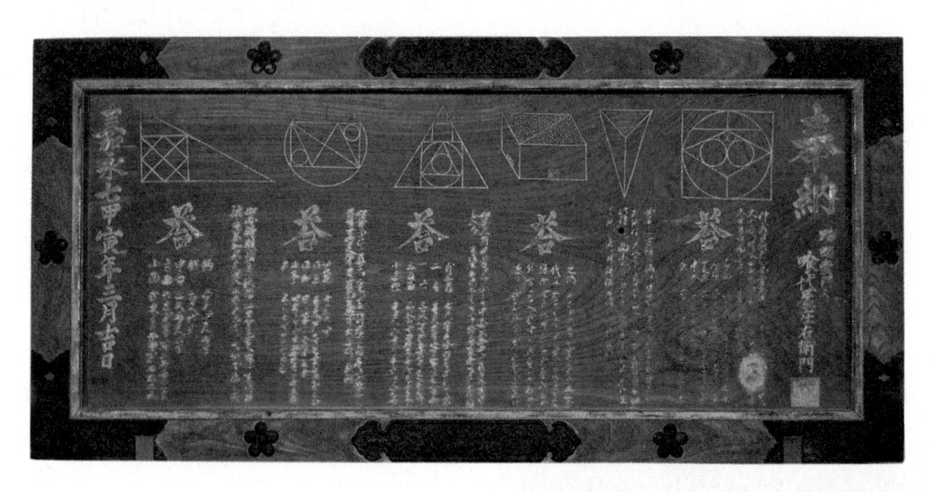

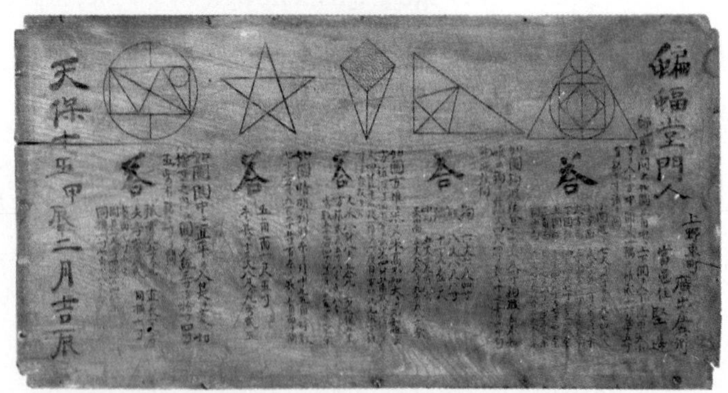

These colorful tablets, some with gold engraving, usually contain geometry problems.
Photographs by Hiroshi Umeoka.

LESSON

12.5

Problem Solving with Trigonometry

One ship drives east and
another drives west
With the self-same winds
that blow,
'Tis the set of the sails and
not the gales
Which tells us the way to go.

ELLA WHEELER WILCOX

There are many practical applications of trigonometry. Some of them involve vectors. In earlier vector activities, you used a ruler or a protractor to measure the size of the resulting vector or the angle between vectors. Now you will be able to calculate the resulting vectors with the Law of Sines or the Law of Cosines.

EXAMPLE

Rowing instructor Calista Thomas is in a stream flowing north to south at 3 km/hr. She is rowing northeast at a rate of 4.5 km/hr. At what speed is she moving? What direction (bearing) is she actually moving?

▶ Solution

First, sketch and label the vector parallelogram. The resultant vector, r, divides the parallelogram into two congruent triangles. In each triangle you know the lengths of two sides and the measure of the included angle. Use the Law of Cosines to find the length of the resultant vector or the speed that it represents.

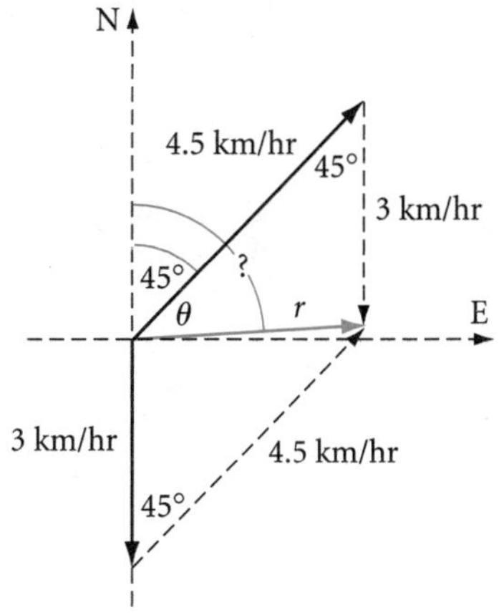

$$r^2 = 4.5^2 + 3^2 - 2(4.5)(3)(\cos 45°)$$

$$r^2 = 4.5^2 + 3^2 - 2(4.5)(3)(\cos 45°)$$

$$r \approx 3.2$$

Calista is moving at a speed of approximately 3.2 km/hr.

To find Calista's bearing (an angle measured clockwise from north), you need to find θ, and add its measure to 45°. Use the Law of Sines.

$$\frac{\sin\theta}{3} = \frac{\sin 45°}{3.2}$$

$$\sin\theta = \frac{3(\sin 45°)}{3.2}$$

$$\theta = \sin^{-1}\left[\frac{3(\sin 45°)}{3.2}\right] \approx 42°$$

Add 42° and 45° to find that Calista is moving at a bearing of 87°.

EXERCISES

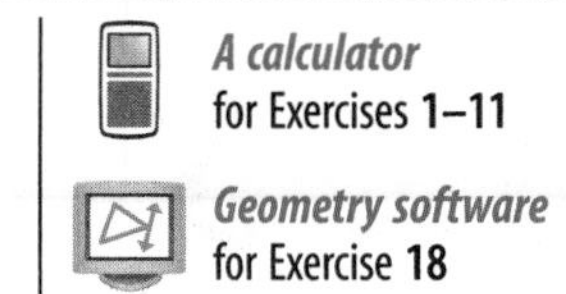

1. **APPLICATION** The steps to the front entrance of a public building rise a total of 1 m. A portion of the steps will be replaced by a wheelchair ramp. By a city ordinance, the angle of inclination for a ramp cannot measure greater than 4.5°. What is the minimum distance from the entrance that the ramp must begin?

2. **APPLICATION** Giovanni is flying his Cessna airplane on a heading as shown. His instrument panel shows an air speed of 130 mi/hr. (Air speed is the speed in still air without wind.) However, there is a 20 mi/hr crosswind. What is the resulting speed of the plane? ⓗ

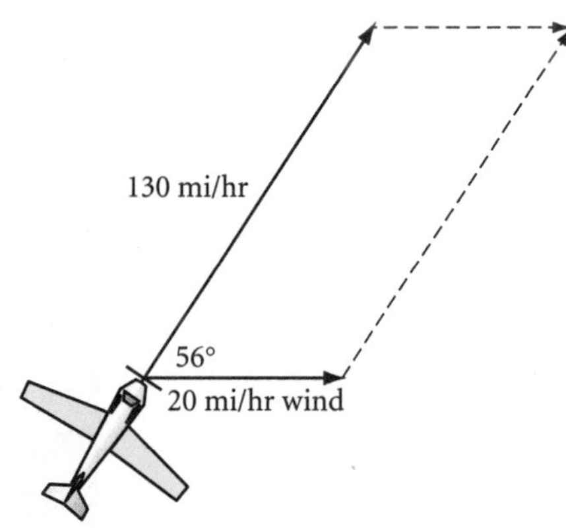

3. **APPLICATION** A lighthouse is east of a Coast Guard patrol boat. The Coast Guard station is 20 km north of the lighthouse. The radar officer aboard the boat measures the angle between the lighthouse and the station to be 23°. How far is the boat from the station?

4. **APPLICATION** The Archimedean screw is a water-raising device that consists of a wooden screw enclosed within a cylinder. When the cylinder is turned, the screw raises water. The screw is very efficient at an angle measuring 25°. If a screw needs to raise water 2.5 meters, how long should its cylinder be?

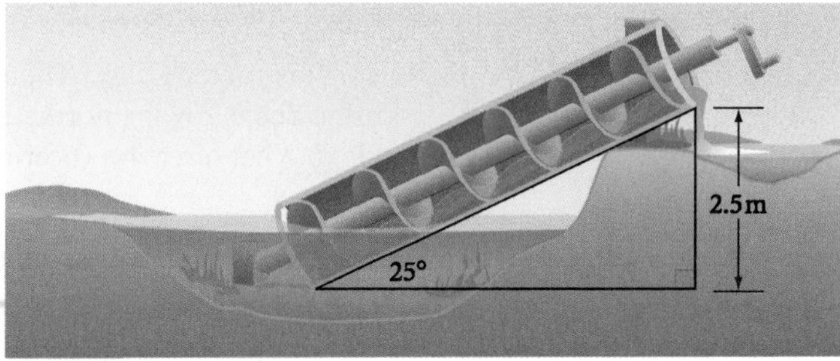

Technology CONNECTION

Used for centuries in Egypt to lift water from the Nile River, the Archimedean screw is thought to have been invented by Archimedes in the third century B.C.E., when he sailed to Egypt. It is also called an Archimedes Snail because of its spiral channels that resemble a snail shell. Once powered by people or animals, the device is now modernized to shift grain in mills and powders in factories.

5. **APPLICATION** Annie and Sashi are backpacking in the Sierra Nevada. They walk 8 km from their base camp at a bearing of 42°. After lunch, they change direction to a bearing of 137° and walk another 5 km. ⓗ
   a. How far are Annie and Sashi from their base camp?
   b. At what bearing must Sashi and Annie travel to return to their base camp?

6. A surveyor at point A needs to calculate the distance to an island's dock, point C. He walks 150 meters up the shoreline to point B such that $\overline{AB} \perp \overline{AC}$. Angle ABC measures 58°. What is the distance between A and C?

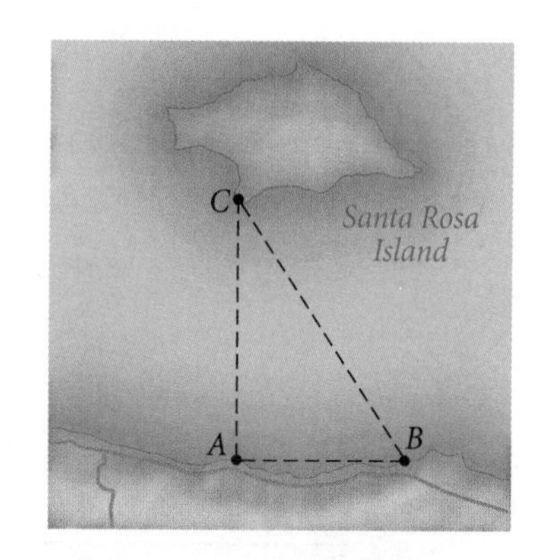

7. During a strong wind, the top of a tree cracks and bends over, touching the ground as if the trunk were hinged. The tip of the tree touches the ground 20 feet 6 inches from the base of the tree and forms a 38° angle with the ground. What was the tree's original height?

8. **APPLICATION** A pocket of matrix opal is known to be 24 meters beneath point A on Alan Ranch. A mining company has acquired rights to mine beneath Alan Ranch, but not the right to bring equipment onto the property. So the mining company cannot dig straight down. Brian Ranch has given permission to dig on its property at point B, 8 meters from point A. At what angle to the level ground must the mining crew dig to reach the opal? What distance must they dig?

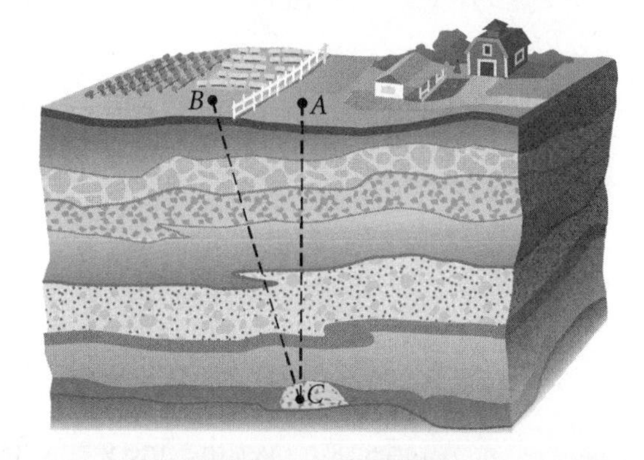

9. Todd's friend Olivia is flying her stunt plane at an elevation of 6.3 km. From the ground, Todd sees the plane moving directly toward him from the west at a 49° angle of elevation. Three minutes later he turns and sees the plane moving away from him to the east at a 65° angle of elevation. How fast is Olivia flying in kilometers per hour? ⓗ

10. A water pipe for a farm's irrigation system must go through a small hill. Farmer Golden attaches a 14.5-meter rope to the pipe's entry point and an 11.2-meter rope to the exit point. When he pulls the ropes taut, their ends meet at a 58° angle. What is the length of pipe needed to go through the hill? At what angle with respect to the first rope should the pipe be laid so that it comes out of the hill at the correct exit point? ⓗ

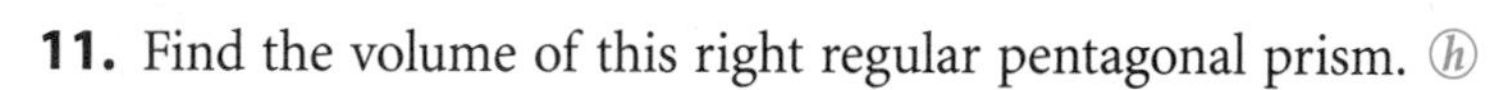

Review

11. Find the volume of this right regular pentagonal prism. ⓗ

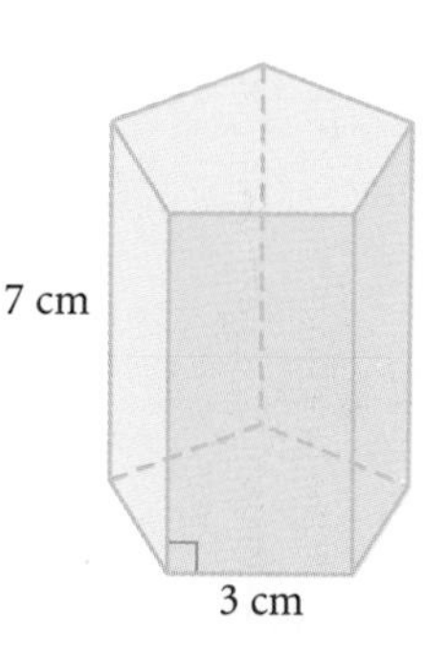

12. A formula for the area of a regular polygon is $A = \frac{ns^2}{4\tan\theta}$, where n is the number of sides, s is the length of a side, and $\theta = \frac{360}{2n}$. Explain why this formula is correct.

13. Find the volume of the largest cube that can fit into a sphere with a radius of 12 cm.

14. How does the area of a triangle change if its vertices are transformed by the rule $(x, y) \rightarrow (-3x, -3y)$? Give an example to support your answer.

15. What's wrong with this picture?

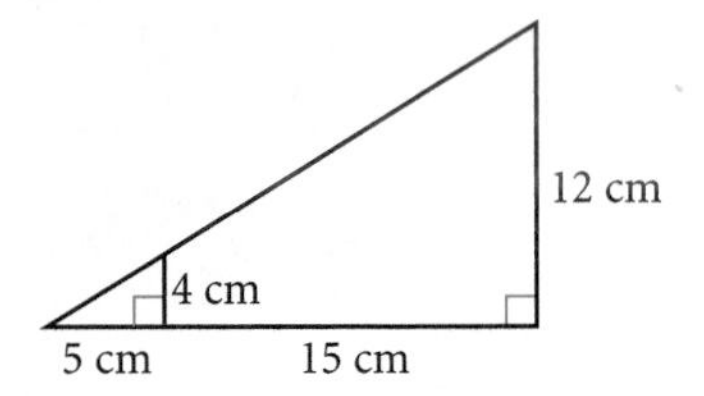

16. As P moves to the right on line ℓ_1, describe what happens to

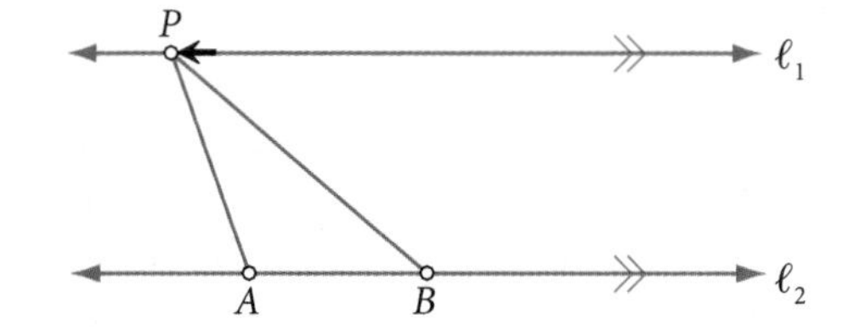

a. PA

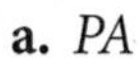

b. Area of $\triangle APB$

17. What single transformation is equivalent to the composition of each pair of functions? Write a rule for each.

a. A reflection over the line $y = x$ followed by a counterclockwise 270° rotation about the origin

b. A rotation 180° about the origin followed by a reflection over the x-axis

18. ***Technology*** Tile floors are often designed by creating simple, symmetric patterns on squares. When the squares are lined up, the patterns combine, often leaving the original squares hardly visible. Use geometry software to create your own tile-floor pattern.

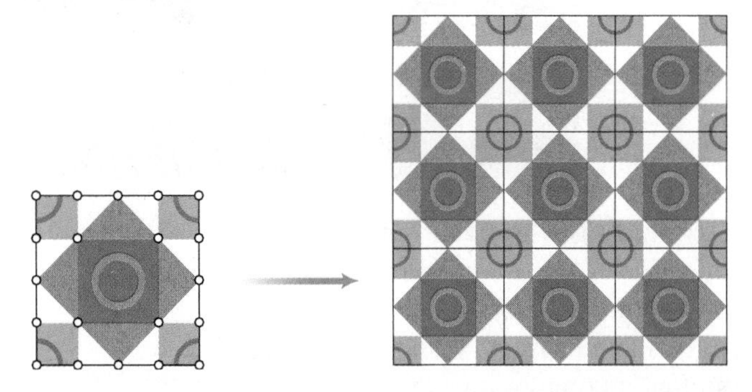

IMPROVING YOUR ALGEBRA SKILLS

Substitute and Solve

1. If $2x = 3y$, $y = 5w$, and $w = \frac{20}{3z}$, find x in terms of z.

2. If $7x = 13y$, $y = 28w$, and $w = \frac{9}{26z}$, find x in terms of z.

Exploration

Trigonometric Ratios and the Unit Circle

In Lesson 12.1, you defined trigonometric ratios in terms of the sides of a right triangle. That limited you to talking about acute angles. But in the coordinate plane, it's possible to define trigonometric ratios for angles with measures less than 0° and greater than 90°. These definitions use a **unit circle**—a circle with center (0, 0) and radius 1 unit.

The height of a seat on a Ferris wheel can be modeled by unit-circle trigonometry. This Ferris wheel, called the *London Eye,* was built for London's year 2000 celebration.

Activity
The Unit Circle

You will need

- The Unit Circle worksheet

In this activity, you will use a Sketchpad construction to explore the unit circle. The first part of this activity will give you some understanding of how a unit circle simplifies the trigonometric ratios for acute angles. Then you will use the unit circle to explore the ratios for all angles, from 0° to 360°, and even negative angle measures.

Step 1 Follow the steps on the worksheet to construct a unit circle with right triangle ADC.

Step 2 In right triangle ADC, write ratios for the sine, cosine, and tangent of $\angle DAC$.

Step 3 What is the length of the hypotenuse, AC, in this unit circle? Use this length to simplify your trigonometric definitions in Step 2.

Step 4 Measure the coordinates of point C. Which coordinate corresponds to the sine of $\angle DAC$? Which coordinate corresponds to the cosine of $\angle DAC$? What parts of your sketch physically represent the sine of $\angle DAC$ and the cosine of $\angle DAC$?

Step 5 How can you use the coordinates of point C to calculate the tangent of $\angle DAC$?

Step 6 Follow the steps on the worksheet to add a physical representation of tangent to your unit circle.

Step 7 Measure the coordinates of point E. Which coordinate corresponds to the tangent of $\angle DAC$? What part of your sketch physically represents the tangent of $\angle DAC$? Use similar triangles ADC and ABE to explain your answers.

So far, you have looked only at right triangle ADC with acute angle DAC. It may seem that you have not gotten any closer to defining trigonometric ratios for angles with measures less than 0° or greater than 90°. That is, even if you move point C into another quadrant, $\angle DAC$ would still be an acute angle.

In order to modify the definition of trigonometric ratios in a unit circle, you need to measure $\angle BAC$ instead. That is, measure the amount of rotation from $\overline{AB}$ to $\overline{AC}$.

You may recall from Chapter 6 that a point on a rolling tire traces a curve called a cycloid. Cycloids are defined by trigonometry.

Step 8 Select point B, point A, and point C, in that order. Measure $\angle BAC$.

Step 9 Move point C around the circle and watch how the measure of $\angle BAC$ changes. Summarize your observations.

Step 10 When point C is in the first quadrant, how is the measure of $\angle DAC$ related to the measure of $\angle BAC$? How about when point C is in the second quadrant? The third quadrant? The fourth quadrant?

Step 11 Use Sketchpad's calculator to calculate the sine, cosine, and tangent of $\angle BAC$. Compare these trigonometric ratios to the coordinates of point C and point E. Do the values support your answers to Steps 4 and 7?

Step 12 | Move point C around the circle and watch how the trigonometric ratios change for angle measures between $-180°$ and $180°$. Answer these questions.

a. How does the sine of $\angle BAC$ change as the measure of the angle goes from $0°$ to $90°$ to $180°$? From $-180°$ to $-90°$ to $0°$?

b. How does the cosine of $\angle BAC$ change?

c. If the sine of one angle is equal to the cosine of another angle, how are the angles related to each other?

d. How does the tangent of $\angle BAC$ change? What happens to the tangent of $\angle BAC$ as its measure approaches $90°$ or $-90°$? Based on the definition of tangent and the side lengths in $\triangle ADC$, what value do you think the tangent of $90°$ equals?

You can add an interesting animation that will graph the changing sine and tangent values in your sketch. You can construct points that will trace curves that algebraically represent the functions $y = \sin(x)$ and $y = \tan(x)$.

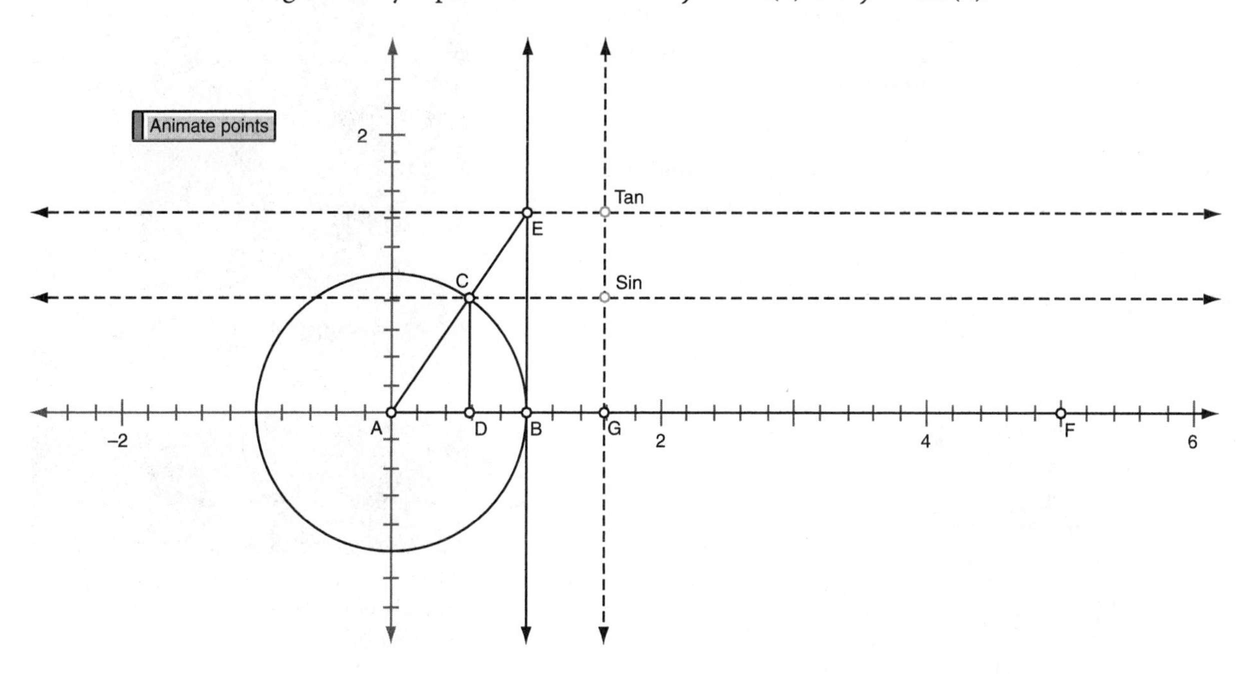

Step 13 | Follow the steps on the worksheet to add an animation that will trace curves representing the sine and tangent functions.

Step 14 | Measure the coordinates of point F, then locate it as close to $(6.28, 0)$ as possible. Move point G to the origin and move point C to $(1, 0)$. Press the Animation button to trace the sine and tangent functions.

The high and low points of tides can be modeled with trigonometry. This tide table from the Savannah River in Fort Jackson, Georgia, shows a familiar pattern in its data.

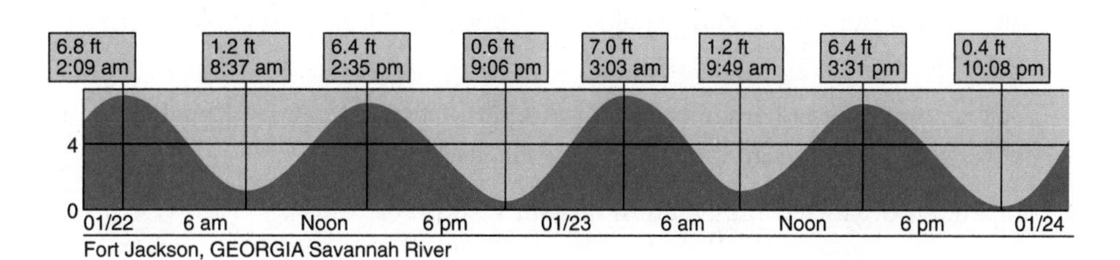

Step 15 | What's special about 6.28 as the x-coordinate of point F? Try other locations for point F to see what happens. Use what you know about circles to explain why 6.28 is a special value for the unit circle.

You will probably learn much more about unit-circle trigonometry in a future mathematics course.

TRIGONOMETRIC FUNCTIONS

You've seen many applications where you can use trigonometry to find distances or angle measures. Another important application of trigonometry is to model **periodic** phenomena, which repeat over time.

In this project you'll discover characteristics of the graphs of trigonometric functions, including their periodic nature. The three functions you'll look at are

$y = \sin(x)$

$y = \cos(x)$

$y = \tan(x)$

These functions are defined not only for acute angles but also for angles with measures less than 0° and greater than 90°.

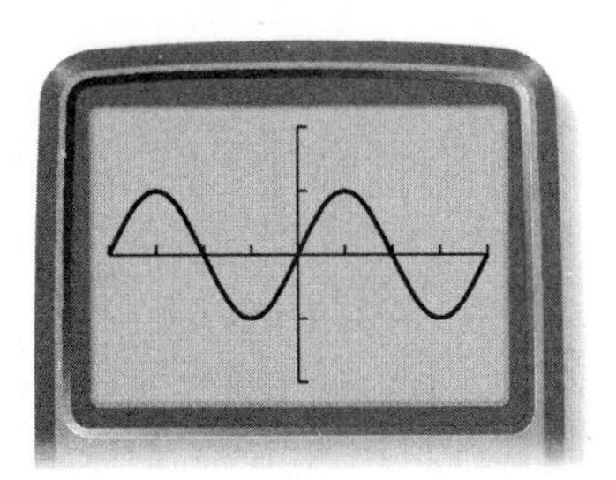

The swinging motion of a pendulum is an example of periodic motion that can be modeled by trigonometry.

Set your calculator in *degree* mode and set a window with an x-range of -360 to 360 and a y-range of -2 to 2. One at a time, graph each trigonometric function. Describe the characteristics of each graph, including maximum and minimum values for y and the **period**—the horizontal distance after which the graph starts repeating itself. Use what you know about the definitions of sine, cosine, and tangent to explain any unusual occurrences.

Try graphing pairs of trigonometric functions. Describe any relationships you see between the graphs. Are there any values in common?

Prepare an organized presentation of your results.

Exploration

Three Types of Proofs

In previous explorations, you learned four forms of valid reasoning: *Modus Ponens* (MP), *Modus Tollens* (MT), the Law of Syllogism (LS), and the Law of Contrapositive (LC). You can use these forms of reasoning to make logical arguments, or proofs. In this exploration you will learn the three basic types of proofs: direct proofs, conditional proofs, and indirect proofs.

In a **direct proof,** the given information or premises are stated, then valid forms of reasoning are used to arrive directly at a conclusion. Here is a direct proof given in two-column form. In a **two-column proof,** each statement in the argument is written in the left column, and the reason for each statement is written directly across in the right column.

Direct Proof

Premises: $P \rightarrow Q$

$R \rightarrow P$

$\sim Q$

Conclusion: $\sim R$

1. $P \rightarrow Q$	**1.** Premise
2. $\sim Q$	**2.** Premise
3. $\sim P$	**3.** From lines 1 and 2, using MT
4. $R \rightarrow P$	**4.** Premise
5. $\therefore \sim R$	**5.** From lines 3 and 4, using MT

A **conditional proof** is used to prove that a $P \rightarrow Q$ statement follows from a set of premises. In a conditional proof, the first part of the conditional statement, called the **antecedent,** is assumed to be true. Then logical reasoning is used to demonstrate that the second part, called the **consequent,** must also be true. If this process is successful, it's demonstrated that *if* P *is true, then* Q *must be true.*

In other words, a conditional proof shows that the antecedent implies the consequent. Here is an example.

Conditional Proof

Premises: $P \to R$

$S \to \sim R$

Conclusion: $P \to \sim S$

1. P	**1.** Assume the antecedent
2. $P \to R$	**2.** Premise
3. R	**3.** From lines 1 and 2, using MP
4. $S \to \sim R$	**4.** Premise
5. $\sim S$	**5.** From lines 3 and 4, using MT

Assuming P is true, the truth of $\sim S$ is established.

$\therefore P \to \sim S$

An **indirect proof** is a clever approach to proving something. To prove indirectly that a statement is true, you begin by assuming it is *not* true. Then you show that this assumption leads to a contradiction. For example, if you are given a set of premises and are asked to show that some conclusion P is true, begin by assuming that the opposite of P, namely $\sim P$, is true. Then show that this assumption leads to a contradiction of an earlier statement. If $\sim P$ leads to a contradiction, it must be false and P must be true. Here is an example.

Indirect Proof

Premises: $R \to S$

$\sim R \to \sim P$

P

Conclusion: S

1. $\sim S$	**1.** Assume the opposite of the conclusion
2. $R \to S$	**2.** Premise
3. $\sim R$	**3.** From lines 1 and 2, using MT
4. $\sim R \to \sim P$	**4.** Premise
5. $\sim P$	**5.** From lines 3 and 4, using MP
6. P	**6.** Premise

But lines 5 and 6 contradict each other. It's impossible for both P and $\sim P$ to be true.

Therefore, $\sim S$, the original assumption, is false. If $\sim S$ is false, then S is true.

$\therefore S$

Many logical arguments can be proved using more than one type of proof. For instance, you can prove the argument in the example above by using a direct proof. (Try it!) With practice you will be able to tell which method will work best for a particular argument.

Activity

Prove It!

Step 1

Copy the direct proof below, including the list of premises and the conclusion. Provide each missing reason.

Premises: $P \rightarrow Q$

$Q \rightarrow \sim R$

R

Conclusion: $\sim P$

1. $Q \rightarrow \sim R$	**1.** ?
2. R	**2.** ?
3. $\sim Q$	**3.** ?
4. $P \rightarrow Q$	**4.** ?
5. $\therefore \sim P$	**5.** ?

Step 2

Copy the conditional proof below, including the list of premises and the conclusion. Provide each missing statement or reason.

Premises: $\sim R \rightarrow \sim Q$

$T \rightarrow \sim R$

$S \rightarrow T$

Conclusion: $S \rightarrow \sim Q$

1. S	**1.** ?
2. $S \rightarrow T$	**2.** ?
3. T	**3.** From lines 1 and 2, using ?
4. $T \rightarrow \sim R$	**4.** ?
5. $\sim R$	**5.** ?
6. ?	**6.** ?
7. ?	**7.** ?

Assuming S is true, the truth of $\sim Q$ is established.

$\therefore$?

Step 3

Copy the indirect proof below and at the top of page 658, including the list of premises and the conclusion. Provide each missing statement or reason.

Premises: $P \rightarrow Q \rightarrow R$

$Q \rightarrow \sim R$

Q

Conclusion: $\sim P$

1. P	**1.** Assume the ? of the ?
2. $P \rightarrow (Q \rightarrow R)$	**2.** ?
3. $Q \rightarrow R$	**3.** ?
4. Q	**4.** ?

Step 3 (continued)

5. R	**5.** ?
6. ?	**6.** ?
7. ?	**7.** From lines ? and ?, using ?

But lines ? and ? contradict each other.

Therefore, P, the assumption, is false.

$\therefore \sim P$

Step 4

Provide the steps and reasons to prove each logical argument. You will need to decide whether to use a direct, conditional, or indirect proof.

a. Premises: $P \rightarrow Q$
$Q \rightarrow \sim R$
$T \rightarrow R$
Conclusion: $T \rightarrow \sim P$

b. Premises: $(R \rightarrow S) \rightarrow P$
$T \rightarrow Q$
$\sim T \rightarrow \sim P$
$\sim Q$
Conclusion: $\sim(R \rightarrow S)$

c. Premises: $S \rightarrow Q$
$P \rightarrow S$
$\sim R \rightarrow P$
$\sim Q$
Conclusion: R

d. Premises: $P \rightarrow Q$
$\sim P \rightarrow S$
$R \rightarrow \sim S$
$\sim Q$
Conclusion: $\sim R$

Step 5

Translate each argument into symbolic terms, then prove it is valid.

a. If all wealthy people are happy, then money can buy happiness. If money can buy happiness, then true love doesn't exist. But true love exists. Therefore, not all wealthy people are happy.

b. If Clark is performing at the theater today, then everyone at the theater has a good time. If everyone at the theater has a good time, then Lois is not sad. Lois is sad. Therefore, Clark is not performing at the theater today.

c. If Evette is innocent, then Alfa is telling the truth. If Romeo is telling the truth, then Alfa is not telling the truth. If Romeo is not telling the truth, then he has something to gain. Romeo has nothing to gain. Therefore, if Romeo has nothing to gain, then Evette is not innocent.

CHAPTER 12 REVIEW

Trigonometry was first developed by astronomers who wanted to map the stars. Obviously, it is hard to directly measure the distances between stars and planets. That created a need for new methods of indirect measurement. As you've seen, you can solve many indirect measurement problems by using triangles. Using sine, cosine, and tangent ratios, you can find unknown lengths and angle measures if you know just a few measures in a right triangle. You can extend these methods to any triangle using the Law of Sines or the Law of Cosines.

What's the least you need to know about a right triangle in order to find all its measures? What parts of a nonright triangle do you need to know in order to find the other parts? Describe a situation in which an angle of elevation or depression can help you find an unknown height.

EXERCISES

You will need

A calculator for Exercises **1–3, 7–28, 51,** and **53**

For Exercises 1–3, use a calculator to find each trigonometric ratio accurate to four decimal places.

1. $\sin 57°$

2. $\cos 9°$

3. $\tan 88°$

For Exercises 4–6, find each trigonometric ratio.

4. $\sin A = \underline{?}$
$\cos A = \underline{?}$
$\tan A = \underline{?}$

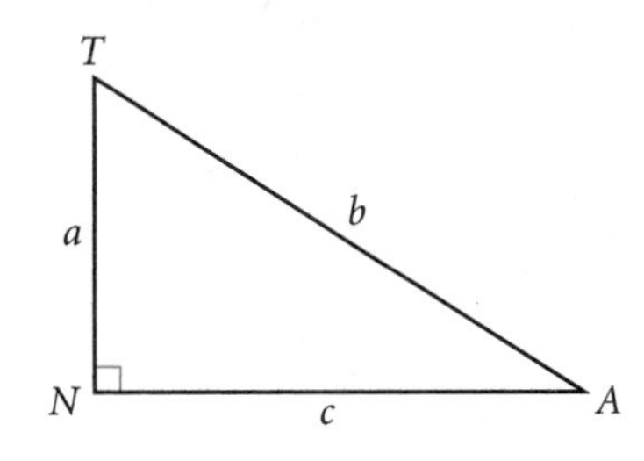

5. $\sin B = \underline{?}$
$\cos B = \underline{?}$
$\tan B = \underline{?}$

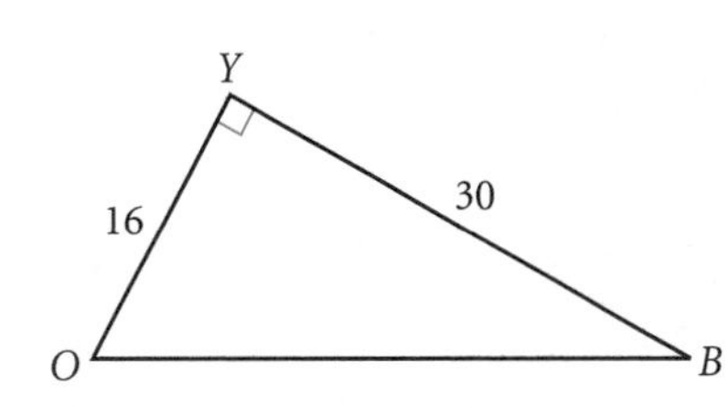

6. $\sin \phi = \underline{?}$
$\cos \phi = \underline{?}$
$\tan \phi = \underline{?}$

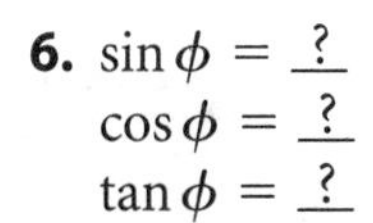

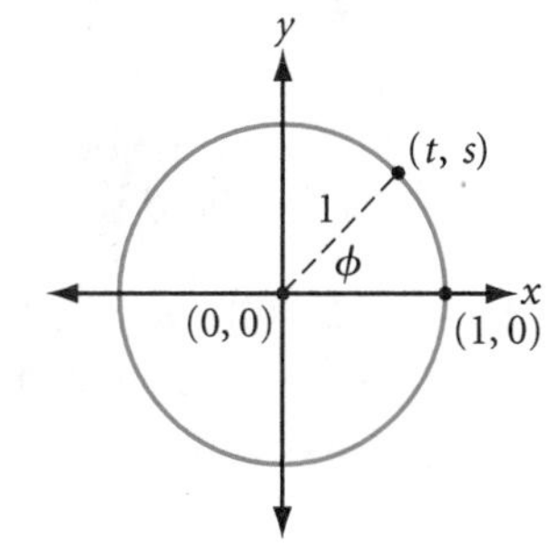

For Exercises 7–9, find the measure of each angle to the nearest degree.

7. $\sin A = 0.5447$

8. $\cos B = 0.0696$

9. $\tan C = 2.9043$

10. Shaded area $\approx \underline{?}$

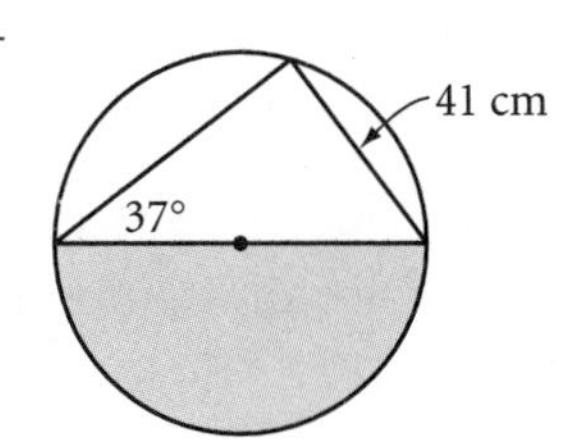

11. Volume $\approx \underline{?}$

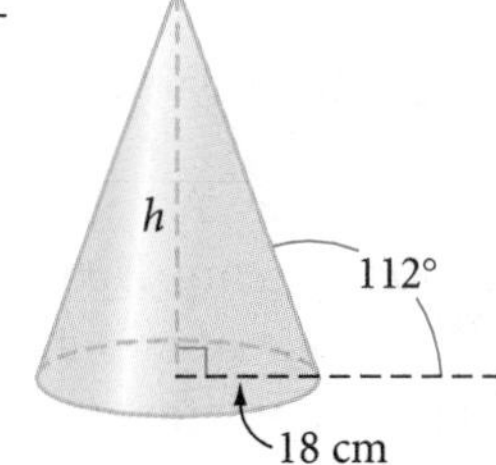

12. **APPLICATION** According to the Americans with Disabilities Act, enacted in 1990, the slope of a wheelchair ramp must be less than $\frac{1}{12}$ and there must be a minimum 5-by-5 ft landing for every 2.5 ft of rise. These dimensions were chosen to accommodate handicapped people who face physical barriers in public buildings and at work. An architect has submitted the orthographic plan shown below. Does the plan meet the requirements of the act? What will be the ramp's angle of ascent?

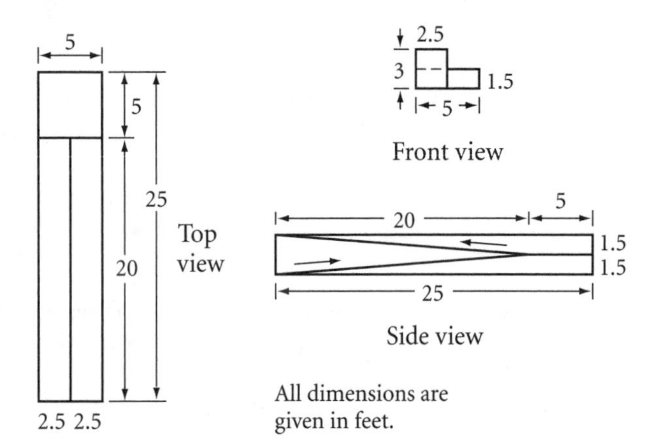

The Axis Dance Company includes performers in wheelchairs. Increased tolerance and accessibility laws have broadened the opportunities available to people with disabilities.

13. **APPLICATION** A lighthouse is east of a sailboat. The sailboat's dock is 30 km north of the lighthouse. The captain measures the angle between the lighthouse and the dock and finds it to be 35°. How far is the sailboat from the dock?

14. **APPLICATION** An air traffic controller must calculate the angle of descent (the angle of depression) for an incoming jet. The jet's crew reports that their land distance is 44 km from the base of the control tower and that the plane is flying at an altitude of 5.6 km. Find the measure of the angle of descent.

15. **APPLICATION** A new house is 32 feet wide. The rafters will rise at a 36° angle and meet above the center line of the house. Each rafter also needs to overhang the side of the house by 2 feet. How long should the carpenter make each rafter?

16. **APPLICATION** During a flood relief effort, a Coast Guard patrol boat spots a helicopter dropping a package near the Florida shoreline. Officer Duncan measures the angle of elevation to the helicopter to be 15° and the distance to the helicopter to be 6800 m. How far is the patrol boat from the point where the package will land?

17. At an air show, Amelia sees a jet heading south away from her at a 42° angle of elevation. Twenty seconds later the jet is still moving away from her, heading south at a 15° angle of elevation. If the jet's elevation is constantly 6.3 km, how fast is it flying in kilometers per hour?

For Exercises 18–23, find each measure to the nearest unit or to the nearest square unit.

18. Area = $\underline{?}$

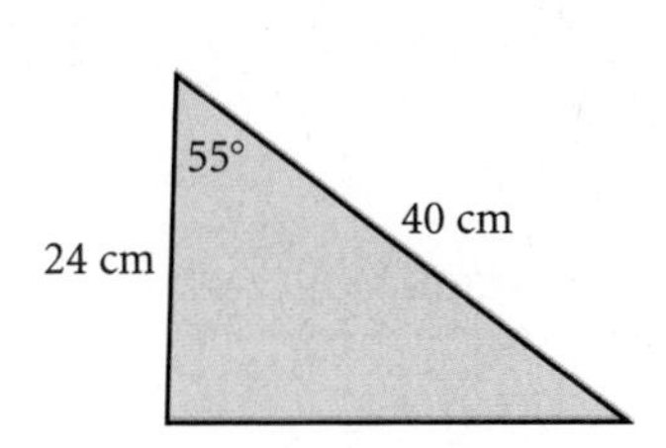

19. $w = \underline{?}$

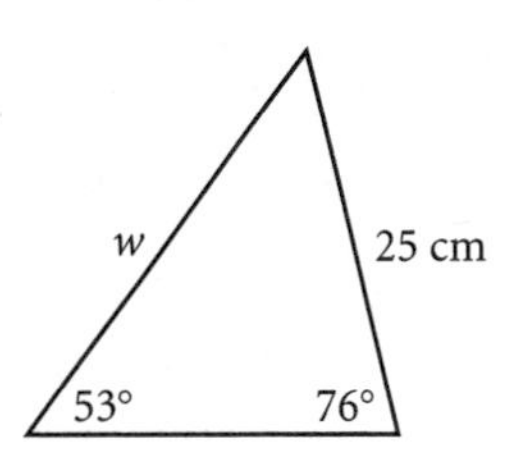

20. $\triangle ABC$ is acute.
$m\angle A = \underline{?}$

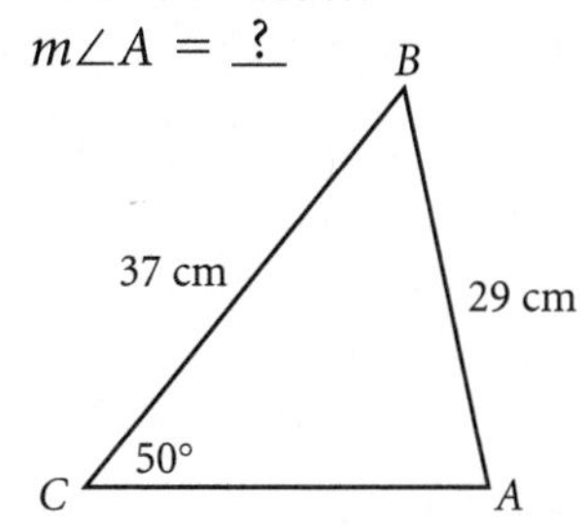

21. $x = \underline{?}$

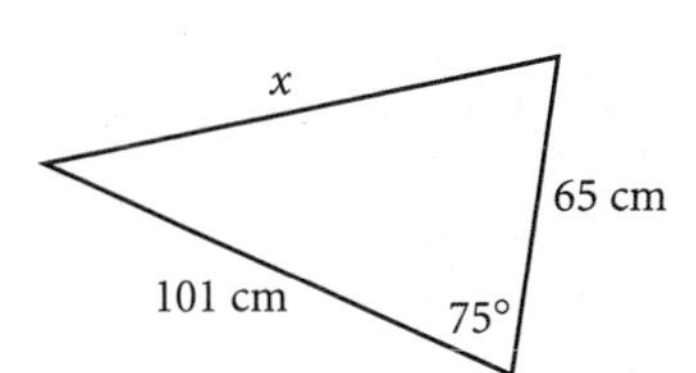

22. $m\angle B = \underline{?}$

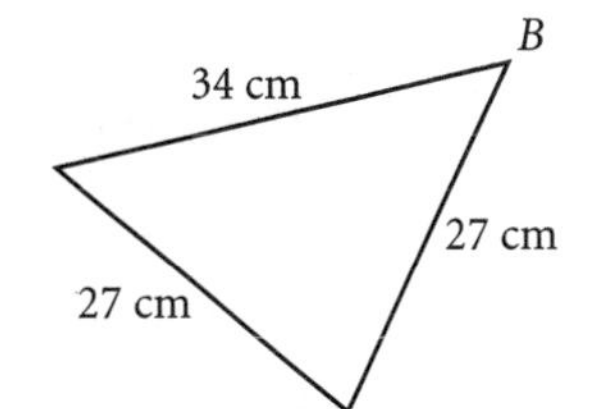

23. Area = $\underline{?}$

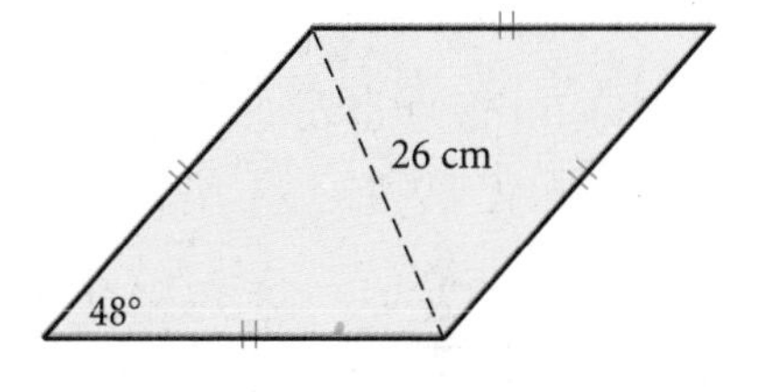

24. Find the length of the apothem of a regular pentagon with a side measuring 36 cm.

25. Find the area of a triangle formed by two 12 cm radii and a 16 cm chord in a circle.

26. A circle is circumscribed about a regular octagon with a perimeter of 48 cm. Find the diameter of the circle.

27. A 16 cm chord is drawn in a circle of diameter 24 cm. Find the area of the segment of the circle. ⓗ

28. Leslie is paddling his kayak at a bearing of 45°. In still water his speed would be 13 km/hr but there is a 5 km/hr current moving west. What is the resulting speed and direction of Leslie's kayak? ⓗ

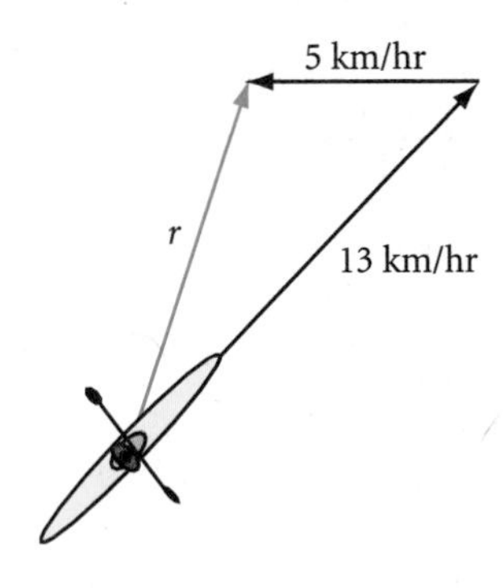

MIXED REVIEW

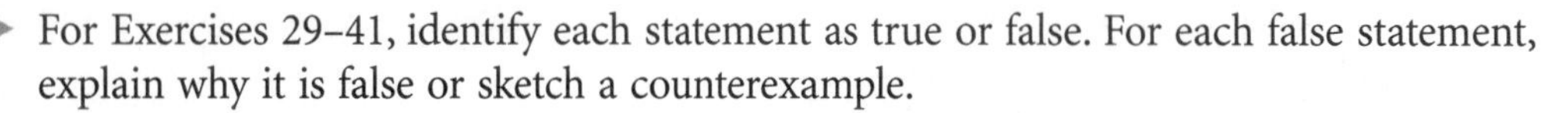

For Exercises 29–41, identify each statement as true or false. For each false statement, explain why it is false or sketch a counterexample.

29. An octahedron is a prism that has an octagonal base.

30. If the four angles of one quadrilateral are congruent to the four corresponding angles of another quadrilateral, then the two quadrilaterals are similar.

31. The three medians of a triangle meet at the centroid.

32. To use the Law of Cosines, you must know three side lengths or two side lengths and the measure of the included angle.

33. If the ratio of corresponding sides of two similar polygons is $\frac{m}{n}$, then the ratio of their areas is $\frac{m}{n}$.

34. The measure of an angle inscribed in a semicircle is always 90°.

35. If $\angle T$ is an acute angle in a right triangle, then

$$\text{tangent of } \angle T = \frac{\text{length of side adjacent to } \angle T}{\text{length of side opposite } \angle T}$$

36. If C^2 is the area of the base of a pyramid, and C is the height of the pyramid, then the volume of the pyramid is $\frac{1}{3}C^3$.

37. If two different lines intersect at a point, then the sum of the measures of at least one pair of vertical angles will be equal to or greater than 180°.

38. If a line cuts two sides of a triangle proportionally, then it is parallel to the third side.

39. If two sides of a triangle measure 6 cm and 8 cm and the angle between the two sides measures 60°, then the area of the triangle is $12\sqrt{3}$ cm^2.

40. A nonvertical line ℓ_1 has slope m and is perpendicular to line ℓ_2. The slope of ℓ_2 is also m.

41. If two sides of one triangle are proportional to two sides of another triangle, then the two triangles are similar.

For Exercises 42–53, select the correct answer.

42. The diagonals of a parallelogram

i. Are perpendicular to each other.

ii. Bisect each other.

iii. Form four congruent triangles.

A. i only **B.** ii only **C.** iii only **D.** i and ii

43. What is the formula for the volume of a sphere?

A. $V = 4\pi r^2$ **B.** $V = \pi r^2 h$ **C.** $V = \frac{4}{3}\pi r^3$ **D.** $V = \frac{1}{3}\pi r^2 h$

44. For the triangle at right, what is the value of $(\cos L)^2 + (\sin L)^2$?

A. About 0.22 **B.** About 1.41

C. 1 **D.** Cannot be determined

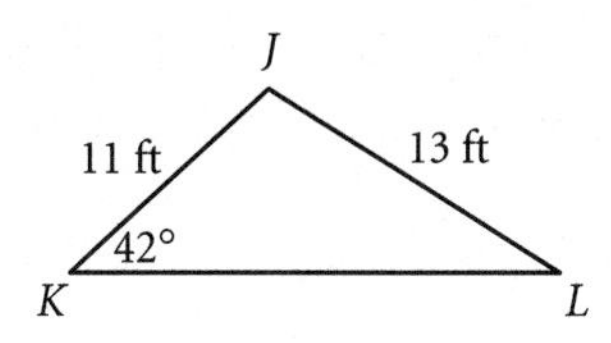

45. The diagonals of a rhombus

i. Are perpendicular to each other.

ii. Bisect each other.

iii. Form four congruent triangles.

A. i only **B.** iii only **C.** i and ii **D.** All of the above

46. The ratio of the surface areas of two similar solids is $\frac{4}{9}$. What is the ratio of the volumes of the solids?

A. $\frac{2}{3}$ **B.** $\frac{8}{27}$ **C.** $\frac{64}{729}$ **D.** $\frac{16}{81}$

47. A cylinder has height T and base area K. What is the volume of the cylinder?

A. $V = \pi K^2 T$ **B.** $V = KT$ **C.** $V = 2\pi KT$ **D.** $V = \frac{1}{3}KT$

48. Which of the following is *not* a similarity shortcut?

A. SSA **B.** SSS **C.** AA **D.** SAS

49. If a triangle has sides of lengths a, b, and c, and C is the angle opposite the side of length c, which of these statements must be true?

A. $a^2 = b^2 + c^2 - 2ab\cos C$ **B.** $c^2 = a^2 + b^2 - 2ab\cos C$

C. $c^2 = a^2 + b^2 + 2ab\cos C$ **D.** $a^2 = b^2 + c^2$

50. In the drawing at right, $\overline{WX} \parallel \overline{YZ} \parallel \overline{BC}$. What is the value of m?

A. 6 ft **B.** 8 ft

C. 12 ft **D.** 16 ft

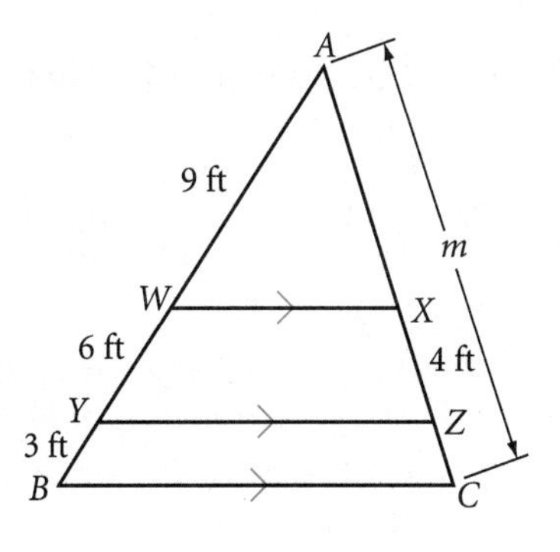

51. When a rock is added to a container of water, it raises the water level by 4 cm. If the container is a rectangular prism with a base that measures 8 cm by 9 cm, what is the volume of the rock?

A. 4 cm³ **B.** 32 cm³

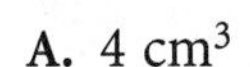

C. 36 cm³ **D.** 288 cm³

52. Which law could you use to find the value of v?

A. Law of Supply and Demand

B. Law of Syllogism

C. Law of Cosines

D. Law of Sines

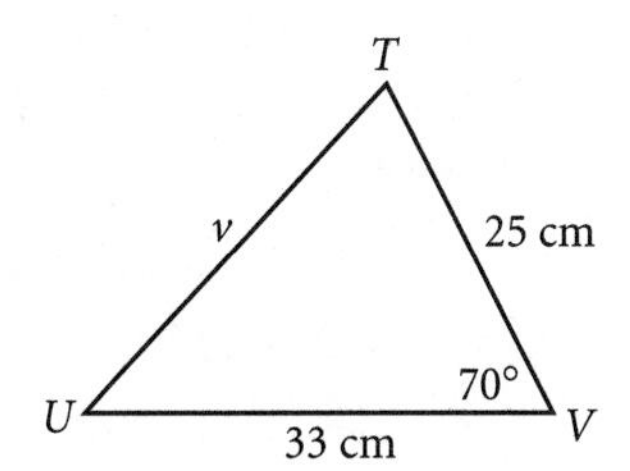

53. A 32-foot telephone pole casts a 12-foot shadow at the same time a boy nearby casts a 1.75-foot shadow. How tall is the boy?

A. 4 ft 8 in. **B.** 4 ft 6 in. **C.** 5 ft 8 in. **D.** 6 ft

Exercises 54–56 are portions of cones. Find the volume of each solid.

54.

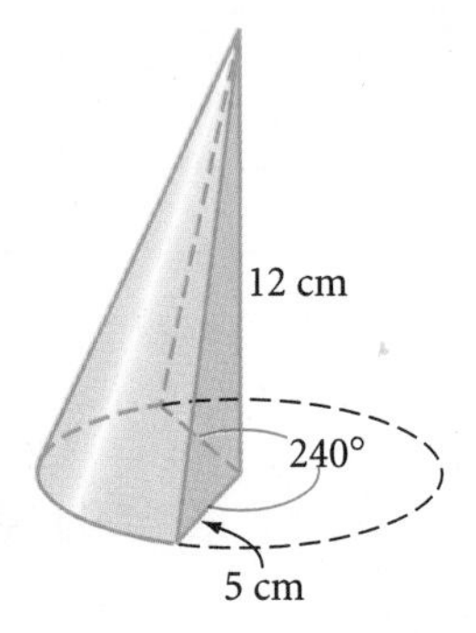

55.

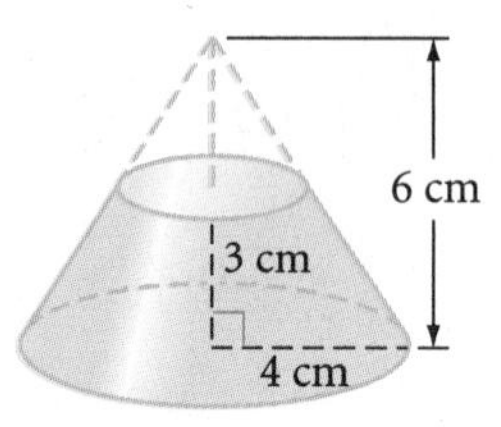

56.

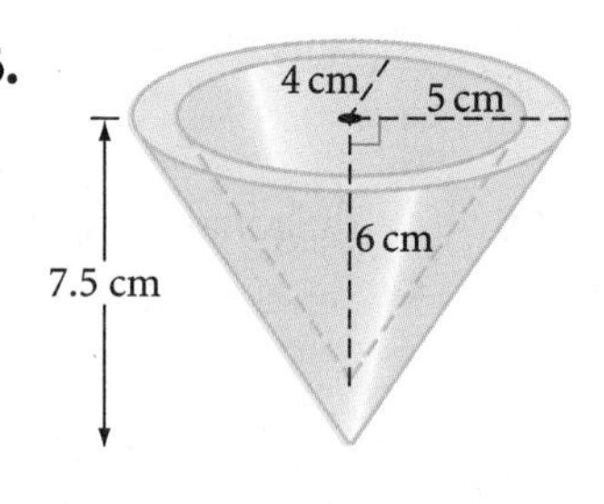

57. Each person at a family reunion hugs everyone else exactly once. There were 528 hugs. How many people were at the reunion?

58. Triangle *TRI* with vertices $T(-7, 0)$, $R(-5, 3)$, and $I(-1, 0)$ is translated by the rule $(x, y) \rightarrow (x + 2, y - 1)$. Then its image is translated by the rule $(x, y) \rightarrow (x - 1, y - 2)$. What single translation is equivalent to the composition of these two translations?

59. $\triangle LMN \sim \triangle PQR$. Find w, x, and y.

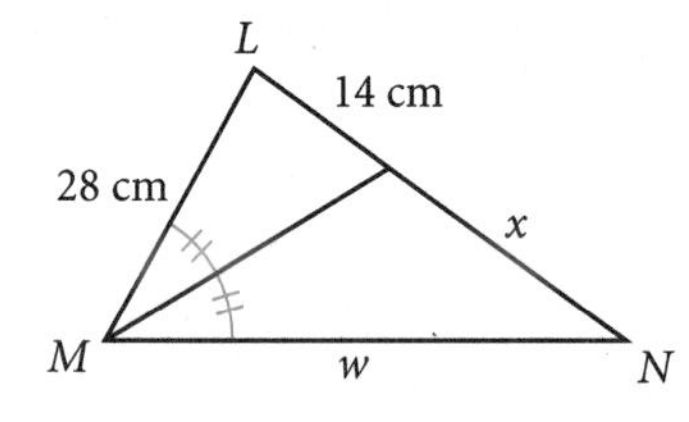

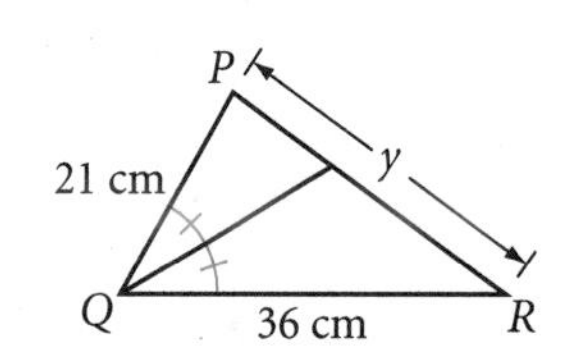

60. Find x.

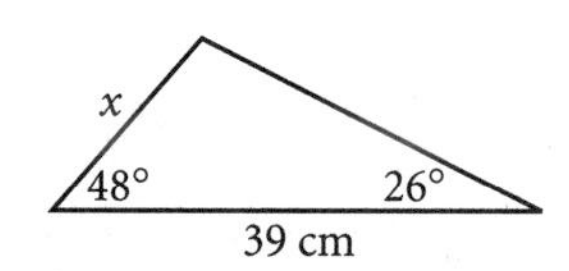

61. The diameter of a circle has endpoints $(5, -2)$ and $(5, 4)$. Find the equation of the circle.

62. Explain why a regular pentagon cannot create a monohedral tessellation.

63. Archaeologist Ertha Diggs uses a clinometer to find the height of an ancient temple. She views the top of the temple with a 37° angle of elevation. She is standing 130 meters from the center of the temple's base, and her eye is 1.5 meters above the ground. How tall is the temple?

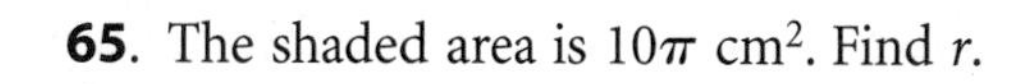

64. In the diagram below, the length of $\overset{\frown}{HK}$ is 20π ft. Find the radius of the circle.

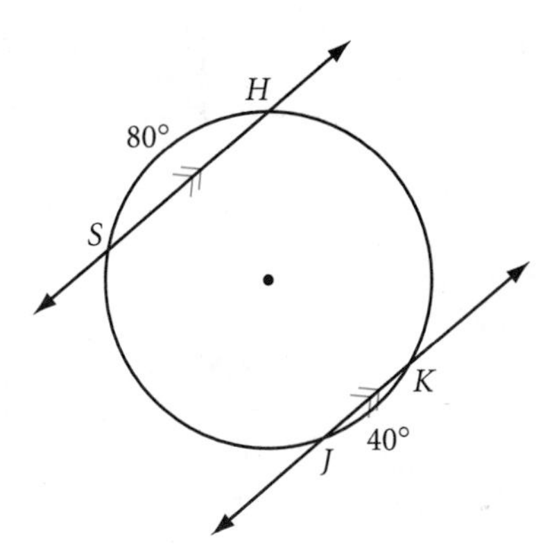

65. The shaded area is 10π cm^2. Find r.

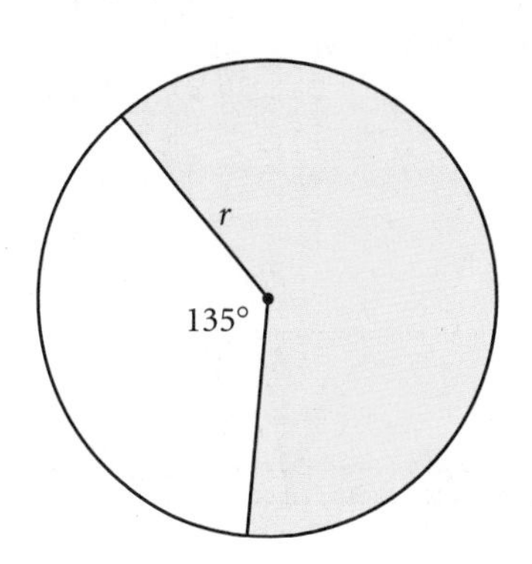

66. Triangle ABC is isosceles with $\overline{AB}$ congruent to $\overline{BC}$. Point D is on $\overline{AC}$ such that $\overline{BD}$ is perpendicular to $\overline{AC}$. Make a sketch and answer these questions.

a. $m\angle ABC = \underline{\ ?\ }\ m\angle ABD$

b. What can you conclude about $\overline{BD}$?

TAKE ANOTHER LOOK

1. You learned the Law of Sines as

$$\frac{\sin A}{a} = \frac{\sin B}{b} = \frac{\sin C}{c}$$

Use algebra to show that

$$\frac{a}{\sin A} = \frac{b}{\sin B} = \frac{c}{\sin C}$$

2. Recall that SSA does not determine a triangle. For that reason, you've been asked to find only acute angles using the Law of Sines. Take another look at a pair of triangles, $\triangle AB_1C$ and $\triangle AB_2C$, determined by SSA. How is $\angle CB_1A$ related to $\angle CB_2A$? Find $m\angle CB_1A$ and $m\angle CB_2A$. Find the sine of each angle. Find the sines of another pair of angles that are related in the same way, then complete this conjecture: for any angle θ, $\sin\theta = \sin(\underline{\ ?\ })$.

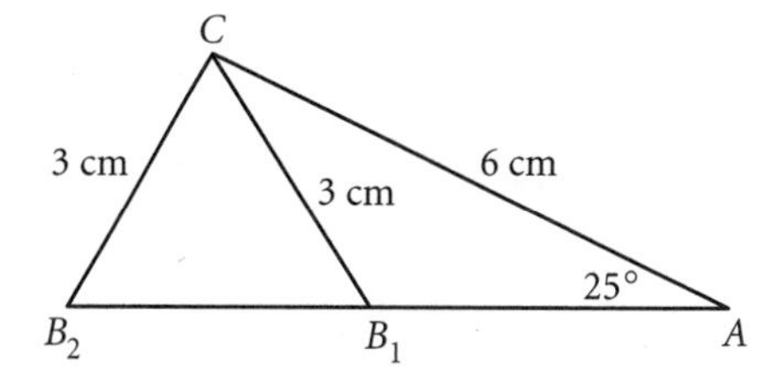

3. The Law of Cosines is generally stated using $\angle C$.

$c^2 = a^2 + b^2 - 2ab\cos C$

State the Law of Cosines in two different ways, using $\angle A$ and $\angle B$.

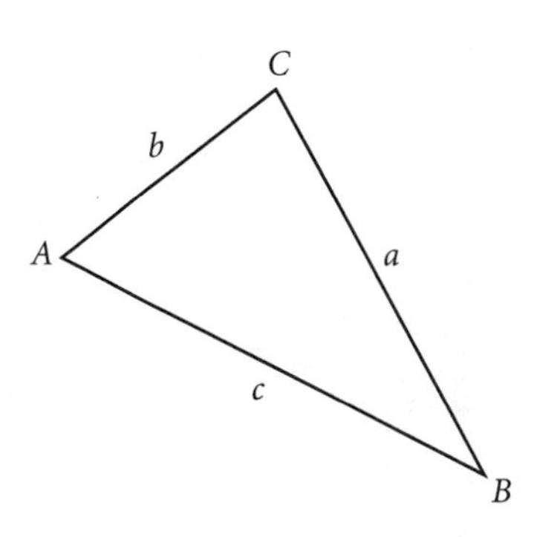

4. Derive the Law of Cosines.

5. Is there a relationship between the measure of the central angle of a sector of a circle and the angle at the vertex of the right cone formed when rolled up?

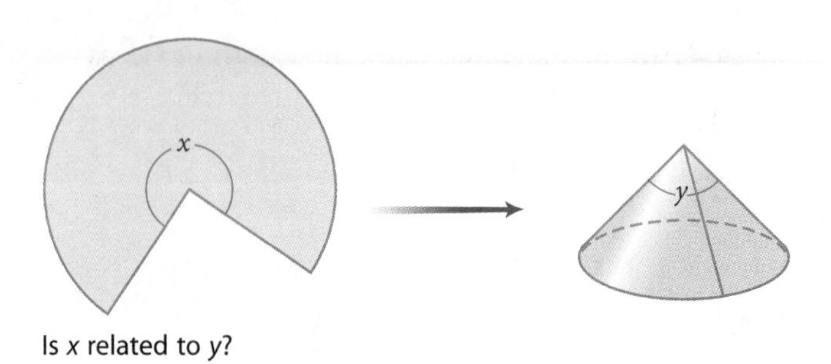

Is x related to y?

Assessing What You've Learned

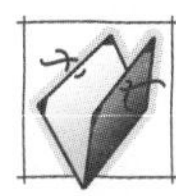

UPDATE YOUR PORTFOLIO Choose a real-world indirect measurement problem that uses trigonometry, and add it to your portfolio. Describe the problem, explain how you solved it, and explain why you chose it for your portfolio.

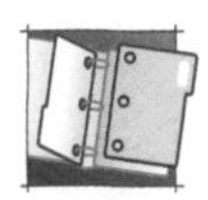

ORGANIZE YOUR NOTEBOOK Make sure your notebook is complete and well organized. Write a one-page chapter summary. Reviewing your notes and solving sample test items are good ways to prepare for chapter tests.

GIVE A PRESENTATION Demonstrate how to use an angle-measuring device to make indirect measurements of actual objects. Use appropriate visual aids.

WRITE IN YOUR JOURNAL The last five chapters have had a strong problem-solving focus. What do you see as your strengths and weaknesses as a problem solver? In what ways have you improved? In what areas could you improve or use more help? Has your attitude toward problem solving changed since you began this course?

CHAPTER

13 Geometry as a Mathematical System

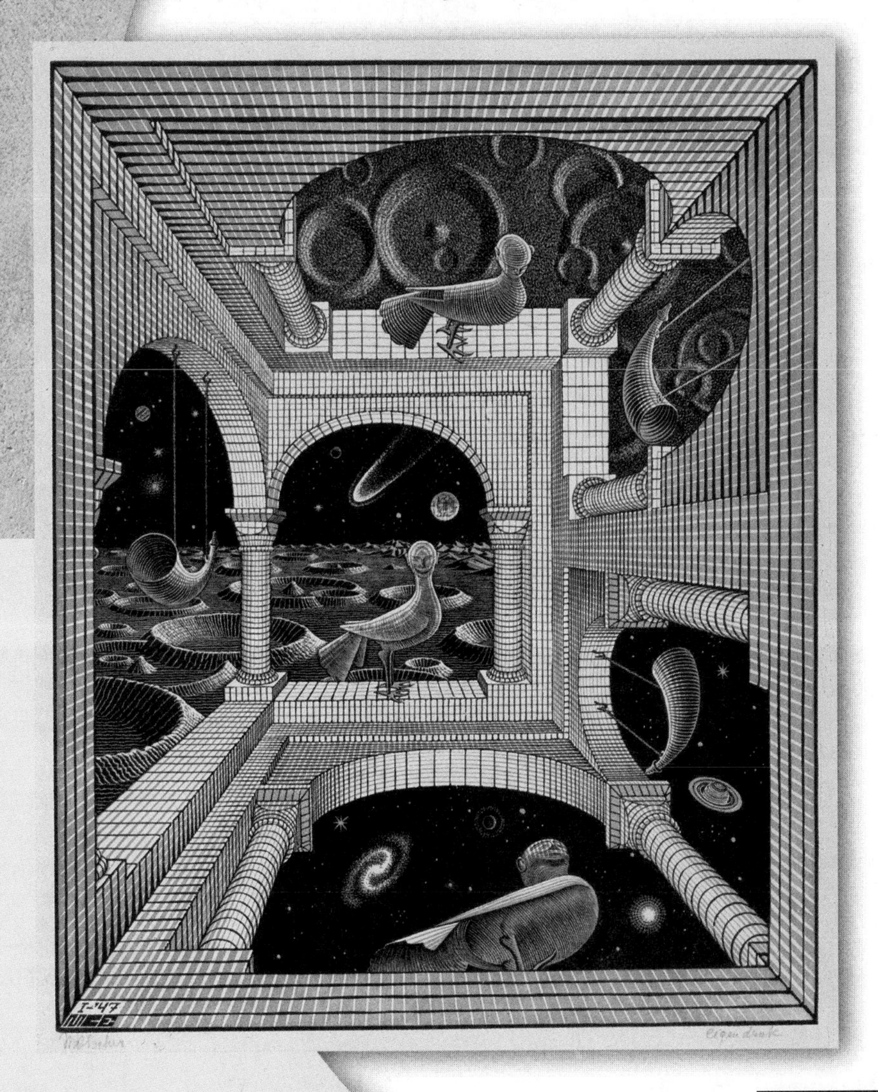

This search for new possibilities, this discovery of new jigsaw puzzle pieces, which in the first place surprises and astonishes the designer himself, is a game that through the years has always fascinated and enthralled me anew.

M. C. ESCHER

Another World (Other World), M. C. Escher, 1947

OBJECTIVES

In this chapter you will

- look at geometry as a mathematical system
- see how some conjectures are logically related to each other
- review a number of proof strategies, such as working backward and analyzing diagrams

LESSON

13.1

The Premises of Geometry

Geometry is the art of correct reasoning on incorrect figures.

GEORGE POLYA

As you learned in previous chapters, for thousands of years Babylonian, Egyptian, Chinese, and other mathematicians discovered many geometry principles and developed procedures for doing practical geometry.

By 600 B.C.E., a prosperous new civilization had begun to grow in the trading towns along the coast of Asia Minor (present-day Turkey) and later in Greece, Sicily, and Italy. People had free time to discuss and debate issues of government and law. They began to insist on reasons to support statements made in debate. Mathematicians began to use logical reasoning to deduce mathematical ideas.

This map detail shows Sicily, Italy, Greece, and Asia Minor along the north coast of the Mediterranean Sea. The map was drawn by Italian painter and architect Pietro da Cortona (1596–1669).

History CONNECTION

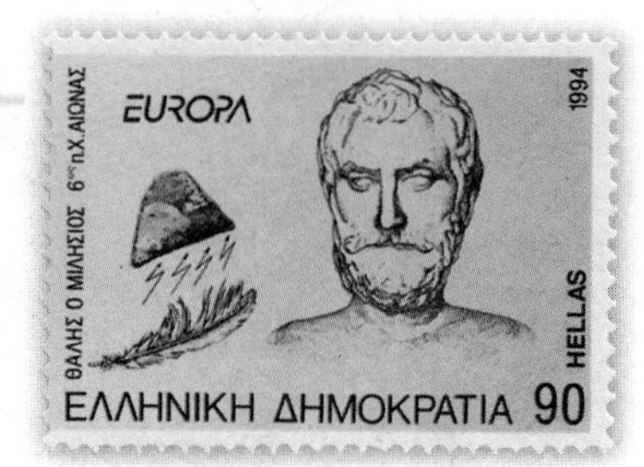

Greek mathematician Thales of Miletus (ca. 625–547 B.C.E.) made his geometry ideas convincing by supporting his discoveries with logical reasoning. Over the next 300 years, the process of supporting mathematical conjectures with logical arguments became more and more refined. Other Greek mathematicians, including Thales' most famous student, Pythagoras, began linking chains of logical reasoning. The tradition continued with Plato and his students. Euclid, in his famous work about geometry and number theory, *Elements,* established a single chain of deductive arguments for most of the geometry known then.

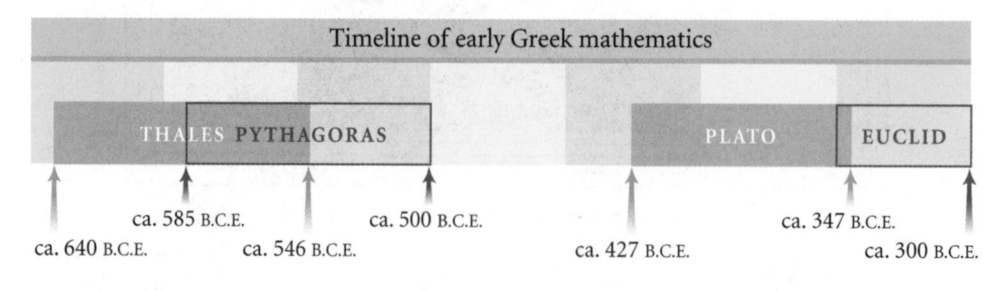

You have learned that Euclid used geometric constructions to study properties of lines and shapes. Euclid also created a **deductive system**—a set of **premises,** or accepted facts, and a set of logical rules—to organize geometry properties. He started from a collection of simple and useful statements he called **postulates.** He then systematically demonstrated how each geometry discovery followed logically from his postulates and his previously proved conjectures, or **theorems.**

Up to now, you have been discovering geometry properties inductively, the way many mathematicians have over the centuries. You have studied geometric figures and have made conjectures about them. Then, to explain your conjectures, you turned to deductive reasoning. You used informal proofs to explain why a conjecture was true. However, you did not prove every conjecture. In fact, you sometimes made critical assumptions or relied on unproved conjectures in your proofs. A conclusion in a proof is true if and only if your premises are true and all your arguments are valid. Faulty assumptions can lead to the wrong conclusion. Have all your assumptions been reliable?

Inductive reasoning process

Deductive reasoning process

In this chapter you will look at geometry as Euclid did. You will start with premises: definitions, properties, and postulates. From these premises you will systematically prove your earlier conjectures. Proved conjectures will become theorems, which you can use to prove other conjectures, turning them into theorems, as well. You will build a logical framework using your most important ideas and conjectures from geometry.

Premises for Logical Arguments in Geometry

1. Definitions and undefined terms
2. Properties of arithmetic, equality, and congruence
3. Postulates of geometry
4. Previously proved geometry conjectures (theorems)

You are already familiar with the first type of premise on the list: the undefined terms—point, line, and plane. In addition, you have a list of basic definitions in your notebook.

You used the second set of premises, properties of arithmetic and equality, in your algebra course.

Properties of Arithmetic

For any numbers a, b, and c:

Commutative property of addition

$a + b = b + a$

Commutative property of multiplication

$ab = ba$

Associative property of addition

$(a + b) + c = a + (b + c)$

Associative property of multiplication

$(ab)c = a(bc)$

Distributive property

$a(b + c) = ab + ac$

These Mayan stone carvings, found in Tikal, Guatemala, show the glyphs, or symbols, used in the Mayan number system. Learn more about Mayan numerals at www.keymath.com/DG .

Properties of Equality

For any numbers a, b, c, and d:

Reflexive property (also called the identity property)

$a = a$

Any number is equal to itself.

Transitive property

If $a = b$ and $b = c$, then $a = c$. (This property often takes the form of the **substitution property,** which says that if $b = c$, you can substitute c for b.)

Symmetric property

If $a = b$, then $b = a$.

Addition property

If $a = b$, then $a + c = b + c$. (Also, if $a = b$ and $c = d$, then $a + c = b + d$.)

Subtraction property

If $a = b$, then $a - c = b - c$. (Also, if $a = b$ and $c = d$, then $a - c = b - d$.)

Multiplication property

If $a = b$, then $ac = bc$. (Also, if $a = b$ and $c = d$, then $ac = bd$.)

Division property

If $a = b$, then $\frac{a}{c} = \frac{b}{c}$ provided $c \neq 0$. (Also, if $a = b$ and $c = d$, then $\frac{a}{c} = \frac{b}{d}$ provided that $c \neq 0$ and $d \neq 0$.)

Square root property

If $a^2 = b$, then $a = \pm\sqrt{b}$.

Zero product property

If $ab = 0$, then $a = 0$ or $b = 0$ or both a and $b = 0$.

Whether or not you remember their names, you've used these properties to solve algebraic equations. The process of solving an equation is really an algebraic proof that your solution is valid. To arrive at a correct solution, you must support each step by a property. The addition property of equality, for example, permits you to add the same number to both sides of an equation to get an equivalent equation.

EXAMPLE Solve for x: $5x - 12 = 3(x + 2)$

▶ ***Solution***

$5x - 12 = 3(x + 2)$	Given.
$5x - 12 = 3x + 6$	Distributive property.
$5x = 3x + 18$	Addition property of equality.
$2x = 18$	Subtraction property of equality.
$x = 9$	Division property of equality.

Why are the properties of arithmetic and equality important in geometry? The lengths of segments and the measures of angles involve numbers, so you will often need to use these properties in geometry proofs. And just as you use equality to express a relationship between numbers, you use congruence to express a relationship between geometric figures.

Definition of Congruence

If $AB = CD$, then $\overline{AB} \cong \overline{CD}$, and conversely, if $\overline{AB} \cong \overline{CD}$, then $AB = CD$.

If $m\angle A = m\angle B$, then $\angle A \cong \angle B$, and conversely, if $\angle A \cong \angle B$, then $m\angle A = m\angle B$.

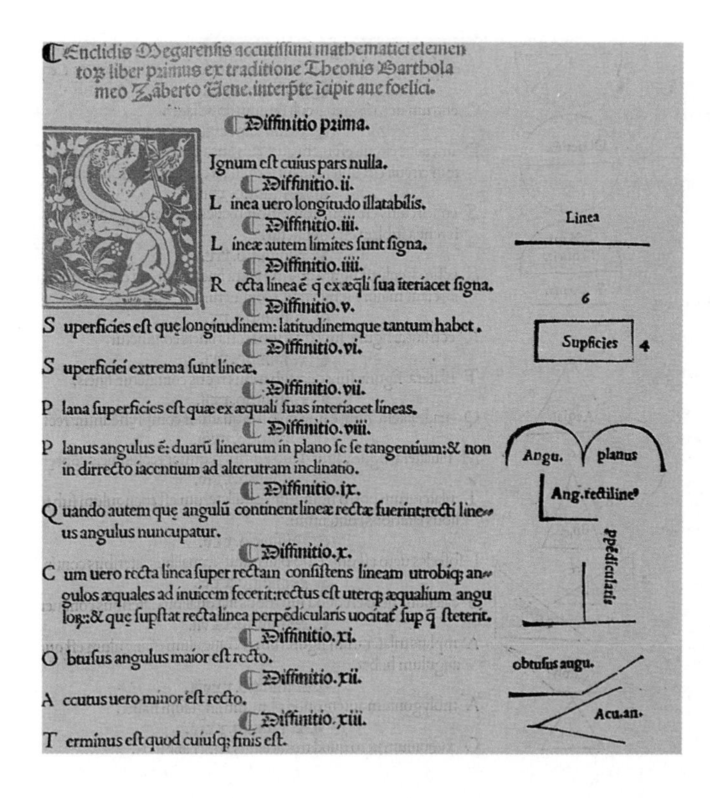
Euclidis Megarensis accutissimi mathematici elementorum liber primus ex traditione Theonis Bartholameo Zaberto Vene. interprete icipit aue foelici.

Diffinitio prima.

Signum est cuius pars nulla.

Diffinitio. ii.

Linea uero longitudo illatabilis.

Diffinitio. iii.

Lineæ autem limites sunt signa.

Diffinitio. iiii.

Recta linea ē q̄ ex æq̄li sua iteriacet signa.

Diffinitio. v.

Superficies est quę longitudinem: latitudinemque tantum habet.

Diffinitio. vi.

Superficiei extrema sunt lineæ.

Diffinitio. vii.

Plana superficies est quæ ex æquali suas interiacet lineas.

Diffinitio. viii.

Planus angulus ē: duarū linearum in plano se se tangentium: & non in dirrecto iacentium ad alterutram inclinatio.

Diffinitio. ix.

Quando autem quę angulū continent lineæ rectæ fuerint: recti lineus angulus nuncupatur.

Diffinitio. x.

Cum uero recta linea super rectam consistens lineam utrobiq; angulos æquales ad inuicem fecerit: rectus est uterq; æqualium angulorum: & quę supstat recta linea perpēdicularis uocitat̄ sup q̄ steterit.

Diffinitio. xi.

Obtusus angulus maior est recto.

Diffinitio. xii.

Accutus uero minor est recto.

Diffinitio. xiii.

Terminus est quod cuiusq; finis est.

A page from a Latin translation of Euclid's *Elements.* Which of these definitions do you recognize?

Congruence is defined by equality, so you can extend the properties of equality to a **reflexive property of congruence,** a **transitive property of congruence,** and a **symmetric property of congruence.** This is left for you to do in the exercises.

The third set of premises is specific to geometry. These premises are traditionally called postulates. Postulates should be very basic. Like undefined terms, they should be useful and easy for everyone to agree on, with little debate.

As you've performed basic geometric constructions in this class, you've observed some of these "obvious truths." Whenever you draw a figure or use an auxiliary line, you are using these postulates.

Postulates of Geometry

Line Postulate You can construct exactly one line through any two points.

Line Intersection Postulate The intersection of two distinct lines is exactly one point.

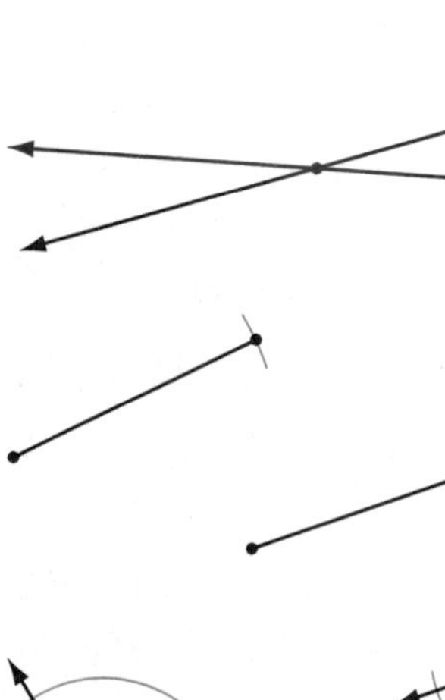

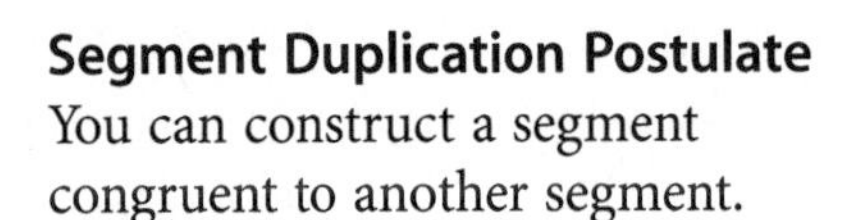

Segment Duplication Postulate You can construct a segment congruent to another segment.

Angle Duplication Postulate You can construct an angle congruent to another angle.

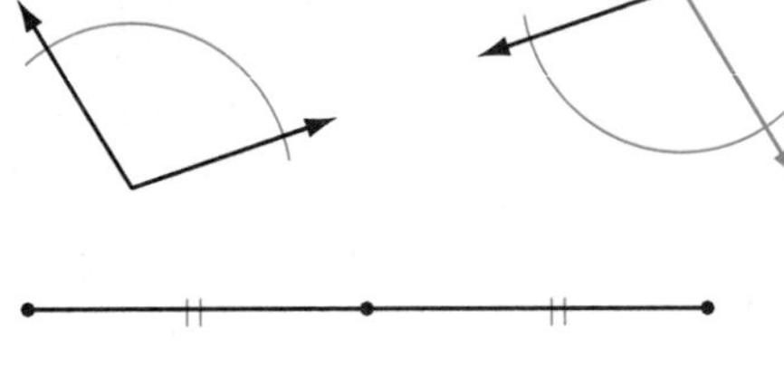

Midpoint Postulate You can construct exactly one midpoint on any line segment.

Angle Bisector Postulate You can construct exactly one angle bisector in any angle.

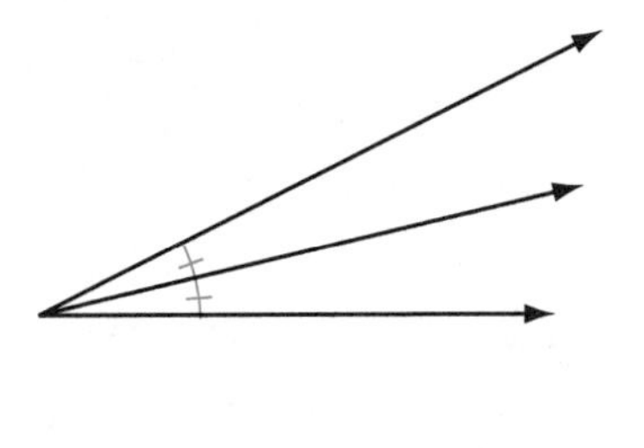

There are certain rules that everyone needs to agree on so we can drive safely! What are the "road rules" of geometry?

Parallel Postulate Through a point not on a given line, you can construct exactly one line parallel to the given line.

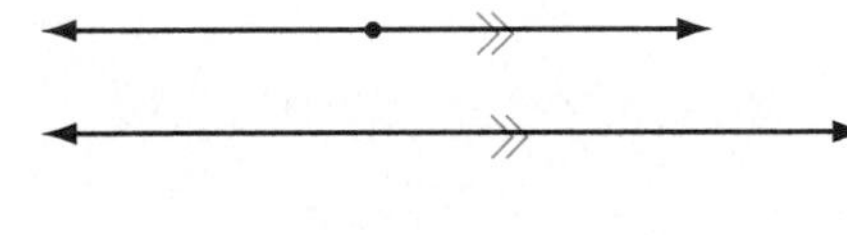

Perpendicular Postulate Through a point not on a given line, you can construct exactly one line perpendicular to the given line.

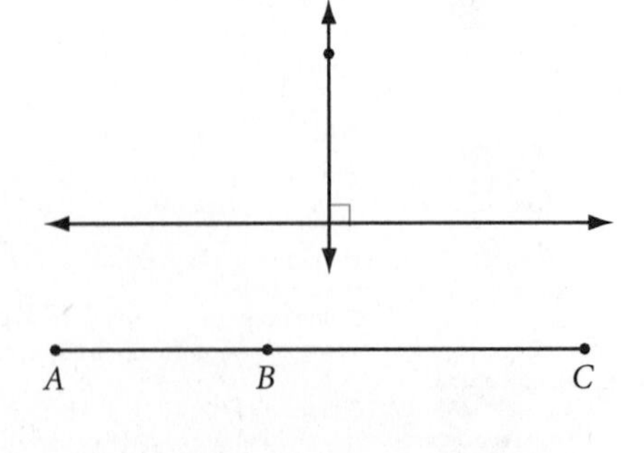

Segment Addition Postulate If point B is on $\overline{AC}$ and between points A and C, then $AB + BC = AC$.

A B C

Angle Addition Postulate If point D lies in the interior of $\angle ABC$, then $m\angle ABD + m\angle DBC = m\angle ABC$.

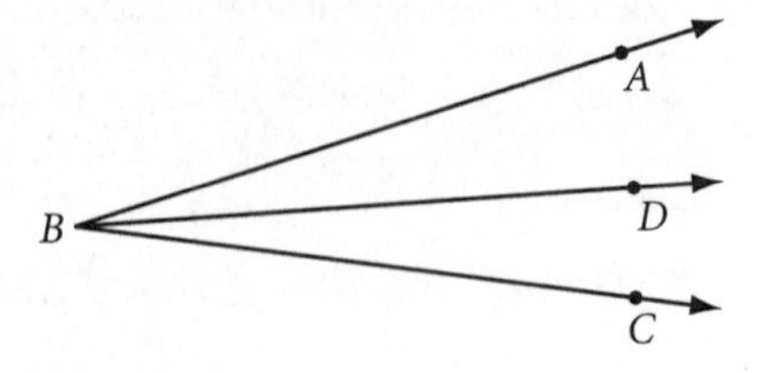

Linear Pair Postulate If two angles are a linear pair, then they are supplementary. (Previously called the Linear Pair Conjecture.)

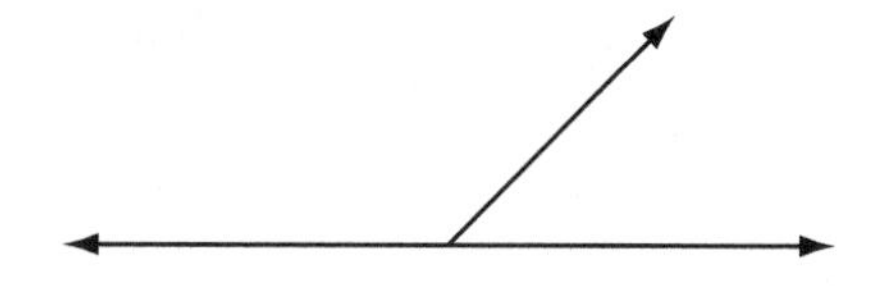

Corresponding Angles Postulate (CA Postulate) If two parallel lines are cut by a transversal, then the corresponding angles are congruent. Conversely, if two lines are cut by a transversal forming congruent corresponding angles, then the lines are parallel. (Previously called the CA Conjecture.)

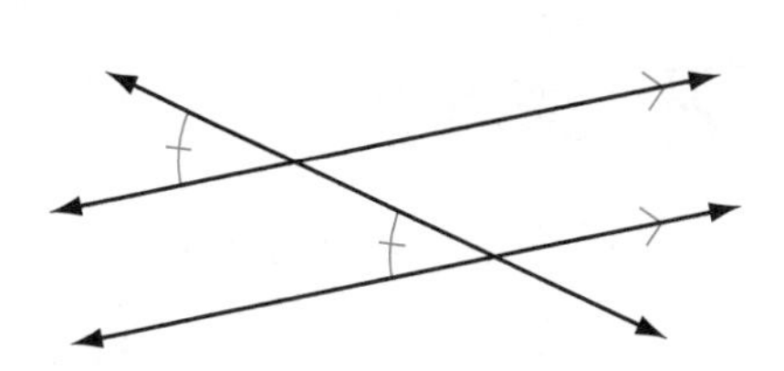

SSS Congruence Postulate If the three sides of one triangle are congruent to three sides of another triangle, then the two triangles are congruent. (Previously called the SSS Congruence Conjecture.)

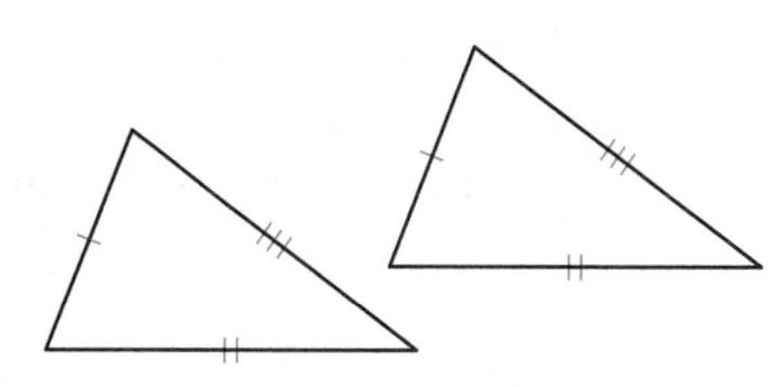

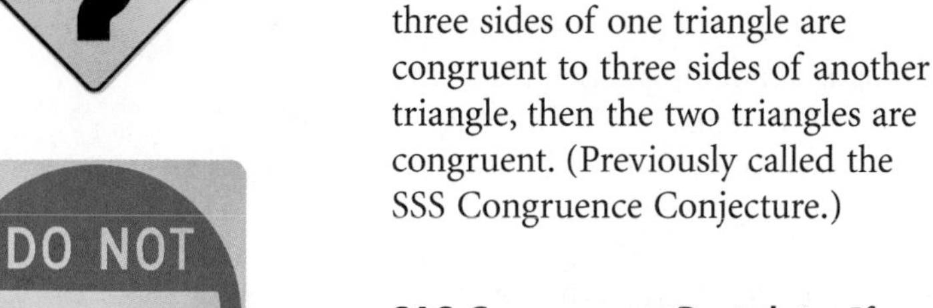

SAS Congruence Postulate If two sides and the included angle in one triangle are congruent to two sides and the included angle in another triangle, then the two triangles are congruent.

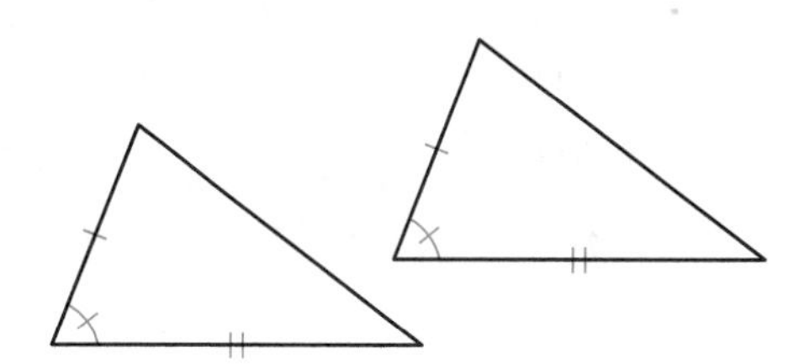

ASA Congruence Postulate If two angles and the included side in one triangle are congruent to two angles and the included side in another triangle, then the two triangles are congruent.

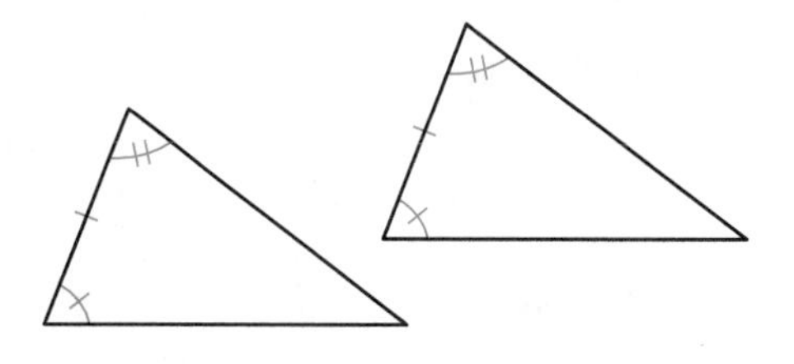

Mathematics CONNECTION

Euclid wrote 13 books covering, among other topics, plane geometry and solid geometry. He started with definitions, postulates, and "common notions" about the properties of equality. He then wrote hundreds of propositions, which we would call conjectures, and used constructions based on the definitions and postulates to show that they were valid. The statements that we call postulates were actually Euclid's postulates, plus a few of his propositions.

To build a logical framework for the geometry you have learned, you will start with the premises of geometry. In the exercises, you will see how these premises are the foundations for some of your previous assumptions and conjectures. You will also use these postulates and properties to see how some geometry statements are logical consequences of others.

EXERCISES

1. What is the difference between a postulate and a theorem?

2. Euclid might have stated the addition property of equality (translated from the Greek) in this way: "If equals are added to equals, the results are equal." State the subtraction, multiplication, and division properties of equality as Euclid might have stated them. (You may write them in English—extra credit for the original Greek!)

3. Write the reflexive property of congruence, the transitive property of congruence, and the symmetric property of congruence. Add these properties to your notebook. Include a diagram for each property. Illustrate one property with congruent triangles, another property with congruent segments, and another property with congruent angles. (These properties may seem ridiculously obvious. This is exactly why they are accepted as premises, which require no proof!)

4. When you state $AC = AC$, what property are you using? When you state $\overline{AC} \cong \overline{AC}$, what property are you using? ⓗ

5. Name the property that supports this statement: If $\angle ACE \cong \angle BDF$ and $\angle BDF \cong \angle HKM$, then $\angle ACE \cong \angle HKM$.

6. Name the property that supports this statement: If $x + 120 = 180$, then $x = 60$.

7. Name the property that supports this statement: If $2(x + 14) = 36$, then $x + 14 = 18$.

In Exercises 8 and 9, provide the missing property of equality or arithmetic as a reason for each step to solve the algebraic equation or to prove the algebraic argument.

8. Solve for x: $7x - 22 = 4(x + 2)$ ⓗ

Solution:	
$7x - 22 = 4(x + 2)$	Given.
$7x - 22 = 4x + 8$	? property.
$3x - 22 = 8$	? property of equality.
$3x = 30$	? property of equality.
$x = 10$	? property of equality.

9. Conjecture: If $\frac{x}{m} - c = d$, then $x = m(c + d)$, provided that $m \neq 0$. ⓗ

Proof:	
$\frac{x}{m} - c = d$	?
$\frac{x}{m} = d + c$	?
$x = m(d + c)$	?
$x = m(c + d)$	?

In Exercises 10–17, identify each statement as true or false. Then state which definition, property of algebra, property of congruence, or postulate supports your answer.

10. If M is the midpoint of $\overline{AB}$, then $AM = BM$.

11. If M is the midpoint of $\overline{CD}$ and N is the midpoint of $\overline{CD}$, then M and N are the same point. ⓗ

12. If $\overrightarrow{AB}$ bisects $\angle CAD$, then $\angle CAB \cong \angle DAB$.

13. If $\overrightarrow{AB}$ bisects $\angle CAD$ and $\overrightarrow{AF}$ bisects $\angle CAD$, then $\overrightarrow{AB}$ and $\overrightarrow{AF}$ are the same ray.

14. Lines ℓ and m can intersect at different points A and B.

15. If line ℓ passes through points A and B and line m passes through points A and B, lines ℓ and m do not have to be the same line.

16. If point P is in the interior of $\angle RAT$, then $m\angle RAP + m\angle PAT = m\angle RAT$.

17. If point M is on $\overline{AC}$ and between points A and C, then $AM + MC = AC$.

18. The Declaration of Independence states "We hold these truths to be self-evident . . . ," then goes on to list four postulates of good government. Look up the Declaration of Independence and list the four self-evident truths that were the original premises of the United States government. You can find links to this topic at www.keymath.com/DG .

Arthur Szyk (1894–1951), a Polish American whose propaganda art helped aid the Allied war effort during World War II, created this patriotic illustrated version of the Declaration of Independence.

19. Copy and complete this flowchart proof. For each reason, state the definition, the property of algebra, or the property of congruence that supports the statement.

Given: $\overline{AO}$ and $\overline{BO}$ are radii

Show: $\triangle AOB$ is isosceles

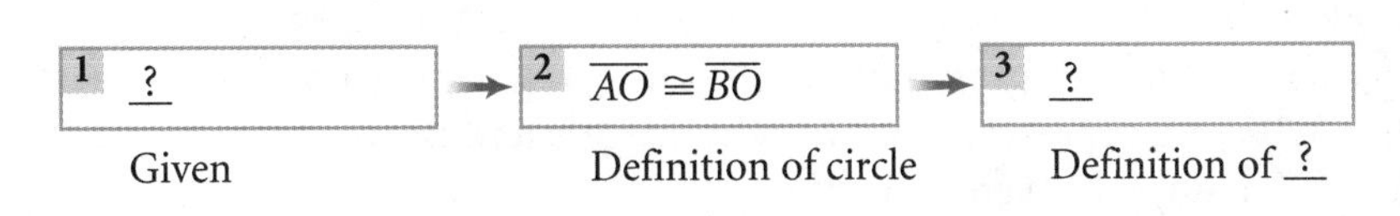

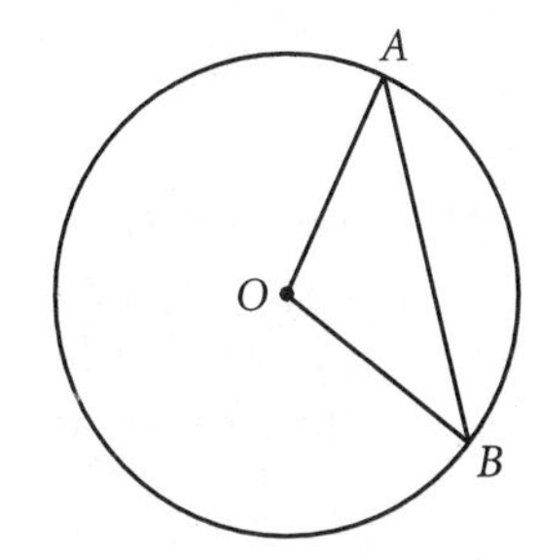

For Exercises 20–22, copy and complete each flowchart proof.

20. Given: $\angle 1 \cong \angle 2$

Show: $\angle 3 \cong \angle 4$

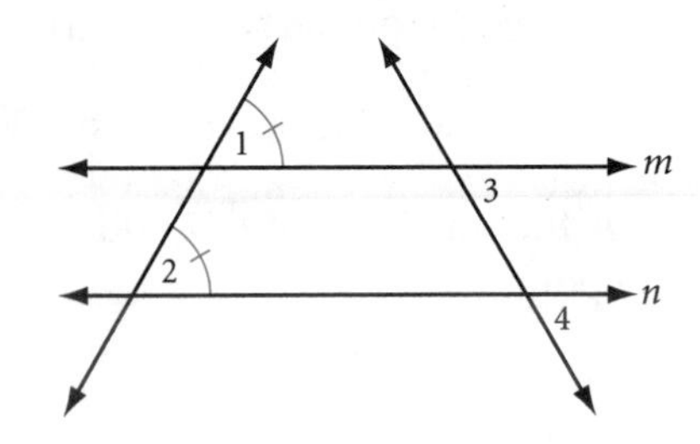

1 $\angle 1 \cong$ _?_ → 2 _?_ → 3 $\angle 3 \cong$ _?_

1: _?_ 2: Corresponding Angles Postulate 3: _?_

21. Given: $\overline{AC} \cong \overline{BD}$, $\overline{AD} \cong \overline{BC}$

Show: $\angle D \cong \angle C$

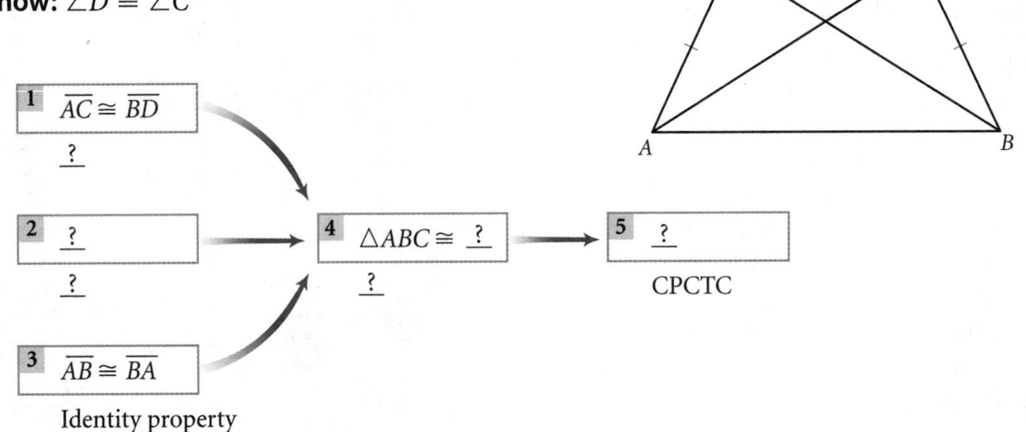

22. Given: Isosceles triangle ABC with $\overline{AB} \cong \overline{BC}$

Show: $\angle A \cong \angle C$ ⓗ

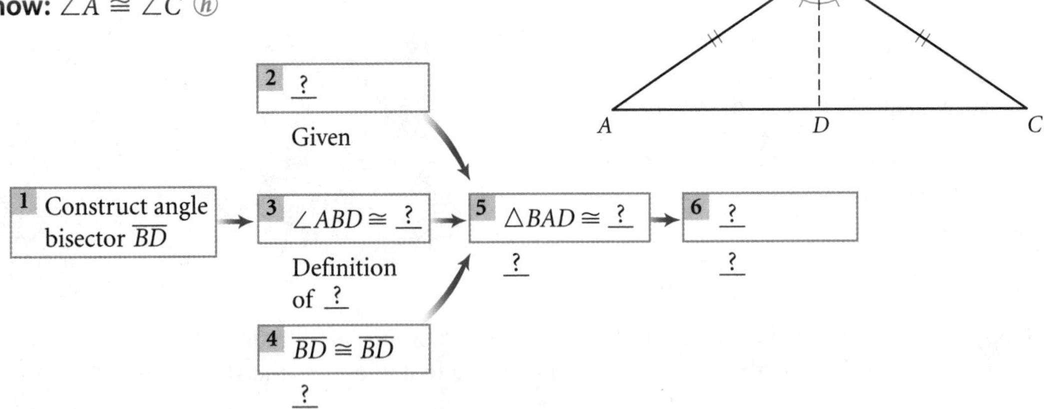

23. You have probably noticed that the sum of two odd integers is always an even integer. The rule $2n$ generates even integers and the rule $2n - 1$ generates odd integers. Let $2n - 1$ and $2m - 1$ represent any two odd integers and prove that the sum of two odd integers is always an even integer.

24. Let $2n - 1$ and $2m - 1$ represent any two odd integers and prove that the product of any two odd integers is always an odd integer.

25. Show that the sum of any three consecutive integers is always divisible by 3. ⓗ

Review

26. Shannon and Erin are hiking up a mountain. Of course, they are packing the clinometer they made in geometry class. At point A along a flat portion of the trail, Erin sights the mountain peak straight ahead at an angle of elevation of 22°. The level trail continues 220 m straight to the base of the mountain at point B. At that point, Shannon measures the angle of elevation to be 38°. From B the trail follows a ridge straight up the mountain to the peak. At point B, how far are they from the mountain peak?

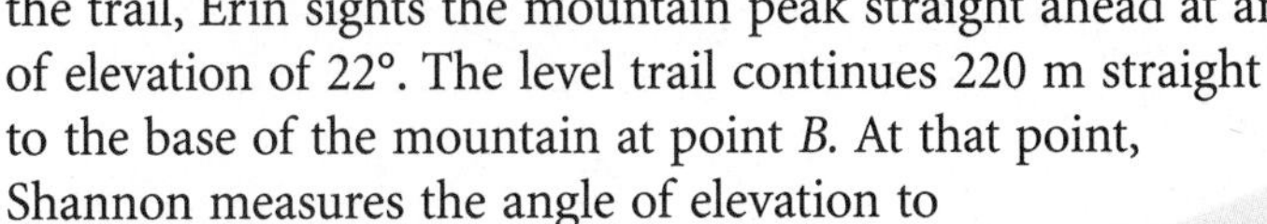

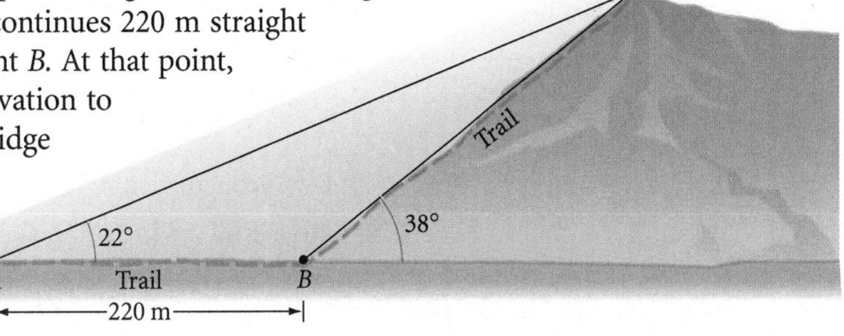

27.

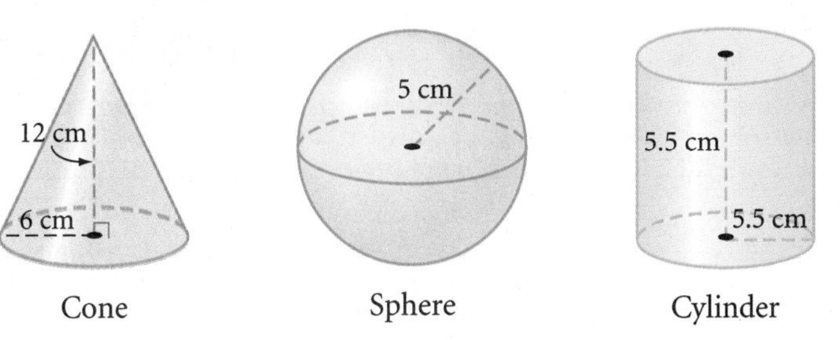

Arrange the names of the solids in order, greatest to least.

Volume:	?	?	?
Surface area:	?	?	?
Length of the longest rod that will fit inside:	?	?	?

28. Two communication towers stand 64 ft apart. One is 80 ft high and the other is 48 ft high. Each has a guy wire from its top anchored to the base of the other tower. At what height do the two guy wires cross?

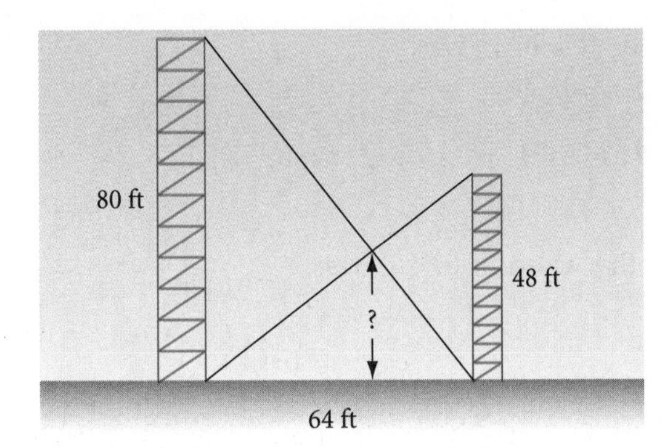

In Exercises 29 and 30, all length measurements are given in meters.

29. What's wrong with this picture?

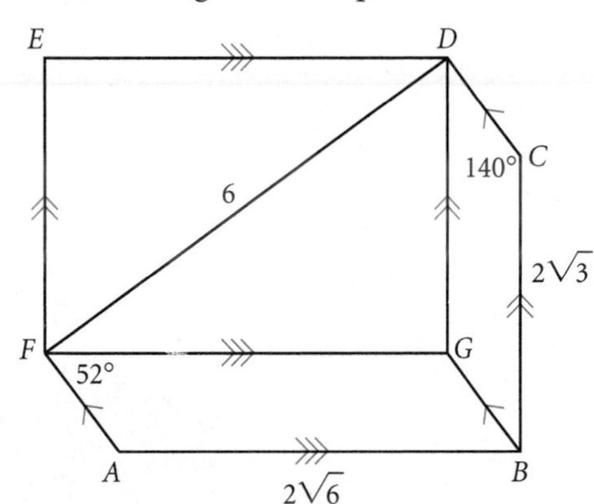

30. Find angle measures x and y, and length a.

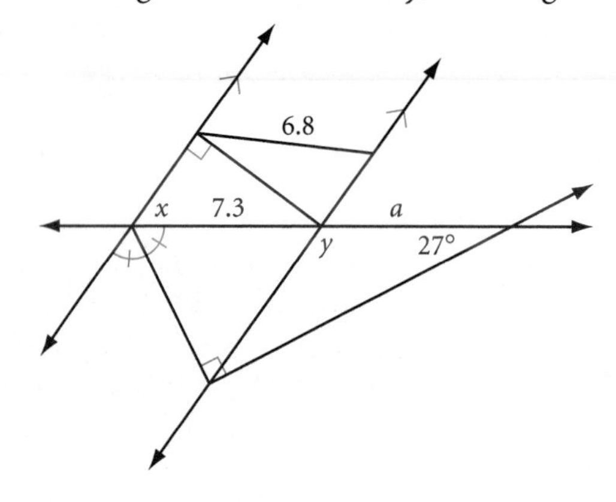

31. Each arc is a quarter of a circle with its center at a vertex of the square.
Given: Each square has side length 1 unit **Find:** The shaded area

a.

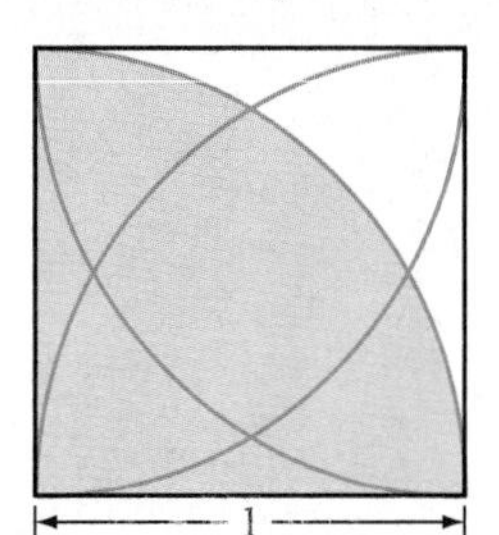

b.

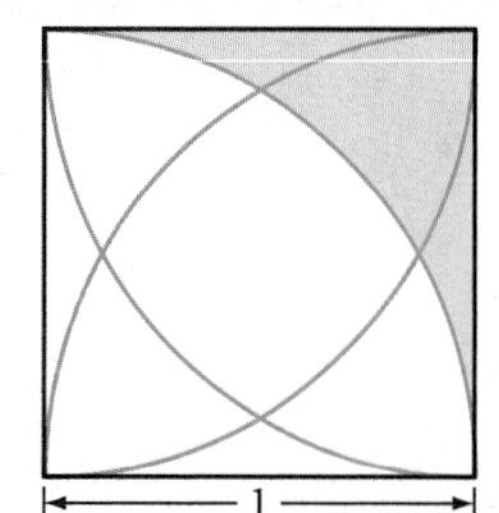

c.

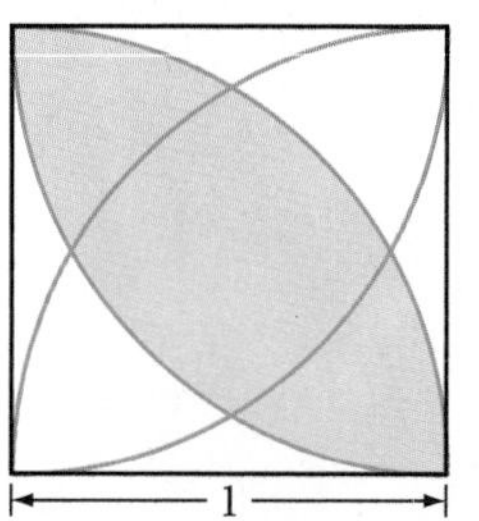

IMPROVING YOUR REASONING SKILLS

PUZZLE

Logical Vocabulary

Here is a logical vocabulary challenge. It is sometimes possible to change one word to another of equal length by changing one letter at a time. Each change, or move, you make gives you a new word. For example, DOG can be changed to CAT in exactly three moves.

DOG ⇒ DOT ⇒ COT ⇒ CAT

Change MATH to each of the following words in exactly four moves.

1. MATH ⇒ _?_ ⇒ _?_ ⇒ _?_ ⇒ ROSE

2. MATH ⇒ _?_ ⇒ _?_ ⇒ _?_ ⇒ CORE

3. MATH ⇒ _?_ ⇒ _?_ ⇒ _?_ ⇒ HOST

4. MATH ⇒ _?_ ⇒ _?_ ⇒ _?_ ⇒ LESS

5. MATH ⇒ _?_ ⇒ _?_ ⇒ _?_ ⇒ LIVE

Now create one of your own. Change MATH to another word in four moves.

LESSON

13.2

Planning a Geometry Proof

What is now proved was once only imagined.

WILLIAM BLAKE

A proof in geometry consists of a sequence of statements, starting with a given set of premises and leading to a valid conclusion. Each statement follows from one or more of the previous statements and is supported by a reason. A reason for a statement must come from the set of premises that you learned about in Lesson 13.1.

In earlier chapters you informally proved many conjectures. Now you can formally prove them, using the premises of geometry. In this lesson you will identify for yourself what is given and what you must show, in order to prove a conjecture. You will also create your own labeled diagrams.

"I THINK YOU SHOULD BE MORE EXPLICIT HERE IN STEP TWO."

As you have seen, you can state many geometry conjectures as conditional statements. For example, you can write the conjecture "Vertical angles are congruent" as a conditional statement: "If two angles are vertical angles, then they are congruent." To prove that a conditional statement is true, you assume that the first part of the conditional is true, then logically demonstrate the truth of the conditional's second part. In other words, you demonstrate that the first part implies the second part. The first part is what you assume to be true in the proof; it is the *given* information. The second part is the part you logically demonstrate in the proof; it is what you want to *show.*

Given	**Show**
Two angles are vertical angles	They are congruent

Next, draw and label a diagram that illustrates the given information. Then, use the labels in the diagram to restate graphically what is given and what you must show.

Once you've created a diagram to illustrate your conjecture and you know where to start and where to go, make a plan. Use your plan to write the proof. Here's the complete process.

Writing a Proof

Task 1 From the conditional statement, identify what is given and what you must show.

Task 2 Draw and label a diagram to illustrate the given information.

Task 3 Restate what is given and what you must show in terms of your diagram.

Task 4 Plan a proof. Organize your reasoning mentally or on paper.

Task 5 From your plan, write a proof.

In Chapter 2, you proved the Vertical Angles Conjecture using conjectures that have now become postulates.

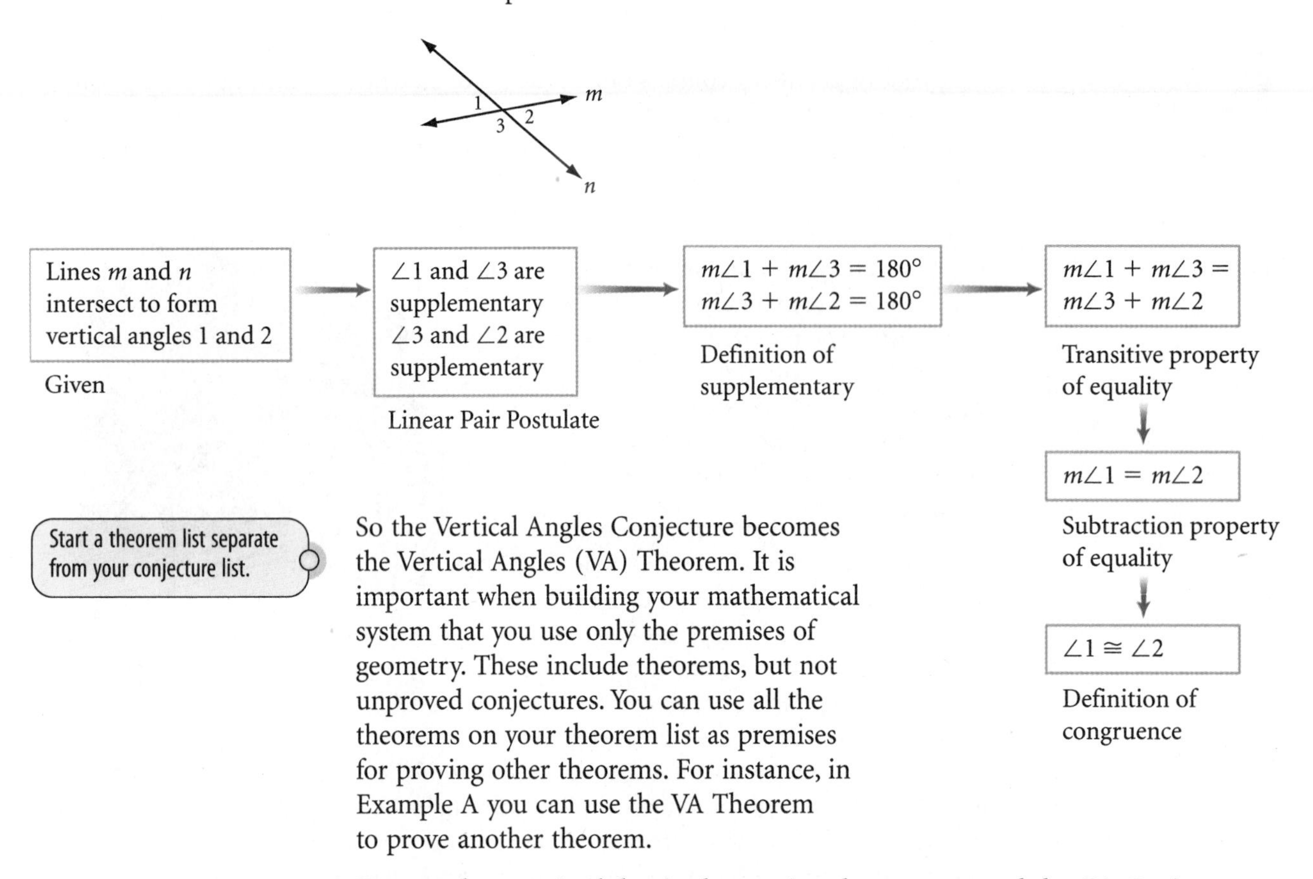

Start a theorem list separate from your conjecture list.

So the Vertical Angles Conjecture becomes the Vertical Angles (VA) Theorem. It is important when building your mathematical system that you use only the premises of geometry. These include theorems, but not unproved conjectures. You can use all the theorems on your theorem list as premises for proving other theorems. For instance, in Example A you can use the VA Theorem to prove another theorem.

You may have noticed that in the previous lesson we stated the CA Conjecture as a postulate, but not the AIA Conjecture or the AEA Conjecture. In this first example you will see how to use the five tasks of the proof process to prove the AIA Conjecture.

EXAMPLE A

Prove the Alternate Interior Angles Conjecture: If two parallel lines are cut by a transversal, then the alternate interior angles are congruent.

▶ **Solution**

For Task 1 identify what is given and what you must show.

Given: Two parallel lines are cut by a transversal

Show: Alternate interior angles formed by the lines are congruent

For Task 2 draw and label a diagram.

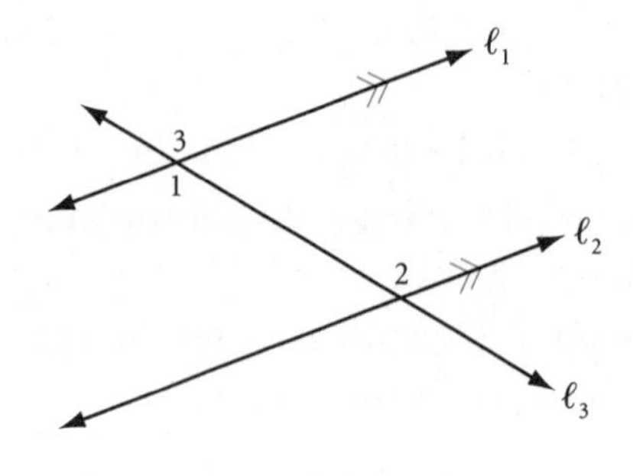

For Task 3 restate what is given and what you must show in terms of the diagram.

Given: Parallel lines ℓ_1 and ℓ_2 cut by transversal ℓ_3 to form alternate interior angles $\angle 1$ and $\angle 2$

Show: $\angle 1 \cong \angle 2$

For Task 4 plan a proof. Organize your reasoning mentally or on paper.

Plan:

I need to show that $\angle 1 \cong \angle 2$.
Looking over the postulates and theorems, the ones that look useful are the CA Postulate and the VA Theorem.
From the CA Postulate, I know that $\angle 2 \cong \angle 3$ and from the VA Theorem, $\angle 1 \cong \angle 3$.
If $\angle 2 \cong \angle 3$ and $\angle 1 \cong \angle 3$, then by substitution $\angle 1 \cong \angle 2$.

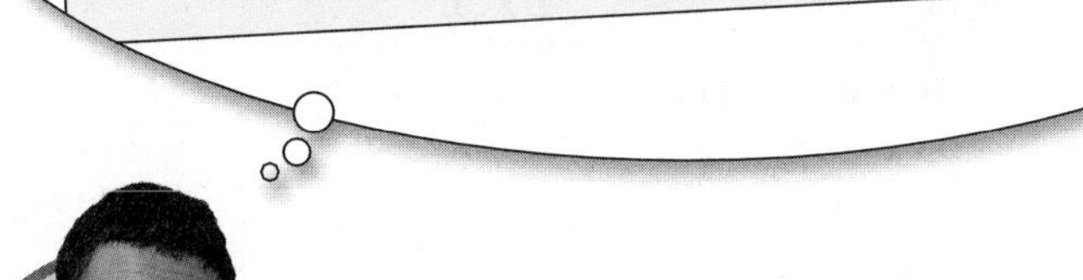

For Task 5 create a proof from your plan.

Flowchart Proof

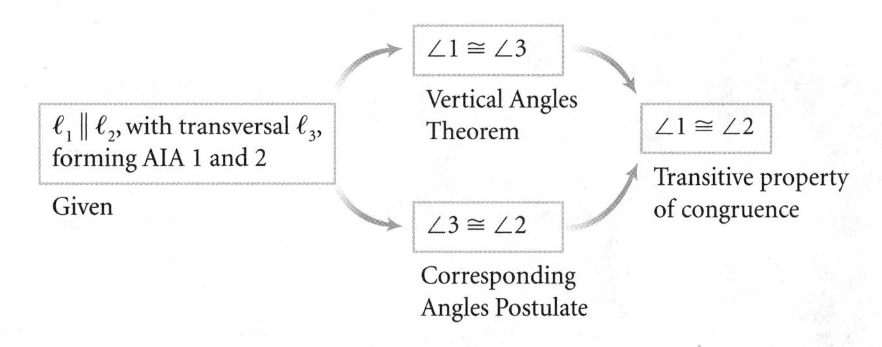

So the AIA Conjecture becomes the AIA Theorem. Add this theorem to your theorem list.

In Chapter 4, you informally proved the Triangle Sum Conjecture. The proof is short, but clever, too, because it required the construction of an auxiliary line. All the steps in the proof use properties that we now designate as postulates. Example B shows the flowchart proof. For example, the Parallel Postulate guarantees that it will always be possible to construct an auxiliary line through a vertex, parallel to the opposite side.

EXAMPLE B Prove the Triangle Sum Conjecture: The sum of the measures of the angles of a triangle is 180°.

▶ Solution

Given: $\angle 1$, $\angle 2$, and $\angle 3$ are the three angles of $\triangle ABC$

Show: $m\angle 1 + m\angle 2 + m\angle 3 = 180°$

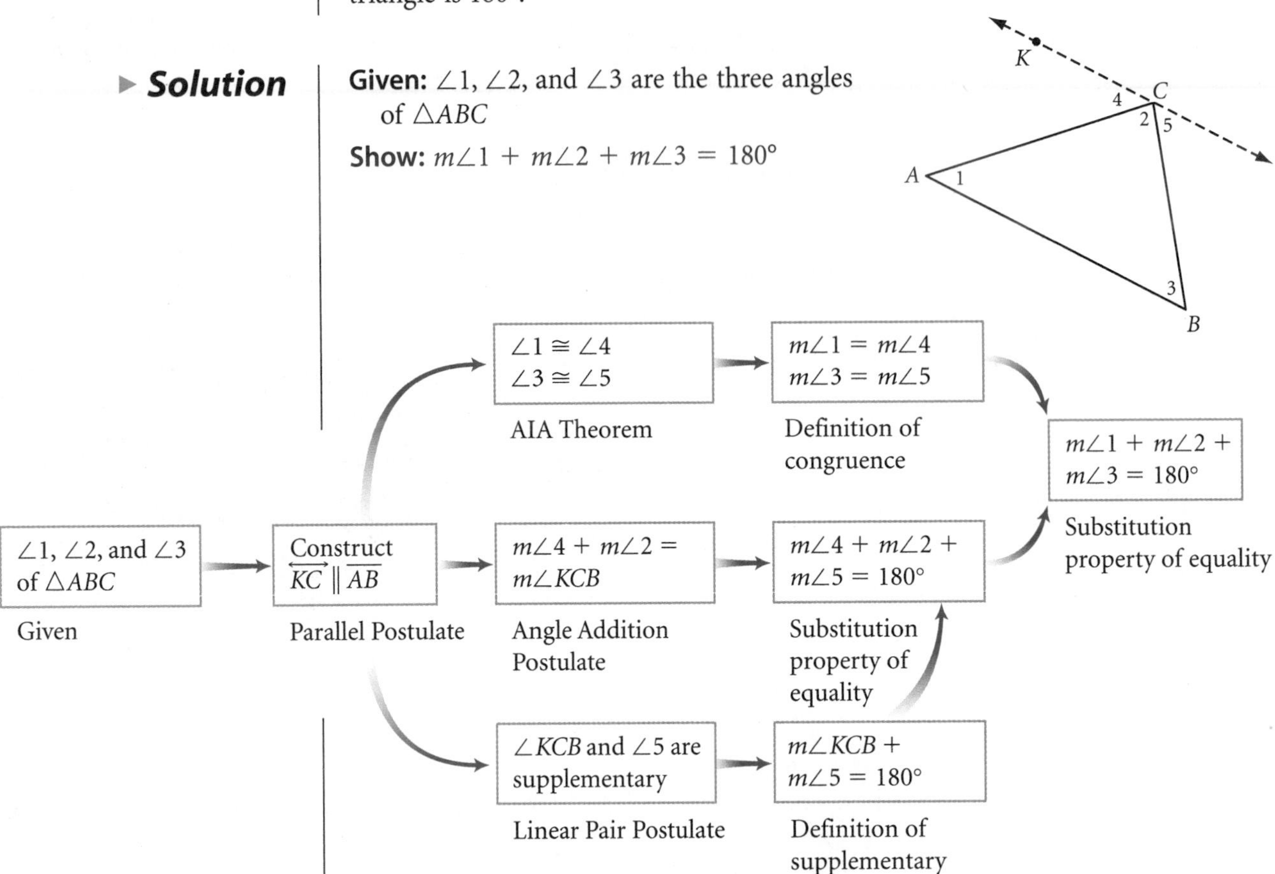

So the Triangle Sum Conjecture becomes the Triangle Sum Theorem. Add it to your theorem list. Notice that each reason we now use in a proof is a postulate, theorem, definition, or property.

When you trace back your family tree, you include the names of your parents, the names of their parents, and so on.

To make sure a particular theorem has been properly proved, you can also check the "logical family tree" of the theorem. When you create a family tree for a theorem, you trace it back to all the postulates that the theorem relied on. You don't need to list all the definitions and properties of equality and congruence: list only the theorems and postulates used in the proof. For the theorems that were used in the proof, what postulates and theorems were used in *their* proofs, and so on. In Chapter 4, you informally proved the Third Angle Conjecture. Let's look again at the proof.

Third Angle Conjecture: If two angles of one triangle are congruent to two angles of a second triangle, then the third pair of angles are congruent.

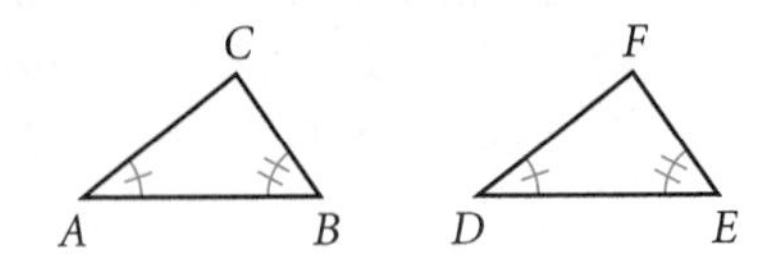

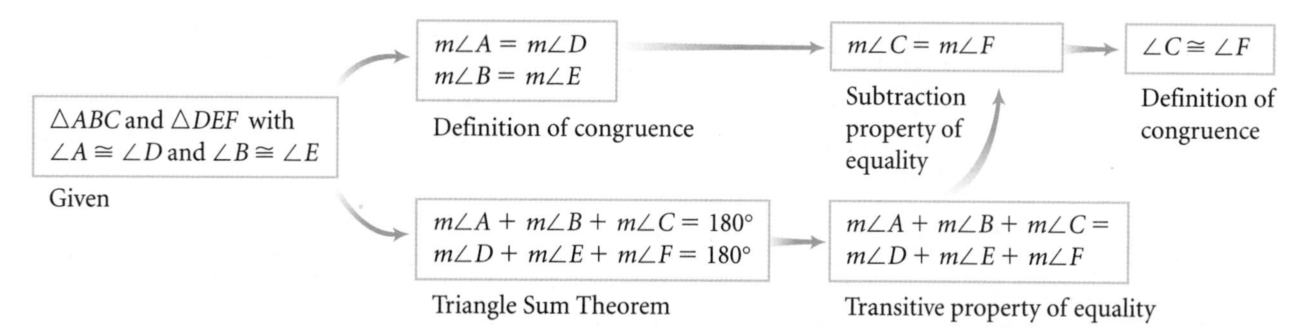

What does the logical family tree of the Third Angle Theorem look like?

You start by putting the Third Angle Theorem in a box. Find all the postulates and theorems used in the proof. The only postulate or theorem used was the Triangle Sum Theorem. Put that box above it.

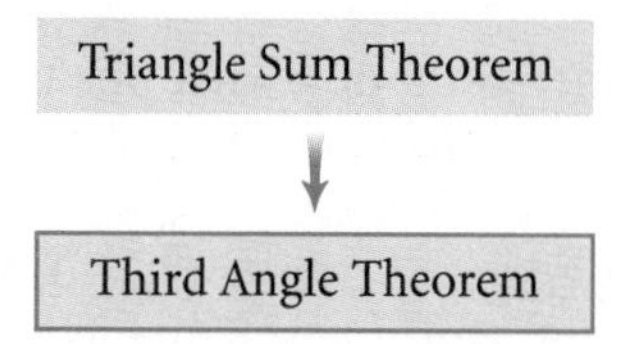

Next, locate all the theorems and postulates used to prove the Triangle Sum Theorem. That proof used the Parallel Postulate, the Linear Pair Postulate, the Angle Addition Postulate, and the AIA Theorem. You place these postulates in boxes above the Triangle Sum Theorem. Connect the boxes with arrows showing the logical connection. Now the family tree looks like this:

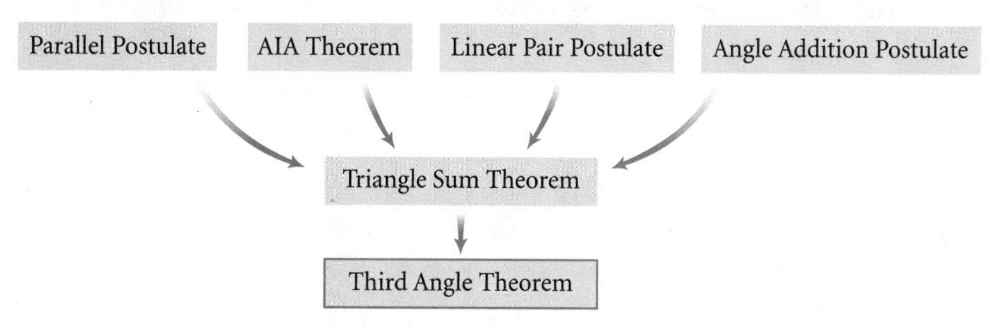

To prove the AIA Theorem, we used the CA Postulate and the VA Theorem, and we used the Linear Pair Postulate to prove the VA Theorem. Notice that the Linear Pair Postulate is already in the family tree, but we move it up so it's above both the Triangle Sum Theorem and the VA Theorem. The completed family tree looks like this:

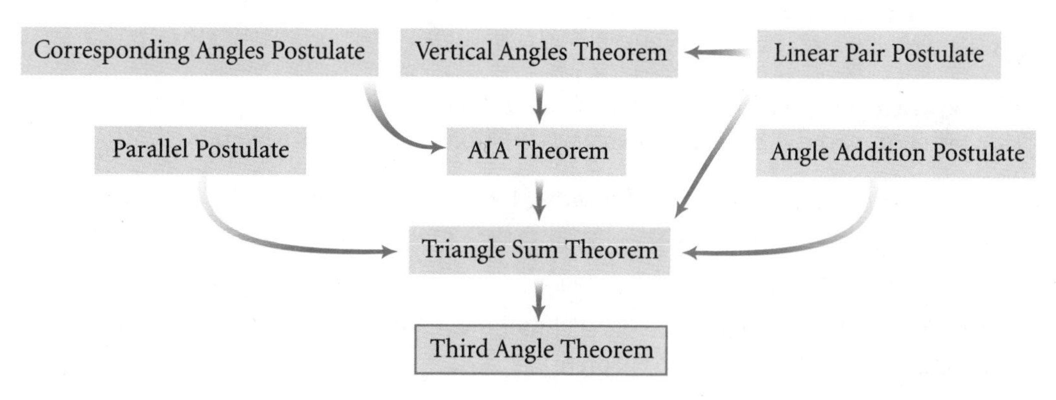

The family tree shows that, ultimately, the Third Angle Theorem relies on the Parallel Postulate, the CA Postulate, the Linear Pair Postulate, and the Angle Addition Postulate. You might notice that the family tree of a theorem looks similar to a flowchart proof. The difference is that the family tree focuses on the premises and traces them back to the postulates.

EXERCISES

1. What postulate(s) does the VA Theorem rely on?

2. What postulate(s) does the Triangle Sum Theorem rely on?

3. If you need a parallel line in a proof, what postulate allows you to construct it?

4. If you need a perpendicular line in a proof, what postulate allows you to construct it?

In Exercises 5–14, write a paragraph proof or a flowchart proof of the conjecture. Once you have completed their proofs, add the statements to your theorem list.

5. If two angles are both congruent and supplementary, then each is a right angle. (Congruent and Supplementary Theorem)

6. Supplements of congruent angles are congruent. (Supplements of Congruent Angles Theorem)

7. All right angles are congruent. (Right Angles Are Congruent Theorem)

8. If two lines are cut by a transversal forming congruent alternate interior angles, then the lines are parallel. (Converse of the AIA Theorem)

9. If two parallel lines are cut by a transversal, then the alternate exterior angles are congruent. (AEA Theorem)

10. If two lines are cut by a transversal forming congruent alternate exterior angles, then the lines are parallel. (Converse of the AEA Theorem)

11. If two parallel lines are cut by a transversal, then the interior angles on the same side of the transversal are supplementary. (Interior Supplements Theorem)

12. If two lines are cut by a transversal forming interior angles on the same side of the transversal that are supplementary, then the lines are parallel. (Converse of the Interior Supplements Theorem)

13. If two lines in the same plane are parallel to a third line, then they are parallel to each other. (Parallel Transitivity Theorem)

14. If two lines in the same plane are perpendicular to a third line, then they are parallel to each other. (Perpendicular to Parallel Theorem)

15. Draw a family tree of the Converse of the Alternate Exterior Angles Theorem.

Review

16. Suppose the top of a pyramid with volume 1107 cm^3 is sliced off and discarded. The remaining portion is called a **truncated pyramid.** If the cut was parallel to the base and two-thirds of the distance to the vertex, what is the volume of the truncated pyramid?

17. Abraham is building a dog house for his terrier. His plan is shown at right. He will cut a door and a window later. After he builds the frame for the structure, can he complete it using one piece of 4-by-8-foot plywood? If the answer is yes, show how he should cut the plywood. If no, explain why not.

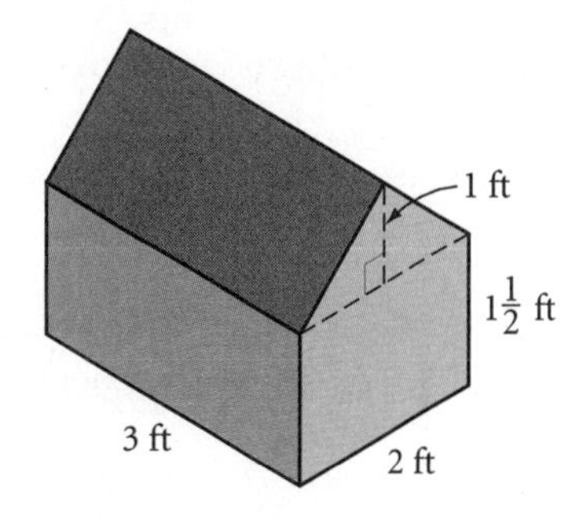

18. Find x.

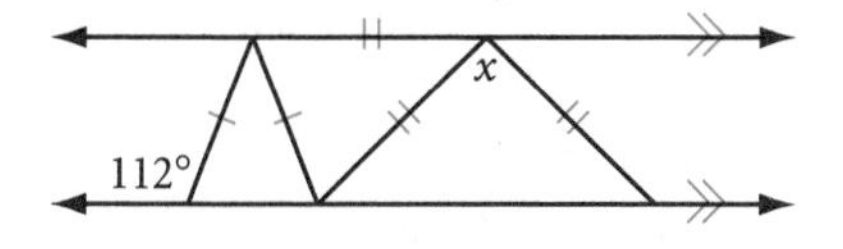

19. M is the midpoint of $\overline{AC}$ and $\overline{BD}$. For each statement, select always (A), sometimes (S), or never (N).

a. $\angle BAD$ and $\angle ADC$ are supplementary.

b. $\angle ADM$ and $\angle MAD$ are complementary.

c. $AD + BC < AC$

d. $AD + CD < AC$

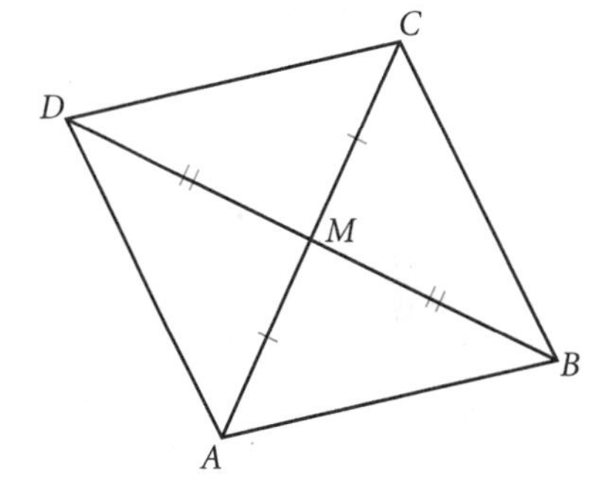

IMPROVING YOUR REASONING SKILLS

Calculator Cunning

Using the calculator shown at right, what is the largest number you can form by pressing the keys labeled 1, 2, and 3 exactly once each and the key labeled y^x at most once? You cannot press any other keys.

LESSON

13.3

Triangle Proofs

The most violent element in our society is ignorance.

EMMA GOLDMAN

Now that the theorems from the previous lesson have been proved, you should add them to your theorem list. They will be useful to you in proving future theorems.

Triangle congruence is so useful in proving other theorems that we will focus next on triangle proofs. You may have noticed that in Lesson 13.1, three of the four triangle congruence conjectures were stated as postulates (the SSS Congruence Postulate, the SAS Congruence Postulate, and the ASA Congruence Postulate). The SAA Conjecture was not stated as a postulate. In Lesson 4.5, you used the ASA Conjecture (now the ASA Postulate) to explain the SAA Conjecture. The family tree for SAA congruence looks like this:

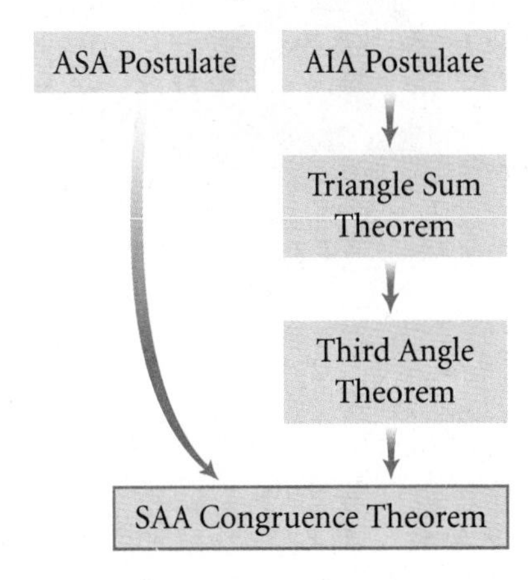

So the SAA Conjecture becomes the SAA Theorem. Add this theorem to your theorem list. This theorem will be useful in some of the proofs in this lesson.

Let's use the five-task proof process and triangle congruence to prove the Angle Bisector Conjecture.

EXAMPLE

Prove the Angle Bisector Conjecture: Any point on the bisector of an angle is equidistant from the sides of the angle.

▶ ***Solution***

Given: Any point on the bisector of an angle

Show: The point is equidistant from the sides of the angle

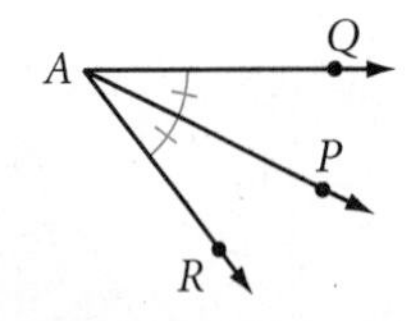

Given: $\overrightarrow{AP}$ bisecting $\angle QAR$

Show: P is equally distant from sides $\overrightarrow{AQ}$ and $\overrightarrow{AR}$

Plan: The distance from a point to a line is measured along the perpendicular from the point to the line. So I begin by constructing $\overline{PB} \perp \overrightarrow{AQ}$ and $\overline{PC} \perp \overrightarrow{AR}$ (the Perpendicular Postulate permits me to do this). I can show that $\overline{PB} \cong \overline{PC}$ if they are corresponding parts of congruent triangles. $\overline{AP} \cong \overline{AP}$ by the identity property of congruence, and $\angle QAP \cong \angle RAP$ by the definition of an angle bisector. $\angle ABP$ and $\angle ACP$ are right angles and thus they are congruent.

So $\triangle ABP \cong \triangle ACP$ by the SAA Theorem. If the triangles are congruent, then $\overline{PB} \cong \overline{PC}$ by CPCTC.

Based on this plan, I can write a flowchart proof.

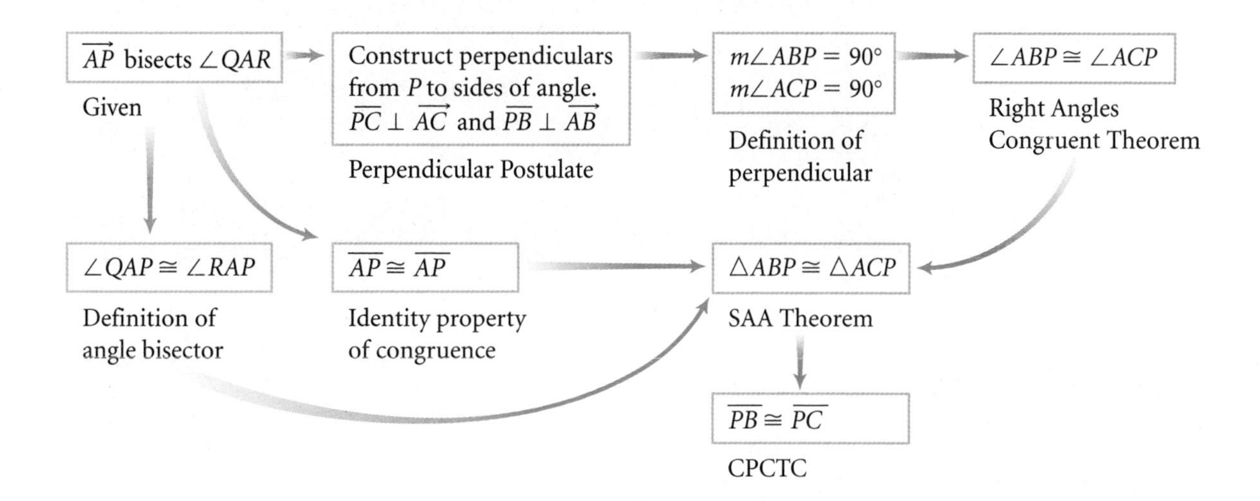

Thus the Angle Bisector Conjecture becomes the Angle Bisector Theorem.

As our own proofs build on each other, flowcharts can become too large and awkward. You can also use a two-column format for writing proofs. A **two-column proof** is identical to a flowchart or paragraph proof, except that the statements are listed in the first column, each supported by a reason (a postulate, definition, property, or theorem) in the second column.

Here is the same proof from the example above, following the same plan, presented as a two-column proof. Arrows link the steps.

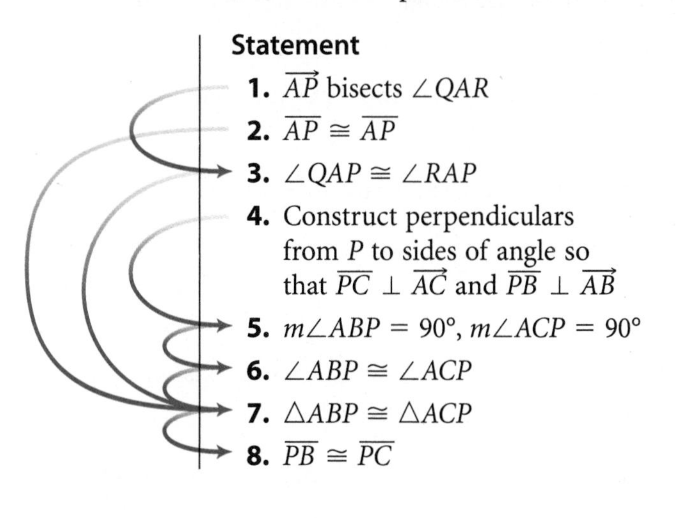

Statement	Reason
1. $\overrightarrow{AP}$ bisects $\angle QAR$	**1.** Given
2. $\overline{AP} \cong \overline{AP}$	**2.** Identity property of congruence
3. $\angle QAP \cong \angle RAP$	**3.** Definition of angle bisector
4. Construct perpendiculars from P to sides of angle so that $\overline{PC} \perp \overrightarrow{AC}$ and $\overline{PB} \perp \overrightarrow{AB}$	**4.** Perpendicular Postulate
5. $m\angle ABP = 90°$, $m\angle ACP = 90°$	**5.** Definition of perpendicular
6. $\angle ABP \cong \angle ACP$	**6.** Right Angles Congruent Theorem
7. $\triangle ABP \cong \triangle ACP$	**7.** SAA Theorem
8. $\overline{PB} \cong \overline{PC}$	**8.** CPCTC

Compare the two-column proof you just saw with the flowchart proof in Example A. What similarities do you see? What are the advantages of each format?

No matter what format you choose, your proof should be clear and easy for someone to follow.

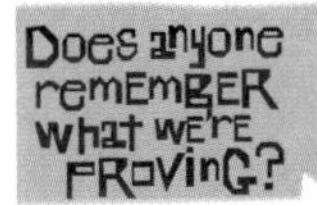

EXERCISES

You will need

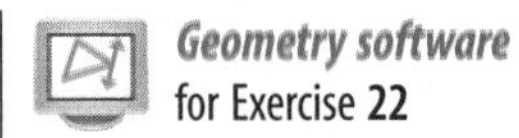
Geometry software for Exercise 22

In Exercises 1–13, write a proof of the conjecture. Once you have completed the proofs, add the theorems to your list.

1. If a point is on the perpendicular bisector of a segment, then it is equally distant from the endpoints of the segment. (Perpendicular Bisector Theorem)

2. If a point is equally distant from the endpoints of a segment, then it is on the perpendicular bisector of the segment. (Converse of the Perpendicular Bisector Theorem) ⓗ

3. If a triangle is isosceles, then the base angles are congruent. (Isosceles Triangle Theorem)

4. If two angles of a triangle are congruent, then the triangle is isosceles. (Converse of the Isosceles Triangle Theorem)

5. If a point is equally distant from the sides of an angle, then it is on the bisector of the angle. (Converse of the Angle Bisector Theorem) ⓗ

6. The three perpendicular bisectors of the sides of a triangle are concurrent. (Perpendicular Bisector Concurrency Theorem)

7. The three angle bisectors of the sides of a triangle are concurrent. (Angle Bisector Concurrency Theorem)

8. The measure of an exterior angle of a triangle is equal to the sum of the measures of the two remote interior angles. (Triangle Exterior Angle Theorem)

9. The sum of the measures of the four angles of a quadrilateral is 360°. (Quadrilateral Sum Theorem)

10. In an isosceles triangle, the medians to the congruent sides are congruent. (Medians to the Congruent Sides Theorem)

11. In an isosceles triangle, the angle bisectors to the congruent sides are congruent. (Angle Bisectors to the Congruent Sides Theorem)

12. In an isosceles triangle, the altitudes to the congruent sides are congruent. (Altitudes to the Congruent Sides Theorem)

13. In Lesson 4.8, you were asked to complete informal proofs of these two conjectures:

The bisector of the vertex angle of an isosceles triangle is also the median to the base.

The bisector of the vertex angle of an isosceles triangle is also the altitude to the base.

To demonstrate that the altitude to the base, the median to the base, and the bisector of the vertex angle are all the same segment in an isosceles triangle, you really need to prove three theorems. One possible sequence is diagrammed at right.

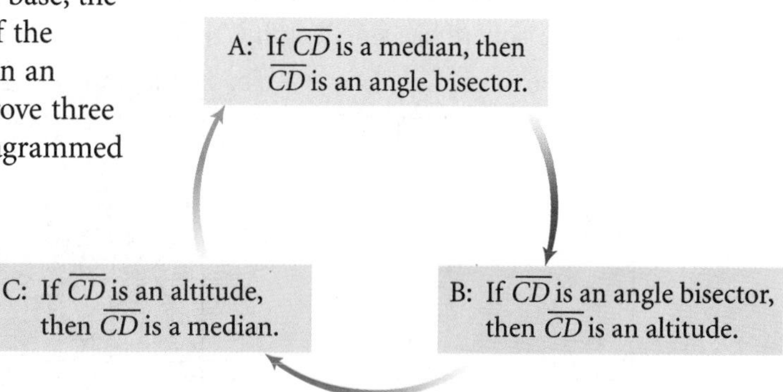

Prove the three theorems that confirm the conjecture, then add it as a theorem to your theorem list. (Isosceles Triangle Vertex Angle Theorem)

Review

14. Find x and y. ⓗ

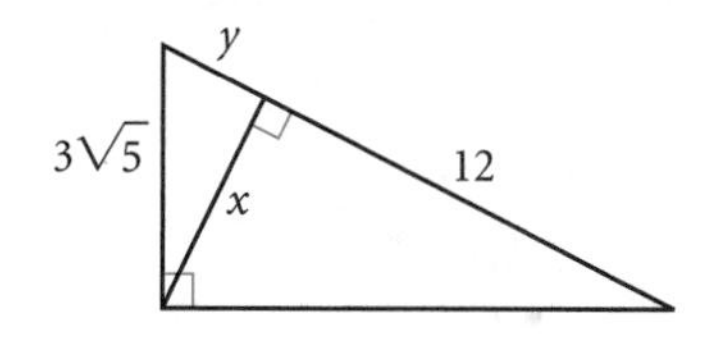

15. Two bird nests, 3.6 m and 6.1 m high, are on trees across a pond from each other, at points P and Q. The distance between the nests is too wide to measure directly (and there is a pond between the trees). A birdwatcher at point R can sight each nest along a dry path. $RP = 16.7$ m and $RQ = 27.4$ m. $\angle QPR$ is a right angle. What is the distance d between the nests?

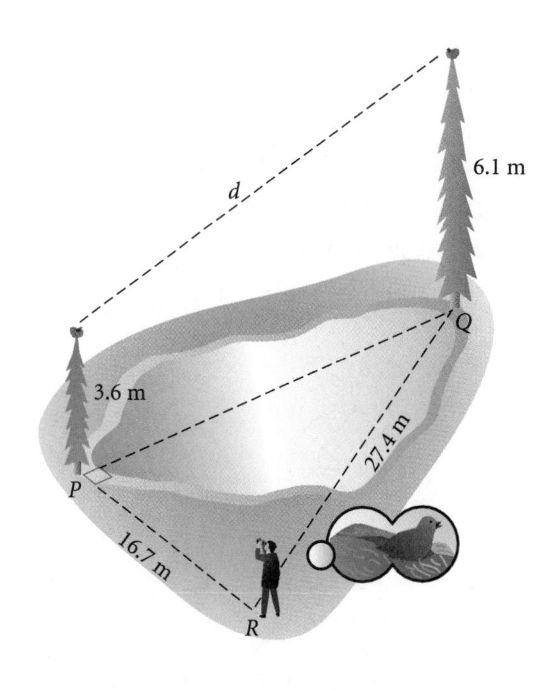

16. Apply the glide reflection rule twice to find the first and second images of the point $A(-2, 9)$. ⓗ

Glide reflection rule: A reflection across the line $x + y = 5$ and a translation $(x, y) \rightarrow (x + 4, y - 4)$.

17. Explain why $\angle 1 \cong \angle 2$.

Given:

B, G, F, E are collinear

$m\angle DFE = 90°$

$BC = FC$

$\overline{AF} \parallel \overline{BC}$

$\overline{BE} \parallel \overline{CD}$

$\overline{AB} \parallel \overline{FC} \parallel \overline{ED}$

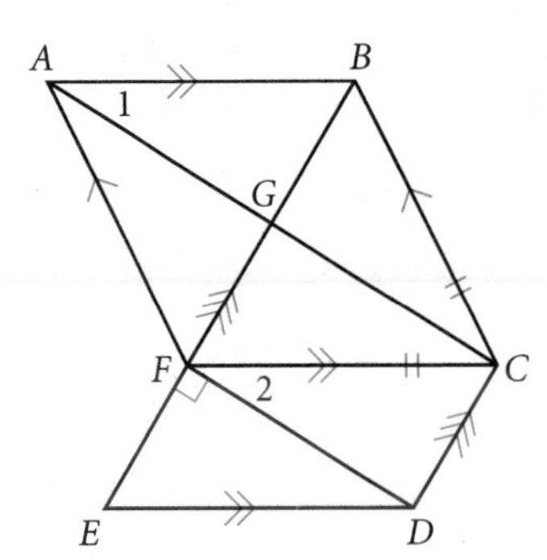

18. Each arc is a quarter of a circle with its center at a vertex of the square.

Given: The square has side length 1 unit **Find:** The shaded area

a. Shaded area = _?_

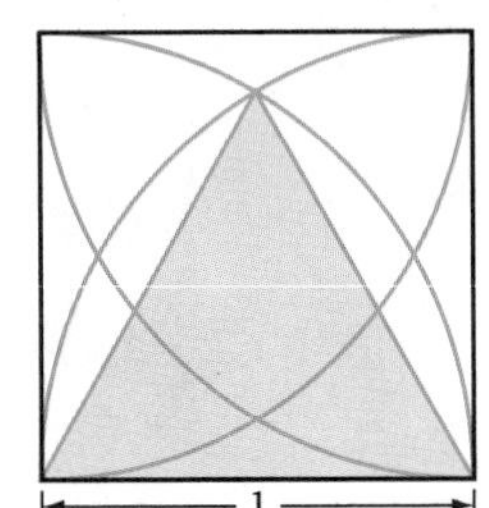

b. Shaded area = _?_

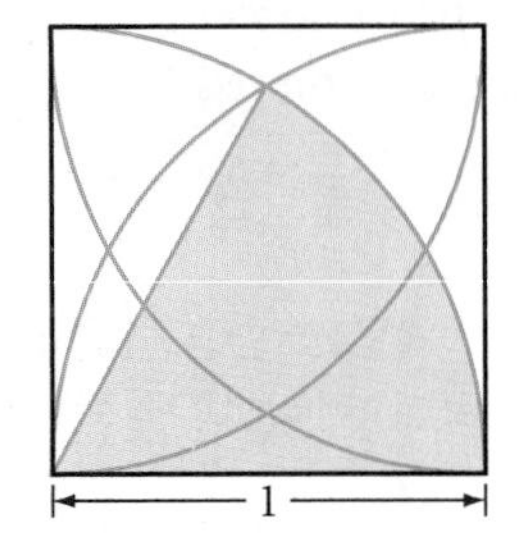

c. Shaded area = _?_

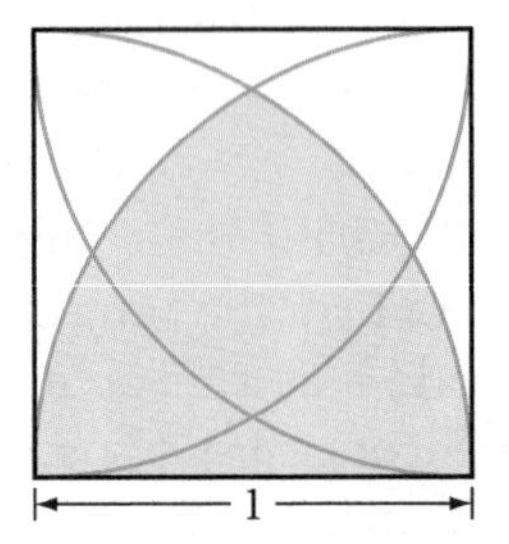

19. Given an arc of a circle on patty paper but not the whole circle or the center, fold the paper to construct a tangent at the midpoint of the arc.

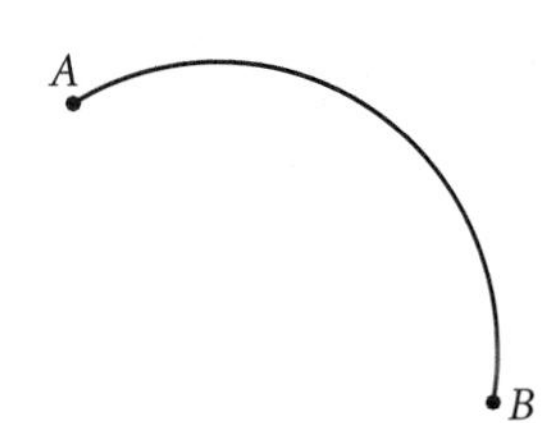

20. Find $m\angle BAC$ in this right rectangular prism.

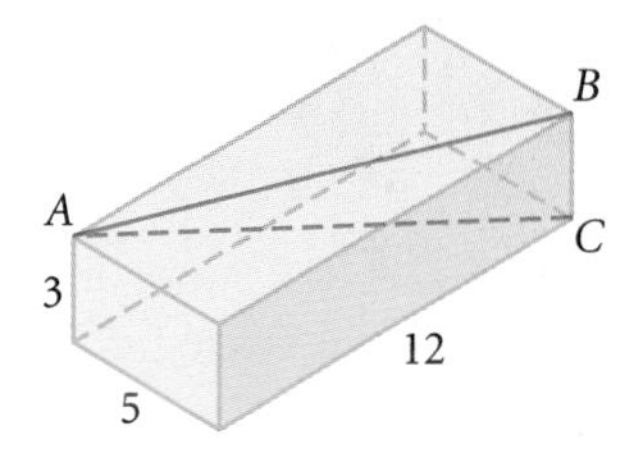

21. Choose **A** if the value of the expression is greater in Figure A.

Choose **B** if the value of the expression is greater in Figure B.

Choose **C** if the values are equal for both figures.

Choose **D** if it cannot be determined which value is greater.

Y, O, X are collinear.

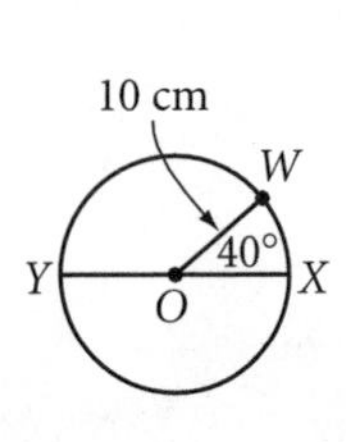

Figure A

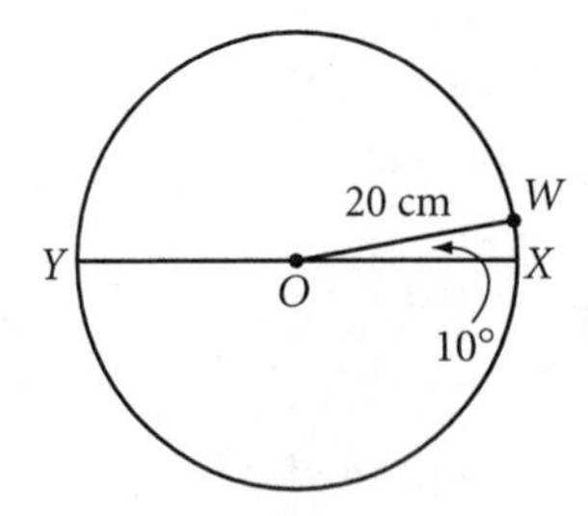

Figure B

a. Perimeter of $\triangle WXY$

b. Area of $\triangle XOW$

22. *Technology* Use geometry software to construct a circle and label any three points on the circle. Construct tangents at those three points to form a circumscribed triangle and connect the points of tangency to form an inscribed triangle.

a. Drag the points and observe the angle measures of each triangle. What relationship do you notice between x, a, and c? Is the same true for y and z?

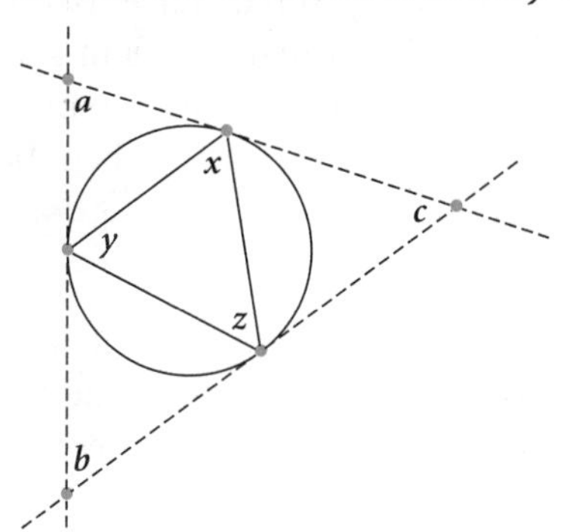

b. What is the relationship between the angle measures of a circumscribed quadrilateral and the inscribed quadrilateral formed by connecting the points of tangency?

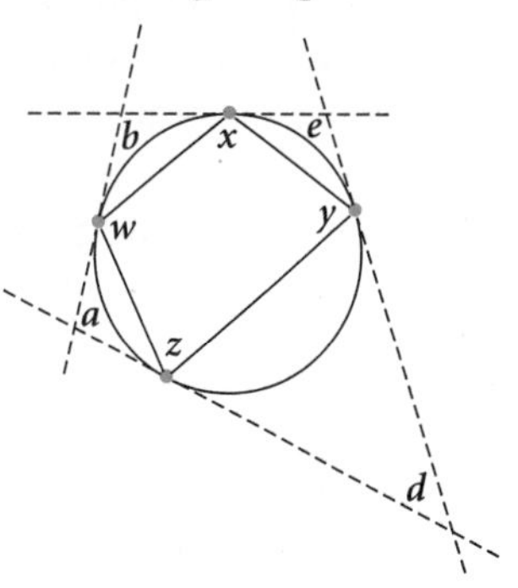

IMPROVING YOUR VISUAL THINKING SKILLS

Picture Patterns III

Sketch the figure that goes in box 12 below.

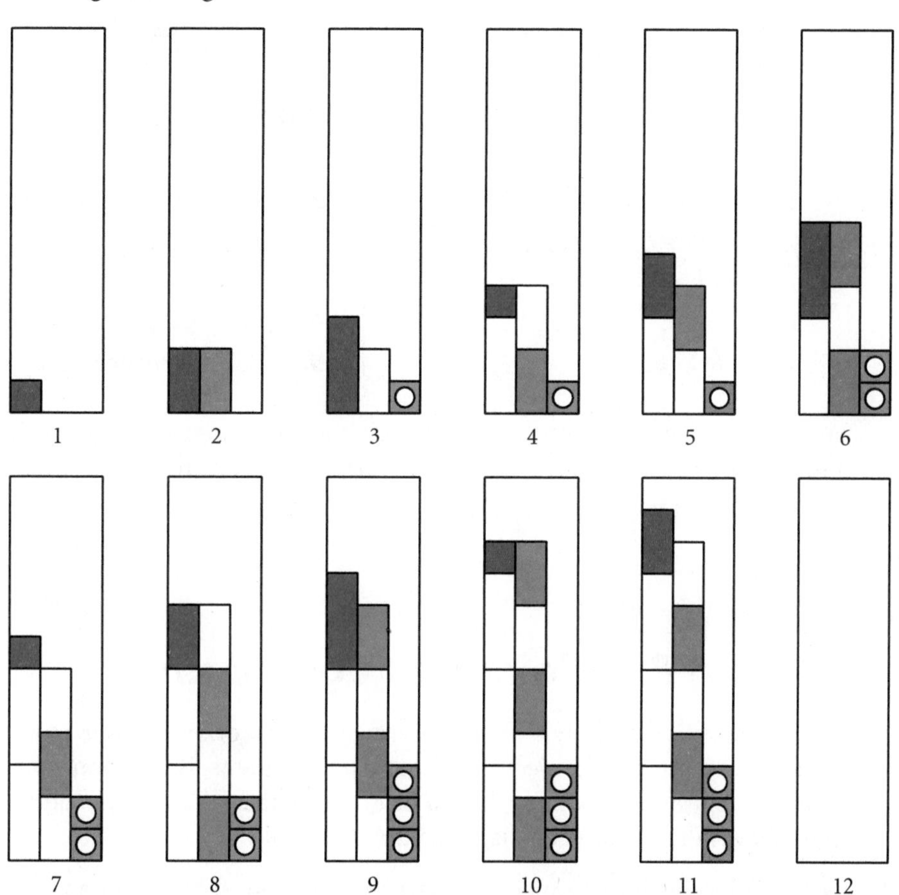

LESSON

13.4

Quadrilateral Proofs

All geometric reasoning is, in the last result, circular.

BERTRAND RUSSELL

In Chapter 5, you discovered and informally proved several quadrilateral properties. As reasons for the statements in some of these proofs, you used conjectures that are now postulates or that you have proved as theorems. So those steps in the proofs are valid. Occasionally, however, you may have used unproven conjectures as reasons. In this lesson you will write formal proofs of some of these quadrilateral conjectures, using only definitions, postulates, and theorems. After you have proved the theorems, you'll create a family tree tracing them back to postulates and properties.

You can prove many quadrilateral theorems by using triangle theorems. For example, you can prove some parallelogram properties by using the fact that a diagonal divides a parallelogram into two congruent triangles. In the example below, we'll prove this fact as a **lemma.** A lemma is an auxiliary theorem used specifically to prove other theorems.

EXAMPLE

Prove: A diagonal of a parallelogram divides the parallelogram into two congruent triangles.

Solution

Given: Parallelogram $ABCD$ with diagonal $\overline{AC}$

Prove: $\triangle ABC \cong \triangle CDA$

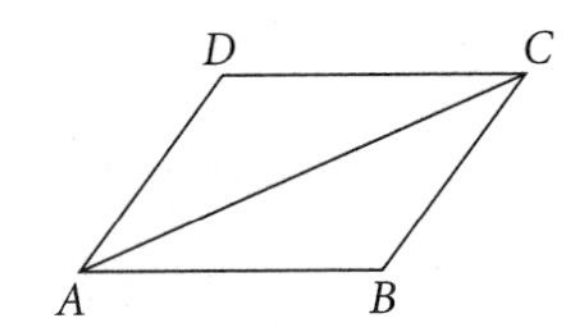

Two-column Proof

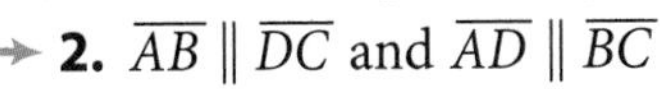

Statement	Reason
1. $ABCD$ is a parallelogram	1. Given
2. $\overline{AB} \parallel \overline{DC}$ and $\overline{AD} \parallel \overline{BC}$	2. Definition of parallelogram
3. $\angle CAB \cong \angle ACD$ and $\angle BCA \cong \angle DAC$	3. AIA Theorem
4. $\overline{AC} \cong \overline{AC}$	4. Identity property of congruence
5. $\triangle ABC \cong \triangle CDA$	5. ASA Congruence Postulate

We'll call the lemma proved in the example the Parallelogram Diagonal Lemma. You can now use it to prove other parallelogram conjectures in the investigation.

Investigation

Proving Parallelogram Conjectures

This investigation is really a proof activity. You will work with your group to prove three of your previous conjectures about parallelograms. Before you try to prove each conjecture, remember to draw a diagram, restate what is given and what you must show in terms of your diagram, and then make a plan.

Step 1 The Opposite Sides Conjecture states that the opposite sides of a parallelogram are congruent. Write a two-column proof of this conjecture.

Step 2 | The Opposite Angles Conjecture states that the opposite angles of a parallelogram are congruent. Write a two-column proof of this conjecture.

Step 3 | State the converse of the Opposite Sides Conjecture. Then write a two-column proof of this conjecture.

After you have successfully proved the parallelogram conjectures above, you can call them theorems and add them to your theorem list.

Step 4 | Create a family tree that shows the relationship among these theorems in Steps 1–3 and that traces each theorem back to the postulates of geometry.

EXERCISES

In Exercises 1–12, write a two-column proof or a flowchart proof of the conjecture. Once you have completed the proofs, add the theorems to your list.

1. If the opposite angles of a quadrilateral are congruent, then the quadrilateral is a parallelogram. (Converse of the Opposite Angles Theorem) ⓗ

2. If one pair of opposite sides of a quadrilateral are parallel and congruent, then the quadrilateral is a parallelogram. (Opposite Sides Parallel and Congruent Theorem)

3. Each diagonal of a rhombus bisects two opposite angles. (Rhombus Angles Theorem)

4. The consecutive angles of a parallelogram are supplementary. (Parallelogram Consecutive Angles Theorem)

5. If a quadrilateral has four congruent sides, then it is a rhombus. (Four Congruent Sides Rhombus Theorem)

6. If a quadrilateral has four congruent angles, then it is a rectangle. (Four Congruent Angles Rectangle Theorem)

7. The diagonals of a rectangle are congruent. (Rectangle Diagonals Theorem)

8. If the diagonals of a parallelogram are congruent, then the parallelogram is a rectangle. (Converse of the Rectangle Diagonals Theorem)

9. The base angles of an isosceles trapezoid are congruent. (Isosceles Trapezoid Theorem)

10. The diagonals of an isosceles trapezoid are congruent. (Isosceles Trapezoid Diagonals Theorem)

11. If a diagonal of a parallelogram bisects two opposite angles, then the parallelogram is a rhombus. (Converse of the Rhombus Angles Theorem)

12. If two parallel lines are intersected by a second pair of parallel lines that are the same distance apart as the first pair, then the parallelogram formed is a rhombus. (Double-Edged Straightedge Theorem)

13. Create a family tree for the Parallelogram Consecutive Angles Theorem.

14. Create a family tree for the Double-Edged Straightedge Theorem.

Review

15. Find the length and the bearing of the resultant vector $\vec{V}_1 + \vec{V}_2$.

$\vec{V}_1$ has length 5 and a bearing of 40°.
$\vec{V}_2$ has length 9 and a bearing of 90°.

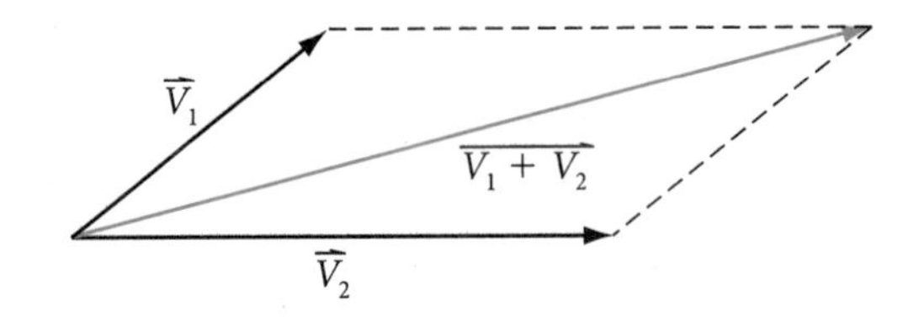

16. A triangle has vertices $A(7, -4)$, $B(3, -2)$, and $C(4, 1)$. Find the coordinates of the vertices after a dilation with center $(8, 2)$ and scale factor 2. Complete the mapping rule for the above dilation: $(x, y) \rightarrow (\underline{?}, \underline{?})$. ⓗ

17. Yan uses a 40 ft rope to tie his horse to the corner of the barn to which a fence is attached. How many square feet of grazing, to the nearest square foot, does the horse have?

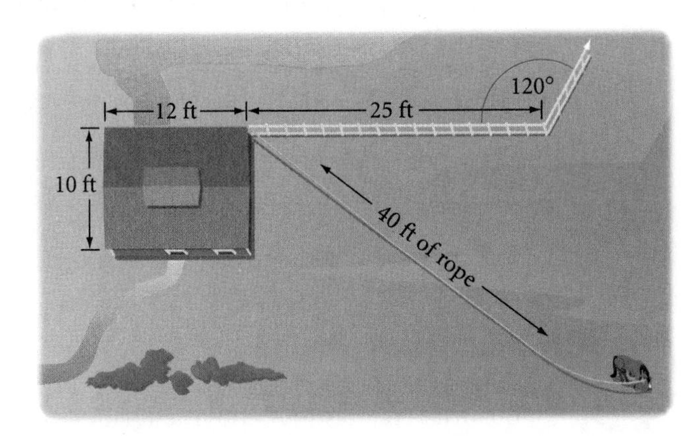

18. Complete the following chart with the symmetries and names of each type of special quadrilateral: parallelogram, rhombus, rectangle, square, kite, trapezoid, and isosceles trapezoid.

Name	Lines of symmetry	Rotational symmetry
	none	
trapezoid		
	1 diagonal	
		4-fold
	2 ⊥ bisectors of sides	
rhombus		
		none

19. Consider the rectangular prisms in Figure A and Figure A.
Choose **A** if the value of the expression is greater in Figure A.
Choose **B** if the value of the expression is greater in Figure B.
Choose **C** if the values are equal in both rectangular prisms.
Choose **D** if it cannot be determined which value is greater.

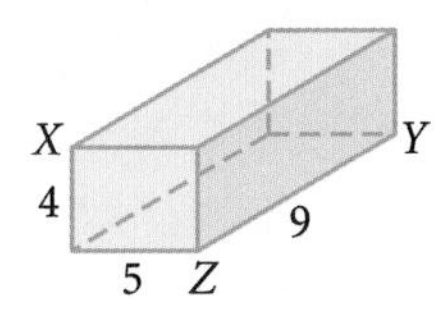

Figure A

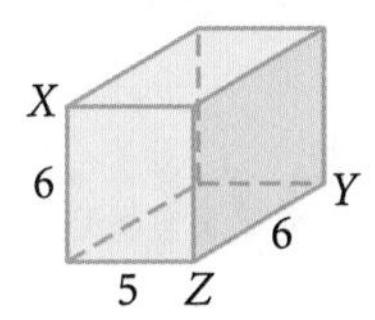

Figure B

a. Measure of $\angle XYZ$

b. Shortest path from X to Y along the surface of the prism

20. Given:

A, O, D are collinear

$\overleftrightarrow{GF}$ is tangent to circle O at point D

$m\angle EOD = 38°$

$\overline{OB} \parallel \overline{EC} \parallel \overline{GF}$

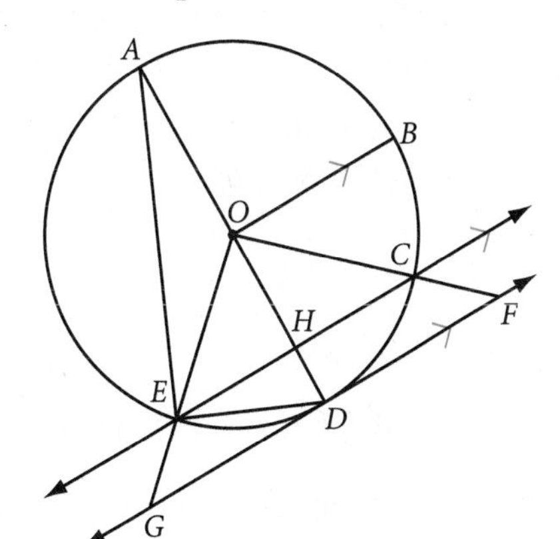

Find:

a. $m\angle AEO$

b. $m\angle DGO$

c. $m\angle BOC$

d. $m\overset{\frown}{EAB}$

e. $m\angle HED$

IMPROVING YOUR VISUAL THINKING SKILLS

Mental Blocks

In the top figure at right, every cube is lettered exactly alike. Copy and complete the two-dimensional representation of one of the cubes to show how the letters are arranged on the six faces.

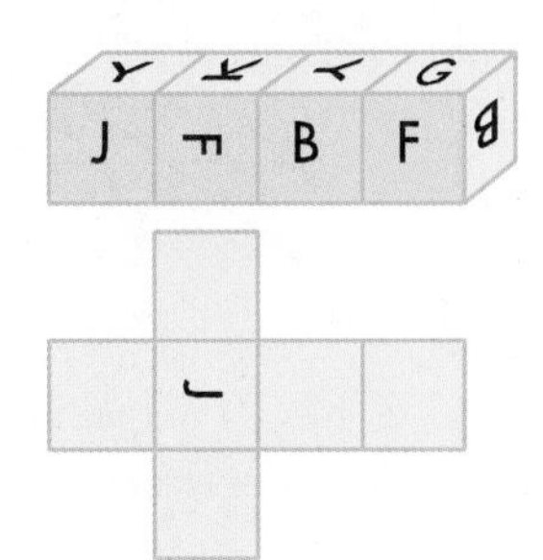

Exploration

Proof as Challenge and Discovery

So far, you have proved many theorems that are useful in geometry. You can also use proof to explore and possibly discover interesting properties. You might make a conjecture, and then use proof to decide whether or not it is always true.

These activities have been adapted from the book *Rethinking Proof with The Geometer's Sketchpad,* 1999, by Michael deVilliers.

Activity

Exploring Properties of Special Constructions

Use Sketchpad to construct these figures. Drag them and notice their properties. Then prove your conjectures.

Parallelogram Angle Bisectors

Construct a parallelogram and its angle bisectors. Label your sketch as shown.

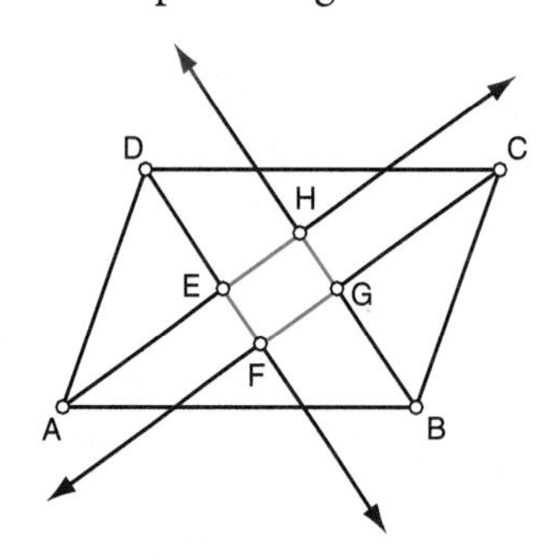

Step 1 Is *EFGH* a special quadrilateral? Make a conjecture. Drag the vertices around. Are there cases when *EFGH* does not satisfy your conjecture?

Step 2 Drag so that *ABCD* is a rectangle. What happens?

Step 3 Drag so that *ABCD* is a rhombus. What happens?

Step 4 Prove your conjectures for Steps 1–3.

Step 5 Construct the angle bisectors of another polygon. Investigate and write your observations. Make a conjecture and prove it.

Parallelogram Squares

Construct parallelogram *ABCD* and a square on each side. Construct the center of each square, and label your sketch as shown.

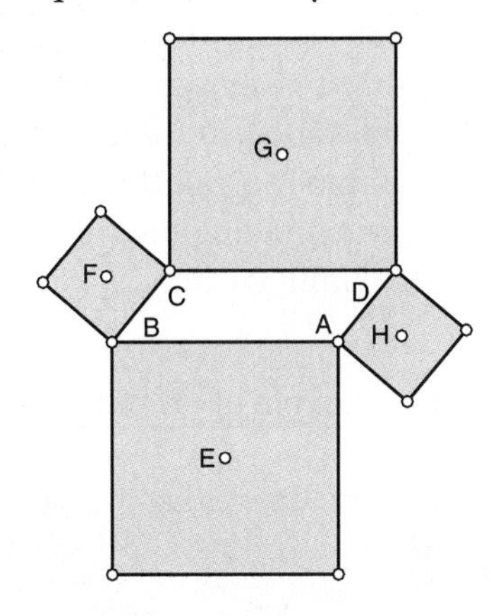

Step 6 Connect *E, F, G,* and *H* with line segments. (Try using a different color.) Drag the vertices of *ABCD.* What do you observe? Make a conjecture. Drag your sketch around. Are there cases when *EFGH* does not satisfy your conjecture?

Step 7 Drag so that *A, B, C,* and *D* are collinear. What happens to *EFGH*?

Step 8 Prove your conjectures for Steps 6 and 7.

Step 9 Investigate other special quadrilaterals and the shapes formed by connecting the centers of the squares on their sides. Write your observations. Make a conjecture and prove it.

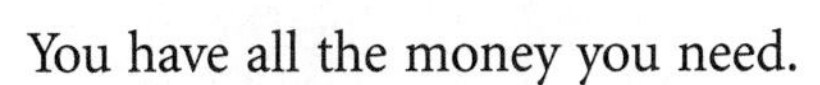

IMPROVING YOUR ALGEBRA SKILLS

PUZZLE

A Precarious Proof

You have all the money you need.

Let h = the money you have.
Let n = the money you need.

Most people think that the money they have is some amount less than the money they need. Stated mathematically, $h = n - p$ for some positive p.

If $h = n - p$, then

$$h(h - n) = (n - p)(h - n)$$
$$h^2 - hn = hn - n^2 - hp + np$$
$$h^2 - hn + hp = hn - n^2 + np$$
$$h(h - n + p) = n(h - n + p)$$

Therefore $h = n$.

So the money you have is equal to the money you need!

Is there a flaw in this proof?

LESSON

13.5

Indirect Proof

How often have I said to you that when you have eliminated the impossible, whatever remains, however improbable, must be the truth?

SHERLOCK HOLMES IN *THE SIGN OF THE FOUR* BY SIR ARTHUR CONAN DOYLE

In the proofs you have written so far, you have shown *directly,* through a sequence of statements and reasons, that a given conjecture is true. In this lesson you will write a different type of proof, called an indirect proof. In an **indirect proof,** you show something is true by eliminating all the other possibilities. You have probably used this type of reasoning when taking multiple-choice tests. If you are unsure of an answer, you can try to eliminate choices until you are left with only one possibility.

This mystery story gives an example of an indirect proof.

Detective Sheerluck Holmes and three other people are alone on a tropical island. One morning, Sheerluck entertains the others by playing show tunes on his ukulele. Later that day, he discovers that his precious ukulele has been smashed to bits. Who could have committed such an antimusical act? Sheerluck eliminates himself as a suspect because he knows he didn't do it. He eliminates his girlfriend as a suspect because she has been with him all day. Colonel Moran recently injured both arms and therefore could not have smashed the ukulele with such force. There is only one other person on the island who could have committed the crime. So Sheerluck concludes that the fourth person, Sir Charles Mortimer, is the guilty one.

For a given mathematical statement, there are two possibilities: either the statement is true or it is not true. To prove indirectly that a statement is true, you start by assuming it is not true. You then use logical reasoning to show that this assumption leads to a contradiction. If an assumption leads to a contradiction, it must be false. Therefore, you can eliminate the possibility that the statement is not true. This leaves only one possibility—namely, that the statement is true!

EXAMPLE A

Conjecture: If $m\angle N \neq m\angle O$ in $\triangle NOT$, then $NT \neq OT$.

Given: $\triangle NOT$ with $m\angle N \neq m\angle O$

Show: $NT \neq OT$

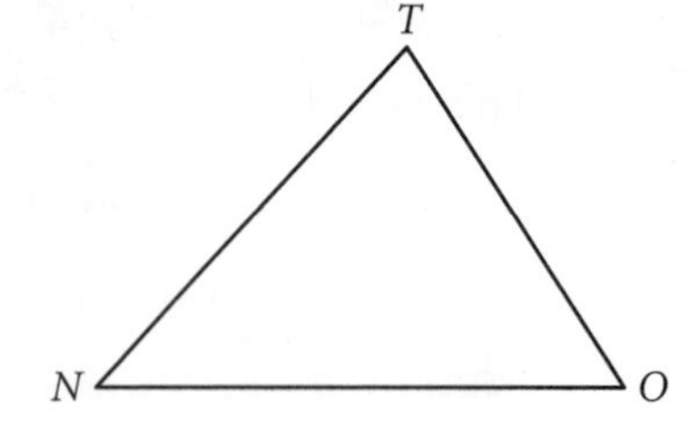

▶ Solution

To prove indirectly that the statement $NT \neq OT$ is true, start by assuming that it is *not* true. That is, assume $NT = OT$. Then show that this assumption leads to a contradiction.

Paragraph Proof

Assume $NT = OT$. If $NT = OT$, then $m\angle N = m\angle O$ by the Isosceles Triangle Theorem. But this contradicts the given fact that $m\angle N \neq m\angle O$. Therefore, the assumption $NT = OT$ is false and so $NT \neq OT$ is true. ■

Here is another example of an indirect proof.

EXAMPLE B

Conjecture: The diagonals of a trapezoid do not bisect each other.

Given: Trapezoid *ZOID* with parallel bases $\overline{ZO}$ and $\overline{ID}$ and diagonals $\overline{DO}$ and $\overline{IZ}$ intersecting at point Y

Show: The diagonals of trapezoid *ZOID* do not bisect each other; that is, $DY \neq OY$ and $ZY \neq IY$

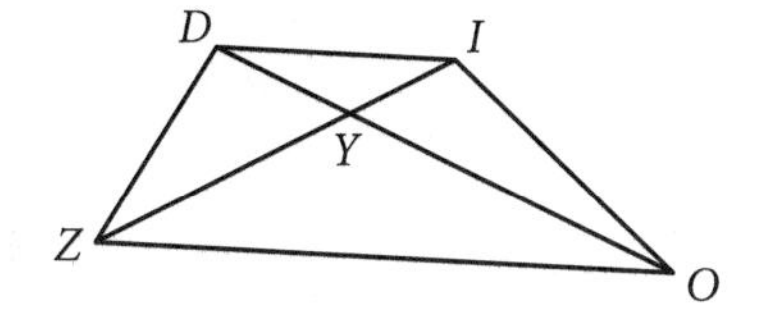

▶ Solution

Paragraph Proof

Assume that at least one of the diagonals of trapezoid *ZOID* *does* bisect the other. Then $\overline{DY} \cong \overline{OY}$. Also, by the AIA Theorem, $\angle DIY \cong \angle OZY$, and $\angle IDY \cong \angle YOZ$. Therefore, $\triangle DYI \cong \triangle OYZ$ by the SAA Theorem. By CPCTC, $\overline{ZO} \cong \overline{ID}$. It is given that $\overline{ZO} \parallel \overline{ID}$. In Lesson 13.4, you proved that if one pair of opposite sides of a quadrilateral are parallel and congruent, then the quadrilateral is a parallelogram. So, *ZOID* is a parallelogram. Thus, *ZOID* has two pairs of opposite sides parallel. But because it is a trapezoid, it has exactly one pair of parallel sides. This is contradictory. So the assumption that its diagonals bisect each other is false and the conjecture is true. ■

In the investigation you'll write an indirect proof of the Tangent Conjecture from Chapter 6.

Investigation
Proving the Tangent Conjecture

This investigation is really a group proof activity. Copy the information and diagram below, then work with your group to complete an indirect proof of the Tangent Conjecture.

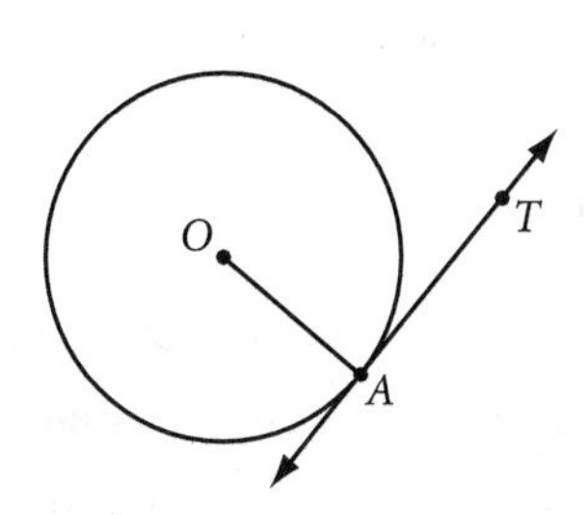

Conjecture: A tangent is perpendicular to the radius drawn to the point of tangency.

Given: Circle O with tangent $\overleftrightarrow{AT}$ and radius $\overline{AO}$

Show: $\overline{AO} \perp \overleftrightarrow{AT}$

Paragraph Proof

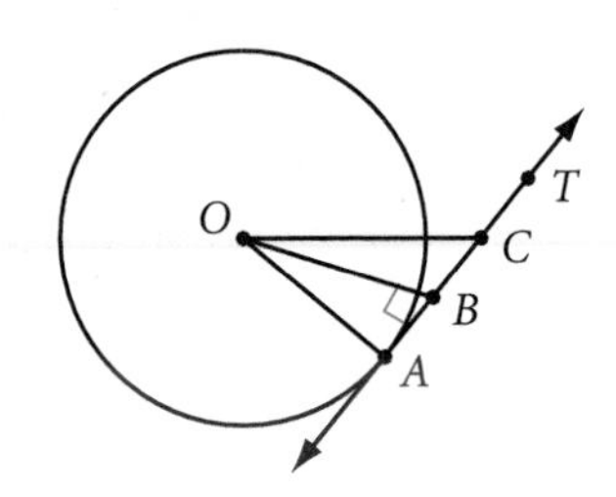

Step 1 Assume $\overline{AO}$ is *not* perpendicular to $\overleftrightarrow{AT}$. Construct a perpendicular from point O to $\overleftrightarrow{AT}$ and label the intersection point B ($\overline{OB} \perp \overleftrightarrow{AT}$). Which postulate allows you to do this?

Step 2 Select a point C on $\overleftrightarrow{AT}$ so that B is the midpoint of $\overline{AC}$. Which postulate allows you to do this?

Step 3 Next, construct $\overline{OC}$. Which postulate allows you to do this?

Step 4 $\angle ABO \cong \angle CBO$. Why?

Step 5 $\overline{AB} \cong \overline{BC}$. What definition tells you this?

Step 6 $\overline{OB} \cong \overline{OB}$. What property of congruence tells you this?

Step 7 Therefore, $\triangle ABO \cong \triangle CBO$. Which congruence shortcut tells you the triangles are congruent?

Step 8 If $\triangle ABO \cong \triangle CBO$, then $\overline{AO} \cong \overline{CO}$. Why?

Step 9 C must be a point on the circle (because a circle is the set of *all* points in the plane at a given distance from the center and points A and C are both the same distance from the center). Therefore, $\overleftrightarrow{AT}$ intersects the circle in *two* points (A and C) and thus $\overleftrightarrow{AT}$ is not a tangent. But this leads to a contradiction. Why? ■

Step 10 Discuss the steps with your group. What was the contradiction? What does it prove? Use the steps above to write a two-column indirect proof of the Tangent Conjecture.

Rename the Tangent Conjecture as the Tangent Theorem and add it to your list of theorems.

EXERCISES

For Exercises 1 and 2, the correct answer is one of the choices listed. Determine the correct answer by indirect reasoning, explaining how you eliminated each incorrect choice.

1. Which is the capital of Mali?

A. Paris **B.** Tucson **C.** London **D.** Bamako

2. Which Italian scientist used a new invention called the telescope to discover the moons of Jupiter?

A. Sir Edmund Halley **B.** Julius Caesar **C.** Galileo Galilei **D.** Madonna

3. Is the proof in Example A claiming that if two angles of a triangle are not congruent, then the triangle is not isosceles? Explain.

4. Is the proof in Example B claiming that if one diagonal of a quadrilateral bisects the other, then the quadrilateral is not a trapezoid? Explain.

5. Fill in the blanks in the indirect proof below.

Conjecture: No triangle has two right angles.

Given: $\angle ABC$

Show: No two angles are right angles

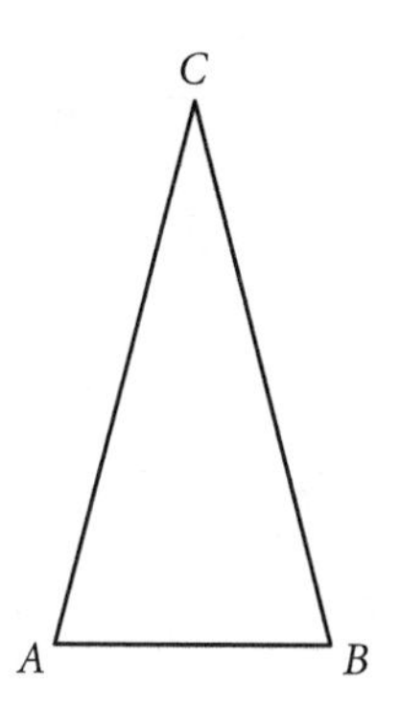

Two-column proof

Statement	Reason
1. Assume $\angle ABC$ has two right angles (Assume $m\angle A = 90°$ and $m\angle B = 90°$ and $0° < m\angle C < 180°$.)	**1.** ?
2. $m\angle A + m\angle B + m\angle C = 180°$	**2.** ?
3. $90° + 90° + m\angle C = 180°$	**3.** ?
4. $m\angle C =$?	**4.** ?

But if $m\angle C = 0$, then the two sides $\overline{AC}$ and $\overline{BC}$ coincide, and thus there is no angle at *C*. This contradicts the given information. So the assumption is false. Therefore, no triangle has two right angles.

6. Write an indirect proof of the conjecture below.

Conjecture: No trapezoid is equiangular.

Given: Trapezoid *ZOID* with bases $\overline{ZO}$ and $\overline{ID}$

Show: *ZOID* is not equiangular

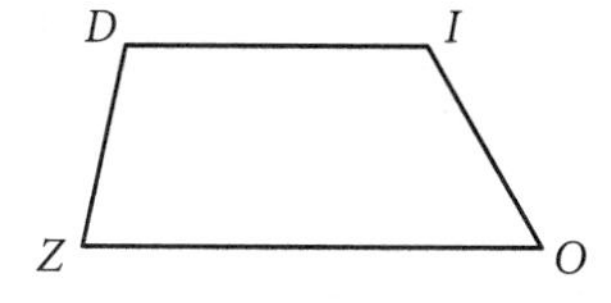

7. Write an indirect proof of the conjecture below.

Conjecture: In a scalene triangle, the median cannot be the altitude.

Given: Scalene triangle *ABC* with median $\overline{CD}$

Show: $\overline{CD}$ is not the altitude to $\overline{AB}$

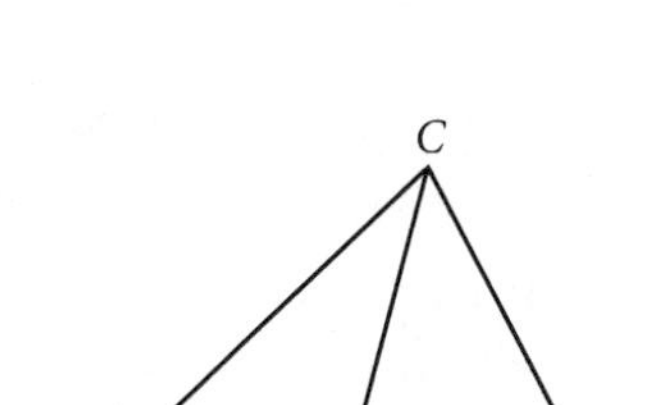

8. Write an indirect proof of the conjecture below.

Conjecture: The bases of a trapezoid have unequal lengths.

Given: Trapezoid *ZOID* with parallel bases and $\overline{ZO}$ and $\overline{ID}$

Show: $ZO \neq ID$

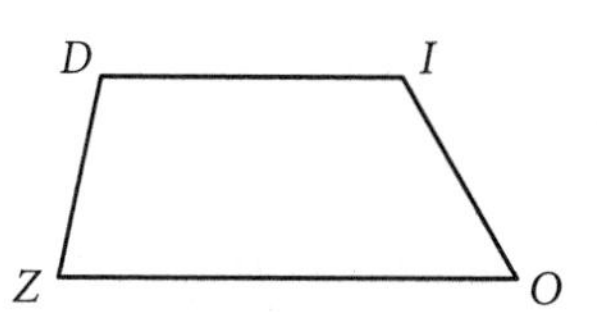

9. Write the "given" and the "show," and then plan and write the proof of the Perpendicular Bisector of a Chord Conjecture: The perpendicular bisector of a chord passes through the center of the circle.

Review

10. Find a, b, and c.

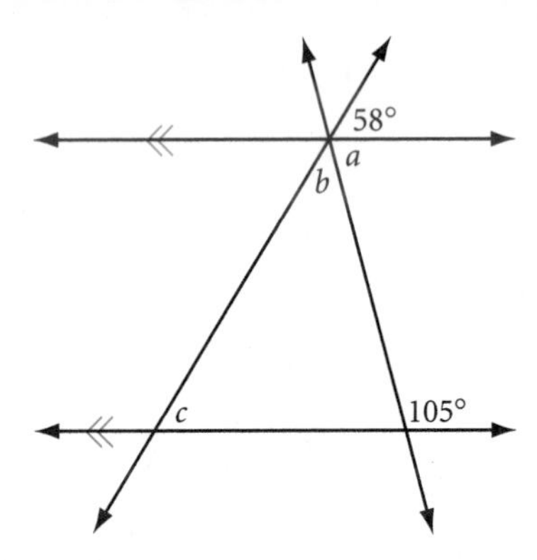

11. A clear plastic container is in the shape of a right cone atop a right cylinder, and their bases coincide. Find the volume of the container.

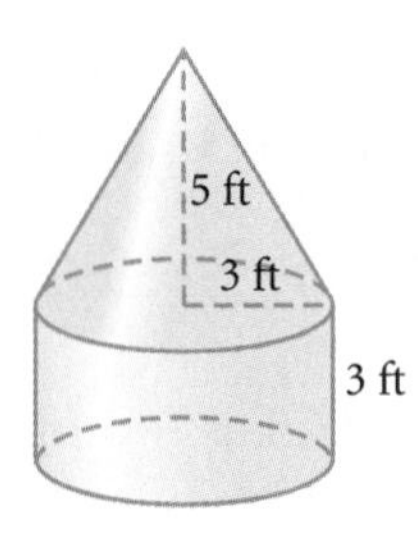

12. Each arc is a quarter of a circle with its center at a vertex of the square.

Given: The square has side length 1 unit **Find:** The shaded area

a. Shaded area = ?

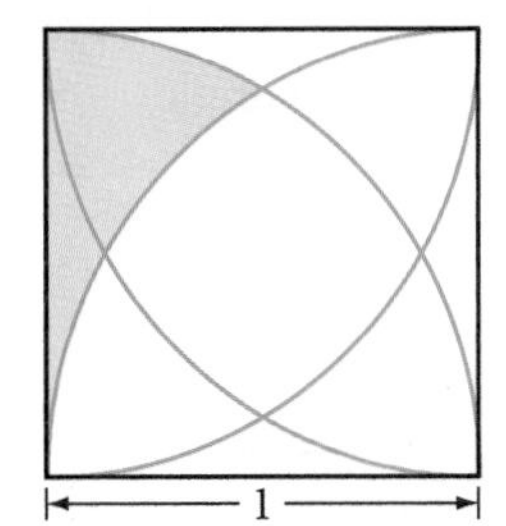

b. Shaded area = ?

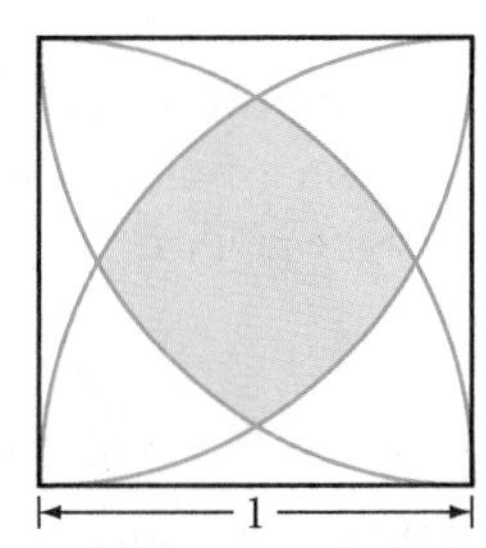

13. For each statement, select always (A), sometimes (S), or never (N).

a. An angle inscribed in a semicircle is a right angle.

b. An angle inscribed in a major arc is obtuse.

c. An arc measure equals the measure of its central angle.

d. The measure of an angle formed by two intersecting chords equals the measure of its intercepted arc.

e. The measure of the angle formed by two tangents to a circle equals the supplement of the central angle of the minor intercepted arc.

IMPROVING YOUR REASONING SKILLS

Symbol Juggling

If $V = \frac{1}{3}BH$, $B = \frac{1}{2}h(a + b)$, $h = 2x$, $a = 2b$, $b = x$, and $Hx = 12$, find the value of V in terms of x.

LESSON

13.6

Circle Proofs

When people say, "It can't be done," or "You don't have what it takes," it makes the task all the more interesting.

LYNN HILL

In Chapter 6, you completed the proof of the three cases of the Inscribed Angle Conjecture: The measure of an inscribed angle in a circle equals half the measure of its intercepted arc. There was a lot of algebra in the proof. You may not have noticed that the Angle Addition Postulate was used, as well as a property that we called *arc addition.* Arc Addition is a postulate that you need to add to your list.

Arc Addition Postulate

If point B is on $\overset{\frown}{AC}$ and between points A and C, then $m\overset{\frown}{AB} + m\overset{\frown}{BC} = m\overset{\frown}{AC}$.

Is the proof of the Inscribed Angle Conjecture now completely supported by the premises of geometry? Can you call it a theorem? To answer these questions, trace the family tree.

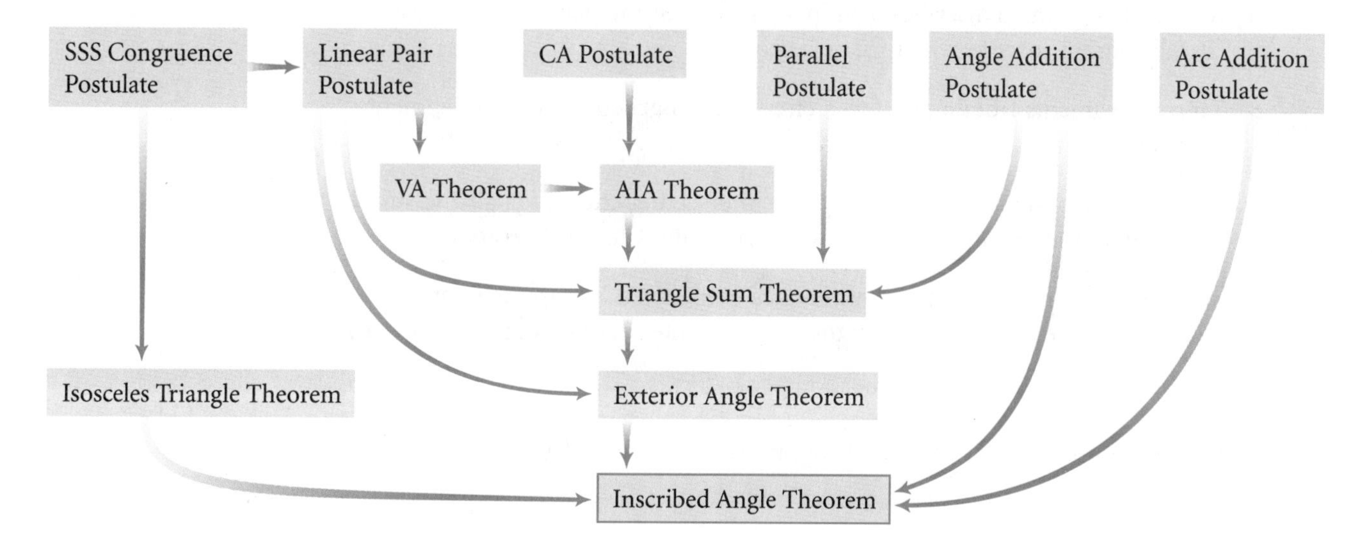

So the Inscribed Angle Conjecture is completely supported by premises of geometry; therefore you can call it a theorem and add it to your theorem list.

A double rainbow creates arcs in the sky over Stonehenge near Wiltshire, England. Built from bluestone and sandstone from 3000 to 1500 B.C.E., Stonehenge itself is laid out in the shape of a major arc.

In the exercises, you will create proofs or family trees for many of your earlier discoveries about circles.

EXERCISES

You will need

Construction tools for Exercise **16**

Geometry software for Exercise **14**

In Exercises 1–7, set up and write a proof of each conjecture. Once you have completed the proofs, add the theorems to your list.

1. Inscribed angles that intercept the same or congruent arcs are congruent. (Inscribed Angles Intercepting Arcs Theorem)

2. The opposite angles of an inscribed quadrilateral are supplementary. (Cyclic Quadrilateral Theorem)

3. Parallel lines intercept congruent arcs on a circle. (Parallel Secants Congruent Arcs Theorem)

4. If a parallelogram is inscribed within a circle, then the parallelogram is a rectangle. (Parallelogram Inscribed in a Circle Theorem)

5. Tangent segments from a point to a circle are congruent. (Tangent Segments Theorem)

6. The measure of an angle formed by two intersecting chords is half the sum of the measures of the two intercepted arcs. (Intersecting Chords Theorem)

7. Write and prove a theorem about the arcs intercepted by secants intersecting outside a circle, and the angle formed by the secants. (Intersecting Secants Theorem)

8. Create a family tree for the Tangent Segments Theorem.

9. Create a family tree for the Parallelogram Inscribed in a Circle Theorem.

Review

10. Find the coordinates of A and P to the nearest tenth.

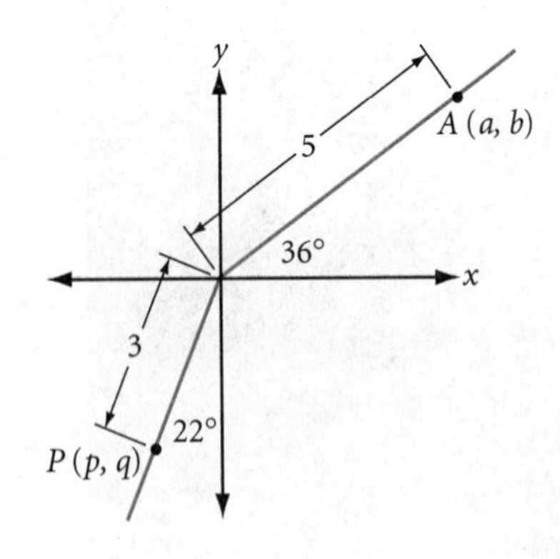

11. Given: $AX = 6$, $XB = 2$, $BC = 4$, $ZC = 3$
Find: $BY = \underline{?}$, $YC = \underline{?}$, $AZ = \underline{?}$

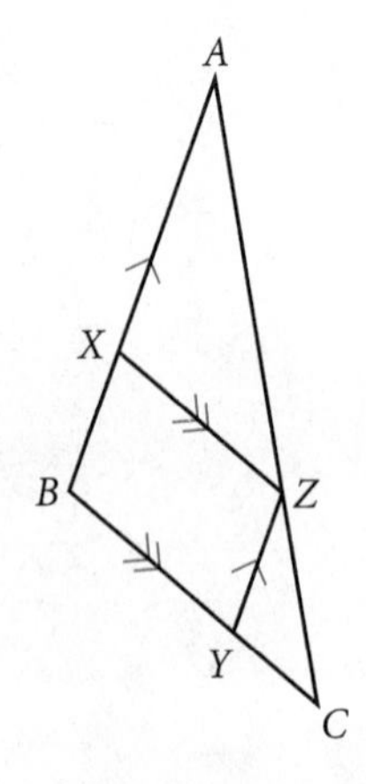

12. List the five segments in order from shortest to longest.

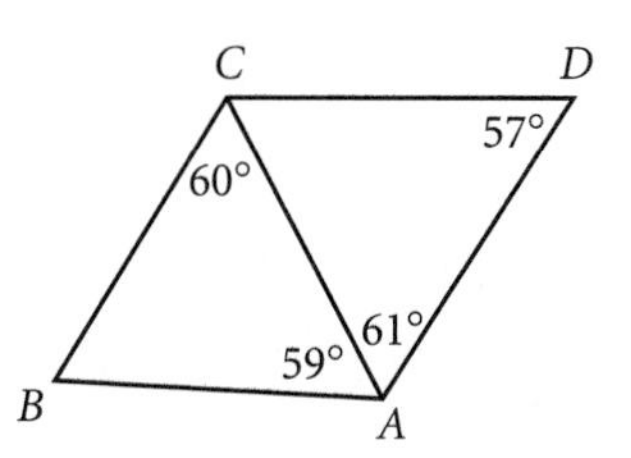

13. $\overline{AB}$ is a common external tangent. Find the length of $\overline{AB}$ (to a tenth of a unit).

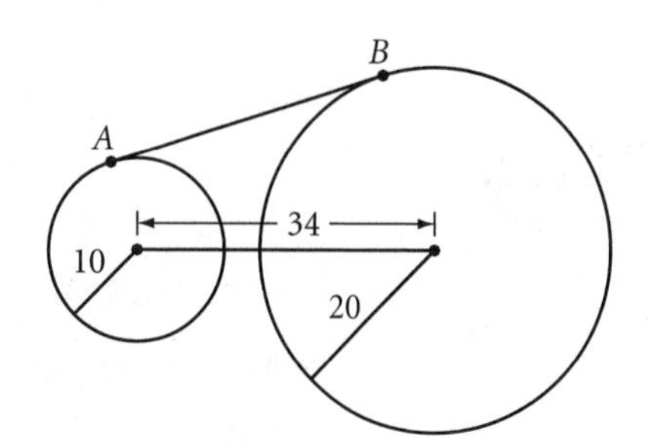

14. *Technology* P is any point inside an equilateral triangle. Is there a relationship between the height h and the sum $a + b + c$? Use geometry software to explore the relationship and make a conjecture. Then write a proof of your conjecture. ⓗ

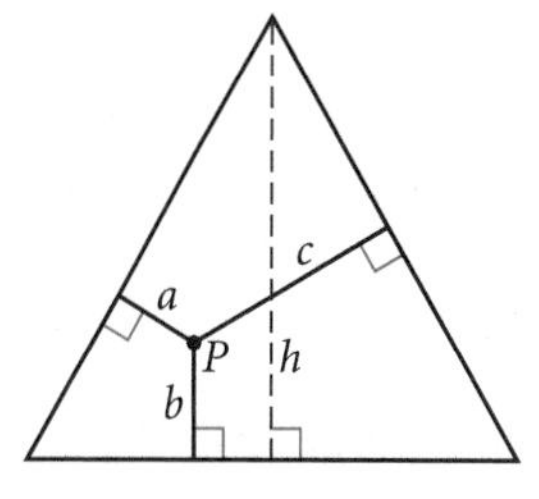

15. Find each measure or conclude that it "cannot be determined."

a. $m\angle P$ **b.** $m\angle QON$

c. $m\angle QRN$ **d.** $m\angle QMP$

e. $m\angle ONF$ **f.** $m\overset{\frown}{MN}$

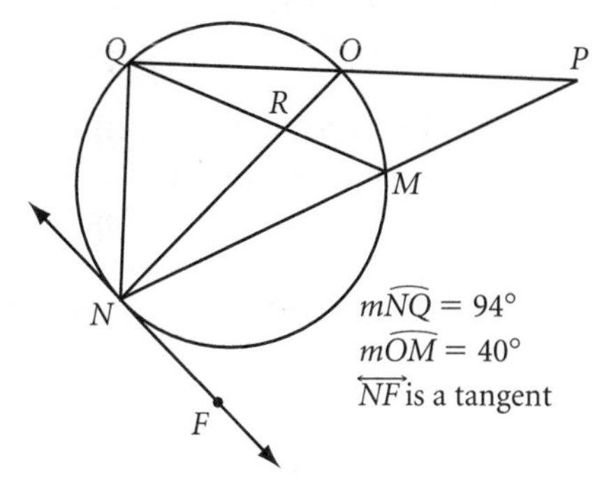

16. *Construction* Use a compass and straightedge to construct the two tangents to a circle from a point outside the circle. ⓗ

IMPROVING YOUR REASONING SKILLS

Seeing Spots

The arrangement of green and yellow spots at right may appear to be random, but there is a pattern. Each row is generated by the row immediately above it. Find the pattern and add several rows to the arrangement. Do you think a row could ever consist of all yellow spots? All green spots?

Could there ever be a row with one green spot? Does a row ever repeat itself?

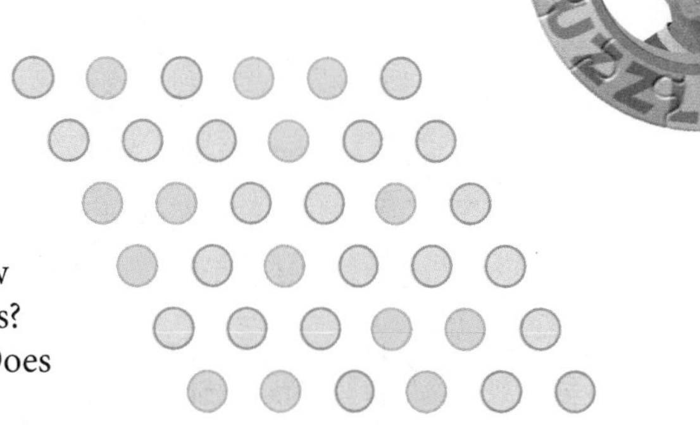

LESSON

13.7

Similarity Proofs

Mistakes are part of the dues one pays for a full life.

SOPHIA LOREN

To prove conjectures involving similarity, we need to extend the properties of equality and congruence to similarity.

Properties of Similarity

Reflexive property of similarity

Any figure is similar to itself.

Symmetric property of similarity

If Figure A is similar to Figure B, then Figure B is similar to Figure A.

Transitive property of similarity

If Figure A is similar to Figure B and Figure B is similar to Figure C, then Figure A is similar to Figure C.

The AA Similarity Conjecture is actually a similarity postulate.

AA Similarity Postulate

If two angles of one triangle are congruent to two angles of another triangle, then the two triangles are similar.

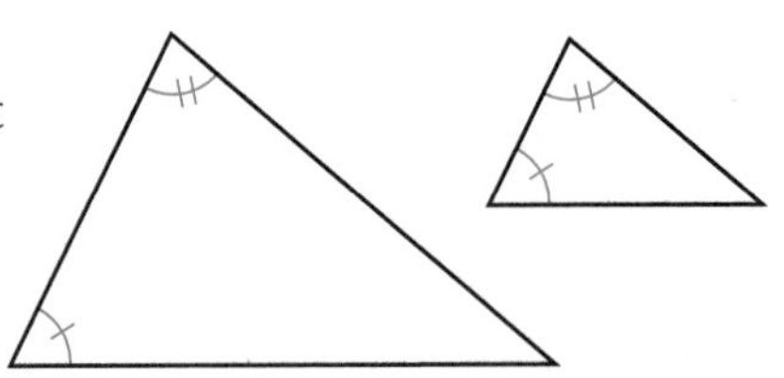

In Chapter 11, you also discovered the SAS and SSS shortcuts for showing that two triangles are similar. In the example that follows, you will see how to use the AA Similarity Postulate to prove the SAS Similarity Conjecture, making it the SAS Similarity Theorem.

EXAMPLE

Prove the SAS Similarity Conjecture: If two sides of one triangle are proportional to two sides of another triangle and the included angles are congruent, then the two triangles are similar.

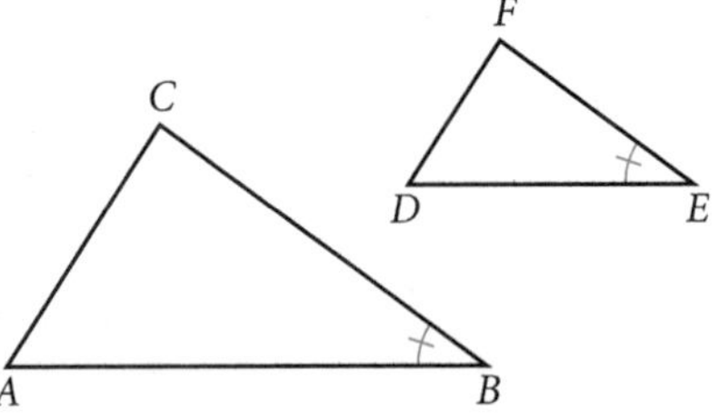

▶ ***Solution***

Given: $\triangle ABC$ and $\triangle DEF$ so that $\frac{AB}{DE} = \frac{BC}{EF}$ and $\angle B \cong \angle E$

Show: $\triangle ABC \sim \triangle DEF$

Plan: The only shortcut for showing that two triangles are similar is the AA Postulate, so you need to find another pair of congruent angles. One way of getting two congruent angles is to find two congruent triangles. You can draw a triangle within $\triangle ABC$ that is congruent to $\triangle DEF$. The Segment Duplication Postulate allows you to locate a point P on $\overline{AB}$ so that $PB = DE$. The Parallel Postulate allows you to construct a line $\overleftrightarrow{PQ}$ parallel to $\overline{AC}$. Then $\angle A \cong \angle QPB$ by the CA Postulate.

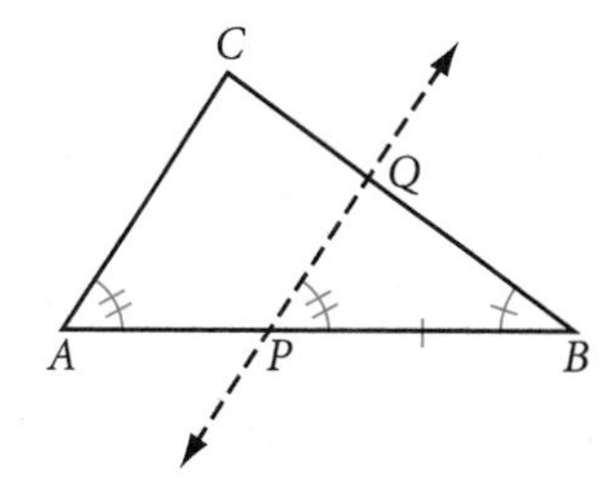

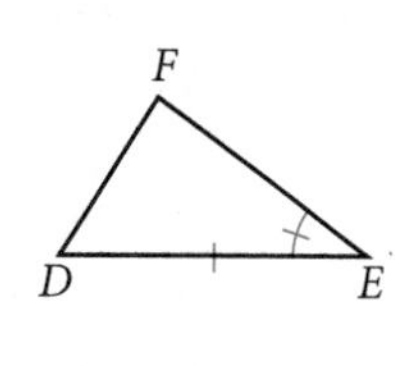

Now, if you can show that $\triangle PBQ \cong \triangle DEF$, then you will have two congruent pairs of angles to prove $\triangle ABC \sim \triangle DEF$. So, how do you show that $\triangle PBQ \cong \triangle DEF$? If you can get $\triangle ABC \sim \triangle PBQ$, then $\frac{AB}{PB} = \frac{BC}{BQ}$. It is given that $\frac{AB}{DE} = \frac{BC}{EF}$, and you constructed $PB = DE$. With some algebra and substitution, you can get $EF = BQ$. Then the two triangles will be congruent by the SAS Congruence Postulate.

Here is the two-column proof.

Statement	Reason
1. Locate P so that $PB = DE$	**1.** Segment Duplication Postulate
2. Construct $\overline{PQ} \parallel \overline{AC}$	**2.** Parallel Postulate
3. $\angle A \cong \angle QPB$	**3.** CA Postulate
4. $\angle B \cong \angle B$	**4.** Identity property of congruence
5. $\triangle ABC \sim \triangle PBQ$	**5.** AA Similarity Postulate
6. $\frac{AB}{PB} = \frac{BC}{BQ}$	**6.** Corresponding sides of similar triangles are proportional (CSSTP)
7. $\frac{AB}{DE} = \frac{BC}{BQ}$	**7.** Substitution
8. $\frac{AB}{DE} = \frac{BC}{EF}$	**8.** Given
9. $\frac{BC}{BQ} = \frac{BC}{EF}$	**9.** Transitive property of equality
10. $BQ = EF$	**10.** Algebra operations
11. $\angle B \cong \angle E$	**11.** Given
12. $\triangle PBQ \cong \triangle DEF$	**12.** SAS Congruence Postulate
13. $\angle QPB \cong \angle D$	**13.** CPCTC
14. $\angle A \cong \angle D$	**14.** Substitution
15. $\triangle ABC \sim \triangle DEF$	**15.** AA Similarity Postulate

This proves the SAS Similarity Conjecture.

The proof in the example above may seem complicated, but it relies on triangle congruence and triangle similarity postulates. Reading the plan again can help you follow the steps in the proof.

You can now call the SAS Similarity Conjecture the SAS Similarity Theorem and add it to your theorem list.

In this investigation you will use the SAS Similarity Theorem to prove the SSS Similarity Conjecture.

Investigation

Can You Prove the SSS Similarity Conjecture?

Similarity proofs can be challenging. Follow the steps and work with your group to prove the SSS Similarity Conjecture: If the three sides of one triangle are proportional to the three sides of another triangle, then the two triangles are similar.

Step 1 Identify the given and show.

Step 2 Restate what is given and what you must show in terms of this diagram.

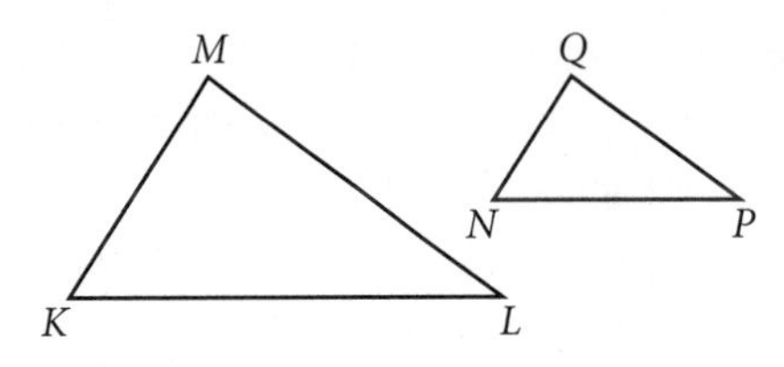

Step 3 Plan your proof. (Hint: Use an auxiliary line like the one in the example.)

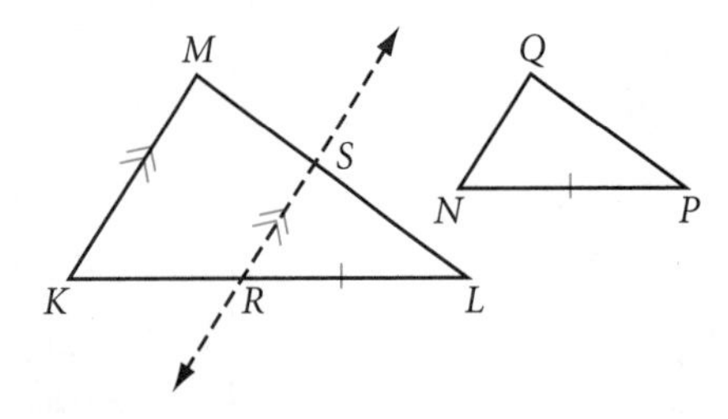

Step 4 Copy the first ten statements and provide the reasons. Then write the remaining steps and reasons necessary to complete the proof.

Statement	Reason
1. Locate R so that $RL = NP$	**1.** ? Postulate
2. Construct $\overleftrightarrow{RS} \parallel \overline{KM}$	**2.** ? Postulate
3. $\angle SRL \cong \angle K$	**3.** ? Postulate
4. $\angle RSL \cong \angle M$	**4.** ? Postulate
5. $\triangle KLM \sim \triangle RLS$	**5.** ?
6. $\frac{KL}{RL} = \frac{LM}{LS} = \frac{MK}{SR}$	**6.** ?
7. $\frac{KL}{NP} = \frac{LM}{LS}$	**7.** ?
8. $\frac{KL}{NP} = \frac{LM}{PQ}$	**8.** ?
9. $\frac{KL}{NP} = \frac{MK}{SR}$	**9.** ?
10. $\frac{KL}{NP} = \frac{MK}{QN}$	**10.** ?
⋮	⋮

Step 5 Draw arrows to show the flow of logic in your two-column proof.

When you have completed the proof, you can call the SSS Similarity Conjecture the SSS Similarity Theorem and add it to your theorem list.

EXERCISES

You will need

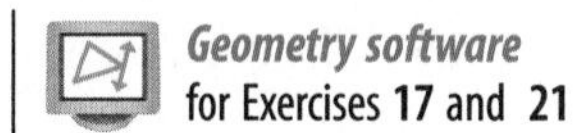

In Exercises 1 and 2, write a proof and draw the family tree of each theorem. If the family tree is completely supported by theorems and postulates, add the theorem to your list.

1. If two triangles are similar, then corresponding altitudes are proportional to the corresponding sides. (Corresponding Altitudes Theorem)

2. If two triangles are similar, then corresponding medians are proportional to the corresponding sides. (Corresponding Medians Theorem)

In Exercises 3–10, write a proof of the conjecture. Once you have completed the proofs, add the theorems to your list. As always, you may use theorems that have been proved in previous exercises in your proofs.

3. If two triangles are similar, then corresponding angle bisectors are proportional to the corresponding sides. (Corresponding Angle Bisectors Theorem)

4. If a line passes through two sides of a triangle parallel to the third side, then it divides the two sides proportionally. (Parallel/Proportionality Theorem) ⓗ

5. If a line passes through two sides of a triangle dividing them proportionally, then it is parallel to the third side. (Converse of the Parallel/Proportionality Theorem) ⓗ

6. If you drop an altitude from the vertex of a right angle to its hypotenuse, then it divides the right triangle into two right triangles that are similar to each other and to the original right triangle. (Three Similar Right Triangles Theorem) ⓗ

7. The length of the altitude to the hypotenuse of a right triangle is the geometric mean of the lengths of the two segments on the hypotenuse. (Altitude to the Hypotenuse Theorem) ⓗ

8. The Pythagorean Theorem ⓗ

9. Converse of the Pythagorean Theorem ⓗ

10. If the hypotenuse and one leg of a right triangle are congruent to the hypotenuse and one leg of another right triangle, then the two right triangles are congruent. (Hypotenuse Leg Theorem) ⓗ

11. Create a family tree for the Parallel/Proportionality Theorem.

12. Create a family tree for the SSS Similarity Theorem.

13. Create a family tree for the Pythagorean Theorem.

This monument in Wellington, New Zealand, was designed by Maori architect Rewi Thompson. How would you describe the shape of the monument? How might the artist have used geometry in planning the construction?

Review

14. A circle with diameter 9.6 cm has two parallel chords with lengths 5.2 cm and 8.2 cm. How far apart are the chords? Find two possible answers.

15. Choose **A** if the value is greater in the regular hexagon.

Choose **B** if the value is greater in the regular pentagon.

Choose **C** if the values are equal in both figures.

Choose **D** if it cannot be determined which value is greater.

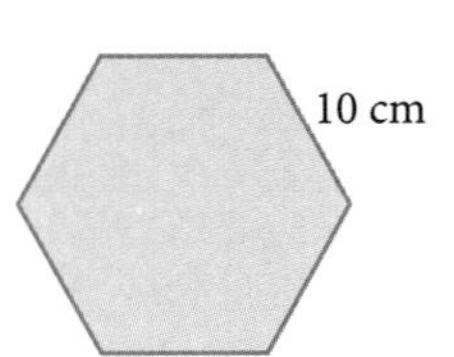

12 cm

Regular hexagon Regular pentagon

a. Perimeter **b.** Apothem **c.** Area

d. Sum of interior angles **e.** Sum of exterior angles

16. ***Mini-Investigation*** Cut out a small nonsymmetric concave quadrilateral. Label the vertices *A, B, C, D*. Use your cut-out as a template to create a tessellation. Trace about 10 images that fit together to cover part of the plane. Number the vertices of each image to match the numbers on your cut-out.

a. Draw two different translation vectors that map your tessellation onto itself. How do these two vectors relate to your original quadrilateral?

b. Pick a quadrilateral in your tessellation. What transformation will map the quadrilateral you picked onto an adjacent quadrilateral? With that transformation, what happens to the rest of the tessellation?

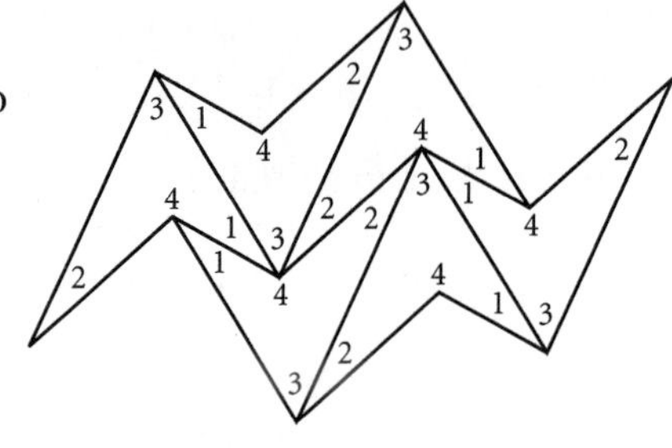

17. ***Technology*** Use geometry software to draw a small nonsymmetric concave quadrilateral.

a. Describe the transformations that will make the figure tessellate.

b. Describe two different translation vectors that map your tessellation onto itself.

18. Given:

Circles P and Q

$\overline{PS}$ and $\overline{PT}$ are tangent to circle Q

$m\angle SPQ = 57°$

$m\overset{\frown}{GS} = 118°$

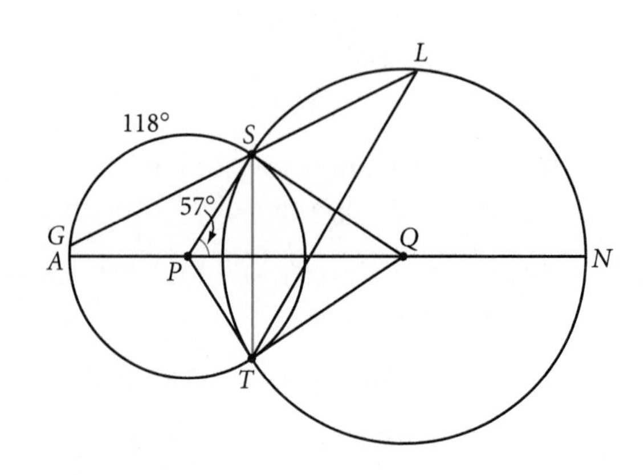

Find:

a. $m\angle GLT$ **b.** $m\angle SQT$

c. $m\angle TSQ$ **d.** $m\overset{\frown}{SL}$

e. Explain why *PSQT* is cyclic.

f. Explain why $\overline{SQ}$ is tangent.

In the figures for Exercises 19 and 20, each arc is a quarter of a circle with its center at a vertex of the square.

Given: The square has side length 1 unit **Find:** The shaded area

19. Shaded area = $\underline{?}$ ⓗ

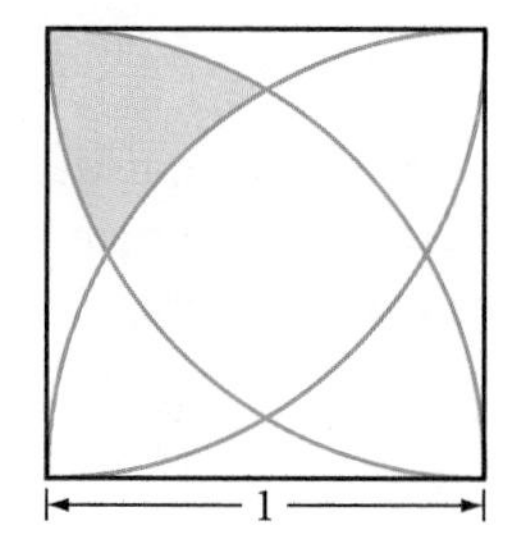

20. Shaded area = $\underline{?}$ ⓗ

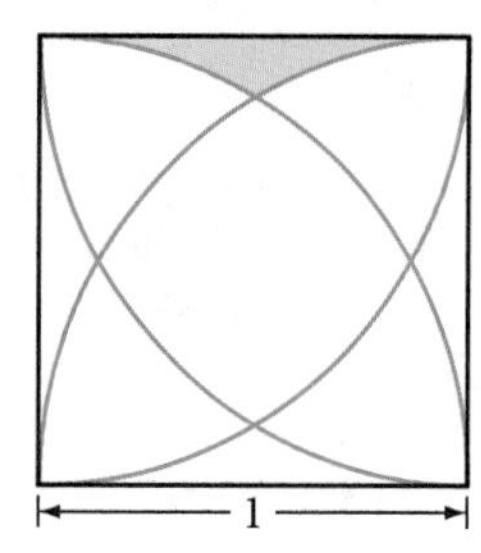

21. *Technology* The diagram below shows a scalene triangle with angle bisector $\overline{CG}$, and perpendicular bisector $\overline{GE}$ of side $\overline{AB}$. Study the diagram.

a. Which triangles are congruent?

b. You can use congruent triangles to prove that $\triangle ABC$ is isosceles. How?

c. Given a scalene triangle, you proved that it is isosceles. What's wrong with this proof?

d. Use geometry software to re-create the construction. What does the sketch tell you about what's wrong?

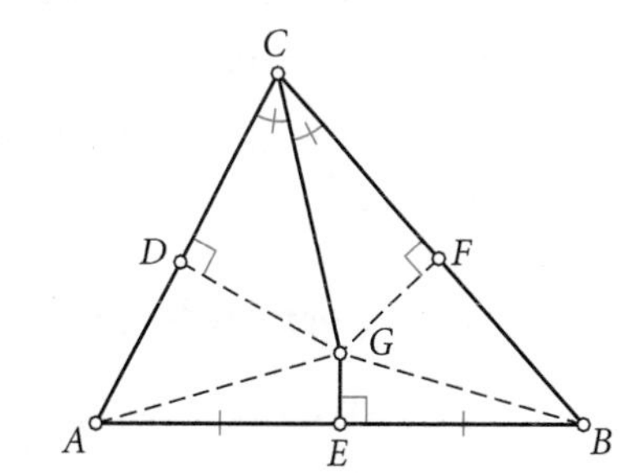

22. Dakota Davis is at an archaeological dig where he has uncovered a stone voussoir that resembles an isosceles trapezoidal prism. Each trapezoidal face has bases that measure 27 cm and 32 cm, and congruent legs that measure 32 cm each. Help Dakota determine the rise and span of the arch when it was standing, and the total number of voussoirs. Explain your method.

IMPROVING YOUR ALGEBRA SKILLS

The Eye Should Be Quicker Than the Hand

How fast can you answer these questions?

1. If $2x + y = 12$ and $3x - 2y = 17$, what is $5x - y$?

2. If $4x - 5y = 19$ and $6x + 7y = 31$, what is $10x + 2y$?

3. If $3x + 2y = 11$ and $2x + y = 7$, what is $x + y$?

Coordinate Proof

You can prove conjectures involving midpoints, slope, and distance using analytic geometry. When you do this, you create a **coordinate proof.** Coordinate proofs rely on the premises of geometry, and these three properties from algebra.

Coordinate Midpoint Property

If (x_1, y_1) and (x_2, y_2) are the coordinates of the endpoints of a segment, then the coordinates of the midpoint are $\left(\frac{x_1 + x_2}{2}, \frac{y_1 + y_2}{2}\right)$.

Parallel Slope Property

In a coordinate plane, two distinct lines are parallel if and only if their slopes are equal.

Perpendicular Slope Property

In a coordinate plane, two nonvertical lines are perpendicular if and only if their slopes are negative reciprocals of each other.

For coordinate proofs, you also use the coordinate version of the Pythagorean Theorem, the distance formula.

Distance Formula

The distance between points $A(x_1, y_1)$ and $B(x_2, y_2)$ is given by $AB^2 = (x_2 - x_1)^2 + (y_2 - y_1)^2$ or $AB = \sqrt{(x_2 - x_1)^2 + (y_2 - y_1)^2}$.

The process you use in a coordinate proof contains the same five tasks that you learned in Lesson 13.2. However, in Task 2, you draw and label a diagram on a coordinate plane. Locate the vertices and other points of your diagram such that they reflect the given information, yet their coordinates should not restrict the generality of your diagram. In other words, do not assume any extra properties for your figure, besides the ones given in its definition.

EXAMPLE A Write a coordinate proof of the Square Diagonals Conjecture: The diagonals of a square are congruent and are perpendicular bisectors of each other.

▶ ***Solution***

Task 1

Given: A square with both diagonals

Show: The diagonals are congruent and are perpendicular bisectors of each other

Task 2

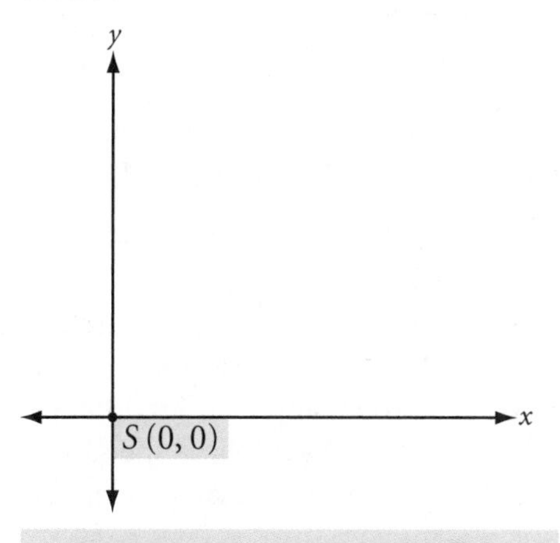

1. Placing one vertex at the origin will simplify later calculations because it is easy to work with zeros.

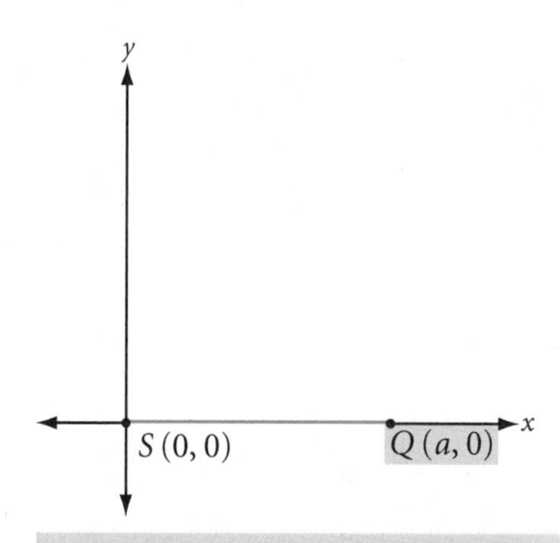

2. Placing the second vertex on the x-axis also simplifies calculations because the y-coordinate is zero. To remain general, call the x-coordinate a.

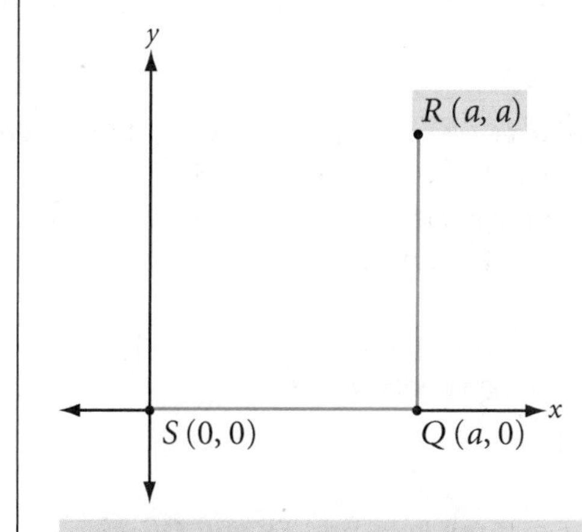

3. $\overline{RQ}$ needs to be vertical to form a right angle with $\overline{SQ}$, which is horizontal. $\overline{RQ}$ also needs to be the same length. So R is placed a units vertically above Q.

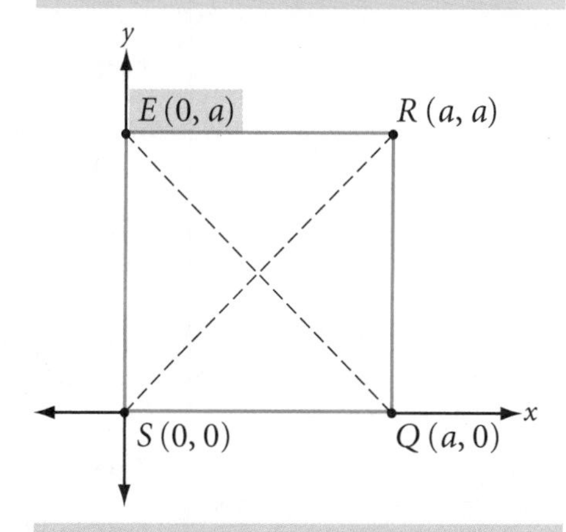

4. The last vertex is placed a units above S.

You can check that $SQRE$ fits the definition of a square—an equiangular, equilateral parallelogram.

$$\text{Slope of } \overline{SQ} = \frac{0-0}{a-0} = \frac{0}{a} = 0$$

$$SQ = \sqrt{(a-0)^2 + (0-0)^2} = \sqrt{a^2} = a$$

$$\text{Slope of } \overline{QR} = \frac{a-0}{a-a} = \frac{a}{0} \text{ (undefined)}$$

$$QR = \sqrt{(a-a)^2 + (a-0)^2} = \sqrt{a^2} = a$$

$$\text{Slope of } \overline{RE} = \frac{a - a}{0 - a} = \frac{0}{-a} = 0$$

$$RE = \sqrt{(0 - a)^2 + (a - a)^2} = \sqrt{a^2} = a$$

$$\text{Slope of } \overline{ES} = \frac{0 - a}{0 - 0} = \frac{-a}{0} \text{ (undefined)}$$

$$ES = \sqrt{(0 - 0)^2 + (0 - a)^2} = \sqrt{a^2} = a$$

Opposite sides have the same slope and are therefore parallel, so $SQRE$ is a parallelogram. Also, from the slopes, $\overline{SQ}$ and $\overline{RE}$ are horizontal and $\overline{QR}$ and $\overline{ES}$ are vertical, so all angles are right angles and the parallelogram is equiangular. Lastly, all the sides have the same length, so the parallelogram is equilateral. $SQRE$ is an equiangular, equilateral parallelogram and is a square by definition.

Task 3

Given: Square $SQRE$ with diagonals $\overline{SR}$ and $\overline{QE}$
Show: $\overline{SR} \cong \overline{QE}$, $\overline{SR}$ and $\overline{QE}$ bisect each other, and $\overline{SR} \perp \overline{QE}$

Task 4

To show that $\overline{SR} \cong \overline{QE}$, you must show that both segments have the same length. To show that $\overline{SR}$ and $\overline{QE}$ bisect each other, you must show that the segments share the same midpoint. To show that $\overline{SR} \perp \overline{QE}$, you must show that the segments have negative reciprocal slopes. Because you know the coordinates of the endpoints of both $\overline{SR}$ and $\overline{QE}$, you can do the necessary calculations to use the distance formula, the coordinate midpoint property, and the perpendicular slope property.

Task 5

Use the distance formula to find SR and QE.

$$SR = \sqrt{(a - 0)^2 + (a - 0)^2} = \sqrt{2a^2} = a\sqrt{2}$$

$$QE = \sqrt{(a - 0)^2 + (0 - a)^2} = \sqrt{2a^2} = a\sqrt{2}$$

So, by the definition of congruence, $\overline{SR} \cong \overline{QE}$ because both segments have the same length.

Use the coordinate midpoint property to find the midpoints of $\overline{SR}$ and $\overline{QE}$.

$$\text{Midpoint of } \overline{SR} = \left(\frac{0 + a}{2}, \frac{0 + a}{2}\right) = (0.5a, 0.5a)$$

$$\text{Midpoint of } \overline{QE} = \left(\frac{0 + a}{2}, \frac{a + 0}{2}\right) = (0.5a, 0.5a)$$

So, $\overline{SR}$ and $\overline{QE}$ bisect each other because both segments have the same midpoint.

Finally, compare the slopes of $\overline{SR}$ and $\overline{QE}$.

$$\text{Slope of } \overline{SR} = \frac{a-0}{a-0} = 1$$

$$\text{Slope of } \overline{QE} = \frac{a-0}{0-a} = -1$$

So, $\overline{SR} \perp \overline{QE}$ by the perpendicular slope property because the segments have negative reciprocal slopes.

Therefore, the diagonals of a square are congruent and are perpendicular bisectors of each other.

Add the Square Diagonals Theorem to your list.

Here's another example. See if you can recognize how the five tasks result in this proof.

EXAMPLE B

Write a coordinate proof of this conditional statement: If the diagonals of a quadrilateral bisect each other, then the quadrilateral is a parallelogram.

▸ Solution

Given: Quadrilateral *ABCD* with diagonals $\overline{AC}$ and $\overline{BD}$ that bisect each other (common midpoint *M*)

Show: *ABCD* is a parallelogram

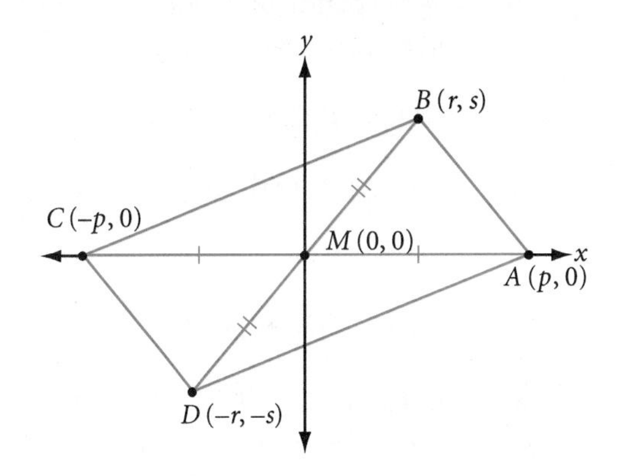

Proof

$$\text{Slope of } \overline{AB} = \frac{s-0}{r-p} = \frac{s}{r-p}$$

$$\text{Slope of } \overline{BC} = \frac{0-s}{-p-r} = \frac{-(s)}{-(p+r)} = \frac{s}{p+r}$$

$$\text{Slope of } \overline{CD} = \frac{-s-0}{-r-(-p)} = \frac{-(s)}{-(r-p)} = \frac{s}{r-p}$$

$$\text{Slope of } \overline{DA} = \frac{0-(-s)}{p-(-r)} = \frac{s}{p+r}$$

Opposite sides $\overline{AB}$ and $\overline{CD}$ have equal slopes of $\frac{s}{r-p}$. Opposite sides $\overline{BC}$ and $\overline{DA}$ have equal slopes of $\frac{s}{p+r}$. So each pair is parallel by the parallel slope property. Therefore, quadrilateral *ABCD* is a parallelogram by definition. Add this theorem to your list.

It is clear from these examples that creating a diagram on a coordinate plane is a significant challenge in a coordinate proof. The first seven exercises will give you some more practice creating these diagrams.

EXERCISES

In Exercises 1–3, each diagram shows a convenient general position of a polygon on a coordinate plane. Find the missing coordinates.

1. Triangle *ABC* is isosceles.

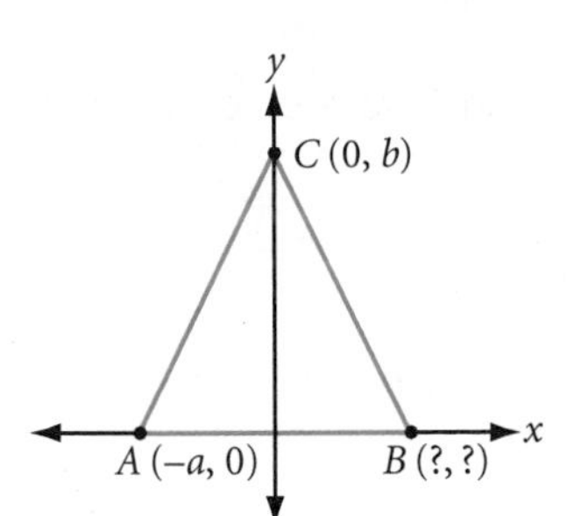

2. Quadrilateral *ABCD* is a parallelogram.

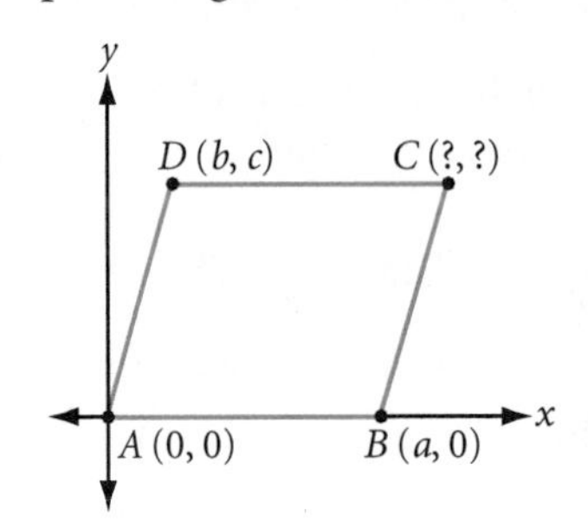

3. Quadrilateral *ABCD* is a rhombus.

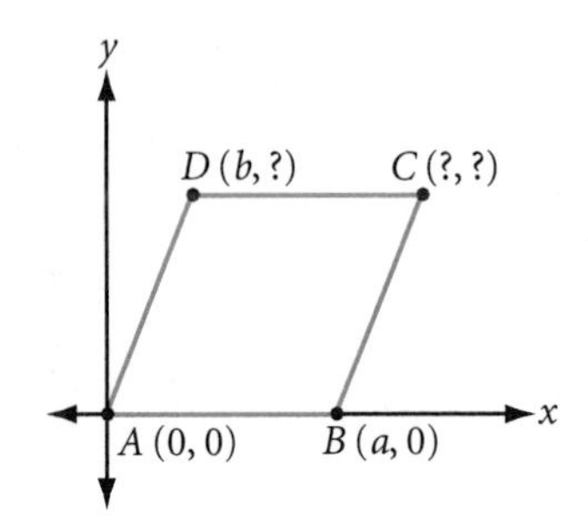

In Exercises 4–7, draw each figure on a coordinate plane. Assign general coordinates to each point of the figure. Then use the coordinate midpoint property, parallel slope property, perpendicular slope property, and/or the distance formula to check that the coordinates you have assigned meet the definition of the figure.

4. Rectangle *RECT*

5. Triangle *TRI* with its three midsegments

6. Isosceles trapezoid *TRAP*

7. Equilateral triangle *EQU*

In Exercises 8–13, write a coordinate proof of each conjecture. If it cannot be proven, write "cannot be proven."

8. The diagonals of a rectangle are congruent.

9. The midsegment of a triangle is parallel to the third side and half the length of the third side.

10. The midsegment of a trapezoid is parallel to the bases.

11. If only one diagonal of a quadrilateral is the perpendicular bisector of the other diagonal, then the quadrilateral is a kite.

12. The figure formed by connecting the midpoints of the sides of a quadrilateral is a parallelogram.

13. The quadrilateral formed by connecting the midpoint of the base to the midpoint of each leg in an isosceles triangle is a rhombus.

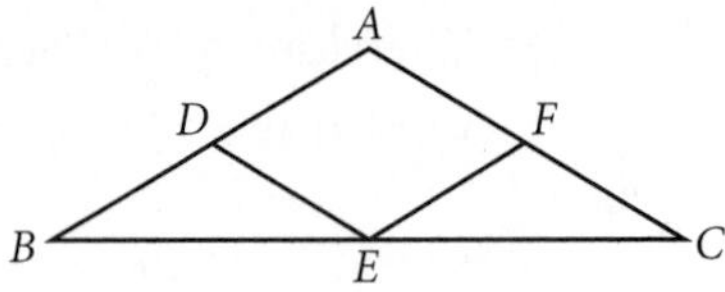

E is the midpoint of base $\overline{BC}$.
D and *F* are the midpoints of the legs.

project

SPECIAL PROOFS OF SPECIAL CONJECTURES

In this project your task is to research and present logical arguments in support of one or more of these special properties.

1. Prove that there are only five regular polyhedra.

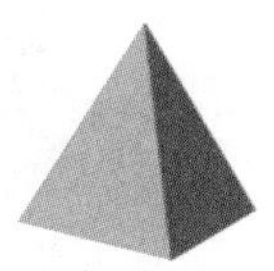

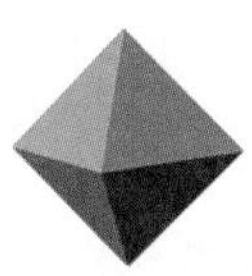
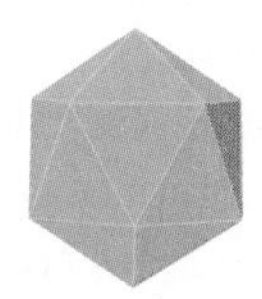
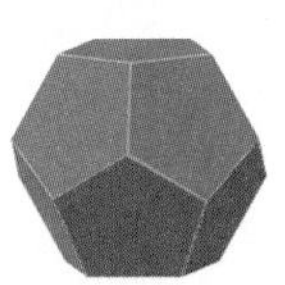

2. You discovered Euler's rule for determining whether a planar network can or cannot be traveled. Write a proof defending Euler's formula for networks.

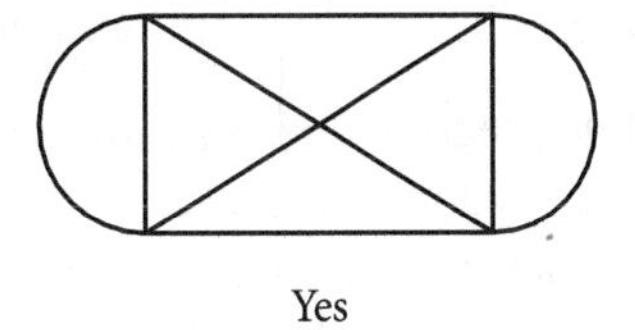
Yes

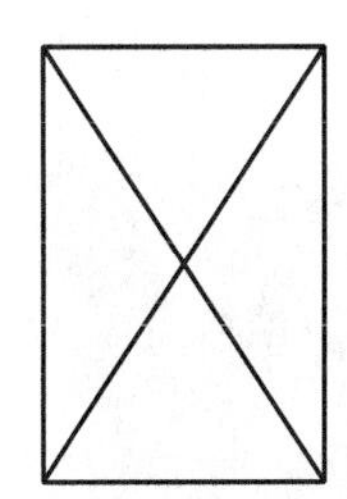
No

3. You discovered that the formula for the sum of the measures of the interior angles of an n-gon is $(n - 2)180°$. Prove that this formula is correct.

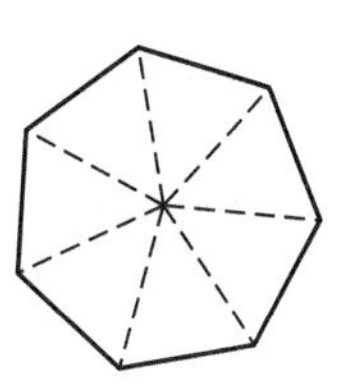

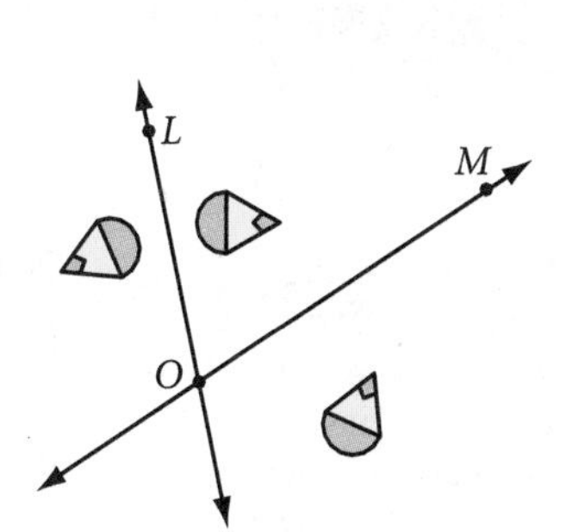

4. You discovered that the composition of two reflections over intersecting lines is equivalent to one rotation. Prove that this always works.

5. The coordinates of the centroid of a triangle are equal to the average of the coordinates of the triangle's three vertices. Prove that this is always true.

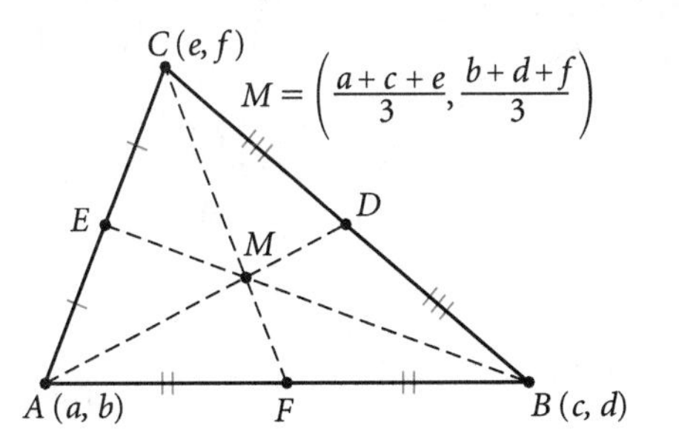

6. Prove that $\sqrt{2}$ is irrational.

7. When you explored all the 1-uniform tilings (Archimedean tilings), you discovered that there are exactly 11 Archimedean tilings of the plane. Prove that there are exactly 11.

Exploration

Non-Euclidean Geometries

Have you ever changed the rules of a game? Sometimes, changing just one simple rule creates a completely different game. You can compare geometry to a game whose rules are postulates. If you change even one postulate, you may create an entirely new geometry.

Euclidean geometry—the geometry you learned in this course—is based on several postulates. A postulate, according to the contemporaries of Euclid, is an obvious truth that cannot be derived from other postulates.

Hungarian mathematician János Bolyai, one of the discoverers of hyperbolic geometry, said, "I have discovered such wonderful things that I was amazed . . . out of nothing I have created a strange new Universe."

The list below contains the first five of Euclid's postulates.

Postulate 1: You can draw a straight line through any two points.

Postulate 2: You can extend any segment indefinitely.

Postulate 3: You can draw a circle with any given point as center and any given radius.

Postulate 4: All right angles are equal.

Postulate 5: Through a given point not on a given line, you can draw exactly one line that is parallel to the given line.

The fifth postulate, known as the Parallel Postulate, does not seem as obvious as the others. In fact, for centuries, many mathematicians did not believe it was a postulate at all and tried to show that it could be proved using the other postulates. Attempting to use indirect proof, mathematicians began by assuming that the fifth postulate was false and then tried to reach a logical contradiction.

If the Parallel Postulate is false, then one of these assumptions must be true.

Assumption 1: Through a given point not on a given line, you can draw *more than one line* parallel to the given line.

Assumption 2: Through a given point not on a given line, you can draw *no line* parallel to the given line.

Interestingly, neither of these assumptions contradict any of Euclid's other postulates. Assumption 1 leads to a new deductive system of non-Euclidean geometry, called **hyperbolic geometry.** Assumption 2 leads to another non-Euclidean system, called **elliptic geometry.**

One model of elliptic geometry applies to lines and angles on a sphere. On Earth, if you walk in a "straight line" indefinitely, what shape will your path take? Theoretically, if you walk long enough, you will end up back at the same point, after walking a complete circle around Earth! (Find a globe and check it!) So, on a sphere, a "straight line" is not a line at all, but a circle.

Hyperbolic geometry is confined to a circular disk. The edges of the disk represent infinity so lines curve and come to an end at the edge of the circle. This may sound like a strange model, but it fits physicists' theory that we live in a closed universe.

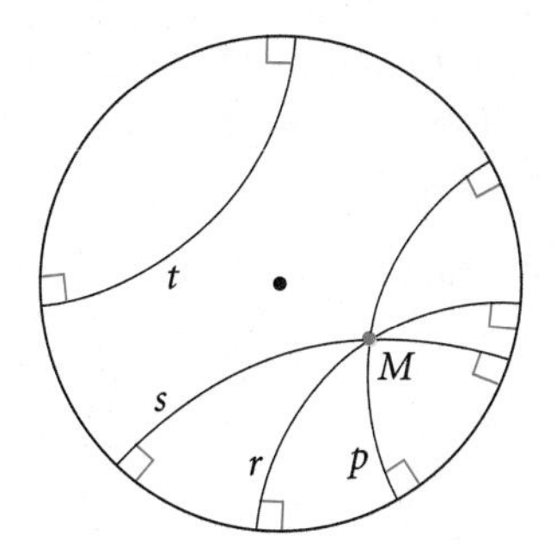

In hyperbolic geometry, many lines can be drawn through a point parallel to another line. Lines *p*, *r*, and *s* all pass through point *M*, and are parallel to line *t*.

In this activity, you will explore elliptic geometry.

Activity
Elliptic Geometry

You will need

- a sphere that you can draw on
- a compass

You can use the surface of a sphere as a model to explore elliptic geometry. Of course, you can't draw a straight line on a sphere. On a plane, the shortest distance between two points is measured along a line. On a sphere, the shortest distance between two points is measured along a great circle. Recall that a **great circle** is a circle on the surface of a sphere whose diameter passes through the center of the sphere. So, in this elliptic-geometry model, a "segment" is an arc of a great circle.

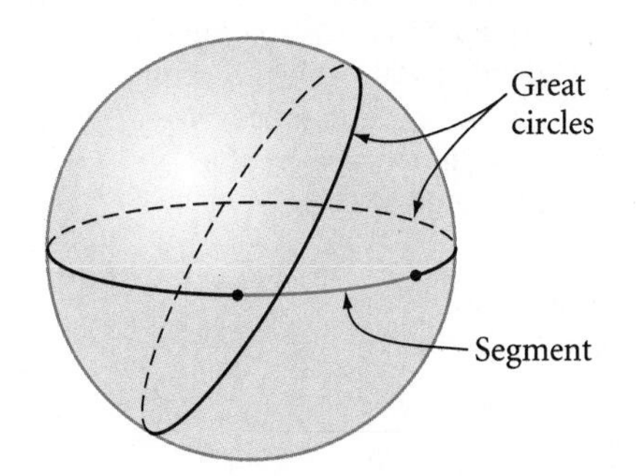

In elliptic geometry, "lines" (that is, great circles) never end; however, their length is finite! Because all great circles have the same diameter, all "lines" have the same length.

A model for elliptic geometry must satisfy the assumption that, through a given point not on a given line, there are *no* lines parallel to the given line. Simply put, there are no parallel lines in elliptic geometry. All great circles intersect, so the spherical model of elliptic geometry supports this assumption.

Step 1 Write a set of postulates for elliptic geometry by rewriting Euclid's first five postulates. Replace the word *line* with the words *great circle*.

Step 2 In Euclidean geometry, two lines that are perpendicular to the same line are parallel to each other. This is not true in elliptic geometry. On your sphere, draw an example of two "lines" that are perpendicular to the same "line" but that are not parallel to each other.

Step 3 On your sphere, show that two points do not always determine a unique "line."

Step 4 Draw an isosceles triangle on your sphere. (Remember, the "segments" that form the sides of a triangle must be arcs of great circles.) Does the Isosceles Triangle Theorem appear to hold in elliptic geometry?

Step 5 In elliptic geometry, the sum of the measures of the three angles of a triangle is always *greater than* 180°. Draw a triangle on your model and use it to help you explain why this makes sense.

Step 6 In Euclidean geometry, no triangle can have two right angles. But in elliptic geometry, a triangle can have three right angles. Find such a triangle and sketch it.

Japanese *temari* balls, colorful balls made of thread or scrap material, are embroidered with geometric designs derived from nature, like flowers or trees. Also called "princess balls," they originated in 700 C.E., when young nobility made them from silk and gave them as gifts. Notice that each "line segment" in the design of a *temari* ball is actually an arc of a great circle.

CHAPTER 13 REVIEW

In this course you have discovered geometry properties and made conjectures based on inductive reasoning. You have also used deductive reasoning to explain why some of your conjectures were true. In this chapter you have focused on geometry as a deductive system. You learned about the premises of geometry. Starting fresh with these premises, you built a system of theorems.

By discovering geometry, and then examining it as a mathematical system, you have been following in the footsteps of mathematicians throughout history. Your discoveries gave you an understanding of how geometry works. Proofs gave you the tools for taking apart your discoveries and understanding why they work.

EXERCISES

In Exercises 1–7, identify each statement as true or false. For each false statement, sketch a counterexample or explain why it is false.

1. If one pair of sides of a quadrilateral are parallel and the other pair of sides are congruent, then the quadrilateral is a parallelogram.
2. If consecutive angles of a quadrilateral are supplementary, then the quadrilateral is a parallelogram.
3. If the diagonals of a quadrilateral are congruent, then the quadrilateral is a rectangle.
4. Two exterior angles of an obtuse triangle are obtuse.
5. The opposite angles of a quadrilateral inscribed within a circle are congruent.
6. The diagonals of a trapezoid bisect each other.
7. The midpoint of the hypotenuse of a right triangle is equidistant from all three vertices.

In Exercises 8–12, complete each statement.

8. A tangent is _?_ to the radius drawn to the point of tangency.
9. Tangent segments from a point to a circle are _?_.
10. The perpendicular bisector of a chord passes through _?_.
11. The three midsegments of a triangle divide the triangle into _?_.
12. A lemma is _?_.
13. Restate this conjecture as a conditional: The segment joining the midpoints of the diagonals of a trapezoid is parallel to the bases.

14. Sometimes a proof requires a construction. If you need an angle bisector in a proof, what postulate allows you to construct one?

15. If an altitude is needed in a proof, what postulate allows you to construct one?

16. Describe the procedure for an indirect proof.

17. **a.** What point is this anti-smoking poster trying to make?

b. Write an indirect argument to support your answer to part a.

In Exercises 18–23, identify each statement as true or false. If true, prove it. If false, give a counterexample or explain why it is false.

18. If the diagonals of a parallelogram bisect the angles, then the parallelogram is a square.

19. The angle bisectors of one pair of base angles of an isosceles trapezoid are perpendicular.

20. The perpendicular bisectors to the congruent sides of an isosceles trapezoid are perpendicular.

21. The segment joining the feet of the altitudes on the two congruent sides of an isosceles triangle is parallel to the third side.

22. The diagonals of a rhombus are perpendicular.

23. The bisectors of a pair of opposite angles of a parallelogram are parallel.

In Exercises 24–27, devise a plan and write a proof of each conjecture.

24. Refer to the figure at right.

Given: Circle O with chords $\overline{PN}$, $\overline{ET}$, $\overline{NA}$, $\overline{TP}$, $\overline{AE}$

Show: $m\angle P + m\angle E + m\angle N + m\angle T + m\angle A = 180°$

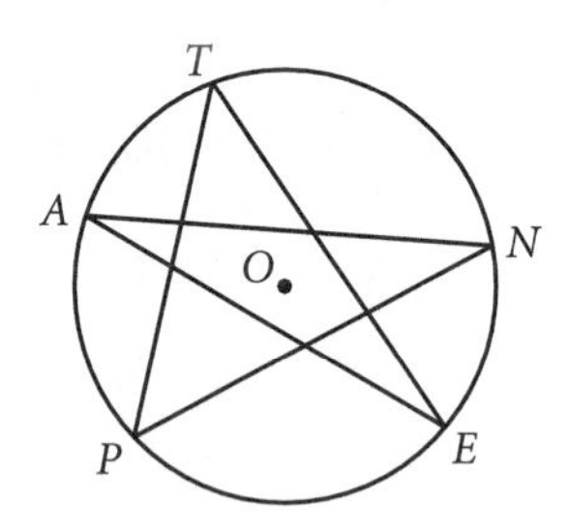

25. If a triangle is a right triangle, then it has at least one angle whose measure is less than or equal to 45°.

26. Prove the Triangle Midsegment Conjecture. ⓗ

27. Prove the Trapezoid Midsegment Conjecture.

In Exercises 28–30, use construction tools or geometry software to perform each mini-investigation. Then make a conjecture and prove it.

28. ***Mini-Investigation*** Construct a rectangle. Construct the midpoint of each side. Connect the four midpoints to form another quadrilateral.

a. What do you observe about the quadrilateral formed? From a previous theorem, you already know that the quadrilateral is a parallelogram. State a conjecture about the type of parallelogram formed.

b. Prove your conjecture.

29. *Mini-Investigation* Construct a rhombus. Construct the midpoint of each side. Connect the four midpoints to form another quadrilateral.

a. You know that the quadrilateral is a parallelogram, but what type of parallelogram is it? State a conjecture about the parallelogram formed by connecting the midpoints of a rhombus.

b. Prove your conjecture.

30. *Mini-Investigation* Construct a kite. Construct the midpoint of each side. Connect the four midpoints to form another quadrilateral.

a. State a conjecture about the parallelogram formed by connecting the midpoints of a kite.

b. Prove your conjecture.

31. Prove this theorem: If two chords intersect in a circle, the product of the segment lengths on one chord is equal to the product of the segment lengths on the other chord.

Assessing What You've Learned

WRITE IN YOUR JOURNAL How does the deductive system in geometry compare to the underlying organization in your study of science, history, and language?

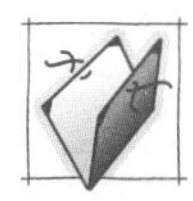

UPDATE YOUR PORTFOLIO Choose a project or a challenging proof you did in this chapter to add to your portfolio.

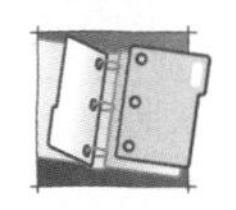

ORGANIZE YOUR NOTEBOOK Review your notebook to be sure it's complete and well organized. Be sure you have all the theorems on your theorem list. Write a one-page summary of Chapter 13.

PERFORMANCE ASSESSMENT While a classmate, friend, family member, or teacher observes, demonstrate how to prove one or more of the theorems proved in this chapter. Explain what you're doing at each step.

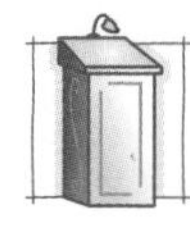

GIVE A PRESENTATION Give a presentation on a puzzle, exercise, or project from this chapter. Work with your group, or try presenting on your own.

Hints for Selected Exercises

. . . there are no answers to the problems of life in the back of the book.

SØREN KIERKEGAARD

You will find hints below for exercises that are marked with an ⓗ in the text. Instead of turning to a hint before you've tried to solve a problem on your own, make a serious effort to solve the problem without help. But if you need additional help to solve a problem, this is the place to look.

CHAPTER 0

LESSON 0.1

7. Here's the title of the sculpture.

Early morning calm
knotweed stalks
pushed into lake bottom
made complete by their own reflection

Derwent Water, Cumbria, 20 February & 8–9 March, 1988

LESSON 0.2

3. Design 1 Do one-half of the Astrid four times.

Design 2 Connect the midpoints of the sides of an equilateral triangle with line segments and leave the middle triangle empty. Repeat the process on the three other triangles. Then, repeat the rule again.

Design 3 On each side of an equilateral hexagon, mark the point that is one-third of the length of the side. Connect the points and repeat the process.

LESSON 0.3

2. Use isometric dot paper, or draw an equilateral equiangular hexagon with sides of length 2. Each vertex of the hexagon will be the center of a circle with radius 1. Fit the seventh small circle inside the other six. The large circle has the same center as the seventh small circle and has radius 3.

LESSON 0.6

5. Draw two identical squares, one rotated $\frac{1}{8}$ turn, or 45°, from the other. Where is the center of each arc located?

CHAPTER 1

LESSON 1.1

3. Because a line is infinitely long in two directions, it doesn't matter where the point used to name the line lies on the line. There are three possible ways to name the line if you don't count switching the letters as two different ways.

21. Because a ray is infinitely long in one direction, it doesn't matter which point you use for the second letter as long as it's not the endpoint.

LESSON 1.2

13. Find $m\angle CQA$ and $m\angle BQA$ and subtract.

26. Don't forget that at half past the hour, the hour hand will be halfway between the 3 and the 4.

36. A $\frac{1}{4}$ rotation $= \frac{1}{4} \cdot 360°$. Subtract the sum of 15° and 21° from that result.

LESSON 1.3

15. Do not limit your thinking to just two dimensions.

24. The measure of the incoming angle equals the measure of the outgoing angle (just as in pool).

25. Use trial and error.

LESSON 1.4

18. The order of the letters matters.

31. Draw your diagram on patty paper or tracing paper. Test your diagram by folding your paper along the line of reflection to see if the two halves coincide. Your diagram should have only one pair of parallel sides.

34. Label the width of each small rectangle as w and its length as l. Therefore the perimeter of the large rectangle is $5w + 4l = 198$ cm.

38. Only one of these is impossible.

LESSON 1.5

24. There are four possible locations for R. The slope of $\overline{CL} = \frac{1}{5}$. Therefore the slope of the perpendicular segment $= \frac{-5}{1}$ or $\frac{5}{-1}$.

31. Locate the midpoint of each rod. Draw the segment that contains all the midpoints. This segment is called the median of the triangle.

LESSON 1.6

19. This ordered pair rule tells you to double the abscissa and double the ordinate. (abscissa, ordinate)

LESSON 1.7

3. Make a large graph (Quadrant I only). Label the vertical axis "feet" and the horizontal axis "days." Or try a number line.

6. The vertical distance from the top of the pole to the lowest point of the cable is 15 feet. Compare that distance with the length of the cable.

7. Draw a diagram. Draw two points, A and B, on your paper. Locate a point that appears to be equally spaced from points A and B. The midpoint of $\overline{AB}$ is only one such point; find others. Connect the points into a line. For points in space, picture a plane between the two points.

11. Copy trapezoid $ABCD$ onto patty paper or tracing paper. Rotate the tracing paper 90°, or $\frac{1}{4}$ turn, counterclockwise. Point A on the tracing paper should coincide with point A on the diagram in the book.

15. The number of hexagons is increasing by one each time, but the perimeter is increasing by four.

34. Because $\overline{PA}$, $\overline{PB}$, $\overline{QA}$, and $\overline{QB}$ are all radii, quadrilateral $PAQB$ is a rhombus. $\overline{AB}$ and $\overline{PQ}$ are the diagonals of rhombus $PAQB$.

LESSON 1.8

8. The biggest face is 3 m by 4 m. Your diagram will look similar to the diagram for Step 4 on page 80, except that the shortest segment will be vertical.

9. The number of boxes that will fit in the solid equals the volume, which is found by $l \times w \times h$.

18. Cut out a rectangle like the one shown and tape it to your pencil. Rotate your pencil to see what shape the rotating rectangle forms.

20. Imagine slicing an orange in half. What shape is revealed?

24. Do not limit your thinking to two dimensions. This situation can be modeled by using three pencils to represent the three lines.

CHAPTER 1 REVIEW

37. Here is one possible method. Draw a circle and one diameter. Draw another diameter perpendicular to the first. Draw two more diameters so that eight 45° angles are formed. Draw the regular octagon formed by the endpoints of the diameters.

45. A clock forms 12 central angles that each measure $30° \left(\frac{360°}{12}\right)$. The angle formed by the hands is $3\frac{1}{2}$ of those central angles.

54. Cut out a semicircle and tape it to your pencil. Rotate your pencil to see what shape the rotating semicircle forms.

57. Rotate your book so that the red line is vertical.

CHAPTER 2 • CHAPTER 2 CHAPTER 2 • CHAPTER

LESSON 2.1

1. Conjectures are statements that generalize from a number of instances to "all." Therefore, Stony is saying "All _?_."

4. Change all fractions to the same denominator.

7. $1 + 1 = 2, 1 + 2 = 3, 2 + 3 = 5,$
$3 + 5 = 8, \ldots$

8. $1^2, 2^2, 3^2, 4^2, \ldots$

13. Add another row with one more rectangle.

14. Each segment branches off into two segments.

15. Connect the midpoints of the sides of the triangle. Make the triangle formed in the middle white. Connect the midpoints of the sides of the shaded triangles. Make the triangles formed in the middle white.

17. Substitute 1 for n and evaluate the expression to find the first term. Substitute 2 for n to find the second term, and so on.

21. For example, "I learned by trial and error that you turn wood screws clockwise to screw them into wood and counterclockwise to remove them."

26. Imagine folding the square up and "wrapping" the two rectangles and the other triangle around the square.

27. Turn your book so that the red line is vertical. Imagine rotating the figure so that the part jutting out is facing back to the right.

28. Cut out a quarter-circle and tape it to your pencil. Rotate your pencil to see the figure formed.

42. Remember that a kite is a quadrilateral with two pairs of consecutive, congruent sides. One of the diagonals does bisect the angles of the kite, the other does not.

LESSON 2.2

8. What is the smallest possible size for an obtuse angle?

9. Compare how many 1's there are in the numbers that are multiplied with the middle digit of the answer; then compare both quantities with the row number.

12. Look for a constant difference among terms.

16. The number of rows increases by one, and the number of columns increases by one.

17. The number of rows increases by two, and the number of columns increases by one.

LESSON 2.3

1. Look for a constant difference, then adjust the rule for the first term.

4. See below.

7. Draw the polygons and all possible diagonals from one vertex. Fill in the table and look for a pattern. You should be able to see the pattern by the time you get to a hexagon.

LESSON 2.4

4. Compare this exercise with the Investigation Party Handshakes, and with Exercise 3. What change can you make to each of those functions to fit this pattern?

5. This is like Exercise 4 except that you add the number of sides to the number of diagonals.

6. In other words, each time a new line is drawn, it passes through all the others. This also gives the maximum number of intersections.

7. Exercises 5 and 6 have the same rule. Every term in the sequence for Exercise 4 is n less than the corresponding term in the sequences for Exercises 5 and 6. This is because in Exercise 5 you count the sides of the polygons.

9. Let a point represent each team, and let the segments connecting the points represent *one* game played between them. What do you then have to multiply this answer by?

10. Use the model from Exercise 5.

14. The tricky part to this problem is that points A and B could be on the same side of point E.

---

4. (*Lesson 2.3*)

Figure number	1	2	3	4	5	6	n
Number of tiles	$1 \cdot 8$	$2 \cdot 8$	$3 \cdot 8$	$4 \cdot 8$	$? \cdot ?$	$? \cdot ?$	$? \cdot ?$

LESSON 2.5

4. Here is how to begin:
$a = 60°$ because of the Vertical Angles Conjecture. $c = 120°$ because of the Linear Pair Conjecture.

23. Refer to Lesson 2.4, Exercise 5.

24. Refer to Lesson 2.4, Exercise 6.

25. Refer to Lesson 2.4, Exercise 5, but subtract the number of couples from each term because the couples don't shake hands.

27. Refer to Lesson 2.4, Exercise 4. Then use "guess and check."

LESSON 2.6

4. Because quadrilateral *TUNA* is a parallelogram, $\overline{TU} \parallel \overline{AN}$ and $\overline{TA} \parallel \overline{UN}$. Form $\overrightarrow{NA}$ by extending side $\overline{NA}$. Place a point Q on $\overrightarrow{NA}$, to the left of A. $\angle T \cong \angle QAT$ by the Alternate Interior Angles Conjecture. $\angle N \cong \angle QAT$ by the Corresponding Angles Conjecture. Because $\angle T$ and $\angle N$ are both congruent to $\angle QAT$, $\angle N \cong \angle T$.

6.

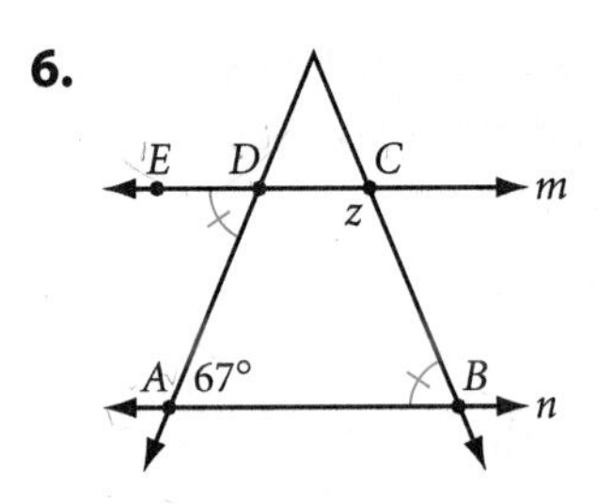

$\angle DAB$ and $\angle EDA$ are congruent by the Alternate Interior Angles Conjecture. $\angle EDA$ and $\angle CDA$ are a linear pair. Now every angle in quadrilateral *ABCD* is known except the one labeled z. The sum of the angles of a quadrilateral is 360°.

7. Measures a, b, c, and d are all related by parallel lines. Measures e, f, g, h, i, j, k, and s are also all related by parallel lines.

13. Squares, rectangles, rhombuses, and kites are eliminated because they have reflectional symmetry.

16. Graph the original triangle and the new triangle on separate graphs. Cut one out and lay it on top of the other to see if they are congruent.

20. Proceed in an organized way.
Number of 1-by-1 squares = ?
Number of 2-by-2 squares = ?
Number of 3-by-3 squares = ?
Number of 4-by-4 squares = ?
Then add.

21. You can add the first two function rules to find the third function rule.

CHAPTER 2 REVIEW

5. Try the alphabet backward and the powers of 2 forward.

7. For the pattern of the letters, number each letter of the alphabet and find the pattern of their differences.

9. Here is how to begin:
$f(1) = 2^{1-1} = 2^0 = 1$
$f(2) = 2^{2-1} = 2^1 = 2$

11. The diagrams are alternating net and solid. The number of sides of the base increases by one.

14. Refer to Lesson 2.4, the Investigation Party Handshakes, discussion of triangular numbers.

15. Exercises 14 and 15 are closely related. Let's examine the rule for Exercise 14.

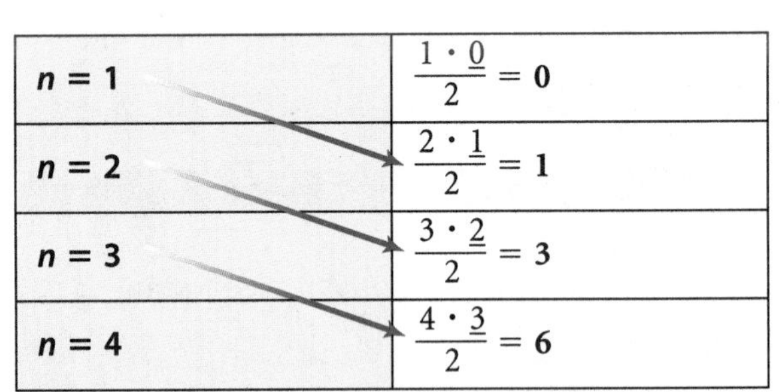

$n = 1$	$\frac{1 \cdot \underline{0}}{2} = 0$
$n = 2$	$\frac{2 \cdot \underline{1}}{2} = 1$
$n = 3$	$\frac{3 \cdot \underline{2}}{2} = 3$
$n = 4$	$\frac{4 \cdot \underline{3}}{2} = 6$

We want 1 to be the first term, not 0. Now look at the pattern for the rest of the terms. The underlined numbers are the n, and the numbers in front of the n are one higher.

17. See below.

18.

$1 = 1 = 1^2$

$1 + 3 = 4 = 2^2$

$1 + 3 + 5 = 9 = 3^2$

$1 + 3 + 5 + 7 = 16 = 4^2$

And so on . . .

20. How many vertical interior segments are there? How many horizontal?

23. Refer to Lesson 2.4, Exercise 6. Then use "guess and check."

CHAPTER 3 • CHAPTER 3 CHAPTER 3 • CHAPTER

LESSON 3.1

2. Copy the first segment onto a ray. Copy the second segment immediately after the first.

10. You duplicated a triangle in Exercise 7. You can think of the quadrilateral as two triangles stuck together (they meet at the diagonal).

14. Fold the paper so that the two congruent sides of the triangle coincide.

LESSON 3.2

2. Bisect, then bisect again.

3. Construct one pair of intersecting arcs, then change your compass setting to construct a second pair of intersecting arcs on the same side of the line segment as the first pair.

4. Bisect $\overline{CD}$ to get the length $\frac{1}{2}CD$. Subtract this length from $2AB$.

5. The average is the sum of the two lengths divided by the number of segments (two). Construct a segment of length $AB + CD$. Bisect the segment to get the average length. Or take half of each, then add them.

8. Construct the median from the vertex to the midpoint.

LESSON 3.3

3. Construct the perpendicular through point B to $\overrightarrow{TO}$.

4. Does your method from Investigation 1 still work? Can you modify it?

5. Construct right angles at Q and R.

13. Look at two columns as a "group."

1st rectangle is 2 groups of 1

2nd rectangle is 3 groups of 3

3rd rectangle is 4 groups of 5

4th rectangle is 5 groups of 7

. . .

nth rectangle is $n + 1$ groups of $2n - 1$

17. (*Chapter 2 Review*)

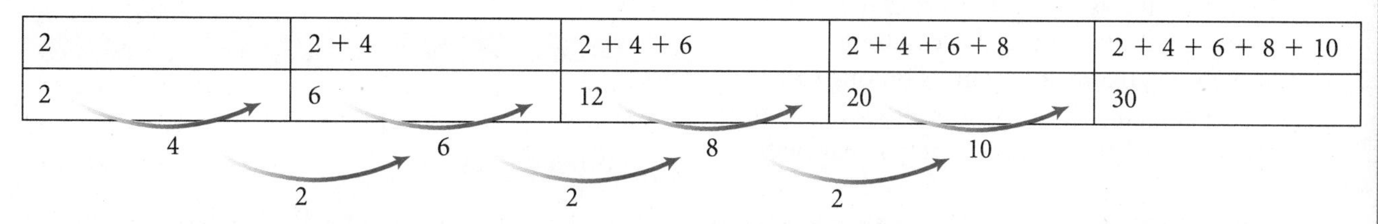

2	2 + 4	2 + 4 + 6	2 + 4 + 6 + 8	2 + 4 + 6 + 8 + 10
2	6	12	20	30

The differences between terms aren't constant, so it's not a linear pattern, but the second set of differences is a constant. Sequences of this type have two linear factors, as the table below shows, and are called *quadratic sequences.*

n	1	2	3	4	5
Term	2	6	12	20	30
Factors	1×2	2×3	3×4	4×5	5×6

17. Each angle of a regular pentagon is 108°.

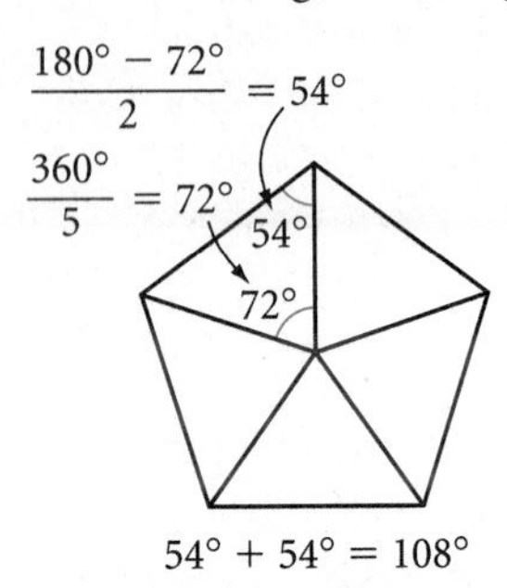

(Or, divide a circle into five congruent arcs, and join the endpoints.)

LESSON 3.4

14. Use your protractor and measure off eight rays 45° apart about a point. Use your compass and swing a circle at the point of intersection.

15. Construct two lines perpendicular to each other, and then bisect the right angles.

16. Because the angles of a triangle add to 180°, the problem could be restated as "Draw a second triangle with a 40° angle, an 80° angle, and a side between the given angles measuring 8 cm."

LESSON 3.5

3. If the perimeter is z, then each side has length $\frac{1}{4}z$. Construct the perpendicular bisector of z to get $\frac{1}{2}z$. Construct the perpendicular bisector of $\frac{1}{2}z$ to get $\frac{1}{4}z$.

10. According to the Perpendicular Bisector Conjecture, if a point is on the perpendicular bisector, then it is equidistant from the endpoints of the segment (which in this case represent fire stations). Therefore, if a point is on one side of the perpendicular bisector, it is closer to the fire station on that side of the perpendicular bisector.

LESSON 3.6

1. Construct a segment, $\overline{MS}$. Draw an arc with radius AS from point S and an arc with radius MA from point M. Connect the point at which the arcs intersect to points M and S to form your triangle.

5. Duplicate $\angle A$ and $\overline{AB}$ on one side of $\angle A$. Open the compass to length BC. If you put the compass point at point B, you'll find two possible locations to mark arcs for point C.

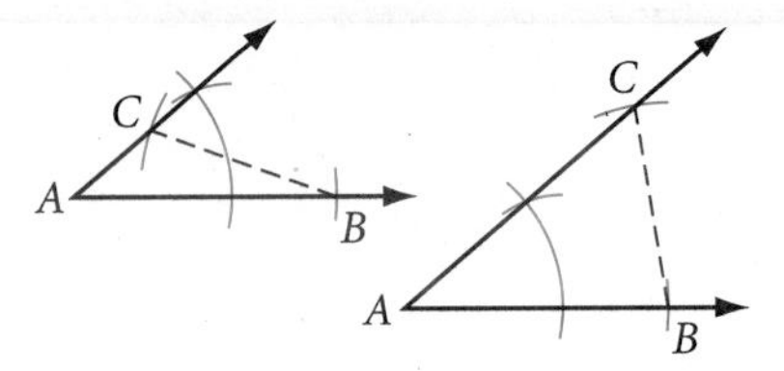

6. $y - x$ is the sum of the two equal sides. Find this length and bisect it to get the length of the other two legs of your triangle.

LESSON 3.7

6. Find the incenter.

7. Find the circumcenter.

8. Draw a slightly larger circle on patty paper and try to fit it inside the triangle.

9. Draw a slightly smaller circle on patty paper and try to fit it outside the triangle.

16. Start by finding points whose coordinates add to 9, such as (3, 6) and (7, 2). Try writing an equation and graphing it.

17. One way is to construct the incenter by bisecting the two given angles. Then find two points on the unfinished sides, equidistant from the incenter and in the same direction (both closer to the missing point.) Now find a point equidistant from those two points. Draw the missing angle bisector through that point and the incenter.

LESSON 3.8

2. If $CM = 16$, then $UM = \frac{1}{2}(16) = 8$. If $TS = 21$, then $SM = \frac{1}{3}(21) = 7$.

8. A quadrilateral can be divided into two triangles in two different ways. How can you use the centroids of these triangles to find *the* centroid?

15. Construct the altitudes for the two other vertices. From the point where the two altitudes meet, construct a line perpendicular to the southern boundary of the triangle.

16. How many people does each person greet? Don't count any greeting twice. There are 60 people. If everyone greets each other, the number of greetings would be $\frac{60 \cdot 59}{2}$. But dorm members aren't greeting their guests, so the number of greetings would be $\frac{60 \cdot 59}{2} - 40$.

CHAPTER 3 REVIEW

28. Draw a long segment and use your compass to add y plus y plus x, bisect z, then subtract it from the sum.

CHAPTER 4 • CHAPTER  CHAPTER 4 • CHAPTER

LESSON 4.1

4. Ignore the 100° angle and the line that intersects the larger triangle. Find the three interior angles of the triangle. Then find z.

6. The total measure of the three angles is 3 times 360° minus the sum of the interior angles of the triangle.

7. The sum of a linear pair of angles is 180°. So the total measure of the three angles is 3 times 180° minus the sum of the interior angles of the triangle.

8. You can find a by looking at the large triangle that has 40° and 71° as its other measures. You can find b because it forms a linear pair with the 133° angle. Continue on your own.

15. $m\angle A + m\angle B + x = 180°$ and $m\angle D + m\angle E + y = 180°$. Then use substitution to prove $x = y$.

LESSON 4.2

1. $m\angle H + m\angle O = 180° - 22°$ and $m\angle H = m\angle O$.

7. Notice that $d = e$ and $d + e + e + 66° = 180°$. Next, find the alternate interior angle to c.

8. Notice that all the triangles are isosceles!

12. The sides do not have to be congruent.

13. Make $GK < MP$.

15. Find the slopes.

21. Move each point right 5 units and down 3 units.

LESSON 4.3

5. Find the value of the unmarked angle and use the Side-Angle Inequality Conjecture.

9. Use the Side-Angle Inequality Conjecture to find an inequality of sides for each triangle, then combine the inequalities.

11. Any side must be smaller than the sum of the other two sides. What must it be larger than?

13. Try using the Triangle Sum Conjecture.

14. Use the Triangle Exterior Angle Conjecture.

17. You need to show that $x = a + b$. From the Triangle Sum Conjecture, you know that $a + b + c = 180°$. Also, $\angle BCA$ and $\angle BCD$ are a linear pair, so $x + c = 180°$.

21. All corresponding sides and angles are congruent. Can you see why? (And remember that the ordering of the points is important in correctly stating the answer.)

LESSON 4.4

1. Rotate one triangle 180°.

2. The shared side is congruent to itself.

9. Match congruent sides.

12. Take a closer look. Are congruent parts corresponding?

15. $UN = YA = 4$, $RA = US = 3$, $m\angle A = m\angle U = 90°$

LESSON 4.5

3. Flip one triangle over.

19. The sides do not have to be congruent.

LESSON 4.6

1. Don't forget to mark the shared segment as congruent to itself.

2. Use $\triangle CRN$ and $\triangle WON$.

4. Use $\triangle ATI$ and $\triangle GTS$.

5. Draw $\overline{UF}$.

7. Draw $\overline{UT}$.

10. Count the lengths of the horizontal and vertical segments, and label the right angles congruent.

17. Make the included angles different.

LESSON 4.7

7. Use $\triangle ABD$ and $\triangle CBD$.

8. Don't forget about the shared side and the Side-Angle Inequality Conjecture.

10. When you look at the larger triangles, ignore the marks for $\overline{OS} \cong \overline{RS}$. When you look at the smaller triangles, ignore the right-angle mark and the given statement $\overline{PO} \cong \overline{PR}$.

11. First, mark the vertical angles. In $\triangle ADM$, the side is included between the two angles. In $\triangle CRM$, the side is not included between the two angles.

14. Review incenter, circumcenter, orthocenter, and centroid.

17. See if your teacher has 14 cubes. Build the figure shown, take away the indicated blocks, and draw what is left.

LESSON 4.8

1. $AB + BC + AC = 48$

$AD = \frac{1}{2}AB$

8. Use the vertex angle bisector as your auxiliary line segment.

12. At 3:15 the hands have not yet crossed each other. At 3:20 the hands have already crossed each other, because the minute hand is on the 4 but the hour hand is only one-third of its way from the 3 toward the 4. So the hands overlap sometime between 3:15 and 3:20.

14. Make a table and look for a pattern.

17. How many H's branch off each C?

CHAPTER 4 REVIEW

19. Look for alternate interior angles.

21. Not enough information is given. The two angles at point H may look the same, but you just don't know.

23. There are actually two isosceles triangles in the figure, and there are three possible answers. It may help to redraw the triangles so that they don't overlap.

25. Calculate the missing angle measure.

32. Look for congruent triangles. Then look for congruent angles to show that lines are parallel.

35. Vertical angles are congruent.

CHAPTER 5 • CHAPTER 5 CHAPTER 5 • CHAPTER

LESSON 5.1

2. All angles are equal in measure.

6. $d + 44° + 30° = 180°$

7. $3g + 117° + 108° = 540°$

13. $(n - 2) \cdot 180° = 2700°$

14. $\frac{(n-2) \cdot 180°}{n} = 156°$

17. Use the Triangle Sum Conjecture and angle addition.

LESSON 5.2

6. First, find the measure of an angle of the equiangular heptagon. Then, find c by the Linear Pair Conjecture.

10. $24° = \frac{360°}{n}$

12. An obtuse angle measures greater than 90°, and the sum of exterior angles of a polygon is always 360°.

15. Look at $\triangle RAC$ and $\triangle DCA$.

16. Draw $\overline{AT}$.

LESSON 5.3

11. Construct $\angle I$ and $\angle W$ at the ends of $\overline{WI}$. Construct $\overline{IS}$. Construct a line through point S parallel to $\overline{WI}$.

13. The nonshared endpoints of the consecutive congruent segments are equidistant from the shared endpoint. So the shared endpoint is on the perpendicular bisector of one diagonal.

16. Look at $\triangle AFG$ and $\triangle BEH$.

LESSON 5.4

2. $PO = \frac{1}{2}RA$

4. Use the Three Midsegments Conjecture.

9. Draw a diagonal of the original quadrilateral. Note that it's parallel to two other segments.

LESSON 5.5

4. $VN = \frac{1}{2}VF$ and $NI = \frac{1}{2}EI$

8. With the midpoint of the longer diagonal as center and using the length of half the shorter diagonal as radius, construct a circle.

10. Complete the parallelogram with the given vectors as sides. The resultant vector is the diagonal of the parallelogram. Refer to the diagram right before the Exercises.

11. $PR = a$. Therefore $MA = a$. $\underline{?} + a = b$. Solve for $\underline{?}$. The height of A is c. Therefore the height of M is c.

LESSON 5.6

1. Consider the parallelogram below.

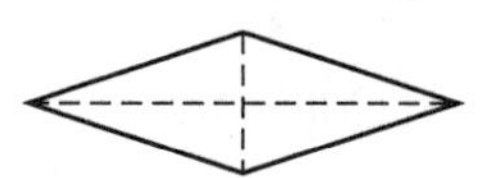

12. $\angle P$ and $\angle PEA$ are supplementary.

17. The diagonals of a square are equal in length and are perpendicular bisectors of each other.

18. Construct $\angle B$, then bisect it. Mark off the length of diagonal $\overline{BK}$ on the angle bisector. Then construct the perpendicular bisector of $\overline{BK}$.

23. The diagonal of a rhombus bisects the angle.

24. Construct a rhombus with your segment as the diagonal. The other diagonal will be the perpendicular bisector.

31. Use the Quadrilateral Sum Conjecture to prove the measure of each angle is 90°. If the consecutive interior angles are supplementary, then the lines are parallel.

LESSON 5.7

1. If you start from square 100 and work backward, the problem becomes much easier.

7. Look at $\triangle YIO$ and $\triangle OGY$.

8. Break up the rectangle into four triangles ($\triangle EAR$, etc.), and show that they are congruent triangles.

12. Imagine what happens to the rectangles when you pull them at their vertices.

13. Calculate the measures of the angles of the regular polygons. Remember that there are 360° around any point.

14. Look at the alternate interior angles.

15. You miss 5 minutes out of 15 minutes.

16. The container is $\frac{8}{12} = \frac{2}{3}$ full. It will be $\frac{2}{3}$ full no matter which face it rests on.

CHAPTER 5 REVIEW

20. Refer to the figure at the end of the Investigation Four Parallelogram Properties for a similar example. Let 1 cm = 100 km be your scale.

23. Construct $\overline{LP}$ and copy $\angle L$. Mark off $\overline{LN}$. At point N, construct a line parallel to $\overline{LP}$.

LESSON 6.1

4. $y + y + y + 72° = 360°$

18. Calculate the slope and midpoint of $\overline{AB}$. Recall that slopes of perpendicular lines are negative reciprocals.

20.

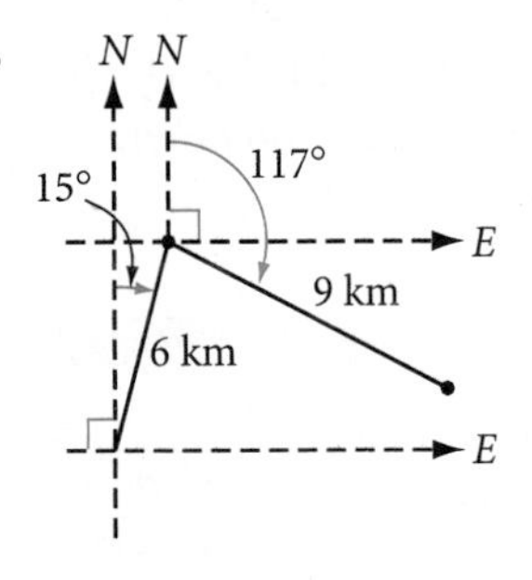

LESSON 6.2

1. $130° + 90° + w + 90° = 360°$

2. $x + x + 70° = 180°$

5. $CP = PA = AO = OR, CT = TD = DS = SR$

8. From the Tangent Conjecture, you know that the tangent is perpendicular to the radius at the point of tangency.

15. Draw a diameter. Then bisect it repeatedly to find the centers of the circles.

16. Look at the angles in the quadrilateral formed.

23. Draw the triangle formed by the three light switches. The center of the circumscribed circle would be equidistant from the three points.

24. Make a list of the powers of 3, beginning with $3° = 1$. Look for a pattern in the units digit.

25. Use a protractor and a centimeter ruler to make a careful drawing. Let 1 cm represent 1 mile.

LESSON 6.3

3. $c + 120° = 2(95°)$

4. Draw in the radius to the tangent to form a right triangle.

14. The measure of each of the five angles is half the measure of its intercepted arc. But the five arcs add up to the complete circle (360°).

15. $a = \frac{1}{2}(70°), b = \frac{1}{2}(80°), y = a + b$

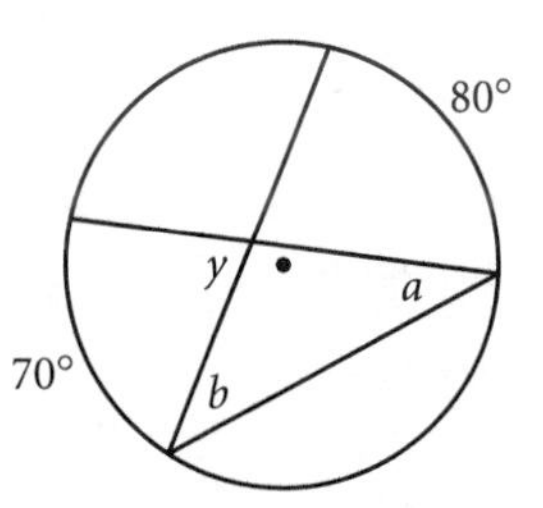

19. Draw the altitude to the point where the side of the triangle (extended if necessary) intersects the circle.

20. One possible location for the camera to get all students in the photo

Students lined up for the photo

23. Show congruent right triangles inside congruent isosceles triangles.

24. Start with an equilateral triangle whose vertices are the centers of the three congruent circles. Then locate the incenter/circumcenter/orthocenter/centroid to find the center of the larger circle.

LESSON 6.4

1. Use angle addition, arc addition, and Case 1.

4. Note that $m\widehat{YLI} + m\widehat{YCI} = 360°$.

5. $\angle 1 \cong \angle 2$ by AIA, and $\angle 1$ and $\angle 2$ are inscribed angles.

6. Apply the Cyclic Quadrilateral Conjecture.

7. Draw one diagonal and use the Inscribed Angle Conjecture.

9. Think about the pair of angles that form a linear pair and the isosceles triangle.

10. Number the points in the grid 1–9. Make a list of all the possible combinations of three numbers. Do this in a logical manner.

Order doesn't matter, so the list beginning with 2 will be shorter than the list beginning with 1. The 3 list will be shorter than the 2 list, and so on. See how many of the possibilities are collinear, and divide that by the total number of possibilities.

11. Make an orderly list. Here is a beginning:

$\overline{RA}$ to $\overline{AL}$ to $\overline{LG}$

$\overline{RA}$ to $\overline{AN}$ to $\overline{NG}$

LESSON 6.5

9. $C = 2\pi r$, $44 \approx 2(3.14)r$

12. The diameter of the circle is 6 cm.

17. Use the Inscribed Angle Conjecture and the Triangle Exterior Angle Conjecture.

LESSON 6.6

1. $\text{speed} = \frac{\text{distance}}{\text{time}} = \frac{\text{circumference}}{12 \text{ hours}}$

$= \frac{2\pi(2000 + 6400) \text{ km}}{12 \text{ hours}}$

8. Calculate the distance traveled in one revolution, or the circumference, at each radius. Multiply this by the rpm to get the distance traveled in 1 minute. Remember, your answer is in inches. You may want to divide your answer by 12 and then by 60 to change its units into feet per second.

10. Use the Inscribed Angle Conjecture and the Triangle Exterior Angle Conjecture.

LESSON 6.7

3. $\frac{210}{360}\pi(24)$

8. $m\widehat{AR} + 70° + m\widehat{AR} + 146° = 360°$,

$\frac{m\widehat{AR}}{360°}(2\pi r) = 40\pi$

9. The length of one lap is equal to $(2 \cdot 100) + (2 \cdot 20\pi)$. The total distance covered in 6 minutes is 4 laps.

11. $\frac{1}{9}(2\pi r) = 12$ meters

15. The midsegment of a trapezoid is parallel to the bases, and the median to the base of an isosceles triangle is also the altitude.

16. The overlaid figure consists of two pairs of congruent equilateral triangles. The length of the side of the smaller pair is half the length of the side of the larger pair. All of the arcs use the lengths of the sides of the triangles as radii.

17. It is not 180°. What fraction of a complete cycle has the minute hand moved since 10:00? Hasn't the little hand moved that same fraction of the way from 10:00 to 11:00?

CHAPTER 6 REVIEW

5. Draw in the radius to the tangent to form a right triangle.

8. See Lesson 6.5, Exercises 16 and 17.

10. The supplement of 88° is 92°. $\frac{1}{2}(118° + f) = 92°$.

12. $C = \pi d$, $132 = \pi d$, $d = \frac{132}{\pi}$

13. Arc length of $\widehat{AB}$ is $\left(\frac{100°}{360°}\right)\pi(54)$.

14. To find the length of $\widehat{DC}$, first find the degree measure of $\widehat{DL}$. $\frac{m\widehat{DL} - 60°}{2} = 50°$.

29. Here is how to calculate 1 nautical mile near a pole: $\frac{2\pi \cdot 6357}{360 \cdot 60}$.

30. The locus of possible locations for Dmitri is a circle with radius $5 \cdot 1100$ ft $= 5500$ ft, and the locus for Tara is a circle with radius $7 \cdot 1100$ ft $= 7700$ ft. How many times can the two circles intersect?

31. The circumference of the table is 2×100. Calculate the diameter.

LESSON 7.2

2. All positive y's become negative y's; therefore, the figure is reflected over the x-axis.

7. Compare the ordered pairs for V and V', R and R', and Y and Y'.

LESSON 7.3

7. Connect a pair of corresponding points with a segment. Construct two perpendiculars to the segment with half the distance between the two given figures between them.

16. Find the midpoint of the segment connecting the two points. Connect the midpoint to one of the endpoints with a curve. Copy the curve onto patty paper and rotate it about the midpoint.

LESSON 7.4

13.

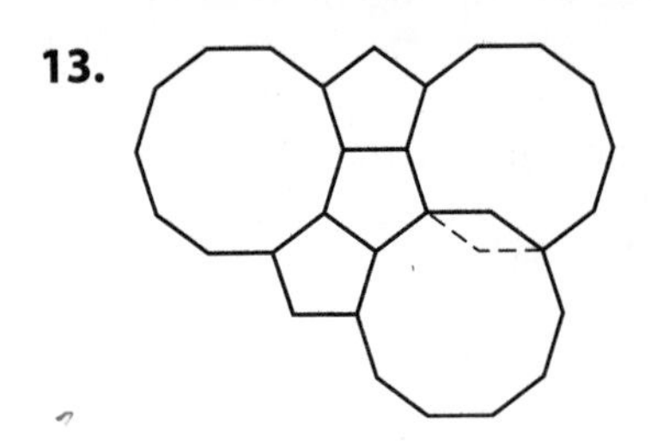

18. Work backward. Reflect a point of the 8-ball over the S cushion. Then reflect this image over the N cushion. Aim at this second image.

LESSON 7.5

2. Connect centers across the common side.

LESSON 7.6

9. If you are still unsure, use patty paper to trace the steps in the examples ("Pegasus" and *Monster Mix*).

LESSON 7.8

5. If you are still unsure, use patty paper to trace the four steps in the Escher *Horseman* example and the Escher *Symmetry Drawing E108* example.

LESSON 8.1

10. Factor 48 in two different ways.

19. Convert inches to fractions of a foot.

LESSON 8.2

8. $50 = \frac{1}{2}h(7 + 13)$

12. The area of the triangle can be calculated in three different ways, but each should give the same area: $\frac{1}{2}(5)y = \frac{1}{2}(15)x = \frac{1}{2}(6)(9)$.

22. Refer to the diagram below.

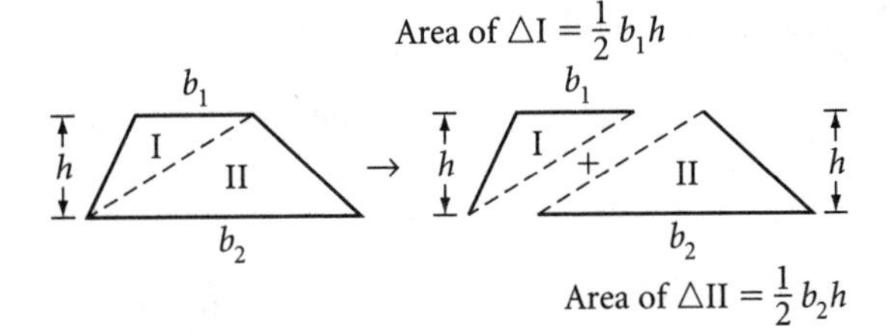

29. Draw the prism unfolded.

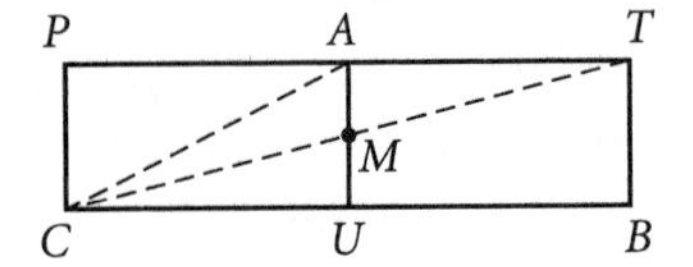

LESSON 8.3

4. Total cost is $\$20/\text{yd}^2 = \$20/9\ \text{ft}^2$; $A_{\text{carpet}} = 17 \cdot 27 - (6 \cdot 10 + 7 \cdot 9)$; 1 yd = 3 ft

8. First, find the area of all the vertical rectangles (walls). Notice that the area of the front and back triangles are the same.

LESSON 8.4

9. Construct a circle with radius 4 cm. Mark off six 4 cm chords around the circle.

10. Draw a regular pentagon circumscribed about a circle with radius 4 cm. Use your protractor to create five 72° angles from the center. Use your protractor to draw five tangent segments.

13. Find the area of the large hexagon and subtract from it the area of the small hexagon. Because they are regular hexagons, the distance from the center to each vertex equals the length of each side.

15. Divide the quadrilateral into two triangles (A and B). Find the areas of the two triangles and add them.

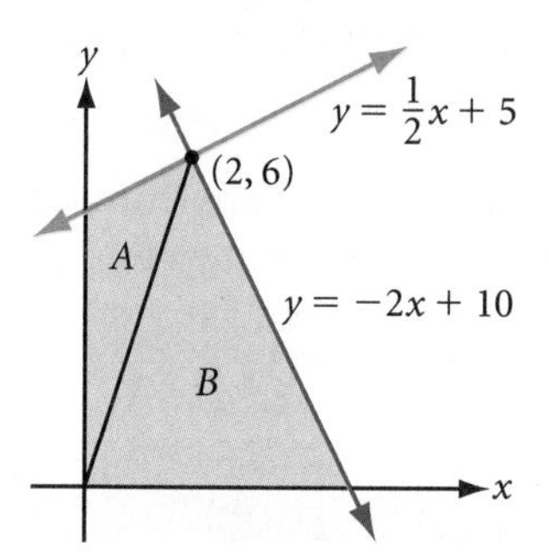

LESSON 8.5

15. Calculate the area of two circles, one with radius 3 cm and one with radius 6 cm. Compare the two areas.

16.

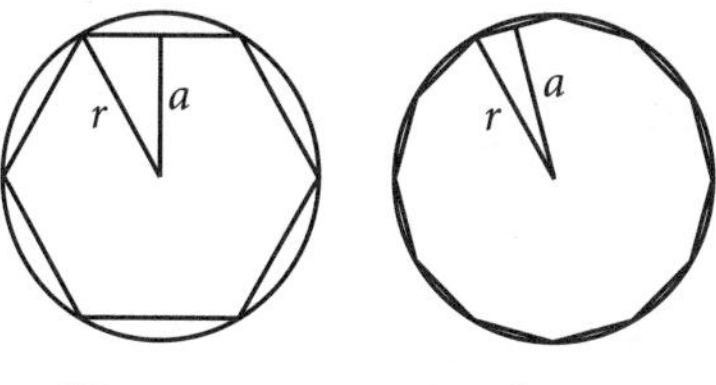

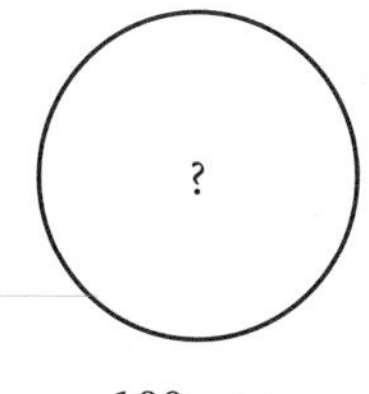

100-gon

LESSON 8.6

6. The shaded area is equal to the area of the whole circle less the area of the smaller circle.

12. $10\pi = \frac{x}{360} \cdot \pi(10^2 - 8^2)$

18. $A = \frac{1}{2} \cdot h(b_1 + b_2)$. Because the length of the midsegment is $\frac{1}{2}(b_1 + b_2)$, the formula can be rewritten $A = \textit{midsegment} \cdot \textit{height}$.

LESSON 8.7

7. Use the formula for finding the area of a regular hexagon to find the area of each base.

To find the area of the six lateral faces, imagine unwrapping the six rectangles into one rectangle. The lateral area of this "unwrapped" rectangle is the height times the perimeter.

9.

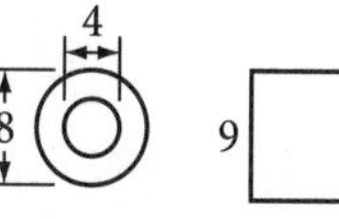

Top and bottom

$C = 8\pi$

9

Outer surface

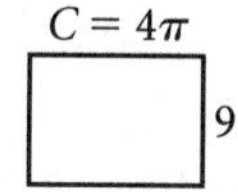

$C = 4\pi$

Inner surface

CHAPTER 8 REVIEW

37. Refer to Lesson 8.2 Project Maximizing Area.

39.

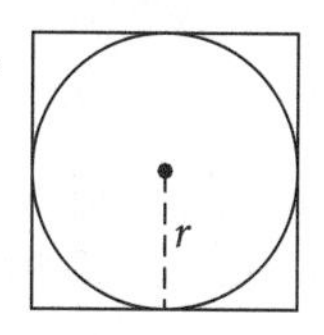

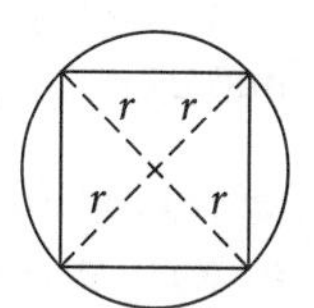

43a.

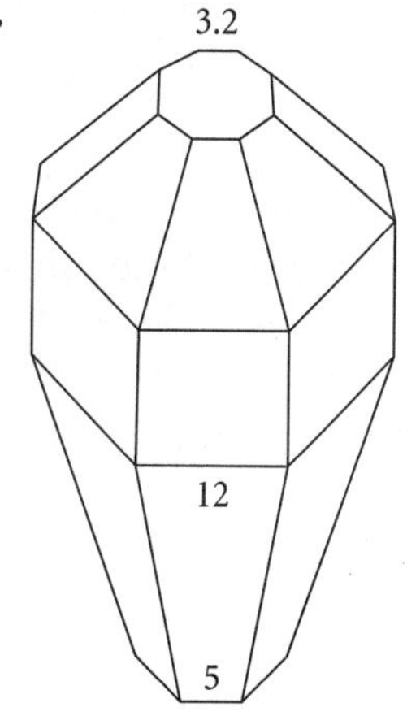

43b. Total surface area = area of octagon + 8 · area of small trapezoid + 8 · area of large trapezoid + 8 · area of square.

LESSON 9.1

6. $6^2 + 6^2 = c^2$

11. The radius of the circle is the hypotenuse of the right triangle.

13. Let s represent the length of the side of the square. Then $s^2 + s^2 = 32^2$.

15. Three consecutive integers can be written algebraically as n, $n + 1$, and $n + 2$.

17. Show that the area of the entire square (c^2) is equal to the sum of the areas of the four right triangles and the area of the smaller square.

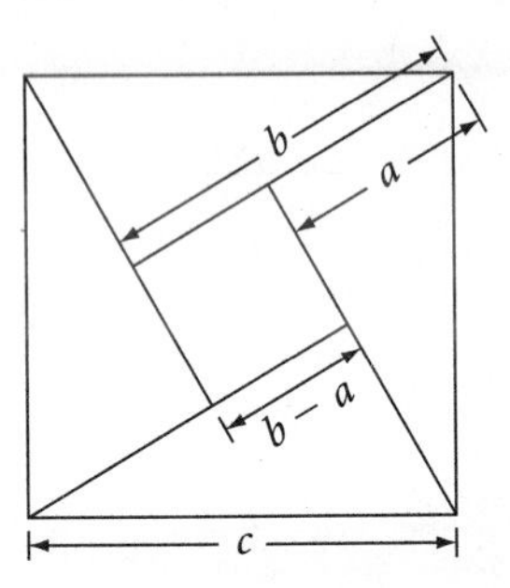

LESSON 9.2

6. $a^2 + b^2$ must exactly equal c^2.

10. Drop a perpendicular from the ordered pair to the x-axis to form a right triangle.

12. Check the list of Pythagorean triples in the beginning of this lesson for a right triangle that has three consecutive even integers.

22. Because a radius is perpendicular to a tangent, $m\angle DCF = 90°$. Because all radii in a circle are congruent, $\triangle DCE$ is isosceles.

LESSON 9.3

2. $b = \text{hypotenuse} \div \sqrt{2}$

4. $d = \frac{1}{2} \cdot 20,\ c = d \cdot \sqrt{3}$

7. Draw diagonal $\overline{DB}$ to form a right triangle on the base of the cube and another right triangle in the interior of the cube.

9. Divide by $\sqrt{2}$ for the length of the leg.

12. This is one way to show the relationship. Draw three 30°-60°-90° triangles with sides of lengths 1, $\sqrt{3}$, and 2. $6^2 - 3^2 = 27$, so $3\sqrt{3} = \sqrt{27}$.

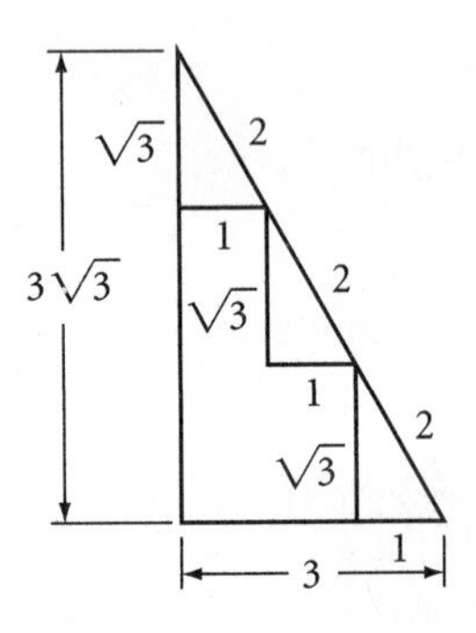

16. Draw an altitude of the equilateral triangle to form two 30°-60°-90° triangles.

19. Construct an isosceles right triangle with legs of length a; construct an equilateral triangle with sides of length $2a$ and construct an altitude; and construct a right triangle with legs of length $a\sqrt{2}$ and $a\sqrt{3}$.

22. Make the rays that form the right angle into lines. OR: Draw an auxiliary line parallel to the other parallel lines through the vertex of the right angle.

LESSON 9.4

1. The length of the hypotenuse is $(36 - x)$. Solve for x.

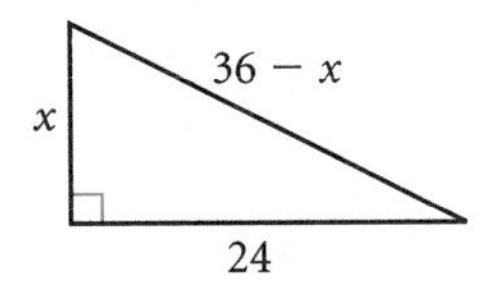

3. Average speed $= \dfrac{d}{4 \text{ hours}}$

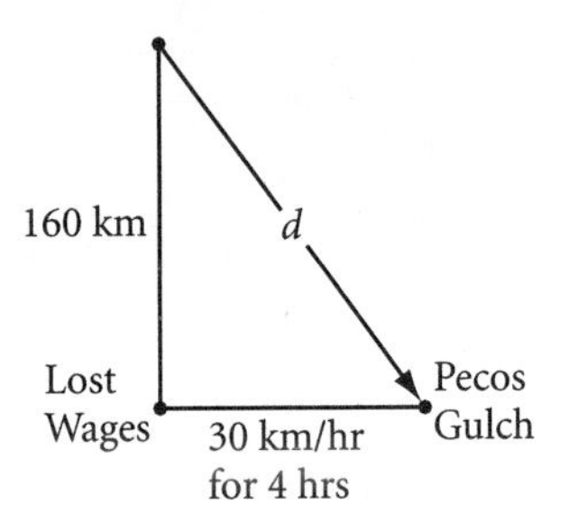

4. This is a two-step problem, so draw two right triangles. Find h, the height of the first triangle. The height of the second is 4 ft less than the height of the first. Then find x.

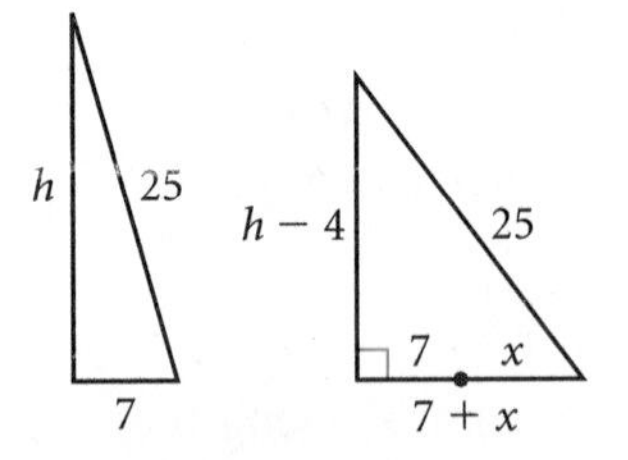

5.

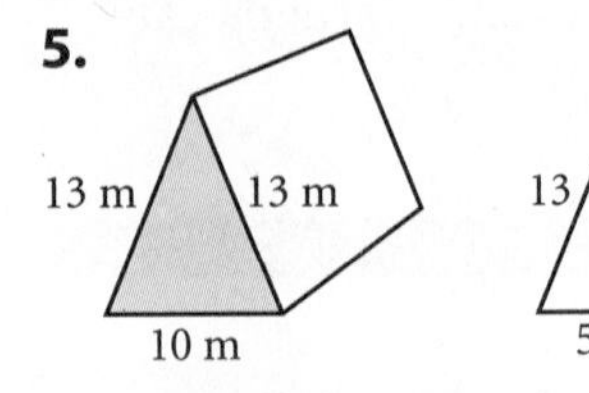

6. Find the apothem of the hexagon.

16. Use the Reflection Line Conjecture and special right triangles.

LESSON 9.5

11. Use the distance formula to find the length of the radius.

12. This is the same as finding the length of the space diagonal of the rectangular prism.

14. In the 45°-45°-90° triangle, $m = \sqrt{3} \cdot \sqrt{2}$; in the 30°-60°-90° triangle, $m = k\sqrt{3}$.

16. $(x + 8)^2 = 40^2 + x^2$

LESSON 9.6

1. Find $m\angle DOB$ using the Quadrilateral Sum Conjecture.

3. See below.

5. When $\overline{OT}$ and $\overline{OA}$ are drawn, they form a right triangle, with $OA = 15$ (length of hypotenuse) and $OT = 12$ (length of leg).

8.

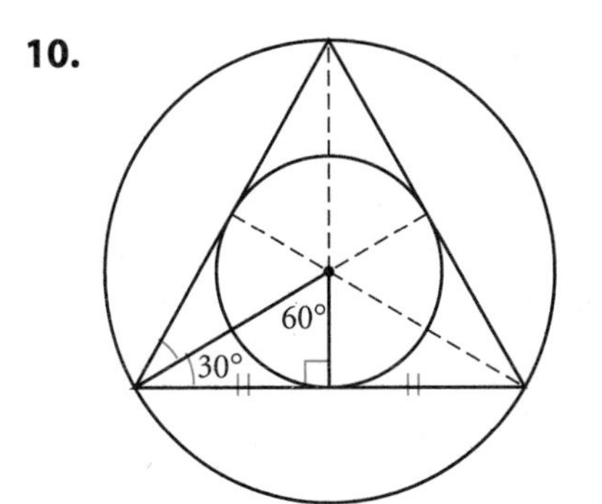

10.

14. The arc length of $\widehat{AC}$ is $\frac{80}{360}[2\pi(9)] = 4\pi$. Therefore, the circumference of the base of the cone is 4π. From this you can determine the radius of the base. The radius of the sector (9) becomes the slant height (the distance from the tip of the cone to the circumference of the base). The radius of the base, the slant height, and the height of the cone form a right triangle.

19. Rearrange the equation:

$$x^2 - 2x + 1 + y^2 = 100$$
$$(\underline{?})^2 + y^2 = 100$$

CHAPTER 9 REVIEW

10. The diameter of the semicircle is the longer leg of a 7-$\underline{?}$-25 right triangle.

12. Each half of the shaded area is equal to a quarter of a circle less the area of the isosceles right triangle.

15. $(45 \cdot 2)^2 + (60 \cdot 2)^2 = d^2$

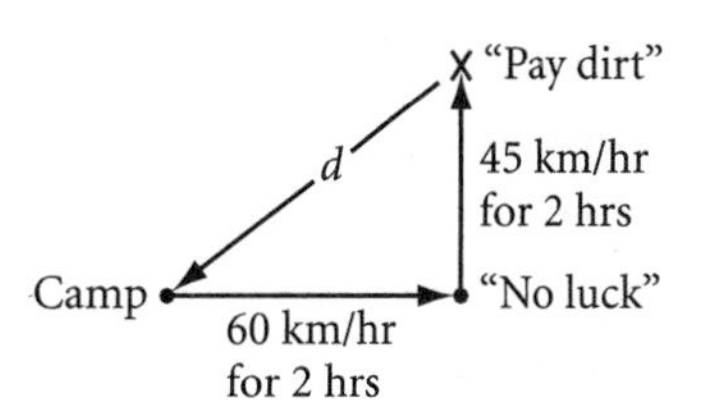

16. What will be the length of the diagonal if the shape is a rectangle?

LESSON 10.1

29. Think of a prism as a stack of thin copies of the bases.

31.

3.

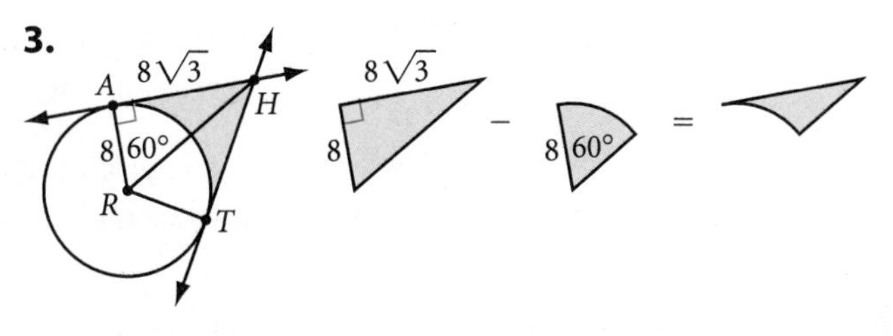

LESSON 10.2

2. $V = BH = \left(\frac{1}{2}bh\right)H$

5. You have only $\frac{1}{2}$ of a cylinder.
$V = BH = \left(\frac{1}{2}\pi r^2\right)H$

6. $\frac{90}{360}$, or $\frac{1}{4}$, of the cylinder is removed. Therefore, you need to find $\frac{3}{4}$ of the volume of the whole cylinder.

7a. What is the difference between this prism and the one in Exercise 2? Does it make a difference in the formula for the volume?

9. Cutie pie!

26. *SOTA* is a square, so the diagonals, $\overline{ST}$ and $\overline{OA}$, are congruent and are perpendicular bisectors of each other and $SM = OM = 6$. Use right triangle *SMO* to find *OS*, which also equals *OP*. Find *PA* with the equation $PA = OA - OP$.

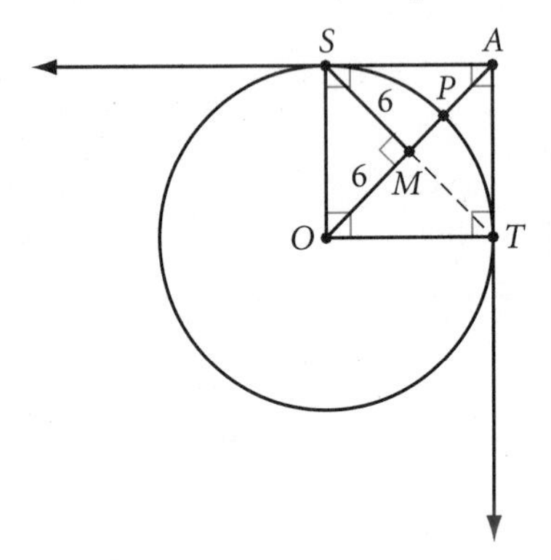

LESSON 10.3

3. $V = \frac{1}{3}BH = \frac{1}{3}\left(\frac{b_1 + b_2}{2} \cdot h\right)H$

6. $V = V_{\text{cylinder}} - V_{\text{cone}}$
$= BH - \frac{1}{3}BH$

10a. What is *B*, the area of the triangular base?

15.

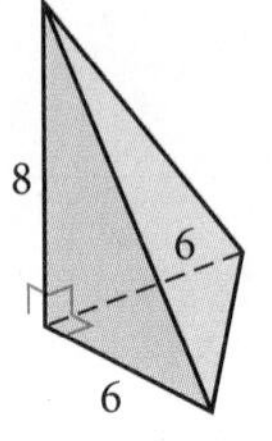

18. The swimming pool is a pentagonal prism resting on one of its lateral faces. The area of the pentagonal base can be found by dividing it into a rectangular region and a trapezoidal region.

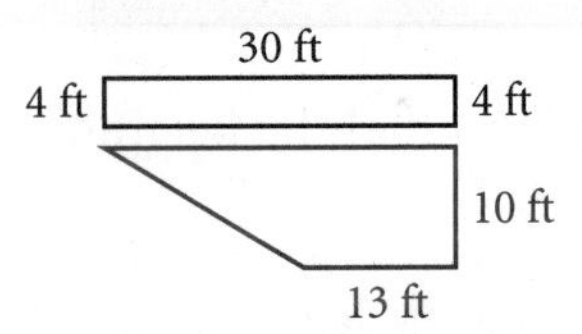

LESSON 10.4

4. $V = \frac{1}{3}BH$; $3168 = \frac{1}{3}\left[\frac{1}{2}(20 + 28)h\right](36)$

9. $V_{\text{ring}} = V_{\text{larger prism}} - V_{\text{missing prism}} =$
$\left[\left(\frac{1}{2}\right)\left(3\sqrt{3}\right)(36)\right](2) - \left[\left(\frac{1}{2}\right)\left(2\sqrt{3}\right)(24)\right](2)$
Then compare V_{ring} to $V_{\text{missing prism}}$.

10. First, change 8 inches to $\frac{2}{3}$ foot.

LESSON 10.5

6. $0.97 = \frac{145.5}{V_{\text{displacement}}}$ and $V_{\text{displacement}} = (10)(10)H$

8. $V_{\text{displacement}} = \frac{7}{8}\left(V_{\text{block of ice}}\right)$ or
$\frac{8}{7}V_{\text{displacement}} = V_{\text{block of ice}}$

LESSON 10.6

4. $V_{\text{capsule}} = 2 \cdot V_{\text{hemisphere}} + V_{\text{cylinder}} =$
$2\left[\left(\frac{2}{3}\right)\pi(6)^3\right] + \left[\pi(6)^2(12)\right]$

6. $\frac{40}{360}$, or $\frac{1}{9}$, of the hemisphere is missing. What fraction is still there?

12. $972\pi = \frac{4}{3}\pi r^3$

LESSON 10.7

1. $V = \frac{4}{3}\pi r^3$, $S = 4\pi r^2$

3. The surface area is technically the curved hemisphere *and* the circular bottom.

4. The surface area of a sphere is how many times the area of a circle with the same radius?

CHAPTER 10 REVIEW

5. $B = (12)(12) - (4)(6)$

12. One-fourth of the hemisphere is missing.

19. $V_{\text{one pipe}} = V_{\text{outer cylinder}} - V_{\text{inner cylinder}} = \pi(3)^2(160) - \pi(2.5)^2(160)$. Remember, a truck cannot carry a fraction of a pipe.

26. The volume of the hollow ball is its weight divided by its density.

27. You'll need to find the area of a segment. Also, look for a familiar triangle.

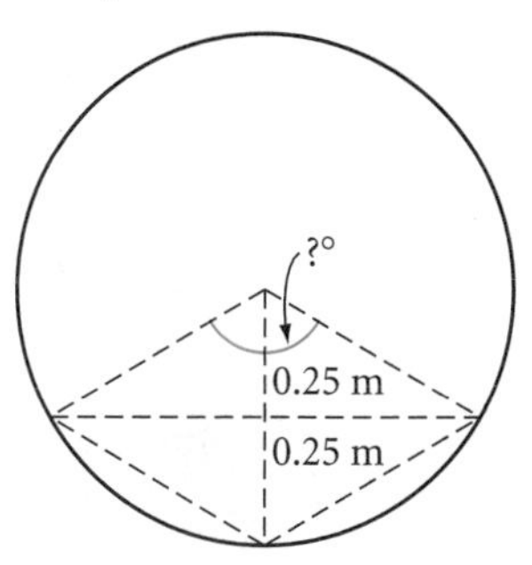

CHAPTER 11 • CHAPTER **11** CHAPTER 11 • CHAPTER

LESSON 11.1

8. All the corresponding angles are congruent. (Why?) Are all these ratios equal? $\frac{150}{165} = \underline{?}$, $\frac{120}{128} = \underline{?}$, $\frac{140}{154} = \underline{?}$, $\frac{180}{192} = \underline{?}$

13. Because the segments are parallel, $\angle B \cong \angle AED$ and $\angle C \cong \angle ADE$.

25. During a rotation, each vertex of the triangle will trace the path of a circle. So, if you connect any vertex of the original figure to its corresponding vertex in the image, you will get a chord. Now, recall that the perpendicular bisector of a chord passes through the center of a circle.

LESSON 11.2

3. It helps to rotate $\triangle ARK$ so that you can see which sides correspond.

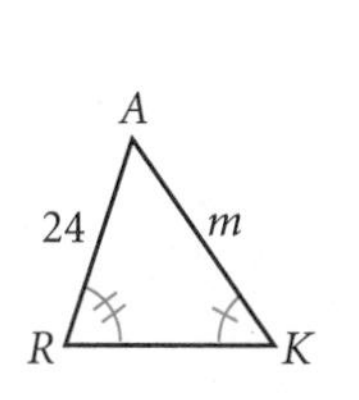

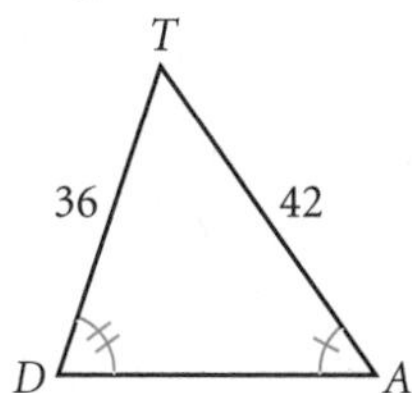

8.

11. Because $\angle H$ and $\angle D$ are two angles inscribed in the same arc, they are congruent. $\angle T$ and $\angle G$ are congruent for the same reason.

15. Draw a perpendicular segment from each point to the x-axis. Then use similar triangles.

16.

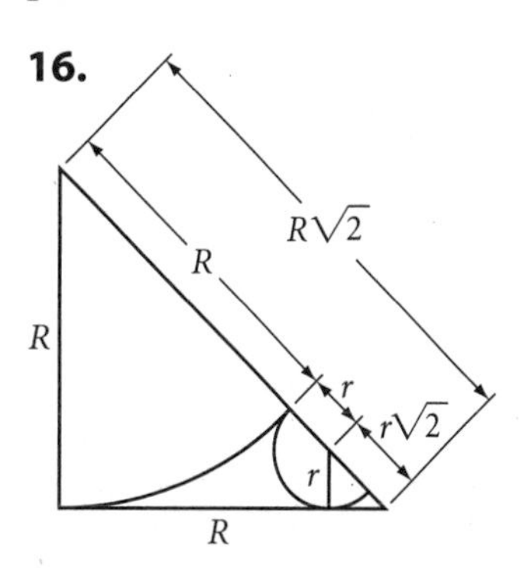

LESSON 11.3

3. Find the height of the flagpole first.

7. Because $\triangle PRE \sim \triangle POC$, then $\frac{PR}{RE} = \frac{PO}{OC}$. Let $x = PR$. Then $\frac{x}{60} = \frac{x+45}{90}$.

8.

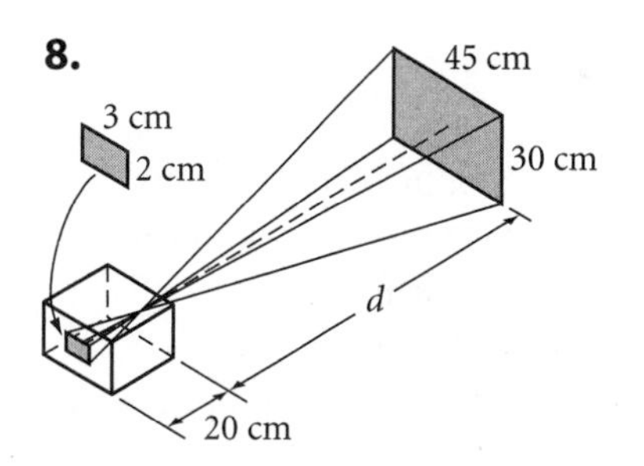

9. Think about what Juanita did in Exercise 5.

11. Draw the large and small triangles separately to label and see them more clearly.

13.

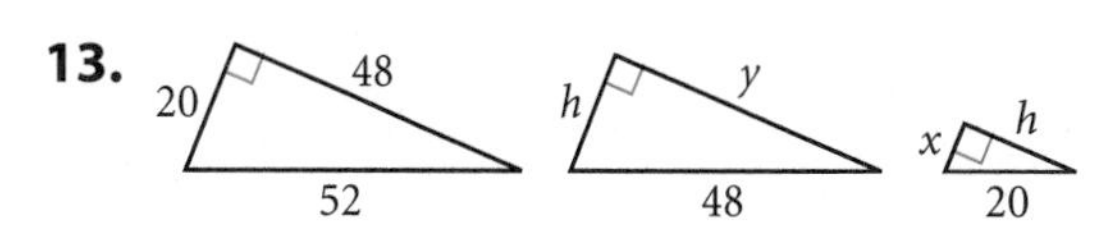

17. The golden ratio is $\frac{2}{\sqrt{5}-1}$. Let $AB = 2$ units. How would you construct the length $\sqrt{5}$? How would you construct the length $\sqrt{5} - 1$?

LESSON 11.4

3. $\frac{IC}{IE} = \frac{CS}{PE}, \frac{IC}{IE} = \frac{CL}{SE}$

9. $\frac{12}{15} = \frac{x}{10 - x}$

12. You don't know the length of the third side, but you do know the ratio of its two parts is $\frac{2}{3}$.

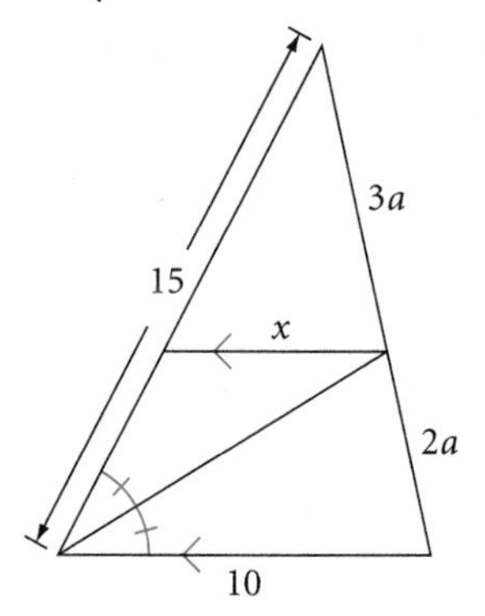

16. Bisect the angle between the sides of lengths $2x$ and $3x$.

17. Use the AA Similarity Conjecture for this proof. One pair of corresponding angles will be a nonbisected pair. The other pair of corresponding angles will consist of one-half of each bisected angle.

20. $\frac{a}{b} = \frac{c}{d}$, then $\frac{a}{b} + 1 = \frac{c}{d} + 1$, then $\frac{a}{b} + \frac{b}{b} = \frac{c}{d} + \frac{d}{d}$.

26. Copy the diagram on your own paper, and connect the centers of two large circles and the center of the small circle to form a 45°-45°-90° triangle. Each radius of the larger circles will be $\frac{s}{4}$. Use the properties of special right triangles and algebra to find the radius of the smaller circle.

LESSON 11.5

1. $\left(\frac{1}{2}\right)^2 = \frac{\text{Area of } \triangle MSE}{72}$

3. If $\frac{\text{Area of } ZOID}{\text{Area of } TRAP} = \frac{16}{25}$, then the ratio of the lengths of corresponding sides is $\frac{4}{5}$.

7.

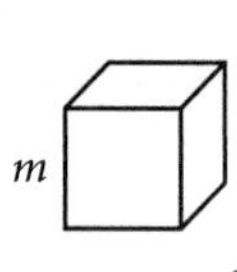

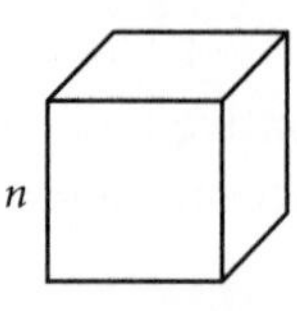

12. $\left(\frac{h}{H}\right)^2 = \frac{9}{25}; \frac{\text{Volume of large prism}}{\text{Volume of small prism}} = \left(\frac{h}{H}\right)^3$

17. Volume of large warehouse $= 2.5^3 \cdot$ (Volume of small warehouse)

26. What kind of triangle is this?

27. A cross section is a section that is perpendicular to the axis.

LESSON 11.6

1. $\frac{4}{4 + 12} = \frac{a}{20}$

3. $\frac{60}{40} = \frac{c + 60}{70}$

4. If $\frac{24}{14} = \frac{36}{d}$, then $\frac{12}{7} = \frac{36}{d}$.

6. $\frac{15}{36} \stackrel{?}{=} \frac{25}{55}$

12. $a + b = \sqrt{(12 - 3)^2 + (0 - 9)^2}$. Once you have solved for $a + b$, then $\frac{4}{12} = \frac{a}{a + b}$.

17. $\frac{x}{10} = \frac{x + 12}{16}$; x is the height of the small missing cone.

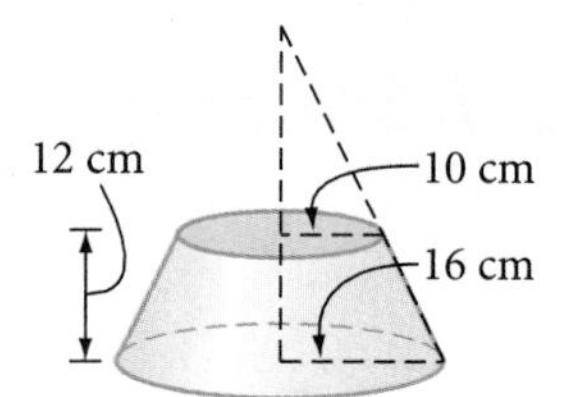

CHAPTER 11 REVIEW

9. Divide $\overline{KL}$ into seven equal lengths.

10. $\triangle ABE \sim \triangle ADC, \frac{AB}{BE} = \frac{AD}{CD}$

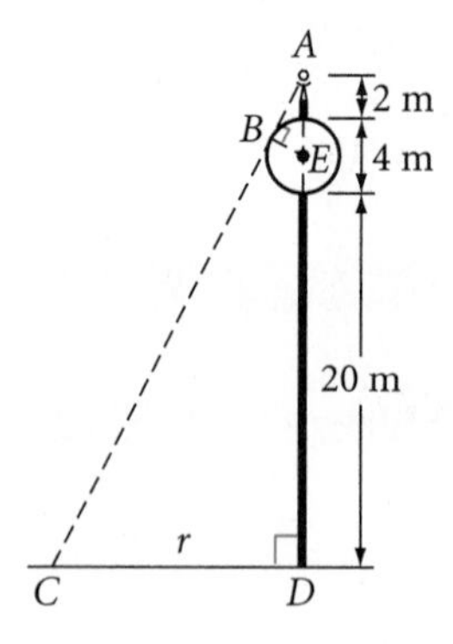

19. Because you are concerned with ratios, it doesn't make any difference what lengths you choose. So, for convenience, assign the square a side of length 2 before calculating the areas and volumes.

CHAPTER 12 • CHAPTER **12** CHAPTER 12 • CHAPTER

LESSON 12.1

7. The length of the side opposite $\angle A$ is s; the length of the side adjacent to $\angle A$ is r; the length of the hypotenuse is t.

10. Use your calculator to find $\sin^{-1}(0.5)$.

14. $\tan 30° = \frac{20}{a}$

21. Use $\sin 35° = \frac{b}{85}$ to find the length of the base, and then use $\cos 35° = \frac{h}{85}$ to find the height.

27. First, find the length of the radius of the circle and the length of the segment between the chord and the center of the circle. Use the Pythagorean Theorem to find the length of half of the chord.

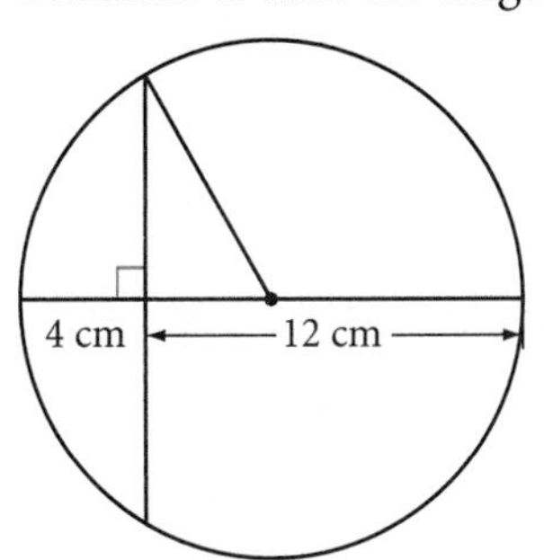

LESSON 12.2

6a. $\sin 44° = \frac{h}{1400}$, where h is the height of the balloon *above* Wendy's sextant.

7.

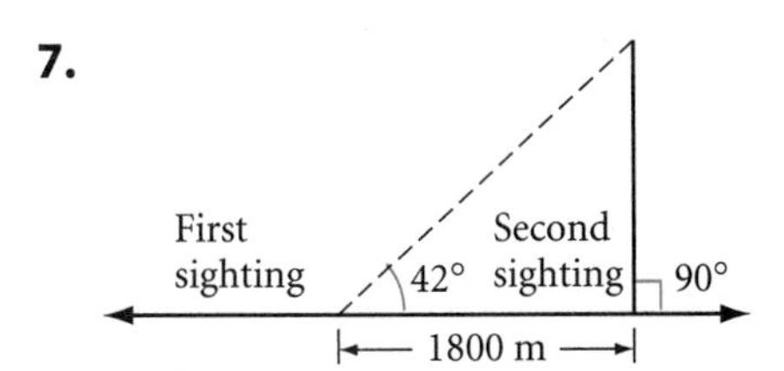

12. $\frac{\left(\frac{1}{2}\right)d_1}{20} = \cos 56°$

15. $\tan\beta = \frac{10}{17}$

LESSON 12.3

3. Sketch a diagonal connecting the vertices of the unmeasured angles. Then find the area of the two triangles.

4. Divide the octagon into eight isosceles triangles. Then use trigonometry to find the area of each triangle.

5. $\frac{\sin 52°}{28°} = \frac{\sin 79°}{w}$

13. $\sin 16° = \frac{a}{18}$

$\cos 16° = \frac{b}{18}$

$\tan 68° = \frac{c}{b}$

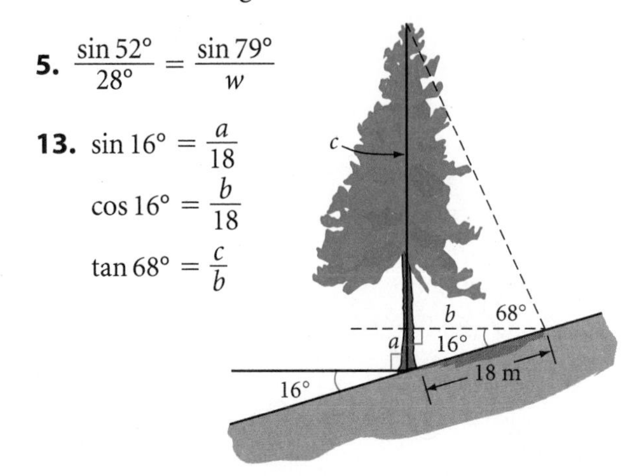

LESSON 12.4

1. $w^2 = 36^2 + 41^2 - 2(36)(41)\cos 49°$

4. $42^2 = 34^2 + 36^2 - 2(34)(36)\cos A$

8. The smallest angle is opposite the shortest side.

10. One approach divides the triangle into two right triangles, where $x = a + b$. When you have the straight-line distance, compare that time against the detour.

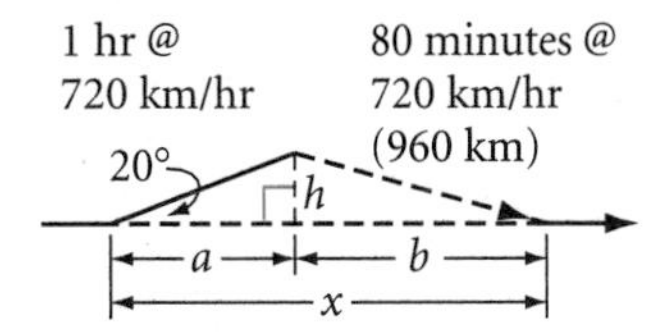

12.

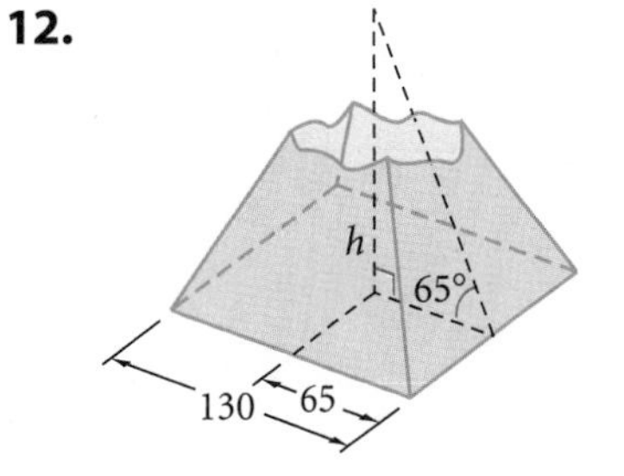

Hints for Selected Exercises

16. The midpoint of the base can be used to create three equilateral triangles.

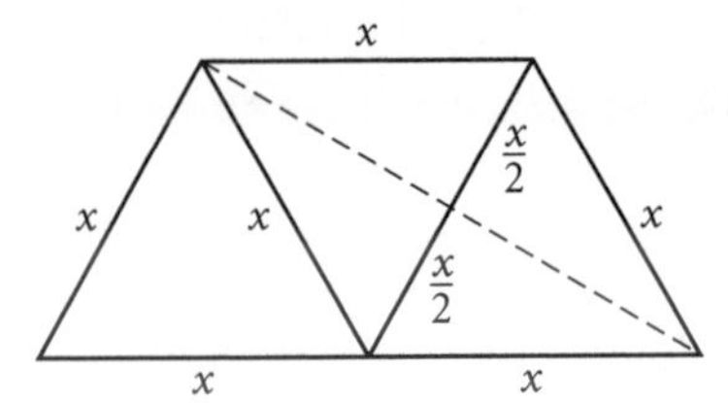

LESSON 12.5

2.

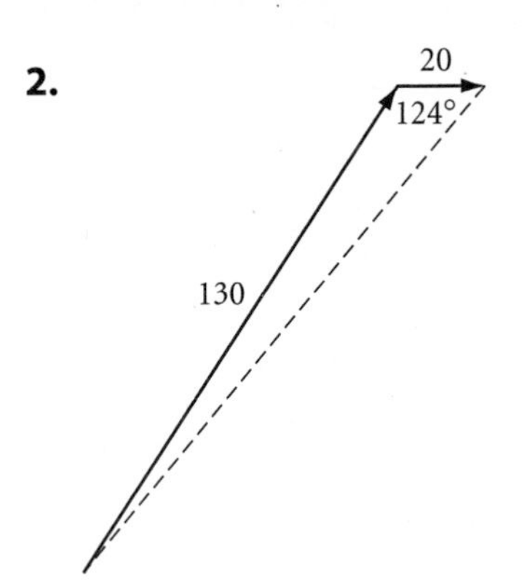

5. First find θ.

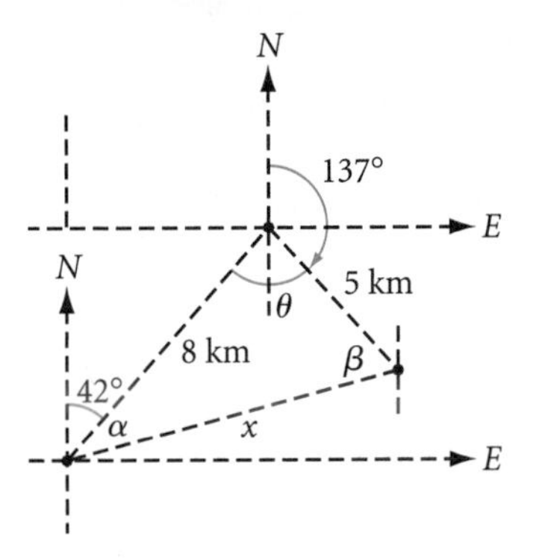

9.

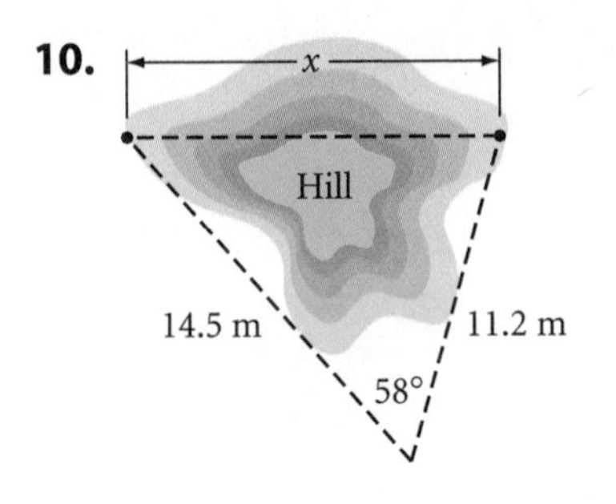

10.

11. Divide the pentagon into five congruent triangles. What is the measure of the interior angle of each triangle at the vertex in the center?

CHAPTER 12 REVIEW

27. $A_{segment} = A_{sector} - A_{triangle}$. You found part of the solution in Exercise 25.

28.

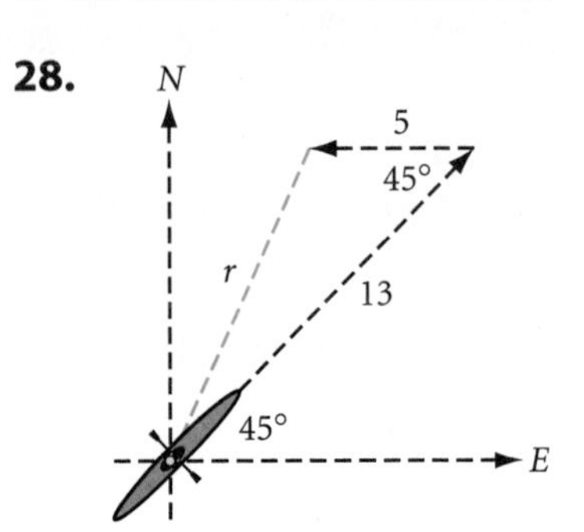

CHAPTER 13 • CHAPTER 13 CHAPTER 13 • CHAPTER

LESSON 13.1

4. It's also called the identity property.

8. Distributive property, subtraction property of equality, _?_, _?_

9. _?_, _?_, multiplication property of equality, _?_

11. The Midpoint Postulate says that a segment has exactly one midpoint.

22. Reason for Box 1: Angle Bisector Postulate

25. Two consecutive integers can be written as n and $n + 1$.

LESSON 13.3

2. Draw an auxiliary line from the point to the midpoint of the segment.

5. Draw an auxiliary line that creates two isosceles triangles. Use isosceles triangle properties and angle subtraction.

14. Use similarity of triangles to set up two proportions.

16. Graph point A and the line. Fold the graph paper along the line to see where point A reflects.

LESSON 13.4

1. Use the Quadrilateral Sum Theorem.

16.

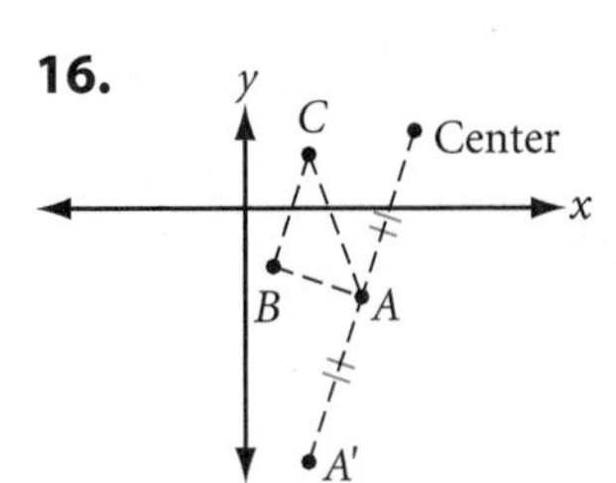

LESSON 13.6

14. Divide the triangle into three triangles with common vertex *P*.

16. Draw a segment from the point to the center of the circle. Using this segment as a diameter, draw a circle. How does this help you find the points of tangency?

LESSON 13.7

4.

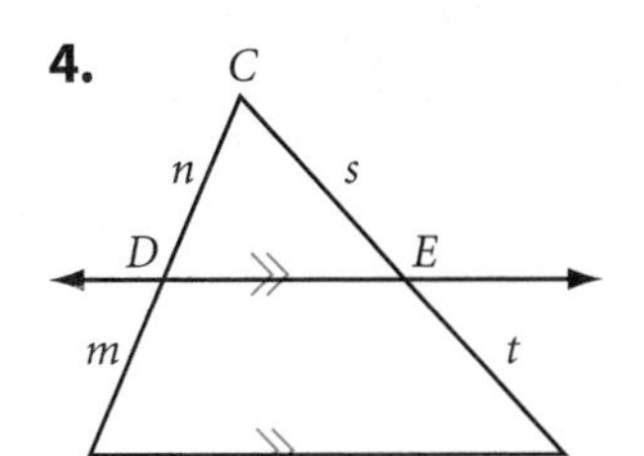

$$\frac{m+n}{n} = \frac{t+s}{s}$$

$$\frac{m}{n} + \frac{n}{n} = \frac{t}{s} + \frac{s}{s}$$

$$\frac{m}{n} + 1 = \frac{t}{s} + 1$$

$$\therefore \frac{m}{n} = \frac{t}{s}$$

5. Try revising the steps of the proof for the Parallel Proportionality Theorem.

6. In $\triangle ATH$, let $m\angle A = a$, $m\angle T = t$, then $a + t = 90$. In $\triangle LTH$, $m\angle T = t$, $m\angle H = h$, then $h + t = 90$. Therefore, $h + t = a + t$ and so $h = a$. Therefore, $\triangle ATH \sim \triangle HTL$ by AA. In like manner you can demonstrate that $\triangle ATH \sim \triangle AHL$, and thus by the transitive property of similarity all three are similar.

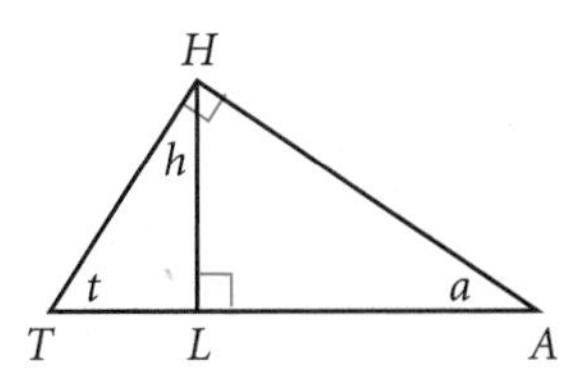

7. If *a* is the geometric mean of *b* and *c*, then $\frac{b}{a} = \frac{a}{c}$, or $a^2 = bc$.

8. Construct $\overline{AD}$ so that it's perpendicular to $\overline{AB}$ and intersects $\overleftrightarrow{BC}$ at point *D*.

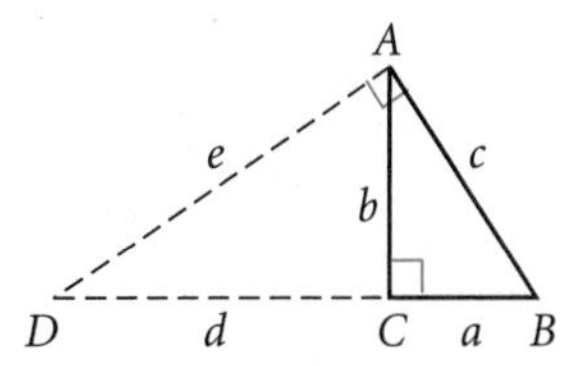

Use the Three Similar Right Triangles Theorem to write proportions and solve.

9.

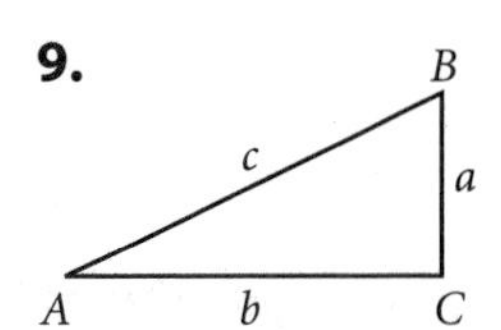

Plan: Construct a right triangle *DEF* with legs of lengths *a* and *b* and hypotenuse of length *x*. Then, $x^2 = a^2 + b^2$ by the Pythagorean Theorem. It is given that $c^2 = a^2 + b^2$. Therefore, $x^2 = c^2$, or $x = c$.

If $x = c$, then $\triangle DEF \cong \triangle ABC$.

10. See Exercise 18 in Lesson 9.1.

19.

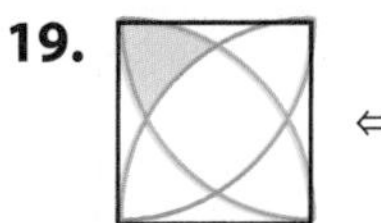
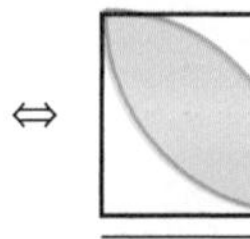
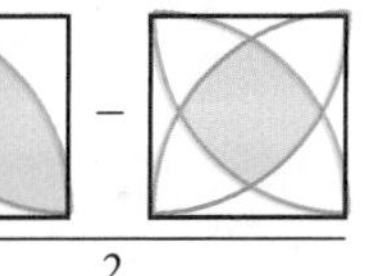

20.

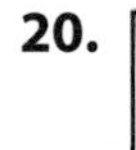
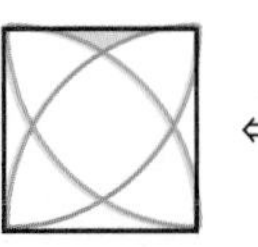
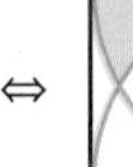
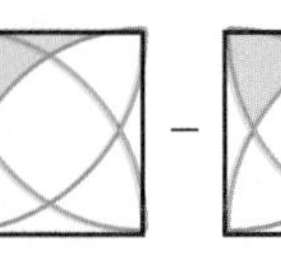

CHAPTER 13 REVIEW

26.

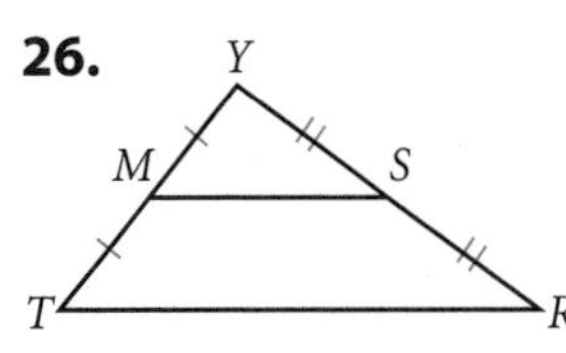

Given: $\triangle TRY$ with midsegment $\overline{MS}$

Show: $\overline{MS} \parallel \overline{TR}$ and $MS = \left(\frac{1}{2}\right)TR$

Hints for Selected Exercises

Index

A

AA Similarity Conjecture, 572
AA Similarity Postulate, 706
AAA (Angle-Angle-Angle), 219, 572, 574
Acoma, 62
Activities
Activity: Boxes in Space, 173–175
Activity: Calculating Area in Ancient Egypt, 453–454
Activity: Chances Are, 86–87
Activity: Convenient Sizes, 599–602
Activity: Dilation Creations, 578–580
Activity: Dinosaur Footprints and Other Shapes, 431–432
Activity: Elliptic Geometry, 719–720
Activity: Exploring Properties of Special Constructions, 696–697
Activity: Exploring Star Polygons, 264–265
Activity: Isometric and Orthographic Drawings, 541
Activity: It's Elementary!, 552–553
Activity: Modeling the Platonic Solids, 528–530
Activity: Napoleon Triangles, 247–248
Activity: Prove It!, 657–658
Activity: The Right Triangle Fractal, 481
Activity: The Sierpiński Triangle, 136–137
Activity: Symbolic Proofs, 613
Activity: Three Out of Four, 189–190
Activity: Toothpick Polyhedrons, 512–513
Activity: Traveling Networks, 118–119
Activity: Turning Wheels, 346–347
Activity: The Unit Circle, 651–654
Activity: Using a Clinometer, 632–633
Activity: Where the Chips Fall, 442–444
acute angle, 49
acute triangle, 60, 636, 641–642
addition of angles, 672
addition, properties of, 670
addition property of equality, 670
adjacent interior angles, 215–216
adjacent side, 620
AEA Conjecture, 127
Africa, 16, 18, 51, 124, 364
agriculture, 70, 435, 452, 453, 548, 648, 649
AIA Conjecture, 127, 129
AIA Theorem, 681
AIDS Memorial Quilt, 519
aircraft travel, 280–281, 337, 338, 340, 392, 423, 570, 644, 648, 649, 660
Alexander the Great, 18
algebra
properties of, in proofs, 670–671, 712
See also **Improving Your Algebra Skills; Using Your Algebra Skills**
Algeria, 567
Almagest (Ptolemy), 620
alternate exterior angles, 126–129
Alternate Exterior Angles Conjecture (AEA Conjecture), 127
alternate interior angles, 126–129, 681
Alternate Interior Angles Conjecture (AIA Conjecture), 127, 129
Alternate Interior Angles Theorem (AIA Theorem), 681
altitude
of a cone, 508
of a cylinder, 407
of a parallelogram, 412
of a prism, 506
of a pyramid, 506
of a triangle, 154, 177, 401, 586
See also orthocenter
Altitude Concurrency Conjecture, 177
analytic geometry. *See* coordinate geometry
Andrade, Edna, 24
angle(s)
acute, 49
addition of, 672
and arcs, 308, 319–320
base. *See* base angles
bearing, 312
bisectors of, 40, 101, 157–158, 587–588
central. *See* central angle
complementary, 50
congruent, 50
consecutive. *See* consecutive angles
corresponding, 126, 128, 673
definitions of, 38, 49–50
duplicating, 144
exterior. *See* exterior angles
included, 219
incoming, 41
inscribed, 307, 319–320, 325–326, 492
interior. *See* interior angles
linear, 50, 120–122
measure of, 39–40
naming of, by vertex, 38
obtuse, 49, 101
outgoing, 41
picture angle, 323
proving conjectures about, 680–681, 686–687
right, 49
sides of, 38
supplementary, 50
symbols for, 38
tessellations and total measure of, 379–380
transversals forming, 126–129
vertical, 50, 121, 679–680
See also polygon(s); *specific polygons listed by name*
Angle Addition Postulate, 672
Angle-Angle-Angle (AAA) case, 219, 572, 574
angle bisector(s)
of an angle, 40, 101, 157–158, 587–588
incenter, 177–179
of a triangle, 176–179, 586, 587–588
of a vertex angle, 242–243
Angle Bisector Concurrency Conjecture, 176
Angle Bisector Conjecture, 157
Angle Bisector/Opposite Side Conjecture, 588
Angle Bisector Postulate, 672
Angle Bisector Theorem, 686–687
Angle Duplication Postulate, 672
angle of depression, 627
angle of elevation, 627
angle of rotation, 359
Angle-Side-Angle (ASA), 219, 225–226, 227, 574
Angles Inscribed in a Semicircle Conjecture, 320
Angola, 18, 124
angular velocity, 344
annulus, 437
Another World (Other World) (Escher), 667

antecedent, 655–656
antiprism(s), 510
Apache, 3
aperture, 264
apothem, 426
applications
agriculture, 70, 435, 452, 453, 548, 648, 649
aircraft travel, 280–281, 337, 338, 340, 392, 423, 570, 644, 648, 649, 660
archaeology, 258, 276, 301, 311, 493, 638, 644, 664
architecture, 6, 9, 12, 15, 23, 25, 55, 65, 70, 83, 134, 159, 174, 198, 204, 271, 278, 319, 345, 379, 386, 405, 414, 428, 445, 448, 449, 505, 520, 539, 548, 564, 610, 631, 660
astronomy, 40, 101, 341, 618, 620
biology, 15, 18, 268, 326, 435, 490, 597, 599–601
botany, 334, 338
business, 212, 570, 595, 609
chemistry, 9, 95, 117, 124, 187, 246, 504, 520, 536, 537, 545, 617
computers, 235
construction and maintenance, 76, 123, 134, 222, 245, 259, 278, 291, 298, 301, 318, 335, 407, 410, 414, 420, 422, 423, 424, 425, 440, 451, 457, 459, 470, 483, 495, 519, 526, 548, 556, 615, 645, 660, 685
consumer awareness, 286, 338, 458, 459, 482, 497, 518, 532, 533, 544, 555, 596, 626, 722
cooking, 458, 498, 556, 609
design, 24, 63, 70, 116, 131, 152, 175, 179, 180, 187, 193, 302, 339, 344, 351, 355, 383, 386, 387, 388, 389–391, 406, 419, 423, 424, 428, 458, 465, 483, 494, 525, 526, 533, 538, 544, 545, 549, 562, 569, 570, 608, 648, 650
distance calculations, 77, 91, 104, 150, 163, 223, 233, 240, 246, 276, 277, 318, 344, 351, 371, 391, 397, 482, 497, 498, 525, 562, 582–583, 584, 614, 621, 627, 628, 629, 638–639, 644, 645, 648, 649, 660, 677, 689
earth science, 549
engineering, 283, 291, 493, 607
environment, 519, 526
flags, 4, 26
forestry, 633
geography, 351, 442, 525, 548
government, 675
landscape architecture, 424
law and law enforcement, 83, 100, 414
manufacturing, 34, 220, 458
maps and mapping, 36, 311, 567
medicine and health, 39, 110
meteorology and climatology, 101, 334, 483, 628
moviemaking, 563, 601
music, 339
navigation, 217, 351, 628, 629, 630, 645, 648, 660
oceanography, 576, 653
optics, 45, 52, 271, 326
photography, 263, 323, 583
physics, 45, 46, 52, 185, 271, 314, 483–484, 520, 535–537, 556, 601–602
public transit, 87
resources, consumption of, 423, 596, 649
seismology, 229
sewing and weaving, 9, 10, 14, 49, 51, 56, 60, 289, 298, 299, 415
shipping, 400, 420, 644
sports, 185, 186, 286, 315, 345, 363, 367–368, 369–370, 371, 383, 406, 435, 443, 498, 501, 562
surveying land, 36, 94, 453, 469, 649
technology, 52, 112, 131, 235, 263, 271, 283, 314, 317, 323, 339, 351, 423, 435, 498, 583, 607, 628, 630
telecommunications, 112, 435, 498
velocity and speed calculations, 134, 293, 302, 337, 338, 340, 344, 345, 351, 392, 483, 497, 660, 661
zoology and animal care, 15, 435, 536, 576, 694
approximately equal to, symbol for, 40
Arbus, Diane, 373
arc(s)
addition of, 703
and angles, 308, 319–320
and chords, 308
defined, 68
endpoints of, 68
inscribed angles intercepting, 320
length of, and circumference, 341–343
major, 68, 308
measure of, 68, 319–321, 341–343
minor, 68
nautical mile and, 351
parallel lines intercepting, 321
semicircles, 68, 320, 492
symbol, 68
Arc Addition Postulate, 703
Arc Length Conjecture, 342
archaeology, 258, 276, 301, 311, 493, 638, 644, 664
arches, 271, 278, 319
Archimedean screw, 648
Archimedean tilings, 381
Archimedes, 381, 535, 536, 616
architecture, 6, 9, 12, 15, 23, 25, 55, 65, 70, 83, 134, 159, 174, 198, 204, 271, 278, 319, 345, 379, 386, 405, 414, 428, 445, 448, 449, 505, 520, 539, 548, 564, 610, 631, 660
Archuleta-Sagel, Teresa, 265
area
ancient Egyptian formula for, 453
ancient Greek formula for, 454
of an annulus, 437
of a circle, 433–434, 437–438, 492
defined, 410
of irregularly shaped figures, 411, 422, 430–432
of a kite, 418
maximizing, 421
of a parallelogram, 412–413
Pick's formula for, 430–432
of a polygon, 411
probability and, 442–444
proportion and, 592–593
of a rectangle, 411
of a regular polygon, 426–427
of a sector of a circle, 437–438
of a segment of a circle, 437–438
surface. *See* surface area
of a trapezoid, 417–418
of a triangle, 411, 417, 454, 634–635
units of measure of, 413
Arendt, Hannah, 634
Around the World in Eighty Days (Verne), 337
art
circle designs, 10–11
Dada Movement, 260
geometric patterns in, 2–4, 9, 19
Islamic tile designs, 20–22
knot designs, 2, 16–17, 21, 396
line designs, 7–8
mandalas, 25, 576
murals, 571
op art, 3, 13–14, 66
perspective, 172
proportion and, 577, 592, 597
symmetry in, 3–4, 5
See also drawing
ASA (Angle-Side-Angle), 219, 225–226, 227, 574
ASA Congruence Conjecture, 225
ASA Congruence Postulate, 673
Asian art/architecture. *See* China; India; Japan

Assessing What You've Learned
Giving a Presentation, 304, 356, 408, 460, 502, 558, 618, 666, 723
Keeping a Portfolio, 26, 92, 140, 196, 254, 304, 356, 408, 460, 502, 558, 618, 666, 723
Organize Your Notebook, 82, 140, 196, 254, 304, 356, 408, 460, 502, 558, 618, 666, 723
Performance Assessment, 196, 254, 304, 356, 460, 480, 558, 723
Write in Your Journal, 140, 196, 254, 304, 356, 408, 460, 502, 558, 666, 723
Write Test Items, 254, 356, 460, 502, 558
associative property
of addition, 670
of multiplication, 670
astronomy, 40, 101, 341, 618, 620
Australia, 388
Austria, 65
auxiliary line, 200
Axis Dance Company, 660
axis of cylinder, 507
Aztecs, 25

B

Babylonia, 94, 463, 469
Bacall, Lauren, 266
Bakery Counter (Thiebaud), 514
Bankhead, Tallulah, 426
base(s)
of a cone, 508
of a cylinder, 507
of an isosceles triangle, 62
of a prism, 445, 506
of a pyramid, 445, 506
of a rectangle, 411
of solids, 445
of a trapezoid, 267
base angles
of an isosceles trapezoid, 269
of an isosceles triangle, 62, 205–206
of a trapezoid, 267
Bauer, Rudolf, 177
Baum, L. Frank, 475
bearing, 312
"Behold" proof, 466
Belvedere (Escher), 619
Benson, Mabry, 299
Berra, Yogi, 73, 437
Bhaskara, 466
biconditional statements, 243
bilateral symmetry, 3
billiards, geometry of, 41–42, 45
binoculars, 271
biology, 15, 18, 268, 326, 435, 490, 597, 599–601
bisector
of an angle, 40, 101, 157–158, 587–588
of a kite, 267
perpendicular. *See* perpendicular bisectors
of a segment, 31, 147–149
of a triangle, 176–179, 586, 587–588
of a vertex angle, 242–243
Blake, William, 679
body temperature, 599–600
Bolyai, János, 718
Bookplate for Albert Ernst Bosman (Escher), 73
Borromean Rings, 18
Borromini, Francesco, 174
botany, 334, 338
Botswana, 364
Boulding, Kenneth, 54
Boy With Birds (Driskell), 303
Braun, Wernher von, 620
Breezing Up (Homer), 629
Brewster, David, 378
Brickwork, Alhambra (Escher), 389
Bruner, Jerome, 384
Buck, Pearl S., 398
buoyancy, 537
business, 212, 570, 595, 609

C

CA Conjecture (Corresponding Angles Conjecture), 127
CA Postulate (Corresponding Angles Postulate), 673
carbon molecules, 504
Carroll, Lewis, 47, 366, 612
Carson, Rachel, 522
Cartesian coodinate geometry. *See* coordinate geometry
ceiling of cloud formation, 628
Celsius conversion, 210
Celtic knot designs, 2, 16
center
of a circle, 67, 309, 325–326
of a sphere, 507
center of gravity, 46, 184–185
Center of Gravity Conjecture, 185
center of rotation, 359
central angle
of a circle, 68, 307–308
of a regular polygon, 272
centroid, 183–185, 189–190
finding, 402
Centroid Conjecture, 184
Cézanne, Paul, 504
Chagall, Marc, 55
chemistry, 9, 95, 117, 124, 187, 246, 504, 520, 536, 537, 545, 617
Chichén Itzá, 639
China, 10, 30, 36, 316, 319, 333, 463, 479, 484, 502, 520, 567
Chokwe, 18
Cholula pyramid, 527
chord(s)
arcs and, 308
center of circle and, 309
central angles and, 307–308
defined, 69
perpendicular bisectors of, 309–310
properties of, 307–310
Chord Arc Conjecture, 308
Chord Central Angles Conjecture, 308
Chord Distance to a Center Conjecture, 309
Christian art and architecture, 12, 159, 174, 198, 345, 405
circle(s)
annulus of concentric, 437
arcs of. *See* arc(s)
area of, 433–434, 437–438, 492
center of, 309, 325–326
central angles of, 68, 307–308
chords of. *See* chord(s)
circumference of, 331–333
circumscribed, 178, 179
concentric, 68
congruent, 68
cycloid, 347–348
defined, 67
diameter of. *See* diameter
epicycloid, 348
equations of, 488–489
externally tangent, 315
inscribed, 179
inscribed angles of, 307, 319–320, 325–326
internally tangent, 315
in nature and art, 2, 10–11
proofs involving, 699–700, 703–704
proving properties of, 325–326
and Pythagorean Theorem, 488–489, 492
radius of, 67
rectifying a, 440
secants of, 321
sectors of, 352, 437–438
segments of, 437–438
tangent, 315
tangents to. *See* tangent(s)
trigonometry using, 651–654
Circle Area Conjecture, 434
circumcenter
defined, 177
finding, 329–330

properties of, 177–178, 189–190
in right triangle, 180
Circumcenter Conjecture, 177–178
circumference, 331–333
Circumference Conjecture, 332
circumference/diameter ratio, 331–333
circumscribed circle, 178, 179
circumscribed triangle, 71, 179
Clark, Karen Kaiser, 572
Claudel, Camille, 597
clinometer, 632–633
clockwise rotation, 359
Clothespin (Oldenburg), 615
coinciding patty papers, 147
Colette, Sidonie Gabriella, 319
collinear points, 30
commutative property
of addition, 670
of multiplication, 670
compass, 7, 143
sector, 608
complementary angles, 50
composition of isometries, 374–376
computer programming, 235
computers, 235
concave kite. *See* dart
concave polygon, 54
concentric circles, 68
concept map. *See* tree diagram; Venn diagram
conclusion, 100, 102
concurrency, point of. *See* point(s) of concurrency
concurrent lines, 115, 176–179
conditional proof, 655–656
conditional statement(s), 551–552
converse of, 122, 611
forming, 679
inverse of, 611
and Law of the Contrapositive, 611–612
proving. *See* logic; proof(s)
symbol for, 552
cone(s)
altitude of, 508
base of, 508
defined, 508
drawing, 81, 82
frustrums of, 451
height of, 508
radius of, 508
right, 508
similar, 593
surface area of, 448–449
vertex of, 508
volume of, 522–524
congruence
of angles, 50
ASA, 225, 227, 673
of circles, 68
CPCTC (corresponding parts of congruent triangles are congruent), 230–231
defined, 671
diagramming of, 59
of polygons, 55
as premise of geometric proofs, 671
properties of, 671
SAA, 226, 227, 686
SAS, 221, 227, 673
of segments, 31
SSS, 220, 227, 673
symbols for, and use of, 31, 40, 59
of triangle(s), 168–169, 219–222, 225–227, 230–231
conjectures
data quality and quantity and, 96
defined, 94
proving. *See* proof(s)
See also postulates of geometry; *specific conjectures listed by name*
Connections
architecture, 9, 23, 198, 204, 271
art, 180, 345, 362, 415, 563
career, 34, 220, 235, 289, 359, 407, 422, 424, 428, 483, 519, 539, 570
consumer, 4
cultural, 25, 341, 364, 386, 414
geography, 351
history, 36, 245, 331, 360, 452, 463, 466, 469, 536, 567, 577, 638, 639, 668
language, 94, 166
literature, 337, 612
mathematics, 74, 142, 333, 381, 385, 440, 673
recreation, 63, 217, 339, 484
science, 18, 40, 52, 101, 117, 185, 271, 326, 334, 484, 504, 520, 620, 628
technology, 263, 283, 314, 498, 630, 648
consecutive angles
defined, 54
of a parallelogram, 280
of a trapezoid, 268
consecutive sides, 54
consecutive vertices, 54
consequent, 655–656
constant difference, 106–108
Constantine the Great, 577
construction, 142–144
of altitudes, 154
of angle bisectors, 157–158
of angles, 144
of auxiliary lines, proofs and, 200
with compass and straightedge, 143
defined, 142–143
of a dilation design, 578–580
of an equilateral triangle, 143
Euclid and, 142
of Islamic design, 156
of an isosceles triangle, 205
of a line segment, 143
of medians, 149
of midsegments, 149
of parallel lines, 161–162
with patty paper, 143, 147
of a perpendicular bisector, 147–149
of perpendiculars, 152–154
of Platonic solids, 529–530
of points of concurrency, 176–179, 183–184
of a regular hexagon, 11
of a regular pentagon, 530
of a rhombus, 287–288
of a triangle, 143, 168–169, 205
See also drawing
construction and maintenance applications, 76, 123, 134, 222, 245, 259, 278, 291, 298, 301, 318, 335, 407, 410, 414, 420, 422, 423, 424, 425, 440, 451, 457, 459, 470, 483, 495, 519, 526, 548, 556, 615, 645, 660, 685
construction exercises, 24, 145, 149–150, 154–155, 158–159, 162, 169–170, 180, 181, 186, 192–193, 202, 208, 228–229, 232, 245, 252, 270, 276, 278, 281–282, 291–293, 302, 311, 312, 316, 324, 340, 345, 350–351, 355, 363, 372, 386, 400, 427, 440, 451, 479, 495, 499, 534, 545, 570, 584, 590, 591, 608, 615, 640, 645, 705
consumer awareness, 286, 338, 458, 459, 482, 497, 518, 532, 533, 544, 555, 596, 626, 722
contrapositive, 611
Contrapositive, Law of the, 612–613
converse of a statement, 122, 611
Converse of Parallel/Proportionality Conjecture, 609
Converse of the Equilateral Triangle Conjecture, 243
Converse of the Isosceles Triangle Conjecture, 206
Converse of the Parallel Lines Conjecture, 128–129
Converse of the Perpendicular Bisector Conjecture, 149
Converse of the Pythagorean Theorem, 468–470
convex polygons, 54
cooking, 458, 498, 556, 609
coordinate geometry
as analytic geometry, 166
centroid, finding, 402
circumcenter, finding, 329–330

dilation in, 566–567, 617
distance in, 486–489
linear equations and, 210–211
ordered pair rules, 366
orthocenter, finding, 401
proofs with, 712–715
reflection in, 374–375, 467
slope and, 165–166
systems of, 401–402
translation in, 359, 366, 373
trigonometry and, 651–654
Coordinate Midpoint Property, 36, 712
Coordinate Transformations Conjecture, 367
coplanar points, 30
corresponding angles, 126, 128, 673
Corresponding Angles Conjecture (CA Conjecture), 127
Corresponding Angles Postulate (CA Postulate), 673
corresponding parts of congruent triangles are congruent (CPCTC), 230–231
cosine (cos), 621–622
Cosines, Law of, 641–643, 647
counterclockwise rotation, 359
counterexamples, 47–48
CPCTC (corresponding parts of congruent triangles are congruent), 230–231
Crawford, Ralston, 154
Crazy Horse Memorial, 569
cubic units, 514
Curie, Marie, 176
Curl-Up (Escher), 305
cyclic quadrilateral, 321
Cyclic Quadrilateral Conjecture, 321
cycloid, 347–348
cylinder(s)
altitude of, 507
axis of, 507
bases of, 507
defined, 507
drawing, 81
height of, 507
oblique, 507, 516–517
radius of, 507
right, 507, 515–517
surface area of, 446–447, 459
volume of, 515–517
Czech Republic, 316

D

Dada Movement, 260
daisy designs, 11
Dali, Salvador, 6
dart
defined, 294
nonperiodic tessellations with, 388
properties of, 294–295
data quantity and quality, conjecture and, 96
Day and Night (Escher), 407
decagon, 54
Declaration of Independence, 675
deductive reasoning, 100–102
defined, 100
geometric. *See* proof(s)
inductive reasoning compared with, 101–102
logic as. *See* logic
deductive system, 668
definitions, 30
imprecise, for basic concepts, 30
writing, tips on, 47–51
degree measure, 39–40
degrees, 39
dendroclimatology, 334
density, 535–536
design, 24, 63, 70, 116, 131, 152, 175, 179, 180, 187, 193, 302, 339, 344, 351, 355, 383, 386, 387, 388, 389–391, 406, 419, 423, 424, 428, 458, 465, 483, 494, 525, 526, 533, 538, 544, 545, 549, 562, 569, 570, 608, 648, 650
deVilliers, Michael, 696
diagonal(s)
of a kite, 267
of a parallelogram, 280
of a polygon, 54
of a rectangle, 289
of a rhombus, 288
of a square, 290
of a trapezoid, 269, 699
diagrams
assumptions possible and not possible with, 59–60
geometric proofs with, 679
problem solving with, 73–75, 482
tree diagrams, 78
vector diagrams, 280–281
Venn diagrams, 78
See also drawing
diameter
defined, 67, 69
ratio to circumference, 331–333
as term, use of, 67
dice, probability and, 86
dilation(s), 566–567
constructing design with, 578–580
Dilation Similarity Conjecture, 567
direct proof, 655
disabilities, persons with, 483, 660
displacement, 535–536
dissection, 462
distance
in coordinate geometry, 486–489
defined, from point to line, 154
of a translation, 358
distance calculations, 77, 91, 104, 150, 163, 223, 233, 240, 246, 276, 277, 318, 344, 351, 371, 391, 397, 482, 497, 498, 525, 562, 582–583, 584, 614, 621, 627, 628, 629, 638–639, 644, 645, 648, 649, 660, 677, 689
distance formula, 486–488, 712
distributive property, 670
division property of equality, 670
dodecagon, 54, 380
dodecahedron, 505
Dodgson, Charles Lutwidge, 612
See also Carroll, Lewis
Doren, Albert Van, 198
Double-edged Straightedge Conjecture, 287
double negation, 552
doubling the angle on the bow, 217
Doyle, Sir Arthur Conan, 698
drawing
circle designs, 10–11
congruence markings in, 31, 59
daisy designs, 11
defined, 142
equilateral triangle, 142
geometric solids, 80–82
isometric, 80–82, 539–541
knot designs, 16
line designs, 8
mandalas, 25
op art, 13–14, 66
orthographic, 539–541
perspective, 172–175
polygons, regular, 272
to scale, 566–567
tessellations, 21, 389–391, 393–396
See also construction
Drawing Hands (Escher), 141
Driskell, David C., 303
dual of a tessellation, 382
Dudeney, Henry E., 490
Dukes, Pam, 315
duplication of geometric figures, 142–144

E

earth science, 549
Ebner-Eschenbach, Marie von, 80
edge, 505
Edison, Thomas, 339
Egypt, 2, 36, 94, 245, 333, 386, 453, 463, 469, 519, 583, 648
Einstein, Albert, 393
Elements (Euclid), 142, 668, 671
elimination method of solving equation systems, 285–286, 401–402
elliptic geometry, 719–720

Emerging Order (Hoch), 260
endpoints
of an arc, 68
of a line segment, 31
engineering, 283, 291, 493, 607
England, 38, 703
environment, 519, 526
epicycloid, 348
equal symbol, use of, 31
equal to, symbol for, 40
equality, properties of, 670–671
equations
of a circle, 488–489
linear. *See* linear equations
equator, 351, 507
equiangular polygon, 55, 261
Equiangular Polygon Conjecture, 261
equiangular triangle. *See* equilateral triangle(s)
Equiangular Triangle Conjecture, 243
equidistant, defined, 148
Equilateral/Equiangular Triangle Conjecture, 242–243
equilateral polygon, 55
equilateral triangle(s)
angle measures of, 242
constructing, 143
defined, 61
drawing, 142
as equiangular, 242–243
as isosceles triangle, 205
Napoleon's theorem and, 247–248
properties of, 205–206
tessellations with, 379–381, 394–395
and 30°-60°-90° triangles, 476
Eratosthenes, 344
Escher, M. C., 1, 27, 93, 141, 197, 255, 305, 357, 409, 462, 503, 559, 619, 667
dilation and rotation designs of, 578
knot designs of, 17
and Napoleon's Theorem, 248
and tessellations, 389, 393–395, 398–399, 407
Euclid, 142, 463, 668, 671, 673, 718
Euclidean geometry, 142, 668, 668–674, 718, 720
Euler, Leonhard, 118, 189, 336, 512
Euler line, 189–190
Euler Line Conjecture, 189
Euler segment, 189–190
Euler Segment Conjecture, 190
Euler's Formula for Networks, 118–119
Euler's Formula for Polyhedrons, 512–513
Euler's magic square, 336
European art and architecture, 2, 5, 25, 38, 55, 62, 65, 159, 174, 278, 316, 405, 440, 703
Evans, Minnie, 404
Explorations
Alternative Area Formulas, 453–454
Constructing a Dilation Design, 578
Cycloids, 346–348
The Euler Line, 189–190
Euler's Formula for Polyhedrons, 512–513
The Five Platonic Solids, 528–530
Geometric Probability I, 86–87
Geometric Probability II, 442–444
Indirect Measurement, 632–633
Ladder Climb, 491
Napoleon's Theorem, 247–248
Non-Euclidean Geometries, 718–720
Orthographic Drawing, 539–541
Patterns in Fractals, 135–137
Perspective Drawing, 172–175
Pick's Formula for Area, 430–432
Proof as Challenge and Discovery, 696–697
A Pythagorean Fractal, 480–481
The Seven Bridges of Königsberg, 118–119
Sherlock Holmes and Forms of Valid Reasoning, 551–553
Star Polygons, 264–265
Three Types of Proofs, 655–658
Trigonometric Ratios and the Unit Circle, 651–654
Two More Forms of Valid Reasoning, 611–613
Why Elephants Have Big Ears, 599–602
Extended Parallel/Proportionality Conjecture, 606, 609
Exterior Angle Sum Conjecture, 261
exterior angles
alternate, 126–129
of a polygon, 260–261
sum of, 261
transversal forming, 126
of a triangle, 215–216
externally tangent circles, 315
Exxon *Valdez,* 519

F

face(s)
and area, 445
lateral, 445, 506
naming of, 505
of a polyhedron, 505
Fahrenheit conversion, 210
Fallingwater (Wright), 9
Fathom projects
Best-Fit Lines, 111
Different Dice, 444
In Search of the Perfect Rectangle, 598
Needle Toss, 336
Random Rectangles, 416
Random Triangles, 218
Ferber, Edna, 586
Fermat, Pierre de, 74
Fermat's Last Theorem, 74
flags, 4, 26
flowchart, 235
flowchart proofs, 235–236, 687
circle conjectures, procedures for, 325–326
quadrilateral conjectures, procedures for, 294–295
forestry, 633
45°-45°-90° triangle, 475–476
fractals
defined, 135
Pythagorean, 480–481
Sierpiński triangle, 135–137
as term, 137
fractions, 560
See also ratio
France, 25, 62, 278, 345, 405, 440, 445
Franklin, Benjamin, 627
frustum of a cone, 451
Fuller, Buckminster, 504
function rule, 106–108
functions
defined, 106
linear, 108
trigonometric. *See* trigonometry

G

Galileo Galilei, 28, 639
Gandhi, Mohandas K., 542
Gardner, Martin, 385
Garfield, James, 463
Gaudí, Antoni, 15
geoglyphs, 565
geography, 351, 442, 525, 548
Geometer's Sketchpad explorations
Alternative Area Formulas, 453–454
Constructing a Dilation Design, 578
Cycloids, 346–348
Is There More to the Orthocenter?, 190
Napoleon's Theorem, 247–248
Patterns in Fractals, 135–137
Proof as Challenge and Discovery, 696–697
A Pythagorean Fractal, 480–481

Star Polygons, 264–265
Trigonometric Ratios and the Unit Circle, 651–654
Geometer's Sketchpad projects, Is There More to the Orthocenter?, 190
geometric figures
congruent. *See* congruence
constructing. *See* construction
drawing. *See* drawing
numbers corresponding to, 115
rectifying, 440
relationships of, 78
similar. *See* similarity
sketching, 142
See also polygon(s); *specific figures listed by name*
geometric proofs. *See* proof(s)
geometric solids. *See* solid(s)
geometry
coordinate. *See* coordinate geometry
Euclidean, 142, 668, 668–674, 718, 720
non-Euclidean, 718–720
as word, 94
Gerdes, Paulus, 124
Germain, Sophie, 74
Giovanni, Nikki, 325
Girodet, Anne-Louis, 247
given. *See* antecedent
glide reflection, 376, 398–399
glide-reflectional symmetry, 376
Goethe, Johann Wolfgang von, 486
golden cut, 585
golden ratio, 585, 598, 610
golden rectangle, 610
golden spiral, 610
Goldman, Emma, 686
Goldsworthy, Andy, 5
Golomb, Solomon, 53
Gordian knot, 18
gou gu, 479, 502
Graphing Calculator projects
Drawing Regular Polygons, 272
Line Designs, 132
Lines and Isosceles Triangles, 248
Maximizing Area, 421
Maximizing Volume, 538
Trigonometric Functions, 654
graphs and graphing
intersections of lines found by, 211
of linear equations, 210–211
of linear functions, 108
of system of equations, 285
of trigonometric functions, 654
See also **Graphing Calculator projects**
gravity and air resistance, 601–602
gravity, center of, 46, 184–185
great circle(s), 351, 507, 719–720
Great Pyramid of Khufu, 519, 527, 683
Greeks, ancient, 18, 30, 36, 142, 233, 344, 381, 425, 454, 463, 528, 610, 616, 620, 668
Greve, Gerrit, 306
Guatemala, 7, 670

Haldane, J. B. S., 592
Hand with Reflecting Sphere (Escher), 93
Hansberry, Lorraine, 581
Harlequin (Vasarely), 13
Harryhausen, Ray, 601
Hawaii, 452
height
of a cone, 508
of a cylinder, 507
of a parallelogram, 412
of a prism, 506
of a pyramid, 447, 506
of a rectangle, 411
of a trapezoid, 417
of a triangle, 154, 634–635
Hein, Piet, 2, 183, 511, 563
hemisphere(s)
defined, 507
drawing, 81, 82
volume of, 542–543
heptagon, 54, 426
heptahedron, 505
Hero, 425, 454
Hero's formula for area, 454
Hesitate (Riley), 13
hexagon(s), 54
in nature and art, 2, 21, 379
regular. *See* regular hexagon(s)
hexagonal prism, 506
hexagonal pyramid, 506
hexahedron, 505
Hill, Lynn, 703
Hindus, 25, 333, 466
Hmong, 363
Hoch, Hannah, 260
Hockney, David, 540
Hoffman, Lisa, 337
Holmes, Sherlock, 537, 551–552
Homer, Winslow, 629
Hopper, Grace Murray, 106
horizon line, 172
Horseman (Escher), 398
Hot Blocks (Andrade), 24
Hubbard, Ellen, 126
Hundertwasser-House, 65
Hungary, 718
Hurston, Zora Neale, 273
hyperbolic geometry, 719
hypotenuse, 462

ice, 520, 537
icosahedron, 528
identity property, 670
if-and-only-if statement, 55, 243
if-then statements. *See* conditional statement(s)
image, 358
Impossible Structures–111 (Roelofs), 396
Improving Your Algebra Skills
Algebraic Magic Squares I, 12
Algebraic Magic Squares II, 562
Algebraic Sequences I, 229
Algebraic Sequences II, 312
Algebraic Sequences III, 472
The Difference of Squares, 365
The Eye Should Be Quicker Than the Hand, 711
Fantasy Functions, 400
Number Line Diagrams, 125
Number Tricks, 246
A Precarious Proof, 697
Pyramid Puzzle I, 9
Pyramid Puzzle II, 146
Substitute and Solve, 650
Improving Your Reasoning Skills
Bagels, 15
Bert's Magic Hexagram, 534
Calculator Cunning, 685
Checkerboard Puzzle, 72
Chew on This for a While, 372
Code Equations, 441
Container Problem I, 212
Container Problem II, 224
The Dealer's Dilemma, 188
How Did the Farmer Get to the Other Side?, 293
Hundreds Puzzle, 209
Logical Liars, 397
Logical Vocabulary, 678
Pick a Card, 240
Puzzling Patterns, 99
Reasonable 'rithmetic I, 495
Reasonable 'rithmetic II, 550
Scrambled Arithmetic, 383
Seeing Spots, 705
Spelling Card Trick, 171
Symbol Juggling, 702
Think Dinosaur, 324
Improving Your Visual Thinking Skills
Build a Two-Piece Puzzle, 577
Coin Swap I and II, 46
Coin Swap III, 160
Colored Cubes, 318
Connecting Cubes, 610
Constructing an Islamic Design, 156

Cover the Square, 454
Dissecting a Hexagon I, 203
Dissecting a Hexagon II, 263
Equal Distances, 85
Folding Cubes I, 151
Folding Cubes II, 474
Fold, Punch, and Snip, 485
Four-Way Split, 425
Hexominoes, 79
Mental Blocks, 695
Moving Coins, 452
Mudville Monsters, 479
Net Puzzle, 259
Painted Faces I, 403
Painted Faces II, 602
Patchwork Cubes, 545
Pentominoes I, 58
Pentominoes II, 117
Pickup Sticks, 6
Picture Patterns I, 340
Picture Patterns II, 387
Picture Patterns III, 691
Piet Hein's Puzzle, 511
Polyominoes, 53
The Puzzle Lock, 182
A Puzzle Quilt, 284
Puzzle Shapes, 633
Random Points, 436
Rolling Quarters, 328
Rope Tricks, 640
Rotating Gears, 105
The Spider and the Fly, 490
The Squared Square Puzzle, 429
3-by-3 Inductive Reasoning Puzzle I, 392
3-by-3 Inductive Reasoning Puzzle II, 626
TIC-TAC-NO!, 585
Visual Analogies, 164
incenter, 177–179
Incenter Conjecture, 178–179
inclined plane, 483–484, 660
included angle, 219
included side, 219
incoming angle, 41
India, 6, 25, 333, 466, 616
indirect measurement
with similar triangles, 581–582
with string and ruler, 584
with trigonometry, 621
indirect proof, 656, 698–700
inductive reasoning, 94–96
deductive reasoning compared to, 101–102
defined, 94
figurate numbers, 115
finding *n*th term, 106–108
mathematical modeling, 112–115
inequalities, triangle, 213–216
inscribed angle(s), 307, 319–320, 325–326, 492
Inscribed Angle Conjecture, 319, 325–326
Inscribed Angle Theorem, 703
Inscribed Angles Intercepting Arcs Conjecture, 320
inscribed circle, 179
inscribed quadrilateral, 321
inscribed triangle, 71, 179
integers
Pythagorean triples, 468–469
rule generating, 106–108
See also numbers
intercepted arc, 320, 321
interior angles
adjacent, 215–216
alternate, 126–129, 681
of polygons, 256–257, 261
remote, 215–216
transversal forming, 126
interior design, 428
internally tangent circles, 315
intersection of lines
finding, 211
reflection images and, 375
See also perpendicular lines; point(s) of concurrency
Interwoven Patterns–V (Roelofs), 396
inverse cosine ($\cos^{-1}$), 624
inverse of a conditional statement, 611
inverse sine ($\sin^{-1}$), 624
inverse tangent ($\tan^{-1}$), 624
Investigations
Angle Bisecting by Folding, 157
Angle Bisecting with Compass, 158
Angles Inscribed in a Semicircle, 320
Arcs by Parallel Lines, 321
Are Medians Concurrent?, 183–184
Area Formula for Circles, 433–434
Area Formula for Kites, 418
Area Formula for Parallelograms, 412
Area Formula for Regular Polygons, 426–427
Area Formula for Trapezoids, 417–418
Area Formula for Triangles, 417
Area of a Triangle, 634–635
Area Ratios, 592–593
Balancing Act, 184–185
Base Angles in an Isosceles Triangle, 205
The Basic Property of a Reflection, 360–361
Can You Prove the SSS Similarity Conjecture?, 708
Chords and the Center of the Circle, 309
Chords and Their Central Angles, 307–308
Concurrence, 176–177
Constructing Parallel Lines by Folding, 161
Copying a Segment, 143–144
Copying an Angle, 144
Corresponding Parts, 586
Cyclic Quadrilaterals, 321
Defining Angles, 49–50
Defining Circle Terms, 69
Dilations on a Coordinate Plane, 566–567
The Distance Formula, 486–487
Do All Quadrilaterals Tessellate?, 385
Do All Triangles Tessellate?, 384
Do Rectangle Diagonals Have Special Properties?, 289
Do Rhombus Diagonals Have Special Properties?, 288
The Equation of a Circle, 488
Extended Parallel/Proportionality, 606
Exterior Angles of a Triangle, 215–216
Finding a Minimal Path, 367–368
Finding the Arcs, 342
Finding the Right Bisector, 147–148
Finding the Right Line, 152–153
Finding the Rule, 106–107
The Formula for the Surface Area of a Sphere, 546–547
The Formula for the Volume of a Sphere, 542
Four Parallelogram Properties, 279–280
Going Off on a Tangent, 313
How Do We Define Angles in a Circle?, 307
Incenter and Circumcenter, 177–178
Inscribed Angle Properties, 319
Inscribed Angles Intercepting the Same Arc, 320
Is AA a Similarity Shortcut?, 572
Is ASA a Congruence Shortcut?, 225
Is SAS a Congruence Shortcut?, 221
Is SAS a Similarity Shortcut?, 574
Is SSS a Congruence Shortcut?, 220
Is SSS a Similarity Shortcut?, 573
Is the Converse True?, 128–129, 206, 468–469
Is There a Polygon Sum Formula?, 256–257
Is There an Exterior Angle Sum?, 260–261

Isosceles Right Triangles, 475
The Law of Sines, 635
The Linear Pair Conjecture, 120
Mathematical Models, 29
Mirror, Mirror, 581
Opposite Side Ratios, 588
Overlapping Segments, 102
Parallels and Proportionality, 604–605
Party Handshakes, 112–114
Patty-Paper Perpendiculars, 153
Perpendicular Bisector of a Chord, 309–310
Proving Parallelogram Conjectures, 692–693
Proving the Tangent Conjecture, 699–700
A Pythagorean Identity, 641
Reflections over Two Intersecting Lines, 375
Reflections over Two Parallel Lines, 374
Right Down the Middle, 148–149
The Semiregular Tessellations, 380–381
Shape Shifters, 96
Solving Problems with Area Formulas, 422
Space Geometry, 82
Surface Area of a Cone, 449
Surface Area of a Regular Pyramid, 448
The Symmetry Line in an Isosceles Triangle, 242
Tangent Segments, 314
A Taste of Pi, 332
30°-60°-90° Triangles, 476
The Three Sides of a Right Triangle, 462–463
Transformations of a Coordinate Plane, 367
Trapezoid Midsegment Properties, 274–275
Triangle Midsegment Properties, 273–274
The Triangle Sum, 199–200
Triangles and Special Quadrilaterals, 60–61
Trigonometric Tables, 622–623
Vertical Angles Conjecture, 121
Virtual Pool, 41–42
The Volume Formula for Prisms and Cylinders, 515–516
The Volume Formula for Pyramids and Cones, 522
Volume Ratios, 594
What Are Some Properties of Kites?, 266–267
What Are Some Properties of Trapezoids?, 268–269
What Can You Draw with the Double-Edged Straightedge?, 287
What Is the Shortest Path from *A* to *B*?, 214
What Makes Polygons Similar?, 564
Where Are the Largest and Smallest Angles?, 215
Which Angles Are Congruent?, 126–128
Iran, 358, 448
irrational numbers, 333
Islamic art and architecture, 2, 6, 10, 20–23, 60, 156, 207, 298, 379, 389, 448, 520
isometric drawing, 80–82, 539–541
isometry
composition of, 373–376
on coordinate plane, 359, 366–367, 373, 467
defined, 358
properties of, 366–370
types of. *See* reflection; rotation; translation
isosceles right triangle, 475–476
Isosceles Right Triangle Conjecture, 475
isosceles trapezoid(s), 268–269
Isosceles Trapezoid Conjecture, 269
Isosceles Trapezoid Diagonals Conjecture, 269
isosceles triangle(s), 204–206
base angles of, 62, 205–206
base of, 62
construction of, 205
defined, 61
legs of, 204
proofs involving, 698–699
right, 475–476
vertex angle bisector, 242–243
vertex angle of, 62
See also equilateral triangle(s)
Isosceles Triangle Conjecture, 205
converse of, 206
Israel, 20, 505
Italy, 18, 159, 174, 463, 525, 549, 639, 668

Jainism, 616
Jamaica, 4
James Fort, 638
James III, Richard, 385
Japan, 4, 7, 14, 17, 19, 83, 299, 386, 387, 414, 525, 646, 720
Jefferson, Thomas, 256
Jewish art, 405
Jimenez, Soraya, 435
Jiuzhang suanshu, 479
John Paul I (pope), 260
journal, defined, 140
Joyce, Bruce, 94
Joyce, James, 603
Juchi, Hajime, 14

kaleidoscopes, 378
Kamehameha I, 452
Keller, Helen, 445
Kickapoo, 548
Kim, Scott, 383
King (Evans), 404
King Kong, 601
King, Martin Luther, Jr., 152
kite(s)
area of, 418
concave. *See* dart
defined, 63
nonperiodic tessellations with, 388
nonvertex angles of, 266
properties of, 266–267
tessellations with, 398
vertex angles of, 266
Kite Angle Bisector Conjecture, 267
Kite Angles Conjecture, 267
Kite Area Conjecture, 418
Kite Diagonal Bisector Conjecture, 267
Kite Diagonals Conjecture, 267
kites (recreational), 63, 419
Klee, Paul, 433
knot designs, 2, 16–17, 21, 396
koban (Japanese architecture), 83, 414
Königsberg, 118

La Petite Chatelaine (Claudel), 597
Lahori, Ustad Ahmad, 6
landscape architecture, 424
Laos, 363
lateral edges
oblique, and volume, 516
of a prism, 506
of a pyramid, 506
lateral faces
of a prism, 445, 506
of a pyramid, 445
and surface area, 445
latitude, 507
Lauretta, Sister Mary, 147
law, 100
law enforcement, 83, 414
Law of Cosines, 641–643, 647
Law of Sines, 634–637, 647

Law of Syllogism, 611, 612–613
Law of the Contrapositive, 612–613
Lec, Stanislaw J., 13
legs
of an isosceles triangle, 204
of a right triangle, 462
lemma, 692
L'Engle, Madeleine, 100
length
of an arc, 341–343
symbols to indicate, 31
Leonardo da Vinci, 331, 360, 440, 463
LeWitt, Sol, 61
light, angles of, 45, 52
Lin, Maya, 130
line(s)
auxiliary, 200
concurrent, 115, 176–179
description of, 28, 29–30
diagrams of, 59
equations of. *See* linear equations
intersecting. *See* intersection of lines
models of, 28–29
naming of, 28
in non-Euclidean geometries, 718–720
parallel. *See* parallel lines
perpendicular. *See* perpendicular lines
postulates of, 672
segments of. *See* line segment(s)
skew, 48
slope of. *See* slope
symbol for, 28
tangent. *See* tangent(s)
transversal, 126
line designs, 7–8
Line Intersection Postulate, 672
line of best fit, 111
line of reflection, 360–361
line of symmetry, 3, 361
Line Postulate, 672
line segment(s)
bisectors of, 31, 147–149
and concurrency. *See* point(s) of concurrency
congruent, 31
construction of, 143
defined, 31
divided by parallel lines, proportion of, 604–607
duplicating, 143
endpoints of, 31
of Euler, 189–190
golden cut on, 585
measure of, notation, 31
midpoint of, 31–32, 36–37, 148–149, 672, 712
overlapping, 102
postulates of, 672
similarity and, 603–607
symbol for, 31
tangent segments, 314
line symmetry. *See* reflectional symmetry
linear equations, 210–211
intersections, finding with, 211
slope-intercept form, 210–211
systems of, solving, 285–286, 401–402
linear functions, 108
linear pair(s) of angles, 120–122
defined, 50
Linear Pair Conjecture, 120–122
Linear Pair Postulate, 673
locus of points, 75
logic
and flowchart proofs, 235–236
forms of, 551–553, 611–613
history of, 668
language of, 551–552
Law of Syllogism, 611, 612–613
Law of the Contrapositive, 612–613
Modus Ponens, 552–553, 612–613
Modus Tollens, 552–553, 613
and paragraph proofs, 235
proofs of, approaches to, 655–658
symbols and symbolic form of, 551–553, 611, 612, 613
See also proof(s)
logical argument, 551
longitude, 507
Loomis, Elisha Scott, 463
Loren, Sophia, 706
lusona patterns, 18, 124

M

Magic Mirror (Escher), 357
Magritte, René, 510
major arc, 68, 308
Malcolm X, 468
mandalas, 25, 576
Mandelbrot, Benoit, 137
Mandelbrot set, 137
manufacturing, 34, 220, 458
Maori, 709
maps and mapping, 36, 311, 567
Martineau, Harriet, 213
Maslow, Abraham, 168
mathematical models, 112–115
defined, 112
mathematics journal, defined, 140
Mattingly, Gene, 313
maximizing area, 421
maximizing volume, 538
maximum volume, 538
Maya, 341, 639, 670
mean, and the centroid, 402–403
measure
of an angle, 39–40
of arcs, defined, 68
indirect. *See* indirect measurement
median(s) of a triangle, 149, 183–185, 402, 586–587
See also centroid
Median Concurrency Conjecture, 183
Medici crest, 549
medicine and health, 39, 110
metals, densities of, 535–536
meteorology and climatology, 101, 334, 483, 628
Mexico, 25, 62, 341, 527, 639
Michelson, Albert Abraham, 52
Michelson Interferometer, 52
midpoint of a line segment, 31–32, 36–37, 148–149, 672, 712
Midpoint Postulate, 672
midsegment(s)
of a trapezoid, 273, 274–275
of a triangle, 149, 273–274, 275
Mini-Investigations, 103, 111, 163, 181, 335, 339, 364, 479, 489, 590, 710, 722, 723
minimal path, 367–370
Minimal Path Conjecture, 368
minor arc, 68
mirror line. *See* line of reflection
mirror symmetry. *See* reflectional symmetry
mirrors, in indirect measurement, 581
Mitchell, Maria, 7
models, mathematical, 112–115
Modus Ponens, 552–553, 612–613
Modus Tollens, 552–553, 613
monohedral tiling, 379–380, 384–385
Morgan, Julia, 198
Morocco, 22
moviemaking, 563, 601
Mozambique, 124
multiplication of radical expressions, 474
multiplication, properties of, 670
multiplication property of equality, 670
murals, 571
music, 339
My Blue Vallero Heaven (Archuleta-Sagel), 265

N

Nadelstern, Paula, 284
Nagaoka, Kunito, 19
NAMES Project AIDS Memorial Quilt, 519
Napoleon Bonaparte, 247
Napoleon's Theorem, 247–248
Native Americans, 3, 49, 62, 361, 449, 548, 569
nature
fractals in, 135

geometry in, 2
op art in, 15
symmetry in, 3, 4
tessellation in, 379
nautical mile, 351
Navajo, 49, 361
navigation, 217, 351, 628, 629, 630, 645, 648, 660
Nazca Lines, 567
negation, 552
nets, 78, 528–530
networks, 118–119
New Zealand, 709
n-gon, 54, 257
See also polygon(s)
Nigeria, 16
nonagon, 54
non-Euclidean geometries, 718–720
nonperiodic tiling, 388
nonrigid transformation, 358, 566–567, 578–580
nonvertex angles, 266
notebook, defined, 92
*n*th term, finding, 106–108
numbers
integers. *See* integers
irrational, 333
rectangular, 115
rounding, 423
square roots, 473–474
triangular, 115
numerical name for tiling, 380

O

Oates, Joyce Carol, 10
oblique cylinder, 507, 516–517
oblique prism, 506, 516, 517
oblique triangular prism, 506
obtuse angle, 49, 101
obtuse triangle, 60, 641–642
oceanography, 576, 653
octagon, 54, 380
octahedron, 528
odd numbers
O'Hare, Nesli, 41
oil spills, 519, 526
Oldenburg, Claes, 615
100 Cans (Warhol), 507
one-point perspective, 173
op art, 3, 13–14, 66
opposite angle, 215, 620
opposite angles of a parallelogram, 279, 692–693
opposite side, trigonometry, 620
opposite sides of a parallelogram, 280, 693–694
optics, 45, 52, 271, 326
ordered-pair rules, 366
orthocenter
defined, 177
finding, 401
properties of, 189–190
in right triangle, 180
orthographic drawing, 539–541
outgoing angle, 41
overlapping angles property, 102
overlapping segments property, 111

P

Palestinian art, 10
paragraph proofs, 122
circle conjectures, procedures for, 326
quadrilateral conjectures, procedures for, 295
parallel lines
arcs on circle and, 321
construction of, 161–162
defined, 48, 161
diagrams of, 59
proofs involving, 680–681
properties of, 127–129, 672
proportional segments by, 603–607
and reflection images, 374
slope of, 165–166, 712
symbol for, 48
Parallel Lines Conjecture, 127–128
converse of, 128–129
Parallel Lines Intercepted Arcs Conjecture, 321
Parallel/Proportionality Conjecture, 605
converse of, 606, 617
Extended Parallel/Proportionality, 606, 609
Parallel Slope Property, 165, 712
parallelogram(s)
altitude of, 412
area of, 412–413
defined, 63
height of, 412
proofs involving, 692–693, 696–697
properties of, 279–280, 287–290
relationships of, 78
special, 287–290
tessellations with, 399
in vector diagrams, 280–281
See also specific parallelograms listed by name
Parallelogram Area Conjecture, 412
Parallelogram Consecutive Angles Conjecture, 280
Parallelogram Diagonal Lemma, 692
Parallelogram Diagonals Conjecture, 280
Parallelogram Opposite Angles Conjecture, 279, 692–693
Parallelogram Opposite Sides Conjecture, 280, 693–694
Parallel Postulate, 672, 718–719
partial mirror, 52
Path of Life I (Escher), 559, 578
patterns. *See* inductive reasoning
patty papers, 142, 143, 147
Pei, I. M., 204, 445
Pelli, Cesar, 23
Penrose, Sir Roger, 388
Penrose tilings, 388
pentagon(s)
area of, 426
congruent, 55
naming of, 54
in nature and art, 2
sum of angles of, 256–257
tessellations with, 379, 385, 386
Pentagon Sum Conjecture, 256
performance assessment, 196
period, 654
periodic curve, 348
periodic phenomena, trigonometry and, 654
periscope, 131
perpendicular bisector(s)
of a chord, 309–310
constructing, 147–149
defined, 147
linear equations for, 211
of a rhombus, 288
of a triangle, 149, 176–178
See also bisector; circumcenter; perpendiculars
Perpendicular Bisector Concurrency Conjecture, 177
Perpendicular Bisector Conjecture, 148
Converse of, 149
Perpendicular Bisector of a Chord Conjecture, 310
perpendicular lines
defined, 48
diagrams of, 59
slope of, 165–166, 712
symbol for, 48
Perpendicular Postulate, 672
Perpendicular Slope Property, 165, 712
Perpendicular to a Chord Conjecture, 309
perpendiculars
constructing, 152–154
defined, 152
postulate of, 672
perspective, 172–175
Peru, 567
Phelps, Elizabeth Stuart, 204
photography, 263, 323, 583
physics, 45, 46, 52, 185, 271, 314, 483–484, 520, 535–537, 556, 601–602
pi, 331–333, 337
symbol for, 331

Picasso, Pablo, 341
Pick, Georg Alexander, 430
Pickford, Mary, 531
Pick's formula, 430–432
picture angle, 323
pinhole camera, 583
plane(s)
concurrent, 176
naming of, 28
as undefined term, 28–30
plane figures. *See* geometric figures
Plato, 528, 668
Platonic solids (regular polyhedrons), 505, 528–530
plumb level, 245
point(s)
collinear, 30
coplanar, 30
description of, 28, 29–30
diagrams of, 59
locus of, 75
models of, 28–29
naming of, 28
symbol for, 28
point(s) of concurrency, 176–179, 183–185
centroid, 183–185, 189–190, 402
circumcenter. *See* circumcenter
construction of, 176–179, 183–184
coordinates for, finding, 329–330, 401, 402
defined, 176
incenter, 177–179
orthocenter. *See* orthocenter
point of tangency, 69
point symmetry, 361
pollution, 519, 526
Polya, George, 234, 668
polygon(s), 54–55
area of, 411
circumscribed circles and, 179
classification of, 54
concave, 54
congruent, 55
consecutive vertices of, 54
convex, 54
defined, 54
diagonals of, 54
drawing, 272
equiangular, 55
equilateral, 55
exterior angles of, 260–261
inscribed circles and, 179
interior angles of, 256–257, 261
naming of, 54
regular. *See* regular polygon(s)
sides of, 54
similar. *See* similarity
sum of angle measures, 256–257, 260–261
symmetries of, 362
tesellations with, 379–381, 384–385
vertex of a, 54
See also specific polygons listed by name
Polygon Sum Conjecture, 257
polyhedron(s)
defined, 505
edges of, 505
Euler's formula for, 512–513
faces of, 505
models of, building, 512–513, 528–530
naming of, 505
nets of, 78, 528–530
regular, 505
similar, 593
vertex of a, 505
See also solid(s); *specific polyhedrons listed by name*
Polynesian navigation, 630
pool, geometry of, 41–42, 45
portfolio, defined, 26
postulates of geometry, 668, 671–673, 703, 706, 718–719
premises of geometric proofs, 668, 669–674, 680, 712
presentations, giving, 304
Print Gallery (Escher), 1
prism(s)
altitude of, 506
antiprisms, 510
bases of, 445, 506
defined, 506
drawing, 81
height of, 506
inverted images and, 271
lateral faces of, 445, 506
oblique, 506, 516, 517
right, 506, 515–517, 593
similarity and, 593
surface area of, 446, 459
volume of, 515–517
Prism-Cylinder Volume Conjecture, 516
probability
area and, 442–444
defined, 86
geometric problems using, 86–87
problem solving
acting it out, 112–113
with diagrams, 73–75, 482
mathematical modeling, 112–115
projects
Best-Fit Lines, 111
Building an Arch, 278
Creating a Geometry Flip Book, 467
Different Dice, 444
Drawing Regular Polygons, 272
Drawing the Impossible, 66
In Search of the Perfect Rectangle, 598
Is There More to the Orthocenter?, 190
Japanese Puzzle Quilts, 299
Japanese Temple Tablets, 646
Kaleidoscopes, 378
Light for All Seasons, 631
Line Designs, 132
Lines and Isosceles Triangles, 248
Making a Mural, 571
Maximizing Area, 421
Maximizing Volume, 538
Needle Toss, 336
Penrose Tilings, 388
Photo or Video Safari, 23
Polya's Problem, 234
Racetrack Geometry, 345
Random Rectangles, 416
Random Triangles, 218
The Soma Cube, 521
Special Proofs of Special Conjectures, 717
Spiral Designs, 35
Symbolic Art, 19
Trigonometric Functions, 654
The World's Largest Pyramid, 527
See also **Fathom projects; Geometer's Sketchpad projects; Graphing Calculator projects**
Promenades of Euclid, The (Magritte), 501
proof(s)
of angle conjectures, 680–684, 686–688
auxiliary lines in, 200
of circle conjectures, 699–700, 703–704
conditional, 655–656
with coordinate geometry, 712–715
direct, 655
flowchart. *See* flowchart proofs
indirect, 656, 698–700
lemmas used in, 692
logical family tree used in, 682–684
paragraph. *See* paragraph proofs
planning and writing of, 294–295, 679–684, 687–688
postulates of, 668, 671–673, 703, 706, 718–719
premises of, 668, 669–674, 680, 712
of the Pythagorean Theorem, 463–464, 466
of quadrilateral conjectures, 294–295, 692–693, 696–697, 699, 712–715
of similarity, 706–709

symbols and symbolic form of, 551–553, 611, 612, 613
of triangle conjectures, 681–684, 686–688, 706–709
two-column, 655, 687–688
See also logic
proportion, 560–561
with area, 592–593
of corresponding parts of triangles, 586–588
defined, 560
indirect measurement and, 581–582
of segments by parallel lines, 603–607
similarity and, 560–561, 565–566, 586–588
with volume, 593–594
See also ratio; trigonometry
Proportional Areas Conjecture, 593
proportional dividers, 609
Proportional Parts Conjecture, 586
Proportional Volumes Conjecture, 594
protractor
angle measure with, 39, 142
and clinometers, 632–633
construction as not using, 143
defined, 47
Ptolemy, Claudius, 620
public transit, 87
Pushkin, Aleksandr, 38
puzzles. *See* **Improving Your Algebra Skills; Improving Your Reasoning Skills; Improving Your Visual Thinking Skills**
pyramid(s)
altitude of, 506
base of, 445, 506
defined, 506
drawing, 81
height of, 447, 506
lateral faces of, 445
slant height of, 447
surface area of, 446, 447–448
truncated, 685
vertex of, 506
volume of, 522–524
Pyramid-Cone Volume Conjecture, 522
pyramids (architectural structures), 62, 204, 245, 445, 447, 519, 527, 583, 639, 644
Pythagoras of Samos, 463, 668
Pythagorean Identity, 641
Pythagorean Proposition (Loomis), 463
Pythagorean Theorem, 462–464
and circle equations, 488–489
circles and, 492
converse of, 468–470
cultural awareness of principle of, 463, 469
and distance formula, 486–488, 712
fractal based on, 480–481
and isosceles right triangle, 475–476
and Law of Cosines, 641–642
picture representations of, 462, 480
problem solving with, 482
proofs of, 463–464, 466
Pythagorean identities and, 641
and similarity, 480
statement of, 463
and 30°-60°-90° triangle, 476–477
trigonometry and, relationship of, 641
Pythagorean triples, 468–469

Q

Q.E.D., defined, 295
quadrilateral(s)
area of, ancient Egyptian formula for, 453
congruent, 55
cyclic, 321
definitions of, 62–64
linkages of, 283
naming of, 54
proofs involving, 294–295, 692–693, 696–697, 699, 712–715
proving properties of, 294–295
special, 62–64
sum of angles of, 256–257
tessellations with, 379, 385
See also specific quadrilaterals listed by name
Quadrilateral Sum Conjecture, 256
quilts and quiltmaking, 9, 56, 284, 299, 415, 519

R

racetrack geometry, 345
radical expressions, 473–474
radiosonde, 628
radius
and arc length, 342–343
of a circle, 67
and circumference formula, 332
of a cone, 508
of a cylinder, 507
of a sphere, 507
tangents and, 313–314
as term, use of, 67
ratio
circumference/diameter, 331–333
defined, 560
equal. *See* proportion
of Euler segment, 189–190
golden, 585, 610
probability, 86
slope. *See* slope
trigonometric. *See* trigonometry
See also similarity
ray(s)
angles defined by, 38
concurrent, 176
defined, 32
naming of, 32
symbol of, 32
reasoning
deductive. *See* logic
inductive. *See* inductive reasoning
record players, 339
rectangle(s)
area of, 411
base of, 411
defined, 63
golden, 610
height of, 411
properties of, 289
sum of angles of, 289
Rectangle Area Conjecture, 411
Rectangle Diagonals Conjecture, 289
rectangular numbers, 115
rectangular prism, 80, 446, 506
rectangular solid(s), drawing, 80
rectifying shapes, 440
recursive rules, 135–137
Red and Blue Puzzle (Benson), 299
reflection
composition of isometries and, 374–376
defined, 360
glide, 376, 398–399
line of, 360–361
minimal path and, 367–370
as type of isometry, 358
Reflection Line Conjecture, 361
reflectional symmetry, 3, 4, 361
Reflections over Intersecting Lines Conjecture, 375
Reflections over Parallel Lines Conjecture, 374
reflexive property of congruence, 671
reflexive property of equality, 670
reflexive property of similarity, 706
regular dodecagon
regular heptagon, 426
tessellations with, 380
regular dodecahedron, 505, 528–530
regular hexagon(s), 11
area of, 426

tessellations with, 379–381, 390, 393–394
regular hexahedron, 528–530
regular icosahedron, 528–530
regular octagon, 380
regular octahedron, 528–530
regular pentagon, 362, 379, 426, 530
regular polygon(s)
apothem of, 426
area of, 426–427
defined, 55
drawing, 272
symmetries of, 362
tessellations with, 379–381
Regular Polygon Area Conjecture, 427
regular polyhedrons (Platonic solids), 505, 528–530
regular tessellation, 380
regular tetrahedron, 528–530
remote interior angles, 215–216
Renaissance, 172
Reptiles (Escher), 394
resources, consumption of, 423, 596, 649
resultant vector, 281
revolution, solid of, 84
rhombus(es)
construction of, 287–288
defined, 63
properties of, 287–288
Rhombus Angles Conjecture, 288
Rhombus Diagonals Conjecture, 288
Rice, Marjorie, 385, 386
right angle, 49
right cone, 508, 593
right cylinder, 507, 515–517
right pentagonal prism, 506
right prism, 506, 515–517, 593
right triangle(s)
adjacent side of, 620
defined, 60
hypotenuse of, 462
isosceles, 475–476
legs of, 462
opposite side of, 620
properties of. *See* Pythagorean Theorem; trigonometry
similarity of, 590–591
30°-60°-90° type, 476–477
rigid transformation. *See* isometry
Riley, Bridget, 13
Roelofs, Rinus, 396
Romans, ancient, 36, 271, 278
Roosevelt, Eleanor, 492
Rosten, Leo, 287
rotation
defined, 359
model of, 359
spiral similarity, 580
tessellations by, 393–395
as type of isometry, 358
rotational symmetry, 3–4, 361
rounding numbers, 423
Rounds and Triangles (Bauer), 177
ruler, 7, 142, 143, 584
Rumi, Jalaluddin, 20
Russell, Bertrand, 692
Russia, 17, 118

S

SAA Congruence Conjecture, 226
SAA Congruence Theorem, 686
sailing. *See* navigation
sangaku, 646
Sarton, May, 379
SAS Congruence Conjecture, 221
SAS Congruence Postulate, 673
SAS Similarity Conjecture, 574
SAS Similarity Theorem, 706–707
SAS Triangle Area Conjecture, 635
satellites, 314, 317
Savant, Marilyn vos, 76
scale drawings, 566–567
scale factor, 566
scalene triangle, 384
Schattschneider, Doris, 385, 580
sculpture, geometry and, 577, 592, 597, 616
secant, 321
section, 84
sector compass, 608
sector of a circle, 352
area of, 437–438
segment(s)
of a circle, 437–438
in elliptic geometry, 719
line. *See* line segment(s)
Segment Addition Postulate, 672
segment bisector, 147
Segment Duplication Postulate, 672
seismology, 229
self-similarity, 135
semicircle(s), 68, 320, 492
semiregular tessellation, 380–381
sewing and weaving, 9, 10, 14, 49, 51, 56, 60, 289, 298, 299, 415
shadows, indirect measurement and, 581–582
Shah Jahan, 6
Shintoism, 646
shipping, 400, 420, 644
Shortest Distance Conjecture, 153
show. *See* consequent
Sicily, 668
side(s)
adjacent, 620
of an angle, 38
included, 219
opposite, 620
of a parallelogram, 280, 293–294
of a polygon, 54
Side-Angle-Angle (SAA), 219, 226, 227, 574
Side-Angle Inequality Conjecture, 215
Side-Angle-Side (SAS), 219, 220, 227, 574, 642
Side-Side-Angle (SSA) case, 219, 221–222, 574, 636
Side-Side-Side (SSS), 219, 220, 227, 573, 642
Sierpiński tetrahedron, 204
Sierpiński triangle, 135–137
Sills, Beverly, 482
similarity, 563–566
and area, 592–593
defined, 565
dilation and, 566–567, 578–580
indirect measurement and, 581–582
mural making and, 571
and polygons, 563–567
proofs involving, 706–709
properties of, 706
and Pythagorean fractal, 480
and Pythagorean Theorem, 480
ratio and proportion in, 560–561, 565–566, 586–588
segments and, 603–607
of solids, 593
spiral, 580
symbol for, 565
of triangles, 200–201, 572–574, 586–588
trigonometry and, 620
and volume, 593–594
simplification of radical expressions, 473
sine (sin), 621–622
Sines, Law of, 634–637, 647
Sioux, 449
sketching, 142
See also construction; drawing
skew lines, 48
slant height, 447
slope
calculating, 133–134
defined, 133
formula, 133
of parallel lines, 165–166, 712
of perpendicular lines, 165–166, 712
slope-intercept form, 210–211
slope-intercept form, 210–211
slope triangle, 133
smoking cigarettes, 722
Snakes (Escher), 17
solar technology, 423
solid(s)
bases of, 445
with curved surfaces, 506–508
drawing, 80–82
faces of, 505

lateral faces of, 445, 506
nets of, 78, 528–530
Platonic, 505, 528–530
of revolution, 84
section, 84
similar, 593–594
surface area of. *See* surface area
See also polyhedron(s); *specific solids listed by name*
solid of revolution, 84
South America, 101
space, 80–82
defined, 80
See also solid(s)
Spain, 15, 20
speed. *See* velocity and speed calculations
Spenger, Sylvia, 18
sphere(s)
center of, 507
coordinates on, 719–720
defined, 507
drawing, 81, 82
elliptic geometry and, 719–720
great circles of, 351, 507, 719–720
hemisphere. *See* hemisphere
radius of, 507
surface area of, 546–547
volume of, 542–543
Sphere Surface Area Conjecture, 547
Sphere Volume Conjecture, 542
spherical coordinates, 719–720
spiral(s)
golden, 610
in nature and art, 35, 65
spiral similarity, 580
sports, 185, 186, 286, 315, 345, 363, 367–368, 369–370, 371, 383, 406, 435, 443, 498, 501, 562
Sproles, Judy, 279
square(s)
defined, 47–48, 64
proofs involving, 712–715
properties of, 290
symmetry of, 362
tessellations with, 379–381, 389
Square Diagonals Conjecture, 290
Square Diagonals Theorem, 712–715
Square Limit (Escher), 409
square pyramid, 506
square root property of equality, 670
square roots, 473–474
square units, 413
Sri Lanka, 451
SSS Congruence Conjecture, 220
SSS Congruence Postulate, 673
SSS Similarity Conjecture, 573
SSS Similarity Theorem, 708–709
star polygons, 264–265
steering linkage, 283
Steinem, Gloria, 225
Still Life and Street (Escher), 255
Still Life With a Basket (Cézanne), 504
Stonehenge, 703
straightedge, 7, 143
double-edged, 287–288
substitution method of solving equation systems, 285–286, 402
subtraction property of equality, 670
Sumners, DeWitt, 18
Sunday Morning Mayflower Hotel (Hockney), 540
supplementary angles, 50
surface area
of a cone, 448–449
of a cylinder, 446–447, 459
defined, 445
of a prism, 446, 459
of a pyramid, 445, 447–448
of a sphere, 546–547
and volume, relationship of, 599–602
surfaces, area and, 445
surveying land, 36, 94, 453, 469, 649
Swenson, Sue, 422
Syllogism, Law of, 611, 612–613
Symbolic Logic, Part I (Dodgson), 612
symbols
angle, 38
approximately equal to, 40
arc, 68
conditional statement, 552
congruence, 31, 40, 59
equals, use of, 31
glide reflection, 398
image point label, 358
line, 28
line segment, 31
of logic, 551–553, 611, 612, 613
measure, 31, 40
negation (logic), 552
parallel, 48
perpendicular, 48
pi, 331
plane, 28
point, 28
ray, 32
same measure, 31
similarity, 565
slant height, 447
therefore, 552
triangle, 54
symmetric property of congruence, 671
symmetric property of equality, 670
symmetry, 3–4, 361–362
bilateral, 3
glide-reflectional, 376
in an isosceles trapezoid, 269
in an isosceles triangle, 242
in a kite, 266
line of, 3, 361
of polygons, 362
reflectional, 3, 4, 361
rotational, 3–4, 361
Symmetry Drawing E25 (Escher), 393
Symmetry Drawing E99 (Escher), 395
Symmetry Drawing E103 (Escher), 197
Symmetry Drawing E105 (Escher), 389
Symmetry Drawing E108 (Escher), 399
systems of linear equations, 285–286, 401–402
Szent-Györgyi, Albert, 120
Szyk, Arthur, 675

T

Taj Mahal, 6
Take Another Look
angle measures, 253
area, 459
circumference, 355
congruence shortcuts, 253
cyclic quadrilaterals, 355
dilation, 617
polygon conjectures, 303–304
Pythagorean Theorem, 501–502
quadrilateral conjectures, 253
similarity, 617–618
tangents, 355
triangle conjectures, 253, 303
trigonometry, 665–666
volume, 557–558
Talmud, 417
Tan, Amy, 67
tangent(s)
defined, 69
external, 315
internal, 315
point of, 69
proofs involving, 699–700
properties of, 313–315
and Pythagorean Theorem, 492
radius and, 313–314
segments, 314
as term, use of, 69
tangent circles, 315
Tangent Conjecture, 313, 314
tangent ratio (tan), 620, 621–622, 624
tangent segments, 314
Tangent Segments Conjecture, 314
Tangent Theorem, 699–700
tangram puzzle, 484
Taoism, 316
tatami, 386
technology
applications, 52, 112, 131, 235,

263, 271, 283, 314, 317, 323, 339, 351, 423, 435, 498, 583, 607, 628, 630
exercises, 123, 145, 150, 160, 170, 180, 181, 186, 201, 259, 284, 316, 318, 323, 345, 382–383, 397, 427, 429, 436, 550, 585, 610, 645, 648, 650, 691, 710, 711
telecommunications, 112, 435, 498
temari balls, 720
temperature conversion, 210
term, *n*th, 106–108
Tessellating Quadrilaterals Conjecture, 385
Tessellating Triangles Conjecture, 384
tessellation (tiling)
creation of, 21, 388, 389–391, 393–396
defined, 20
dual of, 382
glide reflection, 398–399
monohedral, 379–380, 384–385
nonperiodic, 388
with nonregular polygons, 384–385
regular, 380
rotation, 393–395
semiregular, 380–381
translation, 389–390
vertex arrangement, 380, 381
test problems, writing, 254
tetrahedron, 505
Thales of Miletus, 233, 583, 668
theodolite, 628
Theorem of Pythagoras. *See* Pythagorean Theorem
theorem(s), 668
defined, 463
logical family tree of, 682–684
proving. *See* proof(s)
See also specific theorems listed by name
therefore, symbol for, 552
Thiebaud, Wayne, 514
thinking backward, 294
Third Angle Conjecture, 200–201
Third Angle Theorem, 682–684
30°-60°-90° triangle, 476–477
30°-60°-90° Triangle Conjecture, 476–477
Thomas, Calista, 647
Thompson, Rewi, 709
Three Midsegments Conjecture, 273
3-uniform tiling, 381
Three Worlds (Escher), 27
Tibet, 576
tiling. *See* tessellation
transformation(s), 358
nonrigid, 358, 566–567, 578–580
rigid. *See* isometry
transitive property of congruence, 671
transitive property of equality, 670
transitive property of similarity, 706
translation, 358–359
and composition of isometries, 373–374, 376
defined, 358
direction of, 358
distance of, 358
tessellations by, 389–390
as type of isometry, 358
vector, 358
transversal line, 126
trapezium, 268
trapezoid(s)
arch design and, 271
area of, 417–418
base angles of, 267
bases of, 267
defined, 62
diagonals of, 269, 699
height of, 417
isosceles, 268–269
linkages of, 283
midsegments of, 273, 274–275
proofs involving, 699
properties of, 268–269
Trapezoid Area Conjecture, 418
Trapezoid Consecutive Angles Conjecture, 268
Trapezoid Midsegment Conjecture, 275
tree diagrams, 78
triangle(s)
acute, 60, 636, 641–642
adjacent interior angles of, 215–216
altitudes of, 154, 177, 401, 586
angle bisectors of, 176–179, 586, 587–588
area of, 411, 417, 454, 634–635
centroid of, 183–185, 189–190, 402
circumcenter. *See* circumcenter
circumscribed, 71, 179
congruence of, 168–169, 219–222, 225–227, 230–231
constructing, 143, 168–169, 205
definitions of, 60–61
determining parts of, 168
drawing, 134
elliptic geometry and, 720
equiangular. *See* equilateral triangle(s)
equilateral. *See* equilateral triangle(s)
exterior angles of, 215–216
height of, 154, 634–635
incenter, 177–179
included angle, 219
included side, 219
inequalities, 213–216
inscribed, 71, 179
interior angles of, 215–216
isosceles. *See* isosceles triangle(s)
medians of, 149, 183–185, 402, 586–587
midsegments of, 149, 273–274, 275
naming of, 54
obtuse, 60, 641–642
orthocenter. *See* orthocenter
parallel lines and proportions of obtuse, 641–642
perpendicular bisectors of, 149, 176–178
points of concurrency of. *See* point(s) of concurrency
proofs involving, 681–684, 686–688, 706–709
relationships of, 78
remote interior angles of, 215–216
right. *See* right triangle(s)
scalene, 384
similarity of, 200–201, 572–574, 586–588
sum of angles of, 199–200
symbol for, 54
tessellations with, 379–381, 384, 394–395
vertex angle, 62, 242–243
Triangle Area Conjecture, 417
Triangle Exterior Angle Conjecture, 216
Triangle Inequality Conjecture, 214
Triangle Midsegment Conjecture, 274, 275
Triangle Sum Conjecture, 198–201
Triangle Sum Theorem, 681–682
triangular numbers, 115
triangular prism, 506
triangular pyramid, 506
triangulation, 229
trigonometry, 620
adjacent side, 620
cosine (cos), 621–622
graphs of functions, 654
inverse cosine ($\cos^{-1}$), 624
inverse sine ($\sin^{-1}$), 624
inverse tangent ($\tan^{-1}$), 624
Law of Cosines, 641–643, 647
Law of Sines, 634–637, 647
opposite side, 620
and periodic phenomena, 654
problem solving with, 627, 647
ratios, 620–624
sine (sin), 621–622
tables and calculators for, 622–624, 654

tangent (tan), 620, 621–622, 624
unit circle and, 651–654
vectors and, 647
truncated pyramid, 685
Tsiga series (Vasarely), 3
Turkey, 379, 668
Twain, Mark, 59, 104
two-column proof, 655, 687–688
two-point perspective, 174–175
2-uniform tiling, 381
Tyson, Cicely, 157

U

undecagon, 54
unit circle, 651–654
units
area and, 413
nautical mile, 351
not stated, 31
volume and, 514
Using Your Algebra Skills
Coordinate Proof, 712–717
Finding the Circumcenter, 329–330
Finding the Orthocenter and Centroid, 401–403
Midpoint, 36–37
Proportion and Reasoning, 560–561
Radical Expressions, 473–474
Slope, 133–134
Slopes of Parallel and Perpendicular Lines, 165–166
Solving Systems of Linear Equations, 285–286
Writing Linear Equations, 210–211
Uzbekistan, 60

V

VA Theorem (Vertical Angles Theorem), 679–680
valid argument, 100, 102, 551
valid reasoning. *See* logic
vanishing point(s), 172, 173, 174
Vasarely, Victor, 3, 13
vector(s)
defined, 280
diagrams with, 280–281
resultant, 281
translation, 358
trigonometry with, 647
vector sum, 281
velocity and speed calculations, 134, 293, 302, 337, 338, 340, 344, 345, 351, 392, 483, 497, 660, 661
velocity vectors, 280–281
Venn diagram, 78
Venters, Diane, 56
Verblifa tin (Escher), 503
Verne, Jules, 337
vertex (vertices)
of a cone, 508
consecutive, 54
defined, 38
naming angles by, 38
of a polygon, 54
of a polyhedron, 505
of a pyramid, 506
tessellation arrangement, 380, 381
vertex angle(s)
bisector of, 242–243
of an isosceles triangle, 62, 242–243
of a kite, 266
Vertex Angle Bisector Conjecture, 242
vertex arrangement, 380, 381
vertical angles, 50, 121, 679–680
Vertical Angles Conjecture, 121–122, 129
Vertical Angles Theorem (VA Theorem), 679–680
vintas, 144
Vichy-Chamrod, Marie de, 142
Vietnam Veterans Memorial Wall, 130
volume
of a cone, 522–524
of a cylinder, 515–517
defined, 514
displacement and density and, 535–536
of a hemisphere, 542–543
maximizing, 538
of a prism, 515–517
problems in, 531
proportion and, 593–594
of a pyramid, 522–524
of a sphere, 542–543
and surface area, relationship of, 599–602
units used to measure, 514
Vries, Jan Vredeman de, 172

W

Walker, Mary Willis, 546
Wall Drawing #652 (LeWitt), 61
Warhol, Andy, 507
water
and buoyancy, 537
and volume, 520, 537
Water Series (Greve), 306
Waterfall (Escher), 461
Weyl, Hermann, 358
Wick, Walter, 66
wigwams, 548
Wilcox, Ella Wheeler, 647
Wilde, Oscar, 462
Wiles, Andrew, 74
Williams, William T., 79
woodworking, 34
work, 484
World Book Encyclopedia, 26
Wright, Frank Lloyd, 9
Wright, Steven, 514
writing test problems, 254

y-intercept, 210
yin-and-yang symbol, 316

Z

zero product property of equality, 670
Zhoubi Suanjing, 502
zillij, 22
zoology and animal care, 15, 435, 536, 576, 694

Index

Photo Credits

Abbreviations: top (T), center (C), bottom (B), left (L), right (R).

Cover

Background image: Doug Wilson/Corbis; Construction image: Sonda Dawes/The Image Works; All other images: Ken Karp Photography.

Front Matter

v (T): Ken Karp Photography; **v (C):** Cheryl Fenton; **v (B):** Cheryl Fenton; **vi (T):** Ken Karp Photography; **vii (T):** Ken Karp Photography; **vii (C):** Courtesy, St. John's Episcopal Church; **vii (B):** Hillary Turner; **viii (T):** Ken Karp Photography; **viii (B):** Corbis/Stockmarket; **ix (T):** Cheryl Fenton; **ix (B):** Cheryl Fenton; **x (T):** Ken Karp Photography; **xi (T):** Ken Karp Photography; **xii (T):** Ken Karp Photography; **xii (B):** Perry Collection/Photo by Cheryl Fenton; **xiii:** Cheryl Fenton.

Chapter 0

1: *Print Gallery,* M. C. Escher, 1956/©2002 Cordon Art B. V.–Baarn–Holland. All rights reserved.; **2 (B):** ©1993 Metropolitan Museum of Art, Bequest of Edward C. Moore, 1891 (91.1.)2064 **2 (C):** Cheryl Fenton; **2 (TL):** NASA; **3 (BL):** Christie's Images; **3 (T):** *Tsiga I,II,III* (1991), Victor Vasarely, Courtesy of the artist.; **4 (CR):** Hillary Turner; **4 (TL):** Cheryl Fenton; **4 (TR):** Cheryl Fenton; **5 (B):** ©Andy Goldsworthy, Courtesy of the artist and Galerie Lelong; **5 (C):** Cheryl Fenton; **5 (CL):** Cheryl Fenton; **5 (TC):** Cheryl Fenton; **5 (TL):** Cheryl Fenton; **6:** Corbis; **7 (BL):** Dave Bartruff/Stock Boston; **7 (BR):** Robert Frerck/Woodfin Camp & Associates; **7 (CL):** Rex Butcher/Bruce Coleman Inc.; **7 (CR):** Randy Juster; **9:** Schumacher & Co./Frank Lloyd Wright Foundation; **10 (R):** Sean Sprague/Stock Boston; **10 (TC):** Christie's Images/Corbis; **12:** W. Metzen/Bruce Coleman Inc.; **13 (L):** *Hesitate,* Bridget Riley/Tate Gallery, London/Art Resource, NY; **13 (R):** *Harlequin* by Victor Vasarely, Courtesy of the artist.; **15 (C):** National Tourist Office of Spain; **15 (T):** Tim Davis/Photo Researchers; **16:** Cheryl Fenton; **17:** *Snakes,* M. C. Escher, 1969/©2002 Cordon Art B. V.–Baarn–Holland. All rights reserved.; **18:** Will & Deni McIntyre/ Photo Researchers Inc.; **19:** *SEKI/PY XVIII* (1978), Kunito Nagaoka/ Courtesy of the artist.; **19 (L):** Cheryl Fenton; **19 (R):** Cheryl Fenton; **20 (B):** Corbis; **20 (T):** Nathan Benn/Corbis; **22 (B):** Ken Karp Photography; **22 (C):** Peter Sanders Photography; **22 (TL):** Peter Sanders Photography; **22 (TR):** Peter Sanders Photography; **23:** Photo Researchers Inc.; **24:** *Hot Blocks* (1966–67) ©Edna Andrade, Philadelphia Museum of Art, Purchased by Philadelphia Foundation Fund; **25 (L):** Comstock; **25 (R):** Scala/Art Resource; **29 (T):** George Lepp/Photo Researchers Inc.

Chapter 1

27: *Three Worlds,* M. C. Escher, 1955/©2002 Cordon Art B. V.–Baarn–Holland. All rights reserved.; **28 (B):** Spencer Grant/Photo Researchers Inc.; **28 (C):** Cheryl Fenton; **28 (T):** Hillary Turner; **29:** By permission of Johnny Hart and Creators Syndicate, Inc.; **30:** Bachman/Photo Researchers Inc.; **32:** S. Craig/Bruce Coleman Inc.; **33 (R):** Grafton Smith/Corbis Stock Market; **33 (L):** Michael Daly/Corbis Stock Market; **34:** Addison Geary/Stock Boston; **35:** Bob Stovall/Bruce Coleman Inc.; **36:** Archivo Iconografico, S. A./Corbis; **38 (TL):** Bruce Coleman Inc.; **38 (TR):** David Leah/Getty Images; **39 (B):** Hillary Turner; **39 (BR):** Cheryl Fenton; **39 (BR):** Comstock; **39 (T):** Ken Karp Photography; **41:** Pool & Billiard Magazine; **47:** Illustration by John Tenniel; **48:** Osentoski & Zoda/Envision; **49:** Christie's Images; **50:** Corbis; **51:** Hillary Turner; **54:** Cheryl Fenton; **55:** Ira Lipsky/International Stock Photography; **56:** Quilt by Diane Venters/*More Mathematical Quilts*; **56 (C):** Cheryl Fenton; **56 (TCL):** Hillary Turner; **56 (TCR):** Hillary Turner; **56 (TL):** Hillary Turner; **56 (TR):** Hillary Turner; **59:** Spencer Swanger/Tom Stack & Associates; **60:** Gerard Degeorge/Corbis; **61:** Sol LeWitt—Wall Drawing #652—On three walls, continuous forms with color ink washes superimposed, Color in wash. Collection: Indianapolis Museum of Art, Indianapolis, IN. September, 1990. Courtesy of the Artist.; **62 (C):** Michael Moxter/Photo Researchers Inc.; **62 (L):** Larry Brownstein/Rainbow; **62 (R):** Stefano Amantini/Bruce Coleman Inc.; **63 (B):** Cheryl Fenton; **63 (T):** Cheryl Fenton; **65:** Friedensreich ©Erich Lessing/Art Resource, NY; **66:** From WALTER WICK'S OPTICAL TRICKS. Published by Cartwheel books, a division of Scholastic Inc. ©1998 by Walter Wick. Reprinted by permission.; **67 (BL):** Joel Tribhout/ Agence Vandystadt/Getty Images; **67 (BR):** Terry Eggers/Corbis Stock Market; **67 (T):** By permission of Johnny Hart and Creators Syndicate Inc.; **68 (L):** Cheryl Fenton; **68 (R):** Getty Images; **70 (CL):** Corbis; **70 (CR):** Alfred Pasieka/Photo Researchers Inc.; **70 (T):** Corbis; **73:** *Bookplate for Albert Ernst Bosman,* M. C. Escher, 1946/©2002 Cordon Art B. V.–Baarn–Holland. All rights reserved.; **74 (BC):** Stock Montage, Inc.; **74 (BL):** Stock Montage, Inc.; **74 (BR):** AP/Wide World; **79:** William Thomas Williams, *"DO YOU THINK A IS B,"* Acrylic on Canvas, 1975–77, Fisk University Galleries, Nashville, Tennessee; **80 (C):** Cheryl Fenton; **80 (L):** Cheryl Fenton; **80 (R):** Cheryl Fenton; **81:** Cheryl Fenton; **82:** Cheryl Fenton; **83:** Courtesy of Kazumata Yamashita, Architect; **84 (B):** Ken Karp Photography; **84 (C):** Mike Yamashita/Woodfin Camp & Associates; **85:** T. Kitchin/Tom Stack & Associates; **86 (C):** Cheryl Fenton; **86 (T):** Ken Karp Photography; **88:** Paul Steel/Corbis-Stock Market; **90:** Cheryl Fenton.

Chapter 2

93: *Hand with Reflecting Sphere (Self-Portrait in Spherical Mirror),* M. C. Escher/©2002 Cordon Art B. V.–Baarn–Holland. All rights reserved.; **94 (B):** Barry Rosenthal/FPG; **94 (T):** Ken Karp Photography; **95:** Andrew McClenaghan/Photo Researchers Inc.; **100:** Bob Daemmrich/Stock Boston; **101:** California Institute of Technology and Carnegie Institution of Washington; **104:** NASA; **106:** Drabble reprinted by permission of United Feature Syndicate, Inc.; **112 (L):** National Science Foundation Network; **112 (R):** Hank Morgan/Photo Researchers Inc.; **113:** Hillary Turner; **115:** Cheryl Fenton; **118 (B):** Ken Karp Photography; **118 (T):** Culver Pictures; **121:** Ken Karp Photography; **126 (B):** Ken Karp Photography; **126 (T):** Alex MacLean/Landslides; **128:** Hillary Turner; **130:** James Blank/Bruce Coleman Inc.; **134 (L):** Photo Researchers; **134 (R):** Mark Gibson/Index Stock; **135 (B):** Ted Scott/Fotofile, Ltd.; **135 (C):** Ken Karp Photography; **137:** Art Matrix.

Chapter 3

141: *Drawing Hand,* M. C. Escher, 1948/©2002 Cordon Art B. V.–Baarn–Holland. All rights reserved.; **142 (L):** Hillary Turner; **142 (R):** Hillary Turner; **142 (T):** Bettmann/Corbis; **143 (L):** Hillary Turner; **143 (R):** Hillary Turner; **144:** Travel Ink/Corbis; **154:** *Overseas Highway* by Crawford Ralston, 1939 by Crawford Ralston/Art Resource, NY; **156:** Rick Strange/Picture Cube; **159:** Corbis; **172 (BL):** Ken Karp Photography; **172 (BR):** Dover Publications; **172 (T):** Greg Vaughn/Tom Stack & Associates; **174:** Art Resource; **175:** Timothy Eagan/Woodfin Camp & Associates; **176:** Ken Karp Photography; **177:** *Rounds and Triangles* by Rudolf Bauer /Christie's Images; **180:** Corbis; **185 (B):** Corbis; **185 (T):** Ken Karp Photography; **186:** Keith Gunnar/Bruce Coleman Inc.; **189:** Ken Karp Photography; **191:** Victoria & Albert Museum, London/Art Resource, NY.

Chapter 4

197: *Symmetry Drawing E103,* M. C. Escher,1959/©2002 Cordon Art B. V.–Baarn–Holland. All rights reserved.; **198 (B):** Courtesy, St. John's Presbyterian Church; **198 (C):** Jim Corwin/Stock Boston; **198 (TL):** Cheryl Fenton; **198 (TR):** Cheryl Fenton; **199:** Ken Karp Photography; **202:** The Far Side® by Gary Larson ©1987 FarWorks, Inc. All rights reserved. Used with permission.; **203:** Hillary Turner;

L): David L. Brown/Picture Cube; **204 (R):** Joe Sohm/The Image Works; **207:** Art Stein/Photo Researchers Inc.; **213:** Robert k/Woodfin Camp & Associates; **217:** Judy March/Photo Researchers Inc.; **220:** Adamsmith Productions/Corbis; **222:** Carolyn Schaefer/Gamma Liaison; **227:** Cheryl Fenton; **234 (B):** Ken Karp Photography; **234 (T):** AP/Wide World; **240:** Ken Karp Photography; **241:** Eric Schweikardt/The Image Bank; **247:** Archivo Iconografico, SA/Corbis; **249 (T):** Ken Karp Photography; **251:** Ken Karp Photography; **252 (L):** Cheryl Fenton; **252 (R):** Cheryl Fenton; **253:** Cheryl Fenton.

Chapter 5

255: *Still Life and Street,* M. C. Escher, 1967–68/©2002 Cordon Art B. V.–Baarn–Holland. All rights reserved.; **256:** Ken Karp Photography; **256 (T):** Cheryl Fenton; **260:** *Emerging Order* by Hannah Hoch/Christie's Images/Corbis; **263:** Richard Megna/Fundamental Photographs; **265:** Courtesy of the artist, Teresa Archuleta-Sagel; **266:** Steve Dunwell/The Image Bank; **268 (C):** Lindsay Hebberd/Woodfin Camp & Associates; **268 (T):** Photo Researchers Inc.; **271:** Cheryl Fenton; **278:** Malcolm S. Kirk/Peter Arnold Inc.; **284:** ©Paula Nadelstern/Photo by Bobby Hansson; **286:** Corbis; **289:** Ken Karp Photography; **290:** Cheryl Fenton; **291:** Bill Varie/The Image Bank; **292:** Bill Varie/The Image Bank; **292:** Alex MacLean/Landslides; **298:** Cheryl Fenton; **299:** *Red and Blue Puzzle* (1994) Mabry Benson/Photo by Carlberg Jones; **302:** Ken Karp Photography; **303:** *Boy With Birds* by David C. Driskell/Collection of Mr. and Mrs. David C. Driskell.

Chapter 6

305: *Curl-Up,* M. C. Escher,1951/©2002 Cordon Art B. V.–Baarn–Holland. All rights reserved.; **306 (B):** Stephen Saks/Photo Researchers, Inc.; **306 (C):** Corbis; **308:** Tom Sanders/Corbis Stock Market; **311:** NASA; **312:** Tony Freeman/PhotoEdit; **313 (L):** Stephen A. Smith; **313 (R):** Mark Burnett/Stock Boston; **314:** NASA; **315:** Gray Mortimore/Getty Images; **316 (L):** Corbis; **316 (L):** Jim Zuckerman/Corbis; **319:** Xinhua/Gamma Liaison; **328 (L):** Cheryl Fenton; **328 (R):** Cheryl Fenton; **331 (B):** Cheryl Fenton; **331 (C):** Ken Karp Photography; **332:** Hillary Turner; **334 (B):** Don Mason/Corbis Stock Market; **334 (C):** David R. Frazier/Photo Researchers Inc.; **334 (T):** Ken Karp Photography; **337:** Photofest; **339:** Calvin and Hobbes ©Watterson. Reprinted with permission of UNIVERSAL PRESS SYNDICATE. All rights reserved.; **340:** Guido Cozzi/Bruce Coleman Inc.; **341:** Grant Heilman Photography; **344 (B):** Nick Gunderson/Corbis; **344 (T):** Charles Feil/Stock Boston; **345 (B):** Bob Daemmrich/Stock Boston; **345 (T):** Cathedrale de Reims; **346:** Gamma Liaison; **349 (B):** Ken Karp Photography; **349 (T):** Cheryl Fenton; **351:** Ken Karp Photography; **352:** The Far Side® by Gary Larson ©1990 FarWorks, Inc. All rights reserved. Used with permission.; **353:** David Malin/Anglo-Australian Telescope Board; **355:** Private collection, Berkeley, California/Ceramist, Diana Hall.

Chapter 7

357: *Magic Mirror,* M. C. Escher,1946/©2002 Cordon Art B. V.–Baarn–Holland. All rights reserved.; **358:** Giraudon/Art Resource; **358 (B):** Ken Karp Photography; **360:** Corbis; **361:** Grant Heilman Photography; **363 (BC):** Denver Museum of Natural History; **363 (BL):** Michael Lustbader/Photo Researchers; **363 (BR):** Richard Cummins/Corbis; **363 (TL):** Phil Cole/Getty Images; **363 (TR):** Corbis; **364:** Oxfam; **369:** Corbis; **372:** By Holland ©1976 Punch Cartoon Library; **373:** Gina Minielli/Corbis; **378:** Ken Karp Photography; **379 (L):** W. Treat Davison/Photo Researchers; **379 (R):** Ancient Art and Architecture Collection; **381:** Springer-Verlag; **385:** Courtesy Doris Schattschneider; **386:** Lucy Birmingham/Photo Researchers Inc.; **388 (B):** Carleton College/photo by Hilary N. Bullock; **388 (T):** Ashton, Ragget, McDougall Architects; **389 (C):** *Symmetry Drawing E105,* M. C. Escher 1960/©2002 Cordon Art B. V.–Baarn–Holland. All rights reserved.; **389 (T):** *Brickwork,* Alhambra, M. C. Escher/©2002 Cordon Art B. V.–Baarn–Holland. All rights reserved.; **393:** *Symmetry Drawing E25,* M. C. Escher, 1939/©2002 Cordon Art B. V.–Baarn–Holland. All rights reserved.; **394:** *Reptiles,* M. C. Escher, 1943/©2002 Cordon Art B. V.–Baarn–Holland. All rights reserved.; **395 (T):** *Symmetry Drawing E99,* M. C. Escher, 1954/©2002 Cordon Art B. V.–Baarn–Holland. All rights reserved.; **396 (L):** *Interwoven patterns–V–structure 17* by Rinus Roelofs/Courtesy of the artist & ©2002 Artist Rights Society (ARS), New York/Beeldrecht, Amsterdam; **396 (R):** *Impossible structures–III–structure 24* by Rinus Roelofs/Courtesy of the artist & ©2002 Artist Rights Society (ARS), New York/Beeldrecht, Amsterdam; **398 (B):** *Horseman sketch,* M. C. Escher/©2002 Cordon Art B. V.–Baarn–Holland. All rights reserved.; **398 (T):** *Horseman,* M. C. Escher, 1946/©2002 Cordon Art B. V.–Baarn–Holland. All rights reserved.; **399:** *Symmetry Drawing E108,* M. C. Escher, 1967/©2002 Cordon Art B. V.–Baarn–Holland. All rights reserved.; **404:** *King* by Minne Evans/Corbis & Luise Ross Gallery; **405 (C):** Paul Schermeister/Corbis; **405 (TR):** Comstock; **406:** Ken Karp Photography; **407 (B):** Starbuck Goldner; **407 (C):** *Day and Night,* M. C. Escher, 1938/©2002 Cordon Art B. V.–Baarn–Holland. All rights reserved.

Chapter 8

409: *Square Limit,* M. C. Escher, 1964/©2002 Cordon Art B. V.–Baarn–Holland. All rights reserved.; **410:** Spencer Grant/Stock Boston; **414 (B):** Karl Weatherly/Corbis; **414 (T):** Courtesy Naoko Hirakura Associates; **421:** Ken Karp Photography; **422:** Joseph Nettis/Photo Researchers Inc.; **423 (B):** NASA; **423 (C):** Ken Karp Photography; **424:** Phil Schermeister/Corbis; **428:** Cheryl Fenton; **434:** Cheryl Fenton; **435 (B):** Reuters NewMedia Inc./Corbis; **435 (C):** Georg Gerster/Photo Researchers Inc.; **437:** Shahn Kermani/Gamma Liaison; **440:** Eye Ubiquitous/Corbis; **442:** Ken Karp Photography; **444:** Cheryl Fenton; **445 (BL):** Stefano Amantini/Bruce Coleman Inc.; **445 (BR):** Alain Benainous/Gamma Liaison; **445 (CB):** Joan Iaconetti/Bruce Coleman Inc.; **445 (CT):** Bill Bachmann/The Image Works; **446:** Greg Pease/Corbis; **447 (L):** Sonda Dawes/The Image Works; **447 (R):** John Mead/Photo Researchers Inc.; **449:** Library of Congress; **451:** Bryn Campbell/Getty Images; **452:** Douglas Peebles; **458:** Ken Karp Photography.

Chapter 9

461: *Waterfall,* M. C. Escher, 1961/©2002 Cordon Art B. V.–Baarn–Holland. All rights reserved.; **462:** FUNKY WINKERBEAN by Batiuk. Reprinted with special permission of North American Syndicate; **463:** Corbis; **463 (L):** Corbis; **464:** Will & Deni McIntyre/Photo Researchers Inc.; **467 (C):** Corbis; **467 (T):** Ken Karp Photography; **468:** Ken Karp Photography; **469:** Rare Book and Manuscript Library/Columbia University; **470:** John Colett/Stock Boston; **472:** Christie's Images/©2002 Sol LeWitt/Artists Rights Society (ARS), New York; **475:** Turner Entertainment ©1939. All rights reserved.; **482:** FUNKY WINKERBEAN by Batiuk. Reprinted with special permission of North American Syndicate; **483:** Bob Daemmrich/The Image Works; **484 (B):** Private collection, Berkeley, California/Photo by Cheryl Fenton; **484 (T):** Ken Karp Photography; **490:** David Muench/Corbis; **494:** Bruno Joachim/Gamma Liaison; **495:** Ken Karp Photography; **498:** Ken Karp Photography.

Chapter 10

503: *Verblifa tin,* M. C. Escher, 1963/©2002 Cordon Art B. V.–Baarn–Holland. All rights reserved.; **504 (BL):** J. Bernholc, etal./North Carolina State/Photo Researchers Inc.; **504 (CB):** Ken Eward/Photo Researchers Inc.; **504 (CT):** Runk-Schoenberger/Grant Heilman Photography; **504 (CT):** Ken Eward/Photo Researchers; **504 (L):** Archivo Iconografico, S. A. /Corbis; **504 (T):** Cheryl Fenton; **505:** Esaias Baitel/Gamma Liaison; **506:** Cheryl Fenton; **507:** Burstein Collection/Corbis/© Andy Warhol/Artists Rights Society (ARS), New York; **510:** ©1996 C. Herscovici, Brussels, Artists Rights Society (ARS) NY/Photo courtesy Minneapolis Institute of

Art; **512:** Ken Karp Photography; **514:** Ken Karp Photography; **514:** Christie's Images/Corbis; **516:** Hillary Turner; **519 (B):** Jeff Tinsly/Names Project; **519 (C):** Larry Lee Photography/ Corbis; **520 (B):** Thomas Kitchin/Tom Stack & Associates; **520 (C):** Eye Ubiquitous/Corbis; **521 (B):** Cheryl Fenton; **521 (C):** Cheryl Fenton; **522:** Ken Karp Photography; **525:** Masao Hayashi/Photo Researchers Inc.; **527:** Charles Lenars/Corbis; **528 (B):** Ken Karp Photography; **528 (T):** Fitzwilliam Museum, Cambridge; **532 (T):** David Sutherland/Stone/Getty Images; **533:** Parson's School of Design; **537:** David Morris/Gamma Liaison; **539 (B):** C. J. Allen/Stock Boston; **539 (T):** Ken Karp Photography; **540:** David Hockney, *Sunday Morning Mayflower Hotel, N. Y.,* Nov. 28, 1982/Photographic collage, ED: 20/50″ × 77″ ©David Hockney; **541:** Ken Karp Photography; **542 (B):** Ken Karp Photography; **544 (B):** The Far Side® by Gary Larson ©1986 FarWorks, Inc. All rights reserved. Used with permission.; **544 (T):** Culver Pictures/ Picture Quest; **546 (C):** John Cooke/Comstock; **546 (T):** NASA; **548:** Larry Lefevre/Grant Heilman Photography; **549:** Arte & Immagini/Corbis; **551 (L):** Ken Karp Photography; **551 (T):** Illustration by Sidney Paget from The Strand Magazine, 1892; **555:** Piranha Club by B. Grace. Reprinted with special permission of King Features Syndicate; **558:** Douglas Peebles/Corbis.

Chapter 11

559: *Path of Life I,* M. C. Escher, 1958/©2002 Cordon Art B. V.–Baarn–Holland. All rights reserved.; **564 (R):** J. Greenberg/The Image Works; **564 (R):** Chromosohn/Photo Researchers Inc.; **567 (B):** National Geographic Society; **567 (C):** Giaudon/Art Resource, NY; **567 (T):** Erich Lessing/Art Resource, NY; **569:** John Elk III/Stock Boston; **570:** Daniel Sheehan/Black Star Publishing/PictureQuest; **571 (C):** Ken Karp Photography; **571 (T):** Michael S. Yamashita/Corbis; **575:** Patricia Lanza/Bruce Coleman Inc.; **576 (B):** Thomas Dove/Douglas Peebles; **576 (L):** Rossi & Rossi; **576 (R):** Rossi & Rossi; **577:** Dave Bartruff/Corbis; **583:** Ken Karp Photography; **586:** Ken Karp Photography; **592:** Alon Reininger/Contact Press Images/PictureQuest; **594:** Ken Karp Photography; **595:** David Fraser/Photo Researchers Inc.; **597 (B):** Dr. Paul A. Zahl/Photo Researchers Inc.; **597 (T):** *La Petite Chatelaine, Version a La Natte Courbe* by Camille Claudel/Christie's Images/© 2002 Artists Rights Society (ARS), New York/ADAGP, Paris; **599:** Ken Karp Photography; **600 (B):** ©1996 Demart Pro Arte Geneva/© Salvador Dali, Gala-Salvador Dali Foundation/ARS NY; **600 (T):** Corbis; **601 (C):** Kurt Krieger/Corbis; **601 (T):**Bettmann/ Corbis; **607:** Charles Feil/Stock Boston; **611 (B):** Ken Karp Photography; **612:** *Alice in Wonderland*; **615 (B):** Robert Holmes/ Corbis; **615 (T):** Dan McCoy/Rainbow; **616 (L):** Chris Lisle/Corbis; **616 (R):** Alex Webb/Magnum; **617:** Underwood & Underwood/ Corbis; **618:** NASA.

Chapter 12

619: *Belvedere*, M. C. Escher, 1958/©2002 Cordon Art B. V.–Baarn–Holland. All rights reserved.; **620:** Bettmann/Corbis; **621 (B):** Perry Collection; **621 (T):** Gary Braasch/Woodfin Camp & Associates; **623:** Cheryl Fenton; **627:** Ken Karp Photography; **628 (B):** Ecoscene/ Corbis; **628 (C):** H. Reinhard/Photo Researchers; **629:** *Breezing Up* by Winslow Homer/Photo by Francis G. Mayer/Corbis; **630:** Dennis Marsico/Corbis; **631 (L):** Wendell Metzen/Bruce Coleman Inc.; **631 (R):** Corbis; **632 (C):** Ken Karp Photography; **633:** Greg Rynders; **638 (C):** Courtesy of Association for the Preservation of Virginia Antiquities; **639 (B):** Stevan Stefanovic/Photo Researchers Inc.; **639 (T):** Steve Owlett/Bruce Coleman Inc.; **644 (B):** Grant Heilman Photography; **646:** Photograph by Hiroshi Umeoka; **647:** Sonda Dawes/The Image Works; **649:** James A. Sugar/Black Star Publishing/Picture Quest; **651:** Angelo Hornak/Corbis; **652:** Walter Hodges/Corbis; **654:** Courtesy California Academy of Sciences; **655:** Ken Karp Photography; **660:** © Marty Sohl; **661:** B. Christensen/Stock Boston.

Chapter 13

667: *Another World (Other World),* M. C. Escher/©2002 Cordon Art B. V.–Baarn–Holland. All rights reserved.; **668:** Springer-Verlag; **668:** Araldo de Luca/CORBIS; **669 (C):** Ken Karp Photography; **669 (R):** Ken Karp Photography; **670:** Charles & Josette Lenars/ Corbis; **671:** Art Resource; **675:** Library of Congress; **679:** ©1977 by Sidney Harris, *American Scientist Magazine*; **682:** Ken Karp Photography; **684:** Ken Karp Photography; **703:** M. Dillon/Corbis; **709:** Paul A. Souders/Corbis; **718 (L):** Springer-Verlag; **718 (R):** Ken Karp Photography; **720:** Collection of Suzanne Summer/Cheryl Fenton Photography.